DISEASES AND THE ORGANISMS THAT CAUSE THEM *(CONTINUED)*

BACTERIAL DISEASES

Disease	Organism	Type*	Page
typhus, scrub (tsutsugamushi disease)	*Rickettsia tsutsugamushi*	CB, NA	684
verruga peruana (bartonellosis)	*Bartonella bacilliformis*	coccoid, −	686
vibriosis	*Vibrio parahaemolyticus*	R, −	643
whooping cough (pertussis)	*Bordetella pertussis*	CB, −	AppB, 606–607
yersiniosis	*Yersinia enterocolitica*	R, −	646–647

Disease	Organism
*Key to types:	
C = coccus I = irregular VAR = Gram-variable	
CB = coccobacillus − = Gram-negative A-F = acid-fast	
R = rod + = Gram-positive NA = not applicable	
S = spiral	
†Species	

VIRAL DISEASES

Disease	Virus	Reservoir	Page	Disease	Virus	Reservoir	Page
aplastic crisis in sickle cell anemia	erythrovirus (B19)	humans	260, 691–692	herpes, oral	usually herpes simplex type 1, sometimes type 2	humans	261, 264, 589
bronchitis, rhinitis	parainfluenza	humans, some other mammals	600, 601	herpes, genital	usually herpes simplex type 2, sometimes type 1	humans	261, 264, 588
Burkitt's lymphoma	Epstein-Barr	humans	689	HIV disease, AIDS	human immunodeficiency virus (HIV)	humans	260, 518–523
chickenpox	varicella-zoster	humans	261, 265, 544–546	infectious mononucleosis	Epstein-Barr	humans	688–689
coryza (common cold)	rhinovirus	humans	260, 262,	influenza	influenza	swine, humans (type A)	260, 263, 616–621
	coronavirus	humans	603–604			humans (typeB)	260, 263, 616–621
cytomegalic inclusion disease	cytomegalovirus	humans	590–591			humans (type (type C)	616–621
Dengue fever	Dengue	humans	314, 667–668				
encephalitis	Colorado tick fever	mammals	314, 691	Lassa fever	arenavirus	rodents	691
	Eastern equine encephalitis	birds	262–263, 405, 706–707	measles (rubeola)	measles	humans	260, 543–544
	St. Louis encephalitis	birds	706–707	meningoencephalitis	herpes	humans	589, 707
	Venezuelan equine encephalitis	rodents	262–263, 706–707	molluscum contagiosum	poxvirus group	humans	547
	Western equine encephalitis	birds	262–263, 314, 405, 706–707	mumps	paramyxovirus	humans	637
epidemic keratoconjunctivitis	adenovirus	humans	555	pneumonia	adenoviruses, respiratory syncytial virus	humans	AppB, 608–609
fifth disease (erythema infectiosum)	erythrovirus (B19)	humans	265–267, 692	poliomyelitis	poliovirus	humans	260, 713–715
hantavirus pulmonary syndrome	bunyavirus	rodents	263, 622	rabies	rabies	all warm-blooded animals	260, 263, 416 704–706
hemorrhagic fever, Bolivian	arenavirus	rodents and humans	691	respiratory infections	adenovirus polyomavirus	humans none	622 707–708
hemorrhagic fever, Korean	bunyavirus (Hantaan)	rodents	263, 691	Rift Valley fever	bunyavirus (phlebovirus)	humans, sheep, cattle	691
hemorrhagic fever	Ebola virus (filovirus)	humans (?)	260, 263, 690	rubella (German measles)	rubella	humans	260, 541–543
	Marburg virus (filovirus)	humans (?)	260, 263, 690	shingles	varicella-zoster	humans	261, 264, 544
hepatitis A (infectious hepatitis)	hepatitis A	humans	260, 651–652	smallpox	variola (major and minor)	humans	261, 265, 546–547
hepatitis B (serum hepatitis)	hepatitis B	humans	261, 651–653	viral enteritis	rotavirus	humans	649–651
hepatitis C (non-A, non-B)	hepatitis C	humans	651–653	warts, common (papillomas)	human papillomavirus	humans	261, 547–549
hepatitis D (delta hepatitis)	hepatitis D	humans	651–653	warts, genital (condylomas)	human papillomavirus	humans	261, 547–548
hepatitis E (enterically transmitted non-A, non-B, non-C)	hepatitis E	humans	651–653	yellow fever	yellow fever	monkeys, humans, mosquitoes	260, 263, 266 314, 688

The tables of fungal and parasitic diseases appear on the following page.

DISEASES AND THE ORGANISMS THAT CAUSE THEM *(CONCLUDED)*

FUNGAL DISEASES

Disease	Organism	Page	Disease	Organism	Page
aspergillosis	Aspergillus sp.	551–2, 625	histoplasmosis	Histoplasma capsulatum	624
blastomycosis	Blastomyces dermatitidis	550–1	Pneumocystis pneumonia	Pneumocystis carinii	625
candidiasis	Candida albicans	551	ringworm (tinea)	various species of Epidermophyton, Trichophyton, Microsporum	549–50
coccidioidomycosis (San Joaquin valley fever)	Coccidioides immitis	623–4			
cryptococcosis	Filobasidiella neoformans	624	sporotrichosis	Sporothrix schenckii	550
ergot poisoning	Claviceps purpurea	757	zygomycosis	Rhizopus sp., Mucor sp.	552

PARASITIC DISEASES

Disease	Organism	Type	Page	Disease	Organism	Type	Page
Acanthamoeba keratitis	Acanthamoeba culbertsoni	protozoan	413	malaria	Plasmodium sp.	protozoan	71, 291, 294–5, 418–20, 693–96,
African sleeping sickness (trypanosomiasis)	Trypanosoma brucei gambiense and T. brucei rhodesiense	protozoan	294, 313, 314 369, 718–9	pediculosis (lice infestation)	Pediculus humanus	louse	560–1
amoebic dysentery	Entamoeba histolytica	protozoan	291, 294, 655	pinworm	Enterobius vermicularis	roundworm	369, 376, 662
ascariasis	Ascaris lumbricoides	roundworm	660–1	river blindness (onchocerciasis)	Onchocerca volvulus	roundworm	369, 556
babesiosis	Babesia microti	protozoan	696–8				
balantidiasis	Balantidium coli	protozoan	655–6	scabies (sarcoptic mange)	Sarcoptes scabiei	mite	560
Chagas' disease	Trypanosoma cruzi	protozoan	314, 383 369, 720–1	schistosomiasis	Schistosoma sp.	flatworm	306, 552, 670–2
chigger dermatitis	Trombicula sp.	mite	560	sheep liver fluke (fascioliasis)	Fasciola hepatica	flatworm	657
chigger infestation	Tunga penetrans	sandflea	560				
Chinese liver fluke	Clonorchis sinensis	flatworm	658	strongyloidiasis	Strongyloides stercoralis	roundworm	662
crab louse	Phthirus pubis	louse	561				
cryptosporidiosis	Cryptosporidium sp.	protozoan	656	swimmer's itch	Schistosoma sp.	flatworm	552
dracunculiasis (Guinea worm)	Dracunculus medinensis	roundworm	552	tapeworm infestation (taeniasis)	Hymenolepsis nana (dwarf tapeworm)	flatworm	658–9
elephantiasis (filariasis)	Wuchereria bancrofti	roundworm	308, 309 672, 673		Taenia saginata (beef tapeworm)	flatworm	658–9
fasciolopsiasis	Fasciolopsis buski	flatworm	658		Taenia solium (pork tapeworm)	flatworm	658–9
giardiasis	Giardia intestinalis	protozoan	654–5				
heartworm disease	Dirofilaria immitis	roundworm	290, 305, 369		Diphyllobothrium latum (fish tapeworm)	flatworm	658–9
hookworm	Ancylostoma duodenale (Old World hookworm)	roundworm	305, 308, 660		Echinococcus granulosus (dog tapeworm)	flatworm	658–9
	Necator americanus (New World hookworm)	roundworm	308, 660	toxoplasmosis	Toxoplasma gondii	protozoan	291, 295, 696
leishmaniasis kala azar oriental sore	Leishmania braziliensis L. donovani L. tropica	protozoan	291, 313, 314, 692–3	trichinosis	Trichinella spiralis	roundworm	308, 334, 659–60
				trichomoniasis	Trichomonas vaginalis	protozoan	573–4
				trichuriasis (whipworm)	Trichuris trichiura	roundworm	661–2
liver/lung fluke (paragonimiasis)	Paragonimus westermani	flatworm	306, 657, 658	visceral larva migrans	Toxocara sp.	roundworm	661
loaiasis	Loa loa	roundworm	313, 556–7				

Microbiology

Jacquelyn G. Black

MARYMOUNT UNIVERSITY, ARLINGTON, VIRGINIA

Jacquelyn Black received her B.A., B.S., and M.S. from the University of Chicago and her Ph.D. from Catholic University of America. She has been teaching microbiology to undergraduates since 1970. She is a member of the American Society for Microbiology, and she has received grants for conducting teacher-training programs.

In addition to her extensive teaching experience, Dr. Black has engaged in fieldwork and studies throughout the globe. Her travels have taken her from the interior of Iceland to Belgium and Portugal to the barrier reef of Belize.

Dr. Black describes herself as an "incorrigible snoop" who is interested in all the various aspects and applications of microbiology. This natural curiosity, coupled with her classroom and laboratory experience, make her uniquely qualified to author an introductory microbiology textbook. This book conveys her sense of excitement for microbiology and offers the most current information on developments and applications within this field.

Dedication
To Laura—for sharing her mother and much of her childhood with "the book."

Microbiology

Principles and Explorations

Jacquelyn G. Black

Marymount University

Prentice Hall
Upper Saddle River, New Jersey 07458

Library of Cataloging -in-Publication Data

Black, Jacquelyn G.
    Microbiolgy : Principles and explorations / Jacquelyn G. Black — 4th ed
    Includes index.
    ISBN 0-13-920711-2
    1. Medical microbiology. 2. Microbiology I. Title
    QR46.C877 1999
    616'.01—dc21

                                                                    98-46806
                                                                    CIP

Senior Editor: Linda Schreiber
Executive Editor: David Kendric Brake
Editor in Chief: Paul F. Corey
Editorial Director: Tim Bozik
Development Editor: Betsy Dilernia
Assistant Vice President of Production & Manufacturing: David W. Riccardi
Executive Managing Editor: Kathleen Schiaparelli
Assistant Managing Editor: Lisa Kinne
Production Editor: Debra A. Wechsler
Marketing Manager: Jennifer Welchans
Creative Director: Paula Maylahn
Associate Creative Director: Amy Rosen
Art Director: Heather Scott
Assistant to Art Director: John Christiana
Art Manager: Gus Vibal
Art Editor: Karen Branson
Interior Design: Douglas & Gayle; Anne Flanagan
Illustrators: William C. Ober, M.D. & Claire W. Garrison, R.N., Medical and Scientific Illustration; Imagineering
Cover Designer: Donna Aiello
Cover Photo Research/Interior Photo Research: Mary Teresa Giancoli
Cover Photo Credits: *Earth*—Photodisc; *bacteria on a pin*—Dr. Tony Brain and David Parker/Science Photo
    Library/Photo Researchers, Inc.; *witch doctor*—Christine Osborne/Corbis Corporation Media; *catfish*—
    Agricultural Research Service, United States Department of Agriculture; *panning for gold*—Corbis-Bettmann
Photo Editors: Lorinda Morris-Nantz; Melinda Reo
Manufacturing Buyer: Trudy Pisciotti
Editor in Chief, Development: Ray Mullaney
Associate Editor in Chief, Development: Carol Trueheart
Copy Editor: Jane Loftus
Editorial Assistant: Damian Hill

Photo credits begin on page PC-1 and constitute a continuation of the copyright page.

 1999, 1996, 1993, 1990 by Prentice-Hall, Inc.
Simon & Schuster/A Viacom Company
Upper Saddle River, New Jersey 07458

Printed in the United States of America
10 9 8 7 6 5 4 3 2 1

ISBN 0-13-920711-2

Prentice-Hall International (UK) Limited, *London*
Prentice-Hall of Australia Pty. Limited, *Sydney*
Prentice-Hall Canada Inc., *Toronto*
Prentice-Hall Hispanoamericana, S.A., *Mexico*
Prentice-Hall of India Private Limited, *New Delhi*
Prentice-Hall of Japan, Inc., *Tokyo*
Simon & Schuster Asia Pte, Ltd., *Singapore*
Editora Prentice-Hall do Brasil, Ltda., *Rio de Janeiro*

OWNER'S MANUAL

Dear Student,

You've probably never bought a textbook that came with an owner's manual. The idea of an owner's manual for a text might sound like a strange concept, but if you are going to be successful in your study of microbiology you'll need to make the most of your lectures, your labs, and your text. You wouldn't think it odd to ask your instructor how to make the most of lectures, so why shouldn't an author provide you with instructions on how to use her book?

On the following pages I've tried to explain how the features of this book can help you learn microbiology more efficiently. Some students race through their texts, highlighting things that they feel might be important, but they never really assimilate the information into their long-term memory. Learning microbiology requires you first to learn a new vocabulary. The second phase of learning involves understanding basic concepts and relating those concepts to other concepts. The third phase of learning involves applying concepts to the world around you. After all, what good is a knowledge of microbiology if you cannot use it to understand the world around you? Whether your future career requires a microbiology course or not, this course will help you understand that microbiology is everywhere. And because it is everywhere, it can be explored. Thus, don't think of microbiology as just a course, think of it as an adventure!

Knowing how to use this book will help you enjoy the adventure. Please read the description of each feature carefully, and ask your instructor should you have questions about its use. Using the features of this book properly may take a little more time than merely racing through each chapter, but I promise that you'll remember more of what you read if you use these features.

And now for the most important advice that I can give you: enjoy your study of microbiology. I hope that I have been able to transmit in the pages of this book a sense of the excitement that is microbiology. The longer I am a microbiologist, the more excited I become. It never grows old or boring—it is constantly new and vital. It really is an adventure, and I'm delighted you've decided to join me!

Warmest Regards,

Jackie Black

Jackie Black

Questions We'll Explore

Conveniently listed at the beginning of each chapter, these questions foreshadow the territory ahead. As you begin the chapter don't try to answer these questions; instead, use them to preview the concepts that will be introduced. When you've finished reading the chapter, however, you *should* attempt to answer these questions. You may even want to jot down your answers and keep them with your lecture notes. Or, say your answers out loud to a friend, roommate or relative. The goal is to fix these questions and their answers into your long-term memory.

Background Basics

Making connections between chapters is critical to your success in this course. I've included a list of important concepts at the beginning of each chapter, concepts that were discussed in earlier chapters and will be relevant in the current chapter. Make an attempt at saying the concept out loud. Try to remember everything you can about the concept. Check your lecture notes. You can also go back to that section of the book and review the concept. (I've included chapter references to help you find it quickly.)

Characteristics of Prokaryotic and Eukaryotic Cells

Overwhelmed by all these doors? Which one do you open? Where will it take you? Will you be able to pass through it? Maybe they're all locked. What then?

Perhaps these are questions you'd ask if you were a relatively large molecule trying to get into or out of a cell. Just as there is a wonderful diversity of doors, there is also a diversity of protein transport molecules found in cell membranes. These protein transport molecules control what enters and leaves the cell. They can let some molecules in while keeping others out. They might require that one molecule go out at the same time as a different molecule is coming in. In their various forms, protein transport molecules help control traffic across the cell membrane.

Background Basics

Gram stain .... (Chapter 3)
acid-fast stain .... (Chapter 3)
polar bonding .... (Chapter 2)
covalent bonding .... (Chapter 2)
steroid .... (Chapter 2)
protein .... (Chapter 2)
ion .... (Chapter 2)
DNA .... (Chapter 2)

We have considered the chemical principles that apply to cells and how to use microscopes and stains to observe cells. We can now look at the structure and function of the cells themselves.

Basic Cell Types

All living cells can be classified as prokaryotic, from the Greek words *pro* (before) and *karyon* (nucleus) or eukaryotic, from *eu* (true) and *karyon* (nucleus). **Prokaryotic** (pro-kar″e-ot′ik) **cells** lack a nucleus and other membrane-enclosed structures, whereas **eukaryotic** (u-kar″e-ot′ik) **cells** have such structures.

All prokaryotes are single-celled organisms, as exemplified by the bacteria. Most of this book will be devoted to the study of prokaryotes. Eukaryotes include all plants, animals, fungi, and protists (organisms such as *Amoeba, Paramecium,* and the malaria parasite). We will also spend some time studying eukaryotes, especially the fungi and various parasites, plus the interactions of eukaryotic cells and prokaryotes.

Prokaryotic and eukaryotic cells are *similar* in several ways. Both are surrounded by a *cell membrane,* or *plasma membrane.* Although some cells have structures that extend beyond this membrane or surround it, the membrane defines the boundaries of the living cell. Both prokaryotic and eukaryotic cells also encode genetic information in DNA molecules.

These two types of cells are *different* in other, important ways. In eukaryotic cells, DNA is in a nucleus surrounded by a *nuclear envelope,* but in prokaryotic cells, DNA is in a nuclear region not surrounded by a membrane. Eukaryotic cells also have a variety of internal structures called **organelles** (or-ga-nelz′), or "little organs," that are

Questions We'll Explore

A What are the characteristics of prokaryotic and eukaryotic cells?

B How do prokaryotic cells differ in size, shape, and arrangement?

C How are structure and function related in bacterial cell walls and cell membranes?

D How are structure and function related in other bacterial components?

E How are structure and function related in eukaryotic plasma membranes?

F How are structure and function related in other eukaryotic components?

G How do passive transport processes function, and why are they important? How does active transport function, and why is it important?

H How do exocytosis and endocytosis occur, and why are they important?

71

ble for "fifth disease" (erythema infectiosum), which causes a deep red rash on children's cheeks and ears and both a rash and arthritis in adults. B19 can cross the placenta and damage blood-forming cells in the fetus, leading to anemia, heart failure, and even fetal death.

Viral Replication

General Characteristics of Replication

In general, viruses go through the following five steps in their **replication cycles** to produce more virions:

1. **Adsorption,** the attachment of viruses to host cells.
2. **Penetration,** the entry of virions (or their genome) into host cells.
3. **Synthesis,** the synthesis of new nucleic acid molecules, capsid proteins, and other viral components within host cells while using the metabolic machinery of those cells.
4. **Maturation,** the assembly of newly synthesized viral components into complete virions.
5. **Release,** the departure of new virions from host cells. Release generally, but not always, kills (lyses) host cells.

Replication of Bacteriophages

Bacteriophages, or simply *phages,* are viruses that infect bacterial cells. Phages were first observed in 1915 by Frederic Twort in England and in 1917 by Felix d'Herelle in France. D'Herelle named them bacteriophages, which means "eaters of bacteria." Over the years, attempts have been made to use phages to fight bacterial infections. These attempts, though not successful until recently (see the box "Viruses Instead of Drugs," page 000), did provide useful information about phages. Phages have been studied in great detail because it is much easier to manipulate bacterial cells and their viruses in the laboratory than to work with viruses that have multicellular hosts.

Properties of Bacteriophages

Like other viruses, bacteriophages can have their genetic information in the form of either double-stranded or single-stranded RNA or DNA. They can be relatively simple or complex in structure. To understand phage replication, we will examine the *T-even phages.* These phages, designated T2, T4, and T6 (T stands for "type"), are complex but well-studied naked phages that have dsDNA as their genetic material. The most widely studied is the T4 phage, an obligate parasite of the common enteric bacterium *Escherichia coli.* T4 has a distinctly shaped capsid made of a head, collar, and tail (➤ Figure 10.6; Table 10.4). The DNA is packaged in the polyhedral head, which is attached to a helical tail.

Replication of T-Even Phages

Infection with and replication of new T4 phages occurs in the series of steps illustrated in ➤ Figure 10.7.

Adsorption If T4 phages collide in the correct orientation with host cells, the phages will attach to, or adsorb onto, the host cell surface. Adsorption is a chemical attraction; it requires specific protein recognition factors found in the phage tail fibers that bind to specific receptor sites on the host cells. The fibers bend and allow the pins to touch the cell surface. Although many phages, including T4, attach to the cell wall, other phages can adsorb to flagella or to pili.

Penetration The enzyme *lysozyme,* which is present within phage tails, weakens the bacterial cell wall. When the tail sheath contracts, the hollow tube (core) in the tail is forced to penetrate the weakened cell wall and come into contact with the bacterial cell membrane. The viral DNA then moves from the head through the tube into the bacterial cell. It is not clear whether the DNA is introduced directly into the cytoplasm; according to recent evidence, T4 phages introduce their DNA into the periplasmic space, between the cell membrane and the cell wall. Either way, the phage capsid remains outside the bacterium.

✓ On what basis are viruses classified today?

✓ Distinguish between positive and negative sense RNA.

✓ What is the difference between dsDNA and ssDNA?

Seawater can hold 100 million bacteriophages/ml.

Viral Sunscreens

We're not the only ones who should stay out of the sun. UV light also damages viruses like bacteriophages. So how do these viruses survive in the ocean where they are exposed to direct sunlight? Ironically, the very bacteria they are trying to destroy end up protecting them. UV light stimulates bacteria to produce photolyase, an enzyme that repairs damaged bacterial DNA. This process is called photoreactivation. Healthy bacteria thus protect the virus' nucleic acid, allowing the virus to replicate and eventually lyse the bacterial cell that protected it.

Concept Checks

Do you hate to be interrupted while you're reading? Well, I'd like to interrupt you from time to time in every chapter. Whenever you see a set of red check marks, stop and check your understanding of the material you've just read. Some of the questions ask you to recall basic facts or concepts while others ask you to apply what you've just learned. Hopefully, you'll learn to like a few interruptions!

Factoid Notes

From time to time I'll give you facts, tidbits, and examples that relate to the discussion in the running text. These fun, little "sticky notes" contain additional thoughts or information that are not significant concepts by themselves but that add some flavor and dimension to the text. Nobody ever said that a textbook couldn't be fun!

Figure Reference Locators

Wherever figures are called out in the text, you'll find a red arrowhead which functions as a place-marker, making it easier for you to toggle between an illustration and the running narrative associated with that illustration.

Application Boxes

You don't really begin to absorb and understand microbiology until you've seen it in action. I've tried to help you explore microbiology by including a number of Application Boxes that introduce you to an interesting aspect of the field. I've been told students don't read boxes; if that's true, they're missing a real opportunity to cement critical concepts into their minds. Read these boxes and ask yourself how the information applies to the concepts and facts you're learning in the chapter. Hopefully you'll find many of the boxes very interesting; you might find other boxes less so, but try to connect the content of the boxes to their related concepts.

Concept Links

Because connecting concepts is so important to your success in this course, I've included numerous Concept Links in each chapter. These blue arrows with chapter and page references link you to relevant concepts covered earlier in the text. Think of them as an extension of the Background Basics; they'll link you to material that might be helpful for understanding new concepts.

causes cholera, are exotoxins that act on tissues of the gut. Many exotoxins can act as *antigens,* foreign substances against which the immune system reacts. Antigenic exotoxins inactivated by treatment with chemical substances such as formaldehyde are called *toxoids*. A **toxoid** (*-oid* is Latin for "like") is an altered toxin that has lost its ability to cause harm but that retains antigenicity. Toxoids can be used to stimulate the development of immunity without causing disease. For example, when you get a tetanus booster shot, you are receiving the tetanus toxoid. It stimulates your body to produce immunity so that if you are exposed to active tetanus toxin through a cut or puncture of the skin, you will not get tetanus. The effects of bacterial exotoxins in human disease are summarized in Table 14.5. The specific effects of such diseases are discussed in later chapters.

How Viruses Cause Disease

Viruses can replicate only after they have attached to cells and then penetrated specific host cells. In tissue culture systems, once inside a cell, viruses cause observable changes collectively called the **cytopathic effect (CPE).** ➡ (Chapter 10, p. 279) CPE can be cytocidal when the viruses kill the cell and noncytocidal when they do not. Cytocidal viruses can kill cells by causing enzymes from cellular lysosomes to be released or by diverting the host cell's synthetic processes, thereby stopping the synthesis of host proteins and other essential macromolecules. CPE

➤ Table 14.5 Effects of exotoxins

Bacterium	Name of toxin or disease	Action of toxin	Host symptoms
Bacillus anthracis	Anthrax (cytotoxin)	Increases vascular permeability	Hemorrhage and pulmonary edema
Bacillus cereus	Enterotoxin	Causes excessive loss of water and electrolytes	Diarrhea
Clostridium botulinum	Botulism (eight serological types; neurotoxins)	Blocks release of acetylcholine at nerve endings	Respiratory paralysis, double vision
Clostridium perfringens	Gas gangrene (alpha toxin, a hemolysin)	Breaks down lecithin in cell membranes	Cell and tissue destruction
	Food poisoning (enterotoxin)	Causes excessive loss of water and electrolytes	Diarrhea
Clostridium tetani	Tetanus (lockjaw) (neurotoxin)	Inhibits antagonists of motor neurons of brain; 1 ng can kill 2 tons of cells	Violent skeletal muscle spasms, respiratory failure
Corynebacterium	Diphtheria; produced by virus-	Inhibits protein synthesis	Heart damage can cause death

---

Chapter Summary: Retracing Our Steps

Some years ago I drove a little two-lane road across the state of Nebraska, stopping at all the historical markers and many of the small towns that dot that long gray highway. In an effort to remember details and facts about the Cornhusker State, I made notes and observations on a big multifold map of the state. To this day, I can look at any map of the state of Nebraska and recall places, people, and historical markers that I encountered on my journey. What does that have to do with a Chapter Summary? A lot actually. The Chapter Summary is like a map detailing the concepts and facts that we visited in the chapter. Review it carefully and ask yourself questions about the information being summarized. Try to recall what you read, and mentally connect the points of the summary outline to things your instructor said in lecture. Use the outline in this way and you'll find yourself remembering a lot more than you thought possible.

Retracing Our Steps

Epidemiology

What Is Epidemiology?

- **Epidemiology** is the study of factors and mechanisms in the spread of diseases in a population.

- In describing infectious diseases, **epidemiologists** use **incidence** to refer to the number of new cases in a specific period, **prevalence** to refer to the number of people infected at any one time, **morbidity rate** to indicate the number of cases as a proportion of the population, and **mortality rate** to indicate the number of deaths as a proportion of the population.

Diseases in Populations

- In a **sporadic disease**, several isolated cases appear in a population. In an **endemic** disease, many cases appear in a population, but the harm to patients is not sufficient to create a public health problem. In an **epidemic** disease, many cases appear in a population, and patients are sufficiently harmed to create a public health problem. Diseases that spread by **common-source outbreaks** originate with a single contaminated substance, such as a water supply. In **propagated epidemics**, diseases spread by person-to-person contact. A **pandemic** disease is an epidemic disease that has spread over an exceptionally wide geographic area or several geographic areas.

Epidemiologic Studies

- The purpose of **epidemiologic studies** is to learn more about the spread of diseases in populations and how to control those diseases.

- The methods of epidemiologic studies are **descriptive, analytical** (retrospective and prospective), and **experimental.**

Reservoirs of Infection

- **Reservoirs of infection** include humans, other animals, and nonliving sources from which infectious diseases can be transmitted.

- Among human reservoirs, **carriers** often transmit diseases. They are intermittent carriers if they periodically release pathogens.

- Diseases in animal reservoirs can be transmitted by direct con-

water, air, and food. **Vectors** of transmission are usually arthropods, which can transmit disease mechanically or biologically.

- Transmission by carriers, transmission of STDs, and transmission of zoonoses pose special epidemiologic problems.

Disease Cycles

- Some diseases occur in cycles—there are just a few cases for several years, and then many cases suddenly appear.

Herd Immunity

- **Herd immunity**, or group immunity, refers to immunity enjoyed by a large proportion of a population that reduces disease transmission among nonimmune individuals.

- A drop in herd immunity can lead to the sudden appearance of cases of a cyclic disease.

Controlling Disease Transmission

- Methods used to control communicable diseases include isolation, quarantine, immunization, and vector control.

- **Isolation** procedures are summarized in Table 15.2. **Quarantine** is rarely used but can prevent exposed individuals from infecting others. Active immunization prevents many infections. Vector control is effective where vectors can be identified and eradicated.

Public Health Organizations

- Public health organizations exist at city, county, state, federal, and world levels. They help establish and maintain health standards, cooperate in the control of infectious diseases, collect and disseminate information, and assist with professional and public education.

Notifiable Diseases

- **Notifiable diseases** are listed in Table 15.3.

Nosocomial Infections

- A **nosocomial infection** is an infection acquired in a hospital or other medical facility.

Terminology Check

aerosol (p. 413)
analytical study (p. 408)
carrier (p. 409)
common-source outbreak (p. 405)
compromised host (p. 426)
contact transmission (p. 413)
descriptive study (p. 407)
direct contact transmission (p. 413)
direct fecal–oral transmission (p. 413)
droplet nucleus (p. 413)
droplet transmission (p. 413)
endemic (p. 403)
endogenous infection (p. 425)
epidemic (p. 403)
epidemiologic study (p. 405)
epidemiologist (p. 402)

epidemiology (p. 402)
etiology (p. 402)
exogenous infection (p. 425)
experimental study (p. 408)
fomite (p. 413)
herd immunity (p. 416)
horizontal transmission (p. 413)
inapparent infection (p. 409)
incidence (p. 403)
index case (p. 408)
indirect contact transmission (p. 413)
indirect fecal–oral transmission (p. 413)
isolation (p. 417)
morbidity rate (p. 403)
mortality rate (p. 403)
nosocomial infection (p. 423)

notifiable disease (p. 422)
pandemic (p. 405)
placebo (p. 408)
portal of entry (p. 410)
portal of exit (p. 412)
prevalence (p. 403)
propagated epidemic (p. 405)
quarantine (p. 418)
reservoir of infection (p. 408)
sporadic disease (p. 405)
subclinical infection (p. 409)
Universal Precautions (p. 427)
vector (p. 415)
vehicle (p. 413)
vertical transmission (p. 413)
zoonosis (p. 409)

CHAPTER 15 FIFTEEN

See You on the Web

Don't forget I am expecting you on the web. The following is a preparedness checklist for your web explorations. If you have questions on any of these items, you can return to the end of Chapter 1 for more details.

- Questions We'll Explore
- Concept Check questions
- Flashcards
- Study Guide

If you are ready, come visit me at

http://www.prenhall.com/black

It will be a journey you'll never forget.

Terminology Check

Speaking the language of microbiology is half the battle. Take ownership of the language by knowing each term, using it properly, and understanding how it relates to a concept in the chapter. Use the Terminology Check to review the key vocabulary from the chapter. Say the word out loud along with your best shot at its definition. Try to recall how the term fits into the fabric of the whole chapter; in other words, think about it in the proper context. If you can't recall much about the word, or you just want a little extra context, use the handy page reference and go to where the word is first introduced in that chapter. If you are looking for the definition of a particular term, use the Glossary/Index to find both a definition and a page reference.

See You on the Web

Thanks to the growing popularity of the World Wide Web you can travel throughout the world without ever leaving your computer. And it is a great way to enhance your study of microbiology. Visit our Web site and you'll find self-grading quizzes for every chapter. You'll also find links to relevant destinations on the Web, and access to search engines that will help you explore the world of microbiology using key terms from each chapter. Once you've read the chapter, you'll definitely want to see me on the Web.

BRIEF CONTENTS

CONTENTS

LIST OF BOXES

For some years now I've been telling people that microbiology is not just a course, but it's an adventure! The fourth edition of this text has taken that slogan to heart. The first place you'll notice it is in the subtitle: "Principles and Explorations." Microbiology is so much more than mere facts, terms, and concepts. Microbiology needs to be explored, and the world at large is a marvelous place to start. We've also tried to invoke this slogan by way of a design change in the book. Whereas the book's previous editions made use of a two-column format, we've opened the design up by using a major column/minor column format that allows the placement of application boxes and brief "factoid" notes in the margin. These elements in the margin allow me to maintain a running dialogue with students; so, instead of giving them just the facts and concepts, I am able to add some applications, thoughts and trivia, as though I were their tour guide for this adventure. My goal certainly is not to replace their instructor—the real tour guide on this journey—but to bring some relevance and excitement to a book. We've also taken a big leap forward by developing a web site that allows you to travel beyond the book. All of this has been done to help students understand the roles microbes play in this world.

To understand the roles microbes play in our lives, including the interplay between microorganisms and humans, we must examine, learn about, and study their world—the world of microbiology. Microorganisms are everywhere. They exist in a range of environments from mountains and volcanoes to deep-sea vents and hot springs. Microorganisms can be found in the air we breathe, in the food we eat, and even within our own body. In fact, we come in contact with countless numbers of microorganisms every day. Although some microbes can cause disease, most are not disease-producers; rather, they play a critical role in the processes that provide energy and make life possible. Some even prevent disease, and others are used in attempts to cure disease.

Because microorganisms play diverse roles in the world, microbiology continues to be an exciting and critical discipline of study. And because microbes affect our everyday lives, microbiology provides many challenges and offers many rewards. Look at your local newspaper, and you will find items concerning microbiology: to mention a few, reports on diseases such as AIDS, tuberculosis, and cancer; the resurgence of malaria or "new" diseases such as those caused by the Ebola virus and the hantavirus; a bacterium that can cause ulcers; technologies designed to increase food production; genetically engineering tomatoes that resist rotting; microorganisms used to clean up toxic wastes and oil spills; and the Human Genome Project, which will identify the complete set of genetic instructions within the cells of the human body. Yes, microbiology represents a broad field of endless challenges and exciting discoveries.

A theme that permeates this book, then, is that microbiology is a current, relevant, exciting central science that affects all of us. In countless areas—from agriculture to evolution, from ecology to dentistry—microbiology is contributing to scientific knowledge as well as solving human problems. As you read this text, you will get a sense of the history of this science, its methodology, its many contributions to humanity, and the many ways in which it continues to be on the cutting edge of scientific advancement.

This book meets the needs both of students of the health sciences and of students majoring in biology. The book is designed to serve both audiences—in part by using an abundance of clinically important information to illustrate the general principles of microbiology and in part by offering a wide variety of additional applications.

Style and Currency

In countless areas, microbiological studies are contributing both to the knowledge base of biological science and to solving human problems. However, the rapid advances being made in microbiology make teaching about—and learning about—microorganisms challenging. Therefore, every effort has been made in the fourth edition of *Microbiology: Principles and Explorations* to ensure that the writing is simple, straightforward, and functional; that microbiological concepts and methodologies are clearly and thoroughly described; and that the information presented is as accessible as possible to students.

Furthermore, students who enjoy a course are more likely to retain far more of its content for a longer period of time than do those who take the course like a dose of medicine. Students should experience microbiology as an exciting, dynamic, rapidly changing field that is important to human welfare. The development of microbiology—from Leeuwenhoek's astonished observations of "animalcules," to Pasteur's first use of rabies vaccine on a human, to Fleming's discovery of penicillin, to today's race to develop an AIDS vaccine—has been one of the most dramatic stories in the history of science. Whether the topic is the eradication of smallpox, the identification of microorganisms responsible for Lyme disease, or viruses responsible for some forms of cancer, there is no reason for a text to be any less interesting than its subject. Because students find courses interesting when they can relate topics to their everyday life or to career goals, I have emphasized the connection between microbiological knowledge and student experiences.

In a field that changes so quickly—with new research, new drugs, and even new diseases—it is essential that a text be as up-to-date as possible. This book incorporates the latest information on all aspects of microbiology, including clinical practice. Special attention has been paid to such important, rapidly evolving topics as genetic engineering; viruses; drug resistance, especially in tuberculosis; and immunology.

Organization of the Fourth Edition

The organization of the fourth edition combines logic with flexibility. The chapter sequence will be useful in most microbiology courses as they are usually taught. Nevertheless, it is not essential that the chapters be assigned in their present order; it should be possible to use this book in courses organized along quite different lines. **Concept link symbols (➡)** are used throughout the chapters to key the student to material that was previously discussed in the text.

The Art Program

The use of clear, attractive drawings and carefully chosen photographs can significantly contribute to the student's understanding of a scientific subject. In response to the comments of reviewers and users who requested additions and changes to some of the art in the previous edition, new figures have been added to the fourth edition. Throughout, color has been used not just decoratively but for its pedagogic value. For example, every effort has been made to color similar molecules and structures the same way each time they appear, making them easier to recognize. The availability of full color also makes it possible to present many subjects, such as staining reactions, as they are seen in the laboratory.

We are very pleased to once again have had the services of William C. Ober, M.D., and Claire W. Garrison, R.N., who have expertly executed many of the new and redrawn illustrations for this edition of *Microbiology: Principles and Explorations*. Bill and Claire combine technical skill with scientific understanding to a degree rarely found in biological illustration. They have received numerous awards, including the Award of Excellence from the Association of Medical Illustrators, the Certificate of Excellence from the American Institute of Graphic Arts, the art and design award of the Chicago Book Clinic, and the Award of Excellence from the Printing Industries of America. Bill and Claire have also received recognition from Bookbuilders West and are recipients of the Art Directors Award.

Special Pedagogic Features

This book is designed not simply as a vehicle for transmitting information but also as a tool for learning. The features that have made *Microbiology: Principles and Explorations* so popular have been retained in the fourth edition. But I've also added several features that should help students assimilate facts and concepts of microbiology. These features are highlighted in the Owner's Manual section.

Instructor's Edition

A special **Instructor's Edition** of this text is available to enhance the usefulness of the text as a teaching tool. For each chapter, the Instructor's Edition provides a wide range of supplemental material designed to assist the instructor in teaching from the text and to allow for deeper and/or more extensive treatment of various topics. Included in the material for each chapter are the following: (1) A **Chapter Overview** explains the organization of the chapter and summarizes its main themes. Its purpose is to place the material of the chapter into a broader context—to focus the attention on the forest rather than on the trees. (2) **Chapter Objectives** are related to the focus questions at the start of each text chapter but are more numerous and detailed. Couched in behavioral terms, they direct the instructor's attention to what students should *be able to do* as well as *know* after successfully completing each chapter. This feature is convenient for instructors who must prepare a course syllabus with chapter-by-chapter goals. (3) A complete **Chapter Outline** includes all levels of headings. (4) **Keys to Exploration** which include: **Teaching Tips, Discussion Topics,** and a **World Wide Web** supplement suitable for student papers or oral reports.

Ancillaries

To supplement the needs of both students and instructors, the following materials are available to qualified adopters. The ancillary package has been carefully crafted by talented microbiology instructors, many of whom have taught from previous editions of this textbook. Their efforts to meet the needs of the instructor and the student are reflected in materials whose coordination with the text will be obvious to the adopters who use them.

For the Instructor

Instructor's Edition Supplemented with an Instructor's Manual that is bound in as an integral part of the book, this textbook is transformed into an indispensable component for the instructor who is serious about his or her students' success. To help ensure this success, a number of pedagogical suggestions have been inserted into each chapter's outline. For easy reference, these suggestions have been linked to one of four icons, each of which represents one of the following categories: Teaching Tips, Discussion Topics, Track It Downs, and Web Destinations. (ISBN 0-13-920729-5)

Test Item File Offering twenty percent more questions than it did in its previous edition, this edition of the Test Item File is filled with a variety of question styles, including multiple choice, matching, true/false, short answer, and essay. To facilitate the construction of tests that take into account your students' level, each question's difficulty is rated as easy, moderate, or challenging. Each question is also classified by its content and the kind of knowledge that it tests, thus providing even greater potential for test customization. (ISBN 0-13-920737-6)

Prentice Hall Custom Test Available for Windows and Macintosh, the Prentice Hall Custom Test offers complete question editing capabilities of Test Item File questions. It also has a 2-track design, one that includes the Full test option for those who wish to incorporate their own questions or create their own graphics. Once created, these exams can be administered online. Results can then be transferred electronically for evaluation. (Windows: ISBN 0-13-920745-7, Macintosh: ISBN 0-13-920752-X)

Transparency Acetates This set of 250 full-color transparencies includes key illustrations from the textbook. (ISBN 0-13-920760-0)

Prentice Hall Presentational CD-ROM for Microbiology Usable on either a Macintosh or from within Windows, this image bank provides an ideal opportunity to bring multimedia into the classroom. Featuring over 600 pieces of art, this CD-ROM's image resources also include dynamic animations that illustrate some of microbiology's most important topics. These animations, like the images by which they are accompanied, can be searched by title, caption, or key term. The same software that makes these searches possible also allows you to preview selected images to whch you can add custom overlays. (ISBN 0-13-920778-3)

Companion Web Site Plus (http://www.prenhall.com/black) This on-line study guide also provides an abundant supply of support resources for instructors, including an on-line syllabus creation and management utility that can be accessed quickly by students for their course assignments. Features not included on Companion Web Sites, but to which CW+ subscribers have access, include integrated messaging for private class communication, as well as access to administrative tools that can be used to monitor results from student activities at the site. (ISBN 0-13-920802-X)

For the Student

Study Guide Featuring all-new introductions and expanded study outlines, this edition of the Study Guide promotes clearly an understanding of the basic facts and concepts that contribute to the development of problem-solving skills. These skills are developed further in the Study Guide's Self-Tests, which include the following question styles: true/false, multiple choice, matching, and fill-in-the-blank. A new feature in this edition of the Study Guide that is also included among the Self-Tests is the labeling exercise. These Self-Tests culminate in their critical thinking questions. (ISBN 0-13-920794-5)

Companion Web Site Plus (http://www.prenhall.com/black) This on-line resource supports and augments the textbook through quizzes from which you can obtain instant results, hyperlinks to eclectic and exciting microbiological resources, and "updates" with embedded hyperlinks and thought-provoking questions that challenge you to take a position on some of microbiology's hottest topics.

A Note on Pronunciation

The scheme used for pronunciation is simple:

- a single accent mark (´) is used for the main accent in a word;
- a double accent mark (´´) is used for secondary accents, if any;
- any vowel not followed by a consonant is assumed to be long;
- syllables are separated by either a hyphen or an accent mark.

Acknowledgments

A textbook requires a large cast of characters to produce it effectively and on time. I would like to thank the many people at Prentice Hall who worked to make this edition the best yet. In particular, I would like to acknowledge the efforts of Debra Wechsler, Production Editor. Debra worked tirelessly to make this book a reality. There are many others at Prentice Hall who made this book their life for several months as we raced toward publication; you can find their names on the copyright page of this book. Words alone cannot express my appreciation for their professionalism and collective effort. Thank you all. The development and production of this book was an adventure.

I wish to acknowledge the many Prentice Hall reviewers for their thoughtful suggestions.

First Edition Reviewers:

Oswald G. Baca, *University of New Mexico*
David L. Balkwill, *Florida State University*
Keith Bancroft, *Southeastern Louisiana University*
Thomas R. Corner, *Michigan State University*
Monica A. Devanas, *Rutgers University*
Lawrence W. Hinck, *Arkansas State University*
Wallace L. Jones, *De Kalb College*
Donald G. Lehman, *Wright State University*
William C. Matthai, *Tarrant County Junior College*
Raymond B. Otero, *Eastern Kentucky University*
Kimberley Pearlstein, *Adelphi University*
Robert A. Pollack, *Nassau Community College*
Dennis J. Russell, *Seattle Pacific University*
Gordon D. Schrank, *St. Cloud State University*
Alan J. Sexstone, *West Virginia University*
Deborah Simon-Eaton, *Santa Fe Community College*
Robert E. Sjogren, *University of Vermont*
Larry Stearns, *Central Piedmont Community College*
Michael R. Yeaman, *University of New Mexico*

Second Edition Reviewers:

D. Andy Anderson, *Utah State University*
Dan C. DeBorde, *University of Montana*
Monica A. Devanas, *Rutgers University*
Von Dunn, *Tarrant County Junior College*
David L. Filmer, *Purdue University*
Eugene Flaumenhaft, *University of Akron*
Denise Y. Friedman, *Hudson Valley Community College*
William R. Gibbons, *South Dakota State University*
Ronald E. Hurlbert, *Washington State University*
Robert J. Janssen, *University of Arizona*
Thomas R. Jewell, *University of Wisconsin-Eau Claire*
Harvey Liftin, *Broward Community College*
William C. Matthai, *Tarrant County Junior College*
Russell A. Normand, *Northeast Louisiana University*
Joseph M. Sobek, *University of Southwestern Louisiana*
Bernice C. Stewart, *Prince George's Community College*
James E. Urban, *Kansas State University*
John Zak, *Texas Tech University*
Thomas E. Zettle, *Illinois Central College*

Third Edition Reviewers:

Ronald W. Alexander, *Tompkins Cortland Community College*
R. L. Bernstein, *San Francisco State University*
David L. Berryhill, *North Dakota State University*
Richard D. Bliss, *Yuba College*
Edward A. Botan, *New Hampshire Technical College*
Sally DeGroot, *St. Petersburg Junior College*
John G. Dziak, *Community College of Allegheny County*
Pamela B. Fouche, *Walters State Community College*
Christine L. Frazier, *Southeast Missouri State University*
Denise Y. Friedman, *Hudson Valley Community College*
William R. Gibbons, *South Dakota State University*
Pamela L. Hanratty, *Indiana University*
Janet Hearing, *State University of New York, Stony Brook*
John J. Iandolo, *Kansas State University*
John W. Kimball, *Harvard University*
Timothy A. Kral, *University of Arkansas*
Caleb Makukutu, *Kingwood College*
Alexsandria Manrov, *Tidewater Community College, Virginia Beach*
Judy D. Marsh, *Emporia State University*
Rosemarie Marshall, *California State University, Los Angeles*
John Martinko, *Southern Illinois University*
William C. Matthai, *Tarrant County Junior College*
Chris H. Miller, *Indiana University*
Rajeev Misra, *Arizona State University*
Christian C. Nwamba, *Wayne State University*
Curtis Pantle, *Community College of Southern Nevada*
Robin K. Pettit, *State University of New York, Potsdam*
Jeff Pommerville, *Glendale Community College*
Russell Robbins, *Drury College*
Richard A. Robison, *Brigham Young University*
Quentin Reuer, *University of Alaska, Anchorage*
Brian R. Shmaefsky, *Kingwood College*
J. Glenn Songer, *University of Arizona*
K. T. Shanmugam, *University of Florida*

Paul M. Steldt, *St. Philips College*
Gerald Stine, *University of Florida*
Teresa Thomas, *Southwestern College*
James E. Urban, *Kansas State University*
Phyllis K. Williams, *Sinclair Community College*
George A. Wistreich, *East Los Angeles College*

Fourth Edition Reviewers

Burke L. Brown, *University of South Alabama*
Michael Dennis, *Montana State University at Billings*
David Filmer, *Purdue University*
Mike Griffin, *Angelo State University*
Richard Hanke, *Rose State College*
Ali Hekmati, *Mott Community College*
Timothy Kral, *University of Arkansas*
Jeff Lodge, *Rochester Institute of Technology*
Rosemarie Marshall, *California State University—Los Angeles*
Anne Mason, *Yavapai College*
Robert McLean, *Southwest Texas State University*
Karen Messley, *Rock Valley College*
Timothy Nealon, *St. Philip's College*
Cathy Oliver, *Manatee Community College*
C.O. Patterson, *Texas A&M University*
Robert W. Phelps, *San Diego Mesa College*
Leodocia Pope, *University of Texas at Austin*
Jane Repko, *Lansing Community College*
Richard Robison, *Brigham Young University*
Jay Sperry, *University of Rhode Island*
Jack Turner, *University of Southern California—Spartanburg*
Shawn Wright, *Albuquerque Technical-Vocational Institute*

Comments and suggestions about the book are most welcome. Please feel free to contact me through my editor, Linda Schreiber, whose email address is linda_schreiber@prenhall.com.

Jacquelyn Black
Arlington, Virginia

Scope and History of Microbiology

When you think about microbiology, you might think about nasty little organisms that are so small they are invisible to the naked eye. You might think about petri dishes and microscopes and people in white lab coats. Perhaps you think about germs and you hear your mother's voice calling from your childhood, "Don't put that in your mouth. You don't know where it's been, and it has germs all over it!" When you think about microbiology you probably DON'T think about our planet Earth revolving in space. But you should.

Without microbes Earth as we know it would not exist, and you would not exist. Microbes are relevant to virtually all aspects of human life. You cannot eat a hamburger, sit in a hot tub, or even brush your teeth without encountering microbes. They are everywhere, on you and around you. They're even on this page. Furthermore, microbes are not all bad; many of them actually improve the quality of life on Earth.

So next time you look at a photo of our blue planet, celebrate the fact that it has "germs" (and other microbes) all over it. And by the way, you might want to mention this to your mother.

Why Study Microbiology?

Microbes (bacteria, viruses, fungi, protozoa, and some algae) live in us, on us, and nearly everywhere around us (➤ Figure 1.1). They have a major impact on our health and environment. They play an important role in many of the foods we eat and the medicines we take. They also provide useful models for many of the life processes that all organisms experience.

Microbes Have a Major Impact on Human Health

Though fewer than 1 percent of all microbes cause disease, identifying how those disease-causing microbes (called pathogens) cause and transmit disease, as well as discovering how those diseases can be treated, is of critical importance to our well-being as a species. In 1962, the U.S. Surgeon General, William H. Stewart, emphatically stated, "The war against infectious diseases has been won." In the subsequent decades, emerging diseases and increasing antibiotic resistance have made it clear that in reality not only is the war far from won, its outcome is no longer certain. It will be our health science professionals who will be on the frontlines of that war into the foreseeable future.

Microbes Help Maintain the Balance of Nature

Many aquatic microorganisms capture energy from sunlight and store it in molecules that other oganisms use as food. Microorganisms decompose dead organisms and waste material from living organisms, and they can decompose some kinds of industrial wastes. Through this decomposition, they make nitrogen available to plants. Certain microorganisms that reside in the digestive tracts of grazing animals play an important role in those animals' ability to digest grass.

Questions We'll Explore

A Why is the study of microbiology important?

B What is the scope of microbiology?

C What are some major events in the early history of microbiology?

D What is the germ theory of disease, and what historical developments led to its formulation?

E What events mark the emergence of immunology, virology, chemotherapy, genetics, and molecular biology as branches of microbiology?

Almost one-half of children under the age of 10 died of infectious disease prior to this century.

➤ Figure 1.1 **A simple experiment shows that microorganisms are almost everywhere in our environment.** Soil was added to nutrient agar, a culture medium (dish on top); another dish with agar was exposed to air (bottom left); and a tongue print was made on an agar surface (bottom right). After 3 days of incubation under favorable conditions, abundant microbial growth is easily visible in all three dishes.

These are just a few examples of the integral nature of microorganisms in our environment. The vast majority of microbes are directly or indirectly beneficial to other organisms, humans included. They form essential links in many food chains that produce the plans and animals that people eat.

Microorganisms Are Essential to Many Human Endeavors

The biochemical reactions carried out by microorganisms have been harnessed by humans for our own ends. The food industry uses these reactions in the preparations of pickles, sauerkraut, yogurt, fructose, and aspartame, the artificial sweetener. Fermentation reactions are exploited in the creation of beer and wine, as well as in the baking of bread. One of the most significant benefits that microorganisms provide is their ability to synthesize **antibiotics,** substances derived from one organism that kill or inhibit the growth of other microorganisms.

These are all contributions that microorganisms make in their natural state. The advent of genetic engineering has created a whole new suite of benefits. Several substances important to humans, such as interferon and growth hormones, can be economically produced by genetically altered microbes. New organisms have been engineered to degrade oil spills, to remove toxic materials from the soil, and to digest explosives. As we move into the future, it is safe to say that genetically engineered microorganisms will play an increasing role in many aspects of human affairs whether they be medical, environmental, or economic.

Microbiology Provides Insight Into Life Processes in All Life Forms

Biologists in many different disciplines use both the ideas from microbiology and the organisms themselves. Ecologists draw on principles of microbiology to understand how matter is decomposed and made available for continuous recycling. Biochemists use microbes to study metabolic pathways—sequences of chemical reactions in living organisms. Geneticists study microorganisms to investigate the processes of gene transmission.

Microorganisms are particularly attractive to many researchers because:

1. Compared to other organisms, microbes have relatively simple structures. It is easier to study most life processes in simple unicellular organisms than in complex multicellular ones.

2. Large numbers of microorganisms can be used in an experiment to obtain statistically reliable results at a reasonable cost. Growing a billion bacteria costs less than maintaining 10 rats. Experiments with large numbers of microorganisms give more reliable results than do those with small numbers of organisms with individual variations.

A bacterium may weigh approximately 0.00000000001 gram.

3. Because microorganisms reproduce very quickly, they are especially useful for studies involving the transfer of genetic information. Some bacteria can undergo three cell divisions in an hour, so the effects of gene transfer can quickly be followed through many generations.

By studying microbes, scientists have achieved remarkable success in understanding life processes and disease control. For example, within the last few decades, vaccines have nearly eradicated several dreaded childhood diseases—including measles, polio, German measles, and mumps. Recently a vaccine was developed for chickenpox. Smallpox, which once accounted for 1 out of every 10 deaths in Europe, has not been reported anywhere on the planet since 1978. Much has also been learned about genetic changes that lead to antibiotic resistance and about how to manipulate genetic information in bacteria. Much more remains to be learned. For examples, how can vaccines be made available on a worldwide basis? How can the development of new antibiotics keep pace with genetic changes in microorganisms? How will increased jet-age world travel affect the spread of infections? Will the continued encroachment of humans into virgin forests result in new emerging diseases? Can a vaccine or effective treatment for AIDS be developed? Therein lie some of the challenges for the next generation of biologists and health scientists.

Scope of Microbiology

Microbiology is the study of **microbes,** organisms so small that a microscope is needed to study them. We consider two dimensions of the scope of microbiology: (1) the variety of kinds of microbes and (2) the kinds of work microbiologists do.

The Microbes

The major groups of organisms studied in microbiology are bacteria, algae, fungi, viruses, and protozoa (➤ Figure 1.2a-e). All are widely distributed in nature. For example, a recent study of bee bread (a pollen-derived nutrient eaten by worker bees) showed it to contain 188 kinds of fungi and 29 kinds of bacteria. Most microbes consist of a single cell. (Cells are the basic units of structure and function in living things; they are discussed in Chapter 4.) Viruses, tiny acellular entities on the borderline between the living and the nonliving, behave like living organisms when they gain entry to cells. They, too, are studied in microbiology. Microbes range in size from small viruses 20 nm in diameter to large protozoans 5 mm or more in diameter. In other words, the largest microbes are as much as 250,000 times the size of the smallest ones! (Refer to Appendix A for a review of metric units.)

Among the great variety of microorganisms that have been identified, bacteria probably have been the most thoroughly studied. The majority of **bacteria** (singular: *bacterium*) are single-celled organisms with spherical, rod, or spiral shapes, but a few types form filaments. Most are so small they can be seen with a light microscope only under very high magnification. Although bacteria are cellular, they do not have a cell nucleus, and they lack the membrane-enclosed intracellular structures found in most other cells. Many bacteria absorb nutrients from their environment, but some make their own nutrients by photosynthesis or other synthetic processes. Some are stationary, and others move about. Bacteria are widely distributed in nature, for example, in aquatic environments and in decaying matter. And some occasionally cause diseases.

In contrast to bacteria, several groups of microorganisms consist of larger, more complex cells that have a cell nucleus. They include algae, fungi, and protozoa, all of which can easily be seen with a light microscope.

Many **algae** (al′je; singular: *alga*) are single-celled microscopic organisms, but some marine algae are large, relatively complex, multicellular organisms. Unlike bacteria, algae have a clearly defined cell nucleus and numerous membrane-enclosed intracellular structures. All algae photosynthesize their own food as plants do, and many can move about. Algae are widely distributed in both fresh water and oceans. Because they are so numerous and because they capture energy from sunlight in the food they make, algae are an important source of food for other organisms. Algae are of little medical importance; only one species has been found to cause disease in humans.

We Are Not Alone

"We are outnumbered. The average human contains about 10 trillion cells. On that average human are about 10 times as many microorganisms, or 100 trillion microscopic beings. . . . As long as they stay in balance and where they belong, [they] do us no harm. . . . In fact, many of them provide some important services to us. [But] most are opportunists, who if given the opportunity of increasing growth or invading new territory, will cause infection."

—Robert J. Sullivan, 1989

500 bacteria, each 1 μm (1/1000 of a millimeter) long, would fit end-to-end across the dot above the letter "i."

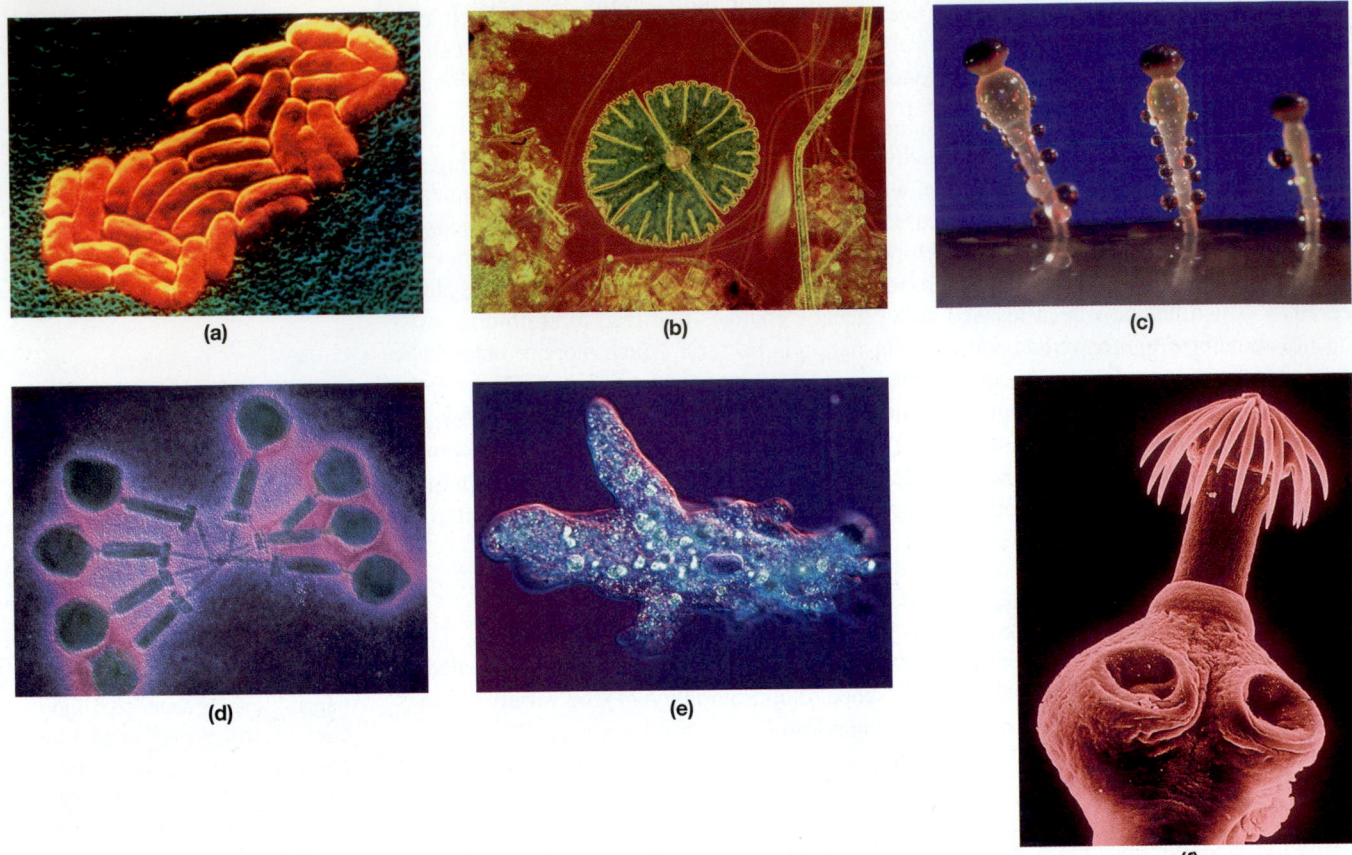

➤ **Figure 1.2   Typical microorganisms.** (a) Several *Klebsiella pneumoniae* cells (magnified 1120X), a bacterium that can cause pneumonia in humans. (b) *Micrasterias,* a type of green algae that lives in fresh water. (c) Fruiting bodies of the fungus *Philobolus crystallinus* with black spore packets on top that will be shot into the air to colonize new areas. (d) Bacteriophages (viruses that infect bacteria; 35,500X). (e) *Amoeba,* a protozoan (175X). (f) Head of the tapeworm *Acanthrocirrus retrirostris* (170X). At the top of the head are hooks and suckers that the worm uses to attach to a host's intestinal tissues.

Like algae, many **fungi** (fun'ji; singular: *fungus*), such as yeasts and some molds, are single-celled microscopic organisms. Some, such as mushrooms, are multicellular, macroscopic organisms. Fungi also have a cell nucleus and intracellular structures. All fungi absorb ready-made nutrients from their environment. Some fungi form extensive networks of branching filaments, but the organisms themselves generally do not move. Fungi are widely distributed in water and soil as decomposers of dead organisms. Some are important in medicine either as agents of disease or as sources of antibiotics.

Viruses are acellular entities too small to be seen with a light microscope. They are composed of specific chemical substances—a nucleic acid and a few proteins (Chapter 2). Indeed, some viruses can be crystallized and stored, but they retain the ability to invade cells. Viruses replicate themselves and display other properties of living organisms only when they have invaded cells. Many viruses can invade human cells and cause disease.

Protozoa (pro-to-zo'ah; singular: *protozoan*) also are single-celled, microscopic organisms with at least one nucleus and numerous intracellular structures. A few species of amoebae are large enough to be seen with the naked eye, but we can study their structure only with a microscope. Many protozoa obtain food by engulfing or ingesting smaller microorganisms. Most protozoa can move, but a few, especially those that cause human disease, cannot. Protozoa are found in a variety of water and soil environments.

In addition to organisms properly in the domain of microbiology, in this text we consider some macroscopic *helminths* (worms) (Figure 1.2f) and *arthropods* (insects and similar organisms). The helminths have microscopic stages in their life cycles that can cause disease, and the arthropods transmit these stages.

We will learn more about the classification of microorganisms in Chapter 9. For now it is important to know only that cellular organisms are referred to by two names: their *genus* and *species* names. For example, a bacterial species commonly found in the human gut is called *Escherichia coli,* and a protozoan species that can cause severe diarrhea is called *Giardia intestinalis*. The naming of viruses is less precise. Some viruses, such as herpesviruses, are named for the group to which they belong. Others, such as polioviruses, are named for the disease they cause.

Disease-causing organisms and the diseases they cause in humans are discussed in detail in Chapters 19–24. Hundreds of infectious diseases are known to medical science. Some of the most important—those diseases that physicians should report to the U.S. Centers for Disease Control and Prevention (CDC)—are listed in Table 1.1 according to the kind of causative organism. The CDC is a federal agency that collects data about diseases and about developing ways to control them.

The Microbiologists

Microbiologists study many kinds of problems that involve microbes. Some study microbes mainly to find out more about a particular type of organism—the life stages of a particular fungus, for example. Other microbiologists are interested in a particular kind of function, such as the metabolism of a certain sugar or the action of a specific gene. Still others focus directly on practical problems, such as how to purify or synthesize a new antibiotic or how to make a vaccine against a particular disease. Quite often the findings from one project are useful in another, as when agricultural scientists use information from microbiologists to control pests and improve crop yields, or when environmentalists attempt to maintain natural food chains and prevent damage to the environment. Some fields of microbiology are described in Table 1.2.

➤ Table 1.1   Reportable diseases caused by microorganisms and parasites[a]

Bacterial diseases	Bacterial diseases	Bacterial diseases	Viral diseases	Algal diseases
Anthrax	Hemolytic uremic syndrome	Rocky Mountain spotted fever	AIDS (symptomatic cases)	None
Bacterial meningitis	Legionnaires' disease	Salmonellosis	Arbovirus infection	
Botulism	Leprosy (Hansen's disease)	Shigellosis	Encephalitis	**Fungal Diseases**
Brucellosis	Listeriosis	Streptococcal disease, invasive, Group A	Hantavirus pulmonary syndrome	None
Campylobacteriosis	Lyme disease	Streptococcal pneumonia, drug-resistant invasive disease	Hepatitis A, B, and C	
Chancroid	Meningitis		Hepatitis (unspecified)	**Protozoan Diseases**
Chlamydial infections	Paratyphoid fever A, B, and C	Syphilis	HIV infection, adult	
Cholera	Pertussis (whooping cough)	Tetanus	HIV infection, pediatric	Giardiasis
Coccidiomycosis	Plague	Toxic shock syndrome	Influenza	Malaria
Cryptosporidiosis	Psittacosis	Trachoma	Measles (rubeola)	
Diphtheria	Q fever	Tuberculosis	Mumps	**Helminth Diseases**
Food poisoning	Relapsing fever	Typhoid fever	Poliomyelitis	
Gonorrhea		Typhus	Rabies (animal and human)	Trichinosis
Haemophilus influenzae infections (invasive)			Rubella (German measles)	
			Yellow fever	

[a]Infectious-disease reporting varies by state. This table lists most of the diseases commonly reported to the U.S. Centers for Disease Control and Prevention.

> **Table 1.2**   **Fields of microbiology**

Field (pronunciation)	Examples of What Is Studied
Microbial taxonomy	Classification of microorganisms
Fields according to organisms studied	
Bacteriology (bak″ter-e-ol′o-je)	Bacteria
Phycology (fi-kol′o-je)	Algae (*phyco*, seaweed)
Mycology (mi-kol′o-je)	Fungi (*myco*, a fungus)
Protozoology (pro″to-zo-ol′o-je)	Protozoa (*proto*, first; *zoo*, animal)
Parasitology (par″a-si-tol′o-je)	Parasites
Virology (vi-rol′o-je)	Viruses
Fields according to processes or functions studied	
Microbial metabolism	Chemical reactions that occur in microbes
Microbial genetics	Transmission and action of genetic information in microorganisms
Microbial ecology	Relationships of microbes with each other and with the environment
Health-related fields	
Immunology (im″u-nol′o-je)	How host organisms defend themselves against microbial infection
Epidemiology (ep-i-de-me-ol′o-je)	Frequency and distribution of diseases
Etiology (e-te-ol′-o-je)	Causes of disease
Infection control	How to control the spread of nosocomial (nos-o-ko′me-al), or hospital-acquired, infections
Chemotherapy	The development and use of chemical substances to treat diseases
Fields according to applications of knowledge	
Food and beverage technology	How to protect humans from disease organisms in fresh and preserved foods
Environmental microbiology	How to maintain safe drinking water, dispose of wastes, and control environmental pollution
Industrial microbiology	How to apply knowledge of microorganisms to the manufacture of fermented foods and other products of microorganisms
Pharmaceutical microbiology	How to manufacture antibiotics, vaccines, and other health products
Genetic engineering	How to use microorganisms to synthesize products useful to humans

Microbiologists work in a variety of settings (➤ Figure 1.3). Some work in universities, where they are likely to teach, do research, and train students to do research. Microbiologists in both university and commercial laboratories are helping to develop the microorganisms used in genetic engineering. Some law firms are hiring microbiologists to help with the complexities of patenting new genetically engineered organisms. These organisms can be used in such important ways as cleaning up the environment (*bioremediation*), controlling insect pests, improving foods, and fighting disease. Many microbiologists work in health-related positions. Some work in clinical laboratories, performing tests to diagnose diseases or determining which antibiotics will cure a particular disease. A few microbiologists develop new clinical tests. Others work in industrial laboratories to develop or manufacture antibiotics, vaccines, or similar biological products. Still others, concerned with controlling the spread of infections and related public health matters, work in hospitals or government labs.

From the point of view of health scientists, today's research is the source of tomorrow's new technologies. Research in *immunology* is greatly increasing our knowledge of how microbes trigger host responses and how the microbes escape these responses. It also is contributing to the development of new vaccines and to the treatment of immunologic disorders. Research in *virology* is improving our understanding of how viruses cause infections and how they are involved in cancer. Research in *chemotherapy* is increasing the number of drugs available to treat infections and is also improving our knowledge of how these drugs work. Finally, research in

(a)   (b)   (c)

(d)   (e)

➤ Figure 1.3 **Microbiology is used in diverse careers.** These careers include such activities as (a) using genetically engineered bacteria to investigate how diet influences the risk of developing cancer; (b) inspecting plastics made with as much as 40 percent starch (pieces inside baskets) for signs that aquatic microbes are degrading them; (c) using bacteria to decontaminate toxic wastes; (d) using beating nets to survey for ticks that can spread disease to livestock and humans; (e) keeping pets and domestic animals healthy, as well as improving their productivity, by means of advances in veterinary science.

- ✓ List three reasons to study microbiology.
- ✓ What is the difference between microbiology and bacteriology?
- ✓ What is the difference between etiology and epidemiology?
- ✓ List 5 bacterial diseases; 5 viral diseases.

genetics is providing new information about the transfer of genetic information and, especially, about how genetic information acts at the molecular level.

Historical Roots

Many of the ancient Mosaic laws found in the Bible about basic sanitation have been used through the centuries and still contribute to our practices of preventive medicine. In Deuteronomy, Chapter 13, Moses instructed the soldiers to carry spades and bury solid waste matter. The Bible also refers to leprosy and to the isolation of lepers. Although in those days the term *leprosy* probably included other infectious and noninfectious diseases, isolation did limit the spread of the infectious diseases.

The Greeks anticipated microbiology, as they did so many things. The Greek physician Hippocrates, who lived around 400 B.C., set forth ethical standards for the practice of medicine that are still in use today. Hippocrates was wise in human relations and also a shrewd observer. He associated particular signs and symptoms with certain illnesses and realized that diseases could be transmitted from one person to another by clothing or other objects. At about the same time, the Greek historian Thucydides observed that people who had recovered from the plague could take care of plague victims without danger of getting the disease again.

The Romans also contributed to microbiology, as early as the first century B.C. The scholar and writer Varro proposed that tiny invisible animals entered the body through the mouth and nose to cause disease. Lucretius, a philosophical poet, cited "seeds" of disease in his *De Rerum Natura (On the Nature of Things)*.

Bubonic plague, also called the Black Death, appeared in the Mediterranean region around A.D. 542, where it reached epidemic proportions and killed millions. In 1347 the plague invaded Europe along the caravan routes and sea lanes from central Asia, affecting Italy first, then France, England, and finally northern Europe. Although no accurate records were kept at that time, it is estimated that tens of millions of people in Europe died during this and successive waves of plague over the next 300 years. The Black Death was a great leveler—it killed rich and poor alike (➤ Figure 1.4). The wealthy fled to isolated summer homes but carried plague-infected fleas with them in unwashed hair and clothing.

One group that escaped the plague's devastation was the Jewish population. Ancient Hebrew laws regarding sanitation offered some protection to those who practiced them. The relatively clean Jewish ghettos harbored fewer rats to spread the disease. When Jews did fall ill,

➤ Figure 1.4   **A portion of "The Triumph of Death" by Pieter Brueghel the Elder.** The picture, painted in the mid-sixteenth century, a time when outbreaks of plague were still common in many parts of Europe, dramatizes the swiftness and inescapability of death for people of all social and economic classes.

they were carefully nursed and treated with herbal remedies rather than by strenuous purging or excessive bleedings with dirty instruments. As a result, a smaller proportion of Jews than gentiles died of the disease. Ironically, some gentiles regarded the Jews' higher survival rates as proof that Jews were the source of the epidemic.

In his *Diary,* the English writer Samuel Pepys gave a vivid, firsthand account of the plague in London in the 1660s.

> The streets mighty empty all the way now even in London, which is a sad sight. . . . Poor Will, that used to sell us ale, . . . his wife and three children died, all I think in a day. . . . [H]ome to draw over anew my will, which I had bound myself by oath to dispatch by tomorrow night, the town growing so unhealthy that a man cannot depend upon living two days to an end. In the City died this week 7,496, and of them 6,102 of the plague. But it is feared that the true number of the dead is near 10,000; partly from the poor that cannot be taken notice of through the greatness of the number. . . . I saw a dead corps in a coffin lie in the Close unburied; and a watch is constantly kept there night and day to keep the people in, the plague making us cruel as doggs one to another.

Until the seventeenth century, the advance of microbiology was hampered by the lack of appropriate tools to observe microbes. Around 1665, the English scientist Robert Hooke built a compound microscope (one in which light passes through two lenses) and used it to observe thin slices of cork. He coined the term *cell* to describe the orderly arrangement of small boxes that he saw because they reminded him of the cells (small, bare rooms) of monks. However, it was Anton van Leeuwenhoek (➤ Figure 1.5), a Dutch cloth merchant and amateur lens grinder, who first made and used lenses to observe living microorganisms. The lenses Leeuwenhoek made were of excellent quality; some gave magnifications up to 300× and were remarkably free of distortion. Making these lenses and looking through them were the passions of his life. Everywhere he looked he found what he called "animalcules." He found them in stagnant water, in sick people, and even in his own mouth.

Over the years Leeuwenhoek observed all the major kinds of microorganisms—protozoa, algae, yeast, fungi, and bacteria in spherical, rod, and spiral forms. He once wrote, "For my part I judge, from myself (howbeit I clean my mouth like I've already said), that all the people living in our United Netherlands are not as many as the living animals that I carry in my own mouth this very day." Starting in the 1670s he wrote numerous letters to the Royal Society in

➤ Figure 1.5 **Anton van Leeuwenhoek (1632–1723).** He is shown holding one of his simple microscopes.

AIDS

Suppose that several close relatives, two of your best friends, and many of your neighbors are suffering from painful illnesses and will soon die. There are no available beds at the hospital, and most of the doctors and nurses have quit work because they fear becoming infected. The local television station has started broadcasting daily the latest figures on deaths and new outbreaks, like the stock market prices or the weather report. Nearly every time you go shopping, you meet a funeral procession.

Over the centuries, many people have found themselves in situations that are similar to this fictional modern epidemic. Again and again, large proportions of the human population have been devastated by infectious diseases such as typhus, smallpox, and bubonic plague. In the mid-fourteenth century (1347–1351), plague alone wiped out 25 million people—one-fourth the population of Europe and neighboring regions—in just 5 years.

We who live in technologically advanced countries tend to think that outbreaks of this sort are a thing of the past. Yet following World War I, within the memory of many people now living, worldwide outbreaks of influenza claimed 20 million lives. Today, acquired immune deficiency syndrome (AIDS) threatens to kill great numbers of people after they have suffered a long and painful illness. Is AIDS in any way comparable to the great killer diseases of the past? More alarmingly, does its appearance mean that all our "triumphs" over infectious disease were an illusion? Will the past few decades, during which vaccines and

A model of the virus that causes AIDS.

antibiotics have largely kept contagious disease in check and even have wiped out certain ancient curses such as smallpox, prove to have been just a brief and atypical episode in an endless war that can't be won?

Despite the fears AIDS has aroused, there are many differences between AIDS and the epidemic diseases mentioned above. For one thing, AIDS is not nearly so easily communicable as the epidemic killers of the past. It is not spread by casual contact but largely by certain behaviors, most of which people can learn to avoid (as we will see in Chapter 19).

More important, however, is the state of our knowledge. During past epidemics, people were terrified and demoralized because they had no idea how the disease was caused and spread or how to fight it. Today, we understand far more about the nature of the enemy. The study of AIDS has occupied virologists, chemotherapists, and immunologists from among the world's most talented microbiologists. They have shown that AIDS is caused by a virus that infects cells of the body's immune system. The virus alters genetic information so that instead of fighting the infection, infected cells make viruses and then die. Scientists have determined the precise structure of many of the components of the virus, and they have learned how it attaches itself to its target cells. They have developed some drugs that are being tested to treat AIDS, and they are working on a vaccine. Despite those efforts, however, AIDS remains one of the most threatening infectious diseases and one of the greatest challenges microbiologists have ever faced.

The AIDS Name Project, a memorial to victims of this disease, showing individual quilts on display in Washington, D.C.

London and pursued his studies until his death in 1723 at the age of 91. Leeuwenhoek refused to sell his microscopes.

After Leeuwenhoek's death, microbiology did not advance for more than a century. Eventually microscopes became more widely available, and progress resumed. Several workers discovered ways to stain microorganisms with dyes to make them more visible. The Swedish botanist Carolus Linnaeus developed a general classification system for all living organisms. The German botanist Matthias Schleiden and the German zoologist Theodor Schwann formulated the **cell theory,** which states that cells are the fundamental units of life and carry out all the basic functions of living things. Today this theory still applies to all cellular organisms, but not to viruses.

The Germ Theory of Disease

The **germ theory of disease** states that microorganisms (germs) can invade other organisms and cause disease. Although this is a simple idea and is generally accepted today, it was not widely accepted when formulated in the mid-nineteenth century. Many people believed that broth, left standing, turned cloudy because of something about the broth itself. Even after it was shown that microorganisms in the broth caused it to turn cloudy, people believed that the microorganisms, like the "worms" (fly larvae, or maggots) in rotting meat, arose from nonliving things, a concept known as **spontaneous generation**. Widespread belief in spontaneous generation, even among scientists, hampered further development of the science of microbiology and the acceptance of the germ theory of disease. As long as they believed that microorganisms could arise from nonliving substances, scientists saw no purpose in considering how diseases were transmitted or how they could be controlled. Dispelling the belief in spontaneous generation took years of painstaking effort.

> The Roman writer, Varro, suggested in the first century. B.C. that people might get sick from inhaling "tiny animals."

Early Studies

For as long as humans have existed, some probably have believed that living things somehow originated spontaneously from nonliving matter. Aristotle's theories about his four "elements"—fire, earth, air, and water—seem to have suggested that nonliving forces somehow contributed to the generation of life. Even some naturalists believed that rodents arose from moist grain, beetles from dust, and worms and frogs from mud. As late as the nineteenth century, it seemed obvious to most people that rotting meat gave rise to "worms."

In the late seventeenth century, the Italian physician Francesco Redi devised a set of experiments to demonstrate that if pieces of meat were covered with gauze so that flies could not reach them, no "worms" appeared in the meat, no matter how rotten it was (➤ Figure 1.6). Maggots did, however, hatch from fly eggs laid on top of the gauze. Despite proof that maggots did not arise spontaneously, some scientists, such as the English clergyman John Needham, still believed in spontaneous generation—at least that of microorganisms. Lazzaro Spallanzani, an Italian cleric and scientist, was more skeptical. He boiled broth infusions containing organic (living or previously living) matter and sealed the flasks to demonstrate that no organisms would de-

➤ Figure 1.6 **Redi's experiments refuting the spontaneous generation of maggots in meat.** When meat is exposed in an open jar, flies lay their eggs on it, and the eggs hatch into maggots (fly larvae). In a sealed jar, however, no maggots appear. If the jar is covered with gauze, maggots hatch from eggs that the flies lay on top of the gauze, but still no maggots appear in the meat.

➤ Figure 1.7   A "swan-necked"
flask that Pasteur used in refuting
the theory of spontaneous generation.
Although air could enter the flasks,
microbes became trapped in the curved
necks and never reached the contents.
The contents, therefore, remained
sterile—and still are—despite their
exposure to the air.

velop spontaneously in them. Critics did not accept this as disproof of spontaneous generation. They argued that boiling drove off oxygen (which they thought all organisms required) and that sealing the flasks prevented its return.

Several scientists tried different ways of introducing air to counter this criticism. Schwann heated air before introducing it into flasks, and other scientists filtered air through chemicals or cotton plugs. All these methods prevented the growth of microorganisms in the flasks. But the critics still argued that altering the air prevented spontaneous generation.

Even nineteenth-century scientists of some stature continued to argue vociferously in favor of spontaneous generation. They believed that an organic compound previously formed by living organisms contained a "vital force" from which life sprang. The force, of course, required air, and they believed that all the methods of introducing air somehow changed it so that it could not interact with the force.

The proponents of spontaneous generation were finally defeated, mainly by the work of the French chemist Louis Pasteur and the English physicist John Tyndall. When the French Academy of Science sponsored a competition in 1859 "to try by well-performed experiment to throw new light on the question of spontaneous generation," Pasteur entered the competition.

During the years Pasteur worked in the wine industry, he had established that alcohol was produced in wine only if yeast was present, and he learned a lot about the growth of microorganisms. Pasteur's experiment for the competition involved his famous "swan-necked" flasks (➤ Figure 1.7). He boiled *infusions* (broths of foodstuffs) in flasks, heated the glass necks, and drew them out into long, curved tubes open at the end. Air could enter the flasks without being subjected to any of the treatments that critics had claimed destroyed its effectiveness. Airborne microorganisms could also enter the necks of the flasks, but they became trapped in the curves of the neck and never reached the infusions. The infusions from Pasteur's experiments remained sterile unless a flask was tipped so that the infusion flowed into the neck and back into the flask. This manipulation allowed microorganisms trapped in the neck to wash into the infusion, where they could grow and cause the infusion to become cloudy. In another experiment Pasteur filtered air through three cotton plugs. He then immersed the plugs in sterile infusions, demonstrating that growth occurred in the infusions from organisms trapped in the plugs.

Tyndall delivered another blow to the idea of spontaneous generation when he arranged sealed flasks of boiled infusion in an airtight box. After allowing time for all dust particles to settle to the bottom of the box, he carefully removed the covers from the flasks. These flasks, too, remained sterile. Tyndall had shown that air could be sterilized by settling, without any treatment that would prevent the "vital force" from acting.

Both Pasteur and Tyndall were fortunate that the organisms present in their infusions at the time of boiling were destroyed by heat. Others who tried the same experiments observed that the infusions became cloudy from growth of microorganisms. We now know that the infusions in which growth occurred contained heat-resistant or spore-forming organisms, but at the time, the growth of such organisms was seen as evidence of spontaneous generation. Still, the works of Pasteur and Tyndall successfully disproved spontaneous generation to most scientists of the time. Recognition that microbes must be introduced into a medium before their growth can be observed paved the way for further development of microbiology—especially for development of the germ theory of disease.

Pasteur's Further Contributions

Louis Pasteur (➤ Figure 1.8) was such a giant among nineteenth-century scientists working in microbiology that we must consider some of his many contributions. Born in 1822, the son of a sergeant in Napoleon's army, Pasteur worked as a portrait painter and a teacher before he began to study chemistry in his spare time. Those studies led to posts in several French universities as professor of chemistry and to significant contributions to the wine and silkworm industries. He discovered that carefully selected yeasts made good wine, but that mixtures of other microorganisms competed with the yeast for sugar and made wine taste oily or sour. To combat this problem, Pasteur developed the technique of pasteurization (heating wine to 56°C in the absence of oxygen for 30 minutes) to kill unwanted organisms. While studying silkworms, he identified three different microorganisms, each of which caused a different disease.

Pasteur was not a physician and had never practiced medicine when he was asked to treat his first case of human exposure to rabies.

➤ Figure 1.8 **Louis Pasteur in his laboratory.** The first rabies vaccine, developed by Pasteur, was made from the dried spinal cords of infected rabbits.

His association of specific organisms with particular diseases, even though in silkworms rather than in humans, was an important first step in proving the germ theory of disease.

Despite personal tragedy—the deaths of three daughters and a cerebral hemorrhage that left him with permanent paralysis—Pasteur went on to contribute to the development of vaccines. The best known of his vaccines is the rabies vaccine, made of dried spinal cord from rabbits infected with rabies, which was tested in animals. When a 9-year-old boy who had been severely bitten by a rabid dog was brought to him, Pasteur administered the vaccine. The boy, who had been doomed to die, survived and became the first person to be immunized against rabies.

In 1894 Pasteur became director of the Pasteur Institute, which was built for him in Paris. Until his death in 1895, he guided the training and work of other scientists at the Institute. Today the Pasteur Institute is a thriving research center—an appropriate memorial to its founder.

Robert Koch disproved the idea that tuberculosis was inherited.

Koch's Contributions

Robert Koch (➤ Figure 1.9), a contemporary of Pasteur's, finished his medical training in 1872 and worked as a physician in Germany throughout most of his career. After he bought a microscope and photographic equipment, he spent most of his time studying bacteria, especially those that cause disease. Koch identified the bacterium that causes anthrax, a highly contagious and lethal disease in cattle and sometimes in humans. He recognized both actively dividing cells and dormant cells (spores) and developed techniques for studying them *in vitro* (outside a living organism).

Koch also found a way to grow bacteria in *pure cultures*—cultures that contained only one kind of organism. He tried streaking bacterial suspensions on potato slices and then on solidified gelatin. But gelatin melts at incubator (body) temperature; even at room temperature, some microbes liquefy it. Finally, Angelina Hesse, the American wife of one of Pasteur's colleagues, suggested that Koch add agar (a thickener used in cooking) to his bacteriological media. This created a firm surface over which microorganisms could be spread very thinly— so thinly that some individual organisms were separated from all others. Each individual organism then multiplied to make a colony of thousands of descendants. Koch's technique of preparing pure cultures is still used today.

Koch's outstanding achievement was the formulation of four postulates to associate a particular organism with a specific disease. **Koch's postulates,** which provided scientists with a method of establishing the germ theory of disease, are as follows:

1. The specific causative agent must be found in every case of the disease.
2. The disease organism must be isolated in pure culture.
3. Inoculation of a sample of the culture into a healthy, susceptible animal must produce the same disease.
4. The disease organism must be recovered from the inoculated animal.

Implied in Koch's postulates is his one organism–one disease concept. The postulates assume that an infectious disease is caused by a single organism, and they are directed toward estab-

➤ Figure 1.9 **Robert Koch.** Koch formulated four postulates for linking a given organism to a specific disease.

What's in the Last Drop?

During the nineteenth century, French and German scientists were fiercely competitive. One area of competition was the preparation of pure cultures. Koch's reliable method of preparing pure cultures from colonies on solid media allowed German microbiologists to forge ahead. The French microbiologists' method of broth dilution, though now often used to count organisms (Chapter 6), hampered their progress. They added a few drops of a culture to fresh broth, mixed it, and added a few drops of the mixture to more fresh broth. After several successive dilutions, they assumed that the last broth that showed growth of microbes had contained a single organism. Unfortunately, the final dilution often contained more than one organism, and sometimes the organisms were of different kinds. This faulty technique led to various fiascos, such as inoculating animals with deadly organisms instead of vaccinating them.

✓ What similarities and differences do you see when comparing past epidemics of plague with today's AIDS epidemic?

✓ State the germ theory of disease. Try to think of an explanation of disease causation which would be contrary to the germ theory.

✓ How did Pasteur's "swan-necked" flasks experiment disprove the theory of spontaneous generation?

✓ Why was the French microbiologists' method of broth dilution inadequate for obtaining pure cultures of organisms?

Pocahontas died of smallpox in 1617 in England.

lishing that fact. This concept also was an important advance in the development of the germ theory of disease.

After obtaining a laboratory post at Bonn University in 1880, Koch was able to devote his full time to studying microorganisms. He identified the bacterium that causes tuberculosis and developed a complex method of staining this organism. He also guided the research that led to the isolation of *Vibrio cholerae,* the bacterium that causes cholera.

In a few years Koch became professor of hygiene at the University of Berlin, where he taught a microbiology course believed to be the first ever offered. He also developed *tuberculin,* which he hoped would be a vaccine against tuberculosis. Because he underestimated the difficulty of killing the organism that causes tuberculosis, use of tuberculin resulted in several deaths from that disease. Although tuberculin was unacceptable as a vaccine, its development laid the groundwork for a skin test to diagnose tuberculosis. After the vaccine disaster, Koch left Germany. He made several visits to Africa, at least two visits to Asia, and one to the United States.

In the remaining 15 years of his life, his accomplishments were many and varied. He conducted research on malaria, typhoid fever, sleeping sickness, and several other diseases. His studies of tuberculosis won him the Nobel Prize for physiology or medicine in 1905, and his work in Africa and Asia won him great respect on those continents.

Work Toward Controlling Infections

Like Koch and Pasteur, two nineteenth-century physicians, Ignaz Philipp Semmelweis of Austria and Joseph Lister of England, were convinced that microorganisms caused infections (➤ Figure 1.10). Semmelweis recognized a connection between autopsies and puerperal (childbed) fever. Many physicians went directly from performing autopsies to examining women in labor without so much as washing their hands. When Semmelweis attempted to encourage more sanitary practices, he was ridiculed and harassed until he had a nervous breakdown and was sent to an asylum. Ultimately, he suffered the curious irony of succumbing to an infection caused by the same organism that produces puerperal fever. In 1865, Lister, who had read of Pasteur's work on pasteurization and Semmelweis's work on improving sanitation, initiated the use of dilute carbolic acid on bandages and instruments to reduce infection. Lister, too, was ridiculed, but with his imperturbable temperament, resolute will, and tolerance of hostile criticism, he was able to continue his work. His methods, the first *aseptic techniques,* were proven effective by the decrease in surgical wound infections in his surgical wards. At age 75, some 37 years after he introduced the use of carbolic acid, Lister was awarded the Order of Merit for his work in preventing the spread of infection. He is considered the father of antiseptic surgery.

Emergence of Special Fields of Microbiology

Pasteur, Koch, and most other microbiologists considered to this point were generalists interested in a wide variety of problems. Certain other contributors to microbiology had more specialized interests, but their achievements were no less valuable. In fact, those achievements helped establish the special fields of immunology, virology, chemotherapy, and microbial genetics—fields that are today prolific research areas. Selected fields of microbiology are defined in Table 1.2.

Immunology

Disease depends not only on microorganisms invading a host but also on the host's response to that invasion. Today, we know that the host's response is in part a response of the immune system.

The ancient Chinese knew that a person scarred by smallpox would not again get the disease. They took dried scabs from lesions of people who were recovering from the disease and ground them into a powder that they sniffed. As a result of inhaling weakened organisms, they acquired a mild case of smallpox but were protected against subsequent infection.

Smallpox was unknown in Europe until the Crusaders carried it back from the Near East in the twelfth century. By the seventeenth century it was widespread. In 1717 Lady Ashley Montagu, wife of the British ambassador to Turkey, introduced a kind of immunization to England. A thread was soaked in fluid from a smallpox vesicle (blister) and drawn through a small incision in the arm. This technique, called *variolation,* was used at first by only a few prominent people, but eventually it became widespread.

In the late eighteenth century, Edward Jenner (➤ Figure 1.11) realized that milkmaids who got cowpox did not get smallpox, and he inoculated his own son with fluid from a cowpox blister. He later similarly inoculated an 8-year-old and subsequently inoculated the same child with smallpox. The child remained healthy. The word *vaccinia* (*vacca,* the Latin name for cow) gave rise both to the name of the virus that causes cowpox and to the word *vaccine.* In the early 1800s Jenner received grants amounting to a total of 30,000 British pounds to extend his work on vaccination. Today, those grants would be worth more than $1 million. They may have been the first grants for medical research.

Pasteur contributed significantly to the emergence of immunology with his work on vaccines for rabies and cholera. In 1879, when Pasteur was studying chicken cholera, his assistant accidentally used an old chicken cholera culture to inoculate some chickens. The chickens did not develop disease symptoms. When the assistant later inoculated the same chickens with a fresh chicken cholera culture, they remained healthy. Although he had not planned to use the old culture first, Pasteur did realize that the chickens had been immunized against chicken cholera. He reasoned that the organisms must have lost their ability to produce disease but retained their ability to produce immunity. This finding led Pasteur to look for techniques that would have the same effect on other organisms. His development of the rabies vaccine was a successful attempt.

Along with Jenner and Pasteur, the nineteenth-century Russian zoologist Elie Metchnikoff was a pioneer in immunology (➤ Figure 1.12). In the 1880s many scientists believed immunity was due to noncellular substances in the blood. Metchnikoff discovered that certain cells in the body could ingest microbes. He named those cells *phagocytes,* which literally means "cell-eating." The identification of phagocytes as cells that defend the body against invading microorganisms was a first step in understanding immunity. Metchnikoff also developed several vaccines. Some were successful, but unfortunately some infected the recipients with the microorganisms against which they were supposedly being immunized. A few of his subjects acquired gonorrhea and syphilis from his vaccines.

Virology

The science of virology emerged after that of bacteriology because viruses could not be recognized until certain techniques for studying and isolating larger particles such as bacteria had been developed. When Pasteur's collaborator Charles Chamberland developed a porcelain filter to remove bacteria from water in 1884, he had no idea that any kind of infectious agent could pass through the filter. But researchers soon realized that some filtrates (materials that passed through the filters) remained infectious even after the bacteria were filtered out. The Dutch microbiologist Martinus Beijerinck determined why such filtrates were infectious and

(a)

(b)

➤ **Figure 1.10   Two nineteenth-century pioneers in the control of infections.** (a) Ignaz Philipp Semmelweis, who died in an asylum before his innovations were widely accepted, depicted on a 1965 Austrian postage stamp; (b) Joseph Lister, who successfully carried on Semmelweis's work.

➤ **Figure 1.11   Edward Jenner vaccinating a child against smallpox.**

Calves, shaved, inoculated with cowpox, and covered with lesions, were led house to house by entrepreneurs offering vaccination during colonial times.

A "Thorny" Problem

Metchnikoff's personal life played a role in his discovery of phagocytes. Becoming despondent after the death of his first wife in 1873, he tried to kill himself with an overdose of opium. He did not due, but married again in 1875. The prenuptial agreement for his second marriage included taking his wife's dozen younger brothers and sisters into his home. On one occasion while Metchnikoff was at lunch, the young pranksters poked rose thorns into a starfish he was studying. Upon return, he found cells of the starfish gathered around the thorns. Study of these cells showed that they could devour foreign substances (phagocytosis)—an important body defense. When his second wife became very ill with typhoid fever in 1880, he again tried to take his own life. This time his method of suicide was an experiment—self-injection with relapsing fever to find out if the disease was transmissible by blood. He was awarded the Nobel Prize in 1908 for his work on immunology and phagocytosis.

➤ Figure 1.12 **Elie Metchnikoff.** Metchnikoff was one of the first scientists to study the body's defenses against invading microorganisms.

was thus the first to characterize viruses. The term *virus* had been used earlier to refer to poisons and to infectious agents in general. Beijerinck used the term to refer to specific *pathogenic* (disease-causing) molecules incorporated into cells. He also believed that these molecules could borrow for their own use existing metabolic and replicative mechanisms of the infected cells, known as *host cells*.

Further progress in virology required development of techniques for isolating, propagating, and analyzing viruses. The American scientist Wendell Stanley crystallized tobacco mosaic virus in 1935, showing that an agent with properties of a living organism also behaved as a chemical substance (➤ Figure 1.13). The crystals consisted of protein and ribonucleic acid (RNA). The nucleic acid was soon shown to be important in the infectivity of viruses. Viruses were first observed with an electron microscope in 1939. From that time both chemical and microscopic studies were used to investigate viruses.

By 1952 the American biologists Alfred Hershey and Martha Chase had demonstrated that the genetic material of some viruses is another nucleic acid, deoxyribonucleic acid (DNA). In 1953 the American postdoctoral student James Watson and the English biophysicist Francis

RNA

Protein subunits

(a)   (b)

➤ Figure 1.13 **The tobacco mosaic virus.** (a) Electron micrograph of tobacco mosaic virus (magnification approx. 617,000X). (b) The structure of the tobacco mosaic virus. A helical core of RNA is surrounded by a coat that consists of repeating protein units. The structure of the particles is so regular that the viruses can be crystallized.

Crick determined the structure of DNA. The stage was set for rapid advances in understanding how DNA functions as genetic material both in viruses and in cellular organisms. Since the 1950s hundreds of viruses have been isolated and characterized. Although much remains to be learned about viruses, tremendous progress has been made in understanding their structure and how they function.

Chemotherapy

The Greek physician Dioscorides compiled *Materia Medica* in the first century A.D. This five-volume work listed a number of substances derived from medicinal plants still in use today—digitalis, curare, ephedrine, and morphine—along with a number of herbal medications. Credit for bringing herbal medicine to the United States is given to many groups of settlers, but Native Americans used many medicinal plants before the arrival of Europeans in the Americas. Many so-called primitive peoples still use herbs extensively, and some pharmaceutical companies finance expeditions into the Amazon Basin and other remote areas to investigate the uses the natives make of the plants around them.

During the Middle Ages virtually no advances were made in the use of chemical substances to treat diseases. Early in the sixteenth century the Swiss physician Aureolus Paracelsus used metallic chemical elements to treat diseases—antimony for general infections and mercury for syphilis. In the mid-seventeenth century Thomas Sydenham, an English physician, introduced cinchona tree bark to treat malaria. This bark, which we now know contains quinine, had been used to treat fevers in Spain and South America. In the nineteenth century morphine was extracted from the opium poppy and used medicinally to alleviate pain.

Paul Ehrlich, the first serious researcher in the field of chemotherapy (➤ Figure 1.14), received his doctoral degree from the University of Leipzig, Germany, in 1878. His discovery that certain dyes stained microorganisms but not animal cells suggested that the dyes or other chemicals might selectively kill microbial cells. This led him to search for the "magic bullet," a chemical that would destroy specific bacteria without damaging surrounding tissues. Ehrlich coined the term *chemotherapy* and headed the world's first institute concerned with the development of drugs to treat disease.

Early in the twentieth century the search for the magic bullet continued, especially among scientists at Ehrlich's institute. After testing hundreds of compounds (and numbering each compound), Ehrlich found compound 418 (arsenophenylglycine) to be effective against sleeping sickness and compound 606 (Salvarsan) to be effective against syphilis. For 40 years Salvarsan remained the best available treatment for this disease. In 1922 Alexander Fleming, a Scottish

Informed Consent

During the Panama Canal construction in 1905, yellow fever struck the workers as they struggled in the swamps. Yellow fever was an often fatal disease. Because the entire canal project was jeopardized, the American physician, Walter Reed, was asked to control the disease. Reed listened to Cuban physician, Carlos Finlay y Barres, who claimed that yellow fever was carried by mosquitoes, and ignored those who insisted that yellow fever was due to swamp air.

Reed designed experiments to determine the role of mosquitoes or swamp air in the disease's transmission. Volunteers were bitten by mosquitoes known to have bitten yellow fever patients, were injected with blood from patients, or slept in sealed huts on sheets soaked with patients' vomitus. Many of those bitten by mosquitoes or injected with patient's blood died from yellow fever. Those who slept on the vomit-soaked sheets, wore the clothing of the dead, and ate from their dishes did not become ill. It is not known how Reed got volunteers for such a deadly experiment. Why do you think these men volunteered for this deadly task?

➤ Figure 1.14 **Paul Ehrlich.** Ehrlich was a pioneer in the development of chemotherapy for infectious disease.

physician, discovered that lysozyme, an enzyme found in tears, saliva, and sweat, could kill bacteria. Lysozyme was the first body secretion shown to have chemotherapeutic properties.

The development of antibiotics began in 1917 with the observation that certain bacteria (actinomycetes) stopped the growth of other bacteria. In 1928 Fleming (➤ Figure 1.15) observed that a colony of *Penicillium* mold contaminating a culture of *Staphylococcus* bacteria had prevented growth of bacteria adjacent to itself. Although he was not the first to observe this phenomenon, Fleming did recognize its potential for countering infections. However, purification of sufficient quantities of the substance he called *penicillin* proved to be very difficult. The great need for such a drug during World War II, money from the Rockefeller Institute, and the hard work of the German biochemist Ernst Chain, the Australian pathologist Howard Florey, and researchers at Oxford University accomplished the task. Penicillin became available as a safe and versatile chemotherapeutic agent for use in humans.

While this work was going on, sulfa drugs were being developed. In 1935, prontosil rubrum, a reddish dye containing a sulfonamide chemical group, was used in treating streptococcal infections. Further study showed that sulfonamides were converted in the body to sulfanilamides; much subsequent work was devoted to developing drugs containing sulfanilamide. The German chemist Gerhard Domagk played an important role in this work, and one of the drugs, prontosil, saved the life of his daughter. In 1939 he was awarded a Nobel Prize for his work, but Hitler refused to allow him to make the trip to receive it. Extensions of Domagk's work led to the development of isoniazid, an effective agent against tuberculosis. Both sulfa drugs and isoniazid are still used today.

The development of antibiotics resumed with the work of Selman Waksman, who was born in Ukraine and moved to the United States in 1910. Inspired by the 1939 discovery, by the French microbiologist Rene Dubos of tyrothricin, an antibiotic produced by soil bacteria, Waksman examined soil samples from all over the world for growth-inhibiting microorganisms or their products. He coined the term *antibiotic* in 1941 to describe actinomycin and other products he isolated. Both tyrothricin and actinomycin proved to be too toxic for general use as antibiotics. After repeated efforts, Waksman isolated the less toxic drug streptomycin in 1943. Streptomycin constituted a major breakthrough in the treatment of tuberculosis. In the same decade Waksman and others isolated neomycin, chloramphenicol, and chlortetracycline.

Examining soil samples proved to be a good way to find antibiotics, and explorers and scientists still collect soil samples for analysis. The more common antibiotic-producing organisms are rediscovered repeatedly, but the possibility of finding a new one always remains. Even the sea has yielded antibiotics, especially from the fungus *Cephalosporium acremonium*. The Italian microbiologist Giuseppe Brotzu noted the absence of disease organisms in seawater where sewage entered, and he determined that an antibiotic must be present. Cephalosporin was subsequently purified, and a variety of cephalosporin derivatives are now available for treating human diseases.

➤ Figure 1.15 **Alexander Fleming.** Fleming discovered the antibacterial properties of penicillin.

How Microbiologists Investigate Problems

Like other scientists, microbiologists investigate problems by designing and carrying out experiments. Such experiments have provided the information health scientists apply to solve medical problems. Much of this text is devoted to presenting information obtained from experiments and to showing how that information is used in understanding infectious diseases. However, we believe that all science students will be interested in knowing how scientific problems are investigated.

First, a scientific problem must concern some aspect of the natural world because scientific methods can deal only with natural conditions and events. Microbiological problems deal with natural conditions and events involving microbes. Second, scientific problems must be clearly defined and sufficiently limited in scope so that a hypothesis and a prediction can be formulated. A **hypothesis** is a tentative explanation to account for an observed condition or event. The hypothesis in a particular experiment must be (1) an explanation for the defined problem and (2) testable. A testable hypothesis is one for which evidence can be collected to support or refute the hypothesis. A **prediction** is an outcome or consequence that will result if the hypothesis is true. Before beginning a scientific experiment, one must define the problem and make a hypothesis and a prediction.

A good hypothesis is one that offers the most reasonable explanation and the simplest solution to a problem. The purpose of scientific experiments is to test hypotheses by determining the correctness of predictions derived from the hypotheses. Scientific progress is made by making and testing hypotheses.

For example, suppose that a microbiologist has isolated an organism in pure culture and wants to know the effects of temperature on its growth. On the basis of information from prior research, he or she might (1) hypothesize that the organism's growth rate increases with temperature and (2) predict that the rate of increase in the number of organisms in a culture is proportional to the increase in temperature. After making the hypothesis and a prediction, the investigator designs an experiment to test the hypothesis. The experiment must be designed specifically to test the hypothesis and to collect evidence to determine whether the prediction is true.

To design a good experiment, an investigator must consider all variables that might affect the outcome. A **variable** is anything that can change for the purposes of an experiment. An experiment should have only one **experimental variable,** the factor that is purposely changed for the experiment. For example, in the study of the effects of tempera-ture on the growth of an organism, temperature is the experimental variable. The hypothesis and prediction are related to the experimental variable. All other variables are **control variables,** factors that *can* change but that are prevented from changing for the duration of the experiment. In our example, the control variables include the number and characteristics of the organism, the quantity and properties of the medium, and all environmental factors except temperature.

When all variables have been identified, the investigator establishes the procedures for carrying out the experiment. Once the experiment has been designed, it must be carried out exactly as planned. All observations must be made and recorded accurately and precisely. If problems or unusual situations are encountered, they must be noted carefully. For example, should an incubator fail to maintain certain cultures at the proper temperature for the appropriate length of time, this failure should be noted and taken into consideration in interpreting the experiment. When the experiment is completed, the researcher analyzes and interprets the results in light of the hypothesis and prediction. The analysis of the results of an experiment often involves preparation of tables and graphs and usually compares results obtained under experimental and control conditions.

The goal of an experiment is to draw conclusions as to whether the prediction is true. If the experimental results, when analyzed, do not support the hypothesis, they may nevertheless suggest a better alternative hypothesis. The experimenter might wish to design further experiments to test this new hypothesis. Often, it is unexpected experimental results that lead to the most interesting discoveries.

In this textbook, you will find some boxes, such as A Winter Dilemma (p. 31) and What Grows in Your Health and Beauty Aids? (p. 554) that suggest projects that you might try yourselves. Even if you are not able to try these projects, at least make a mental plan of what hypotheses and steps you could use to investigate such a problem. Perhaps you can discuss your experimental designs during lecture or in lab. What other questions occur to you? How would you go about forming and testing hypotheses for them? In addition to performing experiments, scientists also should report their results so that other scientists can verify and use the information. Scientific knowledge increases by the sharing of information. This allows other scientists to repeat experiments and determine whether the results are reproducible. It also allows them to develop new experiments that build on existing information.

The fact that many antibiotics have been discovered does not stop the search for more. As long as there are untreatable infectious diseases, the search will continue. Even when effective treatment becomes available, it is always possible that a better, less toxic, or cheaper treatment can be found. Of the many chemotherapeutic agents currently available, none can cure viral infections. Consequently, much of today's drug research is focused on developing effective antiviral agents.

Genetics and Molecular Biology

Modern genetics began with the rediscovery in 1900 of Gregor Mendel's principles of genetics. Even after this significant event, for nearly three decades little progress was made in understanding how microbial characteristics are inherited. For this reason, microbial genetics is the youngest branch of microbiology. In 1928 the British scientist Frederick Griffith discovered that previously harmless bacteria could change their own nature so as to become capable of causing disease. The remarkable thing about this discovery was that live bacteria were shown to acquire heritable traits from dead ones. During the early 1940s, Oswald Avery, Maclyn McCarty, and Colin MacLeod of the Rockefeller Institute in New York City demonstrated that the change was produced by DNA. After that finding came the crucial discovery of the structure of DNA by James Watson and Francis Crick. This breakthrough ushered in the modern era of molecular genetics.

About the same time, the American geneticists Edward Tatum and George Beadle used genetic variations in the mold *Neurospora* to demonstrate how genetic information controls metabolism. In the early 1950s the American geneticist Barbara McClintock discovered that some genes (units of inherited information) can move from one location to another on a chromosome. Before McClintock's work, genes were thought to remain stationary. Her revolutionary discovery forced geneticists to revise their thinking about genes.

More recently, scientists have discovered the genetic basis that underlies the human body's ability to make an enormous diversity of *antibodies,* molecules that the immune system produces to combat invading microbes and their toxic products. Within cells of the immune system, genes are shuffled about and spliced together in various combinations, allowing the body to make millions of different antibodies, including some that can protect us from threats that the body has never previously encountered.

Tomorrow's History

Today's discovery is tomorrow's history. In an active research field such as microbiology, it is impossible to present a complete history. Some of the microbiologists omitted from this discussion are listed in Table 1.3. The period represented there, 1874–1917, is called the Golden Age of Microbiology. Many terms used to describe these scientists' accomplishments will be unfamiliar, but you will become familiar with them as you pursue the study of microbiology. Since 1900, Nobel Prizes have been awarded annually to outstanding scientists, many of whom were in the fields of physiology or medicine (Table 1.4). In some years the prize has been shared by several scientists, although the scientists may have made independent contributions. Refer to Tables 1.3 and 1.4 as you begin to study each new area of microbiology.

You can see from Table 1.4 that microbiology has been in the forefront of research in medicine and biology for several decades, and probably never more so than today. One reason is the renewed focus on infectious disease brought about by the advent of AIDS. Another is the dramatic progress in genetic engineering that has been made in the past two decades. Microorganisms have been and continue to be an essential part of the genetic engineering revolution. Most of the key discoveries that led to our present understanding of genetics emerged from research with microbes. Today scientists are attempting to redesign microorganisms for a variety of purposes (as we will see in Chapter 8). Bacteria have been converted into factories that produce drugs, hormones, vaccines, and a variety of biologically important compounds. And microbes, viruses in particular, are often the vehicle by which scientists insert new genes into

➤ Table 1.3   **The Golden Age of microbiology: Early microbiologists and their achievements**

Year	Investigator	Achievement
1874	Billroth	Discovery of round bacteria in chains
1876	Koch	Identification of *Bacillus anthracis* as causative agent of anthrax
1878	Koch	Differentiation of staphylococci
1879	Hansen	Discovery of *Mycobacterium leprae* as causative agent of leprosy
1880	Neisser	Discovery of *Neisseria gonorrhoeae* as causative agent of gonorrhea
1880	Laveran and Ross	Identification of life cycle of malarial parasites in red blood cells of infected humans
1880	Eberth	Discovery of *Salmonella typhi* as causative agent of typhoid fever
1880	Pasteur and Sternberg	Isolation and culturing of pneumonia cocci from saliva
1881	Koch	Animal immunization with attenuated anthrax bacilli
1882	Leistikow and Loeffler	Cultivation of *Neisseria gonorrhoeae*
1882	Koch	Discovery of *Mycobacterium tuberculosis* as causative agent of tuberculosis
1882	Loeffler and Schutz	Identification of actinobacillus that causes the animal disease glanders
1883	Koch	Identification of *Vibrio cholerae* as causative agent of cholera
1883	Klebs	Identification of *Corynebacterium diphtheriae* and toxin as causative agent of diphtheria
1884	Loeffler	Culturing of *Corynebacterium diphtheriae*
1884	Rosenbach	Pure culturing of streptococci and staphylococci
1885	Escherich	Identification of *Escherichia coli* as a natural inhabitant of the human gut
1885	Bumm	Pure culturing of *Neisseria gonorrhoeae*
1886	Flugge	Staining to differentiate bacteria
1886	Fraenckel	*Streptococcus pneumoniae* related to pneumonia
1887	Weichselbaum	*Neisseria meningitidis* related to meningitis
1887	Bruce	Identification of *Brucella melitensis* as causative agent of brucellosis in cattle
1887	Petri	Invention of culture dish
1888	Roux and Yersin	Discovery of action of diphtheria toxin
1889	Charrin and Roger	Discovery of agglutination of bacteria in immune serum
1889	Kitasato	Discovery that *Clostridium tetani* produces tetanus toxin
1890	Pfeiffer	Identification of Pfeiffer bacillus, *Haemophilus influenzae*
1890	von Behring and Kitasato	Immunization of animals with diphtheria toxin
1892	Ivanovski	Discovery of filterability of tobacco mosaic virus
1894	Roux and Kitasato	Identification of *Yersinia pestis* as causative agent of bubonic plague
1894	Pfeiffer	Discovery of bacteriolysis in immune serum
1895	Bordet	Discovery of alexin (complement) and hemolysis
1896	Widal and Grunbaum	Development of diagnostic test based on agglutination of typhoid bacilli by immune serum
1897	van Ermengem	Discovery of *Clostridium botulinum* as causative agent of botulism
1897	Kraus	Discovery of precipitins
1897	Ehrlich	Formulation of side-chain theory of antibody formation
1898	Shiga	Discovery of *Shigella dysenteriae* as causative agent of dysentery
1898	Loeffler and Frosch	Discovery of filterability of virus that causes foot-and-mouth disease
1899	Beijerinck	Discovery of intracellular reproduction of tobacco mosaic virus
1901	Bordet and Gengou	Identification of *Bordetella pertussis* as causative agent of whooping cough; development of complement fixation test
1901	Reed and colleagues	Identification of virus that causes yellow fever
1902	Portier and Richet	Work on anaphylaxis
1903	Remlinger and Riffat-Bey	Identification of virus that causes rabies
1905	Schaudinn and Hoffmann	Identification of *Treponema pallidum* as causative agent of syphilis
1906	Wasserman, Neisser, and Bruck	Development of Wasserman reaction for syphilis antibodies
1907	Asburn and Craig	Identification of virus that causes dengue fever
1909	Flexner and Lewis	Identification of virus that causes poliomyelitis
1915	Twort	Discovery of viruses that infect bacteria
1917	d'Herelle	Independent rediscovery of viruses that infect bacteria (bacteriophages)

➤ Table 1.4   **Nobel prize awards for research involving microbiology**

Year of prize	Prize winner	Topic studied
1901	von Behring	Serum therapy against diphtheria
1902	Ross	Malaria
1905	Koch	Tuberculosis
1907	Laveran	Protozoa and the generation of disease
1908	Ehrlich and Metchnikoff	Immunity
1913	Richet	Anaphylaxis
1919	Bordet	Immunity
1928	Nicolle	Typhus exanthematicus
1939	Domagk	Antibacterial effect of prontosil
1945	Fleming, Chain, and Florey	Penicillin
1951	Theiler	Vaccine for yellow fever
1952	Waksman	Streptomycin
1954	Enders, Weller, and Robbins	Cultivation of polio virus
1958	Lederberg	Genetic mechanisms
	Beadle and Tatum	Transmission of hereditary characteristics
1959	Ochoa and Kornberg	Chemical substances in chromosomes that play a role in heredity
1960	Burnet and Medawar	Acquired immunological tolerance
1962	Watson, Crick, and Wilkins	Structure of deoxyribonucleic acid
1965	Jacob, Lwoff, and Monod	Regulatory mechanisms in microbial genes
1966	Rous	Viruses and cancer
1968	Holley, Khorana, and Nirenberg	Genetic code
1969	Delbruck, Hershey, and Luria	Mechanism of virus infection in living cells
1972	Edelman and Porter	Structure and chemical nature of antibodies
1975	Baltimore, Temin, and Dulbecco	Interactions between tumor viruses and genetic material of the cell
1976	Blumberg and Gajdusek	New mechanisms for the origin and dissemination of infectious diseases
1978	Smith, Nathans, and Arber	Restriction enzymes for cutting DNA
1980	Benacerraf, Snell, and Dausset	Immunological factors in organ transplants
1980	Berg	Recombinant DNA
1984	Milstein, Köhler, and Jerne	Immunology
1987	Tonegawa	Genetics of antibody diversity
1988	Black, Elion, and Hitchings	Principles of drug therapy
1989	Bishop and Varmus	Genetic basis of cancer
1990	Murray, Thomas, and Corey	Transplant techniques and drugs
1993	Mullis	Polymerase chain reaction method to amplify (copy) DNA
1993	Smith	Method to splice foreign components into DNA
1993	Sharp and Roberts	Genes can be discontinuou

other organisms. Such techniques are beginning to enable us to produce improved varieties of plants and animals such as pest-resistant crops (➤ Figure 1.16) and may even enable us to correct genetic defects in human beings.

In September 1990, a 4-year-old girl became the first gene-therapy patient. She had inherited a defective gene that crippled her immune system. Doctors at the National Institutes of Health (NIH) inserted a normal copy of the gene into some of her white blood cells in the laboratory and then injected these gene-treated cells back into her body, where, it was hoped, they will restore her immune system. Critics were worried that a new gene randomly inserted into her white blood cells could damage other genes and cause cancer. The experiment was a success and she enjoys good health today.

New information is constantly being discovered and sometimes supersedes earlier findings. Occasionally, new discoveries lead almost immediately to the development of medical appli-

➤ Figure 1.16 **Plant scientists using knowledge gained from research.** Microorganisms can be used to produce superior agricultural crops, such as strains that resist insect pests or that have greater productivity due to soil microorganisms that help the plants grow better. The poinsettia on the right has been inoculated with a combination of fungi and bacteria that help the plant grow better.

cations, as occurred with penicillin and as will most certainly occur when a cure or vaccine for AIDS is discovered. However, old ideas such as spontaneous generation and old practices such as unsanitary measures in medicine can take years to replace. Many new bioethics problems will require considerable thought. Decisions regarding AIDS testing and reporting, transplants, environmental cleanup, and related issues will not come easily or quickly. Because of the wealth of prior knowledge, it is likely that you will learn more about microbiology in a single course than many pioneers learned in a lifetime. Yet, those pioneers deserve great credit because they worked with the unknown and had few people to teach them.

Human Genome Project

Microbial genetic techniques have made possible the undertaking of a colossal and controversial scientific plan, the Human Genome Project. At a cost of approximately $3 billion over a period of about 15 years, this project is making progress in identifying the location and chemical sequence of all the genes in the human genome—that is, all the genetic material in the human species—now estimated to consist of about 3 billion base pairs. The project is expected to be completed by the year 2005. When finished, it will be like having the "owner's manual" for humankind. It will make possible an incredible array of manipulations of human genetics and functions. Researchers are now locating and sequencing genes via simple microbes at first, such as *Escherichia coli,* a common organism found in the colon and feces. They have completely sequenced *Haemophilus influenzae,* a cause of ear infections; *Helicobacter pylori,* the cause of stomach ulcers; *Borrelia burgdorferi,* which causes Lyme disease; plus several other bacteria, viruses, and a yeast. Knowing the genetics of pathogens will help us discover means of combating their subterfuges for evading our immune systems. And knowing about good microbes will help us to make them even more useful such as in cleaning up toxic and nuclear waste sites, manufacturing fertilizer, and producing new drugs. Within a few years we'll probably have the complete sequences for 50 to 100 microbes. Now investigators are shifting to human genes as techniques have become more efficient.

In 1991, less than 2000 human genes were known. Then Craig Ventner and his colleagues devised more efficient techniques, and in a few months doubled that number. Today there is some controversy over who should finish sequencing the human genome, and whether patenting of these genes should be allowed (e.g., by pharmaceutical houses). In any case, the project will doubtless finish ahead of its original deadline of the year 2005. Says Ventner, "The secrets of life are all spelled out for us in the genome, we just have to learn how to read it."

✓ What were the scientific contributions of: Jenner, Metchnikoff, Ehrlich, Fleming, McClintock, and Ventner?

✓ When was the Golden Age of Microbiology? What types of discoveries were mostly made during this period?

✓ What is the Human Genome Project? How has microbiology been associated with it?

Retracing Our Steps

Why Study Microbiology?

- Microorganisms are part of the human environment and are therefore important to human health and activities.

- The study of microorganisms provides insight into life processes in all forms of life.

Scope of Microbiology

The Microbes

- **Microbiology** is the study of all **microorganisms (microbes)** in the microscopic range. These include **bacteria, algae, fungi, viruses,** and **protozoa**.

The Microbiologists

- Immunology, virology, chemotherapy, and genetics are especially active research fields of microbiology.

- Microbiologists work as researchers or teachers in university, clinical, and industrial settings. They do basic research in the biological sciences; help to perform or devise diagnostic tests; develop and test antibiotics and vaccines; work to control infection, protect public health, and safeguard the environment; and play important roles in the food and beverage industries.

Historical Roots

- The ancient Greeks, Romans, and Jews all contributed to early understandings of the spread of disease.

- Diseases such as bubonic plague and syphilis caused millions of deaths because of the lack of understanding of how to control or treat the infections.

- The development of high-quality lenses by Leeuwenhoek made it possible to observe microorganisms and later to formulate the **cell theory**.

The Germ Theory of Disease

- The **germ theory of disease** states that microorganisms (germs) can invade other organisms and cause disease.

Early Studies

- Progress in microbiology and acceptance of the germ theory of disease required that the idea of **spontaneous generation** be refuted. Redi and Spallanzani demonstrated that organisms did not arise from nonliving material. Pasteur, with his swan-necked flasks, and Tyndall, with his dust-free air, finally dispelled the idea of spontaneous generation.

Pasteur's Further Contributions

- Pasteur also studied wine making and disease in silkworms and developed the first rabies vaccine. His association of particular microbes with certain diseases furthered the establishment of the germ theory.

Koch's Contributions

- Koch developed four postulates that aided in the definitive establishment of the germ theory of disease. **Koch's postulates** are as follows:

  1. The specific causative agent must be found in every case of the disease.

  2. The disease organism must be isolated in pure culture.

  3. Inoculation of a sample of the culture into a healthy, susceptible animal must produce the same disease.

  4. The disease organism must be recovered from the inoculated animal.

- Koch also developed techniques for isolating organisms, identified the bacillus that causes tuberculosis, developed tuberculin, and studied various diseases in Africa and Asia.

Work Toward Controlling Infections

- Lister and Semmelweis contributed to improved sanitation in medicine by applying the germ theory and using aseptic technique.

Emergence of Special Fields of Microbiology

Immunology

- Immunization was first used against smallpox; Jenner used fluid from cowpox blisters to immunize against it.

- Pasteur developed techniques to weaken organisms so they would produce immunity without producing disease.

Virology

- Beijerinck characterized viruses as pathogenic molecules that could take over a host cell's mechanisms for their own use.

- Reed demonstrated that mosquitoes carry the yellow fever agent, and several other investigators identified viruses in the early twentieth century. The structure of DNA, the genetic material in many viruses and in all cellular organisms, was discovered by Watson and Crick.

- New techniques for isolating, propagating, and analyzing viruses were developed. Viruses could then be observed and in many cases crystallized, and their nucleic acids could be studied.

Chemotherapy

- Substances derived from medicinal plants were virtually the only source of chemotherapeutic agents until Ehrlich began a systematic search for chemically defined substances that would kill bacteria.

- Fleming and his colleagues developed penicillin, and Domagk and others developed sulfa drugs.

- Waksman and others developed streptomycin and other antibiotics derived from soil organisms.

Genetics and Molecular Biology

- Griffith discovered that previously harmless bacteria could change their nature so as to become capable of causing disease. This genetic change was shown by Avery, McCarty, and MacLeod to be due to DNA. Tatum and Beadle studied biochemical mutants of *Neurospora* to show how genetic information controls metabolism.

Tomorrow's History

- Microbiology has been at the forefront of research in medicine and biology, and microorganisms continue to play a critical role in genetic engineering and gene therapy.

Human Genome Project

- The Human Genome Project will identify the location and sequence of all bases in the human genome. Microbes and microbiological techniques have contributed to this work.

Terminology Check

algae   *(p. 3)*
antibiotics   *(p. 2)*
bacteria   *(p. 3)*
cell theory   *(p. 11)*
control variable   *(p. 19)*
experimental variable   *(p. 19)*

fungi   *(p. 4)*
germ theory of disease   *(p. 11)*
hypothesis   *(p. 19)*
Koch's postulates   *(p. 13)*
microbe   *(p. 3)*
microbiology   *(p. 3)*

microorganism   *(p. 1)*
prediction   *(p. 19)*
protozoa   *(p. 4)*
spontaneous generation   *(p. 11)*
variable   *(p. 19)*
viruses   *(p. 4)*

Meet Me on the Web

If you feel like you've got the lay of the land in this chapter, go check out the web for futher adventures. But before you do, make sure you are up to the challenge by taking at least some of these steps:

- Go back to the beginning of the chapter and make sure you can answer all of the questions I posed. Better yet, explain your answers to a classmate, roommate, friend, parent, or stranger on the bus. You'll be amazed at how your understanding of the material will be enhanced by explaining it to someone else.

- Review the concept check questions in the chapter. Answer them again to ensure that you have retained your understanding.

- Photocopy the key illustrations from this chapter, and use them to make your own flashcards. I think you'll find that actually making the flashcards is at least as beneficial as using them for review.

- Work through this same chapter in the companion Study Guide.

Once you are on the web, you will get one more chance to make sure you are on top of everything by taking a quiz for this chapter. It will challenge you with many different types of questions and then score your responses.

There are many other exciting features on the website, guaranteed to excite the explorer in you. To check them out, meet me at

http://www.prenhall.com/black

Fundamentals of Chemistry

Free-falling and weightless, you climb toward the others. You link hands with your skydiving partners to form the pattern you discussed just moments ago in the safety of the airplane. Suspended between heaven and earth you consider the significance of the connections that bind you.

Just as the hands of these skydivers hold them in a complex pattern, chemical bonds between atoms hold together complex molecular patterns. The shapes of molecules are very important—when shapes change, the properties of the molecules themselves change. In an organism, this change can mean the difference between life and death.

All living and nonliving things, including microbes, are composed of matter. Thus, it is not surprising that all properties of microorganisms are determined by the properties of matter.

Why Study Chemistry?

Chemistry is the science that deals with the basic properties of matter. Therefore we need to know some chemistry to begin to understand microorganisms. Chemical substances undergo changes and interact with one another in *chemical reactions*. Metabolism, the use of nutrients for energy or for making the substance of cells, consists of many different chemical reactions. This is true regardless of whether the organism is a human or a microorganism. Thus, understanding the basic principles of chemistry is essential to understanding metabolic processes in living things. A microbiologist uses chemistry to understand the structure and function of microorganisms themselves and to understand how they affect humans in disease processes as well as how they affect all life on earth.

Chemical Building Blocks and Chemical Bonds

Chemical Building Blocks

Matter is composed of very small particles that form the basic chemical building blocks. Over the years, chemists have observed matter and deduced the characteristics of these particles. Just as the alphabet can be used to make thousands of words, the chemical building blocks can be used to make thousands of different substances. The complexity of chemical substances greatly exceeds the complexity of words. Words rarely contain more than 20 letters, whereas some complex chemical substances contain as many as 20,000 building blocks!

The smallest chemical unit of matter is the **atom.** Many different kinds of atoms exist. Matter composed of one kind of atom is called an **element.** Each element has specific properties that distinguish it from other elements. Carbon is an element; a pure sample of carbon consists of a vast number of carbon atoms. Oxygen and nitrogen also are elements; they are found as gases in the earth's atmosphere. Chemists use one- or two-letter symbols to designate elements—such as C for carbon, O for oxygen, N for nitrogen, and Na for sodium (from its Latin name, *natrium*).

Atoms combine chemically in various ways. Sometimes atoms of a single element combine with each other. For example, carbon atoms form long chains that are im-

Questions We'll Explore

A Why is knowledge of basic chemistry necessary to the understanding of microbiology?

B What terms describe the organization of matter, and which elements are found in living organisms?

C What are the properties of chemical bonds and chemical reactions?

D Which properties of water, solutions, colloidal dispersions, acids, and bases make them important in living things?

E What is organic chemistry, and what are the major functional groups of organic molecules?

F How do the structures and properties of carbohydrates contribute to their roles in living things?

G How do the structures and properties of fats, phospholipids, and steroids contribute to their roles in living things?

H How do the structures and properties of proteins, including enzymes, contribute to their roles in living things?

I How do the structures and properties of nucleotides contribute to their roles in living things?

portant in the structure of living things. Both oxygen and nitrogen form paired atoms, O_2 and N_2. More often, atoms of one element combine with atoms of other elements. Carbon dioxide (CO_2) contains one atom of carbon and two atoms of oxygen; water (H_2O) contains two atoms of hydrogen and one atom of oxygen. (The subscripts in these formulas indicate how many atoms of each element are present.)

When two or more atoms combine chemically, they form a **molecule.** Molecules can consist of atoms of the same element, such as N_2, or atoms of different elements, such as CO_2. Molecules made up of atoms of two or more elements are called **compounds.** Thus, CO_2 is a compound, but N_2 is not. The properties of compounds are different from those of their component elements. For example, in their elemental state, both hydrogen and oxygen are gases at ordinary temperatures. They can combine to form water, however, which is a liquid at ordinary temperatures.

Living things consist of atoms of relatively few elements, principally carbon, hydrogen, oxygen, and nitrogen, but these are combined into highly complex compounds. A simple sugar molecule, $C_6H_{12}O_6$, contains 24 atoms. Many molecules found in living organisms contain thousands of atoms.

The Structure of Atoms

Although the atom is the smallest unit of any element that retains the properties of that element, atoms do contain even smaller particles that together account for those properties. Physicists study many such subatomic particles, but we discuss only **protons, neutrons,** and **electrons.** Three important properties of these particles are atomic mass, electrical charge, and location in the atom (Table 2.1). *Atomic mass* is measured in terms of *atomic mass units (AMU).* The mass of a proton or a neutron is almost exactly equal to 1 AMU; electrons have a much smaller mass. With respect to electrical charge, electrons are negatively ($-$) charged, and protons are positively ($+$) charged. Neutrons are neutral, with no charge. Atoms normally have an equal number of protons and electrons and so are electrically neutral. The heavy protons and neutrons are densely packed into the tiny, central *nucleus* of the atom, whereas the lighter electrons move around the nucleus in what have commonly been described as orbits.

The atoms of a particular element always have the same number of protons; that number of protons is the **atomic number** of the element. Atomic numbers range from 1 to over 100. The numbers of neutrons and electrons in the atoms of many elements can change, but the number of protons—and therefore the atomic number—remains the same for all atoms of a given element.

Protons and electrons are oppositely charged. Consequently, they attract each other. This attraction keeps the electrons near the nucleus of an atom. The electrons are in constant, rapid motion, forming an electron cloud around the nucleus. Because some electrons have more energy than others, chemists use a model with concentric circles, or *electron shells,* to suggest different energy levels. Electrons with the least energy are located nearest the nucleus, and those with more energy are farther from the nucleus. Each energy level corresponds to an electron shell (➤ Figure 2.1).

An atom of hydrogen has only one electron, which is located in the innermost shell. An atom of helium has two electrons in that shell; two is the maximum number of electrons that can be found in the innermost shell. Atoms with more than two electrons always have two electrons in the inner shell and up to eight additional electrons in the second shell. The inner shell is filled before electrons are found in the second shell; the second shell is filled before electrons are found in the third shell, and so on. Very large atoms have several more electron shells of larger capacity, but in elements found in living things, the outer shell is chemically stable if

➤ Table 2.1		**Properties of atomic particles**	
Particle	**Atomic mass**	**Electrical charge**	**Location**
Proton	1	+	Nucleus
Neutron	1	None	Nucleus
Electron	1/1836	−	Orbiting the nucleus

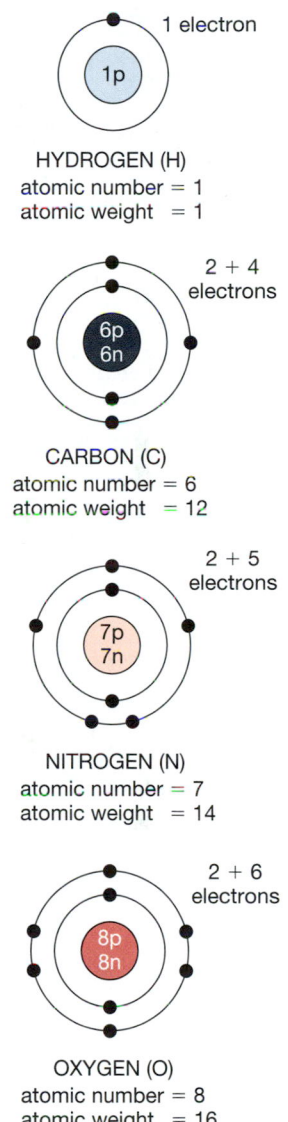

HYDROGEN (H)
atomic number = 1
atomic weight = 1

CARBON (C)
atomic number = 6
atomic weight = 12

NITROGEN (N)
atomic number = 7
atomic weight = 14

OXYGEN (O)
atomic number = 8
atomic weight = 16

➤ Figure 2.1   **The structure of four biologically important atoms.** Hydrogen, the simplest element, has a nucleus consisting of a single proton and a single electron in the first shell. In carbon, nitrogen, and oxygen the first shell is filled with two electrons and the second shell is partly filled. Carbon, with six protons in its nucleus, has six electrons, four of them in the second shell. Nitrogen has five electrons, and oxygen six electrons in the second shell. It is the electrons in the outermost shell that take part in chemical bonding.

**Sodium atom
Na**

(a)

**Chlorine atom
Cl**

**Formation of
ions**

**Sodium ion
Na⁺**

(b)

**Chloride ion
Cl⁻**

**Formation of
ionic bond**

**Sodium chloride
NaCl**

➤ **Figure 2.2**   **The formation of ions, or electrically charged atoms.** (a) When a neutral sodium atom loses the single electron in its outermost shell, the result is a sodium ion, Na⁺. When a neutral chlorine atom gains an extra electron in its outer shell, the result is a chloride ion, Cl⁻. (b) Oppositely charged ions attract one another. Such attraction creates an ionic bond and results in the formation of an ionic compound, in this case sodium chloride (NaCl).

Nitrogen is called stickstoff in German and azoto in italian, but its symbol is N in every country of the world.

it contains eight electrons. This principle, known as the **rule of octets,** is important for understanding chemical bonding, which we will discuss shortly.

Atoms whose outer electron shells are nearly full (containing six or seven electrons) or nearly empty (containing one or two electrons) tend to form ions. An **ion** is a charged atom produced when an atom gains or loses one or more electrons (➤ Figure 2.2a). When an atom of sodium (atomic number 11) loses the one electron in its outer shell without losing a proton, it becomes a positively charged ion, called a **cation** (kat'i-on). When an atom of chlorine (atomic number 17) gains an electron to fill its outer shell, it becomes a negatively charged ion, called an **anion** (an'i-on). In the ionized state, chlorine is referred to as chloride. Ions of elements such as sodium or chlorine are chemically more *stable* than atoms of these same elements because the ions' outer electron shells are full. Many elements are found in microorganisms or their environments as ions (Table 2.2). Those with one or two electrons in their outer shell tend to lose electrons and form ions with +1 or +2 charges, respectively; those with seven electrons in their outer shell tend to gain an electron and form ions with a charge of −1. Some ions, such as the hydroxyl (hi-drok'sil) ion (OH⁻), are compounds—they contain more than one element.

Although all atoms of the same element have the same atomic number, they may not have the same atomic weight. **Atomic weight** is the sum of the number of protons and neutrons in an atom. Many elements consist of atoms with differing atomic weights. For example, carbon usually has six protons and six neutrons; it has an atomic weight of 12. But some naturally occurring carbon atoms have one or two extra neutrons, giving these atoms an atomic weight of 13 or 14. In addition, laboratory techniques are available to create atoms with different numbers of neutrons. Atoms of a particular element that contain different numbers of neutrons are called **isotopes.** The superscript to the left of the symbol for the element indicates the atomic weight of the particular isotope. For example, carbon with an atomic weight of 14, which is often used to date fossils, is written ¹⁴C. The atomic weight of an element that has naturally occurring isotopes is the average atomic weight of the natural mixture of isotopes. Thus, atomic weights are not always whole numbers, even though any particular atom contains a specific number of whole neutrons and protons.

A **gram molecular weight,** or **mole,** is the weight of a substance in grams (g) equal to the sum of the atomic weights of the atoms in a molecule of the substance. For example, a mole of glucose, $C_6H_{12}O_6$, weighs 180 grams: [6 carbon atoms × 12 (atomic weight)] + [12 hydrogen atoms × 1 (atomic weight)] + [6 oxygen atoms × 16 (atomic weight)] = 180 grams. The mole is defined such that one mole of any substance always contains 6.023×10^{23} particles. Table 2.3 summarizes properties of elements found in living things.

➤ **Table 2.2**   **Some common ions**

Ion	Name	Brief description
Na⁺	Sodium	Contributes to salinity of natural bodies of water and body fluids of multicellular organisms.
K⁺	Potassium	Important ion that maintains cell turgor.
H⁺	Hydrogen	Responsible for the acidity of solutions and commonly regulates motility.
Ca²⁺	Calcium	Often acts as a chemical messenger.
Mg²⁺	Magnesium	Commonly required for chemical reactions to occur.
Fe²⁺	Ferrous iron	Carries electrons to oxygen during some chemical reactions that produce energy. Can prevent growth of some microbes that cause human disease.
NH₄⁺	Ammonium	Found in animal wastes and degraded by some bacteria.
Cl⁻	Chloride	Often found with a positively charged ion, where it usually neutralizes charge.
OH⁻	Hydroxyl	Usually present in excess in basic solutions where H⁺ is depleted.
HCO₃⁻	Bicarbonate	Often neutralizes acidity of bodies of water and body fluids.
NO₃⁻	Nitrate	A product of the action of certain bacteria that convert nitrite (NO₂⁻) into a form plants can use.
SO₄²⁻	Sulfate	Component of sulfuric acid in atmospheric pollutants and acid rain.
PO₄³⁻	Phosphate	Can be combined with certain other molecules to form high-energy bonds where energy is stored in a form living things can use.

| ➤ Table 2.3 | Some properties of important elements found in living organisms (in order of abundance and importance) |

Element	Symbol	Atomic number	Atomic weight	Electrons in outer shell	Biological occurrence
Oxygen	O	8	16.0	6	Component of biological molecules; required for aerobic metabolism
Carbon	C	6	12.0	4	Essential atom of all organic compounds
Hydrogen	H	1	1.0	1	Component of biological molecules; H^+ released by acids
Nitrogen	N	7	14.0	5	Component of proteins and nucleic acids
Calcium	Ca	20	40.1	2	Found in bones and teeth; regulator of many cellular processes
Phosphorus	P	15	31.0	5	Found in nucleic acids, ATP, and some lipids
Sulfur	S	16	32.0	6	Found in proteins; metabolized by some bacteria
Iron	Fe	26	55.8	2	Carries oxygen; metabolized by some bacteria
Potassium	K	19	39.1	1	Important intracellular ion
Sodium	Na	11	23.0	1	Important extracellular ion
Chlorine	Cl	17	35.4	7	Important extracellular ion
Magnesium	Mg	12	24.3	2	Needed by many enzymes
Copper	Cu	29	63.6	1	Needed by some enzymes; inhibits growth of some microorganisms
Iodine	I	53	126.9	7	Component of thyroid hormones
Fluorine	F	9	19.0	7	Inhibits microbial growth
Manganese	Mn	25	54.9	2	Needed by some enzymes
Zinc	Zn	30	65.4	2	Needed by some enzymes; inhibits microbial growth

Some isotopes are stable, and others are not. The nuclei of unstable isotopes tend to emit subatomic particles and radiation. Such isotopes are said to be *radioactive* and are called **radioisotopes.** Emissions from radioactive nuclei can be detected by radiation counters. Such emissions can be useful in studying chemical processes, but they also can harm living things.

Chemical Bonds

Chemical bonds form between atoms through interactions of electrons in their outer shells. Energy associated with these bonding electrons holds the atoms together, forming molecules. Three kinds of chemical bonds commonly found in living organisms are ionic, covalent, and hydrogen bonds.

Ionic bonds result from the attraction between ions that have opposite charges. For example, sodium ions, with a positive charge (Na^+) combine with chloride ions, with a negative charge (Cl^-) (➤ Figure 2.2b).

Many compounds, especially those that contain carbon, are held together by **covalent bonds.** Instead of gaining or losing electrons as in ionic bonding, carbon and some other atoms in covalent bonds share pairs of electrons (➤ Figure 2.3). One carbon atom, which has four electrons in its outer shell, can share an electron with each of four hydrogen atoms. At the same time, each of the four hydrogen atoms shares an electron with the carbon atom. Four pair of electrons are shared, each pair consisting of one electron from carbon and one electron from hydrogen. Such mutual sharing makes a carbon atom stable with eight electrons in its outer shell and a hydrogen atom stable with two electrons in its outer shell. Equal sharing produces *nonpolar compounds*—compounds with no charged regions. Sometimes a carbon atom and an atom such as an oxygen atom share two pair of electrons to form a double bond. The octet rule still applies, and each atom has eight electrons in its outer shell and is therefore stable. In structural formulas, chemists use a single line to represent a single pair of shared electrons and a double line to represent two pair of shared electrons (➤ Figure 2.3).

Atoms of four elements, carbon, hydrogen, oxygen, and nitrogen, commonly form covalent bonds that fill their outer electron shells. Carbon shares four electrons, hydrogen one electron, oxygen two electrons, and nitrogen three electrons. Unlike many ionic bonds, covalent bonds are stable and thus are important in molecules that form biological structures.

Covalent bonds: methane
(four single bonds)
CH_4

(a)

Formaldehyde
(double bond)
CH_2O

(b)

➤ **Figure 2.3   Covalent bonds are formed by sharing electrons.** (a) In methane, a carbon atom, with four electrons in its outermost shell, shares pairs of electrons with four hydrogen atoms. In this way all five atoms acquire stable, filled outer shells. Each shared electron pair constitutes a single covalent bond. (b) In formaldehyde, a carbon atom shares pairs of electrons with two hydrogen atoms and also shares two pairs of electrons with an oxygen atom, forming a double covalent bond.

Hydrogen bonds, though weaker than ionic and covalent bonds, are important in biological structures and are typically present in large numbers. The atomic nuclei of oxygen and nitrogen attract electrons very strongly. When hydrogen is covalently bonded to oxygen or nitrogen, the electrons of the covalent bond are shared unevenly—they are held closer to the oxygen or nitrogen than to the hydrogen. The hydrogen atom then has a partial positive charge, and the other atom has a partial negative charge. In this case of unequal sharing, the molecule is called a **polar compound** because of its oppositely charged regions. The weak attraction between such partial charges is called a hydrogen bond.

Polar compounds such as water often contain hydrogen bonds. In a water molecule, electrons from the hydrogen atoms stay closer to the oxygen atom, and the hydrogen atoms lie to one side of the oxygen atom (➤ Figure 2.4). Thus, water molecules are polar molecules that have a positive hydrogen region and a negative oxygen region. Covalent bonds between the hydrogen and oxygen atoms hold the atoms together. Hydrogen bonds between the hydrogen and oxygen regions of different water molecules hold the molecules in clusters.

Hydrogen bonds also contribute to the structure of large molecules such as proteins and nucleic acids, which contain long chains of atoms. The chains are coiled or folded into a three-dimensional configuration that is held together in part by hydrogen bonds.

Chemical Reactions

Chemical reactions in living organisms typically involve the use of energy to form chemical bonds and the release of energy as chemical bonds are broken. For example, the food we eat consists of molecules that have much energy stored in their chemical bonds. During **catabolism** (ka-tab'o-lizm), the breakdown of substances, food is degraded and some of that stored energy is released. Microorganisms use nutrients in the same general way. A catabolic reaction can be symbolized as follows:

$$X—Y \rightarrow X + Y + \text{energy}$$

where X—Y represents a nutrient molecule and where energy was originally stored in the bond between X and Y.

Catabolic reactions are **exergonic**—that is, they release energy. Conversely, energy is used to form chemical bonds in the synthesis of new compounds. In **anabolism** (a-nab'o-lizm), the buildup, or *synthesis,* of substances, energy is used to create bonds. An anabolic reaction can be symbolized as follows:

$$X + Y + \text{energy} \rightarrow X—Y$$

where energy is stored in the new substance X—Y. Anabolic reactions occur in living cells when small molecules are used to synthesize large molecules. Cells can store small amounts of

➤ **Figure 2.4   Polar compounds and hydrogen bonding.** Water molecules are polar—they have a region with a partial positive charge (the hydrogen atoms) and a region with a partial negative charge (the oxygen atom). Hydrogen bonds, created by the attraction between oppositely charged regions of different molecules, hold the water molecules together in clusters.

Hydrogen
bonding

energy for later use or can expend energy to make new molecules. Most anabolic reactions are **endergonic**—that is, they require energy.

Water and Solutions

Water, one of the simplest of chemical compounds, is also one of the most important to living things. It takes part directly in many chemical reactions. Numerous substances dissolve in water or form mixtures called colloidal dispersions. Acids and bases exist and function principally in water mixtures.

Water

Water is so essential to life that humans can live only a few days without it. Many microorganisms die almost immediately if removed from their normal aqueous environments, such as lakes, ponds, oceans, and moist soil. Yet, others can survive for several hours or days without water, and spores formed by a few microorganisms survive for many years away from water. Several bacteria find the moist, nutrient-rich secretions of human skin glands to be an ideal environment.

Water has several properties that make it important to living things. Because water is a polar compound and forms hydrogen bonds, it can form thin layers on surfaces and can act as a *solvent,* or dissolving medium. Water is a good solvent for ions because the polar water molecules surround the ions. The positive region of water molecules is attracted to negative ions, and the negative region of water molecules is attracted to positive ions. Many different kinds of ions can therefore be distributed evenly through a water medium, forming a *solution* (➤ Figure 2.5).

Water forms thin layers because it has a high surface tension. **Surface tension** is a phenomenon in which the surface of water acts as a thin, invisible, elastic membrane (➤ Figure 2.6). The polarity of water molecules gives them a strong attraction for one another but no attraction for gas molecules in air at the water's surface. Therefore, surface water molecules cling together, forming hydrogen bonds with other molecules below the surface. In living cells this feature of surface tension allows a thin film of water to cover membranes and to keep them moist.

Water has a high *specific heat,* that is, it can absorb or release large quantities of heat energy with little temperature change. This property of water helps to stabilize the temperature of living organisms, which are composed mostly of water, as well as bodies of water where many microorganisms live.

Finally, water provides the medium for most chemical reactions in cells, and it participates in many of these reactions. Suppose, for example, that substance X can gain or lose H^+ and that

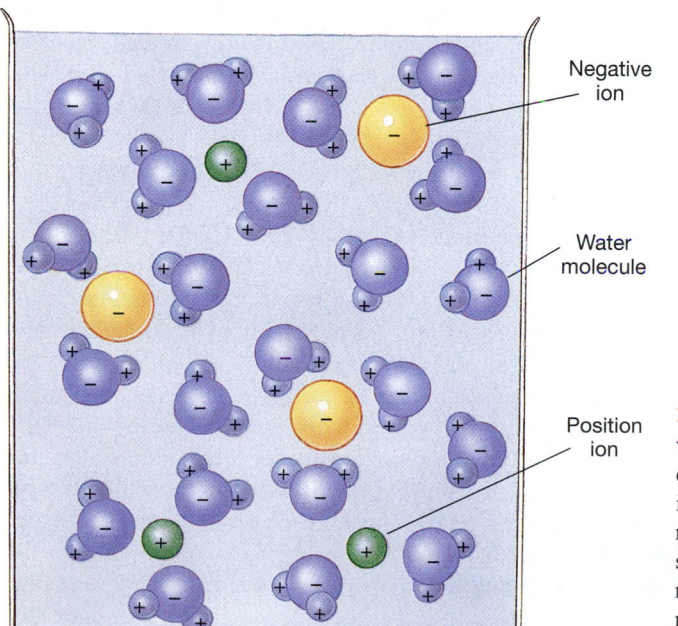

> Figure 2.5 **Polarity and water molecules.** Polarity enables water to dissolve many ionic compounds. The positive regions of the water molecules surround negative ions, and the negative regions of the water molecules surround positive ions, holding the ions in solution.

Negative ion

Water molecule

Position ion

✓ Which number tells us the identity of an atom?

✓ Is it possible to have a molecule of an element? A molecule of a compound? Give examples.

✓ If isotopes can be thought of as "twins," "triplets," "sextuplets," and so on, in what ways are they identical? Different?

✓ What type of bonding is produced by the equal sharing of an electron pair between two atoms? By unequal sharing?

A Winter Dilemma

Picture it—a harsh winter, snow and ice, roads covered with salt and sand. Now it's spring. However, there's something wrong with the trees and other plants growing along the roads. Has excessive release of chemicals from winter salt runoff affected soil chemistry or the soil's ability to support microbes?

As an environmental scientist with the State Laboratories, it's your job to investigate. You test the soil and runoff water for chemicals used during the winter. Are there high concentrations of de-icing chemicals near the affected roads? How far have they spread? Where is normal plant growth again observed? Do collected soils demonstate typical and divergent microbial populations? Are there any unusual pathologies associated with the plants growing near the affected area?

Place yourself in this scenario and use the scientific method to design an experiment to show what is happening here.

> Figure 2.6   **Surface tension.**
(a) Hydrogen bonding between water molecules creates surface tension, which causes the surface of water to behave like an elastic membrane.
(b) Surface tension is strong enough to support the weight of the insects known as water striders.

(a)

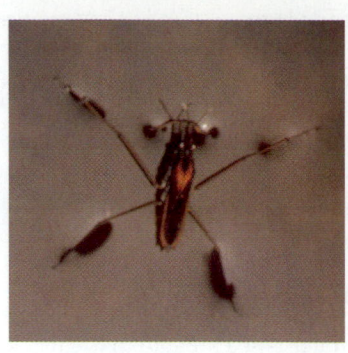

(b)

substance Y can gain or lose OH^-. The substances that enter a reaction are called **reactants.** In an anabolic reaction, the components of water (H^+ and OH^-) are removed from the reactants to form a larger product molecule:

$$X—H + HO—Y \rightarrow X—Y + H_2O$$

This kind of reaction, called **dehydration synthesis,** is involved in the synthesis of complex carbohydrates, some lipids (fats), and proteins. Conversely, in many catabolic reactions, water is added to a reactant to form simpler products:

$$X—Y + H_2O \rightarrow X—H + HO—Y$$

This kind of reaction, called **hydrolysis,** occurs in the breakdown of large nutrient molecules to release simple sugars, fatty acids, and amino acids.

Solutions and Colloids

Solutions and colloidal dispersions are examples of mixtures. Unlike a chemical compound, which consists of molecules whose atoms are present in specific proportions, a **mixture** consists of two or more substances that are combined in any proportion and are not chemically bound. Each substance in a mixture contributes its properties to the mixture. For example, a mixture of sugar and salt could be made using any proportions of the two ingredients. The degree of sweetness or saltiness of the mixture would depend on the relative amounts of sugar and salt present, but both sweetness and saltiness would be detectable.

A **solution** is a mixture of two or more substances in which the molecules of the substances are evenly distributed and ordinarily will not separate out upon standing. In a solution the medium in which substances are dissolved is the **solvent.** The substance dissolved in the solvent is the **solute.** Solutes can consist of atoms, ions, or molecules. In cells and in the medium in which cells live, water is the solvent in nearly all solutions. Typical solutes include the sugar glucose, the gases carbon dioxide and oxygen, and many different kinds of ions. Many smaller proteins also can act as solutes in true solutions.

Few living things can survive in highly concentrated solutions. We make use of this fact in preserving several kinds of foods. Can you think of foods that are often kept unrefrigerated and unsealed for long periods of time? Jellies, jams, and candies do not readily spoil because most microorganisms cannot tolerate the high concentration of sugar. Salt-cured meats are too salty to allow growth of most microorganisms, and pickles are too acidic for most microbes.

Particles too large to form true solutions can sometimes form *colloidal dispersions,* or **colloids.** Gelatin dessert is a colloid in which the protein gelatin is dispersed in a watery medium. Similarly, colloidal dispersions in cells usually are formed from large protein molecules dispersed in water. The fluid or semifluid substance inside living cells is a complex colloidal system. Large particles are suspended by opposing electrical charges, layers of water molecules around them, and other forces. Media for growing microorganisms sometimes are solidified with agar; these media are colloidal dispersions. Some colloidal systems have the ability to change from a semisolid state, such as gelatin that has "set," to a more fluid state, such as gelatin that has melted. Amoebae seem to move, in part, by the ability of the colloidal material within them to change back and forth between semisolid and fluid states.

Bacterial Acids Are Eating *The Last Supper*

Past civilizations built their temples, tombs, and monuments out of stone to last forever. Indeed, structures such as the Egyptian pyramids have lasted for thousands of years. But recently, the ancient stone has begun to crumble into dust. Experts first believed that the destruction of stone was due to the gaseous pollutants spewed into the air from automobiles and industrial chimneys. Now, however, experts have found that a whole new world—a microsystem of microbes—is living in the stone itself, interacting with the pollutant gases and wreaking chemical havoc.

The prime villain devouring ancient marble is *Thiobacillus thioparis,* a bacterium that converts the pollutant sulfur dioxide (SO_2) gas into sulfuric acid (H_2SO_4). The sulfuric acid acts on the calcium carbonate ($CaCO_3$) in marble, forming carbon dioxide (CO_2) and the salt calcium sulfate ($CaSO_4$), which is a form of plaster. The bacteria use the CO_2 as their source of carbon. The plaster produced by this process is soft and washes away by rain or just crumbles and falls off. On many buildings, such as the Parthenon in Greece, this bacterial "epidemic" has already eroded a 2-in.-thick layer of marble surface into plaster. In fact, more marble has been destroyed in the last 35 years than had been destroyed in the previous 300 years.

Not all the damage, however, is due to chemical processes. Some is due to the physical activities of fungi pushing their threadlike growth (hyphae) into the rock, splitting and reducing it to powder. Dozens of kinds of bacteria, yeast, filamentous fungi, and even algae produce acids that eventually eat into the stone. Even paintings aren't immune to the effects of microbial growth. Frescoes painted on walls, such as Leonardo da Vinci's *Last Supper,* are also being consumed, pieces flaking off and brilliant colors being dulled.

Is there a cure for these microbial invasions? A treatment plan identifies which microorganisms are attacking the stone and then determines which antibiotics will kill them most effectively. Administering antibiotics to a building is a tricky business. A spray gun must be used instead of a hypodermic needle or pills. But some antibiotics cannot be

A close-up of Leonardo da Vinci's *Last Supper* shows deterioration due to microbial action.

applied by spraying, so it is difficult to find a suitable one. Sometimes disinfectants such as isothiazolinone chloride are used, but such chemicals must be used carefully. In Angkor, Cambodia, for instance, unskilled Army workers armed with hard brushes and fungicidal solutions have scrubbed away ancient temple paintings in their zeal to remove fungal growth.

Unfortunately, antibiotic treatment does not reverse damage already done. Statues cannot grow a new skin, as you might do after recovering from an infection. Therefore, researchers are working on a process to harden the plaster layer formed by microbes by baking the affected pieces at high temperatures. This method, however, is hardly feasible for an entire statue or a cathedral. And how do we protect these vulnerable structures as long as the pollutants remain in the atmosphere? Stopgap measures have included taking some statues inside and building protective domes over others. The real—but elusive—answer lies in cleaning up our environment.

Acids, Bases, and pH

In chemical terms, most living things exist in relatively neutral environments, but some microorganisms live in environments that are *acidic* or *basic (alkaline)*. Understanding acids and bases is important in studying microorganisms and in studying their effects on human cells. An **acid** is a hydrogen ion (H^+) donor. (A hydrogen ion is a proton.) An acid donates H^+ to a solution. The acids found in living organisms usually are weak acids such as acetic acid (vinegar), although some are strong acids such as hydrochloric acid. Acids release H^+ when carboxyl groups (—COOH) ionize to COO^- and H^+. A **base** is a proton acceptor, or a hydroxyl ion donor. It accepts H^+ from, or donates OH^- (hydroxyl ion) to, the solution. The bases found in living organisms usually are weak bases such as the amino (NH_2) group, which accepts H^+ to form NH_3^+.

Hydrochloric acid in your stomach digests chicken proteins as well as the *Salmonella* bacteria that may be on the chicken.

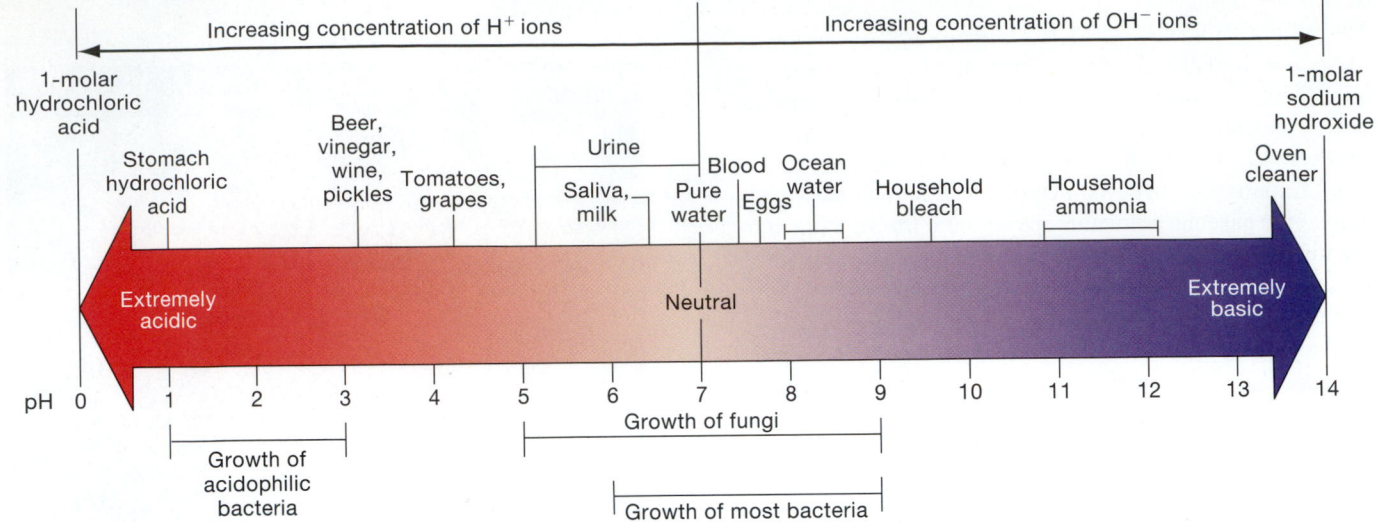

> Figure 2.7   **The pH values of some common substances.** Each unit of the pH scale represents a tenfold increase or decrease in the concentration of hydrogen ions. Thus, vinegar, for example, is 10,000 times more acidic than pure water.

✓ What properties of a water molecule enable it to act as a good solvent for ionic molecules?

✓ Why is a higher number, such as pH 11, a stronger base than pH 9, whereas a higher number, such as pH 5, is a weaker acid than pH 3?

Chemists have devised the concept of pH to specify the acidity or alkalinity of a solution. The **pH** scale (➤ Figure 2.7), which relates proton concentrations to pH, is a logarithmic scale. This means that the concentration of hydrogen ions (protons) changes by a factor of 10 for each unit of the scale. The practical range of the pH scale is from 0 to 14. A solution with a pH of 7 is **neutral**—neither acidic nor **alkaline** (basic). Pure water has a pH of 7 because the concentrations of H^+ and OH^- in it are equal. ➤ Figure 2.7 shows the pH of some body fluids, selected foods, and other substances.

Complex Organic Molecules

The basic principles of general chemistry also apply to **organic chemistry,** the study of compounds that contain carbon. The study of the chemical reactions that occur in living systems is the branch of organic chemistry known as **biochemistry.** Early in the 1800s it was believed that molecules from living things were filled with a supernatural "vital force" and therefore could not be explained by the laws of chemistry and physics. It was considered impossible to make *organic compounds* outside of living systems. That idea was disproved in 1828 when the German scientist Friedrich Wohler synthesized the organic compound urea, a small molecule excreted as a waste material by many animals. Since that time thousands of organic compounds—plastics, fertilizers, and medicines—have been made in the laboratory. Organic compounds such as carbohydrates, lipids, proteins, and nucleic acids occur naturally in living things and in the products or remains of living things. The ability of carbon atoms to form covalent bonds and to link up in long chains makes possible the formation of an almost infinite number of organic compounds.

The simplest carbon compounds are the *hydrocarbons,* chains of carbon atoms with their associated hydrogen atoms. The structure of the hydrocarbon propane, C_3H_8, for example, is as follows:

$$
\begin{array}{ccccccc}
 & H & & H & & H & \\
 & | & & | & & | & \\
H - & C & - & C & - & C & - H \\
 & | & & | & & | & \\
 & H & & H & & H & \\
\end{array}
$$

Carbon chains can have not only hydrogen but other atoms such as oxygen and nitrogen bound to them. Some of these atoms form functional groups. A **functional group** is a part of a molecule that generally participates in chemical reactions as a unit and that gives the molecule some of its chemical properties.

Some Like It Sour

Most natural environments have pH values between 5 and 9, and microbes living in those locations grow within that pH range. The known species of bacteria that grow at pH values below 2 or greater than 10 often have unique properties that we are able to use to our advantage. Certain bacteria that live at low pH, acidophiles, are used to leach economically important metals from low-yielding ores. Low-grade copper ore, dumped into a large pile, is treated with a liquid containing dilute sulfuric acid and the bacterium *Thiobacillus ferrooxidans*. The presence of *Thiobacillus ferrooxidans* increases the rate of copper sulfite oxidation and the production of copper sulfate. Because copper sulfate is extremely water-soluble, it is possible to economically extract and precipitate copper from the liquid collected from the leach dump.

Ketone

Alcohol   Aldehyde   Organic acid

Reduced ◄─────────────────────────► **Oxidized**

➤ **Figure 2.8   Four classes of organic compounds that incorporate oxygen.** Alcohols contain one or more hydroxyl groups (—OH); aldehydes and ketones contain carbonyl groups (—C=O); and organic acids contain carboxyl groups (—COOH).

Four significant groups of compounds-alcohols, aldehydes, ketones, and organic acids—have functional groups that contain oxygen (➤ Figure 2.8). An alcohol has one or more hydroxyl groups (— OH). An aldehyde has a carbonyl group (— CO) at the end of the carbon chain; a ketone has a carbonyl group within the chain. An organic acid has one or more carboxyl groups (—COOH). One key functional group that does not contain oxygen is the amino group (— NH_2). Found in amino acids, amino groups account for the nitrogen in proteins.

The relative amount of oxygen in different functional groups is significant. Groups with little oxygen, such as alcohol groups, are said to be *reduced;* groups with relatively more oxygen, such as carboxyl groups, are said to be *oxidized* (➤ Figure 2.8). As we shall see in Chapter 5, *oxidation* is the addition of oxygen or the removal of hydrogen or electrons from a substance. Burning is an example of oxidation. *Reduction* is the removal of oxygen or the addition of hydrogen or electrons to a substance. In general, the more reduced a molecule, the more energy it contains. Hydrocarbons, such as gasoline, have no oxygen and thus represent the extreme in energy-rich, reduced molecules. They make good fuels because they contain so much energy. Conversely, the more oxidized a molecule, the less energy it contains. Carbon dioxide (CO_2) represents the extreme in an oxidized molecule because no more than two oxygen atoms can bond to a single carbon atom. As we shall see, oxidation releases energy from molecules.

Let us now consider the major classes of large, complex biochemical molecules of which all living things, including microbes, are composed.

Carbohydrates

Carbohydrates serve as the main source of energy for most living things. Plants make carbohydrates, including structural carbohydrates such as cellulose and energy-storage carbohydrates such as starch. Animals use carbohydrates as food, and many, including humans, store energy in a carbohydrate called *glycogen.* Many microorganisms use carbohydrates from their environment for energy and also make a variety of carbohydrates. Carbohydrates in the membranes of cells can act as markers that make a cell chemically recognizable. Chemical recognition is important in immunological reactions and other processes in living things.

All carbohydrates contain the elements carbon, hydrogen, and oxygen, generally in the proportion of two hydrogen atoms to each carbon and oxygen atom. There are three groups of carbohydrates: monosaccharides, disaccharides, and polysaccharides. **Monosaccharides** consist of a carbon chain or ring with several alcohol groups and one other functional group, either an aldehyde group or a ketone group. Several monosaccharides, such as glucose and fructose, are **isomers**—they have the same molecular formula, $C_6H_{12}O_6$, but different structures and different properties (➤ Figure 2.9). Thus, even at the chemical level we can see that structure and function are related.

Glucose, the most abundant monosaccharide, can be represented schematically as a straight chain, a ring, or a three-dimensional structure. The chain structure, in ➤ Figure 2.10a, clearly shows a carbonyl group at carbon 1 (the first carbon in the chain, at the top in this orientation)

Glucose
($C_6H_{12}O_6$)

Fructose
($C_6H_{12}O_6$)

➤ **Figure 2.9   Isomers.** Glucose and fructose are isomers: They contain the same atoms, but they differ in structure.

➤ Figure 2.10   **Three ways of representing the glucose molecule.** (a) In solution, the straight-chain form is rarely found. (b) Instead, the molecule bonds to itself, forming a six-membered ring. The ring is conventionally depicted as a flat hexagon. (c) The actual three-dimensional structure is more complex. The spheres in this depiction represent carbon atoms.

Candy-Coated Chemistry

Liquid centers, such as those found in chocolate-covered cherries, are made from invert sugar. Invert sugar is used to make many different confections including jelly, candy, and prepared frostings. Invert sugar is beet sugar (sucrose) that has been treated with the enzyme *invertase*. The enzyme breaks down sucrose into its two components, glucose and fructose, thereby reducing the size of the sugar crystals. Invert sugar produces a smoother product as a result of its fine crystal structure. In nature, invertase is associated with many tuber vegetables, such as beets and potatoes, and is synthesized by many bacteria. What do you think the role of invertase might be in nature? in bacteria?

and alcohol groups on all the other carbons. ➤ Figure 2.10b shows how a glucose molecule in solution rearranges and bonds to itself to form a closed ring. The three-dimensional projection in ➤ Figure 2.10c more closely approximates the actual shape of the molecule. In studying structural formulas, it is important to imagine each molecule as a three-dimensional object.

Monosaccharides can be reduced to form deoxy sugars and sugar alcohols (➤ Figure 2.11). The deoxy sugar *deoxyribose,* which has a hydrogen atom instead of an —OH group on one of its carbons, is a component of DNA. Certain sugar alcohols, which have an additional alcohol group instead of an aldehyde or ketone group, can be metabolized by particular microorganisms. Mannitol and other sugar alcohols are used to identify some microorganisms in diagnostic tests.

Disaccharides are formed when two monosaccharides are connected by the removal of water and the formation of a **glycosidic bond,** a sugar alcohol/sugar linkage (➤ Figure 2.12a). Sucrose, common table sugar, is a disaccharide made of glucose and fructose. **Polysaccharides** are formed when many monosaccharides are linked by glycosidic bonds (➤ Figure 2.12b). Polysaccharides such as starch, glycogen, and cellulose are **polymers**—long chains of repeating units—of glucose. However, the glycosidic bonds in each polymer are arranged differently. Plants and most algae make starch and cellulose. Starch serves as a way to store energy, and cellulose is a structural component of cell walls. Animals make and store glycogen, which they can break down to glucose as energy is needed. Microorganisms contain several other important polysaccharides, as we shall see in later chapters.

Table 2.4 summarizes the types of carbohydrates.

(a) Deoxyribose   (b) Ribose   (c) Glycerol   (d) Mannitol

➤ Figure 2.11   **Deoxy sugars and sugar alcohols.** (a) "Deoxy" indicates one less oxygen atom. One of the carbon atoms of the deoxy sugar deoxyribose lacks a hydroxyl group that (b) ribose has. (c) Glycerol is a three-carbon sugar alcohol that is a component of fats. (d) Mannitol is a sugar alcohol used in diagnostic tests for certain microbes.

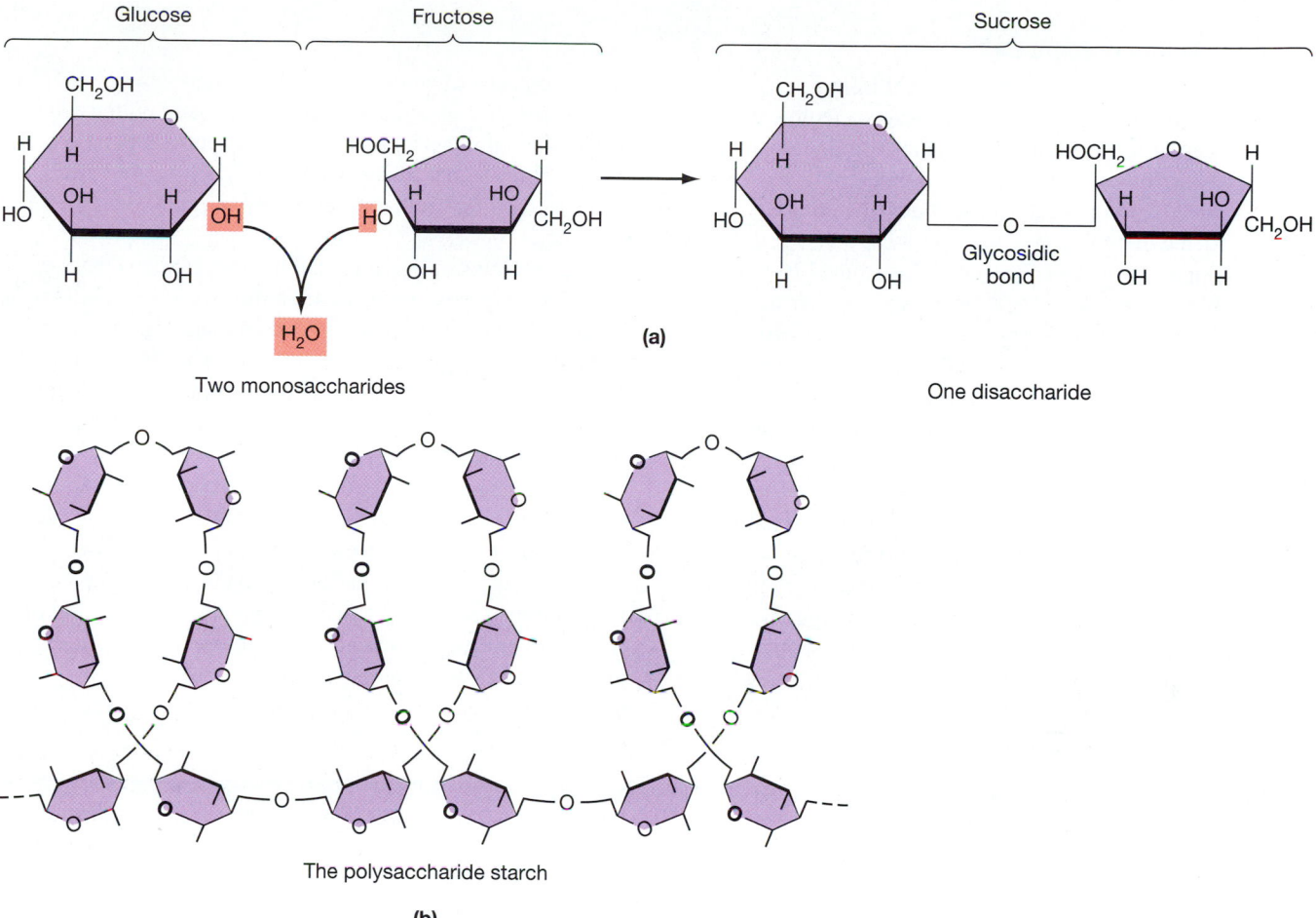

(a)

Two monosaccharides   One disaccharide

The polysaccharide starch

(b)

➤ Figure 2.12   **Disaccharides and polysaccharides.** (a) Two monosaccharides are joined to form a disaccharide by dehydration synthesis and the formation of a glycosidic bond. (b) Polysaccharides such as starch are formed by similar reactions that link many monosaccharides into long chains.

➤ Table 2.4	Types of carbohydrates	
Class of carbohydrates	**Examples**	**Description and occurrence**
Monosaccharides	Glucose	Sugar found in most organisms
	Fructose	Sugar found in fruit
	Galactose	Sugar found in milk
	Ribose	Sugar found in RNA
	Deoxyribose	Sugar found in DNA
Disaccharides	Sucrose	Glucose and fructose; table sugar
	Lactose	Glucose and galactose; milk sugar
	Maltose	Two glucose units; product of starch digestion
Polysaccharides	Starch	Polymer of glucose stored in plants, digestible by humans
	Glycogen	Polymer of glucose stored in animal liver and skeletal muscles
	Cellulose	Polymer of glucose found in plants, not digestible by humans; digested by some microbes

Ticks' body walls are impervious to most pesticides thanks to chitin, a polysaccharide, also found in fungal cell walls.

Can a Cow Actually Explode?

Cows can derive a good deal of nourishment from grass, hay, and other fibrous vegetable matter that are inedible to humans. We can't digest cellulose, the chief component of plants. If you had to live on hay, you would probably starve to death. How, then, do cows and other hooved animals manage on such a diet?

Oddly enough, cows can't digest cellulose either. But they don't need to—it's done for them. Cows and their relatives harbor in their stomachs large populations of microorganisms that do the work of breaking down cellulose into sugars that the animal can use. The same is true of termites: If it weren't for microbes in their guts that help them to digest cellulose, they couldn't dine on the wooden beams in your house.

Cellulose is very similar to starch—both consist of long chains of glucose molecules. The bonds between these molecules, however, are slightly different in their geometry in the two substances. As a result, the enzymes that animals use to break down a starch molecule into its component glucose units have no effect on cellulose. In fact, very few organisms produce enzymes that can attack cellulose. Even the protists (unicellular organisms with a nucleus) that live

in the stomachs of cows and termites cannot always do it by themselves. Just as cows and termites depend on the protists in their stomachs, the protists frequently depend on certain bacteria that reside permanently within them. It is these bacteria that actually make the essential digestive enzymes.

The activities of the intestinal microorganisms that perform these digestive services are a mixed blessing, both to the cows and to the humans who keep them. The bacteria also produce methane gas, CH_4—as much as 190 to 380 l per day from a single cow. (Methane production can be so rapid that a cow's stomach may rupture if the cow can't burp. Some ingenious inventors have actually patented cow safety valves to release the gas buildup directly through the animal's side.) When this gas eventually makes its way out of the cow by one route or another, it rises to the upper atmosphere. There it is suspected of contributing to the "greenhouse effect," trapping solar heat and causing an overall warming of the earth's climate (Chapter 25). Scientists have estimated that the world's cows release 50 million metric tons of methane annually; that's not counting the sheep, goats, antelope, water buffalo, and other grass eaters.

Lipids

Lipids constitute a chemically diverse group of substances that includes fats, phospholipids, and steroids. They are relatively insoluble in water but are soluble in nonpolar solvents such as ether and benzene. Lipids form part of the structure of cells, especially cell membranes, and many can be used for energy. Generally, lipids contain relatively more hydrogen and less oxygen than carbohydrates and therefore contain more energy than carbohydrates.

Fats contain the three-carbon alcohol glycerol and one or more fatty acids. A **fatty acid** consists of a long chain of carbon atoms with associated hydrogen atoms and a carboxyl group at one end of the chain. The synthesis of a fat from glycerol and fatty acids involves removing water and forming an ester bond between the carboxyl group of the fatty acid and an alcohol group of glycerol (➤ Figure 2.13a). A **triacylglycerol,** formerly called a *triglyceride,* is a fat formed when three fatty acids are bonded to glycerol. *Monoacylglycerols* (monoglycerides) and *diacylglycerols* (diglycerides) contain one and two fatty acids, respectively, and usually are formed from the digestion of triacylglycerols.

Fatty acids can be saturated or unsaturated. A **saturated fatty acid** contains all the hydrogen it can have; that is, it is saturated with hydrogen (➤ Figure 2.13b). An **unsaturated fatty acid** has lost at least two hydrogen atoms and contains a double bond between the two carbons that have lost hydrogen atoms (➤ Figure 2.13c). "Unsaturated" thus means not completely saturated with hydrogen. Oleic acid is an unsaturated fatty acid. *Polyunsaturated fats,* many of which are vegetable oils that remain liquid at room temperature, contain many unsaturated fatty acids.

Some lipids contain one or more other molecules in addition to fatty acids and glycerol. For example, **phospholipids,** which are found in all cell membranes, differ from fats by the substitution of phosphoric acid (H_3PO_4) for one of the fatty acids (➤ Figure 2.14a). The charged phosphate group (— HPO_4^-) typically attaches to another charged group. Both can mix with

Higher fat content of food causes slower movement of feces through the bowels. There, bacteria convert the undigested fats into cancer-causing compounds.

Evidence from Babylonian excavations indicate that soapmaking was known as early as 2800 B.C.

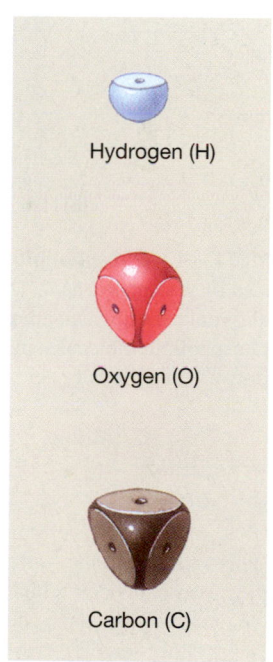

Hydrogen (H)

Oxygen (O)

Carbon (C)

O
‖
R—C—OH   HO—C—H
O   H
‖
R—C—OH   HO—C—H
O
‖
R—C—OH   HO—C—H
H

O   H
‖
R—C—O—C—H
O
‖
R—C—O—C—H   + 3H₂O
O
‖
R—C—O—C—H
H

Ester bond

Three fatty   +   Glycerol
acids

Triacylglycerol

(a)

H₃C   CH₂   CH₂   CH₂   CH₂   CH₂   CH₂   CH₂   CH₂   O
CH₂   CH₂   CH₂   CH₂   CH₂   CH₂   CH₂   CH₂   C
OH

(b)

CH₂   CH₂   O
CH₃   CH₂   CH₂   CH₂   C
CH₂   CH₂   CH₂   CH₂   OH
CH₂   CH₂   CH₂   CH₂
CH₂   **Double bond**   CH₂
HC ══ CH

(c)

➤ Figure 2.13   **The structure of fats.** (a) Three fatty acids combine with glycerol to form a molecule of triacylglycerol, a type of fat. The group designated R is a long hydrocarbon chain that varies in length in different fatty acids. It may be saturated or unsaturated. (b) Saturated fatty acids have only single covalent bonds between carbon atoms in their carbon chains and can therefore accommodate the maximum possible number of hydrogens. (c) Unsaturated fatty acids, such as oleic acid, have one or more double bonds between carbons and thus contain fewer hydrogens. The double bond causes a bend in the carbon chain. In (b) and (c), both structural formulas and space-filling models are shown.

water, but the fatty acid end cannot (➤ Figure 2.14b). Such properties of phospholipids are important in determining the characteristics of cell membranes (Chapter 4).

Steroids have a four-ring structure (➤ Figure 2.15a) and are quite different from other lipids. They include cholesterol, steroid hormones, and vitamin D. Cholesterol (➤ Figure 2.15b) is insoluble in water and is found in the cell membranes of animal cells and the group of bacteria called mycoplasmas. Steroid hormones and vitamin D are important in many animals.

Soap's effectiveness is reduced in hard water when mineral salts, like calcium, magnesium, and iron, react with the soap and form an insoluble precipitate (soap scum).

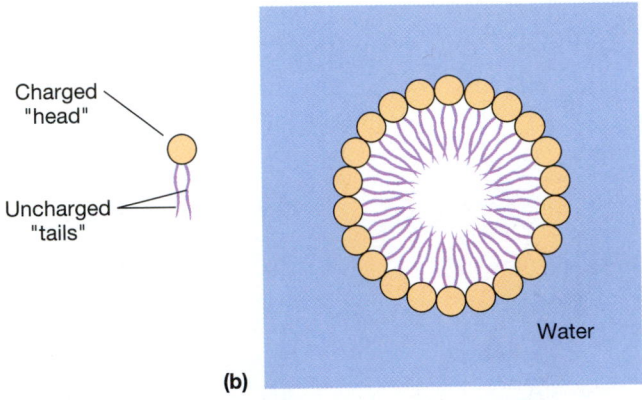

$CH_3 - CH_2 - CH_2 - CH_2 - CH_2 - CH_2 - CH_2 - CH_2 - CH_2 - CH_2 - CH_2 - CH_2 - CH_2 - C$

$CH_3 - CH_2 - CH_2 - CH_2 - CH_2 - CH_2 - CH_2 - CH_2 - CH = CH - CH_2 - CH_2 - CH_2 - CH_2 - CH_2 - C$

Uncharged fatty acid chains

Other charged group

Charged phosphate group

Glycerol portion

(a)

Charged "head"

Uncharged "tails"

Water

(b)

➤ **Figure 2.14   Phospholipids.** (a) In phospholipids, one of the fatty acid chains of a fat molecule is replaced by phosphoric acid. The charged phosphate group and another attached group can interact with water molecules, which are polar, but the two long, uncharged fatty acid tails cannot. (b) As a result, phospholipid molecules in water tend to form globular structures with the phosphate groups facing outward and the fatty acids in the interior.

Down the Drain

You have washed enough dishes to know that soaps get oily substances off the dishes and into the water. Although you probably don't think about how soap works, soap chemistry is an application of what you have learned in Chapter 2. Because water is a polar molecule, it has a high surface tension and beads on clean surfaces. To make water "wetter" it is necessary to reduce the surface tension by adding surfactants. Soaps are anionic surfactants made from fats and oils treated with strong alkali. This process produces a complex molecule with a charged carboxyl group at one end and a nonionized saturated hydrocarbon at the other. The saturated hydrocarbon portion of the soap molecule mixes with the fats on dishes while the charged carboxyl group mixes with the dishwater. The chemical interaction between the grease, soap, and water loosens the food particles off your dishes and down the drain.

(a)

(b)

➤ **Figure 2.15   Steroids.** (a) Steroids are lipids with a characteristic four-ring structure. The specific chemical groups attached to the rings determine the properties of different steroids. (b) One of the most biologically important steroids is cholesterol, a component of the membranes of animal cells and one group of bacteria.

Proteins

Properties of Proteins and Amino Acids

Among the molecules found in living things, proteins have the greatest diversity of structure and function. **Proteins** are composed of building blocks called **amino acids,** which have at least one amino ($-NH_2$) group and one acidic carboxyl ($-COOH$) group. The general structure of an amino acid and some of the 20 amino acids found in proteins are shown in ➤ Figure 2.16. Each amino acid is distinguishable by a different chemical group, called an **R group,** attached to the central carbon atom. Because all amino acids contain carbon, hydrogen, oxygen, and nitrogen, and some contain sulfur, proteins also contain these elements.

A protein is a polymer of amino acids joined by **peptide bonds**—that is, covalent bonds that link an amino group of one amino acid and a carboxyl group of another amino acid (➤ Figure 2.17). Two amino acids linked together make a *dipeptide,* three make a *tripeptide,* and many make a **polypeptide.** In addition to the amino and carboxyl groups, some amino acids have an R group called a *sulfhydryl group* ($-SH$). Sulfhydryl groups in adjacent chains of amino acids can lose hydrogen and form *disulfide linkages* ($-S-S-$) from one chain to the other.

The Structure of Proteins

Proteins have several levels of structure. The **primary structure** of a protein consists of the specific sequence of amino acids in a polypeptide chain (➤ Figure 2.18a). The **secondary structure** of a protein consists of the folding or coiling of amino acid chains into a particular pattern, such as a helix or pleated sheet (➤ Figure 2.18b). Hydrogen bonds are responsible for such patterns. Further bending and folding of the protein molecule into globular (irregular spherical) shapes or fibrous threadlike strands produces the **tertiary structure** (➤ Figure 2.18c). Some large proteins such as hemoglobin have **quaternary structure,** formed by the association of several tertiary-structured polypeptide chains (➤ Figure 2.19). Tertiary and quaternary structures are maintained by disulfide linkages, hydrogen bonds, and other forces between R groups of amino acids. The three-dimensional shapes of protein molecules and the nature of sites at which other molecules can bind to them are extremely important in determining how proteins function in living organisms.

Several conditions can disrupt hydrogen bonds and other weak forces that maintain protein structure. They include highly acidic or basic conditions and temperatures above 50°C. Such disruption of secondary, tertiary, and quatenary structures is called **denaturation.** Sterilization and disinfection procedures often make use of heat or chemicals that kill microorganisms by denaturing their proteins. Also, the cooking of meat tenderizes it by denaturing proteins. Therefore, microbes and cells of larger organisms must be maintained within fairly narrow ranges of pH and temperature to prevent disruption of protein structure.

Classification of Proteins

Most proteins can be classified by their major functions as structural proteins or enzymes. **Structural proteins,** as the name implies, contribute to the three-dimensional structure of cells, cell parts, and membranes. Certain proteins, called *motile proteins,* contribute both to

➤ **Figure 2.16   Amino acids.** (a) The general structure of an amino acid, and (b) six representative examples. All amino acids have four groups attached to the central carbon atom: an amino ($-NH_2$) group, a carboxyl ($-COOH$) group, a hydrogen atom, and a group designated R that is different in each amino acid. The R group determines many of the chemical properties of the molecule—for example, whether it is nonpolar, polar, acidic, or basic.

➤ **Figure 2.17   Peptide linkage.** Two amino acids are joined by the removal of a water molecule (dehydration synthesis) and the formation of a peptide bond between the —COOH group of one and the —NH$_2$ group of the other.

(a) Primary structure

(b) Secondary structure

(c) Tertiary structure

➤ Figure 2.18   **Three levels of protein structure.** (a) Primary structure is the sequence of amino acids (A1, A2, . . .) in a polypeptide chain. Imagine it as a straight telephone cord. (b) Polypeptide chains, especially those of structural proteins, tend to coil or fold into a few simple, regular, three-dimensional patterns called secondary structure. Imagine the telephone cord as a coiled cord. (c) Polypeptide chains of enzymes and other soluble proteins may also exhibit secondary structure. In addition, the chains tend to fold up into complex, globular shapes that constitute the protein's tertiary structure. Imagine the knot formed when a coiled telephone cord tangles.

(a) Hemoglobin molecule

(b) Keratin fiber

➤ Figure 2.19   **Quaternary protein structure.** (a) Many large proteins such as hemoglobin, which carries oxygen in human red blood cells, are made up of several polypeptide chains. The arrangement of these chains makes up the protein's quaternary structure. (b) Some structural proteins such as keratin, a component of human skin and hair, also consist of several polypeptide chains and so have quaternary structure.

structure and to movement. They account for the contraction of animal muscle cells and for some kinds of movement in microbes. **Enzymes** are protein *catalysts*—substances that control the rate of chemical reactions in cells. A few proteins are neither structural proteins nor enzymes. They include proteins that form receptors for certain substances on cell membranes and antibodies that participate in the body's immune responses (Chapter 17).

Enzymes

Enzymes increase the rate at which chemical reactions take place within living organisms in the temperature range compatible with life. We discuss enzymes in more detail in Chapter 5 but summarize their properties here. In general, enzymes speed up reactions by decreasing the energy required to start reactions. They also hold reactant molecules close together in the proper orientation for reactions to occur. Each enzyme has an **active site,** which is the site at which it combines with its **substrate,** the substance on which an enzyme acts. Enzymes have **specificity**—that is, each enzyme acts on a particular substrate or on a certain kind of chemical bond.

Like catalysts in inorganic chemical reactions, enzymes are not permanently affected or used up in the reactions they initiate. Enzyme molecules can be used over and over again to catalyze a reaction, although not indefinitely. Because enzymes are proteins, they are denatured by extremes of temperature and pH.

Nucleotides and Nucleic Acids

The chemical properties of *nucleotides* allow these compounds to perform several essential functions. One key function is storage of energy in **high-energy bonds**—bonds that, when broken, release more energy than do most covalent bonds. Nucleotides joined to form *nucleic acids* are, perhaps, the most remarkable of all biochemical substances. They store information that directs protein synthesis and that can be transferred from parent to progeny.

A **nucleotide** consists of three parts: (1) a nitrogenous base, so named because it contains nitrogen and has alkaline properties; (2) a five-carbon sugar; and (3) one or more phosphate groups, as ➤ Figure 2.20a shows for the nucleotide *adenosine triphosphate (ATP)*. The sugar and base alone make up a *nucleoside* (➤ Figure 2.20b).

The nucleotide ATP is the main source of energy in cells because it stores chemical energy in a form cells can use. The bonds between phosphates in ATP that are high-energy bonds are designated by wavy lines (➤ Figure 2.20c). They contain more energy than most covalent bonds, in that more energy is released when they are broken. Enzymes control the forming and breaking of high-energy bonds so that energy is released as needed within cells. The capture, storage, and use of energy is an important component of cellular metabolism (Chapter 5).

Nucleic acids consist of long polymers of nucleotides, called **polynucleotides.** They contain genetic information that determines all the heritable characteristics of a living organism, be it a microbe or a human. Such information is passed from generation to generation and di-

➤ **Figure 2.20   Nucleotides.** (a) A nucleotide consists of a nitrogenous base, a five-carbon sugar, and one or more phosphate groups. (b) A nucleoside is comprised of the sugar and base without the phosphates. (c) The nucleotide adenosine triphosphate (ATP), the immediate source of energy for most activities of living cells. In ATP the base is adenine, and the sugar is ribose. Adding a phosphate group to adenosine diphosphate greatly increases the energy of the molecule; removal of the third phosphate group releases energy that can be used by the cell.

rects protein synthesis in each organism. By directing protein synthesis, nucleic acids determine which structural proteins and enzymes an organism will have. The enzymes determine which other substances the organism can make and which other reactions it can carry out.

The two nucleic acids found in living organisms are **ribonucleic acid (RNA)** and **deoxyribonucleic acid (DNA).** Except in a few viruses, RNA is a single polynucleotide chain, and DNA is a double chain of polynucleotides arranged as a double helix. In both nucleic acids, the phosphate and sugar molecules form a sturdy but inert "backbone" from which nitrogenous bases protrude. In DNA each chain is connected by hydrogen bonds between the bases, so the whole molecule resembles a ladder with many rungs (➤ Figure 2.21a).

DNA and RNA contain somewhat different building blocks (Table 2.5). RNA contains the sugar ribose, whereas DNA contains deoxyribose, which has one less oxygen atom than ribose. Three nitrogenous bases, adenine, cytosine, and guanine, are found in both DNA and RNA. In addition, DNA contains the base thymine, and RNA contains the base uracil. Of these bases, adenine and guanine are **purines,** nitrogenous base molecules that contain double-ring structures, and thymine, cytosine, and uracil are **pyrimidines,** nitrogenous base molecules that contain a single-ring structure (➤ Figure 2.22). All cellular organisms have both DNA and RNA. Viruses have either DNA or RNA but not both.

The two nucleotide chains of DNA are held together by hydrogen bonds between the bases and by other forces. The hydrogen bonds always connect adenine to thymine and cytosine to guanine, as shown in ➤ Figure 2.21b. This linking of specific bases is called **complemen-**

➤ **Figure 2.21   Nucleic and structure.** Nucleic acids consist of a backbone of alternating sugar and phosphate groups to which nitrogenous bases are attached. (a) RNA is usually single-stranded. DNA molecules typically consist of two chains held together by hydrogen bonds between bases. (b) The complementary base pairs in DNA, showing how hydrogen bonds are formed.

➤ Table 2.5	Components of DNA and RNA		
Component		**DNA**	**RNA**
	Phosphoric acid	X	X
Sugars	Ribose		X
	Deoxyribose	X	
	Adenine	X	X
	Guanine	X	X
Bases	Cytosine	X	X
	Thymine	X	
	Uracil		X

Purines   **Pyrimidines**

Adenine   Thymine   Cytosine
                 (in DNA only)

Guanine   Uracil
                 (in RNA only)

➤ **Figure 2.22** **The five bases found in nucleic acids.** DNA contains the purines adenine and guanine and the pyrimidines cytosine and thymine. In RNA, thymine is replaced by the pyrimidine uracil.

tary base pairing. It is determined by the sizes and shapes of the bases. The same kind of complementary base pairing also occurs when information is transmitted from DNA to RNA at the beginning of protein synthesis (Chapter 7). In that situation, adenine in DNA base pairs with uracil in RNA.

DNA and RNA chains contain hundreds or thousands of nucleotides with bases arranged in a particular sequence. This sequence of nucleotides, like the sequence of letters in words and sentences, contains information that determines what proteins an organism will have. As noted earlier, an organism's structural proteins and enzymes, in turn, determine what the organism is and what it can do. Like changing a letter in a word, changing a nucleotide in a sequence can change the information it carries. The number of different possible sequences of bases is almost infinite, so DNA and RNA can contain a great many different pieces of information.

The functions of DNA and RNA are related to their ability to convey information. DNA is transmitted from one generation to the next. It determines the heritable characteristics of the new individual by supplying the information for the proteins its cells will contain. In contrast, RNA carries information from the DNA to the sites where proteins are manufactured in cells. There it directs and participates in the actual assembly of proteins. The functions of nucleic acids are discussed in more detail in Chapters 7 and 8.

✓ Why are starch, DNA, and RNA all considered to be polymers?

✓ Distinguish among primary, secondary, tertiary, and quaternary levels of protein structure.

Retracing Our Steps

Why Study Chemistry?

- A knowledge of basic chemistry is needed to understand how microorganisms function and how they affect humans and our environment.

Chemical Building Blocks and Chemical Bonds

Chemical Building Blocks

- The smallest chemical unit of matter is an **atom.** An **element** is a fundamental kind of matter, and the smallest unit of an element is an atom. An element is composed of only one type of atom. A **molecule** consists of two or more atoms chemically combined, and a **compound** consists of two or more different kinds of atoms chemically combined.
- The most common elements in all forms of life are carbon (C), hydrogen (H), oxygen (O), and nitrogen (N).

The Structure of Atoms

- Atoms consist of positively charged **protons** and neutral **neutrons** in the atomic nucleus and very small, negatively charged **electrons** orbiting the nucleus.
- The number of protons in an atom is equal to its **atomic number**. The total number of protons and neutrons determines the element's **atomic weight.**
- **Ions** are atoms that have gained or lost one or more electrons.
- **Isotopes** are atoms of the same element that contain different numbers of neutrons; some may be **radioisotopes.**

Chemical Bonds

- Atoms of molecules are held together by **chemical bonds.**
- **Ionic bonds** involve attraction of oppositely charged ions. In **covalent bonds,** atoms share pairs of electrons. **Hydrogen bonds** are weak attractions between polar regions of hydrogen atoms and oxygen or nitrogen atoms.

Chemical Reactions

- Chemical reactions involve breaking or forming chemical bonds and associated energy changes.
- **Catabolism,** the breaking down of molecules, releases energy. **Anabolism,** the synthesis of larger molecules, requires energy.

Water and Solutions

Water

- Water is a **polar compound,** acts as a solvent, and forms thin layers because it has high **surface tension.**
- Water also has high specific heat, and it serves as a medium for and participates in many chemical reactions.

Solutions and Colloids

- **Solutions** consist of **mixtures** with one or more **solutes** evenly distributed throughout a **solvent.**
- **Colloids** contain particles too large to form true solutions.

Acids, Bases, and pH

- In most solutions containing acids or bases, **acids** release H^+ ions, and **bases** accept H^+ ions (or release OH^- ions).
- The **pH** of a solution is a measure of its acidity or alkalinity. A pH of 7 is neutral, below 7 is acidic, and above 7 is basic, or **alkaline.**

Complex Organic Molecules

- **Organic chemistry** is the study of carbon-containing compounds.
- Organic compounds such as alcohols, aldehydes, ketones, organic acids, and amino acids can be identified by their **functional groups.**

Carbohydrates

- **Carbohydrates** consist of carbon chains in which most of the carbon atoms have an associated alcohol group and one carbon has either an aldehyde or a ketone group.
- The simplest carbohydrates are **monosaccharides,** which can combine to form **disaccharides** and **polysaccharides.** Long chains of repeating units are called **polymers.**
- The body uses carbohydrates primarily for energy.

Lipids

- All **lipids** are insoluble in water but soluble in nonpolar solvents.
- **Fats** consist of glycerol and **fatty acids.**
- **Phospholipids** contain a phosphate group in place of a fatty acid.
- **Steroids** have a complex four-ring structure.

Proteins

- **Proteins** consist of chains of **amino acids** linked by **peptide bonds.**
- Proteins form part of the structure of cells, act as enzymes, and contribute to other functions such as motility, transport, and regulation.
- **Enzymes** are biological catalysts of great **specificity** that increase the rate of chemical reactions in living organisms. Each enzyme has an **active site** to which its **substrate** binds.

Nucleotides and Nucleic Acids

- A **nucleotide** consists of a nitrogenous base, a sugar, and one or more phosphates.
- Some nucleotides contain **high-energy bonds.**
- **Nucleic acids** are important information-containing molecules that consist of chains of nucleotides. The nucleic acids that occur in living organisms are **ribonucleic acid (RNA)** and **deoxyribonucleic acid (DNA).**

Terminology Check

acid   (p. 337)
active site   (p. 42)
alkaline   (p. 34)
amino acid   (p. 41)
anabolism   (p. 30)
anion   (p. 28)
atom   (p. 26)
atomic number   (p. 27)
atomic weight   (p. 28)
base   (p. 33)
biochemistry   (p. 34)
carbohydrate   (p. 35)
catabolism   (p. 30)
cation   (p. 28)
chemical bond   (p. 29)
colloid   (p. 32)
complementary base pairing   (p. 44)
compound   (p. 27)
covalent bond   (p. 29)
dehydration synthesis   (p. 32)
denaturation   (p. 41)
deoxyribonucleic acid   (DNA)   (p. 44)
disaccharide   (p. 36)
electron   (p. 27)
element   (p. 26)
endergonic   (p. 31)
enzyme   (p. 42)
exergonic   (p. 30)

fat   (p. 38)
fatty acid   (p. 38)
functional group   (p. 34)
glycosidic bond   (p. 36)
gram molecular weight (p. 28)
high-energy bond   (p. 42)
hydrogen bond   (p. 30)
hydrolysis   (p. 32)
ion   (p. 28)
ionic bond   (p. 29)
isomer   (p. 35)
isotope   (p. 28)
lipid   (p. 38)
mixture   (p. 32)
mole   (p. 28)
molecule   (p. 27)
monosaccharide   (p. 35)
neutral   (p. 34)
neutron   (p. 27)
nucleic acid   (p. 43)
nucleotide   (p. 43)
organic chemistry   (p. 34)
peptide bond   (p. 41)
pH   (p. 34)
phospholipid   (p. 38)
polar compound   (p. 30)
polymer   (p. 36)
polynucleotide   (p. 42)

polypeptide   (p. 41)
polysaccharide   (p. 36)
primary structure   (p. 41)
protein   (p. 41)
proton   (p. 27)
purine   (p. 44)
pyrimidine   (p. 44)
quaternary structure   (p. 41)
radioisotope   (p. 29)
reactant   (p. 32)
R group   (p. 41)
ribonucleic acid (RNA) (p. 44)
rule of octets   (p. 28)
saturated fatty acid   (p. 38)
secondary structure   (p. 41)
solute   (p. 32)
solution   (p. 32)
solvent   (p. 32)
specificity   (p. 42)
steroid   (p. 39)
structural protein   (p. 41)
substrate   (p. 42)
surface tension   (p. 31)
tertiary structure   (p. 41)
triacylglycerol   (p. 38)
unsaturated fatty acid   (p. 38)

See You on the Web

Don't forget I am expecting you on the web. The following is a preparedness checklist for your web explorations. If you have questions on any of these items, you can return to the end of Chapter 1 for more details.

- Questions We'll Explore
- Concept Check questions
- Flashcards
- Study Guide

If you are ready, come visit me at

http://www.prenhall.com/black

It will be a journey you'll never forget.

Microscopy and Staining

Come closer. What do you see? Probably not a lot more than you saw from farther away. The unaided eye can see dust particles as small as 20 micrometers (1 micrometer = 1/1000 of a millimeter) glimmering in a shaft of sunlight, but what's amazing is how much you aren't seeing! Most bacteria are only 1/2 to 2 micrometers long and therefore remain invisible without the help of a microscope.

How many bacteria can fit on a pinpoint? Look at the picture here, and start counting. Each gold rod is a bacterium. If you were to stick yourself with a "clean" pin, how many bacteria might be left behind in your finger? How many are necessary to start an infection? Sometimes as few as 10 bacteria are enough.

Background Basics

nature of the electron . . . . (Chapter 2)
acidic, basic . . . . (Chapter 2)
ionization . . . . (Chapter 2)

Historical Microscopy

Anton van Leeuwenhoek (1632–1723), living in Delft, Holland, was almost certainly the first person to see individual microorganisms. He constructed simple microscopes capable of magnifying objects 100 to 300 times. These instruments were unlike what we commonly think of as microscopes today. Consisting of a single tiny lens, painstakingly ground, they were actually very powerful magnifying glasses (➤ Figure 3.1). It was so difficult to focus one of Leeuwenhoek's microscopes that instead of changing specimens, he built a new microscope for each specimen, leaving the previous specimen and microscope together. When foreign investigators came to Leeuwenhoek's laboratory to look through his microscopes, he made them keep their hands behind their backs to prevent them from touching the focusing apparatus!

In a letter to the Royal Society of London in 1676, Leeuwenhoek described his first observations of bacteria and protozoa in water. He kept his techniques secret, however. Even today we are not sure of his methods of illumination, although it is likely that he used indirect lighting, with light bouncing off the sides of specimens rather than passing through them. Leeuwenhoek was also unwilling to part with any of the 419 microscopes he made. It was only near the time of his death that his daughter, at his direction, sent 100 of them to the Royal Society.

Following Leeuwenhoek's death in 1723, no one came forward to continue the work of perfecting the design and construction of microscopes, and the progress of microbiology slowed. Still, he had taken the first steps. Through Leeuwenhoek's letters to the Royal Society in the mid-1670s, the existence of microbes was revealed to the scientific community. And in 1683 he described bacteria taken from his own mouth. However, Leeuwenhoek could see very little detail of their structure. Further study required the development of more complex microscopes, as we shall soon see.

Questions We'll Explore

A How is the evolution of microscopy instruments related to progress in microbiology?

B Which metric units are most useful for the measurement of microbes?

C What are the relationships among wavelength, resolution, and numerical aperture?

D How are the following properties of light related to microbiology: transmission, absorption, fluorescence, luminescence, phosphorescence, reflection, refraction, and diffraction?

E What is the function of each part of a compound microscope, and what is total magnification?

F What are the special uses and adaptations of bright-field, dark-field, phase-contrast, differential interference contrast, and fluorescence (UV) microscopes?

G What are the principles of transmission and scanning electron microscopy? How do the advantages and limitations of electron microscopy compare with those of light microscopy?

H Which techniques are used to prepare and heighten contrast in specimens to be viewed with a light microscope?

I What are the uses of the common types of microbial stains?

J What are the functions and results of each of the steps in the Gram staining procedure?

(a)

(b)

The Secret Kingdom Uncovered
"I have had several gentlewomen in my house, who were keen on seeing the little eels in vinegar; but some of 'em were so disgusted at the spectacle that they vowed they'd never use vinegar again. But what if one should tell such people in [the] future that there are more animals living in the scum on the teeth in a man's mouth, than there are men in a whole kingdom?"

—Anton van Leeuwenhoek, 1683

➤ **Figure 3.1** **Leeuwenhoek's investigations.** (a) A drawing of one of Leeuwenhoek's microscopes. This simple microscope, really a very powerful magnifying glass, made use of a single, tiny, almost spherical lens set into the metal plate. The specimen was mounted on the needlelike end of the vertical shaft and examined through the lens from the opposite side. The various screws were used to position the specimen and bring it into focus—a very difficult process. (b) The vinegar eels (nematodes) that so upset Leeuwenhoek's friends.

Principles of Microscopy

Metric Units

Microscopy is the technology of making very small things visible to the human eye. Because microorganisms are so small, the units used to measure them are likely to be unfamiliar to beginning students used to dealing with a macroscopic world (Table 3.1).

The **micrometer** (μm), formerly called a micron (μ), is equal to 0.000001 m. A micrometer can also be expressed as 10^{-6} m. The second unit, the **nanometer** (nm), formerly called a millimicron (mμ), is equal to 0.000000001 m. It also is expressed as 10^{-9} m. A third unit, the **angstrom** (Å), is found in much of the current and older literature but no longer has any official recognition. It is equivalent to 0.0000000001 m, 0.1 nm, or 10^{-10} m. ➤ Figure 3.2 shows a scale that summarizes the metric system unit equivalents, the ranges of sizes that can be detected by the unaided human eye and by various types of microscopy, and examples of where various organisms fall on this scale.

Properties of Light: Wavelength and Resolution

Light has a number of properties that affect our ability to visualize objects, both with the unaided eye and (more crucially) with the microscope. Understanding these properties will allow you to improve your practice of microscopy.

The earliest known publication of microscopic illustrations appeared in Uppsala, Sweden, in 1555.

➤ Table 3.1	Some commonly used units of length		
Unit (abbreviation)	**Prefix**	**Metric Equivalent**	**English Equivalent**
meter (m)			3.28 ft
centimeter (cm)	*centi* = one hundredth	0.01 m = 10^{-2} m	0.39 in.
millimeter (mm)	*milli* = one thousandth	0.001 m = 10^{-3} m	0.039 in.
micrometer (μm)	*micro* = one millionth	0.000001 m = 10^{-6} m	0.000039 in.
nanometer (nm)	*nano* = one billionth	0.000000001 m = 10^{-9} m	0.000000039 in.
angstrom (Å)		0.0000000001 m = 10^{-10} m	0.0000000039 in.

> **Figure 3.2   Relative sizes of objects.** Various organisms and their relation to metric units of measurement are shown; red terms indicate organisms studied in microbiology. Chlamydia and Rickettsia are groups of bacteria that are much smaller in size than other bacteria. The range of effective use for various instruments is also depicted.

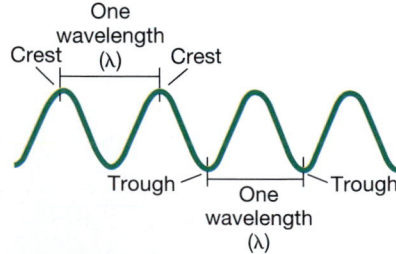

> **Figure 3.3   Wavelength.** The distance between two adjacent crests or two adjacent troughs of any wave is defined as one wavelength, designated by the Greek letter lambda (λ).

One of the most important properties of light is its **wavelength,** or the length of a light ray
> (Figure 3.3). Represented by the Greek letter lambda (λ), wavelength is equal to the distance between two adjacent crests or two adjacent troughs of a wave. The sun produces a continuous *spectrum* of electromagnetic radiation with waves of various lengths (> Figure 3.4). Visible light rays as well as ultraviolet and infrared rays constitute particular parts of this spectrum. White light is the combination of all colors of visible light. Black is the absence of visible light.

The wavelength used for observation is crucially related to the resolution that can be obtained. **Resolution** refers to the ability to see two items as separate and discrete units (> Figure 3.5a) rather than as a fuzzy, overlapped single image (> Figure 3.5b). We can magnify objects, but if the objects cannot be resolved, the magnification is useless. Light must pass between two objects for them to be seen as separate things. If the wavelength of the light by which we see the objects is too long to pass between them, they will appear as one. The key to resolution is to get light of a short-enough wavelength to fit between the objects we want to see separately.

To visualize this phenomenon, imagine a target with a foot-high letter *E* hanging in front of a white background. Suppose that you throw at the target ink-covered objects with diameters corresponding to various wavelengths (> Figure 3.6, page 52). If one object has a diameter smaller than the distance between the "arms" of the letter *E,* the object will pass between the arms, and the arms will be distinguishable as separate structures. First, imagine tossing bas-

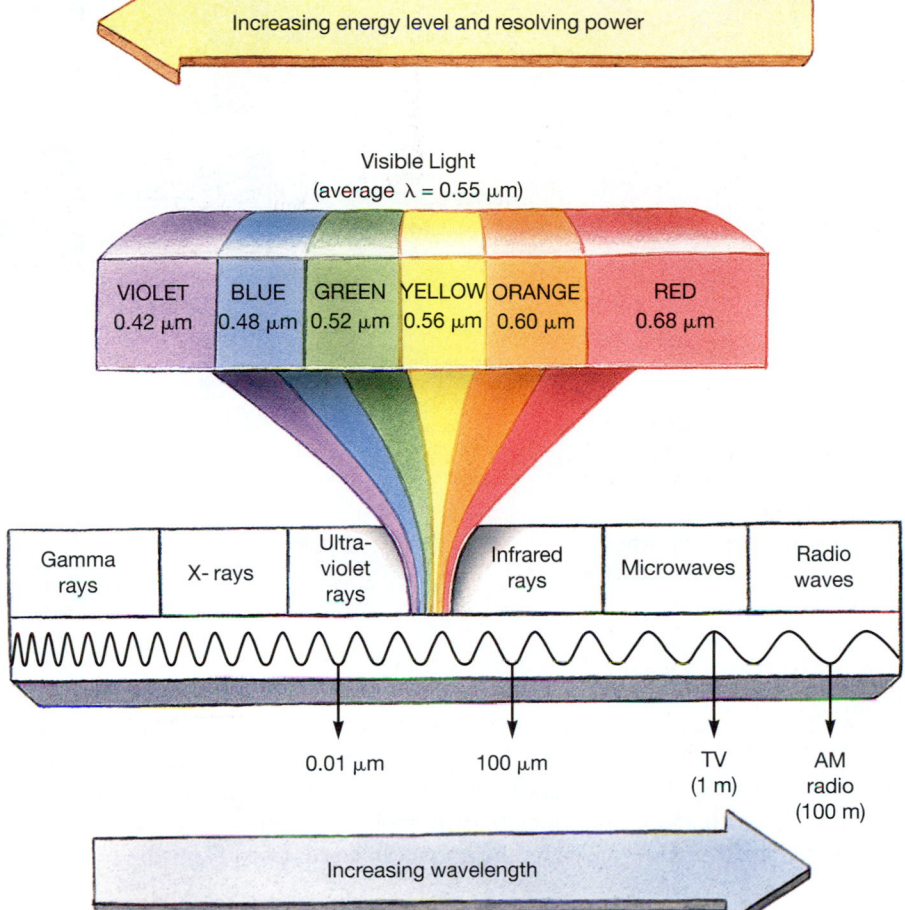

Increasing energy level and resolving power

Visible Light
(average λ = 0.55 μm)

VIOLET	BLUE	GREEN	YELLOW	ORANGE	RED
0.42 μm	0.48 μm	0.52 μm	0.56 μm	0.60 μm	0.68 μm

Gamma rays	X-rays	Ultra-violet rays	Infrared rays	Microwaves	Radio waves

0.01 μm   100 μm   TV   AM
(1 m)   radio
(100 m)

Increasing wavelength

➤ **Figure 3.4   The electromagnetic spectrum.** Only a narrow range of wavelengths—those of visible and ultraviolet light—are used in light microscopy. The shorter the wavelength used, the greater the resolution that can be attained. White light is the combination of all colors of visible light.

ketballs. Because they cannot fit between the arms, light rays of that size would give poor resolution. Next toss tennis balls at the target. The resolution will improve. Then try jelly beans and, finally, tiny beads. With each decrease in the diameter of the object thrown, the number of such objects that can pass between the arms of the *E* increases. Resolution improves, and the shape of the letter is revealed with greater and greater precision.

Microscopists use shorter and shorter wavelengths of electromagnetic radiation to improve resolution. Visible light, which has an average wavelength of 550 nm, cannot resolve distances less than 220 nm. Ultraviolet light, which has a wavelength of 100 to 400 nm, can resolve distances as small as 110 nm. Thus, microscopes that used ultraviolet light instead of visible light allowed researchers to find out more about the details of cellular structures. But the invention of the electron microscope, which uses electrons rather than light, was the major step in increasing the ability to resolve objects. Electrons behave both as particles and as waves. Their wavelength is about 0.005 nm, which allows resolution of distances as small as 0.2 nm.

The **resolving power (RP)** of a lens is a numerical measure of the resolution that can be obtained with that lens. The smaller the distance between objects that can be distinguished, the greater the resolving power of the lens. We can calculate the RP of a lens if we know its **numerical aperture (NA)**, a mathematical expression relating to the extent that light is concentrated by the condenser lens and collected by the objective. The formula for calculating resolving power is RP = λ/2NA. As this formula indicates, the smaller the value of λ and the larger the value of NA, the greater the resolving power of the lens.

(a)

(b)

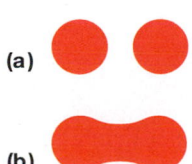

➤ **Figure 3.5   Resolution.** (a) The two dots are resolved—that is, they can clearly be seen as separate structures. (b) These two dots are not resolved—they appear to be fused.

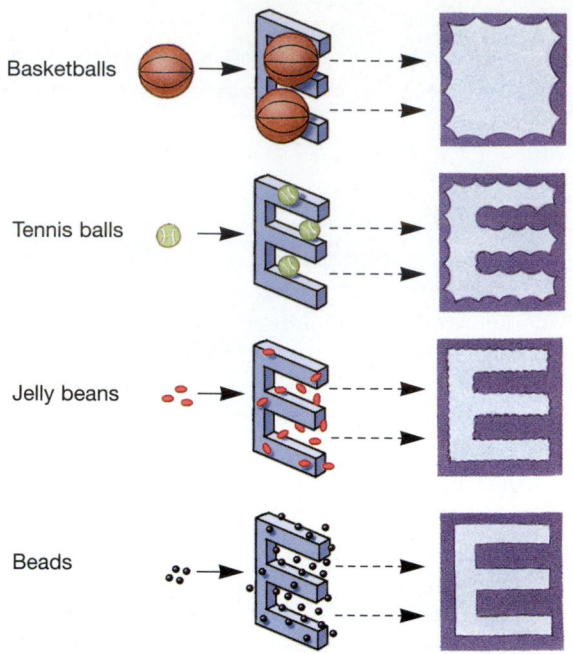

Basketballs

Tennis balls

Jelly beans

Beads

➤ Figure 3.6   **An analogy for the effect of wavelength on resolution.** Smaller objects (corresponding to shorter wavelengths) can pass more easily between the arms of the letter *E,* defining it more clearly and producing a sharper image.

The NA values of lenses differ in accordance with the power of magnification and other properties. The NA value is engraved on the side of each objective lens (the lens nearest the stage) of a light microscope. Look at the NA values on the microscope you use the next time you are in the laboratory. Typical values for the objective lenses commonly found on modern light microscopes are 0.25 for low power, 0.65 for high power, and 1.25 for the oil immersion lens. The higher the NA value, the better the resolution that can be obtained.

Properties of Light: Light and Objects

Various things can happen to light as it travels through a medium such as air or water and strikes an object (➤ Figure 3.7). Let us look at some of those things now and consider how they can affect what we see through a microscope.

Reflection

If the light strikes an object and bounces back (giving the object color), we say that **reflection** has occurred. For example, light rays in the green range of the spectrum are reflected off the surfaces of the leaves of plants. Those reflected rays are responsible for our seeing the leaves as green.

➤ Figure 3.7   **Various interactions of light with an object it strikes.** (a) Light may be reflected back from the object. The particular wavelengths reflected back to the eye determine the perceived color of the object. (b) Light may be transmitted directly through the object. (c) Light may be absorbed, or taken up, by the object. In some cases, the absorbed light rays are reemitted as longer wavelengths, a phenomenon known as fluorescence. (d) Light passing through the object may be refracted, or bent, by it.

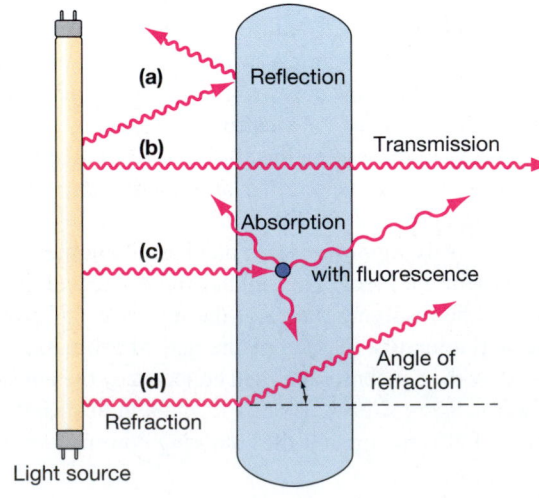

(a) Reflection

(b) Transmission

Absorption

(c) with fluorescence

Angle of refraction

(d) Refraction

Light source

Transmission

Transmission refers to the passage of light through an object. You cannot see through a rock because light cannot pass through it, as it does through a glass window. In order for you to see objects through a microscope, light must either be reflected from the objects or transmitted through them. Most of your observations of microorganisms will make use of transmitted light.

Absorption

If light rays neither pass through nor bounce off an object but are taken up by the object, **absorption** has occurred. Energy in absorbed light rays can be used in various ways. For example, all wavelengths of the sun's light rays except those in the green range are absorbed by a leaf. Some of the energy in these other light rays is captured in photosynthesis and used by the plant to make food. Energy from absorbed light can also raise the temperature of an object. A black object, which reflects no light, will gain heat much faster than a white object, which reflects all light rays.

In some cases, absorbed light rays, especially ultraviolet light rays, are changed into longer wavelengths and reemitted. This phenomenon is known as **luminescence.** If luminescence occurs only during irradiation (when light rays are striking an object), the object is said to **fluoresce.** Many fluorescent dyes are important in microbiology, especially in the field of immunology, because they help us visualize immune reactions and internal processes in microorganisms. If an object continues to emit light when light rays no longer strike it, the object is **phosphorescent.** Some bacteria that live deep in the ocean are phosphorescent.

Refraction

Refraction is the bending of light as it passes from one medium to another of different density. The bending of the light rays gives rise to an *angle of refraction,* the degree of bending (➤ Figure 3.7d). You have probably seen how the underwater portion of a pole sticking out of water or a drinking straw in a glass of water seems to bend (➤ Figure 3.8). When you remove the object from the water, it is clearly straight. It looks bent because light rays deviate, or bend, when they pass from the water into the air as their speed changes across the water-air interface. The **index of refraction** of a material is a measure of the speed at which light passes through the material. When two substances have different indices of refraction, light will bend as it passes from one material into the other.

Light passing through a glass microscope slide, through air, and then through a glass lens is refracted each time it goes from one medium to another. This causes loss of light and a blurred image. To avoid this problem, microscopists use **immersion oil,** which has the same index of refraction as glass, to replace the air. The slide and the lens are joined by a layer of oil; there is no refraction to cause the image to blur (➤ Figure 3.9). If you forget to use oil with

➤ Figure 3.8 **Refraction.** The refraction of light rays passing from water into air causes the pencil to appear bent.

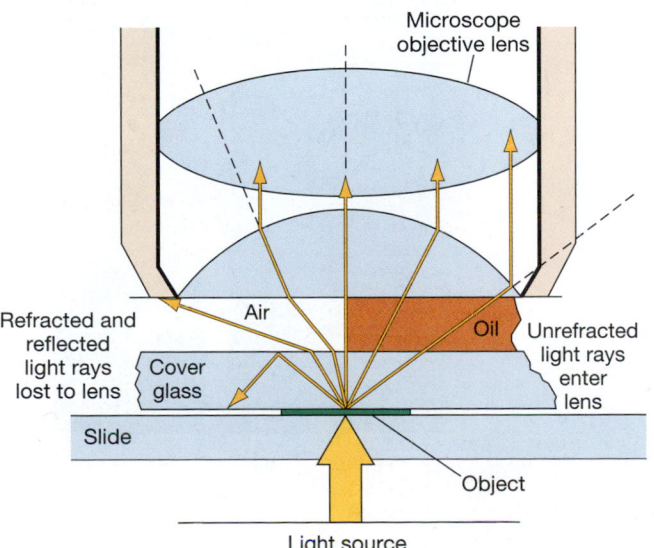

➤ Figure 3.9 **Immersion oil.** Immersion oil is used to prevent the loss of light that results from refraction. The focusing of as much light as possible adds to the clarity of the image. Immersion oil may also be added between the top of the condenser and the bottom of the slide to eliminate another site for refraction.

A Life of Crime

If you want to take some diamonds through customs without declaring them here's how. Obtain an oil with the same refractive index as the diamonds. Pour it into a clear glass bottle labeled "Baby Oil," and drop the diamonds in. The diamonds are invisible, if their surfaces are clean. This trick works because light is not bent when it passes from one medium to another with the same index of refraction. Thus, the boundary between the diamonds and the oil is not apparent.

If larceny is not in your heart or the price of diamonds not in your pocket, try this entertaining, but legal, activity. Clean a glass rod and dip it in and out of a bottle of immersion oil. Watch it disappear and reappear. This experiment will help you understand what is happening when you use the oil immersion lens.

✓ When using red (wavelength 0.68 μm) or blue (wavelength 0.48 μm) visible light, which would give you better resolution? Why?

✓ If you built a light microscope having a total magnification of 5000X, would it give you better, worse, or the same resolution as one that has a magnification of 1000X? Why?

✓ Why would radio waves and microwaves be unsuitable for examining microbes?

✓ Would standard immersion oil placed on a plastic slide prevent refraction? Why or why not?

Leewenhoek's best microscope probably magnified about 270X.

the oil immersion lens of a microscope, it will be impossible to focus clearly on a specimen. Staining (dying) a specimen increases differences in indices of refraction, making it easier to observe details.

Diffraction

As light passes through a small opening, such as a hole, slit, or space between two adjacent cellular structures, the light waves are bent around the opening. This phenomenon is **diffraction.** ➤ Figure 3.10 shows diffraction patterns formed when light passes through a small aperture or around the edge of an object. Similar patterns occur when water passes through an opening in, or around the back of, a breakwater.

Diffraction is a problem for microscopists because the lens acts as a small aperture through which the light must pass. A blurry image results. The higher the magnifying power of a lens, the smaller the lens must be, and therefore the greater the diffraction and blurring it causes. The oil immersion (100X) lens, with its total magnification capacity of about 1000X (when combined with a 10X ocular lens), represents the limit of useful magnification with the light microscope. The small size of higher-power lenses causes such severe diffraction that resolution is impossible.

Light Microscopy

Light microscopy refers to the use of any kind of microscope that uses visible light to make specimens observable. The modern light microscope is a descendant not of Leeuwenhoek's single lenses but of Hooke's compound microscope—a microscope with more than one lens. ➥ (Chapter 1, p. 9) Single lenses produce two problems: They cannot bring the entire field into focus simultaneously, and there are colored rings around objects in the field. Both problems are solved today by the use of multiple correcting lenses placed next to the primary magnifying lens (➤ Figure 3.11). Used in the objectives and eyepieces of modern compound microscopes, correcting lenses give us nearly distortion-free images.

Over the years, several kinds of light microscopes have been developed, each adapted for making certain kinds of observations. We look first at the standard light microscope and then at some special kinds of microscopes.

The Compound Light Microscope

The **optical microscope,** or *light microscope,* has undergone various improvements since Leeuwenhoek's time and essentially reached its current form shortly before the turn of the twentieth century. This microscope is a **compound light microscope**—that is, it has more than

Direction of travel of light

(a)   (b)   (c)

➤ Figure 3.10   **Diffraction.** Light waves are diffracted as they pass (a) around the edge of an object and (b) through a small aperture. (c) Water waves being diffracted as they pass through an opening in a breakwater.

one lens. The parts of a modern compound microscope and the path light takes through it are shown in ➤ Figure 3.12. A compound microscope with a single eyepiece (*ocular*) is said to be **monocular;** one with two eyepieces is said to be **binocular.**

Light enters the microscope from a source in the **base** and often passes through a blue filter, which filters out the long wavelengths of light, leaving the shorter wavelengths and improving resolution. It then goes through a **condenser,** which converges the light beams so that they pass through the specimen. The **iris diaphragm** controls the amount of light that passes through the specimen and into the objective lens. The higher the magnification, the greater the amount of light needed to see the specimen clearly. The **objective lens** magnifies the image before it passes through the **body tube** to the ocular lens in the eyepiece. The **ocular lens** further magnifies the image. A **mechanical stage** allows precise control of moving the slide, which is especially useful in the study of microbes.

The focusing mechanism consists of a **coarse adjustment** knob, which changes the distance between the objective lens and the specimen fairly rapidly, and a **fine adjustment** knob, which changes the distance very slowly. The coarse adjustment knob is used to locate the specimen. The fine adjustment knob is used to bring it into sharp focus.

Compound microscopes have up to six interchangeable objective lenses that have different powers of magnification.

The **total magnification** of a light microscope is calculated by multiplying the magnifying power of the objective lens (the lens used to view your specimen) by the magnifying power of the ocular lens (the lens nearest your eye). Typical values for a 10X ocular are:

- scanning (3X) × (10X) = 30X magnification
- low power (10X) × (10X) = 100X magnification
- high "dry" (40X) × (10X) = 400X magnification
- oil immersion (100X) × (10X) = 1000X magnification

Most microscopes are designed so that when the microscopist increases or decreases the magnification by changing from one objective lens to another, the specimen will remain very nearly in focus. Such microscopes are said to be **parfocal** (*par* means equal). The development of parfocal microscopes greatly improved the efficiency of microscopes and reduced the amount of damage to slides and objective lenses. Most student-grade microscopes are parfocal today.

Some microscopes are equipped with an **ocular micrometer** for measuring objects viewed. This is a glass disk with a scale marked on it that is placed inside the eyepiece between its

➤ **Figure 3.11   Cutaway view of a modern microscope objective.** What we refer to as a single objective lens is really a series of several lenses, which are necessary to correct aberrations of color and focus. The best objectives may have as many as a dozen or more elements.

In 1867, a microscopic flea, enlarged to the size of a "small pig," was exhibited in Paris, leading to the idea of using enlarged photographs of microscopic objects for the purpose of education.

➤ **Figure 3.12   The compound light microscope.** Yellow indicates the path of light through the microscope.

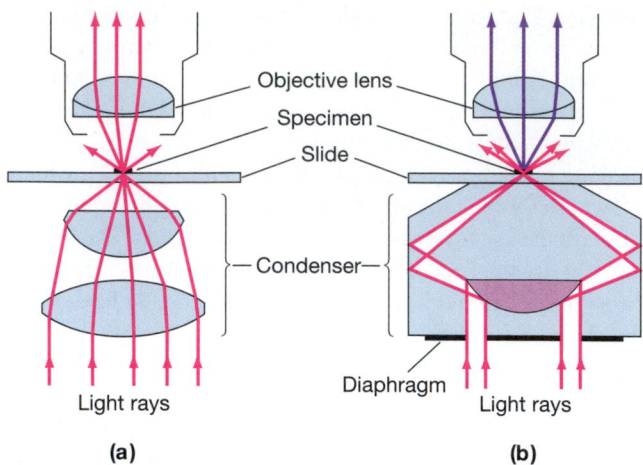

> Figure 3.13 **A comparison of the illumination in bright-field and dark-field microscopy.** (a) The condenser of the bright-field microscope concentrates and transmits light directly through the specimen. (b) The dark-field condenser deflects light rays so that they reflect off the specimen at an angle before they are collected and focused into an image.

lenses. This scale must first be calibrated with a stage micrometer, which has metric units engraved on it. When these units are viewed through the microscope at various magnifications, the microscopist can determine the corresponding metric values of the divisions on the ocular micrometer for each objective lens. Thereafter, he or she needs only to count the number of divisions covered by the observed object and multiply by the calibration factor for that lens in order to determine the actual size of the object.

Dark-Field Microscopy

The condenser used in an ordinary light microscope causes light to be concentrated and transmitted directly through the specimen, as shown in ➤ Figure 3.13a. This gives **bright-field illumination** (➤ Figure 3.14a). In some cases, however, it is more useful, especially with light-sensitive organisms, to examine specimens that would lack contrast with their background in a bright field under other illumination. Live spirochetes (spi'ro-kets), spiral-shaped bacteria that cause syphilis and other diseases, are just such organisms. In this situation **dark-field illumination** is used. A microscope adapted for dark-field illumination has a condenser that prevents light from being transmitted through the specimen but instead causes the light to reflect off the specimen at an angle (Figure 3.13b). When these rays are gathered and focused into an image, a light object is seen on a dark background (➤ Figure 3.14b).

(a)

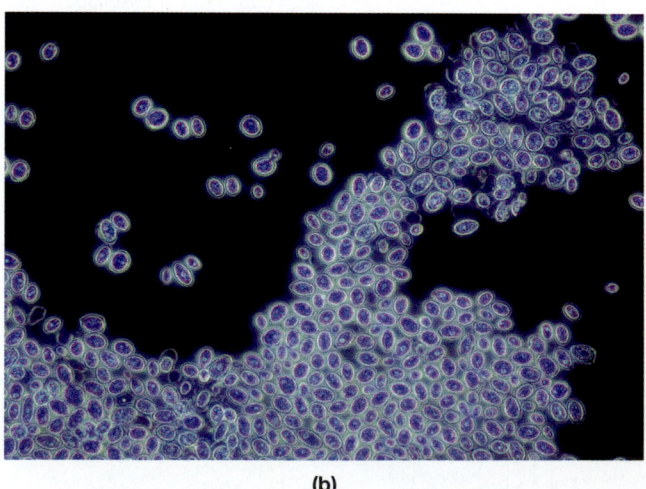
(b)

> Figure 3.14 **A comparison of bright-field and dark-field images.** (a) Bright-field and (b) dark-field microscope views of *Saccharomyces cerevisiae* (brewer's yeast, magnified 975X). Dark-field illumination provides an enormous increase in contrast.

Phase-Contrast Microscopy

Most living microorganisms are difficult to examine because they cannot be stained by coloring them with dyes—stains usually kill the organisms. To observe them alive and unstained requires the use of **phase-contrast microscopy.** A phase-contrast microscope has a special condenser and objective lenses that accentuate small differences in the refractive index of various structures within the organism. Light passing through objects of different refractive indices is slowed down and diffracted. The changes in the speed of light are seen as different degrees of brightness (➤ Figure 3.15).

Nomarski (Differential Interference Contrast) Microscopy

Nomarski microscopy, like phase-contrast microscopy, makes use of differences in refractive index to visualize cells and structures. However, the microscope used, a *differential interference contrast microscope,* produces much higher resolution than the standard phase-contrast microscope. It has a very short *depth of field* (the thickness of specimen that is in focus at any one time) and can produce a nearly three-dimensional image (➤ Figure 3.16).

Fluorescence Microscopy

In **fluorescence microscopy** ultraviolet light is used to excite molecules so that they release light of a longer wavelength than that originally striking them (see ➤ Figure 3.7c). The different wavelengths produced are often seen as brilliant shades of orange, yellow, or yellow green. Some organisms, such as *Pseudomonas,* fluoresce naturally when irradiated with ultraviolet light. Other organisms, such as *Mycobacterium tuberculosis* and *Treponema pallidum* (the cause of syphilis), must be treated with a fluorescent dye called a *fluorochrome.* They then stand out sharply against a dark background (➤ Figure 3.17). Acridine orange is a fluorochrome that binds to nucleic acids, and it colors cells bright green, orange green, or yellow, depending on the filter system used with the fluorescence microscope. It is sometimes used to screen samples for microbial growth, with live cells showing up in bright orange or green.

Fluorescent antibody staining is now widely used in diagnostic procedures to determine whether an *antigen* (a foreign substance such as a microbe) is present. *Antibodies*—molecules produced by the body as an immune response to an invading antigen—are found in many clinical specimens such as blood and serum. If a patient's specimen contains a particular antigen, that antigen and the antibodies specifically made against it will clump together. However, this reaction is ordinarily not visible. Therefore, fluorescent dye molecules are attached to the antibody molecules. If the dye molecules are retained by the specimen, the antigen is presumed to be present, and a positive diagnosis can be made. Thus, if fluorescent dye–tagged antibodies against syphilis organisms are added to a specimen containing spirochetes and are seen to bind to the tagged organisms, those organisms can be identified as the cause of syphilis. This technique is especially important in *immunology,* in which the reactions of antigens and antibodies are studied in great detail (see Chapters 17 and 18, especially Figure 18.33 on the technique of fluorescent antibody staining). Diagnoses can often be made in minutes rather than the hours or days it would take to isolate, culture, and identify organisms.

➤ Figure 3.18 shows the images produced by four different microscopic techniques.

Electron Microscopy

The light microscope opened doors to the world of microbes. However, because it could not resolve distances smaller than 0.2 μm, the view was limited to observations at the level of whole cells and their arrangements. Few subcellular structures could be seen; neither could viruses. The advent of the **electron microscope (EM)** allowed these small structures to be visualized and studied. The EM was developed in 1932 and was in use in many laboratories by the early 1940s.

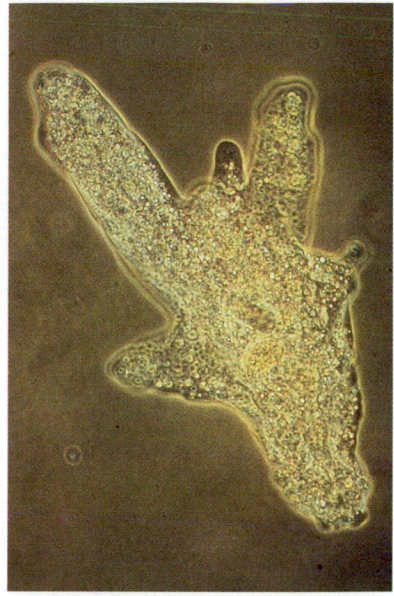

➤ **Figure 3.15   A phase-contrast image.** *Amoeba,* a protozoan.

➤ **Figure 3.16   A Nomarski image.** The protozoan *Paracineta* is attached by a long stalk to the green alga *Spongomorpha* (magnified 400X).

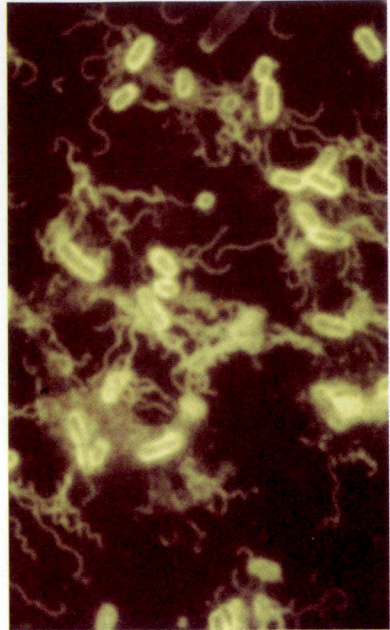

➤ Figure 3.17   **Fluorescent antibody staining.** The fluorescent dye–tagged antibodies clearly show the bacterial cells and their wavy flagella.

The EM uses a beam of electrons instead of a beam of light, and electromagnets rather than glass lenses are used to focus the beam (➤ Figure 3.19). The electrons must travel through a vacuum because collisions with air molecules would scatter the electrons and result in a distorted image. Electron microscopes are much more expensive than light microscopes. They also take up much more space and require additional rooms for preparation of specimens and for processing of photographs. Photographs taken on any microscope are called *micrographs;* those taken on an electron microscope are called **electron micrographs.** Nothing else can show the great detail of minute biological structures that EMs can (➤ Figure 3.20).

The two most common types of electron microscopes are the transmission electron microscope and the scanning electron microscope. Both are used to study various life forms, including microbes. The more advanced scanning tunneling microscope and atomic force microscope let us see actual molecules and even individual atoms.

Transmission Electron Microscopy

The **transmission electron microscope** (**TEM**) gives a better view of the internal structure of microbes than do other types of microscopes. Because of the very short wavelength of illumination (electrons) on which the TEM operates, it can resolve objects as close as 1 nm and magnify microbes (and other objects) up to 500,000X. To prepare specimens for transmission electron microscopy, a specimen may be embedded in a block of plastic and cut with a glass or diamond knife to produce very thin slices (*sections*). These sections are placed on thin wire grids for viewing so that a beam of electrons will pass directly through the section. The section must be exceedingly thin (70–90 nm) because electrons cannot penetrate very far into materials. The specimens can also be treated with special preparations that contain heavy metal elements. The heavy metals scatter electrons and contribute to forming an image.

(a)

(b)

(c)

(d)

➤ Figure 3.18   **Images of the same organism (*Paramecium*, 600X) produced by four different techniques.** (a) Bright-field microscopy, (b) dark-field microscopy, (c) phase-contrast microscopy, (d) Nomarski microscopy. One microscope can have the optics for all four techniques.

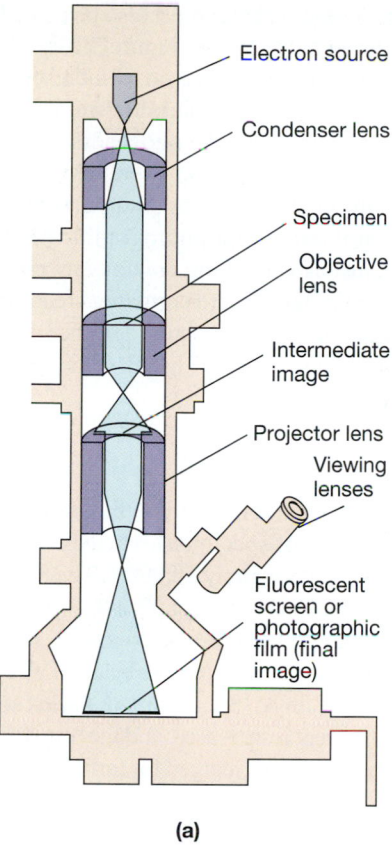

- Electron source
- Condenser lens
- Specimen
- Objective lens
- Intermediate image
- Projector lens
- Viewing lenses
- Fluorescent screen or photographic film (final image)

(a)

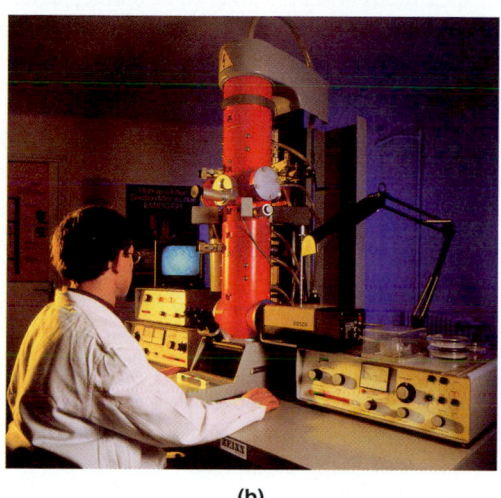

(b)

➤ Figure 3.19   **Electron microscopy.** (a) A cross-sectional diagram of an electron microscope, showing the pathways of the electron beam as it is focused by electromagnetic lenses. (b) A modern scanning electron microscope in use.

Very small specimens, such as molecules or viruses, can be placed directly on plastic-coated grids. Then a heavy metal such as gold or platinum is sprayed at an angle onto the specimen, a technique known as **shadow casting.** A thin layer of the metal is deposited. Areas behind the specimen that did not receive a coating of metal appear as "shadows," which can give a three-dimensional effect to the image (➤ Figure 3.21). Electron beams are deflected by the densely coated parts of the specimen but, for the most part, pass through the shadows.

It is also possible to view the interior of a cell with a TEM by a technique called **freeze-fracturing.** In this technique the cell is frozen and then fractured with a knife. The cleaving of a specimen reveals the surfaces of structures inside the cell (➤ Figure 3.22a). **Freeze-etching,** which involves the evaporation of water from the frozen and fractured specimen, can then ex-

Paramecium

Didinium

(a)

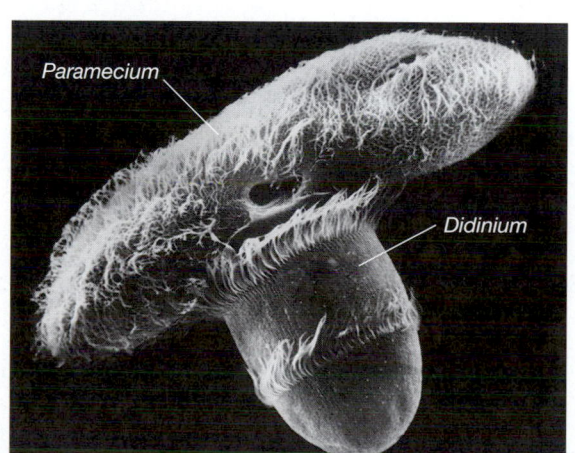

Paramecium

Didinium

(b)

➤ Figure 3.20   **Light and electron microscopy images compared.** (a) Light and (b) electron microscope images of a *Didinium* eating a *Paramecium* (425X). Notice how much more detail is revealed by the scanning electron micrograph.

> ➤ **Figure 3.21   Shadow casting.** Spraying a heavy metal (such as gold or platinum) at an angle over a specimen leaves a "shadow," or darkened area, where metal is not deposited. This technique, known as *shadow casting,* produces images with a three-dimensional appearance, as in this photograph of polio viruses (magnified 324,000X). You can calculate the height of the organisms from the length of their shadows if you know the angle of the metal spray.

pose additional surfaces for examination (➤ Figure 3.22b). These surfaces must also be coated with a heavy metal layer. This layer, called a *replica,* is viewed by TEM (➤ Figure 3.23).

The image formed by the electron beam is made visible as a light image on a fluorescent screen, or monitor. (The actual image made by the electron beam is not visible and would burn your eyes if you tried to view it directly.) The electrons are used to excite the phosphors (light-generating compounds) coating the screen. However, the electron beam will eventually burn through the specimen. Therefore, electron micrographs are made, either by photographing the image on the video screen or by replacing the screen itself with a photographic plate (➤ Figure 3.24a). Electron micrographs can be enlarged, just as you would enlarge any photograph, to obtain an image magnified 20 million times! The micrographs are permanent records of specimens observed and can be studied at leisure. The study of electron micrographs has provided much of our knowledge of the internal structure of microbes. The "M" in TEM and SEM (below) can refer to either microscope or micrograph.

Scanning Electron Microscopy

The **scanning electron microscope (SEM)** is a more recent invention than the transmission electron microscope and is used to create images of the surfaces of specimens. The SEM can resolve objects as close as 20 nm, giving magnifications up to approximately 50,000X. The SEM gives us wonderful three-dimensional views of the exterior of cells (➤ Figure 3.24b).

Preparing a specimen for the SEM involves coating it with a thin layer of a heavy metal, such as gold or palladium. The SEM is operated by scanning, or sweeping, a very narrow beam of electrons (an electron probe) back and forth across a metal-coated specimen. Secondary, or backscattered, electrons leaving the specimen surface are collected, the current is increased, and the resulting image is displayed on a screen. Photographs of the image can be made and enlarged for further study.

Views of the three-dimensional world of microbes, as shown in ➤ Figure 3.25, are breathtakingly beautiful.

Scanning Tunneling Microscopy

In 1980, Gerd Binnig and Heinrich Rohrer invented the first of a series of rapidly improving **scanning tunneling microscopes (STMs),** also called scanning probe microscopes. Five years later, they received the Nobel Prize for their discovery.

> ➤ **Figure 3.22   Freeze-fracturing and freeze-etching.** (a) In freeze-fracturing, a specimen is frozen in a block of ice and broken apart with a very sharp knife. The fracture reveals the interiors of cellular structures and typically passes through the center of membrane bilayers, exposing their inner faces. (b) In freeze-etching, water is evaporated directly from the ice and frozen cytoplasm of the specimen, uncovering additional surfaces for observation.

➤ **Figure 3.23   A freeze-etch preparation.** The toxic cyanobacterium *Microcystis aeruginosa* (magnified 18,000X), showing details of large spherical gas vacuoles.

(a)

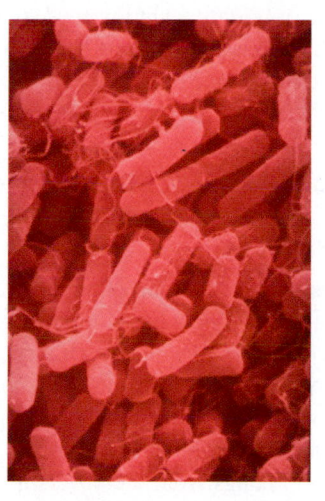

(b)

➤ **Figure 3.24   TEM and SEM compared.** Colorized electron micrographs of *Escherichia coli* produced by (a) transmission electron microscopy and (b) scanning electron microscopy.

(a)

(b)

(c)

➤ **Figure 3.25   Colorized SEM photos of representative microbes.** (a) The fungus *Aspergillus,* a cause of human respiratory disease; (b) *Stella, a star-shaped bacterium;* (c) the diatom *Cyclotella meneghiniana* (1800X), one of many that carry on photosynthesis and form the base of many aquatic food chains.

✓ What does the electron microscope use instead of light beams and glass lenses? What does the fluorescence microscope use?

✓ How can you distinguish between TEM and SEM micrographs? Which type is Figure 3.23?

✓ Why are colored photos of electron micrographs referred to as "colorized" or "false color"?

✓ Rank the following types of microscopes according to the wavelength of illuminating beam they use, beginning with the longest wavelength: fluorescent, TEM, bright-field. What effect does wavelength have on the resolving power of these microscopes?

➤ **Figure 3.26   Scanning tunneling microscopy.** Individual atoms of the element xenon can be clearly distinguished using a scanning tunneling microscope (STM). This chain of seven xenon atoms was built by IBM scientists moving each atom into position, one at a time. The atoms are 20 billionths of an inch, or 0.5 nm, apart. The atoms are bonded together; moving an end atom will relocate up to three of them at a time.

A thin wire probe made of platinum and viridium is used to trace over the surface of a substance, much as you would use your finger to feel the bumps while reading braille. Electron clouds (regions of electron movement) from the surfaces of the probe and the specimen overlap, producing a kind of pathway through which electrons can "tunnel" into one another's clouds. This tunneling sets up an observable current. The stronger the current, the closer the top of the atom is to the probe. Running the probe across in a straight line reveals the highs and lows of individual molecules or atoms in a surface (➤ Figure 3.26).

Even movies can be made using this technique. The first one ever produced showed individual fibrin molecules coming together to form a blood clot. STM also works well under water, and it can be used to examine *live* specimens, such as virus-infected cells exploding and releasing newly formed viruses.

The **atomic force microscope (AFM),** a more advanced member of this family of microscopes, allows three-dimensional imaging and measurement of structures from atomic size to about 1 μm. AFM has been very useful in studying DNA, because it enables investigators to distinguish between bases, such as adenine and guanine, from differences in their electron density states.

The various types of microscopy and their uses are summarized in Table 3.2.

Techniques of Light Microscopy

Microscopes are of little use unless the specimens for viewing are prepared properly. Here we explain some important techniques used in light microscopy.

Although resolution and magnification are important in microscopy, the degree of contrast between structures to be observed and their backgrounds is equally important. Nothing can be seen without contrast, so special techniques have been developed to enhance contrast.

Preparation of Specimens for the Light Microscope

Wet Mounts

Wet mounts, in which a drop of medium containing the organisms is placed on a microscope slide, can be used to view living microorganisms. The addition of a 2 percent solution of carboxymethyl cellulose, a thick, syrupy solution, helps to slow fast-moving organisms so they can be studied. A special version of the wet mount, called a **hanging drop,** often is used with dark-field illumination (➤ Figure 3.27). A drop of culture is placed on a coverslip that is encircled with petroleum jelly. The coverslip and drop are then inverted over the well of a depression slide. The drop hangs from the coverslip, and the petroleum jelly forms a seal that prevents evaporation. This preparation gives good views of microbial motility.

Smears

Smears, in which microorganisms from a loopful of medium are spread onto the surface of a glass slide, can be used to view killed organisms. Although they are living when placed on the slide, the organisms are killed by the techniques used to fix (attach) them to the slide. Smear preparation often is difficult for beginners. If you make smears too thick, you will have trouble seeing individual cells; if you make them too thin, you may find no organisms. If you stir the drop of medium too much as you spread it on the slide, you will disrupt cell arrangements. You may see organisms that normally appear in tetrads (groups of four) as single or double organisms. Such variations lead some beginners to imagine that they see more than one kind of organism when, in fact, the organisms are all of the same species.

After a smear is made, it is allowed to air-dry completely. Then it is quickly passed three or four times through an open flame. This process is called **heat fixation.** Heat fixation accomplishes three things: (1) It kills the organisms; (2) it causes the organisms to adhere to the slide; and (3) it alters the organisms so that they more readily accept stains (dyes). If the slide is not completely dry when you pass it through the flame, the organisms will be boiled and destroyed. If you heat-fix too little, the organisms may not stick and will wash off the slide in subsequent steps. Any cells remaining alive will stain poorly. If you heat-fix too much, the organisms may be incinerated, and you will see distorted cells and cellular remains. Certain structures, such

➤ Table 3.2	Comparison of types of microscopy			
Type	**Special features**		**Appearance**	**Uses**
Bright-Field	Uses visible light; simplest to use, least expensive		Colored or clear specimen on light background	Observation of dead stained organisms or live ones with sufficient natural color contrast
Dark-Field	Uses visible light with a special condenser that causes light rays to reflect off specimen at an angle		Bright specimen on dark background	Observation of unstained living or difficult-to-stain organisms. Allows one to see motion
Phase-Contrast	Uses visible light plus phase-shifting plate in objective with a special condenser that causes some light rays to strike specimen out of phase with each other		Specimen has different degrees of brightness and darkness	Detailed observation of internal structure of living unstained organisms
Nomarski	Uses visible light out of phase; has higher resolution than standard phase-contrast microscope		Produces a nearly three-dimensional image	Observation of finer details of internal structure of living unstained organisms
Fluorescence	Uses ultraviolet light to excite molecules to emit light of different wavelengths, often brilliant colors; because UV can burn eyes, special lens materials are used		Bright, fluorescent, colored specimen on dark background	Diagnostic tool for detection of organisms or antibodies in clinical specimens or for immunologic studies
Transmission Electron	Uses electron beam instead of light rays and electromagnetic lenses instead of glass lenses; image is projected on a video screen; very expensive; preparation requires considerable time and practice		Highly magnified, detailed image; not three-dimensional except with shadow casting	Examination of thin sections of cells for details of internal structure, exterior of cells, and viruses, or surfaces when freeze-fracturing is used
Scanning Electron	Uses electron beam and electromagnetic lenses; expensive; preparation requires considerable time and practice		Three-dimensional view of surfaces	Observation of exterior surfaces of cells or of internal surfaces

as the capsules found on some microbes, are destroyed by heat-fixing, so this step is omitted and these microbes are affixed to the slide just by air-drying.

Principles of Staining

A **stain**, or dye, is a molecule that can bind to a cellular structure and give it color. Staining techniques make the microorganisms stand out against their backgrounds. They are also used to help investigators group major categories of microorganisms, examine the structural and chemical differences in cellular structures, and look at the parts of the cell.

(a)

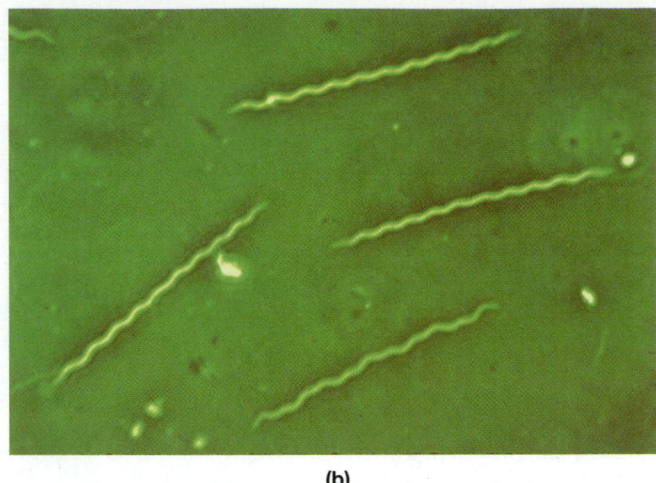

(b)

➤ Figure 3.27   **The hanging-drop technique.** (a) A drop of culture is placed on a coverslip, ringed with petroleum jelly, that is then inverted and placed over the well in a depression slide. The petroleum jelly forms a seal to prevent evaporation. (b) Dark-field micrograph of a hanging-drop preparation (2500X) showing the spiral bacterium *Treponema pallidum,* the cause of syphilis.

In microbiology the most commonly used dyes are **cationic** (positively charged), or **basic, dyes,** such as methylene blue, crystal violet, safranin, and malachite green. These dyes are attracted to any negatively charged cell components. The cell membranes of most bacteria have negatively charged surfaces and thus attract the positively charged basic dyes. Other stains, such as eosin and picric acid, are **anionic** (negatively charged), or **acidic, dyes.** They are attracted to any positively charged cell materials.

Two main types of stains, simple stains and differential stains, are used in microbiology. They are compared in Table 3.3. A **simple stain** makes use of a single dye and reveals basic cell shapes and cell arrangements. Methylene blue, safranin, carbolfuchsin, and crystal violet are commonly used simple stains. A **differential stain** makes use of two or more dyes and distinguishes between two kinds of organisms or between two different parts of an organism. Common differential stains are the Gram stain, the Ziehl-Neelsen acid-fast stain, and the Schaeffer-Fulton spore stain.

The Gram Stain

The **Gram stain,** probably the most frequently used differential stain, was devised by a Danish physician, Hans Christian Gram, in 1884. Gram was testing new methods of staining biopsy and autopsy materials, and he noticed that with certain methods some bacteria were stained differently than the surrounding tissues. As a result of his experiments with stains, the highly useful Gram stain was developed. In Gram staining, bacterial cells take up crystal violet. Iodine is then added; it acts as a **mordant,** a chemical that helps retain the stain in certain cells. Those structures that cannot retain crystal violet are decolorized with 95 percent ethanol or an ethanol-acetone solution, rinsed, and subsequently stained (counterstained) with safranin. The steps in the Gram staining procedure are shown in ➤ Figure 3.28.

Four groups of organisms can be distinguished with the Gram stain: (1) *Gram-positive* organisms, whose cell walls retain crystal violet stain; (2) *Gram-negative* organisms, whose cell walls do not retain crystal violet stain; (3) *Gram-nonreactive* organisms, which do not stain or which stain poorly; and (4) *Gram-variable* organisms, which stain unevenly. The differentiation between Gram-positive and Gram-negative organisms reveals a fundamental difference in the nature of the cell walls of bacteria, as is explained in Chapter 4. Furthermore, the reactions of bacteria to the Gram stain have helped in distinguishing Gram-positive, Gram-negative, and Gram-nonreactive groups that belong to radically different taxonomic groups (Chapter 9).

Gram-variable organisms have somehow lost their ability to react distinctively to the Gram stain. Organisms from cultures over 48 hours old (and sometimes only 24 hours old) are often

---

Are You Positive?—or, No Strings Attached

The Gram stain is not foolproof. Some anaerobic Gram-positive organisms decolorize easily and may falsely appear to be Gram-negative. Gram-negative organisms such as *Streptobacillus moniliformis* can stain Gram-positive. Is there some way to be sure of the organism's correct Gram reaction? There are several ways. One is the potassium hydroxide (KOH) test.

Place 2 drops of a 3 percent solution of KOH on a slide. Remove an inoculating loopful of the organism in question from a pure colony. Add it to the KOH on the slide, and mix continuously for 30 seconds. As you stir, occasionally lift the loop up 1 or 2 cm from the surface to see if "strings" of gooey material hang down. If the organism is truly Gram-negative, the KOH will break down its cell walls, releasing its DNA and forming strings. Gram-positive organisms will not form strings.

➤ Table 3.3   Comparison of staining techniques

Type	Examples	Result	Uses
Simple stain			
Use a single dye; do not distinguish organisms or structures by different staining reactions	Methylene blue Safranin Crystal violet ➔	Uniform blue stain Uniform red stain Uniform purple stain	Shows sizes, shapes, and arrangements of cells
Differential stains			
Use two or more dyes that react differently with various kinds or parts of bacteria, allowing them to be distinguished	Gram stain	Gram +: purple with crystal violet Gram −: red with safranin counterstain Gram-variable: inter-mediate or mixed colors (some stain + and some − on same slide) Gram-nonreactive: stain poorly or not at all	Distinguish Gram +, Gram −, Gram-variable, and Gram-nonreactive organisms
	Ziehl-Neelsen acid-fast stain	Acid-fast bacteria retain carbolfuchsin and appear red. Non-acid-fast bacteria accept the methylene blue counterstain and appear blue.	Distinguishes members of the genera *Mycobacterium* and *Nocardia* from other bacteria
	Negative stain	Capsules appear clear against a dark background.	Allows visualization of organisms with structures that will not accept most stains, such as capsules
Special stains			
Identify various specialized structures	Flagellar stain	Flagella appear as dark lines with silver, or red with carbolfuchsin.	Indicates presence of flagella by building up layers of stain on their surface
	Schaeffer-Fulton spore stain	Endospores retain malachite green stain. Vegetative cells accept safranin counterstain and appear red.	Allows visualization of hard-to-stain bacterial endospores, as for members of genera *Clostridium* and *Bacillus*

Gram-variable, probably because of changes in the cell wall with aging. Therefore, to determine the reaction of an organism to the Gram stain, you should use organisms from cultures 18 to 24 hours old.

The Ziehl-Neelsen Acid-Fast Stain

The **Ziehl-Neelsen acid-fast stain** is a modification of a staining method developed by Paul Ehrlich in 1882. It can be used to detect tuberculosis- and leprosy-causing organisms of the genus *Mycobacterium* (➤ Figure 3.29). Slides of organisms are covered with carbolfuchsin and heated, rinsed, decolorized with 3 percent hydrochloric acid (HCl) in 95 percent ethanol, rinsed, and then stained with Loeffler's methylene blue. Most genera of bacteria will lose the red car-bolfuchsin stain when decolorized. However, those that are *"acid-fast"* retain the bright red color. The lipid components of their walls responsible for this characteristic are discussed in Chapter 4. Bacteria that are not acid-fast lose the red color and can therefore be stained blue with the Loeffler's methylene blue counterstain.

(a) Crystal violet (1 minute)

All purple

Drain, rinse

(b) Iodine (1 minute)

All purple;
iodine acts as mordant
to set stain

Drain, rinse

(e)

(c) Decolorize with alcohol
(one quick rinse);
immediately after, rinse with water

Gram+ cocci = purple
Gram− rods = clear

(d) Safranin (30–60 seconds)

Gram+ cocci = purple
Gram− rods = red (pink)

Drain, rinse, blot

(f)

➤ Figure 3.28   **The Gram stain.** (a)–(d) Steps in Gram staining. (e) Gram-positive cells retain the purple color of crystal violet, whereas (f) Gram-negative cells are decolorized with alcohol and subsequently pick up the red color of the safranin counterstain.

➤ Figure 3.29   **The Ziehl-Neelsen acid-fast stain.** This stain produces vivid red color in acid-fast organisms such as *Mycobacterium leprae* (magnified 1025X), the cause of leprosy.

Special Staining Procedures

Negative Staining   **Negative stains** are used when a specimen—or a part of it, such as the capsule—resists taking up a stain. The *capsule* is a layer of polysaccharide material that surrounds many bacterial cells and can act as a barrier to host defense mechanisms. It also repels stains. In negative staining, the background around the organisms is filled with a stain, such as India ink, or an acidic dye, such as nigrosin. This process leaves the organisms themselves as clear, unstained objects that stand out against the dark background. A second simple or differential stain can be used to demonstrate the presence of the cell inside the capsule. Thus, a typical slide will show a dark background and clear, unstained areas of capsular material, inside of which are purple cells stained with crystal violet (➢ Figure 3.30) or blue cells stained with methylene blue.

Flagellar Staining   *Flagella,* appendages that some cells have and use for locomotion, are too thin to be seen easily with the light microscope. When it is necessary to determine their presence or arrangement, **flagellar stains** are painstakingly prepared to coat the surfaces of the flagella with dye or a metal such as silver. These techniques are very difficult and time-consuming and so are usually omitted from the beginning course in microbiology. (See Figure 4.12 for some examples of stained flagella.)

Endospore Staining   A few types of bacteria produce resistant cells called *endospores*. Endospore walls are very resistant to penetration of ordinary stains. When a simple stain is used, the spores will be seen as clear, glassy, easily recognizable areas within the bacterial cell. Thus, strictly speaking, it is not absolutely necessary to perform an endospore stain to see the spores. However, the differential **Schaeffer-Fulton spore stain** makes spores easier to visualize (➢ Figure 3.31). Heat-fixed smears are covered with malachite green and then gently heated until they steam. Approximately 5 minutes of such steaming causes the endospore walls to become more permeable to the dye. The slide is then rinsed with water for 30 seconds to remove the green dye from all parts of the cell except for the endospores, which retain it. Then a counterstain of safranin is placed on the slide to stain the non-spore-forming, or vegetative, areas of the cells. Cells of cultures without endospores appear red; those with endospores have green spores and red vegetative cells.

Although microscopy and staining techniques can offer valuable information about microorganisms, these methods are usually not enough to permit identification of most microbes. Many species look identical under the microscope—after all, there are only a limited number of basic shapes, arrangements, and staining reactions but thousands of kinds of bacteria. This means that biochemical and genetic characteristics usually must be determined before an identification can be made (Chapter 9).

➢ Figure 3.30   **Negative staining.** Negative staining for capsules reveals a clear area (the capsule, which does not accept stain) in a dark pink background of India ink and crystal violet counterstain. The cells themselves are stained deep purple with the counterstain. The bacteria are *Streptococcus pneumoniae* (3300X), which are arranged in pairs.

➤ **Figure 3.31** **The Schaeffer-Fulton spore stain.** Endospores of *Bacillus megaterium* (1000X) are visible as green, oval structures inside and outside the rod-shaped cells. Vegetative cells, which represent a non-spore-forming stage, and cellular regions without spores stain red.

Retracing Our Steps

Historical Microscopy

- The existence of microorganisms was unknown until the invention of the microscope. Leeuwenhoek, probably the first to see microorganisms (in the 1600s), set the stage for **microscopy,** the technology of making very small things visible to the human eye.

- Leeuwenhoek's simple microscopes could reveal little detail of specimens. Today, multiple-lens, compound microscopes give us nearly distortion-free images, enabling us to delve further into the study of microbes.

Principles of Microscopy

Metric Units

- The three units most used to describe microbes are the **micrometer** (μm), formerly called a micron, which is equal to 0.000001 m, also written as 10^{-6} m; the **nanometer** (nm), formerly called a millimicron (mμ), which is equal to 0.000000001 m, or 10^{-9} m; and the **angstrom** (Å), which is equal to 0.0000000001 m, 0.1 nm, or 10^{-10} m, but is no longer officially recognized.

Properties of Light: Wavelength and Resolution

- The **wavelength,** or the length of light rays, is the limiting factor in resolution.

- **Resolution** is the ability to see two objects as separate, discrete entities. Light wavelengths must be small enough to fit between two objects in order for them to be resolved.

- **Resolving power** can be defined as RP = λ/2NA, where λ = wavelength of light. The smaller the value of λ and the larger the value of NA, the greater the resolving power of the lens.

- **Numerical aperture** (NA) relates to the extent to which light is concentrated by the condenser and collected by the objective. Its value is engraved on the side of each objective lens.

Properties of Light: Light and Objects

- If light strikes an object and bounces back, **reflection** (which gives an object its color) has occurred.

- **Transmission** is the passage of light through an object. Light must either be reflected from or transmitted through an object for it to be seen with a light microscope.

- **Absorption** of light rays occurs when they neither bounce off nor pass through an object but are taken up by that object. Absorbed light energy is used in performing photosynthesis or in raising the temperature of the irradiated body.

- Reemission of absorbed light as light of longer wavelengths is known as **luminescence.** If reemission occurs only during irradiation, the object is said to **fluoresce.** If reemission continues after irradiation ceases, the object is said to be **phosphorescent.**

- **Refraction** is the bending of light as it passes from one medium to another of different density. **Immersion oil,** which has the same **index of refraction** as glass, is used to replace air and to prevent refraction at a glass-air interface.

- **Diffraction** occurs when light waves are bent as they pass through a small opening, such as a hole, a slit, a space between two adjacent cellular structures, or a small, high-powered, magnifying lens in a microscope. The bent light rays distort the image obtained and limit the usefulness of the light microscope.

Light Microscopy

The Compound Light Microscope

The major parts of a compound light microscope and their functions are as follows:

- **Base** Supporting structure that generally contains the light source.

- **Condenser** Converges light beams to pass through the specimen.

- **Iris diaphragm**   Controls the amount of light passing through the specimen.
- **Objective lens**   Magnifies image.
- **Body tube**   Conveys light to the ocular lens.
- **Ocular lens**   Magnifies the image from the objective. A microscope with one ocular lens (eyepiece) is **monocular;** a microscope with two oculars is **binocular.**
- **Mechanical stage**   Allows precise control in moving the slide.
- **Coarse adjustment**   Knob used to locate specimen.
- **Fine adjustment**   Knob used to bring specimen into sharp focus.
- The **total magnification** of a light microscope is calculated by multiplying the magnifying power of the objective lens by the magnifying power of the ocular lens. Increased magnification is of no value unless good resolution can also be maintained.

Dark-Field Microscopy
- **Bright-field illumination** is used in the ordinary light microscope, with light passing directly through the specimen.
- **Dark-field illumination** utilizes a special condenser that causes light to reflect off the specimen at an angle rather than pass directly through it.

Phase-Contrast Microscopy
- **Phase-contrast microscopy** uses microscopes with special condensers that accentuate small differences in refractive index of structures within the cell, allowing live, unstained organisms to be examined.

Nomarski (Differential Interference Contrast) Microscopy
- **Nomarski microscopy** uses microscopes that operate essentially like phase-contrast microscopes but with a much greater resolution and a very short depth of field. They produce a nearly three-dimensional image.

Fluorescence Microscopy
- **Fluorescence microscopy** uses ultraviolet light instead of white light to excite molecules within the specimen or dye molecules attached to the specimen. These molecules emit different wavelengths, often of brilliant colors.

Electron Microscopy
- The **electron microscope (EM)** uses a beam of electrons instead of a beam of light and electromagnets instead of glass lenses for focusing. They are much more expensive and difficult to use but give magnifications of up to 500,000X and a resolving power of less than 1 nm. Viruses can be seen only by using EMs.

- Advanced types of electron microscopes can visualize actual molecules and individual atoms.

Transmission Electron Microscopy
- For the **transmission electron microscope (TEM)**, very thin slices (sections) of a specimen are used, revealing the internal structure of microbial and other cells.

Scanning Electron Microscopy
- For the **scanning electron microscope (SEM)**, a specimen is coated with a metal. The electron beam is scanned, or swept, over this coating to form a three-dimensional image.

Scanning Tunneling Microscopy
- The **scanning tunneling microscope (STM)** can produce three-dimensional images of individual molecules and atoms, as well as movies.

Techniques of Light Microscopy
Preparation of Specimens for the Light Microscope
- **Wet mounts** are used to view living organisms. The **hanging-drop** technique is a special type of wet mount, often used to determine whether organisms are motile.
- **Smears** of appropriate thickness are allowed to air-dry completely and are then passed through an open flame. This process, called **heat fixation,** kills the organisms, causing them to adhere to the slide and more readily accept stains.

Principles of Staining
- A **stain,** or dye, is a molecule that can bind to a structure and give it color.
- Most microbial stains are **cationic** (positively charged), or **basic, dyes,** such as methylene blue. Because most bacterial surfaces are negatively charged, these dyes are attracted to them.
- **Simple stains** use one dye and reveal basic cell shapes and arrangements. **Differential stains** use two or more dyes and distinguish various properties of organisms. The **Gram stain,** the **Schaeffer-Fulton spore stain,** and the **Ziehl-Neelsen acid-fast stain** are examples.
- **Negative stains** color the background around cells and their parts, which resist taking up stain.
- **Flagellar stains** add layers of dye or metal to the surfaces of flagella to make those surfaces visible.
- In the Schaeffer-Fulton spore stain, endospores stain green due to the uptake of malachite green, whereas vegetative cells stain red due to safranin uptake.

Terminology Check

electron micrograph   *(p. 58)*
electron microscope (EM)   *(p. 57)*
fine adjustment   *(p. 55)*
flagellar stain   *(p. 67)*
fluoresce   *(p. 53)*
fluorescence microscopy *(p. 57)*
fluorescent antibody staining   *(p. 57)*
freeze-etching   *(p. 59)*
freeze-fracturing   *(p. 59)*
Gram stain   *(p. 64)*
hanging drop   *(p. 62)*
heat fixation   *(p. 62)*
immersion oil   *(p. 53)*
index of refraction   *(p. 53)*
iris diaphragm   *(p. 55)*
light microscopy   *(p. 54)*
luminescence   *(p. 53)*
mechanical stage   *(p. 55)*

micrometer   *(p. 49)*
microscopy   *(p. 49)*
monocular   *(p. 55)*
mordant   *(p. 64)*
nanometer   *(p. 49)*
negative stain   *(p. 67)*
Nomarski microscopy   *(p. 57)*
numerical aperture (NA)   *(p. 51)*
objective lens   *(p. 55)*
ocular lens   *(p. 55)*
ocular micrometer   *(p. 55)*
optical microscope   *(p. 54)*
parfocal   *(p. 55)*
phase-contrast microscopy   *(p. 57)*
phosphorescent   *(p. 53)*
reflection   *(p. 52)*
refraction   *(p. 53)*
resolution   *(p. 50)*

resolving power (RP)   *(p. 51)*
scanning electron microscope (SEM)
   (p. 60)
scanning tunneling microscope (STM)
   (p. 60)
Schaeffer-Fulton spore stain   *(p. 67)*
shadow casting   *(p. 59)*
simple stain   *(p. 64)*
smear   *(p. 62)*
stain   *(p. 63)*
total magnification   *(p. 55)*
transmission   *(p. 53)*
transmission electron microscope (TEM)
   (p. 58)
wavelength   *(p. 50)*
wet mount   *(p. 62)*
Ziehl-Neelsen acid-fast stain   *(p. 65)*

See You on the Web

Don't forget I am expecting you on the web. The following is a preparedness checklist for your web explorations. If you have questions on any of these items, you can return to the end of Chapter 1 for more details.

- Questions We'll Explore
- Concept Check questions
- Flashcards
- Study Guide

If you are ready, come visit me at

http://www.prenhall.com/black

It will be a journey you'll never forget.

Characteristics of Prokaryotic and Eukaryotic Cells

Overwhelmed by all these doors? Which one do you open? Where will it take you? Will you be able to pass through it? Maybe they're all locked. What then?

Perhaps these are questions you'd ask if you were a relatively large molecule trying to get into or out of a cell. Just as there is a wonderful diversity of doors, there is also a diversity of protein transport molecules found in cell membranes. These protein transport molecules control what enters and leaves the cell. They can let some molecules in while keeping others out. They might require that one molecule go out at the same time as a different molecule is coming in. In their various forms, protein transport molecules help control traffic across the cell membrane.

Background Basics

Questions We'll Explore

A What are the characteristics of prokaryotic and eukaryotic cells?

B How do prokaryotic cells differ in size, shape, and arrangement?

C How are structure and function related in bacterial cell walls and cell membranes?

D How are structure and function related in other bacterial components?

E How are structure and function related in eukaryotic plasma membranes?

F How are structure and function related in other eukaryotic components?

G How do passive transport processes function, and why are they important? How does active transport function, and why is it important?

H How do exocytosis and endocytosis occur, and why are they important?

We have considered the chemical principles that apply to cells and how to use microscopes and stains to observe cells. We can now look at the structure and function of the cells themselves.

Basic Cell Types

All living cells can be classified as prokaryotic, from the Greek words *pro* (before) and *karyon* (nucleus) or eukaryotic, from *eu* (true) and *karyon* (nucleus). **Prokaryotic** (pro-kar″e-ot′ik) **cells** lack a nucleus and other membrane-enclosed structures, whereas **eukaryotic** (u-kar″e-ot′ik) **cells** have such structures.

All prokaryotes are single-celled organisms, as exemplified by the bacteria. Most of this book will be devoted to the study of prokaryotes. Eukaryotes include all plants, animals, fungi, and protists (organisms such as *Amoeba, Paramecium,* and the malaria parasite). We will also spend some time studying eukaryotes, especially the fungi and various parasites, plus the interactions of eukaryotic cells and prokaryotes.

Prokaryotic and eukaryotic cells are *similar* in several ways. Both are surrounded by a *cell membrane,* or *plasma membrane*. Although some cells have structures that extend beyond this membrane or surround it, the membrane defines the boundaries of the living cell. Both prokaryotic and eukaryotic cells also encode genetic information in DNA molecules.

These two types of cells are *different* in other, important ways. In eukaryotic cells, DNA is in a nucleus surrounded by a *nuclear envelope,* but in prokaryotic cells, DNA is in a nuclear region not surrounded by a membrane. Eukaryotic cells also have a variety of internal structures called **organelles** (or-ga-nelz′), or "little organs," that are surrounded by one or more membranes. Prokaryotic cells generally lack organelles that are membrane-enclosed. We take advantage of some of the differences between eukaryotic human cells and prokaryotic bacterial cells when we try to control disease-causing bacteria without harming the human host.

In this chapter we examine the similarities and differences of prokaryotic cells and eukaryotic cells, as summarized in Table 4.1. (Refer to this table each time you learn about a new cellular structure.) Viruses do not fit in either category, as they are acellular. However, some viruses infect prokaryotic cells, while other viruses infect eukaryotes. Chapter 10 will examine viruses in detail.

> ➤ **Table 4.1**   **Similarities and differences between prokaryotic and eukaryotic cells**

Characteristic	Prokaryotic cells	Eukaryotic cells
Genetic structures		
Genetic material (DNA)	Found in single chromosome	Typically found in paired chromosomes
Location of genetic information	Nuclear region (nucleoid)	Membrane-enclosed nucleus
Nucleolus	Absent	Present
Histones	Absent	Present
Extrachromosomal DNA	In plasmids	In organelles, such as mitochondria and chloroplasts, and in plasmids
Intracellular structures		
Mitotic spindle	Absent	Present during cell division
Plasma membrane	Fluid-mosaic structure lacking sterols	Fluid-mosaic structure containing sterols
Internal membranes	Only in photosynthetic organisms	Numerous membrane-enclosed organelles
Endoplasmic reticulum	Absent	Present
Respiratory enzymes	Cell membrane	Mitochondria
Chromatophores	Present in photosynthetic bacteria	Absent
Chloroplasts	Absent	Present in some
Golgi apparatus	Absent	Present
Lysosomes	Absent	Present
Peroxisomes	Absent	Present
Ribosomes	70S	80S in cytoplasm and on endoplasmic reticulum, 70S in organelles
Cytoskeleton	Absent	Present
Extracellular structures		
Cell wall	Peptidoglycan found on most cells	Cellulose, chitin, or both found on plant and fungal cells
External layer	Capsule or slime layer	Pellicle, test, or shell in certain protists
Flagella	When present, consist of fibrils of flagellin	When present, consist of complex membrane-enclosed structure with "9 + 2" microtubule arrangement
Cilia	Absent	Present as structures shorter than, but similar to, flagella in some eukaryotic cells
Pili	Present as attachment or conjugation pili in some prokaryotic cells	Absent
Reproductive process		
Cell division	Binary fission	Mitosis and/or meiosis
Sexual exchange of genetic material	Not part of reproduction	Meiosis
Sexual or asexual reproduction	Only asexual reproduction	Sexual or asexual reproduction

Prokaryotic Cells

Detailed studies of cells have revealed that prokaryotes differ enough to be split into two large groups called *domains*. A relatively new concept in classification, domain is the highest taxonomic category, higher even than kingdom. Three domains exist: two prokaryotic and one eukaryotic:

- Archaea (archaeobacteria) (from *archae,* ancient)
- Bacteria (eubacteria)
- Eukarya

All members of Archaea and Bacteria are prokaryotes and have traditionally been called types of bacteria. A problem of terminology arises over the use of a capital versus a lower-case b in the word *bacteria*. All bacteria (lowercase b) are prokaryotes, but not all prokaryotes belong to the domain Bacteria (capital B). The differences between Archaea and Bacteria are not so much structural as molecular. Therefore, most of what we have to say about "bacteria" in this chapter applies to both Archaea and Bacteria. (We will discuss Archaea further in Chapter 11.

Most bacteria on this planet, both in the environment and living in and on humans, are members of the domain Bacteria. As yet, we know of no disease-causing Archaea. However, they are very important in the ecology of our planet, especially in extreme environments, such as in deep-sea hydrothermal vents where sulfur-laden water, at temperatures exceeding the boiling point of water, gushes out from openings in the ocean floor.

Size, Shape, and Arrangement

Size

Prokaryotes are among the smallest of all organisms. Most prokaryotes range from 0.5 to 2.0 μm in diameter. For comparison, a human red blood cell is about 7.5 μm in diameter. Keep in mind, however, that although we often use diameter to specify cell size, many cells are not spherical in shape. Some spiral bacteria have a much larger diameter, and some cyanobacteria are 60 μm long. Because of their small size, bacteria have a large surface-to-volume ratio. For example, spherical bacteria with a diameter of 2 μm have a surface area of about 12 μm^2 and a volume of about 4 μm^3. Their surface-to-volume ratio is 12:4, or 3:1. In contrast, eukaryotic cells with a diameter of 20 μm have a surface area of about 1200 μm^2 and a volume of about 4000 μm^3. Their surface-to-volume ratio is 1200:4000, or 0.3:1—only one-tenth as great. The large surface-to-volume ratio of bacteria means that no internal part of the cell is very far from the surface and that nutrients can easily and quickly reach all parts of the cell.

Shape

Typically, bacteria display three basic shapes—spherical, rodlike, and spiral (➤ Figure 4.1)—but variations abound. A spherical bacterium is called a **coccus** (kok′us; plural: *cocci* [kok′se]), and a rodlike bacterium is called a **bacillus** (ba-sil′us; plural: *bacilli* [bas-il′e]). Some bacteria, called *coccobacilli,* are short rods intermediate in shape between cocci and bacilli. Spiral bacteria have a variety of curved shapes. A comma-shaped bacterium is called a **vibrio** (vib′re-o); a rigid, wavy-shaped one, a **spirillum** (spi-ril′um; plural: *spirilla*); and a corkscrew-shaped one, a **spirochete** (spi′ro-ket). Some bacteria do not fit any of the preceding categories but rather have spindle shapes or irregular, lobed shapes. Square bacteria were discovered on the shores of the Red Sea in 1981. They are 2 to 4 μm on a side and sometimes aggregate in wafflelike sheets.

Even bacteria of the same kind sometimes vary in size and shape. When nutrients are abundant in the environment and cell division is rapid, rods are often twice as large as those in an environment with only a moderate supply of nutrients. Although variations in shape within a single species of bacteria are generally small, there are exceptions. Some bacteria vary widely in form even within a single culture, a phenomenon known as **pleomorphism.** Moreover, in aging cultures where organisms have used up most of the nutrients and have deposited wastes, cells not only are generally smaller, but they often display a great diversity of unusual shapes.

How Is a Bacteria Like a Lunch Meat?

Bacterial cell walls, organelles, and eukaryotic cells are diagrammed in your textbook to demonstrate their three-dimensional arrangements. Are these illustrations a figment of the imagination, or is there some way to construct complex three-dimensional structures of extremely small sub-macroscopic objects? Using serial sections (i.e., examining adjacent slices, one right after the next) is one method used to elucidate the three-dimensional relationships between microscopic cell parts. The determination of complex shapes using serial sections is not unlike determining the shape and size of the fat globules and peppercorns that lace a good salami. After making many orderly slices through the salami, the geometry of the salami inclusions is revealed as the shape from each section is superimposed over the preceding section. Using serial sections, in conjunction with chemical and biophysical data, makes it possible to produce the three-dimensional cell structures that illustrate your text.

Coccus

Coccobacillus

Vibrio

Bacillus

Spirillum

Spirochete

➤ Figure 4.1 **The most common bacterial shapes.**

(a) Diplococcus

Streptococci

(b) Tetrad

(c) Sarcinae

(d) Staphylococci

(e) Rosette

2 µ

(f) Star-shaped bacteria

(g) Square bacteria

2 µ

➤ Figure 4.2   **Arrangements of bacteria.** (a) Cocci arranged in pairs (diplococci of *Neisseria*) and in chains (*Streptococcus*), formed by division in one plane. (b) Cocci arranged in a tetrad (*Merisopedia,* 100X), formed by division in two planes. (c) Cocci arranged in a sarcina (*Sarcina lutea,* 16,000X), formed by division in three planes. (d) Cocci arranged in a cluster (*Staphylococcus,* 5400X), formed by division in many planes. (e) Bacilli arranged in a rosette (*Caulobacter,* 2400X), attached by stalks to a substrate. (f) Star-shaped bacteria (*Stella*). (g) Square-shaped bacteria.

Arrangement

In addition to characteristic shapes, many bacteria also are found in distinctive arrangements of groups of cells (➤ Figure 4.2). Such groups occur when cells divide without separating. Cocci can divide in one or two planes, or randomly. Division in one plane produces cells in pairs (indicated by the prefix *diplo-*) or in chains (*strepto-*). Division in two planes produces cells in *tetrads* (four cells arranged in a cube). Division in three planes produces *sarcinae* (singular: *sarcina;* eight cells arranged in a cube). Random division planes produce grapelike clusters (*staphylo-*). Bacilli divide in only one plane, but they can produce cells connected end-to-end (like train cars) or side-by-side. Spiral bacteria are not generally grouped together.

Prokaryotes divide by binary fission, rather than by mitosis or meiosis. New cell wall material grows, and the cell pinches in half through this area. Inside, the chromosome has duplicated, and one is found in each daughter cell.

An Overview of Structure

Structurally, bacterial cells (➤ Figure 4.3) consist of the following:

1. A cell membrane, usually surrounded by a cell wall and sometimes by an additional outer layer.
2. An internal cytoplasm with ribosomes, a nuclear region, and in some cases granules and/or vesicles.
3. A variety of external structures, such as capsules, flagella, and pili.

Let us look at each of these kinds of structures in some detail.

The Cell Wall

The semirigid **cell wall** lies outside the cell membrane in nearly all bacteria. It performs two important functions. First, it maintains the characteristic shape of the cell. If the cell wall is digested away by enzymes, the cell takes on a spherical shape. Second, it prevents the cell from bursting when fluids flow into the cell by *osmosis* (described later in this chapter). Although the cell wall surrounds the cell membrane, in many cases it is extremely porous and does not play a major role in regulating the entry of materials into the cell.

Components of Cell Walls

Peptidoglycan   **Peptidoglycan** (pep″ti-do-gly′-kan), also called *murein,* is the single most important component of the bacterial cell wall. It is a polymer so large that it can be thought of as one immense, covalently linked molecule. It forms a supporting net around a bacterium that resembles multiple layers of chain-link fence (➤ Figure 4.4). In the peptidoglycan poly-

✓ Are viruses prokaryotes? Eukaryotes? Why or why not?

✓ Compare surface-to-volume ratios of prokaryotes and eukaryotes. Of what importance is this difference?

✓ Explain how sex is *not* related to reproduction in prokaryotes. What then *is* its purpose?

✓ Prokaryotes lack mitochondria. What structure *does* perform the functions of mitochondria in prokaryotes?

Square bacteria were not discovered until 1980; triangular ones in 1986.

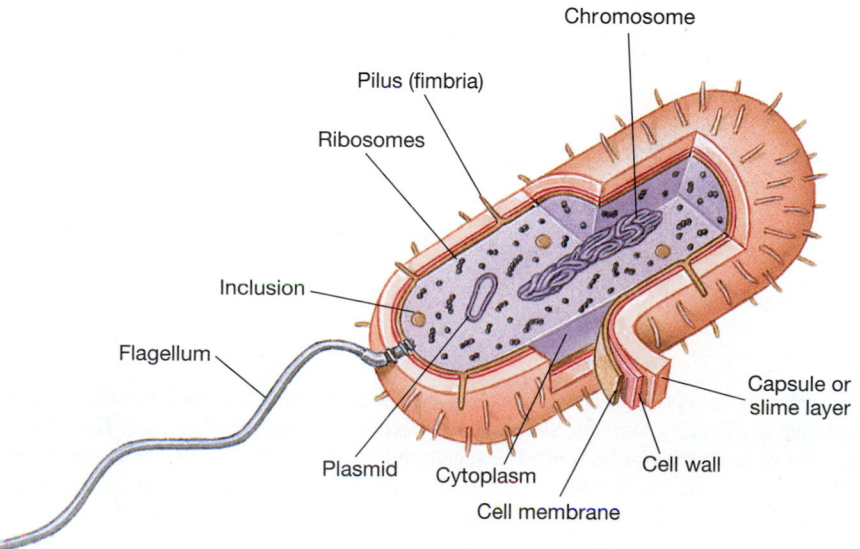

➤ **Figure 4.3   A typical prokaryotic cell.** The cell depicted is a bacillus with a polar flagellum (a flagellum at one end).

➤ **Figure 4.4   Peptidoglycan.** (a) A two-dimensional view of peptidoglycan for the Gram-negative bacterium *Escherichia coli;* a polymer of two alternating sugar units (purple), *N*-acetylglucosamine and *N*-acetylmuramic acid, both of which are derivatives of glucose. The sugars are joined by short peptide chains (tetrapeptides) that consist of four amino acids (red). The sugars and tetrapeptides are cross-linked by a simple peptide bond. (b) A three-dimensional view of peptidoglycan for the Gram-positive bacterium *Staphylococcus aureus.* Compare the components with those in (a). Different organisms can have different amino acids in the tetrapeptide chain as well as different cross-links.

mer, molecules of *N*-acetylglucosamine (gluNAc) alternate with molecules of *N*-acetylmuramic acid (murNAc). These molecules are cross-linked by tetrapeptides, chains of four amino acids. In most Gram-positive organisms, the third amino acid is lysine; in most Gram-negative organisms, it is diaminopimelic acid. Amino acids, like many other organic compounds, have *stereoisomers*—structures that are mirror images of each other, just as a left hand is a mirror image of a right hand. Some of the amino acids in the tetrapeptide chains are mirror images of those amino acids most commonly found in living things. Those chains are not readily broken down because most organisms lack enzymes that can digest stereoisomeric forms.

Cell walls of Gram-positive organisms have an additional molecule, teichoic acid. **Teichoic (tie-ko′ik) acid,** which consists of glycerol, phosphates, and the sugar alcohol ribitol, occurs in polymers up to 30 units long. These polymers extend beyond the rest of the cell wall, even beyond the capsule in encapsulated bacteria. Although its exact function is unclear, teichoic acid furnishes attachment sites for bacteriophages (viruses that infect bacteria).

Outer Membrane   The **outer membrane,** found primarily in Gram-negative bacteria, is a bilayer membrane (discussed in the next section). It forms the outermost layer of the cell wall and is attached to the peptidoglycan by an almost continuous layer of small lipoprotein molecules (proteins combined with a lipid). The lipoproteins are embedded in the outer membrane and covalently bonded to the peptidoglycan. The outer membrane acts as a coarse sieve and exerts little control over the movement of substances into and out of the cell. However, it does control the transport of certain proteins from the environment. Proteins called porins form channels through the outer membrane. The outer surface of the outer membrane has surface antigens and receptors. Some receptors bind viruses and thereby help them to infect the bacterium.

Lipopolysaccharide (LPS), also called **endotoxin,** is an important part of the outer membrane and can be used to identify Gram-negative bacteria. It is an integral part of the cell wall and is not released until the cell walls of dead bacteria are broken down. LPS consists of polysaccharides and **lipid A** (➤ Figure 4.5). The polysaccharides are found in repeating side chains that extend outward from the organism. It is these repeating units that are used to identify different Gram-negative bacteria. The lipid A portion is responsible for the toxic properties that make any Gram-negative infection a potentially serious medical problem. It causes fever and dilates blood vessels, so the blood pressure drops precipitously. Because bacteria release endotoxin mainly when they are dying, killing them may increase the concentration of this very toxic substance.

Periplasmic Space   Another distinguishing characteristic of many bacteria is the presence of a gap between the cell membrane and the cell wall. The gap is most easily observed by electron microscopy of Gram-negative bacteria. In these organisms the gap is called the **periplasmic (per′e-plaz″mik) space.** It represents a very active area of cell metabolism. This space contains not only the cell wall peptidoglycan but also many digestive enzymes and transport proteins that destroy potentially harmful substances and transport metabolites into the bacterial

➤ **Figure 4.5   Lipopolysaccharide.** Lipopolysaccharide (LPS), also called endotoxin, is an important component of the outer membrane in Gram-negative cell walls. The lipid A portion of the molecule consists of a backbone of alternating pyrophosphate units (POP, linked phosphate groups) and glucosamine (G, a glucose derivative), to which long fatty acid side chains are attached. Lipid A is a toxic substance that contributes to the danger of infection by Gram-negative bacteria. Polysaccharide side chains extending outward from the glucosamine units make up the remainder of the molecule.

cytoplasm, respectively. The *periplasm* consists of the peptidoglycan, protein constituents, and metabolites found in the periplasmic space.

Periplasmic spaces are rarely observed in Gram-positive bacteria. However, such bacteria must accomplish many of the same metabolic and transport funtions that Gram-negative bacteria do. At present most Gram-positive bacteria are thought to have only periplasms—not periplasmic spaces—where metabolic digestion occurs and new cell wall peptidoglycan is attached. The periplasm in Gram-positive cells is thus part of the cell wall.

Distinguishing Bacteria by Cell Walls

Certain properties of cell walls result in different staining reactions. Gram-positive, Gram-negative, and acid-fast bacteria can be distinguished on the basis of these reactions (Table 4.2 and ➢ Figure 4.6).

Gram-Positive Bacteria   The cell wall in Gram-positive bacteria has a relatively thick layer of peptidoglycan 20 to 80 nm across. The peptidoglycan layer is closely attached to the outer surface of the cell membrane. Chemical analysis shows that 60 to 90 percent of the cell wall of a Gram-positive bacterium is peptidoglycan. Except for streptococci, most Gram-positive cell walls contain very little protein. If peptidoglycan is digested from their cell walls, Gram-positive bacteria become **protoplasts,** or cells with a cell membrane but no cell wall. Protoplasts shrivel or burst unless they are kept in an *isotonic* solution—a solution that has the same pressure as that inside the cell.

The thick cell walls of Gram-positive bacteria retain such stains as the crystal violet-iodine dye in the cytoplasm, but yeast cells, many of which have thick walls, also retain them. Thus, retention of Gram stain seems to be directly related to wall thickness and not to peptidoglycan. Physiological damage or aging can make a Gram-positive cell wall leaky, so the dye complex escapes. Such organisms can become Gram-variable or even Gram-negative as they age. Therefore, Gram staining must be performed on cultures less than 24 hours old.

Gram-positive bacteria lack both an outer membrane and a periplasmic space. Therefore, digestive enzymes not retained in the periplasm are released into the environment, where they sometimes become so diluted that the organisms derive no benefit from them.

Gram-Negative Bacteria   The cell wall of a Gram-negative bacterium is thinner but more complex than that of a Gram-positive bacterium. Only 10 to 20 percent of the cell wall is peptidoglycan; the remainder consists of various polysaccharides, proteins, and lipids. The cell wall contains an outer membrane, which constitutes the outer surface of the wall, leaving only a very narrow periplasmic space. The inner surface of the wall is separated from the cell membrane by a wider periplasmic space. Toxins and enzymes remain in the periplasmic space in sufficient concentrations to help destroy substances that might harm the bacterium, but they do not harm the organism that produced them. If the cell wall is digested away, Gram-negative bacteria become **spheroplasts,** which have both a cell membrane and most of the outer membrane. Gram-negative bacteria fail to retain the crystal violet-iodine dye during the decolorizing

---

Gram-positive cell walls may have as many as 40 layers of peptidoglycan.

The Heat Is On

You may notice that many prepackaged hospital products, in addition to being labeled "sterile," are also labeled "pyrogen free". In a sense this really does mean that the product will not ignite the patient. Pyrogens, the lipopolysaccharide component of the outer membrane portion of the Gram-negative cell wall, are released into the patient's circulatory system when Gram-negative bacteria die or are killed with antibiotics. Although sterile products are free of living microbes, the presence of Gram-negative cell wall residue can cause serious problems. Pyrogens induce the production of interleukin-1, which in turn affects the hypothalamus, the temperature control center, in the brain. A patient exposed to pyrogens, in addition to fever, may also experience the formation of small blood clots, shock, and sometime death.

➢ Table 4.2	Characteristics of the cell walls of Gram-positive, Gram-negative, and acid-fast bacteria		
Characteristic	**Gram-positive bacteria**	**Gram-negative bacteria**	**Acid-fast bacteria**
Peptidoglycan	Thick layer	Thin layer	Relatively small amount
Teichoic acid	Often present	Absent	Absent
Lipids	Very little present	Lipopolysaccharide	Mycolic acid and other waxes and glycolipids
Outer membrane	Absent	Present	Absent
Periplasmic space	Absent	Present	Absent
Cell shape	Always rigid	Rigid or flexible	Rigid or flexible
Results of enzyme digestion	Protoplast	Spheroplast	Difficult to digest
Sensitivity to dyes and antibiotics	Most sensitive	Moderately sensitive	Least sensitive

(a) Gram-positive bacteria

(b) Gram-negative bacteria

(c) Acid-fast bacteria

➤ Figure 4.6   **The bacterial cell wall.** Schematic drawings, paired with TEM photos of representative bacteria. (a) Gram-positive (*Bacillus fastidosus*). (b) Gram-negative (*Azomonas insignis*). (c) Acid-fast (*Mycobacterium phlei*).

procedure partly because of their thin cell walls and partly because of the relatively large quantities of lipoproteins and lipopolysaccharides in the walls.

Acid-Fast Bacteria   The cell wall of *acid-fast bacteria,* the mycobacteria, is thick, like that of Gram-positive bacteria. It is approximately 60 percent lipid and contains much less peptidoglycan. In the acid-fast staining process, carbolfuchsin binds to cytoplasm and resists removal by an acid-alcohol mixture. ➡ (Chapter 3, p. 65) The lipids make acid-fast organisms impermeable to most other stains and protect them from acids and alkalis. The organisms grow slowly because the lipids impede entry of nutrients into cells, and the cells must expend large quantities of energy to synthesize lipids. Acid-fast cells can be stained by the Gram stain method; they stain as Gram-positive.

Controlling Bacteria by Damaging Cell Walls   Some methods of controlling bacteria are based on properties of the cell wall. For example, the antibiotic penicillin blocks the final stages of peptidoglycan synthesis. If penicillin is present when bacterial cells are dividing, the cells cannot form complete walls, and they die. Similarly, the enzyme lysozyme, found in tears and other human body secretions, digests peptidoglycan. This enzyme helps prevent bacteria from entering the body and is the body's main defense against eye infections (➤ Figure 19.4).

The Cell Membrane

The **cell membrane,** or *plasma membrane,* is a living membrane that forms the boundary between a cell and its environment. Also known as the cytoplasmic membrane, this dynamic, constantly changing membrane is not to be confused with the cell wall. The latter is a static structure external to the cell membrane.

Bacterial cell membranes have the same general structure as the membranes of all other cells. Such membranes, formerly called *unit membranes,* consist mainly of phospholipids and proteins. The **fluid-mosaic model** (➤ Figure 4.7) represents the current understanding of the structure of such a membrane. The model's name is derived from the fact that phospholipids in the membrane are in a fluid state and that proteins are dispersed among the lipid molecules in the membrane, forming a mosaic pattern.

Membrane phospholipids form a *bilayer,* or two adjacent layers. In each layer, the phosphate ends of the lipid molecules extend toward the membrane surface, and the fatty acid ends extend inward. The charged phosphate ends of the molecules are **hydrophilic** (water-loving) and thus can interact with the watery environment (➤ Figure 4.7a). The fatty acid ends, consisting largely of nonpolar hydrocarbon chains, are **hydrophobic** (water-fearing) and form a barrier between the cell and its environment. Some membranes also contain other lipids. The membranes of mycoplasmas, bacteria that lack a cell wall, include lipids, that add rigidity, called *sterols.*

Interspersed among the lipid molecules are protein molecules (➤ Figure 4.7b). Some extend through the entire membrane and act as carriers or form pores or channels through which materials enter and leave the cell. Proteins on the outer surface include those that make the cell identifiable as a particular organism. Others are embedded in, or loosely attached to, the inner or outer surface of the membrane. Proteins on the inner surface are usually enzymes.

Cell membranes are dynamic, constantly changing entities. Materials constantly move through pores and through the lipids themselves, although selectively. Also, both the lipids and the proteins in membranes are continuously changing positions.

The main function of the cell membrane is to regulate the movement of materials into and out of a cell by transport mechanisms, which are discussed later in this chapter. In bacteria this membrane also performs some functions carried out by other structures in eukaryotic cells. It synthesizes cell wall components, assists with DNA replication, secretes proteins, carries on respiration, and captures energy as ATP. It also contains bases of appendages called *flagella;* the actions of the bases cause the flagella to move. Finally, some proteins in the cell membrane respond to chemical substances in the environment.

Internal Structure

Bacterial cells typically contain *ribosomes,* a *nucleoid,* and a variety of *vacuoles* within their *cytoplasm.* Figure 4.3 shows the locations of these structures in a generalized prokaryotic cell. Certain bacteria sometimes contain *endospores* as well.

✓ Compare peptidoglycan and teichoic acid as to location and function.

✓ What takes place in the periplasmic space? Which organisms have such a space?

✓ Relate the following terms: cell wall, outer membrane, lipopolysaccharide, endotoxin, lipid A, death of cell.

✓ Compare the cell walls of Gram-positive, Gram-negative, and acid-fast bacteria.

➤ **Figure 4.7   The fluid-mosaic model of the cell membrane.** (a) The basic structural component of the membrane is the phospholipid molecule. A phospholipid has two long fatty acid "tails" of hydrocarbon. The tails are very hydrophobic—they do not interact with water and form an oily barrier to most water-soluble substances. The "head" of the molecule consists of a charged phosphate group, usually joined to a charged nitrogen-containing group. The head is very hydrophilic—it interacts with water. (b) The fluid-mosaic model of membrane structure. The phospholipids form a bilayer in which the hydrophobic tails form the central core and the hydrophilic heads form the surfaces that face both the interior of the cell and the outside environment. In this fluid bilayer, proteins float like icebergs. Some extend through the bilayer; others are anchored to the inner or outer surface. Proteins and membrane lipids to which carbohydrate chains are attached are called *glycoproteins* and *glycolipids,* respectively. A few bacteria, such as mycoplasmas, have cholesterol molecules in their cell membranes, as do most eukaryotes. Mycoplasmas lack cell walls; cholesterol molecules add rigidity to the cell membrane.

➤ Figure 4.8   **A centrifuge.** Suspended particles in tubes of liquid are whirled around at high speeds, causing them to settle to the bottom of the tubes or to form bands at different levels. The rate of settling or the locations of the bands can be used to determine the size, weight, and shape of the particles.

Cytoplasm

The **cytoplasm** of prokaryotic cells is the semifluid substance inside the cell membrane. Because these cells typically have only a few clearly defined structures, such as a chromosome and some ribosomes, they consist mainly of cytoplasm. Cytoplasm is about four-fifths water and one-fifth substances dissolved or suspended in the water. These substances include enzymes and other proteins, carbohydrates, lipids, and a variety of inorganic ions. Many chemical reactions, both anabolic and catabolic, occur in the cytoplasm. Unlike eukaryotic cytoplasm, that of prokaryotes does not carry out the movement known as "streaming."

Ribosomes

Ribosomes consist of ribonucleic acid and protein. They are abundant in the cytoplasm of bacteria, often grouped in long chains called **polyribosomes.** Ribosomes are nearly spherical, stain densely, and contain a large subunit and a small subunit. Ribosomes serve as sites for protein synthesis (Chapter 7).

The relative sizes of ribosomes and their subunits can be determined by measuring their *sedimentation rates*—the rates at which they move toward the bottom of a tube when the tube is rapidly spun in an instrument called a *centrifuge* (➤ Figure 4.8). Sedimentation rates are expressed in terms of *Svedberg* (S) *units,* which generally vary with molecular size. Whole bacterial ribosomes, which are smaller than eukaryotic ribosomes, have a rate of 70S; their subunits have rates of 30S and 50S. Certain antibiotics, such as streptomycin and erythromycin, bind specifically to 70S ribosomes and disrupt protein synthesis. Because those antibiotics do not affect the 80S ribosomes found in eukaryotic cells, they kill bacteria without harming host cells.

Nuclear Region

One of the key features differentiating prokaryotic cells from eukaryotic cells is the absence of a nucleus bounded by a nuclear membrane. Instead of a nucleus, bacteria have a **nuclear region,** or **nucleoid** (➤ Figure 4.9). The centrally located nuclear region consists mainly of DNA but has some RNA and protein associated with it. The DNA is arranged in one large, circular chromosome. Some bacteria also contain small circular molecules of DNA called *plasmids*. Genetic information in plasmids supplements information in the chromosome (Chapter 8).

Internal Membrane Systems

Photosynthetic bacteria and cyanobacteria contain internal membrane systems, sometimes known as **chromatophores** (➤ Figure 4.10). The membranes of the chromatophores, derived from the cell membrane, contain the pigments used to capture light energy for the synthesis of sugars. Nitrifying bacteria, soil organisms that convert nitrogen compounds into forms usable by green plants, also have internal membranes. They house the enzymes used in deriving energy from the oxidation of nitrogen compounds (Chapter 5).

Electron micrographs of bacterial cells often show large infoldings of the cell membrane, called *mesosomes*. Although these were originally thought to be structures present in living

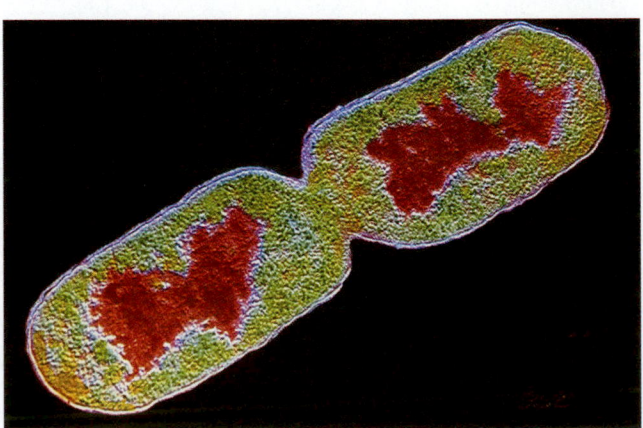

➤ Figure 4.9   **The bacterial nuclear region.** A colorized TEM of a thin section of *Escherichia coli* with the DNA outlined.

Carbohydrate granule

Photosynthetic membrane

➤ Figure 4.10 **Internal membrane systems.** TEM of the cyanobacterium *Coccochloris elabens* (47,000X). The cell's outer regions are filled with photosynthetic membranes. The dark spots between the membranes are granules where carbohydrates produced by photosynthesis are stored.

cells, they have now been proven to be artifacts: That is, they were created by the processes used to prepare specimens for electron microscopy.

Inclusions

Bacteria can have within their cytoplasm a variety of small bodies collectively referred to as **inclusions.** Some are called *granules;* others are called *vesicles.*

Granules, although not bounded by membrane, contain substances so densely compacted that they do not easily dissolve in cytoplasm. Each granule contains a specific substance, such as glycogen or polyphosphate. *Glycogen,* a glucose polymer, is used for energy. *Polyphosphate,* a phosphate polymer, supplies phosphate for a variety of metabolic processes. Polyphosphate granules are called **volutin** (vo-lu′tin), or **metachromatic granules,** because they display **metachromasia.** That is, although most substances stained with a simple stain such as methylene blue take on a uniform solid color, metachromatic granules exhibit different intensities of color. Although quite numerous in some bacteria, these granules become depleted during starvation.

Certain bacteria have specialized membrane-enclosed structures called **vesicles** (or *vacuoles*). Some aquatic photosynthetic bacteria and cyanobacteria have rigid gas-filled vacuoles (➤ Figure 3.23). These organisms regulate the amount of gas in vacuoles and, therefore, the depth at which they float to obtain optimum light for photosynthesis. Another type of vesicle, found only in bacteria, contains deposits of poly-β-hydroxybutyrate. These lipid deposits serve as storehouses of energy and as sources of carbon for building new molecules.

Endospores

The properties of bacterial cells just described pertain to **vegetative cells,** or cells that are metabolizing nutrients. However, vegetative cells of some bacteria, such as *Bacillus* and *Clostridium,* produce resting stages called **endospores.** Although bacterial endospores are commonly referred to simply as *spores,* do not confuse them with fungal spores. A bacterium produces a single endospore, which merely helps that organism survive and is not a means of reproduction. A fungus produces numerous spores, which help the organism survive and provide a means of reproduction.

Endospores, which are formed within cells, contain very little water and are highly resistant to heat, drying, acids, bases, certain disinfectants, and even radiation. The depletion of a nutrient will usually induce a large number of cells to produce spores. However, many investigators believe that spores are part of the normal life cycle and that a few are formed even when nutrients are adequate and environmental conditions are favorable. Thus, *sporulation,* or endospore formation, seems to be a means by which some bacteria prepare for the possibility of adverse conditions.

In 1877, F. Cohn first demonstrated bacterial spores.

Living Magnets

Magnetotactic bacteria synthesize magnetite (Fe_3O_4), or lodestone, and store it in membranous vesicles called *magnetosomes*. (Lodestone was the first substance with magnetic properties to be discovered.) The presence of these magnetic inclusions enables these bacteria to respond to magnetic fields. In the Northern Hemisphere, magnetotactic bacteria swim toward the North Pole; in the Southern Hemisphere, they swim toward the South Pole; and near the equator, some swim north and others south. However, they also swim downward in water, because the magnetic force from the earth's poles is deflected through the earth and not over its horizon. This phenomenon is called *magnetotaxis*. It appears to help these anaerobic bacteria move down toward sediments where their food (iron oxide) is abundant and where oxygen, which they cannot tolerate, is deficient.

Magnetotactic bacteria live in mud and brackish waters, and more than a dozen species have been identified. Most have a single flagellum, but *Aquaspirillum magnetotacticum* has two, one at each end, so it can swim forward or backward. When magnetotactic bacteria with one flagellum are placed in the field of an electromagnet, they make U-turns as the poles of the magnet are reversed.

Magnetosomes are nearly constant in size and are oriented in parallel chains like a string of tiny magnets. Experiments are currently under way to use these bacteria in the manufacture of magnets, audiotapes, and videotapes.

Magnetosomes have been found inside meteorites from Mars. They are associated with bacillus-shaped structures, which some scientists believe to be Martian bacterial life forms.

TRY IT: You can easily find your own magnetotactic bacteria. They are very common organisms. Bring back a bucket with a few inches of mud from a pond and enough pondwater to fill the rest of the bucket. Try mud from different locations: fresh, salt, and brackish (mixed fresh and salt) waters. Cover the bucket and store it in the dark (you don't want to grow algae) for about 1 month. Remove a large beaker of the water, and place a magnet against the outside of the glass. Allow this to stand for a day or two, until you see a whitish spot in the water near the magnet's end. Which attracts more organisms, the north or south end of the magnet? With a pipette, remove a sample of the cloudy liquid and examine it under the microscope. When you lay a magnet on the stage, any magnetotactic bacteria that are present will orient themselves to the field. They can be purified by streaking on agar, a process described in Chapter 6. Microbiology doesn't exist only in books or labs. It's everywhere around you in the real world—just look for it!

(a)

(b)

(a) Electron micrograph of the magnetotactic bacterium *Aquaspirillum magnetotacticum*. The numerous dark, square inclusions, called magnetosomes, are composed of iron oxide, (Fe_3O_4). (b) The magnetosomes enable these organisms to orient themselves in a magnetic field.

Structurally, an endospore consists of a *core,* surrounded by a *cortex,* a *spore coat,* and in some species a delicately thin layer called the *exosporium* (➢ Figure 4.11). The core has an outer core wall, a cell membrane, nuclear region, and other cell components. Unlike vegetative cells, endospores contain *dipicolinic acid* and a large quantity of calcium ions (Ca^{2+}). These materials, which are probably stored in the core, appear to contribute to the heat resistance of endospores, as does their very low water content.

Endospores are capable of surviving adverse environmental conditions for long periods of time, some over 10,000 years. Some withstand hours of boiling. When conditions become more favorable, endospores *germinate,* or begin to develop into functional vegetative cells. (The processes of spore formation and germination are discussed in Chapter 6.) Because endospores are so resistant, special methods must be used to kill them during sterilization. Otherwise, they

Endospore
Core
Cortex
Spore coat

➤ **Figure 4.11 Endospores.** A colorized electron micrograph of an endospore within a *Clostridium perfringens* cell.

germinate and grow in media thought to be sterile. Methods to ensure that endospores are killed when bacterial media or foods are sterilized are described in Chapter 13.

External Structure

In addition to cell walls, many bacteria have structures that extend beyond or surround the cell wall. *Flagella* and *pili* extend from the cell membrane through the cell wall and beyond it. *Capsules* and *slime layers* surround the cell wall. These structures have a variety of properties and functions.

Flagella

About half of all known bacteria are *motile,* or capable of movement. They often move with speed and apparent purpose, and they usually move by means of long, thin, helical appendages called **flagella** (singular: *flagellum*). A bacterium can have one flagellum or two or many flagella. Bacteria with a single *polar* flagellum located at one end, or pole, are said to be **monotrichous** (mon-o-trik'-us; ➤ Figure 4.12a); bacteria with two flagella, one at each end, are **amphitrichous** (am-fet-rik'-us; ➤ Figure 4.12b); both types are said to be *polar*. Bacteria with two or more flagella at one or both ends are **lophotrichous** (lo-fot-rik'us; ➤ Figure 4.12c); and those with flagella all over the surface are **peritrichous** (pe-rit-rik' us; ➤ Figure 4.12d). Bacteria without flagella are **atrichous** (a-trik' us).

Cocci rarely have flagella.

(a)

(b)

(c)

(d)

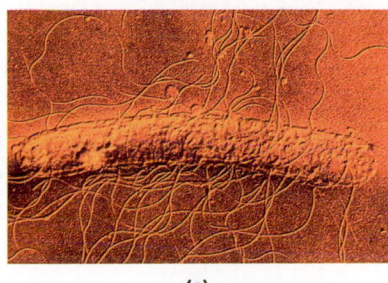

(e)

➤ **Figure 4.12 Arrangements of bacterial flagella.** (a) Polar, monotrichous (single flagellum at one end) *Pseudomonas* (3300X). (b) Polar, amphitrichous (single flagellum at each end) *Spirillum.* (c) Lophotrichous (tuft of flagella at one or both ends) *Spirillum.* (d) Peritrichous (flagella distributed all over) *Salmonella* (1200X). (e) SEM of peritrichous flagella, *Proteus* (29,400X).

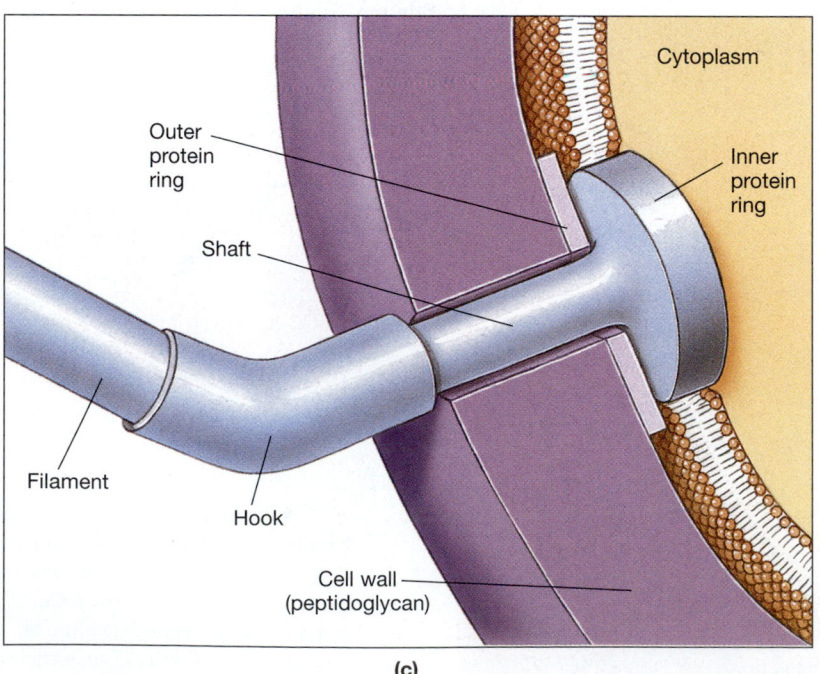

Cell wall

Outer membrane   Peptidoglycan

Cell membrane

Cytoplasm

Filament

Hook

(a)

Outer rings   Rod   Inner rings

Basal body

(b)

Cytoplasm

Outer protein ring

Inner protein ring

Shaft

Filament

Hook

Cell wall (peptidoglycan)

(c)

➤ Figure 4.13   **Structure of two different bacterial flagella.** (a) Drawing and (b) electron micrograph of the basal region of the flagellum of a Gram-negative bacterium. The flagellum has three main parts: a filament, a hook, and a basal body consisting of a rod surrounded by four rings. (c) Gram-positive bacteria have only two rings, one attached to the peptidoglycan of the cell wall and one to the cell membrane.

The diameter of a prokaryote's flagellum is about one-tenth that of a eukaryote's flagellum. It is made of protein subunits called *flagellin*. Each flagellum is attached to the cell membrane by a basal region consisting of a protein other than flagellin (➤ Figure 4.13). The basal region has a hooklike structure and a complex *basal body*. The basal body consists of a central rod or shaft surrounded by a set of rings. Gram-negative bacteria have a pair of rings embedded in the cell membrane and another pair of rings associated with the peptidoglycan and lipopolysaccharide layers of the cell wall. Gram-positive bacteria have one ring embedded in the cell membrane and another in the cell wall.

Most flagella rotate like twirling L-shaped hooks, such as a dough hook on a kitchen mixer or the rotating string on a hand-carried grass trimmer. Motion is thought to occur as energy is used to make one of the rings in the cell membrane rotate with respect to the other. When flagella bundle together (➤ Figure 4.14a), they rotate counterclockwise, and the bacteria *run*, or

(a)

Cell wall

Flagellum

Plasma membrane

Flagellar motors

Flagellar bundle

(b)

Twiddle

Run

Run

Twiddle

(c)

No attractant or repellant

Attractant

Repellant

➤ Figure 4.14   **Chemotaxis.** (a) When all the flagella of a bacterium rotate in a counterclockwise direction, the flagella bundle together and push the bacterium in a fairly straight, forward movement called a *run*. When the flagella reverse into a clockwise direction, the bundle comes apart, each flagellum acts independently, and the cells tumble about in random directions, a movement called a *twiddle*. (b) Peritrichous and lophotrichous flagellated bacteria doing runs and twiddles. Note that the cell swims forward (a run) only when flagella are bundled, and that the bacterium changes direction following a twiddle. (c) When nothing attracts or repels a bacterium, it has frequent twiddles, and short runs, resulting in random movement.

➤ **Figure 4.15   Axial filaments, or endoflagella.** (a) Axial filaments made visible by false coloring are clearly seen as yellow ribbons running along the body of the spirochete *Leptospira interrogans* (50,000X). (b) TEM (cross section) of a spirochete, showing numerous axial filaments (dark circles). Axial filaments lie between the outer sheath and the cell wall.

(a)

Axial filaments

(b)

move in a straight line. When the flagella rotate clockwise, the flagellar bundle comes apart, causing the bacteria to *twiddle,* or tumble randomly. Both runs and twiddles are generally random movements; that is, no one direction of movement is more likely than any other direction. Runs last an average of 1 second, during which the bacteria swim about 10 to 20 times the length of their body. Twiddles last about 0.1 second, and no forward progress is made.

Chemotaxis   Sometimes bacteria move toward or away from substances in their environment by a nonrandom process called **chemotaxis** (➤ Figure 4.14). Concentrations of most substances in the environment vary along a gradient—that is, from high to low concentration. When a bacterium is moving in the direction of increasing concentration of an attractant (such as a nutrient), it tends to lengthen its runs and to reduce the frequency of its twiddles. When it is moving away from the attractant, it shortens its runs and increases the frequency of its twiddles. Even though the direction of the individual runs is still random, the net result is movement toward the attractant, or *positive chemotaxis*. Movement away from a repellent, or *negative chemotaxis,* results from the opposite responses: long runs and few twiddles while the bacterium moves in the direction of lower concentration of the harmful substance; short runs and many twiddles while it moves in the direction of higher concentration. The exact mechanism that produces these behaviors is not fully understood, but certain structures on bacterial cell surfaces can detect changes in concentration over time. *E. coli* cells have at least four different types of receptors (called *transducers*) extending through the cell membrane that dectect chemicals and signal the cells to respond.

Phototaxis   Some bacteria can move toward or away from light; this response is called **phototaxis.** Bacteria that move toward light exhibit *positive phototaxis,* whereas those that move away from light exhibit *negative phototaxis*. This process may be accomplished by means of flagella. Or, in the case of some photosynthetic aquatic bacteria, oil droplet inclusions in their cytoplasm may give them the buoyancy to rise toward the water surface, where light is more available.

Axial Filaments

Spirochetes have **axial filaments,** or **endoflagella,** instead of flagella that extend beyond the cell wall (➤ Figure 4.15). Each filament is attached at one of its ends to an end of the cytoplasmic cylinder that forms the body of the spirochete. Axial filaments cause the rigid spirochete body to rotate like a corkscrew when they twist inside the outer sheath.

Pili

Pili (singular: *pilus*) are tiny, hollow projections. They are used to attach bacteria to surfaces and are not involved in movement. A pilus is composed of subunits of the protein *pilin*. Bacteria can have two kinds of pili (➤ Figure 4.16): (1) long *conjugation pili,* or *F pili* (also called sex pili), and (2) short *attachment pili,* or *fimbriae* (fim'-bre-e; singular: *fimbria*).

Fimbriae

Conjugation pilus

➤ **Figure 4.16   Pili.** Electron micrograph of an *Escherichia coli* cell (13,000X), showing two kinds of pili. The shorter ones are fimbriae, used for attachment to surfaces. The long tube reaching to another cell is a conjugation pilus, perhaps used to transfer DNA.

Conjugation Pili   **Conjugation pili** (or *sex pili*), found only in certain groups of bacteria, attach two cells and may furnish a pathway for the transfer of the genetic material DNA. This transfer process is called *conjugation* (➤ Figures 4.16 and 8.7). Transfer of DNA furnishes genetic variety for bacteria as sexual reproduction does for many other life forms. Such transfers among bacteria cause problems for humans because antibiotic resistance can be passed on with the DNA transfer. Consequently, more and more bacteria acquire resistance, and humans must look for new ways to control the growth of these bacteria.

Attachment Pili   **Attachment pili,** or **fimbriae,** help bacteria adhere to surfaces, such as cell surfaces and the interface of water and air. They contribute to the *pathogenicity* of certain bacteria—their ability to produce disease—by enhancing colonization (the development of colonies) on the surfaces of the cells of other organisms. For example, some bacteria adhere to red blood cells by attachment pili and cause the blood cells to clump, a process called *hemagglutination.* In certain species of bacteria, some individuals have attachment pili and others lack them. In *Neisseria gonorrhoeae,* strains without pili are rarely able to cause gonorrhea, but those with pili are highly infectious because they attach to epithelial cells of the urogenital system. Such pili also allow them to attach to sperm cells and thereby spread to the next individual.

Some aerobic bacteria form a shiny or fuzzy, thin layer at the air-water interface of a broth culture. This layer, called a **pellicle,** consists of many bacteria that adhere to the surface by their attachment pili. Thus, attachment pili allow the organisms to remain in the broth, from which they take nutrients, while they congregate near air, where the oxygen concentration is greatest.

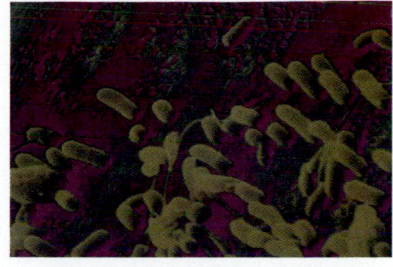

➤ **Figure 4.17   The slime layer.**
SEM of bacteria growing on tooth enamel, to which they initially adhere by means of their slime layer.

Glycocalyx

Glycocalyx is the currently accepted term used to refer to all polysaccharide-containing substances found external to the cell wall, from the thickest *capsules* to the thinnest *slime layers.* All bacteria have at least a thin slime layer.

Capsule   A **capsule** is a protective structure outside the cell wall of the organism that secretes it. Only certain bacteria are capable of forming capsules, and not all members of a species have capsules. For example, the bacterium that causes anthrax, a disease found mainly in cattle, does not produce a capsule when it grows outside an organism but does when it infects an animal. Capsules typically consist of complex polysaccharide molecules arranged in a loose gel. However, the chemical composition of each capsule is unique to the strain of bacteria that secreted it. Anthrax bacteria have a capsule composed of protein. When encapsulated bacteria invade a host, the capsule prevents host defense mechanisms, such as phagocytosis, from destroying the bacteria. If bacteria lose their capsules, they become less likely to cause disease and more vulnerable to destruction.

Slime Layer   A **slime layer** is less tightly bound to the cell wall and is usually thinner than a capsule. When present, it protects the cell against drying, helps trap nutrients near the cell, and sometimes binds cells together. Slime layers allow bacteria to adhere to objects in their environments, such as rock surfaces or the root hairs of plants, so that they can remain near sources of nutrients or oxygen (➤ Figure 4.17). Some oral bacteria, for example, adhere by their slime layers and form dental plaque. The slime layer keeps the bacteria in close proximity to the tooth surface, where they can cause dental caries. Plaque is extremely tightly bound to tooth surfaces. If not removed regularly by brushing, it can be removed only by a dental professional in a procedure called *scaling.*

✓ How do bacteria move in chemotaxis? Distinguish between runs and twiddles, and their frequencies.

✓ What risk would a spore-forming bacterial species run if it did not produce any spores until conditions became adverse?

✓ What are the functions of pili?

✓ Distinguish between nucleus and nucleoid.

Eukaryotic Cells

An Overview of Structure

Eukaryotic cells are larger and more complex than prokaryotic cells. Most eukaryotic cells have a diameter of more than 10 μm, and many are much larger. They also contain a variety of highly differentiated structures. These cells are the basic structural unit of all organisms in the kingdoms Protista, Plantae, Fungi, and Animalia (Chapter 9). Eukaryotic organisms include microscopic protozoa, algae, and fungi and are thus appropriately considered in microbiology. The general structure of the eukaryotic cell is shown diagrammatically in ➤ Figure 4.18.

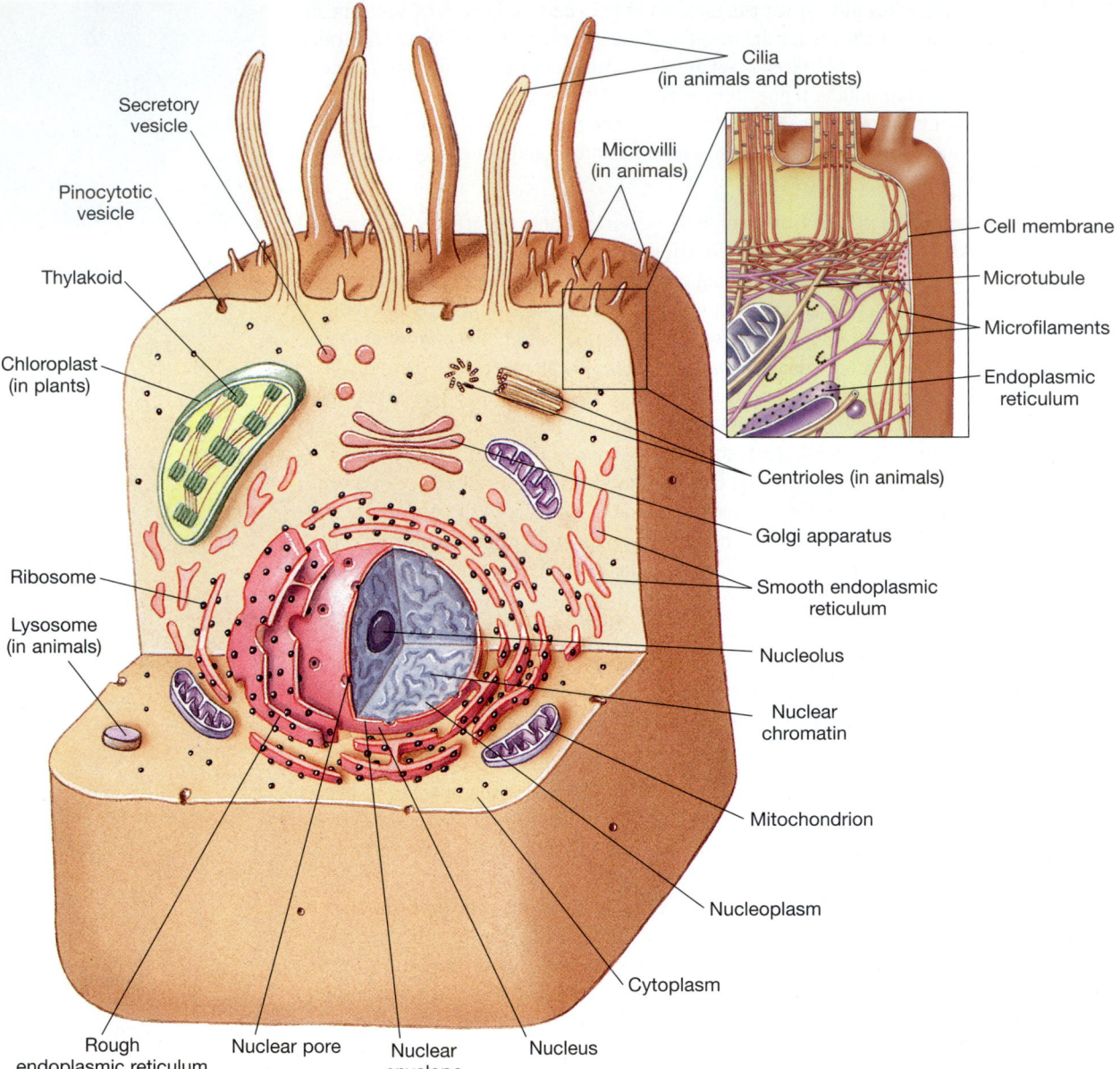

Cilia
(in animals and protists)

Secretory
vesicle

Pinocytotic
vesicle

Microvilli
(in animals)

Thylakoid

Cell membrane

Microtubule

Chloroplast
(in plants)

Microfilaments

Endoplasmic
reticulum

Ribosome

Centrioles (in animals)

Golgi apparatus

Smooth endoplasmic
reticulum

Lysosome
(in animals)

Nucleolus

Nuclear
chromatin

Mitochondrion

Nucleoplasm

Cytoplasm

Rough
endoplasmic reticulum

Nuclear pore

Nuclear
envelope

Nucleus

> Figure 4.18   **A generalized eukaryotic cell.** Most of the features shown are present in nearly all eukaryotic cells, but some (the centrioles, microvilli, and lysosomes) occur only in animal cells, and others (the chloroplast) are found only in cells capable of carrying out photosynthesis.

The Plasma Membrane

The cell membrane, or **plasma membrane,** of a eukaryotic cell has the same fluid-mosaic structure as that of a prokaryotic cell. In addition, eukaryotes also contain several organelles enclosed by membranes that have a similar membrane structure.

Eukaryotic membranes differ from prokaryotic membranes in some respects, especially in the greater variety of lipids they contain. Eukaryotic membranes contain sterols, found among prokaryotes only in the mycoplasmas. Sterols add rigidity to a membrane, and this may be important in keeping membranes intact in eukaryotic cells. Because of their larger size, eukaryotic cells have a much lower surface-to-volume ratio than prokaryotic cells. As the volume of cytoplasm enclosed by a membrane increases, the membrane is placed under greater stress. The sterols in the membrane may help it withstand the stress.

Functionally, eukaryotic plasma membranes are less versatile than prokaryotic ones. They do not have respiratory enzymes that capture metabolic energy and store it in ATP; in the course of evolution, that function has been taken over by mitochondria.

Internal Structure

The internal structure of eukaryotic cells is exceedingly more complex than that of prokaryotic cells. It is also much more highly organized and contains numerous organelles.

Cytoplasm

The cytoplasm makes up a relatively smaller portion of eukaryotic cells than of prokaryotic cells because it contains a *nucleus* and many organelles. Like the cytoplasm of prokaryotic cells, the cytoplasm of eukaryotic cells is a semifluid substance consisting mainly of water with the same substances dissolved in it. In addition, this cytoplasm contains elements of a *cytoskeleton,* a fibrous network that give these larger cells shape and support.

Cell Nucleus

The most obvious difference between eukaryotic and prokaryotic cells is the presence of a nucleus in the eukaryotic cells. The **cell nucleus** (➤ Figure 4.19) is a distinct organelle with a nuclear envelope, nucleoplasm, nucleoli, and (typically paired) chromosomes. The **nuclear**

(a)

(b)

➤ **Figure 4.19   Pores through the cell nucleus.** (a) TEM of a cell nucleus. The dark, granular material is chromatin. The pores in the nuclear membrane allow for entry and exit of materials. (b) Freeze-fracture TEM of a nucleus (compare with Figure 3.23). The many circular structures are nuclear pores.

2N
(Diploid)

DNA replication

Chromosomes
become visible

Dyads line up
on cell equator

Dyads separate,
forming two sets
of chromosomes

Cell division
produces two
diploid cells
identical to
parent cell

(a)

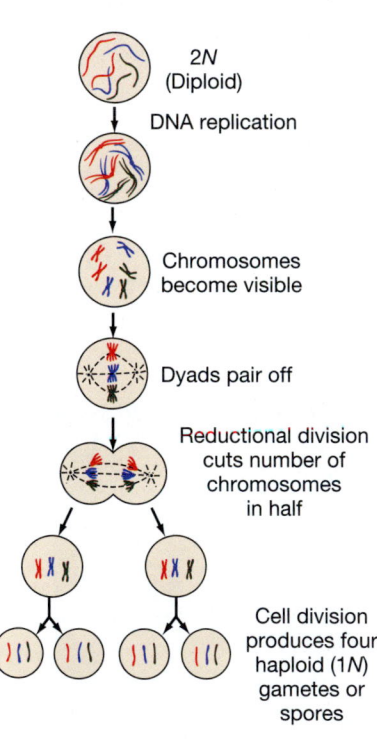

2N
(Diploid)

DNA replication

Chromosomes
become visible

Dyads pair off

Reductional division
cuts number of
chromosomes
in half

Cell division
produces four
haploid (1N)
gametes or
spores

(b)

➤ Figure 4.20 **Mitosis and meiosis compared.** Both processes are preceded by duplication of DNA; soon after, the chromosomes become visible. (a) In mitosis, two identical daughter cells with the same number and kinds of chromosomes are formed. (b) In meiosis, two divisions give rise to four cells, each with half the number of chromosomes as the original parent cell. For this reason, meiosis is sometimes called *reduction division.*

envelope consists of a double membrane, each layer of which is structurally like the plasma membrane. **Nuclear pores** in the envelope allow RNA molecules to leave the semifluid portion of the nucleus, known as **nucleoplasm,** and to participate in protein synthesis. Each nucleus has one or more **nucleoli** (singular: *nucleolus*), which contain a significant amount of RNA and serve as sites for the assembly of ribosomes.

Also present in the nucleus of most eukaryotic organisms are paired **chromosomes,** each of which contains DNA and proteins called **histones.** Histones contribute directly to the structure of chromosomes, and other proteins probably regulate the chromosomes' function. During cell division, the chromosomes are extensively coiled and folded into compact structures. Between divisions, however, the chromosomes are visible only as a tangle of fine threads called **chromatin** that gives the nucleus a granular appearance.

The nuclei of eukaryotic cells divide by the process of **mitosis** (➤ Figure 4.20a). Prior to the actual division of the nucleus, the chromosomes replicate but remain attached, forming **dyads.** In most eukaryotic cells, the nuclear envelope fragments during mitosis, and a system of tiny fibers called the **spindle apparatus** guides the movement of chromosomes. Dyads aggregate in the center of the spindle and separate into single chromosomes as they move along fibers to the poles of the spindle. Each new cell receives one copy of each chromosome that was present in the parent cell. Because the parent cell contained paired chromosomes, the progeny likewise contain paired chromosomes. Cells with paired chromosomes are said to be **diploid** (2N) cells.

During sexual reproduction, the nuclei of sex cells divide by a process called **meiosis** (➤ Figure 4.20b). After the chromosomes replicate, forming dyads, pairs of dyads come together. During the course of two cell divisions, the dyads are distributed to four new cells. Thus, each cell receives only one chromosome from each pair. Such cells are said to be **haploid** (1N) cells. Haploid cells can become gametes or spores. **Gametes** are haploid cells that participate in sexual reproduction; gametes from each of two parent organisms unite to form a diploid **zygote,** the first cell of a new individual. Some **spores** become dormant, whereas others reproduce by mitosis as haploid vegetative cells. Dormant spores allow for survival during adverse environmental conditions. When conditions improve, the spores germinate and begin to divide. Eventually some of these cells produce gametes, which can unite to form zygotes. Thus, the organism alternates between haploid and diploid generations.

Mitochondria and Chloroplasts

Mitochondria (singular: *mitochondrion*), known as the powerhouses of eukaryotic cells, are exceedingly important organelles. They are quite numerous in some cells and can account for up to 20 percent of the cell volume. Mitochondria are complex structures about 1 μm in diameter with an outer membrane, an inner membrane, and a fluid-filled **matrix** inside the inner membrane (➤ Figure 4.21). The inner membrane is extensively folded to form **cristae,** which extend into the matrix. Mitochondria carry out the oxidative reactions that capture energy in adenosine triphosphate (ATP). Energy in ATP is in a form usable by cells for their activities.

Eukaryotic cells capable of carrying out photosynthesis contain **chloroplasts** (➤ Figure 4.22). They too have an outer and an inner membrane. The inner **stroma** of these organelles corresponds structurally with the matrix of mitochondria. Unlike mitochondria, choloroplasts have separate internal membranes, called **thylakoids,** that contain the pigment *chlorophyll,* which captures energy from light during photosynthesis. Both mitochondria and chloroplasts contain DNA and can replicate independently of the cell in which they function. This and other evidence has led many biologists to speculate that these organelles may have originated as free-living organisms.

Ribosomes

Ribosomes of eukaryotic cells, which are larger than those of prokaryotic cells, are about 60 percent RNA and 40 percent protein. They have a sedimentation rate of 80S, and their subunits have sedimentation rates of 60S and 40S. All ribosomes provide sites for protein synthesis, and some are arranged in chains as polyribosomes. Those that are attached to an organelle called the *endoplasmic reticulum* usually make proteins for secretion from the cell; those that are free in the cytoplasm usually make proteins for use in the cell.

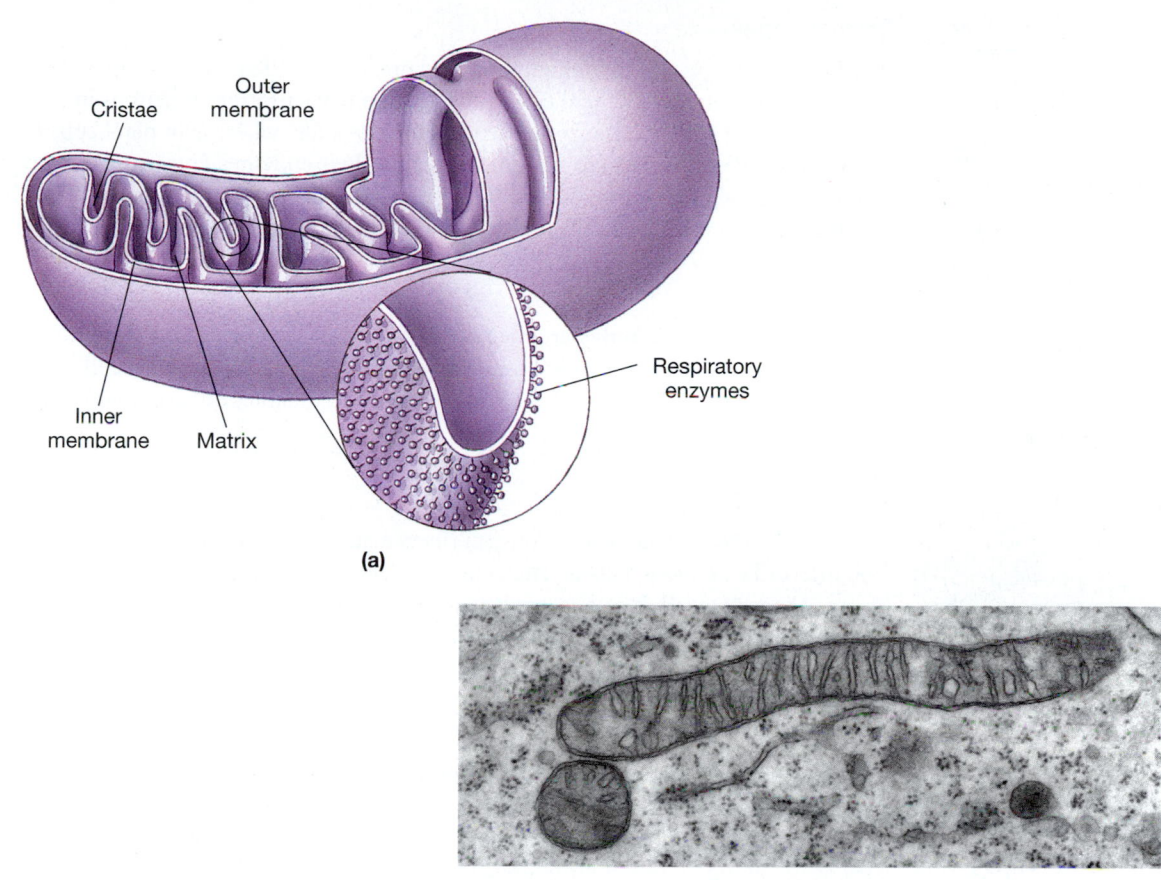

Cristae

Outer membrane

Respiratory enzymes

Inner membrane   **Matrix**

(a)

(b)

> ➤ Figure 4.21   **Mitochondria.** (a) Respiratory enzymes that make ATP are located on the surfaces of the inner membrane and the cristae, which are infoldings of the inner membrane. (b) TEM of two mitochondria, one in longitudinal section and the other in cross section.

Endoplasmic Reticulum

The **endoplasmic reticulum** (➤ Figure 4.18) is an extensive system of membranes that forms numerous tubes and plates in the cytoplasm. Endoplasmic reticulum can be smooth or rough textured. *Smooth endoplasmic reticulum* contains enzymes that synthesize lipids, especially those to be used in making membranes. *Rough endoplasmic reticulum* has ribosomes bound to its surface, which give it a rough texture. Its function is, along with the ribosomes, to manufacture proteins. Vesicles from this membrane system transport to the Golgi apparatus the lipids and proteins synthesized in or on the endoplasmic reticulum membrane.

Golgi Apparatus

The **Golgi** (gol´je) **apparatus** (➤ Figure 4.18) consists of a stack of flattened membrane sacs. The Golgi apparatus receives substances transported from the endoplasmic reticulum, stores the substances, and typically alters their chemical structure. It packages these substances in small segments of membrane called **secretory vesicles.** The secretory vesicles fuse with the plasma membrane and release secretions to the exterior of the cell. The Golgi apparatus also helps to form the plasma membrane and membranes of the lysosomes.

Lysosomes

Lysosomes (➤ Figure 4.18) are extremely small, membrane-covered organelles made by the Golgi apparatus in animal cells. They contain digestive enzymes that could destroy a cell if those enzymes were released into the cytoplasm. Lysosomes fuse with *vacuoles* that form as a cell ingests substances, and they release enzymes that digest the substances in the vacuoles.

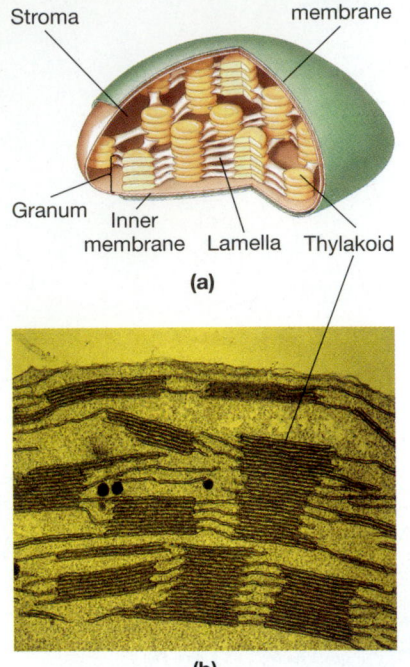

➤ Figure 4.22   **Chloroplasts.**
(a) The thylakoid membranes contain chlorophyll and other pigments and enzymes needed for photosynthesis. Thylakoids occur in stacks called *grana;* grana are joined by membranous flat sheets called *lamellae.*
(b) A colorized TEM of a chloroplast (magnified 24,000X) from a leaf of corn.

Peroxisomes

Peroxisomes are small, membrane-enclosed organelles filled with enzymes. Peroxisomes are found in both plant and animal cells but appear to have different functions in the two kinds of cells. In animal cells their enzymes oxidize amino acids, whereas in plant cells they typically oxidize fats. Peroxisomes are so named because their enzymes convert hydrogen peroxide to water in both plant and animal cells. If hydrogen peroxide were to accumulate in cells, it would kill them just as it kills bacteria when humans use it as an antiseptic.

Vacuoles

In eukaryotic cells **vacuoles** are membrane-enclosed structures that store materials such as starch, glycogen, or fat to be used for energy. Some vacuoles form when cells engulf food particles. As we have already noted, the contents of these vacuoles are eventually digested by lysosomal enzymes.

Cytoskeleton

The **cytoskeleton** is a network of protein fibers made of **microtubules** (which are hollow tubes) and **microfilaments** (which are filamentous fibers). The cytoskeleton supports and gives rigidity and shape to a cell. It also is involved in cell movements, such as those that occur when cells engulf substances or when they make *amoeboid movements* (to be explained in a later section).

External Structure

Like external structures of prokaryotic cells, external structures of eukaryotic cells either assist with movement or provide a protective covering for the plasma membrane. These structures include flagella, cilia, and cell walls and other coverings. Although *pseudopodia* are not, strictly speaking, external structures, they do achieve movement and so are discussed here. Cells of algae and macroscopic green plants have cell walls, and some protozoa have special cell coverings.

Flagella

Flagella in eukaryotes, which are larger and more complex than those in prokaryotes, consist of two central microtubules and nine pairs of peripheral microtubules (a 9 + 2 arrangement) surrounded by a membrane (➤ Figure 4.23). Each fiber is a microtubule made of the protein *tubulin.* One of these microtubules is about the same size as an entire prokaryotic flagellum. Associated with each pair of peripheral microtubules are small molecules of the protein *dynein.* Eukaryotic flagella move like a whip, whereas prokaryotic flagella move like a rotating hook. One mechanism of eukaryotic flagellar movement is a crossbridging among dynein and other flagellar proteins. Through ATP hydrolysis, dynein plays a role in converting chemical energy in ATP to mechanical energy that makes the flagellum move. Microtubules in the flagellum are thought to slide toward or away from the base of the cell in a wavelike manner and thereby cause the whole flagellum to move.

Flagella are most common among protozoa but are found among algae as well. Most flagellated eukaryotes have one flagellum, but some have two or more.

Cilia

Cilia are shorter and more numerous than flagella, but they have the same chemical composition and basic arrangement of microtubules. Cilia are found mainly among ciliated protozoa, which have 10,000 or more cilia distributed over their cell surface (➤ Figure 4.24). Each cilium passes through a stroke-and-recovery cycle as it beats. Together the cilia of an organism beat in a coordinated pattern, which creates a wave that passes from one end of the organism to the other. The large number of cilia and their coordinated beating allow ciliated organisms, such as paramecia, to move much more rapidly than those with flagella. Cilia on some cells can also propel fluids, dissolved particles, bacteria, mucus, and so on past the cell. This function can be of great importance in host defenses against diseases, particularly in the respiratory tract.

Pseudopodia

Pseudopodia (su″do-po′de-a; singular: *pseudopodium*), or "false feet," are temporary projections of cytoplasm associated with **amoeboid movement.** This kind of movement occurs only in cells without walls, such as amoebas and some white blood cells, and only when the cell is resting on a solid surface. When an amoeba first extends a portion of its body to form a pseudopodium, the cytoplasm is much less dense in the pseudopodium than in other areas of the cell (➤ Figure 4.25). As a result, cytoplasm from elsewhere in the organism flows into the pseudopodium by **cytoplasmic streaming.** Amoeboid movement is a slow, inching-along process.

Cell Walls

A number of unicellular eukaryotic organisms have cell walls, none of which contains the peptidoglycan that is characteristic of bacteria. Algal cell walls consist mainly of cellulose, but some contain other polysaccharides. Cell walls of fungi consist of cellulose or chitin, or both. *Chitin* is a structural polysaccharide that is also common in the exoskeletons of arthropods such as insects and crustacea. Regardless of composition, cell walls give cells rigidity and protect them from bursting when water moves into them from the environment.

(a)

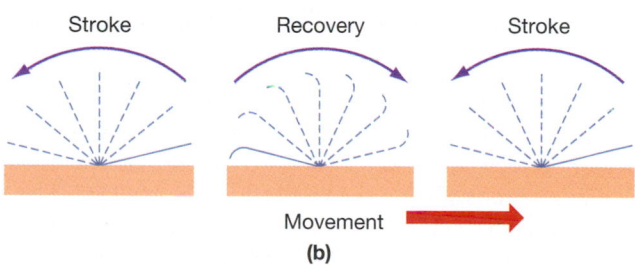

Stroke   Recovery   Stroke

Movement

(b)

Direction of wave

Stroke   Recovery   Stroke

Direction of cell movement

(c)

➤ Figure 4.24   **Cilia.**
(a) SEM of the ciliated protozoan *Oxytricha.* (b) The stroke-and-recovery motion of a cilium.
(c) Cilia on an organism move in a synchronized fashion, creating a wave that propels the organism forward.

Prokaryotic flagellum

(a)

Microtubules

Eukaryotic flagellum

(b)

Flagellar movement

Ciliary movement

(c)

➤ Figure 4.23   **Prokaryotic and eukaryotic flagella compared.** (a) A prokaryotic flagellum; (b) a eukaryotic flagellum. Notice the substantial difference in the diameter of these two structures. (c) The movement of a eukaryotic flagellum and cilium.

Resting amoeba with
cytoplasm distributed evenly.

Newly formed pseudopodium
with less dense cytoplasm.

(a)

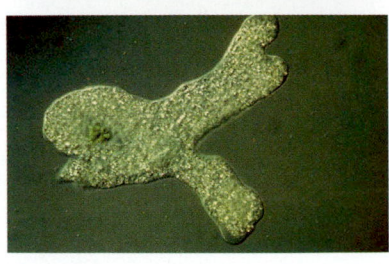

(b)

➤ Figure 4.25 **Pseudopodia.**
(a) The formation of a pseudopodium,
a cytoplasmic extension that allows
organisms such as amoebas to move
and capture food. (b) Micrographs of
an amoeba engulfing food particles
(magnified 160X).

Evolution by Endosymbiosis

Biologists believe that life arose on earth (or perhaps was "seeded in" by meteorites) about 4 billion years ago, in the form of simple organisms much like the prokaryotic organisms of today. However, fossil evidence suggests that eukaryotic organisms arose only about 1 billion years ago. How development from prokaryote to eukaryote took place is unknown, but the **endosymbiotic theory** offers a plausible explanation. As we have seen, the major difference between prokaryotes and eukaryotes is that eukaryotes possess specialized membrane-enclosed organelles, including a true nucleus. According to the endosymbiotic theory, the organelles of eukaryotic cells arose from prokaryotic cells that had developed a *symbiotic* relationship with the eukaryote-to-be. Symbiosis is a relationship between two different kinds of organisms that live in closed contact. If one lives inside the other, the relationship is known as *endosymbiosis.*

It is suggested that the first eukaryotic cell was an amoeba-like cell that somehow had developed a nucleus. Knowing the ease with which bits of cell membrane pinch off to form vesicles, it is fairly easy to imagine that a primitive chromosome might have become surrounded by membrane, thereby creating a rudimentary nucleus. This primitive eukaryote was probably a phagocytic cell that obtained its nutrients by engulfing material from its environment—including, presumably, other cells. Although most engulfed prokaryotic cells were probably digested and used to nourish the phagocyte, some apparently survived and became permanent residents within the cytoplasm, eventually becoming incorporated as organelles. Both organisms benefited from this arrangement. The engulfed prokaryotes were protected by the eukaryote, and the eukaryote acquired some new capabilities through the presence of its symbionts.

Evidence supporting this theory comes from a comparison of the characteristics of eukaryotic organelles with those of prokaryotic organisms:

- Mitochondria and chloroplasts are approximately the same size as prokaryotic cells.
- Unlike other organelles, mitochondria and chloroplasts have their own DNA. Organelle DNA is present in the form of a single circular loop, like the chromosome of a prokaryote (➤ Figure 4.26).
- Organelles have their own 70S ribosomes, which are like prokaryotic ribosomes, in contrast to the 80S ribosomes of eukaryotes.
- Organelle DNA and ribosomes carry out protein synthesis as it occurs in bacteria, rather than as it occurs when directed by nuclear DNA of modern eukaryotes.
- Mitochondria and chloroplasts divide independently of the eukaryotic cell cycle, by means of binary fission.
- The double-membrane structures of mitochondria and chloroplasts strongly resemble the cell membranes of Gram-negative bacteria, even having the same type of pores.
- Chloroplasts strongly resemble the structure of photosynthetic chlorophyll-containing prokaryotic cyanobacteria.

The notion of prokaryotic endosymbionts living in eukaryotes is not mere speculation, however. Examples of such relationships abound in nature. Certain eukaryotes living in low-oxygen environments lack mitochondria, yet they get along quite well, thanks to bacteria that live inside them and serve as "surrogate mitochondria." Protists living symbiotically in the hindgut of termites are, in turn, colonized by symbiotic bacteria similar in size and distribution to mitochondria (➤ Figure 4.27). The bacteria function better in this low-oxygen condition than mitochondria would. They oxidize food and provide energy in the form of ATP for their protist partner. Some primitive eukaryotes today still lack mitochondria. *Giardia,* a parasitic protist that causes diarrhea, is an example of a eukaryote that probably never acquired any mitochondria.

Living in the mud at the bottom of ponds is a giant amoeba, *Pelomyxa palustris.* It also lacks mitochondria and has at least two kinds of endosymbiotic bacteria. Killing just the bacteria with antibiotics allows lactic acid to accumulate. This suggests that the bacteria oxidize the end-products of glucose fermentation, a function that mitochondria ordinarily perform. What else do mitochondria do? They must perform some task necessary for the formation or functioning of the Golgi apparati. This group includes all prokaryotes. Perhaps integration of bacterial endosymbionts into a cell led to the development of mitochondria and Golgi apparati.

➤ Figure 4.26   **Yeast cells stained to show mitochondrial DNA.** Each mitochondrion of *Saccharomyces cerevisiae* has two to four circular chromosomes that stain blue with the fluorescent dye used.

Dr. Lynn Margulis proposes that eukaryotic flagellae and cilia (she calls them "undulipodia") originated from symbiotic associations of motile bacteria, called spirochetes, with heterotrophic protists. Such associations of present-day species are well known. *Mixotricha paradoxa,* a protist endosymbiont found in the hindgut of the Australian termite *Mastotermes darwiniensis* uses the four flagellae at its front end to steer, but depends on the half-million spirochetes covering its surface for driving power. These spirochetes have a natural tendency to coat living or dead surfaces. Dramatic motion pictures show that once attached, they coordinate their undulations and beat in unison, propelling their host particle along. Margulis hypothesizes that some ancient spirochetes integrated into their host cells to become cilia and flagella. She further suggests that other spirochetes were drawn down inside the cell (a process that can be observed in modern species) and eventually transformed into microtubules.

The spirochetes would have obtained nutrients that leaked from the eukaryote, while giving the eukaryote motility. Giant tube worms (6 feet long) living near hydrothermal vents deep in the ocean lack mouths, anuses, and digestive tracts. What keeps them alive? Prokaryotic endosymbiont bacteria colonize their internal tissues. The bacteria generate energy by metabolizing the hydrogen sulfide spewing forth from the hot vents. Excess energy is transferred to the tube worms. A similar relationship exists between endosymbiont bacteria and giant clams that live at the vents. Endosymbiosis is a common pattern of life.

➤ Figure 4.27   **Endosymbiosis.** In the cytoplasm of *Pyrsonympha,* a protist that lives symbiotically in the hindgut of termites, bacteria (dark ovals) acts as mitochondria for the protist. At lower left, one of the bacteria is dividing.

The Movement of Substances Across Membranes

A living cell, either prokaryotic or eukaryotic, is a dynamic entity. A cell is separated from its environment by a membrane, across which substances constantly move in a carefully controlled manner. Understanding how these movements occur is essential to understanding how a cell functions. Very small polar substances, such as water, small ions, and small water-soluble molecules, probably pass through pores in the membrane. Nonpolar substances, such as lipids and other uncharged particles (molecules or ions), dissolve in and pass through the membrane lipids. Still other substances are moved through the membrane by carrier molecules. Most large molecules are unable to enter cells without the aid of specific carriers.

The mechanisms by which substances move across membranes can be passive or active. In passive transport, the cell expends no energy to move substances down a *concentration gradient,* that is, from higher to lower concentration. Passive processes include *simple diffusion, facilitated diffusion,* and *osmosis.* In active processes, the cell expends energy from ATP, enabling it to transport substances against a concentration gradient. These processes include *active transport.* The processes *endocytosis* and *exocytosis,* which occur only in eukaryotic cells, are separate mechanisms for moving substances across the plasma membrane.

Simple Diffusion

All molecules have kinetic energy; that is, they are constantly in motion and are continuously redistributed. **Simple diffusion** is the net movement of particles from a region of higher to lower concentration (➤ Figure 4.28). Suppose, for example, that you drop a lump of sugar into a cup of coffee. At first a concentration gradient exists, with the sugar concentration greatest at the lump and least at the rim of the cup. Eventually, though, sugar molecules will become evenly distributed throughout the coffee (they reach *equilibrium*) even without stirring.

Diffusion occurs because of random movement of particles. Although particles move at high velocity, they do not travel far in a straight line before they collide with other randomly moving particles. Even so, some particles from a region of high concentration eventually move toward a region of lower concentration. Fewer particles move in the opposite direction for two reasons: (1) There are fewer of them in regions of low concentration to begin with, and (2) they are likely to be repelled by collision with particles from a region of high concentration.

The length of time required for particles to diffuse across a cell increases with cell diameter. Materials can diffuse throughout small prokaryotic cells very quickly and throughout larger eukaryotic cells fast enough to supply nutrients and to remove wastes fairly efficiently. If cells were much larger, diffusion throughout the cell would be too slow to sustain life, so diffusion rates may be responsible in part for limiting the size of cells.

Any membrane severely limits diffusion, but many substances diffuse through the lipids of membranes. Diffusion through the phospholipid bilayer is affected by several factors: (1) the solubility of the diffusing substance in lipid, (2) the temperature, and (3) the difference between the highest and lowest concentration of the diffusing substance. Nonpolar substances such as steroids and gases (CO_2, O_2) cross the membrane rapidly by dissolving in the nonpolar fatty acid tails of the membrane phospholipids.

➤ Figure 4.28   **Simple diffusion.** The random movements of molecules cause them to spread out (diffuse) from an area of high concentration to areas of lower concentration until eventually they are equally distributed throughout the available space (that is, they reach equilibrium).

(a)   (b)   (c)

A few substances also diffuse through pores. Such diffusion is affected by the size and charge of the diffusing particles and the charges on the pore surface. Pores probably have a diameter of less than 0.8 nm, so only water, small water-soluble molecules, and ions such as H^+, K^+, Na^+, and Cl^- pass through them. This is one reason a membrane is said to be **selectively permeable** (*semipermeable*).

Facilitated Diffusion

Facilitated diffusion is diffusion down a concentration gradient and across a membrane with the assistance of special pores or carrier molecules. In fact, membranes contain protein-lined pores for specific ions. These pores have an arrangement of charges that allows rapid passage of a particular ion. The carrier molecules are proteins, embedded in the membrane, that bind to one or to a few specific molecules and assist in their movement. By one possible mechanism for facilitated diffusion, a carrier acts like a revolving door or shuttle that provides a convenient one-way channel for the movement of substances across a membrane (➤ Figure 4.29). Carrier molecules can become saturated, and similar molecules sometimes compete for the same carrier. Saturation occurs when all the carrier molecules are moving the diffusing substance as fast as they can. Under these conditions the rate of diffusion reaches a maximum and cannot increase further. When a carrier molecule can transport more than one substance, the substances compete for the carrier in proportion to their concentrations. For example, if there is twice as much of substance A as substance B, substance A will move across the membrane twice as fast as substance B.

Osmosis

Osmosis is a special case of diffusion-in which water molecules diffuse across a selectively permeable membrane. To demonstrate osmosis, we start with two compartments separated by a membrane permeable only to water. One compartment contains pure water, and the other compartment contains some large, nondiffusible molecules, such as proteins (➤ Figure 4.30a). Water molecules move in both directions, but their net movement is from pure water (concen-

Diffusing substance

Lipid bilayer membrane

Trans-membrane carrier protein

➤ Figure 4.29   **Facilitated diffusion.** Carrier protein molecules aid in the movement of substances through the cell membrane, but only down their concentration gradient (from a region where their concentration is high to one where their concentration is low). This process does not require the expenditure of any energy (ATP) by the cell.

Selectively permeable membrane

1% Sugar solution

Distilled water

Net movement of water

(a)

(b)

Net movement

(c)

➤ Figure 4.30   **Osmosis.** (a) The diffusion of water from an area of higher water concentration (the right side) to an area of lower water concentration (the left side) through a semipermeable membrane. (b) Here the net movement of water is into the sugar solution because the concentration of water there is slightly lower than on the other side of the membrane. (c) As a result of the net movement of water, the column rises on the left.

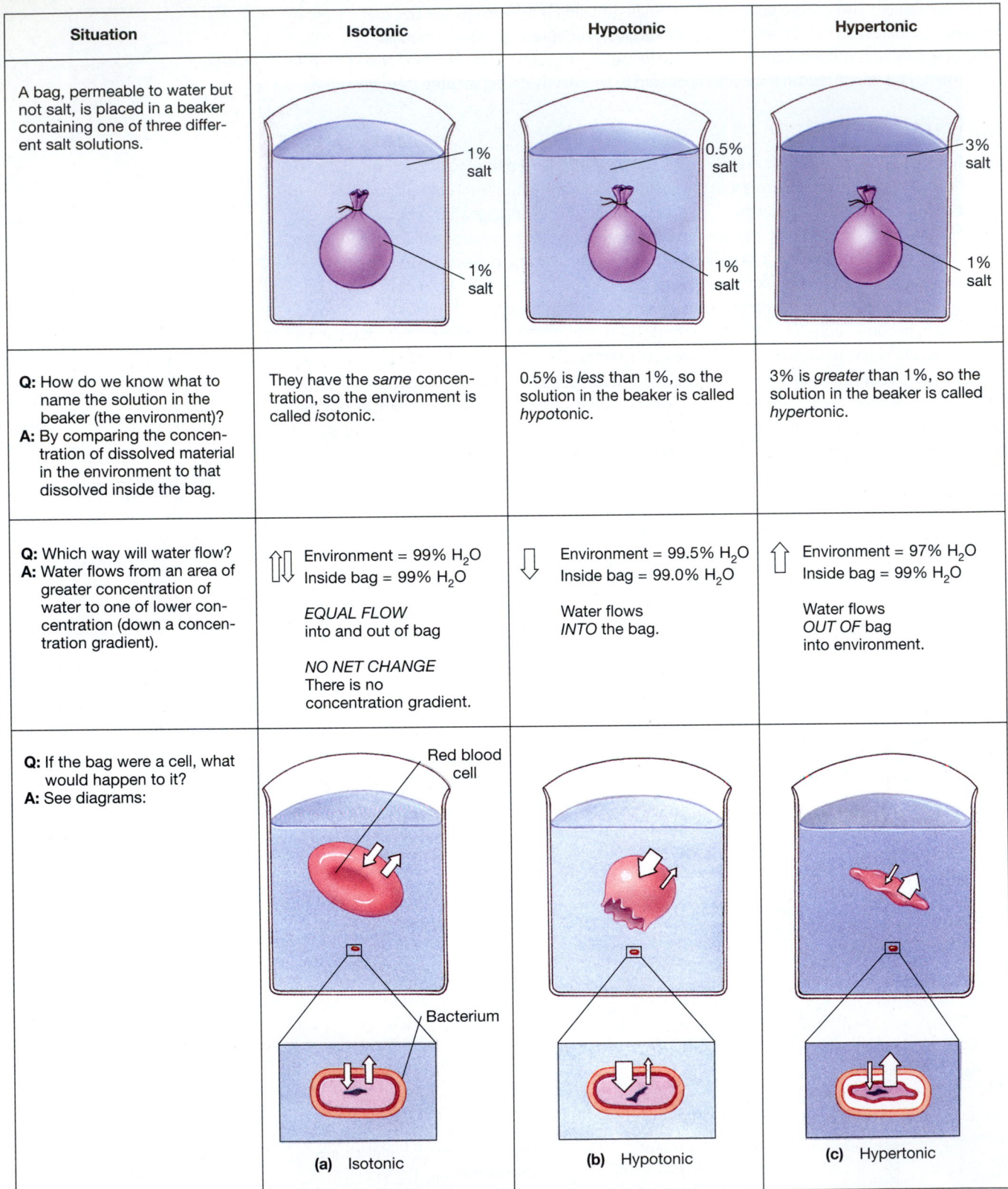

Situation	Isotonic	Hypotonic	Hypertonic
A bag, permeable to water but not salt, is placed in a beaker containing one of three different salt solutions.	1% salt 1% salt	0.5% salt 1% salt	3% salt 1% salt
Q: How do we know what to name the solution in the beaker (the environment)? **A:** By comparing the concentration of dissolved material in the environment to that dissolved inside the bag.	They have the *same* concentration, so the environment is called *iso*tonic.	0.5% is *less* than 1%, so the solution in the beaker is called *hypo*tonic.	3% is *greater* than 1%, so the solution in the beaker is called *hyper*tonic.
Q: Which way will water flow? **A:** Water flows from an area of greater concentration of water to one of lower concentration (down a concentration gradient).	Environment = 99% H_2O Inside bag = 99% H_2O *EQUAL FLOW* into and out of bag *NO NET CHANGE* There is no concentration gradient.	Environment = 99.5% H_2O Inside bag = 99.0% H_2O Water flows *INTO* the bag.	Environment = 97% H_2O Inside bag = 99% H_2O Water flows *OUT OF* bag into environment.
Q: If the bag were a cell, what would happen to it? **A:** See diagrams:	Red blood cell Bacterium **(a)** Isotonic	**(b)** Hypotonic	**(c)** Hypertonic

➤ Figure 4.31   **Experiments that examine the effects of tonicity on osmosis.** (a) A cell in an isotonic environment—one that has the same concentration of dissolved material as the interior of the cell—will experience no net gain or loss of water and will retain its original shape. (b) A cell in a hypotonic environment—one with a lower concentration of dissolved material than the interior of the cell—will gain water and swell. Unlike a bacterial cell, a red blood cell will burst because it lacks a cell wall. (c) A cell in a hypertonic environment—one with a higher concentration of dissolved material than the interior of the cell—will lose water and shrink.

tration 100 percent) toward the water that contains other molecules (concentration less than 100 percent; ➤ Figure 4.30b). Thus, osmosis is the net flow of water molecules from a region of higher concentration of those molecules to a region of lower concentration across a semipermeable membrane (➤ Figure 4.30c).

Osmotic pressure is defined as the pressure required to *prevent* the net flow of water by osmosis. The least amount of hydrostatic pressure required to prevent the movement of water from a given solution into pure water is the osmotic pressure of the solution. The osmotic pressure of a solution is proportional to the number of particles dissolved in a given volume of that solution. Thus, NaCl and other salts that form two ions per molecule exert twice as much osmotic pressure as glucose and other substances that do not ionize, provided each compound is present at the same concentration.

The important thing for a microbiologist to know about osmosis and osmotic pressure is how particles dissolved in fluid environments affect microorganisms in those environments (➤ Figure 4.31). For this purpose, tonicity is a useful concept. *Tonicity* describes the behavior of cells in a fluid environment. The cells are the reference point, and the fluid environments are compared to the cells. The fluid surrounding cells is **isotonic** to the cells when no change in cell volume occurs (➤ Figure 4.31a). The fluid is **hypertonic** to the cells if the cells shrivel or shrink as water moves out of them into the fluid environment (➤ Figure 4.31b); it is **hypotonic** to the cells if the cells swell or burst as water moves from the environment into the cells (➤ Figure 4.31c). Although bacteria become dehydrated and their cytoplasm shrinks away from the cell wall in a hypertonic environment, their cell walls usually prevent them from swelling or bursting in the hypotonic environments they typically inhabit.

Active Transport

In contrast to passive processes, **active transport** moves molecules and ions against concentration gradients from regions of lower concentration to those of higher concentration. This process is analogous to rolling something uphill, and it does require the cell to expend energy from ATP. Active transport is important in microorganisms for moving nutrients that are present in low concentrations in the environment of the cells. It requires membrane proteins that act as both carriers and enzymes (➤ Figure 4.32). These proteins display specificity in that each carrier transports a single substance or a few closely related substances. The results of active transport are to concentrate a substance on one side of a membrane and to maintain that concentration against a gradient. As with facilitated diffusion, active transport carriers also are subject to saturation and competition for binding sites by similar molecules.

Group translocation reactions move a substance from the outside of a bacterial cell to the inside while chemically modifying the substance so that it cannot diffuse out. This process allows molecules such as glucose to be accumulated against a concentration gradient. Because the modified molecule inside the cell is different from those outside, no actual concentration exists. Energy for this process is supplied by phosphoenolpyruvate (PEP), a high-energy phosphate compound. Many eukaryotic cells have a similar active transport mechanism for preventing diffusion.

Endocytosis and Exocytosis

In addition to the processes that move substances directly across membranes, eukaryotic cells move substances by forming membrane-enclosed vesicles. Such vesicles are made from portions of the plasma membrane. If they form by invagination (poking in) and surround substances outside the cell, the process is called **endocytosis.** These vesicles pinch off from the plasma membrane and enter the cell. If vesicles inside the cell fuse with the plasma membrane and extrude their contents from the cell, the process is called **exocytosis.** Both endocytosis and exocytosis require energy, probably to allow contractile proteins of the cell's cytoskeleton to move vesicles.

Endocytosis

Several types of endocytosis occur. In one type, known as *receptor-mediated endocytosis,* a substance outside the cell binds to the plasma membrane, which invaginates and surrounds the substance. The exact mechanisms that trigger binding and invagination depend on specific receptor sites on the plasma membrane. Once the substance is completely surrounded by plasma membrane to form a vesicle, the vesicle pinches off from the plasma membrane.

➤ **Figure 4.32   Active transport.** Carrier protein molecules aid in movement of molecules through the membrane. This process can take place against a concentration gradient and so requires the use of energy (in the form of ATP) by the cell. The accessory protein participates in the carrier protein's function. (P_i is inorganic phosphate, HPO_4^{2-}.)

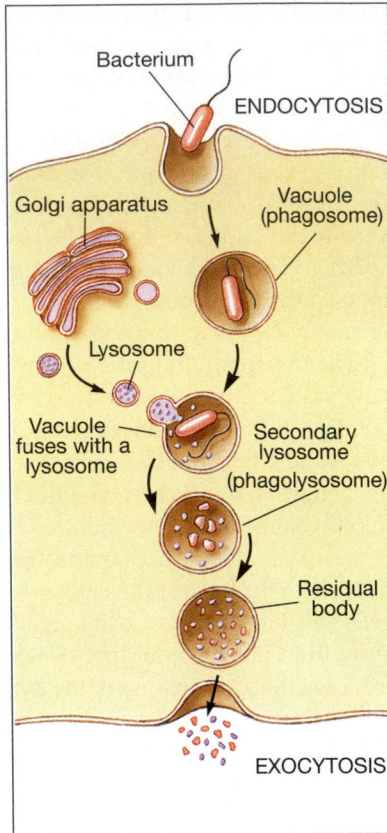

➤ Figure 4.33   **Endocytosis and exocytosis.** Endocytosis is the process of taking materials into the cell; exocytosis is the process of releasing materials from the cell. Material taken in by the form of endocytosis called *phagocytosis* is enclosed in vacuoles known as *phagosomes.* The vacuoles fuse with lysosomes, which release powerful enzymes that degrade the vacuolar contents. Reusable components are absorbed into the cell, and debris is released by exocytosis.

Of all the types of endocytosis, only phagocytosis is of special interest to microbiologists. In **phagocytosis,** large vacuoles called *phagosomes* form around microorganisms and debris from tissue injury. These vacuoles enter the cell, taking with them large amounts of the plasma membrane (➤ Figure 4.33). The vacuole membrane fuses with lysosomes, which release their enzymes into the vacuoles. The enzymes digest the contents of the vacuoles (*phagolysosomes*) and release small molecules into the cytoplasm. Often, undigested particles in residual bodies are returned to and fuse with the plasma membrane. The particles are released from the cell by exocytosis. Certain white blood cells are especially adept at phagocytosis and play an important role in defending the body against infection by microorganisms.

Exocytosis

Exocytosis, the mechanism by which cells release secretions, can be thought of as the opposite of endocytosis. Most secretory products are synthesized on ribosomes or smooth endoplasmic reticulum. They are transported through the membrane of the endoplasmic reticulum, packaged in vesicles, and moved to the Golgi apparatus, where their contents are processed to form the final secretory product. Once secretory vesicles form, they move toward the plasma membrane and fuse with it (➤ Figure 4.33). The contents of the vesicles are then released from the cell.

✓ Most prokaryotes lack sterols in their plasma membranes. What functions do sterols, such as cholesterol, fulfill in eukaryotic plasma membranes?

✓ Compare the number and structure of chromosomes in prokaryotes and eukaryotes.

✓ Give two specific arguments that support the idea that prokaryotes were involved in the evolution of eukaryotes by means of endosymbiosis.

Retracing Our Steps

Basic Cell Types

- Both **prokaryotic cells** and **eukaryotic cells** have membranes that define the bounds of the living cell, and both contain genetic information stored in DNA.
- Prokaryotic cells differ from eukaryotic cells in that they lack a defined nucleus and membrane-enclosed **organelles** (except for a few simple membrane-covered bodies in certain types of prokaryotes).

Prokaryotic Cells

- All prokaryotes are classified in either the domain Archaea or the domain Bacteria.

Size, Shape, and Arrangement

- Prokaryotes are the smallest living organisms.
- Bacteria are grouped by shape: **cocci** (spherical), **bacilli** (rod-shaped), **spirilli** (rigid, wavy), **vibrios** (comma-shaped), and **spirochetes** (corkscrew-shaped).
- Arrangements of bacteria include groupings such as pairs, tetrads, grapelike clusters, and long chains.

An Overview of Structure

- Bacterial cells include a cell membrane, cytoplasm, ribosomes, a nuclear region, and external structures.

The Cell Wall

- The rigid **cell wall** outside the cell membrane is composed mainly of the polymer **peptidoglycan.**
- Cell walls differ in compostition and structure. In Gram-positive bacteria, the cell wall consists of a thick, dense layer of peptidoglycan, with **teichoic acid** in it. In Gram-negative bacteria, the cell wall has a thin layer of peptidoglycan, separated from the cytoplasmic membrane by the **periplasmic space** and enclosed by an **outer membrane** made of **lipopolysaccharide,** or **endotoxin.** In acid-fast bacteria, the cell wall consists mainly of lipids, some of which are true waxes, and some of which are glycolipids.
- Some bacterial cell walls are damaged by penicillin and lysozyme.

The Cell Membrane

- The **cell membrane** has a **fluid-mosaic** structure with phospholipids forming a bilayer and proteins interspersed in a mosaic pattern.
- The main function of the cell membrane is to regulate the movement of materials into and out of cells.
- Bacterial cell membranes also perform functions usually carried out by organelles of eukaryotic cells.

Internal Structure

- The **cytoplasm** is the semifluid substance inside the cell membrane.
- **Ribosomes,** which consist of RNA and protein, serve as sites for protein synthesis.
- The **nuclear region** includes a single, large, circular chromosome, which contains the prokaryotic cell's DNA and some RNA and protein.
- Bacteria contain a variety of **inclusions,** including **granules** that store glycogen or other substances and **vesicles** filled with gas.

- Some bacteria form resistant **endospores.** The core of an endospore contains living material and is surrounded by a cortex, spore coat, and exosporium.

External Structure

- Motile bacteria have one or more **flagella,** which propel the cell by the action of rings in their basal body.
- Much bacterial movement is random, but some bacteria exhibit **chemotaxis** (movement toward attractants and away from repellents) and/or **phototaxis** (movement toward or away from light).
- Some bacteria have **pili: Conjugation pili** allow exchange of DNA, whereas **attachment pili (fimbriae)** help bacteria adhere to surfaces.
- The **glycocalyx** includes all polysaccharides external to a bacterial cell wall. **Capsules** prevent host cells from destroying a bacterium; capsules of any species of bacteria have a specific chemical composition. **Slime layers** protect bacterial cells from drying, trap nutrients, and sometimes bind cells together, as in dental plaque.

Eukaryotic Cells

An Overview of Structure

- Eukaryotic cells, which are generally larger and more complex than prokaryotic cells, are the basic structural unit of microscopic and macroscopic organisms of the kingdoms Protista, Plantae, Fungi, and Animalia.

The Plasma Membrane

- **Plasma membranes** of eukaryotic cells are almost identical to those of prokaryotic cells, except that they contain sterols. The function of eukaryotic plasma membranes, however, is limited primarily to regulating movement of substances into and out of cells.

Internal Structure

- Eukaryotic cells are characterized by the presence of a membrane-enclosed **cell nucleus,** with a **nuclear envelope, nucleoplasm, nucleoli,** and **chromosomes** (typically paired) that contain DNA and proteins called **histones.**
- In cell division by **mitosis,** each cell receives one of each chromosome found in parent cells. In cell division by **meiosis,** each cell receives one member of each pair of chromosomes, and the progeny can be **gametes** or **spores.**
- **Mitochondria,** the powerhouses of eukaryotic cells, carry out the oxidative reactions that capture energy in ATP.
- Photosynthetic cells contain **chloroplasts,** which capture energy from light.
- Eukaryotic ribosomes are larger than those of prokaryotes and can be free or attached to endoplasmic reticulum. Free ribosomes make protein to be used in the cell; those that are attached to endoplasmic reticulum make proteins to be secreted.
- The **endoplasmic reticulum** is an extensive membrane network. Without ribosomes (smooth ER), the endoplasmic reticulum synthesizes lipids; when combined with ribosomes (rough ER), it produces proteins.
- The **Golgi apparatus** is a set of stacked membranes that receive, modify, and package proteins into **secretory vesicles.**

- **Lysosomes,** in animal cells, are organelles that contain digestive enzymes, which destroy dead cells and digest contents of vacuoles.
- **Peroxisomes** are membrane-enclosed organelles that convert peroxides to water and oxygen and sometimes oxidize amino acids and fats.
- **Vacuoles** contain various stored substances and materials engulfed by phagocytosis.
- The **cytoskeleton** is a network of **microfilaments** and **microtubules** that support and give rigidity to cells and provide for cell movements.

External Structure

- Most external components of eukaryotic cells are concerned with movement. Eukaryotic flagella are composed of microtubules; sliding of proteins at their bases causes them to move.
- **Cilia** are smaller than flagella and beat in coordinated waves.
- **Pseudopodia** are projections into which cytoplasm flows, causing a creeping movement.
- Eukaryotic cells of the plant and fungi kingdoms have cell walls, as do the algal protists.

Evolution by Endosymbiosis

- The **endosymbiont theory** holds that organelles of eukaryotic cells arose from prokaryotes that had been engulfed and survived to develop a symbiotic relationship by living inside the larger cell.
- Mitochondria, chloroplasts, flagella, and microtubules are believed to have originated from endosymbiont prokaryotes.
- Many examples exist of modern prokaryotes living endosymbiotically inside eukaryotes.

The Movement of Substances Across Membranes

- All passive processes involved in movement across membranes involve net movement of substances from a region of higher concentration to a region of lower concentration. These processes do not require expenditure of energy by the cell.

Simple Diffusion

- **Simple diffusion** results from the molecular kinetic energy and random movement of particles. The role of diffusion in living cells depends on the size of particles, nature of membranes, and distances substances must move inside cells.

Facilitated Diffusion

- **Facilitated diffusion** uses protein carrier molecules or protein-lined pores in membranes in moving ions or molecules from high to low concentrations.

Osmosis

- **Osmosis** is the net movement of water molecules through a **selectively permeable** membrane from a region of higher to a region of lower concentration. The **osmotic pressure** of a solution is the pressure required to prevent such a flow.

Active Transport

- Active processes involved in movement of substances across membranes generally result in movement from regions of lower concentration to those of higher concentration and require the cell to expend energy.
- **Active transport** requires a protein carrier molecule in a membrane, a source of ATP, and an enzyme that releases energy from ATP.
- Active transport is important in cell functions because it allows cells to take up substances that are in low concentration in the environment and to concentrate those substances within the cell.

Endocytosis and Exocytosis

- **Endocytosis** and **exocytosis,** which occur only in eukaryotic cells, involve formation of vesicles from fragments of plasma membrane and fusion of vesicles with the plasma membrane, respectively.
- In endocytosis the vesicle enters the cell, as in **phagocytosis.**
- In exocytosis the vesicle leaves the cell, as in secretion.
- Endocytosis and exocytosis are important because they allow the movement of relatively large quantities of materials across plasma membranes.

Terminology Check

active transport (p. 101)
amoeboid movement (p. 95)
amphitrichous (p. 85)
atrichous (p. 85)
attachment pilus (p. 89)
axial filament (p. 88)
bacillus (p. 73)
capsule (p. 89)
cell membrane (p. 80)
cell nucleus (p. 91)
cell wall (p. 75)
chemotaxis (p. 88)
chloroplast (p. 92)
chromatin (p. 92)
chromatophore (p. 82)
chromosome (p. 92)

cilium (p. 94)
coccus (p. 73)
conjugation pilus (p. 89)
crista (p. 92)
cytoplasm (p. 82)
cytoplasmic streaming (p. 95)
cytoskeleton (p. 94)
diploid (p. 92)
dyad (p. 92)
endocytosis (p. 101)
endoflagellum (p. 88)
endoplasmic reticulum (p. 93)
endospore (p. 83)
endosymbiotic theory (p. 96)
endotoxin (p. 77)
eukaryotic cell (p. 71)

exocytosis (p. 101)
facilitated diffusion (p. 99)
fimbria (p. 89)
flagellum (p. 85)
fluid-mosaic model (p. 80)
gamete (p. 92)
glycocalyx (p. 89)
Golgi apparatus (p. 93)
granule (p. 83)
group translocation reaction (p. 101)
haploid (p. 92)
histone (p. 92)
hydrophilic (p. 80)
hydrophobic (p. 80)
hypertonic (p. 101)
hypotonic (p. 101)

inclusion   *(p. 83)*
isotonic   *(p. 101)*
lipid A   *(p. 77)*
lipopolysaccharide (LPS)   *(p. 77)*
lophotrichous   *(p. 85)*
lysosome   *(p. 93)*
matrix   *(p. 92)*
meiosis   *(p. 92)*
metachromasia   *(p. 83)*
metachromatic granule   *(p. 83)*
microfilament   *(p. 94)*
microtubule   *(p. 94)*
mitochondrion   *(p. 92)*
mitosis   *(p. 92)*
monotrichous   *(p. 85)*
nuclear envelope   *(p. 91)*
nuclear pore   *(p. 92)*
nuclear region   *(p. 82)*
nucleoid   *(p. 82)*
nucleolus   *(p. 92)*

nucleoplasm   *(p. 92)*
organelle   *(p. 71)*
osmosis   *(p. 99)*
osmotic pressure   *(p. 101)*
outer membrane   *(p. 77)*
pellicle   *(p. 89)*
peptidoglycan   *(p. 75)*
periplasmic space   *(p. 77)*
peritrichous   *(p. 85)*
peroxisome   *(p. 94)*
phagocytosis   *(p. 102)*
phototaxis   *(p. 88)*
pilus   *(p. 88)*
plasma membrane   *(p. 90)*
pleomorphism   *(p. 73)*
polyribosome   *(p. 82)*
prokaryotic cell   *(p. 71)*
protoplast   *(p. 78)*
pseudopodium   *(p. 95)*
ribosome   *(p. 82)*

secretory vesicle   *(p. 93)*
selectively permeable   *(p. 99)*
simple diffusion   *(p. 98)*
slime layer   *(p. 89)*
spheroplast   *(p. 78)*
spindle apparatus   *(p. 92)*
spirillum   *(p. 73)*
spirochete   *(p. 73)*
spore   *(p. 92)*
stroma   *(p. 92)*
teichoic acid   *(p. 77)*
thylakoid   *(p. 92)*
vacuole   *(p. 94)*
vegetative cell   *(p. 83)*
vesicle   *(p. 83)*
vibrio   *(p. 73)*
volutin   *(p. 83)*
zygote   *(p. 92)*

See You on the Web

Don't forget I am expecting you on the web. The following is a preparedness checklist for your web explorations. If you have questions on any of these items, you can return to the end of Chapter 1 for more details.

- Questions We'll Explore
- Concept Check questions
- Flashcards
- Study Guide

If you are ready, come visit me at

http://www.prenhall.com/black

It will be a journey you'll never forget.

Essential Concepts of Metabolism

You're right! I do have my arm inside the cow. It's nice and warm and squishy there inside her rumen, one of the cow's four stomach-like compartments. The grass and hay she's eaten are being digested there. Why can't we digest such low-cost meals? We lack the enzymes needed for the metabolic pathways that digest grass—as does the cow. However, she has billions of microbes, a different mix in each of the four "stomachs," that do metabolize the grass for her. Without them she would starve. I'm going to remove a sample of rumen contents, squeeze out the juice, and examine it under the microscope, plus try to grow some of the fascinating microbes in culture. Microbes are able to do far more types of metabolism than are humans. And remember; most of the methane in the earth's atmosphere comes from rumen microbes. Microbial metabolism keeps our world running.

Background Basics

Until the middle of the nineteenth century, people didn't know what caused a fruit juice to become wine, or milk to sour. Then, in 1857, Louis Pasteur proved that alcoholic fermentation was due to microorganisms. A few years later he identified specific organisms from samples of fermenting juices and souring milk. Pasteur was one of the first to study chemical processes in a living organism. Since his time, much has been learned about such processes.

Metabolism: An Overview

Metabolism is the sum of all the chemical processes carried out by living organisms (➤ Figure 5.1). It includes **anabolism,** reactions that require energy to synthesize complex molecules from simpler ones, and **catabolism,** reactions that release energy by breaking complex molecules into simpler ones that can then be reused as building blocks. Anabolism is needed for growth, reproduction, and repair of cellular structures. Catabolism provides the organism with energy for its life processes, including movement, transport, and the synthesis of complex molecules—that is, anabolism.

All catabolic reactions involve *electron transfer,* which allows energy to be captured in high-energy bonds in ATP and similar molecules. ➡ (Appendix E) Electron transfer is directly related to oxidation and reduction (Table 5.1). **Oxidation** can be defined as the loss or removal of electrons. Although many substances combine with

Questions We'll Explore

A How do the following terms relate to metabolism: autotrophy, heterotrophy, oxidation, reduction, photoautotrophy, photoheterotrophy, chemoautotrophy, chemoheterotrophy, glycolysis, fermentation, aerobic metabolism, and biosynthetic processes?

B What are the characteristics of enzymes, and how do those characteristics contribute to their function?

C What are the main steps and significance of glycolysis and fermentation?

D What are the main steps and significance of the Krebs cycle?

E What are the roles of electron transport and oxidative phosphorylation in energy capture?

F How do microorganisms metabolize fats and proteins for energy?

G What are the main steps and significance of photosynthesis in microbes?

H How do photoheterotrophy and chemoautotrophy differ?

I How do bacteria carry out biosynthetic activities?

J How do bacteria use energy for membrane transport and for movement?

➤ Figure 5.1 **Metabolism, the sum of catabolism and anabolism.**
Large, complex molecules are generally richer in energy than are
small, simple ones. Catabolic reactions break down large molecules
into smaller ones, releasing energy. Organisms capture some of this
energy for their life processes. Anabolic reactions use energy to build
larger molecules from smaller components. The molecules synthesized
in this way are used for growth, reproduction, and repair.

oxygen and transfer electrons to oxygen, oxygen need not be present if another electron acceptor
is available. **Reduction** can be defined as the gain of electrons. When a substance loses electrons, or is oxidized, energy is released, but another substance must gain the electrons, or be reduced, at the same time. For example, during the oxidation of organic molecules, hydrogen
atoms are removed and used to reduce oxygen to form water:

$$2 \ H_2 \ + \ O_2 \ \longrightarrow \ 2 \ H_2O$$
hydrogen oxygen   water

In this reaction, hydrogen is an **electron donor,** or *reducing agent,* and oxygen is an **electron acceptor,** or *oxidizing agent.* Because oxidation and reduction must occur simultaneously,
the reactions in which they occur are sometimes called *redox reactions.*

Among all living things, microorganisms are particularly versatile in the ways in which they
obtain energy. The ways different microorganisms capture energy, and obtain carbon, can be
classified as **autotrophy** (aw-to-trof′e)—"self-feeding"—or **heterotrophy** (het″er-o-trof′e)—
"other-feeding" (➤ Figure 5.2). **Autotrophs** use carbon dioxide (an inorganic substance) to synthesize organic molecules. They include **photoautotrophs,** which obtain energy from light,
and **chemoautotrophs,** which obtain energy from oxidizing simple inorganic substances such
as sulfides and nitrites. **Heterotrophs** use ready-made organic molecules obtained from other
organisms, living or dead. They include **photoheterotrophs,** which obtain chemical energy

*Eyeliner is a favorite
food for the follicle eye
mite because it contains
all the nutrients they
need to survive.*

➤ Table 5.1	Comparison of oxidation and reduction
Oxidation	**Reduction**
Loss of electrons (A)	Gain of electrons (B)
Gain of oxygen	Loss of oxygen
Loss of hydrogen	Gain of hydrogen
Loss of energy (liberates energy)	Gain of energy (stores energy in the reduced compound)
Exothermic; exergonic (gives off heat energy)	Endothermic; endergonic (requires energy, such as heat)

Oxidation

Transfer

A + B → A + B e⁻

A   B   Oxidized   Reduced

Reduction

➤ Figure 5.2   **The main types of energy-capturing metabolism.**

from light, and **chemoheterotrophs,** which obtain chemical energy from breaking down ready-made organic compounds.

Autotrophic metabolism (especially photosynthesis) is important as a means of energy capture in many free-living microorganisms. However, such microorganisms do not usually cause disease. We emphasize metabolic processes that occur in chemoheterotrophs because many microorganisms, including nearly all infectious ones, are chemoheterotrophs. These processes include *glycolysis* (oxidation of glucose to pyruvic acid), *fermentation* (conversion of pyruvic acid to ethyl alcohol, lactic acid, or other organic compounds), and *aerobic respiration* (oxidation of pyruvic acid to carbon dioxide and water). Glycolysis and fermentation (anaerobic processes) do not require oxygen, and only a small amount of the energy in a glucose molecule is captured in ATP. Aerobic respiration does require oxygen as an electron acceptor and captures a relatively large amount of the energy in a glucose molecule in ATP. Complete oxidation of glucose by glycolysis and aerobic respiration is summarized in the following equation:

$$C_6H_{12}O_6 \ + \ 6\,O_2 \ \longrightarrow \ 6\,CO_2 \ + \ 6\,H_2O \ + \ \text{energy}$$

glucose   oxygen   carbon   water
                      dioxide

A large number of microorganisms obtain energy by *photosynthesis,* the use of light energy and hydrogen from water or other compounds to reduce carbon dioxide to an organic substance containing more energy. The overall synthesis of glucose by photosynthesis in cyanobacteria and algae (and green plants) is summarized in the following equation:

$$6\,CO_2 \ + \ 6\,H_2O \ \xrightarrow[\text{chlorophyll}]{\text{light energy}} \ C_6H_{12}O_6 \ + \ 6\,O_2$$

carbon   water   glucose   oxygen
dioxide

(Other photosynthetic bacteria, as we shall see later, use a different version of this process.) Photosynthetic organisms then use the glucose or other carbohydrates made in this way for energy. Notice that the two equations above are the reverse of each other. ➤ Figure 5.3 shows the relationship between respiration and photosynthesis.

➤ **Figure 5.3   Photosynthesis and respiration form a cycle.** In photosynthesis, light energy is used to reduce carbon dioxide, forming energy-rich compounds such as glucose and other carbohydrates. In aerobic respiration, energy-rich compounds are oxidized to carbon dioxide and water, and some of the energy released is captured for use in life processes. (The form of photosynthesis depicted here is carried out by cyanobacteria, algae, and green plants. Green and purple bacteria use compounds other than water as a source of hydrogen atoms to reduce CO_2.)

Like nearly all other chemical processes in living organisms, glycolysis, fermentation, aerobic respiration, and photosynthesis each consist of a series of chemical reactions in which the *product* of one reaction serves as the *substrate* (reacting material) for the next: A ⟶ B ⟶ C ⟶ D ⟶ E, and so on. Such a chain of reactions is called a **metabolic pathway.** Each reaction in a pathway is controlled by a particular enzyme. In this pathway, A is the initial *substrate,* E is the final product, and B, C, and D are *intermediates*.

Metabolic pathways can be catabolic or anabolic (biosynthetic). **Catabolic pathways** capture energy in a form cells can use. **Anabolic pathways** make the complex molecules that form the structure of cells, enzymes, and other molecules that control cells. These pathways use building blocks such as sugars, glycerol, fatty acids, amino acids, nucleotides, and other molecules to make carbohydrates, lipids, proteins, nucleic acids, or combinations such as glycolipids (made from carbohydrates and lipids), glycoproteins (from carbohydrates and proteins), lipoproteins (from lipids and proteins), and nucleoproteins (from nucleic acids and proteins). ATP molecules are the links that couple catabolic and anabolic pathways. Energy released in catabolic reactions is captured and stored in the form of ATP molecules, which are later broken down to provide the energy needed to build up new molecules in biosynthetic pathways. Bacteria transfer approximately 40 percent of the energy in a glucose molecule to ATP during aerobic metabolism and 5 percent during anaerobic fermentation processes. Yields are higher in aerobic processes because their end products are highly oxidized, whereas end products of anaerobic processes are only partially oxidized.

Enzymes

Enzymes are a special category of proteins found in all living organisms. In fact, most cells contain hundreds of enzymes, and cells are constantly synthesizing proteins, many of which are enzymes. Enzymes act as *catalysts*—substances that remain unchanged while they speed up reactions to as much as a million times the uncatalyzed rate, which is ordinarily not sufficient to sustain life. The only other way to speed up the reaction rate would be to increase the temperature: In general, a 10 degree increase in temperature results in a doubling of the reaction rate. However, most cells would die when exposed to such a rise in temperature. Thus, enzymes are necessary for life at temperatures that cells can withstand. To explain how enzymes do these things, we must consider their properties. ➥ (Chapter 2, p. 42)

Properties of Enzymes

In general, chemical reactions that release energy can occur without input of energy from the surroundings. Nevertheless, such reactions often occur at unmeasurably low rates because the molecules lack the energy to start the reaction. For example, although the oxidation of glucose releases energy, that reaction does not occur unless energy to start it is available. The energy required to start such a reaction is called **activation energy** (➤ Figure 5.4). Activation energy can be thought of as a hurdle over which molecules must be raised to get a reaction started. By

Ecologically friendly drain cleaners contain bacteria or bacterial enzymes.

➤ Figure 5.4   **The effect of enzymes on activation energy.** A chemical reaction cannot take place unless a certain amount of activation energy is available to start it. Enzymes lower the amount of activation energy needed to initiate a reaction. They thus make it possible for biologically important reactions to occur at the relatively low temperatures that living organisms can tolerate.

analogy, a rock resting in a depression at the top of a hill would easily roll down the hill if pushed out of the depression. Activation energy is like the energy required to lift the rock out of the depression.

A common way to activate a reaction is to raise the temperature, thereby increasing molecular movement, as you do when you strike a match. Matches ordinarily do not burst into a flame spontaneously. If the energy from friction (striking) is added to the reactants on the match head, the temperature increases, and the match bursts into flame. Such a reaction in cells would raise the temperature enough to denature proteins and evaporate liquids. Enzymes lower the activation energy so reactions can occur at mild temperatures in living cells.

Enzymes also provide a surface on which reactions take place. Each enzyme has a certain area on its surface called the **active site,** a binding site. The active site is the region at which the enzyme forms a loose association with its **substrate,** the substance on which the enzyme acts (➤ Figure 5.5a). Like all molecules, a substrate molecule has kinetic energy, and it collides with various molecules within a cell. When it collides with the active site of its enzyme, an **enzyme–substrate complex** forms (➤ Figure 5.5b). As a result of binding to the enzyme, some of the chemical bonds in the substrate are weakened. The substrate then undergoes chemical change, and the product or products are formed.

Enzymes generally have a high degree of **specificity;** they catalyze only one type of reaction, and most act on only one particular substrate. An enzyme's shape (its tertiary structure; Chapter 2), especially the shape and electrical charges at its active site, accounts for its specificity. ➥ (p. 42) When an enzyme acts on more than one substrate, it usually acts on substrates with the same functional group or the same kind of chemical bond. For example, *proteolytic,* or protein-splitting, enzymes act on different proteins but always act on the peptide bonds in those proteins.

Enzymes are usually named by adding the suffix *-ase* to the name of the substrate on which they act. For example, phosphatases act on phosphates, sucrase breaks down the sugar sucrose, lipases break down lipids, and peptidases break peptide bonds. Enzymes can be divided into two categories on the basis of where they act. **Endoenzymes,** or intracellular enzymes, act within the cell that produced them. **Exoenzymes,** including extracellular enzymes, are synthesized in a cell but cross the cell membrane to act in the periplasmic space or in the cell's immediate environment.

Properties of Coenzymes and Cofactors

Many enzymes can catalyze a reaction only if substances called *coenzymes,* or *cofactors,* are present. Such enzymes consist of a protein portion, called the **apoenzyme,** that must combine with a nonprotein coenzyme or cofactor to form an active **holoenzyme** (➤ Figure 5.6). A **coenzyme** is a nonprotein organic molecule bound to or loosely associated with an enzyme. Many coenzymes are synthesized from *vitamins,* which are essential nutrients precisely because they are required to make coenzymes. For example, *coenzyme A* is made from the vitamin pantothenic acid, and **NAD** *(nicotinamide adenine dinucleotide)* is made from the vitamin

(a)

(b)

➤ **Figure 5.5   The action of enzymes on substrates to yield products.** (a) A computer-generated model of an enzyme (blue) with a substrate molecule (purple) bound to the active site. The active site of the enzyme is a cleft or pocket with a shape and chemical composition that enable it to bind a particular substrate—the molecule on which the enzyme acts. (b) Each substrate binds to an active site, producing an enzyme-substrate complex. The enzyme helps a chemical reaction occur, and one or more products are formed. In this example, the reaction is one that joins two substrate molecules. Other enzyme-catalyzed reactions can involve the splitting of one substrate molecule into two or more parts or the chemical modification of a substrate.

niacin. A **cofactor** is usually an inorganic ion, such as magnesium, zinc, or manganese. Cofactors often improve the fit of an enzyme with its substrate, and their presence can be essential in allowing the reaction to proceed.

Carrier molecules such as coenzymes carry hydrogen atoms or electrons in many oxidative reactions (➤ Figure 5.7). When a coenzyme receives hydrogen atoms or electrons, it is

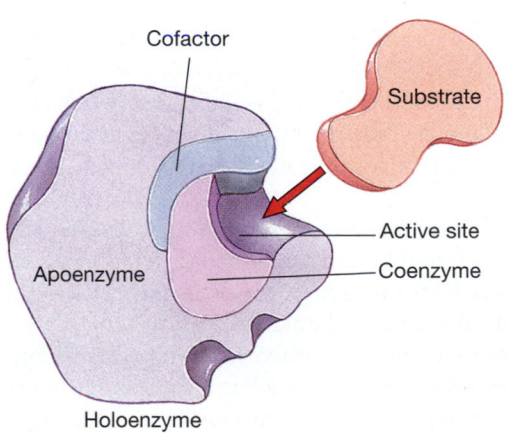

➤ **Figure 5.6   The parts of an enzyme.** Many enzymes consist of a protein apoenzyme that must combine with a nonprotein coenzyme (an organic molecule) or cofactor (an inorganic ion) to form the functional holoenzyme. Can you explain why vitamins and minerals are important in your diet?

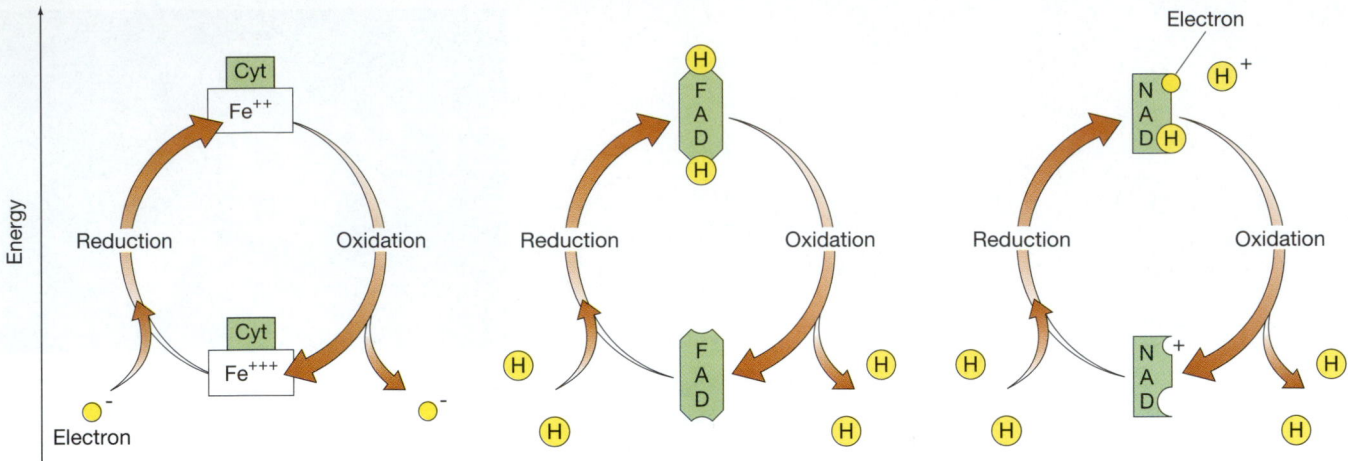

➤ Figure 5.7   **Energy transfer by carrier molecules.** Carrier molecules such as cytochromes and coenzymes carry energy in the form of electrons in many biochemical reactions. Coenzymes such as FAD carry whole hydrogen atoms (electrons together with protons). NAD carries one hydrogen atom and one "naked" electron. When coenzymes are reduced (gain electrons), they increase in energy; when they are oxidized (lose electrons), they decrease in energy.

reduced; when it releases them, it is oxidized. The coenzyme **FAD** *(flavin adenine dinucleotide),* for example, receives two hydrogen atoms to become $FADH_2$ (reduced FAD). The coenzyme NAD has a positive charge in its oxidized state (NAD^+). In its reduced state, NADH, it carries a hydrogen atom and an electron from another hydrogen atom, the proton of which remains in the cellular fluids. In all such oxidation–reduction reactions, the electron carries the energy that is transferred from one molecule to another. Thus, for simplicity, we will refer to *electron transfer* regardless of whether "naked" electrons or hydrogen atoms (electrons with protons) are transferred.

Enzyme Inhibition

No organism can afford to allow continual maximum activity of all its enzymes. Not only is this a waste of materials and energy, but it also may allow harmful quantities of compounds to accumulate, whle others are lacking. Therefore, there must be ways to inhibit enzyme activity, so as to slow or even stop its rate. How, then, are enzymes inhibited? Knowing the answers can help us to control the rates of microbial growth, or the production of certain products they form.

A molecule similar in structure to a substrate can sometimes bind to an enzyme's active site even though the molecule is unable to react. This nonsubstrate molecule is said to act as a **competitive inhibitor** of the reaction because it competes with the substrate for the active site (➤ Figure 5.8). When the inhibitor binds to an active site, it prevents the substrate from binding and thereby inhibits the reaction.

Because the attachment of such a competitive inhibitor is reversible, the degree of inhibition depends on the relative concentrations of substrate and inhibitor. When the concentration of the substrate is high and that of the inhibitor is low, the active sites of only a few enzyme molecules are occupied by the inhibitor, and the rate of the reaction is only slightly reduced. When the concentration of the substrate is low and that of the inhibitor is high, the active sites of many enzyme molecules are occupied by the inhibitor, and the rate of the reaction is greatly reduced.

The *sulfa drugs* (Chapter 13) are competitive inhibitors. Normally, bacterial cells have the enzymes to convert *para-aminobenzoic acid* (*PABA*) to folic acid, an essential vitamin. If sulfa drugs are present, they compete with PABA for the enzymes' active sites. The greater the sulfa drug concentration, the greater the inhibition of folic acid synthesis.

Enzymes also can be inhibited by substances called **noncompetitive inhibitors.** Some noncompetitive inhibitors attach to the enzyme at an **allosteric site,** which is a site other than the active site (➤ Figure 5.9). Such inhibitors distort the tertiary protein structure and alter the shape

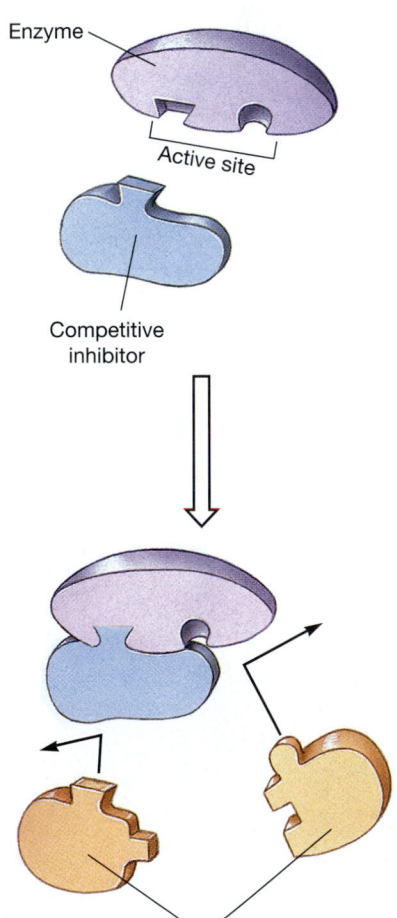

➤ Figure 5.8   **Competitive inhibition of enzymes.** A competitive inhibitor binds to the active site of an enzyme, preventing the normal substrate from reaching it, but cannot take part in the reaction.

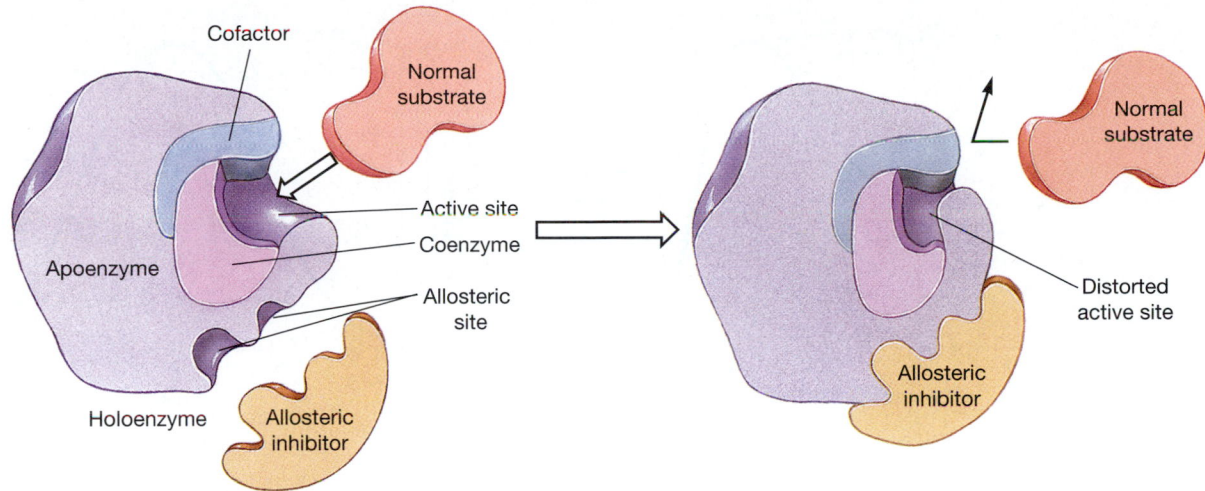

➤ **Figure 5.9   Noncompetitive (allosteric) inhibition of enzymes.** A noncompetitive (allosteric) inhibitor usually binds at a site other than the active site (that is, at an allosteric site). Its presence changes the shape of the enzyme enough to interfere with binding of the normal substrate. Some noncompetitive inhibitors are used in the regulation of metabolic pathways, but others are poisons.

of the active site. Any enzyme molecule thus affected no longer can bind substrate, so it cannot catalyze a reaction. Although some noncompetitive inhibitors bind reversibly, others bind irreversibly and permanently inactivate enzyme molecules, thereby greatly decreasing the reaction rate. In noncompetitive inhibition, increasing the substrate concentration does not increase the reaction rate as it does in the presence of a competitive inhibitor. Lead, mercury, and other heavy metals, although not noncompetitive inhibitors, can bind to other sites on the enzyme molecule and permanently change its shape, thus inactivating it.

Feedback inhibition, a kind of reversible noncompetitive inhibition, regulates the rate of many metabolic pathways. For example, when an end product of a pathway accumulates, the product often binds to and inactivates the enzyme that catalyzes the first reaction in the pathway. Feedback inhibition is discussed in more detail in Chapter 7.

Factors That Affect Enzyme Reactions

Factors that affect the rate of enzyme reactions include:

- Temperature.
- pH.
- Concentrations of substrate, product, and enzyme.

Temperature and pH

Like other proteins, enzymes are affected by heat and by extremes of pH. Even small pH changes can alter the electrical charge on various chemical groups in enzyme molecules, thereby altering the enzyme's ability to bind its substrate and catalyze a reaction.

Most human enzymes have an *optimum temperature,* near normal body temperature, and an *optimum pH,* near neutral, at which they catalyze a reaction most rapidly. Microbial enzymes likewise function best at optimum temperatures and pHs, which are related to an organism's normal environment. The enzymes of microbes that infect humans have approximately the same optimum temperature and pH requirements as human enzymes.

Changes in *enzyme activity,* the rate at which an enzyme catalyzes a reaction, are shown in ➤ Figure 5.10. In this example, enzyme activity increases with temperature up to the enzyme's optimum temperature. Above 40°C, however, the enzyme is rapidly denatured, and its activity decreases accordingly. ➡ (Chapter 2, p. 42) Activity is maximal at an enzyme's optimum pH and decreases as the pH rises or drops from the optimum. Like high temperatures, extremely

How to Ruin an Enzyme

If an enzyme has a vital metabolic function, its inhibitor acts as a poison. A competitive inhibitor temporarily poisons enzyme molecules and slows the reaction. If it binds to the active site of all molecules at one time, it can stop the reaction. The enzyme itself is unharmed, however, and resumes function if the poison is removed. If the poison forms a covalent bond to the enzyme, it is a permanent poison, and the enzyme's ability to function is irreversibly destroyed.

Enzyme inhibition has wide-ranging applications. Some antibiotics, such as penicillin, kill bacteria by damaging their cell wall and causing lysis. Many pesticides and herbicides exert their effects through competitive inhibition. Certain chemotherapeutic agents used to treat cancer inhibit enzymes that are most active in rapidly dividing cells, including malignant cells. Finally, heavy metals inactivate enzymes noncompetitively and permanently and so function as active ingredients in many disinfectants.

> Figure 5.10   **Factors affecting enzyme activity.** (a) Enzymes become more active as the temperature rises. Above about 40°C, however, most enzymes become denatured, and their activity falls off sharply. (b) Most enzymes also have an optimal pH at which they function most effectively. How might adding vinegar (an acid) to food retard spoilage by microbes?

acidic or alkaline conditions also denature enzymes. Such conditions are used to kill or control the growth of microorganisms.

Concentration

To understand the effects of concentrations of substrates and products on enzyme-catalyzed reactions, we must first note that all chemical reactions are, in theory, reversible. Enzymes can catalyze a reaction to go in either direction: AB $\longrightarrow$ A + B or A + B $\longrightarrow$ AB. The concentrations of substrates and products are among several factors that determine the direction of a reaction. A high concentration of AB drives the reaction toward formation of A and B. Use of A and B in other reactions as fast as these products are formed also drives the reaction toward the formation of more A and B. Conversely, use of AB in another reaction so that its concentration remains low drives the reaction toward formation of AB. When neither AB nor A and B are removed from the system, the reaction will ultimately reach a steady state known as **chemical equilibrium.** At equilibrium, no net change in the concentrations of AB, A, or B occurs.

The quantity of enzyme available usually controls the rate of a metabolic reaction. A single enzyme molecule can catalyze only a specific number of reactions per second, that is, can act on only a specific number of substrate molecules. The reaction rate increases with the number of enzyme molecules and reaches a maximum when all available enzyme molecules are working at full capacity. However, if the substrate concentration is too low to keep all enzyme molecules working at capacity, the substrate concentration will determine the rate of the reaction.

With an overview of metabolic processes and an understanding of enzymes and how they work, we are ready to look at metabolic processes in more detail. We begin with glycolysis, fermentation, and aerobic respiration, the processes used by most microorganisms to capture energy.

✓ How are photosynthesis and respiration related to each other?

✓ What is the main difference between chemoautotrophs and chemoheterotrophs? Which of these groups include bacteria? Organisms that cause human disease?

✓ Distinguish between coenzyme and cofactor. How are vitamins related to these?

✓ Why are inhibitors attaching to the allosteric site called noncompetitive inhibitors?

Anaerobic Metabolism: Glycolysis and Fermentation

Glycolysis

Glycolysis (gli-kol′i-sis) is the metabolic pathway used by most autotrophic and heterotrophic organisms, both aerobes and anaerobes, to begin to break down glucose. It does not require oxygen, but it can occur either in the presence or absence of oxygen. ➤ Figure 5.11 shows the 10 steps of the glycolytic pathway, within which four important events occur:

1. Substrate-level phosphorylation.
2. Breaking of a six-carbon molecule (glucose) into two three-carbon molecules.
3. The transfer of two electrons to the coenzyme NAD.
4. The capture of energy in ATP.

Step 1: A phosphate group is transferred to glucose from ATP, forming glucose 6-phosphate.

Steps 2 & 3: The glucose molecule is rearranged to form fructose, and a second phosphate group is added to give fructose 1,6-diphosphate.

Step 4: The six-carbon sugar fructose is split into two different three-carbon sugars.

Step 5: Dihydroxyacetone phosphate is rearranged to form a second molecule of glyceraldehyde 3-phosphate.

Step 6: Another phosphate group is added, and two hydrogen atoms, with their electrons, are transferred to NAD.

Step 7: A phosphate group is transferred to ADP, forming ATP.

Steps 8 & 9: The remaining phosphate group is moved from the end to the middle carbon atom, and a molecule of water is removed.

Step 10: The phosphate group is transferred to ADP, forming more ATP.

> Figure 5.11   The reactions of glycolysis. Note that in steps 1 and 3, two molecules of ATP are used. (A = adenosine) In steps 7 and 10, two molecules of ATP are formed. Because each glucose molecule yields two of the three-carbon sugars that undergo reactions 7 and 10, four molecules of ATP are actually formed, giving a net yield of two ATP per glucose.

Phosphorylation (fos"for-i-la′shun) is the addition of a phosphate group to a molecule, often from ATP. This addition generally increases the molecule's energy. Thus, phosphate groups commonly serve as energy carriers in biochemical reactions. Early in glycolysis, phosphate groups from two molecules of ATP are added to glucose. This expenditure of two ATPs raises the energy level of glucose. It can then participate in subsequent reactions (like the rock pushed out of the depression atop the hill) and is rendered incapable of leaving the cell. The phosphorylated molecule drives the cell's metabolic reactions.

After phosphorylation, glucose is broken into two three-carbon molecules, and each molecule is oxidized as two electrons are transferred from it to NAD. The end products are two molecules of pyruvic acid (called pyruvate in its ionized form), and two molecules of reduced NAD (NADH).

Energy is captured in ATP at the substrate level—that is, in the direct course of glycolysis—in two separate reactions late in the process. With *adenosine diphosphate (ADP)* and *inorganic phosphate (P$_i$)* available in the cytoplasm, the energy released from substrate molecules is used to form high-energy bonds between ADP and P$_i$:

$$ADP + P_i + energy \longrightarrow ATP$$

Glycolysis provides cells with a relatively small amount of energy. Energy is captured in two molecules of ATP during the metabolism of each three-carbon molecule, and a total of four ATPs are formed as one six-carbon glucose molecule is metabolized by glycolysis to two molecules of pyruvic acid. (See Appendix E for a more detailed account of glycolysis.) Because energy from two ATPs was used in the initial phosphorylations, glycolysis results in a net energy capture of only two ATPs per glucose molecule. When atmospheric oxygen is present and the organism has the enzymes to carry out aerobic respiration, electrons from reduced NAD are transferred to oxygen during biological oxidation, as is explained later.

Besides glycolysis, many microorganisms have one or two other metabolic pathways for glucose oxidation. For example, many bacteria, including *Escherichia coli* and *Bacillus subtilis,* have a *pentose phosphate pathway.* This pathway, which can function at the same time as glycolysis, breaks down not only glucose but also five-carbon sugars (pentoses). (Appendix E outlines this pathway.) In a few species of bacteria, including *Pseudomonas,* enzymes carry out the Entner-Doudoroff pathway, which replaces the glycolytic and pentose phosphate pathways. In this pathway, glucose goes through a short series of reactions, one intermediate (glyceraldehyde 3-phosphate) of which goes through the last five steps of a typical glycolysis and produces two ATP molecules in the process of forming pyruvic acid.

Two additional features illustrate principles that apply to metabolic pathways in general:

1. Each reaction is catalyzed by a specific enzyme. Although enzyme names have been omitted in our discussion for simplicity, it is important to remember that each reaction in a metabolic pathway is catalyzed by an enzyme.

2. When electrons are removed from intermediates in metabolic pathways, they are transferred to one of two coenzymes—NAD or NADP *(nicotinamide adenine dinucleotide phosphate).* In a reduced form (NADH or NADPH), these coenzymes store a cell's *reducing power.* For example, in glycolysis, oxidized NAD becomes reduced NAD (NADH). As explained next, electrons are removed from reduced NAD during fermentation, freeing it to remove more electrons from glucose and keep glycolysis operating. Because cells contain limited quantities of both enzymes and coenzymes, the rate at which the reactions of glycolysis and other pathways occur is limited by the availability of these important molecules.

Although glucose is the main nutrient of some microorganisms, other microbes can obtain energy from other sugars. Such organisms usually have specific enzymes to convert a sugar into an intermediate in the glycolytic pathway. Once the sugar has entered glycolysis, it is metabolized to pyruvic acid and then fermented or metabolized aerobically by processes to be described later.

Fermentation

The metabolism of glucose or another sugar by glycolysis is a process carried out by nearly all cells. One process by which pyruvic acid is subsequently metabolized in the absence of oxygen is **fermentation.** Fermentation is the result of the need to recycle the limited amount of NAD by passing the electrons of reduced NAD off to other molecules. It occurs by many different

➤ Figure 5.12   **Fermentation pathways.** Many different fermentation pathways are found among microorganisms. Would two different microbes, each fermenting a quantity of identical materials necessarily produce the same products or by-products? Flavor?

pathways (➤ Figure 5.12). Two of the most important and commonly occurring pathways are homolactic acid fermentation and alcoholic fermentation. Neither captures energy in ATP from the metabolism of pyruvic acid, but both pathways remove electrons from reduced NAD so that it can continue to act as an electron acceptor. Thus, they indirectly foster energy capture by keeping glycolysis going.

Homolactic Acid Fermentation

The simplest pathway for pyruvic acid metabolism is **homolactic acid fermentation,** in which only (homo-) lactic acid is made (➤ Figure 5.13). Pyruvic acid is converted directly to lactic acid, using electrons from reduced NAD. Unlike other fermentations, this type produces no gas. It occurs in some types of the bacteria called lactobacilli, in streptococci, and in mammalian muscle cells. This pathway in lactobacilli is used in making some cheeses.

Live yeast cells and lactic acid-producing bacteria are fed to calves and lactating cows to protect the anaerobic rumen from being colonized by pathogens.

➤ Figure 5.13   **Homolactic acid fermentation.** Pyruvic acid is reduced to lactic acid by the NAD from step 6 of glycolysis (Figure 5.11).

Involuntary Drunkenness

A man was arrested in Virginia for drunk driving. He offered a most unusual defense: involuntary drunkenness—due to yeast fermenting in food in his stomach, thereby producing alcohol that was absorbed into his bloodstream. The judge didn't think much of his plea and found him guilty. The blood alcohol level present in this particular defendant was considerably higher than that ordinarily found in people with such infections. However, there are documented cases in Japan and in the United States of people who were unable to remain sober due to stomach infections of peculiar strains of the yeast *Candida albicans. Candida* is found in various parts of the digestive tract, where it ordi-

narily causes no problems. But those odd strains convert any meal or drink containing carbohydrates into alcohol, although usually not enough to raise blood alcohol levels to the legal limits of intoxication (unless a big meal was eaten). Fortunately the infection can be cured, and the victim returned to sobriety. However, until this occurs, it would seem prudent for people with this problem to refrain from driving.

The yeasts that leaven bread also produce alcohol. Why, then, don't you get drunk from eating your dinner rolls? The reason is that what little alcohol does form evaporates in the oven during baking.

Light of My Death

The culture surrounding death and funeral traditions has undergone many changes. In early America, and through the early 1900s, the corpse was often kept at home for several days before burial. Later, when mortuaries and funeral homes became popular, the body was displayed for an extended time. Before bodies were stored in refrigerated mortuaries, gas build -up and bloating was a practical and cosmetic problem that undertakers had to control. To prevent distortion of the corpse due to the accumulation of bacterial gasses, undertakers would prick tiny holes in the body and briefly hold a candle to the openings. Long blue flames would appear, fed by the gasses escaping from the corpse. These flames would last three or four days until all the gas had been consumed. How does this evidence explain the observation that 19 times as much energy is captured in aerobic metabolism as in fermentation.

Alcoholic Fermentation

In **alcoholic fermentation** (➤ Figure 5.14), carbon dioxide is released from pyruvic acid to form the intermediate acetaldehyde, which is quickly reduced to ethyl alcohol by electrons from reduced NAD. Alcoholic fermentation, although rare in bacteria, is common in yeasts and is used in making bread and wine. Chapter 26 deals extensively with these topics.

Other Kinds of Fermentation

The other kinds of fermentation summarized in ➤ Figure 5.12 are performed by a great variety of microorganisms. One of the most important things about these processes is that they occur in certain infectious organisms, and their products are used in diagnosis. For example, the Voges-Proskauer test for acetoin, an intermediate in butanediol fermentation, helps detect the bacterium *Klebsiella pneumoniae,* which can cause pneumonia. Anaerobic butyric-butylic fermentation occurs in *Clostridium* species that cause tetanus and botulism. The production of butyric acid by *Clostridium perfringens* is what causes the severe tissue damage of gangrene. This fermentation also produces the unpleasant odors of rancid butter and cheese.

The ability to ferment sugars other than glucose forms the basis of other diagnostic tests. One such test (➤ Figure 5.15) uses the sugar mannitol and the pH indicator phenol red. The pathogenic bacterium *Staphylococcus aureus* ferments mannitol and produces acid, which causes the phenol red in the medium to turn yellow. The nonpathogenic bacterium *Staphylococcus epidermidis* fails to ferment mannitol and does not change the color of the medium.

➤ **Figure 5.14   Alcoholic fermentation** In this two-step process, a molecule of carbon dioxide is first removed from pyruvic acid to form acetaldehyde. Acetaldehyde is then reduced to ethyl alcohol by NAD.

> Figure 5.15   **A positive (yellow) mannitol-fermentation test.** This test distinguishes the pathogenic *Staphylococcus aureus* (right) from most nonpathogenic *Staphylococcus* species. *S. aureus* ferments mannitol, producing acid that turns the pH indicator (phenol red) in the medium to yellow. The medium before inoculation (left) is light red.

Many of the products formed by these and other fermentations, such as acetic acid, acetone, and glycerol, are of commercial value. Chapter 26 discusses industrial, pharmaceutical, and food products produced by microbial fermentation. Some of these may allow us to reduce our dependence on costly petrochemicals.

Yeast and bacteria on berries can ferment them to alcohol. Birds eating these berries can become quite drunk, as is obvious from their flight patterns.

Aerobic Metabolism: Respiration

As we have noted, most organisms obtain some energy by metabolizing glucose to pyruvate by glycolysis. Among microorganisms, both anaerobes and aerobes carry out these reactions. **Anaerobes** are organisms that do not use oxygen; they include some that are killed by exposure to oxygen. **Aerobes** are organisms that *do* use oxygen; they include some that must have oxygen. In addition, a significant number of species of microorganisms are *facultative anaerobes* (Chapter 6), which use oxygen if it is available but can function without it. Although aerobes obtain some of their energy from glycolysis, they use glycolysis chiefly as a prelude to a much more productive process, one that allows them to obtain far more of the energy potentially available in glucose. This process is **aerobic respiration** via the *Krebs cycle* and *oxidative phosphorylation*.

The Krebs Cycle

The **Krebs cycle,** named for the German biochemist Hans Krebs, who identified its steps in the late 1930s, metabolizes two-carbon units called *acetyl groups* to CO_2 and H_2O. It also is called the **tricarboxylic acid (TCA) cycle,** because some molecules in the cycle have three carboxyl (COOH) groups, or the **citric acid cycle,** because citric acid is an important intermediate.

Before pyruvic acid (the product of glycolysis) can enter the Krebs cycle, it must first be converted to *acetyl-CoA*. This complex reaction involves the removal of one molecule of CO_2, transfer of electrons to NAD, and addition of coenzyme A (➤ Figure 5.16). In prokaryotes, these reactions occur in the cytoplasm; in eukaryotes, they occur in the matrix of mitochondria.

The Krebs cycle is a sequence of reactions in which acetyl groups are oxidized to carbon dioxide. Hydrogen atoms are also removed, and their electrons are transferred to coenzymes that

Pyruvic acid

Acetyl

Coenzyme A

Acetyl-CoA

CO_2

> Figure 5.16   **The doorway into the Krebs cycle.** Pyruvic acid loses a molecule of CO_2 and is oxidized by NAD. The resulting two-carbon acetyl group is attached to coenzyme A, forming acetyl-CoA.

serve as electron carriers (➤ Figure 5.17). (The hydrogens, as we will see, are eventually combined with oxygen to form water.) Each reaction in the Krebs cycle is controlled by a specific enzyme, and the molecules are passed from one enzyme to the next as they go through the cycle. The reactions form a cycle because oxaloacetic acid (oxaloacetate), a first reactant, is regenerated at the end of the cycle. As one acetyl group is metabolized, oxaloacetate combines with another to form citric acid and goes through the cycle again. (Appendix E gives more detail on the Krebs cycle.)

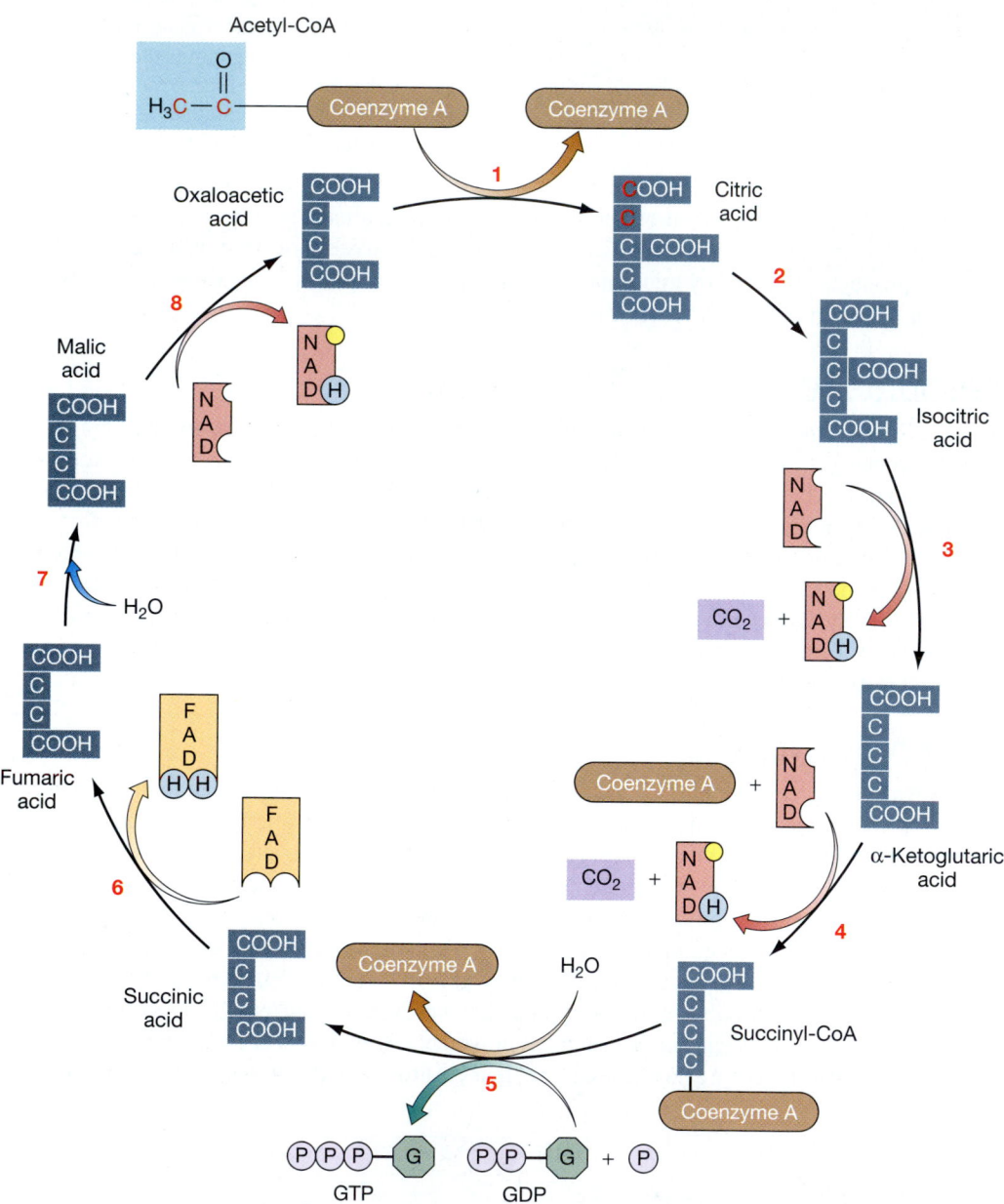

➤ **Figure 5.17** **The reactions of the Krebs cycle.** The intermediates are simplified to show only the number of carbon atoms and carboxyl groups for each. A two-carbon acetyl group enters the cycle as acetyl-CoA in step 1, and two carbon atoms leave the cycle as molecules of CO_2 in steps 3 and 4. Energy is captured in guanosine triphosphate (GTP) in step 5 and eventually transferred to adenosine triphosphate (ATP). In addition, electrons are removed by coenzymes in steps 3, 4, 6, and 8. More energy will be extracted from these electrons when they are subsequently fed into the electron transport chain.

Certain events in the Krebs cycle are of special significance:

- The oxidation of carbon.
- The removal of electrons to coenzymes.
- Substrate-level energy capture.

As each acetyl group goes through the cycle, two molecules of carbon dioxide arise from the complete oxidation of its two carbons. Four pairs of electrons are transferred to coenzymes: three pairs to NAD, and one pair to FAD. Much energy is derived from these electrons in the next phase of respiration, as we will soon see. Finally, some energy is captured in a high-energy bond in guanosine triphosphate (GTP). This reaction takes place at the substrate level. That is, it occurs directly in the course of a reaction of the Krebs cycle. Energy in GTP is easily transferred to ATP. Note that because each glucose molecule produces two molecules of acetyl-CoA, the quantities of the products just given must be doubled to represent the yield from metabolism of a single glucose molecule.

Electron Transport and Oxidative Phosphorylation

Electron transport and oxidative phosphorylation can be likened to a series of waterfalls in which the water makes many small descents and three larger ones (➤ Figure 5.18). In most electron transfers (the small descents), only small amounts of energy are released. At three points (the larger descents), more energy is released, some of which is used to form ATP by the addition of P_i to ADP.

Electron transport, the process leading to the transfer of electrons from substrate to O_2, begins during one of the energy-releasing dehydrogenation reactions of catabolism. Two hydrogen atoms (each consisting of one electron and one proton) are transferred to NAD, forming reduced NAD. The resulting compound in turn transfers the pair of atoms to one of a series of other carrier compounds embedded in the cell membrane of bacteria or in the inner membrane of mitochondria. These carrier compounds form an **electron transport chain,** which is often called the *respiratory chain* (➤ Figure 5.19). Through a series of oxidation-reduction reactions, the electron transport chain performs two basic functions: (1) accepting electrons from an electron donor and transferring them to an electron acceptor, and (2) conserving for ATP synthesis some of the energy released during the electron transfer. The large amounts of energy obtained from respiration result from the transfer of electrons through the electron transport chain, from a level of high energy to one of low energy, with the formation of ATP (➤ Figure 5.19). Energy is captured in high-energy bonds as P_i combines with ADP to form ATP. This process is known as

➤ Figure 5.18   **The electron transport chain modeled as a waterfall.** As the electrons are passed from carrier to carrier in the chain, they decrease in energy, and some of the energy they lose is harnessed to make ATP.

➤ Figure 5.19   **The electron transport chain.** The carriers are NAD; FMN (flavin mononucleotide), a derivative of riboflavin (vitamin B_2); iron sulfide (FeS); coenzyme Q (CoQ); and cytochromes a, b, and c. Each carrier is reduced (*red*) as it picks up electrons and is oxidized (*ox*) as it
passes them on to the next carrier in the chain. Eventually, the electrons combine with the final electron acceptor, oxygen, to form water. Note that FAD feeds its electrons into the chain at a lower point
than does NAD. Because of the arrangement of the various carriers in the cell membrane of prokaryotes or inner mitochondrial membrane of eukaryotes, and because some carriers accept protons (H^+)
whereas others carry only electrons, protons are pushed from the bacterial cytoplasm to outside the
bacterial cell (or from the matrix into the space between the two mitochondrial membranes) during
electron transport. The proton gradient created in this way is used to make ATP (Figure 5.20).

oxidative phosphorylation. Each member of the chain becomes reduced as it picks up electrons;
then upon giving up electrons to the next member in line, it is oxidized. In aerobic respiration,
oxygen is the final electron acceptor and becomes reduced to water (➤ Figure 5.19)

   Several types of enzyme complexes are involved in electron transport. These include NADH
dehydrogenase, cytochrome reductase, and cytochrome oxidase. The electron carriers include
flavoproteins (such as FAD and flavin mononucleotide, FMN), iron-sulfur (FeS) proteins, and
cytochromes, proteins with an iron-containing ring called *heme.* A group of nonprotein, lipid-

soluble electron carriers known as the **quinones,** or *coenzymes Q,* are also found in electron transport systems.

All electron transport chains are not alike; they differ from organism to organism. However, they all have compounds such as flavoproteins and quinones, which accept only hydrogen atoms, and compounds such as cytochromes, which accept only electrons. Unless electrons are continuously transferred from reduced NAD and FAD to oxygen via the electron transport chain, these enzymes cannot accept more electrons from the Krebs cycle, and the entire process will grind to a halt.

From the metabolism of a single glucose molecule, 10 pairs of electrons are transported by NAD (2 pairs from glycolysis, 2 pairs from the pyruvic acid to acetyl-CoA conversion, and 6 pairs from the Krebs cycle). Two additional pairs are transported by FAD (from the Krebs cycle). All these electrons are passed to other electron carriers in the electron transport chain.

In our waterfall analogy, we can think of water entering the falls at two sites, one higher up the mountain than the other. Water from the higher site falls farther than water entering lower down the mountain. In bacteria, electrons entering the electron transport chain at NAD start at the top, and their descent releases enough energy to make three ATPs. Electrons entering at FAD start partway down the chain and contribute only enough energy to make two ATPs. Thus, during aerobic metabolism of a glucose molecule, the 10 pairs of electrons from NAD produce 30 ATPs, and 2 pairs from FAD produce 4 ATPs, for a total of 34 ATPs. Including the 2 ATP molecules from glycolysis and the 2 GTP molecules (= 2 ATPs) from the Krebs cycle gives a total yield of 38 ATPs per glucose molecule.

Oxidative phosphorylation, when compared with fermentation, generates the greatest amount of energy available from glucose. Fermentation, through the substrate-level production of ATP during glycolysis, yields only about 5 percent as much. The net gain of ATP molecules from fermentation is 2.

Chemiosmosis

Electrons for the hydrogen atoms removed from the reactions of the Krebs cycle are transferred through the electron transport system, which generates the high-energy bonds of ATP. The conversion of ADP to ATP occurs via a large ATP-synthesizing complex called *ATP synthase* (or *ATPase*) in a process known as **chemiosmosis** (kem″e-os-mo′sis). This process is the result of a series of chemical reactions that occur in and around a membrane. Although the mechanism for the process, first proposed by the British biochemist Peter Mitchell in 1961, took a number of years to become fully accepted, it is now recognized as a major contribution to the understanding of how ATP is formed during electron transport. Mitchell was awarded the Nobel Prize in 1978 for the development of the chemiosmosis hypothesis.

Chemiosmosis occurs in the cell membrane of prokaryotes such as *Escherichia coli* and in the inner mitochondrial membrane of eukaryotes (➤ Figure 5.20). As electrons are transferred along the electron transport chain, protons are pumped outside the membrane, so the ions' concentration is higher outside the membrane than inside. This process causes a lowering of the proton concentration on the inside and the development of a force that drives the protons back into the cell or mitochondrial matrix to equalize their concentration on both sides of the membrane. Any concentration gradient naturally tends to equalize itself.

In addition to the proton concentration gradient across the membrane, there is also an electrochemical gradient, which makes the membrane a type of biological battery that can power the formation of ATP. The H^+ excess on one side of the membrane gives that side a positive charge compared with the other side. The power generated by the gradient is called the *proton motive force.* The protons flow through special channels within the synthase complex. Energy is thus released and used to form ATP from ADP and inorganic phosphate (P_i).

The Significance of Energy Capture

In glycolysis and fermentation, as we noted earlier, a net of 2 ATPs is usually produced for every glucose molecule metabolized anaerobically. When glycolysis is followed by aerobic respiration, each glucose molecule produces an additional 2 ATPs at the substrate level in the Krebs cycle and 34 ATPs by oxidative phosphorylation. Thus, a glucose molecule yields 38 ATPs

Fueling the Race to Reproduce

Two students, working in the laboratory, observed the growth of yeast in a sugar solution on microscopic slides. One student focused her microscope near the edge of the coverslip, where oxygen levels were sufficient for aerobic respiration. The other student focused his microscope on yeast growing under the center of the coverslip. They kept the same fields in focus for the duration of the class period, and they counted the number of yeast cells in the field every 20 to 30 minutes.

Can you predict their results? What would explain these results? With your instructor's consent, you could easily try this experiment while completing your assigned laboratory exercises and then share your data with the rest of the class. Do you think this experiment would give the same results with all types of organisms? Why?

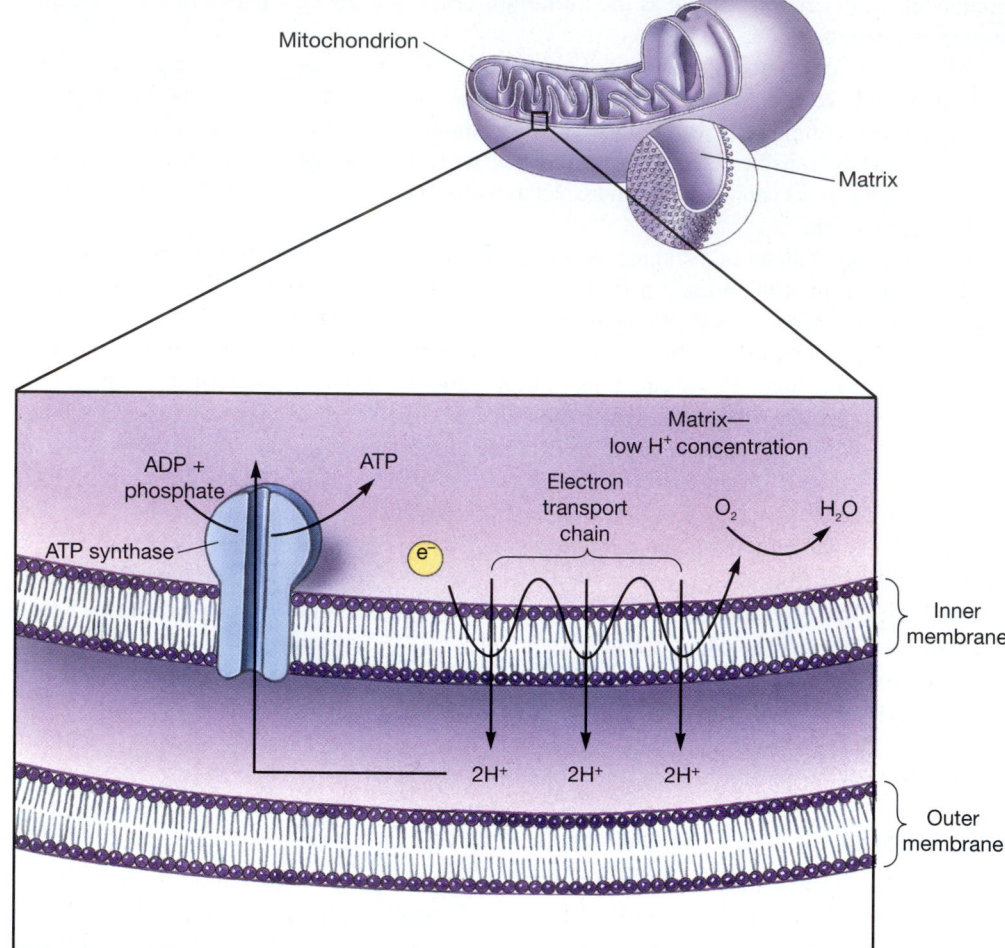

Mitochondrion

Matrix

Matrix—
low H$^+$ concentration

ADP +
phosphate

ATP

Electron
transport
chain

O_2 H_2O

ATP synthase

e$^-$

Inner
membrane

2H$^+$ 2H$^+$ 2H$^+$

Outer
membrane

➤ **Figure 5.20   Chemiosmosis.** Energy capture by chemiosmosis in a mitochondrion is shown here. Protons "pumped" out of the matrix during electron transport pass back in through channels in the ATP synthases that phosphorylate ADP to make ATP. The flow of protons provides the energy to drive this reaction.

✓ If oxygen is present, will it stop glycolysis? Fermentation? The Kreb cycle? The electron transport chain?

✓ If four actual molecules of ATP are produced for each glucose molecule going through glycolysis, why do we say glycolysis yields only two ATPs?

✓ What are the functions of NAD (NADH) and FAD (FADH)?

✓ Where does the electron transport chain function in prokaryotes? In eukaryotes?

by aerobic metabolism but only 2 ATPs by glycolysis and fermentation (Table 5.2). Thus 19 times as much energy is captured in aerobic metabolism as in fermentation! Therefore, aerobic microorganisms in environments with ample oxygen generally grow more rapidly than anaerobes. But aerobes will die if oxygen is depleted, unless they can switch to fermentation. Table 5.3 summarizes the metabolic processes we have studied thus far.

➤ Table 5.2 | **Energy captured in ATP molecules from a glucose molecule by anaerobic and aerobic metabolism in prokaryotes**

Prokaryotic metabolic process	*Number of ATP Molecules*	
	Anaerobic conditions	**Aerobic conditions**
Glycolysis		
Substrate level	4	4
Hydrogen to NAD	0	6
Pyruvate to acetyl-CoA		
Hydrogen to NAD	0	6
Krebs cycle		
Substrate level	0	2
Hydrogen to NAD	0	18
Hydrogen to FAD	0	4
Less energy for phosphorylation	-2	-2
Total	2	38

> ➤ Table 5.3   **A comparison of metabolic processes**

	Glycolysis	**Fermentation**	**Krebs cycle[a]**	**Electron transport chain**
Location	In cytoplasm	In cytoplasm	Prokaryotes: in cytoplasm Eukaryotes: in the mitochondrial matrix	Prokaryotes: in cell membrane Eukaryotes: in inner mitochondrial membranes
Oxygen Conditions	Anaerobic, oxygen is not required; does not stop, however, if oxygen is present	Without O_2; presence of oxygen will cause it to stop	Aerobic	Aerobic
Starting Molecule(s)	1 glucose (6C)	Various substrate molecules go through glycolysis, yielding 2 pyruvic acid	2 pyruvic acid	6 O_2
Ending Molecules	2 pyruvic acid (3C) 2 NADH	Various, depending on which form of fermentation occurs, e.g., ethanol, lactic acid, CO_2, acetic acid	6 CO_2 8 NADH 2 FADH	6 H_2O
Amount of ATP Produced	4 ATP (net 2 ATP)	Various, depending on which form of fermentation occurs, usually 2 or 3 ATP; always far less than is produced in aerobic respiration	2 GTP (= 2 ATP)	34 ATP

[a]Includes the pyruvic acid ⟶ acetyl-CoA step.

The Metabolism of Fats and Proteins

For most organisms, including microorganisms, glucose is a major source of energy. However, for almost any organic substance, we can find a type of microorganism that can degrade that substance for energy. This attribute of microorganisms, and the fact that microbes are found almost everywhere on our planet, accounts for their ability to degrade dead and decaying remains and wastes of all organisms.

Fat Metabolism

Most microorganisms, like most animals, can obtain energy from lipids. The following examples give a general idea of how such processes occur. Fats are hydrolyzed to glycerol and three fatty acids. The glycerol is metabolized by glycolysis. The fatty acids, which usually have an even number of carbons (16, 18, or 20), are broken down into two-carbon pieces by a metabolic pathway called **beta oxidation.** In this process a fatty acid first combines with coenyzme A. Oxidation of the beta carbon (second carbon from the carboxyl group) of the fatty acid results in the release of acetyl-CoA and the formation of a fatty acid shorter by two carbon atoms. The process is then repeated, and another acetyl-CoA molecule is released. The newly formed acetyl Co-A is then oxidized via the Krebs cycle to obtain additional energy (➤ Figure 5.21).

Protein Metabolism

Proteins also can be metabolized for energy (➤ Figure 5.22). They are first hydrolyzed into individual amino acids by *proteolytic* (protein-digesting) *enzymes*. Then the amino acids are *deaminated*—that is, their amino groups are removed. The resulting deaminated molecules enter glycolysis, fermentation, or the Krebs cycle. The metabolism of all major nutrients (fats, carbohydrates, and proteins) for energy is summarized in ➤ Figure 5.23.

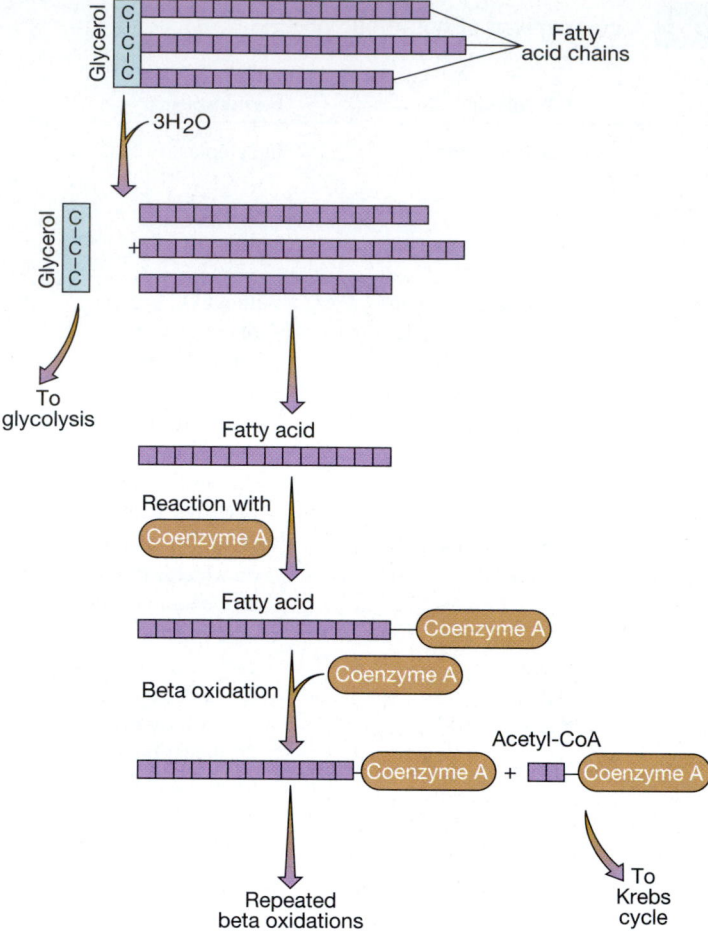

➤ Figure 5.21   **The catabolism of fats.** Triglycerides are hydrolyzed into glycerol and fatty acids. The glycerol is broken down via glycolysis. The fatty acids are broken down into two-carbon units and fed into the Krebs cycle, where they are metabolized to produce additional energy.

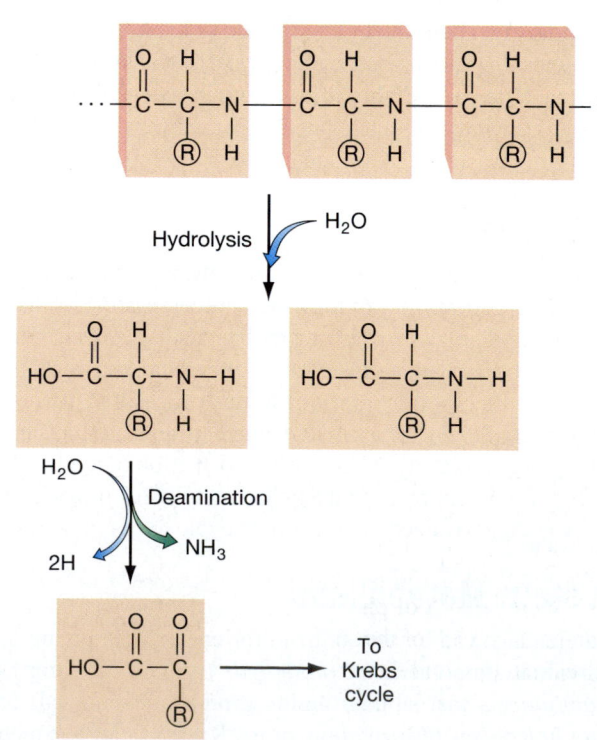

➤ Figure 5.22   **The catabolism of proteins.** Polypeptides are hydrolyzed to amino acids. The amino acids are deaminated, and the resulting molecules enter pathways leading to the Krebs cycle.

➤ **Figure 5.23   Metabolism of the major classes of biomolecules: a summary.** This can be thought of as a giant funnel that eventually channels all three types of nutrients into the Krebs cycle.

Other Metabolic Processes

Having considered energy capture in chemoheterotrophs, we will now briefly consider energy capture in photoautotrophs, photoheterotrophs, and chemoautotrophs.

Photoautotrophy

Organisms called photoautotrophs carry out **photosynthesis,** the capture of energy from light and the use of this energy to manufacture carbohydrates from carbon dioxide. Photosynthesis occurs in green and purple bacteria, in cyanobacteria, in algae, and in higher plants. Photosynthetic bacteria, which probably evolved early in the evolution of living organisms, perform their own version of photosynthesis in the absence of O_2. However, algae and green plants make much more of the world's carbohydrate supply, so we will consider the process in those organisms first and then see how it is modified in green and purple bacteria.

In green plants, algae, and cyanobacteria, photosynthesis occurs in two parts—the "photo" part, or the *light reactions,* in which light energy is converted to chemical energy, and the "syn-

Microbial Clean-up

A few species of bacteria, such as some members of the genus *Pseudomonas,* can use crude oil for energy. They can grow in seawater with only oil, potassium phosphate, and urea (a nitrogen source) as nutrients. These organisms can clean up oil spills in the ocean acting as "bioremediators." They have also proven useful in degrading oil that remains in the water carried by tankers as ballast after unloading their oil. Then the water pumped from the tankers into the sea in preparation for a new cargo of oil does not pollute. A detergentlike substance has been isolated recently from these organisms. When the detergent is added to a quantity of oil sludge, it converts 90 percent of the sludge into usable petroleum in about 4 days, thereby reducing waste and providing a convenient means of cleaning oil-fouled tanks.

(a)

(b)

(c)

(a) The 1989 Exxon *Valdez* oil spill left great quantities of pooled oil on sites in the Gulf of Alaska, such as on Green Island. Bioremediation in 1989 (b) by the application of nutrients (nitrogen and phosphorus) to the shoreline accelerated the bacterial biodegradation of the oil into carbon dioxide and water. (c) In 1991, the area was surveyed and found to be mostly cleared of oil, with no further treatment recommended.

thesis" part, or the *dark reactions,* in which chemical energy is used to make organic molecules. Each part involves a series of steps.

In the **light reactions,** light strikes the green pigment chlorophyll *a* in thylakoids of chloroplasts. ➡ (Chapter 4, p. 94) Electrons in the chlorophyll become excited—that is, raised to a higher energy level. These electrons participate in generating ATP in cyclic photophosphorylation and in noncyclic photoreduction (➤ Figure 5.24). In **cyclic photophosphorylation,** excited electrons from chlorophyll are passed down an electron transport chain. As they are

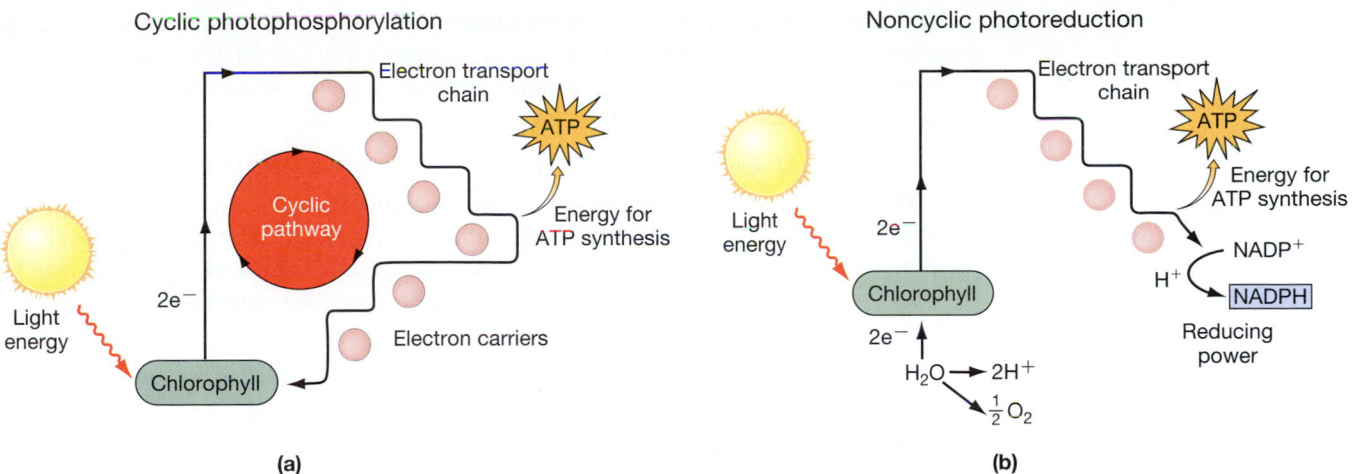

Cyclic photophosphorylation

Noncyclic photoreduction

(a)

(b)

➢ **Figure 5.24   The light reactions of photosynthesis, performed by cyanobacteria, algae, and green plants.** Electrons in chlorophyll receive a boost in energy from light, and their extra energy is used to make ATP. In the pathway of cyclic photophosphorylation, the electrons return to chlorophyll and thus can be used over and over again. In noncyclic photoreduction, the electrons receive a second boost that gives them enough energy to reduce NADP. The electrons are replaced by the splitting of water.

transferred, energy is captured in ATP by chemiosmosis (as described previously in connection with oxidative phosphorylation). When the electrons return to the chlorophyll, they can be excited over and over again, so the process is said to be cyclic.

In **noncyclic photoreduction,** energy also is captured by chemiosmosis. In addition, membrane proteins and energy from light are used to split water molecules into protons, electrons, and oxygen molecules, a process called **photolysis** (fo-tol′eh-sis). The electrons replace those lost from chlorophyll, which are thus freed to reduce the coenzyme NADP. ATP and reduced NADP (NAPDH)—the products of the light reaction—and atmospheric CO_2 subsequently participate in the dark reactions.

The **dark reactions,** or *carbon fixation,* occur in the stroma of chloroplasts. Carbon dioxide is reduced by electrons from NADPH in a process known as the *Calvin-Benson cycle.* (See Appendix E.) Energy from ATP and electrons from NADPH are required in this synthetic process. Various carbohydrates, chiefly glucose, are the products of the dark reactions (➢ Figure 5.25).

➢ **Figure 5.25   The relation between the light and dark reactions.** In the dark reactions, ATP and NADPH (the products of the light reactions) are used to reduce carbon dioxide, forming carbohydrates such as glucose. The dark reactions do not require darkness; they are so named because they can take place in the dark, so long as the products of the light reactions are available.

Photosynthesis in green and purple sulfur bacteria differs from that in green plants, algae, and cyanobacteria in ways related to the evolution of organisms. The first photosynthetic organisms probably were purple and green bacteria, which evolved in an atmosphere containing much hydrogen but no oxygen. They differ from green plants, algae, and cyanobacteria as follows:

1. Bacterial chlorophyll absorbs slightly longer wavelengths of light than does chlorophyll *a*.

2. They use hydrogen compounds such as hydrogen sulfide (H_2S), rather than water (H_2O), for reducing carbon dioxide. Electrons from their pigments reach an energy level high enough to split H_2S but not high enough to split H_2O and to generate an H^+ gradient for ATP synthesis. (Some purple and green bacteria produce elemental sulfur as a by-product; a few produce strong sulfuric acid.)

3. They are usually strict anaerobes and can live only in the absence of oxygen. They do not release oxygen as a product of photosynthesis, as green plants do.

Characteristics of the groups of bacteria that carry out this primitive form of photosynthesis are summarized in Table 5.4.

The cyanobacteria also are photosynthetic, but they probably evolved after the purple and green bacteria. Although prokaryotic, the cyanobacteria release oxygen during photosynthesis, as do green plants and algae. In fact, cyanobacteria are probably responsible for the addition of oxygen to the primitive atmosphere.

Photoheterotrophy

Photoheterotrophs are a small group of bacteria that can use energy from light but require organic substances such as alcohols, fatty acids, or carbohydrates as carbon sources. These organisms include the nonsulfur, purple or green bacteria.

Chemoautotrophy

✓ Lipids are broken down into glycerol and fatty acids. How are each of these then further metabolized?

✓ What is returned to chlorophyll in cyclic photophosphorylation that is not returned in noncyclic photoreduction?

✓ What were probably the first kind of photosynthetic organisms on our planet? Were they aerobes or anaerobes? Did they release oxygen gas as a product of photosynthesis?

✓ What type of metabolism is characteristic of nitrifying bacteria? Why are they important organisms?

Chemoautotrophic bacteria (also called *chemolithotrophs*) are unable to carry out photosynthesis but can oxidize inorganic substances for energy. With this energy and carbon dioxide as a carbon source, these bacteria can synthesize a great variety of substances, including carbohydrates, fats, proteins, nucleic acids, and substances that are required as vitamins by many organisms.

The ability to oxidize, and therefore extract energy from, inorganic substances is probably the most outstanding characteristic of chemoautotrophs, but these bacteria have other noteworthy attributes. The nitrifying bacteria are especially important because they increase the quantity of usable nitrogen compounds available to plants and replace nitrogen that plants remove from the soil. *Thiobacillus* and some other sulfur bacteria produce sulfuric acid by oxidizing elemental sulfur or hydrogen sulfide. Acidity lower than pH 1 has been produced by sulfur bacteria. Sulfur is sometimes added to alkaline soil to acidify it, a practice that works because of the numerous thiobacilli present in most soils. Finally, some chemoautotrophic Archaeobacteria have been found near volcanic vents in the ocean floor, where they grow at extremely high temperatures and sometimes under very acidic conditions. Characteristics of chemoautotrophs are summarized in Table 5.5.

➢ Table 5.4	**Characteristics of photosynthetic bacteria**	
Group	**Family and representative genus**	**Pigments**
Green sulfur bacteria	Chlorobiaceae *Chlorobium*	Bacterial chlorophyll
Purple sulfur bacteria	Chromaticeae *Chromatium*	Bacterial chlorophyll and red and purple carotenoid pigments

> Table 5.5 | **Characteristics of chemoautotrophic bacteria**

Group and representative genus/genera	Source of energy	Products after oxidizing reaction
Nitrifying bacteria		
Nitrobacter	HNO_2	HNO_3
Nitrosomonas	NH_3	$HNO_2 + H_2O$
Nonphotosynthetic sulfur bacteria		
Thiothrix	H_2S	$H_2O + 2S$
Thiobacillus	S	H_2SO_4
Iron bacteria		
Siderocapsa	Fe^{2+}	$Fe^{3+} + OH^-$
Hydrogen bacteria		
Alcaligenes	H_2	H_2O

The Uses of Energy

Microorganisms use energy for such processes as biosynthesis, membrane transport, movement, and growth. Here we will summarize some biosynthetic activities and some mechanisms for membrane transport and movement. We will consider microbial growth in Chapter 6.

Biosynthetic Activities

Microorganisms share many biochemical characteristics with other organisms. All organisms require the same building blocks to make proteins and nucleic acids. Many of these building blocks (amino acids, purines, pyrimidines, and ribose) can be derived from intermediate products of energy-yielding pathways (➢ Figure 5.26). When the energy-yielding pathways were first discovered, they were thought to be purely catabolic. Now that many of their intermediates are known to be involved in biosynthesis, they are more properly called **amphibolic** (am-fe-bol′ik) **pathways** (*amphi-*, either) because they can yield either energy or building blocks for synthetic reactions.

Some biosynthetic pathways are quite complex. For example, synthesis of amino acids in those organisms that can make them often requires many reactions, with an enzyme for each reaction. Tyrosine synthesis requires no fewer than 10 enzymes, and tryptophan synthesis needs at least 13. The synthetic pathways for making purines and pyrimidines also are complex. The absence of a single enzyme in a synthetic pathway can prevent the synthesis of a substance. Any essential substance that an organism cannot synthesize must be accessible in the environment, or the organism will die. Missing enzymes thus increase the nutritional needs of organisms.

Microorganisms of many different types also synthesize a variety of carbohydrates and lipids. The rate at which they are synthesized also varies and depends on the availability and activity of enzymes. Some organisms, such as the aerobe *Acetobacter,* synthesize cellulose, which is ordinarily found in plants. As strands of cellulose reach the cell surface, they form a mat that traps carbon dioxide bubbles and keeps the cell afloat. Because these organisms must have oxygen, the mat contributes to their survival by keeping them near the surface, where oxygen is plentiful.

Many bacteria synthesize peptidoglycan, lipopolysaccharide, and other polymers associated with cell walls. ➡ (Chapter 4, p. 75) Some bacteria form capsules, especially in media that contain serum or large amounts of sugar. Capsules usually consist of polymers of one or more monosaccharides. However, in *Bacillus anthracis,* the bacterium that causes anthrax, the capsule is a polypeptide of glutamic acid. The biosynthetic processes in microorganisms are summarized in ➢ Figure 5.27.

The Swamp of Eternal Stench

It is possible to convert a lovely lake into a "Swamp of Eternal Stench" because such a delicate environmental balance exists between aerobic respiration and fermentation. When excess nutrients, such as carbon, nitrogen, or phosphorous are present, there is an increase in both plant and bacterial growth. The aerobic bacteria will deplete the lake of dissolved oxygen. As more fish, plants, and animals die, the anaerobic bacteria get the ecological advantage. Fermentation rules! Our bucolic lake is now a bubbling stew emitting methane and hydrogen sulfide gasses that are spiced with putrescine and cadaverine—the odors of death.

Codependent on Cometabolism

Environmental scientists observe that hazardous-waste sites that support the growth of highly diverse microbial populations are much easier to bioremediate than sites inhabited by a few microbial species. Sites inhabited by diverse populations remediate faster and with fewer problems associated with the production and accumulation of toxic metabolic by-products. The reason for this is cometabolism—an advantageous situation where an organism, in the process of oxidizing a particular substrate, oxidizes a second substrate as well. The second metabolite is not a source of nutrients or energy for the oxidizing organism, but is a nutrient for the second organism. Hazardous-waste sites where cometabolism is encouraged can efficiently mineralize organic wastes to carbon dioxide and water.

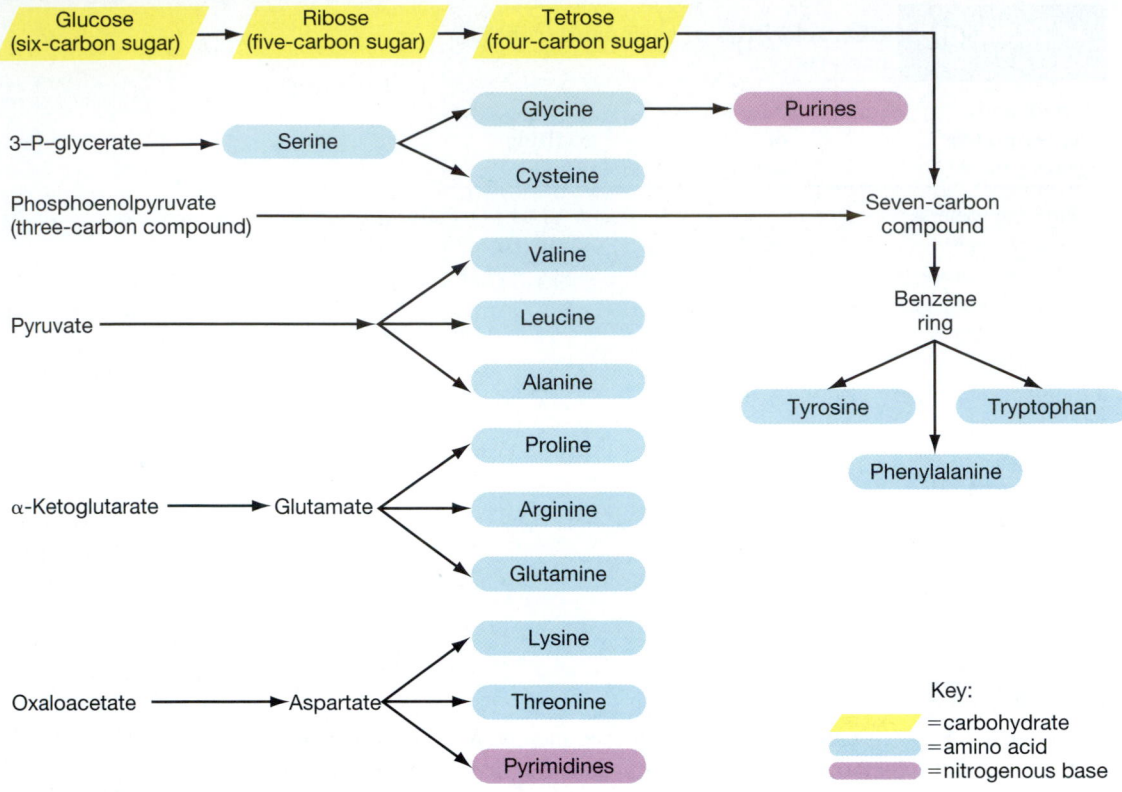

➤ Figure 5.26   **Some biosynthetic pathways.** This flow diagram shows how amino acids, nucleic acid bases, and ribose are made from intermediates in glycolysis and from the Krebs cycle.

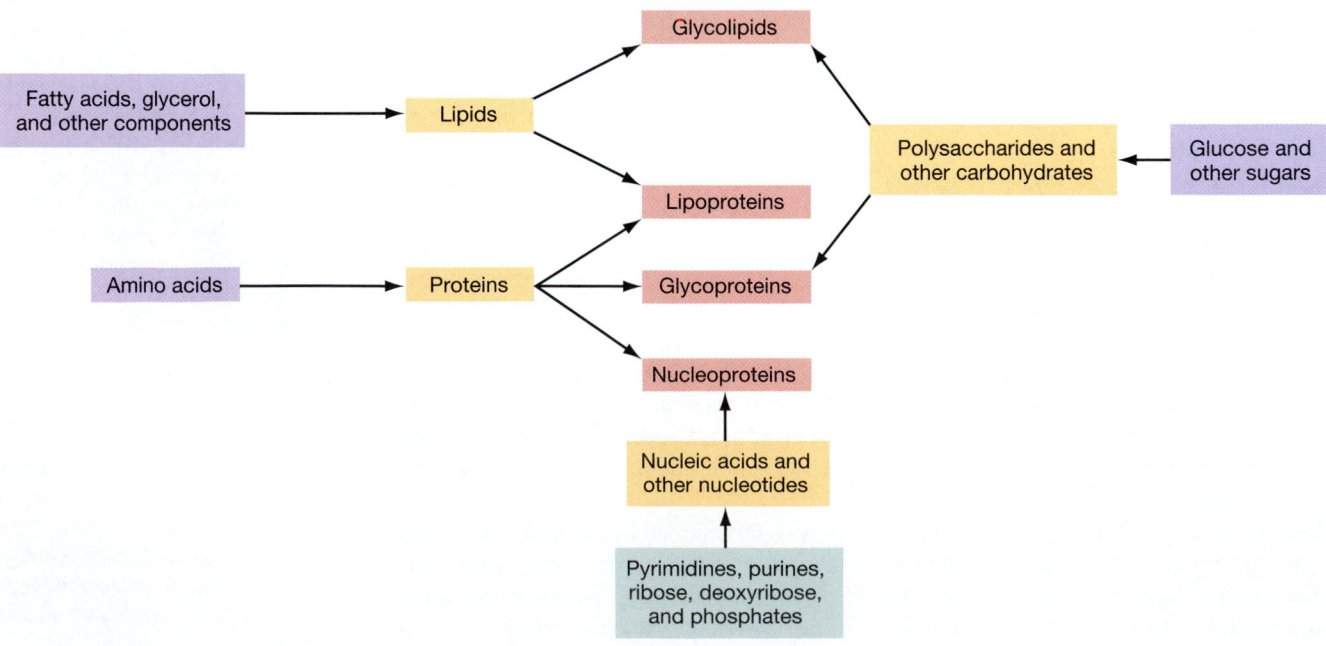

➤ Figure 5.27   **The formation of complex biomolecules from simpler components.**

Membrane Transport and Movement

In addition to using energy for biosynthetic processes, microorganisms also use energy for transporting substances across membranes and for their own movement. These energy uses are as important to the survival of the organisms as are their biosynthetic activities.

Membrane Transport

Microbes use energy to move most ions and metabolites across cell membranes against concentration gradients. For example, bacteria can transport a sugar or an amino acid from a region of low concentration outside the cell to a region of higher concentration inside the cell. This means that they accumulate nutrients within cells in concentrations a hundred to a thousand times the concentration outside the cell. They also concentrate certain inorganic ions by the same means.

Two mechanisms exist in bacteria for concentrating substances inside cells, and both require energy. One active transport mechanism is specific to Gram-negative bacteria, such as *E. coli.* Such bacteria have two membranes—the cell membrane, which surrounds the cell's cytoplasm, and the outer membrane, which forms part of the cell wall. ➥ (Chapter 4, p. 75) Transmembrane carrier proteins called **porins** form channels through the outer membrane. Porins allow entry of ions and small hydrophilic metabolites via *facilitated diffusion.* ➥ (Chapter 4, p. 99) After entering the periplasmic space, a specific periplasmic protein combines with one of the diffusing ions or metabolites. The periplasmic protein then facilitates the transport of the substance into the cytoplasm via a specific carrier protein in the cell membrane. Such substances generally gain entry by active transport. Through ATP hydrolysis, the carrier protein changes shape, allowing the metabolite into the cytoplasm (see Figure 4.32).

Another mechanism, present in all bacteria, is called the **phosphotransferase system (PTS)**. It consists of sugar-specific enzyme complexes called **permeases** (per'me-a-sez), which form a transport system through the cell membrane. The PTS uses energy from the high-energy phosphate molecule phosphoenolpyruvate (PEP). When PEP is present in the cytoplasm, it can provide energy and a phosphate group to a permease in the membrane. Then the permease transfers the phosphate to a sugar molecule and at the same time moves the sugar across the membrane. A phosphorylated sugar is thus transported inside the cell and is prepared to undergo metabolism. This *group translocation* was discussed in Chapter 4. ➥ (p. 101)

Movement

Most motile bacteria move by means of flagella, but some move by gliding or creeping or in a corkscrew motion. Flagellated bacteria move by rotating their flagella. ➥ (Chapter 4, p. 85) The mechanism for rotation, though not fully understood, appears to involve a proton gradient as in chemiosmosis. As the protons move down the gradient, they drive the rotation. Gliding bacteria move only when in contact with a solid surface, such as decaying organic matter. Rotation of the cell on its own axis often occurs with gliding. A number of mechanisms have been proposed to explain gliding, but the mechanism that propels the gliding bacterium *Myxococcus* is best understood. This organism uses energy to secrete a substance called a **surfactant** (ser-fak'tant), which lowers surface tension at the bacterium's posterior end. The difference in surface tension between the anterior and posterior ends (a passive phenomenon) causes *Myxococcus* to glide.

Spirochetes expend energy for both creeping and thrashing motions. On a solid surface they creep along like an inchworm by alternately attaching front and rear ends. Suspended in a liquid medium, they thrash (twist and turn). Both creeping and thrashing motions probably occur by waves of contraction within the cell substance that exert force against axial filaments.

Bioluminescence

Bioluminescence, the ability of an organism to emit light, appears to have evolved as a byproduct of aerobic metabolism. Bacteria of the genera *Photobacterium* and *Achromobacter,* fireflies, glowworms, and certain marine organisms living at great depths in the ocean exhibit bioluminescence (➤ Figure 5.28). Many light-emitting organisms have the enzyme *luciferase* (lu-sif'er-ace), along

(a)

(b)

➤ **Figure 5.28 Microbial bioluminescence.** (a) Bioluminescent bacteria in the Petri dish produce enough light to read by. (b) Angler fish lights up the dark deep-ocean depths with bioluminescent bacteria living symbiotically in its long "lure" that attracts prey to within reach of its jaws.

with other components of the electron transport system. (Luciferase derives its name from Lucifer, which means "morning star.") Luciferase catalyzes a complex reaction in which molecular oxygen is used to oxidize a long-chain aldehyde or ketone to a carboxylic acid. At the same time, $FMNH_2$ from the electron transport chain is oxidized to an excited form of *flavin mononucleotide* (FMN), a carrier molecule derived from riboflavin (vitamin B_2), that emits light as it returns to its unexcited state. In this process, phosphorylation reactions are bypassed, and no ATP is generated. Instead, energy is released as light.

Luminescent microorganisms often live on the surface of marine organisms such as some squids and fish. More than 300 years ago, the Irish chemist Robert Boyle observed that the familiar glow of the skin of dead fish lasted only as long as oxygen was available. At that time the electron transport system and the role of oxygen in it were not understood.

Bioluminescence exhibited by larger organisms has survival value. It is the sole light source for marine creatures that live at great depths, and it helps land organisms such as fireflies find mates. How bioluminescence came to be established among microorganisms is less clear. One hypothesis is that early in the evolution of living things, bioluminescence served to remove oxygen from the atmosphere as it was produced by some of the first photosynthetic organisms. Although this is not an advantage to aerobes, it is an advantage to strict anaerobes. Because most of the microorganisms in existence at that time were anaerobes susceptible to the toxic effects of oxygen, bioluminescence would have been beneficial to them. Today, many bioluminescent microbes are beneficiaries of symbiotic relationships with their hosts. They provide light in return for a shelter and nutrients.

Scientists have found a way to put bioluminescent bacteria to work. In the Microtox Acute Toxicity Test, bioluminescent bacteria are exposed to a water sample to determine if the sample is toxic. Any change—positive or negative—in the bacteria's growth is observable as a change in their light output. The sample's toxicity is calculated by comparing before-and-after readings of the light levels. The brainchild of Microbics Corp. of Carlsbad, California, this toxicity test takes only minutes to perform.

The Microtox Acute Toxicity Test is useful for testing the quality of drinking water and for numerous other industrial applications. For example, waste-water treatment plants use it to determine quickly whether their treated effluent will be able to pass government toxicity compliance tests. Paper mills use the test to determine how much disinfectant is needed to rid their equipment of the microbial growth that slows down the manufacturing process and affects product quality. Makers of household cleansers, shampoos, or cosmetics use the test in place of controversial animal testing in which drops of the products are put into the eyes of rabbits to determine the products' irritancy levels. And unlike cell-culturing techniques, the test requires little skill to perform and to interpret. Bioluminescence could prove to be a very important process to industry in the future.

Retracing Our Steps

Metabolism: An Overview

- **Metabolism** is the sum of all the chemical processes in a living organism. It consists of **anabolism,** reactions that require energy to synthesize complex molecules from simpler ones, and **catabolism,** reactions that release energy by breaking complex molecules into simpler ones.

- **Autotrophs,** which use carbon dioxide to synthesize organic molecules, include **photoautotrophs** (which carry on photosynthesis) and **chemoautotrophs.**

- **Heterotrophs,** which use organic molecules made by other organisms, include **chemoheterotrophs** and **photoheterotrophs.**

- For growth, movement, and other activities, **metabolic pathways** use energy captured in the **catabolic pathways.**

Enzymes

Properties of Enzymes

- **Enzymes** are proteins that catalyze chemical reactions in living organisms by lowering the **activation energy** needed for a reaction to occur.

- Enzymes have an **active site,** the binding site to which the **substrate** (the substance on which the enzyme acts) attaches to form an **enzyme–substrate complex.** Enzymes typically exhibit a high degree of **specificity** in the reactions they catalyze.

Properties of Coenzymes and Cofactors

- Some enzymes require **coenzymes,** nonprotein organic molecules that can combine with the **apoenzyme,** the protein portion of the enzyme, to form a **holoenzyme.** Some enzymes also require inorganic ions as **cofactors.**

Enzyme Inhibition

- Enzyme activity can be reduced by **competitive inhibitors,** molecules that compete with the substrate for the enzyme's active site, or by **noncompetitive inhibitors,** molecules that bind to an **allosteric site,** a site other than the active site.

Factors That Affect Enzyme Reactions

- Factors that affect the rate of enzyme reactions include temperature, pH, and concentrations of substrate, product, and enzyme.

Anaerobic Metabolism: Glycolysis and Fermentation

Glycolysis

- **Glycolysis** is a metabolic pathway by which glucose is oxidized to pyruvic acid.

- Under anaerobic conditions, glycolysis yields a net of two ATPs per molecule of glucose.

Fermentation

- **Fermentation** refers to the reactions of metabolic pathways by which NADH is oxidized to NAD.

- Six pathways of fermentation are summarized in Figure 5.12. **Homolactic acid** and **alcoholic fermentations** are two of the most important and commonly occurring fermentation pathways.

Aerobic Metabolism: Respiration

- **Anaerobes** do not use oxygen; **aerobes** use oxygen and obtain energy chiefly via **aerobic respiration.**

The Krebs Cycle

- The **Krebs cycle** involves the metabolism of two-carbon compounds to CO_2 and H_2O, the production of one ATP directly from each acetyl group, and the transfer of hydrogen atoms to the electron transport system.

- In energy production the Krebs cycle processes acetyl-CoA so that (in the electron transport chain) hydrogen atoms can be oxidized for energy.

Electron Transport and Oxidative Phosphorylation

- **Electron transport** is the transfer of electrons to oxygen (the final electron acceptor).

- **Oxidative phosphorylation** involves the **electron transport chain** for ATP synthesis and is a membrane-regulated process not directly related to the metabolism of specific substrates.

- The theory of **chemiosmosis** explains how energy is used to synthesize ATP.

The Significance of Energy Capture

- In prokaryotes, aerobic (oxidative) metabolism captures 19 times as much energy as does anaerobic metabolism.

The Metabolism of Fats and Proteins

- Most organisms get energy mainly from glucose. But for almost any organic substance, there is some microorganism that can metabolize it.

Fat Metabolism

- Fat metabolism involves hydrolysis and the enzymatic formation of glycerol and free fatty acids. Fatty acids are in turn oxidized by **beta oxidation,** which results in the release of acetyl-CoA. Acetyl-CoA then enters the Krebs cycle.

Protein Metabolism

- The metabolism of proteins involves the breakdown of proteins to amino acids, the deamination of the amino acids, and their subsequent metabolism in glycolysis, fermentation, or the Krebs cycle.

Other Metabolic Processes

Photoautotrophy

- **Photosynthesis** is the use of light energy to synthesize carbohydrates: (1) The **light reactions** can involve **cyclic photophosphorylation** or **photolysis** accompanied by **noncyclic photoreduction** of NADP; (2) the **dark reactions** involve the reduction of CO_2 to carbohydrate.

- Photosynthesis in cyanobacteria and algae provides a means of making nutrients, as it does in green plants; however, photosynthetic bacteria generally use some substances besides water to reduce carbon dioxide.

Photoheterotrophy

- **Photoheterotrophy** is the use of light as a source of energy. It requires organic compounds as sources of carbon.

Chemoautotrophy

- **Chemoautotrophs,** or **chemolithotrophs,** oxidize inorganic substances to obtain energy. Chemolithotrophs require only carbon dioxide as a carbon source.

The Uses of Energy

Biosynthetic Activities

- An **amphibolic pathway** is a metabolic pathway that can capture energy or synthesize substances needed by the cell.

- Figure 5.26 summarizes the intermediate products of energy-yielding metabolism and some of the building blocks for synthetic reactions that can be made from them.

- Bacteria synthesize a variety of cell wall polymers.

Membrane Transport and Movement

- Membrane transport uses energy derived from the ATP-producing electron transport system in the membrane to concentrate substances against a gradient. It occurs by active transport and by the **phosphotransferase system.**

- Movement in bacteria can be by flagella, by gliding or creeping, or by axial filaments.

Terminology Check

activation energy   *(p. 109)*
active site   *(p. 110)*
aerobe   *(p. 119)*
aerobic respiration   *(p. 119)*
alcoholic fermentation   *(p. 118)*
allosteric site   *(p. 112)*
amphibolic pathway   *(p. 131)*
anabolic pathway   *(p. 109)*
anabolism   *(p. 106)*
anaerobe   *(p. 119)*
apoenzyme   *(p. 110)*
autotroph   *(p. 107)*
autotrophy   *(p. 107)*
beta oxidation   *(p. 125)*
catabolic pathway   *(p. 109)*
catabolism   *(p. 106)*
chemical equilibrium   *(p. 114)*
chemiosmosis   *(p. 123)*
chemoautotroph   *(p. 107)*
chemoheterotroph   *(p. 108)*
citric acid cycle   *(p. 119)*
coenzyme   *(p. 110)*
cofactor   *(p. 111)*

competitive inhibitor   *(p. 112)*
cyclic photophosphorylation   *(p. 128)*
cytochrome   *(p. 122)*
dark reactions   *(p. 129)*
electron acceptor   *(p. 107)*
electron donor   *(p. 107)*
electron transport   *(p. 121)*
electron transport chain   *(p. 121)*
endoenzyme   *(p. 110)*
enzyme   *(p. 109)*
enzyme–substrate complex   *(p. 110)*
exoenzyme   *(p. 110)*
FAD   *(p. 112)*
feedback inhibition   *(p. 113)*
fermentation   *(p. 116)*
flavoprotein   *(p. 122)*
glycolysis   *(p. 114)*
heterotroph   *(p. 107)*
heterotrophy   *(p. 107)*
holoenzyme   *(p. 110)*
homolactic acid fermentation   *(p. 117)*
Krebs cycle   *(p. 119)*

light reactions   *(p. 128)*
metabolic pathway   *(p. 109)*
metabolism   *(p. 106)*
NAD   *(p. 110)*
noncompetitive inhibitor   *(p. 112)*
noncyclic photoreduction   *(p. 129)*
oxidation   *(p. 106)*
oxidative phosphorylation   *(p. 122)*
permease   *(p. 133)*
phosphorylation   *(p. 116)*
phosphotransferase system (PTS)   *(p. 133)*
photoautotroph   *(p. 107)*
photoheterotroph   *(p. 107)*
photolysis   *(p. 129)*
photosynthesis   *(p. 127)*
porin   *(p. 133)*
quinone   *(p. 123)*
reduction   *(p. 107)*
specificity   *(p. 110)*
substrate   *(p. 110)*
surfactant   *(p. 133)*
tricarboxylic acid (TCA) cycle   *(p. 119)*

See You on the Web

Don't forget I am expecting you on the web. The following is a preparedness checklist for your web explorations. If you have questions on any of these items, you can return to the end of Chapter 1 for more details.

- Questions We'll Explore
- Concept Check questions
- Flashcards
- Study Guide

If you are ready, come visit me at

http://www.prenhall.com/black

It will be a journey you'll never forget.

Growth and Culturing of Bacteria

I stood in awe near the geyser in Iceland, completely oblivious to the grandeur of my surroundings. The entire reason for my expedition undulated gently in the current of the runoff stream before me. The long, waving filaments of sulfur bacteria looked like long blond hair blowing in a light breeze. They were magnificent! Finally I was able to see with my own eyes, bacteria I had been reading about for years.

My excitement overcame me, and despite the warning steam billowing off the water, I plunged my hand into the water. I just wanted to find out what the strands felt like, but I guess I will never know. The near boiling water scalded my hand immediately. Later, as I nursed my blisters and wounded pride, I pondered the phenomena that allows the bacteria to thrive in an environment that is so hostile to most life forms (including me).

Background Basics

Questions We'll Explore

A How is growth defined in bacteria?

B How does cell division occur in microorganisms?

C What are the phases of growth in a bacterial culture?

D How is bacterial growth measured?

E How do physical factors affect bacterial growth?

F How do biochemical factors affect bacterial growth?

G What occurs in sporulation, and what is its significance?

H What methods are used to obtain a pure culture of an organism for study in the laboratory?

I How are different nutritional requirements supplied by various media?

In this chapter, we will use what we learned in Chapter 5 about energy in microorganisms to study how to grow them in the laboratory. Bacterial growth, which has been more thoroughly studied than growth in other microorganisms, is affected by a variety of physical and nutritional factors. Knowing how these factors influence growth is useful in culturing organisms in the laboratory and in preventing their growth in undesirable places. Furthermore, growing the microbes in pure cultures is essential in performing diagnostic tests that are used to identify a number of disease-causing organisms.

Growth and Cell Division

Microbial Growth Defined

In everyday language, growth refers to an increase in size. We are accustomed to seeing children, other animals, and plants grow. Unicellular organisms also grow, but as soon as a cell, called the **mother** (or *parent*) **cell,** has approximately doubled in size and duplicated its contents, it divides into two **daughter cells.** Then the daughter cells grow, and subsequently they also divide. Because individual cells grow larger only to divide into two new individuals, **microbial growth** is defined not in terms of cell size but as the increase in the number of cells, which occurs by cell division.

Cell Division

Cell division in bacteria, unlike cell division in eukaryotes, usually occurs by *binary fission* or sometimes by *budding*. In **binary fission,** a cell duplicates its components and divides into two cells (➤ Figure 6.1a). The daughter cells become independent when a *septum* (partition) grows between them and they separate (➤ Figure 6.1b). Unlike eukaryotic cells, prokaryotic cells do not have a cell cycle with a specific period

Each square centimeter of skin hosts an average of 100,000 organisms. Bacteria reproduce so quickly, their population is restored within hours of washing.

➤ Figure 6.1   **Binary fission.** (a) The stages of binary fission in a bacterial cell. (b) A colorized TEM of a thin section of the bacterium *Staphylococcus* undergoing binary fission. (c) The nucleoid of a bacterial cell.

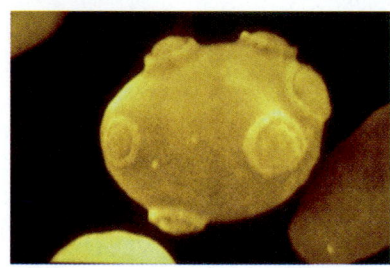

➤ Figure 6.2   **Budding in yeast.**

of DNA synthesis. Instead, in continuously dividing cells, DNA synthesis also is continuous and replicates the single bacterial chromosome shortly before the cell divides. The chromosome is attached to the cell membrane, which grows and separates the replicated chromosomes. Replication of the chromosome is completed before cell division, when the cell may temporarily contain two or more nucleoids. In some species, incomplete separation of the cells produces linear chains (linked bacilli), **tetrads** (cuboidal groups of four cocci), **sarcinae** (singular: *sarcina;* groups of eight cocci in a cubical packet), or grapelike clusters (staphylococci) ➡ (➤ Figure 4.2). Some bacilli always form chains or filaments; others form them only under unfavorable growth conditions. Streptococci form chains when grown on artificial media but exist as single or paired cells when isolated from a rapidly growing lesion in an infected human host.

Cell division in yeast and a few bacteria occurs through **budding.** In that process, a small, new cell develops from the surface of an existing cell and subsequently separates from the parent cell (➤ Figure 6.2).

Phases of Growth

Consider a population of organisms introduced into a fresh, nutrient-rich **medium** (plural: *media*), a mixture of substances on or in which microorganisms grow. Such organisms display four major phases of growth: (1) the lag phase, (2) the log (logarithmic) phase, (3) the stationary phase, and (4) the decline phase, or death phase. These phases form the **standard bacterial growth curve** (➤ Figure 6.3).

The Lag Phase

In the **lag phase,** the organisms do not increase significantly in number, but they are metabolically active—growing in size, synthesizing enzymes, and incorporating various molecules from the medium. During this phase the individual organisms increase in size, and they produce large quantities of energy in the form of ATP.

The length of the lag phase is determined in part by characteristics of the bacterial species and in part by conditions in the media—both the medium from which the organisms are taken and the one to which they are transferred. Some species adapt to the new medium in an hour or two; others take several days. Organisms from old cultures, adapted to limited nutrients and

➤ Figure 6.3   **A standard bacterial growth curve.**

large accumulations of wastes, take longer to adjust to a new medium than do those transferred from a relatively fresh, nutrient-rich medium.

The Log Phase

Once organisms have adapted to a medium, population growth occurs at an **exponential,** or **logarithmic (log), rate.** When the scale of the vertical axis is logarithmic, growth in this **log phase** appears on a graph as a straight diagonal line, which represents the size of the bacterial population. (On the base-10 logarithmic scale, each successive unit represents a 10-fold increase in the number of organisms.) ➥ (Appendix A) During the log phase, the organisms divide at their most rapid rate—a regular, genetically determined interval called the **generation time.** The population of organisms doubles in each generation time. For example, a culture containing 1000 organisms per milliliter with a generation time of 20 minutes would contain 2000 organisms per milliliter after 20 minutes, 4000 organisms after 40 minutes, 8000 after 1 hour, 64,000 after 2 hours, and 512,000 after 3 hours. Such growth is said to be *exponential,* or *logarithmic.*

The generation time for most bacteria is between 20 minutes and 20 hours and is typically less than 1 hour. Some bacteria, such as those that cause tuberculosis and leprosy, have much longer generation times. Some organisms take slightly longer than others to go from the lag phase to the log phase, and they do not all divide precisely together. If they divided together and the generation time was exactly 20 minutes, the number of cells in a culture would increase in a stair-step pattern, exactly doubling every 20 minutes—a hypothetical situation called **synchronous growth.** In an actual culture, each cell divides sometime during the 20-minute generation time, with about 1/20 of the cells dividing each minute—a natural situation called **nonsynchronous growth.** Nonsynchronous growth appears as a straight line, not as steps, on a logarithmic graph (➤ Figure 6.4).

Organisms in a tube of culture medium can maintain logarithmic growth for only a limited period of time. As the number of organisms increases, nutrients are used up, metabolic wastes accumulate, living space may become limited, and aerobes suffer from oxygen depletion. Generally, the limiting factor for logarithmic growth seems to be the rate at which energy can be produced in the form of ATP. As the availability of nutrients decreases, the cells become less able to generate ATP, and their growth rate decreases. The decrease in growth rate is shown in ➤ Figure 6.3 by a gradual leveling off of the growth curve (the curved segment to the right of the log phase).

> Under ideal conditions, one bacterium can multiply to 2,097,152 within 7 hours.

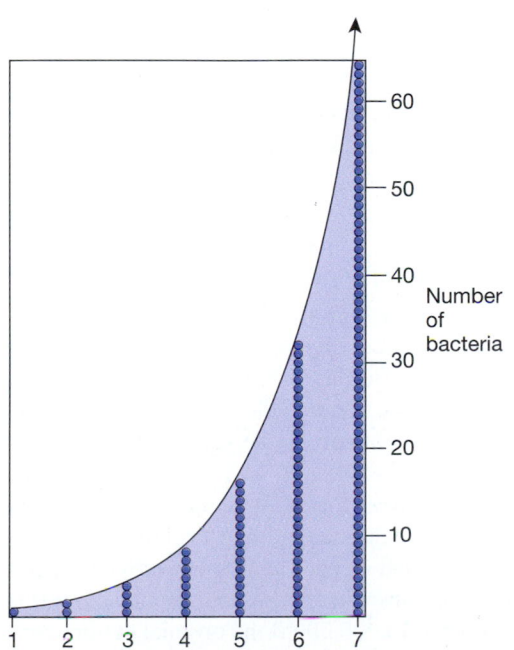

➤ **Figure 6.4 Nonsynchronous growth.** A growth curve for an exponentially increasing population, plotted logarithmically (dashed line) and arithmetically (solid line).

➤ Figure 6.5   **Microbes growing in a chemostat.** A chemostat constantly renews nutrients in a culture, making it possible to grow organisms continuously in the log phase.

Leveling off of growth is followed by the stationary phase unless fresh medium is added or organisms are transferred to fresh medium. Logarithmic growth can be maintained by a device, much like a thermostat, called a **chemostat** (➤ Figure 6.5), which has a growth chamber and a reservoir from which fresh medium is continuously added to the growth chamber as old medium is withdrawn. Alternatively, organisms from a culture in the stationary phase can be transferred to a fresh medium. After a brief lag phase, such organisms quickly reenter the log phase of growth.

The Stationary Phase

When cell division decreases to the point that new cells are produced at the same rate as old cells die, the number of live cells stays constant. The culture is then in the **stationary phase,** represented by a horizontal straight line in ➤ Figure 6.3. The medium contains a limited amount of nutrients and may contain toxic quantities of waste materials. Also, the oxygen supply may become inadequate for aerobic organisms, and damaging pH changes may occur.

The Decline (Death) Phase

As conditions in the medium become less and less supportive of cell division, many cells lose their ability to divide, and thus the cells die. In this **decline phase,** or **death phase,** the number of live cells decreases at a logarithmic rate, as indicated by the straight, downward-sloping diagonal line in ➤ Figure 6.3. During the decline phase, many cells undergo *involution*—that is, they assume a variety of unusual shapes, which makes them difficult to identify. In cultures of spore-forming organisms, more spores than vegetative (metabolically active) cells survive. The duration of this phase is as highly variable as the duration of the logarithmic growth phase. Both depend primarily on the genetic characteristics of the organism. Cultures of some bacteria go through all growth phases and die in a few days; others contain a few live organisms after months or even years.

Growth in Colonies

Growth phases are displayed in different ways in colonies growing on a solid medium. Typically, a cell divides exponentially, forming a small **colony**—all the descendants of the original cell. The colony grows rapidly at its edges; cells nearer the center grow more slowly or begin to die because they have smaller quantities of available nutrients and are exposed to more toxic waste products. All phases of the growth curve occur simultaneously in a colony.

Measuring Bacterial Growth

Bacterial growth is measured by estimating the number of cells that have arisen by binary fission during a growth phase. This measurement is expressed as the number of *viable* (living) organisms per milliliter of culture. Several methods of measuring bacterial growth are available.

Serial Dilution and Standard Plate Counts

One method of measuring bacterial growth is the *standard plate count*. This technique relies on the fact that under proper conditions, only a living bacterium will divide and form a visible colony on an agar plate. An *agar plate* is a nutrient medium solidified with **agar,** a complex polysaccharide extracted from certain marine algae. Because it is difficult to count more than 300 colonies on one agar plate, it is usually necessary to dilute the original bacterial culture before you plate (transfer) a known volume of the culture onto the solid plate. *Serial dilutions* accomplish this purpose.

To make **serial dilutions** (➤ Figure 6.6), you start with organisms in liquid medium. Adding 1 ml of this medium to 9 ml of sterile water makes a 1:10 dilution; adding 1 ml of the 1:10 dilution to 9 ml of sterile water makes a 1:100 dilution; and so on. The number of bacteria per milliliter of fluid is reduced by 9/10 in each dilution. Subsequent dilutions are made in ratios of 1:1000, 1:10,000, 1:100,000, 1:1,000,000, or even 1:10,000,000 if the original culture contained an extremely large number of organisms.

1 ml   1 ml   1 ml   1 ml

| 50,000 bacteria/ml (approximate number) | 9 ml water | 9 ml water | 9 ml water | 9 ml water |

1:10 dilution   1:100 dilution   1:1000 dilution   1:10,000 dilution

5,000 bacteria/ml   500 bacteria/ml   50 bacteria/ml   5 bacteria/ml

➤ **Figure 6.6   Serial dilution.** One milliliter is taken from a broth culture and added to 9 ml of sterile water, thereby diluting the culture by a factor of 10. This procedure is repeated until the desired concentration is reached.

From each dilution, usually beginning with the 1:100, 1 ml of the culture is transferred to an agar plate. (One milliliter of the 1:10 dilution typically contains too many organisms to yield countable colonies when transferred to a Petri plate.) The transfer can be done by either the pour plate method or the spread plate method (➤ Figure 6.7). A **pour plate** is made by first adding 1.0 ml of a diluted culture from a serial dilution to 9 ml of melted nutrient agar. After the medium is mixed, it is poured into an empty Petri plate. Once the agar medium cools, solidifies, and is incubated, colonies will develop both within the medium and on its surface. Cells suspended in the melted agar during preparation may be heat-damaged, and then they will not form colonies. Those that do grow inside the agar will form smaller colonies than those growing on the surface. The **spread plate method** eliminates such problems because all cells remain on the surface of the solid medium. The diluted sample is first placed on the center of a solid, cooled agar medium. The sample is then spread evenly over the medium's surface with a sterile, bent glass rod. After incubation, colonies develop on the agar surface.

Wherever a single living bacterium is deposited on an agar plate, it will divide to form a colony. Each bacterium represents a *colony-forming unit* (*CFU*). One or more plates should have a small enough number of colonies such that each one is clearly distinguishable and can be counted. If you have made dilutions properly, you should get plates with a **countable number** of colonies (30 to 300 per plate).

To count the actual number of colonies present, you would place the plate under the magnifying lens of a *colony counter* (➤ Figure 6.8), and colonies on the entire plate are counted. To determine the number of colony-forming units in the original culture, multiply the number of colonies found on a plate by the *dilution factor;* if it is a fraction, use the denominator. A dilution factor of 1000 would be expressed as 1:1000 or 1/1000, and a dilution factor of 10,000 would be expressed as 1:10,000. A typical calculation for an average colony count of 81 produced by plating a 1/100,000 dilution (dilution factor = 100,000 would be as follows:

$$81 \times 100,000 = 8,100,000, \text{ or } 8.1 \times 10^6 \text{ CFU/ml}$$

The accuracy of the serial dilution and plate count method depends on homogeneous dispersal of organisms in each dilution. Error can be minimized by shaking each culture before sampling and making several plates from each dilution. Accuracy is also affected by the death of cells. Because the number of colonies counted represents the number of living organisms, it does not include organisms that may have died by the time plating was done; nor does it include organisms that cannot grow on the chosen medium. Using young cultures in the log phase of growth minimizes this kind of error.

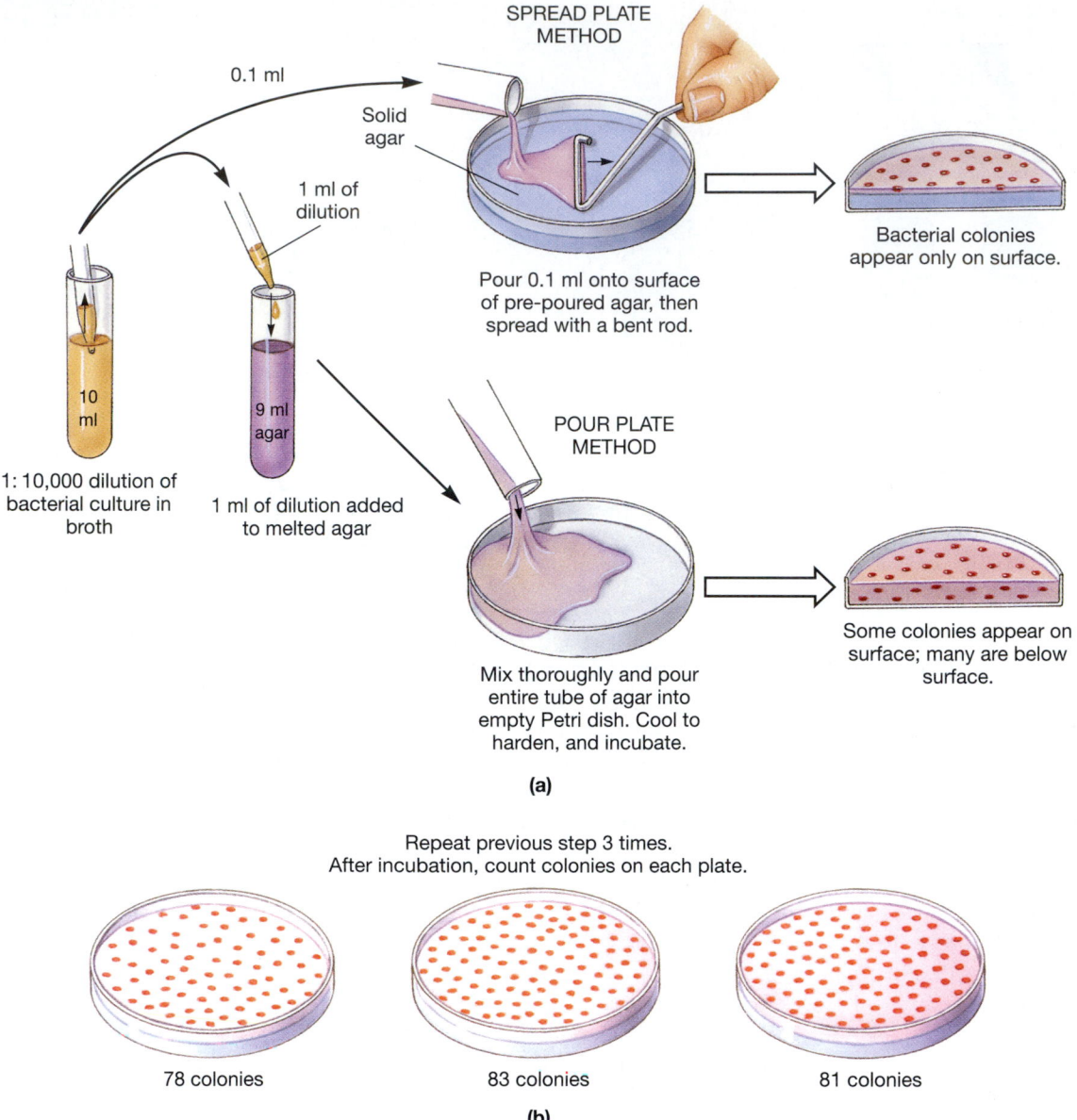

SPREAD PLATE
METHOD

0.1 ml

Solid
agar

1 ml of
dilution

Pour 0.1 ml onto surface
of pre-poured agar, then
spread with a bent rod.

Bacterial colonies
appear only on surface.

10
ml

9 ml
agar

1:10,000 dilution of
bacterial culture in
broth

1 ml of dilution added
to melted agar

POUR PLATE
METHOD

Mix thoroughly and pour
entire tube of agar into
empty Petri dish. Cool to
harden, and incubate.

Some colonies appear on
surface; many are below
surface.

(a)

Repeat previous step 3 times.
After incubation, count colonies on each plate.

78 colonies   83 colonies   81 colonies

(b)

> Figure 6.7   **Calculation of the number of bacteria per milliliter of culture using serial dilution.** (a) One milliliter of a 1:10,000 dilution is mixed with 9 ml of melted agar, which is warm enough to stay liquid but not hot enough to kill the organisms being mixed into it. After thorough mixing, the warm agar is quickly poured into an empty, sterile Petri dish (by the pour plate method). When cooled to hardness, it is incubated. Alternatively, 0.1 ml of a 1:10,000 dilution is poured onto a surface of pre-poured agar and then spread with a bent rod (by the spread plate method). Next it is incubated. (b) The colonies that develop are counted. A single measurement is not very reliable, so the procedure is repeated at least three times, and the results are averaged. The average number of colonies is multiplied by the dilution factor to ascertain the total number of organisms per milliliter of the original culture.

Direct Microscopic Counts

Bacterial growth can be measured by **direct microscopic counts.** In this method a known volume of medium is introduced into a specially calibrated etched glass slide called a *Petroff-Hausser counting chamber* (➤ Figure 6.9). A bacterial suspension is introduced onto the chamber with a calibrated pipette. After the bacteria settle and the liquid currents have slowed, the microorganisms are counted in specific calibrated areas. Their number per unit volume of the original suspension is calculated by using an appropriate formula. The number of bacteria per milliliter of medium can be estimated with a reasonable degree of accuracy. The accuracy of direct microscopic counts depends on the presence of more than 10 million bacteria per milliliter of culture. This is because counting chambers are designed to allow accurate counts only when large numbers of cells are present. An accurate count also requires that the bacteria be homogeneously distributed throughout the culture. This technique has the disadvantage of generally not distinguishing between living and dead cells.

Most Probable Number

When samples contain too few organisms to give reliable measures of population size by the standard plate count method, as in food and water sanitation studies, or when organisms will not grow on agar, the **most probable number (MPN)** method is used. With this method, the technician observes the sample, estimates the number of cells in it, and makes a series of progressively greater dilutions. As the dilution factor increases, a point will be reached at which some tubes will contain a single organism and others, none. A typical MPN test consists of five tubes each of three volumes (10, 1, and 0.1 ml) of a dilution (➤ Figure 6.10). Those that contain an organism will display growth by producing gas bubbles, and or by becoming cloudy, when incubated. The number of organisms in the original culture is estimated from a most probable number table. The values in the table, which are based on statistical probabilities, specify that the number of organisms in the original culture has a 95 percent chance of falling within a particular range. A complete MPN table is given in Appendix A. The more tubes that show growth, especially at greater dilutions, the more organisms were present in the sample.

Filtration

Another method of estimating the size of small bacterial populations makes use of **filtration.** A known volume of water or air is drawn through a filter with pores too small to allow passage of bacteria. When the filter is placed on a solid medium, each colony that grows represents originally one organism. Thus, the number of organisms per liter of water or air can be calculated. (➤ Figure 25.18 shows the filtration process and colonies grown on a filter pad.)

Other Methods

Several other methods of monitoring bacterial growth are available. They include simple observation with or without special instruments, measurement of metabolic products by the detection of gas or acid production, and determination of dry weight of cells.

Turbidity (a cloudy appearance) in a culture tube indicates the presence of organisms (➤ Figure 6.11). Fairly accurate estimates of growth can be obtained by measuring turbidity

(a)

(b)

(c)

➤ **Figure 6.8   Counting colonies.** (a) Using a bacterial colony counter. (b) Bacterial colonies viewed through the magnifying glass against a colony-counting grid. The plate was produced by the pour plate method. How many different colony types can you identify on this plate? (c) Which of these plates would be the correct one to count? Why?

Cover slip

Slide with shallow wells and inscribed grid

Cover slip

Bacterial cells in squares of grid are counted

Bacterial suspension is placed on slide and seeps under cover slip. Suspension fills shallow space of known volume over grid.

➤ Figure 6.9   **The Petroff-Hausser counting chamber.** The volume of suspension filling the narrow space between the grid and the cover slide is known, so the number of bacteria per unit of volume can be calculated.

Volume of Dilution Added	Culture Results					Number of Positive Tubes
10 ml	+	+	+	+	+	5
1 ml	+	−	−	−	+	2
0.1 ml	−	−	−	−	−	0

➤ Figure 6.10   **A most probable number (MPN) test.** Those tubes in which gas bubbles are visible (labeled +) contain organisms.

with a photoelectric device, such as a *colorimeter* or a *spectrophotometer* (➤ Figure 6.12). This method is particularly useful in monitoring the rate of growth without disturbing the culture. Samples with very high cell densities, however, must be diluted to ensure accurate readings. Measures of bacterial growth based on turbidity are likewise especially subject to error when cultures contain fewer than 1 million cells per milliliter. Such cultures can display little or no turbidity even when growth is occurring. Conversely, turbidity can be produced by a high concentration of dead cells in a culture.

Measuring the metabolic products of a population can be used to estimate bacterial growth indirectly. The rate at which metabolic products such as gases and/or acids are formed by a culture reflects the mass of bacteria present. Gas production can be detected (rather than measured) by capturing the gas in small inverted tubes placed inside larger tubes of liquid medium containing bacteria. Acid production can be detected by incorporating *pH indicators*—chemical substances that change color with changes in pH—in a liquid medium containing metabolically active bacteria.

The rate at which a substrate such as glucose or oxygen is used up also reflects cell mass. For example, one method for estimating bacterial mass is the *dye reduction test,* which measures the direct or indirect uptake of oxygen. In this test, a dye such as methylene blue is incorporated into a medium containing milk. Bacteria inoculated into the medium use oxygen as they metabolize the milk. Methylene blue is blue in the presence of oxygen and turns colorless in its absence. Thus, the faster the medium loses color, the faster the oxygen is being used up, and the more bacteria are presumed to be present. The rate at which the dye is decolorized (dye reduction) is a highly indirect approach; it is not an accurate measure of bacterial mass.

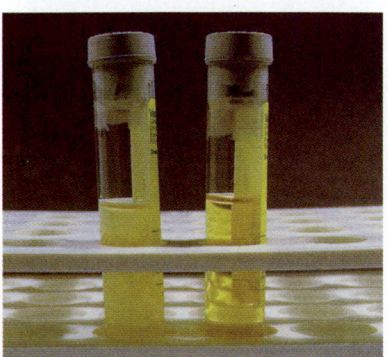

➤ Figure 6.11   **Turbidity.** Turbidity, or a cloudy appearance, is an indicator of bacterial growth in urine in the tube on the left.

Finally, the number of cells in a culture can be determined by *dry weight measurements*. To calculate the dry weight of cells, they must be separated from the medium by some physical means such as filtration or centrifugation. The cells are then dried, and the resulting mass is weighed.

Factors Affecting Bacterial Growth

Microorganisms are found in nearly every environment on earth, including environments in which no other life forms can survive. Microbes can exist in a great many environments because they are small and easily dispersed, occupy little space, need only small quantities of nutrients, and are remarkably diverse in their nutritional requirements. They also have great capacity for adapting to environmental changes. For almost any substance, there is some microbe that can metabolize it as a nutrient; for almost any environmental change, there is some microbe that can survive the change.

As warmblooded, air-breathing, land-dwelling mammals, we tend to forget that 72 percent of our planet's surface is water, that 90 percent of that water is salt water, and that environments containing living organisms have an average temperature of about 5°C. Unlike humans, microorganisms live mostly in water, and many are adapted to temperatures above or below those we consider optimum. The organisms of particular interest in the health sciences account for only a fraction of all microorganisms—those that have adapted to conditions found in or on the human body.

Different species of microorganisms can grow in a wide range of environments—from highly acidic to somewhat alkaline conditions, from Antarctic ice to hot springs, in pure spring water or in salty marshes, in oceans with or without oxygen, and even under great pressure and in boiling steam vents on the ocean floor. Microorganisms use a variety of substances to obtain energy, and some require special nutrients.

The kinds of organisms found in a given environment and the rates at which they grow can be influenced by a variety of factors, both physical and biochemical. **Physical factors** include pH, temperature, oxygen concentration, moisture, hydrostatic pressure, osmotic pressure, and radiation. **Nutritional** (*biochemical*) **factors** include availability of carbon, nitrogen, sulfur, phosphorus, trace elements, and, in some cases, vitamins.

Physical Factors

pH

Remember that the acidity or alkalinity of a medium is expressed in terms of pH. ➡ (Chapter 2, p. 33) Although the pH scale is now widely used in chemistry, it was invented by the Danish chemist Søren Sørenson to describe the limits of growth of microorganisms in various media. Microorganisms have an **optimum pH**—the pH at which they grow best. Their optimum pH is usually near neutrality (pH 7). Most microbes do not grow at a pH more than 1 pH unit above or below their optimum pH.

✓ What are the differences between the lag phase and he log phase of a bacterial growth curve?

✓ How does logarithmic rate of increase differ from arithmetic rate of increase? The following sequence of number of cells is an example of which type of rate of increase? 1, 2, 4, 8, 16, 32.

✓ If a broth culture of bacteria, initially containing 37,000 bacteria/ml, is diluted to a 1:1000 dilution, how many bacteria/ml of the diluted broth would be present on average?

✓ Why does a direct microscopic count of bacteria using a Petroff-Hausser chamber not give you a viable count? How does this differ from the spread plate and pour plate methods?

The total weight for the number of bacteria that live in the soil and underground is estimated to be 10034 trillion tons.

According to their tolerance for acidity or alkalinity, bacteria are classified as:

- acidophiles,
- neutrophiles, or
- alkaliphiles.

However, no single species can tolerate the full pH range of any of these categories, and many tolerate a pH range that overlaps two categories. **Acidophiles** (a-sid′o-filz), or acid-loving organisms, grow best at a pH of 0.1 to 5.4. *Lactobacillus,* which produces lactic acid, is an acidophile, but it tolerates only mild acidity. Some bacteria that oxidize sulfur to sulfuric acid, however, can create and tolerate conditions as low as pH 1. **Neutrophiles** (nu′tro-filz) exist from pH 5.4 to 8.5. Most of the bacteria that cause disease in humans are neutrophiles. **Alkaliphiles** (al′kah-li-filz), or alkali-loving (base-loving) organisms, exist from pH 7.0 to 11.5. *Vibrio cholerae,* the causative agent of the disease Asiatic cholera, grows best at a pH of about 9. *Alcaligenes faecalis,* which sometimes infects humans already weakened by another disease, can create and tolerate alkaline conditions of pH 9 or higher. The soil bacterium *Agrobacterium* grows in alkaline soil of pH 12.

The effects of pH on organisms can in part be related to the concentration of organic acids in the medium and to the protection that bacterial cell walls sometimes provide. *Lactobacillus* and other organisms that produce organic acids during fermentation inhibit their own growth as acids such as lactic acid and pyruvic acid accumulate in the medium. It appears that the acids themselves rather than the hydrogen ions per se inhibit growth. Changes in pH can lead to denaturing of enzymes and other proteins and can interfere with pumping of ions at the cell membrane. Other organisms have relatively impervious cell walls that prevent the cell membrane from being exposed to an extreme pH in the medium. These organisms appear to tolerate environmental acidity or alkalinity because the cell itself is maintained at a nearly neutral pH.

Many bacteria often produce sufficient quantities of acids as metabolic byproducts that eventually interfere with their own growth. To prevent this situation in the laboratory cultivation of bacteria, *buffers* are incorporated into growth media to maintain the proper pH levels. Phosphate salts are commonly used for this purpose.

Temperature

Most species of bacteria can grow over a 30°C temperature range, but the minimum and maximum temperatures for different species vary considerably. Seawater remains liquid below 0°C, and organisms living there can tolerate below-freezing temperatures. According to their growth temperature range, bacteria can be classified as:

- psychrophiles,
- mesophiles, or
- thermophiles.

Most bacteria, however, do not tolerate the whole temperature range of a category, and some tolerate a range that overlaps categories. Within these groups bacteria are further classified as obligate or facultative. **Obligate** means that the organism *must* have the specified environmental condition. **Facultative** means that the organism is *able* to adjust to and tolerate the environmental condition, but it can also live in other conditions.

Psychrophiles (si′kro-filz), or cold-loving organisms, grow best at temperatures of 15° to 20°C, although some live quite well at 0°C. They can be further divided into **obligate psychrophiles,** such as *Bacillus globisporus,* which cannot grow above 20°C, and **facultative psychrophiles,** such as *Xanthomonas pharmicola,* which grows best below 20°C but also can grow above that temperature. Psychrophiles live mostly in cold water and soil. None can live in the human body, but some are known to cause spoilage of refrigerated foods.

Mesophiles (mes′o-filz), which include most bacteria, grow best at temperatures between 25° and 40°C. Human pathogens are included in this category, and most of them grow best near human body temperature (37°C). *Thermoduric* organisms ordinarily live as mesophiles but can withstand short periods of exposure to high temperatures. Inadequate heating during canning or in pasteurization may leave such organisms alive and therefore able to spoil food.

Thermophiles (therm'o-filz), or heat-loving organisms, grow best at temperatures from 50° to 60°C. Many are found in compost heaps, and a few tolerate temperatures as high as 110°C in boiling hot springs. They can be further classified as **obligate thermophiles,** which can grow only at temperatures above 37°C, or **facultative thermophiles,** which can grow both above and below 37°C. *Bacillus stearothermophilus,* which usually is considered an obligate thermophile, grows at its maximum rate at 65° to 75°C but can display minimal growth and cause food spoilage at temperatures as low as 30°C. Thermophilic sulfur bacteria display zones of optimum growth temperatures in the runoff troughs of geysers (➤ Figure 6.13). Different species collect at various locations along the trough. The most heat-tolerant are near the geyser, and those with lesser heat tolerance are distributed in regions where the water has cooled to their optimum temperature. In deep channels the most heat-tolerant species are found at the greatest depths and the least heat-tolerant near the surface, where the water has cooled. Under laboratory conditions that use high pressure to increase water temperature above 100°C, archaeobacteria from deep-sea vents have grown at 115°C (238°F). (Chapter 9 provides more information about these remarkable organisms.)

The temperature range over which an organism grows is determined largely by the temperatures at which its enzymes function. Within this temperature range, three critical temperatures can be identified:

1. The *minimum growth temperature,* the lowest temperature at which cells can divide;

2. The *maximum growth temperature,* the highest temperature at which cells can divide;

3. The *optimum growth temperature,* the temperature at which cells divide most rapidly—that is, have the shortest generation time.

Regardless of the type of bacteria, growth gradually increases from the minimum to the optimum temperature and decreases very sharply from the optimum to the maximum temperature. Furthermore, the optimum temperature is often very near the maximum temperature (➤ Figure 6.14). These growth properties are due to changes in enzyme activity. ➡ (Chapter 5, p. 113) Enzyme activity generally doubles for every 10°C rise in temperature until the high temperature begins to denature all proteins, including enzymes. The sharp decrease in enzyme activity at a temperature only slightly higher than the optimum temperature occurs as enzyme molecules become so distorted by denaturation that they cannot catalyze reactions.

Temperature is important not only in providing conditions for microbial growth, but also in preventing such growth. The refrigeration of food, usually at 4°C, reduces the growth of psychrophiles and prevents the growth of most other bacteria. However, food and other materials, such as blood, can support growth of some bacteria even when refrigerated. For this reason, materials that can withstand freezing are stored at temperatures of −30°C if they are to be kept for long periods of time. High temperatures also can be used to prevent bacterial growth (Chapter 12). Laboratory equipment and media are generally sterilized with heat, and food is frequently preserved by heating and storing in closed containers. Bacteria are more apt to survive

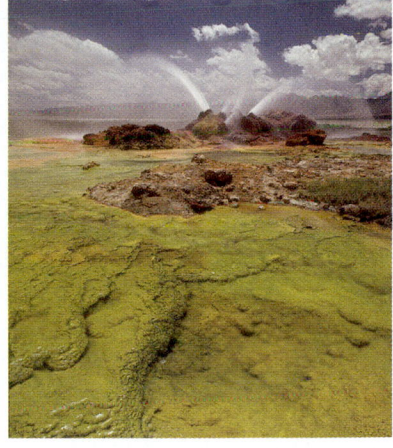

➤ **Figure 6.13   Thermophiles.** Geyser Hot Springs, Black Rock Desert, Nevada. Thermophilic sulfur bacteria can live and grow in the runoff waters from such geysers despite the near-boiling temperatures.

In the laboratory, bacteria isolated from underwater superheated sulfide mounds must be grown at 480°F and at 265 ATM.

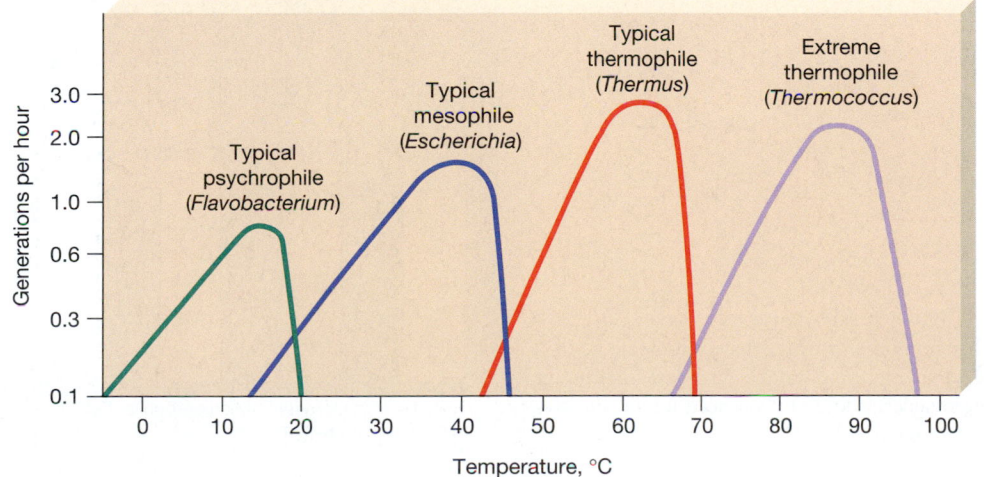

➤ **Figure 6.14   Growth rates of psychrophilic, mesophilic, and thermophilic bacteria.** Notice the overlap of temperature ranges at which these organisms can survive. Growth rates are much lower at the extreme ends of the ranges.

extremes of cold than extremes of heat; enzymes are not denatured by chilling but can be permanently denatured by heat.

Oxygen

Bacteria, especially heterotrophs, can be divided into aerobes, which require oxygen to grow, and anaerobes, which do not require it. ➥ (Chapter 5, p. 119) Among the aerobes, cultures of rapidly dividing cells require more oxygen than do cultures of slowly dividing cells. **Obligate aerobes,** such as *Pseudomonas,* which is a common cause of hospital-acquired infections, must have free oxygen for aerobic respiration, whereas **obligate anaerobes,** such as *Bacteroides* are killed by free oxygen. In a culture tube containing nutrient broth, obligate aerobes grow near the surface, where atmospheric oxygen diffuses into the medium; obligate anaerobes grow near the bottom of the tube, where little or no free oxygen reaches them (➤ Figure 6.15).

For aerobes, oxygen is often the environmental factor that limits growth rate. Oxygen is poorly soluble in water, and a variety of methods are sometimes employed to maintain a high O_2 concentration in cultures, including vigorous mixing or forced aeration by bubbling air through a culture, as is done in a fish tank. This is especially important in such commercial processes as the production of antibiotics and in sewage treatment.

Between the extremes of obligate aerobes and obligate anaerobes are the *microaerophiles,* the *facultative anaerobes,* and the *aerotolerant anaerobes.* **Microaerophiles** (mi″kro-aer′o-filz) appear to grow best in the presence of a small amount of free oxygen. They grow below the surface of the medium in a culture tube at the level where oxygen availability matches their needs. Microaerophiles such as *Campylobacter,* which can cause intestinal disorders, also are **capnophiles,** or carbon dioxide–loving organisms. They thrive under conditions of low oxygen and high carbon dioxide concentration. **Facultative anaerobes** ordinarily carry on aerobic metabolism when oxygen is present, but they shift to anaerobic metabolism when oxygen is absent. *Staphylococcus* and *Escherichia coli* are facultative anaerobes; they often are found in the intestinal and urinary tracts, where only a small amount of oxygen is available. **Aerotolerant anaerobes** can survive in the presence of oxygen but do not use it in their metabolism. *Lactobacillus,* for example, always captures energy by fermentation, regardless of whether the environment contains oxygen.

Compared with other groups of organisms defined according to oxygen requirements, facultative anaerobes have the most complex enzyme systems. They have one set of enzymes that enables them to use oxygen as an electron acceptor and another set that enables them to use another electron acceptor when oxygen is not available. In contrast, the enzymes of the other groups defined here are limited to either aerobic or anaerobic respiration.

Obligate anaerobes are killed not by gaseous oxygen but by a highly reactive and toxic form of oxygen called **superoxide** (O_2^-). Superoxide is formed by certain oxidative enzymes and is

➤ Figure 6.15 **Patterns of oxygen use.** Different organisms incubated for 24 hours in tubes of a nutrient broth accumulate in different regions depending on their need for, or sensitivity to, oxygen.

Bacterial growth

Obligate aerobe · Obligate anaerobe · Microaerophile · Facultative anaerobe converted to molecular oxygen (O_2) and toxic hydrogen peroxide (H_2O_2) by an enzyme called **superoxide dismutase**. Hydrogen peroxide is converted to water and molecular oxygen by the enzyme **catalase.** Obligate aerobes and most facultative anaerobes have both enzymes. Some facultative and aerotolerant anaerobes have superoxide dismutase but lack catalase. Most obligate anaerobes lack both enzymes and succumb to the toxic effects of superoxide and hydrogen peroxide.

Moisture

All actively metabolizing cells generally require a water environment. Unlike larger organisms that have protective coverings and internal fluid environments, single-celled organisms are exposed directly to their environment. Most vegetative cells can live only a few hours without moisture; only the spores of spore-forming organisms can exist in a dormant state in a dry environment.

Hydrostatic Pressure

Water in oceans and lakes exerts **hydrostatic pressure,** pressure exerted by standing water, in proportion to its depth. Such pressure doubles with every 10 m increase in depth. For example, in a lake 50 m deep, the pressure is 32 times the atmospheric pressure. Some ocean valleys have depths in excess of 7000 m, and certain bacteria are the only organisms known to survive the extreme pressure at such depths. Bacteria that live at high pressures, but die if left in the laboratory for only a few hours at standard atmospheric pressure, are called **barophiles.** It appears that their membranes and enzymes do not simply tolerate pressure but require pressure to function properly. The high pressure is necessary to keep their enzyme molecules in the proper three-dimensional configuration. Without it, the enzymes lose their shape and denature, and the organisms die.

Osmotic Pressure

We saw in Chapter 4 that the membranes of all microorganisms are selectively permeable. The cell membrane allows water to move by osmosis between the cytoplasm and the environment. ➥ (➢ Figure 4.31) Environments that contain dissolved substances exert osmotic pressure, and the pressure can exceed that exerted by dissolved substances in cells. Cells in such *hyperosmotic* environments lose water and undergo **plasmolysis** (plas-mol′e-sis), or shrinking of the cell. In microorganisms with a cell wall, the cell or plasma membrane separates from the cell wall. Conversely, cells in distilled water have a higher osmotic pressure than their environment and, therefore, gain water. In bacteria, the rigid cell wall prevents cells from swelling and bursting, but the cells fill with water and become *turgid* (distended).

   Most bacterial cells can tolerate a fairly wide range of concentrations of dissolved substances. Their cell membranes contain transport systems that regulate the movement of dissolved

When the Going Gets Tough, Hide inside a Rock

A few bacteria live in bitterly cold, dry valleys of Antarctica where very few other organisms can survive. The relative humidity is so low that water passes directly from the frozen to the vapor state and is rarely found as a liquid. Organisms that live there nevertheless manage to carry out their metabolic activities, either by using water vapor or by melting tiny amounts of ice with their metabolic heat. But they cannot survive the harsh conditions of the Antarctic atmosphere. The bacteria must hide inside translucent rocks (such as quartz, feldspar, and certain marbles), which allow the sun's rays to penetrate so these bacteria can carry out photo-synthesis. Because they do not produce mineral-dissolving chemicals, these endolithic organisms must colonize only porous rocks. They are usually able to invade several millimeters into the rock, where they find safe refuge until the rock is eroded by wind.

   Mars was originally a warm planet, but it cooled down when it lost its atmosphere. If life had evolved on Mars during its warm phase, would that life have sought shelter inside surface rocks? Examination of Martian meteorites reveals evidence of possible early life forms resembling bacteria entombed in what may have been their last refuge.

In the Pink

The Great Salt Lake is an extremely salty environment that supports bacterial growth. The Great Salt Lake, nearly 10 times saltier than ocean water, supports many varieties of halobacteria. Because all halobacteria lack peptidoglycan in their cell wall, they are Gram-negative. In addition, they are insensitive to most antibiotics, contain unusually large plasmids, and are obligate aerobes. The halobacteria require large amounts of sodium for growth, a requirement that is not satisfied when a similar ion is used in its place. Certain species of the extreme halophiles use a light-mediated mechanism to produce ATP. Unlike green plants, the pigments they use for light-dependent ATP synthesis are the red-orange carotinoids and the red-purple bacterioruberins and bacteriorhodopsins. The brilliant color of these bacteria can be seen when high salinity lakes and holding ponds are photographed from the air—giving the distinct appearance of a pink patchwork quilt.

substances across the membrane. ➡ (Chapter 5, p. 133) Yet, if concentrations outside the cells become too high, water loss can inhibit growth or even kill the cells.

The use of salt as a preservative in curing hams and bacon and in making pickles is based on the fact that high concentrations of dissolved substances exert sufficient osmotic pressure to kill or inhibit microbial growth. The use of sugar as a preservative in making jellies and jams is based on the same principle.

Bacteria called **halophiles** (hal′o-filz), or salt-loving organisms, require moderate to large quantities of salt (sodium chloride). Their membrane transport systems actively transport sodium ions out of the cells and concentrate potassium ions inside them. Two possible explanations for why halophiles require sodium have been proposed. One is that the cells need sodium to maintain a high intracellular potassium concentration so that their enzymes will function. The other is that they need sodium to maintain the integrity of their cell walls.

Halophiles are typically found in the ocean, where the salt concentration (3.5 percent) is optimum for their growth. Extreme halophiles require salt concentrations of 20 to 30 percent (➢ Figure 6.16). They are found in exceptionally salty bodies of water, such as the Dead Sea, and sometimes even in brine vats, where they cause spoilage of pickles being made there.

Radiation

Radiant energy, such as gamma rays and ultraviolet light, can cause mutations (changes in DNA) and even kill organisms. However, some microorganisms have pigments that screen radiation and help to prevent DNA damage. Others have enzyme systems that can repair certain kinds of DNA damage.

(a)

➢ Figure 6.16   **Responses to salt.** (a) Growth rates of halophilic (salt-loving) and nonhalophilic organisms are related to sodium ion concentration. (b) The Great Salt Lake in Utah, an example of an environment in which halophilic organisms thrive. Note the white areas of dried salt around the edges of the lake.

(b)

Nutritional Factors

The growth of microorganisms is affected by nutritional factors as well as by physical factors. Nutrients needed by microorganisms include carbon, nitrogen, sulfur, phosphorus, certain trace elements, and vitamins. Although we are concerned with ways microorganisms satisfy their own nutritional needs, we can note that in satisfying such needs, they also help to recycle elements in the environment. Activities of microbes in the carbon, nitrogen, sulfur, and phosphorus cycles are described in Chapter 26. A few microbes are **fastidious**—that is, they have special nutritional needs that can be difficult to meet in the laboratory. Some fastidious organisms, including those that cause gonorrhea, grow quite well in the human body but still cannot be easily grown in the laboratory on nutrient media.

Carbon Sources

Most bacteria use some carbon-containing compound as an energy source, and many use carbon-containing compounds as building blocks to synthesize cell components. Photoautotrophic organisms reduce carbon dioxide to glucose and other organic molecules. Both autotrophic and heterotrophic organisms can obtain energy from glucose by glycolysis, fermentation, and the Krebs cycle. They also synthesize some cell components from intermediates in these pathways.

Nitrogen Sources

All organisms, including microorganisms, need nitrogen to synthesize enzymes, other proteins, and nucleic acids. Some microorganisms obtain nitrogen from inorganic sources, and a few even obtain energy by metabolizing inorganic nitrogen-containing substances. Many microorganisms reduce nitrate ions (NO_3^-) to amino groups (NH_2) and use the amino groups to make amino acids. Some can synthesize all 20 amino acids found in proteins, whereas others must have 1 or a few amino acids provided in their medium. Certain fastidious organisms require all 20 amino acids and other building blocks in their medium. Many disease-causing organisms obtain amino acids for making proteins and other nitrogenous molecules from the cells of humans and other organisms they invade.

Once amino acids are synthesized or obtained from the medium, they can be used in protein synthesis. Similarly, purines and pyrimidines can be used to make DNA and RNA. The processes by which proteins and nucleic acids are synthesized are directly related to the genetic information contained in a cell. Thus, the synthesis of proteins and of nucleic acids will be discussed in Chapters 7 and 8.

Sulfur and Phosphorus

In addition to carbon and nitrogen, microorganisms need a supply of certain minerals, especially sulfur and phosphorus, which are important cell components. Microorganisms obtain sulfur from inorganic sulfate salts and from sulfur-containing amino acids. They use sulfur and sulfur-containing amino acids to make proteins, coenzymes, and other cell components. Some organisms can synthesize sulfur-containing amino acids from inorganic sulfur and other amino acids. Microorganisms obtain phosphorus mainly from inorganic phosphate ions (PO_4^{3-}). They use phosphorus (as phosphate) to synthesize ATP, phospholipids, and nucleic acids.

Trace Elements

Many microorganisms require a variety of **trace elements,** tiny amounts of minerals such as copper, iron, zinc, and cobalt, usually in the form of ions. Trace elements often serve as cofactors in enzymatic reactions. All organisms require some sodium and chloride, and halophiles require large amounts of these ions. Potassium, zinc, magnesium, and manganese are used to activate certain enzymes. Cobalt is required by organisms that can synthesize vitamin B_{12}. Iron is required for the synthesis of heme-containing compounds (such as the cytochromes of the electron transport system) and for certain enzymes. Although little iron is required, a shortage severely retards growth. Calcium is required by Gram-positive bacteria for synthesis of cell walls and by spore-forming organisms for synthesis of spores.

Vitamins

A **vitamin** is an organic substance that an organism requires in small amounts and that is typically used as a coenzyme. Many microorganisms make their own vitamins from simpler substances. Other microorganisms require several vitamins in their media because they lack the enzymes to synthesize them. Vitamins required by some microorganisms include folic acid, vitamin B_{12}, and vitamin K. Human pathogens often require a variety of vitamins and thus are able to grow well only when they can obtain these substances from the host organism. Growing such organisms in the laboratory requires a complex medium that contains all the nutrients they normally obtain from their hosts. Microbes living in the human intestine manufacture vitamin K, which is necessary for blood clotting, and some of the B vitamins, thus benefiting their host.

Nutritional Complexity

An organism's **nutritional complexity,** the number of nutrients it must obtain to grow, is determined by the kind and number of its enzymes. The absence of a single enzyme can render an organism incapable of synthesizing a specific substance. The organism therefore must obtain the substance as a nutrient from its environment. Microorganisms vary in the number of enzymes they possess. Those with many enzymes have simple nutritional needs because they can synthesize nearly all the substances they need. Those with fewer enzymes have complex nutritional requirements because they lack the ability to synthesize many of the substances they need for growth. Thus, nutritional complexity reflects a deficiency in biosynthetic enzymes.

Picky Eaters

Spiroplasma species, tiny spiral bacteria that lack cell walls, are among the most nutritionally fastidious organisms known. Recently, Kevin Hackett, a U.S. Department of Agriculture (USDA) scientist, devised an exact formula of 80 ingredients, including lipids, carbohydrates, amino acids, salts, vitamins, organic acids, and penicillin (to suppress potential competitors) to meet their needs. In his laboratory, he uses this medium to keep more than 30 species of spiroplasmas alive and well, making it possible for researchers to study them outside the more than 100 insect, tick, and plant species that they normally inhabit. Until now, most *Spiroplasma* species have been impossible to keep alive outside their hosts.

The spiroplasmas are responsible for hundreds of crop and animal diseases. Medical researchers are particularly interested in one species that experimentally causes tumors in animals. Another species kills honey bees, and a third lives harmlessly in the Colorado potato beetle, an insect that damages potato, eggplant, and tomato plants. Scientists hope to alter this last species genetically so that it will kill its potato beetle host.

USDA scientists are now trying to formulate complex media to grow mycoplasmalike organisms, a related group of bacteria that also lack cell walls. These bacteria cause hundreds of crop diseases and millions of dollars in economic losses each year. They are spread from plant to plant by infected insects. Another medium being designed would grow the bacterium *Mycoplasma pneumoniae,* which is the cause of a form of "walking pneumonia" in humans.

USDA scientist Kevin Hackett working on his microbial "witch's brew"—a mix of some 80 ingredients that will support the growth of nutritionally fastidious spiroplasmas outside their hosts.

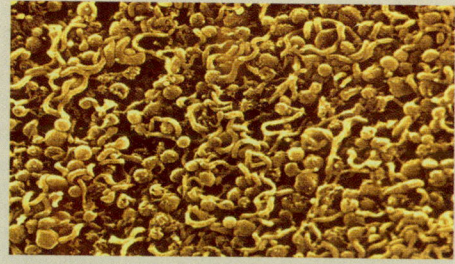

Spiroplasma (magnified 29,000X).

Locations of Enzymes

Most microorganisms move a variety of small molecules across their cell or plasma membranes and metabolize them. These substances include glucose, amino acids, small peptides, nucleosides, and phosphates as well as various inorganic ions. In addition to the endoenzymes that are produced for use within the cell (Chapter 5), many bacteria (and fungi) produce *exoenzymes* and release them through the cell or plasma membrane. ➡ (p.110) These enzymes include **extracellular enzymes,** usually produced by Gram-positive rods, which act in the medium around the organism, and **periplasmic enzymes,** usually produced by Gram-negative organisms, which act in the periplasmic space. Most exoenzymes are hydrolases; they add water as they split large molecules of carbohydrate, lipid, or protein into smaller ones that can be absorbed (Table 6.1). Although microbes cannot move large molecules across membranes, in nature they use large molecules from other organisms by digesting those molecules with exoenzymes before absorbing them.

Adaptation to Limited Nutrients

Microorganisms adapt to limited nutrients in several ways:

1. Some synthesize increased amounts of enzymes for uptake and metabolism of limited nutrients. This allows the organisms to obtain and use a larger proportion of the few nutrient molecules that are available.

2. Others have the ability to synthesize enzymes needed to use a different nutrient. For example, if glucose is in short supply, some microorganisms can make enzymes to take up and use a more plentiful nutrient such as lactose.

3. Many organisms adjust the rate at which they metabolize nutrients and the rate at which they synthesize molecules required for growth to fit the availability of the least plentiful nutrient. Both metabolism and growth are slowed, but no energy is wasted on synthesizing products that cannot be used. Growth is as rapid as conditions will allow.

✓ What does the ending *-phile* mean? Distinguish between the terms obligate and facultative.

✓ What enzymes do most obligate anaerobes lack? How does this cause them to die in the presence of oxygen?

✓ Do fastidious microorganisms have greater or lesser numbers of enzyme types than microbes with simpler nutritional needs? Why?

➢ Table 6.1	**Examples of exoenzymes**
Enzymes	**Action**
Enzymes that act on complex carbohydrates	
Carbohydrases	Break down large carbohydrate molecules into smaller ones
Amylase	Breaks down starch to maltose
Cellulase	Breaks down cellulose to cellobiose
Enzymes that act on sugars	
Sucrase	Breaks down sucrose to glucose and fructose
Lactase	Breaks down lactose to glucose and galactose
Maltase	Breaks down maltose to two glucose molecules
Enzymes that act on lipids	
Lipases	Break down fats to glycerol and fatty acids
Enzymes that act on proteins	
Proteases	Break down proteins to peptides and amino acids
Caseinase	Breaks down milk protein to amino acids and peptides
Gelatinase	Breaks down gelatin to amino acids and peptides

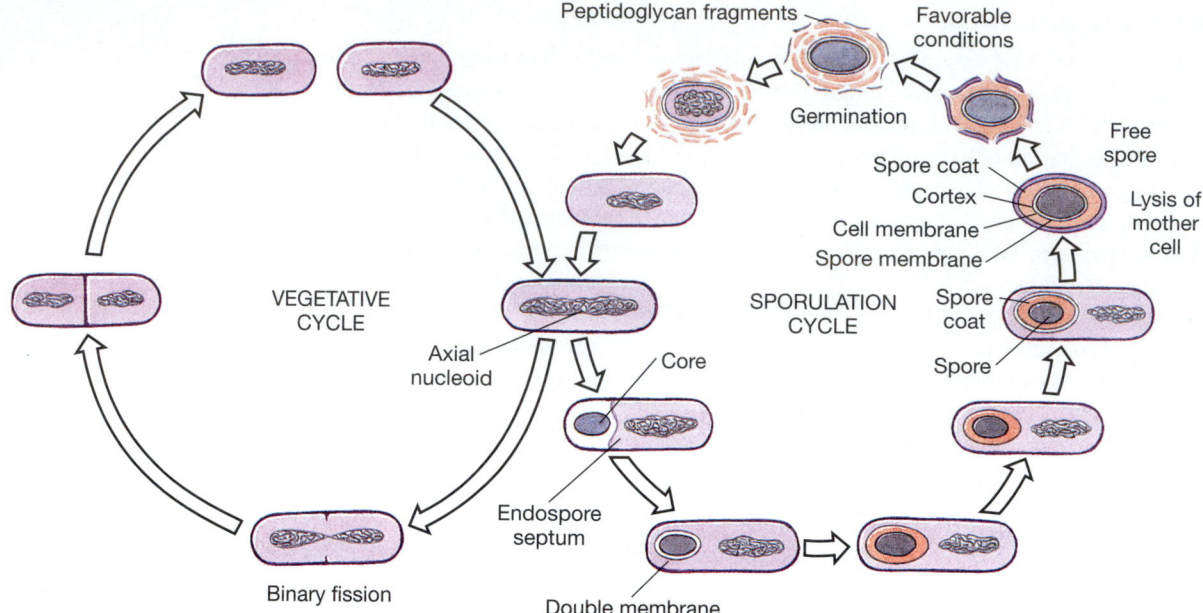

➤ Figure 6.17   **The vegetative and sporulation cycles in bacteria capable of sporulation.**

Sporulation

Sporulation, the formation of endospores, occurs in *Bacillus, Clostridium,* and a few other Gram-positive genera but has been studied most carefully in *B. subtilis* and *B. megaterium.* Bacteria that form endospores generally do so during the stationary phase in response to environmental, metabolic, and cell cycle signals.

When nutrients such as carbon or nitrogen become limiting, highly resistant endospores form inside mother cells. (With very low frequency, some bacteria form endospores even when nutrients are available.) Although endospores are not metabolically active, they can survive long periods of drought and are resistant to killing by extreme temperatures, radiation, and some toxic chemicals. Some endospores can withstand much higher temperatures than vegetative cells can. The endospore itself cannot divide, and the parent cell can produce only one endospore, so sporulation is a protective or survival mechanism, not a means of reproduction.

As endospore formation begins, DNA is replicated and forms a long, compact, *axial nucleoid* (➤ Figure 6.17). The two chromosomes formed by replication separate and move to different locations in the cell. In some bacteria the endospore forms near the middle of the cell, and in others it forms at one end (➤ Figure 6.18). The DNA where the endospore will form directs endospore formation. Most of the cell's RNA and some cytoplasmic protein molecules gather around the DNA to make the **core,** or living part, of the endospore. The core contains **dipicolinic** (di-pik-o-lin´ik) **acid** and calcium ions, which probably contribute to an endospore's heat resistance by stabilizing protein structure. An **endospore septum,** consisting of a cell membrane but lacking a cell wall, grows around the core, enclosing it in a double thickness of cell membrane. Both layers of this membrane synthesize peptidoglycan and release it into the space between the membranes. Thus, a laminated layer called the **cortex** is formed. The cortex protects the core against changes in osmotic pressure, such as those that result from drying. A **spore coat** of keratinlike protein, which is impervious to many chemicals, is laid down around the cortex by the mother cell. Finally, in some endospores an **exosporium,** a lipid-protein membrane, is formed outside the coat by the mother cell. The function of the exosporium is unknown. Under laboratory conditions, sporulation takes about 7 hours.

Once favorable conditions return, an endospore develops into a vegetative cell, which lacks the endospore's resistant properties. **Germination,** in which a spore returns to its vegetative state, occurs in three stages. The first stage, *activation,* usually requires some traumatic agent such as low pH or heat, which damages the coat. Without such damage, some endospores germinate slowly, if at all. The second stage, *germination proper,* requires water and a germination agent (such as the amino acid alanine or certain inorganic ions) that penetrates the damaged coat. During this process,

> Endospores trapped in amber for 25 million years germinate when placed in nutrient media.

much of the cortical peptidoglycan is broken down, and its fragments are released into the medium. The living cell (which occupied the core) now takes in large quantities of water and loses its resistance to heat and staining as well as its *refractility* (ability to bend light rays). Finally, *outgrowth* occurs in a medium with adequate nutrients. Proteins and RNA are synthesized, and in about an hour, DNA synthesis begins. The cell is now a vegetative cell and undergoes binary fission.

Thus, bacterial cells capable of sporulation display two cycles—the *vegetative cycle* and the *sporulation cycle* (➤ Figure 6.17). The vegetative cycle is repeated at intervals of 20 minutes or more, and the sporulation cycle is initiated periodically. Endospores known to be 300 or more years old have been observed to undergo germination when placed in a favorable medium. They are a very good form of insurance against extinction.

Other Sporelike Bacterial Structures

Certain bacteria, such as *Azotobacter,* form resistant **cysts,** or spherical, thick-walled cells, that resemble endospores. Like endospores, cysts are metabolically inactive and resist drying. Unlike endospores, they lack dipicolinic acid and have only limited resistance to high temperatures. Cysts germinate into single cells and therefore are not a means of reproduction.

Some filamentous bacteria, such as *Micromonospora* and *Streptomyces,* form asexually reproduced **conidia** (ko-nid'e-ah), or chains of aerial spores with thick outer walls. These spores are temporarily dormant but are not especially resistant to heat or drying. When the spores, which are produced in large numbers, are dispersed to a suitable environment, they form new filaments. Unlike endospores, these spores do contribute to reproduction of the species.

(a)

Culturing Bacteria

Culturing of bacteria in the laboratory presents two problems. First, a pure culture of a single species is needed to study an organism's characteristics. Second, a medium must be found that will support the growth of the desired organism. Let us look at some of the ways these problems are solved.

Methods of Obtaining Pure Cultures

To study bacteria in the laboratory, it is important to obtain a **pure culture,** a culture that contains only a single species of organism. Prior to the development of pure culture techniques, scientists studied *mixed cultures,* or cultures containing several different kinds of organisms. Researchers could make observations of different shapes and sizes of organisms, but they could find out little about the nutritional needs or growth characteristics of individual species. Today, pure cultures are obtained by isolating the progeny of a single cell.

Simple as it seems now, the technique of isolating pure cultures was difficult to develop. Attempts to isolate single cells by serial dilution were often unsuccessful because two or more organisms of different species were often present in the highest dilutions. Koch's technique of spreading bacteria thinly over a solid surface was more effective because it deposited a single bacterium at some sites. However, he tried several different solid substances. Using the discovery of Angelina Hesse, the wife of an associate, he settled on agar as the ideal solidifying agent. Only a very few organisms digest it, and in 1.5 percent solution it does not melt below 95°C. Furthermore, after being melted, agar remains in the liquid state until it has cooled to about 40°C, a temperature cool enough to allow the addition of nutrients and living organisms that might be destroyed by heat.

(b)

➤ Figure 6.18   **Bacterial endospores in two *Clostridium* species.** (a) Cells with centrally located endospores. (b) A colorized TEM of cells with terminally located endospores, which give the organisms a club-shaped appearance.

The Streak Plate Method

Today, the accepted way to prepare pure cultures is the **streak plate method,** which uses agar plates. Bacteria are picked up on a sterile wire loop, and the wire is moved lightly along the agar surface, depositing streaks of bacteria on the surface. The inoculating loop is flamed, and a few bacteria are picked up from the region already deposited and streaked onto a new region (➤ Figure 6.19). Fewer and fewer bacteria are deposited as the streaking continues, and the loop is flamed after each streaking. Individual organisms are deposited in the region streaked last. After the plate is incubated at a suitable growth temperature for the organism, small colonies—each derived from a single bacterial cell—appear. The wire loop is used to pick up a portion of an isolated colony and transfer it to any appropriate sterile medium for further study. The use of sterile (aseptic) technique assures that the new medium will contain organisms of a single species.

(a)

(b)

➤ **Figure 6.19 The streak plate method of obtaining pure cultures.** (a) A drop or bit of culture on a wire inoculating loop is lightly streaked across the top of the agar in region 1. The loop is flamed, the plate is rotated, and a few organisms are picked up from region 1 and streaked out into region 2. The loop is flamed again, and the process is repeated in region 3. The plate is then incubated. (b) A streak plate of *Serratia marcescens* after incubation. Note the greatly reduced numbers of colonies in each successive region.

The Pour Plate Method

Another way to obtain pure cultures, the **pour plate method,** makes use of serial dilutions (➤ Figure 6.7a). A series of dilutions are made such that the final dilution contains about a thousand organisms. Then 1 ml of liquid medium from the final dilution is placed in 9 ml of melted agar medium (45°C), and the medium is quickly poured into a sterile plate. The resulting pour plate will contain a small number of bacteria, some of which will form isolated colonies on the agar. This method allows some organisms to be embedded in the medium. It is particularly useful for growing microaerophiles that cannot tolerate exposure to oxygen in the air at the surface of the medium.

Culture Media

In nature many species of bacteria and other microorganisms are found growing together in oceans, lakes, and soil and on living or dead organic matter. These materials might be thought of as *natural media*. Although soil and water samples are often brought into the laboratory, organisms from them are typically isolated, and pure cultures are prepared for study.

Growing bacteria in the laboratory requires knowledge of their nutritional needs and the ability to provide the needed substances in a medium. Through years of experience in culturing bacteria in the laboratory, microbiologists have learned what nutrients must be supplied to each of many different organisms. Certain organisms, such as those that cause syphilis and leprosy, still cannot be cultured in laboratory media. They must be grown in cultures that contain living human or other animal cells. Many other organisms whose nutritional needs are reasonably well known can be grown in one or more types of media.

Types of Media

Laboratory media are generally synthetic media, as opposed to the natural media mentioned previously. A **synthetic medium** is a medium prepared in the laboratory from materials of precise or reasonably well-defined composition. A **defined synthetic medium** is one that contains known specific kinds and amounts of chemical substances. Examples of defined synthetic media are given in Tables 6.2 and 6.3. A **complex medium,** or **chemically nondefined medium,** is one that contains certain reasonably familiar materials but that varies slightly in chemical composition from batch to batch. Such media contain blood or extracts from beef, yeasts, soybeans, and other organisms. A common ingredient is **peptone,** a product of enzyme digestion of proteins. It provides small peptides that microorganisms can use. Although the exact concentrations are not known, trace elements and vitamins are present in sufficient quantities in complex media to support the growth of many organisms. Both liquid nutrient broth and solidified agar medium used to culture many organisms are complex media. An example of a complex medium is given in Table 6.4.

Commonly Used Media

Most routine laboratory cultures use media containing peptone from meat or fish in nutrient broth or solid agar medium. Such media are sometimes enriched with **yeast extract,** which contains a number of vitamins, coenzymes, and nucleosides. **Casein hydrolysate** made from milk protein contains many amino acids and is used to enrich certain media. Because blood con-

➤ Table 6.2	A defined synthetic medium for growing *Proteus vulgaris*		
Ingredient	**Amount**	**Ingredient**	**Amount**
Water	1 liter	K_2HPO_4	1 g
$MgSO_4 \cdot 7H_2O$	200 mg	$FeSO_4 \cdot 7H_2O$	10 mg
$CaCl_2$	10 mg	Glucose	5 g
NH_4Cl	1 g	Nicotinic acid	0.1 mg
Trace elements (Mn, Mo, Cu, Co, Zn as inorganic salts, known quantities of 0.02–0.5 mg each)			

Source: Adapted from R. Y. Stanier et al. 1986. *The microbial world.* 5th ed. Englewood Cliffs, NJ: Prentice Hall.

➤ Table 6.3	A defined synthetic medium for growing the fastidious bacterium *Leuconostoc mesenteroides*

Ingredient	Amount	Ingredient	Amount
Water	1 liter		
Energy Source			
Glucose	25 g		
Nitrogen Source			
NH_4Cl	3 g		
Minerals			
KH_2PO_4	600 mg	$FeSO_4 \cdot 7H_2O$	10 mg
K_2HPO_4	600 mg	$MnSO_4 \cdot 4H_2O$	20 mg
$MgSO_4 \cdot 7H_2O$	200 mg	NaCl	10 mg
Organic Acid			
Sodium acetate	20 g		
Amino Acids			
DL-α-Alanine	200 mg	L-Lysine · HCl	250 mg
L-Arginine	242 mg	DL-Methionine	100 mg
L-Asparagine	400 mg	DL-Phenylalanine	100 mg
L-Aspartic acid	100 mg	L-Proline	100 mg
L-Cysteine	50 mg	DL-Serine	50 mg
L-Glutamic acid	300 mg	DL-Threonine	200 mg
Glycine	100 mg	DL-Tryptophan	40 mg
L-Histidine · HCl	62 mg	L-Tyrosine	100 mg
DL-Isoleucine	250 mg	DL-Valine	250 mg
DL-Leucine	250 mg		
Purines and Pyrimidines			
Adenine sulfate · H_2O	10 mg	Uracil	10 mg
Guanine · HCl · $2H_2O$	10 mg	Xanthine · HCl	10 mg
Vitamins			
Thiamine · HCl	0.5 mg	Riboflavin	0.5 mg
Pyridoxine · HCl	1.0 mg	Nicotinic acid	1.0 mg
Pyridoxamine · HCl	0.3 mg	*p*-Aminobenzoic acid	0.1 mg
Pyridoxal · HCl	0.3 mg	Biotin	0.001 mg
Calcium pantothenate	0.5 mg	Folic acid	0.01 mg

Source: H. E. Sauberlich and C. A. Baumann. "A factor required for the growth of *Leuconostoc citrovorum*." *J. Biol. Chem.* 176(1948):166.

Tomato juice agar, which is actually made with tomato juice, is used for the cultivation of lactobacilli.

tains many nutrients needed by fastidious pathogens, **serum** (the liquid part of the blood after clotting factors have been removed), whole blood, and heated whole blood can be useful in enriching media. **Blood agar** is useful in identifying organisms that can cause hemolysis, or breakdown of red blood cells. Sheep's blood is used because its hemolysis is more clearly defined than when human blood is used in the agar medium. **Chocolate agar,** made with heated blood, is so named because it turns chocolate brown. It is used to culture fastidious pathogens.

Selective, Differential, and Enrichment Media

To isolate and identify particular microorganisms, especially those from patients with infectious diseases, *selective, differential,* or *enrichment media* are often used. Such special media are an essential part of modern diagnostic microbiology. Table 6.5 shows some examples of special diagnostic media.

➤ Table 6.4	A complex medium suitable for many heterotrophic organisms

Nutrient Broth Ingredient	Amount
Water	1 liter
Peptone	5 g
Beef extract	3 g
NaCl	8 g
Solidified Medium	
Agar	15 g

Above ingredients in amounts specified.

Don't Leave Home Without Your CO₂!

Transporting specimens from patients to laboratories sometimes presents problems. The organisms must not be subjected to drying conditions or to too much or too little oxygen. Of course, specimen handlers must be protected from infection. Cultures that may contain *Neisseria gonorrhoeae* from patients with gonorrhea pose one such problem—providing an atmosphere relatively high in carbon dioxide. Commercial systems, such as the JEMBEC (John E. Martin Biological Environmental Chamber), are available. This system consists of a small plastic plate of selective medium and a tablet of sodium bicarbonate and citric acid. The plate is inoculated, and the table is placed in it. The plate is then placed in a plastic bag and sealed. Moisture from the medium causes the tablet to release carbon dioxide; an appropriate concentration of 5 to 10 percent is obtained. The culture is incubated to allow growth to begin before shipment to the laboratory.

A **selective medium** is one that encourages the growth of some organisms but suppresses the growth of others. For example, to identify *Clostridium botulinum* in food samples suspected of being agents of food poisoning, the antibiotics sulfadiazine and polymyxin sulfate (SPS) are added to anaerobic cultures of *Clostridium* species. This culture medium is called *SPS agar*. It allows growth of *Clostridium botulinum* while inhibiting growth of most other *Clostridium* species.

A **differential medium** has a constituent that causes an observable change (a color change or a change in pH) in the medium when a particular biochemical reaction occurs. This change allows microbiologists to distinguish a certain type of colony from others growing on the same plate. SPS agar also serves as a differential medium. Colonies of *Clostridium botulinum* formed on this medium are black because of hydrogen sulfide made by the organisms from the sulfur-containing additives.

Many media, such as SPS agar and MacConkey agar, are both selective and differential. *MacConkey agar* contains crystal violet and bile salts, which inhibit growth of Gram-positive bacteria while allowing growth of Gram-negative bacteria. MacConkey agar also contains the sugar lactose plus a pH indicator that turns colonies of lactose fermenters red and leaves colonies of nonfermenters colorless and translucent. Although there are some exceptions, most organisms that are normally found in the human intestines ferment lactose, whereas most pathogens (disease-causing microorganisms) do not.

An **enrichment medium** contains special nutrients that allow growth of a particular organism that might not otherwise be present in sufficient numbers to allow it to be isolated and identified. Unlike a selective medium, an enrichment medium does not suppress others. For example, because *Salmonella typhi* organisms may not be sufficiently numerous in a fecal sample to allow positive identification, they are cultured on a medium containing the trace element selenium, which supports growth of the organism. After incubation in the enrichment medium, the greater numbers of the organisms increase the likelihood of a positive identification.

Controlling Oxygen Content of Media

Obligate aerobes, microaerophiles, and obligate anaerobes require special attention to maintain oxygen concentrations suitable for growth. Most obligate aerobes obtain sufficient oxygen from nutrient broth or on the surface of solidified agar medium, but some need more. Oxygen gas is bubbled through the medium or into the incubation environment with filters between the gas source and the medium to prevent contamination of the culture. Microaerophiles can be incubated in tubes of a nutrient medium or agar plates in a jar in which a candle is lit before the jar is sealed (➤ Figure 6.20). (Scented candles should not be used because oils from them inhibit bacterial growth.) The burning candle uses the oxygen in the jar and adds carbon dioxide to it. When the carbon dioxide extinguishes the flame, conditions are optimum for the growth of microorganisms that require small amounts of carbon dioxide, such as the bacterium *Neisseria gonorrhoeae*, which causes gonorrhea.

To culture obligate anaerobes, all molecular oxygen must be removed and kept out of the medium. Addition of oxygen-binding agents such as thioglycollate, the amino acid cysteine, or sodium sulfide to the medium prevents oxygen from exerting toxic effects on anaerobes. Media can be dispensed in sealed screw-cap tubes, completely filled to exclude air, or in Petri plates. When cultures must be grown in plates so that colonial growth can be studied, special

> **TABLE 6.5**   Selected examples of diagnostic media

Medium	Organism(s) identified	Selectivity and/or differentiation achieved
Brilliant green agar	*Salmonella*	**Selective:** Brilliant green dye inhibits gram-positive bacteria and thus selects Gram-negative ones. **Differential:** Differentiates *Shigella* colonies (which do not ferment lactose or sucrose and are red to white) from other organisms that do ferment one of those sugars and are yellow to green.
Eosin methylene blue agar (EMB)	Gram-negative enterics	**Selective:** Medium partially inhibits gram-positive bacteria. **Differential:** Eosin and methylene blue differentiate among organisms: *Escherichia coli* colonies are purple and typically have a metallic green sheen; *Enterobacter aerogenes* colonies are pink, indicating that they ferment lactose; and colonies of other organisms are light pink, indicating they do not ferment lactose.
MacConkey agar	Gram-negative enterics	**Selective:** Crystal violet and bile salts inhibit Gram-positive bacteria. **Differential:** Lactose and the pH indicator neutral red (red when acidic), identify lactose fermenters as red colonies and nonfermenters as light pink. Most intestinal pathogens are nonfermenters and hence do not produce acid.
Triple sugar-iron agar (TSI)	Gram-negative enterics	**Not Selective.** **Differential:** Used in agar slants (tubes cooled in slanted position), where differentiation is based on both aerobic surface growth and anaerobic growth in agar in base of tube. Medium contains specific amounts of glucose, sucrose, and lactose, sulphur-containing amino acids, iron, and a pH indicator, so relative use of each sugar and H_2S formation can be detected. 1. Uninoculated tube of TSI 2. Inoculated: red slant and red butt=no change; no sugar fermented 3. Yellow slant and yellow butt=lactose and glucose fermented to acid; trapped bubbles in butt indicate fermentation to acid and gas 4. Red slant (lactose not fermented) and yelow butt (glucose fermented to acid); black precipitate=H_2S produced; sometimes obscures yellow butt. Almost all enteric pathogens produce red slant and yellow butt, with or without H_2S and or gas.

Differentiation of Intestinal Bacilli Based on TSI

red slant red butt	yellow slant yellow butt no H_2S	yellow slant yellow butt H_2S produced	red slant yellow butt no H_2S	red slant yellow butt H_2S produced
Pseudomonas *Acinetobacter* *Alcaligenes*	*Escherichia* *Enterobacter* *Klebsiella*	*Citrobacter* *Arizona* some *Proteus* sp.	*Shigella* some *Proteus* sp.	most *Salmonella* *Citrobacter* *Arizona*

➤ Figure 6.20   **Candle jar culture of anaerobes and microaerophiles.** Microaerophiles are growing in culture tubes and on Petri plates in a sealed jar in which a candle burned until it was extinguished by carbon dioxide accumulation in the atmosphere of the jar. A small amount of oxygen remains. Although this method is not used much anymore, years ago it was the main means of growing anaerobes.

✓ Why are endospores not considered to be a form of reproduction? Why would waiting until unfavorable conditions occurred before beginning spore formulation place an organism at greater risk than one that continuously forms spores?

✓ Distinguish between the various kinds of media: synthetic, complex, defined, selective, differential, enrichment, transport. Can a given medium be more than one of these types?

✓ What is the purpose of a stock culture? Why is it never used itself for laboratory tests?

jars are used that can hold both plates and tubes. Agar plates are incubated in sealed jars containing chemical substances that remove oxygen from the air and generate carbon dioxide (➤ Figure 6.21). *Stab cultures* can be made by stabbing a straight inoculating wire coated with organisms into a tube of agar-solidified medium. In laboratories where anaerobes are regularly handled, an *anaerobic transfer chamber* is often used (➤ Figure 6.22). Equipment and cultures are introduced through an air lock, and the technician uses glove ports to manipulate the cultures.

Maintaining Cultures

Once an organism has been isolated, it can be maintained indefinitely in a pure culture called a **stock culture.** When needed for study, a sample from a stock culture is inoculated into fresh medium. The stock culture itself is never used for laboratory studies. However, organisms in stock cultures go through growth phases, deplete nutrients, and accumulate wastes just as those in any culture do. As the culture ages, the organisms may acquire odd shapes or other altered characteristics. Stock cultures are maintained by making subcultures in fresh medium at frequent intervals to keep the organisms growing.

The use of careful aseptic techniques is important in all manipulations of cultures. **Aseptic techniques** minimize the chances that cultures will be contaminated by organisms from the environment or that organisms, especially pathogens, will escape into the environment. Such techniques are especially important in making subcultures from stock cultures. Otherwise an undesirable organism might be introduced, and the stock organism would have to be reisolated. Even with regular transfers of organisms from stock cultures to fresh media, organisms can undergo mutations (changes in DNA) and develop altered characteristics.

Preserved Cultures

To avoid the risk of contamination and to reduce the mutation rate, stock culture organisms also should be kept in a **preserved culture,** a culture in which organisms are maintained in a dormant state. The most commonly used technique for preserving cultures is *lyophilization* (freeze-drying), in which cells are quickly frozen, dehydrated while frozen, and sealed in vials under vacuum (Chapter 12). Such cultures can be kept indefinitely at room temperature.

Because microorganisms frequently undergo genetic changes, reference cultures are maintained. A **reference culture** is a preserved culture that maintains the organisms with the characteristics as originally defined. Reference cultures of all known species and strains of bacteria and many other microorganisms are maintained in the American Type Culture Collection, and many also are maintained in universities and research centers. Then if stock cultures in a particular laboratory undergo change or if other laboratories wish to obtain certain organisms for study, reference cultures are always available.

Methods of Performing Multiple Diagnostic Tests

Many diagnostic laboratories use culture systems that contain a large number of differential and selective media, such as the Enterotube Multitest System or the Analytical Profile Index (API). These systems allow simultaneous determination of an organism's reaction to a variety of carefully chosen diagnostic media from a single inoculation. The advantages of these systems are that they use small quantities of media, occupy little space in an incubator, and provide an efficient and reliable means of making positive identification of infectious organisms.

The Enterotube System® is used to identify enteric pathogens, or organisms that cause intestinal diseases such as typhoid and paratyphoid fevers, shigellosis, gastroenteritis, and some kinds of food poisoning. The causative organisms are all Gram-negative rods indistinguishable from one another without biochemical tests. The Enterotube System consists of a tube with compartments, each of which contains one or more different media, and a sterile inoculating rod (➤ Figure 6.23a). Each compartment is inoculated when the tip of the rod is touched to a colony and the rod is drawn through the tube. After the tube has been incubated for 24 hours at 37°C, the results of 15 biochemical tests can be obtained by observing (1) whether gas was produced and (2) the color of the medium in each compartment. Tests are grouped in sets of three; within each group, tests are assigned a number 1, 2, or 4 (➤ Figure 6.23b). The sum of the numbers of positive tests in each group indicates which tests are positive. The sum 3 means that tests 1 and 2 were positive; the sum 5 means that tests 1 and 4 were positive; the sum 6, that tests 2 and 4 were positive; and the sum 7, that all tests were positive. The single-digit sums for each of the five sets of tests are combined to form a five-digit identification number for a particular organism. For example, 36601 is *Escherichia coli,* and 34363 is *Klebsiella pneumoniae.* A list of identification numbers and the corresponding organisms is provided with the system.

API consists of a plastic tray with 20 microtubes called *cupules,* each containing a different kind of dehydrated medium (➤ Figure 6.24). Each cupule medium is rehydrated and inoculated with a suspension of bacteria from an isolated colony. As with Enterotubes, the tray is incubated, test results are determined, and the values 1, 2, and 4 are summed for sets of three tests. The seven-digit profile number identifies the organism.

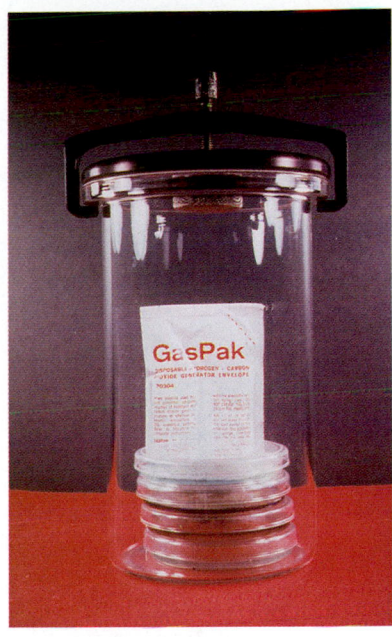

➤ Figure 6.21   **CO$_2$ incubator.** When activated, these chemicals remove oxygen and are enclosed with cultures in a sealed jar to create an anaerobic chamber. These are useful for the small laboratory that has only a few plates needing anaerobic incubation.

➤ Figure 6.22   **Anaerobic transfer.** A large transfer chamber with an air lock for introducing equipment and cultures and with ports to allow manipulation of the cultures.

➤ Figure 6.23   **The Enterotube Multitest System.** (a) Removal of a sterile end cap allows you to "pick" a colony from the surface of a plate. The wire is then drawn through all compartments, thereby inoculating each of them. (b) After inoculation and incubation, compartments with positive test results are assigned a number. The numbers are summed within zones to get a definitive index number that identifies an organism on the list in the coding manual. Any necessary confirmatory tests are also noted there. By numbering each test in a zone with a digit equal to a power of 2 (1, 2, 4, 8, and so on), the sum of any set of positive reactions results in a unique number. A given species may, however, be coded for many different numbers, as individual strains of that species will vary somewhat in their characteristics.

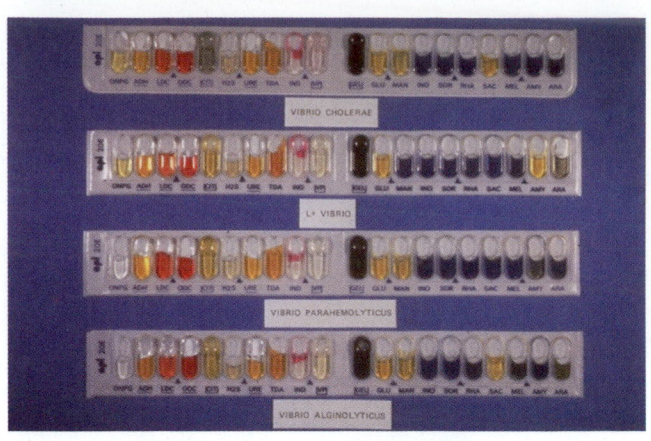

➢ Figure 6.24   **The Analytical Profile Index (API) 20E System.** Various species of the genus *Vibrio* are shown here, with the differences in reactions that enable them to be identified. This system allows identification to species level of 125 Gram-negative intestinal bacilli.

In this brief discussion of diagnostic systems, we have considered only the tip of the iceberg. Of the many other available tests, a large number are based on immunological properties of organisms. We will consider some of them when we discuss immunology or particular infectious agents. Also, much is known about which organisms are likely to infect certain human organs and tissues, and many diagnostic tests are designed to distinguish among organisms found in respiratory secretions, fecal samples, blood, other tissues, and body fluids.

Retracing Our Steps

Growth and Cell Division

Microbial Growth Defined

*   Microbial growth can be defined as the orderly increase in quantity of all cell components and in the number of cells of an organism.

*   Because of limited increase in cell size and the frequency of cell division, growth in microorganisms is measured by increase in cell number.

Cell Division

*   Most cell divisions in bacteria occur by **binary fission,** in which the nuclear body divides and the cell forms a transverse septum that separates the original cell into two cells.

*   Yeast cells and some bacteria divide by **budding,** in which a small, new cell develops from the surface of an existing cell.

Phases of Growth

*   In a nutrient-rich **medium** (a mixture of substances on or in which microorganisms grow), bacteria divide rapidly. The length of time required for one division is called the **generation time.** Such growth is said to occur at an **exponential,** or **logarithmic, rate.**

*   Bacteria introduced into a fresh, nutrient-rich medium display four major phases of growth: (1) In the **lag phase,** the organisms are metabolically active—growing and synthesizing various substances but not increasing in number. (2) In the **log phase,** organisms divide at an exponential, or logarithmic, rate and with a constant generation time. These properties of growth in the log phase can be used to calculate both the number of generations and the generation time. Cultures can be maintained by the use of a **chemostat,** which allows continuous addition of fresh medium. (3) In the **stationary phase,** the number of new cells produced equals the number of cells dying. The medium contains limited amounts of nutrients and may contain toxic quantities of waste materials. (4) In the **decline phase,** or **death phase,** many cells lose their ability to divide and eventually die. A logarithmic decrease in the number of cells results.

*   Growth in colonies parallels that in liquid medium, except that most growth occurs at the edge of the colony, and all phases of growth occur simultaneously somewhere in the colony.

Measuring Bacterial Growth

*   Growth can be measured by **serial dilution,** in which successive 1:10 dilutions of a liquid culture of bacteria are made and transferred onto an **agar plate;** the colonies that arise are counted. Each colony represents one live cell from the original sample.

*   Growth also can be measured by **direct microscopic counts,** the **most probable number** technique, **filtration,** observing or measuring **turbidity,** measuring products of metabolism, and obtaining the dry weight of cells.

Factors Affecting Bacterial Growth

Physical Factors

- Acidity and alkalinity of the medium affect growth, and most organisms have an **optimum pH** range of no more than one pH unit.

- Temperature affects bacterial growth. (1) Most bacteria can grow over a 30°C temperature range. (2) Bacteria can be classified according to growth temperature into three categories: **psychrophiles,** which grow at low temperatures (below 25°C); **mesophiles,** which grow best at temperatures between 25° and 40°C; and **thermophiles,** which grow at high temperatures (above 40°C). (3) The temperature range of an organism is closely related to the temperature at which its enzymes function best.

- The quantity of oxygen in the environment affects the growth of bacteria. (1) **Obligate aerobes** require relatively large amounts of free molecular oxygen to grow. (2) **Obligate anaerobes** are killed by free oxygen and must be grown in the absence of free oxygen. (3) **Facultative anaerobes** can metabolize substances aerobically if oxygen is available or anaerobically if it is absent. (4) **Aerotolerant anaerobes** metabolize substances anaerobically but are not harmed by free oxygen. (5) **Microaerophiles** must have only small amounts of oxygen to grow.

- Actively metabolizing bacteria require some water in their environment.

- Some bacteria, but no other living things, can withstand extreme **hydrostatic pressures** in deep valleys in the ocean.

- Osmotic pressure affects bacterial growth, and water can be drawn into or out of cells according to the relative osmotic pressure created by dissolved substances in the cell and the environment. (1) Active transport minimizes the effects of high osmotic pressure in the environment. (2) Bacteria called **halophiles** require moderate to large amounts of salt and are found in the ocean and in exceptionally salty bodies of water.

Nutritional Factors

- All organisms require a carbon source: (1) Autotrophs use CO_2 as their carbon source and synthesize other substances they need. (2) Heterotrophs require glucose or another organic carbon source from which they obtain energy and intermediates for synthetic processes.

- Microorganisms require an organic or inorganic nitrogen source from which to synthesize proteins and nucleic acids. They also require a source of other elements found within them, including sulfur, phosphorus, potassium, iron, and many **trace elements.**

- Microorganisms that lack the enzymes to synthesize particular **vitamins** must obtain those vitamins from their environment.

- The nutritional requirements of an organism are determined by the kind and number of its enzymes. **Nutritional complexity** reflects a deficiency in biosynthetic enzymes.

- Bioassay techniques use metabolic properties of organisms to determine quantities of vitamins and other compounds in foods and other materials.

- Most microorganisms move substances of low molecular weight across their cell membranes and metabolize them internally. Some bacteria (and fungi) also produce exoenzymes that digest large molecules outside the cell membrane of the organism.

- Microorganisms adjust to limited nutrient supplies by increasing the quantities of enzymes they produce, by making enzymes to metabolize another available nutrient, or by adjusting their metabolic activities to grow at a rate consistent with availability of nutrients.

Sporulation

- **Sporulation,** which occurs in *Bacillus, Clostridium,* and a few other Gram-positive genera, involves the steps summarized in Figure 6.17.

- Sporulation lets the bacterium withstand long periods of dry conditions and extreme temperatures.

- When more favorable conditions are restored, **germination** occurs—endospores begin to develop into vegetative cells.

Culturing Bacteria

Methods of Obtaining Pure Cultures

- The **streak plate method** of obtaining a **pure culture** involves spreading bacteria across a sterile, solid surface such as an agar plate so that the progeny of a single cell can be picked up from the surface and transferred to a sterile medium.

- The **pour plate method** of obtaining a pure culture involves serial dilution, transferring to melted agar a specific volume of the dilution containing a few organisms and picking up cells from a colony on the agar.

Culture Media

- In nature, microorganisms grow on natural media, or the nutrients available in water, soil, and living or dead organic material.

- In the laboratory, microorganisms are grown in **synthetic media:** (1) **Defined synthetic media** consist of known quantities of specific nutrients. (2) **Complex media** consist of nutrients of reasonably well-known composition that vary in composition from batch to batch.

- Most routine laboratory cultures make use of **peptones,** or digested meat or fish proteins. Other substances such as **yeast extract, casein hydrolysate, serum,** whole blood, or heated whole blood are sometimes added.

- Diagnostic media are (1) **selective media** if they encourage growth of some organisms and inhibit growth of others, (2) **differential media** if they allow different kinds of colonies on the same plate to be distinguished from one another, or (3) **enrichment media** if they provide a nutrient that fosters growth of a particular organism.

- Cultures are maintained as **stock cultures** for routine work, as **preserved cultures** to prevent risk of contamination or change in characteristics, and as **reference cultures** to preserve specific characteristics of species and strains.

Methods of Performing Multiple Diagnostic Tests

- These systems allow simultaneous reaction of an organism to a large number of differential and selective media. Analysis of results allows rapid identification of the organism.

- The two most commonly used systems are the Enterotube Multitest System and the Analytic Profile Index (API).

Terminology Check

acidophile   *(p. 146)*
aerotolerant anaerobe *(p. 148)*
agar   *(p. 140)*
alkaliphile   *(p. 146)*
aseptic technique   *(p. 160)*
barophile   *(p. 149)*
binary fission   *(p. 137)*
blood agar   *(p. 157)*
budding   *(p. 138)*
capnophile   *(p. 148)*
casein hydrolysate   *(p. 156)*
catalase   *(p. 149)*
chemically nondefined medium   *(p. 156)*
chemostat   *(p. 140)*
chocolate agar   *(p. 157)*
colony   *(p. 140)*
complex medium   *(p. 156)*
conidium   *(p. 155)*
core   *(p. 154)*
cortex   *(p. 154)*
countable number   *(p. 141)*
cyst   *(p. 155)*
daughter cell   *(p. 137)*
death phase   *(p. 140)*
decline phase   *(p. 140)*
defined synthetic medium   *(p. 156)*
differential medium   *(p. 158)*
dipicolinic acid   *(p. 154)*
direct microscopic count *(p. 142)*
endospore septum   *(p. 154)*
enrichment medium   *(p. 158)*
exosporium   *(p. 154)*

exponential rate   *(p. 139)*
extracellular enzyme   *(p. 153)*
facultative   *(p. 146)*
facultative anaerobe   *(p. 148)*
facultative psychrophile *(p. 146)*
facultative thermophile *(p. 147)*
fastidious   *(p. 151)*
filtration   *(p. 143)*
generation time   *(p. 139)*
germination   *(p. 154)*
halophile   *(p. 150)*
hydrostatic pressure   *(p. 149)*
lag phase   *(p. 138)*
logarithmic rate   *(p. 139)*
log phase   *(p. 139)*
medium   *(p. 138)*
mesophile   *(p. 146)*
microaerophile   *(p. 148)*
microbial growth   *(p. 137)*
most probable number (MPN)   *(p. 143)*
mother cell   *(p. 137)*
neutrophile   *(p. 146)*
nonsynchronous growth   *(p. 139)*
nutritional complexity   *(p. 152)*
nutritional factor   *(p. 145)*
obligate   *(p. 146)*
obligate aerobe   *(p. 148)*
obligate anaerobe   *(p. 148)*
obligate psychrophile   *(p. 146)*
obligate thermophile   *(p. 147)*
optimum pH   *(p. 145)*
peptone   *(p. 156)*

periplasmic enzyme   *(p. 153)*
physical factor   *(p. 145)*
plasmolysis   *(p. 149)*
pour plate   *(p. 141)*
pour plate method   *(p. 156)*
preserved culture   *(p. 160)*
psychrophile   *(p. 146)*
pure culture   *(p. 155)*
reference culture   *(p. 160)*
sarcina   *(p. 138)*
selective medium   *(p. 158)*
serial dilution   *(p. 140)*
serum   *(p. 157)*
spore coat   *(p. 154)*
sporulation   *(p. 154)*
spread plate method   *(p. 141)*
standard bacterial growth curve   *(p. 138)*
stationary phase   *(p. 140)*
stock culture   *(p. 160)*
streak plate method   *(p. 155)*
superoxide   *(p. 148)*
superoxide dismutase   *(p. 149)*
synchronous growth   *(p. 139)*
synthetic medium   *(p. 156)*
tetrad   *(p. 138)*
thermophile   *(p. 147)*
trace element   *(p. 151)*
turbidity   *(p. 143)*
vitamin   *(p. 152)*
yeast extract   *(p. 157)*

See You on the Web

Don't forget I am expecting you on the web. The following is a preparedness checklist for your web explorations. If you have questions on any of these items, you can return to the end of Chapter 1 for more details.

- Questions We'll Explore
- Concept Check questions
- Flashcards
- Study Guide

If you are ready, come visit me at

http://www.prenhall.com/black

It will be a journey you'll never forget.

Microbial Genetics

That should be Chapter 007, because his name is Bond, James Bond. In all of his cinematic exploits you have seen him thwart evil and get the girl(s) with the help of a laboratory of high-tech gadgets and gizmos. There is one gadget, however, that has thus far proven to be beyond the reach of even Q. It is a sapphire-tipped gene gun that can muster 2,000 psi of pressure to shoot an atomized mist of genetic material through the skin and into the targeted specimen.

Background Basics

nucleic acid structure . . . .(Chapter 2)
ribosome structure and function . . . .(Chapter 4)
enzyme structure and function . . . .(Chapter 5)

We have considered many aspects of metabolism and growth, but we have yet to consider the synthesis of nucleic acids and proteins. Synthesis of these complex molecules is the basis of **genetics**, the study of heredity. The genetics of microorganisms is an exciting and active research area, and it is also a rewarding area for microbiologists. Since the inception of the Nobel Prize in physiology or medicine in 1900, more than 30 annual prizes have been awarded in microbiology-related fields, especially microbial genetics. Because of this intensive investigation, much is now known about microbial genetics. We will begin our study of genetics by seeing how bacteria synthesize nucleic acids—DNA and RNA—and how the nucleic acids are involved in the synthesis of proteins. We will also see how *genes* (specific segments of DNA) act, how they are regulated, and how they are altered in mutations. In the next chapter we will discuss the mechanisms by which genetic information is transferred among microorganisms.

An Overview of Genetic Processes

The Basis of Heredity

All information necessary for life is stored in an organism's genetic material, DNA, or, for many viruses, RNA. To explain **heredity**—the transmission of this information from an organism to its progeny (offspring)—we must consider the nature of chromosomes and genes.

A **chromosome** is typically a circular (in prokaryotes) or linear (in eukaryotes) threadlike molecule of DNA. Recall that DNA consists of a double chain of nucleotides arranged in a helix with the nucleotide base pairs held together by hydrogen bonds (➤ Figure 7.1). ➡ (Chapter 2, p. 44) The particular nucleotide sequence in DNA provides information for the synthesis of new DNA and for the synthesis of proteins.

A typical prokaryotic cell contains a single circular chromosome, composed primarily of a single DNA molecule about 1 mm long when fully stretched out—some 1000 times longer than the cell itself. This immense molecule fits compactly into the

Questions We'll Explore

A How are genes, chromosomes, and mutations involved in heredity in prokaryotic organisms?

B How do nucleic acids store and transfer information?

C How is DNA replicated in prokaryotic cells?

D What are the major steps in protein synthesis?

E How do mechanisms that regulate enzyme activity differ from those that regulate gene expression?

F What happens in feedback inhibition, enzyme induction, and enzyme repression?

G What changes in DNA occur as a result of mutations, and how do mutations affect organisms?

H How do spontaneous and induced mutations differ?

I How do the fluctuation test, replica plating, and the Ames test make use of bacteria in studying mutations?

➤ Figure 7.1   **The structure of DNA.** (a) The two upright strands, composed of the sugar deoxyribose (D) and phosphate groups (P), are held together by hydrogen bonding between complementary bases. Adenine (A) always pairs with thymine (T), and guanine (G) always pairs with cytosine (C). Each strand can thus provide the information needed for the formation of a new DNA molecule. (b) The DNA molecule is twisted into a double helix. The two sugar-phosphate strands run in opposite (antiparallel) directions. Each new strand grows from the 5′ end toward the 3′ end.

cell, where it forms the nucleoid ➥ (Chapter 4, p. 82) by twisting tightly around itself, a process known as *supercoiling.* When a prokaryotic cell reproduces by binary fission, the chromosome reproduces, or *replicates,* itself, and each daughter cell receives one of the chromosomes. This mechanism provides for the orderly transmission of genetic information from parent cell to daughter cells.

A **gene,** the basic unit of heredity, is a linear sequence of nucleotides of DNA that form a functional unit of a chromosome or of a plasmid. All information for the structure and function of an organism is coded in its genes. In many cases, a gene determines a single characteristic. However the information in a specific gene, found at a particular **locus** (location) on the chromosome or plasmid, is not always the same. Genes with different information at the same locus are called **alleles** (al-eelz′). Because prokaryotes have a single chromosome, they generally have only one version, or allele, of each gene. (In Chapter 8, we will discover exceptions to this rule.) Many (but not all) eukaryotes have two sets of chromosomes and thus two alleles of each gene, which may be the same or different. For example, in human blood types, any one of three genes—A, B, or O—can occupy a certain locus. Allele A causes red blood cells to have a certain glycoprotein, which we will designate as molecule A, on their surfaces. Allele B causes them to have molecule B, and allele O does not cause them to have any glycoprotein molecule on the cell surfaces. People with type AB blood produce both molecules A and B because they have both alleles A and B.

Heritable variations in the characteristics of progeny can arise from mutations. A **mutation** is a permanent alteration in DNA. Mutations usually change the sequence of nucleotides in DNA and thereby change the information in the DNA. When the mutated DNA is transmitted to a daughter cell, the daughter cell can differ from the parent cell in one or more characteristics. We will see in Chapter 8 that heritable variations in the characteristics of prokaryotic organisms can occur by a variety of mechanisms.

Nucleic Acids in Information Storage and Transfer

Information Storage

All the information for the structure and functioning of a cell is stored in DNA. For example, in the chromosome of the bacterium *Escherichia coli,* each of the paired strands of DNA contains about 5 million bases arranged in a particular linear sequence. The information in those bases is divided into units of several hundred bases each. Each of these units is a gene. Some of the genes and their locations on the chromosome of *E. coli* are shown in ➤ Figure 7.2.

We might think of a gene as a sentence in the language of nucleic acids. Each sentence in this language is constructed from a four-letter alphabet corresponding to the four nitrogenous bases in DNA: adenine (A), thymine (T), cytosine (C), and guanine (G). When these four letters combine to make "sentences" several hundred letters long, the number of possible sentences becomes almost infinite. Likewise, an almost infinite number of possible genes exists. If each gene contained 500 bases, a chromosome containing 5 million bases could contain 10,000 different genes. Thus, the information storage capacity of DNA is exceedingly large!

> *Haemophilus influenzae* is the first microbe to have its genome (1.83 kilobases) completely sequenced. Its sequence was published in *Science,* July 28, 1995.

Information Transfer

Information stored in DNA is used both to guide the replication of DNA in preparation for cell division and to direct protein synthesis. The three ways in which this information is transferred are as follows:

1. *Replication:* DNA makes new DNA.
2. *Transcription:* DNA makes RNA as the first step in protein synthesis.
3. *Translation:* RNA links amino acids together to form proteins.

In both DNA replication and transcription, DNA serves as a **template** (much like a sewing pattern) for the synthesis of a new nucleotide polymer. The sequence of bases in each new polymer is complementary to that in the original DNA. Such an arrangement is accomplished by base pairing. Recall from Chapter 2 that in complementary base pairing in DNA, adenine always pairs with thymine (A–T), and guanine always pairs with cytosine (G–C). Recall also that when DNA serves as a template for synthesis of RNA, the pairing is different: In RNA, thymine is replaced by uracil (U), which pairs with adenine.

In **DNA replication,** the new polymer is also DNA. In protein synthesis, the new polymer is a particular type of RNA called *messenger RNA (mRNA),* which then serves as a second template that dictates the arrangement of amino acids in a protein. For example, some proteins form the structure of a cell, others (enzymes) regulate its metabolism, and still others transport substances across a membrane.

> The *E. coli* genome consists of 4,639,221 base pairs, which code for at least 4288 proteins.

Enemy DNA

How do cells of your immune system identify foreign invaders, such as bacteria and viruses? Until now, immunologists have believed that proteins on the outer surface of invading cells and viruses were the triggers that alerted the body's defense system. Now, however, preliminary studies by Dr. Arthur M. Krieg of the University of Iowa College of Medicine indicate that even before your body responds to these surface proteins, it recognizes bacterial and viral DNA and begins an assault on the invaders. It detects a pattern of bases unique to bacteria and viruses—the frequent occurrence of C–G sequences. This combination of bases is uncommon in mammalian DNA. When it does occur, attached to the cytosine is a methyl group—a group that bacterial and viral C–G sequences do not have.

There is some evidence that people with systemic lupus erythematosus, an autoimmune disease in which the immune system attacks the body's own DNA, may not have normal ability to add methyl groups to their DNA. Thus their DNA may look like foreign DNA to the immune system. Perhaps the cure for this disease lies in increasing the patient's ability to add methyl groups. Another clinical application may lie in administering artificially produced C–G sequences to patients whose immune systems need stimulation. In the laboratory the addition of such C–G sequences to flasks of B cells (immune cells that make antibodies, or proteins that respond to foreign invaders) causes 95 percent of the cells to begin multiplying within a half hour. Further research is needed to see if the same effect occurs in whole organisms.

➤ Figure 7.2   **A partial chromosome map of *E. coli.*** (a) The entire genome of *E. coli* consists of approximately 3000 genes. The outer circle is a simplified representation of the chromosome, with a number of the most commonly studied genes marked on it. It takes about 100 minutes to transfer the entire chromosome from a donor to a recipient cell in conjugation (Chapter 8), a mechanism by which genes are transferred between bacteria. (b) These times are determined by allowing 2 different strains of bacteria to conjugate inside a blender that is turned on after a given number of minutes, causing the pili to break, halting chromosome transfer. Recipient cells are then grown and examined to determine which genes have been transferred to them. By varying the times used, it is possible to discover the order of the genes on the chromosome. The numbers marked inside the circle represent the number of minutes of transfer required to reach that point on the chromosome. The insert is a small segment of the *E. coli* map, enlarged to show some of the additional genes that have been located within that region (after Bachman).

Conjugation pilus

Chromosome crossing from one cell to the other

Blender is turned on after a given number of minutes and causes pilus to break, halting chromosome transfer

(b)

In the overall process of protein synthesis, the synthesis of mRNA from a DNA template is called **transcription,** and the synthesis of protein from information in mRNA is called **translation.** By analogy, transcription transfers information from one nucleic acid to another as you might transcribe handwritten sentences to typewritten sentences in the same language. Translation transfers information from the language of nucleic acids to the language of amino acids as you might translate English sentences into another language. There are even "proofreading" enzymes that try to eliminate any errors that occur, ensuring that a correct copy is passed on.

In the case of viruses that have RNA as their genetic material, scientists were unable to understand how these viruses could make more RNA. Then, the discovery of enzymes for **reverse transcription** revealed a process whereby RNA can make DNA. This DNA can then make more RNA. Such viruses are known as *retroviruses* because of this reverse process. (We will study them in greater detail in Chapter 10.) HIV, the virus that causes AIDS, is a retrovirus. Reverse transcription is a less accurate process than regular transcription. Uncorrected errors

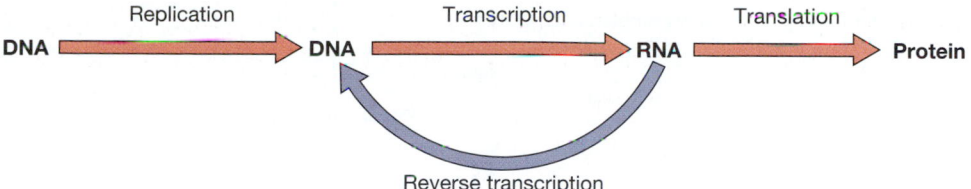

> Figure 7.3   **The transfer of information from DNA to protein.** As we shall see later, certain viruses, such as the one that causes AIDS, can direct synthesis of DNA from their RNA (reverse transcription).

✓ Compare and contrast chromosomes in prokaryotes and eukaryotes.

✓ DNA is not always the genetic material. What are the exceptions?

✓ How could mutations give rise to new alleles of a gene?

✓ How do bacterial DNA and viral DNA differ from human DNA? What disease may be related to this difference?

✓ How does translation differ from transcription?

---

are passed on as mutations, or permanent changes in the genes of an organism. HIV has a mutation rate 500 times higher than that of most organisms, an unfortunate fact for would-be vaccine makers.

DNA replication, transcription, and translation all transfer information from one molecule to another (➤ Figure 7.3). These processes allow information in DNA to be transferred to each new generation of cells and to be used to control the functioning of cells through protein synthesis.

DNA Replication

To understand DNA replication, we need to recall from Chapter 2 that pairs of helical DNA strands are held together by base pairing of adenine with thymine and cytosine with guanine. We also need to know that the ends of each strand are different. At one end, called the 3′ (3 prime) end, carbon 3 of deoxyribose is free to bind to other molecules. At the other end, the 5′ (5 prime) end, carbon 5 of deoxyribose is attached to a phosphate (➤ Figure 7.1). This structure is somewhat analogous to that of a freight train, with the 3′ end the engine and the 5′ end the caboose. When the two strands of a double helix combine by base pairing, they do so in a head-to-tail, or **antiparallel,** fashion. The arrangement of the strands is somewhat like two trains pointed in opposite directions, and base pairing is like passengers in the two trains shaking hands.

DNA replication begins at a specific location in the circular chromosome of a prokaryotic cell and usually proceeds simultaneously in both directions, creating two moving **replication forks,** the points at which the two strands of DNA separate to allow replication of DNA (➤ Figure 7.4). Various enzymes break the hydrogen bonds between the bases in the two DNA strands, unwind the strands from each other, and stabilize the exposed single strands, preventing them from joining back together. Molecules of the enzyme **DNA polymerase** then move along behind each replication fork, synthesizing new DNA strands complementary to the original ones at a speed of approximately 1000 nucleotides per second.

The enzyme DNA polymerase can add nucleotides only to the 3′ end of a growing DNA strand. Consequently, only one strand, the **leading strand** of original DNA, can serve as a template for the synthesis of a continuous new strand going in the 5′ to 3′ direction. Along the other strand, the **lagging strand,** which runs in the 3′ to 5′ direction, the synthesis of new DNA must be *discontinuous;* that is, the polymerase must continually jump ahead and work backward, making a series of short DNA segments called **Okazaki fragments.** The fragments are then joined together by another enzyme called a **ligase.** Ultimately, two separate chromosomes are formed (➤ Figure 7.4), each double helix consisting of one strand of old, or parent, DNA and one strand of new DNA. Such replication is called **semiconservative replication** because one strand is always conserved

Protein Synthesis

Transcription

All cells must constantly synthesize proteins to carry out their life processes: reproduction, growth, repair, and regulation of metabolism. This synthesis involves the accurate transfer of linear information of the DNA strands into a linear sequence of amino acids in proteins. To set the stage for protein synthesis, hydrogen bonds between bases in DNA strands are broken enzymatically in certain regions so that the strands separate. Short sequences of unpaired DNA

If DNA Makes Only Proteins, Who Makes Carbohydrates and Lipids?

If genetic information in DNA is used specifically to determine the structure of proteins, how are the structures of carbohydrates and lipids determined? Stop and think of the kinds of proteins a cell has. Many are enzymes, and, of course, some of those enzymes direct the synthesis of carbohydrates and lipids. The entire cell is controlled by DNA—either directly, in DNA replication and synthesis of structural proteins, or indirectly, by the synthesis of enzymes that in turn control the synthesis of carbohydrates and lipids.

➤ Figure 7.4   **DNA replication in a prokaryote.** DNA strands separate, and replication begins at a replication fork on each strand. As synthesis proceeds, each strand of DNA serves as a template for the replication of its partner. Note the antiparallel arrangement of the complementary strands of the DNA double helix. Because synthesis of new DNA can take place in only one direction, the process must be discontinuous along one strand. Short segments are formed and then spliced together, as the arrows indicate. Each new cell can undergo subsequent replications.

bases are thus exposed to serve as templates in transcription. Only one strand directs the synthesis of mRNA for any one gene; the complementary strand is used as a template during DNA replication or during the transcription of some other gene. Recall that RNA contains the base uracil instead of thymine. ➡ (Chapter 2, p. 44) Thus, when mRNA is transcribed from DNA, uracil pairs with adenine; otherwise, base pairing occurs just as it does in DNA replication. Messenger RNA is formed in the 5′ to 3′ direction.

For transcription to occur, a cell must have sufficient quantities of nucleotides containing high-energy phosphate bonds, which provide energy for the nucleotides to participate in subsequent reactions. The enzyme **RNA polymerase** binds to one strand of exposed DNA. As shown in ➤ Figure 7.5, after an enzyme binds to the first base in DNA (adenine, in this case), the appropriate nucleotide joins the DNA base–enzyme complex. The new base then attaches by base pairing to the template base of DNA. The enzyme moves to the next DNA base, and the appropriate phosphorylated nucleotide joins the complex. The phosphate of the second nucleotide is linked to the ribose of the first nucleotide, and *pyrophosphate* (two attached molecules of phosphate) is released. This forms the first link in a new polymer of RNA. Energy to form this link comes from the hydrolysis of ATP and the release of two more phosphate groups. This process is repeated until the RNA molecule is completed.

In prokaryotes, transcription and translation both take place in the cytoplasm, whereas in eukaryotes, transcription takes place in the cell nucleus. The mRNA of transcription must be completely formed and transported through the nuclear envelope to the cytoplasm before translation can begin. Moreover, the mRNA molecule undergoes additional processing before it is ready to leave the nucleus. In eukaryotic cells the regions of genes that code for proteins are called **exons.** Exons are typically separated within a gene by DNA segments that do not code for proteins. Such noncoding *intervening regions* are called **introns.** In the nucleus, RNA polymerase first forms mRNA from the entire gene, including all exons and introns. The newly formed long mRNA molecule is streamlined by other enzymes, which remove the introns and splice together the exons. The resulting mRNA is ready to direct protein synthesis and to leave the nucleus (➤ Figure 7.6).

Kinds of RNA

Three kinds of RNA—*ribosomal RNA, messenger RNA,* and *transfer RNA*—are involved in protein synthesis. Each RNA consists of a single strand of nucleotides and is synthesized by transcription, using DNA as a template. To complete the story of protein synthesis, we will need more information about these types of RNA.

Ribosomal RNA (rRNA) binds closely to certain proteins to form two kinds of ribosome subunits. A subunit of each kind combines to form a ribosome. Recall that ribosomes are sites of protein synthesis in a cell. ➡ (Chapter 4, p. 82) They serve as binding sites for transfer RNA, and some of their proteins act as enzymes that control protein synthesis. Prokaryotic ribosomes are made of a small (30S) and a large (50S) subunit. (Eukaryotic ribosomes are formed from a 40S and a 60S subunit.) After the two subunits join together around the strand of mRNA (➤ Figure 7.7), peptide synthesis occurs. The newly formed polypeptide chain grows out via a tunnel in the 50S subunit.

➤ **Figure 7.5   The transcription of RNA from template DNA.** The —PPP represents a triphosphate, and PP represents pyrophosphate. In RNA, U (rather than T) pairs with A.

Eukaryotic
DNA

| E | I | E | I | E | I | E |

(A) Entire gene (both introns and exons) is transcribed to RNA by RNA polymerase enzymes

mRNA
transcript

| E | I | E | I | E | I | E |

(B) Processing enzymes remove introns

Nuclear
membrane

Pore in
nuclear
membrane

| E | I | E | I | E | I | E |

| I | | I | | I |

Excised
introns

| E | E | E | E |

Cytoplasm

(C) Exons are spliced together, without introns, forming mRNA that can pass through the nuclear membrane into the cytoplasm and be translated. Note that while introns were transcribed, they will not be translated.

Nucleus

➤ Figure 7.6   **Eukaryotic genes differ in complexity from prokaryotic genes.** Coding sequences of DNA called exons (E) are interrupted by noncoding sequences called introns (I). While both are transcribed into RNA, the introns are removed, leaving only the spliced-together exons to enter the cytoplasm where they will be translated. Prokaryotes lack introns.

➤ Figure 7.7   **Prokaryotic ribosomal structure.** The small (30S) and large (50S) subunits are shown from two different angles. The subunits enfold the mRNA strand. The region of peptide synthesis is the junction of these three components. The growing polypeptide chain passes through a tunnel in the 50S subunit, which can be seen in cross section.

Large 50S
subunit

mRNA

Region of
peptide synthesis

3'

5'

Large (50S)
subunit

Tunnel

Small (30S)
subunit

Growing polypeptide chain

Growing polypeptide chain

Messenger RNA (mRNA) is synthesized in units that contain sufficient information to direct the synthesis of one or more polypeptide chains. One mRNA molecule corresponds to one or more genes, the functional units of DNA. Each mRNA molecule becomes associated with one or more ribosomes. At the ribosome, the information coded in mRNA acts during translation to dictate the sequence of amino acids in the protein.

In translation, each triplet (sequence of three bases) in mRNA constitutes a **codon** (ko′don). Codons are the "words" in the language of nucleic acids. Each codon specifies a particular amino acid or acts as a terminator codon. The first codon in a molecule of mRNA acts as a **start codon.** It always codes for the amino acid methionine, even though the methionine may be removed from the protein later. The last codon to be translated in a molecule of mRNA is a **terminator,** or **stop codon.** It acts as a kind of punctuation mark to indicate the end of a protein molecule. Using a sentence as an analogy, the methionine codon is the capital letter at the beginning of the sentence, and the terminator codon is the punctuation mark at the end.

At least one codon exists for each of the 20 amino acids found in proteins. Several codons exist for some amino acids; for example, six different codons code for leucine. Find these in ➤ Figure 7.8. The relationship between each codon and a specific amino acid constitutes the **genetic code** (➤ Figure 7.8). Those codons that code for an amino acid are called **sense codons.** Early in the study of the genetic code, investigators found a few codons that did not code for any amino acid. Those codons were therefore named **nonsense codons.** It was later found that they were stop codons. Although genetic information is stored in DNA, the genetic code is written in codons of mRNA. Of course, the information in the codons is derived *directly* from DNA by complementary base pairing during transcription.

Comparisons of the codons among different organisms have shown them to be nearly the same in all organisms, from bacteria to humans. This universality of the genetic code allows research on other organisms to be applied to the understanding of information transmission in human cells. Much of what is known about how the genetic code operates has been learned from research on bacteria.

The function of **transfer RNA (tRNA)** is to transfer amino acids from the cytoplasm to the ribosomes for placement in a protein molecule. Many different kinds of tRNAs have been isolated from the cytoplasm of cells. A tRNA molecule consists of 75 to 80 nucleotides and is folded back on itself to form several loops that are stabilized by complementary base pairing (➤ Figure 7.9). Each tRNA has a three-base **anticodon** (an′ti-ko″don) that is complementary to a particular mRNA codon. It also has a binding site for an amino acid—the particular amino acid specified by the mRNA codon. (The mRNA codon, of course, got its information directly from DNA.) Thus, the tRNAs are the link between the codons and the corresponding amino

First position	Second position				Third position
	U	**C**	**A**	**G**	
U	UUU Phe UUC Phe UUA Leu UUG Leu	UCU Ser UCC Ser UCA Ser UCG Ser	UAU Tyr UAC Tyr UAA Stop UAG Stop	UGU Cys UGC Cys UGA Stop UGG Trp	U C A G
C	CUU Leu CUC Leu CUA Leu CUG Leu	CCU Pro CCC Pro CCA Pro CCG Pro	CAU His CAC His CAA Gln CAG Gln	CGU Arg CGC Arg CGA Arg CGG Arg	U C A G
A	AUU Ile AUC Ile AUA Ile AUG Met	ACU Thr ACC Thr ACA Thr ACG Thr	AAU Asn AAC Asn AAA Lys AAG Lys	AGU Ser AGC Ser AGA Arg AGG Arg	U C A G
G	GUU Val GUC Val GUA Val GUG Val	GCU Ala GCC Ala GCA Ala GCG Ala	GAU Asp GAC Asp GAA Glu GAG Glu	GGU Gly GGC Gly GGA Gly GGG Gly	U C A G

➤ Figure 7.8 **The genetic code, with standard three-letter abbreviations for amino acids.** To find the amino acid for which the mRNA codon AGU codes, go down the left column to the block labeled A, move across to the fourth square labeled G at the top of the figure, and find the first line in the square labeled U on the right side of the figure. There you will find Ser, the abbreviation for serine. Stop designates a terminator codon. The Start codon is AUG, which also codes for methionine. Therefore, protein synthesis always begins with methionine. The methionine is usually removed later, however, so not all proteins actually start with methionine. When found in the middle of an mRNA strand, AUG codes for methionine.

(a)

(b)

➤ Figure 7.9   **Transfer RNA.** (a) The two-dimensional structure of the tryptophan transfer RNA. The anticodon end will pair up with a codon on a strand of messenger RNA and deliver the desired amino acid, which is bonded to the acceptor stem at its opposite end. The molecule is maintained in its cloverleaf pattern by hydrogen bonding between strands that form the arms (dashed lines). (b) A tRNA molecule folded into its complex three-dimensional shape, in diagram form and as a computer-generated model.

➤ Table 7.1	**Properties of the different kinds of RNA**
Kind of RNA	**Properties**
Ribosomal	Combines with specific proteins to form ribosomes.
	Serves as a site for protein synthesis.
	Associated enzymes function in controlling protein synthesis.
Messenger	Carries information from DNA for synthesis of a protein.
	Molecules correspond in length to one or more genes in DNA.
	Has base triplets called codons that constitute the genetic code.
	Attaches to one or more ribosomes.
Transfer	Found in the cytoplasm, where they pick up amino acids and transfer them to mRNA.
	Molecules have a cloverleaf shape with an attachment site for a specific amino acid.
	Each has a single triplet of bases called an anticodon, which pairs complementarily with the corresponding codon in mRNA.

acids. Amino acid attachment to specific tRNA molecules is achieved by the action of amino acid-activating enzymes and energy derived from ATP.

The anticodon attaches by complementary base pairing to the appropriate mRNA codon such that its amino acid is aligned for incorporation into a protein. The accuracy of amino acid placement in protein synthesis depends on this precise pairing of codons and anticodons. The properties of the three types of RNA are summarized in Table 7.1.

Translation

Protein synthesis, an important process in bacterial growth, uses 80 to 90 percent of a bacterial cell's energy. Generally, during protein synthesis, the various RNAs and amino acids are available in sufficient quantities. The RNAs can be reused many times before they lose their ability to function. Of the types of RNA, mRNA is produced in the most precise quantity in accordance with the cell's need for a particular protein. ➤ Figure 7.10 shows the three types of RNA and how they function in protein synthesis.

(a)   (b)

➤ Figure 7.10 **Transcription and translation.** (a) Transcription from DNA to RNA. (b) Translation from RNA to protein. Many ribosomes that are connected to and read the same piece of mRNA are called a polyribosome.

> Figure 7.11 **Concurrent transcription and translation in prokaryotes.** A portion of *E. coli* DNA runs horizontally across this electron micrograph. On either side, transcripts of mRNA can be seen. Ribosomes have attached to the mRNA and are synthesizing proteins (which cannot be seen). The mRNA can be seen increasing in length from left to right, indicating the direction of transcription.

✓ Distinguish between leading and lagging strands.

✓ What do 5′ and 3′ refer to? How do they determine the direction of synthesis of new DNA?

✓ Does the genetic code have synonyms?

✓ Contrast the three kinds of RNA. Does DNA make all three kinds?

Once an mRNA molecule has been transcribed and has combined with a ribosome, the ribosome initiates protein synthesis and provides the site for protein assembly. Each ribosome attaches first to the end of the mRNA that corresponds to the beginning of a protein. The length of each polypeptide chain extending from a ribosome corresponds to the amount of mRNA the ribosome has "read." Several ribosomes can be attached at different points along an mRNA molecule to form a **polyribosome** (or *polysome*).

In prokaryotes (unlike eukaryotes), transcription and translation take place in the cytoplasm (➤ Figure 7.11), where all necessary enzymes and ribosomes are present. In eukaryotes, the mRNA formed in the nucleus must pass through the nuclear membrane before it is available to the ribosomes, which carry out protein synthesis.

The main steps in protein synthesis (➤ Figure 7.12) can be summarized as follows: The process begins when a molecule of mRNA becomes properly oriented on a ribosome. As each codon of the mRNA is "read," the appropriate tRNA combines with it and thereby delivers a particular amino acid to the protein assembly site. The location on the ribosome where the first tRNA pairs is called the *P site*. The second codon of the mRNA then pairs with a tRNA that transports the second amino acid to the *A site*, which is next to the P site. Matching of codon and anticodon by base pairing allows coded information in mRNA to specify the sequence of amino acids in a protein. Any tRNAs with nonmatching anticodons simply do not bind to the ribosome. As amino acids are delivered one after another and peptide bonds form between them, the length of the polypeptide chain increases. This process continues until the ribosome recognizes a stop codon. When the ribosome "reads" a stop codon at the A site, it releases the finished protein from the P site.

Any mRNA molecule can direct simultaneous synthesis of many identical protein molecules—one for each ribosome passing along it. Ribosomes, mRNAs, and tRNAs are reusable. The tRNAs shuttle back and forth between the cytoplasm, where they pick up amino acids, and the ribosome, where the amino acids are incorporated into protein.

The Regulation of Metabolism

The Significance of Regulatory Mechanisms

Bacteria use most of their energy to synthesize substances needed for growth. These substances include structural proteins, which form cell parts, and enzymes, which control both energy production and synthetic reactions. The survival of bacteria depends on their ability to grow even when conditions are less than ideal—for example, when nutrients are in short supply. In their evolution, cells of bacteria (and all other organisms) have developed mechanisms to turn reactions on and off in accordance with their needs. Energy and materials are too valuable to waste. Also, the cell has a limited amount of space for storing excesses of materials it synthesizes. Thus, cells use energy to synthesize substances in the amounts needed and shut off these processes before wasteful excesses are produced.

All living organisms are presumed to have control mechanisms that regulate their metabolic activities. However, more research on control mechanisms has been done on bacteria than on all other organisms. Bacteria are ideal for such studies for several reasons:

1. They can be grown in large numbers relatively inexpensively under a variety of controlled environmental conditions.

2. They produce many new generations quickly.

3. Because they reproduce so rapidly, a variety of mutations can be observed in a relatively short time.

Mutant organisms that have an alteration in their control mechanisms can be isolated and studied along with nonmutated organisms to understand better the operation of control mechanisms.

Categories of Regulatory Mechanisms

The mechanisms that control metabolism either regulate enzyme activity directly or regulate enzyme synthesis by turning on or off genes that code for particular enzymes. Of the various mechanisms that regulate metabolism, three have been extensively investigated in bacteria:

- feedback inhibition
- enzyme induction
- enzyme repression

In *feedback inhibition,* enzyme activity is regulated directly, and the control mechansim determines how rapidly enzymes already present will catalyze reactions. In *enzyme induction* and *enzyme repression,* regulation occurs indirectly by enzymes synthesis, and the control mechanism determines which enzymes will be synthesized and in what amounts.

Feedback Inhibition

In **feedback inhibition,** also called **end-product inhibition,** the end product of a biosynthetic pathway directly inhibits the first enzyme in the pathway. This mechanism was discovered when it was observed that the addition of one of several amino acids to a growth medium could cause a bacterium suddenly to stop synthesizing that particular amino acid. Synthesis of the amino acid threonine, for example, is regulated by feedback inhibition. Threonine is made from aspartate, and the allosteric enzyme that acts on aspartate is inhibited by threonine (➤ Figure 7.13). (Aspartate is derived from oxaloacetate formed in the Krebs cycle.) When an inhibitor (threonine) attaches to the allosteric site, it alters the enzyme's shape so the substrate (aspartate) cannot attach to the active site. ➥ (Chapter 5, p. 112) Thus, feedback inhibition occurs when the end product of a reaction sequence binds to the allosteric site of the enzyme for the first step in the sequence.

Feedback inhibition regulates the synthesis of various substances other than amino acids (pyrimidines, for example). This regulatory mechanism also occurs in many organisms other than bacteria. Because feedback inhibition acts quickly and directly on a metabolic process, it allows the cell to conserve energy in two ways:

1. When it is plentiful, the inhibitor (end product) attaches to the enzyme; when it is in short supply, it is released from the enzyme. Thus, the cell expends energy to synthesize the end product only when it is needed.

2. Regulation of enzyme activity requires less energy than the more complex processes involved in the regulation of gene expression.

You Can't Beat the System

Inducible enzyme systems are important in humans as well as in microbes. If diabetes mellitus is suspected, patients are sent for a glucose tolerance test, where they ingest a measured load of glucose. The fluctuations of blood glucose are measured. Failure to remove glucose from the bloodstream at the normal rate *could* indicate diabetes—maybe. Some patients fearing a diagnosis of diabetes plan to "beat" the test by not eating sugar for days before hand. But they also won't have good supplies of enzymes to use the load of sugar. These enzymes are inducible, and the patients haven't been eating inducer (sugar). They can't instantly produce the enzymes, so their blood glucose level will drop slower than normal. A patient should eat at least the amount of sugar in a candy bar each day for the three days preceding the test to induce as much enzyme as possible.

➤ Figure 7.12   **Protein synthesis.** Steps 1–5: The main steps in protein synthesis. Steps 6 and 7: Many ribosomes can "read" the same strand of mRNA simultaneously. The ribosomes are shown moving from left to right.

Enzyme Induction

At one point in the investigation of metabolic regulation, it was discovered that certain organisms always contain active enzymes for glucose metabolism even when glucose is not present in the medium. Such enzymes are called **constitutive enzymes;** they are synthesized continuously regardless of the nutrients available to the organism. The genes that make these enzymes are always active. In contrast, enzymes that are synthesized by genes that are sometimes active and sometimes inactive depending on the presence or absence of substrate are called **inducible enzymes.**

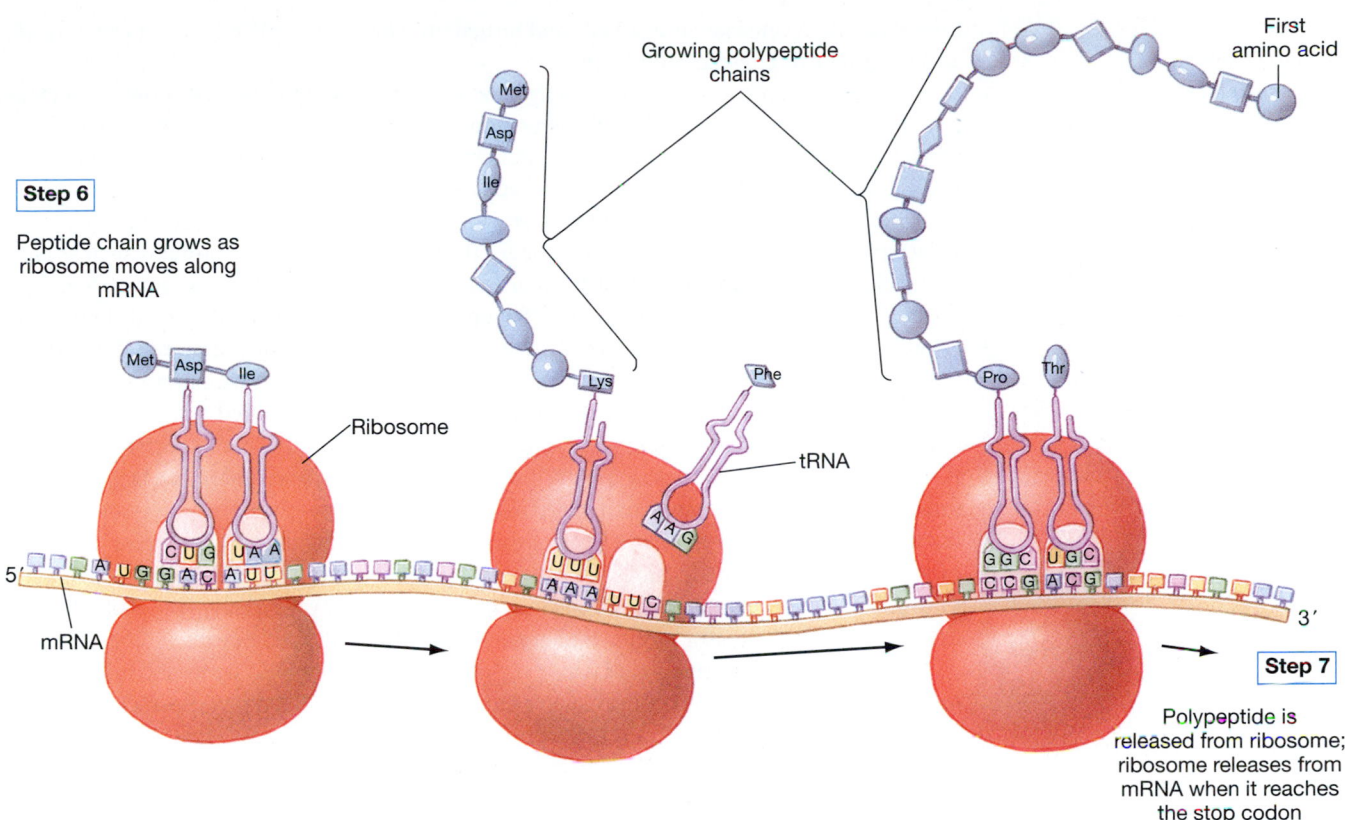

Step 6

Peptide chain grows as ribosome moves along mRNA

Growing polypeptide chains

First amino acid

Met
Asp
Ile

Met Asp Ile

Ribosome

Lys

Phe

Pro Thr

tRNA

AAG

5′ AUGGACAUU

CUG UAA

UUU
AAA UUC

GGC UGC
CCG ACG

3′

mRNA

Step 7

Polypeptide is released from ribosome; ribosome releases from mRNA when it reaches the stop codon

➤ **Figure 7.12** **(continued)**

When bacteria such as *E. coli* are grown on a nutrient medium that contains no lactose, the cells do not make any of the enzymes that they would need to utilize lactose as an energy source. When lactose is present, however, the cells synthesize the enzymes needed for its metabolism. This phenomenon is an example of **enzyme induction.** Enzyme induction controls the breakdown of nutrients as they become available in the growth medium. Such a system is

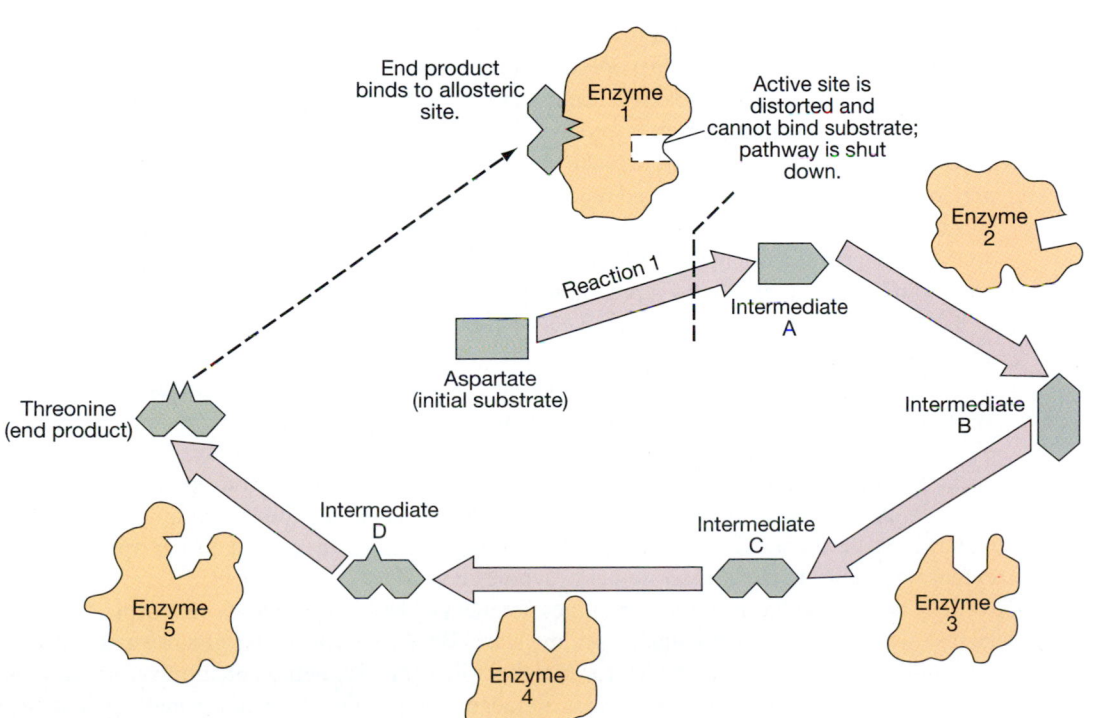

End product binds to allosteric site.

Enzyme 1

Active site is distorted and cannot bind substrate; pathway is shut down.

Enzyme 2

Reaction 1

Intermediate A

Aspartate (initial substrate)

Threonine (end product)

Intermediate B

Intermediate D

Intermediate C

Enzyme 5

Enzyme 4

Enzyme 3

➤ **Figure 7.13** **Feedback inhibition.** The synthesis of threonine involves five enzymatically controlled reactions (arrows), and four intermediate products (A, B, C, and D). Threonine (the end product) inhibits an allosteric enzyme (1) that catalyzes Reaction 1. The allosteric enzyme is functional when its allosteric site is not occupied and is nonfunctional when the end product of a sequence of reactions is bound to that site.

turned on when a nutrient is available and turned off when it is depleted. The nutrient itself acts as an **inducer** of enzyme production.

The *operon* (op'er-on) *theory,* a model that explains the regulation of protein synthesis in bacteria, was proposed in 1961 by French scientists Francois Jacob and Jacques Monod, who received a Nobel Prize in 1965 for their work. Although the theory applies to several operons, we will illustrate it with the *lac* operon, which regulates lactose metabolism. An **operon** is a sequence of closely associated genes that regulate enzyme production. An operon includes one or more **structural genes,** which carry information for the synthesis of specific proteins such as enzyme molecules, and **regulatory sites,** which control the expression of the structural genes. A **regulator** (*i*) **gene** works in conjunction with the operon but may be located some distance from it. In prokaryotes several structural genes are controlled by one operon—a more efficient method than that of eukaryotes, in which each gene is controlled by its own regulatory site.

The *lac* operon (➤ Figure 7.14) consists of regulatory sites called a *promoter* and an *operator,* and three structural genes *Z, Y,* and *A,* which direct synthesis of specific enzymes. An RNA polymerase molecule must bind to the promoter before transcription can begin. When lactose is not present in the medium, the separate *i* gene directs synthesis of a substance called the *lac repressor*. The **repressor** is a protein that binds to the operator and prevents transcription of the adjacent *Z, Y,* and *A* genes. Consequently, the enzymes that metabolize lactose are not synthesized. The *i* gene represents a *constitutive gene*—it is always undergoing protein synthesis to produce more repressor protein and is not controlled by the promoter.

When present in the medium, lactose acts as the inducer by binding to and inactivating the *lac* repressor. The repressor then no longer blocks the operator. The RNA polymerase then binds to the promoter, causing the operator to initiate transcription of the *Z, Y,* and *A* genes as a single long strand of mRNA. This mRNA becomes associated with ribosomes and directs synthesis of three enzymes: β-galactosidase (*Z* gene), permease (*Y* gene), and transacetylase (*A* gene). Permease transports lactose into cells, and β-galactosidase breaks down lactose into glucose and galactose. Although the role of transacetylase is not clear, it is thought to facilitate the escape of galactosides. When the available lactose has been broken down, it is no longer available to bind to the repressor. The active repressor again binds to the operator, and the operon is turned off.

Enzyme Repression

In contrast to enzyme induction, which typically regulates catabolism, **enzyme repression** typically regulates anabolism. It controls processes in which substances needed for growth are synthesized. Synthesis of the amino acid tryptophan, for example, is regulated by enzyme repression through actions of the *trp* operon, which consists of five structural genes.

When tryptophan is available to a bacterial cell, the amino acid binds to an inactive repressor. Binding activates the repressor protein, which can now bind to the promoter and repress synthesis of the enzymes needed to make tryptophan. When tryptophan is not available, the repressor protein remains inactive, and repression does not occur. Structural genes are transcribed, and tryptophan is synthesized. When tryptophan becomes plentiful, it again represses the operon. An even finer control mechanism, called **attenuation,** allows transcription of the *trp* operon to begin but terminates it prematurely by a complex process when sufficient amounts of tryptophan are already present in the cell. Several operons, especially those for amino acid synthesis, have attenuation mechanisms.

A slightly different kind of repression operates in connection with some catabolic pathways. When certain bacteria (*E. coli,* for example) are grown on a nutrient medium containing both glucose and lactose, they grow at a logarithmic rate as long as glucose is available. When the glucose is depleted, they enter a stationary phase but soon begin to grow again at a logarithmic rate, though not quite as rapidly (➤ Figure 7.15). This time the logarithmic growth rate results from the metabolism of lactose. The stationary phase is the period during which the enzymes needed to utilize lactose are being synthesized.

Why was the synthesis of these enzymes not induced before the glucose was depleted, since lactose was present in the medium from the start? Because bacteria use glucose as a nutrient with high efficiency. The enzymes for metabolizing glucose, being constitutive, are always present in the cell. Thus, when glucose is abundant, there is no advantage in making enzymes for metabolizing lactose even if lactose is also available. Consequently, the *lac* operon that we

➤ Figure 7.14 **Enzyme induction.** The mechanisms of operation of the *lac* operon. (a) In the absence of lactose, the repressor is synthesized and binds to the operator, preventing transcription of the genes coding for enzymes used to metabolize lactose. (b) When lactose is present, it binds to the repressor and inactivates it. The structural genes of the operon are transcribed, and enzymes for metabolizing lactose are synthesized. The regulator (*i*) gene may be some distance away from the operon.

described previously is repressed when glucose is present in adequate quantities, an effect known as **catabolite repression.** In this way the cell saves energy by not making enzymes it doesn't need. When glucose supplies fall, the repression is lifted, the *lac* operon genes are transcribed, and the cell is ready to switch over to using lactose. In short, the transcription of the *lac* operon requires both that lactose be present and that glucose *not* be present.

Both enzyme induction and enzyme repression are regulatory mechanisms that control enzyme production by altering gene expression. Although these two mechanisms have different effects, they actually represent two examples of the operation of a single mechanism for turning genes on and off (Table 7.2).

✓ What "feeds back" in feedback inhibition? What does it inhibit? How does it do this?

✓ What is the inducer for the *lac* operon?

✓ Compare enzyme induction and enzyme repression.

➤ Figure 7.15 **Catabolite repression.** The growth curve for bacteria in a medium initially containing both glucose and lactose. When glucose is used up, growth stops temporarily but begins again at a slower rate, using lactose as an energy source.

Mutations

Mutations, or changes in DNA, can now be defined more precisely as heritable changes in the sequence of nucleotides in DNA. Mutations account for evolutionary changes in microorganisms (and larger organisms) and for alterations that produce different strains within species. Here we will consider what kinds of changes occur in DNA during mutations and how these changes affect the organisms.

Types of Mutations and Their Effects

Before we can consider mutations and their effects, we need to distinguish between an organism's genotype and its phenotype. **Genotype** refers to the genetic information contained in the DNA of the organism. **Phenotype** refers to the specific characteristics displayed by the organism. Mutations always change the genotype. Such a change may or may not be expressed in the phenotype, depending on the nature of the mutation.

Two important kinds of mutations are *point mutations,* which affect a single base, and *frameshift mutations,* which can affect more than one base in DNA. Mutations often make an organism unable to synthesize one or more proteins. The absence of a protein often leads to changes in the organism's structure or in its ability to metabolize a particular substance.

A **point mutation** involves base substitution, or nucleotide replacement, in which one base is substituted for another at a specific location in a gene (➤ Figure 7.16). The mutation changes a single codon in mRNA, and it may or may not change the amino acid sequence in a protein. Let's look at some examples.

Suppose a three-base sequence of DNA is changed from AAA to AAT. During transcription the mRNA codon will change from UUU to UUA. (Recall that uracil in RNA pairs with adenine in DNA.) ➡ (Chapter 2, p. 44) When the information in the mRNA is used to synthesize protein, the amino acid leucine will be substituted for phenylalanine in the protein. (To verify this for yourself, refer to the genetic code in ➤ Figure 7.8.) Because of the single amino acid

➤ Table 7.2	Effects of regulatory systems involving an operon		
Regulatory mechanism (example)	**Type of pathway regulated**	**Regulating substance**	**Condition that leads to gene expression**
Enzyme induction (*lac* operon)	Catabolic (degradational) and releases energy	Nutrient (lactose)	Presence of nutrient (lactose)
Enzyme repression (*trp* operon)	Anabolic (biosynthetic) and uses energy	End product (tryptophan)	Absence of end product (tryptophan)

substitution, the new protein will be different from the normal protein. The effects on the phenotype of the organism will be negligible if the new protein functions as well as the original one. They will be significant if the new protein functions poorly or not at all. In rare instances the new protein may function better and produce a phenotype that is better adapted to its environment than the original phenotype.

Should the code in DNA be changed from AAA to AAG, the mRNA code becomes UUC instead of UUU. Because the UUC and UUU codons both code for phenylalanine, the mutation has no effect on the protein being synthesized. In this case, though the genotype has changed, the phenotype is unaffected.

Sometimes the substitution of a single base in DNA produces a terminator codon in mRNA. If the terminator codon is introduced in the middle of a molecule of mRNA destined to produce a single protein, synthesis will be terminated part way through the molecule. A polypeptide that will most likely be unable to function in the cell will be released, and the appropriate protein will not be synthesized. If the missing protein is essential to cell structure or function, the effect can be lethal.

A **frameshift mutation** is a mutation in which there is a **deletion** or an **insertion** of one or more bases (➤ Figure 7.17). Such mutations alter all the three-base sequences beyond the deletion or insertion. When mRNA transcribed from such altered DNA is used to synthesize a protein, many amino acids in the sequence may be altered. (Remember, a ribosome reads an mRNA in codons, sets of three bases.) Such mutations also commonly introduce terminator codons and cause protein synthesis to stop when only a short polypeptide has been made.

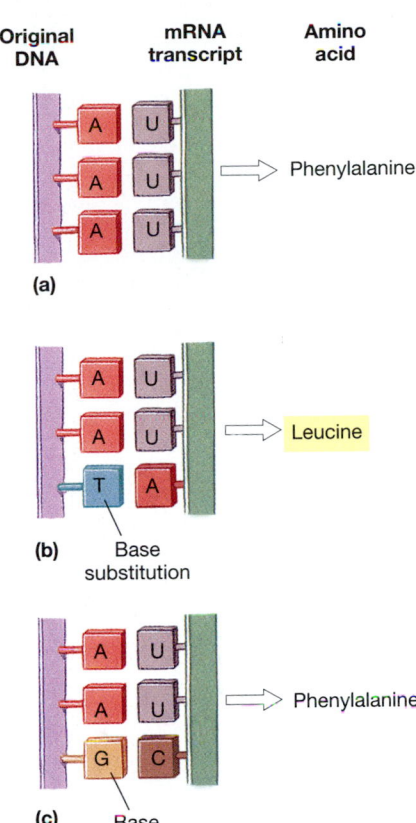

➤ Figure 7.16 **The effects of base substitution (a point mutation).** The resulting protein may or may not be significantly affected, depending on whether the new codon specifies the same amino acid, one with similar properties, or an entirely different one.

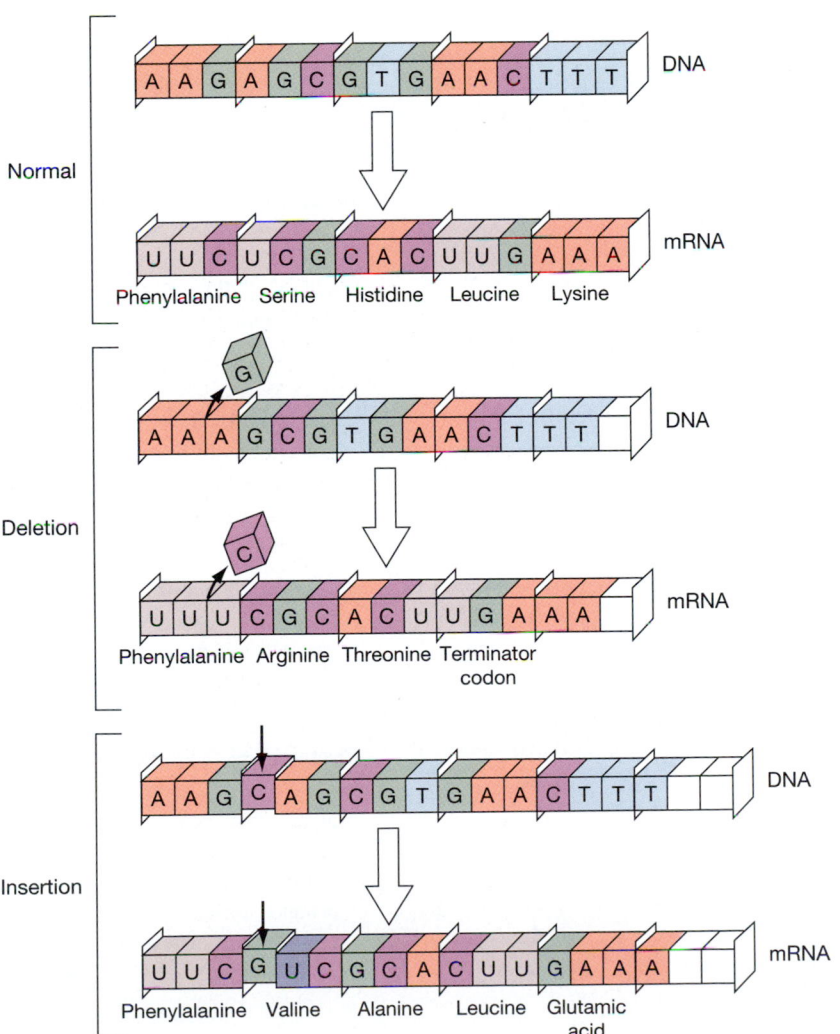

➤ Figure 7.17 **The effects of frameshift mutations.** The addition or deletion of one or more nucleotides changes the amino acid sequence coded for by the entire gene from that point on. (The addition or deletion of three nucleotides might not affect the resulting protein very much. Can you see why?)

Frameshift mutations usually prevent synthesis of a particular protein, and they change both the genotype and the phenotype. Their effect on the organism depends on the role of the missing protein in the organism's function. These types of mutations and their effects are summarized in Table 7.3. If three bases, or a multiple of three bases, were inserted or lost, what would happen? One or more amino acids would be gained or lost.

Phenotypic Variation

Phenotypic variations frequently seen in mutated bacteria include alterations in colony morphology, colony color, or nutritional requirements. For example, instead of being a normal smooth, glossy, raised colony, a colony with mutated DNA may have a flat, rough appearance. The mutation has impaired synthesis of certain cell surface substances. In organisms that typically form capsules, the mutation can prevent synthesis of capsular polysaccharides. Mutations that alter nutritional requirements generally increase the nutritional needs of an organism, usually by impairing the organism's ability to synthesize one or more enzymes. As a result, the organism may require certain amino acids or vitamins in its medium because it can no longer make them itself.

Studies of bacteria that have lost the ability to synthesize a particular enzyme have played an important role in our understanding of metabolic pathways. Such nutritionally deficient mutants are called **auxotrophs** (awks′o-trofs″; *auxo*, increase, and *trophos*, food); they require special substances in their medium to maintain growth. In contrast to auxotrophs, normal nonmutant forms are called **prototrophs** (pro′to-trofs), or *wild types*. Comparisons of characteristics of auxotrophs and prototrophs show the effects of a mutation on metabolism. By observing which metabolites accumulate and which nutrients must be added to the medium

➤ Table 7.3	Types of mutations and their effects on organisms
Types of mutations	**Effects on organisms**
Point mutation	
Single base change in DNA with no change in the amino acid specified by the mRNA codon.	No effect on protein; a "silent" mutation.
Change in DNA with change in the amino acid sequence specified by the mRNA codon.	Change in protein by substitution of one amino acid for another; can significantly alter function of protein.
Change in DNA that creates a terminator codon in mRNA.	Produces polypeptide of no use to organism and prevents synthesis of normal protein.
Frameshift mutation	
Deletion or insertion of one or more bases in DNA.	Changes entire sequence of codons and greatly alters amino acid sequence; can introduce terminator codon and produce useless polypeptides instead of normal proteins.

of auxotrophs, microbiologists have determined the specific steps in the metabolism of certain substances.

Still another type of phenotypic variation of genetic origin is temperature sensitivity. For example, suppose that an organism at one time could grow over a wide range of environmental temperatures. As a result of a mutation, it loses the ability to grow at the higher temperatures of its former range. It can still grow at 25°C, but it can no longer grow at 40°C. This phenomenon may be due to a point mutation that changed a single amino acid in an enzyme. The slightly altered enzyme may function at moderate temperatures but may be easily denatured and inactivated at higher temperatures.

Some phenotypic variations are caused by environmental factors and occur without any change in the genotype (alteration in DNA). For example, large amounts of sugar or irritants in the medium can cause some organisms to form a larger-than-normal capsule. Some organisms, such as the anthrax bacterium, form spores in open air, in spilled blood, or on tissue surfaces but not inside tissues. Variations in environmental temperature can affect pigment synthesis. *Serratia marcescens* usually produces pigment at room temperature but may not do so at higher temperatures. It has the gene for pigment production, but the gene is expressed only at certain temperatures.

Spontaneous and Induced Mutations

Mutations appear to be random or chance events; it is usually impossible to predict when a mutation will occur or which genes will be altered. Although all mutations result from permanent changes in DNA, they can be spontaneous or induced. **Spontaneous mutations** occur in the absence of any agent known to cause changes in DNA. They arise during the replication of DNA and appear to be due to errors in the base pairing of nucleotides in the old and new strands of DNA. Various genes in the DNA of bacteria have different spontaneous *mutation rates,* ranging from 10^{-3} to 10^{-9} per cell division. In other words, one gene might undergo a spontaneous mutation once in every thousand ($1/10^3$) cell divisions, whereas another gene might undergo a spontaneous mutation only once in every billion ($1/10^9$) cell divisions. **Induced mutations** are produced by agents called **mutagens,** which increase the mutation rate above the spontaneous mutation rate. Mutagens include chemical agents and radiation (Table 7.4).

➤ **Table 7.4** **Some mutagens and their effects**

Mutagen	Effects
Chemical agents	
Base analog *Examples:* caffeine, 5-bromouracil	Substitutes "look-alike" molecule for the normal nitrogenous base during DNA replication ⟶ point mutation.
Alkylating agent *Example:* nitrosoguanidine	Adds an alkyl group, such as methyl group ($-CH_3$), to nitrogenous base, resulting in incorrect pairing ⟶ point mutation.
Deaminating agent *Example:* nitrous acid, nitrates, nitrites	Removes an amino group ($-NH_2$) from a nitrogenous base ⟶ point mutation.
Acridine derivative *Examples:* acridine dyes, quinacrine	Inserts into DNA ladder between backbones to form a new rung, distorting the helix ⟶ frameshift mutation.
Radiation	
Ultraviolet	Links adjacent pyrimidines to each other, as in thymine dimer formation, and thereby impairs replication.
X-ray and gamma ray	Ionize and break molecules in cells to form free radicals, which in turn break DNA.

Thymine

5-Bromouracil

➤ **Figure 7.18   Base analogs.** The similarity between the structure of the base analog 5-bromouracil and the structure of the normal base thymine allows it in some cases to be taken up in place of thymine. The bromine (Br) group occupies an area about the same size as the methyl (CH_3) group.

Chemical Mutagens

Chemical mutagens act at the molecular level to alter the sequence of bases in DNA. They include base analogs, alkylating agents, deaminating agents, and acridine derivatives.

A **base analog** is a molecule quite similar in structure to one of the nitrogenous bases normally found in DNA. A cell may incorporate a base analog into its DNA in place of the normal base. For example, 5-bromouracil can be inserted in DNA instead of thymine (➤ Figure 7.18). When DNA containing 5-bromouracil is replicated, the analog can cause an error in base pairing. The 5-bromouracil that replaced thymine may pair with guanine instead of with adenine, which normally pairs with thymine. When DNA is replicated in the presence of a significant quantity of 5-bromouracil, the analog can be incorporated at many sites in the DNA molecule. A mutation occurs wherever the analog causes the insertion of guanine instead of adenine in the subsequent replication. Another purine base analog, caffeine, can cause mutations in an unborn child. For this reason pregnant women are advised to avoid or limit their caffeine intake.

Alkylating agents are substances that add alkyl groups (such as a methyl group, —CH_3) to other molecules. Adding an alkyl group to a nitrogenous base alters the shape of the base and can cause an error in base pairing. For example, the addition of a methyl group to guanine can cause it to pair with thymine instead of cytosine. Such a change can give rise to a point mutation.

Deaminating agents such as nitrous acid (HNO_2) remove an amino group (—NH_2) from a nitrogenous base. Removing an amino group from adenine causes it to resemble guanine, and the deaminated base pairs with cytosine instead of thymine. Nitrates (NO_3^-) and nitrites (NO_2^-) are sometimes added to foods such as hot dogs and cold cuts for coloring, flavoring, or antibacterial action. The hazard of such additives is that, in the body, they form nitrosamines—deaminating agents known to cause birth defects, cancer, and other mutations in laboratory animals.

In contrast to these alterations, which cause point mutations, **acridine derivatives** cause frameshift mutations. The acridine molecule contains one pyrimidine ring and two benzene rings (➤ Figure 7.19). This molecule or one of its derivatives can become inserted in the DNA double helix, displacing both members of a base pair. Such a modification distorts the helix and causes partial unwinding of the DNA strands. The distortion allows one or more bases to be added or deleted, and a frameshift mutation results. The drug quinacrine (Atabrine) is an acridine derivative that was used to treat malaria until other drugs with less unpleasant side effects were developed. It causes mutations in the malarial parasite and possibly in the human host that receives the drug.

Radiation as a Mutagen

Radiation such as X-rays and ultraviolet rays can act as mutagens. Ultraviolet rays affect only the skin of humans because the rays lack energy for deeper penetration, but they have significant effects on microorganisms, which they penetrate easily. Ultraviolet lights are sometimes mounted in hospitals and laboratories to kill airborne bacteria. When ultraviolet rays strike DNA, they can cause pyrimidine dimers to form. A **dimer** (di'mer) consists of two adjacent pyrimidines (two thymines, two cytosines, or thymine and cytosine) bonded together in a DNA strand (➤ Figure 7.20). Binding of pyrimidines to each other prevents base pairing during replication of the adjacent DNA strand, so a gap is produced in the replicated DNA. Transcription of mRNA stops at the gap, and the affected gene fails to transmit information.

X-rays and gamma rays, which are more energetic than ultraviolet radiation, easily break chemical bonds in molecules. ➡ (Chapter 3, p. 51) The product is often a *free radical,* a highly reactive atom, molecule, or ion that in turn attacks other cell molecules, including DNA.

Until recently, microbiologists had no control over which genes underwent mutation when organisms were treated with mutagens. Now certain enzymes are available that greatly facilitate such studies. **Restriction endonucleases** cut DNA at precise base sequences, and **exonucleases** remove segments of DNA. These enzymes allow individual genes to be isolated and mutated at predetermined sites. The mutated gene can be inserted into a host's chromosome, and the effect of the specific mutation studied.

Backbone of DNA (deoxyribose and phosphate groups)

Acridine

Base pairs

➤ **Figure 7.19   Acridine, a chemical mutagen.** Insertion of acridine into a DNA helix can produce a frameshift mutation.

The Repair of DNA Damage

Many bacteria, and other organisms as well, have enzymes that can repair certain kinds of damage to DNA. Two mechanisms, *light repair* and *dark repair,* are known to repair damage caused by dimers.

Light repair, or **photoreactivation,** occurs in the presence of visible light in bacteria previously exposed to ultraviolet light. When organisms containing dimers are kept in visible light, the light activates an enzyme that breaks the bonds between the pyrimidines of a dimer (➤ Figure 7.21a). Thus, mutations that might have been passed along to daughter cells are corrected, and the DNA is returned to its normal state. This mechanism contributes to the survival of the bacteria but creates a problem for microbiologists. Cultures that are irradiated with ultraviolet light to induce mutations must be kept in the dark for the mutations to be retained.

Dark repair, which occurs in some bacteria, requires several enzyme-controlled reactions (➤ Figure 7.21b). First, an endonuclease breaks the defective DNA strand near the dimer. Second, a DNA polymerase synthesizes new DNA to replace the defective segment, using the normal complementary strand as a template. Third, an exonuclease removes the defective DNA segment. Finally, a ligase connects the repaired segment to the remainder of the DNA strand. These reactions were identified in *E. coli* but are now known to occur in many other bacteria. Human cells have similar mechanisms; some human skin cancers, such as xeroderma pigmentosum (➤ Figure 7.22), are caused by a defect in the cellular DNA repair mechanism.

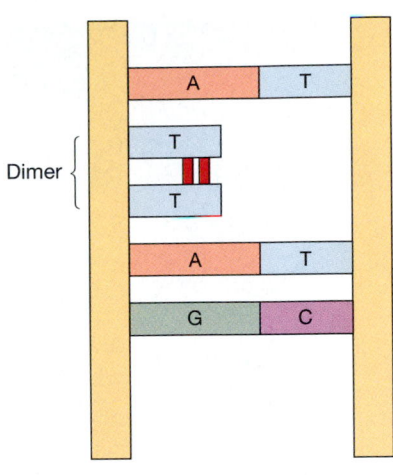

➤ **Figure 7.20   Thymine dimers caused by radiation.** The formation of a dimer prevents the affected bases from pairing with bases in the complementary chain of DNA, impairing replication and preventing transcription.

➤ **Figure 7.21   Thymine dimer repairs.** (a) Light repair of DNA (photoreactivation) removes dimers. (b) In dark repair, a defective segment of DNA is cut out and replaced.

(a)

(b)

➤ Figure 7.22   **The inability to repair UV-caused dimers.** (a) Sunbathers acquire dimers caused by UV radiation, which can cause skin cancer if they are not repaired. (b) Xeroderma pigmentosum is a genetic disease in which the enzymes that normally repair UV damage to DNA are defective, and exposure to sunlight results in multiple skin cancers.

The Study of Mutations

Microorganisms are especially useful in studying mutations because of their short generation time and the relatively small expense of maintaining large populations of mutant organisms for study. Comparisons of normal and mutant organisms have led to important advances in the understanding of both genetic mechanisms and metabolic pathways. Microorganisms continue to be important to researchers attempting to further our knowledge of these processes. However, the study of mutations is not without its problems. Two common problems are (1) distinguishing between spontaneous and induced mutations and (2) isolating particular mutants from a culture containing both mutated and normal organisms. The *fluctuation test* and the technique of *replica plating* are used to distinguish between spontaneous and induced mutations; replica plating also is used to isolate mutants.

Why is it important to differentiate between spontaneous and induced mutations? Making this distinction helps us understand mechanisms in the evolution of microorganisms and presumably other organisms as well. For example, some organisms are penicillin resistant—they grow in the presence of penicillin despite its antibiotic properties. Theoretically, there are two ways organisms can acquire such resistance: Either the penicillin *induces* a change in the organism that enables it to grow in the presence of penicillin, or a mutation occurs *spontaneously* that will allow the organism to grow if it is later exposed to penicillin. In the latter case, penicillin will kill nonresistant organisms, thereby *selecting* for the resistant mutant. Various experiments, two of which are described next, have shown that the second mechanism, the selection of spontaneous mutants, is the primary means of evolution in microorganisms.

The **fluctuation test,** designed by Salvador Luria and Max Delbruck in 1943, is based on the following hypothesis: If mutations that confer resistance occur spontaneously and at random, we would expect great fluctuation in the number of resistant organisms per culture among a large number of cultures. This fluctuation would occur regardless of whether the substance to which resistance develops is present. A mutation might occur early in the incubation period, late in that period, or not at all. Cultures with early mutations would contain many mutated progeny. Those with late mutations would have few mutated progeny, and those without a mutation would have none. An alternative hypothesis is that mutations conferring resistance to a substance occur only in the presence of the substance. Then cultures containing the substance would be expected to have approximately equal numbers of resistant organisms, whereas cultures lacking the substance would have no resistant organisms.

To test these hypotheses, Luria and Delbruck inoculated a large flask of liquid medium with a type of bacterium that was sensitive to the antibiotic streptomycin. At the same time they inoculated 100 small tubes of liquid medium with the same bacteria. No streptomycin was present in either the flask or the tubes. Both the flask and the tubes were allowed to reach maximum growth (10^9 organisms per milliliter). One-milliliter samples were then used to inoculate agar plates containing streptomycin; a plate was made from each tube, and many plates were made from the flask. After 24 hours the colonies on each plate were counted. Each colony represented a resistant mutant that could grow in the presence of streptomycin. There was far greater fluctuation in the number of colonies among the plates inoculated from the tubes than among the plates inoculated from the flask (➤ Figure 7.23). Therefore, mutations must have occurred at different times or not at all in the various tubes. Mutations also must have occurred at different times in the flask, but progeny of mutated organisms became distributed through the medium, so that the number of mutants in each sample did not vary greatly. Luria and Delbruck concluded that resistance was conferred from random mutations occurring at different times among the organisms in the tubes and not from exposure to streptomycin. (Can you predict what results would have been obtained if resistance arose only from exposure to streptomycin?)

The technique of **replica plating,** devised by Joshua and Esther Lederberg in 1952, is also used to study mutations. Based on the same reasoning as the fluctuation test, it hypothesizes that resistance to a substance arises spontaneously and at random without the need for exposure to the substance. In the original replica plating studies (➤ Figure 7.24), bacteria from a liquid culture were evenly spread on a master agar plate and allowed to grow for 4 to 5 hours. A sterile velveteen pad was then gently pressed against the surface of the master plate to pick up organisms from each colony. The tiny fibers of velveteen acted like hundreds of tiny inoculating needles. The pad was carefully kept in the same orientation and used to inoculate an agar plate containing a substance

Ozone Biosensors

Because ozone filters out harmful ultraviolet radiation, the discovery of holes in the ozone layer of the earth's atmosphere has raised concern about how much ultraviolet light reaches the earth's surface. Of particular concern are the questions of how deeply into sea water ultraviolet radiation penetrates and how it affects marine organisms, especially plankton (floating microorganisms) and viruses that attack plankton. Plankton form the base of the marine food chains and are believed to affect our planet's temperature and weather by their uptake of CO_2 for photosynthesis.

Deneb Karentz, a researcher at the Laboratory of Radiobiology and Environmental Health (University of California in San Francisco) has devised a simple method for measuring ultraviolet penetration and intensity. Working in the Antarctic Ocean, she submerged to various depths thin plastic bags containing special strains of *E. coli* that are almost totally unable to repair ultraviolet radiation damage to their DNA. Bacterial death rates in these bags were compared with rates in unexposed control bags of the same organism. The bacterial "biosensors" revealed constant

significant ultraviolet damage at depths of 10 m and frequently at 20 and 30 m. Karentz plans additional studies of how ultraviolet may affect seasonal plankton blooms (growth spurts) in the oceans.

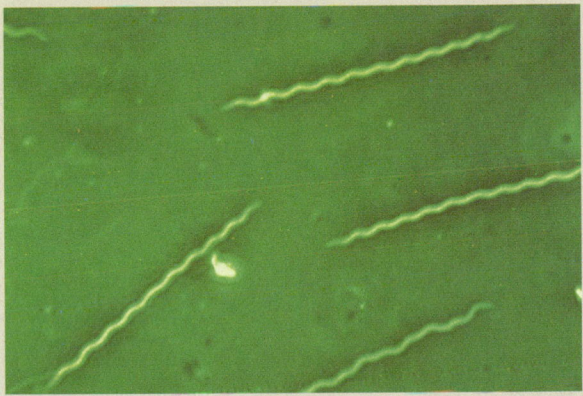

Collecting *E. coli* samples in the Antarctic Ocean to measure UV penetration and intensity.

such as penicillin to which bacteria might be resistant. After incubation, the exact positions of corresponding colonies on the two plates were noted. The bacteria in colonies found on the penicillin plate had resistance to penicillin without ever having been exposed to it.

Replica plating not only demonstrates spontaneity of mutations that confer resistance, it also provides a means of isolating resistant organisms without exposing them to a substance. By keeping the velveteen pad in perfect alignment during the transfer process, colonies on the original plate that contain resistant organisms can be identified by their location relative to colonies on the master penicillin plate.

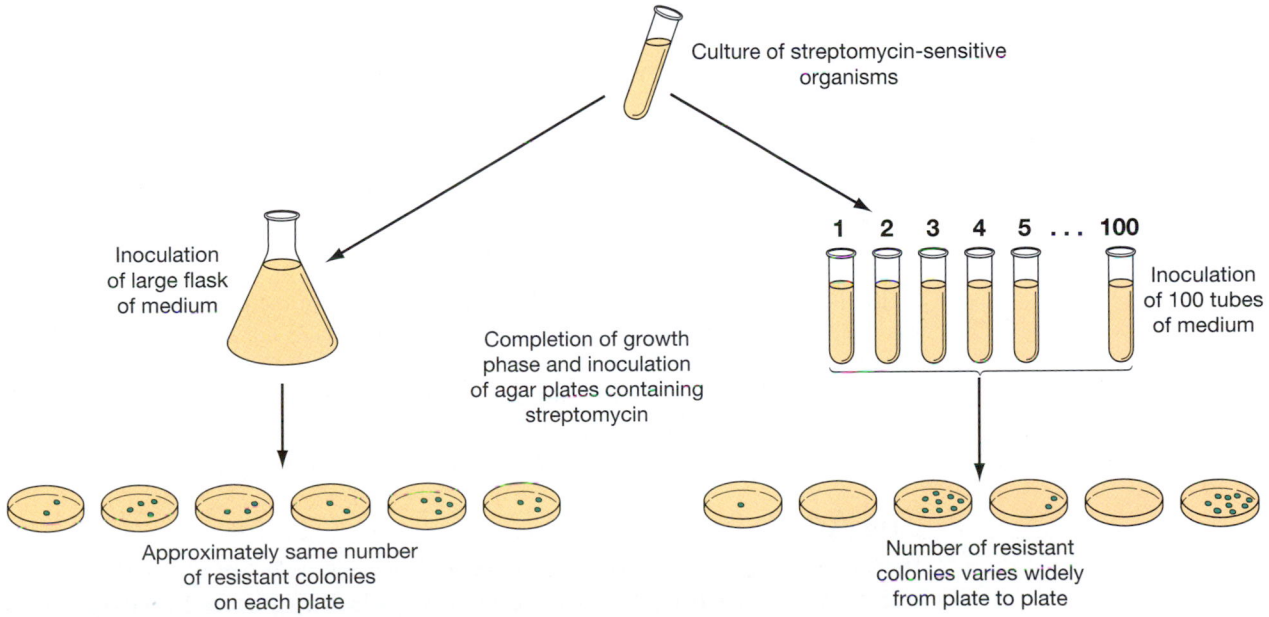

➤ **Figure 7.23 Fluctuation testing.** Luria and Delbruck's fluctuation test proves that mutations conferring antibiotic resistance are random—they are not induced by exposure to the antibiotic.

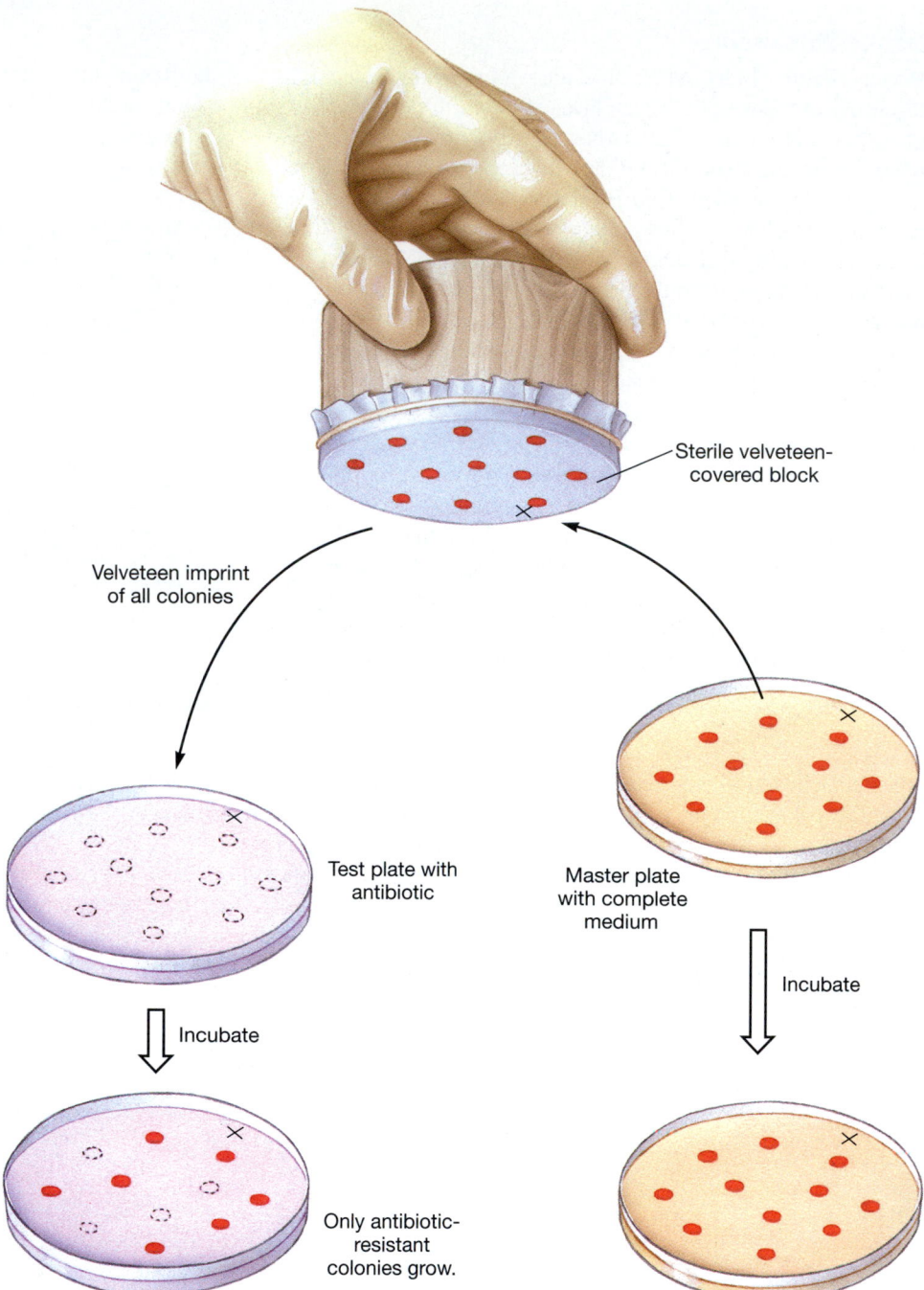

➤ Figure 7.24 **Replica plating.** This technique allows detection of antibiotic-resistant organisms. The × on the side of the plate provides a reference for identifying colonies from the same organism.

Replica plating is now widely used to study changes in the characteristics of many bacteria. The velveteen pads have been replaced by other materials that are easier to sterilize and manipulate. The technique is especially useful for identifying mutants whose nutritional needs have changed. Replicas can be transferred to a variety of different media, each deficient in a particular nutrient. Failure of particular colonies to grow on the deficient medium indicates that a mutation prevented the organism from synthesizing that nutrient.

The Ames Test

Human cancers can be induced by environmental substances that act by altering DNA. Much research effort is now being devoted to determining which substances are **carcinogens** (cancer-producing compounds). Carcinogens tend to be mutagenic, so determining whether a substance is mutagenic is often a first step in identifying it as a carcinogen. Bacteria, being subject to

> Figure 7.25   **The Ames test for mutagenic properties of chemicals.** (a) Plates used in the Ames test.
(b) The test is used to determine whether a substance is a mutagen and therefore a potential carcinogen.

mutation and being easier and cheaper to study than larger organisms, are ideal organisms to use in screening substances for mutagenic properties. Proving that a substance causes mutations in bacteria does not prove that it does so in human cells. Even proving that a substance causes mutations in human cells does not prove that the mutations will lead to cancer. Additional tests, including tests in animals, are necessary to identify carcinogens, but initial screening with bacteria can eliminate some substances from further study. If a substance induces no mutations in a large population of bacteria, most researchers believe that it is not likely to be a carcinogen.

The **Ames test** (➤ Figure 7.25a), devised by the American microbiologist Bruce Ames, is used to test whether substances induce mutations in certain strains of *Salmonella* (auxotrophs) that have lost their ability to synthesize the amino acid histidine. These strains easily undergo another mutation that restores their ability to synthesize histidine. The Ames test is based on the hypothesis that if a substance is a mutagen, it will increase the rate at which these organisms revert to histidine synthesizers (➤ Figure 7.25b). Furthermore, the more powerful a substance's mutagenic capacity, the greater the number of reverted organisms it causes to appear. In practice, the organism is grown in the presence of a test substance. If any organisms regain the ability to synthesize histidine, the substance is suspected of being a mutagen. The larger the number of organisms that regain the synthetic ability, the stronger the substance's mutagenic capacity is likely to be.

✓ If you had to be exposed to a mutagen, would you rather it be one that causes point mutations or one that causes frameshift mutations? Why?

✓ How can a change in genotype fail to produce a change in phenotype? Does a change in phenotype always require a change in genotype?

✓ Why are nitrates and nitrites, which have been added to sausages to retard bacterial growth and thereby prevent food poisoning, a danger to humans?

✓ Does exposure to antibiotics cause bacteria to mutate so as to be resistant to these antibiotics? How was this proven?

Polymerase Chain Reaction (PCR)—Key to Past and Future Worlds of DNA

Is there privacy in the grave? Not anymore. Ancient DNA, sometimes as old as 17 million years, is being recovered, and its bases sequenced. Which genes did past organisms have? How many were handed down to us, and how many mutant versions do we have? DNA has been extracted from the brains of 91 prehistoric Native Americans buried and mummified in Florida peat bogs 7500 years ago. That DNA is now undergoing analysis, thanks to **polymerase chain reaction (PCR),** a technique that first became available in 1985. PCR allows us to produce rapidly (amplify) a billion copies of DNA without needing a living cell. These large quantities are then easily analyzed.

Scientists have applied this tool to many questions concerning the past. For example, they have extracted DNA from fossil flies embedded in amber—and, ironically, did so shortly after the book *Jurassic Park* was written. They have also studied DNA from fossil leaves embedded in Idaho shale 17 million years ago (the DNA is very similar to that of modern magnolias) and from the bloodstains, hair, and bone chips preserved by doctors attending Abraham Lincoln at the time of his assassination. Lincoln is suspected of having had the hereditary disease Marfan's syndrome, which causes weakened arteries that can rupture and cause death. Most people with Marfan's syndrome die before they reach Lincoln's age. Would Lincoln have died soon had he not been assassinated? We can now create a library of Lincoln's DNA. As the Human Genome Project identifies the sequence of various genes (including those for Marfan's), we can match them to Lincoln's DNA and know with certainty which genes he had.

A modern forensic problem has brought the term PCR to the lips of the average American. Attorneys in the O.J. Simpson trial argued about DNA analysis and about the reliability of the PCR techniques in front of millions worldwide. Seldom has a scientific advance entered public awareness so rapidly. Prisoners already incarcerated for many years began asking for DNA analysis of evidence from their trials. PCR made it possible to bring forth evidence that was not available earlier.

DNA analysis can be used to protect and free the innocent as well as to convict the guilty. After PCR and DNA analysis were conducted on semen samples, one man convicted of rape was shown not to have been the rapist. His family had never lost faith in his innocence throughout the 10 years he had been locked up. (Analysis of semen in sexual assault cases has led to the freeing of 30 percent of initial suspects.)

These techniques are powerful forensic tools. Enough DNA can be recovered, and amplified by PCR, from the sweatband of a baseball cap to identify its wearer with extremely high certainty. DNA analysis of saliva on the back of a stamp on an envelope can identify the person who licked the stamp.

Rising from an 11,000-year-old grave in Ohio are cultures of two strains of *Enterobacter cloacae,* recovered alive but frozen in a state of suspended animation, from the intestine of a 4-ton mastodon that had been killed by prehistoric hunters,

(a)   **(b)**

(a) *Enterobacter cloacae.* These bacteria were isolated and cultured from the remains of an 11,000-year-old mastodon fossil found in Newark, Ohio, in 1989. (b) The mastodon's remains. Its digestive tract could be identified as a darker-colored, discrete cylindrical mass bent into the shape of intestinal loops. No organisms were found in areas sampled adjacent to the intestine.

butchered, and then sunk into a peat bog (a primitive form of food preservation) (see the photos). And preserved they are! No mutations have occurred over the past 11,00 years. PCR analysis will reveal how these ancient organisms differ from today's strains of *E. cloacae*. Botanists await the tantalizing analysis of DNA from the mastodon's last meal: pollen, grains stuck in his teeth, swamp grass, mosses, leaves, and a water lily. Chapters of evolutionary history may need to be rewritten.

For the living, PCR-amplified DNA analysis can reveal the presence of organisms that are difficult, dangerous, slow-growing, or require extra skill to culture in standard clinical laboratories. Tuberculosis cultures require 8 weeks to grow; PCR techniques will confirm the presence of the DNA of tuberculosis organisms in just hours. Medical-technology programs will need to train students in these techniques of the future, and current personnel will need to be retrained.

How does PCR amplification of DNA work? A large piece, or mixture of pieces, of DNA is cut up into smaller pieces by restriction endonuclease enzymes. You need to know the base composition of the ends of the exact piece of DNA you wish to replicate. In less than 24 hours, automated synthesizing equipment can make oligonucleotide (*oligos,* few) primer molecules that will pair complementarily with the ends of the desired section of DNA. A *primer* is a molecule that serves as a starting point for DNA synthesis.

The sequence of events in PCR amplification is shown in the figure. The heating process (thermal cycling), which converts newly formed DNA into single strands, is repeated until it has produced billions of copies of the desired piece of DNA. The DNA is then easily detected (as in a clinical diagnostic test) or analyzed for total base sequence. Cutting a large piece of DNA into smaller pieces, sequencing the PCR-amplified quantities of the pieces, and then looking for overlaps at the ends allows us finally to determine the sequence of the entire original DNA piece.

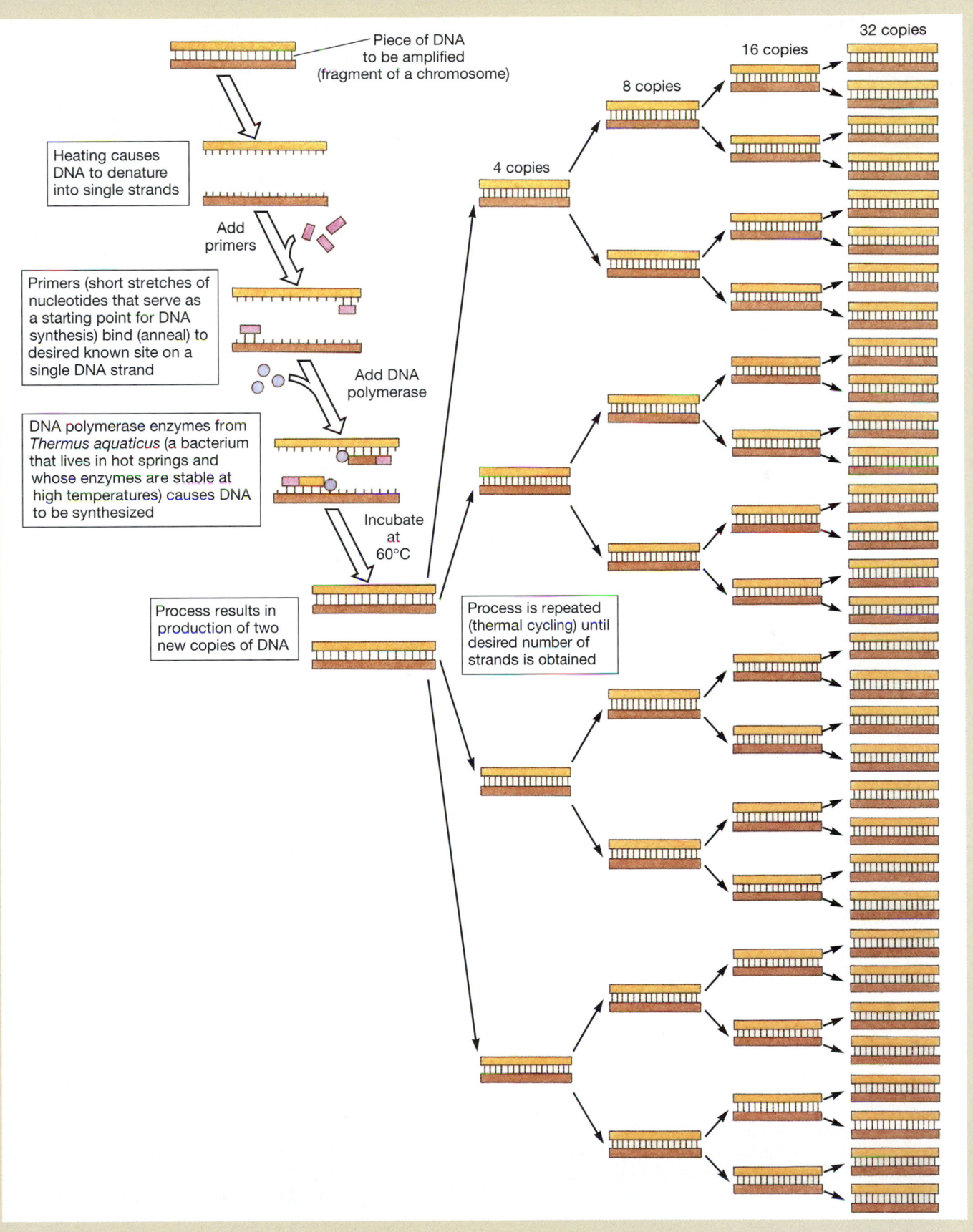

Piece of DNA to be amplified (fragment of a chromosome)

Heating causes DNA to denature into single strands

Add primers

Primers (short stretches of nucleotides that serve as a starting point for DNA synthesis) bind (anneal) to desired known site on a single DNA strand

Add DNA polymerase

DNA polymerase enzymes from *Thermus aquaticus* (a bacterium that lives in hot springs and whose enzymes are stable at high temperatures) causes DNA to be synthesized

Incubate at 60°C

Process results in production of two new copies of DNA

Process is repeated (thermal cycling) until desired number of strands is obtained

4 copies

8 copies

16 copies

32 copies

Retracing Our Steps

An Overview of Genetic Processes

The Basis of Heredity

- **Heredity** involves the transmission of information from an organism to its progeny.

- **Genes** are linear sequences of DNA that carry coded information for the structure and function of an organism.

- Prokaryotic chromosomes are threadlike circular structures made of DNA. Transmission of information in prokaryotes typically occurs during asexual reproduction in which the chromosome is reproduced (replicates), and each daughter cell receives a chromosome like the one in the parent cell.

- **Mutations** (alterations in DNA) transmitted to progeny account for much of the variation in organisms.

Nucleic Acids in Information Storage and Transfer

- All information for the functioning of a cell is stored in DNA in a specific sequence of the nitrogenous bases: adenine, thymine, cytosine, and guanine.

- Information stored in DNA is used for two purposes: (1) to replicate DNA in preparation for cell division and (2) to provide information for protein synthesis. In both of these processes, information is transferred by base pairing.

DNA Replication

- **Replication** of bacterial DNA begins at a specific point in the circular chromosome and usually proceeds in both directions simultaneously. The main steps in DNA replication are summarized in Figure 7.4.

- DNA replication is **semiconservative**—each chromosome consists of one strand of old (parent) DNA and one of newly synthesized DNA.

Protein Synthesis

Transcription

- In **transcription, messenger RNA (mRNA)** is transcribed from DNA, as summarized in Figure 7.5, and serves as a **template** for protein synthesis.

Kinds of RNA

- Besides mRNA, two other kinds of RNA are similarly produced: (1) **ribosomal RNA (rRNA),** which combines with specific proteins to form ribosomes, the sites for protein assembly; and (2) **transfer RNA (tRNA),** which carries amino acids to the assembly site.

Translation

- In the process of **translation,** three-base sequences in mRNA act as **codons** and are matched by base pairing with **anticodons** of tRNA. The mRNA codons constitute the **genetic code**—a code that is essentially the same for all living organisms.

- Once the mRNA and ribosomes are aligned, the process of protein synthesis proceeds as summarized in Figure 7.12 until a **terminator,** or **stop codon,** is reached.

The Regulation of Metabolism

The Significance of Regulatory Mechanisms

- Mechanisms that regulate metabolism turn reactions on and off in accordance with the needs of cells, allowing the cells to use various energy sources and to limit synthesis of substances to the amounts needed.

Categories of Regulatory Mechanisms

- The two basic categories of regulatory mechanisms are: (1) mechanisms that regulate the activity of enzymes already available in the cell and (2) mechanisms that regulate the action of genes, which determine what enzymes and other proteins will be available.

Feedback Inhibition

- In **feedback inhibition,** the end-product of a biochemical pathway directly inhibits the first enzyme in the pathway (Figure 7.13).

- Enzymes subject to such regulation are generally allosteric.

- Feedback inhibition regulates the activity of existing enzymes and is a quick-acting control mechanism.

Enzyme Induction

- In **enzyme induction** (Figure 7.14), the presence of a substrate activates an **operon,** a sequence of closely associated genes that includes **structural genes** and **regulatory sites:** (1) In the absence of lactose, a **repressor,** a product of the **regulator** (*i*) **gene,** attaches to the operator and prevents transcription of the genes of the *lac* operon. (2) When lactose is present, it inactivates the repressor and allows transcription of the genes of the *lac* operon.

Enzyme Repression

- In **enzyme repression,** the presence of a synthetic product inhibits its further synthesis by inactivating an operon: (1) When tryptophan is present, it attaches to the repressor protein and represses genes of the *trp* operon. (2) In the absence of tryptophan, the repressor is not activated, and genes of the *trp* operon are transcribed.

- In **catabolite repression,** the presence of a preferred nutrient (often glucose) represses the synthesis of enzymes that would be used to metabolize some alternative substance.

- Both enzyme induction and enzyme repression regulate by altering gene expression. The effect on enzyme synthesis in both cases depends on the presence or absence of the regulatory substance—lactose, tryptophan, or glucose in the preceding examples.

Mutations

Types of Mutations and Their Effects

- The genetic makeup of an organism is its **genotype;** the physical expression of the genotype is the **phenotype.**

- Mutations cause a change in the genotype, of an organism; the change may or may not be expressed in the phenotype.

- Two major classes of mutations are (Table 7.3): (1) **point mutations,** which consist of changes in a single nucleotide, and (2) **frameshift mutations,** which consist of the **insertion** or **deletion** of one or more nucleotides.

Phenotypic Variation

- Phenotypic variations produced by mutations can involve alterations in colony morphology, nutritional requirements, and temperature sensitivity.

Spontaneous and Induced Mutations

- **Spontaneous mutations** occur in the absence of any known mutagen and appear to be due to errors in base pairing during DNA replication. Various genes have different rates of mutation.

- **Induced mutations** are mutations produced by agents called **mutagens;** mutagens increase the mutation rate.

Chemical Mutagens

- Chemical mutagens include **base analogs, alkylating agents, deaminating agents,** and **acridine derivatives.**

Radiation as a Mutagen

- **Radiation** often causes the formation of **dimers**—two adjacent pyrimidine bases that are bound to each other and interfere with DNA replication.

The Repair of DNA Damage

- Many bacteria have enzymes that can repair certain damages to DNA (Figure 7.21). (1) **Light repair** involves an enzyme that is activated by visible light and that breaks bonds between pyrimidines of a dimer. (2) **Dark repair** involves several enzymes that do not require light for activation; they excise defective DNA and replace it with DNA complementary to the normal DNA strand.

The Study of Mutations

- Microorganisms are useful in studying mutations because many generations can be produced quickly and inexpensively.

- The **fluctuation test** demonstrated that resistance to chemical substances occurs spontaneously rather than being induced.

- **Replica plating** likewise demonstrated the spontaneous nature of mutations; it also can be used for isolating mutants without exposing them to the substance to which they are resistant.

The Ames Test

- The **Ames test** is based on the ability of **auxotrophic** bacteria to mutate by reverting to their original synthetic ability. It is used for screening chemicals for mutagenic properties, which indicate potential **carcinogens.**

Terminology Check

Meet Me on the Web

Don't forget I am expecting you on the web. The following is a preparedness checklist for your web explorations. If you have questions on any of these items, you can return to the end of Chapter 1 for more details.

- Questions We'll Explore

- Concept Check questions

- Flashcards

- Study Guide

If you are ready, come visit me at

http://www.prenhall.com/black

It will be a journey you'll never forget.

Recombinant DNA and Genetic Engineering

I can't even keep my eyes open and you're starting in with a discussion of bacteria. I need a cup of coffee . . . What's that? What do you mean bacteria probably helped make this coffee?

Processes that use water to decaffeinate coffee are slow and costly. Easier processes use organic solvents that leave toxic traces behind in the coffee. Either way, it's an added cost and step in the manufacturing process. Watch for a new type of decaf. Genetic engineers have used microbially developed techniques to remove the gene responsible for caffeine production from some coffee plants. No need to remove what's not even there. No extra cost; no toxic residues. Drink up!

Background Basics

The transfer of genetic material from one organism to another can have far-reaching consequences. In microbes, it provides ways for viruses to introduce genetic information into bacteria and mechanisms for bacteria to increase their disease-causing capabilities or to become resistant to antibiotics. Information obtained from studying the transfer of genetic material between microorganisms can be applied to agricultural, industrial, and medical problems and to the unique problems of the prevention and treatment of infectious diseases. In this chapter we will discuss the mechanisms by which genetic transfers occur, and the significance of such transfers.

The Types and Significance of Gene Transfer

Gene transfer refers to the movement of genetic information between organisms. In most eukaryotes, it is an essential part of the organism's life cycle and usually occurs by sexual reproduction. Male and female parents produce *gametes* (sex cells), which unite to form a zygote, the first cell of a new individual. Because each parent produces many genetically different gametes, many different combinations of genetic material can be transferred to offspring. In bacteria, gene transfer is not an essential part of the life cycle. When it does take place, usually only some of the genes of the *donor* cell are transferred to the other participating, or *recipient,* cell. This combining of genes (DNA) from two different cells is called **recombination,** and the resulting cell is referred to as a *recombinant*. Before the 1920s, bacteria were thought to reproduce only by binary fission and to have no means of genetic transfer comparable to that achieved through sexual reproduction in eukaryotes. Since then, three mechanisms of gene transfer in bacteria have been discovered, none of which is associated with reproduction. We will discuss each mechanism—*transformation, transduction,* and *conjugation*—in this chapter.

Gene transfer is significant because it greatly increases the genetic diversity of organisms. As noted in Chapter 7, mutations account for some genetic diversity, but gene transfer between organisms accounts for even more. ➡ (p. 182) When organisms are subjected to changing environmental conditions, genetic diversity increases the

Questions We'll Explore

A What are the types and significance of gene transfer?

B What are the mechanisms and significance of transformation?

C What are the mechanisms and significance of transduction?

D What are the mechanisms and significance of conjugation?

E What are the characteristics and actions of plasmids?

F How are the following techniques of genetic engineering used: genetic fusion, protoplast fusion, gene amplification, recombinant DNA, and hybridomas?

G Why are scientists concerned about the uses of recombinant DNA?

likelihood that some organisms will adapt to any particular condition. Such diversity leads to evolutionary changes. Organisms with genes that allow them to adapt to an environment survive and reproduce, whereas organisms lacking those genes perish. If all organisms were genetically identical, all would survive and reproduce, or all would die.

In *recombinant DNA technology,* genes from one kind of organism are introduced into the genome of another kind of organism (for example, when human genes are inserted into the cells of a pig).

Transformation

The Discovery of Transformation

Bacterial **transformation,** a change in an organism's characteristics because of the transfer of genetic information, was discovered in 1928 by Frederick Griffith, an English physician, while he was studying pneumococcal infections in mice. Pneumococci with capsules (Chapter 4) produce smooth, glistening colonies. ➡ (p. 89) Those lacking capsules produce rough colonies with a coarse, nonglistening appearance. Only the capsule-producing (encapsulated) pneumococci inoculated into mice were *pathogens*—that is, they had the power to cause disease (pneumonia). One such organism can multiply rapidly and kill a mouse! Mice are said to be "exquisitely" sensitive to pneumococci; therefore, they make excellent test animals. Capsules help prevent molecules produced by the mouse's immune system from reaching the surface of the bacterium. They also make it difficult for white blood cells to engulf the invading bacteria. In other words, the capsule protects the bacteria from the mouse's immune system.

Griffith injected one group of mice with heat-killed smooth pneumococci, a second group with live smooth pneumococci, a third group with live rough pneumococci, and a fourth group with a mixture of live rough and heat-killed smooth pneumococci (➤ Figure 8.1). As expected, mice that received live smooth pneumococci developed pneumonia and died, whereas those that received either heat-killed smooth pneumococci or live rough pneumococci did not develop pneumonia and survived. Surprisingly, those that received the mixture also died of pneumonia. Imagine Griffith's shock when he isolated live smooth organisms from them. He had no way of knowing exactly what had happened, but he realized that some cells had been "transformed" from type R to type S. Moreover, the change was heritable. We now know that the R-type bacteria picked up naked DNA, liberated from disintegrated dead S-type bacteria, and incorporated it into their own DNA. R-type bacteria that picked up DNA that had the genes for capsule production were genetically transformed into S-type organisms.

In subsequent studies of transformation, Oswald Avery discovered that a capsular polysaccharide was responsible for the virulence of pneumococci. In 1944, Avery, Colin MacLeod, and Maclyn McCarty isolated the substance responsible for the transformation of pneumococci and determined that it was DNA. In retrospect, this discovery marked the "birth" of molecular genetics, but at the time DNA was not known to carry genetic information. Researchers working with plant and animal chromosomes had isolated both DNA and protein from them, but they thought the genetic information was in the protein. Only when James Watson and Francis Crick determined the structure of DNA did it become clear how DNA encodes genetic information. After this original work with pneumococci (now called *Streptococcus pneumoniae*), transformation was observed in organisms from a wide variety of genera, including *Acinetobacter, Bacillus, Haemophilus, Neisseria,* and *Staphylococcus,* and in the yeast *Saccharomyces cerevisiae.*

The Mechanism of Transformation

To study the mechanism of transformation, scientists extract DNA from donor organisms by a complex biochemical process that yields hundreds of naked DNA fragments from each bacterial chromosome. (*Naked DNA* is DNA that is not incorporated into chromosomes or other structures.) When the extracted DNA is placed in a medium with organisms capable of incorporating it, most organisms can take up a maximum of about 10 fragments, which is less than 5 percent of the amount of DNA normally present in the organism.

Griffith was actually an Army medical officer. You don't have to be a professional scientist to make important discoveries.

Heat-killed smooth
pneumococci, with capsule

Live, virulent, smooth
pneumococci, with capsule

Live, nonvirulent, rough
pneumococci, no capsule

Mouse lives

Mouse dies

Mouse lives

Heat-killed smooth
pneumococci, with capsule

Live, nonvirulent, rough
pneumococci, without capsule

Mouse dies

Live smooth pneumococci (with capsule) plus live rough pneumococci
(without capsule) isolated from dead mouse

➤ Figure 8.1   **The discovery of transformation:** Griffith's experiment with pneumococcal
infections in mice. When S-type pneumococci (which produce smooth-appearing colonies, due to the
presence of capsules) are injected into mice, the mice die of pneumonia. The mice survive when R-type
pneumococci (which produce rough-appearing colonies, due to the lack of capsules) or heat-killed
S-type pneumococci are injected. But when a mixture of live R-type and heat-killed S-type pneumococci—neither of which is lethal by itself—are injected, the mice die, and live S-type organisms as
well as R-types are recovered from the dead animals.

Uptake of DNA occurs only at a certain stage in a cell's growth cycle, probably prior to the
completion of cell wall synthesis. In this stage, a protein called **competence factor** is released
into the medium and apparently facilitates the entry of DNA. When competence factor from
one culture is used to treat a culture that lacks it, cells in the treated culture become *competent*
to receive DNA—they can now take up DNA fragments. However, not all bacteria can become
competent; thus not all bacteria can be transformed. DNA entry depends on such factors as modifications of the cell wall and the formation of specific receptor sites on the plasma membrane
that can bind DNA. Receptor sites seem to recognize different sources of DNA. They bind

1. Naked DNA fragments from disintegrated cells in the area of a potential recipient cell. This cell must be of the correct genus and be in a state of competence, a proper physiologic condition, to permit entry of the DNA fragments.

2. Entry of naked DNA into competent cell

3. Recombination

Some DNA fragments replace (recombine with) original host cell DNA. The resultant recombinant cell is said to have been genetically transformed and will now express the foreign genes it has received and pass them on to all its offspring.

DNA that has not recombined is broken down by enzymes.

➤ Figure 8.2   **The mechanism of bacterial transformation.**

DNA from the same genus and species, or from closely related genera and species, but generally reject DNA from distantly related organisms (➤ Figure 8.2).

Once DNA reaches the entry sites, endonucleases cut double-stranded DNA into units of 7000 to 10,000 nucleotides. The strands separate, and only one strand enters the cell. Single-stranded DNA is vulnerable to attack by various nucleases and can enter a cell only if the nucleases on the cell surface somehow have been inactivated. Inside the cell, the donor single-stranded DNA must combine by base pairing with a portion of the recipient chromosome immediately or else be destroyed. In transformation, as well as in other mechanisms of gene transfer, the donor single-stranded DNA is positioned alongside the recipient DNA so that identical loci are next to one another. Splicing of a DNA strand involves breaking the strand, removing a segment, inserting a new segment, and attaching the ends. Enzymes in the recipient cell excise (cut out) a portion of the recipient's DNA and replace it with the donor DNA, which now becomes a permanent part of the recipient's chromosome. The leftover recipient DNA is subsequently broken down, so the number of nucleotides in the cell's DNA remains constant.

The Significance of Transformation

Although transformation has been observed mainly in the laboratory, it occurs in nature. It probably follows the breakdown of dead organisms in an environment where live ones of the same or a closely related species are present. However, the degree to which transformation contributes to the genetic diversity of organisms in nature is not known. In the laboratory, researchers induce transformation in order to study the effects of DNA that differs from the DNA that the organism already has. Transformation also can be used to study the locations of genes on a chromosome and to insert DNA from one species into that of another species, thereby producing *recombinant* DNA.

Transduction

The Discovery of Transduction

Transduction, like transformation, is a method of transferring genetic material from one bacterium to another. Unlike transformation, in which naked DNA is transferred, in transduction DNA is carried by a **bacteriophage** (bak-ter′-e-o-faj)—a virus that can infect bacteria. The phenomenon of transduction was originally discovered in *Salmonella* in 1952 by Joshua Lederberg and Norton Zinder and has since been observed in many different genera of bacteria.

The Mechanisms of Transduction

To understand the mechanisms of transduction, we need to describe briefly the properties of bacteriophages, also called **phages** (faj´ez). Phages, which are described in more detail in Chapter 10, are composed of a core of nucleic acid covered by a protein coat. They infect bacterial cells (*hosts*) and reproduce within them, as shown in ➤ Figure 8.3. A phage capable of infecting a bacterium attaches to a receptor site on the cell wall. The phage nucleic acid enters the bacterial cell after a phage enzyme weakens the cell wall. The protein coat remains outside, attached to the cell wall. Once the nucleic acid is in the cell, further events follow one of two pathways, depending on whether the phage is *virulent* or *temperate*.

A **virulent phage** is capable of causing infection and, eventually, the destruction and death of a bacterial cell. Once the phage nucleic acid enters the cell, phage genes direct the cell to synthesize phage-specific nucleic acids and proteins. Some of the proteins destroy the host cell's DNA, whereas other proteins and the nucleic acids assemble into complete phages. When the cell becomes filled with a hundred or more phages, phage enzymes rupture the cell, releas-

> Human beings are not the only ones infected by viruses. There are viruses that specifically infect bacteria, plants, other animals, fungi, algae, and even protozoa.

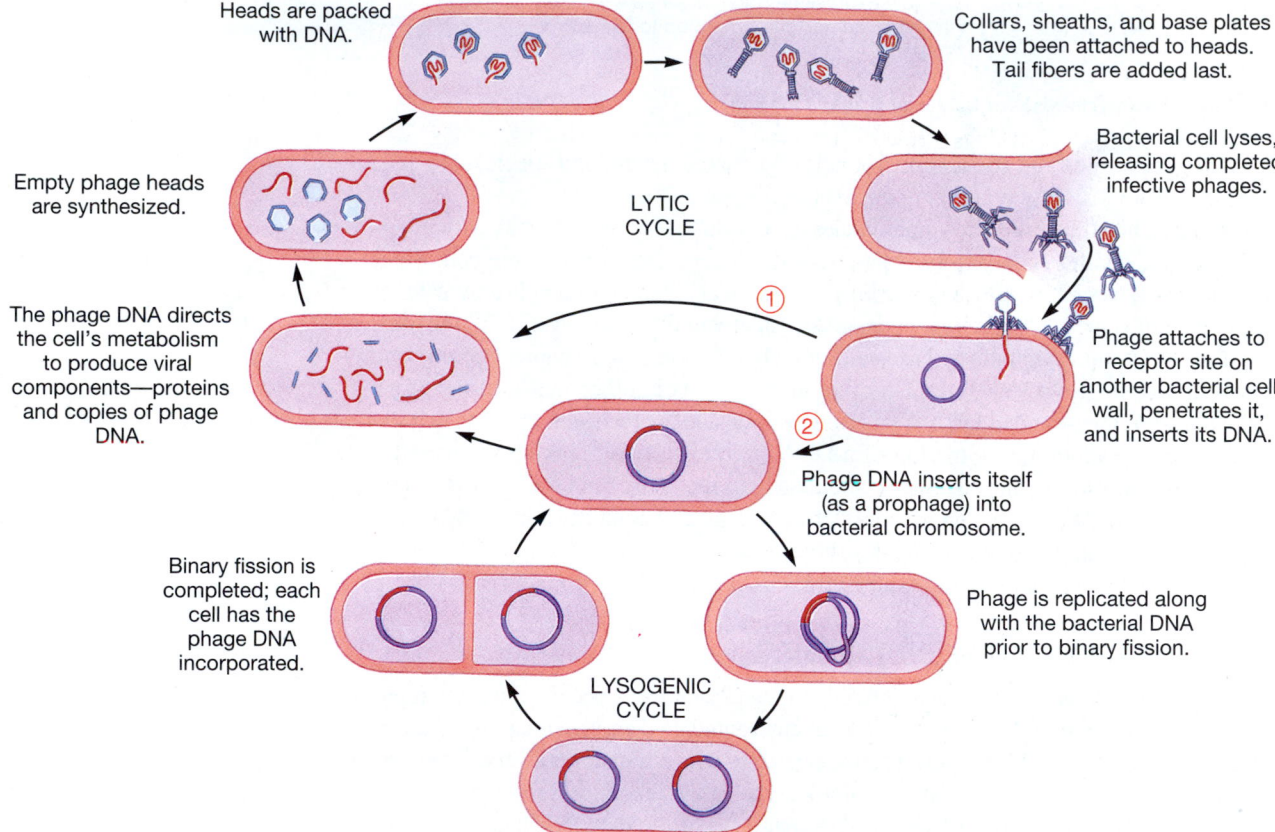

> Figure 8.3   **Bacteriophase life cycles.** When a bacteriophage injects its viral DNA into a host bacterial cell, at least two different outcomes are possible. In the lytic cycle, characteristic of virulent phages, the phage DNA takes control of the cell and ① causes it to synthesize new viral components, which are assembled into whole viral particles. The cell lyses, releasing the infective viruses, which can then enter new host cells. In the lysogenic cycle, the DNA of a temperate phage enters the host cell, ② becomes incorporated into the bacterial chromosome as a prophage, and replicates along with the chromosome through many cell divisions. However, a lysogenic phage can suddenly revert to the lytic life cycle. A prophage is thus a sort of "time bomb" sitting inside the infected cell.

ing newly formed phages, which can then infect other cells. Because this cycle results in **lysis** (li´sis), or rupture, of the infected (host) cell, it is called a **lytic** (lit´ik) **cycle.**

A **temperate phage** ordinarily does not cause a disruptive infection. Instead the phage DNA is incorporated into a bacterium's DNA and is replicated with it. This phage also produces a repressor substance that prevents the destruction of bacterial DNA, and the phage's DNA does not direct the synthesis of phage particles. Phage DNA that is incorporated into the host bacterium's DNA is called a **prophage** (pro´faj). Persistence of a prophage without prophage replication and destruction of the bacterial cell is called **lysogeny** (li-soj´e-ne), and cells containing a prophage are said to be **lysogenic** (li˝so-jen´ik). Several ways to induce such cells to enter the lytic cycle are known, and most involve inactivation of the repressor substance.

Temperate phages can replicate themselves either as a prophage in a bacterial chromosome or independently by assembling into new phages. Temperate phages can carry out both generalized and specialized forms of transduction. In *generalized transduction,* any bacterial gene can be transferred by the phage; in *specialized transduction,* only specific genes are transferred.

Specialized Transduction

Several lysogenic phages are known to carry out specialized transduction, but lambda (λ) phage in *Escherichia coli* has been extensively studied. Phages usually insert at a specific location when they integrate with a chromosome. Lambda phage inserts into the *E. coli* chromosome between the *gal* gene, which controls galactose use, and the *bio* gene, which controls biotin synthesis. The *gal* gene and *bio* gene are operons. ➡ (Chapter 7, p. 180) When cells containing lambda phage are induced to enter the lytic cycle, genes of the phage form a loop and are excised from the bacterial chromosome (➢ Figure 8.4). Lambda phage then directs the synthesis and assembly of new phage particles, and the cell lyses.

In most cases, the new phage particles released contain only phage genes. Occasionally (about one excision in a million) the phage contains one or more bacterial genes that were adjacent to the phage DNA when it was part of the bacterial chromosome. For example, the *gal* gene might be incorporated into a phage particle. When it infects another bacterial cell, the particle transfers not only the phage genes but also the *gal* gene. This process, in which a phage particle transduces (transfers) specific genes from one bacterial cell to another, is called **specialized transduction.** In specialized transduction, the bacterial DNA transduced is limited to one or a few genes lying adjacent to the prophage.

Generalized Transduction

When bacterial cells containing phage DNA enter the lytic cycle, phage enzymes break host cell DNA into many small segments (➢ Figure 8.5). As the phage directs synthesis and assembly of new phage particles, it packages DNA by the "headful" (enough DNA to fill the head of a virus). This allows a bacterial DNA fragment occasionally to be incorporated into a phage particle. Once this phage particle, with its newly acquired bacterial DNA, leaves the infected host, it may infect another susceptible bacterium, thereby transferring the bacterial genes through the process of **generalized transduction.** Each bacterial fragment from the host cell chromosome has an equal chance of accidentally becoming a part of phage particles during the phage's replication cycle.

The Significance of Transduction

Transduction is significant for several reasons. First, it transfers genetic material from one bacterial cell to another and alters the genetic characteristics of the recipient cell. As demonstrated by the specialized transduction of the *gal* genes, a cell lacking the ability to metabolize galactose could acquire that ability. Other characteristics also can be transferred either by specialized or generalized transduction.

Second, the incorporation of phage DNA into a bacterial chromosome demonstrates a close evolutionary relationship between the prophage and the host bacterial cell. The DNA of the prophage and that of the host chromosome must have regions of quite similar base sequences. Otherwise, the prophage would not bind to the bacterial chromosome.

Third, the discovery that a prophage can exist in a cell for a long period of time suggests a similar possible mechanism for the viral origin of cancer. If a prophage can exist in a bacterial

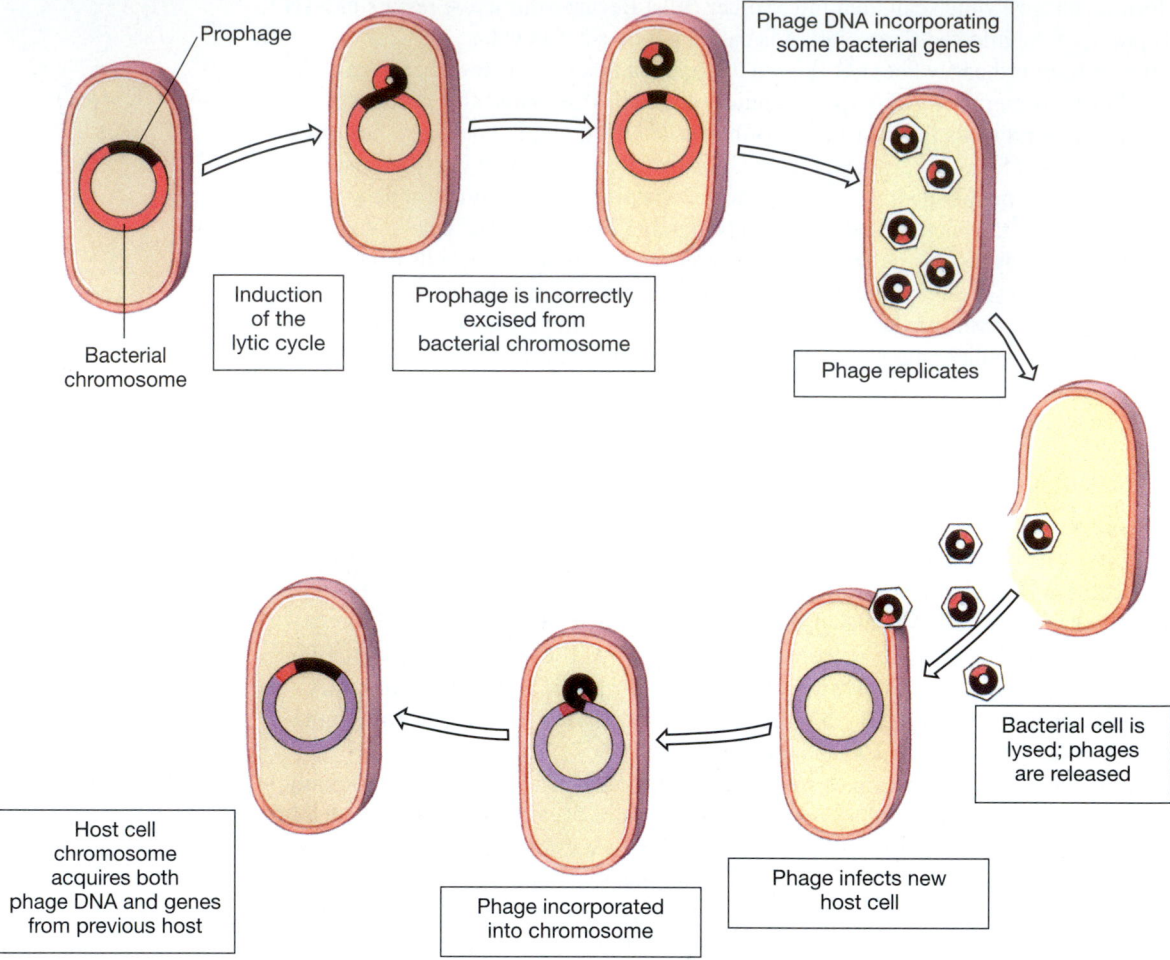

➤ Figure 8.4   **Specialized transduction by lambda phage in *E. coli.*** In this process, phage DNA always inserts itself into the bacterial host chromosome at a particular site. When the phage DNA replicates, it takes bacterial genes from either side of the site and packages them with its own DNA into new phages. Only genes adjacent to the insertion site, not genes from other parts of the host chromosome, are transduced. These genes can then be introduced into the phage's next host cell, where they will confer new genetic traits.

cell and at some point alter the expression of the cell's DNA, this could explain how animal viruses cause malignant changes. For example, viral genes inserted into a human operon could disrupt regulation of that operon, allowing structural genes to be active at the wrong times, continuously, or maybe even not at all. Fetal genes cause rapid proliferation of cells during early development, but growth soon slows down and eventually stops in adulthood. If these fetal genes were turned on again later in life in a group of cells, they could rapidly grow into a tumor. (Viruses and cancer are discussed in Chapter 10.)

Fourth, an interesting thought is that some animal viruses probably brought along genes from their previous host(s) when they infected new human hosts (such as you). These previous hosts were not necessarily humans. In that sense, you may not be entirely human now! You may be "transgenic."

Finally, and most importantly to molecular geneticists, transduction provides a way to study gene linkage. Genes are said to be *linked* when they are so close together on a DNA segment that they are likely to be transferred together. Different phages can be incorporated into a bacterial chromosome, each kind usually entering at a specific site. By studying many different phage transductions, scientists can determine where they were inserted on the chromosome and which adjacent genes they are capable of transferring. The combined findings of many such studies eventually allow identification of the sequence of genes in a chromosome. This procedure is called **chromosome mapping.**

✓ How does transformation differ from transduction?

✓ How does gene transfer in prokaryotes differ from that in eukaryotes?

✓ Which genes are transferred in generalized transduction? In specialized transduction?

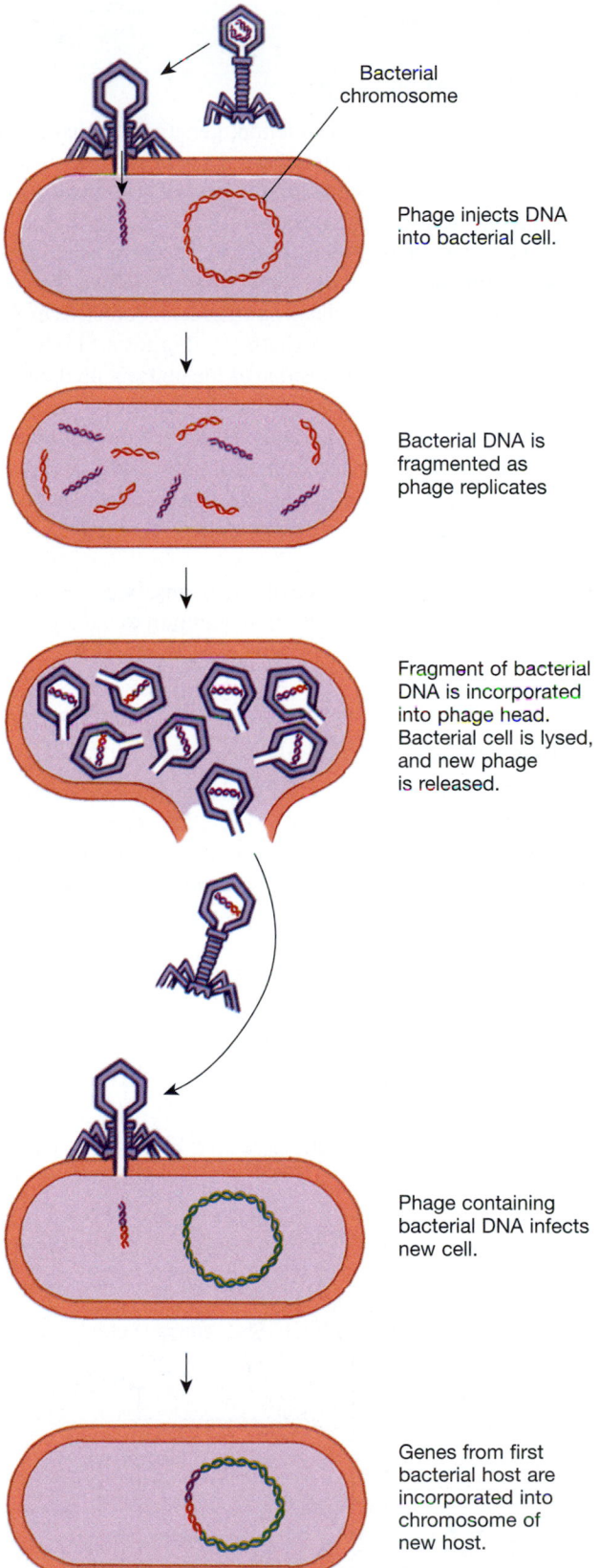

Bacterial chromosome

Phage injects DNA into bacterial cell.

Bacterial DNA is fragmented as phage replicates

Fragment of bacterial DNA is incorporated into phage head. Bacterial cell is lysed, and new phage is released.

Phage containing bacterial DNA infects new cell.

Genes from first bacterial host are incorporated into chromosome of new host.

➤ Figure 8.5 **Generalized transduction.** Bacteriophage infection of a host bacterium initiates the lytic cycle. The bacterial chromosome is broken into many fragments, any of which can be picked up and packaged along with phage DNA into new phage particles. When those particles are released and infect another bacterial cell, the new host acquires the genes that were brought along (transduced) from the previous bacterial host cell.

Conjugation

The Discovery of Conjugation

In **conjugation,** like transformation and transduction, genetic information is transferred from one bacterial cell to another. Conjugation differs from those other mechanisms in two ways: (1) It requires contact between donor and recipient cells, and (2) it transfers much larger quantities of DNA (occasionally whole chromosomes).

Conjugation was first discovered in 1946 by Joshua Lederberg, who at that time was still a medical student. In his experiments, Lederberg used mutated strains of *E. coli* that were unable to synthesize certain substances. He selected two strains, each defective in a different synthetic pathway, and grew them in a nutrient-rich medium (➢ Figure 8.6). He removed cells from each culture and washed them to remove the residue of the nutrient medium. He then attempted to culture cells of each strain on agar plates that lacked the special nutrients needed by the strain. He also mixed cells from the two strains and plated them on the same medium. Whereas cells from the original cultures failed to grow, some from the mixed cultures did grow. The latter must have acquired the ability to synthesize all the substances they needed. Lederberg and others continued to study this phenomenon and eventually discovered many of the details of the mechanism of conjugation.

Lederberg was indeed fortunate in his choice of organisms, because similar studies of other strains of *E. coli* failed to demonstrate conjugation. In addition to the mutations that led to synthetic deficiencies in Lederberg's organisms, other changes had modified their cell surfaces so that conjugation could occur.

The Mechanisms of Conjugation

The mechanisms involved in conjugation were clarified through several important experiments, each of which built on the findings of the preceding one. Of those experiments, we will consider three: transfer of F plasmids, high-frequency recombinations, and transfer of F' plasmids. Recall from Chapter 4 that **plasmids** are small *extrachromosomal DNA* molecules. ➡ (p. 82) Bacterial cells often contain several different plasmids that carry genetic information for various nonessential cell functions.

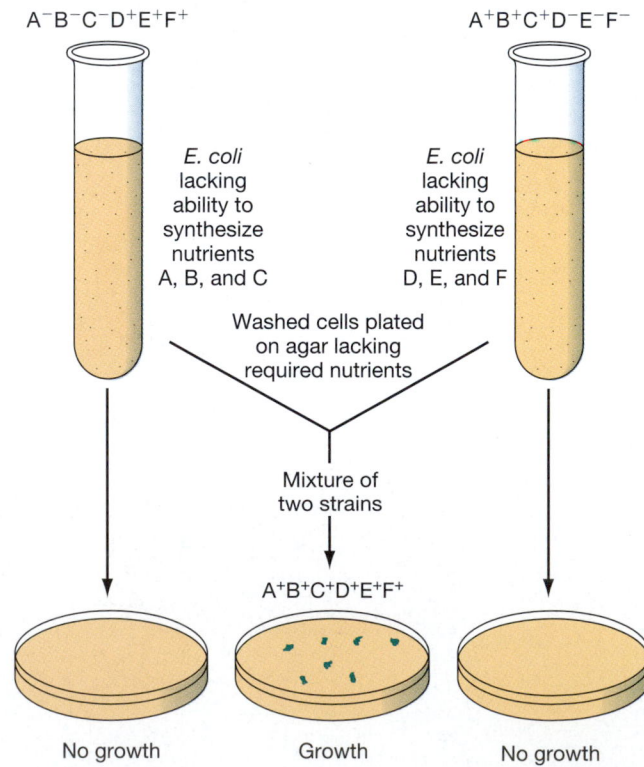

$A^-B^-C^-D^+E^+F^+$

E. coli lacking ability to synthesize nutrients A, B, and C

$A^+B^+C^+D^-E^-F^-$

E. coli lacking ability to synthesize nutrients D, E, and F

Washed cells plated on agar lacking required nutrients

Mixture of two strains

$A^+B^+C^+D^+E^+F^+$

➢ **Figure 8.6** **The discovery of conjugation: Lederberg's experiment.**

No growth   Growth   No growth

> Figure 8.7   **TEM of F pili of E. coli.** (1 F⁺ and 2 F⁻; 18,000X). Phages along the pili make them visible. Unlike the shorter attachment pili (fimbriae), this long type of pilus is used for transfer of genes in conjugation and is often called a sex pilus.

The Transfer of F Plasmids

After Lederberg's initial experiment, an important discovery about the mechanism of conjugation was made. Two types of cells, called F^+ and F^-, were found to exist in any population of *E. coli* capable of conjugating. **F^+ cells** contain extrachromosomal DNA called **F (fertility) plasmids;** **F^- cells** lack F plasmids. (Lederberg coined the term *plasmid* in the 1950s to describe these fragments of DNA.)

Among the genetic information carried on the F plasmid is information for the synthesis of proteins that form F pili. The F^+ cell makes an **F pilus** (or *sex pilus* or *conjugation pilus*), a bridge by which it attaches to the F^- cell when F^+ and F^- cells conjugate (> Figure 8.7). (Chapter 4, p. 89) A copy of the F plasmid is then transferred from the F^+ cell to the F^- cell (> Figure 8.8). F^+ cells are called *donor* or *male* cells, and F^- cells are called *recipient* or *female* cells.

Although the exact transfer process remains unknown, the DNA is transferred as a single strand via a *conjugation bridge.* Because the sex pilus contains a hole that would permit the passage of single-stranded DNA, it is possible that DNA enters the recipient through this channel. However, there is also evidence to suggest that mating cells temporarily fuse, during which time the DNA is transferred. Each cell then synthesizes the complementary strand of DNA, so both have a complete F plasmid. Because all F^- cells in a mixed culture of F^+ and F^- cells receive the F plasmid, the entire population quickly becomes F^+; but in a culture of only F^- cells, no transfer occurs, and cells remain F^- cells.

High-Frequency Recombinations

The mechanisms of conjugation were further clarified when the Italian scientist L. L. Cavalli-Sforza isolated a **clone,** a group of identical cells descended from a single parent cell, from an F^+ strain that could induce more than a thousand times the number of genetic recombinations seen in the F^+ and F^- conjugations. Such a donor strain is called a **high frequency of recombination (Hfr) strain.**

Hfr strains arise from F^+ strains when the F plasmid is incorporated into the bacterial chromosome at one of several possible sites (> Figure 8.9a). When an Hfr cell serves as a donor in conjugation, the F plasmid initiates transfer of chromosomal DNA. Usually, only part of the F plasmid, called the **initiating segment,** is transferred, along with some adjacent chromosomal genes (> Figure 8.9b).

> Figure 8.8   **An $F^+ \times F^-$ mating.** The F^+ cell transfers one strand of DNA from its F plasmid to the F^- cell via the conjugation bridge. As this occurs, the complementary strands of F plasmid DNA are synthesized. Thus, the recipient cell gets a complete copy of the F plasmid, and the donor cell retains a complete copy.

(a) Conversion of F⁺ cell to Hfr condition

(b) Conjugation

➤ Figure 8.9   **High-frequency re-combinations.** (a) Conversion of F⁺ cells to the Hfr condition. Hfr cells arise from F⁺ cells when their F plasmid is incorporated into a bacterial chromosome at one of several possible sites. (b) During conjugation, the (pink) initiating site of the F plasmid and adjacent genes are transferred to a recipient cell. Genes are transferred in linear sequence, and the number of genes transferred depends on the duration of conjugation and whether the DNA strand breaks or remains intact.

F plasmids are circular, double-stranded DNA molecules containing about 100,000 nucleotide pairs (that's the same as about 2 percent of a bacterial chromosome).

In the 1950s the French scientists Elie Wollman and François Jacob studied this process in a series of interrupted mating experiments. They combined cells of an Hfr strain with cells of an F⁻ strain and removed samples of cells at short intervals. Each cell sample was subjected to mechanical agitation through vibration or whirling in a blender to disrupt the conjugation process. Cells from each sample were plated on a variety of media, each of which lacked a particular nutrient, to determine their nutrient requirements. By careful observation of the genetic characteristics of cells from many experiments, the investigators determined that transfer of DNA in conjugation occurred in a linear fashion and according to a precise time schedule. When conjugation was disrupted after 8 minutes, most recipient cells had received one gene. When it was disrupted after 120 minutes, recipient cells had received much more DNA, sometimes an entire chromosome. At intermediate intervals, the number of donor genes transferred was proportional to the length of time conjugation was allowed to proceed. However, because of a tendency of chromosomes to break during transfer, some cells received fewer genes than would have been predicted by the time allowed. Whatever the number of genes transferred, they were always transferred in linear sequence from the initiation site created by the incorporation of the F plasmid.

➤ Table 8.1	Results of selected conjugations		
Donor	**Recipient**	**Molecule(s) transferred**	**Product**
F^+	F^-	F plasmid	F^+ cells
Hfr	F^-	Initiating segment of F plasmid and variable quantity of chromosomal DNA	F^- with variable quantity of chromosomal DNA
F'	F^-	F' plasmid and some chromosomal genes it carries with it	F' cell with some duplicate gene pairs: one on chromosome, one on plasmid

The Transfer of F′ Plasmids

The process of incorporating an F plasmid into a bacterial chromosome is reversible. In other words, DNA incorporated into a chromosome can separate from it and again become an F plasmid. In some cases this separation occurs imprecisely, and a fragment of the chromosome is carried with the F plasmid, creating what is called an **F′** (F prime) **plasmid** (➤ Figure 8.10). Cells containing such plasmids are called *F′ strains*. When F′ cells conjugate with F^- cells, the whole plasmid (including the genes from the chromosome) is transferred. Hence, recipient cells have two of some chromosomal genes—one on the chromosome and one associated with the plasmid. F′ plasmids generally do not become part of the recipient cell's chromosome.

In the transfer of F^+ and F′ plasmids, as in all other transfers during conjugation, the donor cell retains all the genes it had prior to the transfer, including copies of the F plasmid. Single-stranded DNA is transferred, and both donor and recipient cells synthesize a complementary strand for any single-stranded DNA they contain.

The results of conjugation with F^+, Hfr, and F′ transfers are summarized in Table 8.1.

The Significance of Conjugation

Like other mechanisms for gene transfer, conjugation is significant because it contributes to genetic variation. Larger amounts of DNA are transferred in conjugation than in other transfers, so conjugation is especially important in increasing genetic diversity. In fact, conjugation may represent an evolutionary stage between the asexual processes of transduction and transformation and the actual fusion of whole cells (the gametes) that occur during sexual reproduction in eukaryotes. For the microbial geneticist, conjugation is of special significance because precise linear transfer of genes is useful in chromosome mapping.

Plasmids that are self-transmissible—that is, have genes for the formation of an F pilus—can sometimes transfer into species other than their own kind. Those that can are said to be **promiscuous.** Sometimes the species are only distantly related; in other cases, transfer even occurs into eukaryotic cells! Obviously this has important implications for health and evolution.

Some Gram-positive bacteria have self-transmissible plasmids that do not form F pili. Instead, bacteria lacking these plasmids secrete peptide compounds, which stimulate nearby bacteria that do contain the plasmids to mate with them. Once a bacterium acquires the plasmid, it stops producing the attracting peptide for it. This is a neat conservation of energy. However, these cells will still secrete other peptides that will serve as mating lures for other plasmids they have not yet acquired.

➤ **Figure 8.10   The formation and transfer of F′ plasmids.** When the F plasmid in an Hfr cell separates from the bacterial chromosome, it may carry some chromosomal DNA with it. Such an F′ plasmid may then be transferred by conjugation to an F^- cell. The recipient cell will then have two copies of some genes—one on its chromosome and one on the plasmid.

Gene Transfer Mechanisms Compared

The most fundamental differences among the major types of transfers of genetic information concern the quantity of DNA transferred and the mechanism by which the transfer takes place. In transformation, less than 1 percent of the DNA in one bacterial cell is transferred to another, and the transfer involves only chromosomal DNA.

➤ Table 8.2 Summary of the effects of various transfers of genetic information

Kind of transfer	Effects
Transformation	Transfers less than 1 percent of cell's DNA. Requires competence factor. Changes certain characteristics of an organism depending on which genes are transferred.
Transduction	Transfer is effected by a bacteriophage.
Specialized	Only genes near the prophage are transferred to another bacterium.
Generalized	Fragments of host bacterial DNA of variable length and number are packed into the head of a virus.
Conjugation	Transfer is effected by a plasmid.
F$^+$	A single plasmid is transferred.
Hfr	An initiating segment of a plasmid and a linear sequence of bacterial DNA that follows the initiating segment are transferred.
F′	A plasmid and whatever bacterial genes adhere to it when it leaves a bacterium are transferred.

In transduction, the quantity of DNA transferred varies from a few genes to large fragments of the chromosome, and a bacteriophage is always involved in the transfer. In specialized transduction, the phage inserts into a bacterial chromosome and carries a few host genes with it when it separates. In generalized transduction, the phage causes fragmentation of the bacterial chromosome; some of the fragments are accidentally packed into viruses as they are assembled.

In conjugation, the quantity of DNA transferred is highly variable, depending on the mechanism. A plasmid is always involved in the transfer. An F plasmid itself can be transferred, as occurs in F$^+$ to F$^-$ conjugation. An initiating segment of a plasmid and any quantity of chromosomal DNA—from a few genes to the whole chromosome—is transferred in Hfr conjugation. A plasmid and whatever chromosomal genes it has carried with it from the chromosome are transferred in F′ conjugation. These characteristics are summarized in Table 8.2.

Plasmids

Characteristics of Plasmids

The F plasmid just described was the first plasmid to be discovered. Since its discovery, many other plasmids have been identified. Most are circular, double-stranded extrachromosomal DNA. They are self-replicating by the same mechanism that any other DNA uses to replicate itself. Most plasmids have been identified by virtue of some recognizable function that they serve in a bacterium. These functions include the following:

1. F plasmids (fertility factors) direct the synthesis of proteins that self-assemble into conjugation pili.

2. *Resistance (R) plasmids* carry genes that provide resistance to various antibiotics such as chloramphenicol and tetracycline and to heavy metals such as arsenic and mercury.

3. Other plasmids direct the synthesis of bacteriocidal (bacteria-killing) proteins called *bacteriocins.*

4. Virulence plasmids, such as those in *Salmonella,* cause disease signs and symptoms.

5. Tumor-inducing (Ti) plasmids can cause tumor formation in plants.

6. Some plasmids contain genes for catabolic enzymes.

Generally, plasmids carry genes that code for functions not essential for cell growth; the chromosome carries the genes that code for essential functions.

Clostridium tetani neurotoxin genes are carried on plasmids.

Resistance Plasmids

Resistance plasmids, also known as *R plasmids* or *R factors,* were discovered when it was noted that some enteric bacteria—bacteria found in the digestive tract—had acquired resistance to several commonly used antibiotics. We don't know how resistance plasmids arise, but we know that they are not induced by antibiotics. This has been demonstrated by the observation that cultures kept in storage from a time prior to the use of antibiotics exhibited antibiotic resistance on first exposure to the drugs. However, antibiotics contribute to the survival of strains that contain resistance plasmids. That is, when a population of organisms containing both resistant and nonresistant organisms is exposed to an antibiotic, the resistance organisms will survive and multiply, whereas the nonresistant ones will be killed. The resistant organisms are thus said to be *selected* to survive. Such selection is a major force in evolutionary change, as Charles Darwin realized.

According to Darwin, all living organisms are subject to *natural selection,* the survival of organisms on the basis of their ability to adapt to their environment. After studying many different kinds of plants and animals, Darwin drew two important conclusions. First, living organisms have certain heritable—that is, genetic—characteristics that help them adapt to their environment. Second, when environmental conditions change, those organisms with characteristics that allow them to adapt to the new environment will survive and reproduce. Organisms lacking such characteristics will perish and leave no offspring. A change in environmental conditions does not directly cause organisms to change. It merely provides a test of their ability to adapt. Only the organisms that can carry out their life processes under the new conditions will survive.

Resistance plasmids (➤ Figure 8.11) generally contain two components: a **resistance transfer factor (RTF)** and one or more **resistance (R) genes.** The DNA in an RTF is similar to that in F plasmids. The RTF implements transfer by conjugation of the whole resistance plasmid and is essential for the transfer of resistance from one organism to another. Each R gene carries information that confers resistance to a specific antibiotic or to a toxic metal. For antibiotic resistance, such genes usually direct synthesis of an enzyme that inactivates the antibiotic. Some resistance plasmids carry R genes for resistance to four widely used antibiotics: sulfanilamide, chloramphenicol, tetracycline, and streptomycin. Transfer of such a plasmid to any recipient confers resistance to all four antibiotics. Other resistance plasmids carry genes for resistance to one or more of these antibiotics. A few plasmids carry genes for resistance to more than four antibiotics.

The transfer of resistance plasmids from resistant to nonresistant organisms is rapid, so large numbers of previously nonresistant organisms can acquire resistance quickly. Furthermore, transfer of resistance plasmids occurs not only within a species, but also between closely related genera such as *Escherichia, Klebsiella, Salmonella, Serratia, Shigella,* and *Yersinia.* Transfer has even been observed between less closely related genera. Transfer of resistance plasmids is of great medical significance because it accounts for increasingly large populations of resistant organisms and reduces the effective use of antibiotics.

As scientists accumulate information on plasmids and how they confer antibiotic resistance, they become more concerned about the development of resistant strains and their potential danger to public health. As we shall see in Chapter 13, penicillin-resistant strains of *Neisseria gonorrhoeae, Haemophilus influenzae,* and some species of *Staphylococcus* already exist. Other antibiotics must now be used to treat the diseases caused by those strains, and the day may come when no antibiotic will effectively treat them. The more frequently antibiotics are used, the greater the selection is for resistant strains. Therefore, it is extremely important to identify the antibiotic to which an organism is most sensitive before using any antibiotic to treat a disease.

Transposons

In addition to being transferred on resistance plasmids by conjugation, R genes also can move from one plasmid to another in a cell or even become inserted in the chromosome. The ability of a genetic sequence to move from one location to another is called **transposition.** Such a mobile genetic sequence is called a **transposable element.** The simplest type of transposable element, an *insertion sequence,* contains a gene that codes for an enzyme (transposase) needed to transpose the insertion sequence; that gene is flanked on either side by a sequence of 9 to 41 nucleotides called *inverted repeats.* Transposable elements replicate only when in a plasmid or in a chromosome. During transposition, the insertion sequence is copied by the transposase

> Antibiotic resistance can be acquired either by mutation or by gene transfer.

(a)

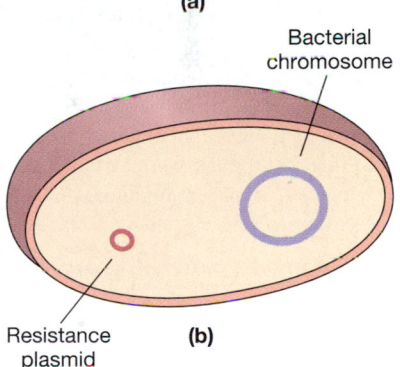

Bacterial chromosome

Resistance plasmid   **(b)**

Resistance Transfer Factor (RTF)

Replication functions

Antibiotic resistance: erythromycin, penicillin, chloramphenicol

Resistance to mercury, lead

Resistance to arsenic, antimony

(c)

➤ Figure 8.11   **Resistance plasmids.** (a) Resistance plasmids (magnified 268,000X). (b) These circular pieces of DNA are much smaller than a bacterial chromosome. (c) A typical resistance plasmid can carry genes for resistance to various antibiotics and to inorganic toxic substances, sometimes used in disinfectants. The resistance transfer factor includes genes needed for the plasmid to undergo conjugation.

Doing It the Hard Way

Geneticists receive respect these days for their research and discoveries, so you might be surprised to see geneticist Barbara McClintock's tiny Cold Spring Harbor laboratory. You might also be surprised to see a tennis racket, a pair of skates, an ironing board, and a hot plate among the Petri dishes and the Bunsen burners. Barbara McClintock worked and lived in tiny laboratory rooms before the days of research grants, huge laboratories, and teams of researchers, in a time when scientific research was considered beyond most women's abilities. Consequently, her mid-1940s reports on transposons were not accepted by most of her scientific peers until the 1970s, when genetic research verified her insights. But a life of being shunned by the scientific community did not stop her from establishing herself among the great geneticists of this century; her work, completed over 50 years ago, is highly esteemed and relevant today.

and cellular enzymes. The copy randomly inserts itself into the bacterial chromosome or into another plasmid; the original insertion element remains in its original position. However, the ability to move among plasmids or to a chromosome greatly increases the ways a transposable element can affect the genetic makeup of a cell. The coding sequence or regulatory regions of any gene into which a transposable element inserts can be disrupted. Transposable elements are known to cause mutations and may be responsible for some spontaneous mutations.

A **transposon** (tranz-pos′on) is a transposable element that contains the genes for transposition and one or more other genes as well (➤ Figure 8.12). Typically, these other genes represent R genes, conferring resistance to antibiotics such as tetracycline, chloramphenicol, or ampicillin. Thus, a transposon can move R genes from one plasmid to another or to the bacterial chromosome. The transposition of a transposon may disrupt gene function.

In 1983 Barbara McClintock won the Nobel Prize for her work on transposons, using corn. Transposons were next found in microorganisms and are now considered a universal phenomenon. Transposition is a relatively rare event and is not easily detected in eukaryotes. It is easier to detect in bacteria because researchers can work with large populations that can be tested more easily for particular characteristics.

Bacteriocinogens

In 1925 the Belgian scientist André Gratia observed that some strains of *E. coli* release a protein that inhibits growth of other strains of the same organism. This allows them to compete more successfully for food and space against the other strains. About 20 such proteins, called **colicins** (ko′leh-sinz), have been identified in *E. coli,* and similar proteins have been identified in many other bacteria. All these growth-inhibiting proteins are now called **bacteriocins** (bak-ter″e-o′sinz). Typically, bacteriocins inhibit growth only in other strains of the same species or in closely related species.

Bacteriocin production is directed by a plasmid called a **bacteriocinogen** (bak-ter″e-o-sin′o-jen). Although in most situations bacteriocinogens are repressed, in some cases the plasmid escapes repression and causes synthesis of its bacteriocin. Ultraviolet radiation can induce the formation and release of bacteriocin. When a bacteriocin is released, it can have a very potent effect on susceptible cells; one molecule of bacteriocin can kill a bacterium.

The mechanisms of action of bacteriocins are quite variable. Some enter a bacterial cell and destroy DNA. Others arrest protein synthesis by disrupting the molecular structure of ribosomes required for protein synthesis. Still others act on cell membranes by inhibiting active transport or by increasing membrane permeability to ions.

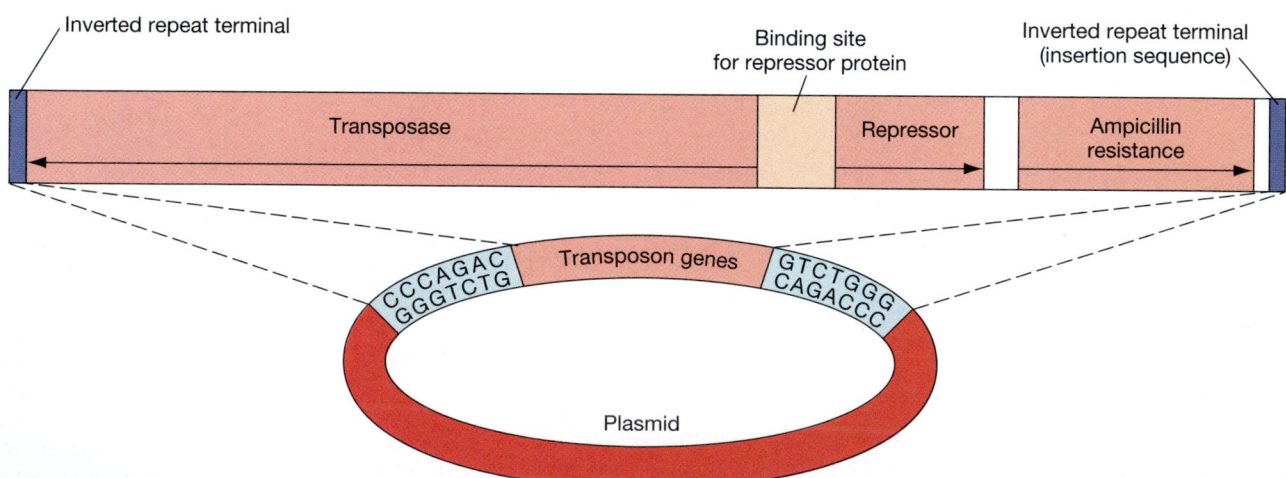

➤ Figure 8.12   **Transposons.** A typical transposon is bounded by inverted repeat terminals—DNA segments with base sequences that are identical when read in opposite directions on different strands. A gene coding for a transposase enzyme that cuts DNA at these insertion sequences allows the transposon to cut itself into and out of plasmids and chromosomes. There is also a gene for a repressor protein that can keep the transposase gene from being transcribed.

Genetic Engineering

Genetic engineering refers to the purposeful manipulation of genetic material to alter the characteristics of an organism in a desired way (Table 8.3). Various methods of genetic manipulation enable microbial geneticists to create new combinations of genetic material in microbes. The transfer of genes between different members of the same species occurs in nature and has been done in the laboratory for several decades. Lederberg's experiment (➤ Figure 8.6) provides one example of such a technique. The transfer of genes between different species also is now possible. Here we will discuss five techniques of genetic engineering: *genetic*

✓ What do the designations F^+, F^-, F', and Hfr mean?

✓ Are resistance plasmids induced by antibiotics?

✓ How are resistance genes moved by transposons?

➤ Table 8.3	Some products and applications of genetic engineering
Pharmaceutical products	**Use**
Human insulin	Treat diabetes
Human growth hormone	Prevent pituitary dwarfism
Blood clotting factor VIII	Treat hemophilia
Erythropoietin	Treat anemia; stimulate formation of new red blood cells
Alpha-, beta-, and gamma-interferon	Treat cancer and viral disease
Tumor necrosis factor	Disintegrate cancer cells
Interleukin-2	Treat cancer and immunodeficiencies
Tissue plasminogen activator	Treat heart attacks, dissolve clots
Taxol	Treat ovarian and breast cancers
Bone growth factor	Heal bone fractures, treat osteoporosis, stimulate bone growth
Epidermal growth factor	Heal wounds
Monoclonal antibodies	Diagnose and treat diseases
Hepatitis A and B vaccines	Prevent hepatitis
AIDS subunit vaccine (in clinical trials)	Incomplete virus vaccine
Human hemoglobin	Blood substitute in emergencies (produced in gene-altered pigs)
Antibiotics	Inhibit or kill microbial growth (increase yields by gene amplification)
Genetic studies	
DNA and RNA probes	Identify organisms, diseases, genetic defects in fetuses and adults
Gene therapy	Insert missing gene, or replace defective gene, in adults or in egg and sperm; treat cystic fibrosis
Gene libraries	Understand gene structure and function, relatedness of organisms, Human Genome Project
Industrial applications	
Oil-eating recombinant bacteria	Clean up oil spills, remove oil residue from empty tankers
Pollutant/toxic materials–degrading recombinant bacteria	Clean up contaminated sites
Enzymes, vitamins, amino acids, industrial chemicals	Various uses (yield increased by gene amplification in producing microbes)
Agricultural applications	
Frostban bacteria (*Pseudomonas syringae*)	Prevent frost damage to strawberry crops
Breeding new types of plants and animals	Provide food, decoration, other uses
Herbicide-resistant crop plants	Allow crop plants to survive weeding done by spraying with herbicides; only crop plants survive
Viruses used as insecticides	Infect and kill insect pests

Bacteria That Don't Deserve a Bad Rap

In the news these days, we hear stories about bacteria causing outbreaks of food poisoning. But did you know that bacteria can actually keep food from spoiling? Naturally occurring proteins from certain bacteria, called bacteriocins, can control the growth of bacteria strains responsible for food spoilage. The U.S. Food and Drug Administration has already approved the use of one bacteriocin, nisin, in pasteurized cheese; it inhibits the growth of *C. botulinum,* a microbial strain that causes food poisoning. Scientists are researching other bacteriocins that kill nearly all of the bacterial cells associated with lean and fat beef tissues. We may soon have bacteria to thank for milk and meat products that are safer and take longer to spoil.

➢ **Figure 8.13   Genetic fusion.** A possible example in which the deletion of a part of a chromosome causes two different adjacent operons to be joined together. The control mechanisms of the first operon will now govern the expression of the genes that were originally part of the second operon.

fusion, protoplast fusion, gene amplification, recombinant DNA technology, and the creation of *hybridomas.*

Genetic Fusion

Genetic fusion allows transposition of genes from one location on a chromosome to another. It can also involve deletion of a DNA segment, resulting in the coupling of portions of two operons. For example, suppose that the *gal* operon, which regulates galactose use, and the *bio* operon, which regulates biotin synthesis, lie adjacent to each other on a chromosome (➢ Figure 8.13). Deletion of the control genes of the *bio* operon and subsequent coupling of the operons would constitute genetic fusion. Such fusion would allow the genes that control the use of galactose to control the entire operon, including the making of the enzymes involved in biotin synthesis.

The major applications of genetic fusion within a species, as just described, are in research studies on the properties of microbes. However, the techniques developed for genetic fusion experiments have been extended and modified in the development of other kinds of genetic engineering.

One application of genetic fusion involves *Pseudomonas syringae,* a bacterium that grows on plants. Genetically altered strains have been developed that increase the resistance of plants, such as potatoes and strawberries, to frost damage. Strains of this bacterium that occur naturally on the leaves of plants produce a protein that forms a nucleus for the formation of ice crystals. The ice crystals damage the plants by causing cracks in the cells and leaves. By removing part of the gene that produces the "ice crystal" protein, scientists have engineered strains of *P. syringae* that cannot make the protein. When organisms of this strain are sprayed on the leaves of plants, they crowd out the naturally occurring strain. The treated plants then become resistant to frost damage at temperatures as low as $-5°C$.

Protoplast Fusion

A **protoplast** is an organism with its cell wall removed. **Protoplast fusion** (➢ Figure 8.14) is accomplished by removing the cell walls of organisms of two strains and mixing the resulting protoplasts. This allows fusion of the cells and their genetic material; that is, material from one strain recombines with that from the other strain before new cell walls are produced. Although genetic recombination occurs in nature in about one in a million cells, it occurs in protoplast fusion in as many as one in five cells. Thus, protoplast fusion simply speeds up a process that occurs in a very limited way in nature.

By mixing two strains, each of which has a desirable characteristic, new strains that have both characteristics can be produced. For example, a slow-growing strain that produces large quantities of a desired substance can be mixed with a fast-growing, poor producer. After protoplast

In protoplast fusion, both organisms' entire genomes, not just individual genes, are transferred, thus avoiding difficult maneuvers like transferring correct promoters.

fusion, some organisms will probably be fast-growing, good producers of the substance. Other organisms that turn out to be slow-growing, poor producers are discarded. Alternatively, two good producers can be mixed to obtain a super producer. This has been done with two strains of the bacterium *Nocardia lactamdurans,* which produce the antibiotic cephalomycin. The new strains produce 10 to 15 percent more antibiotic than the best of the parent strains.

Protoplast fusion works best between strains of the same species. It has been accomplished in molds, however, between two species of the same genus (*Aspergillus nidulans* and *A. rugulosus*) and even between two genera of yeasts (*Candida* and *Endomycopsis*).

Microbiologists are exploring possible applications of protoplast fusion. It offers great promise for the future as procedures are refined and useful strains are developed.

Gene Amplification

Gene amplification is a process by which plasmids, or in some cases bacteriophages, are induced to reproduce within cells at a rapid rate. If the genes required for the production of a substance are in the plasmids or can be moved to them, increasing the number of plasmids will increase production of the substance by the host cells.

Most bacteria and many fungi, including those that produce antibiotics, contain plasmids. Such plasmids, which often carry genes for antibiotic synthesis, provide many opportunities for using gene amplification to increase antibiotic yields. Even when genes concerned with antibiotic production are in the chromosome, scientists can transfer them to plasmids. Increased reproduction of plasmids would then greatly increase the number of copies of genes that act in antibiotic synthesis. This, in turn, would significantly increase the amount of antibiotic such cells could produce.

The possible applications of gene amplification are not limited to increasing antibiotic production. In fact, gene amplification may turn out to be even more effective in increasing production of substances that are synthesized by somewhat simpler pathways. These substances include enzymes and other products such as amino acids, vitamins, and nucleotides.

Rapid reproduction of bacteriophages already can be used to make the amino acid tryptophan. Bacteriophages carrying the *trp* operon (genes that control synthesis of enzymes to make tryptophan) of *E. coli* are induced to reproduce rapidly. Thus, cells containing large numbers of copies of the *trp* operon synthesize large quantities of the enzymes. Subsequent analysis of such cells has shown that half the intracellular proteins are enzymes for tryptophan synthesis.

Recombinant DNA Technology

One of the most useful of all techniques of genetic engineering is the production of **recombinant DNA**—DNA that contains information from two different species of organisms. If these genes integrate permanently into the egg or sperm cells such that the genes can be transferred to offspring, the resulting organism is said to be a **transgenic,** or *recombinant,* organism. Making recombinant DNA involves three processes:

1. The manipulation of DNA *in vitro*—that is, outside cells.

2. The recombination of another organism's DNA with bacterial DNA in a phage or a plasmid.

3. The *cloning,* or production of many genetically identical progeny, of phages or plasmids that carry foreign DNA.

These processes were first carried out in 1972 by Paul Berg and A. D. Kaiser, who inserted other prokaryotic DNA into bacteria, and then by S. N. Cohen and Herbert Boyer, who inserted eukaryotic DNA into bacteria.

DNA from either prokaryotic or eukaryotic cells is removed from the cells and cut into small segments. The donor DNA segments are then incorporated into a **vector,** or self-replicating carrier such as a phage or a plasmid (➤ Figure 8.15). First, a restriction endonuclease is used to cut double-stranded DNA in the vector and donor DNA. The cuts leave overlapping ends. A particular restriction endonuclease always produces the same complementary ends. Then, donor DNA is incorporated into the vector by an enzyme called a *ligase,* which reunites the ends of nucleotide chains. Thus, the vector contains all the original DNA plus a new segment of donor DNA.

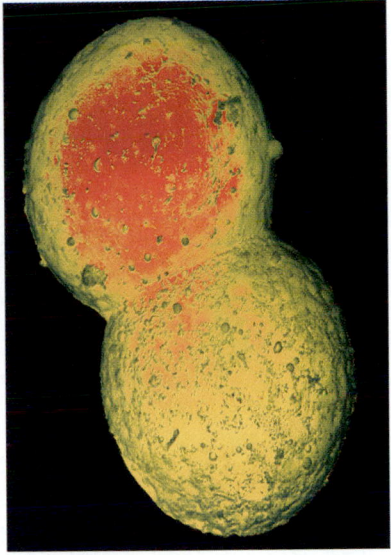

➤ Figure 8.14   **Protoplast fusion.** A colorized TEM of two tobacco plant leaf cells (magnified 775X) undergoing protoplast fusion. Protoplast fusion involves the use of enzymes to digest away the cell walls of cells of organisms from two different strains. When placed together, the cells fuse and develop a new cell wall around the hybrid cell containing the genes of both organisms.

Bacteria, genetically engineered to be bioluminescent, can be inoculated into mice. Their transport and related disease can be followed by monitoring the glow.

➤ Figure 8.15 **Producing recombinant DNA.** The main steps in the procedure shown are (1) obtain the release of DNA containing a gene of interest from mammalian or other cell types; (2) construct recombinant DNA molecules with the use of restriction enzymes (endonucleases), which cut (cleave) DNA at specific sites and into short segments. Cleaving the DNA leaves "sticky ends" that can be joined by hydrogen bonding to similar ends on any DNA segment cleaved with the same endonuclease; (3) introduce the recombinant DNA into a host bacterium; (4) identify and isolate cells that carry the gene of interest. Cloning vectors (plasmids) can carry one or more genes for antibiotic resistance, which are of value in isolating and identifying cells that carry the recombinant DNA.

Once this new segment of DNA is inserted into the vector, it can be introduced into cells such as *E. coli* that have been rendered competent by heating (annealing) in a solution of calcium chloride or by **electroporation.** This technique uses a brief electrical pulse to produce temporary pores in the cell membrane through which the vector can pass. As the *E. coli* cells divide, the vectors in them also are reproduced by cloning. Such vector-containing cells can be identified, grown, and lysed, and the vectors containing a specific cloned segment of donor DNA can be retrieved. In other cases, large quantities of the protein product expressed by the donor gene are required.

Medical Applications of Recombinant DNA

One of the most medically significant applications of recombinant DNA technology is the modification of bacterial cells to make substances useful to humans. To make bacterial cells produce human proteins, a human DNA gene with the information for synthesizing the protein is inserted into the vector. Interferon, a substance used to treat certain viral infections and cancers (Chapter 16), and the hormone insulin were among the first products made with recombinant DNA. Human growth hormone now can be made that way, and new products—vaccines, blood coagulation proteins for people with hemophilia, and enzymes such as cholesterol oxidase to diagnose disorders in cholesterol metabolism—have been developed. Many of the products in Table 8.3 are the result of recombinant DNA technology.

The use of recombinant DNA technology to make substances useful to humans makes certain treatments potentially safer, cheaper, and available to more patients. For example, prior to the use of recombinant DNA to manufacture human insulin, the insulin for diabetic patients came exclusively from slaughtered cattle and pigs. Some patients develop allergies to such insulin, and the number of patients requiring insulin is increasing. Making nonallergenic human insulin and increasing the insulin supply are two important benefits of insulin production by recombinant DNA technology. Most patients in the United States now use a preparation containing genetically engineered insulin.

Likewise, prior to the manufacture of human growth hormone by recombinant DNA technology, the hormone was obtained from the pituitary glands of cadavers at autopsy, and several cadavers were needed to obtain a single dose with which to treat children with the congenital condition of pituitary dwarfism. This treatment was suspended in April 1985 in both the United States and the United Kingdom due to reports of several cases of the fatal, degenerative neurological Creutzfeldt-Jakob disease. But later in 1985, the FDA approved a genetically engineered form of the hormone that produces the same effects as the human cadaver hormone. In addition to correcting congenital disorders, this product of recombinant DNA technology might prove useful in the treatment of delayed wound or fracture healing and in the metabolic problems associated with aging.

The manufacture of certain blood coagulation proteins by recombinant DNA technology makes these substances more readily available to individuals with hemophilia or other blood disorders. It also assures that the recipient will not acquire AIDS or hepatitis B from a contaminated blood product.

Recombinant DNA is being used to make vaccines more economically and in larger quantities than before. In this application, some microorganisms are used to combat the disease-causing capacity of others. Genes that direct the synthesis of specific substances, called *antigens,* from a disease-causing bacterium, virus, or protozoan parasite are inserted into another organism. The organism then makes a pure antigen. When the antigen is introduced into a human, the human immune system makes another specific substance, called an *antibody,* which takes part in the body's defense against the disease-causing organism (Chapter 17).

Recombinant DNA procedures for making vaccines for hepatitis A and B and influenza are already available. The vaccines not only are cheaper than conventional ones but also are purer and more specific, and they cause fewer undesirable side effects. Vaccines have proved to be highly specific and extremely effective against hepatitis A and B.

Many other applications of recombinant DNA techniques are being developed. An especially important one is the diagnosis of genetic defects in a fetus, which can be done by studying enzymes in fetal cells from amniotic fluid. Such defects are detected by using recombinant DNA with a known nucleotide sequence to find errors in the nucleotide sequence in fetal DNA

Recombinant insulin received FDA approval in 1982. The production of recombinant insulin is a $500 million a year business, making it one of the top 200 drugs sold.

Hepatitis B vaccine, licensed in 1981, was the first vaccine for human use to be produced by recombinant DNA technology.

Need a Blood Transfusion? Call on a Genetically Altered Pig

By the year 2000, researchers expect to get government approval of a blood-substitute product composed primarily of human hemoglobin that is produced by transgenic (genetically altered) pigs. The biotechnology firm DNX, located in Princeton, New Jersey, injected thousands of copies of two human hemoglobin genes into 1-day-old pig embryos that had been removed from their mothers' uteri. The embryos were then implanted into a second pig's uterus to grow until delivery. Two days after birth, the piglets were tested to see if they produced human hemoglobin along with pig hemoglobin—that is, if they were transgenic. Only about 0.5 percent of such transfers succeed.

It costs $50,000 to $75,000 to make one transgenic animal. DNX was initially successful in making three such pigs. Then the company bred the one transgenic male with more than 1000 regular females and interbred the offspring of these crosses. DNX now has hundreds of transgenic pigs. Such crossing has continued for four generations. The human hemoglobin genes still function perfectly in the altered pigs, allaying fears that the genes would mutate or disintegrate in their new environment. The transgenic pigs currently produce blood containing more than 30 percent human hemoglobin, about 20 percent hybrid human/pig hemoglobin, and almost 50 percent pig hemoglobin. DNX hopes eventually to reach a blood level of 50 percent human hemoglobin in the pigs.

This pig has genes for the production of human hemoglobin. Such pigs will be bred and bled to collect human hemoglobin to save people's lives by transfusion. This is an example of biotechnological "pharming" of molecules.

To obtain the blood substitute, the pigs are bled and the red blood cells are ruptured. Pure human hemoglobin is separated from hybrid human/pig and pig hemoglobin by a multistep purification process that uses all available forms of chromatography—to safeguard against reliance on a single method that might fail and retain impurities.

The substitute product has several advantages over actual human blood:

1. It has a storage life of months instead of weeks.
2. Because naked hemoglobin does not stimulate the immune system to act against it, as do intact red blood cells containing hemoglobin, it can be transfused into anyone without the need for blood typing and matching.
3. It can ensure safety from human pathogens (including the AIDS virus) that might now contaminate human blood.
4. Because of the oxygen-carrying capacity of hemoglobin, it can serve as an immediate source of oxygen in earthquakes, in accidents, or in times of war, thereby enabling the injured to survive the trip to the hospital.

Human blood costs have recently risen to $300 to $500 per unit. By the year 2000, the blood substitute may cost the same as, or even less than, a unit of blood. Storage will be cheaper, and blood-typing costs would be eliminated.

One drawback to this procedure is that, once transfused, naked hemoglobin lasts only hours or days instead of 6 months, but this might be long enough to treat emergency cases. Another problem with this product is possible contamination with pig molecules or pig pathogens if purification processes fail. One issue that does *not* appear to be a problem is the use of this product by Jewish people who, for religious reasons, do not eat pork. The director of the Rabbinical Council of America has said that there would probably be no religious objection. By Judaic law, pigs may be used for purposes other than eating (such as sources of heart-valve replacements and insulin), and kosher laws are suspended in cases of life or death.

DNX had hoped to have government approval by now, but it has been unable to raise the $50 million needed to do so. In the meantime, it has been working on increasing its herd and techniques for large-scale production. Future use of the hybrid hemoglobin has not been ruled out, but further research is needed to determine how well it would function. DNX is also looking into the production of transgenic pig hearts for transplant into humans.

segments. Such errors in fetal DNA denote genetic defects that can be responsible for absent or defective enzymes. Application of these techniques could greatly improve prenatal diagnosis of many genetic defects. Ultimately, as techniques for preparing recombinant DNA in animal cells improve, it may become possible to insert a missing gene or to replace a defective one in human cells (*gene therapy*). In fact, insertion of functional genes in appropriate cells may have cured the genetic disease severe combined immunodeficiency disease (Chapter 18).

Insertion of such a gene in a defective gamete (egg or sperm) might prevent offspring from inheriting a genetic disease.

Forensic applications of DNA technology are rapidly coming into use in the courtroom. In paternity cases, for example, experts can now determine with about 99 percent certainty that a given man is the father of a particular child, on the basis of DNA comparison. ➡ (p. 192) Likewise, rapists and murderers can be identified by the "DNA fingerprints" they leave behind at the scene of the crime in the form of semen, blood, hair, or tissue under their victim's fingernails. Sufficient DNA can even be collected for analysis from the rim of a used drinking glass or from the sweatband of a hat.

Industrial Applications of Recombinant DNA

Fermentation processes used in making wine, antibiotics, and other substances might be greatly improved by the use of recombinant DNA. For example, addition of genes for the synthesis of amylase to the yeast *Saccharomyces* could allow these organisms to produce alcohol from starch. Malting of grain to make beer would be unnecessary, and wines could be made from juices containing starches instead of sugars. Still other applications might include degradation of cellulose and lignin (plant materials often wasted), manufacture of fuels, clean-up of environmental pollutants, and leaching of metals from low-grade ores. Strains of *Pseudomonas putida,* already known to degrade different components of oil, might be engineered so that one strain degrades all components. Industrial leaching, or extraction, of metals from copper and uranium ores is already carried out by certain bacteria of the genus *Thiobacillus*. If these organisms could be made more resistant to heat and to the toxicity of the metals that they leach, the extraction process could be greatly accelerated.

Agricultural Applications of Recombinant DNA

Some bacteria are being engineered to control insects that destroy crops. The Monsanto Company has recently modified the genetic makeup of a strain of *Pseudomonas fluorescens,* which colonizes the roots of corn. This bacterium has been induced to carry genetic information inserted into it from *Bacillus thuringiensis,* allowing *P. fluorescens* to synthesize a protein that kills insects (➤ Figure 8.16). Toxin made by *B. thuringiensis* has been extracted and used for many years as an insecticide. Now the pseudomonads, applied to the surface of corn seeds, can make the toxin as they grow around the roots of the corn.

If the pseudomonads survive in corn fields as well as they have in greenhouses, they could replace the use of chemical insecticides to control black cutworm and probably other insect larvae that damage crops. With further research, other bacteria normally present on crops might be modified to control additional pests. Some optimistic scientists believe chemical pesticides may be phased out in favor of these safer and cheaper methods of pest control.

Also under development are genetically engineered seeds for crop plants that have high yield and other desirable characteristics and that will resist herbicides that kill weeds. The Environmental Protection Agency has begun steps to approve growth of genetically altered crops to build up supplies of seeds that will produce their own insecticides. Potatoes, corn, and cotton have received genes for the production of insect-killing toxin from *B. thuringiensis*. Their seeds will kill the Colorado potato beetle, the European corn borer and other mothlike insects, and the cotton bollworm, pink bollworm, and tobacco budworm, respectively. Farmers will be able to buy seeds that will solve many cultivation problems. Attempts have also been made to introduce nitrogen-fixing genes into nonleguminous plants. (We will discuss nitrogen fixation in Chapter 25.) This work has been successful in some plants but not yet in any important crop plants. If it can be extended to crop plants, many agricultural crops could be made to satisfy their nitrogen needs and thrive without commercial fertilizers, which are expensive and tend to pollute groundwater. This would be especially beneficial in some developing nations, where famine is an ever-present threat, and money for expensive fertilizers is not available.

Hybridomas

Along with the study of genetic recombinations in microorganisms came studies of such combinations in higher organisms. The first combination useful in industrial microbiology was the fusion of a myeloma (bone marrow cancer) cell with an antibody-producing white blood cell.

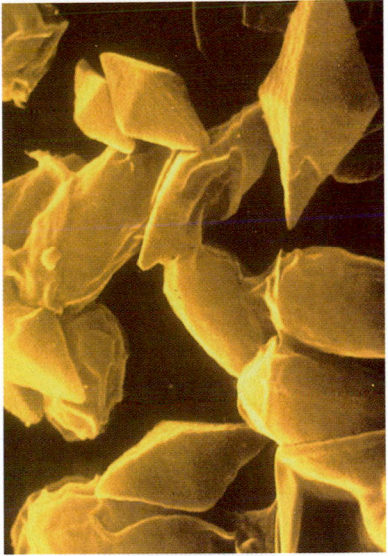

➤ Figure 8.16 **SEM photo (118,750X) of crystals of a substance toxic to many insects.** The genes for production of this toxin are being taken from *Bacillus thuringiensis,* which produces it naturally, and are being incorporated into other organisms through genetic engineering techniques. Imagine the benefits of crop plants that have built-in pesticide: no expense for chemical pesticides, no danger during application, no buildup in soil or water, and no entry or magnification of pesticides in the food chain.

Virus with a Scorpion's Sting

Viruses that cause convulsions, paralysis, and death probably don't sound like viruses you want to keep around. Farmers, though, are glad for the geneticists who created a virus with these characteristics. Under ordinary circumstances, caterpillars that devour valuable crop plants are sometimes infected by a naturally occurring virus harmless to other organisms. The caterpillars develop a long illness that culminates in their climbing to the top of a plant and bursting, releasing a shower of infectious viral particles. But, during their long illness, they still eat a lot of the crop. Genetic engineers have inserted the gene for scorpion venom into this virus. Caterpillars with the recombinant virus manufacture scorpion venom and quickly develop convulsions, paralysis, and death, all without stopping for a bite to eat. Don't worry—any altered virus that remains is destroyed in sunlight, and no chemical pesticide residues enter the food chain.

Redesigning Bacteria

Anyone who's ever tried to grow tomatoes knows how damaging the first frost can be. Beautiful red tomatoes turn into blackened, spongy masses overnight. If that were your main source of food, you'd be in trouble. So the concept of making plants—especially food plants—frost-resistant becomes very attractive.

Many people are surprised to learn that frost is partly a microbiological phenomenon—certain bacteria are largely responsible for the formation of ice crystals. These "ice-plus bacteria" produce a protein that has the same structure as that of water molecules in an ice crystal, and water molecules tend to align themselves on it. As the ice crystals grow, they enter the spaces inside the leaf or flower, causing frost injury. This damage makes the plant susceptible to disease caused by *Pseudomonas syringae,* the predominant frost-causing strain.

Chemicals can control the bacteria that cause frost to form. But such chemicals are expensive, harm the environment, are of limited effectiveness, and may lead to diseases. These chemicals kill beneficial bacteria as well as harmful ones. Hence they destroy the delicate balance of the plant's ecology.

One of the microbiologists who is working on this problem is Trevor Suslow, director of microbial pesticides at the DNA Plant Technology Corporation in Oakland, California. Suslow conducts studies to improve the production of food for developing countries. His research focuses on alternatives to chemical pesticides—on how to use the biological controls that already exist in nature and on evolving new ones. The goal of Suslow's team in controlling frost-causing bacteria was to leave the beneficial bacteria untouched and remove just the ice-plus strains. The strategy was to locate or develop "ice-minus bacteria" to compete successfully with ice-plus strains. Then if the ice-minus bacteria were applied to a frost-susceptible crop, they might protect the plant by keeping the ice-plus bacteria from becoming established—the effect of competitive exclusion.

Suslow's team focused on altering ice-plus bacteria so that they could no longer produce the ice-building protein. Using genetic engineering techniques, the team located the gene that controls this production and transferred it to *Escherichia coli* cells. With the aid of restriction enzymes, about one-third of the gene was moved. This was enough not only to inactivate the gene but also to ensure that no mutation could ever restore it to functional form. Next, the team cloned the defective gene, put it on a plasmid, and reinserted the plasmid into *P. syringae.* A plasmid and a chromosome in the same cell can exchange similar genes that are present on both. So Suslow's team looked for bacteria in which this natural process (called homologous recombination) had taken place, leaving the bacterial cells with a nonfunctional copy of the ice-forming gene in its chromosome. These were ice-minus bacteria. Deletion mutations of this kind occur constantly in nature. But it's more accurate, predictable, and cost effective to perform the genetic surgery in the laboratory. The altered ice-minus strain—whether naturally or ge-

netically engineered—is identical in every other respect to the unaltered ice-plus variety.

Suslow and his team conducted several hundred tests of the ice-minus strain, which they called Frostban. Then it was time to do field trials. Suslow obtained the necessary permits and approvals for field testing but was not prepared for the commotion such tests would cause. Protests came from several quarters. Many organizations opposed the release of the bacteria into the environment. It didn't matter that Frostban was safe. They objected to the genetic engineering of a living organism—any organism. Frostban became the symbol in their fight against a future full of recombinant DNA.

Environmentalists were concerned about the destruction of the ecological balance of an increasingly fragile earth. Local farmers had more immediate concerns. Some simply weren't sure what these bacteria were and what they might do. Others worried about potential risks to their own crops from a product that offered them no benefit. In fact, some California farmers like frost on their crops! A good frost sweetens up certain vegetables, such as carrots and Brussels sprouts. These farmers were concerned that Frostban bacteria would drift over their fields and keep their crops from getting the frost they needed. Research showing that there was virtually no likelihood that Frostban would spread in this way didn't convince the farmers.

Suslow and his team were plagued by an increasing number of lawsuits and even vandalism. The world didn't seem ready for strawberries served with a large dollop of recombinant DNA technology. Discontinuing the field trials, they went back to the laboratory to isolate natural ice-minus bacteria—those in which the ice-building gene is already missing. Although such organisms have the same effect as the bioengineered ones, they're more acceptable to many people because they occur naturally.

Why did the field trials of Frostban cause such controversy? Probably because Frostban was a first—the first commercially available, genetically engineered microbial agent. As such, it attracted much publicity. More fundamentally, people are unaware of the roles that microorganisms play in our lives. If people aren't comfortable with naturally occurring microbes, how can they be comfortable with bioengineered ones? Still, there is tremendous potential for genetically engineered biological controls.

A Frostban field test in Brentwood, California.

A fusion of two such cells is called a **hybridoma** (hi-brid-o'ma), or hybrid cell ($\succ$ Figure 8.17). This particular hybridoma can be grown in the laboratory, and it produces pure specific antibodies, called **monoclonal** (mon-o-klo'nal) **antibodies,** against any antigen to which the white blood cell was previously sensitized. Prior to the production of hybridomas and their monoclonal antibodies, no source of pure antibodies existed. Now many different kinds of monoclonal antibodies are produced commercially, and they represent a major advance in immunology. The production and uses of monoclonal antibodies are discussed in more detail in Chapter 17.

The ability to produce hybridomas may in the future lead to other important advances. Recently, agricultural scientists used this technique to fuse cells from a commercial potato plant with cells from an extremely rare wild strain. The wild strain was selected because it contains the gene for production of a natural insect repellent. The researchers were able to grow the fused cells into a plant that has the necessary commercial properties but also synthesizes the insect repellent in its leaves.

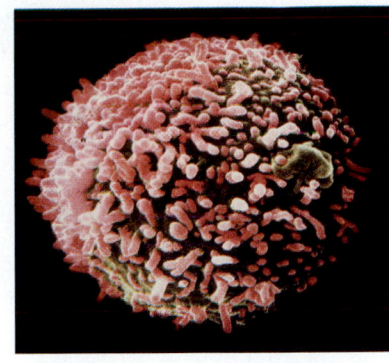

$\succ$ **Figure 8.17 A hybridoma cell.** A colorized SEM (9500×) of a hybridoma, a single cell fused from two other cells. Hybridomas are often made by fusing an antibody-producing plasma cell and a cancer cell. The latter keeps the culture dividing and growing indefinitely; the former causes the hybridoma to produce pure antibody against the antigen to which it was sensitized.

Weighing the Risks and Benefits of Recombinant DNA

Despite the many potential benefits of recombinant DNA research, at first some scientists working with it were concerned about its hazards. They feared that some recombinants might prove to be new and especially virulent pathogens for which humans would have no natural defenses and no effective treatments. In 1974 they called for a moratorium on certain experiments until the hazards could be assessed. From this assessment emerged the idea of biological containment—the practice of making recombinant DNA only in organisms with mutations that prevent them from surviving outside the laboratory.

In 1981 the constraints on recombinant DNA research were relaxed because of the following observations:

1. No illnesses in laboratory workers could be traced to recombinants.

2. The strain of *E. coli* used in the experiments failed to infect humans who voluntarily received large doses.

3. Incorporation of mammalian genes into *E. coli* was observed in nature, and these genes invariably impaired the organism's ability to adapt to the environment. This suggested that if laboratory organisms did escape, they probably would not survive in the natural environment.

4. Mutants of *E. coli* containing recombinant DNA were subject to control by accepted sanitary practices.

Most scientists now agree that, as currently practiced, recombinant DNA techniques offer significant benefits and exceedingly small risks to humans.

Retracing Our Steps

The Types and Significance of Gene Transfer

- **Gene transfer** refers to the movement of genetic information between organisms. It occurs in bacteria by transformation, transduction, and conjugation.

- Gene transfer is significant because it increases genetic diversity within a population, thereby increasing the likelihood that some members of the population will survive environmental changes.

Transformation

The Discovery of Transformation

- Bacterial **transformation** was discovered in 1928 by Griffith, who showed that a mixed culture of live rough and heat-killed smooth pneumococci could produce live smooth pneumococci capable of killing mice.

- Avery later showed that a capsular polysaccharide was responsible for virulence and that DNA was the substance responsible for transformation. Watson and Crick determined the structure of DNA, which led to studies showing that a cell's genetic information is encoded in its nucleic acids.

The Mechanism of Transformation

- Transformation involves the release of naked DNA fragments and their uptake by other cells at a certain stage in their growth cycle: (1) Uptake of DNA requires a protein called **competence factor** to make recipient cells ready to bind DNA. (2) Endonucleases cut double-stranded DNA into units; the strands separate, and only one strand is transferred. (3) Ultimately, donor DNA is spliced into recipient DNA. Leftover recipient DNA is broken down, so a cell's total DNA remains constant.

The Significance of Transformation

- Transformation is significant because (1) it contributes to genetic diversity; (2) it can be used to introduce DNA into an organism, observe its effects, and study gene locations; (3) it can be used to create recombinant DNA.

Transduction

The Discovery of Transduction

- In **transduction,** genetic material is carried by a **bacteriophage (phage)**.

The Mechanisms of Transduction

- Phages can be virulent or temperate. (1) **Virulent phages** destroy a host cell's DNA, direct synthesis of phage particles, and cause lysis of the host cell in the lytic cycle. (2) **Temperate phages** can replicate themselves as a **prophage**—part of a bacterial chromosome—or eventually produce new phage particles and lyse the host cell. Persistence of the phage in the cell without the destruction of the host cell is called **lysogeny.**

- Prophage can be incorporated into the bacterial chromosome, or it can exist as a plasmid, a piece of extrachromosomal DNA. Cells containing a prophage are called **lysogenic** cells because they have the potential to enter the **lytic cycle.**

- Transduction can be specialized or generalized. (1) In **specialized transduction,** the phage is incorporated into the chromosome and can transfer only genes adjacent to the phage. (2) In **generalized transduction,** the phage exists as a plasmid and can transfer any DNA fragment attached to it.

The Significance of Transduction

- Transduction is significant because it transfers genetic material and demonstrates a close evolutionary relationship between prophage and host cell DNA. Also, its persistence in a cell suggests a mechanism for the viral origins of cancer, and it provides a possible mechanism for studying gene linkage.

Conjugation

The Discovery of Conjugation

- In **conjugation** large quantities of DNA are transferred from one organism to another during contact between donor and recipient cells.

- Conjugation was discovered by Lederberg in 1946 when he observed that mixing strains of *E. coli* with different metabolic deficiencies allowed the cells to overcome deficiencies.

- **Plasmids** are extrachromosomal DNA molecules.

The Mechanisms of Conjugation

- Three mechanisms of conjugation have been observed: (1) In the transfer of **F plasmids,** a piece of extrachromosomal DNA (a **plasmid**) is transferred. (2) In high-frequency recombinations, parts of F plasmids that have been incorporated into the chromosome (the **initiating segment**) are transferred along with adjacent bacterial genes. (3) An F plasmid incorporated into the chromosome and subsequently separated becomes an **F' plasmid** and transfers chromosomal genes attached to it.

The Significance of Conjugation

- The significance of conjugation is that it increases genetic diversity, it may represent an evolutionary stage between asexual and sexual reproduction, and it provides a means of mapping genes in bacterial chromosomes.

Gene Transfer Mechanisms Compared

- Genetic transfer mechanisms differ in the quantity of DNA transferred.

Plasmids

Characteristics of Plasmids

- Plasmids are circular, self-replicating, double-stranded extrachromosomal DNA that carry information that is usually not essential for cell growth.

Resistance Plasmids

- **Resistance (R) plasmids** carry genetic information that confers resistance to various antibiotics and to certain heavy metals. They generally consist of a **resistance transfer factor (RTF)** and one or more **resistance (R) genes.**

Transposons

- R genes that move from one plasmid to another in a cell or become inserted in the chromosome are part of a **transposon** because they transpose, or change, their locations.

Bacteriocinogens

- **Bacteriocinogens** are plasmids that produce **bacteriocins,** which are proteins that inhibit growth of other strains of the same species or closely related species.

Genetic Engineering

- **Genetic engineering** is the manipulation of genetic material to alter the characteristics of an organism.

Genetic Fusion

- **Genetic fusion** allows transposition of genes from one location on a chromosome to another, sometimes deleting a portion, thereby causing the joining of genes from two different operons.

Protoplast Fusion

- **Protoplast fusion** combines **protoplasts** (organisms without cell walls) and allows mixing of genetic information.

Gene Amplification

- **Gene amplification** involves the addition of plasmids to microorganisms to increase yield of useful substances.

Recombinant DNA Technology

- **Recombinant DNA** is DNA produced when genes from one kind of organism are introduced into the genome of a different kind of organism. The resulting organism is a **transgenic,** or recombinant, organism.
- Recombinant DNA has proven especially useful in medicine, industry, and agriculture.

Hybridomas

- **Hybridomas** are genetic recombinations involving cells of higher organisms.

Weighing the Risks and Benefits of Recombinant DNA

- When recombinant DNA techniques were first developed, scientists were concerned that virulent pathogens might be created, and they developed containment procedures. As research proceeded and no illnesses caused by recombinants were observed, most scientists came to believe that the benefits of recombinant DNA techniques outweigh the risks.

Terminology Check

bacteriocin   (*p. 210*)
bacteriocinogen   (*p. 210*)
bacteriophage   (*p. 199*)
chromosome mapping   (*p. 202*)
clone   (*p. 205*)
colicin   (*p. 210*)
competence factor   (*p. 198*)
conjugation   (*p. 204*)
electroporation   (*p. 215*)
F^+ cell   (*p. 205*)
F^- cell   (*p. 205*)
F pilus   (*p. 205*)
F plasmid   (*p. 205*)
F′ plasmid   (*p. 207*)
gene amplification   (*p. 213*)
gene transfer   (*p. 196*)
generalized transduction (*p. 201*)

genetic engineering   (*p. 211*)
genetic fusion   (*p. 212*)
high frequency of recombination (Hfr) strain   (*p. 205*)
hybridoma   (*p. 219*)
initiating segment   (*p. 205*)
lysis   (*p. 201*)
lysogenic   (*p. 201*)
lysogeny   (*p. 201*)
lytic cycle   (*p. 201*)
monoclonal antibody   (*p. 219*)
phage   (*p. 200*)
plasmid   (*p. 204*)
promiscuous   (*p. 207*)
prophage   (*p. 201*)
protoplast   (*p. 212*)
protoplast fusion   (*p. 212*)

recombinant DNA   (*p. 213*)
recombination   (*p. 196*)
resistance (R) gene   (*p. 209*)
resistance plasmid   (*p. 209*)
resistance transfer factor   (RTF) (*p. 209*)
specialized transduction (*p. 201*)
temperate phage   (*p. 201*)
transduction   (*p. 199*)
transformation   (*p. 197*)
transgenic   (*p. 213*)
transposable element   (*p. 209*)
transposition   (*p. 209*)
transposon   (*p. 210*)
vector   (*p. 213*)
virulent phage   (*p. 200*)

CHAPTER 8 EIGHT

See You on the Web

Don't forget I am expecting you on the web. The following is a preparedness checklist for your web explorations. If you have questions on any of these items, you can return to the end of Chapter 1 for more details.

- *Questions We'll Explore*
- *Concept Check questions*
- *Flashcards*
- *Study Guide*

If you are ready, come visit me at

http://www.prenhall.com/black

It will be a journey you'll never forget.

An Introduction to Taxonomy: The Bacteria

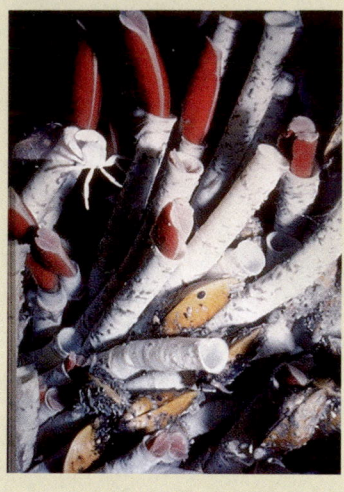

Exploring deep-sea geography, you might notice giant tubeworms (up to 2 m) and other interesting invertebrates living in habitats near geothermal vents. But how do these organisms survive in locations that appear so hostile?

The answer: invertebrates like the giant tubeworm have formed symbiotic relationships with members of Archaea, such that products from the metabolic processes of the microorganisms nourish the macroorganism. For example, chemolithotrophic bacteria have been found in the modified gastrointestinal tract of giant tubeworms (giant tubeworms lack a mouth, gut, or anus). These chemolithotrophs have metabolisms that fix inorganic carbon sources (CO_3^-, HCO_3^-) into organic carbon sources via the same enzymes utilized in the Calvin cycle of certain autotrophs. The tubeworms are than able to use the organic carbon sources in cellular processes.

What do the tubeworms do for the bacteria? The tubeworms have well-vascularized plumes that trap O_2 and H_2S from the thermal vents and transport those substances to the chemolithotrophs. The bacteria use the O_2 and H_2S in their life-sustaining energy reactions. It's an exquisite partnership!

Background Basics

H umans appear to have an innate need to name things. In many primitive societies, a person who knows the true name of an object or of another person is believed to have power over that object or person. Naming helps us to understand our world and to communicate with others about it.

Taxonomy: The Science of Classification

In science accurate and standardized names are essential. All chemists must mean the same thing when they talk about an element or a compound; physicists must agree on terms when they discuss matter or energy; and biologists must agree on the names of organisms, be they tigers or bacteria.

Faced with the great number and diversity of organisms, biologists use the characteristics of different organisms to describe specific forms of life and to identify new ones. The grouping of related organisms together is the basis of *classification*. The most obvious reasons for classification are (1) to establish the criteria for identifying organisms; (2) to arrange related organisms into groups; and (3) to provide important information on how organisms evolved. **Taxonomy** is the science of classification. It provides an orderly basis for the naming of organisms and for placing organisms into a category, or **taxon** (plural: *taxa*).

Another important aspect of taxonomy is that it makes use of and makes sense of the fundamental concepts of unity and diversity among living things. Organisms classified in any particular group have certain common characteristics—that is, they have unity with respect to these characteristics. For example, humans walk upright and have a well-developed brain; *Escherichia coli* cells are rod-shaped and have a Gram-negative cell wall. The organisms within taxonomic groups exhibit diversity as well. Even members of the

Questions We'll Explore

A How are microorganisms named?

B What did Linnaeus contribute to taxonomy?

C How is a dichotomous key used to identify organisms?

D What types of problems and major developments in taxonomy have occurred since Linnaeus founded that science?

E What are the main characteristics of the kingdoms in the five-kingdom system of taxonomy?

F What are the main characteristics of the three domains and their members, and how might they be related to a universal common ancestor?

G How are viruses classified?

H What special methods are needed for determining evolutionary relationships among prokaryotes?

I What criteria are used for classifying bacteria?

J What are the history and significance of *Bergey's Manual of Systematic Bacteriology?*

K What problems are associated with bacterial taxonomy?

same species display variations in size, shape, and other characteristics. Humans vary in height, weight, hair and eye color, and facial features. Certain kinds of bacteria vary somewhat in shape and in their ability to form specific structures, such as endospores.

A basic principle of taxonomy is that members of higher-level groups share fewer characteristics than those in lower-level groups. Like all other vertebrates, humans have backbones, but humans share fewer characteristics with fish and birds than with other mammals. Likewise, nearly all bacteria have a cell wall, but in some the wall is Gram-positive and in others it is Gram-negative.

Linnaeus, the Father of Taxonomy

The eighteenth-century Swedish botanist Carolus Linnaeus is credited with founding the science of taxonomy (➤ Figure 9.1). He originated **binomial nomenclature,** the system that is still used today to name all living things. In the binomial, or "two-name," system, the first name designates the **genus** (plural: *genera*) of an organism, and its first letter is capitalized. The second name is the **specific epithet,** and it is not capitalized even when derived from the name of the person who discovered it. Together the genus and specific epithet identify the **species** to which the organism belongs. Both words are italicized in print but underlined when handwritten. When there is no danger of confusion, the genus name may be abbreviated to a single letter. Thus, *Escherichia coli* is often written *E. coli,* and humans (*Homo sapiens)* may be identified as *H. sapiens.*

The name of an organism often tells something about it, such as its shape, where it is found, what nutrients it uses, who discovered it, or what disease it causes. Some examples of names and their meanings are shown in Table 9.1.

The members of a species generally have several common characteristics that distinguish that species from all other species. As a rule, members of the species cannot be divided into significantly different groups on the basis of a particular characteristic, but there are exceptions to this rule. Sometimes members of a species are divided on the basis of a small but permanent genetic difference, such as a need for a particular nutrient, resistance to a certain antibiotic, or the presence of a particular antigen. When organisms in one pure culture of a species differ from

➤ **Figure 9.1   Carolus Linnaeus (1707–1778).** Linnaeus is known as the father of taxonomy.

Why taxonomy? Common names are confusing. Passer domesticus is the English sparrow in America, house sparrow in England, gorrion in Spain, musch in Holland, and hussparf in Sweden.

➤ Table 9.1	The meaning of the names of some microorganisms

Name of microorganism	Meaning of name
Entamoeba histolytica	*Ent* = intestinal, *amoebae* = shape and means of movement, *histo* = tissue, *lytic* = lysing, or digesting, tissue
Escherichia coli	Named after Theodor Escherich in 1888; found in the colon
Haemophilus ducreyi	*Hemo* = blood, *phil* = love; named after Augusto Ducrey in 1889
Neisseria gonorrhoeae	Named after Albert L. Neisser in 1879; causes gonorrhea
Saccharomyces cerevisiae	*Saccharo* = sugar, *myco* = mold, *cerevisia* = beer or ale
Staphylococcus aureus	*Staphylo* = cluster, *kokkus* = berry, *aureus* = golden
Lactococcus lactis	*Lacto* = milk, *kokkus* = berry
Shigella etousae	Named after Kiyoshi Shiga in 1898; European Theater of Operations of the U.S. Army (final *e* gives proper Latin ending)

Wouldn't It Be Great?

How would you feel if you were included in textbooks as the Father or Mother of some area in science? Carl Linne, whose name has been Latinized to Carolus Linnaeus, has this honor. He is known as the Father of Taxonomy. This title is well-earned, since his system for naming, ranking, and classifying organisms is still widely used. For example, his tenth edition of *Systema Naturae* (1758) included a classification of over 4000 species of animals. Linnaeus was the first to designate human beings as *Homo sapiens*. His *Species Plantarum* (1753) and *Systema Naturae,* 10th edition, volume 1 (1758), have been accepted, by international agreement, as the official starting points for botanical and zoological nomenclature. Linnaeus's ideas on classification have influenced generations of biologists, both during and after his own lifetime.

the organisms in another pure culture of the same species, the organisms in each culture are designated as strains. A **strain** is a subgroup of a species with one or more characteristics that distinguish it from other subgroups of the same species. Each strain is identified by a name, number, or letter that follows the specific epithet. For example, *E. coli* strain K12 has been extensively studied because of its plasmids and other genetic characteristics, and *E. coli* strain 0157:H7 causes hemorrhagic inflammation of the colon in humans.

In addition to introducing the binomial system of nomenclature, Linnaeus also established a hierarchy of taxonomic ranks: species, genus, family, order, class, phylum or division, and kingdom. At the highest level, Linnaeus divided all living things into two *kingdoms*—plant and animal. In his taxonomic hierarchy, much of which is still used today, each organism is assigned a species name, and species of very similar organisms are grouped into a genus. As we proceed up the hierarchy, several similar genera are grouped to form a *family,* several families to form an *order,* and so on to the top of the hierarchy. Some hierarchies today have additional levels, such as *subphyla.* Also, it has become accepted practice to refer to the first categories within the animal kingdom as *phyla* and to those within other kingdoms (we now have five) as *divisions.* Recently, the five kingdoms have been grouped together into three *domains,* a new category even higher than kingdom. Domains will be discussed later in the chapter. The classifications of a human, dog, wolf, and a bacterium are shown in ➤ Figure 9.2.

Using a Taxonomic Key

Biologists often use a taxonomic *key* to identify organisms according to their characteristics. The most common kind of key is a **dichotomous key,** which has paired statements describing characteristics of organisms. Paired statements present an "either-or" choice, such that only one statement is true. Each statement is followed by directions to go to another pair of statements until the name of the organism finally appears. ➤ Figure 9.3 is a dichotomous key that will identify each of the four most common U.S. coins: quarters, dimes, nickels, and pennies. Read statements 1a and 1b, and decide which statement applies to a given coin. Look at the number to the right of the statement; it tells you which pair of statements to look at next. Continue in this manner until you reach a group designation. If you have followed the key carefully, that designation will name the coin.

Of course, you don't need a taxonomic key to identify something as simple and as familiar as coins. But identifying all the many kinds of bacteria in the world is a more difficult task. Major groups of bacteria can be identified with the key in ➤ Figure 9.4. More detailed keys use staining reactions, metabolic reactions (fermentation of particular sugars or release of different gases), growth at different temperatures, properties of colonies on solid media, and sim-

KINGDOM	Animalia	Animalia	Monera (Prokaryotae)
DIVISION/PHYLUM	Chordata	Chordata	Gracilicutes
SUBPHYLUM	Vertebrata	Vertebrata	
CLASS	Mammalia	Mammalia	Scotobacteria
ORDER	Primates	Carnivores	Spirochaetales
FAMILY	Hominidae	Canidiae	Spirochaetaceae
GENUS	*Homo*	Canis	*Treponema*
SPECIFIC EPITHET	*sapiens*		*pallidum*
SUBSPECIES (STRAIN)		*familiaris* (Dog) / *lupus* (Wolf)	

(Human)
Homo sapiens

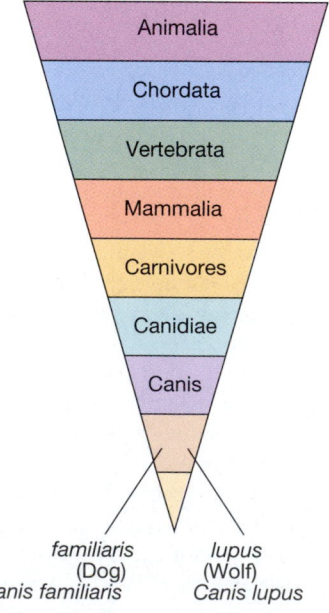
familiaris (Dog)
Canis familiaris
lupus (Wolf)
Canis lupus

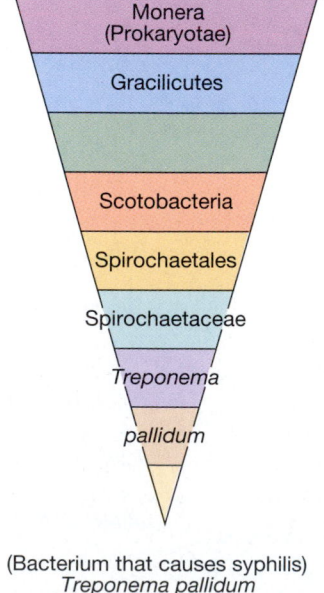
(Bacterium that causes syphilis)
Treponema pallidum

➤ **Figure 9.2** **Classifications of a human, a dog, wolf, and a bacterium.**

| 1a | Smooth-edged | Go to 2 |
| 1b | Rough-edged | Go to 3 |

| 2a | Silver-colored | Nickel |
| 2b | Copper-colored | Penny |

| 3a | Large (about 1-in. diameter) | Quarter |
| 3b | Small (about 3/4-in. diameter) | Dime |

> Figure 9.3   **A dichotomous key for classifying typical U.S. coins.** Why would the word "flat" not be useful in this key?

| 1a | Gram-positive | Go to 2 |
| 1b | Not Gram-positive | Go to 3 |

| 2a | Cells spherical in shape | Gram-positive cocci |
| 2b | Cells not spherical in shape | Go to 4 |

| 3a | Gram-negative | Go to 5 |
| 3b | Not Gram-negative (lack cell wall) | Mycoplasma |

| 4a | Cells rod-shaped | Gram-positive bacilli |
| 4b | Cells not rod-shaped | Go to 6 |

| 5a | Cells spherical in shape | Gram-negative cocci |
| 5b | Cells not spherical in shape | Go to 7 |

| 6a | Cells club-shaped | Corynebacteria |
| 6b | Cells variable in shape | Propionibacteria |

| 7a | Cells rod-shaped | Gram-negative bacilli |
| 7b | Cells not rod-shaped | Go to 8 |

| 8a | Cells helical with several turns | Spirochetes |
| 8b | Cells comma-shaped | Vibrioids |

> Figure 9.4   **A dichotomous key for classifying major groups of bacteria.**

ilar characteristics of cultures. By proceeding step by step through the key, one should be able to identify an unknown organism, or even a strain, if the key is sufficiently detailed.

Problems in Taxonomy

Among the aims of a taxonomic system are organizing knowledge about living things and establishing standard names for organisms so that we can communicate about them. Ideally, we would like to classify organisms according to their **phylogenetic,** or evolutionary, relationships, but this is not always easy. Evolution occurs continuously and at a relatively rapid rate in microorganisms, and our knowledge of the evolutionary history of organisms is incomplete. Taxonomy must change with evolutionary changes and new knowledge. *It is far more important to have a taxonomic system that reflects our current knowledge than to have a system that never changes.*

Creating a taxonomic system that provides an organized overview of all living things and how they are related to each other poses certain problems. Two such problems arise at opposite ends of the taxonomic hierarchy: (1) deciding what constitutes a species, and (2) deciding what constitutes a kingdom or in which domain a kingdom belongs. In the first case, taxonomists try to decide how much diversity can be tolerated within the unity of a species. In the second, taxonomists try to decide how to sort the diverse characteristics of living things into categories that reflect fundamental differences of evolutionary significance. In most advanced organisms, such as plants and animals, species that reproduce sexually are distinguished primarily by their reproductive capabilities. A male and a female of the same species are capable of DNA transfer through mating and producing fertile offspring, whereas members of different species ordinarily either cannot mate successfully or will have sterile offspring. *Morphology* (structural characteristics) and geographic distribution also are considered in defining species.

In bacteria, such criteria normally cannot be used in defining a species, primarily because genetic transfer (genetic recombination) among bacteria is relatively rare and morphological differences are minor. A bacterial species is defined by the similarities found among its members. Properties such as biochemical reactions, chemical composition, cellular structures, genetic characteristics, and immunological features are used in defining a bacterial species. Identifying a species and determining its limits present the most challenging aspects of biological classification—for any type of organism.

Before taxonomists turned their attention to microorganisms, the two-kingdom system of plants and animals worked reasonably well. Anyone can tell plants from animals—for example, trees from dogs. Plants make their own food but cannot move, and animals move but cannot make their own food. Simple enough, or is it? In this scheme, how do you classify *Euglena,* a mobile microorganism that makes its own food? How would you classify jellyfishes and sponges, which are motile or immotile depending on their stage of life? And how do you classify colorless fungi that neither move nor make their own food? Finally, how do you classify slime molds, organisms that can be unicellular or multicellular and mobile or immobile? Obviously, many organisms pose a number of problems when one tries to use a two-kingdom system.

Developments Since Linnaeus's Time

The problem of classifying microorganisms was first addressed by the German biologist Ernst H. Haeckel in 1866 when he created a third kingdom, the Protista. He included among the protists all "simple" forms of life such as bacteria, many algae, protozoa, and multicellular fungi and sponges. Haeckel's original term, Protista, is still used in taxonomic schemes today, but it is now limited mainly to unicellular eukaryotic organisms.

Classification of bacteria has posed taxonomic problems over the centuries and still does. Until recently, many taxonomists regarded bacteria as small plants that lacked chlorophyll. As late as 1957, the seventh edition of *Bergey's Manual of Determinative Bacteriology,* a work devoted to the identification of bacteria, considered bacteria to be unicellular plants. Changes in this viewpoint came as the tools to study bacteria were developed. First, light microscopy and staining techniques were used to describe the basic structure of cells. Second, electron microscopy was used to study the ultrastructure of cells. And third, biochemical techniques were used to study chemical composition and chemical reactions in cells. One of the most important discoveries from these various studies was that DNA looked and behaved differently during cell division in bacteria than in cells whose DNA is organized into chromosomes within a nucleus.

Studies of the structure and function of cells also led to the recognition of two general patterns of cellular organization, prokaryotic and eukaryotic. Basing taxonomy on these two different patterns of cellular organization was proposed as early as 1937. Various taxonomists such as H. F. Copeland, R. Y. Stanier, C. B. van Niel, and R. H. Whittaker, working in the late 1950s, placed bacteria in a separate kingdom of anucleate (lacking a cell nucleus) organisms rather than with organisms that have true nuclei. In 1962, Stanier and van Niel stated, "The distinctive property of bacteria is the prokaryotic nature of their cells."

In 1956, Lynn Margulis and H. F. Copeland proposed a scheme of classifying prokaryotes and eukaryotes by the following four-kingdom system of classification:

1. Monera: all prokaryotes, including true bacteria and blue-green algae

2. Protoctista: all eukaryotic algae, protozoa, and fungi

3. Plantae: all green plants

4. Animalia: all animals derived from a zygote, a cell formed by the union of an egg and a sperm

These taxonomists also proposed that evolution from prokaryotic to eukaryotic life forms had taken place by endosymbiosis. ➡ (Chapter 4, p. 96)

R. H. Whittaker felt that endosymbiosis could not account for all the differences between prokaryotes and eukaryotes. He also felt that a taxonomic system should give more consideration to the methods organisms use to obtain nourishment. Autotrophic nutrition by photosynthesis and heterotrophic nutrition by the ingestion of substances from other organisms had been considered in earlier taxonomies. Absorption as a sole means of acquiring nutrients had been overlooked. To Whittaker, fungi, which acquire nutrients solely by absorption, were sufficiently different from plants to justify placing them in a different kingdom. Also, fungi have certain reproductive processes not shared with any other organisms. Consequently, Whittaker proposed a taxonomic system in 1969 that separated the Protoctista into two kingdoms—Protista (pro-tis'tah) and Fungi—but retained the Monera, Plantae, and Animalia. Finally, through refinements of Whittaker's system by several taxonomists over the past few decades, the five-kingdom system was created.

The Five-Kingdom Classification System

Before we discuss the five-kingdom classification system and how it applies to microorganisms, we must emphasize that all living organisms, regardless of the kingdom to which they are assigned, display certain characteristics that define the unity of life. All organisms are composed of cells, and all carry out certain functions, such as obtaining nutrients and getting rid of wastes. The cell is the basic structural and functional unit of all living things. The fact that viruses are not cells is one reason they are not considered to be living organisms. All cells are bounded by a cell or plasma membrane, carry genetic information in DNA, and have ribosomes where proteins are made. All cells also contain the same kinds of organic compounds—proteins, lipids, nucleic acids, and carbohydrates. They also selectively transport material between their cytoplasm and their environment. Thus, although organisms may be classified in very diverse taxonomic groups, their cells have many similarities in structure and function.

No single classification system is completely accepted by all biologists. One of the most widely accepted is the **five-kingdom system** (➤ Figure 9.5). A major advantage of this system is the clarity with which it deals with microorganisms. It places all **prokaryotes,** microorganisms that lack a cell nucleus, in the kingdom Monera (Prokaryotae). ➡ (Chapter 4,

➤ Figure 9.5   **The five-kingdom system of classification.**

> **Table 9.2** **The five-kingdom system of classification**

	Monera (prokaryotae)	Protista	Fungi	Plantae	Animalia
Cell type	Prokaryotic	Eukaryotic	Eukaryotic	Eukaryotic	Eukaryotic
Cell organization	Unicellular; occasionally grouped	Unicellular; occasionally multicellular	Unicellular or multicellular	Multicellular	Multicellular
Cell wall	Present in most	Present in some, absent in others	Present	Present	Absent
Nutrition	Absorption, some photosynthetic, some chemosynthetic	Ingestion or absorption, some photosynthetic	Absorption	Absorptive, photosynthetic	Ingestion; occasionally in some parasites by absorption
Reproduction	Asexual, usually by binary fission	Mostly asexual, occasionally both sexual and asexual	Both sexual and asexual, often involving a complex life cycle	Both sexual and asexual	Primarily sexual

p. 73) It places most unicellular **eukaryotes,** organisms whose cells contain a distinct nucleus, in the kingdom Protista. (Margulis proposed a very similar five-kingdom system in 1982, but she referred to the kingdom of simple eukaryotes as Protoctista instead of Protista.) The five-kingdom system also places fungi in the separate kingdom Fungi.

The properties and members of each of the five kingdoms are described below and summarized in Table 9.2. A more detailed classification of bacteria is provided in Appendix B.

Kingdom Monera

The kingdom **Monera** (mo-ner′ah) is also called the kingdom **Prokaryotae,** as suggested by the French marine biologist Edouard Chatton in 1937. It consists of all prokaryotic organisms, including the eubacteria ("true bacteria"), the cyanobacteria, and the archaeobacteria (➤ Figure 9.6).

All monerans are unicellular; they lack true nuclei and generally lack membrane-enclosed organelles. Their DNA has little or no protein associated with it. Reproduction in the kingdom Monera occurs mainly by binary fission. Of all monerans, the **eubacteria** (u′bak-ter″e-ah) are of greatest concern in the health sciences and will be considered in detail in several chapters of this book.

The **cyanobacteria** (si″an-o-bak-ter′e-ah), formerly known as blue-green algae, are of special importance in the balance of nature. They are photosynthetic, typically unicellular, organisms, although cells may sometimes be connected to form threadlike filaments. Being autotrophs, cyanobacteria do not invade other organisms, so they pose no health threat to humans, except for toxins (poisons) some release into water.

Cyanobacteria grow in a great variety of habitats, including anaerobic ones, where they often serve as food sources for more complex heterotrophic organisms. Some "fix" atmospheric nitrogen, converting it to nitrogenous compounds that algae and other organisms can use. Certain cyanobacteria also thrive in nutrient-rich water and are responsible for algal blooms—a thick layer of algae on the surface of water that prevents light from penetrating to the water below. Such blooms release toxic substances that can give the water an objectionable odor and even harm fish and livestock that drink the water.

Archaeobacteria (ar′ke-o-bak′ter″e-ah) surviving today are primitive prokaryotes adapted to extreme environments. The methanogens reduce carbon-containing compounds to the gas methane. The extreme halophiles live in excessively salty environments, and the thermoacidophiles live in

Cyanobacteria

Mycoplasma

Spirochetes

Gram-positive
bacteria

Gram-negative
bacteria

Archaeobacteria

Rickettsias
(in a eukaryotic cell)

➤ Figure 9.6 **Some typical monerans.** Monerans are prokaryotic organisms without a cell nucleus and other internal, membrane-enclosed structures.

hot acidic environments, such as volcanic vents in the ocean floor (➤ Figure 9.7). Believed to be of very ancient origin, archaeobacteria have been found to differ from eubacteria in several distinctive ways, including the structure of their cell wall and the structure of their RNA polymerase. These organisms will be discussed in greater detail later in the chapter.

Kingdom Protista

Although the modern protist group is very diverse, it contains fewer kinds of organisms than when first defined by Haeckel. All organisms now classified in the kingdom **Protista** (➤ Figure 9.8) are eukaryotic. Most are unicellular, but some are organized into colonies. Protists have a true membrane-enclosed nucleus and organelles within their cytoplasm, as do other eukaryotes. Many protists live in fresh water, some live in seawater, and a few live in soil. They are distinguished more by what they don't have or don't do than by what they have or do. Protists do not develop from an embryo, as plants and animals do, and they do not develop from distinctive spores, as the fungi do. Yet, among the protists are the algae, which resemble plants; the protozoa, which resemble animals; and the euglenoids, which have both plant and animal charac-

➤ Figure 9.7 **Archaeobacteria: extreme halophiles.** Organisms living at deep-ocean vents, where hot volcanic gases are released from the earth's interior, survive in one of the most extreme environments known. Temperature and pressure are extremely high, yet these vents are the most productive ecosystem on our planet.

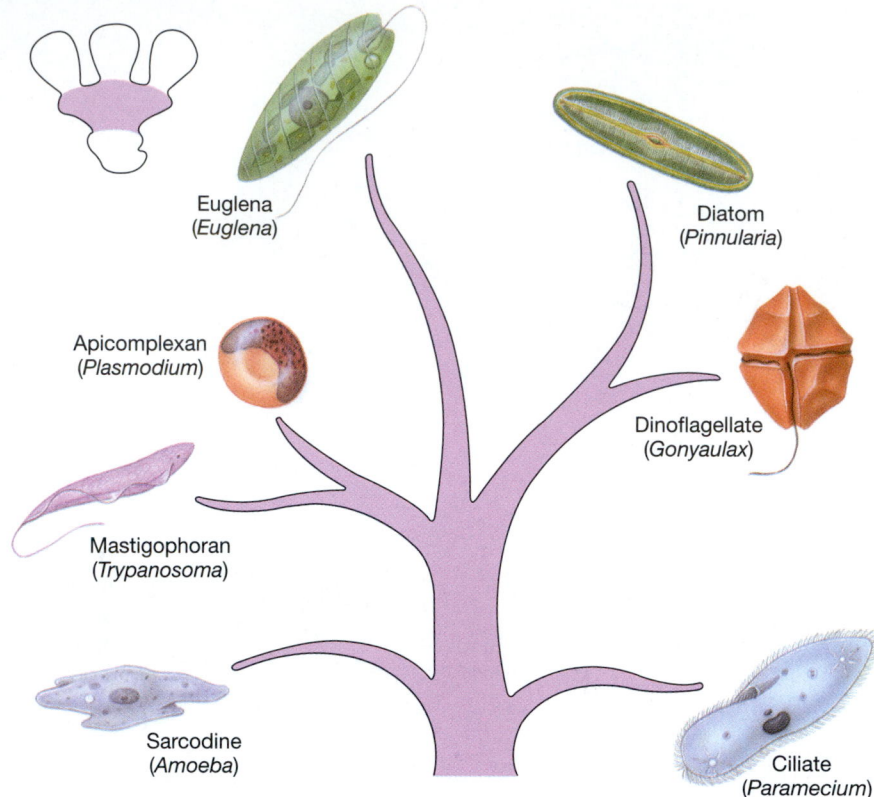

➤ Figure 9.8   **Some typical protists.** Protists are unicellular eukaryotic organisms.

teristics. The protists of greatest interest to health scientists are the protozoa that can cause disease. ➥ (Chapter 11, p. 294)

Kingdom Fungi

The kingdom **Fungi** (➤ Figure 9.9) includes mostly multicellular and some unicellular organisms. Fungi obtain nutrients solely by absorption of organic matter from dead organisms. Even when they invade living tissues, fungi typically kill cells and then absorb nutrients from them. Although the fungi have some characteristics in common with plants, their structures are much simpler in organization than true leaves or stems. Fungi form spores but do not form seeds. Many fungi pose no threat to other living things, but some attack plants and animals, even humans. ➥ (Chapter 11, p. 297)

Kingdom Plantae

The placement of most microscopic eukaryotes with the protists leaves only macroscopic green plants in the kingdom **Plantae.** Most plants live on land and contain chlorophyll in organelles called chloroplasts. Plants are of interest to microbiologists only because some contain medicinal substances such as quinine, which has been used to treat microbial infections.

Kingdom Animalia

The kingdom **Animalia** includes all animals derived from zygotes (a cell formed by the union of two gametes, such as an egg and a sperm). Although nearly all members of this kingdom are macroscopic and therefore of no concern to microbiologists, several groups of animals live in or on other organisms, and some serve as carriers of microorganisms (➤ Figure 9.10).

Certain *helminths* (worms) are parasitic in humans and other animals. Helminths include flukes, tapeworms, and roundworms, which live inside the body of their host. They also include leeches, which live on the surface of their hosts. Microbiologists often need to identify both microscopic and macroscopic forms of helminths. ➥ (Chapter 11, p. 304)

Helminthiasis is the most widespread human parasitic infection. Ascaris now infects 1.4 billion people; Trichuris, 1.3 billion; and hookworms 2 billion.

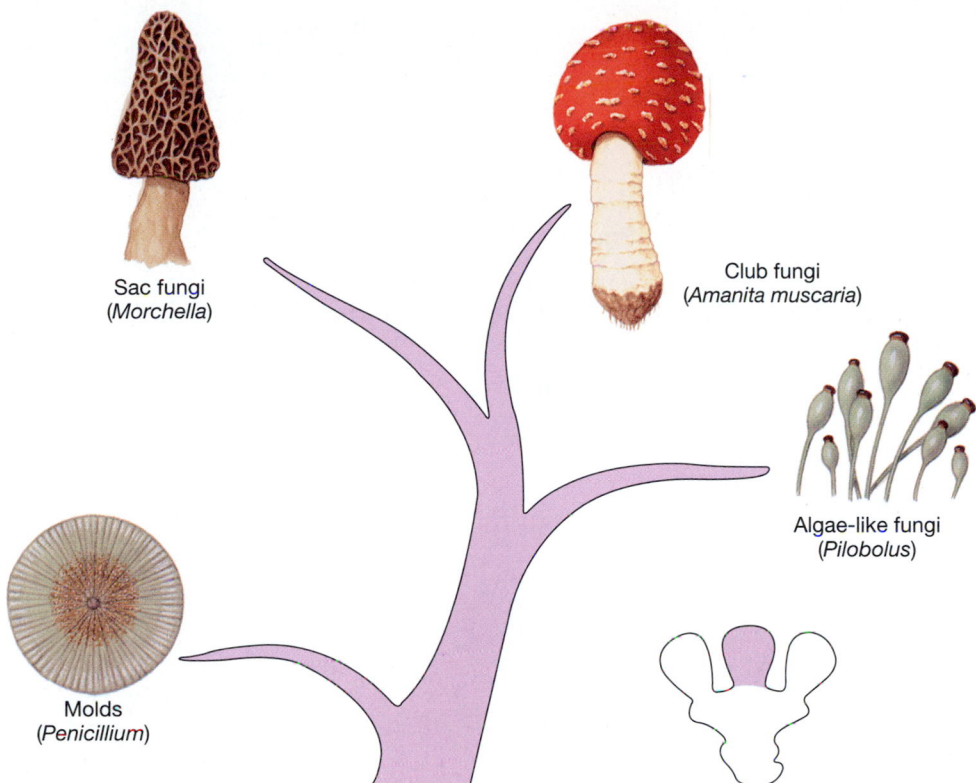

Sac fungi
(*Morchella*)

Club fungi
(*Amanita muscaria*)

Algae-like fungi
(*Pilobolus*)

Molds
(*Penicillium*)

➢ Figure 9.9   **Some typical fungi.**
Fungi are eukaryotic organisms that
have cell walls and do not carry out
photosynthesis. Fungi take their food
from other organic sources (that is, they
are chemoheterotrophs).

Certain *arthropods* live on the surface of their hosts, and some spread disease. Ticks, mites, lice, and fleas are arthropods that live on their hosts for at least part of their lives. Ticks, lice, fleas, and mosquitoes can spread infectious microorganisms from their bodies to those of humans or other animals. ➥ (Chapter 11, p. 312)

The Three-Domain Classification System

Studies of the archaeobacteria in the late 1970s by Carl Woese, G. E. Fox, and others suggested that these organisms represented a third cell type, and they proposed another scheme for the evolution of living things from a universal common ancestor (➢ Figure 9.11). They hypothesized a group of *urkaryotes,* the earliest or original cells, that gave rise to the eukaryotes directly rather than by way of the prokaryotes. They proposed that nucleated urkaryotes became true eukaryotes by acquiring organelles by endosymbiosis from certain eubacteria.

The Evolution of Prokaryotic Organisms

At about the same time that archaeobacteria were first being investigated, studies of stromatolites were also being conducted. **Stromatolites** (stro-mat′o-lites) are fossilized photosynthetic prokaryotes that appear as masses of cells or microbial mats. Commonly found associated with lagoons or hot springs, they are still forming today. Because stromatolites are fossilized prokaryotes, they do not provide any evidence for phylogenetic, or evolutionary, relationships, but can be used to determine the period during which they arose. Studies of stromatolites indicate that life arose nearly 4 billion years ago, and that an "Age of Microorganisms," in which there were no multicellular living organisms, lasted for about 3 billion years. Combined evidence from studies of archaeobacteria and the most ancient stromatolites convinced many scientists that three branches of the tree of life formed during the Age of Microorganisms, and that each branch gave rise to distinctly different groups of organisms.

The three-branch tree was not accepted by all scientists. In 1977, the English taxonomist T. Cavalier-Smith proposed instead that the archaeobacteria arose later than the eubacteria by divergent evolution from a group of Gram-positive bacteria similar to present-day actino-

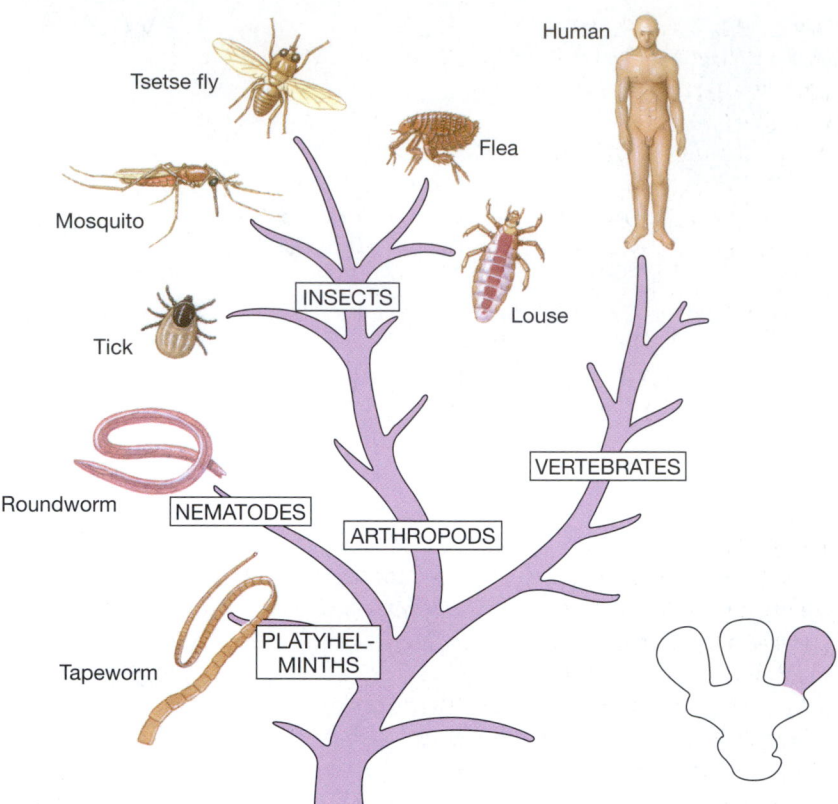

➤ Figure 9.10 **Groups from the kingdom Animalia that are relevant to microbiology.**

mycetes, which were once thought to be fungi. The American taxonomist J. A. Lake proposed in 1988 still another model with two main branches. In Lake's model, one branch gave rise to the eubacteria and to two groups of archaeobacteria—those that live in extremely salty environments and those that release methane. The other branch gave rise to the eukaryotes and to a group of archaeobacteria Lake calls the *eocytes,* which grow in hot, acidic environments.

In 1990, Woese suggested that a new taxonomic category, the **domain,** be erected above the level of kingdom. He based this suggestion on comparative studies of prokaryotes and eukaryotes at the molecular level and of their probable evolutionary relationships. Woese concluded that the archaeobacteria may be more closely related to eukaryotes than to eubacteria.

In 1998, Woese discussed theories about how the three domains may have arisen (➤ Figure 9.12). The standard view is that a universal common ancestor first split into Bacteria and

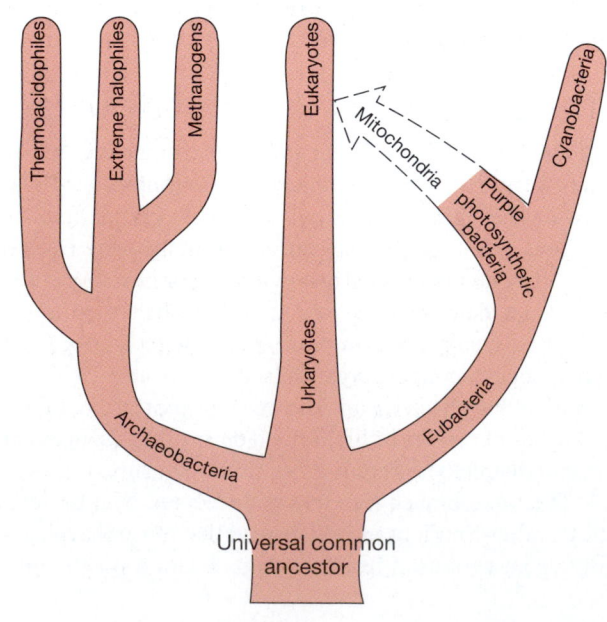

➤ Figure 9.11 **A model of the major evolutionary lines of descent proposed after the discovery of archaeobacteria.**

Archaea, and then the Eukarya branched off from Archaea. A second view holds that all three domains arose simultaneously from a pool of common ancestors that were all able to exchange genes with one another—hence, the universal genetic code. A third view seeks to explain how so many genes are present in Eukarya, but lacking in Archaea and Bacteria. It postulates the existence of a fourth domain that directly contributed genes to the Eukarya, and then became extinct. Thus, we see no modern-day descendants of this group.

The three domains Woese proposes are shown in ➤ Figure 9.13. The domain Eukarya contains all those kingdoms of eukaryotic organisms—the animals, plants, fungi, and protists. The traditional kingdom Monera has been divided into two domains: the domain Bacteria and the domain Archaea. A comparison of the three domains is presented in Table 9.3.

Which, if any, of these models is correct remains to be seen. Further analysis of nucleotide and amino acid sequences promises to provide much better information about relationships among organisms than has ever been available before. It also may help clarify the nature and time of origin of the universal common ancestor.

The Archaea

The Archaea exhibit many differences from the Bacteria. One of the first variations to be noted was that of cell wall structure, and thus far a significant number of variations have been observed (see Table 9.3). However, not all archaeobacteria are the same. Three major groups of archaeobacteria are commonly recognized: methanogens, extreme halophiles, and extreme thermophiles. These groupings are based on physiological characteristics of the organisms and therefore cannot be considered phylogenetic, or evolutionary, classifications. The **methanogens** are strictly anaerobic organisms, having been isolated from such divergent anaerobic environments as waterlogged soils, lake sediments, marshes, marine sediments, and the gastrointestinal tracts of animals, including humans. As members of the anaerobic food chain, they degrade organic molecules to methane. **Extreme halophiles** grow in highly saline environments such as the Great Salt Lake, the Dead Sea, salt evaporation ponds, and the surfaces of salt-preserved foods. Unlike the methanogens, extreme halophiles are generally obligate aerobes. The **extreme thermoacidophiles** occupy unique niches where bacteria are very rarely found, such as hot springs, geothermally heated marine sediments, and submarine hydrothermal vents. With optimum temperatures usually in excess of 80°C, they may be either obligate aerobes, facultative aerobes,

Evolution from unicellular to multicellular organisms took about 3 billion years.

➤ Figure 9.12   **Theories about the three domains.** The standard view is that the universal ancestor split into Bacteria and Archaea, and the Eukarya then branched off from the Archaea. An emerging view is that all three branches evolved independently from the same pool of genes. A third view is that there was a fourth branch, now lost, that contributed genes to the Eukarya. (*Source:* Adapted from Dr. Carl Woese and Dr. Norman R. Pace, *New York Times,* April 14, 1998, p. C1.)

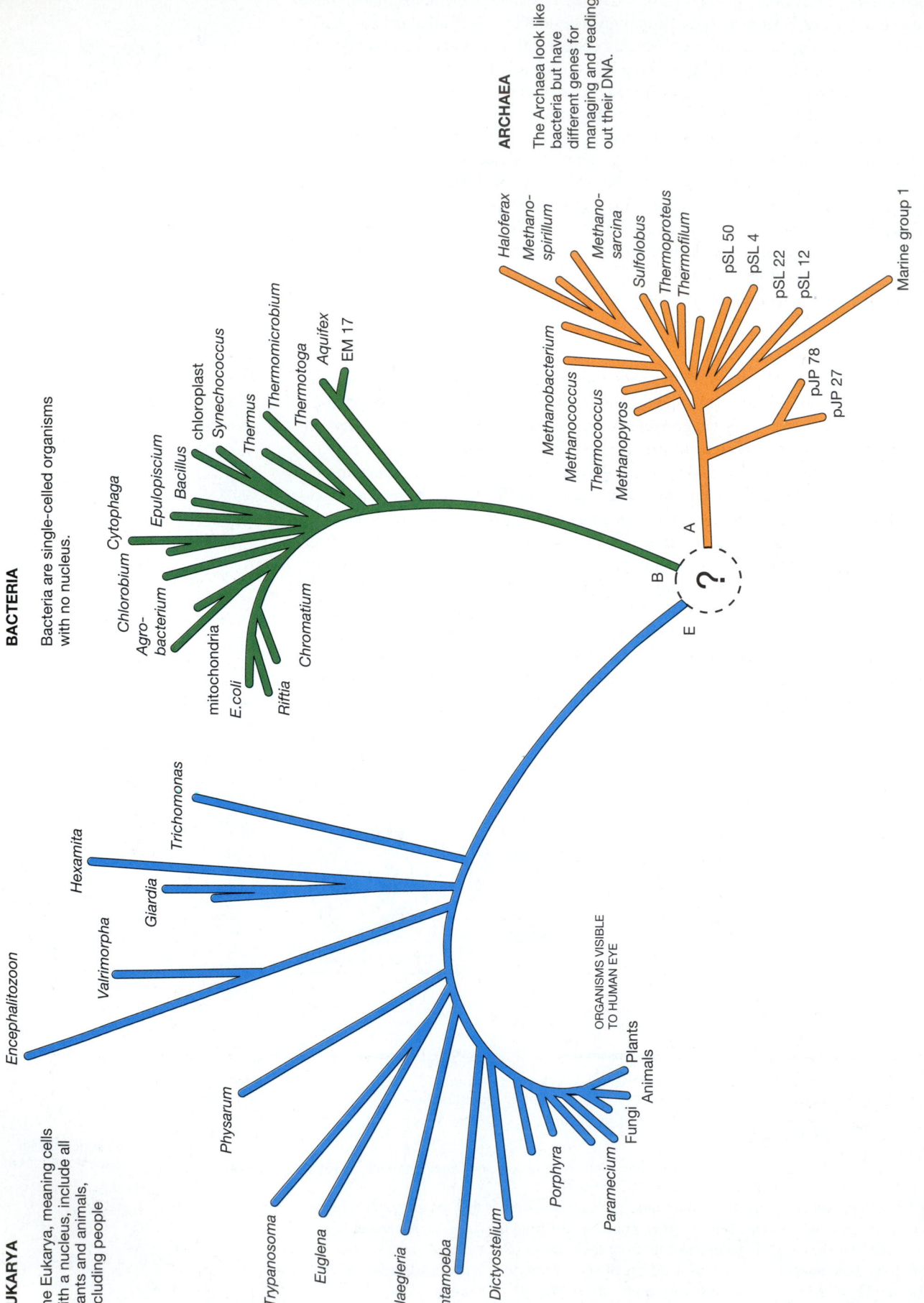

EUKARYA

The Eukarya, meaning cells with a nucleus, include all plants and animals, including people

BACTERIA

Bacteria are single-celled organisms with no nucleus.

ARCHAEA

The Archaea look like bacteria but have different genes for managing and reading out their DNA.

ORGANISMS VISIBLE TO HUMAN EYE

Λ Figure 9.13   **The three-domain system of classification.** Shown here are selected members of the three domains. Lengths of the branches indicate the extent of genetic differences in each organism, based on the similarities of their ribosomal RNA. (*Source:* Adapted from Dr. Carl Woese and Dr. Norman R. Pace, *New York Times,* April 14, 1998, p. C1.)

> **Table 9.3**   **Bacteria, archaea, and eukarya compared**

	Bacteria	Archaea	Eukarya
Cell type	Prokaryotic	Prokaryotic	Eukaryotic
Typical size	0.5–4 µm	0.5–4 µm	> 5 µm
Cell wall	Usually present, contain peptidoglycan	Present, lack peptidoglycan	Absent or made of other materials
Lipids in membranes	Fatty acids present, linked by ester bonds	Isoprenes present, linked by ether bonds	Fatty acids present, linked by ester bonds
Protein synthesis	First amino acid = methionine; impaired by antibiotics such as chloramphenicol	First amino acid = formyl-methionine; not impaired by antibiotics such as chloramphenicol	First amino acid = methionine; most not impaired by antibiotics such as chloramphenicol
Genetic material	Small circular chromosome and plasmids, histones absent	Small circular chromosome and plasmids, histone-like proteins present	Complex nucleus with more than one large linear chromosome, histones present
RNA polymerase	Simple	Complex	Complex
Locomotion	Simple flagella, gliding, gas vesicles	Simple flagella, gas vesicles	Complex flagella, cilia, legs, fins, wings
Habitat	Wide range of environments	Usually only extreme environments	Wide range of environments
Typical organisms	Enteric bacteria, cyanobacteria	Methane-producing bacteria, halo-bacteria, extreme thermophiles	Algae, protozoa, fungi, plants, animals

The Uses of Extremozymes

Various archaeobacteria are able to survive under highly adverse environmental conditions—from freezing waters to deep-sea vents, from concentrated brine to hot sulfur springs. The conditions present in these environments would inactivate, or denature, most enzymes. In order for these organisms to not only survive but to flourish under such conditions, they must possess special adaptations—namely, resistant enzymes. Enzymes that can survive and function under such adverse conditions are called *extremozymes*.

For many years, ordinary microbial enzymes have been used in manufacturing processes, such as the production of artificial sweeteners and "stonewashed" jeans, as well as in PCR and DNA fingerprinting. A big problem has been maintaining appropriate environmental conditions for the action or storage of microbial enzymes. The use of extremozymes would eliminate this concern. In PCR

➡ (Chapter 7, p. 192), the reactions must be cycled between low and high temperatures. The high temperature inactivates ordinary DNA polymerases, which then must be added again as the temperature lowers. The *Taq* DNA polymerase, isolated from the thermophile *Thermus aquaticus,* survives the high-temperature cycling and has enabled a totally automated PCR technology to be developed. An even more heat-resistant DNA polymerase, *Pfu,* has been isolated from the hyperthermophile *Pyrococcus furiosus* ("flaming fireball"). This enzyme works best at 100°C.

Proteases and lipases derived from alkaliphilic bacteria are being used as detergent additives to increase their stain-removal ability. They are also being used to produce the stonewashed appearance of denim. As more archaeobacteria and their extremozymes are discovered, new manufacturing applications are certain to be developed.

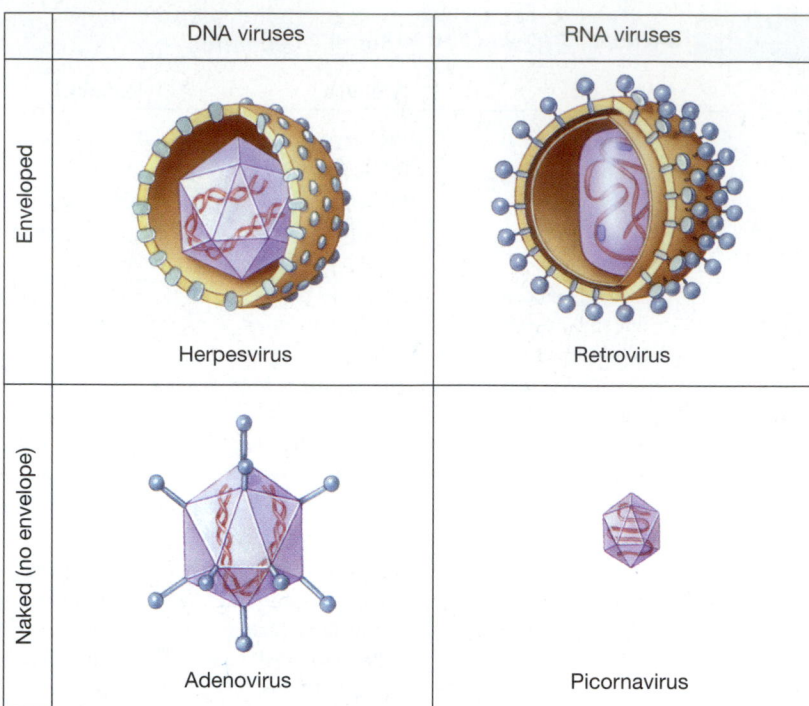

	DNA viruses	RNA viruses
Enveloped	Herpesvirus	Retrovirus
Naked (no envelope)	Adenovirus	Picornavirus

➤ Figure 9.14   **Some categories of viruses.**

✓ What is the difference between a taxon and taxonomy?

✓ What is the difference between species and specific epithet?

✓ What is meant by a system of taxonomy that is phylogenetic? Why does such a system change frequently?

✓ What is the difference between a kingdom and a domain? Name the five kingdoms, the three domains, and the types of organisms contained in each.

✓ Where do viruses, viroids, and prions fit in today's taxonomy?

Who's responsible for naming viruses? This duty is performed by the over 400 participating virologists of the International Committee on Taxonomy of Viruses (ICTV).

or obligate anaerobes. The heat-stable enzymes known as *extremozymes* found in these organisms have become of special interest to scientists (see Box "The Uses of Extremozymes").

Classification of Viruses

Viruses are acellular infectious agents that are smaller than cells. They contain nucleic acid (DNA or RNA) and are coated with protein. They have not been assigned to a kingdom. In fact, they display only a few characteristics associated with living organisms.

Initially viruses were classified according to the hosts they invaded and by the diseases they caused. As more was learned about viruses, the early concept of "one virus, one disease" used in classification was found to be invalid for many viruses. Today viruses are classified by chemical and physical characteristics such as the type and arrangement of their nucleic acids, shape (cubical or tubular), the symmetry of the protein coat that surrounds the nucleic acid, and the presence or absence of such things as a membrane covering (called an envelope), enzymes, tail structures, or lipids (➤ Figure 9.14). These groupings reflect only common characteristics and are not intended to represent evolutionary relationships. A classification of viruses is presented in Appendix B.

The study of viruses, or *virology,* is extremely important in any microbiology course for two reasons: (1) Virology is a recognized branch of microbiology, and techniques to study viruses are derived from microbiological techniques; and (2) viruses are of concern to health scientists because many cause diseases in humans, other animals, plants, and even microorganisms.

The Search for Evolutionary Relationships

Many biologists are interested in how living things evolved and how they are related to one another. In fact, most people have some curiosity about how life originated and gave rise to the diverse assortment of living things we see today. Although the details of the search for evolutionary relationships are of interest mainly to taxonomists, they are of some significance to health scientists. For example, many of the biochemical properties used to establish evolutionary relationships also can be used in identifying microorganisms. Whether it be a symbiotic association (e.g., between nitrogen-fixed bacteria and legumes) or a relationship between an infectious agent and its host, evolutionary relationships generally evolve together. Knowledge of such evolution is useful in order to understand the circumstances under which one organism becomes capable of infecting another, sometimes resulting in a symbiotic relationship and other times in the disease process.

Special Methods Needed for Prokaryotes

The taxonomy of most eukaryotes is based on morphology (structural characteristics) of living organisms, genetic features, and on knowledge of their evolutionary relationships from fossil records. However, morphology and fossil records provide little information about prokaryotes. For one thing, prokaryotes have left few fossil records. As mentioned earlier, stromatolites, fossilized mats of prokaryotes, have been found mainly at sites where the environment millions of years ago allowed the deposition of dense layers of bacteria (➤ Figures 9.15a and b). Stromatolites have provided much of our knowledge of the origin of the Archaea. Unfortunately, most bacteria do not form such mats, so most ancestral prokaryotes have disappeared without a trace.

Some rocks containing fossils of individual cells of cyanobacteria have been discovered (➤ Figure 9.15c), but they have failed to reveal much information about the organisms. Moreover, prokaryotes have few structural characteristics, and these characteristics are subject to rapid change when the environment changes. Large organisms tend to require a fairly long period of time to reproduce, but prokaryotes reproduce rapidly. Assuming the same number of mutations per generation, organisms that reproduce more rapidly will accumulate a greater number of

(a)

(b)

10 μm

(c)

➤ Figure 9.15   **Stromatolites.**
(a) Mats of bacteria, growing as stromatolites, in western Australia. (b) A cross section through fossil stromatolites from Montana, showing horizontal layers of bacterial growth. (c) Filamentous cyanobacteria (*Paleolyngbya*) from the Lakhanda Formation in eastern Siberia. These microfossils date from the late Precambrian and are approximately 950 million years old.

Viroids and Prions

Science changes. For a long time, everyone thought that viruses were the smallest infectious agents. But particles even smaller than viruses have recently been discovered, and some appear to serve as infectious agents. They include viroids (vir′oids) and prions (pre′onz). A viroid is simply a fragment of RNA. The viroid that causes potato spindle tuber disease contains only 359 bases—enough information to specify the location of only 119 amino acids if all the bases function as codons. That's one-tenth the amount of nucleic acid found in the smallest viruses! Prions, or *proteinaceous infections particles,* are only one-tenth the size of a virus and consist of a protein molecule that folds incorrectly as a result of mutation. These particles, which are self-replicating, are responsible for some mysterious brain infections in humans, as well as mad cow disease in cattle.

Yale biologist Timothy Goldsmith has called evolution "one of the most important triumphs of human understanding in the history of science."

Chemical clues found in a manganese mine in South Africa suggest that bacteria, algae, or other simple organisms inhabited the soil there 2 billion years ago.

mutations over a given period of time. Because of this rapid mutational change rate, it is far more difficult to show the relationship between fossilized forms of prokaryotes and current organisms.

Because morphology and evolution are of little use in classifying prokaryotes, metabolic reactions, genetic relatedness, and other specialized properties have been used instead. Health scientists use these properties to identify infectious prokaryotes in the laboratory, but such identification does not necessarily reflect evolutionary relationships among the organisms. The methods described next are of use in exploring evolutionary relationships. Although the methods are particularly appropriate for eukaryotes, they can be used for prokaryotes as well.

Numerical Taxonomy

Numerical taxonomy is based on the idea that increasing the number of characteristics of organisms that we observe increases the accuracy with which we can detect similarities among them. If the characteristics are genetically determined, the more characteristics two organisms share, the closer their evolutionary relationship. Although the idea of numerical taxonomy was developed before computers were available, computers allow us to compare large numbers of organisms rapidly and according to many different characteristics. In a simple example of numerical taxonomy, each characteristic is assigned a value of 1 if present and 0 if not present. Characteristics such as reaction to Gram staining, oxygen requirements, presence or absence of a capsule, properties of nucleic acids and proteins, and the presence or absence of particular enzymes and chemical reactions can be evaluated. Organisms are then compared, and patterns of similarities and differences are detected (➤ Figure 9.16). With the use of numerical taxonomy, no single characteristic is used to arbitrarily divide all organisms into groups. If two organisms match on 90 percent or more of the characteristics studied, they are presumed to belong to the same species. Computerized numerical taxonomy offers great promise for improving our understanding of relationships among all organisms.

Genetic Homology

The discovery of the structure of DNA by James Watson and Francis Crick in 1953 provided new knowledge that was quickly applied by taxonomists, especially those studying taxonomic relationships and the evolution of eukaryotes. These scientists began to study the **genetic homology,** or the similarity of DNA, among organisms. Several techniques for determining genetic homology are now available. Similarities in DNA can be studied directly by determining the base composition of the DNA, by sequencing the bases in DNA or RNA, and by DNA hybridization. Because an organism's proteins are determined by its DNA, similarities in DNA can be studied indirectly by preparing *protein profiles* and by analyzing amino acid sequences in proteins.

Base Composition

Organisms can be grouped by comparing the relative percentages of bases present in the DNA of their cells. DNA contains four bases, abbreviated as A (adenine), T (thymine), G (guanine), and C (cytosine). ➡ (Chapter 2, p. 44) Base pairing occurs only between A and T and between G and C. In making base comparisons, the total amount of G and C in a sample of DNA is determined and expressed as a percentage of total DNA. By subtracting this percentage from 100, we get the percent of total A and T in the sample. For example, if the DNA is 60 percent G–C, then it is 40 percent A–T. The base composition of an organism is generally stated in terms of the percent of guanine plus cytosine and is referred to as the G–C content. Base composition only determines the total amount of each nucleotide base present; it does not give any indication of the sequence of these bases.

Studies of base composition have shown that the G–C content varies from 23 to 75 percent in bacteria. These studies also have shown that certain species of bacteria, such as *Clostridium tetani* and *Staphylococcus aureus,* have very similar DNA compositions, but that *Pseudomonas aeruginosa* has a very different DNA composition. Thus, *C. tetani* and *S. aureus* are probably more closely related to each other than either is to *P. aeruginosa.* Similar percentages of bases do not in themselves prove that the organisms are closely related, because the *sequence* of bases may be quite different. (Human beings and *Bacillus subtilis,* for example, have nearly identi-

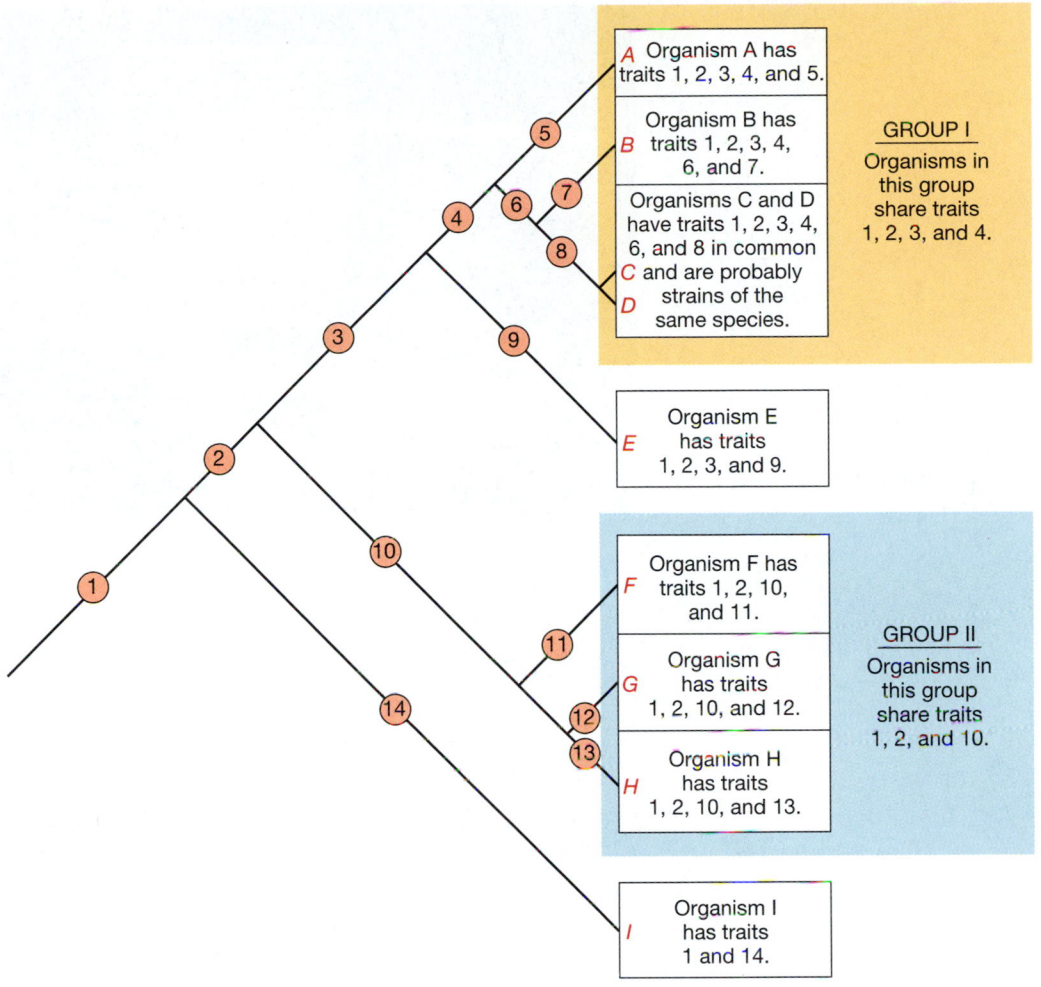

A Organism A has traits 1, 2, 3, 4, and 5.

B Organism B has traits 1, 2, 3, 4, 6, and 7.

Organisms C and D have traits 1, 2, 3, 4, 6, and 8 in common **C** and are probably **D** strains of the same species.

GROUP I
Organisms in this group share traits 1, 2, 3, and 4.

E Organism E has traits 1, 2, 3, and 9.

F Organism F has traits 1, 2, 10, and 11.

G Organism G has traits 1, 2, 10, and 12.

H Organism H has traits 1, 2, 10, and 13.

GROUP II
Organisms in this group share traits 1, 2, and 10.

I Organism I has traits 1 and 14.

➢ Figure 9.16   **A probable evolutionary tree arrived at by numerical taxonomy.** In numerical taxonomy, organisms are compared on the basis of a large number of characteristics. Organisms with a high proportion of characteristics in common are presumed to be closely related.

cal G–C percentages.) We can say, however, that if the percentages in two organisms are quite different, they are not likely to be closely related.

DNA and RNA Sequencing

Automated equipment for identifying the base sequences in DNA or RNA is now available at reasonable cost (➢ Figure 9.17). It is therefore easier than before to search a culture for base sequences known to be unique to certain species. Using PCR techniques and a DNA synthesizer, one can produce a large number of **probes,** single-stranded DNA fragments that have sequences complementary to those being sought. ➡ (Chapter 7, p. 192) A fluorescent dye or a radioactive tag (an indicator molecule) can be attached to the probe. When the probe finds its target DNA, it complementarily binds to it and does not wash off when rinsed. The specimen is then examined for fluorescing dye or for radioactivity. The presence or absence of the unique DNA sequence helps in identification of the specimen.

DNA Hybridization

In **DNA hybridization,** the double strands of DNA of each of two organisms are split apart, and the split strands from the two organisms are allowed to combine (➢ Figure 9.18). The strands from different organisms will anneal (bond to each other) by base pairing—A with T and G with C. The amount of annealing is directly proportional to the quantity of identical base sequences in the two DNAs. A high degree of homology (similarity) exists when both organisms have long

DNA sequencing is now extremely common. For $20 the Macromolecular Structure Analysis Facility will determine the DNA sequence of your sample.

You can use complementary DNA hybridization to detect the presence of DNA sequences from infectious viral agents.

➤ Figure 9.17   **A DNA sequencer.** Automated systems can identify the sequence of nucleotide bases in a piece of DNA.

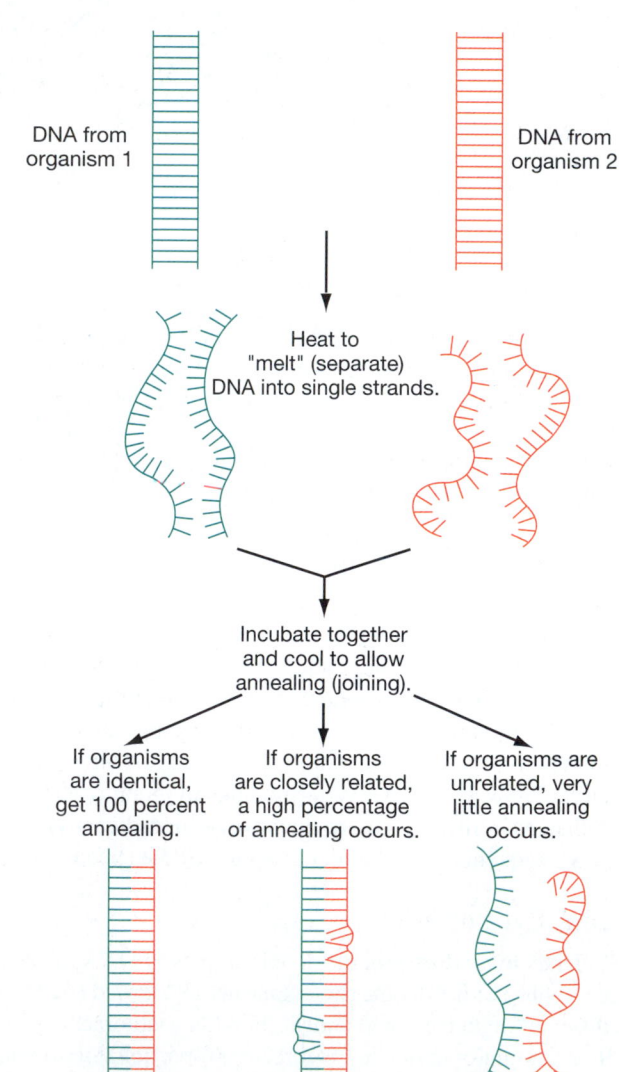

➤ Figure 9.18   **DNA hybridization.** Strands of DNA are separated, and individual strands from two different organisms are allowed to anneal (join by hydrogen bonding at sites where there are many complementary base pairs). The degree of annealing reflects the degree of relatedness between the organisms, on the basis of the assumption that annealing takes place only where genes, or parts of genes, are identical.

identical sequences of bases. Close DNA homology indicates that the two organisms are closely related and that they probably evolved from a common ancestor. A small degree of homology indicates that the organisms are not very closely related. Ancestors of such organisms probably diverged from each other thousands of centuries ago and have since evolved along separate lines.

Protein Profiles and Amino Acid Sequences

Every protein molecule consists of a specific sequence of amino acids and has a particular shape with an assortment of surface charges. Modern laboratory methods allow cells or organisms to be compared according to these properties of their proteins. Although variations in proteins among cells make these techniques difficult to apply to multicellular organisms, they are quite helpful in studying unicellular organisms.

A **protein profile** is a laboratory-prepared pattern of the proteins found in a cell (➤ Figure 9.19a). Because a cell's proteins are the products of its genes, the cells of each species synthesize a unique array of proteins—as distinctive as a fingerprint is for humans. Analysis of the profiles of one or more proteins of different bacterial species provides a reasonable basis for comparisons.

Protein profiles are produced by the **polyacrylamide gel electrophoresis (PAGE)** method, which separates proteins on the basis of molecular size (➤ Figure 9.19b). In this method, samples of protein obtained from lysed cells are dissolved in a detergent and poured into wells (depressions) of a thin slab of polyacrylamide gel. The slab is inserted into a buffer-filled chamber. An electric current is then passed through the gel for a period of time. The current causes the protein molecules to migrate to the opposite end of the gel. Large protein molecules migrate more slowly than do smaller ones. After the smallest proteins have migrated, the current is turned off, and the gel slab is then removed. Next it is stained so that the various proteins show up as separate stained bands in the gel slab. ➥ (Chapter 18, p. 531)

Each band in the profile from one kind of cell represents a different protein in that cell. Bands at the same location in profiles from different kinds of cells indicate that the same protein is present in the different cells.

Determination of amino acid sequences in proteins also identifies similarities and differences among organisms. Certain proteins, such as cytochromes, which contribute to oxidative metabolism in many organisms, are commonly used to study amino acid sequences. The amino acid sequences in the same kind of protein from several organisms are determined. As with DNA hybridization, the extent of matching sequences of amino acids in the proteins indicates the relatedness of the organisms.

The proteins an organism contains are determined directly by the information in that organism's DNA. Thus, both protein profiles and determinations of amino acid sequences are as significant a measure of the relatedness of organisms as are DNA homologies. All are also related to the evolutionary history of the organisms.

Other Techniques

Other techniques for studying evolutionary relatedness include determining properties of ribosomes, immunological reactions, and phage typing.

Properties of Ribosomes

Ribosomes serve as sites of protein synthesis in both prokaryotic and eukaryotic cells. RNA in ribosomes can be separated into several types according to the size of the RNA units. A particular RNA unit, the 16S rRNA component, has proven especially useful in studying evolutionary relationships for several reasons. Because rRNA molecules are easily rendered nonfunctional by even slight alterations in their genetic structure, mutations are rarely tolerated, and therefore ribosomes have evolved very slowly. It is the degree of similarity in 16S rRNA sequences between two organisms that indicates their evolutionary relatedness. If the nucleotide sequences of 16S rRNA molecules from two types of organisms are very similar, those organisms are likely to be quite close evolutionarily. Although direct sequencing of 16S rRNA is used to show evolutionary relationships between species, newer methods, such as PCR, are beginning to replace it. The PCR technique, which is being used to amplify rRNA genes, requires less cell material and is more rapid and convenient for large studies than is direct rRNA sequencing.

Since ribosomal RNAs are functionally constant in bacteria, we may find drugs that target specific bacterial 16S or rRNA sequences to treat different infections.

Step 1: Gel solution is prepared and poured into the electrophoresis apparatus.

Step 2: Cells are broken apart, and a drop of sample solution is placed on the gel.

Step 3: Electric current is run through the gel to separate proteins into bands.

Step 4: Gel is stained to make bands of protein visible.

(a)   (b)

➤ Figure 9.19   **Separation of proteins.** (a) Protein profiles, which provide a "fingerprint" of the proteins present in particular cells, can be used to compare different organisms to determine their degree of relatedness. (b) The PAGE process.

Immunological Reactions

Immunological reactions also are used to identify and study surface structures and the composition of microorganisms, as explained in Chapter 17. As we shall see, one highly specific and sensitive technique involves proteins called *monoclonal antibodies*. Monoclonal antibodies can be created so that they will bind to a specific protein, usually a protein found on a cell surface. If the antibodies bind to the surfaces of more than one kind of organism, the organisms have that protein in common. This technique promises to be particularly useful in identifying specific biochemical properties of microorganisms. In turn, identification of such properties will be extremely useful in determining taxonomic relationships.

➢ Figure 9.20   **Phage typing.**
Receptor sites for bacteriophages are highly specific; certain strains of a species of bacterium are attacked only by particular types of phages. Clear sites (plaques) are left when phages have killed bacterial cells. On the basis of which phages have attacked a bacterial culture, one can determine which strain of that bacterial species is present.

Phage Typing

Phage typing involves the use of bacteriophages, viruses that attack bacteria, to determine similarities among different bacteria. A separate agar plate is inoculated for each bacterium being studied. A sterile cotton swab or bent glass rod is used to spread the inoculum over the agar surface. After incubation, a *lawn,* or continuous sheet, of *confluent* bacterial growth is produced. The underside of the plate is marked with numbered squares so that drops of known phages can be spotted onto specific zones of the lawn and later identified. After a suitable incubation period, zones of lysis (*plaques*) appear in the bacterial lawn (➢ Figure 9.20). Because receptor sites for bacteriophages are highly specific, certain strains of a species of bacterium are attacked only by particular types of phages. By observing which phages cause holes in the lawn, researchers can identify the strain. Strains lysed by the same phages are presumed to be more closely related than strains that show different patterns of lysis by phages.

The Significance of Findings

The main significance of methods of determining evolutionary relationships is that these methods can be used to group closely related organisms and to separate them from less closely related ones. When groups of closely related organisms are identified, it is presumed that they probably had a common ancestor and that small differences among them have arisen by *divergent evolution*. **Divergent evolution** occurs as certain subgroups of a species with common ancestors undergo sufficient mutation to be identified as separate species.

Within the eubacteria, an early divergence gave rise to two important subgroups, the Gram-positive bacteria and the Gram-negative ones. Subsequent divergence within each group has given rise to many modern species of bacteria. Among the Gram-negative bacteria, the purple nonsulfur bacteria gave rise to modern bacteria that inhabit animal digestive tracts. One proposed scheme of divergent evolution in this group is shown in ➢ Figure 9.21.

✓ What are stromatolites? What can they tell us about the evolution of prokaryotes?

✓ Why are genetic homologies more useful than morphology in the study of evolution and classification of microbes?

✓ Why is the study of ribosomal RNA especially useful in studying evolutionary relationships among organisms?

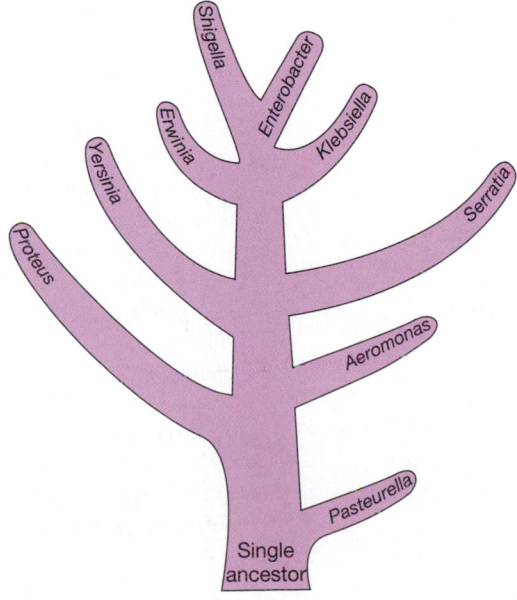

➢ Figure 9.21   **Divergent evolution.**
Shown here is a model of divergent evolution of selected enteric bacteria (most of which live in the intestinal tracts of some animals, including humans).

Bacterial Taxonomy and Nomenclature

Criteria for Classifying Bacteria

Most macroscopic organisms can preliminarily be classified according to observable structural characteristics. But it is more difficult to classify microscopic organisms, especially bacteria, because many of them have similar structures. Separating them according to cell shape, size, and arrangement does not produce a very useful classification system. Nor does the presence of specific structures such as flagella, endospores, or capsules allow identification of particular species. Therefore, other criteria must be used. Staining reactions, especially the Gram stain, were among the first properties other than morphology to be used to classify bacteria. Other properties now in use include features related to growth, nutritional requirements, physiology, biochemistry, genetics, and molecular analysis. These features include properties of DNA and proteins. Important criteria used in classifying bacteria are summarized in Table 9.4, and biochemical tests used in classifying and identifying them are described in Table 9.5.

By using various classification criteria, we can identify an organism as belonging to a particular genus and species. For bacteria, a species is regarded as a collection of strains that share many common features and differ significantly from other strains. A bacterial *strain* consists of descendants of a single isolation in pure culture. Bacteriologists designate one strain of a

<table>
<tr><td colspan="3">➤ Table 9.4   Criteria for classifying bacteria</td></tr>
<tr><th>Criteria</th><th>Examples</th><th>Uses</th></tr>
<tr><td>Morphology</td><td>Size and shape of cells; arrangements in pairs, clusters, or filaments; presence of flagella, pili, endospores, capsules</td><td>Primary distinction of genera and sometimes species</td></tr>
<tr><td>Staining</td><td>Gram-positive, Gram-negative, acid-fast</td><td>Separates eubacteria into divisions</td></tr>
<tr><td>Growth</td><td>Characteristics in liquid and solid cultures, colony morphology, development of pigment</td><td>Distinguishes species and genera</td></tr>
<tr><td>Nutrition</td><td>Autotrophic, heterotrophic, fermentative with different products, energy sources, carbon sources, nitrogen sources, needs for special nutrients</td><td>Distinguish species, genera, and higher groups</td></tr>
<tr><td>Physiology</td><td>Temperature (optimum and range), pH (optimum and range), oxygen requirements, salt requirements, osmotic tolerance, antibiotic sensitivities and resistances</td><td>Distinguish species, genera, and higher groups</td></tr>
<tr><td>Biochemistry</td><td>Nature of cellular components such as cell wall, RNA molecules, ribosomes, storage inclusions, pigments, antigens; biochemical tests</td><td>Distinguish species, genera, and higher groups</td></tr>
<tr><td>Genetics</td><td>Percentage of DNA bases (G + C ratio), DNA hybridization</td><td>Determine relatedness within genera and families</td></tr>
<tr><td>Serology</td><td>Slide agglutination, fluorescent-labeled antibodies</td><td>Distinguish strains and some species</td></tr>
<tr><td>Phage typing</td><td>Susceptibility to a group of bacteriophages</td><td>Identification and distinguishing of strains</td></tr>
<tr><td>Sequence of bases in rRNA</td><td>rRNA sequencing</td><td>Determine relatedness among all living things</td></tr>
<tr><td>Protein profiles</td><td>Separate proteins by two-dimensional PAGE (electrophoresis)</td><td>Distinguish strains</td></tr>
</table>

➤ Table 9.5   Specific biochemical tests sometimes used in identifying and classifying bacteria

Biochemical test	Nature of test
Sugar fermentation	Organism is inoculated into a medium containing a specific sugar; growth and end products of fermentation, including gases, are noted. Anaerobic fermentations can be detected by inoculating organisms via a "stab" culture into solid medium.
Gelatin liquefaction	Organism is inoculated (stabbed) into a solid medium containing gelatin; liquefaction at room temperature or inability to resolidify at refrigerator temperature indicates the presence of proteolytic (protein-digesting) enzymes.
Starch hydrolysis	Organism is inoculated onto an agar medium containing starch; after the plate is flooded with Gram's iodine, clear areas around colonies indicate the presence of starch-digesting enzymes.
Litmus milk	Organism is inoculated into litmus milk medium (10 percent powdered skim milk plus litmus indicator); characteristic changes such as alteration of pH to acid or alkaline, denaturation of the protein casein (curdling), and gas production can be used to help identify specific organisms.
Catalase	Hydrogen peroxide (H_2O_2) is poured over heavy growth of an organism on an agar slant; release of O_2 gas bubbles indicates the presence of catalase, which oxidizes H_2O_2 to H_2O and O_2.
Oxidase	Two or three drops (or a disk) of an oxidase test reagent is added to an organism growing on an agar plate; a color change of the test reagent to blue, purple, or black indicates the presence of cytochrome oxidase.
Citrate utilization	Organism is inoculated into citrate agar medium in which citrate is the sole carbon source; an indicator in the medium changes color if citrate is metabolized; use of citrate indicates the presence of the permease complex that transports citrate into the cell.
Hydrogen sulfide	Organism is inoculated into peptone iron medium; formation of black iron sulfide indicates the organism produces hydrogen sulfide (H_2S).
Indole production	Organism is inoculated into a medium containing the amino acid tryptophan; production of indole, a nitrogenous breakdown product of tryptophan, indicates the presence of a set of enzymes that convert tryptophan to indole.
Nitrate reduction	Organism is inoculated into a medium containing nitrate (NO_3^-); presence of nitrite (NO_2^-) indicates that the organism has the enzyme nitrate reductase; absence of nitrite indicates either absence of nitrate reductase or presence of nitrite reductase (which reduces nitrite to N_2 or NH_3).
Methyl red	Organism is cultured in MR-VP broth; a methyl red indicator is added; presence of acid causes an indicator color change (red).
Voges-Proskauer	Organism is cultured in MR-VP broth; alpha naphthol and KOH-creatine are added; presence of the enzyme cytochrome oxidase causes color change in an indicator (rose color).
Phenylalanine deaminase	Organism is inoculated into a medium containing phenylalanine and ferric ions; formation of phenylpyruvate and its reaction with ferric ions produces a color change that demonstrates the presence of the enzyme phenylalanine deaminase.
Urease	Organism is inoculated into a medium containing urea; production of ammonia, usually detected by an indicator for alkaline pH, indicates the presence of the enzyme urease.
Specific nutrient	Organism is inoculated into a medium containing a specific nutrient, such as a particular amino acid (for example, cysteine) or vitamin (for example, niacin); growth of an organism that fails to grow in media lacking the specific nutrient can be used to identify some auxotrophs.

species as the **type strain.** Usually this is the first strain described. It is the name-bearer of the species and is preserved in one or more type culture collections. The American Type Culture Collection (ATCC), a nonprofit scientific organization established in 1925, collects, preserves, and distributes authenticated type cultures of microorganisms. Many important research studies dealing with classification, identification, and the industrial uses of microorganisms would be seriously hampered without the services of the ATCC.

For many strains of bacteria, scientists are able to determine that they are members of a particular species. For other strains, however, difficult judgments must be made to decide whether the strain belongs to an existing species or differs sufficiently to be defined as a separate species. In recent years, similarities of DNA and proteins among organisms have proved a reliable means of assigning a strain to an existing species or establishing the basis for a new species.

Curiously, assigning bacterial genera to higher taxonomic levels—families, orders, classes, and divisions (or phyla)—can be even more difficult than organizing species and strains *within*

> Figure 9.22   **David H. Bergey, the originator of *Bergey's Manuals*.** The first *Bergey's Manual* was published in 1923. In 1936 he set up an educational trust to which all rights and royalties from the *Manuals* would be transferred for preparing, editing, and publishing future editions, as well as providing funds for research to clarify problems arising in the process. This nonprofit trust ensures that *Bergey's Manual* will be a self-perpetuating publication.

genera. Many macroscopic organisms are classified by establishing their evolutionary relationships to other organisms from fossil records. Efforts are being made to classify bacteria by evolutionary relationships, too, but these efforts are hampered by the incompleteness of the fossil record and by the limited information gleaned from what fossils have been found. Even a complete fossil record might supply only morphological information and would thus be inadequate for determining evolutionary relationships.

The History and Significance of *Bergey's Manual*

The accepted reference on the identification of bacteria is commonly referred to as *Bergey's Manual*. The first edition of *Bergey's Manual of Determinative Bacteriology* was published in 1923 by the American Society for Microbiology; David H. Bergey (➤ Figure 9.22) was chairperson of its editorial board. Since then, eight editions, an abridged version, and several supplements have been published. *Bergey's Manual* has become an internationally recognized reference for bacterial taxonomy. It has also served as a reliable standby for medical workers interested in identifying the causative agents of infections.

Bergey's Manual contains names and descriptions of organisms and diagnostic keys and tables for identifying organisms. It is organized into sections, with each section devoted to a group of similar organisms. The word *section* is used, rather than established taxonomic terms such as *family* and *order,* because evolutionary relationships between sections (and even within sections) are not yet clearly understood.

A new four-volume manual, *Bergey's Manual of Systematic Bacteriology,* has a much broader scope than previous manuals. All the kinds of information found in earlier determinative manuals are included, and information on taxonomy, ecology, cultivation, maintenance, and the preservation of organisms has been expanded. The first volume was published in 1984, the second in 1986, and the final two in 1989. Then *determinative information* (information used to identify bacteria) was collected into a single volume, the ninth edition of *Bergey's Manual of Determinative Bacteriology,* which was published in 1994. And in 1995, *Bergey's Manual* became available in CD-ROM format as well.

Bergey's Manual of Determinative Bacteriology is widely accepted as a reference for identifying bacteria. The four-volume *Bergey's Manual of Systematic Bacteriology* also has that significance, but it is of greater significance to those interested in bacterial taxonomy. An important motivation for publishing the larger work is to stimulate research to improve the understanding of taxonomic relationships among bacteria. However, it is important to remember that, in their current state, both *Bergey's Manuals* do *not* present an accurate picture of evolutionary relationships among bacteria. Rather, they are practical groupings of bacteria that make it easy to identify them. We do not yet have enough information to draw a complete evolutionary tree for bacteria.

Bergey's Manual of Systematic Bacteriology provides many kinds of information, such as descriptions and photographs of species, tests to distinguish among genera and species, DNA relatedness among organisms, and various numerical taxonomy studies (➤ Figure 9.23).

Problems Associated with Bacterial Taxonomy

Despite the tremendous effort spent in classifying bacteria and the wide acceptance of *Bergey's Manual,* no complete classification of bacteria from kingdom to species has been established. The plight of bacteriologists looking at the taxonomy of bacteria might be described as follows: Those looking from the top level down can propose at least plausible divisions of the kingdom Monera, or kingdom Prokaryotae, as it is commonly designated by bacteriologists. The kingdom Prokaryotae has traditionally included the eubacteria, cyanobacteria, and archaeobacteria. Those bacteriologists looking from the bottom up can establish strains, species, and genera and can sometimes assign bacteria to higher-level groups. But too little is known about evolutionary relationships to establish clearly defined taxonomic classes and orders for many bacteria.

Taxonomy from the Top Down

Taxonomists generally agree that true bacteria belong in the kingdom Prokaryotae. Like other prokaryotes, they lack nuclear membranes. Taxonomists do not agree, however, on how to separate the true bacteria into divisions or phyla. In 1968, the bacteriologist R. G. E. Murray proposed four divisions: Gram-negative, Gram-positive, Gram-variable, and organisms lacking a

➤ Figure 9.23 **Samples from *Bergey's Manual*.** Shown here are five representative pages reproduced from the four-volume set of *Bergey's Manual of Systematic Bacteriology*. (a) Description of a genus. (b) Comparison of DNA homology among different species of the same genus. (c) Comparison of immunological relationships of enzymes within a genus. (d) Key characteristics differentiating three closely related genera. (e) DNA relatedness among members of the family Enterobacteriaceae.

cell wall. In 1982, Lynn Margulis grouped bacteria as fermenting heterotrophs, respiring heterotrophs, or autotrophs.

Bergey's Manual of Systematic Bacteriology currently divides the kingdom Prokaryotae into the following four divisions on the basis of cell wall properties:

Division I.	Gracilicutes (gras″ih-lik-yoo′teez)	Prokaryotes with thin cell walls, which are typical of Gram-negative organisms
Division II.	Firmicutes (fer-mik′you-teez)	Prokaryotes with thick cell walls, indicating a Gram-positive type of cell wall
Division III.	Tenericutes (ten″er-ik′yoo-teez)	Prokaryotes of a pliable, soft nature, and lacking cell walls
Division IV.	Mendosicutes (men-doh-sik′yoo-teez)	Prokaryotes having cell walls that lack peptidoglycan

Taxonomy from the Bottom Up

Within each division, bacteria have been classified into 33 groups, called "sections," in *Bergey's Manual*. The sections are based on relatively easily observable characteristics such as shape, staining reactions, the presence or absence of a cell wall, motility, reproduction by budding, and mode of metabolism. Most sections were established many years ago as a practical means of classifying bacteria. The criteria were established before modern techniques for molecular analysis (especially of DNA and RNA) and biochemical studies were available and before any fossil bacteria had been found.

As more and more bacteria have been subjected to modern analyses, certain discrepancies in assignments to sections have become apparent. For example, the cell walls of some genera have unusual properties that prevent their being classified as either Gram-positive or Gram-negative. The organisms assigned to a particular section of *Bergey's Manual* are not necessarily closely related.

The Muddle in the Middle

The difficulties of classifying bacteria are greatly magnified as one proceeds up the taxonomic hierarchy. Although families, orders, and classes can be clearly defined for a few bacteria, too little information is available to do so for many. Until discrepancies in section assignments can be resolved and taxonomic levels between genera and divisions can be more precisely determined, it is practical to continue to use section designations. The medically important sections are summarized in Table 9.6. A complete listing of the sections is provided in Appendix B.

Bacterial Nomenclature

Despite all the taxonomic problems, there is an established nomenclature for bacteria. *Bacterial nomenclature* refers to the naming of species according to internationally agreed-upon rules. Both taxonomy and nomenclature are subject to change as new information is obtained. Organisms are sometimes moved from one category to another, and their official names are sometimes changed. For example, the bacterium that causes tularemia, a fever acquired by handling infected rabbits, was for many years called *Pasturella tularensis*. Its genus name was changed to *Francisella* after DNA hybridization studies revealed that hybridization between its DNA and that of *Pasteurella* species did not occur. It does, however, have a 78 percent match with the DNA of *Francisella novicida*. When considering specific orders and families, we must remember that such names have consistent endings: orders always end in *-ales* and families in *-aceae*.

Bacteria by Section of *Bergey's Manual*

Of the 33 sections of prokaryotes described in *Bergey's Manual of Systematic Bacteriology*, some contain only a few organisms, whereas others contain hundreds. Certain sections contain organisms of medical significance, whereas others are of interest mainly to ecologists, taxonomists, or researchers.

Going Where None Have Gone Before

If you were a microorganism, you'd love to be able to grow in places where competing microorganisms couldn't survive. That's what the archaeobacteria have going for them. At the time of their discovery in 1977, the archaeobacteria were already thought to be quite odd. They lived in brines five times as salty as the oceans, geothermal environments that would cook other organisms to a crisp, and anaerobic habitats where even trace amounts of oxygen couldn't be found. Now they have been proven to be even odder. *Pyrolobus fumarii* holds the current record for life at high temperatures, growing at temperatures as high as a scalding 113°C. Antarctic archaeobacteria thrive at −1.8°C. Archaeobacteria have also been found in rice paddies, terrestrial soils, freshwater lake sediments, and even winery by-products.

➤ Table 9.6   **Characteristics and medically important members of selected sections of bacteria defined in**
Bergey's Manual of Systematic Bacteriology

Section (number)	Medically important members	Diseases or products
Spirochetes (1): Gram-negative, helical, move by axial filaments	*Treponema* *Borrelia* *Leptospira*	Syphilis Relapsing fever, Lyme disease Leptospirosis
Aerobic, Motile, Helical Gram-Negative Bacteria (2): move by flagella, helical or comma-shaped	*Campylobacter* *Helicobacter*	Urogenital and digestive tract infections Peptic ulcers
Gram-Negative Aerobic Rods and Cocci (4): Some contain pigments or oxidase, some have fastidious nutritional requirements, some are obligate parasites.	*Pseudomonas* *Legionella* *Neisseria* *Moraxella* *Brucella* *Bordetella* *Francisella*	Urinary tract infections, burns, and wounds Pneumonia and other respiratory infections Gonorrhea, meningitis, and nasopharyngeal infections Conjunctivitis Brucellosis Whooping cough Tularemia
Facultatively Anaerobic Gram-Negative Rods (5): Some have peritrichous flagella, many can be distinguished by their characteristic fermentation reactions.	*Escherichia* *Shigella* *Salmonella* *Klebsiella* *Enterobacter* *Serratia* *Proteus* *Providencia* *Morganella* *Yersinia* *Vibrio* *Pasteurella* *Haemophilus* *Calymmatobacterium* *Gardnerella* *Eikenella* *Streptobacillus*	Opportunistic infections of colon and other sites Bacillary dysentery Typhoid fever, enteritis, and food poisoning Respiratory and urinary tract infections Opportunistic infections Opportunistic infections Urinary tract infections (especially nosocomial) Wound and burn infections, urinary tract infections Summer diarrhea, opportunistic infections Plague, mesenteric lymphadenitis, septicemia Cholera, acute gastroenteritis Cat- and dog-bite wounds Respiratory infections, meningitis, conjunctivitis, chancroid Granuloma inguinale Vaginitis Wound infections Rat-bite fever
Anaerobic Gram-Negative Rods (6): straight, curved, or helical, motile	*Bacteroides,* *Fusobacterium*	Oral, digestive, respiratory, urogenital infections, wounds, and abscesses
Anaerobic Gram-Negative Cocci (8): nonmotile	*Veillonella*	Oral microbiota and abscesses
Rickettsiae and Chlamydiae (9): intracellular parasites; chlamydiae form reticulate and elementary bodies.	*Rickettsia* *Rochalimaea* *Coxiella* *Bartonella* *Chlamydia*	Typhus, Rocky Mountain spotted fever, rickettsialpox Trench fever Q fever Oroya fever Trachoma, inclusion conjunctivitis, nongonococcal urethritis, lymphogranuloma venereum, parrot fever

➤ Table 9.6 (continued)

Section (number)	Medically important members	Diseases or products
Mycoplasmas (10): lack cell walls, extremely small	*Mycoplasma*	Atypical pneumonia, urogenital infections
	Ureaplasma	Opportunistic urogenital infections
Gram-Positive Cocci (12): aerobic to strictly anaerobic; non-spore-forming; typically pyogenic (pus-forming)	*Staphylococcus*	Skin abscesses, opportunistic infections
	Streptococcus	Strep throat and other respiratory infections, skin and other abscesses, puerperal fever, opportunistic infections
	Peptococcus	Postpartum septicemia, visceral lesions
	Peptostreptococcus	Puerperal fever and various pyogenic infections
Endospore-Forming Gram-Positive Rods and Cocci (13): aerobic to strictly anaerobic; some motile and some nonmotile	*Bacillus*	Anthrax, food poisoning, source of the antibiotic bacitracin
	Clostridium	Tetanus, botulism, gas gangrene, bacteremia
Regular Nonsporing Gram-Positive Rods (14): facultatively or strictly anaerobic; nonmotile	*Lactobacillus*	Microflora of the digestive tract and vagina
	Listeria	Listeriosis
	Erysipelothrix	Erysipeloid
Irregular Nonsporing Gram-Positive Rods (15): club-shaped, pleomorphic, filamentous; aerobic to anaerobic	*Corynebacterium*	Diphtheria and skin opportunists
	Propionibacterium	Wound infections and abscesses
	Eubacterium	Oral and other infections
	Actinomyces	Actinomycoses
Mycobacteria (16): Gram-positive, acid-fast	*Mycobacterium*	Tuberculosis, leprosy, and chronic infections
Nocardioforms (17): Gram-positive, filamentous, some acid-fast	*Nocardia*	Nocardiosis, mycetoma, abscesses
Actinomycetes That Divide in More Than One Plane (27): Gram-positive, filamentous	*Dermatophilus*	Skin lesions
Streptomycetes and Their Allies (29): Gram-positive, filamentous	*Streptomyces*	Produce over 500 different antibiotics

Some sections, such as Section 9, **Rickettsiae** and **Chlamydiae,** contain rather unusual organisms. These two groups are obligate intracellular parasites; that is, they can only grow inside living cells. Chlamydiae have an interesting and complex life cycle (➤ Figure 9.24), rather than dividing by binary fission, as do Rickettsiae and most other bacteria. **Mycoplasmas** (Section 10) lack cell walls and form colonies that look like eggs fried sunny-side up. They have sterols in their cell membranes that give them great flexibility of shape (pleomorphism ➡ (Chapter 4, p. 80). Also in Section 10 are the **Ureaplasmas,** also with unusual cell walls and/or cell membranes. Table 9.7 compares these groups with more typical bacteria and viruses.

Bacterial Taxonomy and You

As a beginning student, you will doubtless find it difficult to remember many characteristics of specific microorganisms that we cover in this course. A four-volume set of *Bergey's Manual* weighs about 21 pounds and costs about $400—not something you could carry to class and back. However, you can use the endpapers, located inside the front and back covers of this textbook. If you wish to find out whether a given organism is Gram-positive or Gram-negative, its shape, the disease(s) it causes, and so on, look it up by name in the back endpapers. Microorganisms are grouped as bacteria, viruses, fungi, and parasites (protozoa and helminths).

New bacteria are constantly being discovered. Scientists estimate the number of as yet unknown species is in the millions.

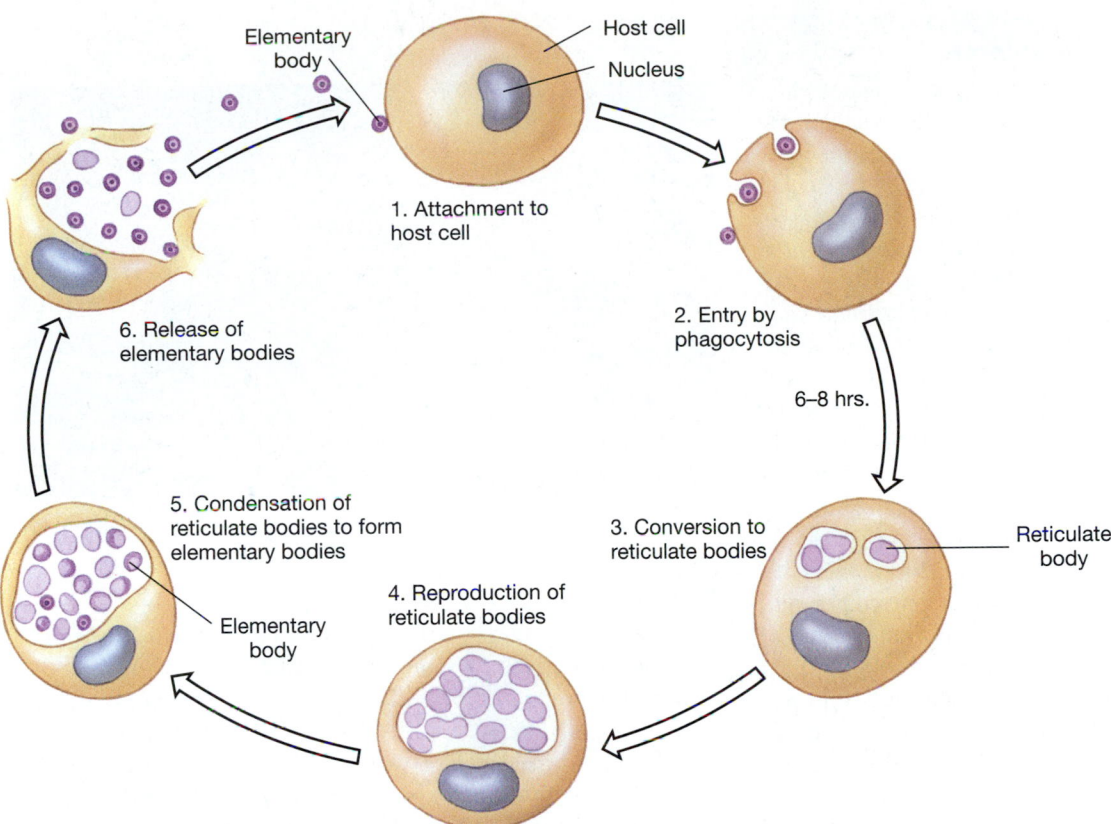

Elementary body

Host cell

Nucleus

1. Attachment to host cell

2. Entry by phagocytosis

6–8 hrs.

3. Conversion to reticulate bodies

Reticulate body

4. Reproduction of reticulate bodies

5. Condensation of reticulate bodies to form elementary bodies

Elementary body

6. Release of elementary bodies

➤ **Figure 9.24**   **The life cycle of a chlamydia.** (Step 1) Small, dark elementary bodies (the only infectious stage of the chlamydial life cycle) attach to a host cell and (2) enter by phagocytosis. (3) The elementary bodies, enclosed within membrane-enclosed vacuoles, lose their thick walls and enlarge to become reticulate bodies. (4) Reticulate bodies reproduce by binary fission, rapidly filling the cell. (5) They condense to form infectious elementary bodies, which (6) are then released by lysis and are free to attach to a new host cell.

Or if you are discussing a disease but can't remember which organism(s) cause it, look in the front endpapers under the name of the disease (again grouped as bacterial, viral, fungal, and parasitic diseases), and you will find the organism's name and some of its characteristics. Page numbers are given to direct you to further information. Thumbing through the book or index can be frustrating when you need some little piece of information. But flipping to the cover of your book is easy, and we encourage you to do so often, until the information gradually becomes more and more familiar.

✓ What is a type strain and a type culture collection? Why is such a collection essential to researchers?

✓ Why is much of *Bergey's Manual* not phylogenetically arranged?

✓ What types of information are contained in the two versions of *Bergey's Manual?*

➤ **Table 9.7**   **Characteristics of typical bacteria, rickettsiae, chlamydiae, mycoplasmas, ureaplasmas, and viruses**

Characteristic	Typical bacteria	Rickettsiae	Chlamydiae	Mycoplasmas	Ureaplasmas	Viruses
Cell wall	Yes	Yes	Yes	No	Sometimes	No
Grow only in cells	No	Yes	Yes	No	No	Yes
Require sterols	No	No	No	Sometimes	Yes	No
Contain DNA and RNA	Yes	Yes	Yes	Yes	Yes	No
Have metabolic system	Yes	Yes	Yes	Yes	Yes	No

The Discovery of New Organisms

Are there any new worlds to discover, or new creatures in them? Yes! In recent years, scientists have discovered living organisms in such diverse environments as submarine hot vents, inside volcanoes, and in deep oil wells. In 1990 a joint U.S. and Soviet team discovered hot vents in fresh water for the first time, complete with an associated community of archaeobacteria, worms, sponges, and other organisms.

The vents lie more than 400 m deep in a most unusual Russian lake, Lake Baikal, which is the deepest lake in the world and holds the greatest quantity of fresh water in the world. Located in central Asia, in Siberia, it lies in a pocket between two continental plates. Asia was formed as a solid mass when several plates collided one after another and remained together. The area of Lake Baikal is being pulled apart, forming a rift valley and eventually a new ocean. This region is comparable to the sea-floor spreading centers (ridges) of the Pacific Ocean, where other vent communities have been found. In both locations, hot materials from deep inside the earth are emerging. Lake Baikal is a unique treasure for studying the evolution of life and microbial forms. Most lakes are only thousands of years old, but Lake Baikal may be 25 million years old. Microbes similar to the early evolving stages of life may still exist in its depths.

What lives inside a volcano? Studies following the 1980 volcanic eruptions of Mount St. Helens have raised some interesting questions. Archaeobacteria, which had previously been known from the deep-sea volcanic vents (black smokers) located 2200 m below the sea surface, have been found living on and in Mount St. Helens at temperatures of 100°C. Where did they come from? Some scientists think that they may have been present deep inside the volcano. For that matter, where do the archaeobacteria in the submarine vents come from? Do they point out a linkage between terrestrial and submarine volcanic activity? We tend to think of life as being present only on the *surface* of the earth, but perhaps there is a whole different range of life that we know nothing about, deep *within* the crust of the earth. Daily, more and more evidence accumulates in favor of the idea of "continuous crustal culture."

Cores taken from the deepest oil wells being bored within the earth reveal archaeobacteria at sites not connected with volcanic activities. Ancient bacteria from the early stages of our planet's development may still be colonizing the hot anaerobic interior of the earth—places with conditions that resemble those formerly on the earth's surface.

Various ecological problems have sent out scientists from universities, government, and industry to hunt for new

Mount St. Helens, shown here erupting in July 1980, is home to archaeobacteria.

microbes with properties that make the organisms useful in cleaning up the environment. Scientists from the Woods Hole Oceanographic Institution, in Massachusetts, took their search to a depth of more than 1800 m in the Gulf of California, where they have discovered anaerobic bacteria that can degrade naphthalene and possibly other hydrocarbons that might be found in oil spills. Sites in need of bioremediation often lack oxygen, making it impossible to utilize aerobic organisms for cleanup; hence the hunt in deep anaerobic environments. General Electric has also found an anaerobic bacterium that it plans to use to destroy polychlorinated biphenyls (PCBs), industrial by-product chemicals that accumulate in animal tissues and cause damage, including cancer and birth defects.

A new bacterium, so far referred to as GS-15, discovered in the Potomac River by U.S. Geological Survey (USGS) scientists, ordinarily changes iron from one form to another. It seems, however, that these bacteria can just as easily feed on uranium, getting twice as much energy in the process and transforming the uranium into an insoluble precipitate. The USGS team plans to use GS-15 to remove uranium from contaminated well and irrigation water found in much of the U.S. West and at uranium mining, processing, and nuclear waste sites.

There are many new microbes yet to be discovered. In addition to the naturally occurring species, new ones will be designed by scientists using genetic engineering techniques—or possibly even discovered on another planet! All these species will need to be classified and named. Clearly, *Bergey's Manual* will never be "finished."

Retracing Our Steps

Taxonomy: The Science of Classification

- Organisms are named according to their characteristics, where they are found, who discovered them, or what disease they cause. **Taxonomy** is the science of classification, and each category is a **taxon.**

Linnaeus, the Father of Taxonomy

- Linnaeus developed the system of **binomial nomenclature,** a two-name identification system for each living organism.

- The **genus** and **specific epithet** of each organism identify the **species** to which it belongs.

- Linnaeus also established the hierarchy of taxonomy and classified organisms into two kingdoms, Plantae and Animalia.

Using a Taxonomic Key

- A **dichotomous key** consists of a series of paired statements presented as either-or choices that describe characteristics of organisms. By selecting appropriate statements to progress through the key, one can classify organisms and, if the key is sufficiently detailed, identify them by genus and species.

Problems in Taxonomy

- Ideally, organisms should be classified by their **phylogenetic,** or evolutionary, relationships.

- Problems in taxonomy include the rapid pace of evolutionary change in microorganisms and the difficulty in deciding what constitutes a kingdom and what constitutes a species.

Developments Since Linnaeus's Time

- Since Linnaeus's time, several taxonomists have proposed three- and four-kingdom systems on the basis of various fundamental characteristics of living things. Whittaker proposed a five-kingdom system in 1969.
- Since 1925, *Bergey's Manual of Determinative Biology* has served as an important tool in identifying bacteria.

The Five-Kingdom Classification System

- The kingdoms of the **five-kingdom system** are **Monera (Prokaryotae)**, **Protista, Fungi, Plantae,** and **Animalia.** The characteristics of members of each kingdom are summarized in Table 9.2.

Kingdom Monera

- All monerans are unicellular **prokaryotes:** They generally lack organelles, have no true nuclei, and their DNA has little or no protein associated with it.

- The **cyanobacteria** are photosynthetic monerans of great ecological importance.

Kingdom Protista

- The protists are a diverse group of mostly unicellular **eukaryotes.**

Kingdom Fungi

- The fungi include some unicellular and many multicellular organisms that obtain nutrients solely by absorption.

Kingdom Plantae

- Most plants live on land and contain chlorophyll in organelles called chloroplasts.

Kingdom Animalia

- All animals are derived from zygotes; most are macroscopic.

The Three-Domain Classification System

- The three **domains** are higher than the category of Kingdom. They include: **Bacteria, Archaea,** and **Eukarya.** Their characteristics are summarized in Table 9.3

Domain Bacteria

- All Bacteria are unicellular prokaryotes and include the eubacteria ("true bacteria").

Domain Archaea

- All Archaea are unicellular prokaryotes, having a cell wall made of materials other than peptidoglycan.

Domain Eukarya

- All are eukaryotic cells, having a true nucleus.

Classification of Viruses

- **Viruses** are acellular infectious agents that share only a few characteristics with living organisms, are not included in any of the five kingdoms. Viruses are classified by their nucleic acids, chemical composition, and their morphology.

The Search for Evolutionary Relationships

Special Methods Needed for Prokaryotes

- Special methods are needed for determining evolutionary relationships among prokaryotes because they have few morphological characteristics and have left only a sparse fossil record.

- Several methods, including numerical taxonomy and genetic homology, are currently used to determine evolutionary relationships among organisms.

Numerical Taxonomy

- In **numerical taxonomy,** organisms are compared on the basis of a large number of characteristics and grouped according to the percentage of shared characteristics.

Genetic Homology

- **Genetic homology** is the similarity of DNA among different organisms, which provides a measure of their relatedness. Several techniques that determine genetic homology are available.

- The relative percentages of bases in DNA are a measure of relatedness. The base composition of DNA is determined, and G–C percentages are compared among organisms.

- Base sequences can be identified by DNA **probes.**

- In **DNA hybridization,** the degree of matching between strands of DNA is compared among organisms.

- **Protein profiles,** made by **polyacrylamide gel electrophoresis (PAGE)** are used to indicate whether the same proteins are present in different organisms.

- The amino acid sequences of related organisms are similar, so the determination of amino acid sequences is another measure of relatedness.

Other Techniques

- Other methods make use of properties of ribosomes, immunological reactions, and **phage typing.**

The Significance of Findings

- Evolutionary relationships can be used to group closely related organisms. Small differences among organisms descended from a common ancestor arise by **divergent evolution.** An early divergence gave rise to the two major subgroups of eubacteria, the Gram-positive bacteria and the Gram-negative ones.

Bacterial Taxonomy and Nomenclature

Criteria for Classifying Bacteria

- The criteria used for classifying bacteria are summarized in Table 9.4. These criteria can be used to classify bacteria into species and even into strains within species.

- For many species a particular strain is designated as the **type strain,** which is preserved in a type culture collection.

The History and Significance of *Bergey's Manual*

- *Bergey's Manual of Determinative Bacteriology* was first published in 1923 and has been revised several times; a ninth edition was published in 1994.

- *Bergey's Manual of Systematic Bacteriology* (a four-volume set) provides definitive information on the identification and classification of bacteria.

Problems Associated with Bacterial Taxonomy

- Taxonomists do not agree on how members of the kingdom Prokaryotae (Monera) should be divided. Many species of bacteria have been grouped into genera and some into families. Four *divisions* (the equivalent of phyla) have been established. Much information is needed to determine evolutionary relationships and establish classes and orders.

Bacteria by Section of *Bergey's Manual*

- The genera of medical significance, their characteristics, and the diseases they cause are summarized in Table 9.6. A complete listing of the sections is provided in Appendix B.

- Groups of important bacteria include the spirochetes, mycoplasmas, rickettsiae, chlamydiae, mycobacteria, and cyanobacteria.

Bacterial Taxonomy and You

- Use the endpaper at the beginning and end of the book to familiarize yourself with taxonomic information.

Terminology Check

Animalia   (*p. 230*)
archaeobacteria   (*p. 228*)
binomial nomenclature (*p. 223*)
Chlamydiae   (*p. 250*)
cyanobacteria   (*p. 228*)
dichotomous key   (*p. 224*)
divergent evolution   (*p. 243*)
DNA hybridization   (*p. 239*)
domain   (*p. 232*)
eubacteria   (*p. 228*)
eukaryote   (*p. 228*)
extreme halophile   (*p. 233*)
extreme thermoacidophile   (*p. 233*)
five-kingdom system (*p. 227*)

Fungi   (*p. 230*)
genetic homology   (*p. 238*)
genus   (*p. 223*)
methanogen   (*p. 233*)
Monera   (*p. 228*)
Mycoplasmas   (*p. 250*)
numerical taxonomy   (*p. 238*)
phage typing   (*p. 243*)
phylogenetic   (*p. 225*)
Plantae   (*p. 230*)
polyacrylamide gel electrophoresis
    (PAGE) (*p. 241*)
probe   (*p. 239*)
Prokaryotae   (*p. 228*)

prokaryote   (*p. 227*)
protein profile   (*p. 241*)
Protista   (*p. 229*)
Rickettsiae   (*p. 250*)
species   (*p. 223*)
specific epithet   (*p. 223*)
strain   (*p. 224*)
stromatolite   (*p. 231*)
taxon   (*p. 222*)
taxonomy   (*p. 222*)
type strain   (*p. 245*)
Ureaplasmas   (*p. 250*)
virus   (*p. 236*)

See You on the Web

Don't forget I am expecting you on the web. The following is a preparedness checklist for your web explorations. If you have questions on any of these items, you can return to the end of Chapter 1 for more details.

- Questions We'll Explore
- Concept Check questions
- Flashcards
- Study Guide

If you are ready, come visit me at

http://www.prenhall.com/black

It will be a journey you'll never forget.

Viruses

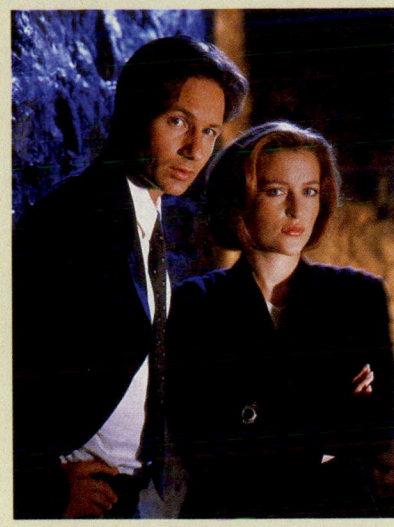

You've been assigned the difficult task of researching the epidemiology of malignant cancers in a certain homeless population. As you proceed, you notice that an uncommon percentage suffer from the same malignancy. From interviews you find that all affected have visited the same mobile clinic. You can't find records of this clinic, but you do enough snooping to find that this clinic is somehow tied to the malignancy.

You're not sure what's going on, but you have to help the victims. Chemotherapy is impractical, but you know a researcher in a government lab who has been developing cancer treatments using genetically engineered viruses. With tissue samples of the cancer, she's able to develop a virus specific for treatment. You infect the victims with this virus; it attacks and kills the cancer cells. You try to go after the mobile clinic, but it seems to have disappeared for good.

Sound like a story-line from TV's *The X-Files*? Possibly. But it's not far from reality. Scientists are working on ways to use viruses to combat cancer and even HIV. What's difficult is making viruses specific to affected cells so that "clean" cells aren't also killed. Maybe soon. Until then it makes a good story.

Background Basics

DNA and RNA structure and function . . . .*(Chapters 2, 7, 8)*
genome . . . .*(Chapter 7)*

T hroughout human history, viral epidemics have caused us to become more aware of the impact microbes have on our lives, and the course of history. Within the current century, we need only look back at the influenza pandemic of 1918 and 1919, polio outbreaks in the 1940s and 1950s, Marburg disease in the 1960s and the related Ebola hemorrhagic disease in the 1970s and 1990s, or HIV in the 1980s and 1990s, to be reminded that virus infections have a significant impact on our society. In fact, we need not look back even that far. In 1996 and 1997, our world experienced outbreaks of hepatitis, dengue fever, yellow fever, Ebola, Lassa fever, and polio. The media is full of both popularized and scientific writings on the "new" or "re-emerging" viruses. What factors are contributing to their increased impact? Are they truly "new" viruses—or are they recurring forms of old viruses? What factors have led to their "re-emergence"? We have seen the eradication of smallpox and read reports that polio is close to eradication in much of the world (in spite of a recent outbreak in the Mediterranean). Hantavirus, dengue fever, and yellow fever are said to be re-emerging. Could smallpox also reappear? Influenza is an annual occurrence, but we haven't seen a major pandemic for many years. Might we once again see a pandemic of the magnitude of that in 1918, which killed half a million Americans in just 10 months? These are just a few of the questions surrounding our knowledge of viral infections today. Current research continues to provide us with a better understanding of viral structure, viral replication, and viral diseases. As we learn more about how viral organisms function, it will increase our ability to develop new methods to control and/or eradicate viral infections. It will also extend our understanding of what the fundamental nature of life is—an important theme for all biologists.

This chapter examines the structure and behavior of viruses and viruslike agents. By the end of this chapter, you will have a better understanding of and appreciation for one of nature's tiniest-sized, but most dangerous, groups of microbes. Indeed, the name *virus* itself comes from the Latin word meaning "poison."

Questions We'll Explore

A What are the general characteristics of viruses?

B How are viruses classified?

C How do bacterial and animal viruses replicate?

D How are animal viruses cultured for experimental studies?

E What is a teratogen, and how do viruses act as teratogens?

F What are the properties of viruslike agents?

G How do viruses cause cancer?

Viral Invaders

Humans may be more similar to insects than you think: Research suggests that insects ward off infections by lining their intestines with a protective protein similar to the protein found in human mucus. Why are scientists excited by this news? Certain viruses that infect insects do so by using enhancin, a molecule that forms large gaps and tears in the insect's intestinal tract. Scientists may soon learn how to prevent viruses from penetrating our intestinal cells, thanks to the insect world.

General Characteristics of Viruses

What Are Viruses?

Viruses are infectious agents that are too small to be seen with a light microscope and that are not cells. They have no cell nucleus, organelles, or cytoplasm. When they invade susceptible host cells, viruses display some properties of living organisms and so appear to be on the borderline between living and nonliving. Viruses can *replicate,* or multiply, only inside a living host cell. As such they are called **obligate intracellular parasites,** a distinction they share with chlamydias and rickettsias. We may have to reconsider the traditional definition of viruses following the announcement in December 1991 by E. Wimmer, A. Molla, and A. Paul that they successfully grew entire polioviruses in test tubes containing ground up human cells, but no live cells. RNA from polioviruses was added to the cell-free extract, and about 5 hours later complete new virus particles began to appear. This has not yet been duplicated with other viruses.

Viruses differ from cells in important ways. Whereas prokaryotic and eukaryotic cells contain both DNA and RNA, individual virus particles contain only one kind of nucleic acid—either DNA or RNA, never both. Cells grow and divide, but viruses do neither. Viral replication requires that a virus particle infect a cell and program the host cell's machinery to synthesize the components required for the assembly of new virus particles. The infected cell may produce hundreds to thousands of new viruses and usually dies. Tissue damage as a result of cell death accounts for the destructive effects seen in many viral diseases.

Components of Viruses

Typical viral components are shown in ➤ Figure 10.1. These components are a nucleic acid core and a surrounding protein coat called a **capsid.** In addition, some viruses have a surrounding lipid bilayer membrane called an **envelope.** A complete virus particle, including its envelope, if it has one, is called a **virion.**

Nucleic Acids

The British immunologist Peter Medawar, who shared the 1960 Nobel Prize in physiology or medicine, once described viruses as "a piece of bad news wrapped up in protein." The nucleic acid is the "bad news" because viruses use their **genome,** their genetic information ➡ (Chap-

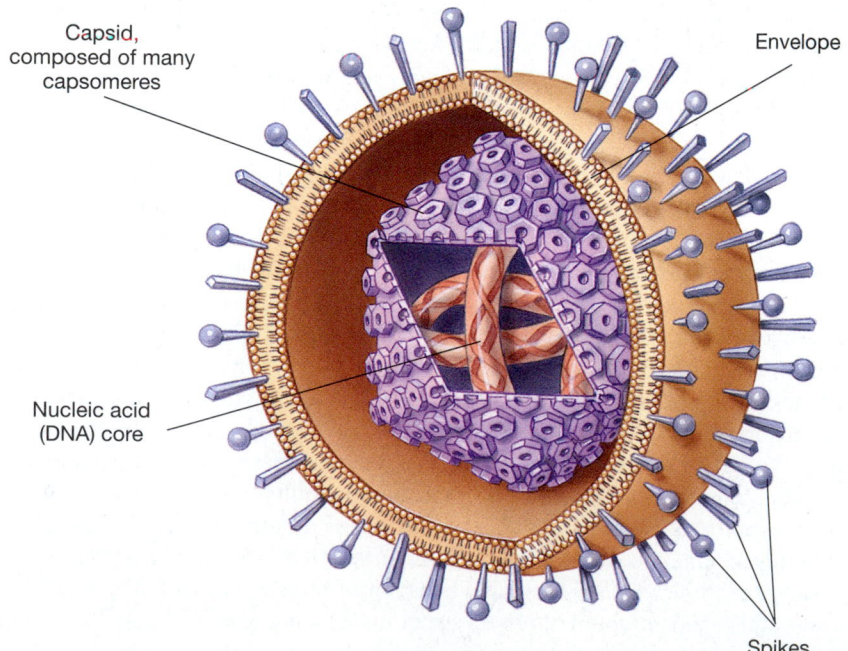

Capsid, composed of many capsomeres

Envelope

Nucleic acid (DNA) core

Spikes

➤ Figure 10.1   **The components of an animal virus (a herpesvirus).**

ter 7), to replicate themselves in host cells. The result is often a disruption of host cellular activities or death of the host. Viral genomes consist of either DNA or RNA. Viral replication depends on the expression of the viral genome for the formation of viral proteins and the replication of new viral genomes within the infected host cell. Viral nucleic acid can be single-stranded or double-stranded, and linear, circular, or segmented (existing as several fragments). All genetic information in RNA viruses is carried by RNA. RNA genomes occur only in viruses and a viruslike agent called the prion.

Capsids

The nucleic acid of an individual virion is in most cases enclosed within a capsid that protects it and determines the shape of the virus. Capsids also play a key role in the attachment of some viruses to host cells. Each capsid is composed of protein subunits called **capsomeres** (➤ Figure 10.1). In some viruses, the proteins found in the capsomere are of a single type. In other viruses several different proteins may be present. The number of proteins and the arrangement of viral capsomeres are characteristic of specific viruses and thus can be useful in virus identification and classification.

Envelopes

Enveloped viruses have a typical bilayer membrane outside their capsids. Such viruses acquire their envelope after they are assembled in a host cell as they *bud,* or move through, one or several membranes. A virion's **nucleocapsid** comprises the viral genome together with the capsid. Viruses with only a nucleocapsid and no envelope are known as **naked,** or nonenveloped, viruses. The composition of an envelope generally is determined by the viral nucleic acid and by the substances derived from host membranes. Combinations of lipids, proteins, and carbohydrates make up most envelopes. Depending on the virus, projections referred to as **spikes** (➤ Figure 10.1) may or may not extend from the viral envelope. These surface projections are **glycoproteins** that serve to attach virions to specific receptor sites on susceptible host cell surfaces. In certain viruses the possession of spikes causes various types of red blood cells to clump, or *hemagglutinate*—a property that is useful in viral identification.

What advantages might envelopes have for viruses? Because envelopes are acquired from and are therefore similar to host cell membranes, viruses may be "hidden" from attack by the host's immune system. Also, envelopes help viruses infect new cells by fusion of the envelope with the host's cell or plasma membrane. Conversely, enveloped viruses are damaged easily. Environmental conditions that destroy membranes—increased temperature, freezing and thawing, pH below 6 or above 8, lipid solvents, and some chemical disinfectants such as chlorine, hydrogen peroxide, and phenol—will also destroy the envelope. Naked viruses generally are more resistant to such environmental conditions.

Host cells are often 1000 times the volume of a virus infecting it.

Sizes and Shapes

Most viruses are too small to be seen with a light microscope, but ➤ Figure 10.2 shows that they have a range of sizes. The largest are the orthopoxviruses, which are about 240 nm by 300 nm—the size of the smallest bacteria or 1/10 the size of a red blood cell. The complex bacteriophages are about 65 nm by 200 nm. Among the smallest viruses known are the enteroviruses, which are less than 30 nm in diameter. But as ➤ Figure 10.2 shows, most viruses are quite small when compared with bacteria or eukaryotic cells. To put things in perspective, consider that typical ribosomes are about 25–30 nm in diameter.

Although some viruses are variable in shape, most viruses have a specific shape that is determined by the capsomeres or the envelope. ➤ Figure 10.2 shows several examples of viral symmetry. A *helical* capsid consists of a ribbonlike protein that forms a spiral around the nucleic acid. The tobacco mosaic virus is a helical virus. ➥ (Chapter 1, p. 16) *Polyhedral* viruses are many-sided. The picornaviruses (pi-kor′na-vi″rus-ez) and the human adenoviruses (ad′e-no-vi″rus-ez) are polyhedral viruses. One of the most common polyhedral capsid shapes is the icosahedron; *icosahedral* viruses have 20 triangular faces. A *complex* capsid is a combination of helical and icosahedral shapes, and some viruses have a *bullet-shaped* capsid.

Another Evil of Tobacco

Keep smokers away from your tomato plants. Cigarette tobacco always contains some tobacco mosaic virus—enough to start an infection in tomato plants when carried there by smokers' hands or cigarette butts. Try an experiment: Is water in which cigarette tobacco has been soaked able to transmit tobacco mosaic disease? Dry tobacco? Cigarette smoke? Washed versus unwashed smokers' fingers?

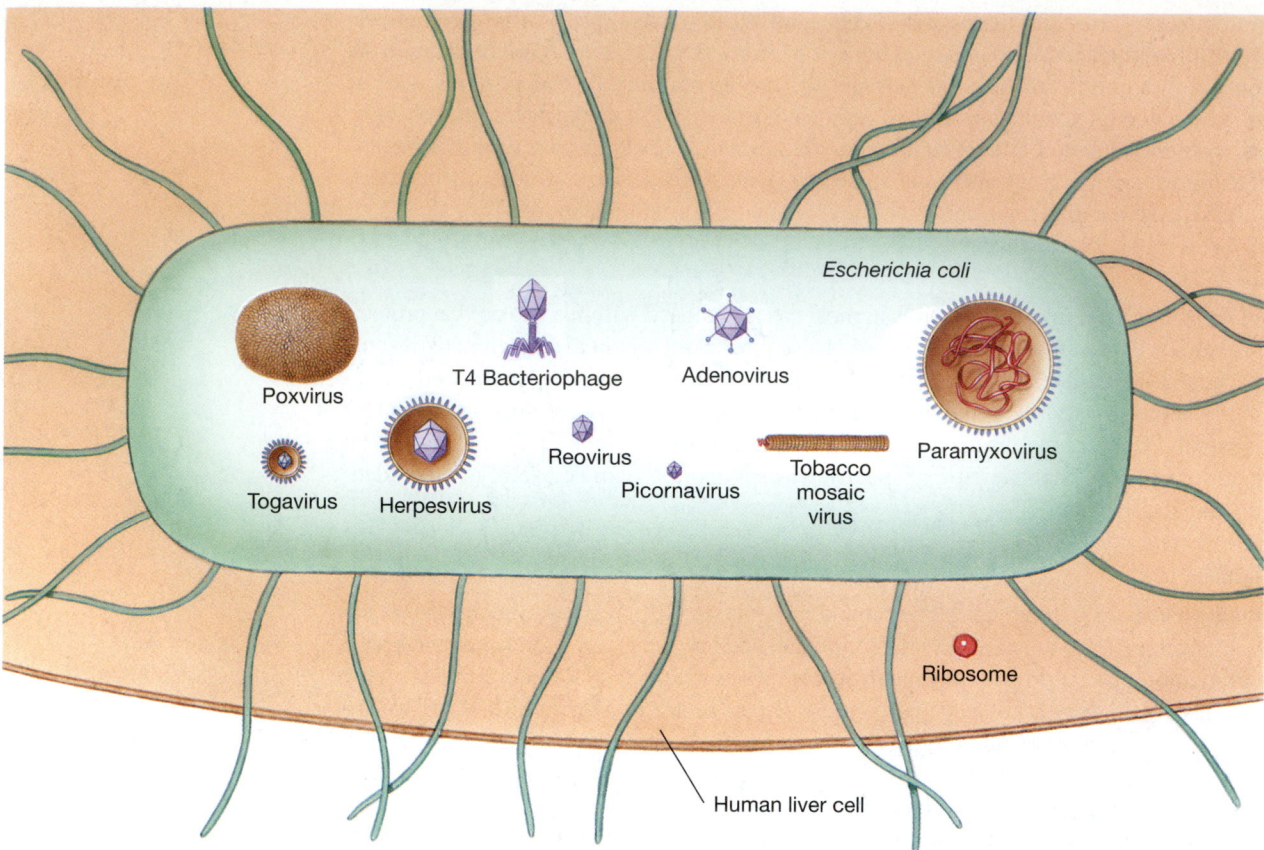

➤ Figure 10.2   **Viral sizes and shapes.** Variations in shapes and sizes of viruses compared with a bacterial cell, an animal cell, and a eukaryotic ribosome.

It's That Pig in Me That's Making Me Sick

Imagine getting extremely sick after a tissue transplant. When you go to your doctor to find out what's wrong, he tells you that infected pig tissues used in your transplant may have caused your illness. "But," you protest, "I thought that viruses are limited to one host!" Not necessarily. A retrovirus in pigs can indeed infect human cells. When the FDA found out about this in October 1997, they ordered a temporary halt to clinical trials involving transplantation of pig tissues into human patients. But you don't have to worry—Carolyn Wilson developed an assay for the pig virus that was so reliable that the FDA lifted their research moratorium by January 1998, as long as companies promised to use the assay to make sure they had virus-free materials. So bring on the pigs!

Viruses with envelopes have a somewhat spherical shape. For example, the herpesvirus shown in ➤ Figure 10.1 has a polyhedral capsid and an envelope.

The poxviruses and many bacterial viruses are called **complex viruses** because they have a more elaborate coat or capsid (➤ Figure 10.2). Many **bacteriophages,** or viruses that infect bacteria, have a complex shape that incorporates specialized structures such as heads, tails, and tail fibers (➤ Figure 10.2). Like spikes, the tail fibers are used by virions to attach to host bacteria. Other specialized bacteriophage structures are used to infect the bacterial cells.

Host Range and Specificity of Viruses

Although viruses are quite small and differ from one another structurally and in their replication strategies, they are capable of infecting all forms of life (hosts). The **host range** of a virus refers to the spectrum of hosts that a virus can infect. Different viruses can infect bacteria, fungi, algae, protozoa, plants, vertebrates, or even invertebrates. However, most viruses are limited to only one host and to only specific cells and/or tissues of that host. Polioviruses, for example, can be grown in the laboratory in monkey kidney cells but have never been observed to cause a natural infection in any animals other than humans. In contrast, the rabies virus can attack the central nervous system of many warmblooded animals. The host range of the rabies virus is much more extensive than that of polioviruses.

Viral specificity, another important property of viruses, refers to the specific kinds of cells a virus can infect. For example, certain papillomaviruses, which cause warts, are so specific in their replication strategy that they infect only skin cells. In contrast, cytomegaloviruses, known for their lethal effects, attack cells of the salivary glands, gastrointestinal tract, liver, lungs, and other organs. They can also cross the placenta and attack fetal tissues, especially those of the central nervous system. The discovery that a given virus can cause varying symptoms in several different body systems made the one virus–one disease concept untenable.

Viral specificity is determined mainly by whether a virus can attach to a cell. Attachment depends on the presence of specific receptor sites on the surfaces of host cells and on specific attachment structures on viral capsids or envelopes. Specificity is also affected by whether appropriate host enzymes and other proteins the virus needs in order to replicate are available inside the cell. Finally, specificity is affected by whether replicated viruses can be released from the cell to spread the infection to other cells.

✓ Why are host cells necessary for viral replication?

✓ Distinguish between capsid and capsomere.

✓ Distinguish between naked and enveloped viruses.

Classification of Viruses

Before they knew much about the structure or chemical properties of viruses, virologists classified viruses by the type of host infected or by the type of host structures infected. Thus, viruses have been classified as bacterial viruses (bacteriophages), plant viruses, or animal viruses. And animal viruses are grouped by the tissues they attack as *dermotropic* if they infected the skin, *neurotropic* if they infected nerve tissue, *viscerotropic* if they infected organs of the digestive tract, or *pneumotropic* if they infected the respiratory system.

As more was learned about the structure of viruses at the biochemical and molecular levels, classification of viruses came to be based on the type and structure of their nucleic acids, method of replication, host range, and other chemical and physical characteristics. And as more viruses were discovered, conflicting classification systems came into use, resulting in much confusion and some bad feelings. The need for a single, universal taxonomic scheme for viruses led to the establishment in 1966 of the International Committee on Taxonomy of Viruses (ICTV). This committee, which meets every 4 years, establishes the rules for classifying viruses. Virus classification is summarized in Appendix B.

Because viruses are so different from cellular organisms, it is difficult to classify them according to typical taxonomic categories—kingdom, phylum, and the like. The family has been the highest taxonomic category used by the ICTV. Viral genera also have been established, but most are new and slow to gain acceptance. Despite advances in classification, the problems of defining and naming *viral species*—a group of viruses that share the same genome and the same relationships with organisms—and of distinguishing between viral species and viral

Virology, as a scientific field of study, is only about 100 years old.

Origins Of Viruses

Viruses are clearly quite different from cellular microbes. Free viruses are incapable of reproduction—they must infect host cells, uncoat their genetic material, and then use the host's machinery to copy or transcribe the viral genetic material. Thus, some debate remains as to whether viruses are living or are nonliving chemical aggregates. Because viruses cannot reproduce or metabolize or perform metabolic functions on their own, some scientists say that they are not living. Other scientists claim that because viruses have the genetic information for replication, and this information is active after infection, they are living. Much of the genetic regulation of viral genes is similar to the regulation of host genes. In addition, viruses use the host's ribosomes for their replication metabolism.

At present, we cannot definitively say whether viruses are living or nonliving. But we can ask, What are the origins of viruses? We do not know that either. However, because viruses cannot replicate without a host cell, it is likely that viruses were not present before primitive cells evolved. One hypothesis proposes that viruses and cellular organisms evolved together, with both viruses and cells originating from self-replicating molecules present in the precellular

world. Another idea is that viruses were once cells that lost all cell functions, retaining only that information to replicate themselves using another cell's metabolic machinery. A third hypothesis proposes that viruses evolved from plasmids, the independently replicating DNA molecules found in many bacterial cells. ➡ (Chapter 8, p. 205) Plasmids are self-replicating and occur in both DNA and RNA forms. They do not, however, have genes to make capsids. In fact, it has been proposed that plasmids evolved from viroids. As some viroids moved from cell to cell, the viroid RNA may have picked up several pieces of genetic information, including the information for making a protein coat.

In trying to understand the origins of viruses, virologists have uncovered some nucleotide sequence relationships common to certain viruses. On the basis of this information, these viruses have been placed into families with similar nucleotide sequences and genetic organization. Thus, it may be possible to predict the potential disease effects of newly discovered viruses by analyzing the nucleotide sequences of their genomes and comparing them with sequences found in other, known viruses.

strains have not yet been resolved completely. Currently, the ICTV requires that the English common name, rather than a Latinized binomial term, be used to designate a viral species. For example, the formal taxonomic designations for the rabies virus and HIV would be as follows: family: Rhabdoviridae; genus: *Lyssavirus;* species: rabies virus; and family: Retroviridae; genus: *Lentivirus;* species: human immunodeficiency virus (HIV).

The names of specific viruses often consist of a group name and a number, such as HIV-1 or HIV-2. Virus families are often distinguished first on the basis of their nucleic acid type, capsid symmetry (shape), envelope, and size (Tables 10.1 and 10.2). As of 1995 the ICTV had assigned more than 4000 member viruses to 71 families, 11 subfamilies, and 175 genera. Many of the virus families contain viruses that cause important infections of humans and some other animals. Additional families contain viruses that infect only other animals, plants, fungi, algae, or bacteria.

Nucleic Acid Classification

Major groups of viruses are distinguished first by their nucleic acid content as either RNA or DNA viruses. Subsequent subdivisions are based largely on other properties of nucleic acids. The RNA viruses can be single-stranded (*ssRNA*) or double-stranded (*dsRNA*), although most are single-stranded (Table 10.1). Because most eukaryotic cells do not have the enzymes to

> **TABLE 10.1** Classification of major groups of RNA viruses that cause human diseases

Family	Envelope and capsid shape	Typical size (nm)	Example (Genus or species)	Infection or disease
(+) *Sense RNA viruses*				
Picornaviridae (1 copy)	Naked, polyhedral	18–30	*Enterovirus* *Rhinovirus* *Hepatovirus*	Polio Common cold Hepatitis A
Togaviridae (1 copy)	Enveloped, polyhedral	40–90	Rubella virus Equine encephalitis virus	Rubella (German measles) Equine encephalitis
Flaviviridae (1 copy)	Enveloped, polyhedral	40–90	*Flavivirus*	Yellow fever
Retroviridae (2 copies)	Enveloped, spherical	100	HTLV-I HIV	Adult leukemia, tumors, AIDS
(–) *Sense RNA viruses*				
Paramyxoviridae (1 copy)	Enveloped, helical	150–200	*Morbillivirus*	Measles
Rhabdoviridae (1 copy)	Enveloped, helical	70–180	*Lyssavirus*	Rabies
Orthomyxoviridae (1 copy in 8 segments)	Enveloped, helical	100–200	*Influenzavirus*	Influenza A and B
Filoviridae (1 copy)	Enveloped, filamentous	80	*Filovirus*	Marburg, Ebola
Bunyaviridae (1 copy in 3 segments)	Enveloped, spherical	90–120	*Hantavirus*	Respiratory distress, hemorrhagic fevers
Double-stranded RNA viruses				
Reoviridae (1 copy in 10–12 segments)	Naked, polyhedral	70	*Rotavirus*	Respiratory and gastrointestinal infections

> **TABLE 10.2** Classification of major groups of DNA viruses that cause human diseases

Family	Envelope and capsid shape	Typical size (nm)	Example (genus or species)	Infection or disease	
Double-stranded DNA viruses					
Adenoviridae (linear DNA)	Naked, polyhedral	75	Human adenoviruses	Respiratory infections	
Herpesviridae (linear DNA)	Enveloped, polyhedral	120–200	_Simplexvirus_	Oral and genital herpes	
			Varicellovirus	Chickenpox, shingles	
Poxviridae (linear DNA)	Enveloped, complex shape	230 X 270	_Orthopoxvirus_	Smallpox, cowpox	
Papovaviridae (circular DNA)	Naked, polyhedral	45–55	Human papilloma-viruses	Warts, cervical and penile cancers	
Hepadnaviridae (circular DNA)	Enveloped, polyhedral	40–45	Hepatitis B virus	Hepatitis B	
Single-stranded DNA viruses					
Parvoviridae (linear DNA)	Naked, polyhedral	22	B19	Fifth disease (erythema infectiosum) in children	

copy viral RNA molecules, the RNA viruses must either carry the enzymes or have the genes for those enzymes as part of their genome. Table 10.1 identifies two types of single-stranded RNA viruses—positive sense and negative sense RNA viruses. Many ssRNA viruses contain **positive (+) sense RNA,** meaning that during an infection the RNA acts like mRNA and can be translated by the host's ribosomes. Other ssRNA viruses have **negative (−) sense RNA.** In such viruses the RNA acts as a template during transcription to make a complementary (+) sense mRNA after a host cell has been entered. ➥ (Chapter 7, p. 175) This strand is translated by host ribosomes. In order to perform the transcription step, (−) sense RNA viruses must carry an RNA polymerase within the virion.

Like RNA viruses, DNA viruses can also occur in single-stranded or double-stranded form (Table 10.2). For example, the human adenoviruses, responsible for some common colds, and the herpesviruses are double-stranded DNA (_dsDNA_) viruses. Only one single-stranded DNA (_ssDNA_) virus is currently known to produce human disease.

With this background, let us briefly examine several families of animal RNA and DNA viruses.

RNA Viruses

General Properties of RNA Viruses

The different families of RNA viruses are distinguished from one another by their nucleic acid content, their capsid shape, and the presence or absence of an envelope (Table 10.1; ➤ Figure 10.3). Most families of RNA viruses contain either one (+) sense RNA or one (−) sense RNA molecule. However, some RNA viruses are placed in separate families if the RNA exists as two complete copies of (+) sense RNA or contains small segments of (−) sense RNA. Finally, one family has segmented dsRNA.

Important Groups of RNA Viruses

Picornaviridae The **picornaviruses** are very small (30 nm in diameter), naked, polyhedral, (+) sense RNA viruses. They include more than 150 species that cause disease in humans. After infection these viruses quickly interrupt all functions of DNA and RNA in the host cell. The Picornaviridae are divided into several groups, including the genera *Enterovirus, Hepatovirus,* and *Rhinovirus.*

Enteroviruses (*entero,* Greek for "intestine") include the polioviruses (➤ Figure 10.3a). These viruses are resistant to many chemical substances and can replicate in and pass through the digestive tract unharmed. Unless inactivated by host defense mechanisms, the viruses invade the blood and lymph, spreading throughout the body but especially into the nervous system. Poor sanitation increases the numbers of enteroviruses, and human overcrowding helps their spread. As a result of early and frequent exposure, most children living in such conditions acquire infection in infancy when paralysis is unlikely to occur, and symptoms are mainly flulike. It is the older child and adult that usually develop paralysis. Thus paralytic polio epidemics are uncommon in developing nations.

Poor sanitation is also responsible for the spread of certain **hepatoviruses** (*hepato,* Greek for "liver"). The hepatitis A virus, for instance, is transmitted via the fecal–oral route, with disease arising from the ingestion of contaminated food or water. The major organ infected is the liver.

The genus *Rhinovirus* (*rhino,* Greek for "nose"), which includes more than 100 types of human **rhinoviruses,** is one of the genera of viruses responsible for the common cold. Human rhinoviruses do not cause digestive tract diseases because they cannot survive the acidic conditions in the stomach. Instead, they enter the body through the mucous membranes of the nasal passages and replicate in the epithelial cells of the upper respiratory tract. Much has recently been learned about the capsids of rhinoviruses (➤ Figure 10.4). Virologists have discovered that these capsids attach to only a few receptors in the nasal mucous membranes. Thus, it may be possible in the future to prevent colds by designing chemicals that cover these receptors so that rhinoviruses cannot attach.

Togaviridae The **togaviruses** are small, enveloped, polyhedral, (+) sense RNA viruses that multiply in the cytoplasm of many mammalian and arthropod host cells. Togaviruses known as arthropodborne viruses are transmitted by mosquitoes and cause several kinds of encephalitis (plural:

> During the 1940s, in New York City, for every case of paralytic polio, 100 more people had symptomless cases of polio infection.

> Rhinoviruses can survive on household objects for up to 3 days.

(a)   (b)   (c)

(d)   (e)

➤ Figure 10.3   **False-color electron micrographs of representative RNA viruses.** (a) picornaviruses (polioviruses; 32,500X); (b) retroviruses (oncovirus; 42,500X); (c) rhabdoviruses (rabies virus; 72,300X); (d) orthomyxoviruses (influenza viruses); (e) reoviruses (respiratory viruses).

encephalitides) in humans and in horses. The rubella virus, which causes German measles (rubella), is in this family but is not transmitted by arthropods; rather, it is spread person to person.

Flaviviridae   The **flaviviruses** are enveloped, polyhedral, (+) sense RNA viruses that are transmitted by mosquitoes and ticks. The viruses produce a variety of encephalitides or fevers in humans. The yellow fever virus is a flavivirus that causes a hemorrhagic fever—in which blood vessels in the skin, mucous membranes, and internal organs bleed uncontrollably. Hepatitis C infection is also caused by a flavivirus.

Retroviridae   The **retroviruses** are enveloped viruses that have two complete copies of (+) sense RNA (➤ Figure 10.3b). They also contain the enzyme **reverse transcriptase,** which uses the viral RNA to form a complementary strand of DNA that is then replicated to form a dsDNA. This reaction is exactly the reverse of the typical transcription step (DNA ⟶ RNA) in protein synthesis. For virus replication to continue, the newly formed DNA must be transcribed into viral RNA that will function as mRNA for viral protein synthesis and be incorporated into new virions. To do so, the DNA must first migrate to the host cell nucleus and become incorporated into chromosomes of host cells. Such integrated viral DNA is known as a **provirus.** Retroviruses cause tumors and leukemia in rodents and birds as well as in humans. The human retroviruses invade immune defense cells called *T lymphocytes* and are referred to as *h*uman *T* cell *l*eukemia *v*iruses (HTLV). HTLV-1 and HTLV-2 are associated with malignancies (leukemia and other tumors), whereas the *h*uman *i*mmunodeficiency *v*irus (HIV-1 and HIV-2 strains) causes *a*cquired *i*mmune *d*eficiency *s*yndrome (AIDS). AIDS is discussed in Chapter 18.

Paramyxoviridae   The **paramyxoviruses** (*para,* Latin for "near"; *myxo,* Greek for "mucus") are medium-sized, enveloped, (−) sense RNA viruses, with a helical nucleocapsid. Different genera of paramyxoviruses are responsible for mumps, measles, viral pneumonia, and bronchitis in children and mild upper respiratory infections in young adults.

Rhabdoviridae   Another (−) sense RNA virus group, the **rhabdoviruses** (*rhabdo,* Greek for "rod") consists of medium-sized, enveloped viruses. Although these viruses have an envelope, the capsid is helical and makes the viruses nearly rod- or bullet-shaped (➤ Figure 10.3c). Rhabdoviridae virions contain an RNA-dependent RNA polymerase that uses the (−) sense strand to form a (+) sense strand. The newly produced strand serves as mRNA and as a template for the synthesis of new viral RNA. Human rabies almost always results from a bite by a rabid animal carrying the rabies virus. Rhabdoviruses also infect other vertebrates, invertebrates, or plants. The Lago virus, which produces disease in bats, and the Mokolo virus, which infects shrews in Africa, are closely related to rabies viruses.

Orthomyxoviridae   The **orthomyxoviruses** (*ortho,* Greek for "straight") are medium-sized, enveloped, (−) sense RNA viruses that vary in shape from spherical to helical (➤ Figure 10.3d). Their genome is segmented into eight pieces. Like the paramyxoviruses, orthomyxoviruses have an affinity for mucus. Influenza virus A, with which we are all too familiar, is an orthomyxovirus that also infects birds, swine, and horses. Influenza virus B appears to be specific to humans.

Filoviridae   The **filoviruses** are enveloped, filamentous, single (−) sense RNA viruses. These viruses can be transmitted from person to person by close contact with blood, semen, or other secretions and by using contaminated needles. The filoviruses include the viruses responsible for Marburg and Ebola diseases, which are hemorrhagic fevers.

Bunyaviridae   The **bunyaviruses** also are enveloped, (−) sense RNA viruses whose genome has three segments. Bunyaviruses can be transmitted by arthropods, but rodents typically are the principal host. The most recently recognized member of the Bunyaviridae is the hantavirus responsible for hantavirus pulmonary syndrome (HPS). Other forms of the genus *Hantavirus* cause hemorrhagic fevers.

Arenaviridae   Like the bunyaviruses, the **arenaviruses** are enveloped, (−) sense RNA viruses, but their genome has only two segments. Arenaviruses are carried by rodents. Human infections occur via aerosols, exposure to infectous urine or feces, or rat bites. Argentinean and Bolivian hemorrhagic fevers and Lassa fever are arenavirus infections.

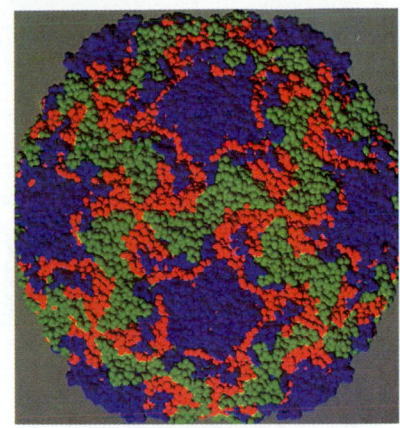

➤ Figure 10.4   **Cold virus.** Computer-generated model of a human rhinovirus, cause of the common cold. The colors represent different capsomeres of the capsid.

Rabies is an ancient disease. It was recognized in Egypt before 2300 B.C., and in ancient Greece, where it was well-described by Aristotle.

Reoviridae   The **reoviruses** have a naked, polyhedral capsid (➤ Figure 10.3e). They are medium-sized dsRNA viruses. They replicate in the cytoplasm and form distinctive inclusions that stain with eosin. Reoviruses include the orthoreoviruses, orbiviruses, and rotaviruses. The rotaviruses are the most common cause of severe diarrhea in infants and in young children under age 2. They also are responsible for minor upper respiratory and gastrointestinal infections in adults. The other reoviruses infect other animals.

DNA Viruses

General Properties of DNA Viruses

Like the RNA viruses, the animal DNA viruses are grouped into families according to their DNA organization (Table 10.2; ➤ Figure 10.5). The dsDNA viruses are further separated into families on the basis of the shape of their DNA (linear or circular), their capsid shape, and the presence or absence of an envelope. Only one family of viruses has ssDNA.

Important Groups of DNA Viruses

Adenoviridae   The **adenoviruses** (*adeno,* Greek for "gland") are medium-sized, naked viruses with linear dsDNA. First identified in adenoid tissue, they are highly resistant to chemical agents and are stable from pH 5 to 9 and from 36° to 47°C. Freezing causes little loss of infectivity. More than 80 different types of adenoviruses have been identified, many being responsible for human respiratory disease. Adenovirus types 40 and 41 cause 10 to 30 percent of all cases of severe diarrhea in babies and young children. Only half the children carrying the virus in their throat actually become ill.

   Diseases caused by adenoviruses are generally acute (that is, have sudden onset and short duration). Soon after entering the body, the virus appears in the blood and a measleslike rash may develop. Sources of adenoviruses are respiratory secretions and feces from infected persons.

Herpesviridae   The **herpesviruses** (*herpes,* Greek for "creeping") are relatively large, enveloped viruses with linear dsDNA (➤ Figure 10.5a). Herpesviruses are widely distributed in nature, and most animals are infected with one or more of the 100 types discovered. These viruses cause a broad spectrum of diseases, which are summarized in Table 10.3.

   The core of the virion contains proteins around which the DNA is coiled. In cells infected with herpesviruses, the viral dsDNA can exist as a provirus. Therefore, a universal property of herpesviruses is **latency,** the ability to remain in host cells, usually in neurons, for long periods and to retain the ability to replicate. For example, a child who has recovered from chickenpox (varicella) will still have the virus in a latent form. Years or decades later the virus may be reactivated as a result of stress and/or physical factors. This adult disease, which can be quite painful and debilitating, is called *shingles* (zoster). Fluids oozing from the vesicles carry the varicella-zoster virus, which can then cause chickenpox in persons who have not previously been infected with the virus.

> *Ingestion of only 10 rotavirus particles is sufficient to cause infection and diarrhea.*

> *Human herpesvirus 8 has been found associated with all cases of Kaposi's sarcoma, a frequent complication of AIDS.*

(a)

(b)

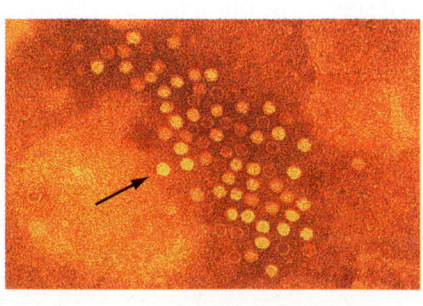

(c)

➤ Figure 10.5   **False-color electron micrographs of representative DNA viruses** (arrows point to one virion): (a) herpesviruses (pink spheres within the cell); (b) papovaviruses (human papilloma-viruses; 13,000X); (c) parvoviruses (70,000X).

> Table 10.3   **Herpesviruses that cause human disease**

Genus	Virus type	Infection or disease
Simplexvirus	Herpes simplex type 1	Oral herpes (sometimes genital and neonatal herpes), encephalitis
	Herpes simplex type 2	Genital and neonatal herpes (sometimes oral herpes), meningo-encephalitis
Varicellovirus	Varicella-zoster	Chickenpox (varicella) and shingles (zoster)
Cytomegalovirus	Cytomegaloviruses (salivary gland virus)	Acute febrile illness; infections in AIDS patients, transplant recipients, and others with reduced immune system function; a leading cause of birth defects
Roseolovirus	Roseola infantum (formerly called human herpesvirus 6)	Exanthema subitum (roseola infantum), a common disease of infancy, featuring rash and fever
Lymphocryptovirus	Epstein-Barr virus	Infectious mononucleosis and Burkitt's lymphoma (cancer of the jaw seen mainly in African children); also linked to Hodgkin's disease (cancer of lymphocytes) and B cell lymphomas, and to nasopharyngeal cancer in Asians

Poxviridae   The **poxviruses,** another group of enveloped, linear dsDNA viruses, are the largest and most complex of all viruses. They are widely distributed in nature; nearly every animal species can be infected by a form of poxvirus. The human poxviruses (orthopoxviruses) are large, enveloped, brick-shaped viruses 250 to 450 nm long and 160 to 260 nm wide. These viruses multiply in specialized portions of the host cell cytoplasm called *viroplasm,* where they can cause skin lesions typical of smallpox, molluscum contagiosum, and cowpox. Other poxviruses, such as monkeypox, can infect humans who have close contact with infected animals.

The smallpox virus represents the first human pathogen purposely eradicated from the face of the earth.

Papovaviridae   The **papovaviruses** (pa-po′va-vi″rus-ez) are named for three related viruses, the *pa*pilloma, *po*lyoma, and *va*cuolating viruses. These are small, naked, polyhedral dsDNA viruses that replicate in the nuclei of their host cells. Papovaviruses are widely distributed in nature; more than 25 human papillomaviruses and 2 human polyomaviruses have been found. Papillomaviruses are frequently found in host cell nuclei without being integrated into host DNA (➤ Figure 10.5b); polyomaviruses are nearly always integrated as a provirus. The papillomaviruses cause both benign and malignant warts in humans, and some papillomaviruses are associated with cervical cancer. The most thoroughly studied vacuolating virus is simian virus 40 (SV-40). This virus has been used by virologists to study the mechanisms of viral replication, integration, and oncogenesis (the development of cancerous cells).

Hepadnaviridae   The **hepadnaviruses** (he′pa-dee-en-ay-vi″rus-ez) are small, enveloped, mostly dsDNA (partially ssDNA) viruses. Their name comes from the infection of the liver—*hepa*titis—by a *DNA* virus. The hepadnaviruses can cause chronic (that is, of long and continued duration) liver infections in humans and other animals, including ducks. In humans, the hepatitis B virus causes hepatitis B, which can progress to liver cancer. We will discuss in Chapter 22 other forms of hepatitis that are caused by viruses.

Parvoviridae   The **parvoviruses** are small, naked, linear ssDNA viruses (➤ Figure 10.5c). Their genetic information is so limited that they must enlist the aid of an unrelated helper virus or a dividing host cell to replicate. Three genera, *Dependovirus, Parvovirus,* and *Erythrovirus,* have been identified in vertebrates. The dependoviruses are often called adeno-associated viruses because, to replicate more virions they require coinfection with adenoviruses (or herpesviruses). No known human disease is associated with this genus. Members of the genus *Parvovirus* can cause disease in rats, mice, swine, cats, and dogs. Rat parvovirus causes congenital defects in the unborn. Canine parvovirus is responsible for severe and sometimes fatal gastroenteritis in dogs and puppies. The only known parvovirus to infect humans (predominantly children) is *Erythrovirus,* which is also called B19. This virus, identified in 1974, is responsi-

Delta Hepatitis: A Defective Virus

Hepatitis delta virus (HDV) was discovered in the mid-1970s. Sequencing of its genome reveals similarity to the viroid and to virusoid RNAs that infect plants. HDV initially was thought to be part of the hepatitis B virus (HBV) because it was never found without the presence of hepatitis B infection. However, it was not present in all cases of hepatitis B, only in especially severe cases. By 1980 it was found to be a separate, defective pathogen that required coinfection with the hepatitis B virus in order to replicate. It can be prevented by vaccinating against HBV, because it cannot infect without its helper virus. HDV has the smallest genome of any known animal virus, with a length of a mere 1679 to 1683 nucleotides. In contrast, HBV is 3000 to 3300. It lacks a well-defined capsid of its own and is surrounded by the portion of the HBV that codes for the surface antigen (formerly called the Australian antigen) of HBV. It is primarily transmitted by blood and blood products.

Emerging Viruses

Viruses have been infecting humans for thousands of years, and the diseases they cause have been responsible for millions of deaths. Microbiologists believe that many recent, unexpected viral diseases have been caused by *emerging viruses*—viruses that were previously *endemic* (low levels of infection in localized areas) or had "crossed species barriers"—that is, expanded their host range to other species. For example, although the poliovirus has been endemic since ancient times, only since 1900 have *pandemics* (high levels of infection worldwide) resulted from this virus, with numerous annual outbreaks. Why the increase in disease?

The poliovirus has not mutated over the centuries into a more pathogenic form. Rather, virologists and epidemiologists suggest that the urban populations that developed after the Industrial Revolution provided an ideal environment for viral spread. The large numbers of nonimmune people who had emigrated from nonendemic areas were exposed to immune people who carried the poliovirus. Thus, polio spread rapidly—and with deadly results. Only through the development of polio vaccines in the 1950s was the epidemic halted. Although there still are areas where polio is endemic, the World Health Organization hopes to eradicate the poliovirus through vaccination programs by the year 2000.

Other endemic viral diseases that are transmitted among humans, such as measles, also are recurring due to changes in human population densities and travel. But several viral diseases involve other animals that act as *reservoirs* (a "healthy" organism harboring an infectious agent that is available to infect another host) or *vectors* (carriers) for a virus. In these cases the virus could cross species barriers if the vector transmitted the virus from a reservoir species to humans. For example, prior to 1930, yellow fever was thought to be carried solely by one species of mosquito, *Aedes aegypti*. By controlling that mosquito in urban areas (through vaccination and spraying with DDT), the disease could be controlled. However, in the late 1950s an outbreak of yellow fever occurred that was not carried by *A. aegypti*. Rather, jungle (sylvan) yellow fever was carried by another mosquito, of the genus *Haemagogus*. In the jungle these mosquitoes transmit the virus among monkeys that live high in the forest canopy. Forest clearing through tree cutting had brought the *Haemagogus* mosquito from the treetops to the forest floor, where it passed the virus to people involved in timber cutting and agricultural activities.

This yellow fever situation represents an excellent example of how viral diseases can remain endemic to those parts of the world where an insect vector lives. However, in tropical areas where once-uninhabited lands are being converted for use in agriculture and farming, contact with the insects (and the viruses they carry) is inevitable.

Of the more than 500 known arboviruses, about 80 cause disease in humans; of these, 20 are considered emerging viruses. The most dangerous are the yellow fever virus, which has been endemic for more than a century—and is reemerging—and the Dengue fever viruses. Both are carried by mosquitoes.

Many virologists believe that a similar event may have occurred in the case of the human immunodeficiency virus (HIV). Retroviruses similar to HIV exist in domesticated cats (feline immunodeficiency virus, FIV) and in monkeys (simian immunodeficiency virus, SIV). It is possible that a mutated form of SIV crossed over to humans from contact with an infected monkey. However, antibodies to SIV have been found in humans. Therefore, SIV itself may initially have infected humans and later mutated. Natural selection could have favored these mutations because they were better adapted than SIV to the new human host.

In the United States one recent emerging virus is the hantavirus, which is transmitted from rodent feces and urine to humans. The virus, which causes hantavirus pulmonary syndrome (HPS), struck first in New Mexico in May 1993. Although we do not know how the virus got into the rodent population, genetic analysis suggests that the virus has been endemic in rodents for years. We do know that conditions were right for an explosion in the rodent population, guaranteeing more contact between rodent feces and humans. Other hantaviruses are known to cause hemorrhagic fever in millions of people worldwide. The first, called the Hantaan virus, was isolated in Korea in 1978 but is believed to have been causing disease since the 1930s.

One of the disease transmission mechanisms of great concern today is air travel. Since the 1950s the annual number of international air travel passengers has risen from 2 million to almost 300 million, and almost 600 million are expected by the year 2000. In the confined air system of a jet airplane (an ideal atmosphere for the rapid spread of disease), a person or mosquito harboring an emerging virus could carry the virus around the world overnight.

So, how do we protect ourselves from such potential viral threats? Many virologists suggest that "virus outposts" be set up to try to detect emerging viruses before they spread. Viruses such as HIV that spread slowly might be hard to detect because they take years to emerge on a global scale. Conversely, the yellow fever virus shows clinical symptoms in nonimmunized individuals within days, if not hours. Perhaps quarantine will be required for people visiting or working in areas suspected of harboring emerging viruses.

Several factors have been proposed the can contribute to the emergence of viral (and other infectious) diseases. These include ecological changes and development (human contact with natural hosts or reservoirs), changes in human demographics (typically due to famine or war), international travel and commerce (allowing rapid introduction of viruses to new habitats and hosts), technology and industry (for example, during processing and rendering of infected animals), microbial adaptations and change (high mutation rates or shifts in genetic composition), and environmental changes (such as extension of vector ranges by global warming).

ble for "fifth disease" (erythema infectiosum), which causes a deep red rash on children's cheeks and ears and both a rash and arthritis in adults. B19 can cross the placenta and damage blood-forming cells in the fetus, leading to anemia, heart failure, and even fetal death.

Viral Replication

General Characteristics of Replication

In general, viruses go through the following five steps in their **replication cycles** to produce more virions:

1. **Adsorption,** the attachment of viruses to host cells.
2. **Penetration,** the entry of virions (or their genome) into host cells.
3. **Synthesis,** the synthesis of new nucleic acid molecules, capsid proteins, and other viral components within host cells while using the metabolic machinery of those cells.
4. **Maturation,** the assembly of newly synthesized viral components into complete virions.
5. **Release,** the departure of new virions from host cells. Release generally, but not always, kills (lyses) host cells.

Replication of Bacteriophages

Bacteriophages, or simply *phages,* are viruses that infect bacterial cells. Phages were first observed in 1915 by Frederic Twort in England and in 1917 by Felix d'Herelle in France. D'Herelle named them bacteriophages, which means "eaters of bacteria." Over the years, attempts have been made to use phages to fight bacterial infections. These attempts, though not successful until recently (see the box "Viruses Instead of Drugs," page 272), did provide useful information about phages. Phages have been studied in great detail because it is much easier to manipulate bacterial cells and their viruses in the laboratory than to work with viruses that have multicellular hosts.

Properties of Bacteriophages

Like other viruses, bacteriophages can have their genetic information in the form of either double-stranded or single-stranded RNA or DNA. They can be relatively simple or complex in structure. To understand phage replication, we will examine the *T-even phages.* These phages, designated T2, T4, and T6 (T stands for "type"), are complex but well-studied naked phages that have dsDNA as their genetic material. The most widely studied is the T4 phage, an obligate parasite of the common enteric bacterium *Escherichia coli.* T4 has a distinctly shaped capsid made of a head, collar, and tail (➤ Figure 10.6; Table 10.4). The DNA is packaged in the polyhedral head, which is attached to a helical tail.

Replication of T-Even Phages

Infection with and replication of new T4 phages occurs in the series of steps illustrated in ➤ Figure 10.7.

Adsorption   If T4 phages collide in the correct orientation with host cells, the phages will attach to, or adsorb onto, the host cell surface. Adsorption is a chemical attraction; it requires specific protein recognition factors found in the phage tail fibers that bind to specific receptor sites on the host cells. The fibers bend and allow the pins to touch the cell surface. Although many phages, including T4, attach to the cell wall, other phages can adsorb to flagella or to pili.

Penetration   The enzyme *lysozyme,* which is present within phage tails, weakens the bacterial cell wall. When the tail sheath contracts, the hollow tube (core) in the tail is forced to penetrate the weakened cell wall and come into contact with the bacterial cell membrane. The viral DNA then moves from the head through the tube into the bacterial cell. It is not clear whether the DNA is introduced directly into the cytoplasm; according to recent evidence, T4 phages introduce their DNA into the periplasmic space, between the cell membrane and the cell wall. Either way, the phage capsid remains outside the bacterium.

✓ On what basis are viruses classified today?

✓ Distinguish between positive and negative sense RNA.

✓ What is the difference between dsDNA and ssDNA?

Seawater can hold 100 million bacteriophages/ml.

Viral Sunscreens

We're not the only ones who should stay out of the sun. UV light also damages viruses like bacteriophages. So how do these viruses survive in the ocean where they are exposed to direct sunlight? Ironically, the very bacteria they are trying to destroy end up protecting them. UV light stimulates bacteria to produce photolyase, an enzyme that repairs damaged bacterial DNA. This process is called photoreactivation. Healthy bacteria thus protect the virus' nucleic acid, allowing the virus to replicate and eventually lyse the bacterial cell that protected it.

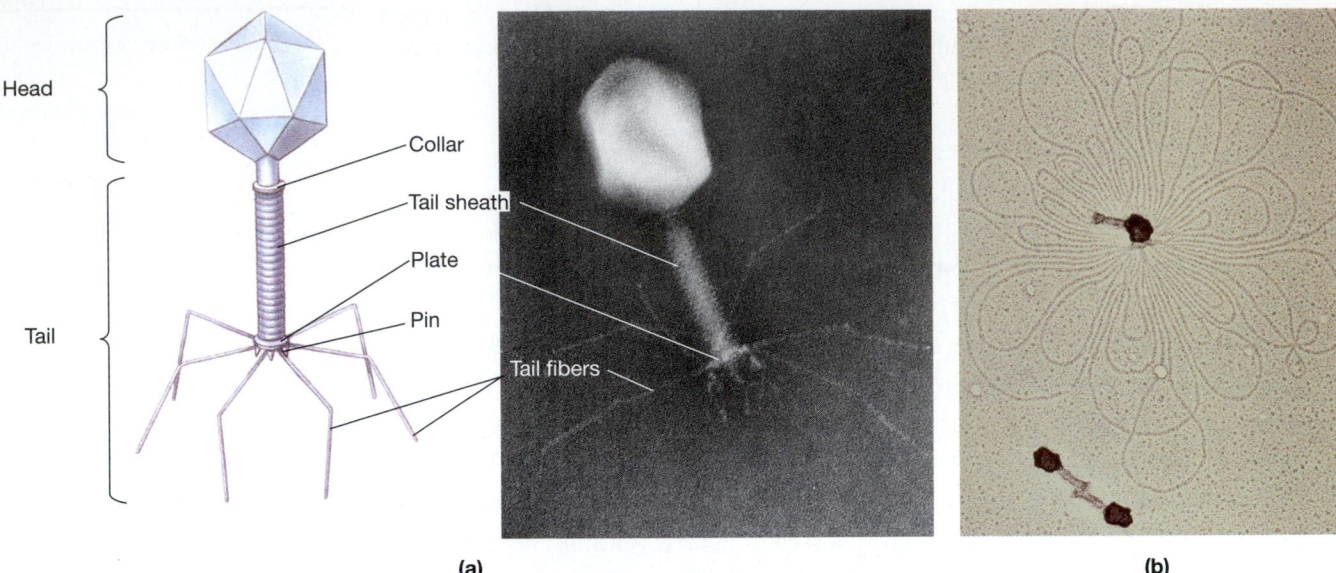

➤ **Figure 10.6   Bacteriophages.** (a) Structure and electron micrograph of a T-even (T4) bacterio-phage (191,500X). (b) DNA normally is packaged into the phage head. Osmotic lysis has released the DNA from this phage, showing the large amount of DNA that must be packaged into a phage (or into an animal or plant virus).

Synthesis   Viral genomes consisting of only a few thousand to 250,000 nucleotides, are too small to contain all the genetic information to replicate themselves. Therefore, they must use the biosynthetic machinery present in host cells. Once the phage DNA enters the host cell, phage genes take control of the host cell's metabolic machinery. Usually, the bacterial DNA is disrupted so that the nucleotides of hydrolyzed nucleic acids can be used as building blocks for new phage. Phage DNA is transcribed to mRNA, using the host cell's machinery. The mRNA, translated on host ribosomes, then directs the synthesis of capsid proteins and viral enzymes. Some of these enzymes are DNA polymerases that replicate the phage DNA. Thus, phage infection directs the host cell to make only viral products—that is, viral DNA and viral proteins.

Maturation   The head of a T4 phage is assembled in the host cell cytoplasm from newly synthesized capsid proteins. Then, a viral dsDNA molecule is packed into each head. At the same time, phage tails are assembled from newly formed base plates, sheaths, and collars. When the head is properly packed with DNA, each head is attached to a tail. Only after heads and tails are attached are the tail fibers added to form mature, infective phages.

Release   The enzyme lysozyme, which is coded for by a phage gene, breaks down the cell wall, allowing viruses to escape. In the process the bacterial host cell is lysed. Thus, phages such as T4

➤ Table 10.6	The functions of bacteriophage structural components
Component	**Function**
Genome	Carries the genetic information necessary for replication of new phage particles
Tail sheath	Retracts so that the genome can move from the head into the host cell's cytoplasm
Plate and tail fibers	Attach phage to specific receptor sites on the cell wall of a susceptible host bacterium

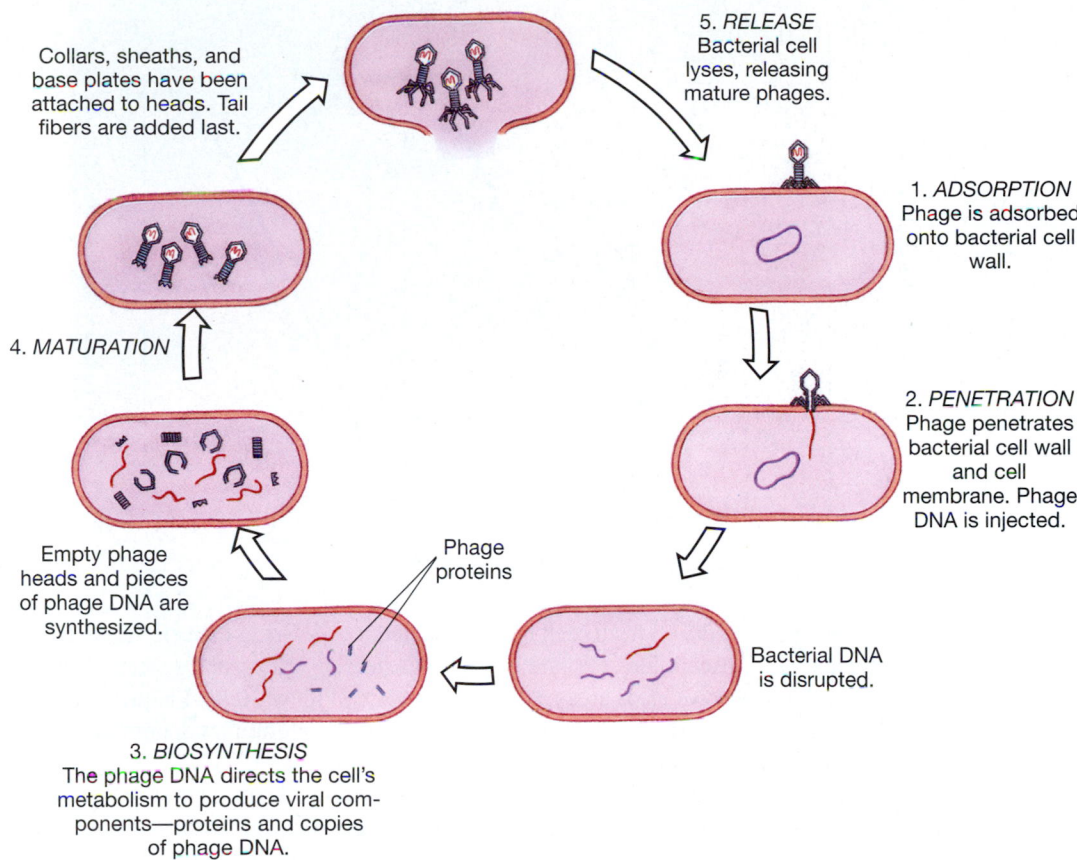

Collars, sheaths, and base plates have been attached to heads. Tail fibers are added last.

5. *RELEASE*
Bacterial cell lyses, releasing mature phages.

1. *ADSORPTION*
Phage is adsorbed onto bacterial cell wall.

4. *MATURATION*

2. *PENETRATION*
Phage penetrates bacterial cell wall and cell membrane. Phage DNA is injected.

Empty phage heads and pieces of phage DNA are synthesized.

Phage proteins

Bacterial DNA is disrupted.

3. *BIOSYNTHESIS*
The phage DNA directs the cell's metabolism to produce viral components—proteins and copies of phage DNA.

➤ **Figure 10.7   Replication of a virulent bacteriophage.** A virulent phage undergoes a lytic cycle to produce new phage particles within a bacterial cell. Cell lysis releases new phage particles that can infect more bacteria.

are called **virulent (lytic) phages** because they lyse and destroy the bacteria they infect. ➡ (Chapter 7, p. 198) The released phages can now infect more susceptible bacteria, starting the infection process all over again. Such infections by virulent phages represent a **lytic cycle** of infection.

The time from adsorption to release is called the **burst time;** it varies from 20 to 40 minutes for different phages. The number of new virions released from each bacterial host represents the **viral yield,** or **burst size.** In phages such as T4, anywhere from 50 to 200 new phages may be released from one infected bacterium.

Phage Growth and the Estimation of Phage Numbers

Like bacterial growth, viral growth (biosynthesis and maturation) can be described by a **replication curve,** which generally is based on observations of phage-infected bacteria in laboratory cultures (➤ Figure 10.8). The replication curve of a phage includes an **eclipse period,** which spans from penetration through biosynthesis. During the eclipse period, mature virions cannot be detected in host cells. The **latent period** spans from penetration up to the point of phage release. As ➤ Figure 10.8 shows, the latent period is longer than—and includes—the eclipse period. The number of viruses per infected host cell rises after the eclipse period and eventually levels off.

If you had a phage suspension in a test tube, how could you determine the number of viruses in the tube? Phages cannot be seen with the light microscope, and it is not feasible to count such viruses from electron micrographs. Therefore, virologists and microbiologists use a different approach to estimate phage number. The viral assay method used is called a **plaque assay.** To perform a plaque assay, virologists start with a suspension of phages. Serial dilutions, like those described for bacteria, are prepared. ➡ (Chapter 6, p. 140) A sample of each dilution is inoculated onto a plate containing a susceptible **bacterial lawn**—a layer of bacteria. Ideally, virologists want a dilution that will permit only one phage to infect one bacterial cell. As a result of in-

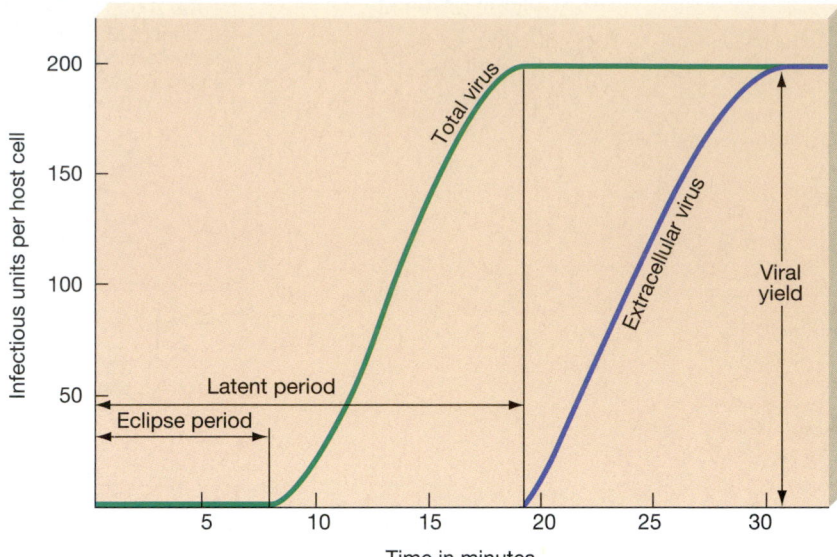

➤ Figure 10.8 **Growth curve for a bacteriophage.** The eclipse period represents the time after penetration through the biosynthesis of mature phages. The latent period represents the time after penetration through the release of mature phages. The number of viruses per infected cell is the viral yield, or burst size.

fection, new phages are produced from each infected bacterial cell, lysing the cell. These phages then infect surrounding susceptible cells and lyse them. After incubation and several rounds of lysis, the bacterial lawn shows clear areas called **plaques** (➤ Figure 10.9). Plaques represent areas where viruses have lysed host cells. In other parts of the bacterial lawn, uninfected bacteria multiply rapidly and produce a turbid growth layer.

Each plaque should represent the progeny from one infectious phage. Therefore, by counting the number of plaques and multiplying that number by the dilution factor, virologists can estimate the number of phages in a milliliter of suspension. Sometimes, however, two phages are deposited so close together that they produce a single plaque. And not all phages are infective. Thus, counting the number of plaques on a plate will approximate, but may not exactly equal, the number of infectious phages in the suspension. Therefore, such counts usually are reported as **plaque-forming units** (pfu) rather than as the number of phages.

Lysogeny

General Properties of Lysogeny

The bacteriophages we have been discussing, virulent phages, destroy their host cells. **Temperate phages,** do not always undergo a lytic cycle. The majority of the time they will exhibit **lysogeny,** a stable long-term relationship between the phage and its host in which the phage nucleic acid becomes incorporated into the host nucleic acid.

Such participating bacteria are called *lysogenic cells*. One of the most widely studied lysogenic phages is the lambda (λ) phage of *Escherichia coli* (➤ Figure 10.10). Lambda phages attach to bacterial cells and insert their linear DNA into the bacterial cytoplasm (➤ Figure 10.11). How-

➤ Figure 10.9 **Plaque assay.** The number of bacteriophages in a sample is assayed by spreading the sample out over a "lawn" of solid bacterial growth. When the phages replicate and destroy the bacterial cells, they leave a clear spot, called a plaque, in the lawn. The number of plaques corresponds roughly to the number of phages that were initially present in the sample. Different kinds of phages produce plaques of different size or shape when replicating in the same bacterial species—in this case, *Escherichia coli.* The upper left-hand plate was inoculated with T2 phage; the upper right-hand plate, with T4 phage; and the lower plate, with lambda phage.

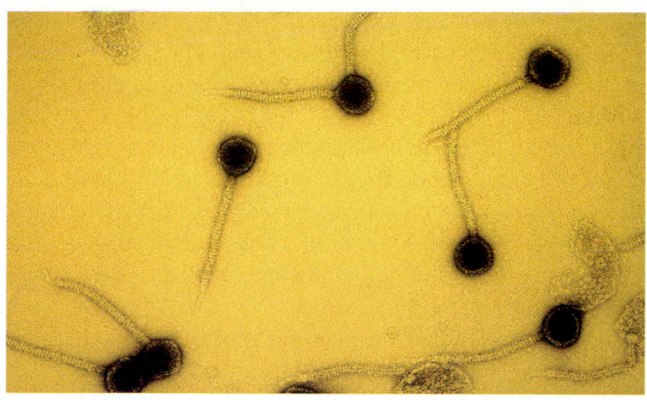

➤ Figure 10.10   **False-color TEM of the temperate lambda phage (84,000ZX).** This virus infects the bacterium *Escherichia coli.*

ever, once in the cytoplasm, the phage DNA circularizes and then integrates into the circular bacterial chromosome at a specific location. This viral DNA within the bacterial chromosome is called a **prophage.** The combination of a bacterium and a temperate phage is called a **lysogen.**

Insertion of a lambda phage into a bacterium alters the genetic characteristics of the bacterium. Two genes present in the prophage produce proteins that repress virus replication. The prophage also contains another gene that provides "immunity" to infection by another phage

4. *MATURATION*
Heads are packed with DNA.

Collars, sheaths, and base plates have been attached to heads. Tail fibers are added last.

Empty phage heads and pieces of phage DNA

LYTIC
CYCLE

5. *RELEASE*
Bacterial cell lyses, releasing completed infective phages.

3. *BIOSYNTHESIS*

The phage DNA directs the cell's metabolism to produce viral components— proteins and copies of phage DNA.

②

①

1. & 2. *ADSORPTION AND PENETRATION*

Phage is adsorbed to receptor site on bacterial cell wall, penetrates it, and inserts its DNA.

Induction

Prophage forms.

Phage DNA inserts itself as a prophage into bacterial chromosome.

Binary fission is completed; each cell has the phage DNA incorporated.

LYSOGENIC CYCLE

Phage is replicated along with the bacterial DNA prior to binary fission.

Indefinite cell divisions

➤ Figure 10.11   **Replication of a temperate bacteriophage.** Following adsorption and penetration, the virus undergoes prophage formation. ① In the lysogenic cycle, temperate phages can exist harmlessly as a prophage within the host cell for long periods of time. Each time the bacterial chromosome is replicated, the prophage also is replicated; all daughter bacterial cells are "infected" with the prophage. Induction involves either a spontaneous or environmentally induced excision of the prophage from the bacterial chromosome. ② A typical lytic cycle, involving biosynthesis and maturation, occurs, and new temperate phages are released.

Viruses Instead of Drugs

Recently, the Polish microbiologist Stefan Slopek and his colleagues successfully used bacteriophages to treat 138 patients with long-term antibiotic-resistant bacterial infections. Each patient received a bacteriophage that specifically attacks the bacterium causing the infection. All patients benefited from the treatment, and 88 percent were completely cured.

Using bacteriophages instead of antibiotics to treat infection has several advantages. The bacteriophages are highly specific. They attack the infectious agent without harming potentially beneficial bacteria that normally inhabit the human digestive tract. They also are cheap and easy to produce, are effective in small doses, and rarely cause side effects. Furthermore, the bacteriophages are effective against bacteria that have developed antibiotic resistance. One potential disadvantage is that bacteria might also develop resistance to bacteriophages.

of the same type. This process, called **lysogenic conversion,** prevents the adsorption or biosynthesis of phages of the type whose DNA is already carried by the lysogen. The gene responsible for such immunity does not protect the lysogen against infection by a different type of temperate phage or by a virulent phage.

Lysogenic conversion can be of medical significance because the toxic effects of some bacterial infections are caused by the prophages they contain. For example, the bacteria *Corynebacterium diphtheriae* and *Clostridium botulinum* contain prophages that have a gene that codes for the production of a toxin. The conversion from non-toxin production to toxin production is largely responsible for the tissue damage that occurs in diphtheria and botulism, respectively. Without the prophages, the bacteria do not cause disease.

Once established as a prophage, the virus can remain dormant for a long time. Each time a bacterium divides, the prophage is copied and is part of the bacterial chromosome in the progeny bacteria. Thus, this period of bacterial growth with a prophage represents a **lysogenic cycle** (➤ Figure 10.11). However, either spontaneously or in response to some outside stimulation, the prophage can become active and initiate a typical lytic cycle. This process, called **induction,** may be due to a lack of nutrients for bacterial growth or the presence of chemicals toxic to the lysogen. The provirus seems to sense that "living" conditions are deteriorating and that it is time to find a new home. Through induction, the provirus removes itself from the bacterial chromosome. The phage DNA then codes for viral proteins to assemble new temperate phages in a manner similar to that used by lytic phages. As a result, new temperate phages mature and are released through cell lysis.

The French microbiologist André Lwoff first described lysogeny in 1950. He also discovered that only a small proportion of lysogens produce phages at any one time. Those that do are lysed as a result of phage release. The remaining lysogens do not undergo induction and, due to lysogenic conversion, remain protected from infection by phages of the same type. In 1965 Lwoff shared the Nobel Prize in physiology or medicine with François Jacob and Jacques Monod.

The majority of bacteriophages undergo lysogeny. The reason may have to do with replication. Remember, virulent phages can move from host to host only by forming new phages that are released from one cell and that infect another. Rather, a lysogenic cycle allows temperate phages to "infect" more bacteria without forming new bacteriophages. As a result of binary fission, a copy of the phage DNA is distributed to each new bacterial cell.

Replication of Animal Viruses

Like other viruses, animal viruses invade and replicate in animal cells by the processes of adsorption, penetration, synthesis, maturation, and release. However, animal viruses perform these processes in ways that differ from those employed by bacteriophages—and in a number of different ways among themselves (Table 10.5). A complete replication cycle for an animal DNA virus is summarized in ➤ Figure 10.12, and two mechanisms for the replication of (+) sense RNA animal viruses are summarized in ➤ Figure 10.13.

Adsorption

As we have seen, bacteriophages have specialized structures for attaching to bacterial cell walls. Although animal cells lack cell walls, animal viruses have ways of attaching to host cells. Specificity involves a combination of virus and host cell recognition.

Naked viruses have attachment sites (proteins) on the surfaces of their capsids that bind to corresponding sites on appropriate host cells. For example, virologists have shown that rhinoviruses have "canyons," or depressions, in their capsids that bind to a specific membrane protein normally involved with cell adhesion (➤ Figure 10.14a, page 274). Conversely, enveloped viruses, such as HIV, have spikes that recognize, in part, a membrane protein receptor on the surface of certain specific immune defense cells (➤ Figure 10.14b).

Penetration

Penetration follows quite quickly after adsorption of the virion to the host's plasma membrane. Unlike bacteriophages, animal viruses do not have a mechanism for injecting their nucleic acid into host cells. Thus, both the nucleic acid and the capsid usually penetrate animal

➤ Table 10.5 **Comparison of bacteriophage and animal virus replication**

Stage	Bacteriophage	Animal virus
Attachment sites	Attachment of tail fibers to cell wall proteins	Attachment of spikes, capsid or envelope to plasma membrane proteins
Penetration	Injection of viral nucleic acid through bacterial cell wall	Endocytosis or fusion
Uncoating	None needed	Enzymatic digestion of viral proteins
Synthesis	In cytoplasm	In cytoplasm (RNA viruses) or nucleus (DNA viruses)
	Bacterial synthesis ceased	Host cell synthesis ceased
	Viral DNA or RNA replicated, formation of viral mRNA	Viral DNA or RNA replicated, formation of viral mRNA
	Viral components synthesized	Viral components synthesized
Maturation	Addition of collar, sheath, base plate and tail fibers to viral nucleic acid containing head	Insertion of viral nucleic acid into capsid
Release	Host cell lysis	Budding (enveloped viruses); cell rupture (nonenveloped viruses)
Chronic Infection	Lysogeny	Latency, chronic infection, cancer

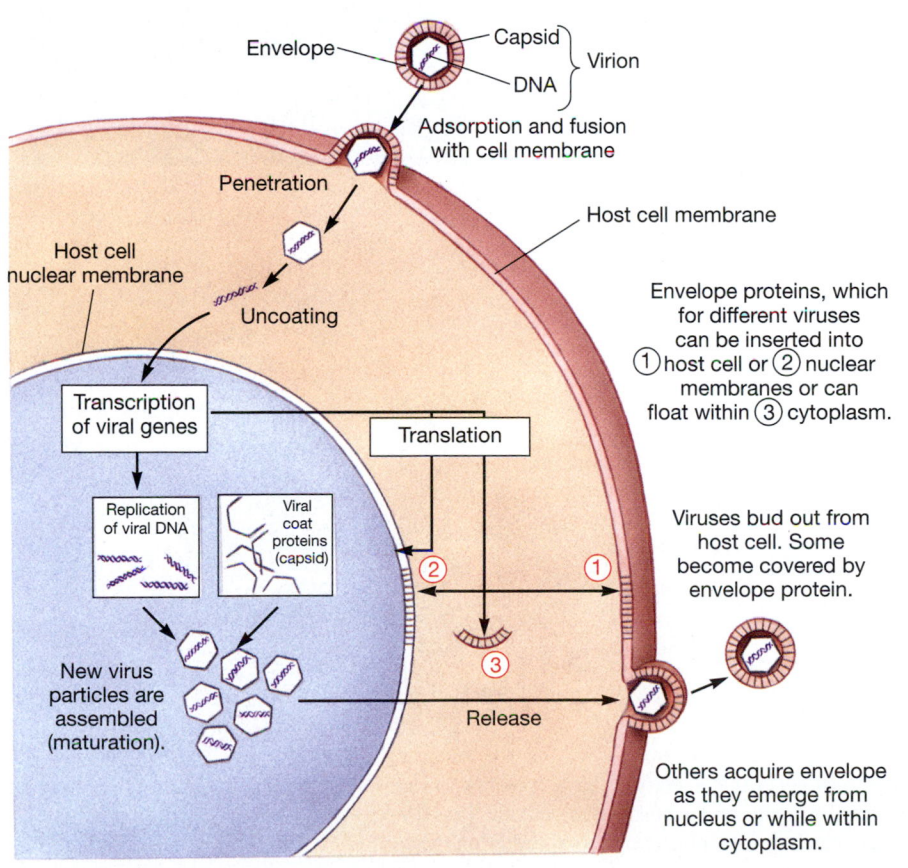

➤ Figure 10.12 **Replication of an enveloped dsDNA animal virus.** Shown is a herpes virus.

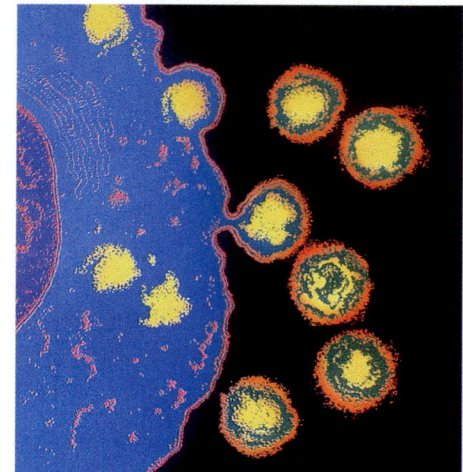

➤ Figure 10.13   **Replication of RNA viruses.** (a) Two of the replication mechanisms used by different (+) sense RNA animal viruses. *Left:* In the poliovirus the viral (+) sense RNA serves as mRNA—it is translated immediately to produce proteins needed for reproduction of the virus. A (−) sense RNA copy is then made, which serves as a template for the production of more viral (+) sense RNA molecules. Mature polioviruses lyse the cell during release. *Right:* In HIV each (+) sense RNA, copied with the help of reverse transcriptase, forms an ssDNA, which serves as template for the synthesis of the complementary strand. The dsDNA is then inserted into the host chromosome, where it can remain for some time. When virus replication occurs, one strand of the DNA becomes the template for the synthesis of viral (+) sense RNA molecules. Mature HIV particles usually do not lyse the cell but rather bud off the cell surrounded by an envelope. (b) HIV viruses budding from a T-4 lymphocyte.

➤ Figure 10.14   **Viral recognition of an animal host cell.** (a) Rhinoviruses have "canyons," or depressions, in the capsid that attach to specific membrane proteins on the host cell membrane. (b) HIV has specific envelope spikes (viral glycoproteins) that attach to a membrane protein receptor on the surface of specific host immune defense cells.

host cells. Most naked viruses enter the cell by endocytosis in which virions are captured by pitlike regions on the surface of the cell and enter the cytoplasm within a membranous vesicle (➤ Figure 10.15). Enveloped viruses may fuse their envelope with the host's plasma membrane or enter by endocytosis. In the latter case the envelope fuses with the vesicle membrane.

Once the animal virus enters the host cell's cytoplasm, the viral genome must be separated from its protein coat (released) through a process called **uncoating.** Naked viruses are uncoated by proteolytic enzymes from host cells or from the viruses themselves. The uncoating of viruses such as the poxviruses is completed by a specific enzyme that is encoded by viral DNA and formed soon after infection. Polioviruses begin uncoating even before penetration is complete.

Synthesis

The synthesis of new genetic material and proteins depends on the nature of the infecting virus.

Synthesis in DNA Animal Viruses   Generally, DNA animal viruses replicate their DNA in the host cell nucleus with the aid of viral enzymes and synthesize their capsid and other proteins in the cytoplasm by using host cell enzymes. The new viral proteins move to the nucleus, where they combine with the new viral DNA to form virions (➤ Figure 10.12). This pattern is typical of adenoviruses, hepadnaviruses, herpesviruses, and papovaviruses. Poxviruses are the only exception; their parts are synthesized in the host cell's cytoplasm.

➤ Figure 10.15   **Animal virus penetration of host cells.** Many naked virions adhere to the cell surface and become trapped in pits of the cell membrane. These pits invaginate to form separate cytoplasmic vesicles. In the electron micrograph, coronaviruses are being taken into the cytoplasm of a host cell.

In dsDNA viruses, replication proceeds in a complex series of steps designated as *early* and *late* transcription and translation. The early events take place before the synthesis of viral DNA and result in the production of the enzymes and other proteins necessary for viral DNA replication. The late events occur after the synthesis of viral DNA and result in the production of structural proteins needed for building new capsids. Compared with bacteriophage replication, synthesis in animal virus replication can take much longer. The capsids of the herpesviruses, for example, contain so many proteins that their synthesis requires 8 to 16 hours.

Some viruses, such as the adenoviruses, contain only ssDNA. Before viral replication can be initiated, the viral DNA must be copied, forming a dsDNA viral genome.

Synthesis in RNA Animal Viruses   Synthesis among RNA animal viruses takes place in a greater variety of ways than is found among DNA animal viruses. In RNA viruses such as the picornaviruses, the (+) sense RNA acts as mRNA, and viral proteins are made immediately after penetration and uncoating (➤ Figure 10.13). The nucleus of the host cell is not involved. Viral proteins also play key roles in the synthesis of these viruses. One protein inhibits synthetic activities of the host cell. For synthesis an enzyme uses the (+) sense RNA as a template to make a (−) sense RNA. This (−) sense RNA in turn acts as a template RNA to replicate many (+) sense RNA molecules for virion formation.

In the retroviruses, such as HIV, the two copies of (+) sense RNA do not act as mRNA. Rather, they are transcribed into ssDNA with the help of reverse transcriptase (➤ Figure 10.13). The ssDNA then is replicated through complementary base pairing to make dsDNA molecules. Once in the cell nucleus, this molecule inserts itself as a provirus into a host cell chromosome. The provirus can remain there for an indefinite period of time. When infected cells divide, the provirus is replicated along with the rest of the host chromosome. Thus, the viral genetic information is passed to progeny host cells.

Unlike prophages, however, the provirus cannot be excised. If an event occurs that activates the provirus, its genes are expressed; that is, the genes are used to make viral mRNA, which directs synthesis of viral proteins. Full-length (+) sense RNA molecules also are transcribed from the prophage. Two copies of the (+) sense RNA are packaged into each virion.

In (−) sense RNA animal viruses, such as the viruses causing measles and influenza A, a packaged transcriptase uses the (−) sense RNA to make (+) sense RNA molecules (mRNA). Prior to assembly, new (−) sense RNA is made from (+) sense RNA templates. The process is essentially the same regardless of whether the viral RNA is in one segment (measles) or in many segments (influenza A).

In the reoviruses the dsRNA codes for several viral proteins. Each strand of the dsRNA acts as a template for its partner. Like DNA replication, RNA replication is semiconservative, so the molecules produced have one strand of old RNA and one strand of new RNA. These viruses have a double-walled capsid that is never completely removed, and replication takes place within the capsid.

Maturation

Once an abundance of viral nucleic acid, enzymes, and other proteins have been synthesized, assembly of components into complete virions starts. This step constitutes maturation or assembly of progeny viruses. The cellular site of maturation varies depending on the virus type. For example, human adenovirus nucleocapsids are assembled in the cell nucleus (➤ Figure 10.12), whereas viruses such as HIV are assembled at the inner surface of the host cell's plasma membrane. The poxviruses, polioviruses, and picornaviruses are assembled in the cytoplasm.

Maturation of enveloped viruses is a longer and more complex process than that of most bacteriophages. As we have seen, both the infecting virus and nucleic acids and enzymes made in the host cell participate in synthesizing components. Among the components destined for the progeny viruses, the proteins and glycoproteins are coded by the viral genome; envelope lipids and glycoproteins are synthesized by host cell enzymes and are present in the host plasma membrane. If the virus is to have an envelope, the virion is not complete until it buds through a host membrane—either the nuclear, endoplasmic reticulum, Golgi, or plasma membrane—depending on the specific virus (➤ Figure 10.12).

Release

The budding of new virions through a membrane may or may not kill the host cell. Human adenoviruses, for example, bud from the host cell in a controlled manner. This *shedding* of new virions does not lyse the host cells. Other types of animal viruses kill the host cell. When an infected animal cell is filled with progeny virions, the plasma membrane lyses and the progeny are released. Lysis of cells often produces the clinical symptoms of the infection or disease. The herpesviruses that cause cold sores and the poxviruses destroy skin cells as a result of virion release. And the polioviruses destroy nerve cells during the release process.

✓ List in correct order the five steps in viral replication.

✓ How do these five stages differ between bacteriophages and animal viruses?

✓ Compare lysogeny with the lytic cycle in bacteriophages.

Latent Viral Infections

Many individuals experience the reoccurrence of skin eruptions commonly called cold-sores or fever blisters. These are caused by the herpes simplex virus, a member of the herpesviruses. As we saw earlier, these are dsDNA viruses that can exhibit a lytic cycle. They can also remain latent within the cells of the host organism throughout the individual's life—not in the skin cells we associate them with, but in the nerve cells. When activated, whether by a cold or fever or by stress or immunosuppression, they once again replicate resulting in cell lysis.

The ability to become latent is held by all of the herpesviuses. Another herpesvirus, that's responsible for chickenpox, can also remain dormant within the central nervous system. When it becomes activated, usually due to changes in cell-mediated immunity, the virus causes a rash to form along the nerve where it lay latent. This reactivation is known as shingles. Many individuals carry these viruses throughout their lives, never exhibiting any symptoms.

Culturing of Animal Viruses

Development of Culturing Methods

Initially, if a virologist wanted to study viruses, the viruses had to be grown in whole animals. This made it difficult to observe specific effects of the viruses at the cellular level. In the 1930s virologists discovered that embryonated (intact, fertilized) chicken eggs could be used to grow herpesviruses, poxviruses, and influenza viruses. Although the chick embryo is simpler in organization than a whole mouse or rabbit, it is still a complex organism. The use of embryos did not completely solve the problem of studying cellular effects caused by viruses. Another problem was that bacteria also grow well in embryos, and the effects of viruses often could not be determined accurately in bacterially contaminated embryos. Virology progressed slowly during these years until techniques for growing viruses in cultures improved.

Two discoveries greatly enhanced the usefulness of cell cultures for virologists and other scientists. First, the discovery and use of antibiotics made it possible to prevent bacterial contamination. Second, biologists found that proteolytic enzymes, particularly trypsin, can free animal cells from the surrounding tissues without injuring the freed cells. After the cells are washed, they are counted and then dispensed into plastic flasks, tubes, or Petri dishes. Cells in such suspensions will attach to the plastic surface, multiply, and spread to form sheets one cell thick, called **monolayers.** These monolayers can be subcultured. **Subculturing** is the process by which cells from an existing culture are transferred to new containers with fresh nutrient media. A large number of separate subcultures can be made from a single tissue sample, thereby assuring a reasonably homogeneous set of cultures with which to study viral effects.

The term **tissue culture** remains in widespread usage to describe the preceding technique, although the term **cell culture** is perhaps more accurate. Today, the majority of cultured cells are in the form of monolayers grown from enzymatically dispersed cells. With a wide variety of cell cultures available and with antibiotics to control contamination, virology entered its "Golden Age." In the 1950s and 1960s, more than 400 viruses were isolated and characterized. Although new viruses are still being discovered, emphasis is now on characterizing the viruses in more detail and on determining the precise steps in viral infection and viral replication.

Plaque assays similar to those used to study phages can be used for animal viruses. For example, cultures of susceptible human cells are grown in cell monolayers and then inoculated with viruses. If the viruses lyse cells, several rounds of infection will produce plaques.

➤ Figure 10.16 **A culture of diploid fibroblasts.** These immature cells, derived from fetal tissue, multiply rapidly and provide an ideal cell culture for many viruses.

➤ Figure 10.17 **Viral culture in eggs.** Some viruses, such as influenza viruses, are grown in embryonated chicken eggs.

Types of Cell Cultures

Three basic types of cell cultures are widely used in clinical and research virology: (1) primary cell cultures, (2) diploid fibroblast strains, and (3) continuous cell lines. **Primary cell cultures** come directly from the animal and are not subcultured. The younger the source animal, the longer the cells will survive in culture. They typically consist of a mixture of cell types, such as muscle and epithelial cells. Although such cells usually do not divide more than a few times, they support growth of a wide variety of viruses.

If primary cell cultures are repeatedly subcultured, one cell type will become dominant, and the culture is called a **cell strain.** In cell strains all the cells are genetically identical to one another. They can be subcultured for several generations with only a very small likelihood that changes in the cells themselves will interfere with the determination of viral effects.

Among the most widely used cell strains are **diploid fibroblast strains** (➤ Figure 10.16). *Fibroblasts* are immature cells that produce collagen and other fibers as well as the substance of connective tissues, such as the dermis of the skin. Derived from fetal tissues, these strains retain the fetal capacity for rapid, repeated cell division. Such strains support growth of a wide range of viruses and are usually free of contaminating viruses often found in cell strains from mature animals. For this reason they are used in making viral vaccines.

The third type of cell culture in extensive use is the continuous cell line. A **continuous cell line** consists of cells that will reproduce for an extended number of generations. The most famous of such cultures is the HeLa cell line, which has been maintained and grown in culture since 1951 and has been used by many researchers worldwide. The original cells of the HeLa cell line came from a woman with cervical cancer and are named from the first letters of her name. In fact, many of the early continuous cell lines used malignant cells because of their capacity for rapid growth. Such immortal cell lines grow in the laboratory without aging, divide rapidly and repeatedly, and have simpler nutritional needs than normal cells. The HeLa cell line, for example, contains two viral genes necessary for its own immortality. Immortal cell lines are heteroploid (have different numbers of chromosomes) and are therefore genetically diverse.

Cell cultures have largely replaced animals and embryonated eggs for studies in animal virology. Yet, the embryonated chicken egg remains one of the best host systems for influenza A viruses (➤ Figure 10.17). In addition, young albino Swiss mice are still used to culture *arboviruses* (*ar*thropod*bo*rne viruses), and other mammalian cell lines—as well as mosquito cell lines—have been used for some time.

The Cytopathic Effect

The visible effect viruses have on cells is called the **cytopathic effect** (CPE). Cells in culture show several common effects, including changes in cell shape and detachment from adjacent cells or the culture container (➤ Figure 10.18). However, CPE can be so distinctive that an experienced virologist often can use it to make a preliminary identification of the infecting virus. For example, human adenoviruses and herpesviruses cause infected cells to swell because of fluid accumulation, whereas picornaviruses arrest cell functions when they enter and lyse cells when they leave. The paramyxoviruses cause adjacent cells in culture to fuse, forming giant, multinucleate cells called **syncytia** (sin-sish′e-a; singular: *syncytium*). Syncytia can contain 4 to 100 nuclei in a common cytoplasm. Another type of CPE produced by some viruses is *transformation:* the conversion of normal cells into malignant ones, which we discuss later in this chapter.

Viruses and Teratogenesis

Teratogenesis is the induction of defects during embryonic development. A **teratogen** (ter′a-to-jen) is a drug or other agent that induces such defects. Certain viruses are known to act as teratogens and can be transmitted across the placenta and infect the fetus. The earlier in pregnancy the embryo is infected, the more extensive the damage is likely to be. During the early stages of embryologic development, when an organ or body system may be represented by only a few cells, viral damage to those cells can interfere with the development of that organ

(a)

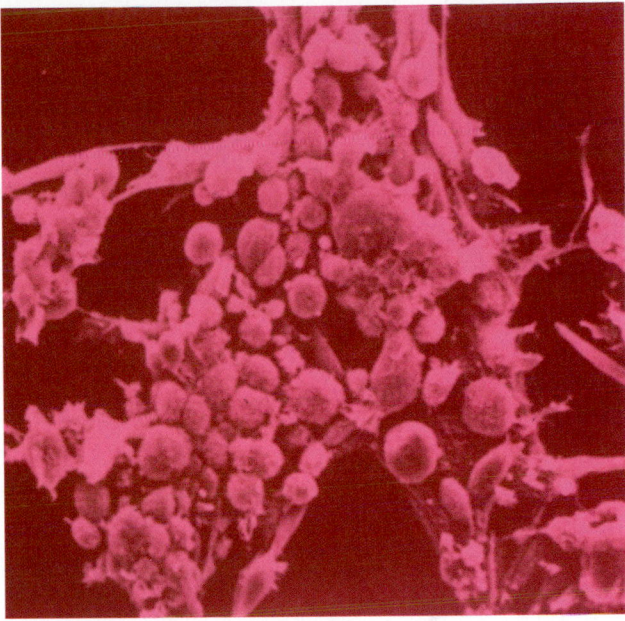

(b)

> Figure 10.18   **Viral transformation of cells.** (a) Normal and (b) transformed (malignant) cells in culture. Such transformation is an example of a cytopathic effect (CPE) caused by infection with the Rous sarcoma virus (RSV). In the transformed state, the cells become rounded and do not adhere to the culture container.

or body system. Viral infections occurring later in development may damage fewer cells and thus have a proportionately smaller effect. That is because, by then, the total cell population in the fetus has greatly increased, and each organ or body system consists of thousands of cells.

Three human viruses—cytomegalovirus (CMV), herpes simplex virus (HSV) types 1 and 2, and rubella—account for a large number of teratogenic effects. Cytomegalovirus (CMV) infections are found in about 1 percent of live births; of those, about 1 in 10 eventually die from the CMV infection. Most of the defects are neurological, and the children have varying degrees of mental retardation. Some also have enlarged spleens, liver damage, and jaundice. HSV infections usually are acquired at or shortly after birth. Infections acquired before birth are rare. In cases of disseminated infections (those that spread through the body), some infants die, and survivors have permanent damage to the eyes and central nervous system.

Rubella virus infections in the mother during the first 4 months of pregnancy are most likely to result in fetal defects referred to as the "rubella syndrome." These defects include deafness, damage to other sense organs, heart and other circulatory defects, and mental retardation. The degree of impairment is highly variable. Some children adapt to their disabilities and live productive lives; in other cases the fetus is so impaired that death and natural abortion occur. Congenital rubella is discussed in Chapter 19.

A series of blood tests often referred to as the **TORCH series** is sometimes used to identify teratogenic diseases in pregnant women and newborn infants. These tests detect antibodies made against _T_oxoplasma (a protozoan), _o_ther disease-causing viruses (including the hepatitis B virus and the varicella, or chickenpox, virus), _r_ubella virus, _C_MV, and _H_SV.

Viruslike Agents: Viroids and Prions

Viruses represent the smallest microbes that, in most cases, have the genetic information to produce new virions in a host cell. The exceptions are those viruses that do not have all their own genetic information to produce new virions. We mentioned earlier that viruses such as the dependoviruses (family Parvoviridae) must use a helper virus to supply the necessary compo-

Some slow-growing viruses, such as cytomegaloviruses, rubella virus, and some adenoviruses, may not produce obvious CPE for 1–4 weeks.

Cytomegalovirus infection during pregnancy is now the major viral cause of congenital abnormalities in the newborn.

nents to produce more virions. However, there are even smaller infectious agents that can cause disease (Table 10.6).

Viroids

In 1971 the plant pathologist T. O. Diener described a new type of infectious agent. He was studying potato tuber spindle disease, which was thought to be caused by a virus. However, no virions could be detected. Rather, Diener discovered molecules of RNA in the nuclei of diseased plant cells (➤ Figure 10.19a). He proposed the concept of a **viroid,** an infectious RNA particle smaller than a virus. Since then viroids have been found to differ from viruses in six ways:

1. Each viroid consists of a single circular RNA molecule of low molecular weight.
2. Viroids exist inside cells as particles of RNA without capsids or envelopes.
3. Unlike viruses such as the parvoviruses, viroids do not require a helper virus.
4. Viroid RNA does not produce proteins.
5. Unlike virus RNA, which may be copied in the host cell's cytoplasm or nucleus, viroid RNA is always copied in the host cell nucleus.
6. Viroid particles are not apparent in infected tissues without the use of special techniques to identify nucleotide sequences in the RNA.

Viroids must disrupt host cell metabolism in some way, but because no protein products are produced, it is not clear how viroids and their RNA cause disease. They may interfere with the cell's ability to process mRNA molecules. Without mature mRNA molecules, proteins cannot be synthesized. If so, cell metabolism would be so disturbed that cell death could result. Although some viroids cause no apparent effect or only mild pathogenic effects in the host, other viroids are known to cause several lethal plant diseases, such as potato spindle tuber disease, Chrysanthemum stunt disease, cucumber pale fruit disease, and tomato apical stunt disease (➤ Figure 10.19b). None of these diseases were recognized before 1922 despite centuries of intense cultivation of these crops. Several other diseases have been recognized recently. Some scientists believe that while isolated plants may have contained viroids for an unknown number of years, modern agricultural methods, such as growing large numbers of the same plant

➤ Table 10.6	Comparison of viruses, viroids, and prions		
	Virus	**Viroid**	**Prion**
Nucleic acid	+ (ssDNA, dsDNA, ssRNA, or dsRNA)	+ (ssRNA)	−
Presence of capsid or envelope	+	−	−
Presence of protein	+	−	+
Need for helper viruses	+/− (Needed by some of the smaller viruses such as the parvoviruses)		
Viewed by	Electron microscopy	Nucleotide sequence identification	Host cell damage
Effected by heat and protein denaturing agents	+	−	−
Effected by radiation of enzymes that digest DNA or RNA	+	+	−
Host	Bacteria, animals, or plants	Plants	Mammals

(a)

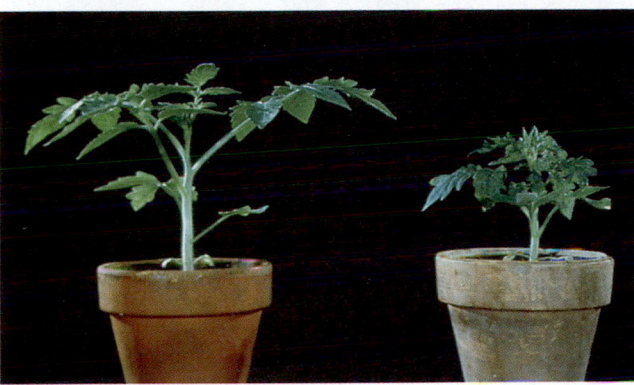

(b)

➢ Figure 10.19 **Viroids and their effects.** (a) Viroid particles that cause potato spindle tuber disease (shown as yellow rods in this artist's rendition of an electron micrograph) are very short pieces of RNA containing only 300 to 400 nucleotides. The much larger (blue and purple) strand is DNA from a T7 bacteriophage. Such comparisons make it easy to see how viroids were overlooked for many years. (b) The tomato plant on the left is normal; the one on the right is infected with a viroid causing tomato apical stunt disease.

in close association and the use of machinery for harvesting, may have allowed viroid diseases to spread, allowing observers to recognize them. The viroid may have entered crop plants from unknown wild plants, an idea supported by the observation that the first viroid-infected crop plants appear along the margins of fields that abut wild plots. No viroid is presently known to infect animals, but there is no reason to suppose they cannot.

At least two hypotheses have been proposed to account for the origin of viroids. One suggests that they originated early in precellular evolution when the primary genetic material probably consisted of RNA. A second suggests that they are relatively new infectious agents that represent the most extreme example of parasitism.

Prions

In the 1920s several cases of a slow but progressive dementing illness in humans were observed independently by Hans Gerhard Creutzfeldt and Alfons Maria Jakob. The disease, now called *Creutzfeldt–Jakob disease* (CJD), is characterized by mental degeneration, loss of motor function, and eventual death. Since that time several similar neurological degenerative diseases have been described (Chapter 23). One is *kuru,* which caused loss of voluntary motor control and eventual death of natives of New Guinea. These deaths were attributed to an infective agent transmitted as a result of cannibalism. In other animals, *scrapie* in sheep and *bovine spongiform encephalopathy* (BSE)—commonly called mad cow disease—in dairy cattle have been observed to cause slow loss of neuronal function that leads to death.

Although some recent research into these similar diseases indicates that they may be caused by viruses, other evidence points to a different type of infective agent. The infective agent may be an exceedingly small *pro*teinaceous *in*fectious particle. In 1982 Stanley Prusiner proposed that such an infectious particle be called a **prion** (pre′on). In 1987, Prusiner received both the

➤ Figure 10.20 **Protein structure model of the two forms of the prion protein (PrP).** The protein helices are represented as spiral ribbons. (a) Harmless form, (b) harmful form.

(a)   **(b)**

Nobel Prize in medicine and the Columbia University Louisa Gross Horwitz Prize for his work with prions. Prions have the following characteristics:

1. Prions are resistant to inactivation by heating to 90°C, which will inactivate viruses.
2. Prion infection is not sensitive to radiation treatment that damages virus genomes.
3. Prions are not destroyed by enzymes that digest DNA or RNA.
4. Prions are sensitive to protein denaturing agents, such as phenol and urea.
5. Prions have direct pairing of amino acids.

Prusiner's research and that of others suggest that prions are normal proteins that become folded incorrectly, possibly as a result of a mutation (➤ Figure 10.20). The harmless, normal proteins are found on the plasma membrane of many mammalian cells, especially brain cells. The prion proteins (*PrP*) are thought to stick together inside cells, forming small fibers, or fibrils. Because the fibrils cannot be organized in the plasma membrane correctly, such aggregations eventually kill the cell.

The most urgent question is to determine how a prion-caused disease spreads. Researchers believe that prion diseases are spread by the prions' causing other copies of the normal protein to fold improperly. In an outbreak of mad cow disease early in the 1990s, the infectious prion originally came from a protein supplement in the feed. This supplement included by-products from scrapie-infected sheep! In fact, experiments show that when mice are injected with prion extracts, disease results.

✓ What are the three most important types of cell culture used for growing viruses?

✓ How are cell lines made immortal?

✓ Define cytopathic effect (CPE), sycytia, and transformation.

✓ Give two examples of viral teratogenesis.

✓ Compare viruses, viroids, and prions.

Viruses and Cancer

Cancer is known as a set of diseases that perturb the normal behavior and functioning of cells. We can define **cancer** as an uncontrolled, invasive growth of abnormal cells—in other words, cancer cells divide repeatedly. In many cases they cannot stop dividing; the result is a **neoplasm,** or localized accumulation of cells known as a **tumor.** A neoplasm can be **benign**—a noncancerous growth. But if the cells invade and interfere with the functioning of surrounding normal tissue, the tumor is **malignant.** Malignant tumors and their cells can **metastasize,** or spread, to other tissues in the body.

That viruses could cause some cancers in animals was discovered in 1911 by F. Peyton Rous. He showed that certain *sarcomas* (neoplasms of connective tissue) in chickens were caused by a virus, named the *Rous sarcoma virus* (RSV). Therefore, it was not surprising to discover that viruses can be associated with cancer in humans as well. Although most human cancers arise from genetic mutations, cellular damages from environmental chemicals, or both, epidemiologists estimate that about 15 percent of human cancers arise from viral infections.

Human Cancer Viruses

After many years of research and testing, we now know of at least six viruses that are associated with human cancers. There probably are many more yet to be identified.

The Epstein–Barr virus (EBV) perhaps is the best understood of the human cancer viruses. This DNA virus is a herpesvirus that was first discovered in African children suffering from Burkitt's lymphoma, a malignant tumor that causes swelling and eventual destruction of the jaw. In fact, evidence points to three other tumors also associated with EBV.

Several of the human papillomaviruses (HPV) have shown a strong correlation with some human cancers. Although some of these DNA viruses cause only benign warts, other types (HPV-8 and HPV-16) lead to a *carcinoma* (neoplasm of epithelial tissues) of the uterine cervix. Another potential cancer-causing DNA virus is hepatitis B virus (HBV). It causes inflammation of the liver and can lead to liver cancer. *Kaposi's sarcoma,* a cancer of the endothelial cells of the blood vessels or lymphatic system, is thought to be associated with a herpesvirus.

The major human cancer viruses discovered so far are dsDNA viruses. However, some (+) sense RNA viruses, specifically the retroviruses, are also associated with cancers. HTLV-I causes *adult T cell leukemia/lymphoma.*

How Cancer Viruses Cause Cancer

Like bacteriophages, some animal viruses that infect animal cells often cause cell death through cell lysis. Other animal viruses can infect cells and form proviruses. In some cases these infections result in physical and genetic changes to the host cells—the CPE discussed earlier. For example, RSV causes cells in culture to detach themselves from the culture flask and round up (➤ Figure 10.18). In the case of **DNA tumor viruses,** which can exist as proviruses, the major CPE is the uncontrollable division of the infected cells. This process, called **neoplastic transformation,** is typical of DNA tumor viruses. Many insert all or part of their DNA at random sites into the host DNA. However, only a few of these viral genes are necessary for transformation.

The papillomaviruses (family Papovaviridae) that cause human cancers infect cells, but their viral DNA remains free in the cytoplasm of the host (➤ Figure 10.21). A few genes of the papillomavirus are active so that the virus can replicate with each cell division. Should the viral DNA accidentally integrate into the host cell DNA, unregulated replication of viral proteins can occur. These proteins cause host cells to divide uncontrollably. Some of these viral proteins block the effects of tumor-suppressor genes, which prevent uncontrolled cell divisions. Without the products of these genes, the host experiences uncontrolled cell divisions—and a tumor develops.

Many of the retroviruses are **RNA tumor viruses.** In fact, they are the only RNA viruses known to cause cancer. Recall that retroviruses use their own reverse transcriptase to transcribe (+) sense RNA into DNA that then integrates as a provirus into the host chromosome. The provirus of HTLV-I codes for proteins that transform the host cells into neoplastic ones. Infection also leads to the production of new virions by budding, which does not kill the infected cell. Thus, RNA tumor viruses can continue to infect other uninfected cells or sex cells. In the latter case, the presence of virus particles ensures transmission of virions to offspring.

Oncogenes

The proteins produced by tumor viruses that cause uncontrolled host cell division come from segments of DNA called **oncogenes** (*onco,* Greek for "mass"). In DNA tumor-causing viruses, not only do oncogenes cause a neoplasm, they also contain the information for synthesizing viral proteins needed for viral replication. The oncogenes in RNA tumor viruses are quite different. Virologists and cell biologists have shown that some RNA tumor viruses pick up "extra" genes from normal host cells during viral replication. These genes, which are similar to oncogenes, are called proto-oncogenes. A **proto-oncogene** is a normal gene that, when under the control of a virus, can cause uncontrolled cell division; that is, it can act as an oncogene. Such oncogenes carried by these viruses are not needed for virus replication.

Many oncogenes have been discovered in oncogenic viruses, and most code for information leading to unlimited cell divisions. Such oncogenes are mutant genes containing deletions or substitutions. ➡ (Chapter 7, p. 183) These mutations cause structural changes in the proteins for which the genes code. Such oncogenes work in one of two ways: (1) The product of the oncogene can disrupt normal cell function, leading to cell divisions. (2) The oncogene is controlled by viral regulators near the site of their integration into the host cell's chromosome. These regulators "turn on" the gene so normal protein is made—but in excessive amounts or at the wrong time in the host cell's life. Again, excessive cell divisions occur. The discovery of oncogenes in viruses has had a major impact on our understanding of cancer. Although there is still much to be learned about cancer in humans, perhaps in the future effective antiviral drugs will prevent virus-induced cancers.

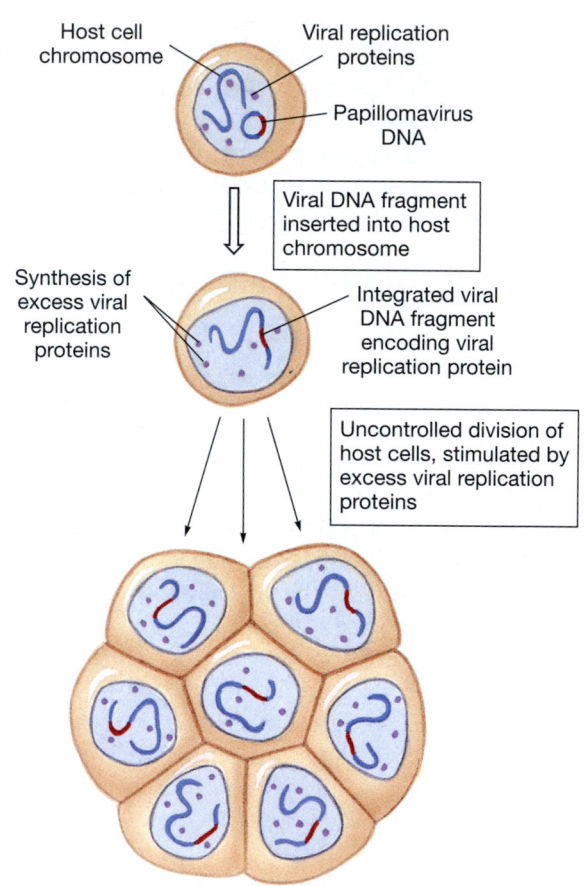

➤ Figure 10.21   **Malignant tumor formation.** This particular tumor is caused by a papillomavirus (DNA tumor virus). Integration of the provirus causes synthesis of viral replication proteins that promote host cell divisions, leading to cancer.

Retracing Our Steps

General Characteristics of Viruses

What Are Viruses?

- **Viruses** are submicroscopic **obligate intracellular parasites**—they replicate only inside a living host cell.

Components of Viruses

- Viruses consist of a nucleic acid core and a protein **capsid.** Some viruses also have a membranous **envelope.**
- Viral genetic information is contained in either DNA or RNA—not both.
- Capsids are made of subunits called **capsomeres.**
- A viral capsid and genome form a **nucleocapsid.** Such viruses are called **naked viruses;** those with a nucleocapsid surrounded by an envelope are **enveloped viruses.**

Sizes and Shapes

- Viruses have polyhedral, helical, binal, bullet, or complex shapes and vary in size from 20 to 300 nm in diameter.

Host Range and Specificity of Viruses

- Viruses vary in **host range** and **viral specificity.** Many viruses infect a specific kind of cell in a single host species; others infect several kinds of cells, several hosts, or both.

Classification of Viruses

- Viruses are classified by the nucleic acid (DNA or RNA) they contain, other chemical and physical properties, their mode of replication, shape, and host range. Some of these characteristics are summarized in Tables 10.1 and 10.2.
- Similar viruses are grouped into genera, and genera are grouped into families. Viruses that share the same genome and relationships with organisms generally constitute a viral species.

RNA Viruses

- Among the (+) sense RNA virus families are the Picornaviridae, which include the poliovirus, the hepatitis A virus and the rhinoviruses; the Togaviridae, which include the virus that causes rubella; the Flaviviridae, which include the yellow fever virus; and the Retroviridae, which cause some cancers and AIDS. The (−) sense RNA viruses include the Paramyxoviridae, which cause measles, mumps, and several respiratory disorders; the Rhabdoviridae, one of which causes rabies; the Orthomyxoviridae, which include the influenza viruses; the Filoviridae, which cause Marburg and Ebola diseases; the Arenaviridae, which cause Lassa fever; and the Bunyaviridae, one of which causes hantavirus pulmonary syndrome. The double-stranded (ds) RNA virus families include the Reoviridae, which cause a variety of upper respiratory and gastrointestinal infections.

DNA Viruses

- The dsDNA virus families include the Adenoviridae, some of which cause respiratory infections; the Herpesviridae, which cause oral and genital herpes, chicken pox, shingles, and infectious mononucleosis; the Poxviridae, which cause smallpox and similar, milder infections. The Papovaviridae cause warts; some papovaviruses are associated with certain cancers. The Hepadnaviridae cause human hepatitis B infection, and the Parvoviridae cause relatively rare human infections.

Viral Replication

General Characteristics of Replication

- Viruses generally go through five steps in the replication process: **adsorption, penetration, synthesis, maturation,** and **release.** These steps are somewhat different in bacteriophages and animal viruses.

Replication of Bacteriophages

- **Bacteriophage** replication has been thoroughly studied in T-even phages, which are **virulent phages.**
- T-even phages have recognition factors that attach to specific receptors on bacterial cell walls during adsorption. Enzymes weaken the bacterial wall so viral nucleic acid can penetrate it.
- During biosynthesis viral DNA directs the making of viral components.
- In the maturation stage, the viral components are assembled into complete virions.
- Release, the final stage, is facilitated by the enzyme lysozyme. **Burst time** is the time from adsorption to release of progeny virions; **burst size** is the number of phage progeny released from one host cell.
- The growth curve of a phage includes an **eclipse period** (the time following penetration through biosynthesis) and a **latent period** (the time after penetration up to release).
- The number of phages produced in an infection can be determined by counting the number of **plaques** produced on a plate of virus-infected bacteria (**plaque assay**). Each plaque represents a **plaque-forming unit.**
- Phages carrying out these stages of replication, leading to host cell destruction, represent a **lytic cycle** of infection.

Lysogeny

- **Lysogeny,** a stable long-term relationship between certain phages and host bacteria, occurs in **temperate phages.** Temperate phage DNA can exist as a **prophage** or revert through **induction** to a lytic cycle.
- Prophages such as the lambda phage insert into a bacterial chromosome at a specific location.

Replication of Animal Viruses

- On the surface of some viruses, proteins are used for attachment to host plasma membranes during adsorption; animal viruses thus gain entry into the cell. **Uncoating** (loss of the capsid) occurs at the plasma membrane or in the cytoplasm.
- Synthesis and maturation differ in DNA and RNA viruses. In most DNA viruses, DNA is synthesized in an orderly sequence in the nucleus, and proteins are synthesized in the cytoplasm of the host cell. In RNA viruses, RNA can act as a template for protein synthesis, for making mRNA, or for making DNA by reverse transcription. Virions are assembled in the cell; sometimes viral DNA is incorporated as a **provirus** into the host cell's chromosome.
- Release can occur through direct lysis of the host cell or by budding through the host membrane.

Culturing of Animal Viruses

Development of Culturing Methods

• The discovery of antibiotics to prevent bacterial contamination of chicken embryos and **cell cultures** and the use of trypsin to separate cells in culture systems into **monolayers** provided an important impetus to the study of virology.

Types of Cell Cultures

• **Primary cell cultures** come directly from animals and are not subcultured.

• All the cells in a **cell strain,** which are derived from subcultured primary cell cultures, are very similar. **Diploid fibroblast strains** from primary cultures of fetal tissues produce stable cultures that can be maintained for years; they are used to produce vaccines.

• **Continuous cell lines,** usually derived from cancer cells, grow in the laboratory without aging, can divide repeatedly, have greatly reduced nutritional needs, and display heteroploidy.

• The visible effects that viruses produce in infected host cells are collectively called the **cytopathic effect.**

Viruses and Teratogenesis

• A **teratogen** is an agent that induces defects during embryonic development.

• Viruses can act as teratogens by crossing the placenta and infecting embryonic cells. The earlier in pregnancy an infection occurs, the more extensive damage is likely to be.

• The rubella virus can be responsible for the death of fetuses and severe birth defects in others; cytomegaloviruses and occasionally herpesviruses also act as teratogens.

Viruslike Agents: Viroids and Prions

Viroids

• **Viroids** are very different from viruses; each viroid is solely a small RNA molecule.

• Viroids may cause plant diseases by interfering with mRNA processing.

Prions

• **Prions** are infectious particles made of protein. Research suggests that prions are normal proteins that become folded incorrectly.

• Prions cause neurological degenerative diseases, including Creutzfeldt-Jakob disease and kuru.

Viruses and Cancer

• **Cancer** is generally an uncontrolled and/or invasive growth of abnormal cells.

Human Cancer Viruses

• **Tumors,** or **neoplasms,** are **benign** (noncancerous) or **malignant** (cancerous). Malignant tumors spread by **metastasis.**

• Several animal viruses are thought to cause some forms of cancer, including the Epstein–Barr virus, certain human papillomaviruses, the hepatitis B virus, and some retroviruses, such as HTLV-I.

How Cancer Viruses Cause Cancer

• **DNA tumor viruses** contain viral genes whose protein products disrupt the activities of normal host cell proteins that control cell division.

• **RNA tumor viruses** contain viral genes used for **neoplastic transformation** and viral replication.

Oncogenes

• **Oncogenes** are viral genes that cause host cells to divide uncontrollably.

• **Proto-oncogenes** are normal genes that, when under the control of a virus, act as oncogenes causing uncontrolled cell division.

• Oncogenes in RNA tumor viruses produce proteins in excessive amounts or produce proteins at the wrong times. In either case, infected host cells start uncontrolled cell division.

Terminology Check

adenovirus   *(p. 264)*
adsorption   *(p. 267)*
arenavirus   *(p. 263)*
bacterial lawn   *(p. 269)*
bacteriophage   *(p. 258)*
benign   *(p. 282)*
bunyavirus   *(p. 263)*
burst size   *(p. 269)*
burst time   *(p. 269)*
cancer   *(p. 282)*
capsid   *(p. 256)*
capsomere   *(p. 257)*
cell culture   *(p. 277)*
cell strain   *(p. 278)*
complex virus   *(p. 258)*
continuous cell line   *(p. 278)*
cytopathic effect   *(p. 278)*
diploid fibroblast strain   *(p. 278)*
DNA tumor virus   *(p. 283)*

eclipse period   *(p. 269)*
enterovirus   *(p. 262)*
envelope   *(p. 256)*
enveloped virus   *(p. 257)*
filoviruses   *(p. 263)*
flavivirus   *(p. 263)*
genome   *(p. 256)*
glycoprotein   *(p. 257)*
hepadnavirus   *(p. 265)*
hepatovirus   *(p. 262)*
herpesvirus   *(p. 264)*
host range   *(p. 258)*
induction   *(p. 272)*
latency   *(p. 264)*
latent period   *(p. 269)*
lysogen   *(p. 271)*
lysogenic conversion   *(p. 272)*
lysogenic cycle   *(p. 272)*
lysogeny   *(p. 270)*

lytic cycle   *(p. 269)*
lytic phage   *(p. 269)*
malignant   *(p. 283)*
maturation   *(p. 267)*
metastasize   *(p. 283)*
monolayer   *(p. 277)*
naked virus   *(p. 257)*
negative (−) sense RNA   *(p. 261)*
neoplasm   *(p. 282)*
neoplastic transformation   *(p. 283)*
nucleocapsid   *(p. 257)*
obligate intracellular parasite   *(p. 256)*
oncogene   *(p. 283)*
orthomyxovirus   *(p. 263)*
papovavirus   *(p. 265)*
paramyxovirus   *(p. 263)*
parvovirus   *(p. 265)*
penetration   *(p. 267)*
picornavirus   *(p. 262)*

plaque   (p. 270)
plaque assay   (p. 269)
plaque-forming unit   (p. 270)
positive (+) sense RNA
   (p. 260)
poxvirus   (p. 265)
primary cell culture   (p. 278)
prion   (p. 281)
prophage   (p. 271)
proto-oncogene   (p. 284)
provirus   (p. 263)
release   (p. 267)
reovirus   (p. 264)

replication curve   (p. 269)
replication cycle   (p. 267)
retrovirus   (p. 263)
reverse transcriptase   (p. 263)
rhabdovirus   (p. 263)
rhinovirus   (p. 262)
RNA tumor virus   (p. 283)
spike   (p. 257)
subculturing   (p. 277)
syncytia   (p. 278)
synthesis   (p. 267)
temperate phage   (p. 270)
teratogen   (p. 278)

teratogenesis   (p. 278)
tissue culture   (p. 277)
togavirus   (p. 262)
TORCH series   (p. 279)
tumor   (p. 282)
uncoating   (p. 275)
viral specificity   (p. 258)
viral yield   (p. 269)
virion   (p. 256)
viroid   (p. 280)
virulent (lytic) phage   (p. 269)
virus   (p. 256)

CHAPTER 10 TEN

See You on the Web

Don't forget I am expecting you on the web. The following is a preparedness checklist for your web explorations. If you have questions on any of these items, you can return to the end of Chapter 1 for more details.

- Questions We'll Explore
- Concept Check questions
- Flashcards
- Study Guide

If you are ready, come visit me at

http://www.prenhall.com/black

It will be a journey you'll never forget.

Eukaryotic Microorganisms and Parasites

Working in their labs, often alone and late at night, searching desperately for the agent responsible for the deaths of over 1 billion fish in North Carolina estuaries, some scientists entered their own horror stories. Scientists and lab workers were struck by a toxin produced by *Pfiesteria piscicida,* a dinoflagellate whose numbers can suddenly skyrocket during an algal bloom.

With the fury of an illness straight from the scariest science fiction, this toxin wields a devastating blow to the immune system and higher-level brain functions. As it starts, the eyes blur, there's difficulty catching the breath, then hours of nausea and vomiting followed by unwarranted fits of rage. After coming out of a narcotic-like delirium, the mind seems to be going, and bleeding lesions appear on the skin. The afflicted can't remember his name, can't form a sentence, can't read. Five to seven years later, nerve damage persists. Shortly after exposure to the toxin, 20–40% of the immune system is destroyed and remains impaired for years.

Background Basics

In our survey of microbes, we have devoted significant attention to bacteria of the kingdom Monera and to viruses. However, some members of eukaryotic kingdoms are also of interest to microbiologists and health scientists. The kingdoms Protista and Fungi contain large numbers of microscopic species, some of which supply food and antibiotics, and some of which cause disease. The kingdom Animalia contains helminths that cause disease and arthropods that cause or transmit diseases. Studying the microscopic eukaryotes, as well as the helminths and arthropods, constitutes a significant part of a health scientist's training. Unless health scientists take a course in parasitology, their only opportunity to learn about helminths and arthropods is in conjunction with the study of microscopic infectious agents.

Principles of Parasitology

A **parasite** is an organism that lives at the expense of another organism, called the **host.** Parasites vary in the degree of damage they inflict on their hosts. Although some cause little harm, others cause moderate to severe damage. Parasites that cause disease are called **pathogens. Parasitology** is the study of parasites.

Although few people realize it, among all living forms, there are probably more parasitic than nonparasitic organisms. Many of these parasites are microscopic throughout their life cycle or at some stage of it. Historically, in the development of the science of biology, parasitology came to refer to the study of protozoa, helminths, and arthropods that live at the expense of other organisms. We will use the term *parasite* to refer to these organisms. Strictly speaking, bacteria and viruses that live at the expense of their hosts also are parasites.

The manner in which parasites affect their hosts differs in some respects from that described in earlier chapters for bacteria and viruses. Special terms also are used to describe parasites and their effects. This introduction to parasitology will make discussions of parasites here and in later chapters more meaningful.

Questions We'll Explore

A What is a parasite, and what are the principles of parasitology?

B What are protists, and why are they important?

C How do groups of protists differ?

D What are fungi, and why are they important?

E How do groups of fungi differ?

F What are parasitic helminths, and why are they important?

G How do groups of parasitic helminths differ?

H What are the characteristics of parasitic and vector arthropods?

I How do groups of parasitic and vector arthropods differ?

The Significance of Parasitism

Parasites have been a scourge throughout human history. In fact, even with modern technology to treat and control parasitic diseases, there are more parasitic infections than there are living humans. It has been estimated that among the 60 million people dying each year, fully one-fourth die of parasitic infections or their complications.

Parasites play an important, though negative, role in the worldwide economy. For example, less than half the world's cultivable land is under cultivation, primarily because parasites endemic to (always present in) those lands prevent humans and domesticated animals from inhabiting some of them. As the world population increases, and the need for food with it, cultivation of such lands will become more important. In some inhabited regions, many people are near starvation and severely debilitated by parasites. Furthermore, parasitic infections in wild and domestic animals provide sources of human infection and cause debilitation and death among the animals, thus preventing the raising of cattle and other animals for food. Given the many human problems created by parasites, all citizens—and especially health scientists—need to understand the problems associated with the control and treatment of parasitic diseases.

Parasites in Relation to Their Hosts

Parasites can be divided into **ectoparasites,** such as ticks and lice, which live on the surface of other organisms, and **endoparasites,** such as some protozoa and worms, which live within the bodies of other organisms. Most parasites are **obligate parasites:** They must spend at least some of their life cycle in or on a host. For example, the protozoan that causes malaria invades red blood cells. A few parasites are **facultative parasites:** They normally are free-living, such as some soil fungi, but they can obtain nutrients from a host, as many fungi do when they cause skin infections. Hosts that are invaded by parasites usually lack effective defenses against them, so such diseases can be serious and sometimes fatal.

Parasites are also categorized according to the duration of their association with their hosts. **Permanent parasites,** such as tapeworms, remain in or on a host once they have invaded it. **Temporary parasites,** such as many biting insects, feed on and then leave their hosts. **Accidental parasites** invade an organism other than their normal host. Ticks that ordinarily attach to dogs or to wild animals sometimes attach to humans; the ticks are then accidental parasites. **Hyperparasitism** refers to a parasite itself having parasites. Some mosquitoes, which are temporary parasites, harbor the malaria parasite or other parasites. Such insects serve as **vectors,** or agents of transmission, of many human parasitic diseases.

An organism that transfers a parasite to a new host is a vector. A vector in which the parasite goes through part of its life cycle is a **biological vector.** The malaria mosquito is both a host and a biological vector. A **mechanical vector** is a vector in which the parasite does not go through any part of its life cycle during transit. Flies that carry parasite eggs, bacteria, or viruses from feces to human food are mechanical vectors.

Hosts are classified as **definitive hosts** if they harbor a parasite while it reproduces sexually; they are said to be **intermediate hosts** if they harbor the parasite during some other developmental stages. The mosquito is the definitive host for the malaria parasite because that parasite reproduces sexually in the mosquito; the human is an intermediate host, even though humans suffer greater damage from the parasite. **Reservoir hosts** are infected organisms that make parasites available for transmission to other hosts. Reservoir hosts for human parasitic diseases typically are wild or domestic animals. **Host specificity** refers to the range of different hosts in which a parasite can mature. Some parasites are quite host specific—they mature in only one host. The malaria parasite matures primarily in *Anopheles* mosquitoes. Other parasites can mature in many different hosts. The worm that causes trichinosis can mature in almost any warmblooded animal, but the parasite is most often acquired by humans from pigs through the consumption of inadequately cooked, contaminated pork.

Over thousands of years of evolution, parasites tend to become less injurious to their hosts. Such an arrangement preserves the host so that the parasites are guaranteed a continuous supply of nutrients. A parasite that destroys its host also destroys its own means of support. The adjustment of parasites and hosts to each other is closely related to the host's defense mecha-

A single tapeworm can live for 30 to 35 years. Its pear-shaped head is about 1 to 2 millimeters in diameter, and it may reach a length of 10 meters.

nisms. Many parasites have one or more of the following mechanisms for evading host defense mechanisms:

1. *Encystment,* the formation of an outer covering that protects against unfavorable environmental conditions. These resistant cyst stages also sometimes provide a site for internal reorganization of the organism and cell division, help attach a parasite to a host, or serve to transmit a parasite from one host to another.

2. Changing the parasite's surface antigens (molecules that elicit immunity) faster than the host can make new antibodies (molecules that recognize and attack antigens).

3. Causing the host's immune system to make antibodies that cannot react with the parasite's antigens.

4. Invading host cells, where the parasites are out of reach of host defense mechanisms.

When parasites successfully evade host defenses, they can cause several kinds of damage. All parasites rob their hosts of nutrients. Some take such a large share of nutrients or damage so much surface area of the host's intestines that the host receives too little nourishment. Many parasites cause significant trauma to host tissues. They cause open sores on the skin, destroy cells in tissues and organs, clog and damage blood vessels, and may even cause internal hemorrhages. Parasites that do not evade defense mechanisms sometimes trigger severe inflammatory and immunological reactions. For example, treatment to rid human hosts of some worm infections effectively kills the worms, but toxins from the dead worms cause more tissue damage than do the living parasites. The dog heartworm, *Dirofilaria immitis,* perforates the heart wall and leaves holes in the heart when the worms die and decay. Therefore, it is important for a veterinarian to test all dogs for the presence of heartworms before administering preventive heartworm medication.

A hallmark of many parasites is their reproductive capability. Parasitism, although an easy life once the parasite is established, is a hazardous existence during transfers from one host to another. For example, many parasites that leave the human body through feces die from desiccation (drying out) before they reach another host. If several hosts are required to complete the life cycle, the hazards are greatly multiplied. Consequently, many parasites have exceptional reproductive capacities. Some parasites, such as certain protozoa, undergo **schizogony** (skiz-og'o-ne), or multiple fission, in which one cell gives rise to many cells, all of which are infective. Others, such as various worms, produce large numbers of eggs. Some worms are **hermaphroditic**—that is, one organism has both male and female reproductive systems. In fact, certain worms, such as tapeworms, lack a digestive tract and consist almost exclusively of reproductive systems.

Protists

Characteristics of Protists

The **protists,** members of the kingdom Protista, are a diverse assortment of organisms that share certain common characteristics. Protists are unicellular (though sometimes colonial), eukaryotic organisms with cells that have true nuclei and membrane-enclosed organelles. Although most protists are microscopic, they vary in diameter from 5 μm to 5 mm.

The Importance of Protists

Protists have captured the fancy of biologists since Leeuwenhoek made his first microscopes. In fact, most of the "animalcules" he observed were protists. Like Leeuwenhoek, many people find protists inherently interesting, and biologists have learned much about life processes from protists.

Protists also are important to humans for other reasons. For instance, they are a key part of food chains. Autotrophic protists capture energy from sunlight. Some heterotrophic protists ingest autotrophs and other heterotrophs. Others decompose, or digest, dead organic matter,

Protozoans are so tiny that some develop in the salivary glands of insects.

In the last 25 years, the number of red tides worldwide has increased significantly—possibly linked to increased pollution.

which then can be recycled to living organisms. Protists also serve as food for higher-level consumers. Ultimately, some energy originally captured by protists reaches humans. For example, energy from the sun is transferred to protists, protists are eaten by oysters, and the oysters are eaten by humans.

Protists can be economically beneficial or detrimental. Certain protists have **tests,** or shells, of calcium carbonate. Carbonate shells deposited in great numbers by such protists that lived in ancient oceans formed the white cliffs of Dover and the limestone used in building the pyramids of Egypt. Because different test-forming protists gained prominence during different geological eras, the identification of the protists in rock layers helps determine the age of the rocks. Certain test-forming protists tend to occur in rock layers near petroleum deposits, so geologists looking for oil are pleased to find them. Some autotrophic protists produce toxins that do not harm the oysters that eat the protists, but the accumulated toxins can cause disease or even death in people who subsequently eat the oysters. Oyster beds infected with such protists can cause great economic losses to oyster harvesters. Other autotrophic protists multiply very rapidly in abundant inorganic nutrients and form a "bloom," a thick layer of organisms over a body of water. This process, called **eutrophication** (u″tro-fi-ka′shun), blocks sunlight, killing plants beneath the bloom and causing fish to starve. Microbes that decompose dead plants and animals use large quantities of oxygen, and the lack of oxygen leads to more deaths. Together these events result in great economic losses in the fishing industry.

Finally, some protists are parasitic. They cause debilitation in large numbers of people, and sometimes death, especially in poor countries that lack the resources to eradicate those protists. Parasitic diseases caused by protozoa include amoebic dysentery, malaria, sleeping sickness, leishmaniasis, and toxoplasmosis. Together, these diseases account for severe losses in human productivity, incalculable human misery, and many deaths.

Classification of Protists

Like all groups of living things, the protists display great variation, which provides a basis for dividing the kingdom Protista into sections and phyla. However, taxonomists do not agree about how these classifications should be made. We can accomplish our main purpose of illustrating diversity and avoid taxonomic problems by grouping protists according to the kingdom of macroscopic organisms they most resemble (Table 11.1). Thus, we speak of protists that resemble plants (➤ Figure 11.1), protists that resemble fungi (➤ Figure 11.2), and protists that resemble animals (➤ Figure 11.3).

The Plantlike Protists

The plantlike protists, or algae, have chloroplasts and carry on photosynthesis. They are found in moist, sunny environments. Most have cell walls and one or two flagella, which allow them to move. The **euglenoids** (u-gle′noidz) usually have a single flagellum and a pigmented eyespot called a *stigma*. The stigma may orient flagellar movement so that the organism moves toward light. A typical euglenoid, *Euglena gracilis* (➤ Figure 11.1a), has an elongated, cigar-shaped, flexible body. Instead of a cell wall, it has a **pellicle,** or outer membranous cover.

➤ Table 11.1	Properties of protists	
Group	**Characteristics**	**Examples**
Plantlike protists	Have chloroplasts; live in moist, sunny environments	Euglenoids, diatoms, dinoflagellates
Funguslike protists	Most are saprophytes; may be unicellular or multicellular	Water molds, plasmodial and cellular slime molds
Animal-like protists	Heterotrophs; most are unicellular, most are free-living, but some are commensals or parasites	Mastigophorans, sarcodines, apicomplexans, and ciliates

Red Tides

Certain species of *Gonyaulax* and some other dinoflagellates produce toxins. When these marine organisms appear seasonally in large numbers, they cause a bloom known as a *red tide.* The toxins accumulate in the bodies of shellfish such as oysters and clams that feed on the protists. Although the toxin does not harm the shellfish, it causes paralytic shellfish poisoning in some fish, and in humans who eat the infected shellfish. Even animals as large as dolphins have been killed in large numbers by this toxin. Inhaling air that contains small quantities of the toxin can irritate respiratory membranes, so sensitive individuals should avoid the sea and its products during red tides.

Keep That Head on Your Beer!

Taxonomists disagree on how to classify eukaryotic algae. They are sometimes classified as protists, sometimes as plants, and sometimes divided between the protist and plant kingdoms. Eukaryotic algae should not be confused with blue-green algae (now called cyanobacteria), which are prokaryotes.

Although both eukaryotic and prokaryotic algae are important as producers in many environments, they are generally not of medical significance. (However, cyanobacteria of the genus *Prototheca,* which have lost their chlorophyll, have been reported to cause skin lesions.) Agar, which is of great importance in the microbiology laboratory, is a product extracted from smaller seaweeds (red algae). Some eukaryotic algae, such as kelps (brown algae), are used as food and in the manufacture of products such as cheese spreads, toothpaste, and mayonnaise, to which they add smoothness and spreadability. They also enable beer to retain a foamy "head."

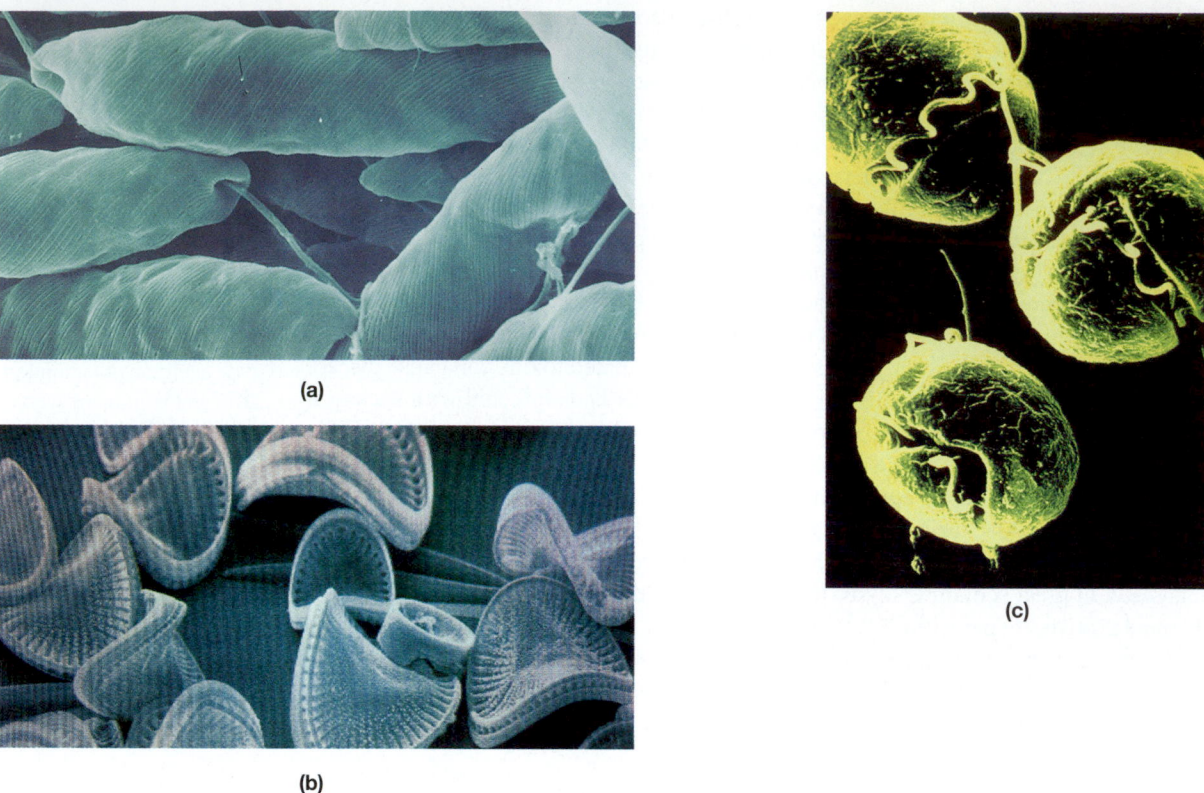

➤ Figure 11.1   **SEMs of representative algae, or plantlike protists.** (a) *Euglena,* a euglenoid. (b) The diatom *Campylodiscus hibernicus* (magnified 250X). (c) *Gonyaulax,* a dinoflagellate that causes red tides.

Euglenoids usually reproduce by binary fission. Most live in fresh water, but a few are found in the soil.

Another group of plantlike protists have other pigments in addition to chlorophyll. These protists usually have cell walls surrounded by a loosely attached secreted test that contains silicon or calcium carbonate. Most reproduce by binary fission. They include the **diatoms** (di´ah-tomz), which lack flagella (➤ Figure 11.1b), and several other groups, which have flagella and are distinguished by their yellow and brown pigments. Diatoms are an especially numerous group and are important as producers in both freshwater and marine environments. Fossil deposits of diatoms, known as diatomaceous earth, are used as filtering agents and abrasives in various industries.

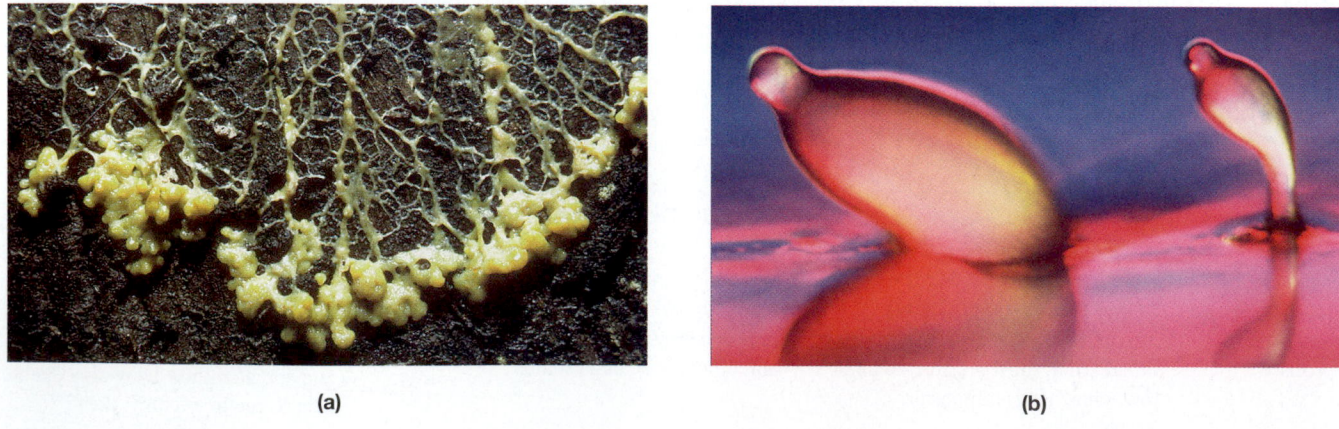

➤ Figure 11.2   **Representative funguslike protists.** (a) A plasmodial slime mold of the genus *Hemitrichia,* on a decaying log. (b) Pseudoplasmodia of a cellular slime mold, *Dictyostelium discoideum.*

(a)   (b)

(c)   (d)

➤ **Figure 11.3   Representative protozoa, or animal-like protists.** (a) *Trichonympha*, a mastigophoran, an endosymbiont from a termite gut. Particles seen inside the body are ingested wood particles. (b) *Amoeba proteus* (350X), a sarcodine, free-living inhabitant of ponds. (c) *Plasmodium vivax* (inside red blood cells), an apicomplexan, one of the parasites that causes malaria. (d) *Paramecium caudatum* (135X), a ciliate.

The **dinoflagellates** (di′no-flaj″el-atz) are plantlike protists that usually have two flagella—one extending behind the organism like a tail, and the other lying in a transverse groove (➤ Figure 11.1c). They are small organisms that may or may not have a cell wall. Some have a *theca*, a tightly affixed secreted layer that typically contains cellulose. Cellulose is an uncommon substance in protists, although it is abundant in plants. Whereas most dinoflagellates have chlorophyll and are capable of carrying on photosynthesis, others are colorless and feed on organic matter. Several dinoflagellates exhibit bioluminescence. The photosynthetic dinoflagellates are second only to the diatoms as producers in marine environments.

The Funguslike Protists

The funguslike protists, or water molds and slime molds, have some characteristics of fungi and some of animals.

Water Molds   The **water molds** and related protists that cause mildew—the **Oomycota**—are sometimes classified as fungi. These molds, mildews, and plant blights produce flagellated spores, called *zoospores,* during asexual reproduction and large motile gametes during sexual reproduction. The most prominent phase of their life cycle consists of diploid cells from the union of gametes. These protists live freely in fresh water or as plant parasites; they cause such diseases as downy mildew on grapes and sugar beets and late blight in potatoes. A member of the Oomycota was responsible for the Irish potato famine in the 1840s. With a few exceptions, water molds are not medically significant to humans. They do, however, cause disease in fish and other aquatic organisms.

Slime Molds   **Slime molds** are commonly found as glistening, viscous masses of slime on rotting logs; they also live in other decaying organic matter or in soil. Most slime molds are **saprophytes** (sap'ro-fitz), or organisms that feed on dead or decaying matter. A few are parasites of algae, fungi, or flowering plants, but not of humans. Slime molds occur as plasmodial slime molds and as cellular slime molds.

Plasmodial slime molds (➤ Figure 11.2a) form a multinucleate, amoeboid mass called a **plasmodium,** which moves about slowly and phagocytizes dead matter. Sometimes a plasmodium stops moving and forms *fruiting bodies*. Each fruiting body develops *sporangia*, sacs that produce spores. When spores are released, they germinate into flagellated gametes. Two gametes fuse, lose their flagella, and form a new plasmodium. As a plasmodium feeds and grows, it can also divide and produce new plasmodia directly.

The **cellular slime molds** (➤ Figure 11.2b) produce pseudoplasmodia, fruiting bodies, and spores with characteristics that are quite different from those of plasmodial slime molds. A **pseudoplasmodium** is a slightly motile aggregation of cells. It produces fruiting bodies, which in turn produce spores. The spores germinate into amoeboid phagocytic cells that divide repeatedly, producing more independent amoeboid cells. Depletion of the food supply causes the cells to aggregate into loosely organized new pseudoplasmodia.

The Animal-like Protists

The animal-like protists, or **protozoa,** are heterotrophic, mostly unicellular organisms, but a few form colonies. Most are free-living. Some are **commensals,** which live in or on other organisms without harming them, and a few are parasites. The parasitic protozoa are of particular interest in the health sciences. Many protozoa live in watery environments and encyst when conditions are not favorable. Some protozoa are protected by a tough outer pellicle. Many are motile and are further classified on the basis of their means of locomotion (➤ Figure 11.3). The protozoa that you will encounter in this book belong to the groups Mastigophora, Sarcodina, Apicomplexa (also known as Sporozoa), or Ciliata (also known as Ciliophora).

Mastigophorans
The **mastigophorans** (mas″ti-gof'or-anz) have flagella. A few species are free-living in either fresh or salt water, but most live in symbiotic relationships with plants or animals. The symbiont *Trichonympha* (➤ Figure 11.3a) lives in the termite gut and contributes enzymes that digest cellulose. Mastigophorans that parasitize humans include members of the genera *Trypanosoma, Leishmania, Giardia,* and *Trichomonas*. Trypanosomes cause African sleeping sickness, leishmanias cause skin lesions or systemic disease with fever, giardias cause diarrhea, and trichomonads cause vaginal inflammation.

Sarcodines
The **sarcodines** are usually amoeboid—they move by means of pseudopodia (➤ Figure 11.3b). ➥ (Chapter 4, p. 96) A few sarcodines have flagella at some stage in their life cycle. They feed mainly on other microorganisms, including other protozoa and small algae. The sarcodines include foraminiferans and radiolarians, which have shells and are found mainly in marine environments, and amoebas, which have no shells and are typically parasites.

Numerous species of amoebas are capable of inhabiting the human intestinal tract. Most form cysts that help them withstand adverse conditions. The more commonly observed genera—*Entamoeba, Dientamoeba, Endolimax,* and *Iodamoeba*—cause amoebic dysenteries of varying degrees of severity. *Entamoeba gingivalis* is found in the mouth. *D. fragilis,* which is unusual in that it has two nuclei and does not form cysts, is found in the large intestine of about 4 percent of the human population. Its means of transmission is unknown. Although usually considered a commensal, it can cause chronic, mild diarrhea.

Apicomplexans
The **apicomplexans** (or sporozoans) are parasitic and immobile (➤ Figure 11.3c). Enzymes present in groups (complexes) of organelles at the tips (apices) of their cells digest their way into host cells, giving the group the name Apicomplexa. These parasites usually have complex life cycles. An important example is the life cycle of the malaria parasite, *Plasmodium,* which requires both a human and a mosquito host (➤ Figure 11.4). (Do not confuse this apicomplexan with the plasmodium form of slime molds.) The parasites, which are present as **sporozoites** (spo-ro-zo'itz) in the salivary glands of an infected mosquito, enter human blood through the mosquito's bite. The sporozoites migrate to the liver and become **merozoites** (meh-ro-zo'itz). After about 10 days, they emerge into the blood, invade red blood cells, and become **trophozoites** (tro-fo-zo'itz). Trophozoites reproduce asexually, producing

People lacking the Duffy blood-group protein are resistant to *Plasmodium vivax* malaria. 90% of West African blacks and 60% of African-Americans lack this molecule.

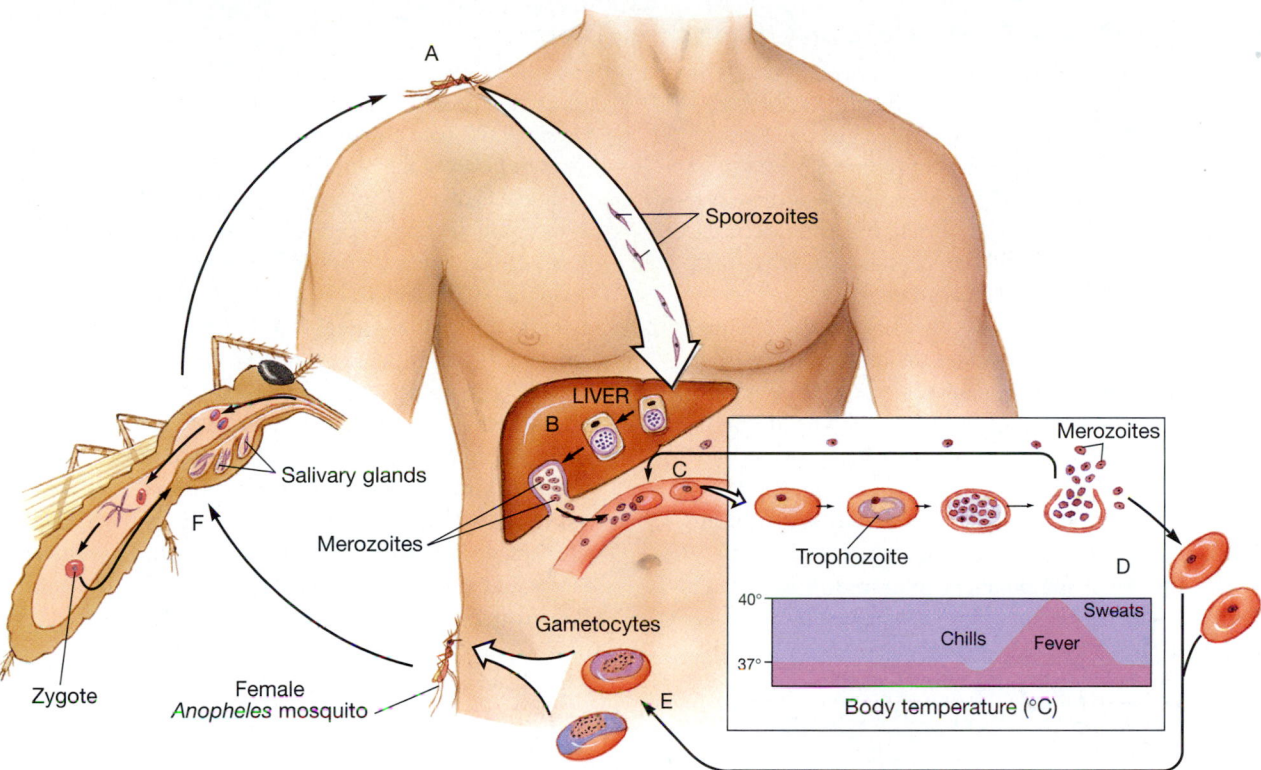

> **Figure 11.4   The life cycle of the malaria parasite *Plasmodium*.** A: Female *Anopheles* mosquito transmits sporozoites from its salivary glands when it bites a human. The sporozoites travel in human blood to the liver. B: In the liver, the sporozoites multiply and become merozoites, which are shed into the bloodstream when liver cells rupture. C: The merozoites enter red blood cells and become trophozoites, which feed and eventually form many more merozoites. D: Merozoites are released by the rupture of the red blood cells, accompanied by chills, high fever (40°C), and sweating. They can then infect other red blood cells. E: After several such asexual cycles, gametocytes (sexual stages) are produced. F: Upon ingestion by a mosquito, the gametocytes form a zygote, which gives rise to more infective sporozoites in the salivary glands. These can then infect other people.

many more merozoites, which are released into the blood by the rupture of red blood cells. Multiplication and release of merozoites is repeated several times during a bout of malaria. Some merozoites enter the sexual reproductive phase and become **gametocytes,** or sex cells. When a mosquito takes a blood meal from an infected human, it also takes in gametocytes, most of which mature and unite to form zygotes in the lining of the mosquito's stomach. Zygotes pass through the stomach wall and produce sporozoites, which eventually make their way to the salivary glands.

Several species of *Plasmodium* cause malaria, and each displays variations in the life cycle just described and in the particular species of mosquito that serves as a suitable host. Another apicomplexan, *Toxoplasma gondii,* causes lymphatic infections and blindness in adults and severe neurological damage to the fetuses of infected pregnant women. Contact with infected domestic cats and their feces, consumption of contaminated raw meat, and failure to wash hands after handling such meat are means of transmitting the parasite.

Ciliates   The largest group of protozoans, the **ciliates,** have cilia over most of their surfaces. Cilia have a basal body near their origin that anchors them in the cytoplasm and enables them to extend from the surface of the cell. Cilia allow the organisms to move, and in some genera, such as *Paramecium* (➢ Figure 11.3d), cilia assist in food gathering. *Balantidium coli,* the only ciliate that parasitizes humans, causes dysentery.

Ciliates have several highly specialized structures. Most ciliates have a well-developed contractile vacuole, which regulates cell fluids. Some have a strengthened pellicle. Others have **trichocysts,** tentacles that can be used to capture prey, or long stalks by which they attach

✓ How do parasites differ from predators?

✓ If a parasite has recently invaded a new population, is it more likely to cause severe or mild symptoms and effects?

✓ What is a "bloom" of protists? Why does it occur? What can its effects be?

The War Against Malaria: Missteps and Milestones

Malaria, caused by protozoa of the genus *Plasmodium,* is one of the most serious parasitic infections that afflicts humans. Despite massive efforts to control the spread of this disease, in any given year it strikes up to 500 million people and claims the lives of 1.5 to 3 million a year, many of them children. Malaria is an ancient human scourge. Written records on Egyptian papyrus from 1550 B.C. describe a disease with high intermittent fever that must have been malaria. They refer to the use of tree oils as mosquito repellents—although not until some 30 centuries later was the mosquito proven to be the vector (or carrier) of the disease. In some low-lying ancient cities, nearly entire populations succumbed to the disease. Only the wealthy could afford to "head for the hills" in summer to escape the heat, mosquitoes, and fevers. Many of the Crusaders of medieval times also died of malaria, and slave trading in later years contributed significantly to its spread. The one bright spot in this dismal history was the discovery in the sixteenth century that malaria could be treated with quinine, a drug from the *Chinchona* tree.

A connection between swamps and fevers had long been recognized but was misinterpreted. Most early investigations of malaria's cause focused on air—indeed, the disease was named for the "bad air" (*mal,* bad) thought to be responsible—and water. The association between the disease and mosquitoes suggested in Egyptian records was ignored. By the 1870s, with the advent of the germ theory, some scientists came to believe that malaria was caused by a bacterium, which they named *Bacillus malariae.*

Alphonse Laveran, a French army physician in North Africa, was not convinced that the malaria organism had been found. With unstained preparations and a poor-quality, low-power microscope, he continued to search for it. He eventually found what we now know to be male sex cells of the malarial parasite in human blood. Most scientists rejected Laveran's findings in favor of the *Bacillus* theory, until he demonstrated the male sex cells to Pasteur in 1884. Only then did the scientific community accept that protozoa of the genus *Plasmodium* cause malaria.

The Italian physiologist Camillo Golgi added to Laveran's work in 1885 by identifying several species of *Plasmodium.* And in 1891 Russian researchers developed the Romanovsky staining procedure (methylene blue and eosin) for malarial blood smears. With slight modifications, it is still used.

Although the causative agent of malaria had been found, its transmission was not yet understood. Ronald Ross, a medical officer in India, spent years searching in his spare time for proof that malaria parasites are carried by mosquitoes. Eventually, he found the parasites in *Anopheles* mosquitoes, though he never fully explained the mode of transmission. His efforts were neither appreciated nor supported by his superiors, but in 1902 he was awarded the Nobel Prize in physiology or medicine.

Once the vector for malaria had been identified, researchers turned to controlling transmission of the disease by controlling vector populations. Credit for developing the first effective mosquito control measures is given to the American physician William Crawford Gorgas, chief medical officer for sanitation during the building of the Panama Canal early in the twentieth century. By draining wet areas and instituting the use of mosquito netting, Gorgas significantly reduced the incidence of both malaria and yellow fever among canal workers.

In the 1930s and 1940s, advances in malarial treatment and control led many people to believe that malaria was no longer a threat. However, when World War I soldiers thought to have been cured with quinine treatment suffered relapses after they had left malaria-infested areas, researchers began a new round of investigations. This time, they searched for sites in which the parasites become sequestered in human tissues. In 1938, S. P. James and P. Tate discovered the parasite outside red blood cells in certain birds; in 1948 H. C. Shortt and P. C. C. Garnham made a similar discovery of *P. vivax* in humans. The parasite is now known to disappear from blood circulation, out of the reach of drugs, by invading cells of the liver and other tissues.

The war against malaria continues, but with limited success. Massive insecticide (DDT) spraying in the 1960s appeared to have eradicated the disease from many regions of the world, but DDT-resistant mosquitoes soon emerged. Incidence of the disease increased, sometimes to epidemic proportions. Some strains of the parasite also became resistant to chloroquine, one of the best drugs for treating malaria.

Another factor contributing to malaria's recent comeback may be environmental changes in developing countries. In the late 1950s, for instance, the inhabitants of Kenya's Karo Plain turned from subsistence farming of maize and cattle grazing to a cash-producing rice culture. Rice must grow in wet conditions, so the dry plains were flooded and the cattle were banished from the area. But along with rice, the Karo Plain soon supported *Anopheles* mosquitoes. Attracted by the water and the increased humidity, malaria-carrying mosquitoes came to outnumber non-malaria-carrying mosquitoes 2 to 1. And because the favorite host of the *Anopheles* mosquitoes—cattle—had been removed from the fields, the mosquitoes turned to humans for their meals. The result of all these environmental changes was a frightening increase in malaria incidence. In fact, global warming is predicted to bring about the spread of *P. falciparum,* the species that causes the most life-threatening form of malaria. Computer models predict it will spread to the eastern United States and Canada, to most of Europe, and to Australia.

Several new methods for controlling malaria are currently being studied. However, a successful vaccine still has not been developed. That means that health education and affordable replacements for chloroquine are paramount in controlling malaria. Mosquito nets impregnated with mosquito-killing chemicals, could also be effective. In Gambia, childhood mortality from malaria fell by an astonishing 63 percent after the introduction of these nets.

themselves to surfaces. Ciliates also undergo **conjugation.** Unlike bacterial conjugation, in which one organism receives genetic information from another, conjugation in ciliates allows exchange of genetic information between two organisms.

Fungi

Characteristics of Fungi

Fungi, studied in the specialized field of **mycology,** are a diverse group of heterotrophs. Many are saprophytes that digest dead organic matter and organic wastes. Some are parasites that obtain nutrients from the tissues of other organisms. Most fungi, such as molds and mushrooms, are multicellular, but yeasts are unicellular.

The body of a fungus is called a **thallus.** The thallus of most multicellular fungi consists of a **mycelium** (my-se´le-um), a loosely organized mass of threadlike structures called **hyphae** (hy´fe; singular: *hypha;* ➤ Figure 11.5). The mycelium is embedded in decaying organic matter, soil, or the tissues of a living organism. Mycelial cells release enzymes that digest the *substratum* (the surface on which the fungus grows) and absorb small nutrient molecules. The cell walls of a few fungi contain cellulose, but those of most fungi contain **chitin** (ki´tin), a polysaccharide also found in the exoskeletons (outer coverings) of arthropods such as ticks and spiders. All fungi have lysosomal enzymes that digest damaged cells and help parasitic fungi to invade hosts. Many fungi synthesize and store granules of the nutrient polysaccharide glycogen. Some fungi, such as yeasts, are known to have plasmids. These plasmids can be used to clone foreign genes into the yeast cells, a technique of great use in genetic engineering. ➡ (Chapter 8, p. 196)

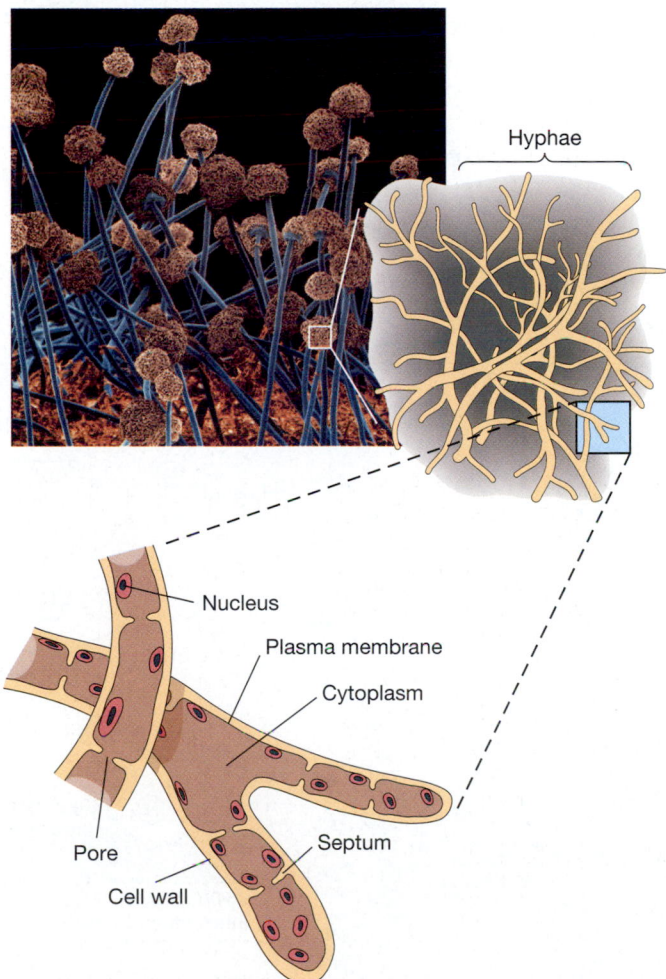

Nucleus

Plasma membrane

Cytoplasm

Pore

Septum

Cell wall

Hyphae

Lichen Housemates

A lichen is not a single organism but a fungus living in symbiosis with either a cyanobacterium or a green alga. Members of the pair can be separated, and each will live a normal life by itself. Presumably there are advantages to living together. Most scientists agree that lichens represent a mutualistic relationship, in which each member of the partnership benefits. The fungus obtains food from the photosynthetic organism while providing it with structure and protection from the elements (especially dehydration).

Lichen-forming organisms take on different and very specific shapes when they grow in association. *Crustose lichens* (pink, purple, and green) grow on surfaces and resemble a crust. *Foliose lichens* (top right) are leaflike, growing in crinkled layers that project up from a substratum, usually a rock or tree trunk. *Fruticose lichens* are the tallest, sometimes looking like miniature forests. Reindeer moss is not a true moss but rather an example of a fruticose lichen.

➤ **Figure 11.5   The mycelium of a typical fungus.** The mold *Aspergillus niger* (85X) consists of filamentous hyphae, the cells of which can be multinucleate and separated by pore-containing septae.

The hyphal cells of most fungi have one or two nuclei, and many hyphal cells are separated by cross-walls called **septa** (singular: *septum*). Pores in septa allow both cytoplasm and nuclei to pass between cells. Some fungi have septa with so many pores that they are sievelike, and a few lack septa entirely. Certain fungi with a single septal pore have an organelle called a *Woronin* (weh-ro´nin) *body*. When a hyphal cell ages or is damaged, the Woronin body moves to and blocks the pore so that materials from the damaged cell cannot enter a healthy cell.

Many fungi reproduce both sexually and asexually, but a few have only asexual reproduction. Asexual reproduction always involves mitotic cell division, which in yeast occurs by budding (➤ Figure 11.6). Sexual reproduction occurs in several ways. In one way, haploid gametes unite, and their cytoplasm mingles in a process called **plasmogamy** (plaz-mog´am-e). However, if the nuclei fail to unite, a **dikaryotic** ("two-nucleus") cell forms; it can persist for several cell divisions. Eventually, the nuclei fuse in a process called **karyogamy** (kar´´-e-og´am-e) to produce a diploid cell. Such cells or their progeny later produce new haploid cells. Some fungi also can reproduce sexually during dikaryotic (diploid) phases of their life cycle. Fungi usually go through haploid, dikaryotic, and diploid phases in their life cycle (➤ Figure 11.7).

Fungi can produce spores both sexually and asexually, and spores can have one or several nuclei (➤ Figure 11.8). Typically, aquatic fungi produce motile spores with flagella, and terrestrial fungi produce spores with thick protective walls. Germinating spores produce either single cells or germ tubes. *Germ tubes* are filamentous structures that break through weakened spore walls and develop into hyphae.

The Importance of Fungi

In ecosystems, fungi are important decomposers. In the health sciences, they are important as facultative parasites—they can obtain nutrients from nonliving organic matter or from living organisms. Fungi are never obligate parasites because all fungi can obtain nutrients from dead

➤ Figure 11.6 **Budding yeast.** Circular scars seen on the surface of the cell on the right represent sites of previous budding (7000X).

Phytophtora infestans, the fungus that caused the great Irish potato famine, led to the death of 1 million Irish and to the emigration of another 2 million.

➤ Figure 11.7 **One method of sexual reproduction in fungi.** Haploid organisms may maintain themselves by asexual spore formation (beige background) or budding. Alternatively (blue background), they may produce gametes that initially undergo plasmogamy (fuse their cytoplasmic portions). After several mitotic divisions of the still-separate nuclei, the two nuclei undergo karyogamy (fuse their nuclei) to form a diploid zygote. The zygote then undergoes meiosis to return to the haploid state and produces reproductive spores.

(a)   (b)

➤ **Figure 11.8   The formation of asexual spores (conidiospores).** (a) Brushlike clusters of fungus *Penicillium* chains of spores (1400X). (b) Spores of the rose rust fungus *Phragmidium* (1000X).

organisms. Even when fungi parasitize living organisms, they kill cells and obtain nutrients as saprophytes. Nearly every form of life is parasitized by some type of fungus. Some fungi produce antibiotics that inhibit the growth of or kill bacteria. Parasitic fungi vary in the damage they inflict. Fungi such as those that cause athlete's foot are nearly always present on the skin and rarely cause severe damage. However, the fungus that causes histoplasmosis can spread through the lymphatic system to cause fever, anemia, and death.

Saprophytic fungi are beneficial as decomposers and as producers of antibiotics. The digestive activities of such fungi provide nutrients not only for the fungi themselves but for other organisms, too. The carbon and nitrogen compounds they release from dead organisms contribute significantly to the recycling of substances in ecosystems. Fungi are essential for decomposing lignins and other woody substances. Some fungi excrete metabolic wastes that are toxic to other organisms, especially soil microorganisms. In the soil, the production of such toxins, which are antibiotics, is called **antibiosis.** These toxins presumably help the species that produce them compete and survive. The antibiotics, when extracted and purified, are used to treat human infections (Chapter 13).

Parasitic fungi can be destructive when they invade other organisms. These fungi have three requirements for invasion: (1) proximity to the host, (2) the ability to penetrate the host, and (3) the ability to digest and absorb nutrients from host cells. Many fungi reach their hosts by producing spores that are carried by wind or water. Other fungi arrive on the bodies of insects or other animals. For example, wood-boring insects spread spores of the fungal Dutch elm disease (➤ Figure 11.9) throughout North America in the decades following World War I, killing almost all elm trees in some parts of the United States. Fungi penetrate plant cells by forming hyphal pegs that press on and push through cell walls. How fungi penetrate animal cells, which lack cell walls, is not fully understood, but lysosomes apparently play an important role. Once fungi have entered cells, they digest cell components and absorb nutrients. As cells die, the fungus invades adjacent cells and continues to digest and absorb nutrients.

Fungal parasites in plants cause diseases such as wilts, mildews, blights, rusts, and smuts and thereby produce extensive crop damage and economic losses. Fungal infections of domestic birds and mammals are also responsible for extensive economic losses. Those fungi that invade humans cause human suffering, decreased productivity, and sometimes long-term medical expenses. Human fungal diseases, or **mycoses,** often are caused by more than one organism. Mycoses can be classified as superficial, subcutaneous, or systemic. *Superficial* diseases affect only keratinized tissue in the skin, hair, and nails. *Subcutaneous* diseases affect skin layers beneath keratinized tissue and can spread to lymph vessels. *Systemic* diseases invade internal organs and cause significant destruction. Some fungi are opportunistic; they do not ordinarily cause disease but can do so in individuals whose defenses are impaired, such as AIDS patients and transplant recipients who are receiving immunosuppressive drugs. More individuals with fungal infections are seeking hospital treatment than ever before.

Fungi and Orchids

When explorers first brought orchids from South America back to England during the nineteenth century, the English were delighted to have such handsome specimens in their conservatories. However, they suffered great disappointment when the plants failed to thrive, no matter how carefully they were potted in fresh soil and new pots. It took the English several years of experimentation, and perhaps the fortuitous importation of a few orchids in their native medium, to learn to grow orchids out of their natural environment. Eventually, it was discovered that orchids require certain fungi to thrive. These fungi form symbiotic associations with the orchid roots; such associations are called *mycorrhizae* (my"ko-ri'ze). When medium in which orchids had grown was used to pot fresh specimens, the orchids and the fungi formed mycorrhizae, and both thrived.

Of the approximately 70,000 known species of fungi, only about 300 of these are pathogenic to humans.

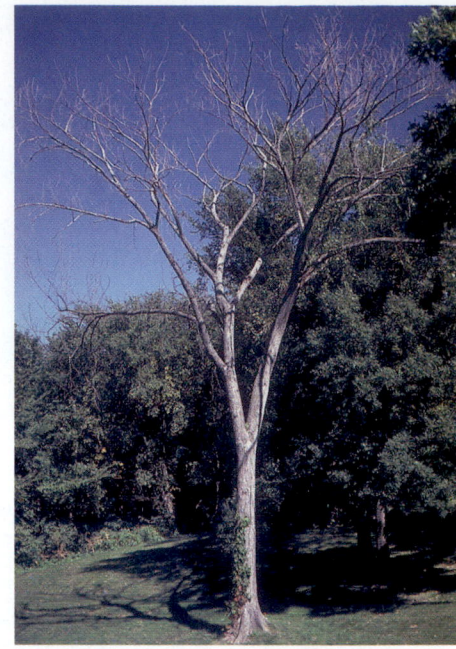

➤ Figure 11.9   **Dutch elm disease.** American elms (*Ulmus americana*) killed by Dutch elm disease.

(a)

(b)

➤ Figure 11.10   **Dimorphism in fungi.** (a) Hyphae of *Mucor.* (b) Yeast form of *Mucor.*

Culturing and identification of the causative agents of mycoses require special laboratory techniques. Acidic, high-sugar media with antibiotics added help prevent bacterial growth and allow fungal growth. The medium Sabouraud agar, which was developed nearly a century ago by a French mycologist, is still used in many laboratories. Under the best conditions, most pathogenic fungi in cultures grow slowly; some may take 2 to 4 weeks to grow as much as bacteria do in 24 hours.

Classification of Fungi

Fungi are classified according to the nature of the sexual stage in their life cycles. Such classification is complicated by two problems: (1) No sexual cycle has been observed for some fungi, and (2) it is often difficult to match the sexual and asexual stages of some fungi. For instance, one researcher may work out an asexual phase and give the fungus a name; another researcher may work out a sexual phase and give the same fungus a different name. Because the relationship between the sexual and asexual phases is not always apparent, a particular species of fungi may have two names until someone discovers that the two phases occur in the same organism. Another problem is that many fungi look quite different when growing in tissues (yeastlike) and when growing in their natural habitats (filamentous). The ability of an organism to alter its structure when it changes habitats is called **dimorphism** (di-mor'fizm) ➤ (Figure 11.10). Dimorphism in fungi has complicated the problem of identifying causative agents in fungal diseases. We will consider bread molds, sac fungi, club fungi, and the so-called Fungi Imperfecti, which are believed to have lost their sexual cycle (Table 11.2).

Bread Molds

The **bread molds, Zygomycota** or conjugation fungi, have complex mycelia composed of hyphae (lacking septa) with chitinous walls. The black bread mold, *Rhizopus* (➤ Figure 11.11), has hyphae that grow rapidly along a surface and into the substratum. Some bread-mold hyphae produce spores that are easily carried by air currents. When the spores reach an appropriate substratum, they germinate to produce new hyphae. Sometimes short branches of the hyphae of two different strains, called plus and minus strains, grow together. This joining of hyphae gave rise to the name conjugation fungi. Chemical attractants are involved in attracting hyphae to each other. Multinucleate cells form where the hyphae join, and many pairs of plus and minus nuclei fuse to form zygotes. Each zygote is enclosed in a **zygospore,** a thick-walled, resistant

> **Table 11.2** **Properties of fungi**

Phylum	Common name	Characteristics	Examples
Zygomycota	Bread molds	Display conjugation	*Rhizopus* and other bread molds
Ascomycota	Sac fungi	Produce asci and ascospores during sexual reproduction	*Neurospora, Penicillium, Saccharomyces,* and other yeasts; *Candida, Trichophyton,* and several other human pathogens
Basidiomycota	Club fungi	Produce basidia and basidiospores	*Amanita* and other mushrooms; *Claviceps* (which produces ergot); *Cryptococcus*
Deuteromycota	Fungi Imperfecti	Sexual stage nonexistent or unknown	Soil organisms, various human pathogens

structure that also produces spores. Genetic information in zygospores comes from two strains, whereas that in hyphal spores comes from a single strain.

Although bread molds interest mycologists and frustrate bacteriologists whose cultures they contaminate, they usually do not cause human disease. *Rhizopus,* however, is an opportunistic human pathogen; it is especially dangerous to people with diabetes mellitus that is not well controlled.

Sac Fungi

The **sac fungi** are a diverse group containing over 30,000 species, including yeasts. Sac fungi have chitin in their cell walls and produce no flagellated spores. With the exception of some yeasts, which do not form hyphae, the hyphae of sac fungi have septa with a central pore. These fungi are properly called **Ascomycota** (as′ko-mi-ko″ta); unlike other fungi, they produce a saclike **ascus** (plural: *asci*) during sexual reproduction (➤ Figure 11.12). Yeasts are included among the ascomycetes, even though most yeasts have no known sexual stage. In species that reproduce both sexually and asexually, the asexual phase forms spores called **conidia** at the ends of modified hyphae. In the sexual phase, one strain has a large *ascogonium,* and an adjacent strain has a smaller *antheridium.* These structures fuse; their nuclei mingle, and hyphal cells with dikaryotic nuclei grow from the fused mass. Eventually, dikaryotic nuclei fuse to form a zygote, and the zygote nucleus divides to form eight nuclei in each ascus. Each ascus forms eight **ascospores,** sometimes releasing them forcefully.

Several sac fungi are of interest in microbiology. *Neurospora* is significant because studies of its ascospores have provided important genetic information. *Penicillium notatum* produces the antibiotic penicillin. *P. roquefortii* and *P. camemberti* are responsible for the color, texture, and flavor of Roquefort and Camembert cheeses. Yeasts, especially those of the genus *Saccharomyces,* release carbon dioxide and alcohol as metabolic products of fermentation and are used to leaven bread and to make alcohol in beer and wine (Chapter 26). A number of sac fungi are human pathogens. *Candida albicans* causes vaginal yeast infections. *Trichophyton* is associated

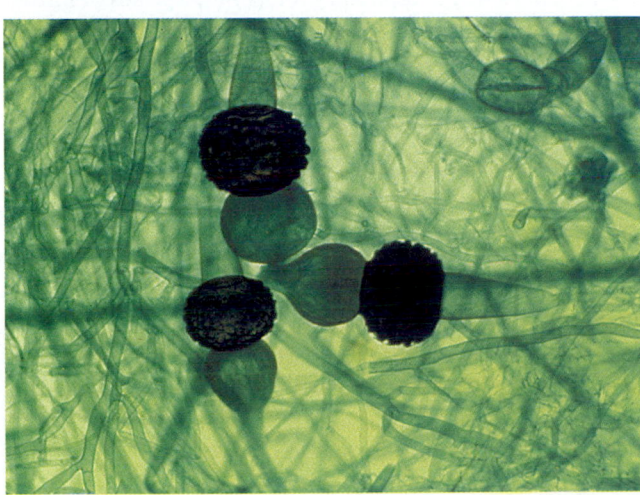

> **Figure 11.11** **The black bread mold, *Rhizopus nigricans.*** Sexual zygospores (black, spiny structures) are the result of the joining and fusion of genetic materials at the tips of special hyphal side branches. The zygospores germinate to produce a sporangium that, in turn, produces many asexual spores (370X).

Spore Prints

Mushroom identification requires knowledge about the spores of your unknown specimen. Some keys for mushroom identification are arranged according to spore color. How do you obtain such information? Try this simple method.

First, collect fresh mushrooms whose caps are just opening or that are fully open. Cut off the stem flush with the bottom of the cap. Place the cap, gill-side down, on a piece of paper. Leave it undisturbed overnight or until dry. Gently lift the cap and see the sunburstlike pattern of spores that have been shed from the surfaces of the gills. If you have two mushrooms of the same variety, use a dark piece of paper under one and white paper under the other before the drying process, because you won't know what color spores to expect. Spores may range in color from black to white, tan, or even pink.

➤ Figure 11.12   **The life cycle of an ascomycete.** In the asexual phase, spores called conidia are formed at the tips of modified hyphae. In the sexual phase, the mycelium that produces conidia also forms the gamete-producing structures, antheridia (male) and ascogonia (female). After cytoplasmic fusion of those structures occurs, dikaryotic hyphal cells develop and interweave into an ascocarp. There saclike asci grow. In each ascus, the dikaryotic nuclei fuse to form a zygote, and the zygote nucleus divides into eight nuclei. From them, eight ascospores form and are forcefully released.

with athlete's foot, and *Aspergillus* with opportunistic respiratory infections. Species of *Blastomyces* and *Histoplasma* cause respiratory infections and can spread throughout the body.

Club Fungi

The **club fungi** include mushrooms, toadstools, rusts, and smuts. The rusts and smuts parasitize plants and cause significant crop damage. In addition to having hyphae aggregated to form mycelia, the club fungi have club-shaped sexual structures called **basidia,** from which the name **Basidiomycota** is derived (➤ Figure 11.13). In a typical basidiomycete life cycle, sexual spores called **basidiospores** germinate to form septate mycelia, and cells of mycelia unite into dikaryotic forms. The dikaryotic mycelium grows and produces basidia, which in turn produce basidiospores. Some mushrooms, such as *Amanita,* produce toxins that can be lethal to humans. *Claviceps purpurea,* a parasite of rye, produces the toxic substance ergot. This substance can be used in small quantities to treat migraine headaches and induce uterine contractions, but in larger quantities it can kill (Chapter 22). The yeast *Cryptococcus* causes opportunistic respiratory infections, which can be fatal if they spread to the central nervous system, causing meningitis and brain infection. This organism is increasingly being seen in AIDS patients.

Are The Biggest and Oldest Organisms On Earth Fungi?

Weighing about 100 tons (more than a blue whale) and extending through nearly 40 acres of soil in a forest near Crystal Falls in upper Michigan is a gigantic individual of the fungus *Armillaria bulbosa*. Between 1500 and 10,000 years ago, most likely at the end of the last Ice Age, a single pair of compatible spores blew in from parent mushrooms, germinated, and mated. They began growth that continues today. The fungus grows primarily under the soil, so it is usually not visible to the casual observer. The hyphae of its mycelium probe through the soil, seeking woody debris to decompose and recycle. Experimental measurements of its growth rate through soil enabled scientists to estimate the time required to reach its present size.

DNA analysis of 12 genes from the organism's fruiting structures—commonly called *button* or *honey mushrooms*—and its stringlike underground colonizing structures—called *rhizomorphs*—revealed the huge fungus to be a giant clone. All parts of the clone are identical in genetic composition. Although there are minor breaks in its continuity, it is still regarded as a single individual.

Despite the massive size of this fungus, its discoverers, Myron Smith and James Anderson of the University of Toronto, and Johann Bruhn of Michigan Technological University, predicted that it might not be the largest organism of its kind. Writing in the journal *Nature* in April 1992, they explained that they found the fungus in a mixed forest containing many kinds of trees. In a single-type forest such as a large stand of birch or aspen, a fungus with a preference for that type of tree could reach even greater size. This one, however, has probably reached its maximum size, as it collides with competing fungi along its borders.

The scientists' prediction quickly proved to be prophetic. About a month after the *Nature* article was published, two forest pathologists—Ken Russell of the State Department of Natural Resources and Terry Shaw of the U.S. Forest Service—announced that they had been studying an even larger fungus near Mount Adams in southwestern Washington. This organism, an individual of *Armillaria ostoyae,* covers 1500 acres (about 2.5 square miles), making it almost 40 times as large as the Michigan fungus. The Washington fungus grows in a region populated largely by a single type of tree—in this case, pine—and therefore enjoys a vast source of nourishment. Although the Washington fungus dwarfs its Michigan counterpart in size, it is actually younger, having an estimated age of 400 to 1000 years. Thus, the Michigan fungus retains the title "oldest" (at least for now) but not "largest."

Will scientists eventually discover fungi that are even bigger than the Washington fungus? Most likely, yes. In fact, in an interview, Shaw referred to an *A. ostoyae* in Oregon that might be larger than the one he discovered in Washington. And still bigger ones may remain to be found. The search for the "biggest and oldest" promises to be an exciting episode in the field of microbiology.

These mushrooms are only a small part of the immense 100-ton fungal organism *Armillaria bulbosa* extending throughout nearly 40 acres of a Michigan forest.

(a)

(b)

> Figure 11.13   **Mushroom spores.** (a) The gills on the bottom of a mushroom (*Leucoagaricus naucinus*) cap have microscopic, club-shaped structures called basidia. (b) Each basidium (of *Psilocybe mexicana,* 225X) produces four balloonlike structures called basidiospores.

Fungi Imperfecti

The **Fungi Imperfecti,** or **Deuteromycota,** are called "imperfect" because no sexual stage has been observed in their life cycles. Without information on the sexual cycle, taxonomists cannot assign them to a taxonomic group. However, by their vegetative characteristics and the production of asexual spores, most of these fungi seem to belong with the sac fungi. Many of the Fungi Imperfecti have recently been placed in other phyla and given new genus names. We have kept the older designations, however, because the new ones are not yet familiar or widely used in clinical work.

Helminths

Characteristics of Helminths

Helminths, or worms, are bilaterally symmetrical—that is, they have left and right halves that are mirror images. A helminth also has a head and tail end, and its tissues are differentiated into three distinct tissue layers: ectoderm, mesoderm, and endoderm. Helminths that parasitize humans include flatworms and roundworms (Table 11.3).

Flatworms

Flatworms (Platyhelminthes) are primitive worms usually no more than 1 mm thick, but some, such as large tapeworms, can be as long as 10 m. Flatworms lack a **coelom** (se′lom), a cavity that lies between the digestive tract and the body wall in higher animals. Most flatworms have a simple digestive tract with a single opening, but some parasitic flatworms, the tapeworms, have lost their digestive tracts. Most flatworms are hermaphroditic, each individual having both male and female reproductive systems. They have an aggregation of neurons in the head end, representing an early stage in the evolution of a brain. Flatworms lack circulatory systems, and most absorb nutrients and oxygen through their body walls.

More than 15,000 species of flatworms have been identified. They include free-living, mostly aquatic organisms such as *planarians* and two groups of parasitic organisms, the **flukes** (*trematodes*) and the **tapeworms** (*cestodes*). Both parasitic groups have highly specialized reproductive systems and suckers or hooks by which they attach to their host. The flukes can be internal or external parasites. *Fasciola hepatica* and several other flukes parasitize humans. Tapeworms parasitize the small intestine of animals almost exclusively but occasionally occur in the eye or brain. The beef tapeworm, *Taenia saginata,* and several other worms parasitize humans.

Roundworms

Roundworms, or **nematodes,** share many characteristics with the flatworms, but they have a **pseudocoelom,** a primitive, fluid-filled body cavity that lacks the complete lining found in higher animals. The roundworms have cylindrical bodies with tapered ends and are covered with

➤ Table 11.3	**Properties of helminths**	
Group	**Characteristics**	**Examples**
Flatworms	Worms live in or on hosts	*Taenia* and other tapeworms are internal parasites; flukes can be internal or external parasites.
Roundworms (nematodes)	Most worms live in the intestine or circulatory system of hosts	Hookworms, pinworms, and several other roundworms live in intestines or lymph system.

a thick, protective cuticle. They vary in length from less than 1 mm to more than 1 m. Contractions of strong muscles in the body wall exert pressure on the fluid in the pseudocoelom and stiffen the body. Pointed ends and stiff bodies allow roundworms to move through soil and tissues easily. Roundworm females are larger than males. Breeding is enhanced by chemical attractants released by females that attract males. Females can lay as many as 200,000 eggs per day. The large number of eggs, well protected by hard shells, ensures that some will survive and reproduce.

Over 80,000 species of roundworms have been described. They occur free-living in soil, fresh water, and salt water and as parasites in every plant and animal species ever studied. A single acre of soil can contain billions of roundworms. Many parasitize insects and plants; only a relatively small number of species infect humans, but they cause significant debilitation, suffering, and death. Most roundworms that parasitize humans, such as hookworms and pinworms, live mainly in the intestinal tract, but a few, such as *Wuchereria,* have larval forms that live in blood or lymph. The effects of roundworms on humans were first recorded in ancient Chinese writings and have been recorded by nearly every civilization since then. (For an account of modern American experiences with sushi and other forms of raw fish, which may harbor roundworms, see the box "Sushi" in Chapter 22.)

Parasitic Helminths

We will concern ourselves only with parasitic helminths and consider four groups: flukes, tapeworms, adult roundworms of the intestine, and roundworm larvae (➤ Figure 11.14). Because helminths have complex life cycles related to their ability to cause diseases, we consider a typical life cycle for each group.

Flukes

Two types of fluke infections occur in humans. One involves tissue flukes, which attach to the bile ducts, lungs, or other tissues; the other involves blood flukes, which are found in blood in some stages of their life cycle. Tissue flukes that parasitize humans include the lung fluke,

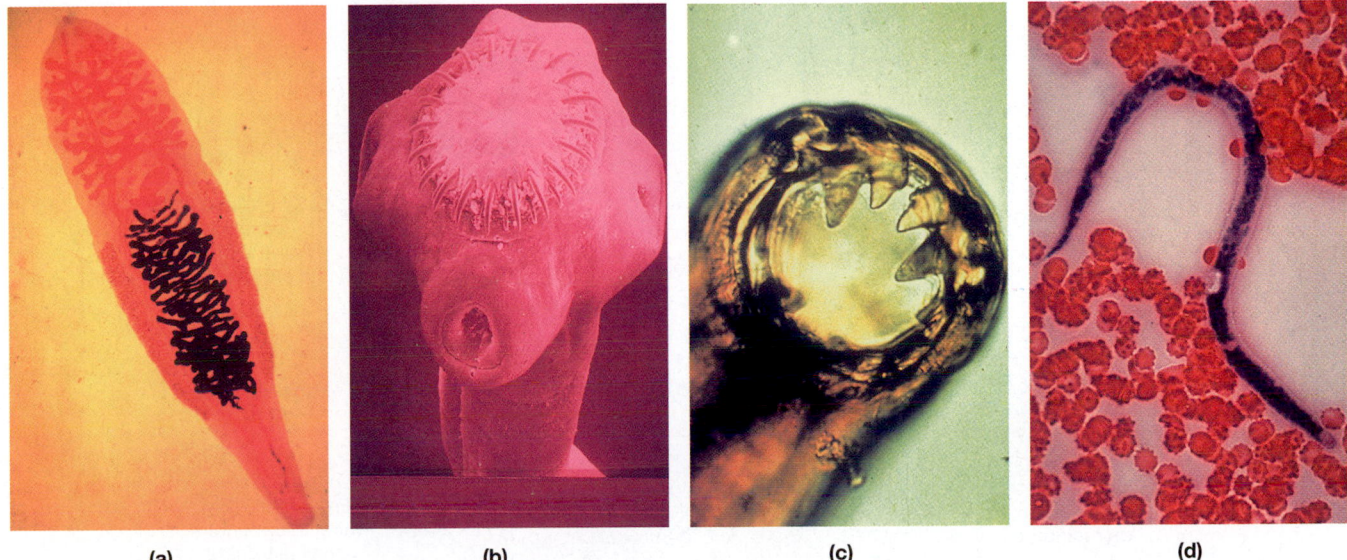

(a)   (b)   (c)   (d)

➤ Figure 11.14   **Representative helminths.** (a) *Clonorchis sinensis,* the Chinese liver fluke, stained to show internal organs. It infests the gallbladder, bile ducts, and pancreatic ducts, where it causes biliary cirrhosis and jaundice. (b) Head (scolex) of a tapeworm (220X). The hooked spines and suckers are used for attachment to intestinal surfaces. (c) Mouth of the Old World hookworm *Ancylostoma duodenale.* The muscular pharynx of this roundworm pumps blood from the intestinal lining of its host. (d) The microfilarial (miniature larval) stage of the heartworm *Dirofilaria immitis,* in a sample of dog blood (370X), is transmitted by mosquito bites. The larger stages live inside the heart and perforate its walls.

Paragonimus westermani, and the liver flukes, *Clonorchis sinensis* (➤ Figure 11.14a) and *Fasciola hepatica*. Blood flukes include various species of the genus *Schistosoma*.

Parasitic flukes have a complex life cycle (➤ Figure 11.15), often involving several hosts. The fusion of male and female gametes produces fertilized eggs that become encased in tough shells during their passage through the female fluke's uterus. The eggs pass from the host with the feces. When the eggs reach water, they hatch into free-swimming forms called **miracidia** (mi″ra-sid′e-ah). The miracidia penetrate a snail or other molluskan host, become **sporocysts,** and migrate to the host's digestive gland. The cells inside the sporocysts typically divide by mitosis to form **rediae** (re′de-e). Rediae, in turn, give rise to free-swimming **cercariae** (ser-ka′re-e), which escape from the mollusk into water. Using enzymes to burrow through exposed skin, cercariae penetrate another host (often an arthropod) and then encyst as **metacercariae.** When this host is eaten by the definitive host, the metacercariae excyst and develop into mature flukes in the host's intestine.

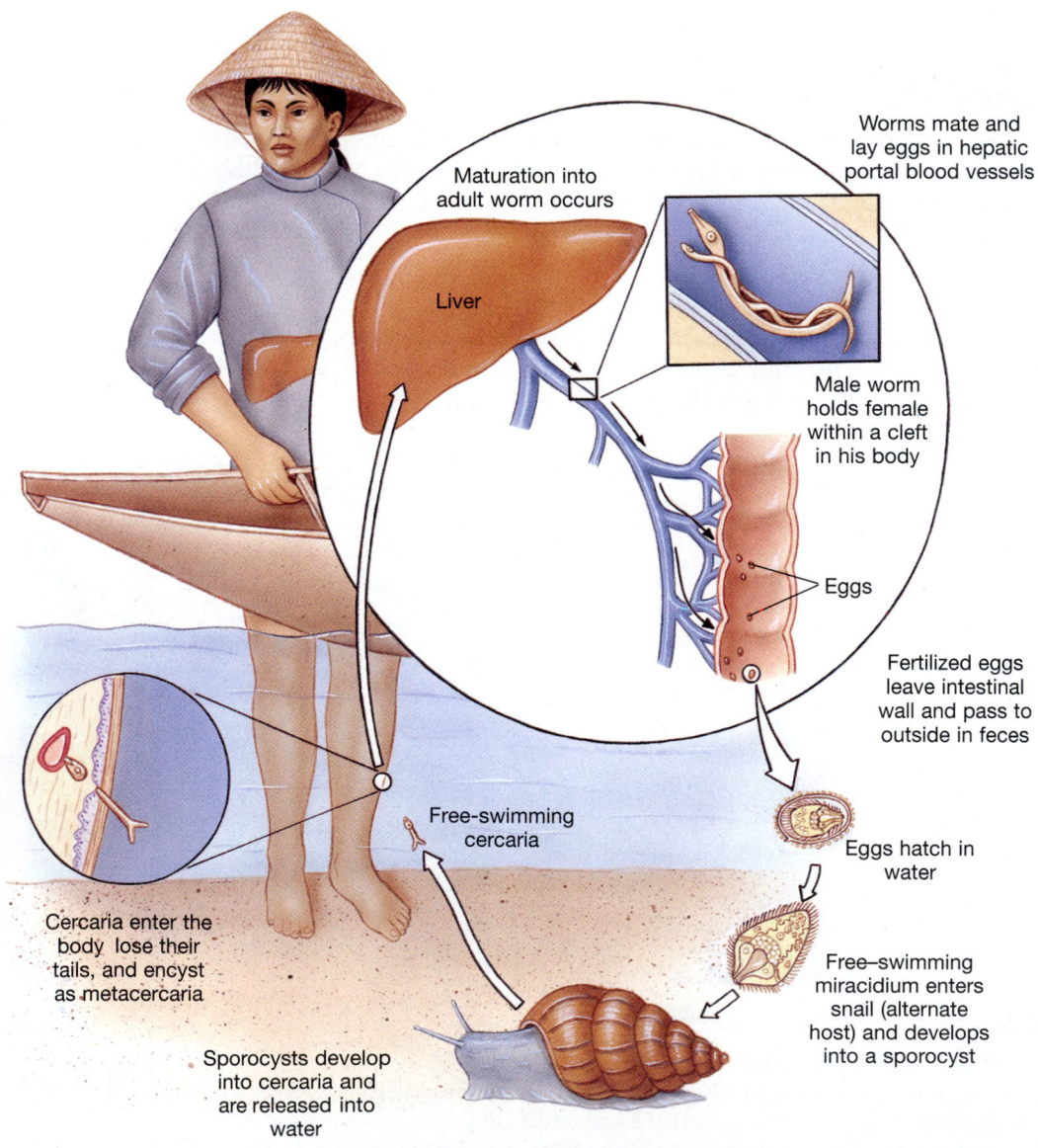

Worms mate and lay eggs in hepatic portal blood vessels

Maturation into adult worm occurs

Liver

Male worm holds female within a cleft in his body

Eggs

Fertilized eggs leave intestinal wall and pass to outside in feces

Eggs hatch in water

Free-swimming cercaria

Free–swimming miracidium enters snail (alternate host) and develops into a sporocyst

Cercaria enter the body lose their tails, and encyst as metacercaria

Sporocysts develop into cercaria and are released into water

➤ **Figure 11.15** **The life cycle of a blood fluke, *Schistosoma japonicum.*** This organism causes schistosomiasis. Unlike some flukes, *S. japonicum* does not have a rediae stage or enter an arthropod host.

Tapeworms

Tapeworms consist of a **scolex** (sko′lex), or head end (➤ Figure 11.14b), with suckers that attach to the intestinal wall, and a long chain of hermaphroditic **proglottids** (pro-glot′tidz), body components that contain mainly reproductive organs of both sexes. New proglottids develop behind the scolex, mature, and fertilize themselves. Old ones disintegrate and release eggs at the rear end. Among the tapeworms that can infect humans are beef and pork tapeworms that are species of *Taenia*, dwarf and rat tapeworms that are species of *Hymenolepis*, the hydatid worm *Echinococcus*, the dog tapeworm *Dipylidium*, and the broad fish tapeworm *Diphyllobothrium*.

Although different species display minor variations, the life cycle of tapeworms (➤ Figure 11.16) usually includes the following stages: Embryos develop inside eggs and are released

Adult birds may temporarily act as host to Trichinella spiralis, but the microbe does not encyst in fowl muscle nor in that of any cold-blooded animal.

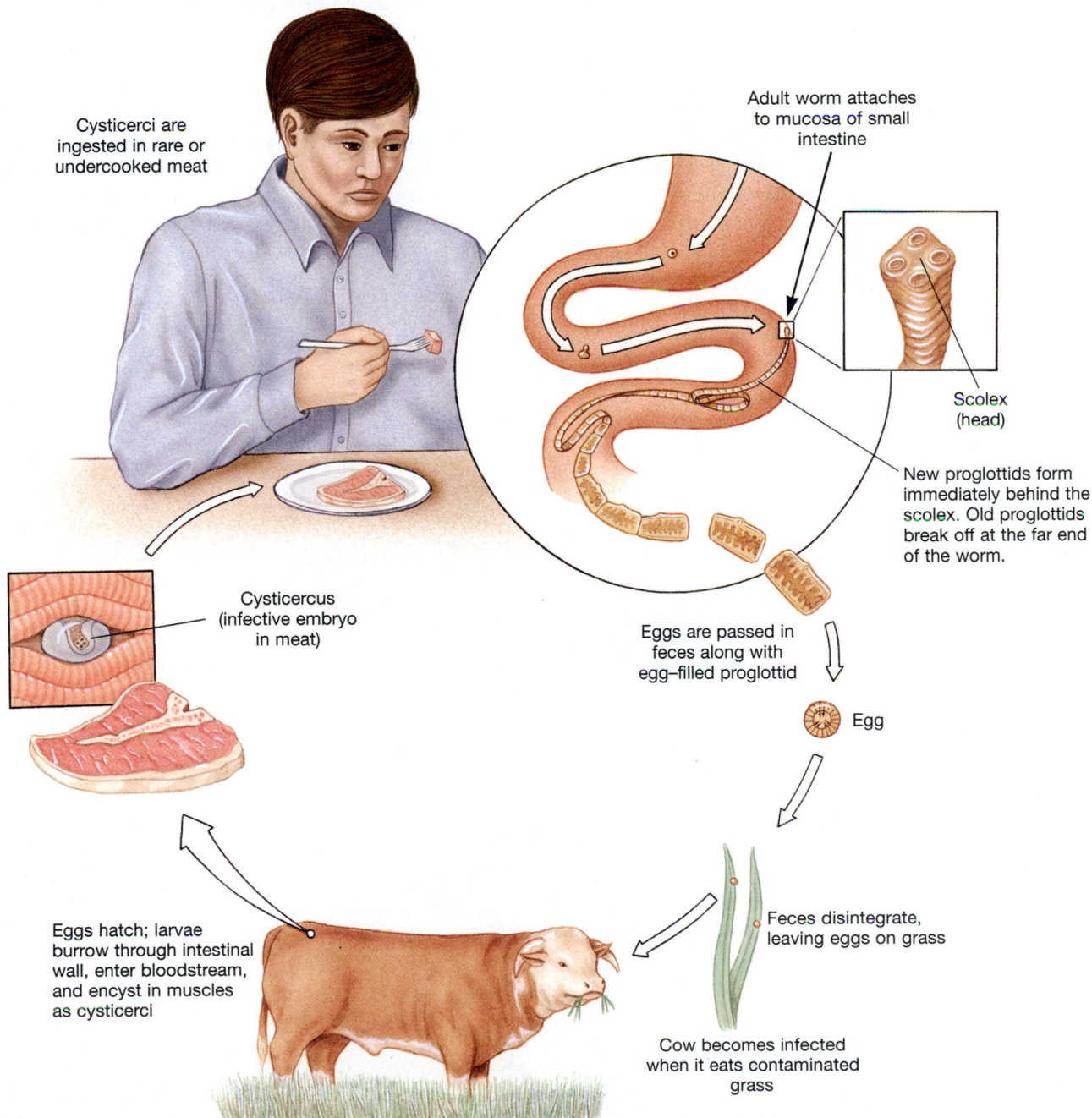

Cysticerci are ingested in rare or undercooked meat

Adult worm attaches to mucosa of small intestine

Scolex (head)

New proglottids form immediately behind the scolex. Old proglottids break off at the far end of the worm.

Cysticercus (infective embryo in meat)

Eggs are passed in feces along with egg–filled proglottid

Egg

Feces disintegrate, leaving eggs on grass

Eggs hatch; larvae burrow through intestinal wall, enter bloodstream, and encyst in muscles as cysticerci

Cow becomes infected when it eats contaminated grass

➤ **Figure 11.16   The life cycle of the beef tapeworm *Taenia saginata*.**

Dracunuculus in the United States! A different species infects raccoons and is sometimes seen as long white "strings" hanging from raccoons' wrists as they wash their food in a pond.

Over 1 billion people in 73 countries are at risk of contracting elephantiasis, and there are over 120 million people already infected.

Adult Loa loa worms migrate through subcutaneous tissue at the rate of 1 inch in 2 minutes. They are particularly troublesome crossing the bridge of the nose.

from proglottids; the proglottids and eggs leave the host's body with the feces. When another animal ingests vegetation or water contaminated with eggs, the eggs hatch into larvae, which invade the intestinal wall and can migrate to other tissues. A larva can develop into a **cysticercus** (sis-ti-ser′kus), or bladder worm, or it can form a cyst. A cysticercus can remain in the intestinal wall or migrate through blood vessels to other organs. A cyst can enlarge and develop many tapeworm heads within it, becoming a **hydatid** (hi-da′tid) **cyst** (Chapter 22). If an animal eats flesh containing such a cyst, each scolex can develop into a new tapeworm.

Adult Roundworms

Most roundworms that parasitize humans live much of their life cycle in the digestive tract. They usually enter the body by ingestion with food or water, but some, such as the hookworm, penetrate the skin. These helminths include the pork roundworm *Trichinella spiralis,* the common roundworm *Ascaris lumbricoides,* the guinea worm *Dracunculus medinensis,* the pinworm *Enterobius vermicularis,* and the hookworms *Ancylostoma duodenale* (➤ Figure 11.14c) and *Necator americanus.*

The life cycles of intestinal roundworms show considerable variation. We use the life cycle of *Trichinella spiralis* as an example (➤ Figure 11.17). These worms enter humans as encysted larvae in the muscle of infected pigs when undercooked pork is eaten. The cyst walls are digested with the meat, and the larvae are released into the intestine. They mature sexually in about 2 days and then mate. Females burrow into the intestinal wall and produce eggs that hatch inside the adult worm and emerge as larvae. The larvae migrate to lymph vessels and are carried to the blood. From the blood the larvae burrow into muscles and encyst. These cysts can

Human ingests cyst in undercooked pork

Stomach acid dissolves cyst cover, releasing worms

Worms mate

Females release larvae into lacteals and vessels

Blood vessel

Larvae migrate to muscle and encyst

Encysted worm in pork

Undercooked pork in garbage

➤ Figure 11.17   **The life cycle of the roundworm *Trichinella spiralis.*** This roundworm causes trichinosis.

remain in muscles for years. The same processes occur in the pigs themselves, so cysts are present in their tissues.

Roundworm Larvae

Whereas most roundworms cause much of their tissue damage as adults in the intestine, some cause their damage mainly as larvae in other tissues. These roundworms include *Wuchereria bancrofti*, which lives in lymphatic tissue and causes elephantiasis; *Loa loa*, which infects the eyes and eye membranes; *Onchocerca volvulus*, which infects both the skin and eyes; and *Dracunculus medinensis* (Guinea worm), which causes symptoms described in the Box relating the work of the Carter Center's efforts to eradicate Guinea worm from the planet.

The life cycles of roundworms that parasitize humans as larvae also require a mosquito host (➤ Figure 11.18). These worms enter the human body as immature larvae called **microfilariae** (mi″kro-fi-lar′e-e) with the bite of an infected mosquito. The microfilariae migrate through the tissues to lymph glands and ducts and mature and mate as they migrate. Females produce large numbers of new microfilariae, which enter the blood (➤ Figure 11.14d), usually at night. The microfilariae are ingested by mosquitoes as they bite infected humans. Any one of several species of mosquitoes can serve as host. When the microfilariae reach the midgut of the mosquito, they penetrate its wall and migrate first to the thoracic muscles and then to the mosquito's mouthparts. There they can be transferred to a new human host, where the cycle is repeated.

✓ How do flukes differ from tapeworms?

✓ Describe the steps in the life cycle of a tapeworm, including cysticercus and hydatid cyst stages.

✓ What are microfilariae? How are they usually transmitted? What do they become?

Microfilariae are produced and migrate to peripheral circulation between 10 p.m. and 4 a.m., remaining in tiny capillaries of lungs during the day.

Mosquito bites human and ingests microfilariae

Microfilariae bore through mosquito stomach and thoracic muscles; in about 10 days they develop into larvae.

Male worm (2.5 - 4 cm long)

Female worm (5 - 10 cm long)

Adult worms mate.

Mosquito bites human; larvae emerge from mouthparts onto skin and enter body through bite wound.

Larvae migrate to lymphatic vessels and lymph nodes. After growing to maturity in about 6 months, the larvae block lymphatic drainage, causing elephantiasis.

➤ **Figure 11.18** **The life cycle of the roundworm *Wuchereria bancrofti*.** This roundworm produces microfilariae and causes elephantiasis (a chronic edema; see Figure 23.4), especially of the legs and scrotum.

Carter Center Expects Eradication of Guinea Worm from World by 2000

Former President Jimmy Carter: Once you've seen a small child with a 2- or 3-foot-long live Guinea worm protruding from her body, right through her skin, you never forget it. I first saw the devastating effects of Guinea worm disease [dracunculiasis] in two villages near Accra, Ghana, in March 1988. In just a few minutes, Rosalynn and I saw nearly 200 victims, including people with worms coming out of their ankles, knees, groins, legs, arms, and other parts of their bodies. One woman, in great agony, was cradling her breast as if it were an infant. On it was an abscess the size of a fist where a Guinea worm was about to emerge. I have seen people with as many as a dozen or more of these worms emerging at the same time. I was shocked to find that this debilitating disease, which strikes nearly 10 million people each year, could be easily prevented. The disease is caused by contaminated drinking water; prevention is simply a matter of showing people how to make their water supply safe.

Former President Jimmy Carter, on a trip to Africa, examining a water hole contaminated with Guinea worms.

We are a compassionate nation, and that is one reason why those people still suffering needlessly from this disease should matter to us. The world is full of difficult problems that we cannot yet solve. This is one we can solve, and we can do it quickly if we put our minds to it. Here in Atlanta, The Carter Center's Global 2000 program is targeting the year 2000 as the date for global eradication of Guinea worm. Dr. Donald Hopkins, Senior Consultant for Global 2000's health programs, directs the Center's Guinea worm eradication efforts. He has helped enlist the support of governments and nonprofit organizations in this fight, and he has helped implement eradication programs in several African countries.

Dr. Donald Hopkins: The international community didn't realize until about 1980 that it was possible to eradicate Guinea worm completely. In fact, they didn't even realize the extent of the problem in their own countries, especially in rural areas. Before 1988, Ghana reported an average of 4500 cases of Guinea worm to the World Health Organization each year. Then in 1989, Ghana conducted its first nationwide search for cases and as a result reported more than 170,000 cases. A UNICEF study in southeastern Nigeria estimated that the rice farmers in that area of 1.6 million people are losing $20 million each year in potential profits because so many of them are crippled by Guinea worm at harvest time. In one heavily affected district of Ghana's northern region, the production of yams reportedly increased by 33 percent during the first 9 months of 1991 because so many farmers were restored to full productivity by the sharp reduction of Guinea worm.

Mr. Carter: The value is not just in terms of money. Healthy mothers can take better care of their children. Children can get a better education. Guinea worm disease causes children to miss an average of 12 weeks of school in a year when they are infected, compared to a total loss of 1 to 2 weeks lost due to all other causes combined. Often children are so badly behind when they return to school, especially if this happens several years in a row, that they become permanent dropouts. Additional income means that people can improve their homes and do all kinds of things they couldn't do before. They become more self-reliant. Tackling the next problem becomes easier when they have pride in the success of this effort.

Guinea worm disease is contracted by drinking from ponds, step wells, cisterns, and other sources of stagnant water that have been contaminated by the worm larvae. Guinea worm, *Dracunculus medinensis,* affects only humans, and it actually uses its human host to further its life cycle. Contaminated water contains water fleas that have eaten immature Guinea worm larvae. The larvae escape when the digestive juices in the person's stomach kill the flea. The larvae penetrate the stomach wall, wander around the abdomen, mature in a few months, and mate, after which the male worms die. It is only the female worm that grows to 2 or 3 feet in length and, about a year later, secretes a toxin that causes a blister on the skin. When the blister ruptures, usually when the infected part of the body is immersed in cool water, the worm starts to emerge. This process can take 30 to 100 days before the worm finally finishes making its way out of the body. When an infected person enters the village pond or watering hole, the worm discharges hundreds of thousands of tiny larvae into the water, beginning the cycle again.

Dr. Hopkins: A small incision made before the emergence blister is raised allows the worm to be wound out gradually, wrapped around a stick. This may have been the origin of the symbol of the medical profession, the caduceus, a serpent coiled around a stick. Many scholars think the Guinea worm is the "fiery serpent" of the Bible. It takes several weeks of daily gentle winding to complete the removal of a worm. If the worm breaks and dies, it will decompose inside the host, causing festering and infection. If a portion of it retracts into the tissues, it can carry tetanus spores back with it, leading to fatal tetanus disease. The local practice in some countries of putting cow dung on the wound makes tetanus especially common. In Upper Volta and Nigeria, Guinea worm is the third leading cause of acquiring tetanus. Other types of microorganisms can also enter the wound, and secondary infections are frequent, even if tetanus is avoided. If the worm emerges near a major joint, permanent scarring leads to stiff and crippled joints. One man died of starvation when a worm came out under his tongue and he couldn't eat.

Although most worms emerge from the lower limbs, they can be found anywhere: scrotum, scalp, chest, face.

Imagine the suspense of living in an area where Guinea worms abound. Will you have worms again this year? How many will you have? Where will they emerge? Will you become crippled, infected, or die of tetanus? Will you be able to go to work or school? Transmission season is only during wet periods. In sub-Saharan Africa these are the months of June, July, and August. The worms emerge about 12 months later during the next wet season. Years when there are droughts help reduce the number of infections.

Mr. Carter: There is no cure, only prevention. No matter how many times you have been infected with Guinea worms, no immunity to them develops. Prevention involves providing safe sources of drinking water such as borehole wells, which people with emerging worms cannot enter to contaminate. It can also be prevented by teaching villages to boil their water (although many cannot afford enough fuel to do so) or to filter their drinking water through a clean, finely woven cloth. The E. I. DuPont Company, in association with Precision Fabrics Group, donated 1.4 million monofilament nylon cloth filters to The Carter Center for use in the eradication campaign. The American Cyanamid Company has donated over $2 million of the larvicide Abate, which is used to treat water safely, killing the water fleas and Guinea worms but harming nothing else. The company has pledged to continue donating Abate until the goal of eradication is reached. Using the nylon filters and Abate, the two villages I visited in Ghana in March 1988 reduced the number of cases of Guinea worm there by more than 90 percent in one year.

Dr. Hopkins: Money is the limiting factor in our fight against the Guinea worm. We have the training and the manpower. Mr. and Mrs. Carter have been very active, making several trips to Africa to talk with leaders and to go into the villages and explain how the worm is transmitted and to teach people how to break that chain of transmission. This is a problem that affects 19 nations and that will probably require $65 million to eradicate. The countries themselves cannot afford the entire cost themselves, so help has to come from wealthier nations. The Japanese have put in 191 wells. The Swedes, Dutch, and Danes are the only European countries helping.

I was part of the international team that eradicated smallpox. It is exciting to be working on another eradication project that is so close to success. We believe we succeeded in totally eliminating Guinea worm from Pakistan at

Female Guinea worm emerging from a blister on the foot of a victim.

the end of 1991. So far we haven't heard of any new cases there.

Mr. Carter: A while back, I was asked what I felt my top presidential achievements were. Of course there were things such as the Panama Canal treaties, the Camp David accord between Israel and Egypt, and normalization of relations with China. But, if in 2000 we find that where there were 10 million cases of Guinea worm there are none, I would say that eradication of the Guinea worm is more important.

Current Status (1998): Ninety-seven percent of all cases of Guinea worm infestation have been eliminated. Most of the last 3 percent of cases are identified and under medical surveillance. Infested persons who step into water sources are fined. Rewards are being offered for reporting new cases. It is hoped that the final eradication of the Guinea worm will occur in time for the new century.

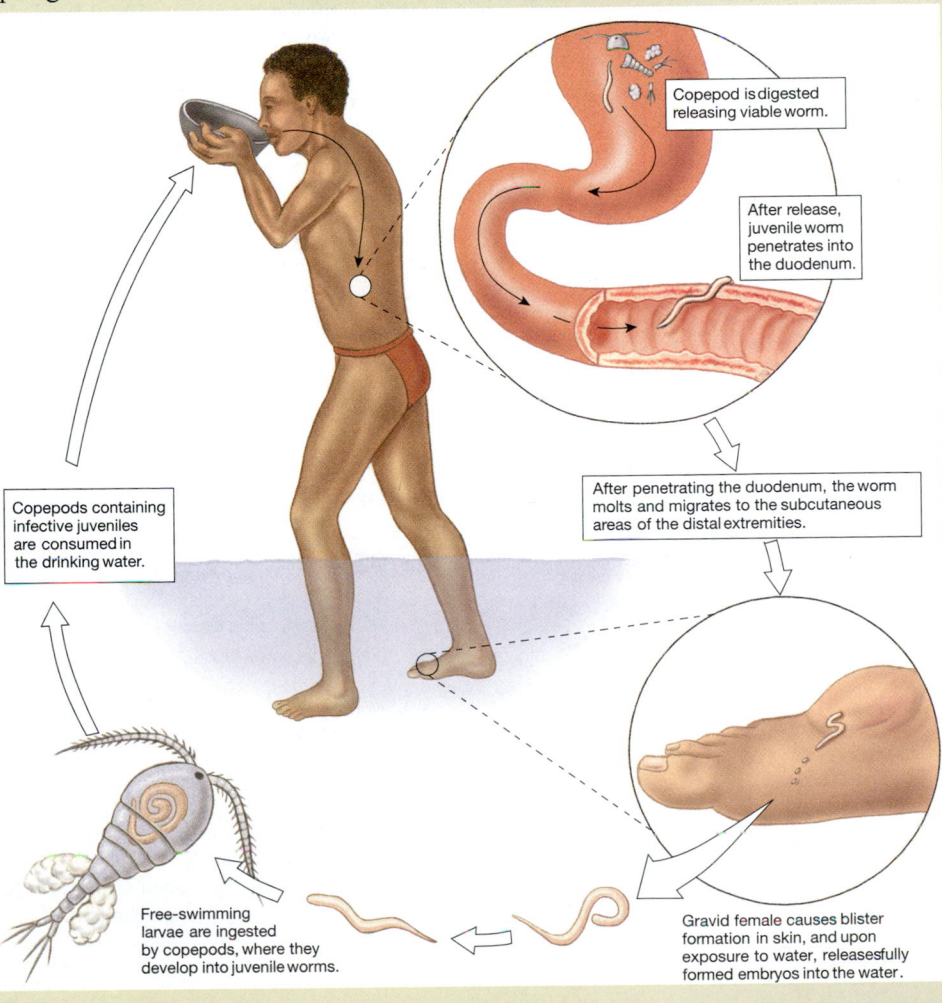

Copepod is digested releasing viable worm.

After release, juvenile worm penetrates into the duodenum.

After penetrating the duodenum, the worm molts and migrates to the subcutaneous areas of the distal extremities.

Copepods containing infective juveniles are consumed in the drinking water.

Gravid female causes blister formation in skin, and upon exposure to water, releasesfully formed embryos into the water.

Free-swimming larvae are ingested by copepods, where they develop into juvenile worms.

Arthropods

Characteristics of Arthropods

Arthropods constitute the largest group of living organisms; as many as 80 percent of all animal species belong to the phylum Arthropoda. Arthropods are characterized by jointed chitinous exoskeletons, segmented bodies, and jointed appendages associated with some or all of the segments. The name arthropod is derived from *arthros,* joint, and *podos,* foot. The exoskeleton both protects the organism and provides sites for the attachment of muscles. These organisms have a true coelom, which is filled with fluid that supplies nutrients, as blood does in higher organisms. Arthropods have a small brain and an extensive network of nerves. Various groups have different structures that extract oxygen from air or from aquatic environments. The sexes are distinct in arthropods, and females lay many eggs. Arthropods are found in nearly all environments—free-living in soil, on vegetation, in fresh and salt water, and as parasites on many plants and animals.

Classification of Arthropods

Certain members of three subgroups (classes) of arthropods, the arachnids, insects, and crustaceans (Table 11.4), are important either as parasites or as disease vectors (➤ Figure 11.19). The diseases transmitted by arthropods are summarized in Table 11.5.

Arachnids

Arachnids have two body regions—a cephalothorax and an abdomen, four pairs of legs, and mouthparts that are used in capturing and tearing apart prey. They include spiders, scorpions, ticks, and mites. Spider bites and scorpion stings can produce localized inflammation and tissue death, and their toxins can produce severe systemic effects. Ticks and mites are external parasites on many animals; some also serve as vectors of infectious agents.

Infected ticks transmit several human diseases. Certain species of *Ixodes* carry viruses that cause encephalitis and the spirochete *Borrelia burgdorferi,* which causes Lyme disease. The common tick *Dermacentor andersoni,* which can cause tick paralysis, can also carry the viruses that cause encephalitis and Colorado tick fever, the rickettsias that cause Rocky Mountain spotted fever, and the bacterium that causes tularemia. Several species of *Amblyoma* ticks also carry the Rocky Mountain spotted fever rickettsia, and *Ornithodorus* ticks transmit the spirochete responsible for relapsing fever. Mites serve as vectors for the rickettsial diseases scrub typhus fever and Q fever.

Insects

Insects have three body regions—head, thorax, and abdomen; three pairs of legs; and highly specialized mouthparts. Some insects have specialized mouthparts for piercing skin and for sucking blood and can inflict painful bites. Insects that can serve as vectors of disease include

> The crab, or pubic, louse dies in 2 days without food. Both sexes suck blood.

➤ Table 11.4	**Properties of three classes of arthropods**

Identifying group	Characteristic	Examples
Arachnids	Have eight legs	Spiders, scorpions, ticks, mites
Insects	Have six legs	Lice, fleas, flies, mosquitoes, true bugs
Crustaceans	A pair of appendages on each body segment	Crabs, crayfish, copepods

(a)   (b)   (c)

(d)   (e)

➤ Figure 11.19   **Representative arthropods that are parasitic or can serve as disease vectors.**
(a) A wood tick, *Dermacentor andersoni*. (b) False-color SEM of the pubic louse, *Phthirus pubis*, also
known as a crab louse, clinging to a human pubic hair (55X). The lice suck blood, feeding about five
times a day. (c) The housefly, *Musca domestica* (4X), can carry microbes on its body. (d) The *Aedes*
mosquito. (e) A flea, *Ctentocephalidis canis* (33X).

all lice and fleas and certain flies, mosquitoes, and true bugs, such as bedbugs and reduviid bugs.
Although we often refer to all insects as "bugs," entomologists—scientists who study insects—
use the term *true bug* to refer to certain insects that typically have thick, waxy wings and sucking, rather than biting, mouthparts.

The body louse is the main vector for the rickettsias that cause typhus and trench fevers and
a spirochete that causes relapsing fever. (This spirochete is a different species of *Borrelia* from
the one carried by ticks.) Epidemics of all louseborne diseases usually occur under crowded,
unsanitary conditions. All louseborne disease agents enter the body when louse feces are
scratched into bite wounds.

The human flea, *Pulex irritans,* lives on other hosts and can transmit plague. However, fleas
that normally parasitize rats and other rodents are more likely to transmit plague to humans.
This bacterial disease still occurs in the United States in individuals who have had contact with
wild rodents and their fleas.

Several kinds of flies feed on humans and serve as vectors for various diseases. The common
housefly, *Musca domestica,* is not part of the life cycle of any pathogens, yet it is an important
carrier of any pathogens found in feces. This fly is attracted to both human food and human excreta, and it leaves a trail of bacteria, vomit, and feces wherever it goes. Other insects, such as
blackflies, serve as vectors for *Onchocerca volvulus*. Sandflies serve as vectors for leishmanias,
for bacteria that cause bartonellosis, and for viruses that cause sandfly fever and several other diseases. Tsetse flies are vectors for trypanosomes that cause African sleeping sickness, and deer flies
are vectors for the worm that causes loaiasis. Eye gnats, which look like tiny houseflies, may be
responsible for transmission of bacterial conjunctivitis and the spirochete that causes yaws.

Head and body lice
interbreed, producing
fertile offspring.

> ## Table 11.5   Diseases transmitted by arthropods

Disease	Causative agents	Principal vectors	Endemic areas
Plague	*Yersinia pestis*	Fleas	Only sporadic in modern times; reservoir of infection maintained in rodents
Tularemia	*Francisella tularensis*	Fleas and ticks	Western United States
Salmonellosis	*Salmonella* species	Flies	Worldwide
Lyme disease	*Borrelia burgdorferi*	Ticks	Parts of United States, Australia, and Europe
Relapsing fever	*Borrelia* species	Ticks and lice	Rocky Mountains and Pacific Coast of United States; many tropical and subtropical regions
Typhus fever	*Rickettsia prowazekii*	Lice	Asia, North Africa, and Central and South America
Tick-borne typhus fever	*Rickettsia conorii*	Ticks	Mediterranean area; parts of Africa, Asia, and Australia
Scrub typhus fever	*Rickettsia tsutsugamushi*	Mites	Asia and Australia
Murine typhus fever	*Rickettsia typhi*	Fleas	Tropical and subtropical regions
Rocky Mountain spotted fever	*Rickettsia rickettsii*	Ticks	United States, Canada, Mexico, and parts of South America
Q fever	*Coxiella burnetii*	Ticks and mites	Worldwide
Trench fever	*Rochalimaea quintana*	Lice	Known only in fighting armies
Viral encephalitis	Togaviruses	Mosquitoes	Worldwide but varies by virus and vector
Yellow fever	Togavirus	Mosquitoes	Tropics and subtropics
Dengue fever	Togavirus	Mosquitoes	India, Far East, Hawaii, Caribbean Islands, and Africa
Sandfly fever	A virus, probably of bunyavirus family	Female sandfly	Mediterranean region, India, and parts of South America
Colorado tick fever	An orbivirus	Ticks	Western United States
Tick-borne encephalitis	Various viruses	Ticks	Europe and Asia
African sleeping sickness	Trypanosomes	Tsetse fly	Africa
Chagas' disease	*Trypanosoma cruzi*	True bug	South America
Kala azar and other leishmaniases	*Leishmania* species	Sandfly	Tropical and subtropical regions

Many species of mosquitoes serve as vectors for diseases. *Culex pipiens,* a common mosquito, breeds in any water and feeds at night. It is a vector for *Wuchereria.* Another mosquito, *C. tarsalis,* breeds in water in sunny locations and also feeds at night. It is a vector for viruses that cause western equine encephalitis (WEE) and St. Louis encephalitis. Although WEE most often causes severe illness in horses, it also can cause severe encephalitis in children and a milder disease with fever and central nervous system infections in adults. (The latter form is sometimes called sleeping sickness, but it should not be confused with African sleeping sickness.) Many species of *Aedes* play a role in human discomfort and disease. *Aedes aegypti* is a vector of a variety of viral diseases, including dengue fever (breakbone fever), yellow fever, and epidemic hemorrhagic fever. Several species of *Anopheles* serve as vectors for malaria. They have a variety of breeding habits, and thus control of them requires the application of several different eradication methods.

Several species of reduviid bugs transmit the parasite that causes Chagas' disease, which is a leading cause of cardiovascular disorders in Central and South America. Bedbugs cause dermatitis and may be responsible for spreading one kind of hepatitis, a liver infection.

Crustaceans

Crustaceans are generally aquatic arthropods that typically have a pair of appendages associated with each segment. Appendages include mouthparts, claws, walking legs, and appendages that aid in swimming or in copulation. Crustaceans that are hosts for disease agents that infect humans include some crayfish, crabs, and smaller crustaceans called copepods.

Retracing Our Steps

Principles of Parasitology

- A **parasite** is an organism that lives at the expense of another organism, the **host. Pathogens** are parasites that cause disease.
- **Parasitology** is the study of parasites, which typically include protozoa, helminths, and arthropods.

The Significance of Parasitism

- Parasites are responsible for much disease and death of humans, plants, and animals and for extensive economic losses.

Parasites in Relation to Their Hosts

- Parasites can live on or in hosts. Parasites can be **obligate** or **facultative** and **permanent, temporary,** or **accidental. Vectors** are agents of parasite transmission.
- Parasites reproduce sexually in **definitive hosts** and spend other life stages in **intermediate hosts. Reservoir hosts** can transmit parasites to humans.
- **Host specificity** refers to the number of different hosts in which a parasite can mature.
- Over time parasites become more adapted to and less destructive of their hosts. Most parasites have mechanisms to evade host defenses and exceptionally adept reproductive capacities.

Protists

Characteristics of Protists

- **Protists** are eukaryotic, and most are unicellular. They can be autotrophic or heterotrophic, and some are parasitic.

The Importance of Protists

- Protists are important in food chains as producers and decomposers; they can be economically beneficial or detrimental.

Classification of Protists

- Protists include plantlike organisms (such as **euglenoids, diatoms,** and **dinoflagellates**), funguslike organisms (the **water molds** and **slime molds**), and animal-like organisms (the **protozoa,** such as **mastigophorans, sarcodines, apicomplexans,** and **ciliates**). The groups of protists are summarized in Table 11.1.
- **Saprophytes** are organisms that feed on dead matter.

Fungi

Characteristics of Fungi

- **Fungi** are saprophytes or parasites that generally have a **mycelium,** a loosely organized mass consisting of threadlike **hyphae.** Most fungi reproduce both sexually and asexually, and their sexual stages are used to classify them.

The Importance of Fungi

- Fungi are important as decomposers in ecosystems and as parasites in the health sciences.

Classification of Fungi

- Fungi include **bread molds, sac fungi, club fungi,** and the **Fungi Imperfecti,** which cannot be classified in another group because either they lack a sexual stage or none has yet been identified. The groups of fungi are summarized in Table 11.2.

Helminths

Characteristics of Helminths

- **Helminths,** or worms, are bilaterally symmetrical and have head and tail ends and differentiated tissue layers.

Parasitic Helminths

- Only two groups of helminths, the flatworms and the roundworms (nematodes), contain parasitic species.
- **Flatworms** lack a **coelom,** have a simple digestive tract with one opening, and are **hermaphroditic.** They include **tapeworms** and **flukes.**
- **Roundworms** have a **pseudocoelom,** separate sexes, and a cylindrical body. They include hookworms, pinworms, and other parasites of the intestinal tract and lymphatics.

Arthropods

Characteristics of Arthropods

- **Arthropods** have jointed, chitinous exoskeletons, segmented bodies, and jointed appendages.

Classification of Arthropods

- Parasitic and vector arthropods include some arachnids and insects; a few crustacea also serve as intermediate hosts for human parasites. Arthropod vectors of disease are summarized in Table 11.5.
- **Arachnids** have eight legs; they include scorpions, spiders, ticks, and mites.
- **Insects** have six legs; they include lice, fleas, flies, mosquitoes, and true bugs.
- **Crustaceans** are generally aquatic arthropods, typically with a pair of appendages on each segment; they include crayfish, crabs, and copepods.

Terminology Check

definitive host   (p. 289)
Deuteromycota   (p. 304)
diatom   (p. 292)
dikaryotic   (p. 298)
dimorphism   (p. 300)
dinoflagellate   (p. 293)
ectoparasite   (p. 289)
endoparasite   (p. 289)
euglenoid   (p. 291)
eutrophication   (p. 291)
facultative parasite   (p. 289)
flatworm   (p. 304)
fluke   (p. 304)
Fungi Imperfecti   (p. 304)
fungus   (p. 297)
gametocyte   (p. 295)
helminth   (p. 304)
hermaphroditic   (p. 290)
host   (p. 288)
host specificity   (p. 289)
hydatid cyst   (p. 308)
hyperparasitism   (p. 289)
hypha   (p. 297)
insect   (p. 312)
intermediate host   (p. 289)

karyogamy   (p. 298)
mastigophoran   (p. 294)
mechanical vector   (p. 289)
merozoite   (p. 294)
metacercaria   (p. 306)
microfilaria   (p. 309)
miracidium   (p. 306)
mycelium   (p. 297)
mycology   (p. 297)
mycosis   (p. 299)
nematode   (p. 304)
obligate parasite   (p. 289)
Oomycota   (p. 293)
parasite   (p. 288)
parasitology   (p. 288)
pathogen   (p. 288)
pellicle   (p. 291)
permanent parasite   (p. 289)
plasmodial slime mold   (p. 294)
plasmodium   (p. 294)
plasmogamy   (p. 298)
proglottid   (p. 307)
protist   (p. 290)
protozoan   (p. 294)
pseudocoelom   (p. 304)

pseudoplasmodium   (p. 294)
redia   (p. 306)
reservoir host   (p. 289)
roundworm   (p. 304)
sac fungus   (p. 301)
saprophyte   (p. 293)
sarcodine   (p. 294)
schizogony   (p. 290)
scolex   (p. 307)
septum   (p. 298)
slime mold   (p. 293)
sporocyst   (p. 306)
sporozoite   (p. 294)
tapeworm   (p. 304)
temporary parasite   (p. 289)
test   (p. 291)
thallus   (p. 297)
trichocyst   (p. 295)
trophozoite   (p. 294)
vector   (p. 289)
water mold   (p. 293)
Zygomycota   (p. 300)
zygospore   (p. 300)

See You on the Web

Don't forget I am expecting you on the web. The following is a preparedness checklist for your web explorations. If you have questions on any of these items, you can return to the end of Chapter 1 for more details.

- Questions We'll Explore
- Concept Check questions
- Flashcards
- Study Guide

If you are ready, come visit me at

http://www.prenhall.com/black

It will be a journey you'll never forget.

Sterilization and Disinfection

"May the saints preserve me!" He should be so lucky! The peasant in this seventeenth-century painting by Flemish artist, David Teniers, has good reason to worry. Before the availability of good antiseptics and antibiotics, infections could quickly get out of hand. Even a small wound such as this could be a death sentence. Who would win: the microbes or the immune system? The microbes that have gained entry to the bloodstream through this cut could send our canvassed friend to his grave within days or weeks, dead of "blood poisoning."

Background Basics

D o you like spicy foods? Perhaps you won't like the original reasons for their popularity. Before modern methods of food preservation, such as canning and refrigeration, were available, control of microbial growth in foods was a difficult problem. Inevitably after a short while, food began to take on the "off" flavors of spoilage. Spices were used to mask these unpleasant tastes. Some spices were also effective as preservatives. The antimicrobial effects of garlic have long been known. Fortunately, we need not eat spoiled food today, and we can use spices solely to enhance our enjoyment of safely preserved foods.

Medical care, especially in the operating room, is also safer today. As we have seen from the work of Ignaz Semmelweis and Joseph Lister, careful washing and the use of chemical agents are effective in controlling many infectious microorganisms. ➡ (Chapter 1, p. 14) In this chapter, we will consider the properties of various chemical and physical agents used to control microorganisms in laboratories, in medical facilities, and in homes.

Principles of Sterilization and Disinfection

Agents called **disinfectants** are typically applied to inanimate objects, and agents called **antiseptics** are applied to living tissue. A few agents are suitable as both disinfectants and antiseptics, although most disinfectants are too harsh for use on delicate skin tissue. *Antibiotics,* though often applied to skin, are considered separately in Chapter 13.

Sterilization is the killing or removal of all microorganisms in a material or on an object. There are no degrees of sterility—**sterility** means that there are *no* living organisms in or on a material. When properly carried out, sterilization procedures ensure that even highly resistant bacterial endospores and fungal spores are killed. Much

Questions We'll Explore

A How do sterilization and disinfection differ, and what terms are used to describe these processes?

B What important principles apply to the processes of sterilization and disinfection?

C What factors affect the potency of antimicrobial chemical agents?

D How is the effectiveness of an antimicrobial chemical agent assessed?

E By what mechanisms do antimicrobial chemical agents act?

F What are the properties of commonly used antimicrobial chemical agents?

G How are dry heat, moist heat, and pasteurization used to control microorganisms?

H How are refrigeration, freezing, drying, and freeze-drying used to control and to preserve microorganisms?

I How are radiation, sonic and ultrasonic waves, filtration, and osmotic pressure used to control microorganisms?

of the controversy regarding spontaneous generation in the nineteenth century resulted from the failure to kill resistant cells in materials that were thought to be sterile. In contrast with sterilization, **disinfection** means reducing the number of pathogenic organisms on objects or in materials so that they pose no threat of disease. Terms related to sterilization and disinfection are defined in Table 12.1.

The Control of Microbial Growth

As explained in the discussion of the growth curves in Chapter 6, both the growth and death of microorganisms occur at logarithmic rates. ➡ (p. 139) Here we are concerned with the death rate and the effects of antimicrobial agents—substances that kill microbes or inhibit their growth—on it.

Organisms treated with antimicrobial agents obey the same laws regarding death rates as those declining in numbers from natural causes. We will illustrate this principle with heat as the agent because its effects have been the most thoroughly studied. When heat is applied to a material, the death rate of the organisms in or on it remains logarithmic but is greatly accelerated. Heat acts as an antimicrobial agent. If 20 percent of the organisms die in the first minute, 20 percent of those remaining alive will die in the second minute, and so on. If, at a different temperature, 30 percent die in the first minute, 30 percent of the remaining ones will die in the second minute, and so on. From these observations we can derive the principle that *a definite proportion of the organisms die in a given time interval.*

➢ Table 12.1	Terms related to sterilization and disinfection
Term	**Definition**
Sterilization	The killing or removal of all microorganisms in a material or on an object.
Disinfection	The reduction of the number of pathogenic microorganisms to the point where they pose no danger of disease.
Antiseptic	A chemical agent that can safely be used externally on living tissue to destroy microorganisms or to inhibit their growth.
Disinfectant	A chemical agent used on inanimate objects to destroy microorganisms. Most disinfectants do not kill spores.
Sanitizer	A chemical agent typically used on food-handling equipment and eating utensils to reduce bacterial numbers so as to meet public health standards. Sanitization may simply refer to thorough washing with only soap or detergent.
Bacteriostatic agent	An agent that inhibits the growth of bacteria.
Germicide	An agent capable of killing microbes rapidly; some such agents effectively kill certain microorganisms but only inhibit the growth of others.
Bactericide	An agent that kills bacteria. Most such agents do not kill spores.
Viricide	An agent that inactivates viruses.
Fungicide	An agent that kills fungi.
Sporocide	An agent that kills bacterial endospores or fungal spores.

Consider now what happens when the number of live organisms that remain becomes small—100, for example. At a death rate of 30 percent per minute, 70 will remain after 1 minute, 49 after 2 minutes, 34 after 3 minutes, and only 1 after 12 minutes. Soon the probability of finding even a single live organism becomes very small. Most laboratories say a sample is sterile if the probability is no greater than one chance in a million of finding a live organism.

The total number of organisms present when disinfection is begun affects the length of time required to eliminate them. We can state a second principle: *The fewer organisms present, the shorter the time needed to achieve sterility*. Thoroughly cleaning objects before attempting to sterilize them is a practical application of this principle. Clearing objects of tissue debris and blood is also important because such organic matter impairs the effectiveness of many chemical agents.

Different antimicrobial agents affect various species of bacteria and their endospores differently. Furthermore, any given species may be more susceptible to an antimicrobial agent at one phase of growth than at another. The most susceptible phase for most organisms is the logarithmic growth phase, because during that phase many enzymes are actively carrying out synthetic reactions, and interfering with even a single enzyme might kill the organism. From these observations, we can state a third principle: *Microorganisms differ in their susceptibility to antimicrobial agents.*

Chemical Antimicrobial Agents

The Potency of Chemical Agents

The potency, or effectiveness, of a chemical antimicrobial agent is affected by time, temperature, pH, and concentration. The death rate of organisms is affected by the length of time the organisms are exposed to the antimicrobial agent, as was explained earlier for heat. Thus, adequate time should always be allowed for an agent to kill the maximum number of organisms. The death rate of organisms subjected to a chemical agent is accelerated by increasing the temperature. Increasing temperature by 10°C roughly doubles the rate of chemical reactions and thereby increases the potency of the chemical agent. Acidic or alkaline pH can increase or decrease the agent's potency. A pH that increases the degree of ionization of a chemical agent often increases its ability to penetrate a cell. Such a pH also can alter the contents of the cell itself. Finally, increasing concentration may increase the effects of most antimicrobial chemical agents. High concentrations may be **bactericidal** (killing), whereas lower concentrations may be **bacteriostatic** (growth-inhibiting).

Both ethyl and isopropyl alcohol are exceptions to the rule regarding increasing concentrations. They have long been believed to be more potent at 70 percent than at higher concentrations, although they are also effective up to 99 percent concentration. Some water must be present for alcohols to disinfect because they act by coagulating (permanently denaturing) proteins, and water is needed for the coagulation reactions. Also, a 70 percent alcohol–water mixture penetrates more deeply than pure alcohol into most materials to be disinfected.

Evaluating the Effectiveness of Chemical Agents

Many factors affect the potency of chemical antimicrobial agents, so evaluation of effectiveness is difficult. No entirely satisfactory method is available. However, we need some way to compare the effectiveness of disinfecting agents, especially as new ones come on the market.

The Phenol Coefficient

Since Lister introduced *phenol* (carbolic acid) as a disinfectant in 1867, it has been the standard disinfectant to which other disinfectants are compared under the same conditions. The result of this comparison is called the **phenol coefficient.** Two organisms, *Salmonella typhi,* a pathogen of the digestive system, and *Staphylococcus aureus,* a common wound pathogen, are typically used to determine phenol coefficients. A disinfectant with a phenol coefficient of 1.0 has the same effectiveness as phenol. A coefficient less than 1.0 means that the disinfectant is less effective than phenol; a coefficient of greater than 1.0 means that it is more effective. Phenol coefficients

	Table 12.2	Phenol coefficients of various chemical agents	
Chemical agent		*Staphylococcus aureus*	*Salmonella typhi*
Phenol		1.0	1.0
Chloramine		133.0	100.0
Cresols		2.3	2.3
Ethyl alcohol		6.3	6.3
Formalin		0.3	0.7
Hydrogen peroxide		—	0.01
Lysol		5.0	3.2
Mercury chloride		100.0	143.0
Tincture of iodine		6.3	5.8

(a)

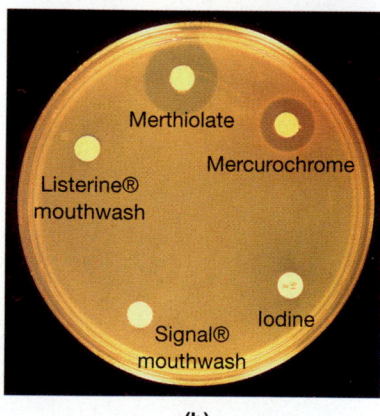

(b)

➤ **Figure 12.1 The filter paper method of evaluating disinfectants and antiseptics.** The difference in response of (a) *Staphylococcus aureus* (Gram-positive) and (b) *Escherichia coli* (Gram-negative) to several common chemical agents is shown here. In both cases, the greatest inhibition of growth is seen near the top of the Petri dish surrounding the merthiolate-soaked paper disk. The various disks were soaked in mercurochrome, iodine, Signal® mouthwash, or Listerine® mouthwash before being placed on the surface of the nutrient medium, which had first been confluently inoculated with one of the test organisms.

are reported separately for the different test organisms (Table 12.2). Lysol, for instance, has a coefficient of 5.0 against *Staphylococcus aureus* but only 3.2 when used on *Salmonella typhi*, whereas ethyl alcohol has a value of 6.3 against both.

The phenol coefficient can be determined by the following steps. Prepare several dilutions of a chemical agent, and place the same volume of each in different test tubes. Prepare an identical set of test tubes, using phenol dilutions. Put both sets of tubes in a 20°C water bath for at least 5 minutes to ensure that the contents of all tubes are at the same temperature. Transfer 0.5 ml of a culture of a standard test organism to each tube. After 5, 10, and 15 minutes, use a sterile loop to transfer a specific volume of liquid from each tube into a separate tube of nutrient broth, and incubate the tubes. After 48 hours, check cultures for cloudiness, and find the smallest concentration (highest dilution) of the agent that killed all organisms in 10 minutes but not in 5 minutes. Find the ratio of this dilution to the dilution of phenol that has the same effect. For example, if a 1:1000 dilution of a chemical agent has the same effect as a 1:100 dilution of phenol, the phenol coefficient of that agent is 10 (1000/100). If you performed this test on a new disinfectant and obtained these results, you would have found a very good disinfectant! The phenol coefficient provides an acceptable means of evaluating the effectiveness of chemical agents derived from phenol, but it is less acceptable for other agents. Another problem is that the materials on or in which organisms are found may affect the usefulness of a chemical agent by complexing with it or inactivating it. These effects are not reflected in the phenol coefficient number.

The Filter Paper Method

The **filter paper method** of evaluating a chemical agent is simpler than determining a phenol coefficient. It uses small filter paper disks, each soaked with a different chemical agent. The disks are placed on the surface of an agar plate that has been inoculated with a test organism. A different plate is used for each test organism. After incubation, a chemical agent that inhibits growth of a test organism is identified by a clear area around the disk where the bacteria have been killed (➤ Figure 12.1). Note: What is effective against one organism may have little or no effect on the others.

The Use-Dilution Test

A third way of evaluating chemical agents, the **use-dilution test,** uses standard preparations of certain test bacteria. A broth culture of one of these bacteria is coated onto small stainless steel cylinders and allowed to dry. Each cylinder is then dipped into one of several dilutions of the chemical agent for 10 minutes, removed, rinsed with water, and placed into a tube of broth. The tubes are incubated and then observed for the presence or absence of growth. Agents that prevent growth at the greatest dilutions are considered the most effective. Many microbiologists feel that this measurement is more meaningful than the phenol coefficient.

Disinfectant Selection

Several qualities should be considered in deciding which disinfectant to use. An ideal disinfectant should

1. Be fast acting even in the presence of organic substances, such as those in body fluids;
2. Be effective against all types of infectious agents without destroying tissues or acting as a poison if ingested;
3. Easily penetrate material to be disinfected without damaging or discoloring the material;
4. Be easy to prepare and stable even when exposed to light, heat, or other environmental factors;
5. Be inexpensive and easy to obtain and use;
6. Not have an unpleasant odor.

No disinfectant is likely to satisfy all these criteria, so the agent that meets the greatest number of criteria for the task at hand is chosen.

In practice, many agents are tested in a wide range of situations and are recommended for use where they are most effective. Thus, some agents are selected for sanitizing kitchen equipment and eating utensils, whereas other agents are chosen for rendering pathogenic cultures harmless. Furthermore, certain agents can be used in dilute concentration on the skin and in stronger concentration on inanimate objects.

✓ Are there degrees of sterility? Why or why not?

✓ Is it worth paying extra for a "sanitized" item? Explain.

✓ Disinfectant A has a phenol coefficient of 0.5; disinfectant B has one of 5.0. How do these two compare to phenol?

Mechanisms of Action of Chemical Agents

Chemical antimicrobial agents kill microorganisms by participating in one or more chemical reactions that damage cell components. Although the kinds of reactions are almost as numerous as the agents, agents can be grouped by whether they affect proteins, membranes, or other cell components.

Reactions That Affect Proteins

Much of a cell is made of protein, and all its enzymes are proteins. Alteration of protein structure is called *denaturation*. ➡ (Chapter 2, p. 41) In denaturation, hydrogen and disulfide bonds are disrupted, and the functional shape of the protein molecule is destroyed. Any agent that denatures proteins prevents them from carrying out their normal functions. When treated with mild heat or with some dilute acids, alkalis, or other agents for a short time, proteins are temporarily denatured. After the agent is removed, some proteins can regain their normal structure. However, most antimicrobial agents are used in a strong enough concentration over a sufficient length of time to denature proteins permanently. Permanent denaturation of a microorganism's proteins kills the organism. Denaturation is bactericidal if it permanently alters the protein such that the protein's normal state cannot be restored. Denaturation is bacteriostatic if it temporarily alters the protein, and the normal structure can be recovered (➢ Figure 12.2).

Reactions that denature proteins include hydrolysis, oxidation, and the attachment of atoms or chemical groups. (Recall that hydrolysis is the breaking down of a molecule by the addition of water and that oxidation is the addition of oxygen to, or the removal of hydrogen from, a molecule.) ➡ (Chapter 2, p. 32) Acids, such as boric acid, and strong alkalis destroy protein by hydrolyzing it. Oxidizing agents (electron acceptors), such as hydrogen peroxide and potassium permanganate, oxidize disulfide linkages (—S—S—) or sulfhydryl groups (—SH). Agents that contain halogens— the elements chlorine, fluorine, bromine, and iodine—also sometimes act as oxidizing agents. Heavy metals, such as mercury and silver, attach to sulfhydryl groups. Alkylating agents, which contain methyl (—CH_3) or similar groups, donate these groups to proteins. Formaldehyde and some dyes are alkylating agents. Halogens can be substituted for hydrogen in carboxyl (—COOH), sulfhydryl, amino (—NH_2), and alcohol (—OH) groups. All these reactions can kill microorganisms.

Reactions That Affect Membranes

Membranes contain proteins and so can be altered by all the preceding reactions. Membranes also contain lipids and thus can be disrupted by substances that dissolve lipids. **Surfactants** (surfak'tantz) are soluble compounds that reduce surface tension, just as soaps and detergents break

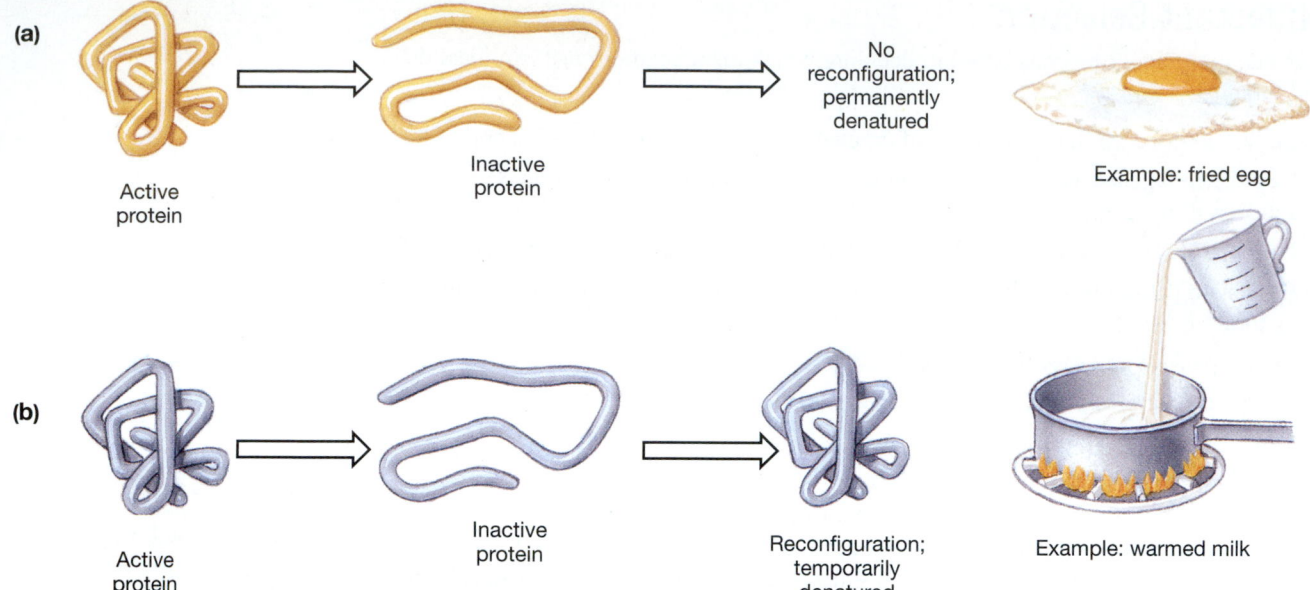

> **Figure 12.2**   **Denaturing proteins.** (a) A permanently denatured protein, like that of a fried egg, cannot return to its original configuration. (b) A temporarily denatured protein, like that in warmed milk, can refold into its original configuration. The protein structure of milk that has been warmed is recovered when the milk is cooled.

up grease particles in dishwater (➤ Figure 12.3). Surfactants include alcohols, detergents, and *quaternary ammonium compounds,* such as benzalkonium chloride, which dissolve lipids. Phenols, which are alcohols, dissolve lipids and also denature proteins. Detergent solutions, also called **wetting agents,** are often used with other chemical agents to help the agent penetrate fatty substances. Although detergent solutions themselves usually do not kill microorganisms, they do help get rid of lipids and other organic materials so that antimicrobial agents can reach the organisms.

Sponges should not be used in the microbiology lab, as soap does not prevent bacterial growth. Sponges spread "bacterial broth" on surfaces. Use paper towels!

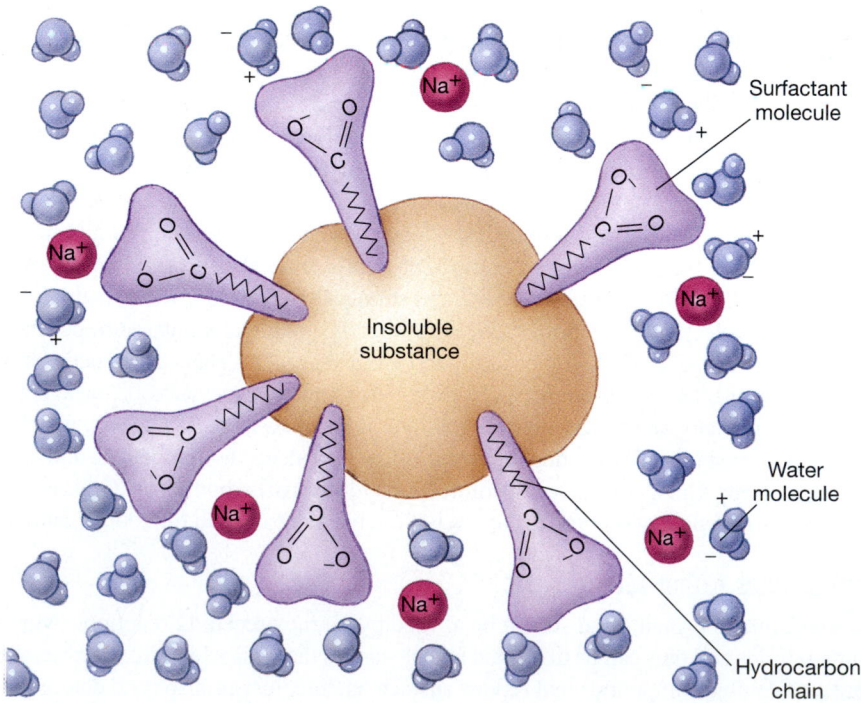

> **Figure 12.3**   **The action of a surfactant.** Here the surfactant molecule has ionized into sodium ions and long hydrocarbon chains, whose zigzagged, covalently bonded tails are able to enter an insoluble substance such as grease. The other end of these molecules has a carboxyl group with a negatively charged oxygen. These negative charges attract the positively charged sides of water molecules, thereby making the attached insoluble substance soluble in the water so the substance can be washed away.

Reactions That Affect Other Cell Components

Other cell components affected by chemical agents include nucleic acids and energy-producing systems. Alkylating agents can replace hydrogen on amino or alcohol groups in nucleic acids. Certain dyes, such as crystal violet, interfere with cell wall formation. Some substances, such as lactic acid and propionic acid (end products of fermentation), inhibit fermentation and thus prevent energy production in certain bacteria, molds, and some other organisms.

Reactions That Affect Viruses

Like many cellular microorganisms, viruses can cause infections and must be controlled. Control of viruses requires that they be inactivated—that is, rendered permanently incapable of infecting or replicating in cells. Inactivation can be effected by destroying either the viruses' nucleic acid or their proteins.

Alkylating agents, such as ethylene oxide, nitrous acid, and hydroxylamine act as chemical mutagens—they alter DNA or RNA. If the alteration prevents the DNA or RNA from directing the synthesis of new viral particles, the alkylating agents are effective inactivators. Detergents, alcohols, and other agents that denature proteins act on bacteria and viruses in the same way. Certain dyes, such as acridine orange and methylene blue, render viruses susceptible to inactivation when exposed to visible light. This process disrupts the structure of the viral nucleic acid.

Viruses sometimes remain infective even after their proteins are denatured, so methods used to rid materials of bacteria may not be as successful with infectious viruses. Also, use of an agent that does not inactivate viruses can lead to laboratory-acquired infections.

Specific Chemical Antimicrobial Agents

Now that we have considered general principles of sterilization and disinfection and the kinds of reactions caused by such agents, we can look at some specific agents and their applications. The structural formulas of some of the most important compounds discussed are shown in ➤ Figure 12.4.

Soaps and Detergents

Soaps and detergents remove microbes, oily substances, and dirt. Mechanical scrubbing greatly enhances their action. In fact, vigorous hand washing is one of the easiest and cheapest means of preventing the spread of disease among patients in hospitals, in medical and dental offices, among employees and patrons in food establishments, and among family members. Unlike surgical scrubs, germicidal soaps usually are not significantly better disinfectants than ordinary soaps.

Soaps contain alkali and sodium and will kill many species of *Streptococcus, Micrococcus,* and *Neisseria* and will destroy influenza viruses. Many pathogens that survive washing with

Soap and Sanitation

Washing and drying clothing in modern public laundry facilities is generally a safe practice because the clothing is almost disinfected if the water temperature is high enough. Soaps, detergents, and bleaches kill many bacteria and inactive many viruses. Agitation of the clothes in the washer provides good mechanical scrubbing. Many microbes that survive this action are killed by heat in the dryer. The use of bar soap in a public washroom is not such a safe practice. The soap may be a source of infectious agents. In a study of 84 samples of bar soap taken from public washrooms, every sample contained microorganisms. More than 100 strains of bacteria and fungi were isolated from the soap samples, and some of the organisms were potential pathogens. Many restaurants and other establishments have installed soap dispensers for this very reason. In fact, many jurisdictions have made the use of bar soap in such facilities illegal.

➤ Figure 12.4   **Structural formulas of some important disinfectants.**

soap can be killed by a disinfectant applied after washing. A common practice after washing and rinsing hands and inanimate objects is to apply a 70 percent alcohol solution. Even these measures do not necessarily rid hands of all pathogens. Consequently, disposable gloves are used where there is a risk that health care workers may become infected or may transmit pathogens to other patients.

Detergents, when used in weak concentrations in wash water, allow the water to penetrate into all crevices and cause dirt and microorganisms to be lifted out and washed away. Detergents are said to be *cationic* if they are positively charged and *anionic* if they are negatively charged. Cationic detergents are used to sanitize food utensils. Although not effective in killing endospores, they do inactivate some viruses. Anionic detergents are used for laundering clothes and as household cleaning agents. They are less effective sanitizing agents than cationic detergents, probably because the negative charges on bacterial cell walls repel them.

Many cationic detergents are **quaternary ammonium compounds,** or **quats,** which have four organic groups attached to a nitrogen atom. A variety of quats are available as disinfecting agents; their chemical structures vary according to their organic groups. One problem with quats is that their effectiveness is decreased in the presence of soap, calcium or magnesium ions, or porous substances such as gauze. An even more serious problem with these agents is that they support the growth of some bacteria of the genus *Pseudomonas* rather than killing them. Zephiran (benzalkonium chloride) was once widely used as a skin antiseptic. It is no longer recommended because it is less effective than originally thought and is subject to the same problems as other quats. Quats are now often mixed with another agent to overcome some of these problems and to increase their effectiveness.

Acids and Alkalis

Soap is a mild alkali, and its alkaline properties help destroy microbes. A number of organic acids lower the pH of materials sufficiently to inhibit fermentation. Several are used as food preservatives. Lactic and propionic acids retard mold growth in breads and other products. Benzoic acid and several of its derivatives are used to prevent fungal growth in soft drinks, catsup, and margarine. Sorbic acid and sorbates are used to prevent fungal growth in cheeses and a variety of other foods. Boric acid, formerly used as an eyewash, is no longer recommended because of its toxicity.

Heavy Metals

Heavy metals used in chemical agents include selenium, mercury, copper, and silver. Even tiny quantities of such metals can be very effective in inhibiting bacterial growth (➢ Figure 12.5). Silver nitrate was once widely used to prevent gonococcal infection in newborn infants. A few drops of silver nitrate solution were placed in the baby's eyes at the time of delivery to protect against infection by gonococci entering the eyes during passage through the birth canal. For a time, many hospitals replaced silver nitrate with antibiotics such as erythromycin. However, the development of antibiotic-resistant strains of gonococci has led some localities to require the use of silver nitrate, to which gonococci do not develop resistance.

Organic mercury compounds, such as merthiolate and mercurochrome, are used to disinfect surface skin wounds. Such agents kill most bacteria in the vegetative state but do not kill spores. They are not effective against *Mycobacterium*. Merthiolate is generally prepared as a **tincture** (tingk'chur), that is, dissolved in alcohol. The alcohol in a tincture may have a greater germicidal action than the heavy metal compound. Thimerosal, another organic mercury compound, can be used to disinfect skin and instruments and as a preservative for vaccines. Phenylmercuric nitrate and mercuric naphthenate inhibit both bacteria and fungi and are used as laboratory disinfectants.

Selenium sulfide kills fungi, including spores. Preparations containing selenium are commonly used to treat fungal skin infections. Shampoos that contain selenium are effective in controlling dandruff. Dandruff, a crusting and flaking of the scalp, is often, though not always, caused by fungi.

Copper sulfate is used to control algal growth. Although algal growth usually is not a direct medical problem, it is a problem in maintaining water quality in heating and air-conditioning

---

For a Clear Aquarium

If you have an aquarium, you probably have contended with water that looks like pea soup because of the large numbers of algae growing in it. This problem can be corrected by placing a few pennies in the tank. Enough copper to inhibit algal growth dissolves from the pennies into the water. For this small investment, you can greatly increase visibility and enjoyment of your fish.

Area of inhibition

➢ **Figure 12.5   Heavy metals inhibit bacterial growth.** The inhibitory effects of silver ions can be seen as clear zones in which no growth has occurred around the silver charm (which has been pushed aside) and silver dime. The nonsilver coin (a copper penny) has not inhibited growth of the organisms as effectively as the silver objects have.

systems and outdoor swimming pools. (The Environmental Protection Agency, however, is evaluating copper sulfate as an environmental hazard.)

Halogens

Hypochlorous acid, formed by the addition of chlorine to water, effectively controls microorganisms in drinking water and swimming pools. It is the active ingredient in household bleach and is used to disinfect food utensils and dairy equipment. It is effective in killing bacteria and inactivating many viruses. However, chlorine itself is easily inactivated by the presence of organic materials. That is why a substance such as copper sulfate is used to control algal growth in water to be purified with chlorine.

Iodine also is an effective antimicrobial agent. Tincture of iodine was one of the first skin antiseptics to come into use. Now *iodophors,* slow-release compounds in which the iodine is combined with organic molecules, are more commonly used. In such preparations, the organic molecules act as surfactants. Betadine and Isodine are used for surgical scrubs and on skin where an incision will be made. These compounds take several minutes to act and do not sterilize the skin. Betadine in concentrations of 3 to 5 percent destroys fungi, amoebas, and viruses as well as most bacteria, but it does not destroy bacterial endospores. Contamination of Betadine with *Pseudomonas cepacia* has been reported.

Bromine is sometimes used in the form of gaseous methyl bromide to fumigate soil that will be used in the propagation of bedding plants. It is also used in some pools and indoor hot tubs because it does not give off the strong odor that chlorine does.

Alcohols

When mixed with water, alcohols denature protein. They are also lipid solvents and dissolve membranes. Ethyl and isopropyl alcohols can be used as skin antiseptics. Isopropyl alcohol is more often used due to legal regulation of ethyl alcohol. It disinfects skin where injections will be made or blood drawn. Alcohol disinfects but does not sterilize skin because it evaporates quickly and stays in contact with microbes for only a few seconds. It also does not penetrate deeply enough into pores in the skin. It kills vegetative microorganisms on the skin surface but does not kill endospores, resistant cells, or cells deep in skin pores.

Phenols

Phenol and phenol derivatives called *phenolics* disrupt cell membranes, denature proteins, and inactivate enzymes. They are used to disinfect surfaces and to destroy discarded cultures because their action is not impaired by organic materials. Amphyl, which contains amylphenol, destroys vegetative forms of bacteria and fungi and inactivates viruses. It can be used on skin, instruments, dishes, and furniture. When used on surfaces, it retains its antimicrobial action for several days. The orthophenylphenol in Lysol gives it similar properties. A mixture of phenol derivatives called *cresols* is found in creosote, a substance used to prevent the rotting of wooden posts, fences, railroad ties, and such. However, because creosote is irritating to skin and a carcinogen, its use is limited. The addition of halogens to phenolic molecules usually increases their effectiveness. Hexachlorophene and dichlorophene, which are halogenated phenols, inhibit staphylococci and fungi, respectively, on the skin and elsewhere. Chlorhexidine gluconate (Hibiclens), which is chlorinated and similar in structure to hexachlorophene, is effective against a wide variety of microbes even in the presence of organic material. It is a good agent for surgical scrubs.

Oxidizing Agents

Oxidizing agents disrupt disulfide bonds in proteins and thus disrupt the structure of membranes and proteins. Hydrogen peroxide (H_2O_2), which forms highly reactive superoxide (O_2^-), is used to clean puncture wounds. When hydrogen peroxide breaks down into oxygen and water, the oxygen kills obligate anaerobes present in the wounds. Hydrogen peroxide is quickly inactivated by enzymes from injured tissues. A recently developed method of sterilization that uses vaporized hydrogen peroxide can now be used for small rooms or areas, such as glove boxes and transfer hoods (➤ Figure 12.6). Another oxidizing agent, potassium permanganate, is used to disinfect instruments and, in low concentrations, to clean skin.

Hexachlorophene

Hexachlorophene is an excellent skion disinfectant. In a 3 percent solution, it kills staphylococci and most other Gram-positive organisms, and its residue on skin is strongly bacteriostatic. Because staphylococcal skin infections can spread easily among newborn babies in hospitals, this antiseptic was used extensively in the 1960s for daily bathing of infants. The unforeseen price paid for controlling infections was permanent brain damage in infants bathed in it over a period of time. Hexachlorophene is absorbed through the skin and travels in the blood to the brain. Baby powder containing hexachlorophene killed 40 babies in France in 1972. Available only by prescription in the United States today, hexachlorophene is used routinely, though very cautiously, in hospital neonatal units because it is still the most effective agent for preventing the spread of staphylococcal infections.

➤ Figure 12.6   **Hydrogen peroxide disinfection.** Recently developed biodecontamination equipment uses vaporized hydrogen peroxide to sterilize small, sealable enclosures such as glove boxes, hoods, or transfer rooms. It is not sufficient, however, to sterilize a larger space such as an operating room.

No More Frogs in Formalin

Formalin, a 37 percent aqueous solution of formaldehyde, was for many years the standard material used to preserve laboratory specimens for dissection. But formaldehyde is toxic to tissues and may cause cancer, so it is rarely used today as a preservative. A variety of other preservatives are now used. Although less toxic to students performing dissections, they are also less effective for long-term preservation. Molds growing on the surface of specimens are now a common problem.

➤ **Figure 12.7** **Ethylene oxide sterilization.** This equipment must be used very carefully, as ethylene oxide is both explosive and carcinogenic. An aerator is used to remove all traces of the gas from sterilized material.

✓ How does a surfactant act?

✓ If bacteria can grow in soap, why do we use it to clean things?

✓ Explain the antimicrobial actions of acridine, mercurochrome, and Lysol.

✓ What are the drawbacks of using ethylene oxide?

Alkylating Agents

Alkylating agents disrupt the structure of both proteins and nucleic acids. Because they can disrupt nucleic acids, these agents may cause cancer and should not be used in situations where they might affect human cells. Formaldehyde, glutaraldehyde, and β-propiolactone are used in aqueous solutions. Ethylene oxide is used in gaseous form.

Formaldehyde inactivates viruses and toxins without destroying their antigenic properties. Glutaraldehyde kills all kinds of microorganisms, including spores, and sterilizes equipment exposed to it for 10 hours. Beta-propiolactone destroys hepatitis viruses, as well as most other microbes, but penetrates materials poorly. It is, however, used to inactivate viruses in vaccines.

Gaseous ethylene oxide has extraordinary penetrating power. Used at a concentration of 500 mg/l at 50°C for 4 hours, it sterilizes rubber goods, mattresses, plastics, and other materials destroyed by higher temperatures. Special equipment used during ethylene oxide sterilization is shown in ➤ Figure 12.7. As will be explained when we discuss autoclaving, an *ampule* (a sealed glass container) of endospores should be processed with ethylene oxide sterilization to check the effectiveness of sterilization.

All articles sterilized with ethylene oxide must be well ventilated with sterile air to remove all traces of this toxic gas, which can cause burns if it reaches living tissues and is also highly explosive. After exposure to ethylene oxide, articles such as catheters, intravenous lines, in-line valves, and rubber tubing must be thoroughly flushed with sterile air. Both the toxicity and flammability of ethylene oxide can be reduced by using it in gas containing 90 percent carbon dioxide. *It is exceedingly important that workers be protected from ethylene oxide vapors, which are toxic to skin, eyes, and mucous membranes and may also cause cancer.*

Dyes

The dye acridine, which interferes with cell replication by causing mutations in DNA (Chapter 7), can be used to clean wounds. ➡ (p. 186) Methylene blue inhibits growth of some bacteria in cultures. Crystal violet (gentian violet) blocks cell wall synthesis, possibly by the same reaction that causes this dye to bind to cell wall material in Gram staining. It effectively inhibits growth of Gram-positive bacteria in cultures and in skin infections. It can be used to treat protozoan (*Trichomonas*) and yeast (*Candida albicans*) infections.

Other Agents

Certain plant oils have special antimicrobial uses. Thymol, derived from the herb thyme, is used as a preservative, and eugenol, derived from oil of cloves, is used in dentistry to disinfect cavities. A variety of other agents are used primarily as food preservatives. They include sulfites and sulfur dioxide, used to preserve dried fruits and molasses; sodium diacetate, used to retard mold in bread; and sodium nitrite, used to preserve cured meats and some cold cuts. Foods containing nitrites should be eaten in moderation because the nitrites are converted during digestion to substances that may cause cancer.

The properties of chemical antimicrobial agents are summarized in Table 12.3.

Physical Antimicrobial Agents

For centuries, physical antimicrobial agents have been used to preserve food. Ancient Egyptians dried perishable foods to preserve them. Scandinavians made holes in the centers of pieces of dry, flat, crisp bread in order to hang them in the air of their homes during the winter; likewise they kept seed grains in a dry place. Otherwise, both flour and grains would have molded during the long and very moist winters. Europeans used heat in the food-canning process 50 years before Pasteur's work explained why heating prevented food from spoiling. Today, physical agents that destroy microorganisms are still used in food preservation and preparation. Such agents remain a crucial weapon in the prevention of infectious disease. Physical antimicrobial agents include various forms of heat, refrigeration, desiccation (drying), irradiation, and filtration.

> **Table 12.3** **Properties of chemical antimicrobial agents**

Agent	Actions	Uses
Soaps and detergents	Lower surface tension, make microbes accessible to other agents	Hand washing, laundering, sanitizing kitchen and dairy equipment
Surfactants	Dissolve lipids, disrupt membranes, denature proteins, and inactivate enzymes in high concentrations; act as wetting agents in low concentrations	Cationic detergents are used to sanitize utensils; anionic detergents to launder clothes and clean household objects; quaternary ammonium compounds are sometimes used as antiseptics on skin.
Acids	Lower pH and denature proteins	Food preservation
Alkalis	Raise pH and denature proteins	Found in soaps
Heavy metals	Denature proteins	Silver nitrate is used to prevent gonococcal infections, mercury compounds to disinfect skin and inanimate objects, copper to inhibit algal growth, and selenium to inhibit fungal growth.
Halogens	Oxidize cell components in absence of organic matter	Chlorine is used to kill pathogens in water and to disinfect utensils; iodine compounds are used as skin antiseptics.
Alcohols	Denature proteins when mixed with water	Isopropyl alcohol is used to disinfect skin; ethylene glycol and propylene glycol can be used in aerosols.
Phenols	Disrupt membranes, denature proteins, and inactivate enzymes; not impaired by organic matter	Phenol is used to disinfect surfaces and destroy discarded cultures; amylphenol destroys vegetative organisms and inactivates viruses on skin and inanimate objects; chlorhexidine gluconate is especially effective as a surgical scrub.
Oxidizing agents	Disrupt disulfide bonds	Hydrogen peroxide is used to clean puncture wounds, potassium permanganate to disinfect instruments.
Alkylating agents	Disrupt structure of proteins and nucleic acids	Formaldehyde is used to inactivate viruses without destroying antigenic properties, glutaraldehyde to sterilize equipment, betapropiolactone to destroy hepatitis viruses, and ethylene oxide to sterilize inanimate objects that would be harmed by high temperatures.
Dyes	May interfere with replication or block cell wall synthesis	Acridine is used to clean wounds, crystal violet to treat some protozoan and fungal infections.

Principles and Applications of Heat Killing

Heat is a preferred agent of sterilization for all materials not damaged by it. It rapidly penetrates thick materials not easily penetrated by chemical agents. Several measurements have been defined to quantify the killing power of heat. The **thermal death point** is the temperature that kills all the bacteria in a 24-hour-old broth culture at neutral pH in 10 minutes. The **thermal death time** is the time required to kill all the bacteria in a particular culture at a specified temperature. The **decimal reduction time,** also known as the **DRT** or **D value,** is the length of time needed to kill 90 percent of the organisms in a given population at a specified temperature. (The temperature is indicated by a subscript: $D_{80°C}$, for example.)

These measurements have practical significance in industry as well as in the laboratory. For example, a food-processing technician wanting to sterilize a food as quickly as possible would determine the thermal death point of the most resistant organism that might be present in the food and would employ that temperature. In another situation it might be preferable to make the food safe for human consumption by processing it at the lowest possible temperature. This could be important in processing foods containing proteins that would be denatured, thereby altering their flavor or consistency. The processor would then need to know the thermal death time at the desired temperature for the most resistant organism likely to be in the food.

Dry Heat, Moist Heat, and Pasteurization

Dry heat probably does most of its damage by oxidizing molecules. *Moist heat* destroys microorganisms mainly by denaturing proteins; the presence of water molecules helps disrupt the hydrogen bonds and other weak interactions that hold proteins in their three-dimensional shapes.

It is impossible to sterilize skin except by burning!

> **Figure 12.8   Hot-air oven used for sterilizing metal and glass items.**

Prions are highly resistant and must be sterilized by longer and higher temperature autoclaving (134°C for 18 minutes).

(Chapter 2, p. 41) Moist heat may disrupt membrane lipids as well. Heat also inactivates many viruses, but those that can infect even after their protein coats are denatured require extreme heat treatment, such as steam under pressure, that will disrupt nucleic acids.

Dry Heat

Dry (oven) heat penetrates substances more slowly than moist (steam) heat. It is usually used to sterilize metal objects and glassware and is the only suitable means of sterilizing oils and powders (> Figure 12.8). Objects are sterilized by dry heat when subjected to 171°C for 1 hour, 160°C for 2 hours or longer, or 121°C for 16 hours or longer, depending on the volume.

An open flame is a form of dry heat used to sterilize inoculating loops and the mouths of culture tubes by incineration and to dry the inside of pipettes. When flaming objects in the laboratory, you must avoid the formation of floating ashes and **aerosols** (droplets released into the air). These substances can be a means of spreading infectious agents if the organisms in them are not killed by incineration as intended. For this reason, specially designed loop incinerators with deep throats are often used for sterilizing inoculating loops.

Moist Heat

Moist heat, because of its penetrating properties, is a widely used physical agent. Boiling water destroys vegetative cells of most bacteria and fungi and inactivates some viruses, but it is not effective in killing all kinds of spores. The effectiveness of boiling can be increased by adding 2 percent sodium bicarbonate to the water. However, if water is heated under pressure, its boiling point is elevated, so temperatures above 100°C can be reached. This is normally accomplished by using an **autoclave** (aw'to-klav), as shown in > Figure 12.9, in which a pressure of 15 lb/in.^{2} above atmospheric pressure is maintained for 15 to 20 minutes, depending on the volume of the load. At this pressure, the temperature reaches 121°C, which is high enough to kill spores as well as vegetative organisms and to disrupt the structure of nucleic acids in viruses. In this procedure it is the increased temperature, and not the increased pressure, that kills microorganisms.

Sterilization by autoclaving is invariably successful if properly done and if two common-sense rules are followed: First, articles should be placed in the autoclave so that steam can easily penetrate them; second, air should be evacuated so that the chamber fills with steam. Wrapping objects in aluminum foil is not recommended because it may interfere with steam penetration. Steam circulates through an autoclave from a steam outlet to an air evacuation port (> Figure 12.10). In preparing items for autoclaving, containers should be unsealed and articles should be wrapped in materials that allow steam penetration. Large packages of dressings and large flasks of media require extra time for heat to penetrate them. Likewise, packing many articles close together in an autoclave lengthens the processing time to as much as 60 minutes to ensure sterility. It is more efficient and safer to run two separate, uncrowded loads than one crowded one.

Several methods are available to ensure that autoclaving achieves sterility. Modern autoclaves have devices to maintain proper pressure and record internal temperature during operation. Regardless of the presence of such a device, the operator should check pressure

> **Figure 12.9   A small counter-top autoclave.**

Exhaust valve

Steam to chamber valve

Safety valve

Pressure gauge

Manual exhaust to atmosphere

Door

Steam

Air

Steam jacket

Sterilization chamber

Thermometer

Traps

Pressure regulator

Steam intake

➤ **Figure 12.10   The autoclave.** Steam is heated in the jacket of an autoclave, enters the sterilization chamber through an opening at the upper rear, and is exhausted through a vent at the bottom front.

periodically and maintain the appropriate pressure. Tapes impregnated with a substance that causes the word "sterile" to appear when they have been exposed to an effective sterilization temperature can be placed on packages. These tapes are not fully reliable because they do not indicate how long appropriate conditions were maintained. Tapes or other sterilization indicators should be placed inside and near the center of large packages to determine whether heat penetrated them. This precaution is necessary because when an object is exposed to heat, its surface becomes hot much more quickly than its center. (When a large piece of meat is roasted, for example, the surface can be well done while the center remains rare.)

The Centers for Disease Control and Prevention recommends weekly autoclaving of a culture containing heat-resistant endospores, such as those of *Bacillus stearothermophilus,* to check autoclave performance. Endospore strips are commercially available to make this task easy (➤ Figure 12.11). The spore strip and an ampule of medium are enclosed in a soft plastic vial. The vial is placed in the center of the material to be sterilized and is autoclaved. Then the inner ampule is broken, releasing the medium, and the whole container is incubated. If no growth appears in the autoclaved culture, sterilization is deemed effective.

In large laboratories and hospitals, where great quantities of materials must be sterilized, special autoclaves, called *prevacuum autoclaves,* are often used (➤ Figure 12.12). The chamber is emptied of air as steam flows in, creating a partial vacuum. The steam enters and heats the chamber much more rapidly than it would without the vacuum, so the proper temperature is reached quickly. The total sterilization time is cut in half, and the costs of sterilization are greatly decreased.

Pasteurization

Pasteurization, a process invented by Pasteur to destroy organisms that caused wine to sour, does not achieve sterility. It does kill pathogens, especially *Salmonella* and *Mycobacterium,* that might be present in milk, other dairy products, and beer. *Mycobacterium* used to cause many cases of tuberculosis among children who drank raw milk. Milk is pasteurized by heating it to 71.6°C for at least 15 seconds in the *flash method* or by heating it to 62.9°C for 30 minutes in the *holding method.* Some years ago certain strains of bacteria of the genus *Listeria* were found in pasteurized milk and cheeses. This pathogen causes diarrhea and encephalitis and can lead to death in pregnant women. A few such infections have prompted questions about the need to revise standard procedures for pasteurization. However, finding these pathogens in pasteurized milk has not become a persistent problem, and no action has been taken.

➤ **Figure 12.11   Checking for sterility.** To check if an autoclave is operating properly, a commercially prepared spore test ampule is placed in the autoclave and run with the rest of the load. Afterward, the vial is crushed to release medium onto a strip containing spores. If the load was truly sterilized, the spores will have been killed, and growth will not occur in the medium. Sometimes an indicator dye is added to the medium, which will turn color if microbial growth occurs, due to the accumulation of acid byproducts. This is faster than waiting for sufficient growth to turn the medium cloudy.

Cap that allows steam to penetrate

Crushable glass ampule

Flexible plastic vial

Nutrient medium

Spore strip

This unit is placed in autoclave.

After autoclaving, flexible vial is squeezed to break ampule and to release medium onto spore strip.

Incubation

If medium remains clear, spores were killed, and sterility was produced.

If medium becomes cloudy, spores were not killed, and sterility was not produced.

➤ **Figure 12.12   A large automatic hospital autoclave.** Recording charts keep records of the actual temperatures and pressures reached during the time of operation. These would alert the operator to some malfunctions.

Although most milk for sale in the United States is pasteurized fresh milk, sterile milk also is available. All evaporated or condensed canned milk is sterile, and some milk packaged in cardboard containers also is sterile. The canned milk is subjected to steam under pressure and has a "cooked" flavor. Sterilized milk in cardboard containers is widely available in Europe and can be found in some stores in the United States. It is subjected to a process that is similar to pasteurization but uses higher temperatures. It too has a "cooked" flavor but can be kept unrefrigerated as long as the container remains sealed. Such milk is often flavored with vanilla, strawberry, or chocolate. **Ultrahigh temperature (UHT) processing** raises the temperature from 74° to 140°C and then drops it back to 74°C in less than 5 seconds. A complex cooling process that prevents the milk from ever touching a surface hotter than itself prevents development of a "cooked" flavor. Some, but not all, small containers of coffee creamer are treated by this method.

Refrigeration, Freezing, Drying, and Freeze-Drying

Cold temperature retards the growth of microorganisms by slowing the rate of enzyme-controlled reactions but does not kill many microbes. Heat is much more effective than cold at killing microorganisms. *Refrigeration* is used to prevent food spoilage. *Freezing, drying,* and *freeze-drying* are used to preserve both foods and microorganisms, but these methods do not achieve sterilization.

Refrigeration

Many fresh foods can be prevented from spoiling by keeping them at 5°C (ordinary refrigerator temperature). However, storage should be limited to a few days because some bacteria and molds continue to grow at this temperature. To convince yourself of this, recall some of the strange things you have found growing on leftovers in the back of your refrigerator. In rare instances strains of *Clostridium botulinum* have been found growing and producing lethal toxins in a refrigerator when the organisms were deep within a container of food, where anaerobic conditions exist.

Freezing

Freezing at −20°C is used to preserve foods in homes and in the food industry. Although freezing does not sterilize foods, it does significantly slow the rate of chemical reactions so that microorganisms do not cause food to spoil. Frozen foods should not be thawed and refrozen. Repeated freezing and thawing of foods causes large ice crystals to form in the foods during slow freezing. Cell membranes in the foods are ruptured, and nutrients leak out. The texture of

Yogurt

Certain foods such as yogurt are made by introducing into milk organisms such as lactobacilli that ferment the milk. The fermented foods are often heat-treated after initial pasteurization to kill the fermenting organisms and to increase the shelf life of the products. The labels of such products indicate whether they contain live fermenting organisms. If you make yogurt at home, be sure to purchase a live-culture brand of yogurt to use as your starter.

Home Canning

Home canning is done in an open water bath or in a pressure cooker. Food must be packed loosely in jars with ample fluid to carry heat to the center of the jar. Space must be left between the jars. Once adequately processed, canned foods will keep indefinitely. Jars of relish in the wreckage of the *Monitor,* an ironclad Civil War ship, were actually sterile after more than 100 years at the bottom of the Atlantic Ocean! The contents would have been edible if poisonous quantities of lead from the lids had not dissolved in the relish.

The water bath reaches a temperature of 100°C and is adequate for preventing spoilage of acidic foods such as fruits and tomatoes. Acid in these foods inhibits the germination of most spores, should some survive the boiling-water treatment. However, meats and alkaline vegetables, such as corn and beans, must be processed in a pressure cooker. Adding onions or green peppers to a jar of tomatoes increases the pH, so such mixtures also must be cooked under pressure. Because acid-tolerant spores do exist, home canning is safest when it is done by pressure cooking. All commercially canned foods are processed in pressurized equipment. A pressure cooker functions like an autoclave. Foods in jars are processed at least 15 minutes at 15 lb/in.^{2} pressure. Any spores that might be present in these foods are killed, so the food is sterile.

Failure to process alkaline foods at the high temperature reached in a pressure cooker can lead to the accumulation of toxin produced by still-living *Clostridium botulinum* bacteria while the food is stored. Even a tiny amount of this toxin can be lethal. Any home-canned or commercially canned foods should be discarded if they have an unpleasant odor or if the lids of the containers bulge, because this indicates that gas is being produced by living organisms inside the container. Unfortunately, toxins can be present even when the lids don't bulge and there is no noticeable odor, so great care must be used in home canning to be sure that adequate pressure is maintained for a long enough time. As a safety precaution, after a jar of home-canned food is opened, 15 to 20 minutes of vigorous boiling should destroy any botulism toxin. However, the best rule to follow is "When in doubt, throw it out."

Canning peaches at home by the hot water bath method.

foods is thus altered, and they become less palatable. It also allows bacteria to multiply while food is thawed, making the food more susceptible to bacterial degradation.

Freezing can be used to preserve microorganisms, but this requires a much lower temperature than that used for food preservation. Microorganisms are usually suspended in glycerol or protein to prevent the formation of large ice crystals (which could puncture cells), cooled with solid carbon dioxide (dry ice) to a temperature of −78°C, and then held there. Alternatively, they can be placed in liquid nitrogen and cooled to −180°C.

Drying

Drying can be used to preserve foods because the absence of water inhibits the action of enzymes. Many foods, including peas, beans, raisins, and other fruits, are often preserved by drying (➤ Figure 12.13). Yeast used in baking also can be preserved by drying. Endospores present on such foods can survive drying, but they do not produce toxins. Dried pepperoni sausage and smoked fish retain enough moisture for microorganisms to grow. Because smoked fish is not cooked, eating it poses a risk of infection. Sealing such fish in plastic bags creates conditions that allow anaerobes such as *Clostridium botulinum* to grow.

Drying also naturally minimizes the spread of infectious agents. Some bacteria, such as *Treponema pallidum,* which causes syphilis, are extremely sensitive to drying and die almost immediately on a dry surface; thus they can be prevented from spreading by keeping toilet seats and other bathroom fixtures dry. Drying of laundry in dryers or in the sunshine also destroys pathogens.

Freeze-Drying

Freeze-drying, or **lyophilization** (li-of″i-li-za′shun), is the drying of a material from the frozen state (➤ Figure 12.14). This process is used in the manufacture of some brands of instant coffee; freeze-dried instant coffee has a more natural flavor than other kinds. Microbiologists use

Home Freezing

Home freezing of foods is probably a more common practice today than home canning. Before freezing fresh fruits and vegetables, they should be blanched, or immersed in boiling water for about a minute. Blanching helps kill microorganisms on the foods, but its main purpose is to denature enzymes in the foods that can cause discoloration or changes in texture even at freezer temperatures. The foods should then be cooled quickly in cold water and placed in clean containers. Finally, they should be placed in the coldest part of the home freezer with space around the containers so that they freeze as quickly as possible.

> Figure 12.13   **Preservation by drying.** Sun drying is an ancient means of preventing the growth of microorganisms. These grapes will remain edible as raisins because microbes need more water than remains inside the dried fruit.

lyophilization for long-term preservation rather than for destruction of cultures of microorganisms. Organisms are rapidly frozen in alcohol and dry ice or in liquid nitrogen and are then subjected to a high vacuum to remove all the water while in the frozen state. Rapid freezing allows only very tiny ice crystals to form in cells, so the organisms survive this process. Organisms so treated can be kept alive for years, stored under vacuum in the freeze-dried state.

Radiation

Four general types of radiation—ultraviolet light, ionizing radiation, microwave radiation, and strong visible light (under certain circumstances)—can be used to control microorganisms and to preserve foods. Refer to the electromagnetic spectrum (➤ Figure 3.4) to review their wavelengths and positions relative to one another along the spectrum.

Ultraviolet Light

Ultraviolet (UV) light consists of light of wavelengths between 40 and 390 nm, but wavelengths in the 200-nm range are most effective in killing microorganisms. UV light is absorbed by the purine and pyrimidine bases of nucleic acids. Such absorption can permanently destroy these important molecules. Ultraviolet light is especially effective in inactivating viruses. However, it kills far fewer bacteria than one might expect because of DNA repair mechanisms. Once DNA is repaired, new molecules of RNA and protein can be synthesized to replace the damaged molecules. ➥ (Chapter 7, p. 187)

Ultraviolet light is of limited use because it does not penetrate glass, cloth, paper, or most other materials, and it does not go around corners or under lab benches. It does penetrate air, effectively reducing the number of airborne microorganisms and killing them on surfaces in operating rooms and rooms that will contain caged animals (➤ Figure 12.15). UV lights lose effectiveness over time and should be monitored often. To help sanitize the air without irradiating humans, these lights can be turned on when the rooms are not in use. Exposure to ultraviolet light can cause burns, as anyone who has had a sunburn knows, and can also damage the eyes; years of exposure of skin can lead to skin cancer. Hanging laundry outdoors on bright, sunny days takes advantage of the ultraviolet light present in sunlight. Although the quantity of UV rays in sunlight is small, these rays may help kill bacteria on clothing, especially diapers.

In some communities, ultraviolet light is replacing chlorine in sewage treatment. When chlorine-treated sewage effluent is discharged into streams or other bodies of water, carcinogenic compounds form and may enter the food chain. The cost of removing chlorine before discharging treated effluent could add as much as $100 per year to the sewage bills of the average American family, and very few sewage plants do this. Running the sewage effluent under ultraviolet light before discharging it can destroy microorganisms without altering the odor, pH, or chemical composition of the water and without forming carcinogenic compounds.

> Figure 12.14   **Freeze-drying (lyophilization) equipment.** (a) A stoppering tray dryer, in which the trays supply the heat needed to remove moisture. After completion of lyophilization (about 24 hours), the device automatically stoppers the vials. (b) A manifold dryer, in which prefrozen samples in vials of different sizes can be attached via ports to the dryer, which supplies a vacuum to remove water. The sample will dry in 4 to 20 hours, depending on its initial thickness. Unlike the tray dryer, this device allows samples to be added and removed.

(a)

(b)

➤ Figure 12.15 **Ultraviolet radiation.** The effects of UV exposure can be seen in this Petri plate of *Serratia marcescens;* the left side was exposed to UV rays while the right side was shielded. Most of the organisms on the left side have been killed.

Ionizing Radiation

X-rays, which have wavelengths of 0.1 to 40 nm, and gamma rays, which have even shorter wavelengths, are forms of *ionizing radiation,* so named because it can dislodge electrons from atoms, creating ions. (Longer wavelengths comprise *nonionizing radiation*.) These forms of radiation also kill microorganisms and viruses. Many bacteria are killed by absorbing 0.3 to 0.4 millirads of radiation; polioviruses are inactivated by absorbing 3.8 millirads. A **rad** is a unit of radiation energy absorbed per gram of tissue; a millirad is one-thousandth of a rad. Humans usually do not become ill from radiation unless they are subjected to doses greater than 50 rad.

Ionizing radiation damages DNA and produces peroxides, which act as powerful oxidizing agents in cells. This radiation can also kill or cause mutations in human cells if it reaches them. It is used to sterilize plastic laboratory and medical equipment and pharmaceutical products. It can be used to prevent spoilage in seafoods by doses of 100 to 250 kilorads, in meats and poultry by doses of 50 to 100 kilorads, and in fruits by doses of 200 to 300 kilorads. (One kilorad equals 1000 rads.) Many consumers in the United States reject irradiated foods for fear of receiving radiation, but such foods are quite safe—free of both pathogens and radiation. In Europe, milk and other foods are often irradiated to achieve sterility.

Irradiation is used by hospitals to sterilize food for immune-compromised patients.

Microwave Radiation

Microwave radiation, in contrast with gamma, X-ray, and ultraviolet radiation, falls at the long-wavelength end of the electromagnetic spectrum. It has wavelengths of approximately 1 mm to 1 m, a range that includes television and police radar wavelengths. Microwave oven frequencies are tuned to match energy levels in water molecules. In the liquid state, water molecules quickly absorb the microwave energy and then release it to surrounding materials as heat. Thus, materials that do not contain water, such as plates made of paper, china, or plastic, remain cool while the moist food on them becomes heated. For this reason the home microwave cannot be used to sterilize items such as bandages and glassware. Conduction of energy in metals leads to problems such as sparking, which makes most metallic items also unsuitable for microwave sterilization. Moreover, bacterial endospores, which contain almost no water, are not destroyed by microwaves. However, a specialized microwave oven has recently become available that can be used to sterilize media in just 10 minutes (➤ Figure 12.16). It has 12 pressure vessels, each of which holds 100 ml of medium. Microwave energy increases the pressure of the medium inside the vessels until sterilizing temperatures are reached.

Caution should be observed in cooking foods in the home microwave oven. Geometry and differences in density of the food being cooked can cause certain regions to become hotter than others, sometimes leaving very cold spots. Consequently, to cook foods thoroughly in a microwave oven, it is necessary to rotate the items either mechanically or by hand. For example, pork roasts must be turned frequently and cooked thoroughly to kill any cysts of the pork round-

> Figure 12.16 **Microwave sterilization.** The MikroClave™ system is specifically designed for rapid sterilization of microbiological media and solutions. Using microwave energy, it can sterilize 1.2 liters of media in 6.5 minutes, or 100 ml in 45 seconds. Agar need not be boiled prior to sterilization.

worm *Trichinella* (Chapter 22). Failure to kill such cysts could lead to the disease trichinosis, in which cysts of the worm become embedded in human muscles and other tissues. All experimentally infected pork roasts, when microwaved without rotation, showed live worms remaining in some portion at the end of standard cooking time.

Strong Visible Light

Sunlight has been known for years to have a bactericidal effect, but the effect is due primarily to ultraviolet rays in the sunlight. Strong visible light, which contains light of wavelengths from 400 to 700 nm (violet to red light), can have direct bactericidal effects by oxidizing light-sensitive molecules such as riboflavin and porphyrins (components of oxidative enzymes) in bacteria. For that reason, bacterial cultures should not be exposed to strong light during laboratory manipulations. The fluorescent dyes eosin and methylene blue can denature proteins in the presence of strong light because they absorb energy and cause oxidation of proteins and nucleic acids. The combination of a dye and strong light can be used to rid materials of both bacteria and viruses.

Sonic and Ultrasonic Waves

Sonic, or sound, waves in the audible range can destroy bacteria if the waves are of sufficient intensity. Ultrasonic waves, or waves with frequencies above 15,000 cycles per second, can cause bacteria to cavitate. **Cavitation** (kav″i-ta′shun) is the formation of a partial vacuum in a liquid—in this case, the fluid cytoplasm in the bacterial cell. Bacteria so treated disintegrate, and their proteins are denatured. Enzymes used in detergents are obtained by cavitating the bacterium *Bacillus subtilis*. The disruption of cells by sound waves is called **sonication** (son″i-ka′shun). Neither sonic nor ultrasonic waves are a practical means of sterilization. We mention them here because they are useful in fragmenting cells to study membranes, ribosomes, enzymes, and other components.

Filtration

Filtration is the passage of a material through a filter, or straining device. Sterilization by filtration requires filters with exceedingly small pores. Filtration has been used since Pasteur's time to separate bacteria from media and to sterilize materials that would be destroyed by heat. Over the years, filters have been made of porcelain, asbestos, diatomaceous earth, and sintered glass (glass that has been heated without melting). *Membrane filters* (➤ Figure 12.17), thin disks with pores that prevent the passage of anything larger than the pore size, are widely used today. They are usually made of nitrocellulose and have the great advantage that they can be manufactured with specific pore sizes from 25 μm to less than 0.025 μm. Particles filtered by various pore sizes are summarized in Table 12.4.

Membrane filters have certain advantages and disadvantages. Except for those with the smallest pore sizes, membrane filters are relatively inexpensive, do not clog easily, and can filter large volumes of fluid reasonably rapidly. They can be autoclaved or purchased already sterilized. A disadvantage of membrane filters is that many of them allow viruses and some my-

(a)   (b)

➤ Figure 12.17   **Sterilization by filtration.** (a) Various types of membrane filters are available to sterilize large or small quantities of liquids. Some can be vacuum-filtered, ensuring that what is forced into the bottle or flask will be sterile. (b) Scanning electron micrograph of *Staphylococcus epidermidis* cells trapped on the surface of a 0.22-μm Millipore membrane filter. Membrane pore size can be selected to allow viruses, but not bacteria, to pass through or to prevent both from passing.

coplasmas to pass through. Other disadvantages are that they may absorb relatively large amounts of the filtrate and may introduce metallic ions into the filtrate.

Membrane filters are used to sterilize materials likely to be damaged by heat sterilization. These materials include media, special nutrients that might be added to media, and pharmaceutical products such as drugs, sera, and vitamins. Some filters can be attached to syringes so that materials can be forced through them relatively quickly. Filtration can also be used instead of pasteurization in the manufacture of beer. When using filters to sterilize materials, it is important to select a filter pore size that will prevent any infectious agent from passing into the product.

In the manufacture of vaccines that require the presence of live viruses, it is important to select a filter pore size that will allow viruses to pass but prevent bacteria from doing so. By selecting a filter with a proper pore size, scientists can separate polioviruses from the fluid and debris in tissue cultures in which they were grown. This procedure simplifies the manufacture

➤ Table 12.4	**Pore sizes of membrane filters and particles that pass through them**
Pore size (in μm)	**Particles that pass through them**
10	Erythrocytes, yeast cells, bacteria, viruses, molecules
5	Yeast cells, bacteria, viruses, molecules
3	Some yeast cells, bacteria, viruses, molecules
1.2	Most bacteria, viruses, molecules
0.45	A few bacteria, viruses, molecules
0.22	Viruses, molecules
0.10	Medium-sized to small viruses, molecules
0.05	Small viruses, molecules
0.025	Only the very smallest viruses, molecules
Ultrafilter	Small molecules

of polio vaccine. Cellulose acetate filters with extremely tiny pores are now available and are capable of removing many viruses (although not the very smallest) from liquids. However, these filters are expensive and clog easily.

Membrane filters used to trap bacteria from air and water samples can be transferred directly to agar plates, and the quantity of bacteria in the sample can be determined. Alternatively, the filters can be transferred from one medium to another, so organisms with different nutrient requirements can be detected. Filtration is also used to remove microorganisms and other small particles from public water supplies and in sewage treatment facilities. This technique, however, cannot sterilize; it merely reduces contamination.

High-efficiency particulate air (HEPA) filters are used in the ventilation systems of areas where microbial control is especially important, such as in operating rooms, burn units, and laminar flow transfer hoods in laboratories. HEPA filters also capture organisms released in rooms occupied by patients with tuberculosis or in laboratories where especially dangerous microbes are studied, such as the maximum containment units shown in ➤ Figure 15.14. These filters remove almost all organisms larger than 0.3 μm in diameter. Used filters are soaked in formalin before they are disposed of.

Osmotic Pressure

✓ Filtering water through paper does not sterilize the water. How, then, can membrane filters sterilize water?

✓ Can a home microwave oven be used to sterilize items? Why or why not?

✓ Are ice cubes safe sources of water in areas with poor water supplies?

✓ Are pasteurized products sterilized?

High concentrations of salt, sugar, or other substances create a hyperosmotic medium, which draws water from microorganisms by osmosis. ➥ (Chapter 4, p. 99) **Plasmolysis** (plaz-mol′i-sis), or loss of water, severely interferes with cell function and eventually leads to cell death. The use of sugar in jellies, jams, and syrups or salt solutions in curing meat and making pickles plasmolyzes most organisms present and prevents growth of new organisms. A few halophilic organisms, however, thrive in these conditions and cause spoilage, especially of pickles, and some fungi can live on the surface of jams.

Properties of physical antimicrobial agents are summarized in Table 12.5.

➤ Table 12.5	Properties of physical antimicrobial agents	
Agent	**Action**	**Use**
Dry heat	Denatures proteins	Oven heat used to sterilize glassware and metal objects; open flame used to incinerate microorganisms.
Moist heat	Denatures proteins	Autoclaving sterilizes media, bandages, and many kinds of hospital and laboratory equipment not damaged by heat and moisture; pressure cooking sterilizes canned foods.
Pasteurization	Denatures proteins	Kills pathogens in milk, dairy products, and beer.
Refrigeration	Slows the rate of enzyme-controlled reactions	Used to keep fresh foods for a few days; does not kill most microorganisms.
Freezing	Greatly slows the rate of most enzyme-controlled reactions	Used to keep fresh foods for several months; does not kill microorganisms; used with glycerol to preserve microorganisms.
Drying	Inhibits enzymes	Used to preserve some fruits and vegetables; sometimes used with smoke to preserve sausages and fish.
Freeze-drying	Dehydration inhibits enzymes	Used to manufacture some instant coffees; used to preserve microorganisms for years.
Ultraviolet light	Denatures proteins and nucleic acids	Used to reduce the number of microorganisms in air in operating rooms, animal rooms, and where cultures are transferred.
Ionizing radiation	Denatures proteins and nucleic acids	Used to sterilize plastics and pharmaceutical products and to preserve foods.
Microwave radiation	Absorbs water molecules, then releases microwave energy to surroundings as heat	Cannot be used reliably to destroy microbes except in special media-sterilizing equipment.
Strong visible light	Oxidation of light-sensitive materials	Can be used with dyes to destroy bacteria and viruses; may help sanitize clothing.
Sonic and ultrasonic waves	Cause cavitation	Not a practical means of killing microorganisms but useful in fractionating and studying cell components.
Filtration microbes	Mechanically removes microbes	Used to sterilize media, pharmaceutical products, and vitamins, in manufacturing vaccines, and in sampling microbes in air and water.
Osmotic pressure	Removes water from microbes	Used to prevent spoilage of foods such as pickles and jellies.

Hospital Sanitation and the Infection Control Practitioner

Maintaining sanitation in a modern hospital is becoming an increasingly complex task for several reasons. First, far more patients with infectious diseases are treated in hospitals today than a decade ago. Second, more techniques are available to assist with maintaining sanitation—disposable equipment and supplies, more complex isolation procedures, and advances in sanitizing agents and equipment. Using these techniques improves sanitation, but it also makes more work for the hospital staff. Third, and most important, increasing numbers of pathogens are becoming resistant to antibiotics, and these organisms are especially common in hospitals, where antibiotics are in continuous use. Hospital-acquired, or *nosocomial,* infections are therefore a constant threat to the lives of already seriously ill patients. In Chapter 15 we discuss the nature of nosocomial infections; here we focus on sterilization and disinfection procedures used to limit the spread of nosocomial and other infections in hospitals.

The hospital environment is a particularly likely place to acquire an infection. Because people with infections come to hospitals for treatment, the density of pathogens is greater in hospitals than in most other environments. The movements of hospital personnel and visitors and even air currents from elevator shafts tend to spread microbes throughout the hospital. The most likely mode of transmission of infections in a hospital is direct contact, as when a health-care worker touches an infected patient and fails to wash his or her hands before touching another patient. Indirect contact, as with contaminated equipment, and airborne transmission of pathogens are less frequent modes of transmission.

Several procedures are used to minimize the spread of infection in a hospital. Each room is disinfected after a patient is discharged and before another patient occupies it. Floors are regularly mopped with disinfectant solutions, and carpets are kept dry and vacuumed often. Linens, especially those from patients with infections, are placed in plastic bags before being dropped into a laundry chute. The linens are then laundered in strong detergents and very hot water (71°C for 25 minutes). Food is heated to an internal temperature of 74°C and kept covered and above 60°C until it is served. Dishes are washed at 60°C for 20 seconds and rinsed at 82°C for 10 seconds. Electronic air filters that ionize airborne microbes are installed in ventilating systems, especially those that serve critical-care units, burn units, and nurseries.

Hospital personnel can minimize the risk of infecting themselves and others in several ways. The single most important way is by thoroughly washing their hands between patient contacts. Another way is by receiving appropriate immunizations. Personnel should be immunized against diphtheria and tetanus and against hepatitis B if they will have contact with blood or other potentially infectious fluids. They should also be immunized against measles and mumps if they have not had those diseases and against influenza if they are susceptible to frequent pulmonary infections. Personnel also can learn and practice good aseptic techniques and carry out recommendations of the hospital's employee health programs.

Hospitals are ethically and legally responsible for patients who acquire nosocomial infections. In fact, to maintain accreditation by the American Hospital Association, hospitals must have a program that includes surveillance of nosocomial infections in both patients and staff, a microbiology laboratory, isolation procedures, accepted procedures for the use of catheters and other instruments, general sanitation procedures, and a nosocomial disease education program for staff members. The program's goal is to engage all hospital personnel in active measures to prevent infections. Most hospitals have an infection-control practitioner (ICP) to manage such a program.

ICP candidates must pass a registry examination offered by the Association for Practitioners in Infection Control (APIC). This specialty requires knowledge of both microbiology and patient-care techniques. Among currently registered specialists, some were first trained as microbiologists and many as nurses. The duties of an ICP include surveillance and identification of infections; supervision of, or collaboration with, the hospital's employee health program; keeping up-to-date on newly available immunizations to determine which ones hospital personnel should receive; assisting with studies of antibiotic use in infection control and detection of resistant organisms; and providing instruction to new staff on aseptic techniques and the hospital's infection-control program, including isolation procedures (Chapter 15).

Some specific and effective means of infection control are as follows: (1) thorough washing of hands with soap and water, which can greatly reduce the risk of physicians, nurses, and other staff members spreading diseases among patients; (2) scrupulous care in sterilizing equipment and maintaining its sterility while inserting and using catheters and other invasive instruments; (3) the use of gloves when drawing blood or handling infectious materials, such as dressings and bedpans.

Other techniques are needed to minimize the development of antibiotic-resistant pathogens. The routine use of antimicrobial agents to prevent infections has turned out to be a misguided effort becaues it contributes to the development of resistant organisms (Chapter 13). Therefore, some hospitals maintain surveillance of antibiotic use. Antibiotics are given for known infections but are given prophylactically (as a preventive measure) only in special situations. Prophylactic antibiotics are justified in surgical procedures such as hysterectomies, colorectal surgeries, and repair of traumatic injuries, in which the surgical field is invariably contaminated with potential pathogens, and in immunosuppressed patients and excessively debilitated patients, whose natural defense mechanisms may fail.

If all the known techniques for preventing nosocomial infections were rigorously practiced, the incidence of such infections could probably be reduced to half the present level.

Retracing Our Steps

Principles of Sterilization and Disinfection

- **Sterilization** refers to the killing or removal of all organisms in any material or on any object.
- **Disinfection** refers to the reduction in numbers of pathogenic organisms on objects or in materials so that the organisms no longer pose a disease threat.
- Important terms related to sterilization and disinfection are defined in Table 12.1.

The Control of Microbial Growth

- Because of the logarithmic death rate of microorganisms, a definite proportion of organisms die in a given time interval.
- The fewer organisms present, the less time will be needed to achieve sterility.
- Microorganisms differ in their susceptibility to antimicrobial agents.

Chemical Antimicrobial Agents

The Potency of Chemical Agents

- The potency, or effectiveness, of a chemical antimicrobial agent is affected by time, temperature, pH, and concentration of the agent.
- Potency increases with the length of time organisms are exposed to the agent, increased temperature, acidic or alkaline pH, and usually increased concentration of the agent.

Evaluating the Effectiveness of Chemical Agents

- Evaluation of effectiveness is difficult, and no entirely satisfactory method is available.
- For agents similar to phenol, the **phenol coefficient** is determined; it is the ratio of the dilution of the agent to the dilution of phenol that will kill all organisms in 10 minutes but not in 5 minutes.

Disinfectant Selection

- A variety of criteria are considered in selecting a **disinfectant.** In practice, most chemical agents are tested in various situations and used in situations where they produce satisfactory results.

Mechanisms of Action of Chemical Agents

- Actions of chemical antimicrobial agents can be grouped according to their effects on proteins, cell membranes, and other cell components.
- Reactions that alter proteins include hydrolysis, oxidation, and attachment of atoms or chemical groups to protein molecules. Such reactions denature proteins, rendering them nonfunctional.
- Membranes can be disrupted by agents that denature proteins and by **surfactants,** which reduce surface tension and dissolve lipids.
- Reactions of other chemical agents damage nucleic acids and energy-producing systems. Damage to nucleic acids is an important means of inactivating viruses.

Specific Chemical Antimicrobial Agents

- Soaps and detergents aid in the removal of microbes, oils, and dirt but do not sterilize.
- Acids are commonly used as food preservatives; alkali in soap helps destroy microorganisms.

- Among the agents containing heavy metals, silver nitrate is used to kill gonococci, and mercury-containing compounds are used to disinfect instruments and skin.
- Among the agents containing halogens, chlorine is used to kill pathogens in water, and iodine is a major ingredient in several skin disinfectants.
- Alcohols are used to disinfect skin.
- Phenol derivatives can be used on skin, instruments, dishes, and furniture, and to destroy discarded cultures; they work well in the presence of organic materials.
- Oxidizing agents are particularly useful in disinfecting puncture wounds.
- Alkylating agents can be used to disinfect or to sterilize a variety of materials, but all are carcinogens.
- Some dyes, plant oils, sulfur-containing substances, and nitrates can be used as disinfectants or food preservatives.

Physical Antimicrobial Agents

Principles and Applications of Heat Killing

- Heat destroys microorganisms by denaturing protein, melting lipids, and, when open flame is used, by incineration.

Dry Heat, Moist Heat, and Pasteurization

- Dry heat is used to sterilize metal objects and glassware.
- Flame is used to sterilize inoculating loops and the mouths of culture tubes.
- The **autoclave,** which uses moist heat under pressure, is a common instrument for sterilization and is very effective when proper procedures are followed.
- **Pasteurization** kills most pathogens in milk, other dairy products, and beer but does not sterilize.

Refrigeration, Freezing, Drying, and Freeze-drying

- Refrigeration, freezing, drying, and freeze-drying can be used to retard the growth of microorganisms.
- **Lyophilization,** drying in the frozen state, can be used for long-term preservation of live microorganisms.

Radiation

- Radiation used to control microorganisms includes ultraviolet light, ionizing radiation, and sometimes microwaves and strong sunlight.

Sonic and Ultrasonic Waves

- Sonic and ultrasonic waves can kill microorganisms but are used mostly for **sonication,** the disruption of cells by sound waves.

Filtration

- **Filtration** can be used to sterilize substances that are destroyed by heat, to separate viruses, and to collect microorganisms from air and water samples.

Osmotic Pressure

- High concentrations of sugar or salt create osmotic pressure that results in the **plasmolysis** of cells (causes them to lose water) and prevents growth of microorganisms in highly sweetened or salted foods.

Terminology Check

aerosol   *(p. 328)*
antiseptic   *(p. 317)*
autoclave   *(p. 328)*
bactericidal   *(p. 319)*
bacteriostatic   *(p. 319)*
cavitation   *(p. 334)*
decimal reduction time (DRT)   *(p. 327)*
disinfectant   *(p. 317)*
disinfection   *(p. 318)*
D value   *(p. 327)*

filter paper method   *(p. 320)*
filtration   *(p. 334)*
lyophilization   *(p. 331)*
pasteurization   *(p. 329)*
phenol coefficient   *(p. 319)*
plasmolysis   *(p. 336)*
quaternary ammonium compound (quat)   *(p. 324)*
rad   *(p. 333)*
sonication   *(p. 334)*

sterility   *(p. 317)*
sterilization   *(p. 317)*
surfactant   *(p. 321)*
thermal death point   *(p. 327)*
thermal death time   *(p. 327)*
tincture   *(p. 324)*
ultrahigh temperature (UHT) processing   *(p. 330)*
use-dilution test   *(p. 320)*
wetting agent   *(p. 322)*

See You on the Web

Don't forget I am expecting you on the web. The following is a preparedness checklist for your web explorations. If you have questions on any of these items, you can return to the end of Chapter 1 for more details.

- Questions We'll Explore
- Concept Check questions
- Flashcards
- Study Guide

If you are ready, come visit me at

http://www.prenhall.com/black

It will be a journey you'll never forget.

Antimicrobial Therapy

Two years is a long time to be pregnant—especially when your reproductive tract allows seawater to leak into it. Infection by the many microbes that enter the uteruslike oviducts of the female dogfish shark (*Squalus acanthias*) would seem a likely end to the mother and/or most of her seven baby "pups." However, the sharks do not develop infections. The secret of their resistance lies in a broad-spectrum antibiotic called squalamine, which is present in all their tissues. It attacks cell membranes of microbes, forming holes in them. Squalamine has been synthesized in the laboratory and is under investigation for possible human use.

Background Basics

bacteriocidal versus bacteriostatic . . . . *(Chapter 12)*
competitive inhibition . . . . *(Chapter 6)*
osmotic pressure . . . . *(Chapter 4)*
nucleic acid synthesis . . . . *(Chapter 2)*
resistance plasmids . . . . *(Chapter 8)*

A s you lie in your sickbed, suffering from some infectious disease, how reassuring it is to be able to reach over, swallow some capsules, and look forward to getting well soon. But people have not always been able to do this. Over the ages, anxious parents have watched their children die of fevers, diarrhea, infected wounds, and other maladies. Whole villages have been wiped out by plagues. During the middle of the fourteenth century, more than a quarter of the population of Europe died of the Black Death (bubonic plague) in just a few years. More people have died of infection in wartime than from swords or bullets. Until relatively recently, the only defenses against infectious diseases were such things as herbal teas and poultices (soaks), or fleeing the disease area. Modern and effective weapons against microbes—antibiotics and sulfa drugs—did not become available until the twentieth century.

Think back through your life. Have there been times when you might have died had it not been for antimicrobial drugs? At what age would you have died? Which of your family members might not be alive now? Happily, we live in a better time with respect to illnesses and deaths caused by infectious organisms. In the United States the life expectancy of a baby born in 1850 was less than 40 years; in 1900, about 50 years; and in 1990, over 70 years. Infectious diseases claimed the lives of about one in every 100 U.S. residents per year as late as 1900 but only about one in every 300 in 1990. Although antimicrobial agents still don't save all patients, they have drastically lowered the death rate from infectious disease. A period of increased infectious diseases could return, however, if patients and the medical community fail to protect the effectiveness of antimicrobial agents. As many pathogens develop resistance to available antimicrobial drugs, our ability to fight infectious diseases is dwindling.

Questions We'll Explore

A What terms are used to discuss chemotherapy and antibiotics, and what do they mean?

B How have chemotherapeutic agents been developed?

C How do the terms selective toxicity, spectrum of activity, and modes of action apply to antimicrobial agents?

D What kinds of side effects are associated with antimicrobial agents?

E What is antibiotic resistance, and how do microorganisms acquire it?

F How are sensitivities of microbes to chemotherapeutic agents determined?

G What are the attributes of an ideal antimicrobial agent?

H What are the properties, uses, and side effects of antibacterial agents?

I What are the properties, uses, and side effects of antifungal agents, antiviral agents, antiprotozoan agents, and antihelminthic agents?

J How do drug-resistant hospital infections arise, and how can they be treated and prevented?

Antimicrobial Chemotherapy

The term **chemotherapy** was coined by the German medical researcher Paul Ehrlich to describe the use of chemical substances to kill pathogenic organisms without injuring the host. Today, chemotherapy refers to the use of chemical substances to treat various aspects of disease—aspirin for headache and inflammation, drugs to regulate heart function, and agents to rid the body of malignant cells. With this broad modern definition of chemotherapy, we describe a **chemotherapeutic agent** as any chemical substance used in medical practice. Such agents also are referred to as **drugs.**

In microbiology we are concerned with **antimicrobial agents,** a special group of chemotherapeutic agents used to treat diseases caused by microbes. Thus, in modern terms, an antimicrobial agent is synonymous with a chemotherapeutic agent as Erhlich originally defined it. In this chapter we consider a variety of antimicrobial agents and a few agents used to treat helminth infections.

Antibiosis literally means "against life." In the 1940s Selman Waksman, the discoverer of streptomycin, defined an **antibiotic** as "a chemical substance produced by microorganisms which has the capacity to inhibit the growth of bacteria and even destroy bacteria and other microorganisms in dilute solution." In contrast, agents synthesized in the laboratory are called **synthetic drugs.** Some antimicrobial agents are synthesized by chemically modifying a substance from a microorganism. More often a synthetic precursor different from the natural one is supplied to a microorganism, which then completes synthesis of the antibiotic. Antimicrobial agents made partly by laboratory synthesis and partly by microorganisms are called **semisynthetic drugs.**

The History of Chemotherapy

Throughout history humans have attempted to alleviate suffering by treating disease—often by taking concoctions of plant substances. Although ancient Egyptians used moldy bread to treat wounds, they had no knowledge of the antibiotics it contained. Extracts of willow bark, now known to contain a compound closely related to aspirin, were used to alleviate pain. Parts of the foxglove plant were used to treat heart disease in the sixteenth century, although the active ingredient, digitalis, had not been identified. Likewise, quinine-containing extracts from the cinchona tree were used to treat malaria.

Despite their reputation for using rituals irrelevant to the cure of disease, traditional healers of primitive societies, especially in the tropics, are quite knowledgeable about medicinal properties of plants. Their knowledge has been passed down from generation to generation. Because these healers are disappearing, pharmaceutical companies are attempting to learn from them and make written records of their treatments as well as test the plants they use.

In Western civilization, the first systematic attempt to find specific chemical substances to treat infectious disease was made by Paul Ehrlich. ➥ (Chapter 1, p. 17) Although his discovery in 1910 of Salvarsan to treat syphilis was of great therapeutic benefit, even more important were the concepts he developed in the new science of chemotherapy. He was interested in the mechanisms by which chemical substances bind to microorganisms and to animal tissues. His studies of chemicals that bind to tissues led to histological (tissue) stains that are still used today.

The next advances in chemotherapy were the nearly concurrent development of sulfa drugs and antibiotics. In 1935 Gerhard Domagk discovered that prontosil, a red dye, inhibits growth of many Gram-positive bacteria. The following year Ernest Fourneau found that the antimicrobial activity was due to the sulfanilamide portion of the prontosil molecule. These discoveries stimulated the development of a group of substances called *sulfonamides,* or *sulfa drugs.* As the number of sulfa drugs grew, it became possible to use them to attack directly a variety of pathogens. However, the usefulness of sulfa drugs is limited. They do not attack all pathogens, and they sometimes cause kidney damage and allergies. But they have saved many lives and continue to do so today.

Alexander Fleming reasoned that the ability of the mold *Penicillium* to inhibit growth of microorganisms might be exploited. This idea led him to identify the inhibitory agent and name it *penicillin.* In 1928 Fleming had observed the contamination of his bacterial cultures with this fungus many times, as had many other microbiologists. However, instead of grumbling about another contaminated culture and tossing it out, Fleming saw the tremendous potential in this accidental finding. If only the substance (penicillin) could be extracted and collected in large quantities, it could be used to combat infection.

Fleming's idea did not come to fruition until the early 1940s, when Ernst Chain and Howard Florey finally isolated penicillin and worked with other researchers to develop methods of mass production. Such mass production occurred during World War II and saved the lives of many people whose wounds became infected. Supplies of the drug were limited, however, and

When Doctors Learned to Cure

"Explanation was the real business of medicine. What the ill patient and his family wanted most was to know the name of the illness, and then if possible, what had caused it, and finally, most important of all, how it was likely to turn out. . . . It gradually dawned on us that we didn't know much that was really useful, that we could do nothing to change the course of the great majority of the diseases we were so busy analyzing. . . . Then came the explosive news of sulfanilamide, and the start of the real revolution in medicine. I remember with astonishment when the first cases of pneumococcal and streptococcal septicemia were treated in Boston in 1937. The phenomenon was almost beyond belief. Here were moribund patients, who would surely have died without treatment, improving in their appearance within a matter of hours of being given the medicine and feeling entirely well within the next day or so."

—Lewis Thomas, 1983

it was not readily available to civilians until after the war. Following the war, research proceeded rapidly, and new antibiotics were discovered one after another.

The introduction of penicillin and sulfonamides in the 1930s can be said to mark the beginning of modern medicine. As the medical writer Lewis Thomas said, "Doctors could now *cure* disease, and this was astonishing, most of all to the doctors themselves."

General Properties of Antimicrobial Agents

Antimicrobial agents share certain common properties. We can learn much about how these agents work and why they sometimes do not work by considering such properties as selective toxicity, spectrum of activity, mode of action, side effects, and resistance of microorganisms to them.

Selective Toxicity

Some chemical substances with antimicrobial properties are too toxic to be taken internally and are used only for topical application—application to the skin's surface. For internal use, an antimicrobial drug must have **selective toxicity**—that is, it must harm the microbes without causing significant damage to the host. Some drugs, such as penicillin, have a wide range between the **toxic dosage level,** which causes host damage, and the **therapeutic dosage level,** which successfully eliminates the pathogenic organism if the level is maintained over a period of time. The relationship between an agent's toxicity to the body and its toxicity to an infectious agent is expressed in terms of its chemotherapeutic index. For any particular agent, the **chemotherapeutic index** is defined as the maximum tolerable dose per kilogram of body weight, divided by the minimum dose per kilogram of body weight, that will cure the disease. Thus an agent with a chemotherapeutic index of 8 would be more effective and less toxic to the patient than an agent with a chemotherapeutic index of 1.

For drugs such as those containing arsenic, mercury, and antimony, the dosage must be calculated very precisely because these substances are highly toxic to human and other animal hosts as well as to pathogens. Treatment of worm infections is especially difficult because what damages the parasite will also damage the host. In contrast, bacterial pathogens often can be treated by interfering with metabolic pathways not shared by the host. For example, penicillin interferes with cell wall synthesis; it is not toxic to human cells, which lack walls, though some patients are allergic to it.

The Spectrum of Activity

The range of different microbes against which an antimicrobial agent acts is called its **spectrum of activity.** Those agents that are effective against a great number of microorganisms from a wide range of taxonomic groups, including both Gram-positive and Gram-negative bacteria, are said to have a **broad spectrum** of activity. Those that are effective against only a small number of microorganisms or a single taxonomic group have a **narrow spectrum** of activity (➤ Figure 13.1). Some common antibiotics are classified according to their spectrum of activity in Table 13.1.

A broad-spectrum drug is especially useful when a patient is seriously ill with an infection caused by an unidentified organism. Using such a drug increases the chance that the organism will be susceptible to it. However, if the identity of the organism is known, a narrow-spectrum drug should be used. Using such a drug minimizes the destruction of the host's *microflora,* or *normal flora*—the indigenous microbes that naturally occur in or on the host—that sometimes compete with and help destroy infectious organisms. The use of narrow-spectrum drugs also decreases the likelihood that organisms will develop drug resistance.

Modes of Action

Like other medicines, antimicrobial agents are sometimes used simply because they work, without our always knowing how they work. Many people's lives have been saved by medicines whose actions at the cellular level have never been understood. However, it is always desirable to know the mode of action of an agent. With that knowledge, effects of actions on patients can be better monitored and controlled, and ways of improving them may be found.

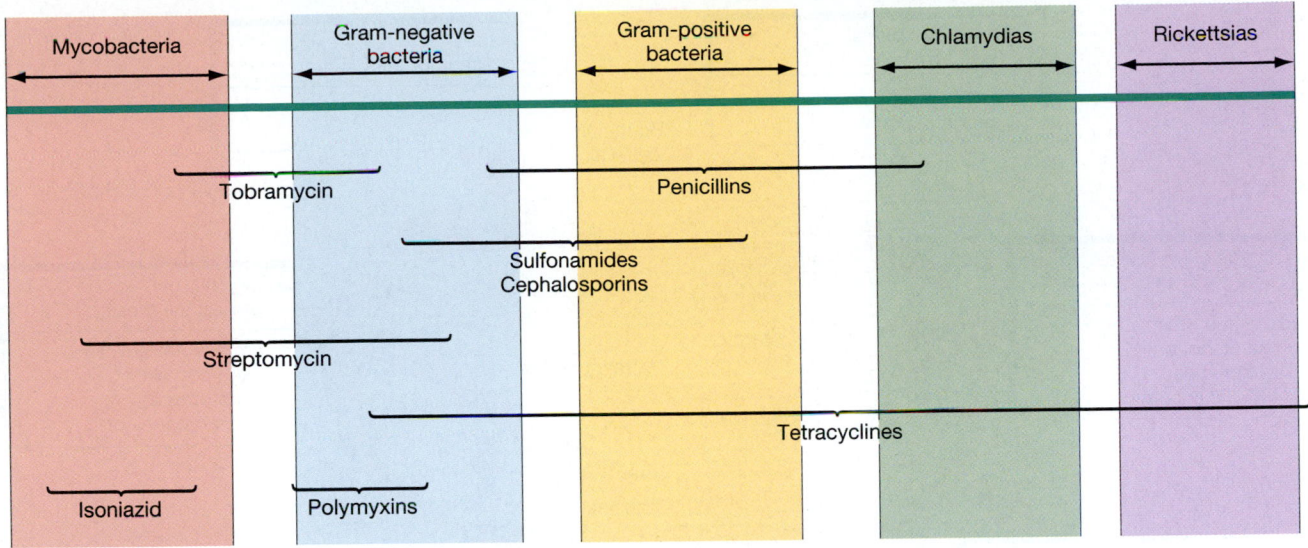

> Figure 13.1 **The spectrum of antibiotic activity.** Broad-spectrum drugs, such as tetracycline, affect a variety of different organisms. Narrow-spectrum drugs, such as isoniazid, affect only a few specific types of organisms.

Antimicrobial drugs generally act on an important microbial structure or function that usually differs from its counterpart in animals. This difference is exploited in exerting a *bactericidal,* or killing, effect or a *bacteriostatic,* or growth-inhibiting, effect on bacteria while having minimal effects on host cells.➡ (Chapter 12, p. 318) However, the host's immune systems or phagocytic defenses must still complete the elimination of the invading microbes.

Five different modes of action of antimicrobials are discussed here: (1) inhibition of cell wall synthesis, (2) disruption of cell membrane function, (3) inhibition of protein synthesis, (4) inhibition of nucleic acid synthesis, and (5) action as antimetabolites (➤ Figure 13.2).

Inhibition of Cell Wall Synthesis

Many bacterial and fungal cells have rigid external cell walls, whereas animal cells lack cell walls. Consequently, inhibiting cell wall synthesis selectively damages bacterial and fungal

> Table 13.1   **The spectrum of activity of selected antimicrobial agents**

Organisms affected	Broad-spectrum agents*	Narrow-spectrum agents
Bacteroides and other anaerobes	Cephalosporins	Lincomycin
Yeasts	Chloramphenicol	Nystatin
Gram-positive bacteria	Gentamicin Ampicillin	Penicillin G Erythromycin
Gram-negative bacteria	Kanamycin	Polymyxins
Streptococci and some Gram-negative bacteria	Tetracyclines	Streptomycin
Staphylococci, enterococci, and some clostridia	Tetracyclines	Vancomycin

*Broad-spectrum agents affect most bacteria.

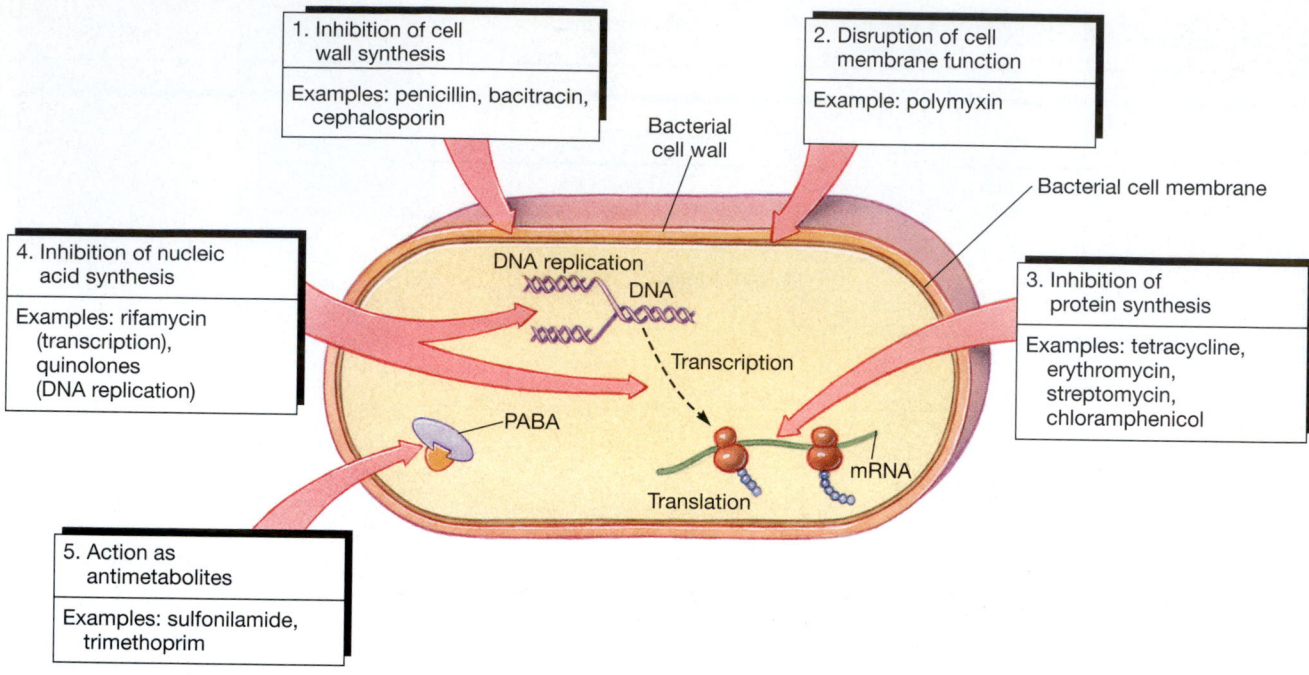

1. Inhibition of cell wall synthesis

Examples: penicillin, bacitracin, cephalosporin

2. Disruption of cell membrane function

Example: polymyxin

Bacterial cell wall

Bacterial cell membrane

4. Inhibition of nucleic acid synthesis

Examples: rifamycin (transcription), quinolones (DNA replication)

DNA replication

DNA

Transcription

3. Inhibition of protein synthesis

Examples: tetracycline, erythromycin, streptomycin, chloramphenicol

PABA

mRNA

Translation

5. Action as antimetabolites

Examples: sulfonilamide, trimethoprim

➤ Figure 13.2   **Modes of action.** Shown here are the five major modes of action by which drugs exert their antimicrobial effects on bacterial cells.

cells. Bacterial cells, especially Gram-positive ones, have a high internal osmotic pressure. Without a normal, sturdy cell wall, these cells burst when subjected to the low osmotic pressure of body fluids. ➡ (Chapter 4, p. 100) Antibiotics such as penicillin and cephalosporin contain a chemical structure called a *β-lactam ring,* which attaches to the enzymes that cross-link peptidoglycans. ➡ (Chapter 4, p. 75) By interfering with the cross-linking of tetrapeptides, these antibiotics prevent cell wall synthesis (➤ Figure 13.3). Fungi, whose cell walls lack peptidoglycan, are unaffected by these antibiotics.

Disruption of Cell Membrane Function

All cells are bounded by a membrane. Although the membranes of all cells are quite similar, those of bacteria and fungi differ sufficiently from those of animal cells to allow selective action of antimicrobial agents. Certain polypeptide antibiotics, such as polymyxins, act as detergents and distort bacterial cell membranes, probably by binding to phospholipids in the membrane. (With this distortion, the membrane is no longer regulated by membrane proteins, and the cytoplasm and cell substances are lost.) These antibiotics are especially effective against Gram-negative bacteria, which have an outer membrane rich in phospholipids. ➡ (Chapter 4, p. 77) Polyene antibiotics, such as amphotericin B, bind to particular sterols present in the membranes of fungal (and animal) cells. Thus, polymyxins do not act on fungi, and polyenes do not act on bacteria.

Inhibition of Protein Synthesis

In all cells, protein synthesis requires not only the information stored in DNA, plus several kinds of RNA, but also ribosomes. Differences between bacterial (70S) and animal (80S) ribosomes allow antimicrobial agents to attack bacterial cells without significantly damaging animal cells—that is, with selective toxicity. Aminoglycoside antibiotics, such as streptomycin, derive their name from the amino acids and glycosidic bonds they contain. They act on the 30S portion of bacterial ribosomes by interfering with the accurate reading (translation) of the mRNA message—that is, the incorporation of the correct amino acids. ➡ (Chapter 7, p. 167) Chloramphenicol and erythromycin act on the 50S portion of bacterial ribosomes, inhibiting the formation of the growing polypeptide. Because animal cell ribosomes consist of 60S and 40S subunits, these antibiotics have little effect on host cells. (Mitochondria, however, which have 70S ribosomes, can be affected by such drugs.)

(a)

(b)

> ➤ **Figure 13.3** **Inhibition of cell wall synthesis by penicillin.** SEM of bacteria (a) before and (b) after exposure to penicillin (magnified 785X). Notice the distortion of the cell shape due to the disruption by penicillin of the cross-linking tetrapeptides in the peptidoglycan layer of the cell wall.

Inhibition of Nucleic Acid Synthesis

Differences between the enzymes used by bacterial and animal cells to synthesize nucleic acids provide a means for selective action of antimicrobial agents. Antibiotics of the rifamycin family bind to a bacterial RNA polymerase and inhibit RNA synthesis. ➡ (Chapter 7, p. 171)

Action as Antimetabolites

The normal metabolic processes of microbial cells involve a series of intermediate compounds called *metabolites* that are essential for cellular growth and survival. **Antimetabolites** are substances that affect the utilization of metabolites and therefore prevent a cell from carrying out necessary metabolic reactions. Antimetabolites function in two ways: (1) by competitively inhibiting enzymes and (2) by being erroneously incorporated into important molecules such as nucleic acids. Antimetabolites are structurally similar to normal metabolites. The actions of antimetabolites are sometimes called **molecular mimicry** because they mimic, or imitate, the normal molecule, preventing a reaction from occurring or causing it to go awry.

In competitive inhibition an enzymatic reaction is inhibited by a substrate that binds to the enzyme's active site but cannot react. ➡ (Chapter 5, p. 112) While this competing substrate occupies the active site, the enzyme is unable to function, and metabolism will slow or even cease if enough enzyme molecules are inhibited. Consider sulfanilamide and *para*-aminosalicylic acid (PAS), which are chemically very similar to *para*-aminobenzoic acid (PABA) (➤ Figure 13.4). They competitively inhibit an enzyme that acts on PABA. Many bacteria require PABA in order to make folic acid, which they use in synthesizing nucleic acids and other metabolic products. When sulfanilamide or PAS instead of PABA is bound to the enzyme, the bacterium cannot make folic acid. Animal cells lack the enzymes to make folic acid and must obtain it from their diets; thus their metabolism is not disturbed by these competitive inhibitors.

Antimetabolites such as the purine analogue vidarabine and the pyrimidine analogue idoxuridine are erroneously incorporated into nucleic acids. These molecules are very similar to the

> ➤ **Figure 13.4** **Competitive inhibition.** (a) *Para*-aminobenzoic acid (PABA), a metabolite required by many bacteria. (b) Sulfanilamide, a sulfa drug. (c) *Para*-aminosalicylic acid (PAS). Sulfanilamide and PAS act as competitive inhibitors to PABA. Notice the similarity in the structures of the three compounds.

normal purines and pyrimidines of nucleic acids (➤ Figure 13.5). When incorporated into a nucleic acid, they garble the information that it encodes because they cannot form the correct base pairs during replication and transcription. Purine and pyrimidine analogues are generally as toxic to animal cells as to microbes because all cells use the same purines and pyrimidines to make nucleotides. These agents are most useful in treating viral infections, because viruses incorporate analogues more rapidly than do cells and are more severely damaged.

Kinds of Side Effects

The side effects of antimicrobial agents on infected persons (hosts) fall into three general categories: (1) toxicity, (2) allergy, and (3) disruption of normal microflora. The development of resistance to antibiotics can also be thought of as a side effect on the microorganisms. As is explained later, resistance produces infections that can be difficult to treat.

Toxicity

By their selective toxicity and modes of action, antimicrobial agents kill microbes without seriously harming host cells. However, some antimicrobials do exert toxic effects on the patients receiving them. The toxic effects of antimicrobial agents are discussed later in connection with specific agents.

Allergy

An *allergy* is a condition in which the body's immune system responds to a foreign substance, usually a protein. For example, breakdown products of penicillins combine with proteins in body fluids to form a molecule that the body treats as a foreign substance. Allergic reactions can be limited to mild skin rashes and itching, or they can be life-threatening. One kind of life-threatening allergic reaction, called *anaphylactic shock* (Chapter 18), occurs when an individual is subjected to a foreign substance to which his or her body has already become sensitized—that is, a substance to which the individual has been exposed and developed antibodies against.

The Disruption of Normal Microflora

Antimicrobial agents, especially broad-spectrum antibiotics, may exert their adverse effects not only on pathogens but also on indigenous microflora—the microorganisms that normally inhabit the skin and the digestive, respiratory, and urogenital tracts. When these microflora are disturbed, other organisms not susceptible to the antimicrobial agent, such as *Candida* yeast, invade the unoccupied areas and multiply rapidly. Invasion by replacement microflora is called **superinfection**. Superinfections are difficult to treat because they are susceptible to few antibiotics.

➤ Figure 13.5   **Base analogues.** Nucleic acid bases and their analogues: molecules so similar in structure that they can be incorporated in place of the correct molecule, thus acting as antimetabolites. (a) Basic structure of a purine. (b) The purine analogue vidarabine. (c) Basic structure of a pyrimidine. (d) The pyrimidine analogue idoxuridine.

Although short-term use of penicillins generally does not severely disrupt normal microflora, oral ampicillin sometimes allows overgrowth of toxin-producing *Clostridia*. Long-term use of penicillin or aminoglycosides can abolish natural microbiota and allow colonization of the gut with resistant Gram-negative bacteria and fungi such as *Candida*. Live-culture yogurt (which contains lactobacilli), or a preparation called Lactinex (which contains normal microflora), can be given to counteract the effects of antibiotics. Oral and vaginal superinfections with species of *Candida* yeasts are common after prolonged use of antimicrobial agents such as cephalosporins, tetracyclines, and chloramphenicol. An old remedy for this problem is douching with dilute suspensions of plain, live-culture yogurt. The risk of serious superinfections is greatest in hospitalized patients receiving broad-spectrum antibiotics, for two reasons. First, patients often are debilitated and less able to resist infection. Second, they are in an environment in which drug-resistant pathogens are prevalent.

The Resistance of Microorganisms

Resistance of a microorganism to an antibiotic means that a microorganism formerly susceptible to the action of the antibiotic is no longer affected by it. An important factor in the development of drug-resistant strains of microorganisms is that many antibiotics are bacteriostatic rather than bactericidal. Unfortunately, the most resilient microbes evade defenses (Chapter 14) and are likely to develop resistance to the antibiotic.

How Resistance Is Acquired

Microorganisms generally acquire antibiotic resistance by genetic changes, but sometimes they do so by nongenetic mechanisms. Nongenetic resistance occurs when microorganisms such as those that cause tuberculosis persist in the tissues out of reach of antimicrobial agents. If the sequestered microorganisms start to multiply and release their progeny, the progeny are still susceptible to the antibiotic. This type of resistance might more properly be called *evasion*. Another type of nongenetic resistance occurs when certain strains of bacteria temporarily change to L forms that lack most of their cell walls. ➡ (Appendix B) For several generations, while the cell wall is lacking, these organisms are resistant to antibiotics that act on cell walls. However, when they revert to producing cell walls, they again become susceptible to the antibiotics.

Genetic resistance to antimicrobial agents develops from genetic changes followed by natural selection (➤ Figure 13.6). ➡ (Chapter 8, p. 209) For example, in most bacterial populations, mutations occur spontaneously at a rate of about 1 per 10 million to 10 billion organisms. Bacteria reproduce so rapidly that billions of organisms can be produced in a short period of time, and among them there will always be a few mutants. If a mutant happens to be resistant to an antimicrobial agent in the environment, that mutant and its progeny will be most likely to survive, whereas the nonresistant organisms will die. After a few generations, most survivors will be resistant to the antimicrobial agent. Antibiotics do *not* induce mutations, but they can create environments that favor the survival of mutant resistant organisms.

Genetic resistance in bacteria, where it is best understood, can be due to changes in the bacterial chromosome or to the acquisition of extrachromosomal DNA, usually in plasmids. (The mechanisms by which genetic changes occur were described in Chapters 7 and 8.) **Chromosomal resistance** is due to a mutation in chromosomal DNA and will usually be effective only against a single type of antibiotic. Such mutations often alter the DNA that directs the synthesis of ribo somal proteins. **Extrachromosomal resistance** is usually due to the presence of particular kinds of **resistance (R) plasmids, or R factors.** ➡ (Chapter 8, p. 209) How R plasmids originated is unknown, but they were first discovered in *Shigella* in Japan in 1959. Since that time many different R plasmids have been identified. Some R plasmids carry as many as six or seven genes, each of which confers resistance to a different antibiotic. R plasmids can also be transferred from one strain or species of bacteria to another. Most transfers occur by transduction (the transfer of plasmid DNA in a bacteriophage), and some occur by conjugation. ➡ (Chapter 8, pp. 201, 204)

Mechanisms of Resistance

Five mechanisms of resistance have been identified, each of which involves the alteration of a different microbial structure. One involves the alteration of the target to which antimicrobial agents bind, a process that generally is caused by a mutation in the bacterial chromosome. The other

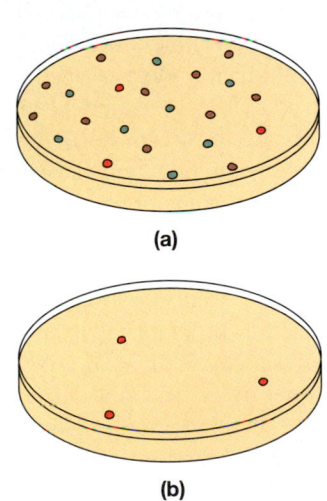

(a)

(b)

➤ Figure 13.6 **A method of detecting genetic resistance.** (a) A mixed population of bacteria of varying resistance to a new antibiotic is present. (b) Antibiotic is added to the Petri plate. Only those organisms with sufficient resistance will survive. Introduction of the antibiotic represents a change in the environment, but it does not create the resistant organisms—they were already there.

Fighting Disease at the Zoo: No Monkey Business
An Interview

Rob Schumaker: As soon as I walked into the monkey house that morning I knew that one of them was really sick. The first thing I noticed was a large amount of liquid stool in the cage. Colobus monkeys have stools like little pellets, so one of them was definitely in trouble.

I looked at the animals to see who it was. We keepers look at each animal every day, you know, and we really get to know them. I realized that it was Emily. Her face was very swollen—so much so that her eyes were nearly swollen shut. Her tail was filthy. I was really concerned. Emily is the oldest female in the group, which makes her the core of the females. Also, we suspected that she was pregnant.

I'm Rob Schumaker. I work as a keeper of monkeys and great apes at the Smithsonian's National Zoo in Washington, D.C. I'm studying for a graduate degree in zoology, and I have a particular interest in microbiology. The monkeys I'm talking about are the black and white colobus: *Colobus abyssinicus*. Their range is Africa, from Ethiopia to Tanzania and Zaire. Their natural diet consists of leaves and fruits. Colobus monkeys live in groups of two to 15 or more. The males are somewhat solitary, but the females have a highly structured society with many rules.

I watched Emily some more. The veterinarians here go on rounds every day, but I was concerned about Emily and didn't want to wait for them to come to us. Even though it could be very stressful for her, we decided to put her in her transport cage and take her to the hospital. We had to find out what this illness was. Our groups had always been okay; up until then we had escaped the difficulties that other monkey colonies have experienced.

Dr. Richard Montali: Those other difficulties were outbreaks of dysentery caused by *Shigella* bacteria. It's a problem in many primate colonies, and even in good zoos, monkeys and apes are often carriers. I'm Dr. Richard Montali, the veterinary pathologist at the National Zoo. Our zoo never had a *Shigella* problem until 1984. Now Dr. Mitchell Bush, the zoo's clinical veterinarian, and I are doing an intensive study of *Shigella* infection. Twice a year, we screen cultures of each animal's feces on three consecutive days. We've found *S. flexneri* and *S. sonnei*. We've found carrier animals. So far none of the keepers has caught it—although one of the veterinarians did.

The *Shigella* could have been introduced by humans, or it could have been brought in by a new animal. We really don't know. However, we are considering the possibility of using human vaccine for the animals. We're talking to Johns Hopkins about a *Shigella* vaccine they're developing that might be applicable to our primates. Walter Reed Army Medical Center has also been helpful. We identified the dysentery bacterium in our own labs, but now they do special cultures for us, to characterize organisms by species and even by strain. It's fortunate that these medical centers are close by.

One of the zoo's black and white colobus monkeys—ruminants that depend on gut microbes to help them digest their food.

We've also identified other pathogens from the zoo animals in our labs: *Salmonella, Campylobacter,* parasites. Timing is very important. For example, to isolate *Yersinia pseudotuberculosis* and *Y. enterocolitica,* carried by rats, we have to take cultures 3 to 5 days in a row. Even then, we may find the pathogen on only one of those days.

It's terrible when a zoo population falls victim to disease like this. We lost two gibbons to *Shigella*. One was old, the other had other underlying problems. But even if the mortality is low, it's upsetting to the staff *and* to the other animals—their whole social structure can be thrown off.

Once Emily was diagnosed as having a *Shigella* infection, the treatment was clear. Colobus monkeys have ruminant-type stomachs, similar to those of cows. That is, the stomach has several compartments and is populated with symbiotic bacteria that help break down the cellulose in the plant material the monkeys eat. [See Chapter 2.] These very special gastrointestinal microorganisms are acquired shortly after birth from handling the feces of other animals, and they are essential for life. A colobus monkey that lacks the proper microbes cannot get enough nutrients from its food to survive.

When these animals are given antibiotics, the bacterial populations that help them digest their food are destroyed. Saving them from possible death by infection involves exposing them to possible illness from malnutrition. So after the *Shigella* infection is cured, we try to repopulate the gut with stomach juices aspirated from other, healthy animals. Unfortunately, this technique doesn't always work. In Emily's case, it proved effective. But Emily was a very weak animal for a long time afterward. She lost the baby she was carrying, and also miscarried in her next two pregnancies. This year, though, she has a healthy baby.

Schumaker: Even though we knew what organism was making Emily so sick, we were still very worried about her. Her separation from her family group could be very stressful for both her and the group, and that stress could be the difference in her recovering or not. Being in a hospital is also stressful for animal patients—they don't understand where they are or why. The hospital keepers are wonderful, but the animals don't know them. And they may not always know how to talk to individual animals or know their specific needs. So the primary keepers go up to the hospital and visit their sick charges. They hand-feed them, if necessary.

Montali: The keepers in the zoo and the veterinary staff work closely together. It's important that the staff be educated, and aware of potential problems. For example, we have learned to avoid mixed exhibits of different species of animals, because some may be endangered by the food of others. We saw some losses in the small mammal house because of such a situation. Some tamarins died from *Streptococcus zooepidemicus,* Group C. This is an opportunistic pathogen in humans and is carried by horses. It is usually present in the respiratory and genital tracts of the horses and can sometimes cause serious infection even in them. In a small mammal, such as a tamarin, infection begins in the lymph nodes and then develops into septicemia.

How did the tamarins get an equine infection? Probably from the food that was given to the armadillos they lived with. We fed the armadillos a commercially sold raw horsemeat product that we received frozen. In a mixed exhibit such as the one we're talking about, the tamarin may have gone over to the armadillo's food dish and tasted the food. Or it may just have handled the meat and then put its fingers in its mouth. Although the pathogen didn't affect the armadillos, it killed several of the tamarins.

It's easy to understand how the horsemeat could be contaminated in this way. Do many *healthy* horses go to slaughter? Probably not. And the quality controls on food intended for animal consumption are much less stringent than those on food for human consumption. Our solution was to switch to beef and to reorganize the mixed exhibits.

Schumaker: The Zoo has developed a cooperative program to reintroduce golden lion tamarins to their native Brazil. They are an endangered species in the wild but breed well in zoos, where a surplus number of animals have accumulated. We run a staging program here in which we keep the tamarins in an outdoor, uncaged setting. This helps them adjust to the conditions they will find in the wild. We want them to survive when they get there.

Montali: And we want to be sure that they don't bring any diseases into the wild population that might kill what remains of the endangered population. In the zoo, animals can pick up parasites and infections not found in their native countries. A cockroach might scurry from the cage of an Asian animal into an adjacent cage, where an animal from South America will eat it, thus acquiring a parasite from a different continent. The German cockroach has been found to carry an intestinal nematode parasite from its indigenous host, the slow lorus, to tamarins. Thus, we examine these tamarins *very* carefully before we send them to Brazil. We also eliminate animals with genetic defects from the program.

Schumaker: We keepers have to be extraordinarily careful to avoid contamination of our animals and of ourselves. Whenever there is illness, we take extra precautions. For example, we always wear waterproof boots around the enclosures—for washing them down and so forth. When Emily got sick, we set up a footbath, containing a diluted bleach solution, at the door to her cage. The keepers stepped in it going into and coming out of the cage. (We didn't want to contaminate anything outside that enclosure.) We kept special tools only for that particular cage. We tended to that cage last and washed all tools in a bleach solution. We took extra care in scrubbing the cage, using a special antiseptic solution.

Those were the precautions we took over and above our regular procedures. We ordinarily clean the inside of the cages daily. They're scrubbed, hosed, and disinfected. We rotate among three products: an antiseptic detergent; bleach, which is a good stain remover; and a deodorizing agent.

As for the keepers, we always wear filter masks so that we don't inhale any aerosol material. We wear gloves, boots, and coveralls on the job, and we don't wear any work clothes out of the building—there are a washer and dryer in the basement of the building that we use. We wear our street clothes in and out.

Just as we check the monkeys to make sure they're healthy on a routine basis, we check ourselves too. All keepers' stools are cultured for three consecutive days twice a year. There are a number of pathogens that can be transmitted back and forth between humans and animals. Early in this century, for example, tuberculosis was a problem among both animals and staff. The keepers gave it to the animals; the animals gave it to the keepers. Even now, mammalian tuberculosis can be a danger to keepers, so we undergo regular TB testing. We have all come to the realization that disease transmission in a zoo can be a two-way problem. And since the animals are totally dependent on us, it is our problem to solve.

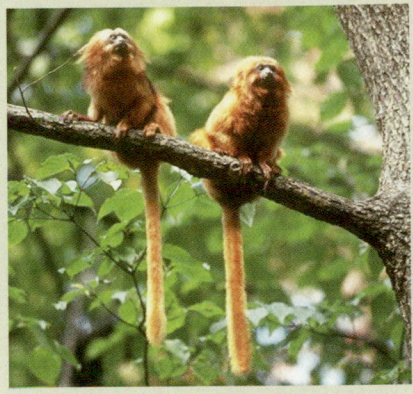

A pair of squirrel-sized golden lion tamarin monkeys becoming acclimated to uncaged life at the National Zoo before being released in Brazil.

Rob Schumaker at the National Zoo in Washington, D.C.

mechanisms involve alterations in membrane permeability, enzymes, or metabolic pathways, which usually are caused by the acquisition of R plasmids. The five mechanisms are explained next.

1. *Alteration of targets*. This mechanism usually affects bacterial ribosomes. The mutation alters the DNA such that the protein produced or target is modified. Antimicrobial agents can no longer bind to the target. Resistance to erythromycin, rifamycin, and antimetabolites has developed by this mechanism.

2. *Alteration of membrane permeability*. This mechanism occurs when new genetic information changes the nature of proteins in the membrane. Such alterations change a membrane transport system or pores in the membrane, so an antimicrobial agent can no longer cross the membrane. In bacteria, resistance to tetracyclines, quinolones, and some aminoglycosides has occurred by this mechanism. The presence of penicillin or cephalosporin can partially overcome such resistance because these agents interfere with cell wall synthesis.

3. *Development of enzymes*. This common cause of resistance can destroy or inactivate antimicrobial agents. One enzyme of this type is β-lactamase. Several β-lactamases exist in various bacteria; they are capable of breaking the β-lactam ring in penicillins and some cephalosporins. Similar enzymes that can destroy various aminoglycosides and chloramphenicol have been found in certain Gram-negative bacteria.

4. *Alteration of an enzyme*. This mechanism allows a formerly inhibited reaction to occur. It is exemplified by a mechanism found among certain sulfonamide-resistant bacteria. These organisms have developed an enzyme that has a very high affinity for PABA and a very low affinity for sulfonamide. Consequently, even in the presence of sulfonamide, the enzyme works well enough to allow the bacterium to function.

5. *Alteration of a metabolic pathway*. This mechanism bypasses a reaction inhibited by an antimicrobial agent that occurs in other sulfonamide-resistant bacteria. These organisms have acquired the ability to use ready-made folic acid from their environment and no longer need to make it from PABA.

First-Line, Second-Line, and Third-Line Drugs

As a strain of microorganism acquires resistance to a drug, another drug must be found to treat resistant infections effectively. If resistance to a second drug develops, a third drug is needed, and so on. Drugs used to treat gonorrhea illustrate this point. Before the 1930s no effective treatment was available for gonorrhea. Then sulfonamides were found to cure the disease. After a few years, sulfonamide-resistant strains developed, but penicillin was soon available as a "second-line" drug. Over several decades penicillin-resistant strains developed but were combatted with very large doses of penicillin. By the 1970s some strains of gonococci developed the ability to produce a β-lactamase enzyme, which completely counteracted the effects of penicillin (➤ Figure 13.7). "Third-line" spectinomycin is now being used. As spectinomycin-resistant strains start to appear, forcing physicians to resort to "fourth-line" drugs, we have to wonder whether the development of new drugs can go on indefinitely.

Drug-resistant organisms have most frequently been encountered within hospitals, where seriously ill patients with lowered resistance to infections serve as convenient hosts. However, more and more resistant organisms are being isolated from infections among the general population, and the risk of acquiring a drug-resistant infection is increasing for everyone. Moreover, many organisms are resistant to multiple antibiotics. Infections with such organisms are particularly difficult to treat. A new use for genetic probes will be to look for resistance genes in organisms, in order to avoid delays in effective treatment.

Cross-Resistance

Cross-resistance is resistance to two or more similar antimicrobial agents via a common mechanism. The action of β-lactamases provides a good example of cross-resistance. In many instances, an enzyme that will break down one β-lactam antibiotic also will break down several other β-lactam antibiotics. The presence of such an enzyme would give a microorganism resistance to all the antibiotics it can break down.

The overall cost of antibiotic resistance in the United States is estimated between $100 million and $30 billion annually.

Microbial Resistance

Microbes resistant to disinfectants, antibiotics, or antiseptics are not created but are already present in microbial populations. It is the inappropriate use of these compounds that tends to select for and therefore promote the growth of resistant organisms in our hospitals and homes. Through the biological mechanisms of microbial competition and genetic exchange, resistant organisms can become the predominant organisms. When antimicrobial compounds are used, they must be used in concentrations that are strong enough to kill the most insensitive organisms present in the target population. If this situation has not been achieved, then the less sensitive bacteria, unencumbered by competition, are able to thrive.

> Figure 13.7   **The effect of β-lactamase on penicillin.** Numerous bacteria (staphylococci, streptococci, and gonococci) produce this enzyme, which inactivates penicillin. The enzyme can be transmitted by plasmids. Cephalosporins, although similar in action to penicillin, have a different cyclic ring structure and are more resistant to the effects of the enzyme.

Limiting Drug Resistance

Although, as we have seen, drug resistance is not induced by antibiotics, it is fostered by environments containing antibiotics. The progress of microbes in acquiring resistance can be thwarted in three ways. First, high levels of an antibiotic can be maintained in the bodies of patients long enough to kill all pathogens, including resistant mutants, or to inhibit them so that body defenses can kill them. This is the reason your doctor admonishes you to be sure to take all of an antibiotic prescription and not to stop taking it once you begin to feel better. The development of resistance when medication is discontinued before all pathogens are killed is illustrated in ➤ Figure 13.8.

Second, two antibiotics can be administered simultaneously so that they can exert an additive effect called **synergism.** For example, when streptomycin and penicillin are combined in therapy, the damage to the cell wall caused by the penicillin allows better penetration by streptomycin. A variation on this principle is the use of one agent to destroy the resistance of microbes to another agent. When clavulanic acid and a penicillin called amoxicillin are given together (Augmentin), the clavulanic acid binds tightly to β-lactamases and prevents them from inactivating the amoxicillin. However, some drugs are less effective when used in combination than when used alone. This decreased effect, called **antagonism,** can be observed when bacteriostatic drugs such as tetracyclines, which inhibit growth, are combined with bactericidal penicillins, which require growth to be effective.

Third, antibiotics can be restricted to essential uses only. For example, most physicians do not prescribe antibiotics for colds and other viral diseases, except in the case of patients at high risk of secondary bacterial infections, because such diseases do not respond to antibiotics. Restrictions on antibiotic use would be particularly valuable in hospitals, where microbes "just waiting to acquire resistance" lurk in antibiotic-filled environments. In addition, the use of antibiotics in animal feeds could be banned; see the box "Antibiotic Resistance: Drugs in Animal Feeds."

According to the CDC, nearly 50 million of the 150 million outpatient prescriptions for antibiotics are unneeded.

✓ Are all antimicrobial compounds properly called antibiotics? Why or why not?

✓ When would you choose a narrow-spectrum, rather than a broad-spectrum, antibiotic? Why?

✓ What is a superinfection? How is one acquired?

✓ If exposure to antibiotics does not cause drug-resistant mutations to occur, why do we see more drug-resistant strains today?

Determining Microbial Sensitivities to Antimicrobial Agents

Microorganisms vary in their susceptibility to different chemotherapeutic agents, and susceptibilities can change over time. Ideally, the appropriate antibiotic to treat any particular infection should be determined before any antibiotics are given. Sometimes an appropriate agent can be prescribed as soon as the causative organism is identified from a laboratory culture. Often tests are needed to show which antibiotic kills the organism. Several methods—disk diffusion, dilution, and automated methods—are available to do this.

The Disk Diffusion Method

In the **disk diffusion method,** or **Kirby-Bauer method,** a standard quantity of the causative organism is uniformly spread over an agar plate. Then several filter paper disks impregnated with specific concentrations of selected chemotherapeutic agents are placed on the agar surface (➤ Figure 13.9a). Finally, the culture with the antibiotic disks is incubated.

During incubation, each chemotherapeutic agent diffuses out from the disk in all directions. Agents with lower molecular weights diffuse faster than those with higher molecular weights. Clear areas called **zones of inhibition** appear on the agar around disks where the agents inhibit

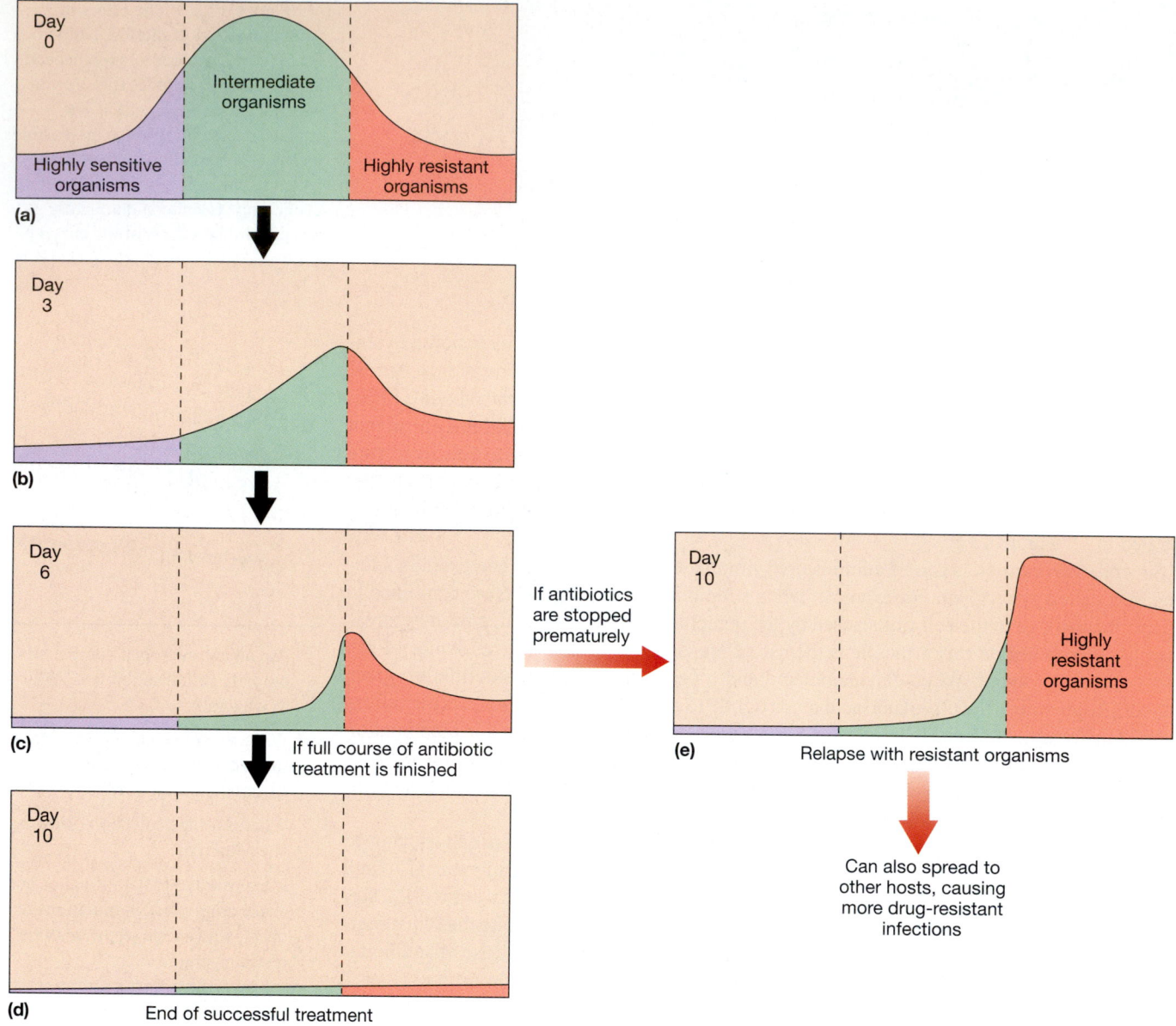

> Figure 13.8   **Effects of premature termination of antibiotic treatment.** (a) Organisms present before treatment begins have differing sensitivities to the antibiotic that will be used. (b) and (c) As treatment progresses, those organisms most sensitive to the antibiotic die off first, leaving mostly resistant ones as survivors. (d) Finishing the full course of treatment finally kills the most resistant organisms. (e) If antibiotics are stopped too early, the remaining resistant organisms will survive and multiply, leading to an infection that will be difficult or impossible to cure using the same antibiotic. This can then spread to other hosts.

the organism. The size of a zone of inhibition is not necessarily a measure of the degree of inhibition because of differences in the diffusion rates of chemotherapeutic agents. An agent of large molecular size might be a powerful inhibitor even though it might diffuse only a small distance and produce a small zone of inhibition. Standard measurements of zone diameters for particular media, quantities of organisms, and drug concentrations have been established and correlated to zone diameters in order to determine whether the organisms are *sensitive, moderately sensitive,* or *resistant* to the drug.

Even when inhibition has been properly interpreted in a disk diffusion test, the most inhibitory chemotherapeutic agent may not cure an infection. The agent will probably inhibit the causative organism, but it may not kill sufficient numbers of the organism to control the in-

(a)

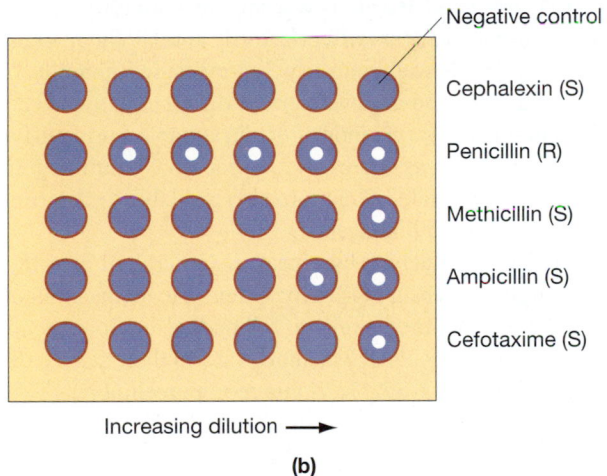

Negative control

Cephalexin (S)

Penicillin (R)

Methicillin (S)

Ampicillin (S)

Cefotaxime (S)

Increasing dilution ⟶

(b)

➢ Figure 13.9 **The disk diffusion (Kirby-Bauer) method of determining microbial sensitivities to various antibiotics.** (a) A Petri plate is prestreaked on an agar medium with the organism to be tested. Paper disks, each containing a measured amount of a particular antibiotic, are placed firmly in contact with the medium and diffuse outward. After the plate is incubated, a clear area of no growth around a disk (a zone of inhibition) represents inhibition of the test organism by the antibiotic. Nonclear areas indicate resistance to that antibiotic. The largest zone of inhibition does not always indicate the most effective antibiotic, as different molecules do not diffuse at the same speed into the medium. Also, some drugs do not behave the same way in living organisms as they do on agar. However, the diameters of zones of inhibition, when compared with measured standards, help indicate whether an organism is sensitive or resistant to a drug. (b) Minimal inhibitory concentration (MIC) microbial susceptibility testing. A standardized microdilution plate with shallow wells that contain increasing dilutions (decreasing concentrations) of selected antibiotics in a broth is inoculated with a test bacterium. The plate is incubated; the lowest concentration that prevents growth (dots in well) is the MIC. The test bacterium on this plate is sensitive (S) to all the antibiotics except penicillin (R). The negative control well contains only broth.

fection. A bactericidal agent is often needed to eliminate an infectious organism, and the disk diffusion method does not assure that a bactericidal agent will be identified. Moreover, results obtained *in vivo* (in a living organism) often differ from those obtained *in vitro* (in a laboratory vessel). Metabolic processes in the body of a living organism may inactivate or inhibit an antimicrobial compound.

The Dilution Method

The **dilution method** of testing antibiotic sensitivity was first performed in tubes of culture broth; it is now performed in shallow wells on standardized plates (➢ Figure 13.9b). In this method a constant quantity of microbial inoculum (specimen) is introduced into a series of broth cultures containing decreasing concentrations of a chemotherapeutic agent. After incubation (for 16 to 20 hours) the tubes or wells are examined, and the lowest concentration of the agent that prevents visible growth (indicated by turbidity or dots of growing organisms) is noted. This concentration is the **minimum inhibitory concentration (MIC)** for a particular agent acting on a specific microorganism. This test can be done for several agents simultaneously by using several sets of tubes or wells, but it is time-consuming and therefore expensive.

Finding an inhibitory agent by the dilution method does no more to prove that it will kill the infectious organism in the patient than finding one by the disk diffusion method. However, the dilution method allows a second test to distinguish between bactericidal agents, which kill microorganisms, and bacteriostatic agents, which merely inhibit their growth. Samples from tubes that show no growth but that might contain inhibited organisms can be used to inoculate broth that contains no chemotherapeutic agent. In this test the lowest concentration of the chemotherapeutic agent that yields no growth following this second inoculation, or *subculturing,* is the **minimum bactericidal concentration (MBC).** Thus, both an effective chemotherapeutic agent and an appropriate concentration to control an infection can be determined. That concentration should be maintained at the sites of infection because it is the minimum concentration that will cure the disease.

Serum Killing Power

Still another method of determining the effectiveness of a chemotherapeutic agent is to measure its **serum killing power.** This test is performed by obtaining a sample of a patient's blood while the patient is receiving an antibiotic. A bacterial suspension is added to a known quantity of the patient's **serum** (blood plasma minus the clotting factors). Growth (turbidity) in the serum after incubation means that the antibiotic is ineffective. Inhibition of growth suggests that the drug is working, and more quantitative determinations can be made to identify the lowest concentration that still provides serum killing power.

Automated Methods

Automated methods (➤ Figure 13.10) are now available to identify pathogenic organisms and to determine which antimicrobial agents will effectively combat them. One such method makes use of prepared trays with small wells into which a measured quantity of inoculum is automatically dispensed. As many as 36 organisms from different patients can be inoculated onto the same tray. Trays containing several kinds of media suitable for identifying members of different groups of organisms—such as Gram-positive bacteria, Gram-negative bacteria, anaerobic bacteria, and yeasts—are available. Trays are also available to determine the sensitivity of organisms to a variety of antimicrobial agents.

The trays are inserted into a machine that measures microbial growth. Some machines do this by using a beam of light to measure turbidity. Others use media containing radioactive carbon. Organisms growing on such media release radioactive carbon dioxide into the air, and a sampling device automatically detects it. Machines vary in their degree of automation and the speed with which results become available. Some require technicians to perform some steps;

(a)   (b)

➤ Figure 13.10   **An automated system for identifying microorganisms and determining their sensitivity to various antimicrobial agents.** (a) A sample containing the organism(s) is automatically inoculated into wells on a thin plastic tray, each containing a specific chemical reagent. (b) Tests are carried out in an incubation chamber, and the results are read and recorded by computer. Trays for a wide variety of different identification and antimicrobial sensitivity tests are available.

others provide a computerized printout of results that is relayed to the patient's chart. Some machines provide results in 3 to 6 hours, and most provide them overnight, although slow-growing organisms may require 48 hours.

Automated methods make laboratory identification of organisms and their sensitivities to antimicrobials more efficient and less expensive. Once the results of laboratory tests are available, the physician can then choose an appropriate drug on the basis of the nature of the pathogen, the location of the infection, and other factors such as the patient's allergies. Automated methods allow physicians to prescribe an appropriate antibiotic early in an infection rather than prescribing a broad-spectrum antibiotic while awaiting laboratory results.

Attributes of An Ideal Antimicrobial Agent

Having considered various characteristics of antimicrobial agents and methods of determining microbial sensitivities to them, we can now list the characteristics of an ideal antimicrobial agent:

1. *Solubility in body fluids.* Agents must dissolve in body fluids to be transported in the body and reach the infectious organisms. Even agents used topically must dissolve in the fluids of injured tissue to be effective; however, they must not bind too tightly to serum proteins.

2. *Selective toxicity.* Agents must be more toxic to microorganisms than to host cells. Ideally, a great difference should exist between the low concentration toxic to microorganisms and the concentration that damages host cells.

3. *Toxicity not easily altered.* The agent should maintain a standard toxicity and not be made more or less toxic by interactions with foods, other drugs, or abnormal conditions such as diabetes and kidney disease in the host.

4. *Nonallergenic.* The agent should not elicit an allergic reaction in the host.

5. *Stability: maintenance of a constant, therapeutic concentration in blood and tissue fluids.* The agent should be sufficiently stable in body fluids to have therapeutic activity over many hours; it should be degraded and excreted slowly.

6. *Resistance by microorganisms not easily acquired.* There should be few, if any, microorganisms with resistance to the agent.

7. *Long shelf life.* The agent should retain its therapeutic properties over a long period of time with a minimum of special procedures such as refrigeration or shielding from light.

8. *Reasonable cost.* The agent should be affordable to patients who need it.

Many antimicrobial agents meet these criteria reasonably well. But few, if any, meet all the criteria for the ideal antimicrobial agents. As long as better drugs might be found, the search for them will continue.

Antibacterial Agents

Most antimicrobial agents are *antibacterial agents,* so we will start our "catalog" of antimicrobial agents with them, keeping in mind that some are effective against other microbes as well. Antibacterial agents can be categorized in several ways; we have chosen to use their modes of action. Another way of grouping antibiotics is by the microorganism that produces them (Table 13.2).

Inhibitors of Cell Wall Synthesis

Penicillins

Natural **penicillins,** such as *penicillin G* and *penicillin V,* are extracted from cultures of the mold *Penicillium notatum.* The discovery in the 1950s that certain strains of *Staphylococcus aureus* are resistant to penicillin provided the impetus to develop semisynthetic penicillins. The first of these was *methicillin,* which is effective against penicillin-resistant organisms because it is not broken down by β-lactamase enzymes. Other semisynthetic penicillins, including *nafcillin, oxacillin, ampicillin, amoxicillin, carbenicillin,* and *ticarcillin,* emerged in rapid succession. Each

Nearly 40% of the antibiotics manufactured in the United States are given to animals. Although some are used to combat infection, most are mixed into feed to promote growth.

Be a Discoverer!

Soil microbes are good sources of antibiotics. Pharmaceutical companies pay people to bring back soil samples from all over the world in hopes of finding new and useful antibiotic-producing microbes. Old cemeteries wildlife refuges, swamps, and airfield landing strips have all been sampled. Try collecting soil samples from some places that interest you. Using aseptic technique, swab the surface of a Petri plate of nutrient agar with a broth culture of an organism such as *E. coli.* Make confluent strokes so that the entire surface is inoculated evenly, like a lawn. You could try different organisms on different plates. Then sprinkle tiny bits of your soil sample over each inoculated plate. Incubate the plates or leave them at room temperature, and examine them daily. Do clear areas develop around some of the colonies that have grown from soil particles? If so, these are areas where natural antibiotics have diffused out into the agar and prevented the background "lawn" bacteria from growing. Congratulations, you've found some antibiotics!

➤ Table 13.2   **Selected microbes that serve as sources of antibiotics**

Microbe	Antibiotic
Fungi	
Aspergillus fumigatus	Fumagillin
Cephalosporium species	Cephalosporins
Penicillium griseofulvum	Griseofulvin
Penicillium notatum and *P. chrysogenum*	Penicillin
Streptomycetes	
Streptomyces nodosus	Amphotericin B
Streptomyces venezuelae	Chloramphenicol
Streptomyces erythreus	Erythromycin
Streptomyces griseus	Streptomycin
Streptomyces kanamyceticus	Kanamycin
Streptomyces fradiae	Neomycin
Streptomyces noursei	Nystatin
Streptomyces antibioticus	Vidarabine
Actinomycetes	
Micromonospora species	Gentamicin
Other bacteria	
Bacillus licheniformis	Bacitracin
Bacillus polymyxa	Polymyxins
Bacillus brevis	Tyrocidin

is synthesized by adding a particular side chain to a penicillin nucleus (➤ Figure 13.11). Both the natural and semisynthetic penicillins are bactericidal.

Penicillin G, the most frequently used natural penicillin, is administered *parenterally*—that is, by some means other than via the gut, such as intramuscularly or intravenously. When administered orally, most of it is broken down by stomach acids. Penicillin is rapidly absorbed into the blood, reaches its maximum concentration, and is excreted unless it is combined with an agent such as procaine, which slows excretion and prolongs activity.

Penicillin G is the drug of choice in treating infections caused by streptococci, meningococci, pneumococci, spirochetes, clostridia, and aerobic Gram-positive rods. It is also suitable for treating infections caused by a few strains of staphylococci and gonococci that are not resistant to it. Because it retains activity in urine, it is suitable for treating some urinary tract infections. Infections caused by organisms resistant to penicillin G can be treated with semisynthetics such as nafcillin, oxacillin, ampicillin, or amoxicillin. Carbenicillin and ticarcillin are especially useful in treating *Pseudomonas* infections. Allergy to penicillin is rare among children but occurs in 1 to 5 percent of adults. Penicillins are generally nontoxic, but large doses can have toxic effects on the kidneys, liver, and central nervous system.

In addition to their use as treatment for infections, penicillins also are used prophylactically—that is, to *prevent* infection. For example, patients with heart defects (in particular, malformed or artificial valves) or heart disease are especially susceptible to endocarditis, an inflammation of the lining of the heart, caused by a bacterial infection. Organisms tend to attack surfaces of damaged valves. To prevent such infections, susceptible patients often receive penicillin before surgery or dental procedures (even cleanings) that could release bacteria into the bloodstream.

Cephalosporins

Natural **cephalosporins** (sef″a-lo-spor′-inz), derived from several species of the fungus *Cephalosporium*, have limited antimicrobial action. Their discovery led to the development of a large number of bactericidal, semisynthetic derivatives of natural cephalosporin C. The nu-

> Figure 13.11   **Penicillins.** A comparison of the penicillin and cephalosporin molecules with β-lactam rings (blue). Cephalosporin differs slightly in the attached ring (red) and has two sites for side-chain attachments (purple) rather than one, as on the penicillin molecule.

cleus of a cephalosporin is quite similar to that of penicillin; both contain β-lactam rings (➤ Figure 13.11). Semisynthetic cephalosporins, like semisynthetic penicillins, differ in the nature of their side chains. Frequently used cephalosporins include *cephalexin* (Keflex), *cephradine,* and *cefadroxil,* all of which are fairly well absorbed from the gut and so can be administered orally. Other cephalosporins, such as *cephalothin* (Keflin), *cephapirin,* and *cefazolin,* must be administered parenterally, usually into muscles or veins.

Although cephalosporins usually are not the first drug considered in the treatment of an infection, they are frequently used when allergy or toxicity prevents the use of other drugs. But because cephalosporins are structurally similar to penicillin, some patients who are allergic to penicillin may also be sensitive to the cephalosporins. Nevertheless, cephalosporins account for one-fourth to one-third of pharmacy expenditures in American hospitals, mainly because they have a fairly wide spectrum of activity, rarely cause serious side effects, and can be used prophylactically in surgical patients. Unfortunately, they are often used when a less expensive and narrower-spectrum agent would be just as effective.

The development of new varieties of cephalosporins seems to be a race against the ability of bacteria to acquire resistance to older varieties. When organisms became resistant to early "first-generation" cephalosporins, new "second-generation" cephalosporins, including *cefuroxime* and *cefaclor,* were produced (➤ Figure 13.11). Now "third-generation" cephalosporins, such as *ceftriaxone* and *cephalexin,* are used against organisms resistant to older drugs. These drugs are especially effective (for now) in dealing with hospital-acquired infections resistant to many antibiotics. They are being tried in patients with AIDS and other immunodeficiencies. (Do not confuse these with second- and third-line drugs, described earlier, which are not derivatives of one another.)

Adverse effects from cephalosporins tend to be local reactions, such as irritation at the injection site or nausea, vomiting, and diarrhea when the drug is administered orally. Four to 15 percent of patients allergic to penicillin also are allergic to cephalosporins. Moreover, newer cephalosporins have little effect on Gram-positive organisms, which can cause superinfections during the treatment of Gram-negative infections.

Other Antibacterial Agents That Act on Cell Walls

Carbapenems (kar′ba-pen-emz) represent a new group of bactericidal antibiotics with two-part structures. *Primaxin,* a typical carbapenem, consists of a β-lactam antibiotic (*imipenem*) that interferes with cell wall synthesis and *cilastatin sodium,* a compound that prevents degradation of the drug in the kidneys. As a group, the carbapenems have an extremely broad spectrum of activity.

Bacitracin, a small bactericidal polypeptide derived from the bacterium *Bacillus licheniformis,* is used only on lesions and wounds of the skin or mucous membranes because it is poorly absorbed and toxic to the kidneys. *Vancomycin* is a large, complex molecule produced by the soil actinomycete *Streptomyces orientalis.* It can be used to treat infections caused by methicillin-resistant staphylococci and enterococci. It is also the drug of choice against antibiotic-induced pseudomembranous colitis (enteritis with the formation of false membranes in stool). Because it is poorly absorbed through the gastrointestinal tract, it must be administered intravenously. Vancomycin is fairly toxic, causing hearing loss and kidney damage, especially in older patients, if the drug is not monitored carefully.

Disrupters of Cell Membranes

Polymyxins

Five **polymyxins,** designated A, B, C, D, and E, have been obtained from the soil bacterium *Bacillus polymyxa.* Polymyxins B and E are the most common clinically. They are usually applied topically, often with bacitracin, to treat skin infections caused by Gram-negative bacteria such as *Pseudomonas.* Used internally, polymyxins can cause numbness in the extremities, serious kidney damage, and respiratory arrest. They are administered by injection when the patient is hospitalized and kidney function can be monitored.

The Role of *Clostridium difficile* in Antibiotic-Associated Intestinal Disease

Antibiotic therapy sometimes has its dark side. Today, *Clostridium difficile* is one of the most important intestinal bacterial pathogens in the developed world, in terms of prevalence and severity of disease. Yet *C. difficile* causes intestinal disease only in patients to whom antibiotics have been administered. (A few exceptions to this rule existed in the preantibiotic era, and no completely convincing explanation is available to explain these cases.)

This pathogen is a Gram-positive anaerobic rod. *C. difficile* was first described in 1935, but it was not definitively associated with disease until 1978. During the 1950s, especially, *Staphylococcus aureus* was blamed for what are now known to have been *C. difficile* infections. *C. difficile* is now understood to be a major nosocomial pathogen and is currently the only organism recognized as a common cause of antibiotic-associated colitis. *C. difficile* is found in 15 to 25 percent of patients with antibiotic-associated diarrhea; 50 to 75 percent of patients with colitis; and more than 90 percent of patients with pseudomembranous colitis (PMC). In individuals with PMC, a fibrous pseudomembrane covers the mucosa of the colon due to fibrin-containing fluid that collects there.

The organism is rarely found in healthy people except for newborns, among whom it occurs frequently but without harm. These newborns can act as reservoirs for *C. difficile* and can easily spread it to others in hospitals or at home. It disappears as babies approach 6 to 12 months of age. Older children rarely develop *C. difficile* problems, despite frequent use of antibiotics. Older adults are most likely to develop disease, but this age factor has not yet been explained. PMC can develop in patients who have undergone abdominal or intestinal surgery and in those whose gastrointestinal microbiota has been changed by the use of antibiotics.

C. difficile is a moderately strict anaerobe. It is widely distributed in nature, being found in soil and in feces of animals such as cows and horses. Once it has established itself in an environment, *C. difficile* is difficult to remove. It may persist in hospital wards or nursing homes for months or

years despite vigorous efforts at eradication. Organisms have been recovered from the floor, bedpans, linens, and even walls of rooms of infected patients.

Clindamycin is the antibiotic most frequently associated with *C. difficile* infections, followed by ampicillin and cephalosporins. These drugs are often administered to patients who will undergo abdominal surgery, in order to decrease normal microbiota. The *C. difficile* organisms are usually very susceptible to these antibiotics but survive by forming endospores. After the drug is discontinued, they regenerate and can overgrow the intestinal tract, but they do not invade its tissues.

Virtually all strains of *C. difficile* produce two toxins, but the strains vary widely in the amount of toxin produced. Toxin A, often called the enterotoxin, has a cytotoxic (cell-killing) effect and is responsible for most of the symptoms observed. Toxin B, called the cytotoxin, is considerably more cytotoxic but is not active in the gastrointestinal tract. The mechanism of action of these toxins is still not completely known.

Symptoms of the colitis include abdominal cramps, diarrhea, fever (up to 106°F, or 41.1°C), electrolyte imbalance, toxic megacolon (an acute dilation of the colon), and even perforation of the colon. The antibiotic vancomycin is the single most effective treatment and is almost 100 percent effective when perforation has not occurred. Fever drops within 24 to 48 hours, and normal bowel action returns in 5 to 7 days. However, relapse is frequent due to the sporulating forms. Vancomycin costs four times as much as an equivalent weight of gold. Cost of the treatment ranges from $200 to $600. Metronidazole is also used due to its anti-anaerobic properties.

Endoscopic observation of pseudomembranes in the colon is considered diagnostic for a severe *C. difficile* disease. Confirmation usually requires isolating the organism from stool specimens and identifying the toxins by ELISA and tissue culture assay. Toxin A is so potent that one molecule is enough to cause the changes seen in cells cultured.

Tyrocidins

Tyrocidins are cyclic polypeptides obtained from *Bacillus brevis*. The first to be discovered was *tyrothricin,* from which *tyrocidine* and *gramicidin* (named for Hans Gram, the originator of the Gram stain) have been derived. All are highly toxic, but they can be used topically to prevent bacterial growth, especially that of Gram-positive cocci.

Inhibitors of Protein Synthesis

Aminoglycosides

Aminoglycosides are obtained from various species of the genera *Streptomyces* and *Micromonospora*. The first, **streptomycin,** was discovered in the 1940s and was effective against a variety of bacteria. Since then, many bacteria have become resistant to it. Moreover, streptomycin can damage kidneys and the inner ear, sometimes causing permanent ringing in the ears and dizziness. Consequently, this compound is now used only in special situations and generally in combination with other drugs. For example, it can be used with tetracyclines to treat plague and tularemia and with isoniazid and rifampin to treat tuberculosis.

Other aminoglycosides, such as *neomycin, kanamycin, amikacin, gentamicin, tobramycin,* and *netilmicin,* also have special uses and display varying degrees of toxicity to the kidneys and inner ear. At lower, less toxic, doses aminoglycosides tend to be bacteriostatic. They are usually administered intramuscularly or intravenously because they are poorly absorbed when given orally.

An important property of aminoglycosides is their ability to act synergistically with other drugs—an aminoglycoside and another drug together often control an infection better than either could alone. For example, gentamicin and penicillin or ampicillin are effective against penicillin-resistant streptococci. In other synergistic actions, gentamicin or tobramycin work with carbenicillin or ticarcillin to control *Pseudomonas* infections, especially in burn patients, and aminoglycosides work with cephalosporins to control *Klebsiella* infections.

Other applications of aminoglycosides include the treatment of bone and joint infections, peritonitis (inflammation of the lining of the abdominal cavity), pelvic abscesses, and many hospital-acquired infections. In bone and joint infections, gentamicin and tobramycin are especially useful because they can penetrate joint cavities. Because peritonitis and pelvic abscesses are severe and are often caused by a mixture of enterococci and anaerobic bacteria, aminoglycoside treatment is usually started before the organisms are identified. Amikacin is especially effective in treating hospital-acquired infections resistant to other drugs. It should not be used in less demanding situations lest organisms become resistant to it, too.

Aminoglycosides often damage kidney cells, causing protein to be excreted in the urine, and prolonged use can kill kidney cells. These effects are most pronounced in older patients and those with preexisting kidney disease. Some aminoglycosides damage the eighth cranial nerve: Streptomycin causes dizziness and disturbances in balance, and neomycin causes hearing loss.

Tetracyclines

Several **tetracyclines** are obtained from species of *Streptomyces,* in which they were originally discovered. Commonly used tetracyclines include tetracycline itself, *chlortetracycline* (Aureomycin), and *oxytetracycline* (Terramycin). Newer semisynthetic tetracyclines include *minocycline* (Minocin) and *doxycycline* (Vibramycin). All are bacteriostatic at normal doses, are readily absorbed from the digestive tract, and become widely distributed in tissues and body fluids (with the exception of cerebrospinal fluid).

The fact that tetracyclines have the widest spectrum of activity of any antibiotics is a two-edged sword. They are effective against many Gram-positive and Gram-negative bacterial infections and are suitable for treating rickettsial, chlamydial, mycoplasmal, and some fungal infections. But because they have such a wide spectrum of activity, they destroy the normal intestinal microflora and often produce severe gastrointestinal disorders. Recalcitrant superinfections of tetracycline-resistant *Proteus, Pseudomonas,* and *Staphylococcus,* as well as yeast infections, also can result.

Tetracyclines can cause a variety of mild to severe toxic effects. Nausea and diarrhea are common, and extreme sensitivity to light is sometimes seen. The drug can also cause pustules to form on the skin. Effects on the liver and kidneys are more serious. Liver damage can be fatal, especially in patients with severe infections or during pregnancy. Kidney damage can lead to acidosis (low blood pH) and to the excretion of protein and glucose. Anemia can occur, but it is rare. Tetracycline can also interfere with the effectiveness of birth control pills.

Staining of the teeth (➤ Figure 13.12) occurs when children under 5 years of age receive tetracycline or when their mothers received it during the last half of their pregnancy. Both deciduous teeth (baby teeth) and permanent teeth will be mottled because the buds of both types of teeth form before birth. Tetracycline taken during pregnancy also can lead to abnormal bone formation in the fetal skull and a permanent abnormal skull shape. The ability of calcium ions to form a complex with tetracycline is responsible for its effects on bones and teeth. Because this reaction destroys the antibiotic effect of the drug, patients should not consume milk or other dairy products with the drug or for a few hours after taking it. Some greens, such as collard greens, are also very high in calcium, and should be avoided when taking tetracyclines.

> Figure 13.12 **Staining of teeth caused by tetracycline.** If the condition results from ingestion of the antibiotic during pregnancy, both the deciduous (baby) and permanent teeth will be affected, as both sets of tooth buds are forming in the fetus at that time.

Chloramphenicol

Chloramphenicol, originally obtained from cultures of *Streptomyces venezuelae,* is now fully synthesized in the laboratory. Like tetracyclines it is bacteriostatic, is rapidly absorbed from the digestive tract, is widely distributed in tissues, and has a broad spectrum of activity. It is used to treat typhoid fever, infections due to penicillin-resistant strains of meningococci and *Haemophilus influenzae,* brain abscesses, and severe rickettsial infections.

Chloramphenicol damages bone marrow in two ways. It causes a dose-related, reversible aplastic anemia, in which bone marrow cells produce too few erythrocytes and sometimes too few leukocytes and platelets as well. Terminating use of the drug usually allows the bone marrow to recover normal function. It also causes a non-dose-related, permanent aplastic anemia due to destruction of bone marrow. Aplastic anemia appears days to months after treatment is discontinued and is most common in newborns. Unless a successful bone marrow transplant can be performed, aplastic anemia is usually fatal. It is seen in only one in 25,000 to 40,000 patients treated with chloramphenicol. Long-term use of chloramphenicol can cause inflammation of the optic and other nerves, confusion, delirium, and mild to severe gastrointestinal symptoms. Since chloramphenicol is sometimes prescribed, or sold without prescription in countries outside the United States, you should always be careful to know the identity of antibiotics that you acquire abroad. In the United States, chloramphenicol is a drug of last choice.

Other Antibacterial Agents That Affect Protein Synthesis

Macrolides Erythromycin (e-rith″ro-mi′sin), a commonly used **macrolide** (large-ring compound), is produced by several strains of *Streptomyces erythreus.* Erythromycin exerts a bacteriostatic effect, is readily absorbed, and reaches most tissues and body fluids (with the exception of cerebrospinal fluid). It is recommended for infections caused by streptococci, pneumococci, and corynebacteria but is also effective against *Mycoplasma* and some *Chlamydia* and *Campylobacter* infections. Erythromycin is most valuable in treating infections caused by penicillin-resistant organisms or in patients allergic to penicillin. Unfortunately, resistance to erythromycin often emerges during treatment. Dual antibiotic treatment—erythromycin and some other drug—is often used on patients with a pneumonia-like disease that might be Legionnaires' disease. Several antibiotics combat other pneumonias, but erythromycin is the only common antibiotic that will combat Legionnaires' disease. Erythromycin is one of the least toxic of commonly used antibiotics. Mild gastrointestinal disturbances are seen in 2 to 3 percent of patients receiving it.

Lincosamides *Lincomycin* is produced by *Streptomyces lincolnensis,* and *clindamycin* is a semisynthetic derivative that is more completely absorbed and less toxic than lincomycin. Both drugs, which are collectively called lincosamides, exert a bacteriostatic effect. Lincomycin can be used to treat a variety of infections but is not significantly better than other widely used antibiotics, and organisms quickly become resistant to it. Clindamycin is effective against *Bacteroides* and other anaerobes, except *Clostridium difficile,* which often becomes established as a superinfection during clindamycin therapy. Toxins from *C. difficile* can cause a severe, and sometimes fatal, colitis (inflammation of the large intestine) unless diagnosed early and treated with oral vancomycin.

Antibiotics and Acne

Low doses of tetracyclines and erythromycin suppress skin bacteria, mostly *Propionibacterium acnes,* and reduce the release of microbial lipases, which contribute to skin inflammation. This therapy is used to treat acne, but its effectiveness has not been proved. Studies intended to assess antibiotic effectiveness in acne therapy have been useless because they lack suitable controls, fail to characterize adequately the type and severity of cases, and employ other concurrent therapies. Some studies have shown that low doses of many antibiotics can lead to the appearance of antibiotic-resistant strains. Are the benefits of antibiotics to acne patients worth the risk of promoting development of resistant organisms?

Inhibitors of Nucleic Acid Synthesis

Rifampin

From among the **rifamycins** produced by *Streptomyces mediterranei,* only the semisynthetic *rifampin* is currently used. Easily absorbed from the digestive tract except when taken directly after a meal, it reaches all tissues and body fluids. Rifampin blocks RNA transcription. Although it is bactericidal and has a wide spectrum of activity, it is approved in the United States only for treating tuberculosis and eliminating meningococci from the nasopharynx of carriers.

Rifampin can cause liver damage but usually does so only when excessive doses are given to patients with preexisting liver disease. It is unusual among antibiotics in its ability to interact with other drugs, and possibilities of such interactions should be considered before the drug is given. Taking rifampin concurrently with oral contraceptives has been implicated in an increased risk of pregnancy and menstrual disorders. Dosages of anticoagulants must be increased while a patient is taking rifampin to achieve the same degree of reduction in blood clotting. Finally, drug addicts receiving methadone sometimes suffer withdrawal symptoms if they are given rifampin without an increase in methadone dosage. One explanation for these diverse effects is that rifampin stimulates the liver to produce greater quantities of enzymes that are involved in the metabolism of a variety of drugs.

Quinolones

Quinolones, a new group of synthetic bactericidal analogues of *nalidixic acid,* are effective against many Gram-positive and Gram-negative bacteria. Quinolones' mode of action is to inhibit bacterial DNA synthesis by blocking DNA gyrase, the enzyme that unwinds the DNA double helix preparatory to its replication. *Norfloxacin, ciprofloxacin,* and *enoxacin* are examples of this group of antibiotics. They are especially effective in the treatment of traveler's diarrhea and in urinary tract infections caused by multiplely resistant organisms.

A recent advance has produced a hybrid class of antibiotics. One of these, a quinolone–cephalosporin combination, is currently being tested. When the β-lactamase enzymes act on the cephalosporin component, the quinolone is released from the hybrid molecule and is available to kill the cephalosporin-resistant organisms. The use of such a dual-acting synergistic antibiotic may also prevent or delay development of antibiotic resistance in organisms.

Antimetabolites and Other Antibacterial Agents

Sulfonamides

The **sulfonamides,** or *sulfa drugs,* are a large group of entirely synthetic, bacteriostatic agents. Many are derived from *sulfanilamide* (sul-fa-nil′a-mid), one of the first sulfonamides (➤ Figure 13.4b). In general, orally administered sulfonamides are readily absorbed and become widely distributed in tissues and body fluids. They act by blocking the synthesis of folic acid, which is needed to make the nitrogenous bases of DNA. Sulfonamides have now been largely replaced by antibiotics because antibiotics are more specific in their actions and less toxic than sulfonamides.

When sulfonamides first came into use in the 1930s, they frequently led to kidney damage. Newer forms of these drugs usually do not damage kidneys, but they do occasionally produce nausea and skin rashes. Certain sulfonamides are still used to suppress intestinal microflora prior to colon surgery. They also are used to treat some kinds of meningitis because they enter cerebrospinal fluid more easily than do antibiotics. *Cotrimoxazole* (Septra), a combination of *sulfamethoxazole* and *trimethoprim,* is used to treat urinary tract infections and a few other infections. Cotrimoxazole is the primary drug of choice to control *Pneumocystis* pneumonia, a common fungal complication of AIDS patients. Unfortunately, both drugs are toxic to bone marrow and may cause nausea and skin rashes.

Isoniazid

Isoniazid (i-so-ni′a-zid) is an antimetabolite for two vitamins—nicotinamide and pyridoxal. It binds to and inactivates the enzyme that converts the vitamins to useful molecules. This bacteriostatic synthetic agent, which has little effect on most bacteria, is effective against the mycobacterium that causes tuberculosis. Isoniazid is completely absorbed from the digestive tract and reaches all tissues and body fluids. Because the mycobacteria present in any such infection usually include some isoniazid-resistant organisms, isoniazid usually is given with another agent such as rifampin or ethambutol (discussed in the following section). Dietary supplements of nicotinamide and pyridoxal also should be given with isoniazid.

Ethambutol

The synthetic agent **ethambutol** is effective against certain strains of mycobacteria that do not respond to isoniazid. Ethambutol is well absorbed and reaches all tissues and body fluids. However, mycobacteria acquire resistance to it fairly rapidly, so it is used with other drugs such as isoniazid and rifampin.

Nitrofurans

Nitrofurans (ni″tro-fyu′ranz) are antibacterial drugs that enter susceptible cells and apparently damage sensitive microbial respiratory systems. Several hundred nitrofurans have been synthesized since the first one was made in 1930. Only a few of these are currently used. Oral doses of *nitrofurantoin* (Furadantin) are bacteriostatic in low doses, easily absorbed, and quickly metabolized. This drug is especially useful in treating acute and chronic urinary infections. The low incidence of resistance to it makes it an ideal prophylactic agent to prevent recurrences. Unfortunately, 10 percent of patients experience nausea and vomiting as a side effect and must then be treated with an antibiotic instead. Another nitrofuran, *nifuratel,* has been shown to have the same clinical advantages as nitrofurantoin with a much lower incidence of gastrointestinal side effects. Unfortunately, this cheap, effective drug is not currently available.

The chemical structures, uses, and side effects of antibacterial agents are summarized in ➤ Figure 13.13.

Antifungal Agents

Antifungal agents are being used with greater frequency because of the emergence of resistant strains and an increase in the number of immunosuppressed patients, especially those with AIDS. Because fungi are eukaryotes and thus similar to human cells, antifungal treatment often causes toxic side effects. At less toxic levels, many systemic fungal infections are slow to respond. Furthermore, laboratory tests are not available to determine appropriate susceptibility and therapeutic levels. Despite these difficulties, numerous effective drugs are now becoming available, many without prescription.

Imidazoles and Triazoles

The **imidazoles** (im″id-az′olz) and *triazoles* comprise a large group of related synthetic fungicides. Several agents, including *clotrimazole, ketoconazole, miconazole,* and *fluconazole,* are currently in use; many are available without prescription. The imidazoles and triazoles appear to affect fungal plasma membranes by disrupting the synthesis of membrane sterols. All these agents are used topically in creams and solutions to control fungal skin infections (dermatomycoses) and *Candida* yeast infections of the skin, nails, mouth, and vagina. Ketoconazole has also been given orally to treat systemic fungal infections, especially when other antifungal agents have not been effective. Some patients, however, have experienced mild to severe skin irritations with the topical agents. Furthermore, potentially severe drug interactions may occur, especially with certain antihistamines and immunosuppresents.

Polyenes

The **polyene** family of antibiotics consists of antifungal agents that contain at least two double bonds. Amphotericin B and nystatin are two of the most common polyene antibiotics.

Amphotericin B   The fungicidal antibiotic *amphotericin* (am″fo-ter′i-sin) *B* [Fungizone] is derived from *Streptomyces nodosus.* This drug binds to plasma membrane ergosterol (a crystallizing sterol) found in fungi and some algae and protozoa but not in human cells. Amphotericin B increases membrane permeability such that glucose, potassium, and other essential

✓ Why is the antibiotic with the largest zone of inhibition not necessarily the best one to use with a given patient?

✓ What harmful side effects can occur when taking tetracycline? What things should be avoided while taking this drug?

✓ What is a β-lactam ring? Where is it found? What importance does it have?

✓ What organisms are effective against isoniazid and ethambutol? Are they antibiotics? What is isoniazid's mode of action?

Agent	Used to Treat	Common Method of Administration*	Side Effects
Agents that inhibit cell wall synthesis			
Penicillin (natural)	Wide variety of infections, mostly of Gram-positive bacteria	IM, O	Relatively few side effects, but allergies do occur
Penicillin (semisynthetic)	Infections resistant to natural penicillin	O, IV	Same as natural penicillin
Cephalosporins	Wide variety of infections when allergy or toxicity make other agents unsuitable	IV, IM, O	Relatively nontoxic but can lead to superinfections
Carbapenems	Mixed infections, nosocomial infections, infections of unknown etiology	IV	Allergic reactions, superinfections, seizures, gastrointestinal disturbances
Bacitracin	Skin infections (topical application)	T	Internal use toxic to kidneys

Imipenem
(a carbapenem)

Bacitracin

Agents that interfere with cell membrane function

Polymyxins	Skin infections (topical application, with bacitracin)	T, IV	Internal use highly toxic
Tyrocidins	Skin infections caused by Gram-positive cocci (topical application)	T, IV	Internal use highly toxic

Polymyxin B

Tyrocidin

Antimetabolites and other agents

Sulfonamides	Some kinds of meningitis and to supress intestinal flora before colon surgery	O, IV	Early forms caused kidney damage, but ones now in use do not
Isoniazid	Tuberculosis (used with ethambutol)	O	May cause pyridoxine deficiency
Ethambutol	Tuberculosis (used with isoniazid)	O	
Nitrofurantoin	Urinary tract infections	O	Nausea and vomiting

Sulfanilamide
(a sulfonamide)

Isoniazid

Ethambutol

Nitrofurantoin

* IM = intramuscular   O = oral
IV = intravenous   T = topical

➤ Figure 13.13   **Selected antibacterial drugs.**

Agent	Used to Treat	Common Method of Administration*	Side Effects
Agents that inhibit protein synthesis			
Streptomycin	Tuberculosis (used with isoniazid and rifampin)	IM, O	Damages kidneys and inner ear
Gentamicin and other aminoglycosides	Antibiotic-resistant and hospital-acquired infections (used synergistically with other drugs)	IM, T (burns)	Varying degrees of kidney and inner ear damage
Tetracyclines	A broad spectrum of bacterial infections and some fungal infections	O	Stain teeth; cause gastrointestinal symptoms; can lead to super-infections
Chloramphenicol	A broad spectrum of bacterial infections, brain abscesses and penicillin-resistant infections	O	Can damage bone marrow and cause aplastic anemia
Erythromycin	Gram-positive bacterial infections, some penicillin-resistant infections, and Legionnaires' disease	O	One of the least toxic of commonly used antibiotics

Erythromycin

Gentamicin

Tetracycline

Chloramphenicol

Streptomycin

Agents that inhibit nucleic acid synthesis			
Rifampin	Tuberculosis and to eliminate meningococci from the nasopharynx	O	Bright orange or red urine, saliva, tears, and skin; liver damage; many disorders when used with other agents
Quinolones	Urinary tract infections, traveller's diarrhea; effective against many drug-resistant organisms	O	Nausea; headaches and other nervous system disturbances

Rifampin

Ciprofloxacin (a quinolone)

substances leak from the cell. The drug is poorly absorbed from the digestive tract and so is given intravenously. Even then, only 10 percent of the dose given is found in the blood. Excretion persists for up to 3 weeks after treatment is discontinued, but it is not known where the drug is sequestered in the meantime.

Amphotericin B is the drug of choice in treating most systemic fungal infections, especially cryptococcosis, coccidioidomycosis, and aspergillosis. Although fungi are not known to develop resistance to this agent, side effects are numerous and sometimes severe. They include abnormal skin sensations, fever and chills, nausea and vomiting, headache, depression, kidney damage, anemia, abnormal heart rhythms, and even blindness. Because some of the fungal infections are fatal without treatment, patients, especially those who are immunocompromised or have AIDS, have little choice but to risk these unfortunate side effects.

Nystatin   The polyene antibiotic *nystatin* (Mycostatin) is produced by *Streptomyces noursei.* This drug has the same mode of action as amphotericin B but is also effective topically in the treatment of *Candida* yeast infections. Because it is not absorbed through the intestinal wall, it can be given orally to treat fungal superinfections in the intestine, which often occur after long-term treatment with antibiotics. Nystatin was named for the New York State Health Department, where it was discovered.

Griseofulvin

Griseofulvin (gris″e-o-ful′vin), originally derived from *Penicillium griseofulvum,* is used primarily for superficial fungal infections. This fungistatic drug is incorporated into new cells that replace infected cells; it interferes with fungal growth, probably by impairing the mitotic spindle apparatus used in cell division. Although griseofulvin (Fulvicin) is poorly absorbed from the intestinal tract, it is given orally and appears to reach the target tissues through perspiration. It is ineffective against bacteria and most systemic fungal agents but is very useful topically in treating fungal infections of the skin, hair, and nails. Most infections are cured within 4 weeks, but recalcitrant infections associated with fingernails and toenails may persist even after a year of treatment. Reactions to griseofulvin are usually limited to mild headaches but can include gastrointestinal disturbances, especially when prolonged treatment is required.

Other Antifungal Agents

Flucytosine is a synthetic drug used in treating infections caused by *Candida* and several other fungi. This fluorinated pyrimidine is transformed in the body to fluorouracil, an analogue of uracil, and thereby interferes with nucleic acid and protein syntheses. The drug can be given orally and is easily absorbed, but 90 percent of the amount given is found unchanged in the urine within 24 hours. Because it is less toxic and causes fewer side effects than amphotericin B, flucytosine should be given instead of amphotericin B whenever possible.

Tolnaftate (Tinactin) is a common topical fungicide that is readily available without prescription. Although its mode of action is still not clear, it is effective in the treatment of various skin infections, including athlete's foot and jock itch.

Terbinafine (Lamisil), a relatively new fungicide, has been approved for topical use in skin infections and cutaneous candidiasis. Because it is absorbed directly through the skin, it reaches therapeutic levels in much less time than do orally administered agents such as griseofulvin.

Antiviral Agents

Until recent years no chemotherapeutic agents effective against viruses were available. One reason for the difficulty in finding such agents is that the agent must act on viruses within cells without severely affecting the host cells. Currently available *antiviral agents* inhibit some phase of viral replication, but they do not kill the viruses.

Purine and Pyrimidine Analogues

Several purine and pyrimidine analogues are effective antiviral agents. All cause the virus to incorporate erroneous information (the analogue) into a nucleic acid and thereby interfere with the replication of viruses. ➡ (Chapter 7, p. 185) The drugs include idoxuridine, vidarabine, ribavirin, acyclovir, and ganciclovir.

Drug-Resistant Viruses

Evidence is accumulating that viruses, like bacteria, can develop resistance to chemotherapeutic agents. Herpesviruses and cytomegaloviruses with resistance to acyclovir have been observed in AIDS patients. Some laboratory strains of the virus that causes AIDS have become resistant to azidothymidine (AZT), the most effective drug currently available to treat the disease. Resistance to chemotherapeutic agents is a greater problem in viruses than in bacteria because so few antiviral agents are available. When a bacterium becomes resistant to one antibiotic, another usually can be found to which the bacterium is susceptible. Unfortunately, this is not the case with viruses, and we must hope that biotechnology can help us battle drug-resistant viruses.

Idoxuridine and *trifluridine,* both analogues of thymine, are administered in eye drops to treat inflammation of the cornea caused by a herpesvirus. They should not be used internally because they suppress bone marrow.

Vidarabine (ARA-A), an analogue of adenine, has been used effectively to treat viral encephalitis, an inflammation of the brain caused by herpesviruses and by cytomegaloviruses. It is not effective against cytomegalovirus infections acquired before birth. Vidarabine is less toxic than either idoxuridine or cytarabine, but it sometimes causes gastrointestinal disturbances.

Ribavirin (Virazole), a synthetic nucleotide analogue of guanine, blocks replication of certain viruses. In an aerosol spray, it can combat influenza viruses; in an ointment, it can help to heal herpes lesions. Although it has low toxicity, it can induce birth defects and should not be given to pregnant women. It has been found to be effective against hantaviruses, such as those that caused the deadly outbreak of respiratory disease on the Navajo reservation in the Four Corners region of the American Southwest in 1993. ➡ (Chapter 21, p. 622) Ribavirin has shown activity against a wide variety of unrelated viruses, raising hopes of finding a broad-spectrum antiviral agent.

Acyclovir (Zovirax), an analogue of guanine, is much more rapidly incorporated into virus-infected cells than into normal cells. Thus, it is less toxic than other analogues. It can be applied topically or given orally or intravenously. It is especially effective in reducing pain and promoting healing of primary lesions in a new case of genital herpes. It is given prophylactically to reduce the frequency and severity of recurrent lesions, which appear periodically after a first attack. It does not, however, prevent the establishment of latent viruses in nerve cells. Acyclovir is more effective than vidarabine against herpes encephalitis and neonatal herpes, an infection acquired at birth, but is not effective against other herpesviruses.

Ganciclovir is an analogue of guanine similar to acyclovir. The drug is active against several kinds of herpesvirus infections and in particular cytomegalovirus eye infections in AIDS cases.

Amantadine

The tricyclic amine **amantadine** prevents influenza A viruses from penetrating cells. Given orally, it is readily absorbed and can be used from a few days before to a week after exposure to influenza A viruses to reduce the incidence and severity of symptoms. Unfortunately, it causes insomnia and ataxia (inability to coordinate voluntary movements), especially in elderly patients, who also are often severely affected by influenza infections. *Rimantadine,* a drug similar to amantadine, may be effective against a wider variety of viruses and be less toxic as well.

The Treatment of AIDS

Several agents are being tested for the treatment of AIDS. New information about AIDS, its complications, and its treatment is becoming available with great rapidity. We consider AIDS, agents used to treat it, and ramifications for health care workers in Chapter 18.

Interferons and Immunoenhancers

Cells infected with viruses produce one or more proteins collectively referred to as *interferons* (Chapter 16). When released, these proteins induce neighboring cells to produce antiviral proteins, which prevent these cells from becoming infected. Thus, interferons represent a natural defense against viral infection. Some interferons are currently being genetically engineered and tested as antiviral agents. Some positive results have been obtained in controlling chronic viral hepatitis and warts and arresting virus-related cancers, such as Kaposi's sarcoma.

Because cells produce interferons naturally, a possible way to combat viruses is to induce cells to produce interferons. Synthetic double-stranded RNA has been shown to increase the quantity of interferon in the blood. Experiments with one such substance in virus-infected monkeys have shown sufficient increase in interferon to prevent viral replication.

Two other agents, *levamisole* and *inosiplex,* appear to stimulate the immune system to resist viral and other infections. Both seem to stimulate activity of leukocytes called T lymphocytes rather than to stimulate interferon release. Levamisole appears to be effective prophylactically in reducing the incidence and severity of chronic upper respiratory infections, which are probably viral in nature. It also reduces symptoms of autoimmune disorders such as rheumatoid arthritis, in which the body reacts against its own tissues. Inosiplex has a more specific action: It stimulates the immune system to resist infection with certain viruses that cause colds and influenza.

Although efforts to improve antiviral therapies by enhancing natural defenses have been somewhat successful, none is yet in widespread use. More research is needed to identify or synthesize effective agents, to determine how they act, and to discover how they can be most effectively used.

Antiprotozoan Agents

Although many protozoa are free-living organisms, a few are parasitic in humans. The parasite that causes malaria invades red blood cells and causes the patient to suffer alternating fever and chills. Other protozoan parasites cause intestinal or urinary tract infections. Several *antiprotozoan agents* have been found that are successful in controlling or even curing most protozoan infections, but some have rather unpleasant side effects.

Quinine

Quinine, from the bark of the cinchona tree (native to Peru and Bolivia, but now cultivated exclusively in Indonesia), was used for centuries to treat malaria. One of the first chemotherapeutic agents to come into widespread use, it is now used only to treat malaria caused by strains of the parasite resistant to other drugs.

Chloroquine and Primaquine

Currently the most widely used antimalarial agents are the synthetic agents **chloroquine** (Aralen) and **primaquine.** Chloroquine appears to interfere with protein synthesis, especially in red blood cells, which it enters more readily than it does other cells. The drug may concentrate in vacuoles within the parasite and prevent it from metabolizing hemoglobin. Chloroquine is used to combat active infections. The malarial parasite persists in red blood cells and can cause relapses when it multiplies and is released into blood plasma. A combination of chloroquine and primaquine can be used prophylactically to protect people who visit or work in regions of the world where malaria occurs and are thus at risk of becoming infected. However, the drugs must be taken both before and after a malarial zone is entered. A new prophylactic agent, *mefloquine* (Lariam), has proved effective against resistant strains.

Metronidazole

The synthetic imidazole **metronidazole** (met-ro-ni′da-zol) is effective in treating *Trichomonas* infections, which typically cause a vaginal discharge and itching. It also is effective against intestinal infections caused by parasitic amoebas and *Giardia*. Although metronidazole (Flagyl) controls these infections, it does not prevent overgrowth of *Candida* yeast infections. It also can cause birth defects and cancer and can be passed to infants in breast milk. Metronidazole sometimes causes an unusual side effect called "black hairy tongue," or "brown furry tongue," because it breaks down hemoglobin and leaves deposits in papillae (small projections) on the surface of the tongue (➤ Figure 13.14).

Other Antiprotozoan Agents

A variety of other organic compounds have been found effective in treating certain infections caused by protozoa. *Pyrimethamine* (pir-i-meth′a-men) interferes with the synthesis of folic

➤ Figure 13.14 **Black hairy tongue, a reaction to the drug metronidazole (Flagyl).** The papilli on the tongue surface become elongated and filled with breakdown products of hemoglobin, which darken the tongue.

acid, which pathogenic protozoa need in greater quantities than do host cells. It is used with sulfanilamide to treat some protozoan infections such as toxoplasmosis. Pyrimethamine (Daraprim) can also be used prophylactically to prevent malaria.

Suramin sodium, a sulfur-containing compound, can be given intravenously to treat African sleeping sickness (trypanosomiasis) and other trypanosome infections. *Nifurtimox,* a nitrofuran, is used against the trypanosomes that cause Chagas' disease. Arsenic and antimony compounds, although very toxic, have been used with some success against stubborn amoebic infections and leishmanias. *Pentamidine isethionate* is used to treat African trypanosomiasis and as a drug of second choice for *Pneumocystis* pneumonia, a fungal complication of AIDS patients.

Antihelminthic Agents

Various helminths can infect humans. A variety of *antihelminthic agents* are available to help rid the body of these unwelcome parasites.

Niclosamide

Niclosamide interferes with carbohydrate metabolism, thereby causing a parasite to release large quantities of lactic acid. This drug may also inactivate products made by the worm to resist digestion by host proteolytic enzymes. It is effective mainly in the treatment of tapeworm infections.

Mebendazole

The imidazole **mebendazole** (Vermox) blocks the uptake of glucose by parasitic roundworms. It is useful in treating whipworm, pinworm, and hookworm infections. However, it can damage a fetus and thus should not be given to pregnant women.

Other Antihelminthic Agents

Piperazine (Antepar), a simple organic compound, is a powerful neurotoxin that paralyzes body wall muscles of roundworms and is useful in treating *Ascaris* and pinworm infections. Although piperazine exerts its effect on worms in the intestine, if absorbed it can reach the human nervous system and cause convulsions, especially in children.

The compound *ivermectin,* originally developed for the treatment of parasitic nematodes in horses (and widely used to prevent heartworm infections in dogs), has been found to be extremely effective against *Onchocerca volvulus* in humans. Infection with this roundworm, widespread in many parts of Africa, causes a progressive loss of sight known as onchocerciasis, or river blindness.

> Figure 13.15 provides the chemical structures, uses, and side effects of antifungal, antiviral, antiprotozoan, and antihelminthic agents.

✓ Do most antifungal agents also kill bacteria, or vice versa?

✓ What difficulties result from the fact that parasitic protozoans and helminths have many of the same biochemical pathways as humans?

Special Problems with Drug-Resistant Hospital Infections

As soon as antibacterial agents became available, resistant organisms began to appear. One of the first successes in treating bacterial infections was the use of sulfanilamide to treat infections caused by hemolytic streptococci. It was then discovered that sulfadiazine is useful in preventing recurrent streptococcal infections of rheumatic fever. Strains of streptococci resistant to sulfonamides soon emerged. Epidemics (mostly in military installations during World War II) caused by resistant strains led to many deaths. These epidemics were brought under control when penicillin became available, but soon penicillin-resistant streptococci were seen.

This chain of events has been repeated again and again. As new antibiotics were developed, strains of streptococci resistant to many of them evolved. Similar events led to the emergence of antibiotic-resistant strains of many other organisms, including staphylococci, gonococci, *Salmonella, Neisseria,* and especially *Pseudomonas. Pseudomonas* infections are now a major problem in hospitals. Many of these organisms are now resistant to several different antibiotics, and new resistant strains are constantly being encountered.

Why are resistant organisms found more often in hospitalized patients than among outpatients? This question can be answered by looking at the hospital environment and the patients likely to

Agent	Used to Treat	Common Method of Administration*	Side Effects
Antifungal agents			
Clotrimazole	Skin and nail infections	O	Skin irritation
Miconazole	Skin infections and systemic infections resistant to other agents	T, IV	Severe itching, nausea, fever, thrombophlebitis
Amphotericin B	Systemic infections	IV	Fever, chills, nausea, vomiting, anemia, kidney damage, blindness
Nystatin	*Candida* yeast infections, intestinal superinfections	T	
Griseofulvin	Infections of skin, hair, and nails	T, O	Mild headaches, nerve inflammations, gastrointestinal disturbances
Flucytosine	*Candida* and some systemic infections	O	Less toxic than many fungal agents

Clotrimazole   Miconazole   Amphotericin B

Nystatin   Griseofulvin   Flucytosine

Antihelminthic agents			
Niclosamide	Tapeworm infections	O	Irritation of gut
Piperazine	Pinworm and *Ascaris* infections	O	Can cause convulsions in children
Mebendazole	Whipworm, pinworm, and hookworm infections	O	Can damage fetus if given to pregnant women
Ivermectin	*Onchocerca volvulus* infections (cause of river blindness), heartworm infections in animals	O	Minimal

Niclosamide   Piperazine   Mebendazole

* IM = intramuscular   O = oral
IV = intravenous   T = topical

➤ Figure 13.15   **Selected antifungal, antihelminthic, antiviral, and antiprotozoan drugs.**

Agent	Used to Treat	Common Method of Administration*	Side Effects
Antiviral agents			
Idoxuridine	Corneal infections	T	Suppresses bone marrow
Ganciclovir	CMV eye infections in AIDS	IV	Suppresses bone marrow
Vidarabine	Viral encephalitis	T, IV	Less toxic than other antiviral agents
Ribavirin	Herpes lesions (topical application), influenza (in aerosol)	T	Can cause birth defects if given to pregnant women
Acyclovir	Herpesvirus infections; lessens severity of symptoms	IV, O, T	Less toxic than other analogues
Amantadine	Infections of influenza A viruses from entering cells (preventive)	O	Insomnia and ataxia

Idoxuridine

Ganciclovir

Vidarabine

Ribavirin

Amantadine

Acyclovir

Agent	Used to Treat	Common Method of Administration*	Side Effects
Antiprotozoan agents			
Quinine	Malaria resistant to other agents	O	
Chloroquine	Malaria	O	Headache, itching
Primaquine	With chloroquine to prevent relapse of malaria	O	Slight nausea and abdominal pain
Pyrimethamine	Various protozoan infections	O	Large doses damage bone marrow
Metronidazole	*Trichomonas, Giardia,* and amoebic infections	O, IV, T	Black hairy tongue

Quinine

Chloroquine

Primaquine

Pyrimethamine

Metronidazole

(a) Single beta-lactam antibiotic against β-lactamase–producing bacteria

Bacteria

Active antibiotic

β-Lactamase

β-Lactam ring

Ring breaks, inactivating β-lactam antibiotic.

Organisms multiply; infection continues.

(b) Double antibiotic therapy against β-lactamase–producing bacteria

β-Lactamase

β-Lactam ring is broken, and active quinolone is released.

Inactive β-lactam antibiotic

Organisms are killed or suppressed.

(c) Double antibiotic therapy against non-β-lactamase bacteria

Organisms are unable to produce β-lactamase and so can't break β-lactam ring.

Active antibiotic

Quinolone remains inactive.

Organisms are killed or suppressed; quinolone is not released, so strains resistant to quinolones do not develop.

➤ Figure 13.16   **The use of double antibiotic therapy to eradicate resistant-strain infections.**

be hospitalized. First, despite efforts to maintain sanitary conditions, a hospital provides an environment where sick people live in close proximity and where many different kinds of infectious agents are constantly present and are easily spread. Second, hospitalized patients tend to be more severely ill than outpatients; many have lowered resistance to infection because of their illnesses or because they have received immunosuppressant drugs. Finally, and most importantly, hospitals typically make intensive use of a variety of antibiotics. Because many infections are being treated and different antibiotics are used, organisms resistant to one or more of the antibiotics are likely to emerge. The resistant strains can readily spread among patients.

Treatment of resistant infections creates a vicious cycle. If an antibiotic can be found to which an organism is susceptible, that drug can be used to treat the infection. However, some strains of the organism that are resistant to the new antibiotic may then proliferate and require treatment with another new drug. A recurrent cycle in which new antibiotics are used and the organisms subsequently develop resistance to them is established.

Preventing infections caused by antibiotic-resistant strains of microorganisms is a difficult task, but several guidelines should be followed. First, the use of antibiotics should be limited to situations in which the patient is unlikely to recover without antibiotic treatment. Second, sensitivity tests should be done, and patients should receive only an antibiotic to which the organism is known to be sensitive. Third, when antibiotics are used, they should be continued until the organism is completely eradicated from the patient's body. Double antibiotic use, as described earlier under quinolones, is especially helpful (➤ Figure 13.16). Finally, any patient with an infectious disease should be isolated from other patients.

Retracing Our Steps

Antimicrobial Chemotherapy

- **Chemotherapy** is the use of any chemical agent in the treatment of disease.
- A **chemotherapeutic agent,** or **drug,** is any chemical agent used in medical practice.
- An **antimicrobial agent** is a chemical agent used to treat a disease caused by a microbe.
- An **antibiotic** is a chemical substance produced by microorganisms that inhibits the growth of or destroys other microorganisms.
- A **synthetic drug** is one made in the laboratory.
- A **semisynthetic drug** is one made partly by microorganisms and partly by laboratory synthesis.

The History of Chemotherapy

- The first chemotherapeutic agents were concoctions from plant materials used by primitive societies.
- Paul Ehrlich's search for the "magic bullet" was the first systematic attempt to find chemotherapeutic agents. Subsequent events included the development of sulfa drugs, penicillin, and many other antibiotics.

General Properties of Antimicrobial Agents

Selective Toxicity

- **Selective toxicity** is the property of antimicrobial agents that allows them to exert greater toxic effects on microbes than on the host.
- The **therapeutic dosage level** of an antimicrobial agent is the concentration over a period of time required to eliminate a pathogen.
- The **chemotherapeutic index** is a measure of the toxicity of an agent to the body relative to its toxicity for an infectious organism.

The Spectrum of Activity

- The **spectrum of activity** of an antimicrobial agent refers to the variety of microorganisms sensitive to the agent. A **broad-spectrum** agent attacks many different organisms. A **narrow-spectrum** agent attacks only a few different organisms.

Modes of Action

- Agents that kill bacteria are bactericidal; those that inhibit bacterial growth are bacteriostatic.
- Agents that inhibit cell wall synthesis allow the membrane of the affected microbe to rupture and release the cell contents.
- Agents that disrupt membrane function dissolve the membrane or interfere with the movement of substances into or out of cells.
- Agents that inhibit protein synthesis prevent the growth of microbes by disrupting ribosomes or otherwise interfering with the process of translation.
- Agents that inhibit nucleic acid synthesis interfere with synthesis of RNA (transcription) or DNA (replication) or disrupt the information these molecules contain.
- Agents that act as **antimetabolites** affect normal metabolites by competitively inhibiting microbial enzymes or by being erroneously incorporated into important molecules such as nucleic acids.

Kinds of Side Effects

- Side effects of antimicrobial agents on the host include toxicity, allergy, and disruption of normal microbiota.
- Allergic reactions to antimicrobial agents occur when the body reacts to the agent as a foreign substance.

- Many antimicrobial agents attack not only the infectious organism but also normal microbiota. **Superinfections** with new pathogens can occur when the defensive capacity of normal microbiota is destroyed.

The Resistance of Microorganisms

- **Resistance** to an antibiotic means that a microorganism formerly susceptible to the action of an antibiotic is no longer affected by it.
- Nongenetic resistance occurs when microorganisms are sequestered from antibiotics or undergo a temporary change, such as the loss of their cell walls, that renders them nonsusceptible to antibiotic action.
- Genetic resistance occurs when organisms survive exposure to an antibiotic because of their genetic capacity to avoid damage by the antibiotic. As susceptible organisms die, the resistant survivors multiply unchecked and increase in numbers.
- **Chromosomal resistance** is due to a mutation in microbial DNA; **extrachromosomal resistance** is due to **resistance (R) plasmids,** or **R factors.**
- Mechanisms of resistance include alterations of receptors, cell membranes, enzymes, or metabolic pathways.
- **Cross-resistance** is resistance against two or more similar antimicrobial agents.
- Drug resistance can be minimized by (1) continuing treatment with an appropriate antibiotic at therapeutic dosage level until all the disease-causing organisms are destroyed; (2) using two antibiotics that exert **synergism,** an additive effect; and (3) using antibiotics only when absolutely necessary.

Determining Microbial Sensitivities to Antimicrobial Agents

- Sensitivity of microbes to chemotherapeutic agents is determined by exposing them to the agents in laboratory cultures.

The Disk Diffusion Method

- In the **disk diffusion (Kirby-Bauer) method,** antibiotic-impregnated filter paper disks are placed on agar plates inoculated with a lawn of the test organism. Sensitivities to the drugs are determined by comparing the size of clear zones around the disks to a table of standard measurements.

The Dilution Method

- In the **dilution method,** a constant inoculum is placed into broth cultures or wells with differing known quantities of chemotherapeutic agents. **The minimum inhibitory concentration (MIC)** of the agent is the lowest concentration in which no growth of the organism is observed. The **minimum bactericidal concentration (MBC)** of the agent is the lowest concentration in which subculturing of broth yields no growth.

Serum Killing Power

- In the **serum killing power** method, a bacterial suspension is added to a patient's **serum** drawn while the patient is receiving an antibiotic, and it is noted whether the organisms are killed.

Automated Methods

- Automated methods allow rapid identification of microorganisms and determination of their sensitivities to antimicrobial agents.

Attributes of an Ideal Antimicrobial Agent

- An ideal antimicrobial agent is soluble in body fluids, selectively toxic, and nonallergenic; can be maintained at a constant therapeutic concentration in blood and body fluids; is unlikely to elicit resistance; has a long shelf life; and is reasonable in cost.

Antibacterial Agents

- Antibacterial agents inhibit cell wall synthesis, disrupt cell membrane functions, inhibit protein synthesis, inhibit nucleic acid synthesis, or act by some other means to kill bacteria.

Antifungal Agents

- Antifungal agents increase plasma membrane permeability, interfere with nucleic acid synthesis, or otherwise impair cell functions.

Antiviral Agents

- Antiviral agents have been difficult to find because they must damage intracellular viruses without severely damaging host cells.
- Most antiviral agents are analogues of purines or pyrimidines.

- Interferon is released by virus-infected cells and stimulates neighboring cells to produce antiviral proteins. Interferons are being made by genetic engineering and being tested in treatment of viral infections and cancer.

Antiprotozoan Agents

- Some antiprotozoan agents interfere with protein synthesis or folic acid synthesis. The mechanism of action of others is not well understood.

Antihelminthic Agents

- Antihelminthic agents interfere with carbohydrate metabolism or act as neurotoxins.

Special Problems with Drug-Resistant Hospital Infections

- Resistant hospital infections are due largely to intensive use of a variety of antibiotics, which fosters the growth of resistant strains. Treatment and prevention of such infections are extremely difficult.

Terminology Check

amantadine (p. 367)
aminoglycoside (p. 359)
antagonism (p. 351)
antibiosis (p. 341)
antibiotic (p. 341)
antimetabolite (p. 345)
antimicrobial agent (p. 341)
broad spectrum (p. 342)
carbapenem (p. 358)
cephalosporin (p. 356)
chemotherapeutic agent (p. 341)
chemotherapeutic index (p. 342)
chemotherapy (p. 340)
chloramphenicol (p. 361)
chloroquine (p. 368)
chromosomal resistance (p. 347)
cross-resistance (p. 350)
dilution method (p. 353)
disk diffusion method (p. 351)
drug (p. 341)
erythromycin (p. 361)

ethambutol (p. 363)
extrachromosomal resistance (p. 347)
griseofulvin (p. 366)
imidazole (p. 363)
isoniazid (p. 362)
Kirby-Bauer method (p. 351)
macrolide (p. 361)
mebendazole (p. 369)
metronidazole (p. 368)
minimum bactericidal concentration
  (MBC) (p. 354)
minimum inhibitory concentration (MIC)
  (p. 353)
molecular mimicry (p. 345)
narrow spectrum (p. 342)
niclosamide (p. 369)
nitrofuran (p. 363)
penicillin (p. 355)
polyene (p. 363)
polymyxin (p. 358)
primaquine (p. 368)

quinine (p. 368)
quinolone (p. 362)
resistance (p. 347)
resistance (R) plasmid (p. 347)
R factor (p. 347)
rifamycin (p. 362)
selective toxicity (p. 342)
semisynthetic drug (p. 341)
serum (p. 354)
serum killing power (p. 354)
spectrum of activity (p. 342)
streptomycin (p. 359)
sulfonamide (p. 362)
superinfection (p. 346)
synergism (p. 351)
synthetic drug (p. 341)
tetracycline (p. 360)
therapeutic dosage level (p. 342)
toxic dosage level (p. 342)
tyrocidin (p. 359)
zone of inhibition (p. 351)

See You on the Web

Don't forget I am expecting you on the web. The following is a preparedness checklist for your web explorations. If you have questions on any of these items, you can return to the end of Chapter 1 for more details.

- Questions We'll Explore
- Concept Check questions
- Flashcards
- Study Guide

If you are ready, come visit me at

http://www.prenhall.com/black

It will be a journey you'll never forget.

Host-Microbe Relationships
and Disease Processes

After having spent the morning exploring your exotic surroundings, you are absolutely starving! Luckily, there's a market in the center of town.

Would you like to shop here? In many parts of the world there's no choice about the cleanliness of your local market. Will flies, rodents and lack of refrigeration spread disease? You bet they will! If you think the melon and peppers look bad, you should see the stalls of meat and fish!

In the face of all this obvious contamination, why isn't everyone sick, dying, or dead? Well, some of them *are* sick, and the average life expectancy in developing countries is not very high. Probability of infection is rather like the 2-pan balances on which the merchants were weighing out the food: host defenses on one side and microbial virulence on the other. When the balance shifts in favor of the microbes, infection occurs. Do you risk it or keep looking, hoping your hunger will pass?

Background Basics

Questions We'll Explore

A What terms are used to define host–microbe relationships, and what do they mean?

B How do Koch's postulates relate to infectious disease?

C What are the major differences between infectious and noninfectious diseases and between communicable and noncommunicable infectious diseases?

D How do microbes cause disease?

E What are the meanings of terms used to describe diseases?

F What stages occur in the course of an infectious disease?

How is it that every now and then, no matter how careful you are, you "catch" an infectious disease? You become ill. With or without antimicrobial agents, you generally recover from the disease. In the process, you *may* develop *immunity*— that is, if you are exposed to the disease agent at another time, you may be protected from contracting the disease again.

Recall from Chapter 11 that a **pathogen** is a parasite capable of causing disease in a host. ➡ (p. 288) The ability of a pathogen to cause a disease in you depends on whether the pathogen or you—the host—wins the battle. Pathogens have certain invasive capabilities, and you have a variety of defenses. For example, in many countries the measles virus is present in a portion of the population at all times. Those who are infected release the virus, and it makes its way into the tissues of susceptible individuals. There, the virus can overcome defenses, invade tissues, and cause disease. But some individuals do not become infected.

If a virus does make its way into your tissues, your immune-system defenses may destroy it before it can cause disease. You could become immune to further exposures without actually becoming sick. Even when your first defenses fail and the disease occurs, you may develop immunity and will not be susceptible to the disease on subsequent exposures.

To begin the study of host–microbe interactions, we will look at a variety of relationships between host and microbe and see how some of these relationships result in disease. We will then characterize diseases and look at the disease process brought on by pathogens.

Host-Microbe Relationships

Microorganisms display a variety of complex relationships with other microorganisms and with larger forms of life that serve as hosts for them. A **host** is any organism that harbors another organism.

The word "parasite" comes from the Greek *parasitos,* meaning "one who eats at another's table."

Symbiosis

Symbiosis is an association between two (or more) species. Meaning "living together" the term *symbiosis* encompasses a spectrum of relationships. These include *mutualism, commensalism,* and *parasitism.*

At one end of the spectrum is **mutualism,** in which both members of the association living together benefit from the relationship (➤ Figure 14.1). For example, the ability of termites to digest wood, or cellulose, depends on protozoa they harbor in their intestines. These protozoa secrete into the intestine enzymes that digest the chewed wood. The protozoa themselves gain a safe, stable environment in which to live. For the termites, the relationship is obligatory; they would starve without their protozoan partners. Similarly, large numbers of *Escherichia coli* live in the large intestine of humans. These bacteria release useful products such as vitamin K, which we use to make certain blood-clotting factors. Although the relationship is not obligatory, *E. coli* does make a modest contribution to satisfying our need for vitamin K. The bacteria, in turn, get a favorable environment in which to live and obtain nutrients.

At the other end of the spectrum is **parasitism,** in which one organism, the parasite, benefits from the relationship, whereas the other organism, the host, is harmed by it (➤ Figure 14.2). By this broad definition of the term **parasite,** bacteria, viruses, protozoa, fungi, and helminths are parasites. (Some biologists use "parasite" to refer only to protozoa, helminths, and arthropods that live on or in their host.) Parasitism encompasses a wide range of relationships, from those in which the host sustains only slight harm to those in which the host is killed. Some parasites obtain comfortable living arrangements by causing only modest harm to their host. Other parasites kill their hosts, thereby rendering themselves homeless. ➡ (Chapter 11, p. 288) The

➤ Figure 14.1 **Many of the bacteria on human skin are mutualistic.** (40,800X) However, most of these organisms are commensals, which indirectly benefit us by competing with harmful organisms for nutrients and preventing those organisms from finding a site to attach to and invade tissues.

➤ Figure 14.2 **Parasite infestation.** Female pinworms leaving the anus of a 5-year-old child to lay eggs on the adjacent skin.

most successful parasites are those that maintain their own life processes without severely damaging their hosts.

Somewhere in the middle lies **commensalism,** in which two species live together in a relationship such that one benefits and the other one neither benefits nor is harmed. For example, many microorganisms live on our skin surfaces and utilize metabolic products secreted from pores in the skin. Because those products are released whether or not they are used by microorganisms, the microorganisms benefit, and ordinarily we are neither benefited nor harmed.

The line between commensalism and mutualism is not always a clear one. By taking up space and utilizing nutrients, microbes that show mutualistic or commensalistic behavior may prevent colonization of the skin by other, potentially harmful, disease-causing microbes—a phenomenon known as *microbial competition.* Hence these symbiotic relationships confer an indirect benefit on the host.

There is also a fine line between parasitism and commensalism. In healthy hosts, many microbes of the large intestine form harmless associations, simply feeding off digested food materials. But a "harmless" microbe could act as a parasite if it gains access to a part of the body where it would not normally exist.

Contamination, Infection, and Disease

Contamination, infection, and disease can be viewed as a sequence of conditions in which the severity of the effects microorganisms have on their hosts increases. **Contamination** means that the microorganisms are present. Inanimate objects and the surfaces of skin and mucous membranes can be contaminated with a wide variety of microorganisms. Commensals do no harm, but parasites have the capacity to invade tissues. **Infection** refers to the multiplication of any parasitic organism within or upon the host's body. (Sometimes the term **infestation** is used to refer to the presence of larger parasites, such as worms or arthropods, in or on the body.) If an infection disrupts the normal functioning of the host, disease occurs. **Disease** is a disturbance in the state of health wherein the body cannot carry out all its normal functions.

Both infection and disease result from interactions between parasites and their hosts. Sometimes an infection produces no observable effect on the host even though organisms have invaded tissues. More often an infection produces observable disturbances in the host's state of health; that is, disease occurs. When an infection causes disease, the effects of the disease range from mild to severe.

Let us look at some examples to understand the differences among contamination, infection, and disease. A health care worker who fails to follow aseptic procedures while dressing a skin wound contaminates her hands with staphylococci. However, after she finishes her task, she washes her hands properly and suffers no ill effects. Although her hands were contaminated, she did not develop an infection. Another worker performing the same task on another patient fails to wash his hands properly after treating the patient, and the organisms gain entrance to the body and infect a small cut. Soon the skin around the cut becomes reddened for a day or so. This worker was contaminated and infected. In a similar situation, a third worker develops a reddened area on her skin; she ignores it and in a few days has a large boil. This worker has experienced contamination, infection, and disease.

Disease, or illness, is characterized by changes in the host that interfere with normal function. These changes can be mild, severe but reversible, or irreversible. For example, if you become infected with one of the viruses that cause the common cold, you may have just a runny nose for a few days. Or you may have a severe cold with a sore throat, cough, fever, and headache, but the disease runs its course in a week or so without any permanent effects. The changes in your state of health are reversible. But if you develop trachoma, a bacterial infection of the eye, without treatment scarring of the cornea can occur, leading to permanent vision impairment and sometimes to blindness. Likewise, if you fail to get proper treatment for streptococcal infections, you might suffer irreversible damage to your heart or kidneys.

Pathogens, Pathogenicity, and Virulence

Pathogens vary in their abilities to disrupt the state of an individual's health—that is, they display different degrees of pathogenicity. **Pathogenicity** is the capacity to produce disease. An organism's pathogenicity depends on its ability to invade a host, multiply in the host, and avoid being damaged by the host's defenses. Some disease agents, such as *Mycobacterium tuberculosis,* frequently cause disease upon entering a susceptible host. Other agents, such as *Staphylococcus epidermidis,* cause disease only in rare instances and usually only in hosts with poor defenses. Most infectious agents exhibit a degree of pathogenicity between these extremes.

An important factor in pathogenicity is the number of infectious organisms that enter the body. If only a small number enter, the host's defenses may be able to eliminate the organisms before they can cause disease. If a large number enter, they may overwhelm the host's defenses and cause disease.

Virulence refers to the intensity of the disease produced by pathogens, and it varies among different microbial species. For example, *Bacillus cereus* causes mild gastroenteritis, whereas the rabies virus causes neurological damage that is nearly always fatal. Virulence also varies among members of the same species of pathogen. For example, organisms freshly discharged from an infected individual tend to be more virulent than those from a carrier, who characteristically shows no signs of disease. The virulence of a pathogen can increase by **animal passage,** the rapid transfer of the pathogen through animals of a species susceptible to infection by the pathogen. As one animal becomes diseased, organisms released from that animal are passed to a healthy animal, which then also gets sick. If this sequence is repeated two or three times, each newly infected animal suffers a more serious case of the disease than the one before it. Presumably the microbe becomes better able to damage the host with each animal passage. Sometimes an infectious disease spreads through human populations in this fashion, and an epidemic of the disease results. Influenza epidemics often proceed in this manner; the first people to become infected have a mild illness, but those infected later have a much more severe form of the disease. This process does not continue forever; the microbe reaches the height of its virulence, and the exposed population acquires immunity.

The virulence of a pathogen can be decreased by **attenuation,** the weakening of the disease-producing ability of the pathogen. Attenuation can be achieved by repeated subculturing on laboratory media or by transposal of virulence. **Transposal of virulence** is a laboratory technique in which a pathogen is passed from its normal host to a new host species and then passed sequentially through many individuals of the new host species. Eventually, the pathogen adapts so completely to the new host that it is no longer virulent for the original host. In other words, virulence has been transposed to another organism. Pasteur made use of transposal of virulence in preparing rabies vaccines. By repeated passage through rabbits, the virus eventually became harmless to humans and was safe to use in a human vaccine. We will see in Chapter 17 that attenuation is an important step in the production of some vaccines in use today.

Normal (Indigenous) Microflora

As we have described, microorganisms found in various symbiotic associations with humans do not necessarily cause disease. An adult human body consists of approximately 10^{13} (10 trillion) eukaryotic cells. It harbors an additional 10^{14} (100 trillion) prokaryotic and eukaryotic microorganisms on the skin surface, on mucous membranes, and in the passageways of the digestive, respiratory, and reproductive systems. Thus, there are 10 times more microbial cells on or in the human body than there are cells making up the body!

Before birth, a fetus exists in a sterile environment. During passage through the birth canal, the fetus acquires certain microorganisms that may become permanently or temporarily associated with it. Organisms that live on or in the body but do not cause disease are referred to collectively as **normal microflora,** or *normal microbiota* (Table 14.1). Many such organisms have well-established associations with humans. Most organisms among the normal microflora are commensals—they obtain nutrients from host secretions—waste substances found on the surfaces of skin and mucous membranes. Two categories of organisms can be distinguished: *resident microflora* and *transient microflora.*

The **resident microflora** (➤ Figure 14.3) comprise microbes that are always present on or in the human body. They are found on the skin and conjunctiva, in the mouth, nose, and throat, in the large intestine, and in passageways of the urinary and reproductive systems, especially near their openings. In each of these body regions, resident microflora are adapted to prevailing conditions. The mouth and the lower part of the large intestine provide warm, moist con-

---

> Attenuated viruses are used to make measles and mumps vaccines.

Can a Bloodhound Find the Correct Identical Twin?

As we walk around, we shed a "dandruff cloud" of skin flakes that lands on the ground or on nearby objects. A bloodhound can follow a trail by sniffing these flakes. Normal microflora on humans metabolize oils and other secretions into by-products with particular odors. Identical twins are not colonized by identical normal microflora, so their skin flakes develop slightly different odors. A bloodhound can distinguish among these odors, leading the dog to the right twin.

> **Table 14.1** | **Major normal microflora (unless otherwise noted, bacteria) of the human body**

Skin
Staphylococcus epidermidis*
Staphylococcus aureus
Lactobacillus species
Propionibacterium acnes*
Pityrosporon ovale (fungus)*

Mouth
Streptococcus salivarius*
Streptococcus pneumoniae
Streptococcus mitis*
Staphylococcus epidermidis*
Staphylococcus aureus
Moraxella catarrhalis
Veillonella alcalescens*
Lactobacillus species*
Klebsiella species
Haemophilus influenzae*
Fusobacterium nucleatum*
Treponema denticola*
Candida albicans (fungus)*
Entamoeba gingivalis (protozoan)*
Trichomonas tenax (protozoan)*

Upper respiratory tract
Staphylococcus epidermidis*
Staphylococcus aureus
Streptococcus mitis*
Streptococcus pneumoniae
Moraxella catarrhalis
Lactobacillus species
Haemophilus influenzae

Intestine
Staphylococcus epidermidis*
Staphylococcus aureus
Streptococcus mitis*
Enterococcus species*
Lactobacillus species*
Clostridium species*
Eubacterium limosum*
Bifidobacterium bifidum*
Actinomyces bifidus
Escherichia coli*
Enterobacter species*
Klebsiella species
Proteus species
Pseudomonas aeruginosa
Bacteroides species*
Fusobacterium species
Treponema denticola
Endolimax nana (protozoan)
Giardia intestinalis (protozoan)

Urogenital tract
Streptococcus mitis*
Streptococcus species*
Staphylococcus epidermidis*
Lactobacillus species*
Clostridium species
Actinomyces bifidus
Candida albicans (fungus)*
Trichomonas vaginalis (protozoan)

*Well-established associations.

ditions and ample nutrients. Mucous membranes of the nose, throat, urethra, and vagina also provide warm, moist conditions, although nutrients are in shorter supply. The skin provides ample nutrients but is cooler and less moist.

Other regions of the body lack resident microflora either because these regions provide conditions unsuitable for microorganisms, are protected by host defenses, or are inaccessible to microorganisms (Table 14.2). For example, conditions in the stomach are too acidic to permit survival of microflora. Under normal conditions the nervous system is inaccessible to microbes. Blood has no resident microflora because it is relatively inaccessible, and host defense mechanisms normally destroy microorganisms before they become established.

Transient microflora are microorganisms that can be present under certain conditions in any of the locations where resident microflora are found. They persist for hours to months, but only as long as the necessary conditions are met. Transient microflora appear on mucous membranes when greater than normal quantities of nutrients are available or on the skin when it is warmer and more moist than usual. Even pathogens can be transient microflora. For example, suppose that you come in contact with a child infected with measles, and some of the viruses enter your nose and throat. You had measles years ago and are immune to the disease, so your body's defenses prevent the viruses from invading cells. But you harbor the viruses as transients for a short time.

Young children's normal microflora of the skin contain a wider variety of species than does an adult's.

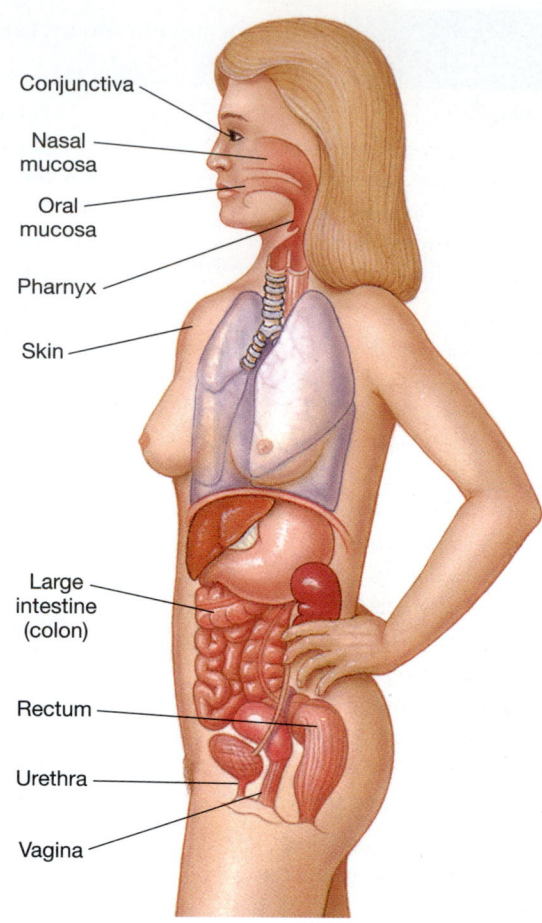

> Figure 14.3   **Locations of resident microflora of the human body.**

Among the resident and transient microflora are some species of organisms that do not usually cause disease but can do so under certain conditions. These organisms are called **opportunists** because they take advantage of particular opportunities to cause disease. Conditions that create opportunities for such organisms include:

1. *Failure of the host's normal defenses.* Individuals with weakened immune defenses are said to be **immunocompromised.** Factors such as advanced malnutrition, the presence of another disease, advanced age, treatment with radiation or immunosuppressive drugs, and physical or mental stress can lead to this state. The failure of host defenses in AIDS patients, for example, allows several different opportunistic infections to develop.

➤ Table 14.2	Body tissues, organs, and fluids that are normally microbe-free
Internal tissues and organs	***Body fluids***
Middle and inner ear	Blood
Sinuses	Cerebrospinal fluid
Internal eye	Saliva prior to secretion
Bone marrow	Urine in kidneys and in
Muscles	bladder
Glands	Semen prior to entry into
Organs	the urethra
Circulatory system	
Brain and spinal cord	
Ovaries and testes	

2. *Introduction of the organisms into unusual body sites.* The bacterium *Escherichia coli* is a normal resident of the human large intestine, but it can cause disease if it gains entrance to unusual sites such as the urinary tract, surgical wounds, or burns.

3. *Disturbances in the normal microflora.* Thriving populations of normal microflora compete with pathogenic organisms and in some instances actively combat their growth, an effect known as **microbial antagonism.** The normal microflora interfere with the growth of pathogens by competing for and depleting nutrients the pathogens need or by producing substances that create environments in which the pathogens cannot grow. As we saw in Chapter 13, antibiotics sometimes destroy or disturb the normal microflora as they bring a pathogen under control. ➡ (p. 346) This disturbance allows other potential pathogens, such as yeasts that are not harmed by the antibiotic, to thrive in the absence of their antagonists, the normal microflora.

Although in later chapters we will focus on microorganisms that cause human disease, we must not lose sight of the importance of the many nonpathogenic microorganisms associated with the human body. In addition, we must remember that disease can result from disturbances in the normal ecological balance between resident populations and the host.

Koch's Postulates

The work of Robert Koch and the role of his postulates in relating causative agents to specific diseases was described briefly in Chapter 1. ➡ (p. 13) Now we can use our understanding of infection and disease to look at those postulates more carefully. For example, we now know that infection with an organism does not necessarily indicate that disease is present. With that knowledge, we can better appreciate the need for all four of **Koch's postulates** to be satisfied in order to prove that a specific organism is the causative agent of a particular disease:

1. The specific causative agent must be observed in every case of a disease.

2. The agent must be isolated from a diseased host and must be grown in pure culture.

3. When the agent from the pure culture is inoculated into healthy, but susceptible, experimental hosts, the agent must cause the same disease.

4. The agent must be reisolated from the inoculated, diseased experimental host and identified as identical to the original specific causative agent.

It is relatively easy today to demonstrate that each postulate is met for a variety of diseases caused by bacteria (➤ Figure 14.4). Some bacteria, however, are difficult to culture because they have fastidious nutritional requirements or other special needs for growth. For example, although the causative agent of syphilis, *Treponema pallidum,* has been known for many years, it has not been successfully grown on artificial media. Moreover, parasites such as viruses and rickettsias cannot be grown in artificial media and must instead be grown in living cells. For some agents that cause disease in humans, no other host has been found. Consequently, in such cases inoculation into a susceptible host is impossible unless human volunteers can be found. There are obvious ethical problems associated with inoculating humans with infectious agents, even if volunteers might be available.

✓ Distinguish among the different types of symbiosis.

✓ Compare and contrast pathogenicity and virulence.

✓ Compare and contrast opportunist and pathogen.

✓ What do Koch's postulates prove?

Kinds of Diseases

Human diseases are caused by infectious agents, structural or functional genetic defects, environmental factors, or any combination of these causes.

Infectious and Noninfectious Diseases

Infectious diseases are diseases caused by infectious agents such as bacteria, viruses, fungi, protozoa, and helminths. Chapters 19 through 24 of this text are devoted to discussions of particular infectious agents and the diseases they cause. **Noninfectious diseases** are caused by any factor other than infectious organisms.

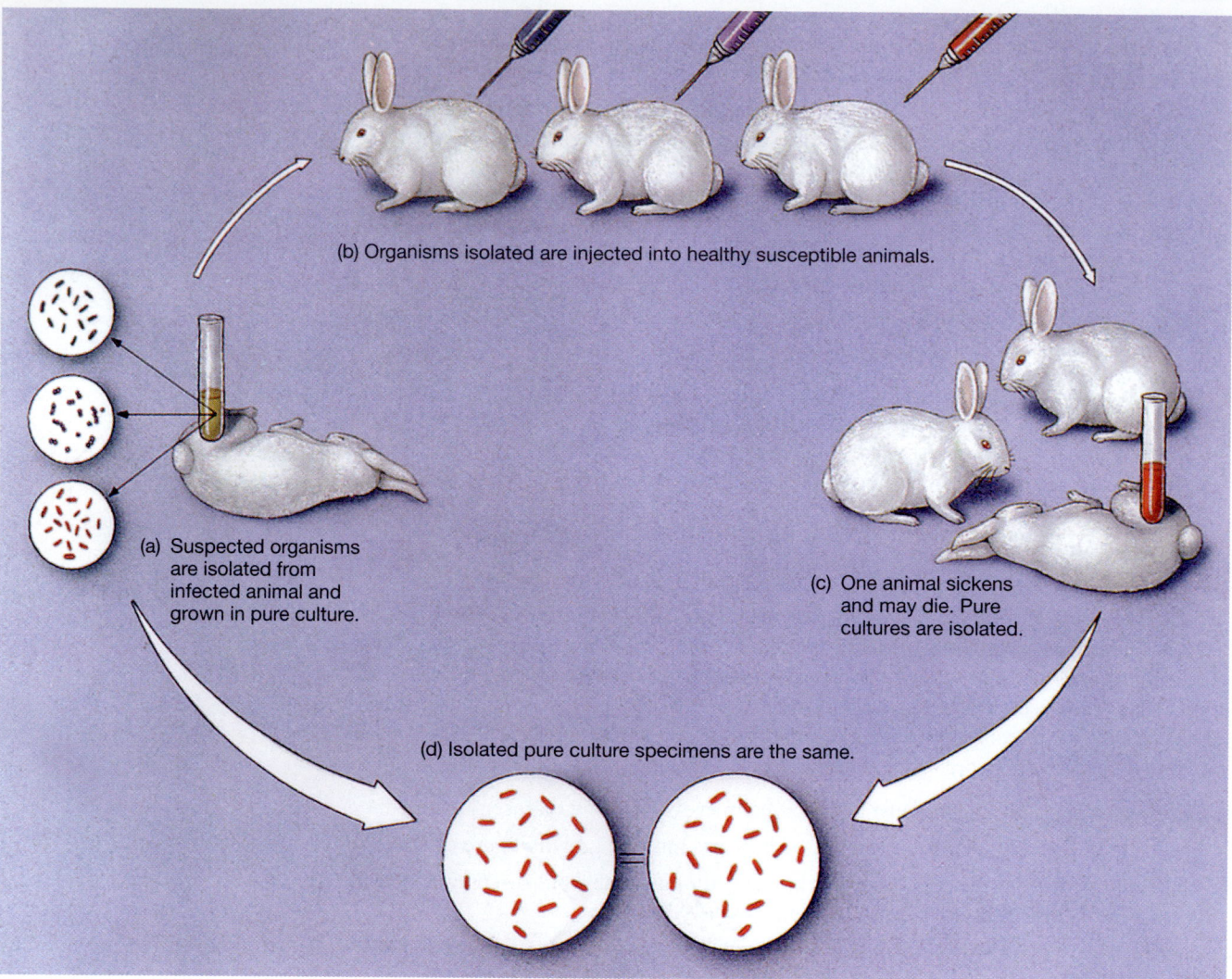

(b) Organisms isolated are injected into healthy susceptible animals.

(a) Suspected organisms are isolated from infected animal and grown in pure culture.

(c) One animal sickens and may die. Pure cultures are isolated.

(d) Isolated pure culture specimens are the same.

➤ Figure 14.4   **Demonstration that a bacterial disease satisfies Koch's postulates.**

Classification of Diseases

Classification of diseases as infectious or noninfectious gives a very limited view of human disease. The following scheme for classifying diseases provides a more comprehensive view. More importantly, it shows that infectious agents can interact with other factors in causing disease.

1. *Inherited diseases* are caused by errors in genetic information. The resulting developmental disorders may be caused by abnormalities in the number and distribution of chromosomes or by the interaction of genetic and environmental factors. Although inherited diseases have a noninfectious cause, some are associated with microbial activities. Sickle-cell anemia weakens patients and makes them more susceptible to infectious diseases. However, sickle-cell patients or carriers of the defect tend to be resistant to malaria. The abnormal hemoglobin S of sickle-cell patients gives up stored oxygen. With less oxygen, red blood cells change into a sickle shape and are removed by the spleen. Malaria-causing parasites that enter red blood cells make the cells sickle and are thereby killed before they complete their life cycle.

2. *Congenital diseases* are structural and functional defects present at birth, caused by drugs, excessive X-ray exposure, or certain infections. When a mother has a rubella (German measles) or a syphilis infection, the infectious agent may cross the placenta and cause congenital defects. Some medicines, such as the antiwrinkle drug retinoid-A and the antibiotic tetracycline, may cause congenital defects when taken by pregnant women.

Armadillo: Culture Vessel for Leprosy

The organism that causes Hansen's disease (leprosy) is difficult to culture. Many different methods had been tested and found unsatisfactory until someone tried inoculating the organism into the footpads of the nine-banded armadillo. There it grows very well; in fact, the organism multiplies faster there than in human tissues. When the organism does infect humans, it can have an incubation period of up to 30 years before disease symptoms appear. Before the armadillo was used to culture the organism, Koch's third postulate could not be satisfied. No one wanted to hold out an arm and say, "Here, try to give me leprosy." Also, 30 years was a long time to wait to determine the results of such an experiment. The use of the armadillo has made it possible to confirm Koch's postulates for *Mycobacterium leprae* as the causative agent of leprosy. This bacterium, seen by Armauer Hansen in 1878, was one of the first infectious agents to be identified and associated with a disease but one of the last to satisfy Koch's postulates. The armadillo was chosen as an experimental host after naturally occurring leprosy infections were found in the armadillo populations of Texas and Louisiana.

DNA studies have shown the strains of *M. leprae* in the armadillo to be identical with those infecting humans. Cases of humans' having acquired infection from armadillos have been confirmed, and we now designate armadillos as a reservoir for the disease in the U.S. Southwest. Cases of leprosy in African chimpanzees and mangabey monkeys have also been found. Injection of *M. leprae* collected from armadillos into mangabeys has caused the monkeys to develop leprosy.

Armadillos are used to culture the organism that causes Hansen's disease (leprosy) in humans.

3. *Degenerative diseases* are disorders that develop in one or more body systems as aging occurs. Patients with degenerative diseases such as emphysema or impaired kidney function are susceptible to infections. Conversely, infectious agents can cause tissue damage that leads to degenerative disease, as occurs in bacterial endocarditis, rheumatic heart disease, and some kidney diseases.

4. *Nutritional deficiency diseases* lower resistance to infectious diseases and contribute to the severity of infections. For example, the bacterium that causes diphtheria (*Corynebacterium diphtheriae*) produces more toxin in people with iron deficiencies than in those with normal amounts of iron. Poor nutrition also increases the severity of measles and contributes to deaths from the disease. Nutritional deficiencies can themselves develop from the action of, for example, helminths that severely damage the intestinal lining.

5. *Endocrine diseases* are due to excesses or deficiencies of hormones. Viral infection has been linked to pancreatic damage that leads to insulin-dependent diabetes.

6. *Mental disease* can be caused by a variety of factors, including those of an emotional, or psychogenic (si-ko-jen′-ik), nature as well as certain infections. For example, psychological stress may give rise to several gastrointestinal disorders, skin irritations, and even breathing difficulties. Mental disease can also result from brain infections such as in cases of neurosyphilis and the prion-caused Creutzfeld-Jakob disease.

7. *Immunological diseases* such as allergies, autoimmune diseases, and immunodeficiencies are caused by malfunction of the immune system. AIDS is a consequence of a viral infection and destruction of certain cells of the immune system.

8. *Neoplastic diseases* involve abnormal cell growth that leads to the formation of various types of generally harmless or cancerous growths or tumors. Causes of such diseases include chemicals, physical agents such as various forms of radiation, and microorganisms, especially viruses. Papillomaviruses, which are known to cause warts, have been associated with the development of cervical cancer, and other viruses are known to cause tumorous growths in plants. ➡ (Chapter 10, p. 283)

9. *Iatrogenic* (i-at″ro-jen′ik) *diseases* (*iatros,* Greek for "physician") are caused by medical procedures and/or treatments. Examples include surgical errors, drug reactions, and infections acquired from hospital treatment. The latter are called *nosocomial infections.* For example, *Staphylococcus aureus* is a common bacterium associated with surgical wound infections. Nosocomial infections are discussed in Chapter 15.

10. *Idiopathic* (id″e-o-path′ik) *diseases* are diseases whose cause is unknown. Some researchers believe that Alzheimer's disease, which causes mental deterioration, has an infectious basis.

Communicable and Noncommunicable Diseases

Some infectious diseases can be spread from one host to another and are said to be **communicable infectious diseases.** Some communicable diseases are more easily spread than others. Rubeola (red measles) and rubella are highly communicable, or **contagious diseases,** especially among young children. Vaccines protect children in developed countries, but nearly all unimmunized children in developing nations still get these diseases. Influenza is highly communicable among adults, especially the elderly. Gonorrhea and genital herpes infections are easily spread among unprotected sexual partners. Although also communicable, certain other diseases such as *Klebsiella* pneumonia are less contagious. Some diseases that normally affect other animals are transmissible to humans (Chapter 15), whereas diseases such as Hansen's disease (leprosy) can also be transmitted from humans to other animals.

Noncommunicable infectious diseases are not spread from one host to another. You cannot "catch" a noncommunicable disease from another person. Such diseases may result from (1) infections caused by an individual's normal microflora, such as an inflammation of the abdominal cavity lining following rupture of the appendix; (2) poisoning following the ingestion of preformed toxins, such as staphylococcal enterotoxin, a common cause of food poisoning; and (3) infections caused by certain organisms found in the environment, such as tetanus, a bacterial infection resulting from spores in the soil gaining access to a wound. Other noncommunicable infectious diseases, such as legionellosis, a form of pneumonia, can spread through contaminated air-conditioning systems.

The Disease Process

How Microbes Cause Disease

Microorganisms act in certain ways that allow them to cause disease. These actions include gaining access to the host, adhering to and colonizing cell surfaces, invading tissues, and producing toxins and other harmful metabolic products. However, host defense mechanisms tend to thwart the actions of microorganisms. The occurrence of a disease depends on whether the pathogen or the host wins the battle; if it is a draw, a chronic, long-lasting disease may result.

Most of the pathogens considered in this text are prokaryotic microorganisms and viruses, which together account for the majority of human disease agents. However, several eukaryotes such as fungi, protozoa, and multicellular parasites (mostly worms) display pathogenicity ➡ (Chapter 11, p. 302) Eukaryotic pathogens can be present in a host without causing disease signs or symptoms, or they can cause severe disease. The extent of damage caused by these pathogens, like that caused by prokaryotic infectious agents, is determined by the properties of the pathogens and by the host's response to them.

How Bacteria Cause Disease

Bacterial pathogens often have special structures or physiological characteristics that improve the chances of successful host invasion and infection. **Virulence factors** are structural or physiological characteristics that help organisms cause infection and disease. These factors include structures such as pili for adhesion to cells and tissues, enzymes that either help in evading host defenses or protect the organism from host defenses, and toxins that can directly cause disease.

Direct Actions of Bacteria   Bacteria can enter the body by penetrating the skin or mucous membranes, by sexual transmission, by being ingested with food, by being inhaled in aerosols, or by transmission on a *fomite* (any inanimate object contaminated with an infectious agent). If the bacteria are immediately swept out of the body in urine or feces or by coughing or sneezing, they cannot initiate an infection.

A critical point in the production of bacterial disease is the organism's **adherence,** or attachment, to a host cell's surface. The occurrence of certain infections depends in part on the interaction between host plasma membranes and bacterial adherence factors. **Adhesins** are proteins or glycoproteins found on attachment pili (fimbriae) and capsules. ➡ (Chapter 4, p. 89) Most adhesins that have been identified permit the pathogen to adhere only to receptors on membranes of certain cells or tissues (Table 14.3). For example, an adhesin on attachment pili of certain strains of *Escherichia coli* attaches to receptors on certain host epithelial cells. (Host leukocytes also have receptors for this adhesin, so the same adhesin that helps the bacterium attach may also help the host destroy it.) However, very often the capsules and attachment pili are also antiphagocytic structures. It is difficult for phagocytic cells to engulf bacteria that have capsules or attachment pili, so these structures make excellent virulence factors.

Attachment to a host cell surface is not enough to cause an infection. The microbes must also be able to colonize the cell's surface or to penetrate it. **Colonization** refers to the growth of microorganisms on epithelial surfaces, such as skin or mucous membranes or other host tissues. For colonization to occur after adherence, the pathogens must survive and reproduce despite host defense mechanisms. For example, pathogenic bacteria on the skin's surface must withstand environmental conditions and bacteriostatic skin secretions. Those on respiratory membranes must escape the action of mucus and cilia. Those on the lining of portions of the digestive tract must withstand peristaltic movements, mucus, digestive enzymes, and acid.

Only a few pathogens cause disease by colonizing surfaces; most have additional virulence factors that enable the pathogen to invade tissues. The degree of **invasiveness** of a pathogen—its ability to invade and grow in host tissues—is related to the virulence factors the pathogen possesses and determines the severity of disease it produces. Some bacteria, such as pneumo-

> Some viruses gain entry to cells by mimicing substances of use to the host cell, e.g., rabies virus mimics the neurotransmitter acetylcholine.

> The syphilis spirochete uses the tail end of its body to "hook" onto host cells.

➤ Table 14.3   Examples of adhesive virulence factors

Bacterium	Disease	Adhesion mechanism
Upper respiratory tract		
Mycoplasma pneumoniae	Atypical pneumonia	Adhesin on cell surface adheres to receptor in respiratory lining
Neisseria meningitidis	Meningitis	Adhesin on pili
Streptococcus pneumoniae	Pneumonia	Surface adhesins attach to carbohydrate on respiratory lining
Mouth		
Streptococcus mutans	Dental caries	Capsule attaches to tooth enamel
Intestinal tract		
Shigella species	Dysentery	Unknown mechanism for attachment to intestinal lining
Escherichia coli	Diarrhea	Adhesins on pili attach to receptor on intestinal lining
Campylobacter jejuni	Diarrhea	Adhesins on flagella attach to intestinal lining
Vibrio cholerae	Cholera	Adhesins on flagella bind to receptors on intestinal lining
Urogenital tract		
Treponema pallidum	Syphilis	Bacterial protein attaches to cells
Neisseria gonorrhoeae	Gonorrhea	Adhesins on pili attach to lining of genital tract

cocci and other streptococci, release digestive enzymes that allow them to invade tissues rapidly and cause severe illnesses. Streptococci produce the enzyme **hyaluronidase,** or *spreading factor*. This enzyme digests hyaluronic acid, a gluelike substance that helps hold the cells of certain tissues together (➤ Figure 14.5a). Digestion of hyaluronic acid allows streptococci to pass between epithelial cells and invade deeper tissues.

In some cases the same pathogen can display varying degrees of invasiveness and pathogenicity in different tissues. Both bubonic plague and pneumonic plague are caused by the bacterium *Yersinia pestis*. In bubonic plague, the organisms enter the body by means of a flea bite, migrate through the blood, and infect many organs and tissues. Untreated, this disease has a mortality rate of about 55 percent. As a victim of pneumonic plague coughs or sneezes, the bacteria are spread by aerosols to other individuals. *Yersinia pestis* can cause a severe infection of the lungs with a mortality rate as high as 98 percent.

Most bacteria that invade tissues damage cells and are found around cells. Thus, enzymes that contribute to tissue damage are another important virulence factor. **Coagulase** is a bacterial enzyme that accelerates the coagulation (clotting) of blood. When blood plasma, the fluid portion of blood, leaks out of vessels into tissues, coagulase causes the plasma to clot. *Staphylococcus aureus* produces coagulase to aid in infection (➤ Figure 14.5b). Coagulase is a two-edged sword: It keeps organisms from spreading but also helps wall them off from immune defenses that might otherwise destroy them. Conversely, the bacterial enzyme **streptokinase**

1. Invasive pathogens reach epithelial surface

2. Pathogens produce hyaluronidase

3. Pathogens invade deeper tissues

1. Pathogens produce coagulase

2. Blood clot forms around pathogens

3. Pathogens produce streptokinase, dissolving clot and releasing bacteria

➤ **Figure 14.5   Enzymatic virulence factors help bacteria invade tissues and evade host defenses. (a)** Hyaluronidase dissolves the "cement" that holds together the cells that line the intestinal tract. Bacteria that produce hyaluronidase can then invade deeper cells within the intestinal tissues. **(b)** Coagulase triggers blood plasma clotting, allowing bacteria protection from immune defenses. Streptokinase dissolves blood clots. Bacteria trapped within a clot can free themselves and spread the infection by producing streptokinase.

dissolves blood clots. Pathogens trapped in blood clots free themselves to spread to other tissues by secreting these virulence factors.

Some bacterial pathogens actually enter cells. The rickettsias, chlamydias, and a few other pathogens must invade cells to grow, reproduce, and produce disease. In other situations, organisms that can survive within host phagocytic cells not only escape destruction by the phagocytes but also obtain free transportation to deeper body tissues. Such organisms include *Mycobacterium tuberculosis* and *Neisseria gonorrhoeae.*

Bacterial Toxins   A **toxin** is any substance that is poisonous to other organisms. Some bacteria produce toxins, which are synthesized inside bacterial cells and are classified according to how they are released. **Exotoxins** are soluble substances secreted into host tissues. **Endotoxins** are part of the cell wall and are released into host tissues—sometimes in large quantities—from Gram-negative bacteria, often when the bacteria die or divide. ➡ (Chapter 4, p. 77) Giving antibiotics that kill such bacteria can release sufficient toxin to cause the patient to die of severely reduced blood pressure (*endotoxic shock*). Let us look at some of the properties and effects of endotoxins and exotoxins (Table 14.4).

Relatively weak (except in large doses), endotoxins are produced by certain Gram-negative bacteria. All endotoxins consist of lipopolysaccharide (LPS) complexes, the components of which vary among genera. They are relatively stable molecules that do not display affinities for particular tissues. Bacterial endotoxins have nonspecific effects such as fever or a sudden drop in blood pressure. They also cause tissue damage in diseases such as typhoid fever and epidemic meningitis (an inflammation of membranes that cover the brain and spinal cord).

Exotoxins are more powerful toxins produced by several Gram-positive and a few Gram-negative bacteria. Most are polypeptides, which are denatured by heat, ultraviolet light, and chemicals such as formaldehyde. Species of *Clostridium, Bacillus, Staphylococcus, Streptococcus,* and several other bacteria produce exotoxins.

Some exotoxins are enzymes. **Hemolysins** were first discovered in cultures of bacteria grown on blood–agar plates. The action of these exotoxins is to lyse (rupture) red blood cells. Two kinds of hemolysins were identified from bacteria grown on blood–agar plates. **Alpha-hemolysin** (α-hemolysin) hemolyze blood cells, partially break down hemoglobin, and produce a greenish ring around colonies; **β-hemolysins** also hemolyze blood cells but completely break down hemoglobin and leave a clear ring around colonies (➤ Figure 14.6). Streptococci and staphylococci produce different hemolysins that are helpful in identifying them in laboratory

Diphtheria toxin molecules have a molecular weight of 62,000, contain 535 amino acids, and are not produced when iron is plentiful.

Hemolysins also attack cells other than red blood cells, but are most easily visualized in blood agar.

➤ Table 14.4 Properties of toxins

Property	Exotoxins	Endotoxins
Organisms producing	Almost all Gram-positive, some Gram-negative	Almost all Gram-negative
Location in cell	Extracellular, excreted into medium	Bound within bacterial cell wall; released upon death of bacterium
Chemical nature	Mostly polypeptides	Lipopolysaccharide complex
Stability	Unstable; denatured above 60°C and by ultraviolet light	Relatively stable; can withstand several hours above 60°C
Toxicity	Among the most powerful toxins known (some are 100 to 1 million times as strong as strychnine)	Weak, but can be fatal in relatively large doses
Effect on tissues	Highly specific; some act as neurotoxins or cardiacmuscle toxins	Nonspecific; ache-all-over systemic effects or local site reactions
Fever production	Little or no fever	Rapid rise in temperature to high fever
Antigenicity	Strong; stimulates antibody production and immunity	Weak; recovery from disease often does not produce immunity
Toxoid conversion and use	By treatment with heat or chemicals; toxoid used to immunize against toxin	Cannot be converted to toxoid; cannot be used to immunize
Examples	Botulism, gas gangrene, tetanus, diphtheria, staphylococcal food poisoning, cholera, enterotoxins, plague	Salmonellosis, tularemia, endotoxic shock

(a)

➤ Figure 14.6   **Types of hemolysis.** (a) Alpha, or partial, hemolysis of red blood cells results in a greenish zone around colonies of *Streptococcus pneumoniae* grown on blood agar. (b) *S. pyogenes* colonies release β-hemolysins, which produce complete breakdown of hemoglobin, causing clear zones to form around colonies grown on blood agar.

(b)

cultures. There is no evidence that red blood cell lysis plays a role in the disease syndrome. Rather, the hemolysins release iron from the hemoglobin molecules in the red blood cells. Iron is a critical element for growth of all cells, both host and microbe. But there is very little free iron within the human body. Most of it is bound in a form such as hemoglobin, and the microbe must enzymatically release it. Bacteria that can produce hemolysins can grow better than those that do not produce these enzymes. Especially in the staphylococci, the hemolysins can damage other types of cells as well. Alpha-hemolysin damages smooth muscle and kills skin cells.

Virulence factors called **leukocidins** are exotoxins produced by many bacteria, including the streptococci and staphylococci. These toxins damage or destroy certain kinds of white blood cells called *neutrophils* and *macrophages*. Leukocidins are most effective when released by microbes that have been engulfed by a neutrophil. Because of the action of leukocidins, the number of white blood cells decreases in certain diseases, although most infections are characterized by an elevated white cell count. A similar substance, called **leukostatin,** interferes with the ability of leukocytes to engulf microorganisms that secrete the exotoxin.

In the preceding examples, the spreading of exotoxins by blood from the site of infection is called **toxemia.** But some diseases caused by microbes are due not to infection and invasion of tissues by pathogens but instead to the ingestion of preformed toxins made by pathogens. For example, botulism food poisoning strikes within hours of ingesting food that contains a significant amount of a toxin produced by *Clostridium botulinum*—too short a time for the microbe to invade tissues and cause disease. The toxins accumulate during the storage of an improperly sterilized jar or can of food and have an immediate and often ultimately lethal effect on the consumer. Diseases that result from the ingestion of a toxin are termed **intoxications** rather than infections.

Many exotoxins have a special attraction for particular tissues. **Neurotoxins,** such as the botulism and tetanus toxins, are exotoxins that act on tissues of the nervous system to prevent muscle contraction (botulism) or muscle relaxation (tetanus). **Enterotoxins,** such as the toxin that

Leukocidins release cell-killing enzymes from lysosomes.

One milligram of botulinum toxin can kill more than 1 million guinea pigs.

causes cholera, are exotoxins that act on tissues of the gut. Many exotoxins can act as *antigens*, foreign substances against which the immune system reacts. Antigenic exotoxins inactivated by treatment with chemical substances such as formaldehyde are called *toxoids*. A **toxoid** (*-oid* is Latin for "like") is an altered toxin that has lost its ability to cause harm but that retains antigenicity. Toxoids can be used to stimulate the development of immunity without causing disease. For example, when you get a tetanus booster shot, you are receiving the tetanus toxoid. It stimulates your body to produce immunity so that if you are exposed to active tetanus toxin through a cut or puncture of the skin, you will not get tetanus. The effects of bacterial exotoxins in human disease are summarized in Table 14.5. The specific effects of such diseases are discussed in later chapters.

How Viruses Cause Disease

Viruses can replicate only after they have attached to cells and then penetrated specific host cells. In tissue culture systems, once inside a cell, viruses cause observable changes collectively called the **cytopathic effect (CPE)**. ➡ (Chapter 10, p. 279) CPE can be cytocidal when the viruses kill the cell and noncytocidal when they do not. Cytocidal viruses can kill cells by causing enzymes from cellular lysosomes to be released or by diverting the host cell's synthetic processes, thereby stopping the synthesis of host proteins and other essential macromolecules. CPE

➣ Table 14.5 **Effects of exotoxins**

Bacterium	Name of toxin or disease	Action of toxin	Host symptoms
Bacillus anthracis	Anthrax (cytotoxin)	Increases vascular permeability	Hemorrhage and pulmonary edema
Bacillus cereus	Enterotoxin	Causes excessive loss of water and electrolytes	Diarrhea
Clostridium botulinum	Botulism (eight serological types; neurotoxins)	Blocks release of acetylcholine at nerve endings	Respiratory paralysis, double vision
Clostridium perfringens	Gas gangrene (alpha toxin, a hemolysin)	Breaks down lecithin in cell membranes	Cell and tissue destruction
	Food poisoning (enterotoxin)	Causes excessive loss of water and electrolytes	Diarrhea
Clostridium tetani	Tetanus (lockjaw) (neurotoxin)	Inhibits antagonists of motor neurons of brain; 1 ng can kill 2 tons of cells	Violent skeletal muscle spasms, respiratory failure
Corynebacterium diphtheriae	Diphtheria; produced by virus-infected (cytotoxin) bacteria	Inhibits protein synthesis	Heart damage can cause death weeks after apparent recovery
Escherichia coli	Traveler's diarrhea (enterotoxin)	Causes excessive loss of water and electrolytes	Diarrhea
Escherichia coli	O157:H7 (enterotoxin)	Hemolytic uremic syndrome	Destroys intestinal lining and causes hemorrhages in kidney; Bleeding and kidney hemorrhage and failure
Pseudomonas aeruginosa	Various infections (exotoxin A)	Inhibits protein synthesis	Lethal, necrotizing lesions
Shigella dysenteriae	Bacillary dysentery (enterotoxin)	Cytotoxic effects; as potent as botulinum toxin	Diarrhea, causes paralysis in rabbits from spinal cord hemorrhage and edema
Staphylococcus aureus	Food poisoning (enterotoxin)	Stimulates brain center that causes vomiting	Vomiting
	Scalded skin syndrome (exfoliatin)	Causes intradermal separation of cells	Redness and sloughing of skin
Streptococcus pyogenes	Scarlet fever (erythrogenic, or red-producing toxin)	Causes vasodilation	Maculopapular (slightly raised, discolored) lesions
Vibrio cholerae	Cholera (enterotoxin)	Causes excessive loss of water (up to 30 l/day) and electrolytes	Diarrhea; can kill within hours

Clinical Use Of Botulinum Toxin

The powerful effects of the neurotoxins produced by *Clostridium botulinum,* best known as a cause of lethal food poisoning, have been harnessed to help victims of dystonia. *Dystonia* refers to a group of neurologic disorders characterized by abnormal, sustained involuntary movements, often twisting, which are of unknown origin. In one form, blepharospasm, the patient's eyes remain tightly closed at all times. Physicians inject small quantities of botulinum toxin (trade name, Oculinum A) at several sites around each eye. The toxin blocks nerve impulses to muscles, thereby relieving the spasms of the eyelids. Injections are needed every 2 to 3 months. Some people receive such treatment for 4 to 5 years without problems; others develop antibodies (immune defenses) against the toxin after many injections. Because there are several different types of the toxin, it is hoped that patients can switch to a different form once they develop antibodies against one form. A round of injections can cost from $400 to $1800.

Other dystonias being helped by botulinum toxin include oromandibular dystonia, in which the patient's jaws are clenched so tightly that the jaw bones may break, causing midfacial collapse. Eating and speaking are difficult, and some patients starve to death. Vocal cord spasms, causing a cracked tremulous voice, and "stenographers' cramp," causing the middle finger to extend rigidly, are also being treated experimentally with the toxin.

A patient with hemifacial spasm before (left) and 2 weeks after (right) treatment with botulinum toxin.

Oculinum A is licensed for treating adults with strabismus (cross-eye, lazy eye). Small amounts are injected into the overcontracted eye muscle, which then relaxes and lengthens. The antagonistic muscles on the other side of the eye contract to take up the slack, and the eye can then look straight ahead.

can be observed in laboratory tissue cultures with a compound microscope (➤ Figure 14.7). CPE can be so distinctive that an experienced clinical virologist can make a tentative identification by looking at infected cells through the microscope, even though further tests are needed to confirm the identification (Table 14.6).

Many viruses produce pathogenic effects in host cells. These include *inclusion bodies,* which consist of nucleic acids and proteins not yet assembled into viruses, masses of viruses, or remnants of viruses. Rabiesviruses make inclusion bodies that are so distinctive they can be used to diagnose rabies. Retroviruses and oncoviruses integrate into host chromosomes and can remain

(a)   (b)

➤ Figure 14.7   **An example of the cytopathic effect (CPE).** (a) Uninfected mouse cells; (b) the same cells 24 hours after infection with vesicular stomatitis virus. A large number have died, and many others have rounded up into abnormal shapes.

> Table 14.6	Examples of changes (cytopathic effects) to virus-infected cells
Virus family	**Cytopathic effect**
Adenoviridae	Cells swell
Herpesviridae	Cells swell
Picornaviridae	Cells swell and lyse
Paramyxoviridae	Cell membranes fuse, and up to 100 nuclei accumulate in a newly formed giant cell
Rhabdoviridae (rabies)	Inclusion bodies called Negri bodies form (site of viral replication or the accumulation of viral antigens)
Orthomyxoviridae	Produce hemagglutinins that cause erythrocytes to agglutinate, or clump together

in cells indefinitely, sometimes leading to the expression of their antigens on host-cell surfaces. Influenza and parainfluenza viruses produce hemagglutinins, which cause agglutination, or clumping together, of erythrocytes. This feature is of value in laboratory testing.

Viral infections can be productive or abortive. A **productive infection** occurs when viruses enter a cell and produce infectious offspring. An **abortive infection** occurs when viruses enter a cell but are unable to express all their genes to make infectious offspring. Productive infections vary in the degree of damage they cause, depending on the kind and number of cells the virus invades. An enterovirus, such as a human rotavirus or human adenovirus, that infects the gut can destroy millions of intestinal epithelial cells. Because these cells are rapidly replaced, the infection causes temporary, though sometimes severe, symptoms such as diarrhea, but no permanent damage. But a poliovirus that infects motor neurons of the central nervous system can destroy these cells. Destroyed neurons cannot be replaced, so permanent paralysis may result. Human papillomaviruses that cause warts are limited to cells in localized areas. In contrast, the measles virus replicates and spreads throughout the body and may damage many tissues.

Latent viral infections are characteristic of herpesviruses. For example, chickenpox infections occur during childhood and usually are brought under control by host immune defenses. However, the virus may retreat into the nervous system and remain inactive, or latent. Later in life, factors such as stress, other infections, or fever can reactivate the virus. A weakened immune system allows the virus to multiply. Whatever the cause, the disease appears as shingles.

Persistent viral infections involve a continued production of viruses over many months or years. The hepatitis B virus (HBV) infects the liver in such a chronic fashion that there may be no outward signs of an infection. However, such persistent infections can lead to cirrhosis of the liver or even liver cancer.

How Fungi, Protozoa, and Helminths Cause Disease

In addition to bacteria and viruses, infectious diseases can also be caused by eukaryotes—specifically fungi, protozoa, and helminths. Even a few algae produce neurotoxins, and one alga (*Prototheca*) directly invades skin cells.

Most fungal diseases result from fungal spores that are inhaled or from fungal cells and/or spores that enter cells through a cut or wound. Fungi damage host tissues by releasing enzymes that attack cells. As the first cells are killed, the fungi progressively digest and invade adjacent cells. Some fungi also release toxins or cause allergic reactions in the host. Certain fungi that parasitize plants produce *mycotoxins,* which cause disease if ingested by humans. Ergot, from a fungus that grows on rye, and aflatoxins, highly carcinogenic compounds that can be found in grains, cereals, even peanut butter made from moldy peanuts, are mycotoxins (see Chapter 22).

Pathogenic protozoans and helminths cause human disease in several ways. Some protozoans, including those that cause malaria, invade and reproduce in red blood cells. ➡ (Chapter 11, p. 294)

✓ Give several examples of noninfectious diseases.

✓ Can a disease be both infectious and noncommunicable?

✓ Compare and contract coagulase and streptokinase.

✓ Compare and contrast infection and intoxication.

(a)

(b)

➤ Figure 14.8   *Giardia intestinalis.* (a) SEM of *Giardia intestinalis* (21,150X). (b) The suction forces of *Giardia's* adhesive disk are so strong that they leave markings behind on the intestinal surface after they drop off. White arrowheads indicate where the suction cup attached, and white arrows point to marks left by the flagella.

The protozoan *Giardia intestinalis* attaches to tissues and ingests cells and tissue fluids of the host. *Giardia's* virulence factor is an *adhesive disc* by which it attaches to cells that line the small intestine (➤ Figure 14.8). While burrowing into the tissue, the parasite uses its flagella to expel tissue fluids. This process creates so strong a suction that the parasite is not disturbed by peristaltic contractions.

Most helminths are extracellular parasites, inhabiting the intestines or other body tissues. Many release toxic waste products and antigens in their excretions that often cause allergic reactions in the host. The outer surface of many helminths is quite tough and resistant to immune attacks.

Signs, Symptoms, and Syndromes

Most diseases are recognized by their signs and symptoms. A **sign** is a characteristic of a disease that can be observed by examining the patient. Signs of disease include such things as swelling, redness, rashes, coughing, pus formation, runny nose, fever, vomiting, and diarrhea. A **symptom** is a characteristic of a disease that can be observed or felt only by the patient. Symptoms include such things as pain, shortness of breath, nausea, sore throat, headache, and malaise (discomfort).

A **syndrome** is a combination of signs and symptoms that occur together and are indicative of a particular disease or abnormal condition. For example, most infectious diseases cause the body to mount an acute inflammatory response. This response, which is discussed in Chapter 16, is characterized by a syndrome of fever, malaise, swollen lymph nodes, and **leukocytosis** (an increase in the numbers of white blood cells circulating in the blood).

In addition to the inflammatory response, many infectious diseases cause other signs and symptoms. Infections of the gut called enteric infections often cause nausea, vomiting, and diarrhea. Upper respiratory infections are usually characterized by coughing, sneezing, sore throat, and runny nose. Unfortunately, the signs and symptoms of diseases caused by different pathogens may be too similar to allow a specific diagnosis to be made. Thus, laboratory tests to identify infectious agents are an important component of modern medicine.

Even after recovery, some diseases leave aftereffects, called **sequelae** (se-kwel'e; singular: *sequela*). Bacterial infections of heart valves often cause permanent valve damage, and poliovirus infections leave permanent paralysis.

Types of Infectious Disease

Infectious diseases vary in duration, location in the body, and other attributes. Several important terms, summarized in Table 14.7, are used to describe these attributes.

An **acute disease** develops rapidly and runs its course quickly. Measles and colds are examples of acute diseases. A **chronic disease** develops more slowly than an acute disease, is

> Table 14.7 | **Terms used to describe infections**

Term	Characteristic of infection
Acute disease	Disease in which symptoms develop rapidly and that runs its course quickly
Chronic disease	Disease in which symptoms develop slowly and disease is slow to disappear
Subacute disease	Disease with symptoms intermediate between acute and chronic
Latent disease	Disease in which symptoms appear and/or reappear long after infection
Local infection	Infection confined to a small region of the body, such as a boil or bladder infection
Focal infection	Infection in a confined region from which pathogens travel to other regions of the body, such as an abscessed tooth or infected sinuses
Systemic infection	Infection in which the pathogen is spread throughout the body, often by traveling through blood or lymph
Septicemia	Presence and multiplication of pathogens in blood
Bacteremia	Presence but not multiplication of bacteria in blood
Viremia	Presence but not multiplication of viruses in blood
Toxemia	Presence of toxins in blood
Sapremia	Presence of metabolic products of saprophytes in blood
Primary infection	Infection in a previously healthy person
Secondary infection	Infection that immediately follows a primary infection
Superinfection	Secondary infection that is usually caused by an agent resistant to the treatment for the primary infection
Mixed infection	Infection caused by two or more pathogens
Inapparent infection	Infection that fails to produce full set of signs and symptoms

usually less severe, and persists for a long, indeterminate period. Tuberculosis and Hansen's disease (leprosy) are chronic diseases. A **subacute disease** is intermediate between an acute and a chronic disease. Gingivitis, or gum disease, can exist as a subacute disease. A **latent disease** is characterized by periods of inactivity either before signs and symptoms appear or between attacks. The herpes simplex virus and several other viral infections produce latent disease.

A **local infection** is confined to a specific area of the body. Boils and bladder infections are local infections. A **focal infection** is confined to a specific area, but pathogens from it, or their toxins, can spread to other areas. Abscessed teeth and sinus infections are focal infections. A **systemic infection,** or *generalized infection,* affects most of the body, and the pathogens are widely distributed in many tissues. Typhoid fever is a systemic infection. When focal infections spread, they become systemic infections. For example, organisms from an abscessed tooth can enter the bloodstream and be carried to other tissues, including the kidneys. The organisms can then infect the kidneys and other parts of the urinary tract.

Pathogens can be present in the blood with or without multiplying there. In **septicemia,** once known as blood poisoning, pathogens are present in and multiply in the blood. In **bacteremia** and **viremia,** bacteria and viruses, respectively, are transported in the blood but do

To Squeeze or Not to Squeeze
There is good cause for the common warning not to squeeze pimples and boils. Left alone, the body's defense mechanisms ordinarily confine these lesions to the skin. However, squeezing them can disperse microorganisms into the blood and cause septicemia—a far worse condition. In septicemia, the organisms are spread throughout the body and can cause severe infections.

not multiply in transit. Such spread of organisms often occurs in cases of injury, such as a cut, abrasion, or even teeth cleaning. As we have seen, some pathogens release toxins into the blood; the presence of toxins in blood is called *toxemia.* Saprophytes feed on dead tissues. Fungi behave as parasites when they destroy cells and as saprophytes when they feed on them or on other dead or decaying matter. They release metabolic products into the blood, thereby causing a condition called **sapremia.**

A **primary infection** is an initial infection in a previously healthy person. Most primary infections are acute. A **secondary infection** follows a primary infection, especially in individuals weakened by the primary infection. A person who catches the common cold as a primary infection, for instance, might come down with a middle-ear infection as a secondary infection. A **superinfection** (Chapter 13) is a secondary infection that results from the destruction of normal microflora and often follows the use of broad-spectrum antibiotics. Although many infections are caused by a single pathogen, **mixed infections** are caused by several species of organisms present at the same time. Dental caries and periodontal disease are due to mixed bacterial infections. An **inapparent,** or **subclinical, infection** is one that fails to produce the full range of signs and symptoms either because too few organisms are present or because host defenses effectively combat the pathogens. People with inapparent infections, such as carriers of the hepatitis B virus, can spread the disease to others.

Stages of an Infectious Disease

At one time or another, all of us have suffered from infectious diseases, such as the common cold, for which there is no cure. We simply must let the disease "run its course." Most diseases caused by infectious agents have a fairly standard course, or series of stages. These stages include the *incubation period,* the *prodromal phase,* the *invasive phase* (which includes the *acme*), the *decline phase,* and the *convalescence period* (➤ Figure 14.9). Even when treatment is available to eliminate the pathogen, the disease process usually passes through most of the stages. Treatment commonly lessens the severity of symptoms because pathogens can no longer multiply. It shortens the duration of the disease and the time required for recovery. The signs and symptoms associated with the stages of an infectious disease and the resulting tissue damage caused by the pathogen are summarized in Table 14.8.

The Incubation Period

The **incubation period** for an infectious disease is the time between infection and the appearance of signs and symptoms. Although the infected person is not aware of the presence of an infectious agent, he or she can spread the disease to others. Each infectious disease has a typical incubation period (➤ Figure 14.10). The length of the incubation period is determined by the properties of the pathogen and the response of the host to the organism.

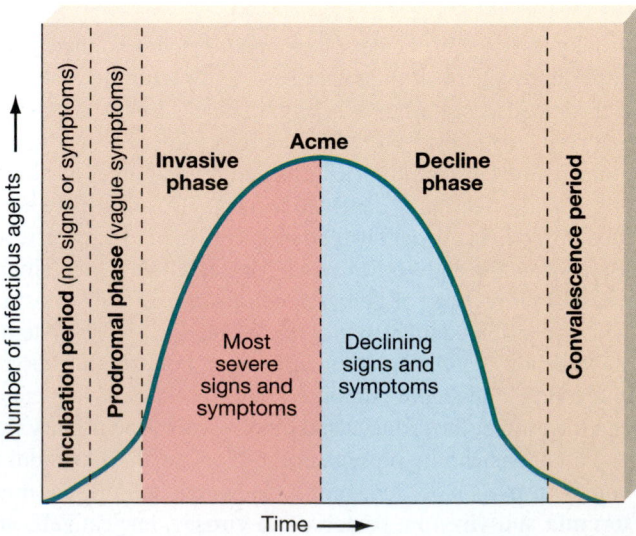

➤ Figure 14.9   **Stages in the course of an infectious disease.**

> Table 14.8 | **Correlation of signs and symptoms with tissue damage**

Signs and symptoms	Probable nature of tissue damage
Incubation period	
None	None
Prodromal phase	
Local redness and swelling	Pathogen has damaged tissue at site of invasion and caused release of chemicals that dilate blood vessels (redness) and allow fluid from blood to enter tissues (swelling).
Headache	Chemicals from tissue injury dilate blood vessels in the brain.
General aches and pains	Chemicals from tissue injury stimulate pain receptors in joints and muscles.
Invasive phase	
Cough	Mucosal cells of respiratory tract have been damaged by pathogens; excess mucus is released, and neural centers in the brain elicit coughing to remove mucus.
Sore throat	Lymphatic tissue of the pharynx is swollen and inflamed by substances released by pathogens and by leukocytes.
Fever	Leukocytes release pyrogens that reset the body's thermostat and cause temperature to rise.
Swollen lymph nodes	Leukocytes release other substances that stimulate cell division and fluid accumulation in lymph nodes; lymph nodes themselves release substances that flow to and affect other lymph nodes; some pathogens multiply in lymph nodes.
Skin rashes	Leukocytes release substances that damage capillaries and allow small hemorrhages; some pathogens invade skin cells and cause pox, vesicles, and other skin lesions.
Nasal congestion	Nasal mucosal cells have been damaged by pathogens (usually viruses) that release fluids and increase mucous secretions.
Pain at specific sites (earache, local pain at a wound site)	Substances from pathogens or leukocytes have stimulated pain receptors; messages are relayed to the brain, where they are interpreted as pain.
Nausea	Toxins from pathogens have stimulated neural centers; you interpret the stimuli as nausea.
Vomiting	Toxins in food have stimulated the brain's vomiting center; vomiting helps rid the body of toxins.
Diarrhea	Toxins in food cause fluids to enter the digestive tract; some pathogens directly injure the intestinal epithelium; both toxins and pathogens stimulate peristalsis; frequent watery stools result.
Acme	
All signs and symptoms. are at peak intensity	Full development of all signs and symptoms
Decline Phase	
Signs and symptoms subside.	Host defense mechanisms (and treatment, if applicable) have contributed to overcoming the pathogen.
Convalescence Period	
Patient regains strength	Tissue repair occurs; substances that caused signs and symptoms no longer released.

Properties that affect the incubation period include the nature of the organism, its virulence, how many organisms enter the body, and where they enter in relation to the tissues they affect. For example, if large numbers of an extremely virulent strain of *Shigella* quickly reach the intestine, profuse diarrhea can appear in a day. In contrast, if only small numbers of a less virulent strain enter the digestive tract with a large quantity of food, the disease will develop more slowly. In fact, host defenses might be able to destroy the small number of organisms such that the disease will not occur at all. In the course of a lifetime, we undoubtedly have many more exposures than infections and more infections than we have overt diseases. As we will see in Chapter 16, host defenses frequently attack pathogens as they start to invade tissues, thus averting potential diseases.

The Prodromal Phase

The **prodromal phase** of disease is a short period during which nonspecific, often mild, symptoms such as malaise and headache sometimes appear. A **prodrome** (*prodromos*, Greek for "forerunner") is a symptom indicating the onset of a disease. You wake up one morning feeling bad, and you know you're coming down with something, but you don't know yet whether you will break out in spots, start to cough, develop a sore throat, or experience other signs or

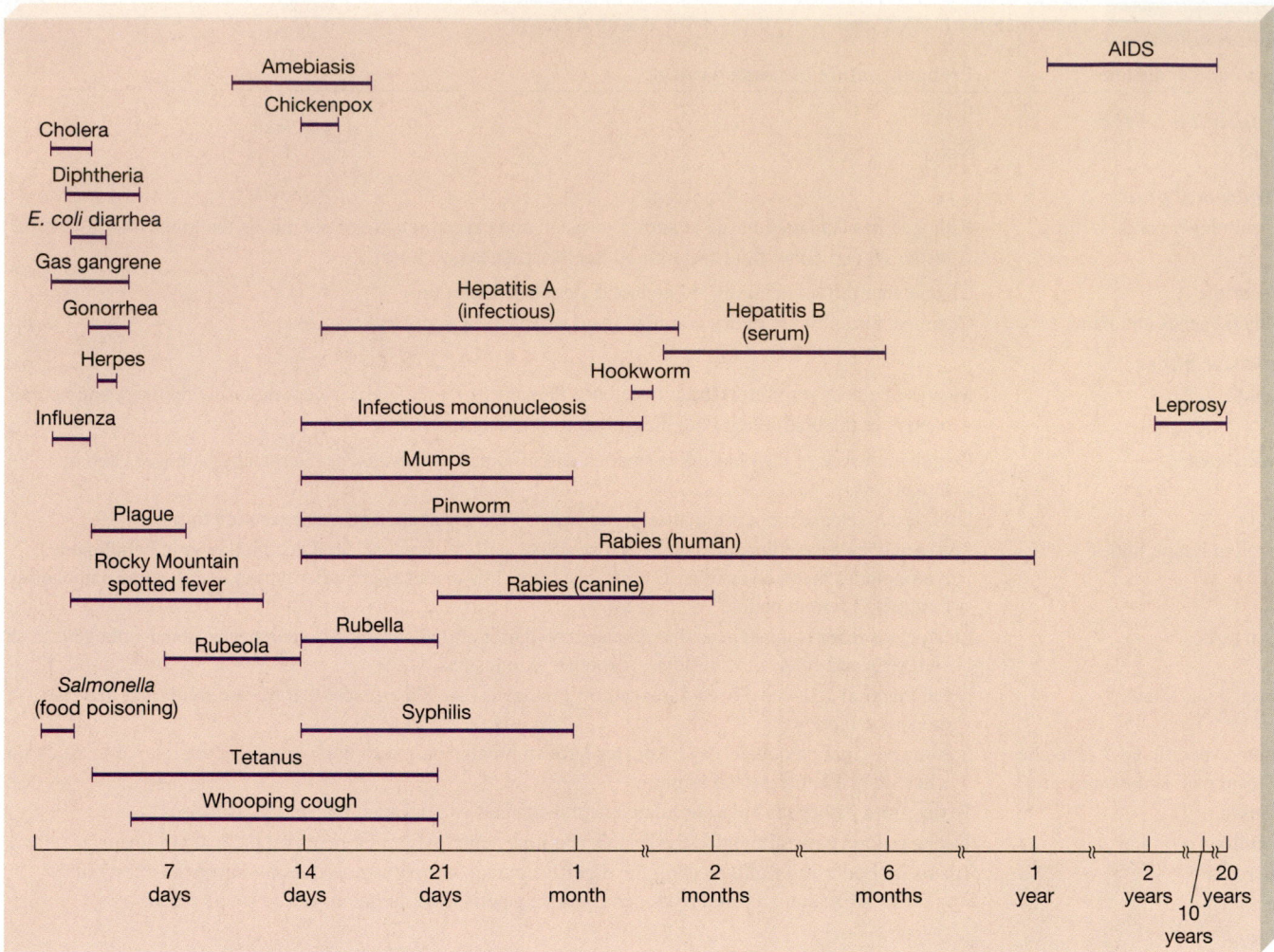

> ➤ Figure 14.10   **Incubation periods of selected infectious diseases.** (Note time axis is not drawn to scale.)

> Fever can be caused by chemicals directly released by pathogens, or liberated from white blood cells destroyed by pathogens.

symptoms. Many diseases lack a prodromal phase and begin with a sudden onset of symptoms such as fever and chills. During the prodromal phase, infected individuals are contagious and can spread the disease to others.

The Invasive Phase

The **invasive phase** is the period during which the individual experiences the typical signs and symptoms of the disease. These may include fever, nausea, headache, rash, and swollen lymph nodes. During this phase, the time when the signs and symptoms reach their greatest intensity is known as the **acme.** During the acme, pathogens invade and damage tissues. In some diseases, such as some kinds of meningitis, this phase is referred to as being **fulminating** (*fulmen,* Latin for "lightning"), or sudden and severe. In other diseases, such as hepatitis B, it can be persistent or chronic or appear gradually with inapparent symptoms. A period of chills followed by fever marks the acme of many diseases. As signs and symptoms appear, the form the infection will take becomes clear. Individuals at this critical stage are still contagious. The battle between pathogens and host defenses is at its height. A pathogen victory could lead to severe impairment of body function; if treatment is not available or provided in time, death can result.

Fever is an important component of the acme of many diseases. Certain pathogens produce substances called **pyrogens** that act on a center in the hypothalamus sometimes referred to as the body's "thermostat." Pyrogens set the thermostat at a higher-than-normal temperature. The body responds with involuntary muscle contraction that generates heat and constriction (narrowing) of blood vessels in the skin to prevent heat loss. Because our bodies function at lower temperatures than the newly set temperature, we feel cold and have chills at this stage. We shiver and get "goose bumps" as muscles contract involuntarily.

As the effects of the pyrogens diminish, the thermostat is reset to the normal, lower temperature, and the body responds to reach and maintain that temperature. This response includes sweating and dilation (widening) of blood vessels in the skin to increase heat loss. Because our bodies are warmer than the newly set temperature, we feel hot and say that we have fever. Our skin gets moist from sweat as more blood circulates near the skin surface. In many infectious diseases, repeated episodes of pyrogen release occur, thereby accounting for bouts of fever and chills. Fever and its function are described in more detail in Chapter 16.

The Decline Phase

As symptoms begin to subside, the disease enters the **decline phase**—the period of illness during which host defenses and the effects of treatment finally overcome the pathogen. The body's thermostat and other body activities gradually return to normal. Secondary infections may occur during this phase.

The Convalescence Period

During the **convalescence period,** tissues are repaired, healing takes place, and the body regains strength and recovers. Individuals no longer have disease symptoms. In some diseases, however, especially those in which scabs form over lesions, persons recovering from the disease can still transmit pathogens to others.

✓ How does a sign differ from a symptom?

✓ How does sapremia differ from bacteremia?

✓ How does the prodromal stage differ from the incubation period?

Infectious Diseases—Past, Present, and Future

In all the centuries of human history up to the twentieth, recovery or death from infectious diseases was determined largely by whether the human host or the pathogen won the war they waged against each other. Assorted potions and palliative (pain-reducing) treatments were available, but none could cure infectious diseases. Sometimes treatment was based on the notion that imbalances in body fluids caused disease. Depending on which fluid was judged to be in excess, efforts were made to remove some of it. Blood was removed by opening a vein or by applying blood-sucking leeches to the patient's skin. In eighteenth-century Europe, patients were bled until they lost consciousness. In 1774 when King Louis XV of France came down with smallpox, his desperate physicians bled him for 3 days in a row, each time removing "four large basinfuls of blood." In other cases, harsh laxatives were given to rid the body of excess bile. In most instances, these treatments failed to rid the patient of infectious agents, which at the time were unknown. At best, such treatments probably reduced suffering by hastening death.

Even after microorganisms came to be recognized as agents of disease, many years of painstaking research were required to relate specific diseases with the agents that caused them. More tedious research was needed to find antimicrobial agents that could cure diseases and to develop vaccines that could prevent them. The effects of these medical advances are clearly reflected in changes in the death rate in the United States. That death rate has decreased from 1560 per 100,000 people in 1900 to 505 per 100,000 people in the 1990s (➤ Figure 14.11). The greatest single factor in the decrease was the control of infectious diseases by better treatment or by immunization. Better sanitation has also helped. ➤ Figure 14.11 shows that in 1900 the proportion of the population dying from infectious diseases was 27 times the current rate. Deaths from typhoid fever, syphilis, and childhood diseases (measles, whooping cough, diphtheria) have nearly been eliminated, and deaths from pneumonia, influenza, and tuberculosis have been greatly reduced.

On a global scale, however, we have failed to eliminate diseases via technology that is already available. Measles kills nearly 1.5 million children each year, mostly in developing nations. Single doses of vaccine that cost less than 12 cents could save most of these lives. In fact, as of 1995 each year 14 million children under the age of 5 die from infectious diseases such as measles, whooping cough, tetanus, diarrhea, and pneumonia—all of which are vaccine-preventable diseases. That is, one child dies every 2 seconds from these diseases.

Look again at ➤ Figure 14.11. Because deaths from infectious diseases have decreased, average lifespans have increased. More people live long enough to develop degenerative diseases, and as they age their likelihood of developing a malignant disease increases. Thus, the

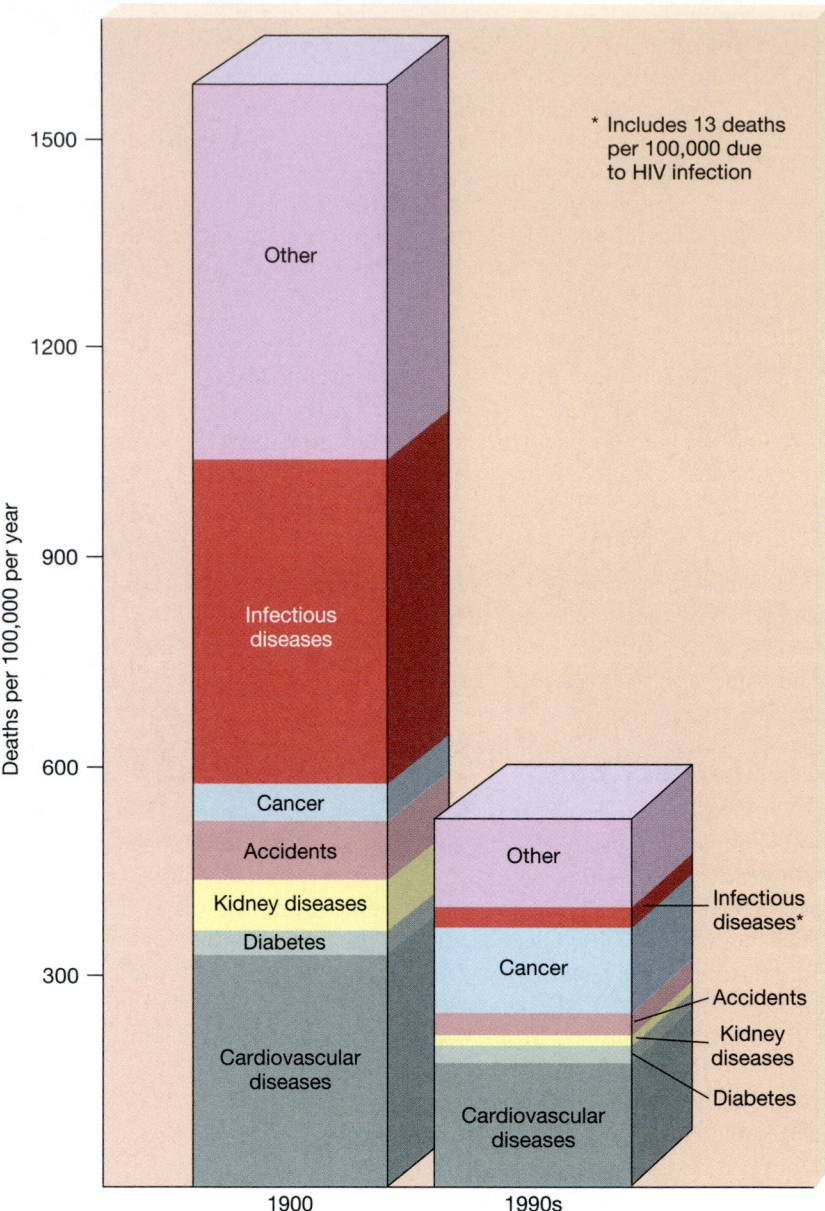

1500

1200

900

600

300

Deaths per 100,000 per year

* Includes 13 deaths per 100,000 due to HIV infection

Other

Infectious diseases

Cancer

Accidents

Kidney diseases

Diabetes

Cardiovascular diseases

Other

Infectious diseases*

Cancer

Accidents

Kidney diseases

Diabetes

Cardiovascular diseases

1900

1990s

➤ Figure 14.11 **Changes in the causes of death in the United States from 1900 to the 1990s.**

death rate for cancer has increased by more than 240 percent—from 55 per 100,000 in 1900 to about 133 per 100,000 in the 1990s.

Past successes in treating infectious diseases suggest that disease eradication should be possible. But at least four factors make eradication difficult:

1. *Available medical expertise is not always applied.* Preventable diseases such as measles and mumps still occur in the United States because parents fail to have their children immunized. Also, some people, both young and old, fail to obtain treatment for curable diseases, a problem that could be solved by improved access to health care.

2. *Infectious agents are often highly adaptable.* Many strains of microorganisms have developed resistance to several of the available antibiotics. The use of antibiotics has prevented so many deaths. However, the misuse and/or the overuse of antibiotics through the years has contributed to the development of mutant drug-resistant bacterial strains. Treating diseases caused by such microbes presents a challenge that will not disappear or be solved quickly.

3. *Previously unknown or rare diseases become significant as a result of changes in human activities and/or social conditions.* The epidemic of legionellosis that marred the festivities of a Bicentennial celebration in 1976 was eventually found to be caused by a mi-

croorganism that was not commonly known and had existed and occasionally caused disease in the past. However, this time it was spread through a hotel air-conditioning system—something that could not have happened before air conditioning was invented. In the early 1980s many cases of toxic shock syndrome (TSS) suddenly appeared. This disease was shown to be caused by a staphylococcal toxin that usually reached the blood from organisms growing in certain rough, high-absorbency tampons used by women during menstruation. TSS was very rare before such tampons were invented; since then they have been modified by the manufacturer so that they no longer pose a health threat.

4. *Immigration and international travel and commerce introduce new or recurrent strains of the pathogen.* The high influx of legal and illegal immigrants often brings specific disease, such as tuberculosis, into areas where the disease had been eradicated. And the ease of international airline travel helps reintroduce previously eradicated diseases.

Earlier in the twentieth century, microbiologists and public health officials thought that the use of antibiotics and vaccines would eliminate infectious disease. Although this dream has been realized for smallpox and probably will be true for polio shortly after the turn of the century, many infectious diseases remain a serious health threat. Tuberculosis and cholera represent diseases again on the rise, mostly due to antibiotic resistance and the absence of sanitary control measures. Other diseases have arisen from new infectious agents; AIDS, caused by HIV, is perhaps the most prominent. Some scientists believe that HIV arose as a mutant strain in another animal species—an African monkey—and has "jumped species" to humans. Such jumps have probably occurred many times before, and with limited mobility of the human population, the virus would have disappeared without escaping into the rest of the world. However, the infection is slow to develop, and so unsuspecting carriers of the disease now take it with them from one part of the world to another, thanks to the ease of international air travel.

In 1992 the Institute of Medicine stated that emerging infectious diseases in the United States should be taken seriously and that public health agencies were not prepared to deal with possible epidemics. In fact, AIDS might be the best example of our unpreparedness—and an indication of other diseases yet to come.

Retracing Our Steps

- A **pathogen** is a parasite capable of causing disease.

Host–Microbe Relationships
- A **host** is an organism that harbors another organism.

Symbiosis
- **Symbiosis** means "living together" and includes **commensalism,** in which one organism benefits and the other neither benefits nor is harmed; **mutualism,** in which both organisms benefit; and **parasitism,** in which one organism (the **parasite**) benefits and the other (the **host**) is harmed.

Contamination, Infection, and Disease
- **Contamination** refers to the presence of microorganisms. In **infection,** pathogens invade the body; in **disease,** pathogens or other factors disturb the state of health such that the body cannot perform its normal functions. **Infestation** refers to the presence of worms or arthropods in or on the body.

Pathogens, Pathogenicity, and Virulence
- **Pathogenicity** is the capacity of a pathogen to produce disease. **Virulence** is the intensity of a disease caused by a pathogen. **Attenuation** is weakening of a pathogen's disease-producing capacity.

Normal (Indigenous) Microflora
- **Normal microflora** (normal flora) are microorganisms found in or on the body that do not normally cause disease.
- **Resident microflora** are those organisms that are always present on or in the body; **transient microflora** are those present temporarily and under certain conditions. **Opportunists** are resident or transient microflora that can cause disease under certain conditions or in certain locations in the body.

Koch's Postulates
- **Koch's postulates** provide a way to link a pathogen with a disease:
  1. A specific causative agent must be observed in every case of a disease.
  2. The agent must be isolated from a host displaying the disease and grown in pure culture.
  3. When the agent from the pure culture is inoculated into an experimental healthy, susceptible host, the agent must cause the disease.
  4. The agent must be reisolated from the inoculated, diseased experimental host and identified as the original specific causative agent.

- When Koch's postulates are met, an organism has been proved to be the causative agent of an infectious disease.

Kinds of Diseases

Infectious and Noninfectious Diseases

- **Infectious diseases** are caused by infectious agents; **noninfectious diseases** are caused by other factors.

Classification of Diseases

- Although many diseases are caused by noninfectious agents, some of these diseases may be associated with an infectious agent.

Communicable and Noncommunicable Diseases

- A **communicable,** or **contagious, infectious disease** can be spread from one host to another. A **noncommunicable infectious disease** cannot be spread from host to host and may be acquired from soil, water, or contaminated foods.

The Disease Process

How Microbes Cause Disease

- Many microbes have **virulence factors** that enable the establishment of infections. These factors include adhesion molecules, enzymes, and toxins.
- Bacteria cause disease by **adhering** to a host, by **colonizing** and/or invading host tissues, and sometimes by invading cells. The ability of a pathogen to invade and grow in host tissues, called **invasiveness,** is related to particular virulence factors. **Hyaluronidase** helps bacteria invade tissues.
- Bacteria release other substances, most of which damage host tissues. **Hemolysins** lyse red blood cells in cultures and may or may not directly cause tissue damage in the host. **Leukocidins** destroy neutrophils. **Coagulase** accelerates blood clotting. **Streptokinase** digests blood clots and helps pathogens spread to body tissues.
- Many bacteria also produce **toxins. Endotoxins** are part of the cell wall of Gram-negative bacteria and are released when cells divide or are killed. **Exotoxins** are produced by and released from bacteria; they are called **neurotoxins** if they affect the nervous system and **enterotoxins** if they affect the digestive system. **Toxoids** are inactivated exotoxins that retain antigenic properties and are used for immunization.

- Viruses damage cells and produce a variety of observable changes called the **cytopathic effect (CPE).** A **productive infection** leads to the release of virus progeny, whereas an **abortive infection** does not produce infectious progeny.

- Pathogenic fungi can invade and progressively digest cells, and some produce toxins.

- Protozoa and helminths damage tissues by ingesting cells and tissue fluids, releasing toxic wastes, and causing allergic reactions.

Signs, Symptoms, and Syndromes

- A **sign** is an observable effect of a disease. A **symptom** is an effect of a disease felt by the infected person. A **syndrome** is a group of signs and symptoms that occur together.

Types of Infectious Disease

- Terms used to describe types of diseases are defined in Table 15.7.

Stages of an Infectious Disease

- The **incubation period** is the time between infection and the appearance of signs and symptoms of a disease.

- The **prodromal phase** is the stage during which pathogens begin to invade tissues; it is marked by early nonspecific symptoms.

- The **invasive phase** is the period during which the individual experiences the typical signs and symptoms of the disease. During this phase the signs and symptoms reach their greatest intensity at the **acme.**

- The **decline phase** is the stage during which host defenses overcome pathogens; signs and symptoms subside during this phase, and secondary infections may occur.

- The **convalescence period** is the stage during which tissue damage is repaired and the patient regains strength. Recovering individuals may still transmit pathogens to others.

Terminology Check

pathogen (*p. 375*)
pathogenicity (*p. 377*)
persistent viral infection (*p. 391*)
primary infection (*p. 394*)
prodromal phase (*p. 395*)
prodrome (*p. 395*)
productive infection (*p. 391*)
pyrogen (*p. 396*)
resident microflora (*p. 378*)
sapremia (*p. 394*)

secondary infection (*p. 394*)
septicemia (*p. 393*)
sequela (*p. 392*)
sign (*p. 392*)
streptokinase (*p. 386*)
subacute disease (*p. 393*)
subclinical infection (*p. 394*)
superinfection (*p. 394*)
symbiosis (*p. 376*)
symptom (*p. 392*)

syndrome (*p. 392*)
systemic infection (*p. 393*)
toxemia (*p. 388*)
toxin (*p. 387*)
toxoid (*p. 389*)
transient microflora (*p. 379*)
transposal of virulence (*p. 378*)
viremia (*p. 393*)
virulence (*p. 378*)
virulence factor (*p. 384*)

See You on the Web

Don't forget I am expecting you on the web. The following is a preparedness checklist for your web explorations. If you have questions on any of these items, you can return to the end of Chapter 1 for more details.

- Questions We'll Explore
- Concept Check questions
- Flashcards
- Study Guide

If you are ready, come visit me at

http://www.prenhall.com/black

It will be a journey you'll never forget.

Epidemiology and Nosocomial Infections

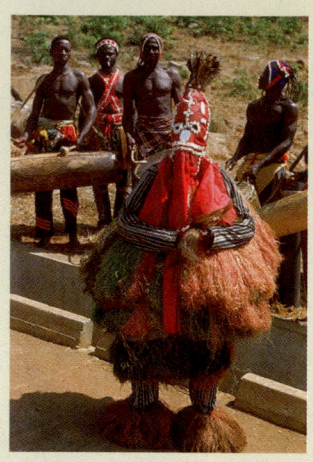

What do you know about witch doctors beyond Hollywood's outdated imagery? The photo here shows a Lassa spiritual healer from Sierra Leone in Africa. The centuries of knowledge that these and other healers hold regarding local diseases and use of local plants to treat these diseases is in danger of dying out. They have the distilled knowledge of countless years of treating illness, and in recent years more Western scientists are becoming apprenticed to these healers to help save this information. Many pre-industrial peoples attributed diseases to "evil spirits" or "bad air"—an approach that differs from that of industrial societies. However, often their keen observations of nature give the clues needed to find the source of a disease. For example, Navajo medicine men noted an unusually high production of piñon pine nuts the year of the hantavirus outbreak in the desert Southwest. Rodents feeding on the nuts spread the virus in their urine.

Background Basics

antibiotic resistance . . . . *(Chapter 13)*

U p to this point in our study of host–microbe interactions, we have examined the characteristics of pathogens that lead to infectious diseases, and we have seen how the disease process occurs in individuals. But people with infectious diseases are members of a population; they acquire infectious diseases and transmit them within a population. Therefore, to further our understanding of such diseases, we must consider their effects on populations, including hospital populations.

Epidemiology

What Is Epidemiology?

Epidemiology (ep″i-de-me-ol′o-je) is the study of factors and mechanisms involved in the frequency and spread of diseases and other health-related problems within populations of humans, other animals, or plants. The term is derived from the Greek words *epidemios,* meaning "among the people," and *logos,* meaning "study." Although **epidemiologists,** scientists who study epidemiology, may consider such health-related problems as automobile accidents, lead poisoning, or cigarette smoking, we will limit our discussion to factors and mechanisms that concern the transmission of infectious diseases as well as their **etiology** (e-te-ol′o-je), or cause, in a population. We will also see how epidemiologists use this information to design ways to control and prevent the spread of infectious agents.

Epidemiology is a branch of microbiology because many diseases that concern epidemiologists, such as AIDS, tuberculosis, and malaria, are caused by microbes. Parasitic worm diseases, such as hookworm disease and ascariasis, also are studied by epidemiologists. Epidemiology includes relationships among pathogens, their hosts, and the environment. It is related to public health because it provides information and methods used to understand and control the spread of disease within the human population. Agricultural and environmental scientists often are concerned with the epidemiology of animal and plant diseases. When such diseases can be transmitted to humans, they, too, become a public health problem.

In following diseases and their spread, epidemiologists are especially interested in the frequencies of diseases within populations.

Questions We'll Explore

A What is epidemiology, and what special terms are used by epidemiologists?

B How are diseases classified according to their spread in populations?

C What are the purposes and methods of epidemiologic studies?

D How do various kinds of reservoirs of infection contribute to human disease?

E What are the roles of portals of entry and exit and modes of transmission in the spread of human disease?

F What is an infectious disease cycle, and how is herd immunity related to disease cycles?

G What methods are used to control disease transmission?

H How do the functions of public health organizations and the reporting of notifiable diseases contribute to public health?

I What are nosocomial infections, and how are they studied epidemiologically?

J How can nosocomial infections be prevented and controlled?

Incidence and Prevalence Rates

The **incidence** of a disease is the number of *new* cases contracted within a set population during a specific period of time (usually expressed as new cases per 100,000 people per year). The **prevalence** of a disease is the *total number* of people infected within the population at any time. Prevalence includes both old and newly diagnosed cases. For example, if epidemiologists conduct weekly surveys regarding a disease that lasts four weeks, an individual infected in week one could be counted as many as four times in a prevalence study but only once in an incidence survey. Thus, incidence data are reliable indicators of the spread of a disease—a drop in incidence suggests a reduction in the spread of the disease. In contrast, prevalence data measure how seriously and how long the disease is affecting a population (➤ Figure 15.1).

Morbidity and Mortality Frequencies

Frequencies also are expressed as proportions of the total population. The **morbidity rate** is the number of individuals affected by a disease during a set period in relation to the total number in the population. It usually is expressed as the number of cases per 100,000 people per year. The **mortality rate** is the number of deaths due to a disease in a population during a specific period in relation to the total population. It is expressed as deaths per 100,000 people per year.

Diseases in Populations

When studying the frequency of diseases in populations, epidemiologists must consider the geographic areas affected and the degree of harm the diseases cause in the population. On the basis of their findings, they classify diseases as *endemic, epidemic, pandemic,* or *sporadic.*

An infectious disease agent is **endemic** if it is present continually in the population of a particular geographic area but both the number of reported cases and the severity of the disease remain too low to constitute a public health problem. For example, mumps is endemic in the entire United States, and valley fever is endemic in the southwestern United States. Chickenpox is an endemic disease with seasonal variation; that is, many more cases are seen from late winter through spring than at other times (➤ Figure 15.2). Endemic diseases also can vary in incidence in different parts of the endemic region.

An **epidemic** arises when a disease suddenly has a higher-than-normal incidence in a population. Then the morbidity rate or the mortality rate or both become high enough to pose a public health problem. Endemic diseases can give rise to epidemics, especially when a particularly virulent strain of a pathogen appears or when most of a population lacks immunity. For example, St. Louis encephalitis, a viral inflammation of the brain, reached epidemic proportions in the United States in 1975 (➤ Figure 15.3). It arose due to the presence of both a large nonimmune bird population that carried the virus and a large mosquito population that transferred the virus from birds to humans.

With the breakup of the former Soviet Union into the New Independent States (NIS), epidemic diphtheria has emerged. Since 1990, overcrowding due to population migrations has resulted in a dramatic increase in the diphtheria incidence rate (➤ Figure 15.4). In 1994, 47,802

> *Over 20 million deaths per year are due to infectious disease.*

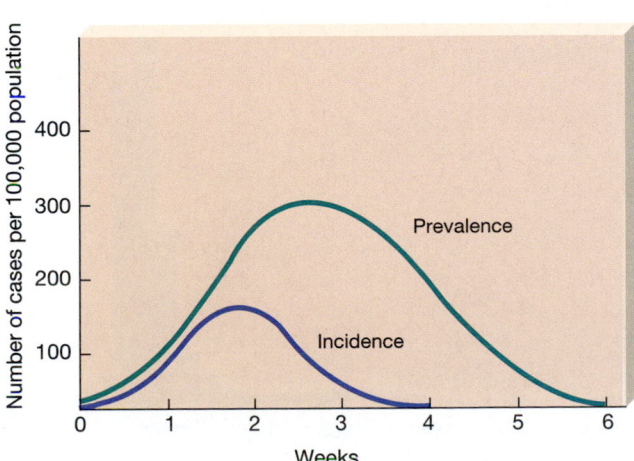

➤ **Figure 15.1   Incidence and prevalence rates.** A hypothetical disease lasting 4 weeks illustrates the difference between the incidence and prevalence numbers.

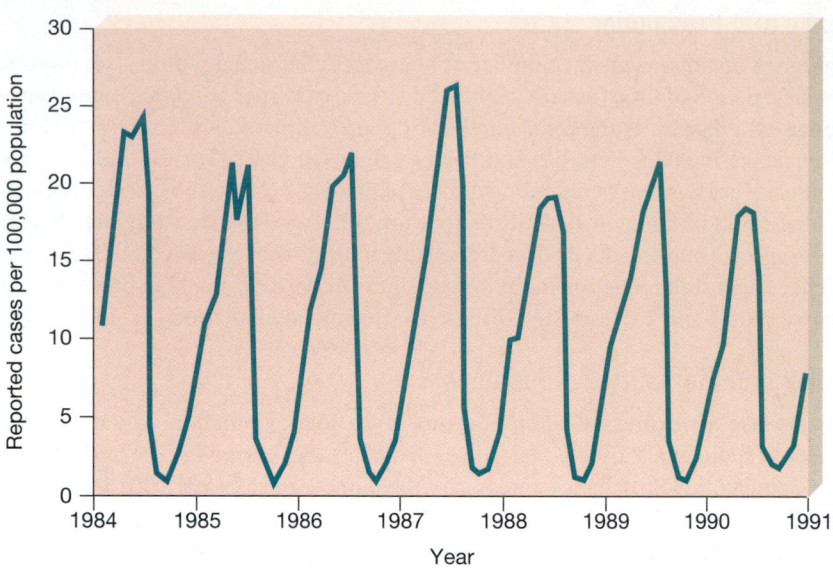

➤ Figure 15.2 **The incidence rate of chickenpox in the United States.** Chickenpox, an endemic disease, shows obvious seasonal variation, with most cases occurring during the spring. (Chickenpox is no longer reportable to the CDC, so current data are not available.)

➤ Figure 15.3 **The incidence rate of St. Louis encephalitis in the United States.** This viral disease showed a major outbreak during late 1975.

➤ Figure 15.4 **Diphtheria cases in the New Independent States, 1965 to 1994.** The number of reported cases increased from 1991 to 1994 because of insufficient health precautions as a result of the breakup of the Soviet Union. (*Source*: World Health Organization)

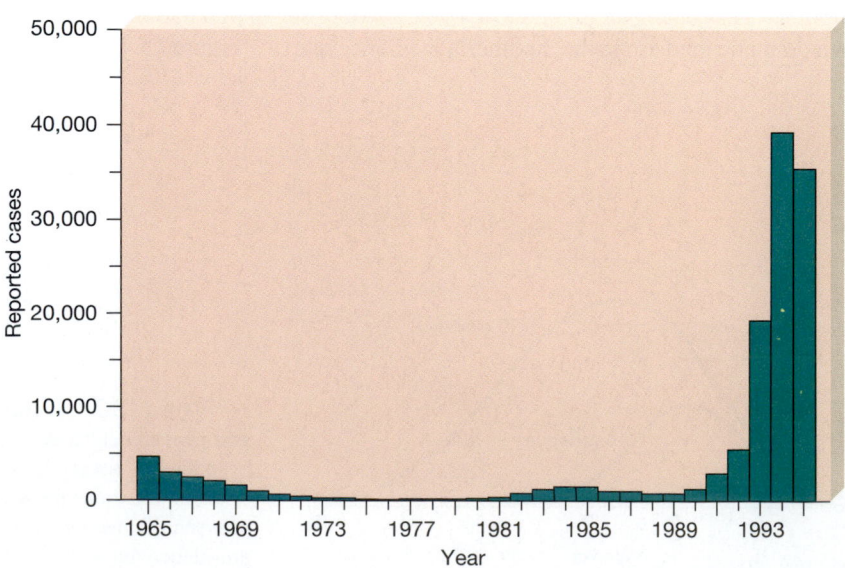

cases were reported, with 1746 deaths. There was also a cultural reluctance to immunize children who were not healthy and robust, as it was considered dangerous for ill or weak children to receive vaccine. World health officials considered the expanding diphtheria epidemic an international health emergency. Diphtheria cases arising in the NIS have been reported in eastern and northern Europe. Health officials outlined a strategy in early 1995 to ensure that all NIS children are vaccinated against diphtheria. Since then, case rates have begun dropping.

A **pandemic** occurs when an epidemic spreads worldwide. In 1918 the swine flu reached pandemic proportions (as we shall see in Chapter 21). Cholera has been responsible for seven pandemics over the centuries. Its spread through the Americas during the epidemic of 1991–1992 is shown in ➤ Figure 15.5.

A **sporadic disease** occurs in a random and unpredictable manner, involving several isolated cases that pose no great threat to the population as a whole. Eastern equine encephalitis (EEE) is a sporadic disease in the Americas. ➤ Figure 15.6 contrasts the sporadic nature of EEE with endemic California encephalitis (CE) in certain parts of the United States and epidemic western equine encephalitis (WEE) in the Americas.

The nature and spread of epidemics can vary according to the source of the pathogen and how it reaches susceptible hosts. A **common-source outbreak** is an epidemic that arises from contact with contaminated substances. The term *outbreak* does not evoke the fear that the word *epidemic* can. The spread of a common-source outbreak can typically be traced to a water supply contaminated with fecal material or to improperly handled food. Many individuals become ill quite suddenly. For example, on a cruise in 1994 from San Pedro, California, to Ensenada, Mexico, 586 of 1589 passengers acquired gastrointestinal illness due to *Shigella flexneri*. Such outbreaks subside quickly once the source of infection is eradicated.

A **propagated epidemic** arises from direct person-to-person contacts (horizontal transmission). The pathogen moves from infected people to uninfected but susceptible individuals. In a propagated epidemic, the number of cases rises and falls more slowly, and the pathogen is more difficult to eliminate, than in a common-source outbreak. Differences between common-source outbreaks and propagated epidemics are illustrated in ➤ Figure 15.7.

Epidemiologic Studies

Collecting frequency data and drawing conclusions is the foundation of any **epidemiologic study.** The British physician John Snow made what may have been the first epidemiologic study. In 1854 Snow investigated the cause of a cholera epidemic then sweeping through London.

City Tales

"Hamburg persisted in postponing costly improvements to its water supply. . . . [It] drew its water from the Elbe without special treatment. Adjacent lay the town of Altona . . . where a solicitous government installed a water filtration plant. In 1892, when cholera broke out in Hamburg, it ran down one side of the street dividing the two cities and spared the other completely. . . . A more clearcut demonstration of the importance of the water supply in defining where the disease struck could not have been devised. Doubters were silenced; and cholera has, in fact, never returned to European cities, thanks to systematic purification of urban water supplies from bacteriological contamination."

—William H. McNeill, 1976

★ Initial epidemics
January 1991

·········· August 1991

— — February 1992

———— December 1992

➤ Figure 15.5 **The spread of cholera.** Cholera spread through South and Central America during the period 1991–1992, beginning in Peru in January 1991. It then moved into Colombia and Ecuador. By late 1992 the epidemic had spread to Venezuela, Bolivia, Chile, and Brazil in South America and to Guatemala, Honduras, Panama, Nicaragua, and El Salvador in Central America. (Map from *Morbidity and Mortality Weekly Record.*)

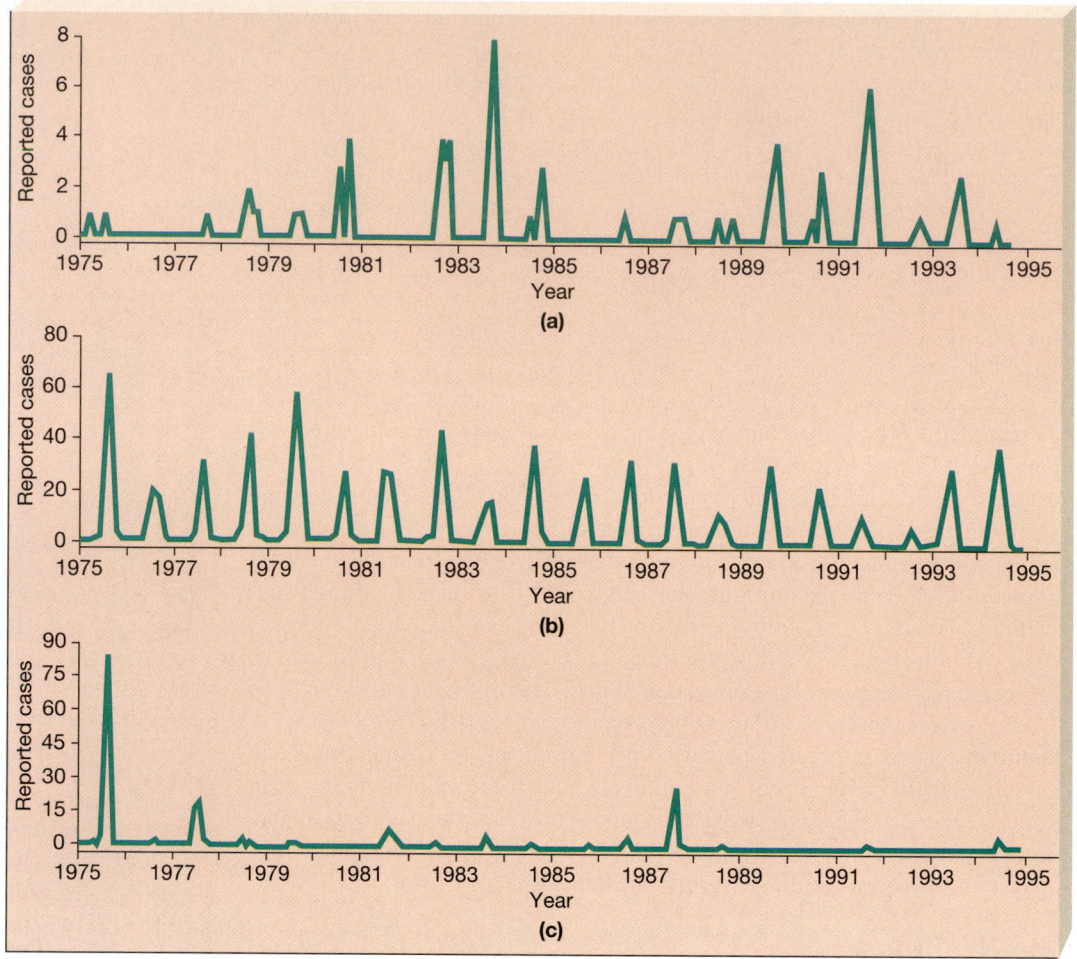

➤ Figure 15.6   **Incidence rates of three different types of encephalitis.** (a) The sporadic pattern
of eastern equine encephalitis is compared with the (b) endemic pattern of California encephalitis and
(c) epidemic outbreaks of western equine encephalitis.

He eventually traced the source of the epidemic to the Broad Street pump in Golden Square
(➤ Figure 15.8). He proved that people became infected by drinking water contaminated with
human feces.

Since Snow's landmark study, many other investigators have conducted epidemiologic studies to learn more about the spread of disease in populations. Such studies can be *descriptive,
analytical,* or *experimental.*

➤ Figure 15.7   **Differences in
incidence patterns of common-source
outbreaks and propagated epidemics.** In
common-source outbreaks, all cases occur
within a fairly short time period after
exposure to the single source and then
stop, whereas in propagated epidemics,
new cases are continually seen.

Pumps   Deaths from Cholera

50   0   50   100   150   200

Yards

> Figure 15.8   **The first epidemiologic study.** In 1854, John Snow, a London physician, recorded the locations of cholera cases in the city. He found that they were clustered around the Broad Street pump, which supplied water to the nearby area. He tracked the epidemic's source to contamination of the pump by sewage. When the handle of the pump was removed, the outbreak subsided. Snow's study was the first modern, systematic, scientific epidemiologic study.

Descriptive Studies   A **descriptive study** is concerned with the physical aspects of an existing disease and disease spread. Such a study records (1) the number of cases of a disease, (2) those segments of the population that were affected, and (3) the locations and time period of the cases. The age, gender, race, marital status, socioeconomic status, and occupation of each patient also are recorded. From careful analysis of data accumulated from several studies, epidemiologists can determine whether people of a certain age group, males or females, or members of a certain race are particularly susceptible to the disease. Data on marital status or sexual behavior can help show whether the disease is transmitted sexually. Data on socioeconomic status might show that a disease is most easily transmitted among the undernourished or among those living in substandard conditions. Noting the occupations of infected individuals can help epidemiologists trace diseases to certain factories, slaughterhouses, or hide-processing plants. If most cases are among veterinarians, for example, the disease probably is transmitted by the animals they handle.

The geographic distribution of cases is also important, as Snow's study showed. Some of Snow's modern counterparts have traced disease outbreaks to contaminated water supplies, to restaurants where workers are infected with viral hepatitis, and to areas where particular infectious agents thrive.

Multiple sclerosis (MS) is a disease in which one's own immune system attacks the myelin sheath surrounding nerve cells in the spinal cord and brain. It eventually leads to a loss of muscle control and paralysis. Epidemiologically, most cases of MS are found in clusters, in temperate climates, and among Caucasians. In the United States, the incidence of MS increases significantly northward from Mississippi to Minnesota. However, an important determining factor seems to be where a person spent the first 14 years of life. For example, for an individual who grew up in a northern region with a high incidence of MS, moving south later would not reduce the likelihood of acquiring the disease. Various studies have led some investigators to propose that an infectious virus may contribute to the development of MS. Studies of the

Faeroe Island, north of Great Britain, showed them to be free of MS until British troops were stationed there during World War II. Since then, there have been several cycles of MS cases. Only people living on those islands where troops lived have developed MS, with one exception—a case in the house next to the dock where British troops changed ships, but never lived on that island. Thus, genetic factors coupled with an infection by one or more specific microbes increases the likelihood that a person will acquire MS later in life. (Chapter 18 discusses MS and other autoimmune disorders.)

Finally, the time period over which cases appear and the season of the year are important considerations in descriptive studies. To study the role of time in an epidemic, epidemiologists define an **index case** as the first case of the disease to be identified. As we have seen, common-source outbreaks can be distinguished from propagated epidemics by the rate at which the number of cases increases and the time required for the epidemic to subside. The season of the year in which epidemics occur may help identify the causative agent. Arthropodborne infections usually occur in relatively warm weather, and certain respiratory infections usually occur in cold weather, when people often are crowded together indoors. The seasonal nature of encephalitis, which is most prevalent in the fall, is shown in ➤ Figure 15.6.

Analytical Studies An **analytical study** focuses on establishing cause-and-effect relationships in the occurrence of diseases in populations. Such studies can be retrospective or prospective. A *retrospective* study takes into account factors that preceded an epidemic. For example, the investigator might ask patients where they had been and what they had done in the month or so before they became ill. The patients are then compared with a *control group*—individuals in the same population who are unaffected by the disease. Thus, if most patients had hiked in a certain wooded area, had contact with horses, or shared another common activity in which the control group did not participate, that activity might provide a clue to the source of the infection. Several investigations of this sort are described in Berton Rouché's fascinating book *The Medical Detectives* (New York: Truman Talley Books/Plume).

A *prospective* study considers factors that occur as an epidemic spreads. Which children in a population get chickenpox, at what age, and under what living conditions, for instance, are factors used to determine susceptibility and resistance to infection. As the 1993 hantavirus outbreak spread in the U.S. Southwest, epidemiologists had to determine what was causing the disease and how to change living conditions so that the infectious agent would stop spreading.

Experimental Studies An **experimental study** designs experiments to test a hypothesis, often about the value of a particular treatment. Such studies are limited to animals or to humans in which participants are not subjected to harm. For example, an investigator might test the hypothesis that a particular treatment will be effective in controlling a disease for which no accepted cure is available. One group (the experimental group) from a population receives the treatment, and another group (the control group) receives a placebo. A **placebo** (pla-se′bo) is a nonmedicinal substance that has no effect on the recipient but that the recipient *believes* is a treatment. From the results of the study, the investigator could learn whether the new treatment was effective.

However, in some circumstances control groups might not be used. In the early days of AIDS research, drug treatments were carried out without a control group because the U.S. government considered it unethical to allow control subjects to be deprived of potentially lifesaving treatment.

Infectious diseases threaten human populations only when the diseases can be spread or transmitted. Factors important to the spread of disease or of infectious agents include (1) reservoirs of infection, (2) portals by which organisms enter and leave the body, and (3) mechanisms of transmission. Let's look at each of these factors in more detail.

✓ Distinguish between morbidity and mortality; between incidence and prevalence; between endemic, epidemic, and pandemic.

✓ How do common-source outbreaks differ from propagated epidemics?

Reservoirs of Infection

Most pathogens infecting humans cannot survive outside the body of a host long enough to serve as a source of infection. Therefore, sites in which organisms can persist and maintain their ability to infect are essential for new human infections to occur. Such sites are called **reservoirs of infection.** Examples are humans, other animals (including insects), plants, and certain nonliving materials, such as water and soil.

Human Reservoirs

Humans with active infections are important reservoirs because they can easily transmit organisms to other humans. **Carriers,** individuals who harbor an infectious agent without having any observable clinical signs or symptoms, also are important reservoirs. Thus, a disease may be transmitted by a person (or animal) with a **subclinical,** or **inapparent, infection**—an infection with signs and symptoms too mild to be recognized, except by special tests. Many cases of whooping cough (pertussis) in adults, for example, are never diagnosed; yet such *"healthy" carriers* harbor and transmit infectious agents. Infectious diseases are said to be *communicable* if they can be transmitted during the incubation period (before symptoms are apparent) and during recovery from the disease. A *chronic carrier* is a reservoir of infection for a long time after he or she has recovered from a disease. An *intermittent carrier* periodically releases organisms.

Depending on the disease, carriers can discharge organisms from the mouth or nose, in urine, or in feces. Diseases commonly spread by carriers include diphtheria, typhoid fever, amoebic and bacillary dysenteries, hepatitis, streptococcal infections, and pneumonia. Our cholera example shows how international jet travel can increase the risk of introducing infectious agents from reservoirs in one region to populations visiting from another region.

Gonorrhea, AIDS, hepatitis, polio, and typhoid fever are examples of diseases that are spread by human carriers.

Animal Reservoirs

About 150 pathogenic microorganisms can infect both humans and some other animals. In such instances, the animals can serve as reservoirs of infection for humans. Animals that are physiologically similar to humans are most likely to serve as reservoirs for human infections. Therefore, monkeys are important reservoirs for malaria, yellow fever, and many other human infections. Once humans are infected, they, too, can serve as reservoirs for the infections.

Diseases that can be transmitted under natural conditions to humans from other vertebrate animals are called **zoonoses** (zo-o-no′ses; singular: *zoonosis*). Selected zoonoses are summarized in Table 15.1. Of these diseases, rabies is perhaps the greatest threat in the United States because of the severity of the disease and because both domestic pets and wild animals can serve as reservoirs for the rabies virus. Where vaccination of dogs and cats is widespread, humans are more likely to acquire rabies from wild animals in which the disease is endemic, such as skunks, raccoons, bats, and foxes.

The larger the animal reservoir, both in the number of species and in the total number of susceptible animals, the more unlikely it is that a disease can be eradicated. This is especially true if the reservoir contains wild animals in which a disease is epidemic. It is impossible to find all the infected animals and to control the disease among them. Even today *Yersinia pestis,* the bacterium responsible for the plague, persists among gophers, ground squirrels, and other wild rodents in the American West and occasionally causes human cases.

Humans, their pets, and other domestic animals similarly serve as reservoirs of infection for wild animals. Distemper, an infectious viral disease in dogs, has spread to and killed many black-footed ferrets. This animal, already an endangered species, is now in even greater jeopardy. Thus, pets and domestic animals should not be allowed to enter wildlife refuges.

Nonliving Reservoirs

Soil and water can serve as reservoirs for pathogens. Soil, for example, is the natural environment of several bacterial species. *Clostridium tetani* (the cause of tetanus) and *C. botulinum* (the cause of botulism) are found everywhere, but especially where animal fecal matter is used as fertilizer. They are part of the normal intestinal microflora of cattle, horses, and some humans. Many fungi, including the organism that causes valley fever, also are common soil inhabitants. Often soil fungi can invade human tissues and cause ringworm, other skin diseases, or systemic infections. Water contaminated by human or animal feces can contain a variety of pathogens. Most of these cause gastrointestinal diseases. Improperly prepared or stored food also can serve as a temporary nonliving reservoir of disease. Poorly cooked contaminated meats can be a source of infection with *Salmonella* species and a variety of helminths. Failure to refrigerate foods can lead to growth of microorganisms and to the production of toxins that cause food poisoning. Even with proper refrigeration, helminth larvae remain infectious unless the foods that contain them are cooked thoroughly.

What's in Dust

Household dust typically contains an amazing assortment of microbes, spores, a few larger organisms, dandruff and other debris from the human body, and other nonliving materials. Spores of *Clostridium perfringens,* which causes gas gangrene in deep wounds, have been found in air conditioner filters. Many different genera of fungi, including *Penicillium, Rhizopus* (bread mold), and *Aspergillus,* which can cause swimmer's ear, have been found in dust. Bacterial endospores, fungal spores, plant pollens, insect parts, and hundreds of mites (small organisms related to spiders; see photo) also abound in dust. Mites feed on skin particles sloughed from our body surfaces; fortunately they do not feed on living skin. Dust also contains large quantities of human and animal hair and sometimes nail clippings as well, all of which have microorganisms on their surfaces. Finally, dust contains nonliving particles from sources as diverse as flaking paint and meteorites.

➤ Table 15.1   Selected zoonoses (with emphasis on those that occur in pets)

Disease	Animals infected	Modes of transmission
Bacterial diseases		
Avian tuberculosis	Birds	Respiratory aerosols
Anthrax	Domestic animals, including dogs and cats	Direct contact with animals, contaminated soil, and hides; ingestion of contaminated milk or meat; inhalation of spores
Brucellosis (undulant fever)	Domestic animals	Direct contact with infected tissues, ingestion of milk from infected animals
Bubonic plague	Rodents	Fleas
Leptospirosis	Primarily dogs; also pigs, cows, sheep, rodents, and other wild animals	Direct contact with urine, infected tissues, and contaminated water
Psittacosis	Parrots, parakeets, and other birds	Respiratory aerosols
Relapsing fever	Rodents	Ticks and lice
Rocky Mountain spotted fever	Dogs, rodents, and other wild animals	Ticks
Salmonellosis	Dogs, cats, poultry, turtles, and rats	Ingestion of contaminated food or water
Viral diseases		
Equine encephalitis (several varieties)	Horses, birds, and other domestic animals	Mosquitoes
Rabies	Dogs, cats, bats, skunks, and wolves	Bites, infectious saliva in wounds, aerosols
Lassa fever, hantavirus pulmonary syndrome, hemorrhagic fevers	Rodents	Urine
Fungal diseases		
Histoplasmosis	Birds	Aerosols of dried infected feces
Ringworm (several varieties)	Cats, dogs, and other domestic animals	Direct contact
Parasitic diseases		
African sleeping sickness	Wild game animals	Tsetse flies
Tapeworms	Cattle, swine, rodents	Ingestion of cysts in meat or via proglottids in feces
Toxoplasmosis	Cats, birds, rodents, and domestic animals	Aerosols, contaminated food and water, and placental transfer

Portals of Entry

To cause an infection, a microorganism must enter body tissues. The sites at which microorganisms can enter the body are called **portals of entry.** Common portals of entry include the skin and the mucous membranes of the digestive, respiratory, and urogenital systems (➤ Figure 15.9). Although intact skin usually prevents the entry of microorganisms, some enter through the ducts of sweat glands, mammary glands, or through hair follicles. Some fungi invade cells on the skin's surface, and a few can pass on to other tissues. Larvae of some parasitic worms such as the hookworm can bore through the skin to enter other tissues.

Openings to the outside of the body, such as the ears, nose, mouth, eyes, anus, urethra, and vagina, allow microbes to enter. Organisms that infect the respiratory system typically enter in inhaled air, on dust particles, or in airborne droplets. Those that infect the digestive system typically enter in food or water but can also enter from contaminated fingers. As a result of fluid discharge from mucous membranes, sexual intercourse provides a portal of entry for urogenital system infections. Some infectious microbes also travel through the urethra and vagina from skin surfaces.

Sometimes microorganisms are introduced directly into the damaged tissues created by bites, burns, injections, and accidental or surgical wounds. *Pseudomonas aeruginosa* infections are especially common in hospitals among burn and surgical patients. Insect bites are a common portal of entry for a variety of parasitic protozoan and helminthic diseases carried by insects.

Finally, a few infectious organisms, mostly viruses, can cross the placenta from an infected mother and cause an infection in the fetus. Congenital infectious diseases such as

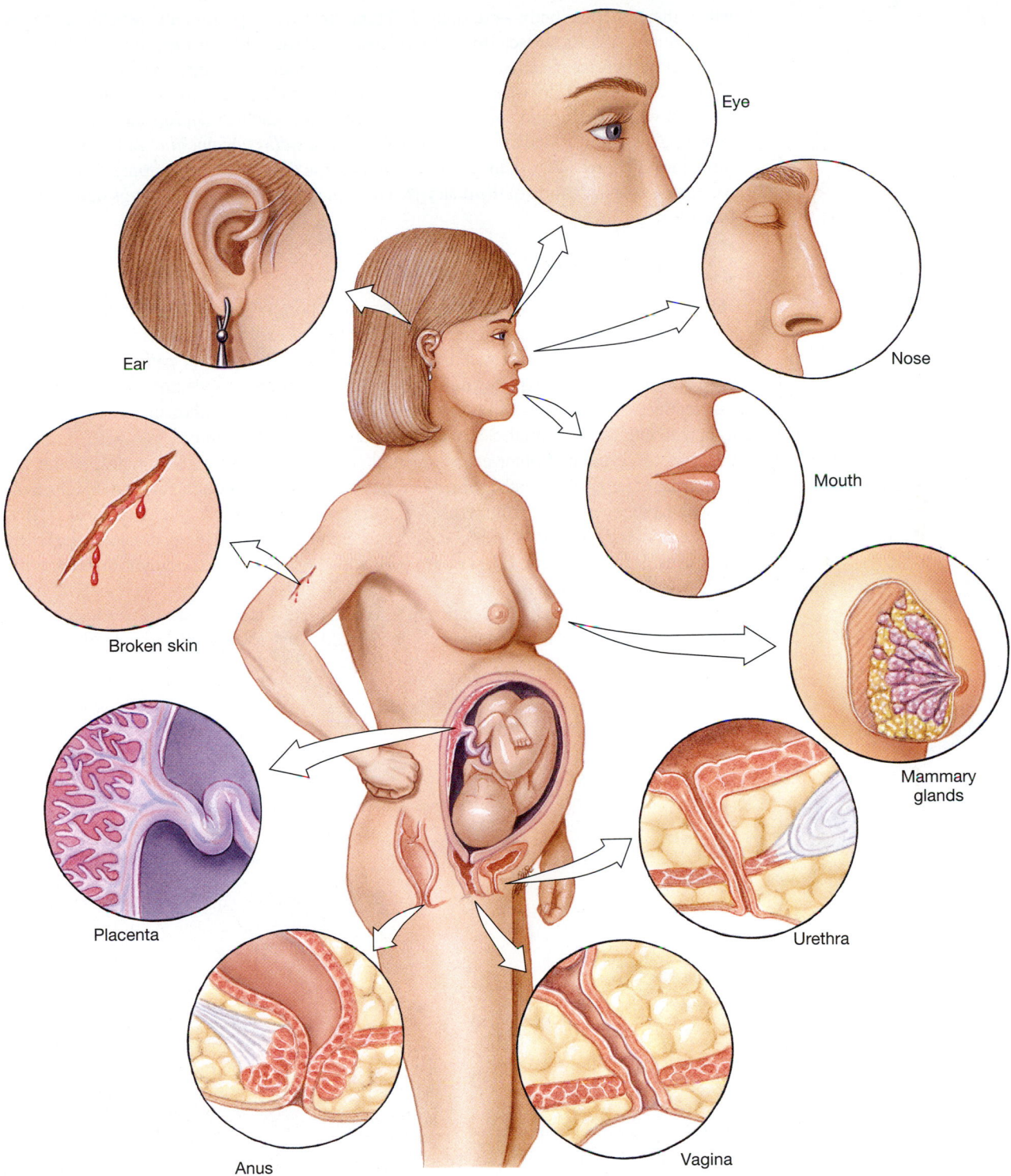

➤ Figure 15.9   **Portals of entry for human pathogens.** Besides those portals common to females and males, additional portals in females include the placenta and the milk ducts of the mammary glands.

cytomegalovirus infection, toxoplasmosis, syphilis, AIDS, and rubella (German measles) are acquired in this way.

Some organisms can enter the body through only a single portal. Others can enter through any of several portals, and their pathogenicity may depend on the portal of entry. For example, many pathogens that cause diseases of the digestive tract cause no disease at all if they happen to enter the respiratory tract. Similarly, most pathogens that cause respiratory diseases do not infect the skin or the tissues of the digestive tract. However, a few organisms cause illnesses no matter where they enter the body—but quite different illnesses, depending on which portal they enter. For example, the plague bacterium (*Yersinia pestis*) causes *bubonic plague,* which has a mortality of about 50 percent if untreated, and is acquired through a flea bite. But when inhaled into the lungs, the same organism causes *pneumonic plague,* which has a mortality of close to 100 percent.

Even when a pathogen enters the body, it may not reach an appropriate site to cause an infection. Almost everyone unknowingly carries *Klebsiella pneumoniae* in the pharynx at some time during winter. Why, then, do so few of us come down with pneumonia? The reason is that we do not get pneumonia of the pharynx. The organism must enter the lungs before it can cause disease.

Portals of Exit

How infectious agents get out of their hosts is important to the spread of diseases. The sites where organisms leave the body are called **portals of exit** (➤ Figure 15.10).

Generally, pathogens exit with body fluids or feces. Respiratory pathogens exit through the nose or mouth in fluids expelled during coughing, sneezing, or speaking. Saliva from dogs, cats, insects, and other animals can transmit infectious organisms. Pathogens of the gastrointestinal tract exit with fecal material. Some of these pathogens are helminth eggs, which are exceedingly resistant to drying and other environmental conditions. Urine and, in males, semen from the urethra carry urogenital pathogens. Semen is an important, though sometimes ignored, means by which pathogens, especially viruses, exit the body. For example, the AIDS virus (HIV) can be carried from the body by white blood cells and sperm present in semen, as can the hepatitis B and C viruses. Hepatitis B and C have now been added to the list of sexually transmitted diseases.

Blood from patients also sometimes contains infectious organisms, such as HIV or hepatitis viruses. As such, blood can be a source of infection for health care workers or others rendering aid to an injured person. Another person's blood should always be considered potentially infectious.

> Breast milk is a significant source of HIV infection in infants born to infected mothers.

> Pasteurization of milk was important in stopping the spread of TB from cows to humans, as was TB testing of herds.

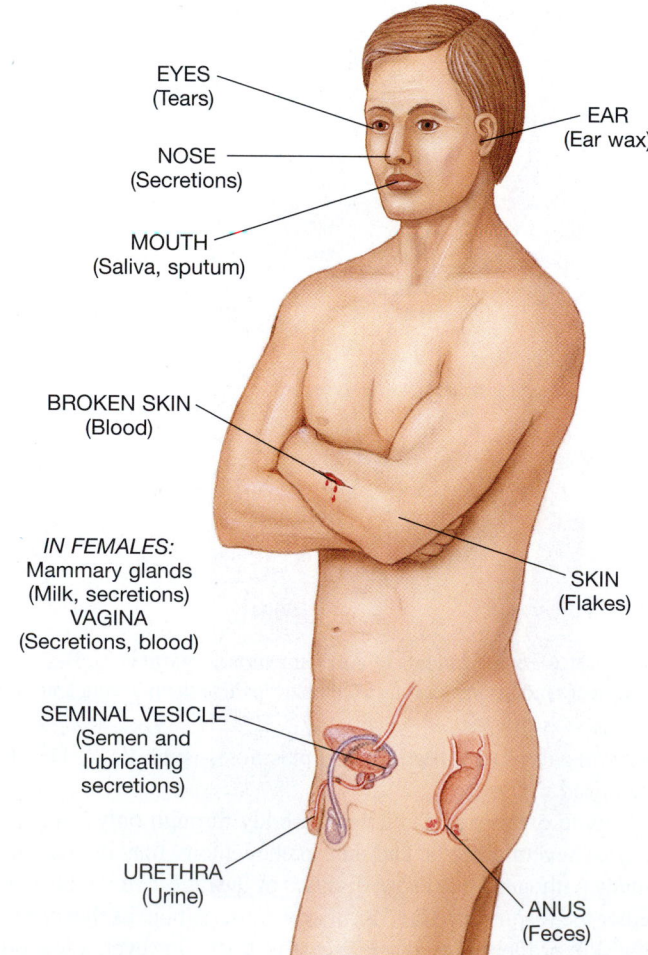

➤ **Figure 15.10** **Portals of exit for human pathogens.** Besides those portals common to males and females, additional portals in males include the seminal vesicles and in females, the mammary glands and vagina.

Modes of Disease Transmission

For new cases of infectious diseases to occur, pathogens must be transmitted from a reservoir or portal of exit to a portal of entry. Transmission can occur by several modes, which we have grouped into three categories: contact transmission, transmission by vehicles, and transmission by vectors. ➤ Figure 15.11 presents an overview of these transmission modes.

Contact Transmission

Contact transmission can be direct, indirect, or by droplets. **Direct contact transmission** requires body contact between individuals. Such transmission can be horizontal or vertical. In **horizontal transmission,** individuals pass on pathogens by shaking hands, kissing, touching sores, or having sexual contact. Pathogens can also be spread from one part of the body to another through unhygienic practices. For example, touching genital herpes lesions and then touching other parts of the body, such as the eyes, can spread the infection. Pathogens from fecal matter also can be spread by unwashed hands to the mouth; this is **direct fecal–oral transmission.** In **vertical transmission,** pathogens are passed from parent to offspring in an egg or sperm, across the placenta, in breast milk, or in the birth canal (as can happen with syphilis and gonorrhea).

    Indirect contact transmission occurs through **fomites** (fo′mi-tez), nonliving objects that can harbor and transmit an infectious agent. Examples of fomites include soiled handkerchiefs, dishes, eating utensils, doorknobs, toys, bar soap, and money. (U.S. paper currency is treated with an antimicrobial agent that reduces transmission of microorganisms.)

    Droplet transmission, a third kind of contact transmission, occurs when a person coughs, sneezes, or speaks near others (➤ Figure 15.12). **Droplet nuclei** consist of dried mucus, which protects microorganisms embedded in it. These particles can be inhaled directly, can collect on the floor with dust particles, or can become airborne. Droplet transmission over a distance of less than 1 m is not considered airborne.

Transmission by Vehicles

A **vehicle** is a nonliving carrier of an infectious agent from its reservoir to a susceptible host. Common vehicles include water, air, and food. Blood, other body fluids, and intravenous fluids also can serve as vehicles of disease transmission.

Waterborne Transmission Although waterborne pathogens do not grow in pure water, some survive transit in water with small quantities of nutrients or in water polluted with fertilizer. Waterborne pathogens usually thrive in and are transmitted in water contaminated with untreated or inadequately treated sewage. Such **indirect fecal–oral transmission** occurs when pathogens from feces of one organism infect another organism. Pathogens have been isolated from public water supplies, semiprivate water supplies (camps, parks, and hotels that have their own water systems), and private water supplies (springs and wells). Polioviruses and enteroviruses as well as several bacteria are waterborne microbes that infect the digestive system and cause gastrointestinal symptoms. Waterborne infections can be prevented by proper treatment of water and sewage, although enteroviruses are especially difficult to eradicate from water.

Airborne Transmission Airborne microorganisms are mainly transients from soil, water, plants, or animals. They do not grow in air, but some reach new hosts through air despite dryness, temperature extremes, and ultraviolet radiation. In fact, dry air actually enhances transmission of many viruses. Pathogens are said to be airborne if they travel more than 1 m through the air. Both airborne pathogens and those suspended in droplets have the best chance of reaching new hosts when people are crowded together indoors. Increased incidence of airborne infections is associated with nearly sealed modern buildings in which temperatures are controlled with heating and air-conditioning systems, and little fresh air enters.

    Airborne pathogens fall to the floor and combine with dust particles or become suspended in aerosols. An **aerosol** is a cloud of tiny water droplets or fine solid particles suspended in air. Microorganisms in aerosols need not come directly from humans; they can also come from dust particles stirred by dry mopping, changing bedding, or even changing clothing. In the microbiology laboratory, flaming a transfer loop full of bacteria can disperse microorganisms into aerosols.

Droplets that travel less than one meter from a sneeze or cough to a host are not classified as airborne.

Are There Amoebas in Your Emergency Eyewash Station?
Emergency eyewash stations are intended to supply large volumes of clean water in case of a chemical accident. The American National Standards Institute recommends weekly flushing to keep them clean. However, often water stands in them for long periods of time at room temperature—ideal conditions for growth of biofilms of bacteria and fungi. These microbes can serve as food for water-dwelling amoebas, such as *Acanthamoeba,* which can cause serious eye infections, leading to blindness. Many species of amoebas supply growth factors and an intracellular home for *Legionella pneumophila,* the cause of Legionnaires' disease. Just 100 cells spread by airborne water droplets can cause infection. *Pseudomonas,* a dangerous pathogen often found, can cause considerable tissue destruction and is difficult to treat.

    In a 1995 study of 30 dual-spray eyewash units connected to building plumbing, 60 percent of the units contained more than one genus of amoebas. Regular flushing could not eliminate them. What would you expect in eyewash bottles that are not connected to plumbing in tap water?

A contaminated whirlpool spa, displayed in a large home improvement center, resulted in 15 cases of Legionnaires' disease and 2 deaths.

Contact transmission

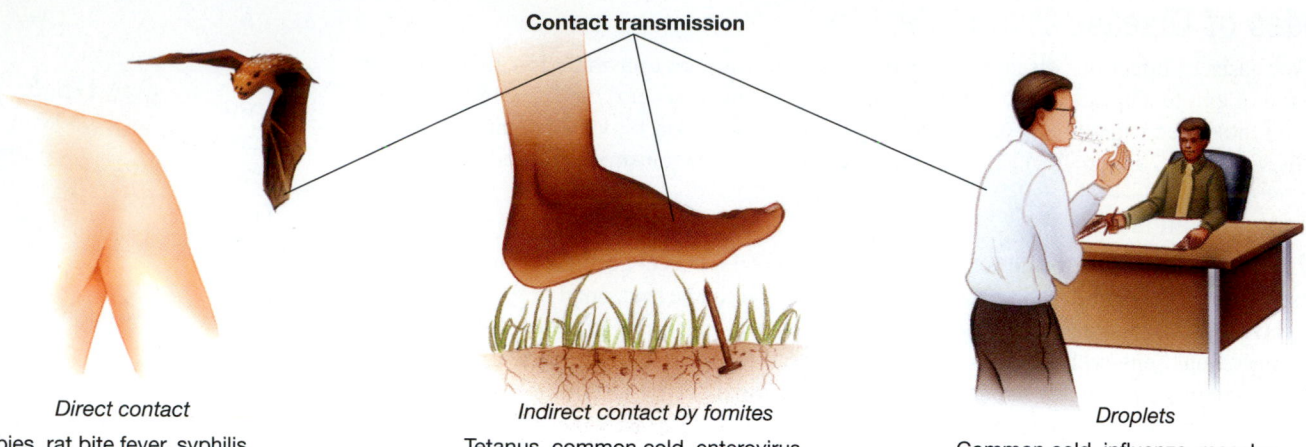

Direct contact

Rabies, rat bite fever, syphilis, gonorrhea, herpes, staphylococcal infections, cutaneous anthrax, genital warts

Indirect contact by fomites

Tetanus, common cold, enterovirus infections, ringworm

Droplets

Common cold, influenza, measles, Q fever, pneumonia, whooping cough

Vehicle transmission

Waterborne

Cholera, shigellosis, leptospirosis, *Campylobacter* infections

Airborne, including dust particles

Chickenpox, tuberculosis, coccidioidomycosis, histoplasmosis, influenza, measles

Foodborne

Intoxication with aflatoxins and botulinum toxin, paralytic shellfish poisoning, staphylococcal food poisoning, typhoid fever, salmonellosis, listeriosis, toxoplasmosis, tapeworms, hepatitis A

Vector transmission

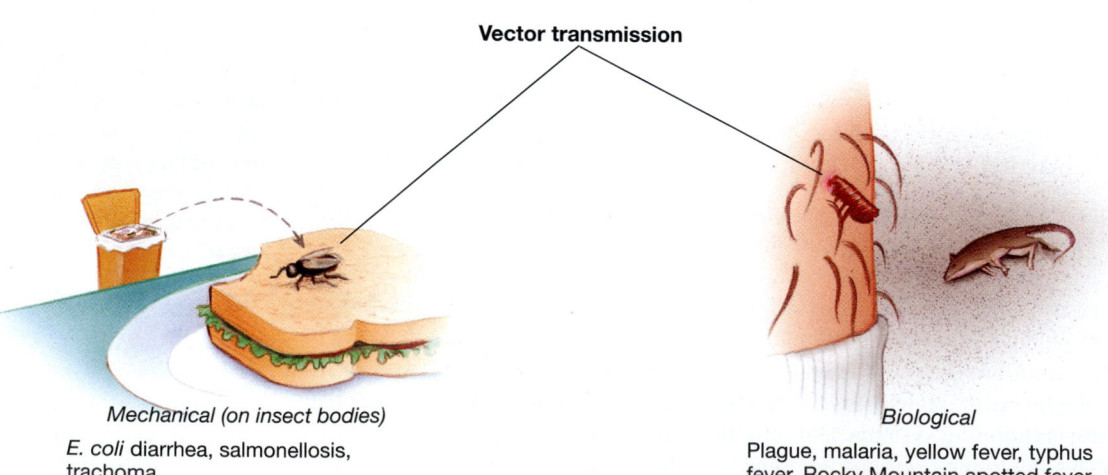

Mechanical (on insect bodies)

E. coli diarrhea, salmonellosis, trachoma

Biological

Plague, malaria, yellow fever, typhus fever, Rocky Mountain spotted fever, Chagas' disease, Lyme disease

➤ Figure 15.11   **Modes of disease transmissions.**

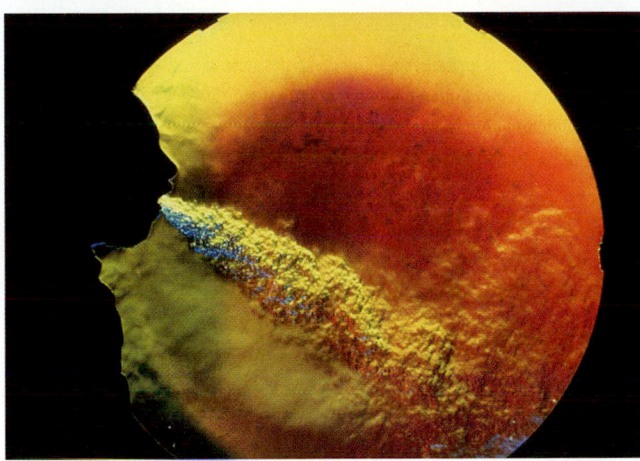

➤ Figure 15.12 **Droplet transmission.** A cough is photographed with the Schlieren technique, which detects air-speed differences as different colored areas. Such dispersal is most important within a radius of about 1 m; however, the smallest particles can be dispersed much farther and kept aloft by air currents. Even a surgical mask will not prevent spread of all droplets.

Dust particles can harbor many pathogens. Bacteria with sturdy cell walls, such as staphylococci and streptococci, can survive for several months in dust particles. Naked viruses as well as bacterial and fungal spores can survive for even longer periods.

Hospitalized patients are at great risk of getting airborne diseases because they often have lowered resistance and because former patients may have left pathogens deposited in dust particles. Cleaning floors with a wet mop, wiping surfaces with a damp cloth, and carefully unfolding bed linens and towels help reduce aerosols. Masks and special clothing are used in operating rooms, burn wards, and other areas where patients are at greatest risk of infection. Some hospitals also use ultraviolet lights and special air flow devices to prevent exposure of patients to airborne pathogens.

Foodborne Transmission   Pathogens are most likely to be transmitted in foods that are inspected improperly, processed unsanitarily, cooked incompletely, or refrigerated poorly. As with waterborne pathogens, foodborne pathogens are most likely to produce gastrointestinal symptoms.

Transmission by Vectors

As you learned in Chapter 11, **vectors** are living organisms that transmit disease to humans. ➡ (p. 289) Most vectors are arthropods such as ticks, flies, fleas, lice, and mosquitoes. However, the mechanism of vector transmission can be mechanical or biological.

Mechanical Vectors   Insects act as *mechanical vectors* when they transmit pathogens passively on their feet and body parts. Houseflies and other insects, for example, frequently feed on animal and, if available, human fecal matter. If they then move on to feed on human fare, they can deposit pathogens in the process. Disease transmission by mechanical vectors does not require that the pathogen multiply on or in the vector.

This method of disease transmission can be prevented simply by keeping these vectors out of areas where food is prepared and eaten. The fly that walked across dog feces in the park should not be allowed to walk across your picnic potato salad. The use of screened areas to keep insects out also reduces disease transmission by mechanical vectors. Unfortunately, in some poverty-ridden areas of the world, screens are lacking on windows—even those that open into hospital operating rooms!

Biological Vectors   Insects act as *biological vectors* when they transmit pathogens actively; that is, the infectious agent must complete part of its life cycle in the vector before the insect can transmit the infective form of the microbe. Compared with direct transmission through animal bites, the transmission of zoonoses through vectors is much more common. In most vector-transmitted diseases, such as malaria and schistosomiasis, a biological vector is the host for some phase of the life cycle of the pathogen. Control of zoonoses transmitted by biological vectors can often be achieved by controlling or eradicating the vectors. The spraying of standing water with oil kills many insect larvae. Spraying breeding grounds with pesticides also can be an effective control, at least until the vectors become resistant to the pesticides.

"Dog Germs!"

When Snoopy kisses Lucy, she always screams, "Dog germs!" However, Snoopy is the one who should be worried. A human mouth is more likely to contain pathogens than is a dog's mouth. Dog saliva is acidic and a less hospitable environment for microorganisms.

Although kissing probably does little harm, swimming with man's best friend may be more dangerous for humans. *Leptospira* spirochetes ordinarily infect animal kidneys and are shed during urination. One man discovered when he went swimming in a river with his dog, that this can lead to human leptospirosis. This disease is characterized by fever, headache, and kidney damage. An infected dog is a hazard in a crowded swimming pool. Although vigilant chlorination of pools and vaccination of pets against leptospirosis might prevent them from transmitting the disease, it is best not to allow pets in pools.

A variation on this occurred when some teenagers drove their swamp buggy through water contaminated with urine from leptospirosis-infected deer. Droplets splashed up by the tires apparently infected the passengers

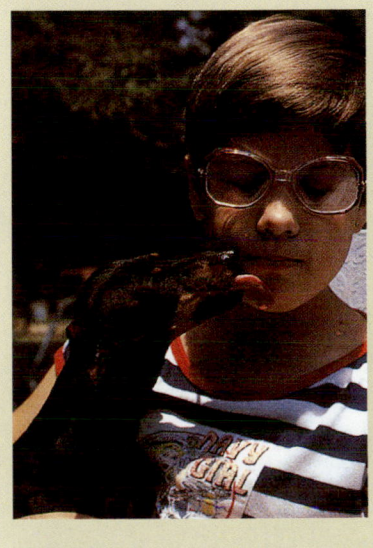

Global warming may result in tropical diseases such as malaria moving into new regions of the world, as their arthropod vectors spread.

Plague in the 1990s

In August 1994, outbreaks of bubonic and pneumonic plague occurred in parts of India where the bacterium was epidemic in rats. Widespread panic erupted, and health care workers worldwide became deeply concerned that the disease could become pandemic. By October 1994 the number of suspected cases of bubonic and pneumonic plague reported in India had reached 693, with 56 deaths. Due to rapid identification and control measures, the epidemic subsided.

With jet travel, it is easy for a disease to become pandemic almost overnight. Thus, the reports from India prompted U.S. health authorities to develop an emergency plan for detecting and managing any suspected cases arriving by plane. In fact, 11 air travelers arriving in New York, suspected of having the plague, were later found to be suffering from other diseases. This time we were lucky. The available vaccine is not completely effective, and many rats have become resistant to the best available rat poisons.

✓ Give examples of a nonliving reservoir and a nonliving vehicle of transmission.

✓ Distinguish between biological and mechanical vectors.

✓ What is herd immunity?

Special Problems in Disease Transmission

Transmission of disease by carriers poses special epidemiologic problems because carriers are often difficult to identify. The carriers themselves usually do not know they are carriers and sometimes cause sudden outbreaks of disease. Depending on the pathogen they carry, carriers can transmit disease by direct or indirect contact or through vehicles such as water, air, or food; they can even be a source of pathogens for vectors.

Another special transmission problem arises with people who have *sexually transmitted diseases (STDs)*. Such diseases are most often transmitted by direct sexual contact, including kissing, but some can be transmitted by oral or anal sex. STDs present epidemiologic problems because infected individuals sometimes have contact with multiple sexual partners. In fact, the incidences of AIDS, genital herpes, genital warts, syphilis, and *Chlamydia* infections are rapidly increasing.

Zoonoses are another epidemiologic problem. They can be transmitted by direct contact, as when humans get rabies from the bite of an infected domestic or wild animal. An oral vaccine for rabies has been developed, to be administered to wildlife.

Disease Cycles

Many diseases occur in cycles. For years or even decades, only a few cases are seen, but then many cases suddenly appear in epidemic or pandemic proportions. Let's look at one example. Bubonic plague—or the Black Death, as it was called—has occurred in pandemic outbreaks followed by recurrent cycles for centuries. Between A.D. 543 and 548, the disease spread from India or Africa, through Egypt, to Constantinople (now Istanbul, Turkey), where it killed 200,000 people in only 4 months. The disease quickly spread via fleas on ship rats to Europe and the Mediterranean basin. About a half century later it appeared in China, with equally devastating results. After the initial outbreaks, it occurred in cycles of 10 to 24 years over the next two centuries.

All of Europe breathed a sigh of relief for about the next 500 years, which were plague-free. But in 1346, a second pandemic, worse than the first, afflicted North Africa, the Middle East, and most of Europe. Nearly one-third of the population of Europe died, and in many cities three-fourths of the population lost their lives to the dreaded disease. Then cyclic recurrences claimed more lives in epidemics of the seventeenth century in England and the eighteenth century in France. Near the beginning of the twentieth century, a pandemic killed more than a million people in India and spread to many parts of the world, including San Francisco.

Thus, cyclic diseases pose special epidemiologic problems. Epidemiologists still cannot predict when one will break out and reach epidemic proportions. It is difficult to be prepared to treat sudden, large increases in the incidence of a disease and nearly impossible to persuade people to be immunized against a disease they have never seen.

Herd Immunity

An important factor in cyclic disease is **herd immunity** (or *group immunity*), which is the proportion of individuals in a community or population who are immune to a particular disease. If herd immunity is high—that is, if most of the individuals in a population are immune to a disease—then the disease can spread only among the small number of susceptible individuals in the population (➤ Figure 15.13). Even when a member of the population becomes infected, the likelihood that that person will transmit the disease to others is small. Thus, a sufficiently high herd immunity protects the entire population, including its susceptible members.

It is easy to see, then, why public health officials want to maintain the highest possible herd immunity, especially against common cyclic diseases. They encourage parents to have children immunized against measles and other communicable diseases. In many cities in the United States, children are required to be immunized against measles before they can start school. As a result, about 95 percent of elementary-age children are immune to measles. Although high school and college students who have neither had measles nor received vaccine are protected by herd immunity, many school systems and some colleges and universities are requiring immunization against measles for all students, whatever their age.

The loss of herd immunity could lead to a reemergence of a disease. The rise in diphtheria incidence in the New Independent States of the former Soviet Union is due in part to a lack of childhood vaccinations. Herd immunity in these populations has dwindled. Some people fear that herd immunity to smallpox is bound to drop now that vaccinations have ceased as a result

The Meningitis Belt

Like many other diseases discussed in Chapter 16, meningococcal meningitis, which is caused by *Neisseria meningitidis,* occurs in roughly 5- to 12-year cycles. During the 1960s, an estimated 3 million cases occurred in China. In mid-April of 1988, patients were entering the hospital in N'Djamena, the capital city of the north central African nation Chad, at a rate of 250 per day. Out in the countryside, thousands of others who were unable to reach medical facilities suffered and died uncounted. In 1989, 40,000 cases were diagnosed in an Ethiopian epidemic; yet the United States does not have such epidemics. Looking specifically at Africa, what epidemiological factors cause meningitis epidemics to sweep across a broad belt of central Africa (see the map) in these cycles?

In considering such multi-year cycles, epidemiologists also must consider the seasonal cycles in Africa. Each year, whether an epidemic year or one of the years in between, meningitis outbreaks occur only during the dry-season months of January through June. Then, as soon as the rains begin, the number of new cases drops to zero—only to climb again at the start of the dry season. However, it can be shown that the causative meningococci are transmitted year-round. How can the seasonality of meningitis outbreaks be explained?

The meningitis belt across central Africa.

1. *Environmental factors*   The environmental factors of low humidity and heat during the dry season may cause nasal and throat membranes to dry out, thus making them more likely to allow meningococci to enter the bloodstream.

2. *Other diseases*   The dry season is also the season of more colds, influenza, and other upper respiratory diseases that may add to tissue vulnerability. Meningitis patients are 23 times more likely than average to have viral infections of the upper respiratory tract. Many meningitis patients are also infected with the bacterium *Mycoplasma hominis.*

3. *Herd immunity*   Why doesn't the seasonal swing erupt into epidemic every year? During an epidemic, most people in a population will develop antibodies against the prevalent strain of causative bacteria. Each year, however, new children are born, and they lack such immunity. Eventually the herd immunity drops so low as to allow another epidemic to occur.

4. *Strain virulence*   During the years between epidemics, mutant strains will arise. If one of these new strains is more virulent than previous strains, it could initiate a new epidemic. Studies of epidemics have shown that usually only one strain of meningococci is responsible for any one epidemic. So virulence of strain is an important factor.

5. *Timing*   The timing of entry of the mutant strain into the population is also crucial. If the mutant strain enters during the dry season, those who are exposed to it are likely to develop meningitis. But if a new strain enters during the rainy season, people will become exposed to it without developing the disease. Instead, they will develop antibodies against the new strain and will be immune to it when the following dry season begins.

Clearly the epidemiology of meningitis epidemics is not a simple matter. Several factors are at work, and each may have a greater or lesser effect on an outbreak of a given epidemic. Unfortunately, the available meningitis vaccine confers only short-term immunity; it does not last from one epidemic to the next. However, it can be used during an epidemic to halt its spread.

of the eradication of the smallpox virus. If the smallpox virus were to appear again, it could be devastating—few people would be immune to the disease.

Controlling Disease Transmission

Several methods are currently available for full or partial control of communicable diseases. They include *isolation, quarantine, immunization,* and *vector control.*

In **isolation,** a patient with a communicable disease is prevented from having contact with the general population. Isolation generally is accomplished in a hospital. There appropriate

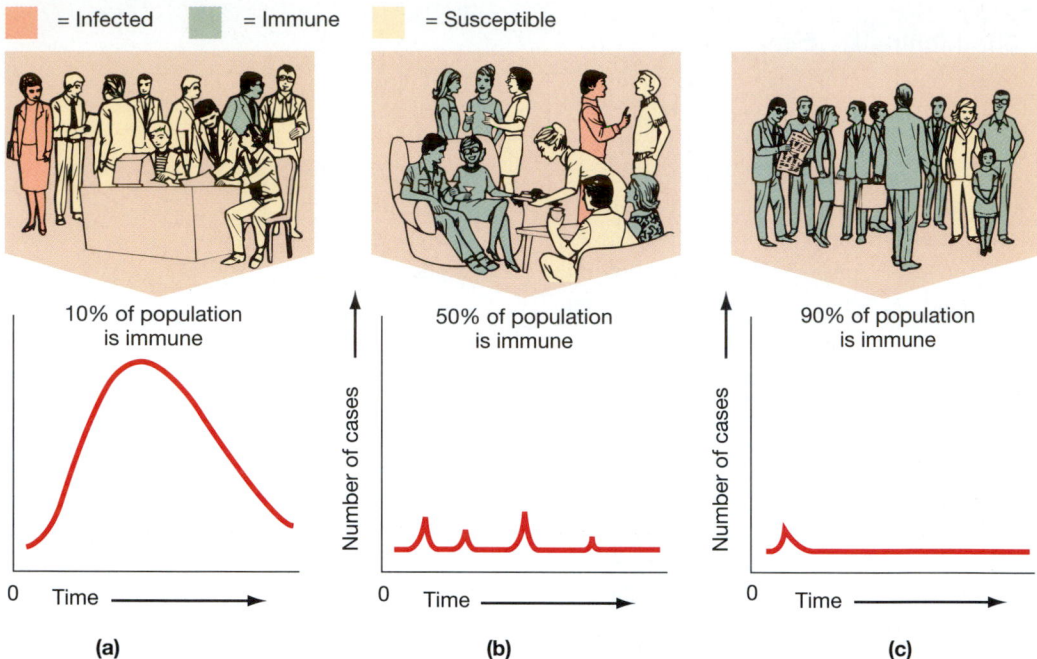

= Infected   = Immune   = Susceptible

10% of population is immune

50% of population is immune

90% of population is immune

(a)   (b)   (c)

➤ **Figure 15.13   Herd immunity.** (a) When the percentage of immune persons in a population is low, it is likely that susceptible individuals will be exposed to the disease. (b and c) As the percentage of immune persons increases, it becomes increasingly less likely that a susceptible individual will be exposed to the disease.

procedures can be carried out to reduce the spread of disease among susceptible individuals and to prevent the spread of disease in the general population. In all, there are seven categories of isolation (Table 15.2). Strict isolation makes use of all available procedures to prevent transmission of organisms or virulent infections to medical personnel and visitors. Even medical researchers working with highly pathogenic strains of microbes must use high levels of isolation and special laboratories equipped to contain such agents (➤ Figure 15.14).

Quarantine is the separation of "healthy" human or animal carriers from the general population when they have been exposed to a communicable disease. Quarantine prevents spread of the disease during its incubation period. Although it is one of the oldest methods of controlling communicable diseases, it is now used mainly for serious diseases such as cholera and yellow fever. Quarantine differs from isolation in two ways: (1) It is applied to healthy people who were exposed to a disease during the incubation period, and (2) it pertains to limiting the movements of such people and not necessarily to precautions during treatment. Quarantine is rarely used today because it is very difficult to carry out. To ensure that no infected persons spread a disease, everyone who had been exposed to it would have to be quarantined for the disease's incubation period. This would mean, for example, that all travelers returning to the United States from a region of the world where cholera is endemic would have to be quarantined for 3 days in accommodations provided at airports and seaports—not likely to be a popular idea.

Large-scale *immunization* programs are an extremely effective means of controlling communicable diseases for which safe vaccines are available. Such programs greatly increase herd immunity and thus greatly decrease human suffering and deaths from infectious diseases. In the United States, immunizations have nearly eradicated polio, measles, mumps, diphtheria, and whooping cough. Unfortunately, as the incidence of these diseases becomes very small, people become complacent about getting immunized. Such complacency can lead to a sufficient decrease in herd immunity, resulting in outbreaks of vaccine-preventable diseases.

Vector control is an effective means of controlling infectious diseases if the vector, such as an insect or rodent, can be identified and its habitat, breeding habits, and feeding behavior determined. Places where a vector lives and breeds can be treated with insecticides or rodenticides. Window screens, mosquito netting, insect repellents, and other barriers can be used to protect humans from becoming victims of the bites of feeding vectors. Unfortunately, vectors have their own defenses. Some escape or become resistant to pesticides or make their way through barriers.

Malaria control in the United States was accomplished mainly by the control of mosquito vectors. Consequently, Americans have little herd immunity to malaria because most have never had the disease. Since the 1940s new cases of malaria in the United States have occurred mainly

Yummy Vaccines

The Oral Vaccination Program uses aerial distribution of an oral vaccine to control rabies in wildlife. A dose of vaccine is inserted in the hollow center of a bait composed of dog food or fish meal. The vaccine is a recombinant vaccine that has been shown to be safe and effective in almost 60 species of mammals and birds. The vaccine, does not cause rabies in human or animals, and produces a protective immune response against the infection. To control rabies in red foxes, a massive bait drop was conducted over several years in Ontario, Canada. The bait program finally broke the rabies cycle in red foxes, and to date only 4 cases of rabies have been found. According to Ontario public health officials, if the trend continues, red fox rabies will be eliminated before the year 2001. Similar programs, targeted for coyotes and gray foxes, are ongoing in the United States.

➢ **Table 15.2** | **A summary of important isolation procedures**

				Isolation Categories		
Strict	**Contact**	**Respiratory**	**Tuberculosis**	**Enteric precautions**	**Drainage/ secretion precautions**	**Blood and body fluid precautions**
Visitors must check in at nursing station before entering patient's room						
Yes	Yes	Yes	Yes	Yes	Yes	Yes
Hands must be washed on entering and on leaving patient's room						
Yes	Yes	Yes	Yes	Yes	Yes	Yes
Gowns must be worn by personnel and visitors						
Yes	Yes	No	Yes	Only for direct patient contact	Only if soiling is likely	Only if soiling is likely
Masks must be worn by personnel and visitors						
Yes	Yes	Unless not susceptible to disease	Only if coughing	No	No	No
Gloves must be worn by personnel and visitors						
Yes	Only for direct patient contact	No	No	Only for direct contact with patient or feces	Only for direct contact with lesion site	Only for direct contact with lesion site
Private room required with door closed						
Yes	Yes	Yes	Yes	Only for children	No	If hygiene is poor
Examples of diseases						
Pneumonic plague, rabies, diphtheria, disseminated herpes zoster, Lassa fever, chickenpox, draining *Staphylococcus aureus* wounds	Severe noninfected dermatitis, noninfected burns	Measles, mumps, rubella, pertussis	Pulmonary tuberculosis	Typhoid fever, cholera, salmonellosis, shigellosis, hepatitis	Bubonic plague, gas gangrene localized herpes, puerperal sepsis	AIDS, hepatitis B

➢ Figure 15.14 **Biosafety level 4 lab.** Scientists who work with very dangerous, and often easily disseminated, microbes do so in an isolation lab. This laboratory worker is in the highest level of isolation, a biosafety level 4 lab. Extreme precautions, such as the use of special ventilation systems and the wearing of "space suits," are required of personnel to avoid contact with microbes and to prevent the escape of the microbes from the facility.

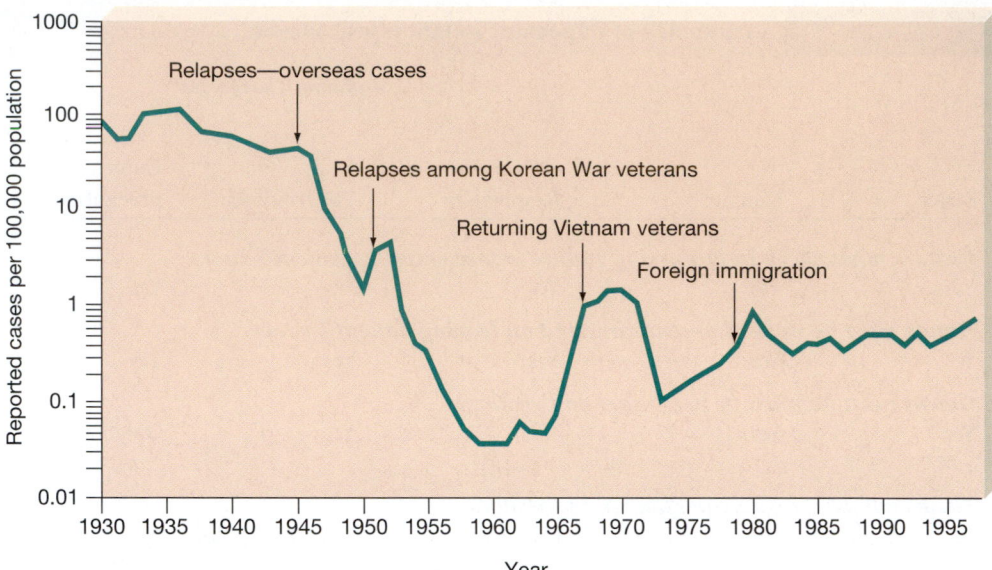

> Figure 15.15   **The incidence rate of malaria, as related to sources of infection into the United States.**

among people infected in other countries (➤ Figure 15.15). As long as infected individuals do not reintroduce the parasite into the mosquito population, a malaria epidemic in the United States is most unlikely despite low herd immunity. In recent years, however, laborers from areas where malaria still is common have brought the parasite northward, and the disease has become endemic in some parts of California.

Although communicable diseases are theoretically preventable, some still have a high incidence in every human population. In countries with relatively high living standards, the common cold and many sexually transmitted diseases occur with great frequency. In countries with lower standards of living—especially those in the tropics—the prevalence of malaria and a variety of other diseases, including some nearly eradicated in other countries, is extremely high. And certainly AIDS has become a worldwide threat that now extends beyond the special high-risk groups with which it was once associated.

Public Health Organizations

In the United States and many other countries, the importance of controlling infectious diseases and reducing other health hazards has led to the creation of public health agencies. City and county health departments provide immunizations, inspect restaurants and food stores, and work with other local agencies to ensure that water and sewage are treated properly. State health departments deal with problems that extend beyond cities and counties. They often do laboratory tests, such as identifying rabies in animals, and hepatitis and toxins in water.

The Centers for Disease Control and Prevention

In the United States, the federal government operates the U.S. Public Health Service (USPHS), which has several branches. Of these branches, the *Centers for Disease Control and Prevention (CDC)* in Atlanta, Georgia (➤ Figure 15.16), has major responsibilities for the control and prevention of infectious diseases and other preventable conditions. Among the microbiology-related activities of the CDC are

1. providing guidelines for occupational health and safety, quarantines, tropical medicine, cooperative activities with national agencies in other countries and with international agencies, and public health education;

2. making recommendations to the medical community regarding the use of antibiotics, especially for the treatment of diseases caused by antibiotic-resistant organisms;

3. storing infrequently used drugs and providing them to physicians who encounter patients with tropical parasitic diseases and other diseases rarely seen in the United States; and

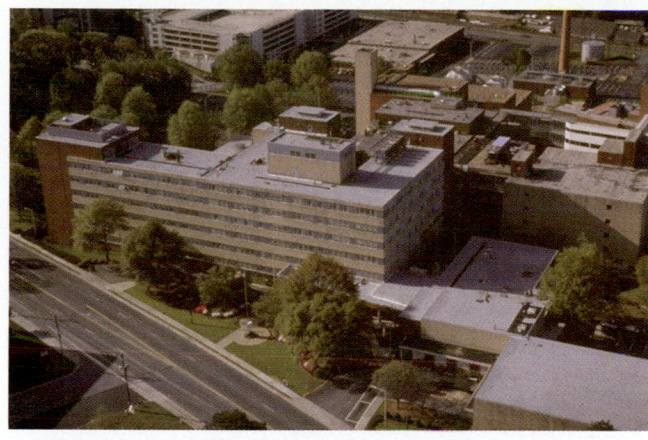

➢ **Figure 15.16   CDC headquarters in Atlanta, Georgia.**

4. making recommendations regarding the administration of vaccines—which should be used, who should receive them, and at what ages.

The CDC carries out epidemiologic studies, which are published as the *Morbidity and Mortality Weekly Report (MMWR)*. This publication provides statistics for specific diseases in various parts of the United States and the world (➢ Figure 15.17). Other CDC periodical reports are *Recommendations and Reports* and *Surveillance Summaries,* which provide in-depth coverage of specific issues, many related to infectious diseases.

The World Health Organization

The *World Health Organization (WHO)* is an international agency based in Geneva, Switzerland, that coordinates and sets up programs to improve health in more than 100 member countries. Its basic objective is that all peoples attain the highest possible level of health. This aim is embodied

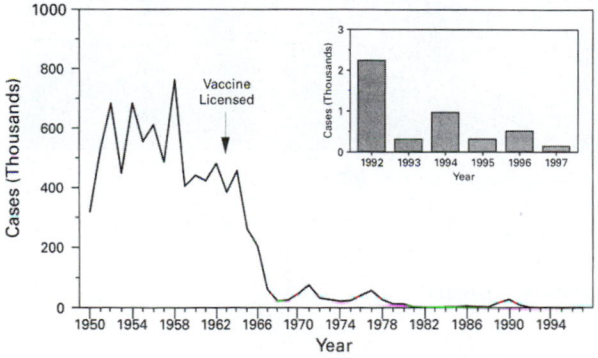

April 17, 1998 / Vol. 47 / No. 14

MORBIDITY AND MORTALITY WEEKLY REPORT

Printed and distributed by the Massachusetts Medical Society, publishers of *The New England Journal of Medicine*

273   Measles — United States, 1997
277   Demographic Characteristics of Persons Without a Regular Source of Medical Care — Selected States, 1995
280   Update: Influenza Activity — United States and Worldwide, 1997–98 Season, and Composition of the 1998–99 Influenza Vaccine
285   Nosocomial *Ralstonia pickettii* Colonization Associated with Intrinsically Contaminated Saline Solution — Los Angeles, California, 1998
287   Notice to Readers

Measles — United States, 1997

During 1997, a provisional total of 138 confirmed measles cases was reported to CDC by local and state health departments, the lowest number of measles cases ever reported in 1 year and a 55% decrease from the previous record low of 309 cases reported in 1995 (Figure 1). This report describes the epidemiology of measles in the United States in 1997, which suggests that no endemic measles virus is circulating in the United States.

Case Classification

Reported measles cases are classified as imported or indigenous based on where transmission of measles virus is likely to have occurred. Cases in persons who traveled outside the United States within 18 days before rash onset are classified as international importations. Indigenous measles cases are classified into three groups:

FIGURE 1. Reported measles cases, by year — United States, 1950–1997

➢ **Figure 15.17   The *Morbidity and Mortality Weekly Report*,** issued weekly by the **CDC.** It reports new and unusual cases and trends. It also lists, by state, the number of cases of notifiable diseases recorded that week, the number recorded for the same week in the previous year, and the cumulative totals. It is available from the Massachusetts Medical Society; C.S.P.O. Box 9120; Waltham, MA 02254-9120.

in WHO's current theme: "Health for All by the Year 2000." Specific activities are carried out by six regional organizations in Africa, the Eastern Mediterranean, Europe, Southeast Asia, the Western Pacific, and the Americas. WHO works closely with the United Nations on population control, management of food supplies, and various other scientific and educational activities.

WHO sets health standards for international disease control; helps developing nations establish effective control and immunization programs; collects, analyzes, and distributes health data; and maintains surveillance of potential epidemics (published in WHO's *Weekly Epidemiological Record*). It also provides training and research programs for health personnel and information for individuals (➤ Figure 15.18). The agency has helped more than 100 countries in immunizing against diphtheria, measles, whooping cough, poliomyelitis, tetanus, and tuberculosis and hopes eventually to eradicate measles worldwide. It conducts research and training to combat widespread tropical diseases such as leprosy, malaria, and several diseases caused by helminths. WHO has been instrumental in coordinating the eradication of smallpox worldwide.

Notifiable Diseases

Cooperation among state and national health organizations in the United States has led to the establishment of a list of **notifiable diseases,** which are infectious diseases that are potentially harmful to the public's health and must be reported by physicians. As of July 1994, 49 infectious diseases were listed as notifiable at the national level (Table 15.3). On the basis of CDC suggestions, each year the

(a)   (b)

(c)   (d)

➤ Figure 15.18 **Some typical activities of the World Health Organization.** (a) Eye tests for the prevention of blindness are given high in the Andes Mountains of Peru. (b) A medical assistant travels by horseback to treat patients in remote villages of China. (c) Ambulance boat transporting a patient to a hospital, Myanmar. (d) Health education classes, Guatemala.

> **Table 15.3** **Infectious diseases designated as notifiable at the national level—United States, 1997**

AIDS	Hepatitis A	Rocky Mountain spotted
Anthrax	Hepatitis B	fever (Typhus fever,
Botulism	Hepatitis, non-A, non-B	tickborne)
Brucellosis	Hepatitis, unspecified	Rubella
Chancroid	HIV infection, adult	Salmonellosis
Chlamydia trachomatis,	HIV infection, pediatric	Shigellosis
genital infection	Legionellosis	Streptococcal disease,
Cholera	Leprosy (Hansen disease)	invasive, Group A
Coccidiomycosis (regional)	Lyme disease	Streptococcal pneumonia,
Congenital rubella syndrome	Malaria	drug-resistant invasive
Diphtheria	Measles	disease
Encephalitis	Meningococcal infection	Syphilis
Escherichia coli O157:H7	Mumps	Syphilis, congenital
Gonorrhea	Pertussis	Tetanus
Hantavirus, pulmonary	Plague	Toxic shock syndrome
syndrome	Poliomyelitis	Trichinosis
Haemophilus influenzae	Psittacosis	Tuberculosis
Hemolytic uremic syndrome,	Rabies, animal	Typhoid fever
post-diarrheal	Rabies, human	Yellow fever

Source: Morbidity and Mortality Weekly Report 43(43):801.

Council of State and Territorial Epidemiologists (CSTE) adds diseases to, or deletes diseases from, the list. If a specific disease shows a decline in incidence, it may be removed from the list.

Although reporting infectious diseases is mandatory only at the state level, the reporting of notifiable diseases at the national level is intended to accomplish two things: (1) to ensure that public health officials learn of diseases that jeopardize the health of populations, and (2) to provide consistency and uniformity in the reporting of those diseases. Various kinds of information about notifiable diseases in the United States are available from the CDC; samples of this information are provided in Tables 15.4A and 15.4B.

Nosocomial Infections

A **nosocomial** (nos-o-ko'me-al) **infection** is an infection acquired in a hospital or other medical facility. The term nosocomial is derived from the Greek words *nosos,* meaning "disease," and *komeo,* meaning "to take care of." Nosocomial diseases are acquired during medical treatment. Although many such infections occur in patients, infections acquired at work by staff members also are considered nosocomial infections.

Among patients admitted to American hospitals each year, about 2 million (10 percent) acquire an infection that increases the risk of death, the duration of the hospital stay, and the cost of treatment. Of these, over 20,000 people per year die of their nosocomial infections. Curiously, nosocomial infections are largely a product of advances in medical treatment. Intravenous, urinary, and other catheters, invasive diagnostic tests, and complex surgical procedures increase the likelihood that pathogens will enter the body. Intensive use of antibiotics contributes to the development of resistant strains of pathogens. And, therapies to reduce the chances of rejection of transplanted organs impair the immune response to pathogens. Despite the risk of nosocomial infections, the medical treatments now available save far more patients than are lost to such infections.

The Epidemiology of Nosocomial Infections

The epidemiology of nosocomial infections, like the epidemiology of diseases acquired in the community, considers sources of infection, modes of transmission, susceptibility to infection, and prevention and control. In addition, it focuses on medical procedures that increase the risk

> Table 15.4A | Provisional cases of selected notifiable diseases, United States, weeks ending May 23, 1998, and May 17, 1997 (20th Week)

Reporting Area	AIDS Cum. 1998*	AIDS Cum. 1997	Chlamydia Cum. 1998	Chlamydia Cum. 1997	Escherichia coli O157:H7 NETSS[†] Cum. 1998	Escherichia coli O157:H7 PHLIS[§] Cum. 1998	Gonorrhea Cum. 1998	Gonorrhea Cum. 1997	Hepatitis C/NA,NB Cum. 1998	Hepatitis C/NA,NB Cum. 1997
UNITED STATES	16,097	22,446	186,752	174,647	357	166	110,456	106,060	1,630	1,121
NEW ENGLAND	489	676	7,230	6,638	40	25	1,871	2,267	17	30
Maine	10	25	364	358	1	–	14	20	–	–
N.H.	14	14	337	297	7	5	32	53	–	3
Vt.	10	18	146	145	–	–	11	22	–	1
Mass.	211	279	3,294	2,710	19	15	792	852	16	24
R.I.	40	55	964	800	3	1	139	193	1	2
Conn.	204	285	2,125	2,328	10	4	883	1,127	–	–
MID. ATLANTIC	4,607	6,832	24,399	21,613	30	9	13,369	13,266	151	123
Upstate N.Y.	545	1,122	N	N	22	–	1,937	2,369	125	94
N.Y. City	2,631	3,293	13,244	11,653	2	4	5,616	5,249	–	–
N.J.	823	1,538	3,628	3,969	6	4	2,398	2,769	–	–
Pa.	608	879	7,527	5,991	N	1	3,418	2,879	26	29
E.N. CENTRAL	1,299	1,639	32,562	27,852	58	19	22,133	16,388	184	270
Ohio	242	348	9,657	8,680	16	3	5,858	5,311	5	5
Ind.	275	301	2,706	3,363	10	7	1,769	2,296	3	6
Ill.	495	505	8,937	4,357	18	–	7,173	2,123	7	43
Mich.	218	394	8,372	7,309	14	4	6,227	4,950	169	201
Wis.	69	91	2,890	4,143	N	5	1,106	1,708	–	15
W.N. CENTRAL	288	449	10,963	11,632	45	26	5,485	5,196	100	25
Minn.	50	79	1,830	2,473	20	12	650	860	–	1
Iowa	14	58	1,639	1,778	3	–	505	453	10	12
Mo.	139	210	4,437	4,431	8	12	3,312	2,962	86	3
N. Dak.	4	3	290	339	1	1	29	23	–	2
S. Dak.	7	2	616	442	1	–	104	41	–	–
Nebr.	32	34	915	724	6	–	331	266	2	1
Kans.	42	63	1,236	1,445	6	1	554	591	2	6
S. ATLANTIC	4,121	5,759	40,137	32,902	31	14	32,533	32,058	71	81
Del.	44	69	992	612	–	1	522	420	–	–
Md.	488	725	3,217	2,768	10	4	3,544	5,072	3	6
D.C.	343	400	N	N	–	–	1,365	1,516	–	–
Va.	284	484	3,500	4,287	N	7	2,336	3,177	1	8
W. Va.	36	38	1,150	1,220	N	–	324	379	3	5
N.C.	273	282	8,753	6,569	7	2	7,366	6,294	10	22
S.C.	283	293	7,320	4,751	1	–	4,725	4,236	–	17
Ga.	501	690	8,863	3,377	2	–	7,308	4,453	8	–
Fla.	1,869	2,778	6,342	9,318	10	–	5,043	6,511	46	23
E.S. CENTRAL	591	628	13,403	12,742	27	7	12,655	12,900	51	136
Ky.	87	61	2,329	2,546	6	–	1,298	1,695	9	6
Tenn.	184	302	4,735	4,845	16	7	3,963	4,070	39	83
Ala.	183	153	3,608	3,088	5	–	4,621	4,329	3	5
Miss.	137	112	2,731	2,263	–	–	2,773	2,806	–	42
W. S. CENTRAL	1,953	2,504	22,753	20,791	24	4	13,471	13,830	452	124
Ark.	71	96	1,228	1,022	1	1	1,108	1,713	1	4
La.	333	428	4,731	3,098	–	–	3,940	2,855	2	77

Table 15.4A (continued)

Reporting Area	AIDS Cum. 1998*	AIDS Cum. 1997	Chlamydia Cum. 1998	Chlamydia Cum. 1997	Escherichia coli O157:H7 NETSS† Cum. 1998	PHLIS§ Cum. 1998	Gonorrhea Cum. 1998	Gonorrhea Cum. 1997	Hepatitis C/NA,NB Cum. 1998	Hepatitis C/NA,NB Cum. 1997
Okla.	106	116	4,052	2,930	3	3	2,179	1,840	1	4
Tex.	1,443	1,864	12,742	13,741	20	–	6,244	7,422	448	39
MOUNTAIN	526	706	6,997	9,822	29	17	2,646	2,939	286	136
Mont.	13	18	415	391	2	–	21	14	4	5
Idaho	12	22	715	578	2	–	63	43	80	19
Wyo.	2	13	268	200	–	–	11	22	128	45
Colo.	91	194	–	1,736	5	4	917	752	10	17
N. Mex.	76	66	1,359	1,322	7	4	267	501	35	27
Ariz.	200	157	3,315	3,848	N	5	1,213	1,212	1	15
Utah	45	46	672	660	9	1	62	87	16	2
Nev.	87	190	253	1,087	4	3	92	308	12	6
PACIFIC	2,223	3,253	28,308	30,655	73	45	6,293	7,216	318	196
Wash.	165	241	4,301	3,575	17	22	718	784	9	10
Oreg.	64	144	2,137	1,807	23	17	292	275	2	2
Calif.	1,947	2,826	20,370	24,092	33	3	5,002	5,803	253	116
Alaska	11	18	726	541	–	–	127	175	1	–
Hawaii	36	24	774	640	N	3	154	179	53	68
Guam	–	2	8	181	N	–	2	24	–	–
P.R.	666	517	U	U	–	U	158	250	–	44
V.I.	15	28	N	N	N	U	–	–	–	–
Amer. Samoa	–	–	–	–	N	U	–	–	–	–
C.N.M.I.	–	–	N	N	N	U	7	15	–	2

N: Not notifiable   U: Unavailable   –: No reported cases   C.N.M.I.: Commonwealth of Northern Mariana Islands
*Updated monthly to the Division of HIV/AIDS Prevention–Surveillance and Epidemiology, National Center for HIV, STD, and TB Prevention, last update April 26; 1998.
†National Electronic Telecommunications System for Surveillance.
§Public Health Laboratory Information System.

of infection, the sites at which infections often occur, and the correlation between procedures and sites of infection.

Sources of Infection

Nosocomial infections can be exogenous or endogenous. **Exogenous infections** are caused by organisms that enter the patient from the environment. The organisms can come from other patients, staff members, or visitors. They can also be passed on by insects (ants, roaches, flies) from fomites (toilet, trash can) to patient. Other inanimate objects, such as equipment used in respiratory or intravenous therapy, catheters, bathroom fixtures and soap, and water systems, also can be a source of exogenous infections. Some nosocomial infections have even been traced to disinfectants such as quaternary ammonium compounds, to which certain organisms are resistant. **Endogenous infections** are caused by opportunists among the patient's own normal microflora. Opportunists are most likely to cause infection if the patient has lowered resistance or if normal microflora that compete with pathogens have been eliminated by antibiotics.

A small group of organisms, including *Escherichia coli, Enterococcus* species, *Staphylococcus aureus,* and *Pseudomonas* species, are responsible for about half of all nosocomial infections (➤ Figure 15.19). These organisms are particularly likely to cause such infections because they are ubiquitous (present everywhere) and can survive outside the body for long periods. In addition, some strains of these organisms are resistant to many antibiotics; methicillin- and vancomycin-resistant staphylococci and carbapenem-resistant *Pseudomonas aeruginosa* are especially problematic.

➤ Table 15.4B	Summary—provisional cases of selected notifiable diseases, United States, cumulative, week ending May 23, 1998 (20th Week)		
	Cum. 1998		**Cum. 1998**
Anthrax	–	Plague	–
Brucellosis	8	Poliomyelitis, paralytic¶	–
Cholera	3	Psittacosis	15
Congential rubella syndrome	2	Rabies, human	–
Cryptosporidiosis*	662	Rocky Mountain spotted fever (RMSF)	32
Diphtheria	1	Streptococcal disease, invasive Group A	893
Encephalitis: California*	–	Streptococcal toxic-shock syndrome*	25
eastern equine*	–	Syphilis, congenital**	64
St. Louis*	–	Tetanus	8
western equine*	–	Toxic-shock syndrome	52
Hansen Disease	45	Trichinosis	5
Hantavirus pulmonary syndrome*†	2	Typhoid fever	109
Hemolytic uremic syndrome, post-diarrheal*	10	Yellow fever	–
HIV infection, pediatric*§	88		

– No reported cases
*Not notifiable in all states.
†Updated weekly from reports to the Division of Viral and Rickettsial Diseases, National Center for Infectious Diseases (NCID).
§Updated monthly to the Division of HIV/AIDS Prevention–Surveillance and Epidemiology, National Center for HIV, STD, and TB Prevention (NCHSTP), last update April 26, 1998.
¶One suspected case of polio with onset in 1998 has also been reported to date.
**Updated from reports to the Division of STD Prevention, NCHSTP.

Susceptibility and Transmission

Compared with the general population, patients in hospitals are much more susceptible to infection; that is, they are **compromised hosts.** Many patients have breaks in the skin from lesions, wounds (surgical and accidental), or bed sores. Some also have breaks in mucous membranes that line the digestive, respiratory, urinary, or reproductive systems. The lack of intact skin and mucous membranes provides easy access for infectious organisms. Also, most patients are debilitated; their resistance to infectious organisms is lower than normal. Patients undergoing organ transplants receive immunosuppressant drugs, and patients with AIDS and other disorders of the immune system also have reduced resistance. Factors that contribute to host resistance are discussed in Chapters 16 and 17.

Theoretically, nosocomial infections can be transmitted by all modes of transmission that occur in the community. However, direct person-to-person transmission between an infected patient, staff member, or visitor and noninfected patients; indirect transmission through equipment, supplies, and hospital procedures; and transmission through air are most common in hospitals (➤ Figure 15.20). Some organisms can be transmitted by more than one route.

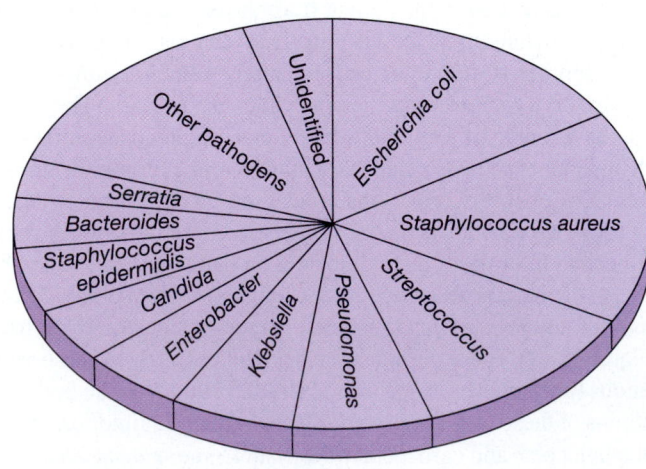

➤ Figure 15.19 **Common causative agents of nosocomial infections.**

Universal Precautions

In 1988, the CDC, concerned about the possibilities that the AIDS virus would be transmitted in the health care setting, issued guidelines to reduce the risks. These guidelines, called **Universal Precautions,** are summarized in Appendix D. Some hospitals and other medical facilities choose to exercise even more caution than the CDC recommends. A shortened list of these precautions is given in Table 15.5. Universal Precautions apply to *all* patients, not just those infected with the viruses that cause AIDS or hepatitis B—hence the term *universal*. Universal Precautions apply to the following body fluids: blood, semen, and vaginal, tissue, cerebrospinal, synovial (joint cavity), pleural, peritoneal, pericardial, and amniotic fluids. The CDC has stated that Universal Precautions do not apply to feces, nasal secretions, sputum, sweat, tears, urine, and vomitus, as long as these do not contain visible blood. This is not to imply that no viruses are present in these fluids but rather that the risk of transmission is either very low or unproved. For example, studies have shown that HIV is present in all tested samples of saliva from AIDS patients. However, the levels are so low that it takes PCR techniques to detect them. ➡ (Chapter 7, p. 192) AIDS patients are often infected with other disease organisms that may be present in these fluids. These organisms include tuberculosis bacilli in sputum, bacteria such as *Salmonella* and *Shigella, Cryptosporidium* protozoa in feces, and herpesvirus in oral secretions. Therefore, some health care facilities require their employees to use Universal Precautions with all body fluids.

Equipment and Procedures That Contribute to Infection

Surgical procedures and the use of equipment such as catheters and respiratory devices are major contributors to nosocomial infections. The smallest abrasion can provide a site of entry for infectious agents. Infections can arise from a contaminated catheter, inadequate cleaning of the site of catheter insertion, or the movement of organisms from leaky connections. In addition, tubing, joints, containers of fluids, and the fluids themselves also can be contaminated.

All surgical procedures expose internal body parts to air, instruments, surgeons, and other operating room personnel, all of which can be contaminated. These procedures can also allow the patient's own microflora to enter sites where they can produce infection. For example, bacteria that cause pneumonia can reach the lungs from the pharynx during surgery.

➤ Figure 15.20   **Some common modes of transmission of nosocomial infections in a hospital setting.**

➤ Table 15.5	Some important universal precautions and recommendations from CDC

1. Wear gloves and gowns if soiling of hands, exposed skin, or clothing with blood or body fluids is *likely.*
2. Wear masks *and* protective eyewear or chin-length plastic faceshields whenever splashing or splattering of blood or body fluids is *likely.* A mask alone is not sufficient.
3. Wash hands before and after patient contact, and after removal of gloves. Change gloves between *each* patient.
4. Use disposable mouthpiece/airway for cardiopulmonary resuscitation.
5. Discard contaminated needles and other sharp items *immediately* into a *nearby,* special puncture-proof container. Needles must *not* be bent, clipped, or recapped.
6. Clean spills of blood or contaminated fluids by: (1) Putting on gloves and any other barriers needed, (2) wiping up with disposable towels, (3) washing with soap and water, and (4) disinfecting with a 1:10 solution of household bleach and water. Allow it to stand on surface for at least 10 minutes. Bleach solution should not have been prepared more than 24 hours beforehand.

Respiratory devices, including nebulizer jets, that administer oxygen or air and medications to expand passageways in the lungs provide a means for disseminating microorganisms deep into the lungs. Organisms can grow in the reservoir pans of both cold mist and warm steam humidifiers, and the organisms can be dispersed in an aerosol as the machines operate. Therefore, all respiratory equipment should be disinfected or sterilized daily and, if not disposable, should be disinfected before being moved from one patient to the next.

Other devices and procedures account for smaller, but significant, numbers of nosocomial infections. For example, hemodialysis, a procedure for removing wastes from blood, provides a variety of means of introducing microorganisms into the body (➤ Figure 15.21). Devices used to monitor blood pressure in the heart or major vessels or cerebrospinal fluid pressure have tubing extending outside the body. Such devices can be contaminated or can allow introduction of organisms from the patient or from the environment. Improperly cleaned gynecological instruments can transmit disease from one patient to another. Endoscopes, which are introduced through body openings and used to examine the linings of organs such as the bladder, large intestine, stomach, and respiratory passageways, are difficult to sterilize and thus can transfer microorganisms from one patient to another.

Another important factor that contributes to nosocomial infections is the intensive use of antibiotics, especially in hospital settings. How antibiotics contribute to the development of antibiotic-resistant pathogens and how these pathogens contribute to nosocomial infections were discussed in Chapter 13.

Sites of Infection

The sites of nosocomial infections, in order from most to least common, are as follows: urinary tract, surgical wounds, respiratory tract, skin (especially burns), blood (bacteremia), gastrointestinal tract, and central nervous system (➤ Figure 15.22).

Preventing and Controlling Nosocomial Infections

The problem of nosocomial infections is widely recognized, and nearly all hospitals now have infection-control programs. In fact, to maintain accreditation by the American Hospital Association, hospitals must have programs that include surveillance of nosocomial infections in both patients and staff, a microbiology laboratory, isolation procedures, accepted procedures for the use of catheters and other instruments, general sanitation procedures, and a nosocomial disease education program for staff members. Most hospitals have an infection-control specialist to manage such a program.

Several techniques are available to prevent the introduction and spread of nosocomial infections. Hand washing is the single most important technique. Physicians, nurses, and other staff members who wash hands thoroughly with soap and water between patient contacts can greatly reduce the risk of spreading diseases among patients. Scrupulous care in obtaining sterile equipment and

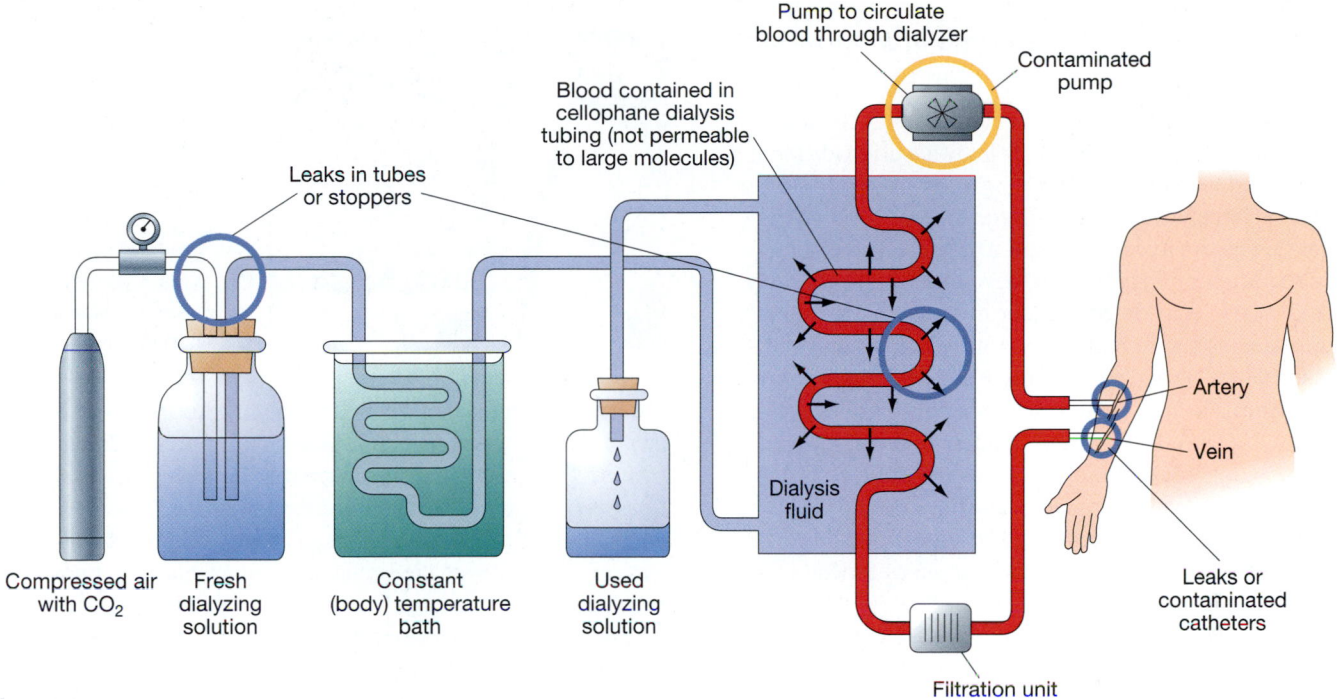

> Figure 15.21   **Possible sites of contamination in hemodialysis equipment.**

maintaining its sterility while in use are also important. In addition, the use of gloves when handling infectious materials such as dressings and bedpans and when drawing blood prevents the spread of infections. And, as mentioned earlier, it is important to prevent insect infestations, as flies, ants, or roaches can easily spread an infectious agent.

Other techniques are needed to reduce the development of antibiotic-resistant pathogens. Routine use of antimicrobial agents to prevent infections has turned out to be a misguided effort because it contributes to the development of resistant organisms. Therefore, some hospitals maintain surveillance of antibiotic use. Antibiotics are given for known infections but are given prophylactically (as preventive measures) only in special situations. Prophylactic antibiotics are justified in surgical procedures such as those involving the intestinal tract and repair of traumatic injuries, in which the surgical field is invariably contaminated with potential pathogens. They also are justified in immunosuppressed and excessively debilitated patients, whose natural defense mechanisms may fail.

If all the known techniques for preventing nosocomial infections were practiced rigorously, the incidence of such infections probably could be reduced to half the present level.

✓ List the Universal Precautions.

✓ What is meant by a nosocomial infection?

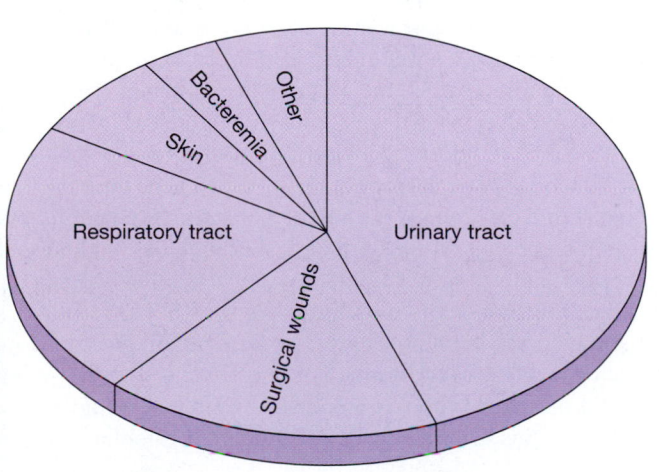

> Figure 15.22   **Relative frequencies of sites of nosocomial infections.**

Controlling Nosocomial Infections in a Burn Unit

An Interview

The Institute of Surgical Research, at Brooke Army Medical Center in Fort Sam Houston, Texas, is known worldwide for its pioneering work in burn treatment. Dr. Arthur D. Mason, Jr., a surgeon, is largely responsible for the research on burn infections and ways to combat them. Dr. Albert McManus, chief of the Microbiology Branch, developed a microbiology laboratory that is considered a leader in research of surgical wound infections and in burn care. Colonel Elisabeth Greenfield is the Chief Nurse of the Institute's Burn Center.

Drs. Arthur Mason, left, and Albert McManus in the laboratory.

Dr. Mason: The treatment of burn victims has changed a great deal in the past half century. When I first entered the field in 1954, two of the major advances in burn care had recently been made. We had learned that skin grafting could be used to replace the skin destroyed by the burn and that shock could be averted by the administration of fluids to replace those lost to the burn.

The advent of effective fluid replacement sharply decreased early loss of life due to burn shock. Thus, a whole new population of burn patients emerged–severely injured people who would not previously have lived for more than a day or two. But, although these patients now lived longer, many still died. In most of these cases, burn wound infection and bacteremia preceded death. Nevertheless, bacteremia was not generally considered to be the cause of death.

Dr. McManus: At the time, bacterial infection of burns was thought to be significant only in so far as it interfered with healing of the burned wound. Organisms such as *Pseudomonas aeruginosa* and other Gram-negative bacteria often were found in the blood and wounds of burn patients. Because these organisms were not recognized as causes of infections in other surgical situations, they were considered nonpathogens. Moreover, the first generations of antibiotics available, such as penicillin, were effective mostly against streptococci and staphylococci—the traditional agents causing wound infections. These antibiotics were of little use against *Pseudomonas,* so treating badly burned patients with these drugs usually made no difference in mortality. Such deaths were considered caused by the burn itself. This fact masked the true role of infection with Gram-negative organisms as an often fatal complication in burn patients.

Dr. Mason: During the 1950s we at the Institute set out on a systematic examination of the process of infection in burns. By the end of the decade we had developed an animal model of *Pseudomonas* sepsis in which contamination of nonlethal burn wounds with *P. aeruginosa* consistently caused death. In this way we were able to convince the medical community that often it wasn't the burn itself that proved lethal, but rather the generalized bacteremia that developed—especially with Gram-negative organisms.

The next step naturally was to explore possible antimicrobial therapies. We soon found that antibiotics in the bloodstream failed in human infections because they did not penetrate the dead tissue of burns, where there was no longer any blood sup-

ply. Even newer antibiotics, such as polymyxin B, that had high activity in the laboratory against strains taken from burned patients did not reach burn wounds in high enough concentrations to be effective. Dead burned tissue was the site of contamination and growth of organisms; by the time the organisms penetrated to the circulatory system, bacterial numbers had become too large for effective antibiotic action.

The logical alternative approach was to try and deliver a drug directly to the surface of the burn site. After much experimentation, we eventually found that sulfamylon, an antibacterial agent used by the German army for wounds during World War II, was highly effective in controlling bacterial growth in burn wounds. The results were remarkable: Bacteremia was much less frequent, and mortality decreased sharply.

But this Utopian result—a single, simple, effective therapy—didn't last; bacteria are much too versatile for that. In 1969 a strain of *Providencia stuartii* that was resistant to sulfamylon appeared in our patients. Within a year or two we again had a serious infection problem. This challenge was ultimately met with another topical agent, but it was clear that the topical approach alone was not a complete answer. It fell to Dr. McManus to develop a more complete method for infection control in these patients.

Dr. McManus: Clearly, microbial adaptation in response to sulfamylon had occurred. These strains were probably being maintained somewhere in the ward environment and being passed to new patients after admission. In addition to organisms already resistant and present in the burn ward, with each transfer of a patient from another hospital to our burn unit there was a chance that other types of resistant nosocomial pathogens would be added to the burn unit flora.

Most microbiology cultures were taken to identify an organism that had produced the clinical symptoms of infection. This type of information was very useful for describing what infections were occurring, but it helped very little in trying to find the sources and mechanisms for perpetuating resistant organisms on the burn unit. We really did not know how many patients were colonized with these same organisms but were not cultured because they were not sick. Without knowing how commonly cross-contamination occurred in such patients, it was essentially impossible to stop the spread. I therefore proposed a culture system that would provide a wider sample of data, reflecting the colonizing as well as infecting organisms in patients of all burn sizes, at admission and throughout their hospitalization.

This Microbial Surveillance System was based on the premise that colonization of the burn wound and other damaged sites occurs soon after injury. Thus, if the patient survives the initial shock, fluid loss, and the burn itself, colonization will occur from the environment or from the patient's own preinjury flora. Because colonization precedes infection, the information would be readily available to help identify an infecting organism at the time of a clinical change if it occurred. So on admission or during transport by our burn flight team to the burn unit, the patient is cultured at the likely places for infection. Samples of wound, sputum, and urine are taken. These sites are sampled three times per week for the duration of care. In addition, the stool is sampled weekly and cultured on selective media to detect resistance to antibiotics that may be used. Organisms are identified, and antibiotic susceptibility tests are completed on potential problem organisms, such as *Pseudomonas aeruginosa* and *Staphylococcus aureus*. The information is entered into a computerized database available in the burn unit. The culture results are updated daily. Microbial colonization can thus be followed for each patient as well as for the entire patient population on the ward. The population data are used by Colonel Greenfield and her staff to maintain and improve patient isolation and infection control.

Col. Greenfield: As Dr. McManus stated, the data provided by the Microbial Surveillance System are extremely useful to us in many respects. They play an important part in our assigning of nursing care teams to specific groups of patients to reduce the potential for cross-contamination. Patients with specific microbial considerations, such as resistant organisms or increased susceptibility to infection, are further isolated and identified as requiring "cohort" isolation. An individual assigned to a cohorted patient may not provide care to any other patient that day. Ideally, we schedule days off for the individuals caring for such a patient before assigning them to other patients.

Microbial surveillance data are also an excellent Quality Improvement mechanism. Daily examination of the data allows for early identification of patient cross-contamination and modification of infection control techniques. These data serve as an objective monitor for many of our infection control policies, such as dress code, handwashing, and traffic control.

Finally, microbial surveillance is a basis for monitoring and revising existing infection control policies or for creating new ones. Any consistent pattern of cross-contamination cues an evaluation of current policies.

Dr. McManus: Of course, knowing what organisms are present or are coming doesn't always guarantee that we have the weapons to deal with them successfully. More than a decade ago, before we had the patient isolation rooms, the Institute was called to evacuate three American merchant seamen who had been burned in a fire in their ship's engine room and were admitted to a hospital in a developing country. They

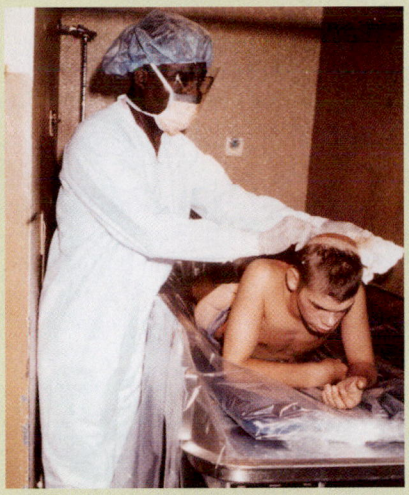

A nurse showering a patient in the Burn Center. Care of burn victims requires much contact between patient and medical staff; cross-contamination of patients is a danger.

were infected with a strain of *Providencia stuartii*. We had done surveillance cultures on the airplane, so we knew the patients had the organism prior to admission at our hospital. Within a few days, all three developed serious infections, and despite the use of infection-control measures, the strain quickly spread to other patients. In addition to being highly transmissible, the organism proved to be resistant not only to many of the antibiotics we could use but also to the disinfectants available for cleaning the unit. The strain harbored a plasmid that was loaded with resistance genes. Within a few weeks, other Gram-negative organisms were isolated with essentially the same antibiotic resistance pattern. Not only were we having an outbreak of *Providencia* infections but antibiotic resistance was being spread to other organisms by a broad-host-range plasmid. (Chapter 8, p. 209).

We finally eliminated this troublesome strain and its plasmid by directing all new admissions into a new single-bed-isolation burn unit. The surveillance system documented that the resistance problem was not transmitted to the new ward. We have not had an outbreak of such an organism since we have been in the single-bed-isolation facility. Not only have endemics of resistant organisms stopped, but we have also documented major reductions in the incidence and outcome of infections. *Pseudomonas aeruginosa,* which was the major burn wound pathogen during the 1960s and 1970s, has become a rare problem with only two documented wound invasions in more than 2000 burn admissions. Mortality associated with infection has also significantly declined.

Col. Greenfield: Single-bed isolation was a giant step forward in reducing the incidence of infections in this institution. A comprehensive infection control program, under the direction of an Infection Control Committee, reinforces these statistics. In addition to staffing and isolation guidelines, the infection control program provides for strict surveillance of traffic policies and dress codes for both staff and visitors, environmental surveillance, equipment selection, and employee health. All staff assigned to the Institution, either permanently or temporarily, undergo occupational health screening and provide proof of hepatitis B vaccination.

Retracing Our Steps

Epidemiology

What Is Epidemiology?

- **Epidemiology** is the study of factors and mechanisms in the spread of diseases in a population.

- In describing infectious diseases, **epidemiologists** use **incidence** to refer to the number of new cases in a specific period, **prevalence** to refer to the number of people infected at any one time, **morbidity rate** to indicate the number of cases as a proportion of the population, and **mortality rate** to indicate the number of deaths as a proportion of the population.

Diseases in Populations

- In a **sporadic disease,** several isolated cases appear in a population. In an **endemic** disease, many cases appear in a population, but the harm to patients is not sufficient to create a public health problem. In an **epidemic** disease, many cases appear in a population, and patients are sufficiently harmed to create a public health problem. Diseases that spread by **common-source outbreaks** originate with a single contaminated substance, such as a water supply. In **propagated epidemics,** diseases spread by person-to-person contact. A **pandemic** disease is an epidemic disease that has spread over an exceptionally wide geographic area or several geographic areas.

Epidemiologic Studies

- The purpose of **epidemiologic studies** is to learn more about the spread of diseases in populations and how to control those diseases.

- The methods of epidemiologic studies are **descriptive, analytical** (retrospective and prospective), and **experimental.**

Reservoirs of Infection

- **Reservoirs of infection** include humans, other animals, and non-living sources from which infectious diseases can be transmitted.

- Among human reservoirs, **carriers** often transmit diseases. They are intermittent carriers if they periodically release pathogens.

- Diseases in animal reservoirs can be transmitted by direct contact with animals or by vectors. Diseases that can be naturally transmitted from animals to humans are called **zoonoses.**

- Diseases in nonliving reservoirs are transmitted by water, soil, or wastes.

Portals of Entry

- **Portals of entry** include skin, mucous membranes that line various body systems, tissues, and the placenta.

Portals of Exit

- **Portals of exit** include the nose, ear, mouth, skin, and openings from which products of the digestive, urinary, and reproductive systems are released. Organisms usually are in body fluids or feces.

Modes of Disease Transmission

- Transmission can be by contact, vehicle, or vector. **Direct contact transmission** includes person-to-person **horizontal transmission** and **vertical transmission** from parent to offspring. **Indirect contact transmission** occurs through **fomites** (inanimate objects) and by droplets. **Vehicles** of transmission include water, air, and food. **Vectors** of transmission are usually arthropods, which can transmit disease mechanically or biologically.

- Transmission by carriers, transmission of STDs, and transmission of zoonoses pose special epidemiologic problems.

Disease Cycles

- Some diseases occur in cycles—there are just a few cases for several years, and then many cases suddenly appear.

Herd Immunity

- **Herd immunity,** or group immunity, refers to immunity enjoyed by a large proportion of a population that reduces disease transmission among nonimmune individuals.

- A drop in herd immunity can lead to the sudden appearance of cases of a cyclic disease.

Controlling Disease Transmission

- Methods used to control communicable diseases include isolation, quarantine, immunization, and vector control.

- **Isolation** procedures are summarized in Table 15.2. **Quarantine** is rarely used but can prevent exposed individuals from infecting others. Active immunization prevents many infections. Vector control is effective where vectors can be identified and eradicated.

Public Health Organizations

- Public health organizations exist at city, county, state, federal, and world levels. They help establish and maintain health standards, cooperate in the control of infectious diseases, collect and disseminate information, and assist with professional and public education.

Notifiable Diseases

- **Notifiable diseases** are listed in Table 15.3.

Nosocomial Infections

- A **nosocomial infection** is an infection acquired in a hospital or other medical facility.

The Epidemiology of Nosocomial Infections

- Nosocomial infections can be **exogenous** (caused by external organisms) or **endogenous** (caused by opportunists in normal microflora). About half are caused by only four types of pathogens, of which many strains are antibiotic resistant.

- Host susceptibility is an important factor in the development of nosocomial infections.

- Medical equipment and procedures, including surgery, are often responsible for infections.

- Modes of nosocomial infections are illustrated in ➤ Figure 15.20.

Preventing and Controlling Nosocomial Infections

- Most hospitals have an extensive infection control program. Hand washing, use of gloves, scrupulous attention to maintaining sanitary conditions and sterility where possible, and surveillance of antibiotic use and other hospital procedures help reduce infections.

- Nosocomial infections could be reduced by half if all known procedures were carefully followed in all medical facilities at all times.

Terminology Check

aerosol (*p. 413*)
analytical study (*p. 408*)
carrier (*p. 409*)
common-source outbreak (*p. 405*)
compromised host (*p. 426*)
contact transmission (*p. 413*)
descriptive study (*p. 407*)
direct contact transmission (*p. 413*)
direct fecal–oral transmission (*p. 413*)
droplet nucleus (*p. 413*)
droplet transmission (*p. 413*)
endemic (*p. 403*)
endogenous infection (*p. 425*)
epidemic (*p. 403*)
epidemiologic study (*p. 405*)
epidemiologist (*p. 402*)

epidemiology (*p. 402*)
etiology (*p. 402*)
exogenous infection (*p. 425*)
experimental study (*p. 408*)
fomite (*p. 413*)
herd immunity (*p. 416*)
horizontal transmission (*p. 413*)
inapparent infection (*p. 409*)
incidence (*p. 403*)
index case (*p. 408*)
indirect contact transmission (*p. 413*)
indirect fecal–oral transmission (*p. 413*)
isolation (*p. 417*)
morbidity rate (*p. 403*)
mortality rate (*p. 403*)
nosocomial infection (*p. 423*)

notifiable disease (*p. 422*)
pandemic (*p. 405*)
placebo (*p. 408*)
portal of entry (*p. 410*)
portal of exit (*p. 412*)
prevalence (*p. 403*)
propagated epidemic (*p. 405*)
quarantine (*p. 418*)
reservoir of infection (*p. 408*)
sporadic disease (*p. 405*)
subclinical infection (*p. 409*)
Universal Precautions (*p. 427*)
vector (*p. 415*)
vehicle (*p. 413*)
vertical transmission (*p. 413*)
zoonosis (*p. 409*)

See You on the Web

Don't forget I am expecting you on the web. The following is a preparedness checklist for your web explorations. If you have questions on any of these items, you can return to the end of Chapter 1 for more details.

- Questions We'll Explore
- Concept Check questions
- Flashcards
- Study Guide

If you are ready, come visit me at

http://www.prenhall.com/black

It will be a journey you'll never forget.

Nonspecific Host Defenses and Host Systems

It's late, dark, everyone has gone. Who is that stranger coming towards me? Defense mechanisms immediately whirl into action: adrenaline, pounding heart, extra sugar released.

What can a plant do when microbial danger threatens? It *can't* run away. All plants do have built-in defense mechanisms. However, they need to be able to trigger them in time.

The signal appears to be an active oxygen burst produced by plant cells in response to the presence of bacteria. This may signal cells to produce lignin, one of the tough components of wood. Lignin then helps retard the movement of invading bacteria through cell walls and within the plant. Or, it might be that the signal causes plant cells to commit suicide, collapse around bacteria, and trap them. Genetic engineers may eventually transfer super-effective defense-signaling genes to important crop plants.

Background Basics

portals of entry and exit . . . . *(Chapter 15)*
phagocytosis . . . . *(Chapter 4)*
the disease process . . . . *(Chapter 14)*

We can look at infectious disease as a battle between the power of infectious agents to invade and damage the body and the body's powers to resist such invasions. In Chapters 14 and 15 we considered how infectious agents enter and damage the body and how they leave the body and spread through populations. In the next three chapters we consider how the body resists invasion by infectious agents.

We begin this chapter by distinguishing between specific and nonspecific defenses. Then we look at the nonspecific defense mechanisms in more detail to see how they function in protecting the body against infectious agents.

Nonspecific and Specific Host Defenses

With potential pathogens ever present, why do we rarely succumb to them in illness or death? The answer is that our bodies have defenses for resisting the attack of many dangerous organisms. Only when our resistance fails do we become susceptible to infection by pathogens.

Host defenses that produce resistance can be specific or nonspecific. **Specific defenses** respond to particular agents called *antigens*. Viruses and pathogenic bacteria are examples of such antigens. Specific defenses then respond to these antigens by producing protein *antibodies*. The human body is capable of making thousands of different antibodies, each effective against a particular antigen. Specific responses also involve the activation of the *lymphocytes,* specific cells of the body's immune system. These antibody and cellular responses are more effective against succeeding invasions by the same pathogen than against initial invasions. Chapter 17 focuses on these specific defenses of the immune system.

In the case of many threats to an individual's well being, specific defenses do not need to be called on because the body is adequately protected by its **nonspecific defenses**—those that act against any type of invading agent. Often such defenses perform their function before specific body defense mechanisms are activated. Nonspecific defenses include the following:

Questions We'll Explore

A How do nonspecific and specific host defenses differ?

B What are the physical barrier defenses?

C What kinds of cells form cellular barrier defenses?

D What are the stages of phagocytosis, and what kinds of cells are involved?

E What is inflammation, and what do the steps in acute inflammatory processes accomplish?

F What are the causes and the effects of chronic inflammation?

G How does fever function as a nonspecific defense mechanism?

H What roles do interferons play in molecular defense mechanisms?

I How do the molecular defense mechanisms involving complement work?

J What is the acute phase response?

1. *Physical barriers,* such as the skin and mucous membranes and the chemicals they secrete.
2. *Chemical barriers,* including antimicrobial substances in body fluids such as saliva, mucus, gastric juices, and the iron limitation mechanisms.
3. *Cellular defenses,* consisting of certain cells that engulf (phagocytize) invading microorganisms.
4. *Inflammation,* the reddening, swelling, and temperature increases in tissues at sites of infection.
5. *Fever,* the elevation of body temperature to kill invading agents and/or inactivate their toxic products.
6. *Molecular defenses,* such as interferon and complement, that destroy or impede invading microbes.

The physical and certain chemical barriers operate to prevent pathogens from entering the body. The other nonspecific defenses (cellular defenses, inflammation, fever, and molecular defenses) act to destroy pathogens or inactivate their toxic products that have gained entry or to prevent the pathogens from damaging additional tissues. The nonspecific defenses serve as the body's *first lines of defense* against pathogens. The specific defenses represent the *second lines of defense.* Let's look at each of the nonspecific defenses now; we will discuss the specific defenses in Chapter 17.

Physical Barriers

The skin and mucous membranes protect your body and internal organs from injury and infectious agents. These two physical barriers are made of cells that line the body surfaces and secrete chemicals, making the surfaces hard to penetrate and inhospitable to pathogens. The **skin,** for example, not only is exposed directly to microorganisms and toxic substances but also is subject to objects that touch, abrade, and tear it. Sunlight, heat, cold, and chemicals can damage the skin. Cuts, scratches, insect and animal bites, burns, and other wounds can disrupt the continuity of the skin and make it vulnerable to infection.

Besides the skin, a **mucous membrane,** or *mucosa,* covers those tissues and organs of the body cavity that are exposed to the exterior. Mucous membranes, therefore, are another physical barrier that makes it difficult for pathogens to invade internal body systems.

The hairs and mucus of the nasal and respiratory system present mechanical barriers to invading microbes. But so do the physical reflex flushing activities of coughing and sneezing. Vomiting and diarrhea similarly act to flush harmful microbes and their chemical products from the digestive tract. Likewise urinary flow is important in removing microbes that enter the urinary tract. Urinary tract infections are especially common among those unable to empty their bladder completely or frequently enough.

Cellular Defenses

Although the physical defense barriers do an excellent job of keeping microbes out of our bodies, we constantly suffer minor breaches of the physical defense barriers. A paper cut, the cracking of dry skin, or even brushing our teeth may temporarily breach the physical defenses and allow some microbes to enter the blood or connective tissue. However, we survive these daily attacks because ever-present cellular defenses can kill invading microbes or remove them from the blood or tissues.

When the skin is broken by any kind of trauma, microorganisms from the environment may enter the wound. Blood flowing out of the wound helps remove the microorganisms. Subsequent constriction of ruptured blood vessels and the clotting of blood help seal off the injured area until more permanent repair can occur. Still, if microorganisms enter blood through cuts in the skin or abrasions in mucous membranes, cellular defense mechanisms come into play.

Defensive Cells

Cellular defense mechanisms use special-purpose cells found in the blood and other tissues of the body. Blood consists of about 60 percent liquid called **plasma** and 40 percent **formed elements** (cells and cell fragments). Formed elements include **erythrocytes** (red blood cells),

Take Two, not Twenty-Two

Do you know someone who is a chronic aspirin or ibuprofen user? These days most people use the "harmless" painkillers freely. But, these little pills can have deadly effects. The problem is that aspirin, ibuprofen, and acetaminophen aren't specific enough. Their beneficial effects come from their ability to permanently block an enzyme that promotes inflammation, pain, and fevers. Unfortunately, the drugs are even more effective at permanently inhibiting a related enzyme that is necessary for the health of the stomach and kidneys. Aspirin also disrupts the body's acid-base balance, which can lead to whole organs—the kidneys, the liver, and the brain—shutting down forever, depending on the amount ingested. Patients can also have seizures and develop heart arrythmias.

Phlegm, Anyone?

Remember that thick, viscous mucus you coughed up last time you had a cold? Pretty gross stuff. And even grosser when you think of the tons of microorganisms your body had trapped with it. With barriers like that, how did those flu organisms manage to infect your respiratory tract in the first place? Some organisms, unfortunately, have evolved ways to get through this mucus barrier. For example, the influenza virus has a surface molecule that allows it to firmly attach itself to cells in the mucous membrane. Cilia can't sweep the attached virus out. As another example, the organism that causes gonorrhea has surface molecules that allow it to bind to mucous membrane cells in the urogential tract. With ingenious microorganisms like these, thank goodness your body has other defenses that lie in wait to attack any organisms that make it past your body's physical barriers.

platelets, and **leukocytes** (white blood cells) (➤ Figure 16.1 and Table 16.1). All are derived from *pluripotent stem cells,* cells that form a continuous supply of blood cells, in the bone marrow. Platelets, which are short-lived fragments of large cells called *megakaryocytes,* are important components of the blood-clotting mechanism.

Leukocytes are defensive cells that are important to both specific and nonspecific host defenses. These cells are divided into two groups—granulocytes and agranulocytes—according to their cell characteristics and staining patterns with specific dyes.

Granulocytes   **Granulocytes** have granular cytoplasm and an irregularly shaped, lobed nucleus. They are derived from *myeloid stem cells* in the bone marrow (*myelos* is Greek for "marrow"). Granulocytes include basophils, mast cells eosinophils, and neutrophils, which are distinguished from one another by the shape of their cell nuclei and by their staining reactions with specific dyes. **Basophils** release *histamine,* a chemical that helps initiate the inflamma-

> A natural antibiotic, human beta-defensin-2, lurks on the human skin and when induced, can kill pathogens by punching holes in the bacterial membranes.

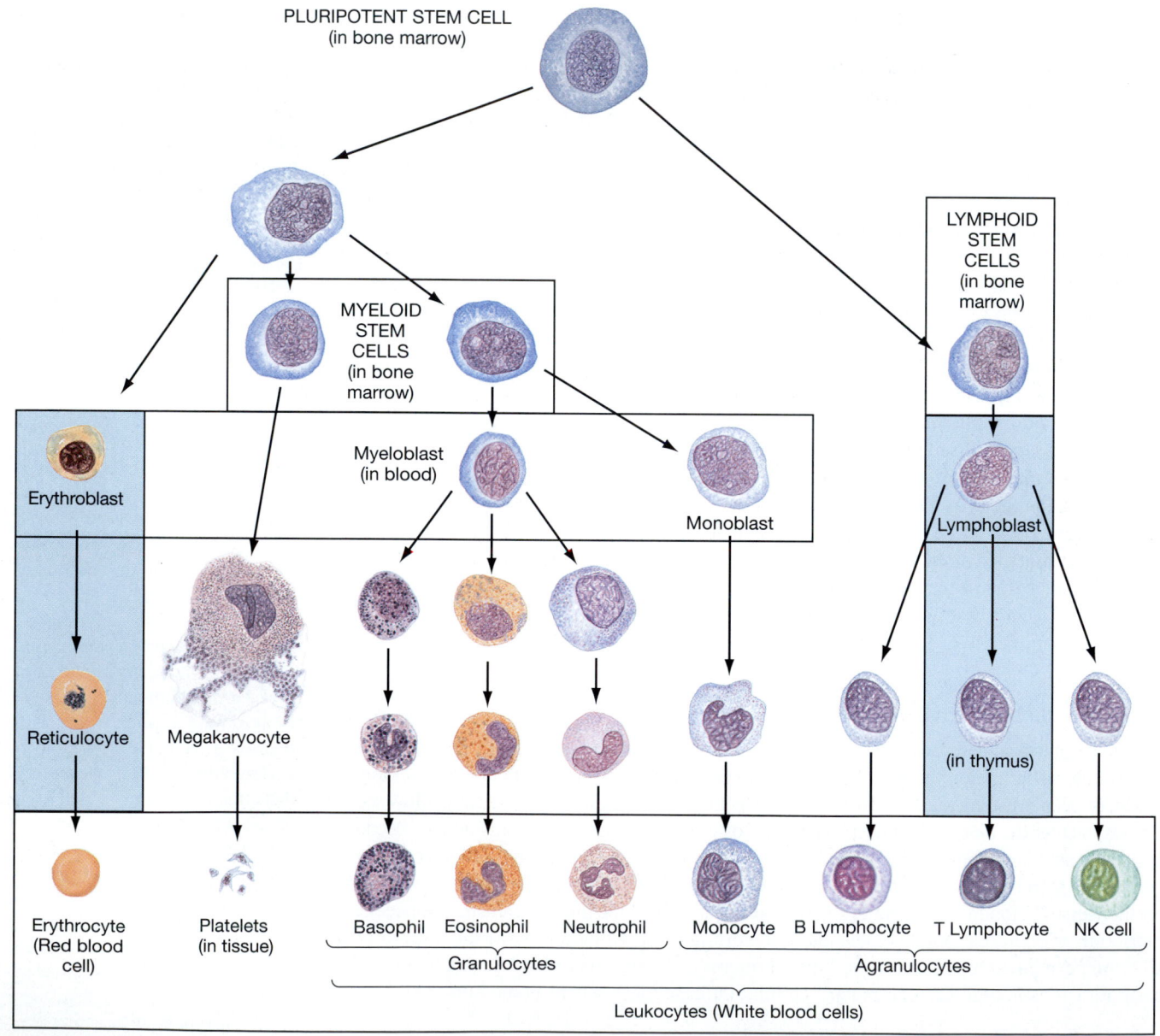

➤ Figure 16.1   **Formed (cellular) elements of the blood.** These elements are derived from pluripotent stem cells (cells that form an endless supply of blood cells) in the bone marrow. The myeloid stem cells differentiate into several kinds of leukocytes, called granulocytes and agranulocytes. Lymphoid stem cells differentiate into B lymphocytes (B cells), T lymphocytes (T cells), and natural killer cells (NK cells).

> ## Table 16.1   Formed elements of the blood in healthy adults

Element	Normal numbers (per microliter*)	Life span	Functions
Erythrocytes		120 days	Transport oxygen gas from lungs to tissues; transport carbon dioxide gas from tissues to lungs
Adult male	4.6 to 6.2 million		
Adult female	4.2 to 5.4 million		
Newborn	5.0 to 5.1 million		
Leukocytes	5000 to 9000	Hours to days	
Granulocytes			
Neutrophils	50–70% of total leukocytes		Phagocytic; contain oxidative chemicals to kill internalized microbes
Eosinophils	1–5% of total leukocytes		Release defensive chemicals to damage parasites (worms); phagocytic
Basophils	0.1% of total leukocytes		Release histamine and other chemicals during inflammation; responsible for allergic symptoms
Agranulocytes			
Monocytes	2–8% of total leukocytes		In tissues, develop into macrophages, which are phagocytic
Lymphocytes	20–40% of total leukocytes		Essential to specific host immune defenses; antibody production
Platelets	250,000 to 300,000	5–9 days	Blood clotting

*1 microliter (μl) = 1 mm^3 = 1/1,000,000 l.

tory response. **Mast cells,** which are prevalent in connective tissue and alongside blood vessels, also release histamine and are associated with allergies. **Eosinophils** (e-o-sin′o-fils) are present in large numbers during allergic reactions (Chapter 18) and worm infections. These cells may also detoxify foreign substances and help turn off inflammatory reactions. **Neutrophils,** also called *polymorphonuclear leukocytes* (PMNLs), guard the skin and mucous membranes against infection. These cells respond quickly wherever tissue injury has occurred.

Agranulocytes   **Agranulocytes** lack granular cytoplasm and have round nuclei. These cells include monocytes and lymphocytes. **Monocytes** are derived from myeloid stem cells, whereas **lymphocytes** are derived from *lymphoid stem cells,* again in the bone marrow. The lymphocytes contribute to specific host immunity. They circulate in the blood and are found in large numbers in the lymph nodes, spleen, thymus, and tonsils.

Neutrophils and monocytes are exceedingly important components of nonspecific host defenses. They are phagocytic cells, or *phagocytes*.

Phagocytes

Phagocytes are cells that literally eat (*phago,* Greek for "eating"; *cyte,* Greek for "cell") or engulf other materials. They patrol, or circulate through the body, destroying dead cells and cellular debris that must be removed constantly from the body as cells die and are replaced. Phagocytes also guard the skin and mucous membranes against invasion by microorganisms. Being present in many tissues, these cells first attack microbes and other foreign material at portals of entry, such as wounds in skin or mucous membranes. If some microbes escape destruction at the portal of entry and enter deeper tissues, phagocytes circulating in blood or lymph mount a second attack on them.

Alcoholics and Pneumonia

The opening to the larynx is covered by the epiglottis, a flap of tissue that prevents food and fluids from entering the larynx. When the epiglottis fails to operate properly, fluids can be drawn or sucked into the lungs. Such failure in alcoholics often leads to pneumonia. When alcoholics are "sleeping off" drunken episodes, the epiglottis gapes and allows pharyngeal secretions to be drawn into the lungs with breathing. If pneumonia-causing organisms are present in those secretions, they can gain access to and infect the lungs. Once in the lungs, it is unlikely that the microbes will be destroyed because alcoholics have "lazy leukocytes." In other words, their macrophages are less efficient in destroying invading microbes than are those of nonalcoholics.

Neutrophils are released into the blood from the bone marrow, circulate for 7–10 hours, and then migrate into the tissues where they have a 3-day life span.

The combined mass of all of the lymphocytes in your body is approximately equal to the mass of your brain or liver.

➤ Table 16.2	Names of fixed macrophages in various tissues
Name of macrophage	**Tissue**
Alveolar macrophage (dust cell)	Lung
Histiocyte	Connective tissue
Kupffer cell	Liver
Microglial cell	Neural tissue
Osteoclast	Bone
Sinusoidal lining cell	Spleen

The neutrophils are released from the bone marrow continuously to maintain a stable circulating population. An adult has about 50 billion circulating neutrophils at all times. If an infection occurs, they are usually first on the scene because they migrate quickly to the site of infection. Being avid phagocytes, they are best at inactivating bacteria and other small particles. They are not capable of cell division and are "programmed" to die after only 1 or 2 days.

The monocytes migrate from the bone marrow into the blood. When these cells move from blood into tissues, they go through a series of cellular changes, maturing into macrophages. **Macrophages** are "big eaters" (*macro,* Greek for "big") that destroy not only microorganisms but also larger particles, such as debris left from neutrophils that have died after ingesting bacteria. Although macrophages take longer than neutrophils to reach an infection site, they arrive in larger numbers than neutrophils.

Macrophages can be fixed or wandering. *Fixed macrophages* remain stationary in tissues and are given different names, depending on the tissue in which they reside (Table 16.2). *Wandering macrophages,* like the neutrophils, circulate in the blood, moving into tissues when microbes and other foreign material are present (➤ Figure 16.2). Unlike neutrophils, macrophages can live for months or years. As we will see in Chapter 17, besides having a nonspecific role in host defenses, macrophages also are critical to specific host defenses.

The Process of Phagocytosis

Phagocytes digest and generally destroy invading microbes and foreign particles by a process called **phagocytosis** (Chapter 4) or by a combination of immune reactions and phagocytosis ➡ (p. 102) If an infection occurs, neutrophils and macrophages use this four-step process to destroy the invading microorganisms. The phagocytic cells must (1) find, (2) adhere to, (3) ingest, and (4) digest the microorganisms.

Chemotaxis

Phagocytes first must find the invading microorganisms. Both the infectious agents and the damaged tissues release specific chemical substances to which phagocytes are attracted. In addi-

> Fixed macrophages are named to reflect their tissue location: in the liver, they are called Kupffer cells; in the connective tissues, histiocytes; in the lung, alveoloar macrophages.

➤ Figure 16.2   **False-color SEM of a macrophage moving over a surface (2150X).** The macrophage has spread out from its normal spherical shape and is using its ruffly cytoplasm to move itself and to engulf particles. Macrophages clear the lungs of dust, pollen, bacteria, and some components of tobacco smoke.

tion, basophils and mast cells release histamine, and phagocytes already at the infection site release chemicals called **cytokines** (si′to-kinz). These chemicals are a diverse group of soluble proteins that have specific roles in host defenses, including the attraction of additional phagocytes to the site of the infection. Phagocytes make their way to this site by **chemotaxis,** the movement of cells toward a chemical stimulus. ➡ (Chapter 4, p. 88) We will discuss cytokines in more depth in Chapter 17.

Some pathogens can escape phagocytes by interfering with chemotaxis. For example, most strains of the bacterium that causes gonorrhea (*Neisseria gonorrhoeae*) remain in the urogenital tract, but some strains escape local cellular defenses and enter the blood. Microbiologists believe the invasive strains fail to release the chemical attractants that bring phagocytes to the infection site.

Adherence and Ingestion

Following chemotaxis and the arrival of phagocytes at the infection site, the infectious agents become attached to the plasma membranes of phagocytic cells. The ability of the phagocyte cell membrane to bind to specific molecules on the surface of the microbe is called **adherence.**

A fundamental requirement for many pathogenic bacteria is to escape phagocytosis. The most common means by which bacteria avoid this defense mechanism is an *antiphagocytic capsule.* The capsules present on bacteria responsible for pneumococcal pneumonia (*Streptococcus pneumoniae*) and childhood meningitis (*Haemophilus influenzae*) make adherence difficult for phagocytes. The cell walls of the bacterium responsible for rheumatic fever (*Streptococcus pyogenes*) contain molecules of *M protein,* which interferes with adherence.

To overcome such resistance to adherence, the host's nonspecific defenses can make microbes more susceptible to phagocytosis. If microbes are first coated with antibodies, or with proteins of the *complement system* (to be discussed later in this chapter), phagocytes have a much easier time binding to the microbes. Because both these mechanisms represent molecular defenses, we will discuss them later in this chapter.

Once captured, phagocytes rapidly ingest (engulf) the microbe. The cell membrane of the phagocyte forms fingerlike extensions, called *pseudopodia,* that surround the microbe (➤ Figure 16.3a). These pseudopodia then fuse, enclosing the microbe within a cytoplasmic vacuole called a **phagosome** (➤ Figure 16.3b).

Complex antigens (substances that the body identifies as foreign), such as whole bacteria or viruses, tend to adhere well to phagocytes and are readily ingested.

Digestion

Phagocytic cells have several mechanisms for digesting and destroying ingested microbes. One mechanism uses the *lysosomes* found in the phagocyte's cytoplasm. ➡ (Chapter 4, p. 102) These organelles, which contain digestive enzymes and small proteins called *defensins,* fuse with the phagosome membrane, forming a **phagolysosome** (➤ Figure 16.3b). (More than 30 different types of antimicrobial enzymes have been identified within lysosomes.) In this way the digestive enzymes and defensins are released into the phagolysosome. The defensins eat holes in the microbes, allowing lysosomal enzymes to digest almost any biological molecule they contact. Thus, lysosomal enzymes rapidly (within 20 minutes) destroy the microbes, breaking them into small molecules (amino acids, sugars, fatty acids) that the phagocyte can use as building blocks for its own metabolic and energy needs.

Macrophages can also use other metabolic products to kill ingested microbes. These phagocytic cells use oxygen to form hydrogen peroxide (H_2O_2), nitric oxide (NO), superoxide ions (O_2^-), and hypochlorite ions (OCl^-). (Hypochlorite is the ingredient in household bleach that accounts for its antimicrobial action.) All these molecules are effective in damaging plasma membranes of the ingested pathogens.

Once the microbes have been destroyed, there may be some indigestible material left over. Such material remains in the phagolysosome, which now is called a *residual body.* The phagocyte transports the residual body to the plasma membrane, where the waste is excreted (➤ Figure 16.3b).

Just as some microbes interfere with chemotaxis and others avoid adherence, some microbes have developed mechanisms to prevent their destruction within a phagolysosome. In fact, a few pathogens even multiply within phagocytes. Some microbes resist digestion by phagocytes in one of three ways:

➤ Figure 16.3 **Phagocytosis of two bacterial cells by a neutrophil.** (a) Extensions of cytoplasm, called pseudopodia, surround the bacteria. Fusion of the pseudopodia forms a cytoplasmic vacuole, called a phagosome, containing the bacteria. (b) Phagocytes find their way to a site of infection by means of chemotaxis. Phagocytes, including macrophages and neutrophils, have proteins in their plasma membranes to which a bacterium adheres. The bacterium is then ingested into the cytoplasm of the phagocyte as a phagosome, which fuses with lysosomes to form a phagolysosome. The bacterium is digested, and any undigested material within the residual body is excreted from the cell.

1. Some bacteria, such as those that cause the plague (*Yersinia pestis*), produce capsules that are not vulnerable to destruction by macrophages. If these bacteria are engulfed by macrophages, their capsule protects them from lysosomal digestion, allowing the bacteria to multiply, even within a macrophage.

2. Other bacteria—such as those that cause Hansen's disease, or leprosy (*Mycobacterium leprae*), and tuberculosis (*M. tuberculosis*)—and the protozoan that causes leishmaniasis (*Leishmania* species) can resist digestion by phagocytes. In the case of *Mycobacterium,* each engulfed bacillus resides in a membrane-enclosed, fluid-filled compartment called a *parasitophorous vacuole* (PV). No lysosomal enzyme activity is associated with the PVs. These organisms' resistance to lysosomal activity is due to the complexity of their acid-fast cell walls ➥ (Chapter 4, p. 80), which consist of wax D and mycolic acids. Lysosomal enzymes are unable to react with and digest these components. As the bacilli reproduce, new PVs arise. For *Leishmania* infections, each PV contains several protozoan cells. Although the lysosomal enzymes are active in these PVs, microbiologists do not understand how the pathogens resist digestion.

3. Still other microbes produce toxins that kill phagocytes by causing the release of the phagocyte's own lysosomal enzymes into its cytoplasm. Examples of such toxins are **leukocidin,** released by staphylococci, and **streptolysin,** released by streptococci.

Thus, some pathogens survive phagocytosis and can even be spread throughout the body in the phagocytes that attempt to destroy them. Because macrophages can live for months, they can provide pathogens with a long-term, stable environment in which they can multiply out of the reach of other host defense mechanisms.

Extracellular Killing

The phagocytic process described previously represents *intracellular killing*—that is, the microbe is degraded within a defensive cell. However, other microbes, such as viruses and parasitic worms, are destroyed without being ingested by a defensive cell; they are destroyed *extracellularly*.

Neutrophils and macrophages are too small to engulf a large parasite such as a worm (helminth). Therefore, another leukocyte, the eosinophil, takes the leading role in defending the body. Although eosinophils can be phagocytic, they are best suited for excreting toxic enzymes such as *major basic protein* (MBP) that can damage or perforate a worm's body. Once such parasites are destroyed, macrophages can engulf the parasite fragments.

Viruses must get inside cells to multiply. ➡ (Chapter 1, p. 254) Therefore, host defenses must eliminate such infectious agents before they can reproduce in the cells they have infected. The leukocytes responsible for killing intracellular viruses are **natural killer** (NK) **cells.** NK cells are a type of lymphocyte whose activity is greatly increased by exposure to interferons and cytokines. Although the exact mechanism of recognition is not known, NK cells probably recognize specific glycoproteins on the cell surface of virus-infected cells. Such recognition does not lead to phagocytosis; rather, the NK cells secrete cytotoxic proteins that trigger the death of the infected cell.

> In humans, Chediak-Higashi syndrome is associated with an absence of natural killer cells and with an increased incidence of lymphomas.

The Lymphatic System

The **lymphatic system,** which is closely associated with the cardiovascular system, consists of a network of vessels, nodes and other lymphatic tissues, and the fluid *lymph* (➤ Figure 16.4). The lymphatic system has three major functions: It (1) collects excess fluid from the spaces between body cells; (2) transports digested fats to the cardiovascular system; and (3) provides many of the nonspecific and specific defense mechanisms against infection and disease.

Lymphatic Circulation

The process of draining excess fluid from the spaces between cells starts with the *lymphatic capillaries* found throughout the body. These capillaries, which are slightly larger in diameter than blood capillaries, collect the excess fluid and plasma proteins that leak from the blood into the spaces between cells. Once in the lymphatic capillaries, this fluid is called **lymph.** Lymphatic capillaries join to form larger **lymphatic vessels.** As fluid moves through the vessels, it passes through **lymph nodes.** Finally, the lymph is returned to the venous blood via the *right* and *left lymphatic ducts,* which drain the fluids into the right and left subclavian veins. There is no mechanism to move or pump lymphatic fluid. Hence, the flow of lymph depends on skeletal muscle contractions, which squeeze the vessels, forcing the lymph toward the lymphatic ducts. Throughout the lymphatic system, there are one-way valves to prevent backflow of lymph.

Lymphoid Organs

Specific organs of the lymphatic system are essential in the body's defense against infectious agents and cancers. These organs include the lymph nodes, thymus, and spleen. Although all lymphatic organs contain numerous lymphocytes, these cells originate in bone marrow and are released into blood and lymph. They live from weeks to years, becoming dispersed to various lymphatic organs or remaining in the blood and lymph. In humans most lymphocytes are either *B lymphocytes* (B cells) or *T lymphocytes* (T cells). B cells differentiate in the bone marrow itself and migrate to the lymph nodes and spleen. Immature T cells from the bone marrow migrate to the thymus, where they mature; they then migrate to the lymph nodes or spleen. We will discuss these cells in more depth in Chapter 17.

At intervals along the lymphatic vessels, lymph flows through lymph nodes distributed throughout the body. They are most numerous in the thoracic (chest) region, neck, armpits,

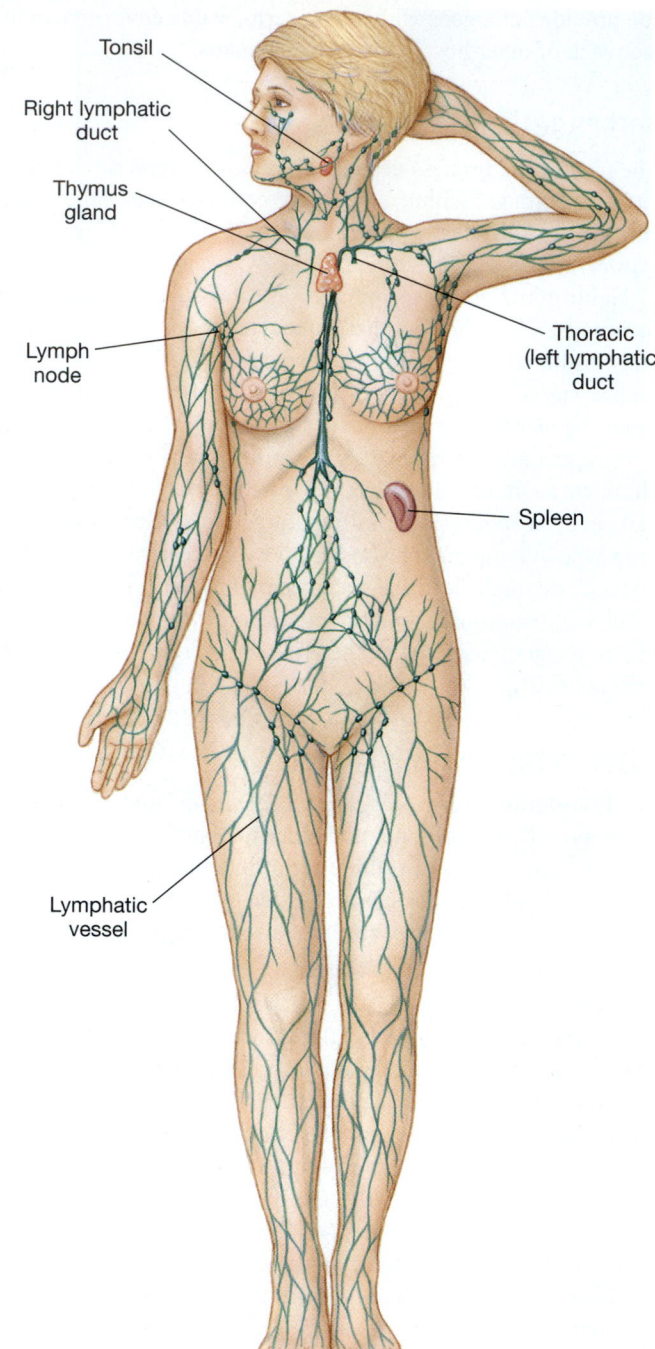

Tonsil

Right lymphatic
duct

Thymus
gland

Lymph
node

Thoracic
(left lymphatic)
duct

Spleen

Lymphatic
vessel

> **Figure 16.4   Structure of the lymphatic system.** The lymphatic system filters out microbes from the fluids surrounding cells. In so doing, it is subject to infections that overrun the ability of the system to destroy the microbes. Lymphocytes are defensive cells commonly found in the lymphatic system.

and groin. The lymph nodes filter out foreign material in the lymph. Most foreign agents passing through a node are trapped and destroyed by the defensive cells present.

Lymph nodes occur in small groups, each group covered in a network of connective tissue fibers called a **capsule** (➤ Figure 16.5). Lymph moves through a lymph node in one direction. Lymph first enters **sinuses,** wide passageways lined with phagocytic cells, in the outer cortex of the lymph node. The *outer cortex* houses large aggregations of B lymphocytes. The lymph then passes through the *deep cortex,* where T lymphocytes exist. The lymph moves through the inner region of a lymph node, the *medulla,* which contains B lymphocytes, macrophages, and plasma cells. Finally, lymph moves through sinuses in the medulla and leaves the lymph node.

This filtration of the lymph is important when an infection has occurred. For example, if a bacterial infection occurs, the bacteria are carried to the lymph nodes. As the lymph passes

Cortical sinus

Afferent lymphatic vessel

Capsule

Cortex

Medulla

Lymphatic nodule

Efferent lymphatic vessel

Medullary sinus

Germinal centers of lymphatic nodules

Lymph node artery and vein

> Figure 16.5   **Structure of a lymph node.** Lymph nodes are centers for removing microbes. These tissues contain phagocytes and lymphocytes. Swollen lymph nodes are usually an indication of a serious infection.

through the nodes, a majority of the bacteria are removed. Macrophages and other phagocytic cells in the nodes bind to and phagocytize the bacterial cells, thereby initiating a specific immune response (Chapter 17).

The **thymus gland** is a multilobed lymphatic organ located beneath the sternum (breast-bone) (➣ Figure 16.4). It is present at birth, grows until puberty, then atrophies (shrinks) and is replaced by fat and connective tissue by adulthood. Around the time of birth, the thymus begins to process lymphocytes and releases them into the blood as T cells. T cells play several roles in immunity: They regulate the development of B cells into antibody-producing cells, and subpopulations of T cells can kill virus-infected cells directly. The thymus also produces at least seven hormones, including cytokines, that stimulate production of T cells in the lymph nodes and spleen.

The **spleen,** located in the upper left quadrant of the abdominal cavity, is the largest of the lymphatic organs (➣ Figure 16.4). Anatomically, the spleen is similar to the lymph nodes. It is encapsulated, lobed, and well supplied with blood and lymphatic vessels. Although it does not filter material, its sinusoids contain many phagocytes that engulf and digest worn-out erythrocytes and microorganisms. It also contains B cells and T cells.

Other Lymphoid Tissues

Earlier, we mentioned the lymphoid masses found in the ileum of the small intestine. These Peyer's patches are **lymphoid nodules,** unencapsulated areas filled with lymphocytes. Collectively, the tissues of lymphoid nodules are referred to as **gut-associated lymphatic tissue** (GALT), which are the main sites of antibody production. Similar nodules are found in the respiratory system, urinary tract, and appendix.

The **tonsils** represent another site for the aggregation of lymphocytes. Although these tissues are not essential for fighting infections, they do contribute to immune defenses, as they contain B cells and T cells.

Although lymphatic tissues contain cells that phagocytize microorganisms, if these cells encounter more pathogens than they can destroy, the lymphatic tissues can become sites of infection. Thus, swollen lymph nodes and tonsillitis are common signs of many infectious diseases.

✓ How do nonspecific and specific defenses differ?

✓ List six categories of nonspecific defenses.

✓ List and describe the steps in phagocytosis.

✓ What are NK cells and how do they function?

✓ What are the parts and functions of the lymphatic system?

As phagocytic cells accumulate at the site of inflammation and begin to ingest bacteria, they release lytic enzymes, which can damage nearby healthy cells.

In summary, lymphoid tissues contribute to nonspecific defenses by phagocytizing microorganisms and other foreign material. They contribute to specific immunity through the activities of their B and T cells, which we will discuss in Chapter 17.

Inflammation

Do you remember the last time you cut yourself? If the cut was not too serious, the bleeding soon stopped. You washed the cut and put on a bandage. A few hours later the area around the cut became warm, red, swollen, and perhaps even painful. It had become *inflamed*.

Characteristics of Inflammation

Inflammation is the body's defensive response to tissue damage from microbial infection. It is also a response to mechanical injury (cuts and abrasions), heat and electricity (burns), ultraviolet light (sunburn), chemicals (phenols, acids, and alkalis), and allergies. But whatever the cause of inflammation, it is characterized by *cardinal signs* or symptoms: (1) calor—an increase in temperature, (2) rubor—redness, (3) tumor—swelling, and (4) dolor—pain at the infected or injured site. What happens in the inflammatory process, and why?

The Acute Inflammatory Process

The duration of inflammation can be either acute (short-term) or chronic (long-term). In **acute inflammation,** the battle between microbes (or other agents of inflammation) and host defenses usually is won by the host. In an infection, acute inflammation functions to (1) kill invading microbes, (2) clear away tissue debris, and (3) repair injured tissue. Let's look at acute inflammation more closely. ➤ Figure 16.6 illustrates the steps described next.

When cells are damaged, the chemical substance histamine is released from basophils and mast cells. **Histamine** diffuses into nearby capillaries and venules, causing the walls of these vessels to dilate (**vasodilation**) and become more permeable. Dilation increases the amount of blood flowing to the damaged area, and it causes the skin around wounds to become red and warm to the touch. Because the vessel walls are more permeable, fluids leave the blood and accumulate around the injured cells, causing **edema** (swelling). The blood delivers clotting factors, nutrients, and other substances to the injured area and removes wastes and some excess fluids. It also brings macrophages, which release cytokines. Some cytokines attract other phagocytes, and another cytokine called *tumor necrosis factor alpha* (TNF-α) additionally causes vasodilation and edema.

All kinds of tissue injury—burns, cuts, infections, insect bites, allergies—cause histamine release. In conjunction with its effects on blood vessels, histamine also causes the red, watery eyes and runny nose of hay fever and the breathing difficulties in certain allergies. The drugs called **antihistamines** alleviate such symptoms by blocking the released histamine from reaching its receptors on target organs.

The fluid that enters the injured tissue carries the chemical components of the blood-clotting mechanism. If the injury has caused bleeding, platelets and clotting factors, such as fibrin, stop the bleeding by forming a blood clot in the injured blood vessel. Because clotting takes place near the injury, it greatly reduces fluid movement around damaged cells and walls off the injured area from the rest of the body. Pain associated with tissue injury is thought to be due to the release of **bradykinin,** a small peptide, at the injured site. How bradykinin stimulates pain receptors in the skin is unknown, but cellular regulators called **prostaglandins** seem to intensify bradykinin's effect.

Inflamed tissues also stimulate **leukocytosis,** an increase in the number of leukocytes in the blood. To do this, the damaged cells release cytokines that trigger the production and infiltration of more leukocytes. Within an hour after the inflammatory process begins, phagocytes start to arrive at the injured or infected site. For example, neutrophils pass out of the blood by squeezing between endothelial cells lining the vessel walls. This process, called **diapedesis** (di-a-pe-de′sis), allows neutrophils to congregate in tissue fluids at the injured region.

Epithelium

1. Cut allows bacteria to get beneath surface of skin.

2. Damaged cells release histamine and bradykinin.

3. Capillaries dilate (vasodilation), bringing more blood to the tissue. Skin becomes reddened and warmer.

4. Capillaries become more permeable, allowing fluids to accumulate and cause swelling (edema).

5. Blood clotting occurs, and scab forms.

6. Bacteria multiply in cut.

7. Phagocytes enter tissue by moving through the walls of blood vessels (diapedesis).

8. Phagocytic cells are attracted to bacteria and tissue debris (chemotaxis) and engulf them.

9. Larger blood vessels dilate, further increasing blood supply to tissue and adding to heat and redness.

10. As dead cells and debris are removed, epithelial cells proliferate and begin to grow under the scab.

11. Scar tissue (connective tissue) replaces cells that cannot replace themselves.

> Figure 16.6   **Steps in the process of inflammation and subsequent healing.**

As we discussed earlier, when phagocytes reach an infected area, they attempt to engulf the invading microbes by phagocytosis. In that process many of the phagocytes themselves die. The accumulation of dead phagocytes, injured or damaged cells, the remains of ingested organisms, and other tissue debris forms the white or yellow fluid called **pus.** Many bacteria, such as *Streptococcus pyogenes,* cause pus formation because of their ability to produce leukocidins that destroy phagocytes. Viruses lack this activity and do not cause pus formation. Pus continues to form until the infection or tissue damage has been brought under control. An accumulation of pus in a cavity hollowed out by tissue damage is called an **abscess.** Boils and pimples are common kinds of abscesses.

Although the inflammatory process is usually beneficial, it can sometimes be harmful. For example, inflammation can cause swelling (edema) of the membranes (meninges) surrounding the brain or spinal cord, leading to brain damage. Swelling, which delivers phagocytes to injured tissue, can also interfere with breathing if it constricts the airways in the lung. Moreover, vasodilation delivers more oxygen and nutrients to injured tissues. Ordinarily this is of greater benefit to host cells than to pathogens, but sometimes it helps the pathogens thrive as well. Even though rapid clotting and the walling off of an injured area prevents pathogens from spreading, it can also prevent natural defenses and antibiotics from reaching the pathogens.

Boils must be lanced before therapeutic drugs can reach them. Attempting to suppress the inflammatory process also can be harmful. Such attempts can allow boils to form when natural defenses might otherwise destroy the bacteria.

In summary, cellular defense mechanisms usually prevent an infection from spreading or from getting worse. However, sometimes these nonspecific defense mechanisms are overwhelmed by sheer numbers of microbes or are inhibited by virulence factors that the microbes possess. The pathogens can then invade other parts of the body. For bacterial infections, medical intervention with antibiotics may inhibit microbial growth in injured tissue and reduce the chance of an infection's spreading. Despite such measures, however, infections do spread. In Chapter 17 we will describe the mechanisms by which various lymphocytes act as agents of specific host immune defenses that help overcome an initial infection and prevent future infections by the same microbe.

Repair and Regeneration

During the entire inflammatory reaction, the healing process is also underway. Once the inflammatory reaction has subsided and most of the debris has been cleared away, healing accelerates. Capillaries grow into the blood clot, and **fibroblasts,** connective tissue cells, replace the destroyed tissue as the clot dissolves. The fragile, reddish, grainy tissue seen at the cut site consists of capillaries and fibroblasts called **granulation tissue.** As granulation tissue accumulates fibroblasts and fibers, it replaces nerve and muscle tissues that cannot be regenerated. New epidermis replaces the part destroyed. In the digestive tract and other organs lined with epithelium, an injured lining can similarly be replaced. Although scar tissue is not as elastic as the original tissue, it does provide a strong durable "patch" that allows the remaining normal tissue to function.

Several factors affect the healing process. The tissues of young people heal more rapidly than those of older people. The reason is that the cells of the young divide more quickly, their bodies are generally in a better nutritional state, and their blood circulation is more efficient. As you might guess from the many contributions of blood to healing, good circulation is extremely important. Certain vitamins also are important in the healing process. Vitamin A is essential for the division of epithelial cells, and vitamin C is essential for the production of collagen and other components of connective tissue. Vitamin K is required for blood clotting, and vitamin E also may promote healing and reduce the amount of scar tissue formed.

Chronic Inflammation

Sometimes an acute inflammation becomes a **chronic inflammation,** in which neither the agent of inflammation nor the host is a decisive winner of the battle. Rather, the agent causing the inflammation continues to produce tissue damage as the phagocytic cells and other host defenses attempt to destroy or at least confine the region of inflammation. In the process, pus may be formed continuously. Such chronic inflammation can persist for years.

Because the cause of inflammation is not destroyed, host defenses attempt to limit or confine the agent so that it cannot spread to surrounding tissue. For example, **granulomatous inflammation** results in granulomas. A **granuloma** is a pocket of tissue that surrounds and walls off the inflammatory agent. The central region of a granuloma contains epithelial cells and macrophages; the latter may fuse to form giant, multinucleate cells. Collagen fibers, which help wall off the inflammatory agent, and lymphocytes surround the core. Granulomas associated with specific disease are sometimes given special names—for example, *gummas* (syphilis), *lepromas* (Hansen's disease), and *tubercles* (tuberculosis).

Tubercles usually contain necrotic (dead) tissue in the central region of the granuloma. As long as necrotic tissue is present, the inflammatory response will persist. If only a small quantity of necrotic tissue is present, the lesions sometimes become hardened as calcium is deposited in them. Calcified lesions are common in tuberculosis patients. When an anti-inflammatory drug such as cortisone is given, the organisms isolated in tubercles may be liberated, and signs and symptoms of tuberculosis reappear (secondary tuberculosis).

Fever

A rise in temperature in infected or injured tissue is one sign of a local inflammatory reaction. **Fever,** a systemic increase in body temperature, often accompanies inflammation. Fever was first studied in 1868, when the German physician Carl Wunderlich devised a method to measure body temperature. He placed a foot-long thermometer in the armpit of his patients and left it in place for 30 minutes! Using this cumbersome technique, he could record human body temperatures during *febrile* (feverish) illnesses.

Normal body temperature is about 37°C (98.6°F), although individual variations in normal temperature within the range 36.1° to 37.5°C (97.0° to 99.5°F) are not uncommon. Fever is defined clinically as an oral temperature above 37.8°C (100.5°F) or rectal temperature of 38.4°C (101.5°F). Fever accompanying infectious diseases rarely exceeds 40°C (104.5°F); if it reaches 43°C (109.4°F), death usually results.

Body temperature is maintained within a narrow range by a temperature-regulating center in the *hypothalamus,* a part of the brain. Fever occurs when the temperature established for this mechanism is reset and raised to a higher temperature. Fever can be caused by many pathogens, by certain immunological processes (such as reactions to vaccines), and by nearly any kind of tissue injury, even heart attacks. Most often, fever is caused by a substance called a **pyrogen** (*pyro,* Greek for "fire"). ➡ (Chapter 14, p. 396) **Exogenous pyrogens** include exotoxins and endotoxins from infectious agents. These toxins cause fever by stimulating the release of an **endogenous pyrogen** from macrophages. The endogenous pyrogen is yet another cytokine, called *interleukin-1* (IL-1), that circulates via the blood to the hypothalamus, where it causes certain neurons to secrete prostaglandins. The prostaglandins then reset the hypothalamus thermostat at a higher temperature, which then causes the body temperature to begin rising within 20 minutes. In such situations, body temperature is still regulated, but the body's "thermostat" is reset at a higher temperature. (The sensation of chills that sometimes accompanies a fever was described in Chapter 14.) ➡ (p. 397)

Fever has several beneficial roles. (1) Fever raises the body temperature above the optimum temperature for growth of many pathogens. This slows their rate of growth, reducing the number of microorganisms to be combated. (2) At the higher temperatures of fever some microbial enzymes or toxins may be inactivated. (3) Fever can heighten the level of immune responses by increasing the rate of chemical reactions in the body. This results in a faster rate at which the body's defense mechanisms attack pathogens, shortening the course of the infection. (4) Fever makes a patient feel ill. In this condition the patient is more likely to rest, preventing further damage to the body and allowing energy to be used to fight the infection.

In an infection, cells also release **leukocyte-endogenous mediator** (LEM). Besides helping to elevate body temperature, LEM decreases the amount of iron absorbed from the digestive tract and increases the rate at which it is moved to iron storage deposits. Thus, LEM lowers the plasma iron concentration. Without adequate iron, growth of microorganisms is slowed. ➡ (Chapter 6, p. 151)

Our current knowledge of the importance of fever has changed the clinical approach to this symptom. In the past, *antipyretics*—fever-reducing drugs such as aspirin—were given almost routinely to reduce fever caused by infections. For the beneficial effects cited above, many physicians now recommend allowing fevers to run their course. Evidence shows that medication can delay recovery. However, if a fever goes above 40°C or if the patient has a disorder that might be worsened by fever, antipyretics are still used. In fact, untreated extreme fever increases the metabolic rate by 20 percent, makes the heart work harder, increases water loss, alters electrolyte concentrations, and can cause convulsions, especially in children. Thus, patients with severe heart disease or fluid and electrolyte imbalances, as well as children subject to convulsions, usually receive antipyretics.

Molecular Defenses

Along with cellular defenses, inflammation, and fever, molecular defenses represent another formidable nonspecific defense barrier. These molecular defenses involve the actions of *interferon* and *complement*.

Sweat It Out, Grandma

When you're in bed with a fever, it's hard to believe that fevers aren't just annoying side effects of being sick. They are actually important in fighting off infections. That's bad news for Grandma and Grandpa, since elderly people have trouble generating fevers. But a researcher at the University of Delaware in Newark found that sick geriatric rats, which also have problems developing fevers, benefited from living in rooms heated to 100°C. That doesn't necessarily mean that humans will benefit from such high temperatures, but if further studies show that they do, then cranking up the thermostat may help Grandma and Grandpa fight off the flu and other infections.

✓ List the cardinal signs or symptoms of inflammation.

✓ What is the role of histamine in the inflammatory process?

✓ Define: diapedesis, pus, edema, granuloma, and pyrogen.

✓ List four benefits of fever.

> ➤ Table 16.3   **Properties of type I and type II human interferons**

Class	Cell source	Subtypes	Stimulated by	Effects
Type I				
Alpha-interferon (INF-α)	Leukocytes	20	Viruses	Production of antiviral proteins in neighboring cells
Beta-interferon (INF-β)	Fibroblasts	1	Viruses	Same as INF-α
Type II				
Gamma-interferon (INF-γ)	T lymphocytes and NK cells	1	Viruses and other antigens	Activates tumor destruction and killing of infected cells

Interferon

As early as the 1930s, scientists observed that infection by one virus prevented for a time infection by another virus. Then, in 1957, a small, soluble protein was discovered that was responsible for this viral interference. This protein, called **interferon** (in-ter-fer'on), "interfered" with virion replication in other cells. Such a molecule suggested to virologists that they might have the "magic bullet" for viral infections, similar to the antibiotics used to treat bacterial infections. As we will see, such hope has dwindled somewhat.

Efforts to purify interferon led to the discovery that many different subtypes of interferon exist in different animal species, and that those produced by one species may be ineffective in other species. For example, interferon produced in a chicken is useful in protecting other chicken cells from viral infection. But chicken interferon is of no use in preventing viral infections in mice or in humans. Different interferons also exist in different tissues of the same animal. In humans there are three groups of interferons, called alpha (α), beta (β), and gamma (γ) (Table 16.3). Analysis of the protein structure and function show *α-interferon* and *β-interferon* to be similar, so they are placed together as *type I interferons*. *Gamma-interferon* is different structurally and functionally and represents the only known *type II interferon*.

Many researchers have tried to determine how these interferons act. The synthesis of α-interferon and β-interferon occurs after a virus infects a cell (➤ Figure 16.7). These interferons do not interfere directly with viral replication. Rather, after viral infection the cell synthesizes and secretes minute amounts of interferon. The interferon then diffuses to adjacent, uninfected cells and binds to their surfaces. Binding stimulates those cells to transcribe specific genes into mRNA molecules, which are then translated to produce many new proteins, most of them enzymes. Together these enzymes are called **antiviral proteins** (AVPs). Although viruses still infect cells possessing the AVPs, many of the proteins interfere with virus replication.

The AVPs are especially effective against RNA viruses. Recall from Chapter 10 that all RNA viruses must either produce dsRNA (Reoviridae) or go through a dsRNA stage during replication of (−) sense or (+) sense RNA. ➥ (p. 261) Two of the AVPs digest mRNA and limit translation of viral mRNA. The result is that the AVPs prevent the formation of new viral nucleic acid and capsid proteins. The infected cell that initially produced the interferon is thus surrounded by cells that can resist the replication of viruses, limiting viral spread.

Gamma-interferon also can block virus replication by AVP synthesis. However, lymphocytes and NK cells do not have to be infected with a virus to synthesize γ-interferon. Rather, it is produced in uninfected lymphocytes and NK cells that are sensitive to specific foreign antigens (viruses, bacteria, tumor cells) present in the body. The exact role of γ-interferon is unclear, but it is known to enhance the activities of lymphocytes, NK cells, and macrophages—the cells needed to attack microbes and tumors. Gamma-interferon (along with tumor necro-

Interferons are usually species-specific but virus non-specific.

Interferons are produced and released in response to viral infections, double-stranded RNA, endotoxins, and many parasitic organisms.

> Figure 16.7   **The mechanism by which interferons α and β act.**

sis factor-α, or TNF-α) also helps infected macrophages rid themselves of pathogens. For example, we mentioned earlier that macrophages can become infected with *Mycobacterium* bacilli. Such infected macrophages can be activated by γ-interferon and TNF-α, which bind to infected macrophages. New bactericidal activity is thereby triggered within the macrophage, usually leading to death of the bacteria and the restoration of normal macrophage function.

Therapeutic Uses of Interferon

Besides having the ability to block virus replication, interferons can also stimulate specific immune defenses. Therefore, interferons provide a potential therapy for viral infections and tumors. Unfortunately, infected animal cells produce very small quantities of interferons. However, today *recombinant interferon* (rINF) can be produced more cheaply and abundantly by using recombinant DNA techniques. ➡ (Chapter 8, p. 213) Manufacture of recombinant interferon starts with the isolation and copying of the interferon gene and its insertion into plasmids. When recombinant plasmids are mixed with appropriate bacterial or yeast cells, some cells will take up the gene-containing plasmid and thereby acquire the human interferon gene. By growing these bacterial or yeast cells in very large vats and extracting the interferon that they produce, pharmaceutical companies can produce relatively significant quantities of recombinant interferon.

The ability to produce recombinant interferons spurred research on therapeutic applications for these proteins. In 1986 α-interferon was approved by the FDA for treating hairy cell leukemia, a very rare blood cancer. Since then, interferons have been approved for treatment of several other viral diseases, including genital warts and cancer. However, in most cases interferon is a treatment, not a cure. Patients must remain on the drug throughout their lives. With hairy cell leukemia, for example, removal of the drug results in a recurrence of the disease in 90 percent of the patients. For hepatitis C virus infection, treatment must be given three times a week for 6 months. Even so, if the patient is taken off treatment, hepatitis C disease will reappear after 6 months in 70 percent of the cases. As of 1994 several clinical trials using interferons for other diseases were in progress.

Other studies have looked at the value of interferons to treat cancer. Tests on one form of bone cancer show that after most of the cancerous tissue is removed by surgery or destroyed by radiation, interferon therapy will reduce the incidence of metastasis (spread). How interferon stops metastasis is not known. Perhaps it interferes with viral replication. Interferon therapy could also prevent growth of the cancer cells through their destruction by macrophages and NK cells.

The therapeutic use of interferons has some drawbacks. When recombinant interferon is injected, it does not remain stable for very long in the body. This makes delivery of the interferons to the site of infection difficult. Injection of interferon (especially α-interferon) also has side effects, including fatigue, nausea, headache, vomiting, weight loss, and nervous system disorders. Whereas fever normally increases interferon production, which helps the body fight viral infections, the injection of interferon *produces* fever as a side effect. High doses can cause toxicity to the liver, kidneys, heart, and bone marrow.

Moreover, some microbes have developed resistance to interferons. Although some DNA viruses, such as the poxviruses, stimulate interferon synthesis, the human adenoviruses have resistance mechanisms to combat antiviral protein activity. In addition, the hepatitis B virus often fails to stimulate adequate interferon production in infected cells.

The therapeutic usefulness of interferon is clearly not the viral magic bullet that was originally envisioned. Nevertheless, interferons are being used to treat life-threatening viral infections and cancers.

Complement

Complement, or the **complement system,** refers to a set of more than 20 large regulatory proteins that play a key role in host defense. They are produced by the liver and circulate in plasma in an inactive form. These proteins account for about 10 percent (by weight) of all plasma proteins. When complement was discovered, it was believed to be a single substance that "complemented," or completed, certain immunological reactions. Although complement can be activated by immune reactions, its effects are nonspecific—it exerts the same defensive effects regardless of which microorganism has invaded the body.

The general functions of the complement system are to (1) enhance phagocytosis by phagocytes, (2) lyse microorganisms, bacteria, and enveloped viruses directly, and (3) generate peptide fragments that regulate inflammation and immune responses. Furthermore, complement goes to work as soon as an invading microbe is detected; the system makes up an effective nonspecific host defense long before specific host immune defenses are mobilized.

The complement system works as a cascade. A **cascade** is a set of reactions that amplify some effect—that is, more product is formed in the second reaction than in the first, still more in the third, and so on. Of the 20 different serum proteins so far identified in the complement system, 13 participate in the cascade itself, and 7 activate or inhibit reactions in the cascade.

Complement Function

Two pathways have been identified in the sequence of reactions carried out by the complement system. They are called the **classical pathway** and the **alternative pathway,** or *properdin pathway* (➤ Figure 16.8a). The classical pathway begins when antibodies bind to antigens such as microbes and involves complement proteins C1, C4, and C2 (*C* stands for complement). The alternative pathway is activated by contact between complement proteins and polysaccharides at the pathogen surface. Complement proteins called factor B, factor D, and factor P (*properdin*) replace C1, C4, and C2 in the initial steps. However, the components of both pathways activate reactions involving C3 through C9. Consequently, the effects of the complement systems are the same regardless of the pathway by which C3 is produced. However, the alternative pathway is activated even earlier in an infection than is the classical pathway.

The contributions of the complement system to nonspecific defenses depend on C3, a key protein in the system. Once C3 is formed, it immediately splits into C3a and C3b, which then participate in three kinds of molecular defenses: opsonization, inflammation, and membrane attack complexes (➤ Figure 16.8b).

Opsonization   Earlier, we mentioned that some bacteria with capsules or surface proteins (M proteins) can prevent phagocytes from adhering to them. The complement system can coun-

The numbers attached to the complement cascade refer to their order of discovery, not the sequence in which they act.

(a)

(b)

➤ Figure 16.8   **The complement system.** (a) Classical and alternative pathways of the complement cascade. Although the two pathways are initiated in different ways, they combine to activate the complement system. (b) Activation of the classical complement pathway. In this cascade each complement protein activates the next one in the pathway. The action of C3b is critical for opsonization and, along with C5b, for formation of membrane attack complexes. C4a, C3a, and C5a also are important to inflammation and phagocyte chemotaxis. (IgG is a class of antibodies that we will discuss in Chapter 17.)

teract these defenses, making possible a more efficient elimination of such bacteria. First, special antibodies called **opsonins** bind to and coat the surface of the infectious agent. C1 binds to these antibodies, initiating the cascade. C1 causes the cleavage of C4 into C4a and C4b. C4b and C1 then cause C2 to split into C2a and C2b. The C4bC2a complex in turn leads to the splitting of C3 into C3a and C3b. C3b then binds to the surface of the microbe. Complement receptors on the plasma membrane of phagocytes recognize the C3b molecules; this recognition stimulates phagocytosis. This process, initiated by opsonins, is called **opsonization,** or *immune adherence*.

Inflammation   The complement system is also potent in initiating and enhancing inflammation. C3a, C4a, and C5a enhance the acute inflammatory reaction by stimulating chemotaxis and thus phagocytosis. These three complement proteins also adhere to the membranes of basophils and mast cells, causing them to release histamine and other substances that increase the permeability of blood vessels.

Membrane Attack Complexes   Another defense triggered by C3b is cell lysis. By a process called **immune cytolysis,** complement proteins produce lesions in the cell membranes of microorganisms and other types of cells. These lesions cause cellular contents to leak out. To cause immune cytolysis, C3b initiates the splitting of C5 into C5a and C5b. C5b then binds C6 and C7, forming a C5bC6C7 complex. This protein complex is hydrophobic ➡ (Chapter 4, p. 80) and inserts into the microbial cell membrane. C8 then binds to C5b in the membrane. Each C5bC6C7C8 complex causes the assembly in the cell membrane of up to 15 C9 molecules (➤ Figure 16.9). By extending all the way through the cell membrane, these proteins form a pore and constitute the **membrane attack complex** (MAC). The MAC is responsible for the direct lysis of invading microorganisms. Importantly, host plasma membranes contain proteins that protect against MAC lysis. These proteins prevent damage by preventing the binding of activated complement proteins to host cells. The MAC forms the basis of *complement fixation,* a laboratory test used to detect antibodies against any one of many microbial antigens. That test is described in Chapter 18.

> The completed MAC has a tubular form and a functional pore size of 70–100 angstroms (1 angstrom = 10^{-10} meters).

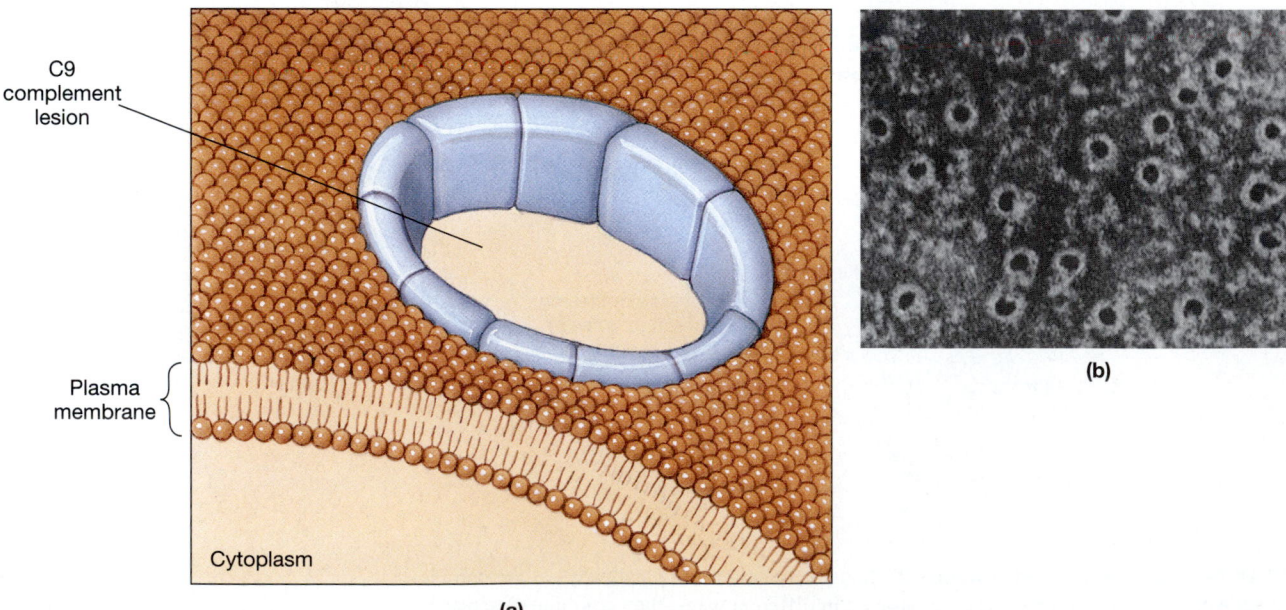

(a)

(b)

C9 complement lesion

Plasma membrane

Cytoplasm

➤ Figure 16.9   **Complement lesions in cell membranes.** (a) Complement lyses a bacterial cell by creating a membrane attack complex (lesion) consisting of 10 to 15 molecules of C9. These protein molecules form a hole in the cell membrane through which the cytoplasmic contents leak out. (b) An EM showing the holes formed in red blood cell membranes by C9.

Table 16.4	Disease states related to complement deficiencies
Disease state	**Complement deficiencies**
Severe recurrent infections	C3
Recurrent infections of lesser severity	C1, C2, C5
Systemic lupus erythematosus (a bodywide immunologic disease)	C1, C2, C4, C5, C8
Glomerulonephritis (an immunological disease of the kidneys)	C1, C8
Gonococcal infections	C6, C8
Meningococcal infections	C6

A great advantage of the complement system to host defenses is that, once it is activated, the reaction cascade occurs rapidly. A very small quantity of an activating substance (microbe) can activate a few molecules of C1. They, in turn, activate large quantities of C3; one C4b2a molecule can split 1000 molecules of C3 into C3a and C3b. Thus, sufficient quantities of C3b are quickly available to cause opsonization and inflammation and to produce membrane attack complexes.

Unfortunately, complement activity can be impaired by the absence of one or more of its protein components. Impaired complement activity makes the host more vulnerable to various diseases (Table 16.4), most of which are acquired or congenital. Acquired diseases result from temporary depletion of a complement protein; they subside when cells again synthesize the protein. Congenital complement deficiencies are due to genetic defects that prevent the synthesis of one or more complement components.

The most significant effect of complement deficiencies is the lack of resistance to infection. Deficiencies in several complement components have been observed. The greatest degree of impaired complement function occurs with a deficiency of C3—which is not surprising, because C3 is the key component in the system. In individuals with C3 deficiencies, chemotaxis, opsonization, and cell lysis are all impaired. Such individuals are especially subject to infection by pyogenic bacteria. A deficiency in MAC components (C5–C9) is associated with recurrent infections, especially by *Neisseria* species. Complement deficiencies are less important in defenses against viruses, although some viruses, such as the Epstein–Barr virus, use complement receptors to invade cells.

Acute Phase Response

Observations of acutely ill patients have led to the characterization of the **acute phase response,** a response to acute illness that involves increased production of specific blood proteins called **acute phase proteins.** In an acute phase response, pathogen ingestion by macrophages stimulates the synthesis and secretion of several cytokines. One, called *interleukin-6* (IL-6), travels through the blood and causes the liver to synthesize and secrete the acute phase proteins into the blood. Thus, acute phase proteins form a nonspecific host defense mechanism distinct from both the inflammatory response and host-specific immune defenses. This mechanism appears to recognize foreign substances before the immune system defenses do and acts early in the inflammatory process, before antibodies are produced.

The best understood acute phase proteins are *C-reactive protein* (CRP) and *mannose-binding protein* (MBP). All humans studied thus far have the capacity to produce CRP and MBP. CRP recognizes and binds to phospholipids, and MBP to mannose sugars, in cell membranes of many bacteria and the plasma membranes of fungi. Once bound, these acute phase proteins

✓ What are interferons? How and where are they produced?

✓ How might interferons be used to treat disease?

✓ Describe the complement system, including the classical and alterntive (properdin) pathways.

✓ What are the results of activating the complement cascade?

✓ What are the functions of acute phase proteins?

Development of the Immune System

Can all organisms defend themselves against attacks by infectious microbes? For vertebrates, the answer is yes. As we saw in this chapter, they have nonspecific defense mechanisms, and as we will see in Chapter 17, they also have well-developed specific immune defenses.

Defenses against infection are not limited to animals. Other biological kingdoms also have host defense mechanisms, usually of a chemical nature. Plants, for example, produce chemical defenses that can wall off areas damaged or infected by bacteria or fungi. In fact, an important determinant of how well a given strain of plant can resist infection after pruning or damage is its chemical and physical defensive abilities. Many fungi are plant pathogens, getting their nutrients by parasitizing certain tissues within the plant. To infect a plant, the fungus must penetrate the plant cell. ➡ (Chapter 11, p. 299) During infection the plant cells produce enzymes that release carbohydrate molecules from the fungal cell walls. These fragments of fungal wall, called *elicitors,* trigger an immunological-like response by the plant. Elicitors cause the plant to produce lipidlike chemicals called *phytoalexins.* Phytoalexins inhibit fungal growth by restricting the infection to a small portion of the plant tissue (see the photo). Plant biotechnologists are trying to "breed" this response into other types of plants that are sensitive to fungal invasion.

Experimentally damaged areas of tree trunk are walled off in trees that survive attack, thus keeping infection from spreading throughout the entire tree.

Invertebrates also have nonspecific defenses for fending off invaders. Phagocytosis is important to invertebrates in obtaining food, but it is also necessary for preventing sedentary organisms permanently fixed to a surface, and living where space is limited, from being overgrown by neighbors. So, phagocytosis is used to defend one's territory. In animals lacking a cardiovascular system, amoebocytes wander through the body, engulfing foreign matter and damaged or aged cells. When your white blood cells phagocytize a bacterium, they are using an ancient mechanism preserved and transformed from simpler life forms.

Opsonization is also observed in invertebrates, made possible by complement-like components of body fluids. For example, fluids in the body cavity of sea urchins share many characteristics with human complement proteins. In fact, complement proteins, like phagocytosis, probably were derived from these early versions in invertebrates. Secretion of antimicrobial enzymes is another means of defense present even in simple protozoa. Thus, nonspecific defense processes, such as phagocytosis and opsonization, are often called a *primitive characteristic* because most animals have these ancient mechanisms.

Almost all invertebrates also can reject grafts of foreign tissue. Vertebrates reject such grafts more vigorously on a second encounter, but invertebrates do not; in fact, the second rejection may be slower than the first. Because invertebrates lack these memory responses, the presence of such specific immune defenses in vertebrates is considered an *advanced characteristic.* These defenses include the B cells, T cells, and antibodies.

Although immune defenses involving the production of specific antibodies are found in all types of fish, the swiftest and most complex immune responses are found in mammals and birds. Birds have a saclike structure, the *bursa of Fabricius,* that is not present in mammals and probably represents a higher state of evolution of the immune system. In chickens, immature B cells in the bone marrow migrate to the bursa of Fabricius. There they are stimulated to mature rapidly and are capable of recognizing foreign substances. In mammals, B cells originate and mature more slowly in the bone marrow. Thus, immune system development culminates in the two-part system of B cells and T cells. In Chapter 17 we will investigate this achievement of specific host defenses.

act like an opsonin: They activate the complement system and immune cytolysis and stimulate phagocyte chemotaxis. If we knew how to enhance CRP and MBP activity, effective therapies could be developed to combat many bacterial and fungal infections.

In summary, the nonspecific defense mechanisms operate regardless of the nature of the invading agent. They constitute the body's first line of defense against pathogens, whereas the specific defense mechanisms (Chapter 17) constitute the second line of defense. ➤ Figure 16.10 reviews the major categories of nonspecific defenses.

➤ Figure 16.10   **A summary of the body's nonspecific defenses.**

Retracing Our Steps

Nonspecific and Specific Host Defenses

- **Nonspecific defenses** operate regardless of the kind of invading agent; they form a first line of defense that is often effective even before specific defenses are activated.
- **Specific defenses** respond to particular invading agents; provided by the immune system, they form a second line of defense against pathogens.

Physical Barriers

- Skin and mucous membranes act as physical barriers to penetration, and secrete chemicals inhospitable to pathogens.
- **Mucous membranes** consist of a thin layer of cells that secrete **mucus.**

Cellular Defenses

Defensive Cells

- **Formed elements,** found in blood but derived from bone marrow, provide a cellular defense barrier to infection.
- Defensive cells include **granulocytes—basophils, mast cells, eosinophils,** and **neutrophils**—and **agranulocytes—monocytes** and **lymphocytes.**

Phagocytes

- A **phagocyte** is a cell that ingests and digests foreign substances.
- Phagocytic cells include neutrophils in the blood and in injured tissues, monocytes in the blood, and fixed and wandering **macrophages.**

The Process of Phagocytosis

- The process of **phagocytosis** occurs as follows: (1) Invading microorganisms are located by **chemotaxis,** which is aided by the release of **cytokines** by phagocytes. (2) Ingestion occurs as the phagocyte surrounds and ingests a microbe or other foreign substance into a **phagosome.** (3) Digestion occurs as lysosomes surround a vacuole and release their enzymes into it, forming a **phagolysosome.** Enzymes and defensins break down the contents of the phagolysosome and produce substances toxic to microbes.
- Some microbes resist phagocytosis by producing capsules or specific proteins, preventing release of lysosomal enzymes, and by producing toxins (**leukocidin** and **streptolysin**).

Extracellular Killing

- Eosinophils defend against parasitic worm infections by secreting cytotoxic enzymes.
- **Natural killer** (NK) **cells** secrete products that kill virus-infected cells and certain cancer cells.

The Lymphatic System

- The **lymphatic system** consists of a network of **lymphatic vessels, lymph nodes** and **lymphoid nodules,** the **thymus gland,** the **spleen,** and **lymph.**
- All lymphatic tissues that filter blood and lymph are susceptible to infection by pathogens they filter when the pathogens overwhelm defenses.
- Nonspecific defenses consist of the actions of phagocytic cells.

Inflammation

Characteristics of Inflammation

- **Inflammation** is the body's response to tissue damage. It is characterized by localized increased temperature, redness, swelling, and pain.

The Acute Inflammatory Process

- **Acute inflammation** is initiated by **histamine** released by damaged tissues, which dilates and increases permeability of blood vessels (**vasodilation**). Activation of cytokines also contributes to initiation of inflammation.
- Dilation of blood vessels accounts for redness and increased tissue temperature; increased permeability accounts for **edema** (swelling).
- Tissue injury also initiates the blood-clotting mechanism.
- **Bradykinin** stimulates pain receptors; **prostaglandins** intensify its effect.
- Inflamed tissues also stimulate an increase in the number of leukocytes in the blood (**leukocytosis**) by releasing cytokines that trigger leukocyte production. Neutrophils and macrophages migrate from the blood to the site of injury (**diapedesis**).
- Leukocytes and macrophages phagocytize microbes and tissue debris.

Repair and Regeneration

- Repair and regeneration occur as capillaries grow into the site of injury and fibroblasts replace the dissolving blood clot. The resulting **granulation tissue** is strengthened by connective tissue fibers (from **fibroblasts**) and the overgrowth of epithelial cells.

Chronic Inflammation

- **Chronic inflammation** is a persistent inflammation in which the inflammatory agent continues to cause tissue injury as host defenses fail to overcome the agent completely.
- **Granulomatous inflammation** is a chronic inflammation in which monocytes, lymphocytes, and macrophages surround necrotic tissue to form a **granuloma.**

Fever

- **Fever** is an increase in body temperature caused by **pyrogens,** which increase the setting (thermostat) of the temperature-regulating center in the hypothalamus.
- **Exogenous pyrogens** (usually pathogens and their toxins) come from outside the body and stimulate a cytokine that acts as an **endogenous pyrogen.**
- Fever and the chemicals associated with it augment the immune response and inhibit the growth of microorganisms by lowering plasma iron concentrations. Fever also increases the rate of chemical reactions, raises the temperature above the optimum growth rate for some pathogens, and makes the patient feel ill (thereby lowering activity).
- Antipyretics are recommended only for high fevers and for patients with disorders that would be exacerbated by fever.

Molecular Defenses

Interferon

- **Interferons** are proteins that act nonspecifically to cause cell killing or to stimulate cells to produce **antiviral proteins.**

- Interferon can be made by recombinant DNA technology and has proved to be therapeutic for certain malignancies; other therapeutic applications are being studied.

Complement

- **Complement** refers to a set of blood proteins that, when activated, produce a **cascade** of protein reactions. The **complement system** can be activated by the **classical pathway** or the **alternative pathway.**

- Action of the complement system is rapid and nonspecific. It promotes opsonization, inflammation, and immune cytolysis through the formation of **membrane attack complexes.** In **opsonization,** invading agents are coated with **opsonins** (antibodies) and C3b complement protein, making the invaders recognizable to phagocytes. In **immune cytolysis,** complement proteins produce lesions on invaders' plasma membranes that cause cell lysis.

- Deficiencies in complement reduce resistance to infection.

Acute Phase Response

- Acutely ill patients increase production of certain blood proteins (**acute phase proteins**). These substances are distinct from those involved in the inflammatory response and act quickly, before antibodies can be made. Such proteins initiate or accelerate inflammation, activate complement, and stimulate chemotaxis of phagocytes.

Terminology Check

abscess (*p. 445*)
acute inflammation (*p. 444*)
acute phase protein (*p. 453*)
acute phase response (*p. 453*)
adherence (*p. 439*)
agranulocyte (*p. 437*)
alternative pathway (*p. 450*)
antihistamine (*p. 444*)
antiviral protein (*p. 448*)
basophil (*p. 436*)
bradykinin (*p. 444*)
capsule (*p. 442*)
cascade (*p. 450*)
chemotaxis (*p. 439*)
chronic inflammation (*p. 446*)
classical pathway (*p. 450*)
complement (*p. 450*)
complement system (*p. 450*)
cytokine (*p. 439*)
diapedesis (*p. 444*)
edema (*p. 444*)
endogenous pyrogen
  (*p. 447*)
eosinophil (*p. 437*)
erythrocyte (*p. 435*)

exogenous pyrogen (*p. 447*)
fever (*p. 447*)
fibroblast (*p. 446*)
formed element (*p. 435*)
granulation tissue (*p. 446*)
granulocyte (*p. 436*)
granuloma (*p. 446*)
granulomatous inflammation (*p. 446*)
gut-associated lymphatic tissue (*p. 443*)
histamine (*p. 444*)
immune cytolysis (*p. 452*)
inflammation (*p. 444*)
interferon (*p. 448*)
leukocidin (*p. 440*)
leukocyte (*p. 436*)
leukocyte-endogenous mediator (*p. 447*)
leukocytosis (*p. 444*)
lymph (*p. 441*)
lymphatic system (*p. 441*)
lymphatic vessel (*p. 441*)
lymph node (*p. 441*)
lymphocyte (*p. 437*)
lymphoid nodule (*p. 443*)
macrophage (*p. 438*)
mast cell (*p. 437*)

membrane attack complex (*p. 452*)
monocyte (*p. 437*)
mucous membrane (*p. 435*)
natural killer cell (*p. 441*)
neutrophil (*p. 437*)
nonspecific defenses (*p. 434*)
opsonin (*p. 452*)
opsonization (*p. 452*)
phagocyte (*p. 437*)
phagocytosis (*p. 438*)
phagolysosome (*p. 439*)
phagosome (*p. 439*)
plasma (*p. 435*)
platelet (*p. 436*)
prostaglandin (*p. 444*)
pus (*p. 445*)
pyrogen (*p. 447*)
sinus (*p. 442*)
skin (*p. 435*)
specific defenses (*p. 434*)
spleen (*p. 443*)
streptolysin (*p. 440*)
thymus gland (*p. 443*)
tonsil (*p. 443*)
vasodilation (*p. 444*)

See You on the Web

Don't forget I am expecting you on the web. The following is a preparedness checklist for your web explorations. If you have questions on any of these items, you can return to the end of Chapter 1 for more details.

- Questions We'll Explore
- Concept Check questions
- Flashcards
- Study Guide

If you are ready, come visit me at

http://www.prenhall.com/black

It will be a journey you'll never forget.

Immunology I: Basic Principles of Specific Immunity and Immunization

How long will it be before you can have your pet goldfish vaccinated? Silly question? Maybe not.

Researchers have developed an effective vaccine against the bacterium, *Edwardsiella ictaluri,* to be given to farm-raised catfish. *E. ictaluri* causes enteric septicemia that costs catfish farmers on the order of $20 million a year. Catfish are big business these days—one acre of catfish farm yields more protein per year than an acre of any other crop. Where does all this catfish meat go? It goes into frozen fish sticks and fast food fish fillet sandwiches, among other foods.

So the next time you find yourself eating a fish fillet sandwich, think about the microbiology that helped provide your lunch.

Background Basics

toxin; toxoid . . . . *(Chapter 14)*
nonspecific host defenses . . . . *(Chapter 16)*
components of the lymphatic system . . . . *(Chapter 16)*

When you were very young you probably received a variety of immunizations against diphtheria, tetanus, whooping cough, polio, and possibly measles, German measles, and mumps as well. Your parents or grandparents, however, probably became immune to both kinds of measles and to mumps by acquiring and then recovering from these diseases. Either being immunized or having a disease may confer specific immunity to the organism that causes that disease. As we saw in the last chapter, nonspecific host defenses protect the host against infections in a general way. This chapter will show how specific host defenses and immunization protect the host against particular infectious agents.

Immunology and Immunity

The word *immune* literally means "free from burden." Used in a general sense, **immunity** refers to the ability of an organism to recognize and defend itself against infectious agents. *Susceptibility,* the opposite of immunity, is the vulnerability of the host to harm by infectious agents.

As we saw in the last chapter, host organisms have many general defenses against invading infectious organisms, regardless of what type of organism invades. ➡ (Chapter 16, p. 434) Immunity produced by such defenses is called **nonspecific immunity.** In contrast, **specific immunity** is the ability of a host to mount a defense against particular infectious agents by physiological responses *specific to that infectious agent.*

Immunology is the study of specific immunity and how the immune system responds to specific infectious agents. The **immune system** consists of various cells, especially lymphocytes,and organs such as the thymus gland, that help provide the host with specific immunity to infectious agents. ➡ (Chapter 16, p. 441)

Types of Immunity

Innate immunity, also called **genetic immunity,** exists because of genetically determined characteristics. One kind of innate immunity is **species immunity,** which is common to all members of a species. For example, all humans have immunity to many

Questions We'll Explore

A What is immunology, and how does immunity differ from susceptibility?

B How do innate immunity, acquired immunity, active immunity, and passive immunity differ?

C What are the properties of antigens and antibodies?

D How do cells and tissues function in immune reactions?

E How do recognition of self, specificity, heterogeneity, and immunological memory function in the immune system?

F How do B cells, plasma cells, and antibodies function in humoral immunity?

G How do primary responses and secondary responses confer immunological memory and immunity to many infectious diseases?

H How does cell-mediated immunity differ from humoral immunity, and how do reactions of cell-mediated immunity occur?

I What factors modify immune responses?

J What are the mechanisms of immunization, recommended immunizations, and the benefits and hazards of immunization?

infectious agents that cause disease in pets and domestic animals, and animals have similar immunity to some human diseases. Humans do not have the appropriate receptor sites and will not become infected with canine distemper no matter how much contact they have with infected puppies. *Mycobacterium avium* causes tuberculosis in birds, but rarely in humans with normal immune systems. (It does often infect people with AIDS.) Some diseases appear only in a few species. Gonococci infect humans and monkeys but usually not other species. *Bacillis anthracis* causes anthrax in all mammals and some birds but not in many other animals.

Acquired Immunity

In contrast to innate immunity, **acquired immunity** is immunity obtained in some manner other than by heredity. It can be naturally acquired or artificially acquired. **Naturally acquired immunity** is most often obtained through having a specific disease. During the course of the disease the immune system responds to molecules called *antigens* on invading infectious agents. It produces molecules called *antibodies* and initiates other specific defenses that protect against future invasions by the same agent. Immunity also can be naturally acquired from antibodies transferred to a fetus across the placenta or to an infant in colostrum and breast milk. **Colostrum** is the first fluid secreted by the mammary glands after childbirth. Although deficient in many nutrients found in milk, colostrum contains large quantities of antibodies that cross the intestinal mucosa and enter the infant's blood.

In contrast, **artificially acquired immunity** is obtained by receiving an antigen by the injection of vaccine or immune serum that produces immunity. Sticking needles full of vaccine or serum into people is not a natural process. Thus, the immunity produced is artificially acquired.

Active and Passive Immunity

Regardless of whether immunity is naturally or artificially acquired, it can be active or passive. **Active immunity** is created when the person's own immune system produces antibodies or other defenses against an infectious agent. It can last a lifetime or for a period of weeks, months, or years, depending on how long the antibodies persist. **Naturally acquired active immunity** is produced when a person is exposed to an infectious agent. **Artificially acquired active immunity** is produced when a person is exposed to a vaccine containing live, weakened, or dead organisms or their toxins. In both types of active immunity the host's own immune system responds specifically to defend the body against an antigen. Furthermore, the immune system generally "remembers" the antigen to which it has responded and will mount another response any time it again encounters the same antigen.

Passive immunity is created when ready-made antibodies are introduced into the body. This immunity is passive because the host's own immune system does not make antibodies. **Naturally acquired passive immunity** is produced when antibodies made by a mother's immune system are transferred to her offspring. New mothers are encouraged to breast-feed for a few days even if they are not planning to continue so that their infants obtain antibodies from **colostrum.** (ko-los'trum), the protein-rich fluid secreted by the mammary glands just after childbirth, before the appearance of true breast milk. **Artificially acquired passive immunity** is produced when antibodies made by other hosts are introduced into a new host. For example, a person who is bitten by a rattlesnake may receive a snake antivenin injection. Antivenins are antibodies produced in another animal, such as horses or rabbits. In this kind of immunity the host's immune system is not stimulated to respond. Ready-made antibodies and the immunity they confer persist for a few weeks to a few months and are destroyed by the host; the host's immune system cannot make new ones.

Relationships among the various types of immunity are shown in ➤ Figure 17.1. The properties of each type of immunity are summarized in Table 17.1

Characteristics of the Immune System

Antigens and Antibodies

Actions of the immune system are triggered by antigens. An **antigen** is a substance the body identifies as foreign and toward which it mounts an immune response—often it is also referred to as an immunogen. Most antigens are large protein molecules with complex structures and

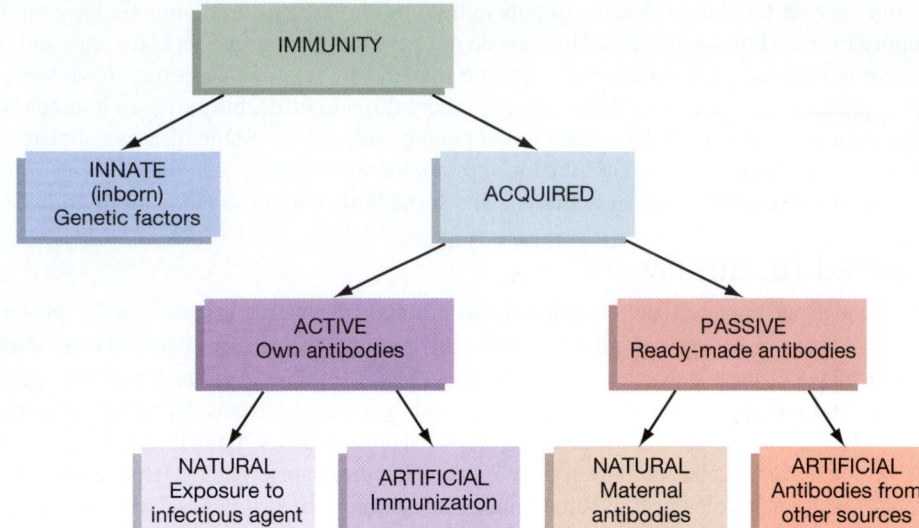

➤ Figure 17.1 **The various types of immunity.** Nonspecific immunity is largely innate or inborn, whereas specific immunity is acquired.

molecular weights greater than 10,000. Some antigens are polysaccharides, and a few are glycoproteins (carbohydrate and protein) or nucleoproteins (nucleic acid and protein). Proteins usually have greater antigenic (immunogenic) strength because they have a more complex structure than polysaccharides. Large, complex proteins can have several **epitopes,** or **antigenic determinants,** areas on the molecule to which antibodies can bind.

Antigens are found on the surface of viruses and all cells, including bacteria, other microorganisms, and human cells. The exact chemical structure of each of a cell's antigens is determined by genetic information in its DNA. Bacteria can have antigens on capsules, cell walls, and even flagella. Many microorganisms have several different antigens somewhere on their surface. Determining how the human body responds to these different antigenic determinants is important in making effective vaccines. As we shall see, antigens on the surfaces of red blood cells determine blood types, and antigens on other cells determine whether a tissue transplanted from another person will be rejected.

In some instances a small molecule called a **hapten** (hap'ten) can act as an antigen if it binds to a larger protein molecule. Haptens act as epitopes on the surfaces of proteins. Sometimes they bind to body proteins and provoke an immune response. Neither the hapten nor the body protein alone acts as an antigen, but in combination they can. For example, penicillin molecules can act as haptens, bind to protein molecules, and elicit an allergic reaction.

One of the most significant responses of the immune system to any foreign substance is to produce antiantigen proteins, or antibodies. An **antibody** is a protein produced in response to

➤ Table 17.1 **Characteristics of types of immunity**

Characteristic	Kind of Immunity		
	Innate	**Actively acquired**	**Passively acquired**
Agent	Genetic and physiological factors	Antibodies elicited by antigens	Ready-made antibodies
Source of antibodies	None	Person immunized	Plasma of other, such as mother
How elicited	Genetic expression	**Natural:** by having disease **Artificial:** by receiving vaccine	**Natural:** by receiving antibodies across placenta or in colostrum **Artificial:** by receiving injection of gamma globulin or immune serum
Time to develop immunity	Always present	5 to 14 days after receiving antigen	Immediately after receiving antibodies
Duration of immunity	Lifetime	Months to lifetime	Days to weeks

an antigen that is capable of binding specifically to the antigen. Each kind of antibody binds to a specific antigenic determinant. Such binding may or may not contribute to inactivation of the antigen. A typical antigen-antibody reaction is shown diagrammatically in ➤ Figure 17.2.

In discussing concentrations of antigens and antibodies, immunologists often refer to titers. A **titer** (ti′ter) is the quantity of a substance needed to produce a given reaction. For example, an antibody titer is the quantity required to bind to and neutralize a particular quantity of an antigen.

Cells and Tissues of the Immune System

Specific immune responses are carried out by lymphocytes, which develop from stem cells as do other white blood cells, red blood cells, and platelets. Early in embryonic development, un-differentiated stem cells from regions in the yolk sac called primitive blood islands proliferate. Later, they migrate into the body through the umbilical cord to various sites where they differentiate into specific types of lymphocytes.

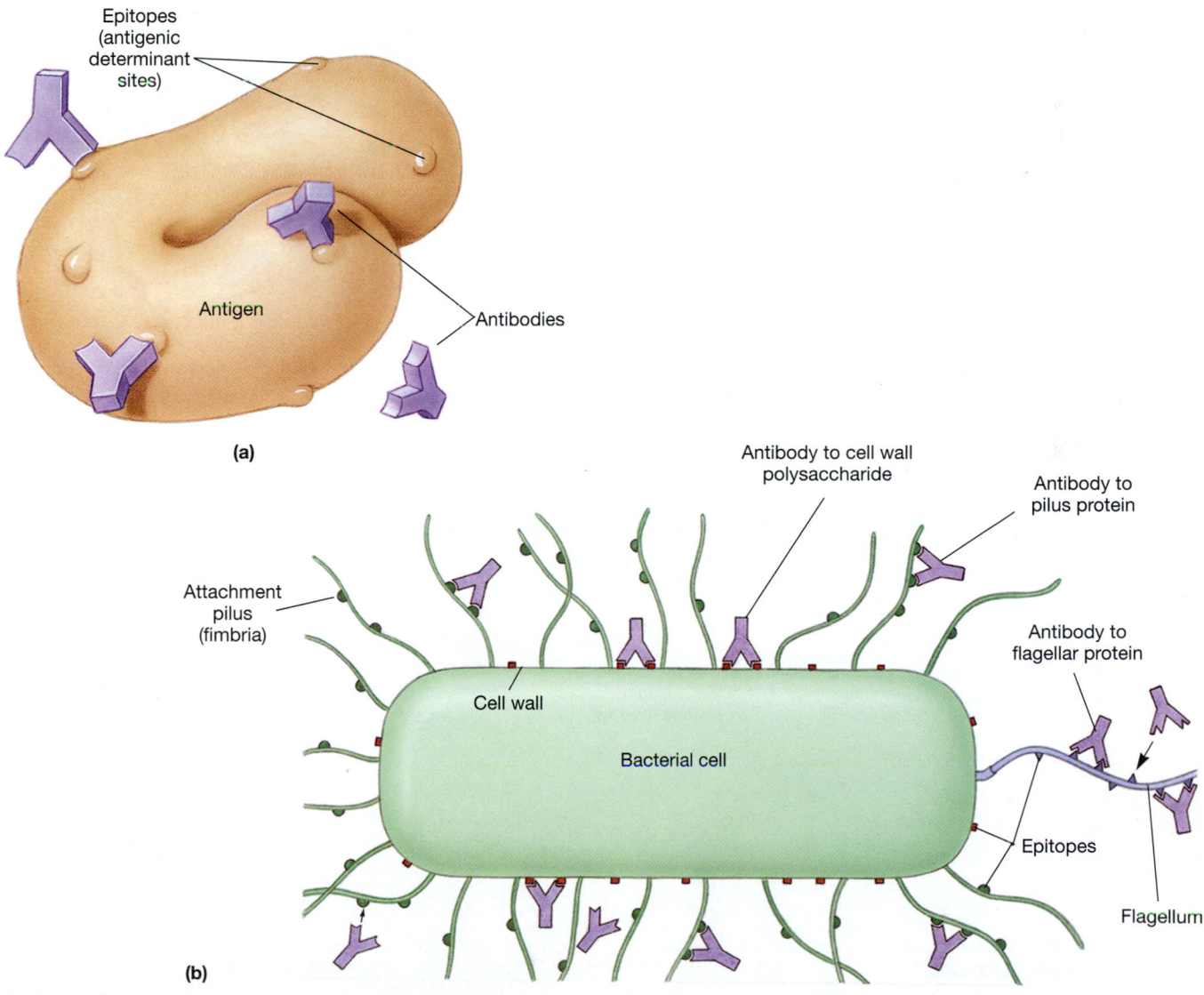

➤ Figure 17.2 **A typical antigen–antibody reaction.** (a) Antibodies bind to specific chemical groups or structures, called epitopes, or antigenic determinants. (b) A Gram-negative bacterial pathogen may have several antigens, or immunogens (for example, for flagella, pili, and cell wall), each with particular epitopes. Large, complex protein molecules may have several different antigenic determinants.

Differentiation of stem cells into lymphocytes is influenced by other tissues of the immune system (➤ Figure 17.3). Lymphocytes that are processed and mature in tissue referred to as bursal-equivalent tissue become **B lymphocytes,** or **B cells.** Differentiation of B cells was first observed in birds, where they are processed in an organ called the *bursa of Fabricius* (Fabris′e-us) (➤ Figure 17.4). Although no site equivalent to the bursa of Fabricius has been identified, B cells are produced in humans. Most investigators think this differentiation takes place in bone marrow or in gut-associated lymphoid tissues (*GALT*)—lymphoid tissues in the digestive tract, including the appendix and Peyer's patches of the small intestine. Regardless of where they differentiate, functional B cells are found in all lymphoid tissues—lymph nodes, spleen, tonsils, adenoids, and gut-associated lymphoid tissues. B cells account for about one-fourth of the lymphocytes circulating in the blood.

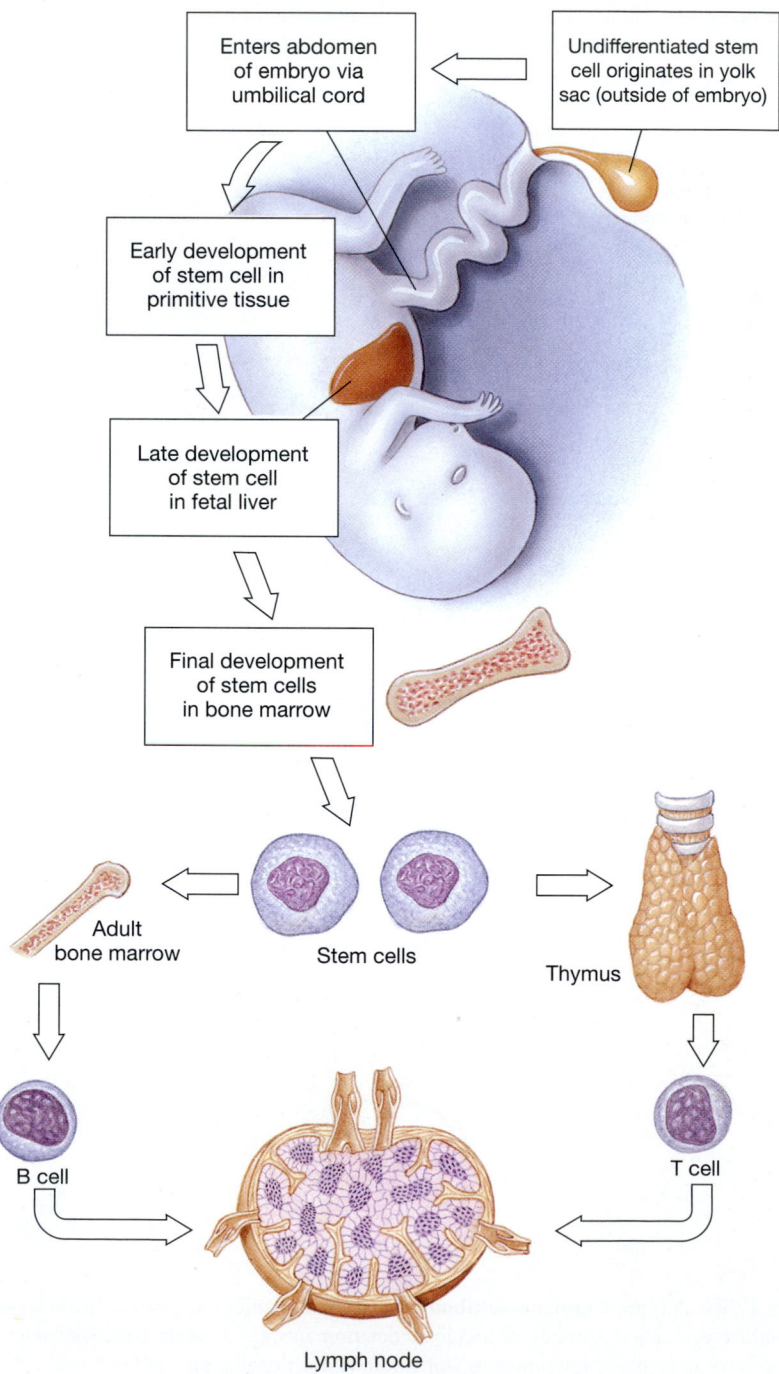

Enters abdomen of embryo via umbilical cord

Undifferentiated stem cell originates in yolk sac (outside of embryo)

Early development of stem cell in primitive tissue

Late development of stem cell in fetal liver

Final development of stem cells in bone marrow

Adult bone marrow

Stem cells

Thymus

B cell

T cell

Lymph node

➤ Figure 17.3 **Differentiation of stem cells into B cells and T cells.** This occurs in the bone marrow and thymus, respectively. The mature lymphocytes then migrate to lymphoid tissues, such as the lymph nodes.

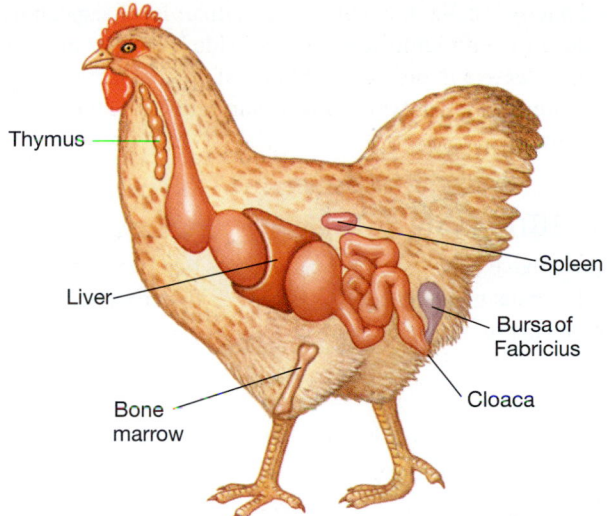

> Figure 17.4   **The bursa of Fabricius.** In chickens this is where B cells develop. It is a pouch located off the cloaca, a chamber into which waste and reproductive materials empty. (Some other organs of importance to the immune system are also shown.)

Labels: Thymus, Liver, Bone marrow, Spleen, Bursa of Fabricius, Cloaca

Other stem cells migrate to the thymus, where they undergo differentiation into thymus-derived cells called **T lymphocytes,** or **T cells.** In adulthood, when the thymus becomes less active, differentiation of T cells is thought to occur in bone marrow or tissues under the influence of hormones from the thymus. ➥ (Chapter 16, p. 443) T cells are found in all tissues that contain B cells and account for about three-fourths of the lymphocytes circulating in the blood. The distribution of B and T cells in lymphatic tissues is summarized in Table 17.2. Subsequent differentiation of T cells produces four different kinds of cells: (1) cytotoxic (killer) T cells, (2) delayed-hypersensitivity T cells, (3) helper T cells, and (4) suppressor T cells. After differentiation these T cells migrate among lymphatic tissues and the blood.

A few lymphocytes that cannot be identified as either B cells or T cells are found in tissues and circulating in blood. These undifferentiated cells are called **null cells** and include the so-called natural killer cells, discussed later in this chapter.

Dual Nature of the Immune System

Lymphocytes give rise to two major types of immune responses, humoral immunity and cell-mediated immunity. However, the presence of a foreign substance in the body often triggers both kinds of responses.

Humoral (hu′mor-al) **immunity** is carried out by antibodies circulating in the blood. When stimulated by an antigen, B lymphocytes initiate a process that leads to the release of antibodies. Humoral immunity is most effective in defending the body against bacterial toxins, bacteria, and viruses before these agents enter cells.

> Table 17.2   **Propotions of B and T lymphocytes in human lymphoid tissues.**[a]

Lymphoid tissue	% B Cells	% T Cells
Peyer's patches and nodules in digestive tract	60	25
Spleen	45	45
Lymph nodes	20	70
Blood	25	75
Thymus	1	99

[a]Where percentages do not add to 100, some lymphocytes are undifferentiated. Based on data from E. J. Moticka. In R. F. Boyd and J. J. Marr, (eds.), 1980. *Medical Microbiology.* New York: Little, Brown & Co.

Humoral Immune Responses: What's in a Name?

The term *humoral* in humoral immune response is derived from the work *humor* (from *umor,* Latin for "liquid"). It originally referred to the four basic body fluids, or "humors"—blood, phlegm, yellow bile, and black bile—which ancient physicians believed must be present in proper proportions for an individual to enjoy good health. If any of these fluids were out of balance a person was side to be "in bad humor" and likely to be diseased. Because this type of acquired immunity involves antibodies that circulate in the blood fluid, "humoral immune response" seemed logical.

Cell-mediated immunity is carried out by T cells. It occurs at the cellular level, especially in situations where antigens are embedded in cell membranes or are inside host cells and are thus inaccessible to antibodies. It is most effective in clearing the body of virus-infected cells, but it also may participate in defending against fungi and other eukaryotic parasites, cancer, and foreign tissues, such as transplanted organs.

General Properties of Immune Responses

Both humoral and cell-mediated responses have certain common attributes that enable them to confer immunity: (1) recognition of self versus nonself, (2) specificity, (3) heterogeneity, and (4) memory. We will look at each in some detail.

Recognition of Self versus Nonself

For the immune system to respond to foreign substances, it must distinguish between host tissues and substances that are foreign to the host. Immunologists refer to normal host substances as **self** and foreign substances as **nonself.** The **clonal** (klo′nal) **selection hypothesis** (➤ Figure 17.5), first proposed by Frank Macfarlane Burnet in the 1950s, explains one way the immune system might distinguish self from nonself. According to this hypothesis, embryos contain many different lymphocytes, each genetically programmed to recognize a particular antigen and make antibodies to destroy it. If a lymphocyte encounters and recognizes that antigen after development is complete, it divides repeatedly to produce a clone, a group of identical progeny cells that make the same antibody. If, during embryological development, it encounters its programmed antigen as part of a normal host substance (self), the lymphocyte is somehow destroyed or inactivated (➤ Figure 17.6). This mechanism removes lymphocytes that can destroy host tissues and thereby creates **tolerance** for self. It also selects for survival lymphocytes that will protect the host from foreign antigens.

Tolerance also can be acquired by irradiation during cancer treatment or the administration of immunosuppressant drugs to prevent rejection of transplanted organs. The host loses the ability to detect and respond to foreign antigens in transplanted organs, but then also fails to respond to infectious organisms.

Specificity

By the time the immune system matures at age 2 to 3, it can recognize a vast number of foreign substances as nonself. Furthermore, it reacts in a different way to each foreign substance. This property of the immune system is called **specificity.** Due to specificity, each reaction is directed toward a specific foreign antigen, and the response to one antigen generally has no effect on other antigens. However, **cross-reactions,** reactions of a particular antibody with very similar antigens, can occur. For example, certain microorganisms, such as the bacterium that causes syphilis, have the same haptens as some human cells, such as heart muscle cells, although the carrier molecules are quite different. This allows antibodies against this particular hapten to react with these otherwise vastly different cells. Cross-reactions also occur between strains of bacteria. For example, if three strains of pneumococci can cause pneumonia, and if each produces a particular antigen, A, B, or C, a person who has recovered from an infection with strain A has anti-A antibodies. The person then may also have some resistance to strains B and C because anti-A antibodies cross-react (that is, they react with antigens B and C).

Heterogeneity

The ability of the immune system to respond specifically allows it to attack particular antigens. But in a lifetime, the human body encounters hundreds of different foreign antigens. The property of **heterogeneity** (versatility, or flexibility) refers to the ability of the immune system to produce many different kinds of antibodies, each of which reacts with a different epitope (antigenic determinant). When a bacterium or other foreign agent has more than one kind of antigenic determinant, the immune system may make a different antibody against each. And it is capable of producing antibodies even against foreign substances, such as newly synthesized molecules never before encountered by any immune system.

> Specificity of each T and B cell is determined by random gene rearrangements that occur during maturation in the bone marrow before it ever contacts an antigen.

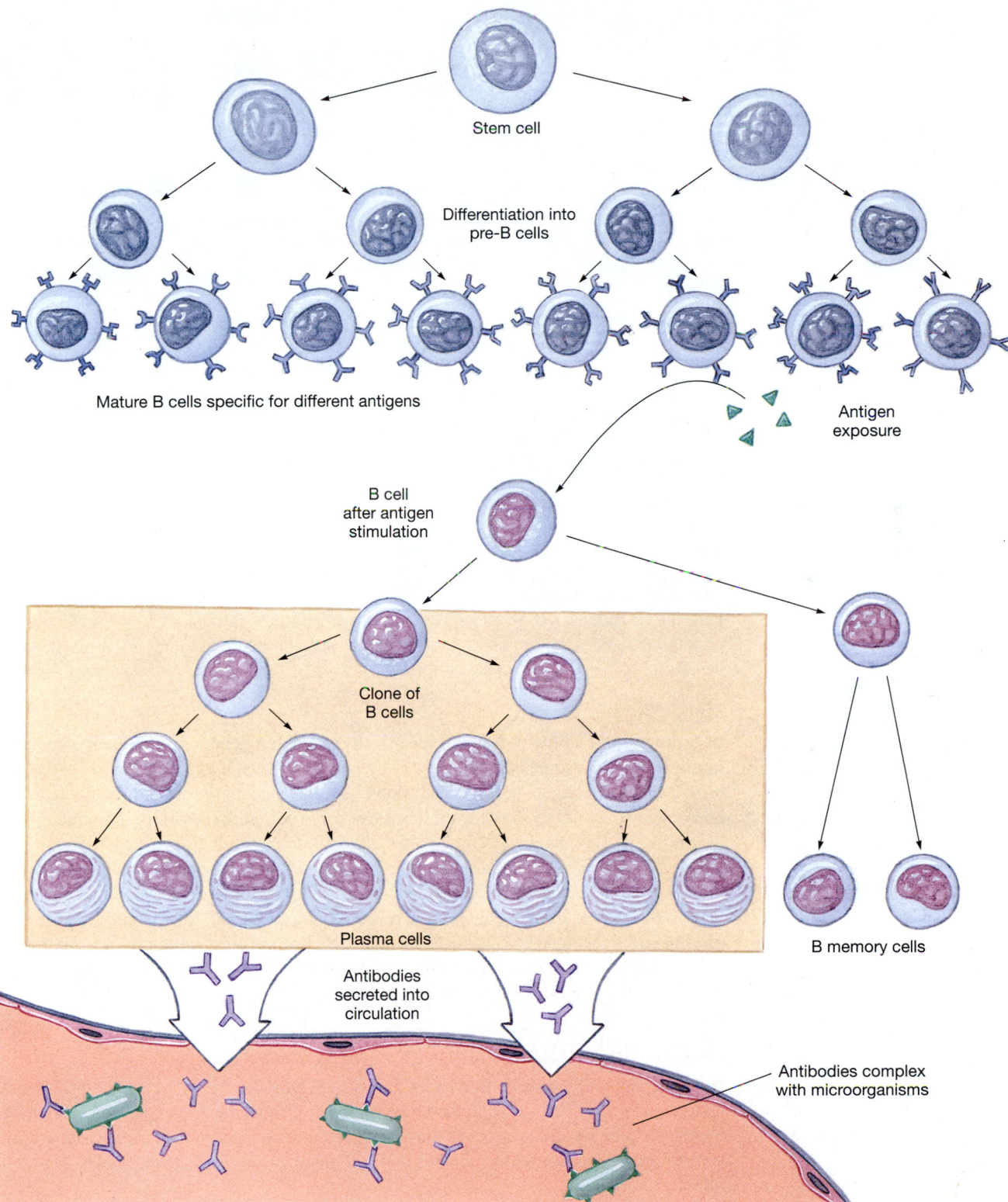

➤ Figure 17.5   **Clonal selection hypothesis.** According to this theory, one of many B cells responds to a particular antigen and begins to divide, thereby producing a large population of identical B cells (a clone). All cells of such a clone produce the same antibody against the original epitope. B memory cells are also produced.

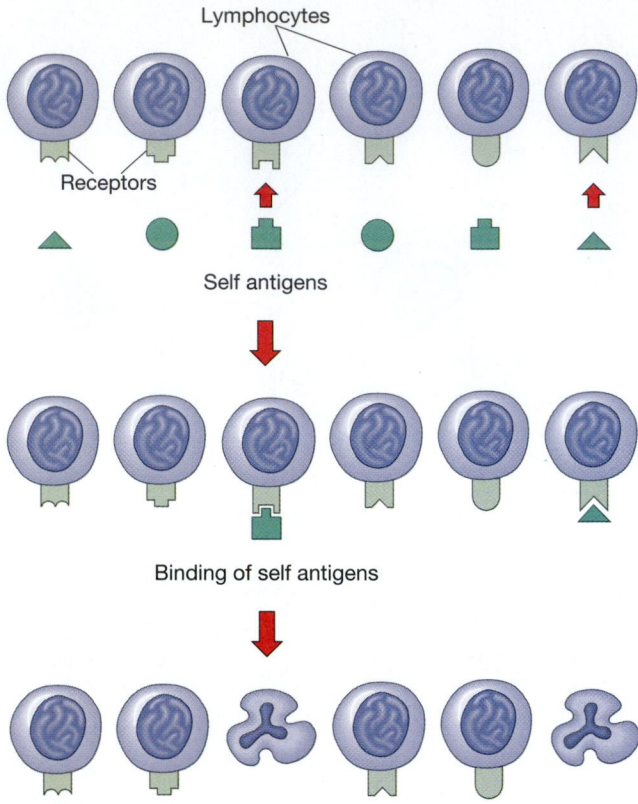

> Figure 17.6 **Clonal deletion.** This is a process that removes those lymphocytes that have receptors for self antigens. When lymphocytes bind to self antigens, clonal deletion occurs; that is, those lymphocytes die as a result of condensation and disintegration of cell nuclei. Lymphocytes lacking self receptors survive.

✓ Distinguish between active and passive immunity. Give examples of each.

✓ Distinguish between innate and acquired immunity. Give examples of each.

✓ What are the differences between antigen, epitope, and hapten?

✓ Distinguish between cellular and humoral immunity.

Memory

In addition to its ability to respond specifically to a heterogeneous assortment of antigens, the immune system also has the property of **memory**—that is, it can recognize substances it has previously encountered. Memory allows the immune system to respond rapidly to defend the body against an antigen to which it has previously reacted. In addition to producing antibodies during its first reaction to the antigen, the immune system also makes **memory cells** that stand ready for years or decades to quickly initiate antibody production. Consequently, the immune system responds to second and subsequent exposures to an antigen much more rapidly than to the first exposure. This prompt response due to "recall" by memory cells is called an **anamnestic** (secondary) **response.** The attributes of specific immunity are summarized in Table 17.3. With these attributes in mind we will now look in more detail at the two kinds of specific immunity, humoral and cell-mediated.

Humoral Immunity

Humoral immunity depends first on the ability of B lymphocytes to recognize specific antigens and second on their ability to initiate responses that protect the body against foreign agents. In most instances the antigens are on the surfaces of infectious organisms. The most common response is the production of antibodies that will inactivate an antigen and lead to destruction of infectious organisms.

Each kind of B cell carries its specific antibody on its membrane and can bind immediately to a specific antigen. The binding of an antigen **sensitizes,** or activates, the B cell and caues it to divide many times. Some of the progeny are memory cells, but most are plasma cells. **Plasma cells** are large lymphocytes that synthesize and release many antibodies like those on their membranes. While it is active a single plasma cell can produce as many as 2000 antibodies per second!

The responses of B cells can be influenced by certain T cells. When B cells encounter very large antigen molecules with many epitopes, they appear to generate plasma cells and antibodies without the aid of T cells. However, most antigens found on infectious agents have relatively

How B Cells Build Diverse Antibodies

How can B cells make antibodies to almost any foreign antigen or foreign substance with which they come in contact? The key to such heterogeneity lies in the immunoglobulin genes within each B cell. When B cells are formed in the bone marrow, each cell randomly pieces together different segments of its antibody genes.

In the embryo, the relatively few gene segments that code for the constant region of each light and heavy chain are not adjacent to the hundreds of gene segments that code for the variable regions. Let's look at how a light chain is built.

Light chains are formed when the DNA that separates a particular variable (V) segment from a constant (C) segment is removed, and the two gene segments are joined by a junction (J) segment. The now-joined segments form one continuous DNA sequence that represents the functional light-chain gene. Heavy chains are formed in a similar manner. Following transcription and translation, light-chain polypeptides are produced. These can be combined with heavy-chain polypeptides to form the functional antibody molecule. Thus, the heterogeneity of antibody-binding sites comes from the random combinations of variable gene segments that join with constant gene segments to form the light and heavy chains.

few epitopes. In such instances helper T cells assist B cells by "processing" antigens (see the following discussion) so that B cells can respond to them. Macrophages also present antigens to B cells. Suppressor T cells can block activity of helper T cells and so may be important in terminating an antigen-antibody reaction. How T cells carry out these functions will be explained later. Current evidence suggests that helper and suppressor T cells secrete helper and suppressor factors, respectively. Their effect on B cells seems to depend on the relative quan-

> ➤ Table 17.3 | **Main attributes of specific immunity**

Attribute	Description
Recognition of self versus nonself	The ability of the immune system to tolerate host tissues while recognizing and destroying foreign substances. Probably due to the destruction (deletion) of clones of lymphocytes during embryonic development.
Specificity	The ability of the immune system to react in a different and particular way to each foreign substance.
Heterogeneity	The ability of the immune system to respond in a specific way to a great variety of different foreign antigens.
Memory	The ability of the immune system to recognize and quickly respond to foreign substances to which it has previously responded.

tities of the two factors present in the region of an immune reaction. In a typical immune re-action, helper T cells facilitate growth and differentiation of plasma cells. After about a week this reaction reaches a peak and then subsides, largely because suppressor T cells inhibit fur-ther antibody production.

Properties of Antibodies (Immunoglobulins)

Antibodies, or **immunoglobulins (Ig)**, are Y-shaped protein molecules composed of four polypeptide chains—two identical **light (L) chains** and two identical **heavy (H) chains** (➤ Figure 17.7). The chains, which are held together by disulfide bonds, have constant regions and variable regions. The chemical structure of the *constant regions* determines the particular

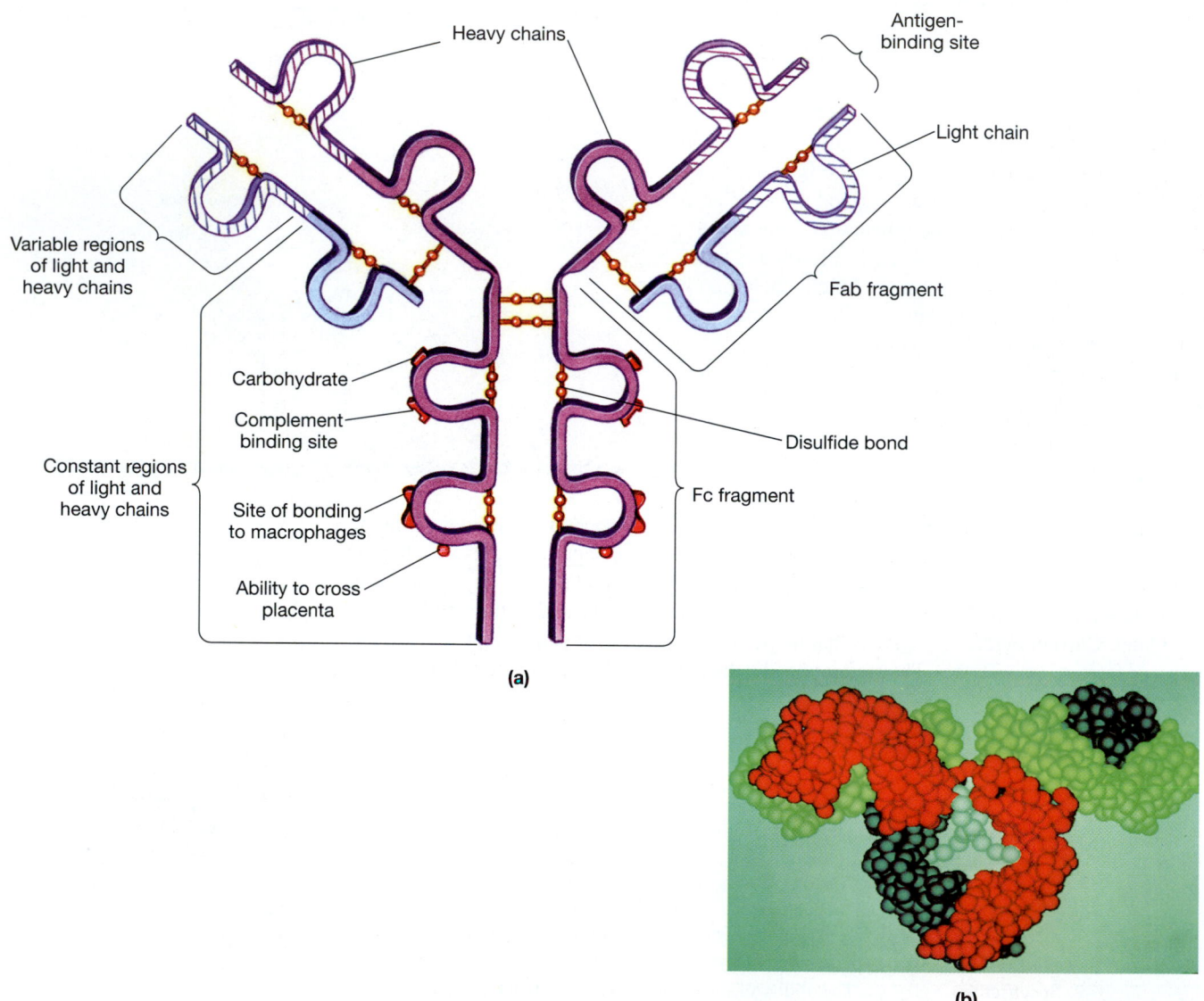

(a)

(b)

➤ Figure 17.7   **Antibody structure.** (a) The basic structure of the most abundant antibody (immunoglobulin) molecule in serum contains two heavy and two light chains, joined by disulfide bonds to form a Y shape. The upper ends of the Y, consisting of variable regions in both the light and heavy chains, differ from antibody to antibody. These variable regions form the two antigen-binding sites (part of the Fab fragment), which are responsible for the specificity of the antibody. The remain-ing part of the molecule consists of constant regions that are similar in all antibodies of a particular class. The Fc fragment determines the role each antibody plays in the body's immune responses. (b) A computer model of antibody structure. The two light chains are depicted in green, one heavy chain in red and the other heavy chain in blue.

class that an immunoglobulin belongs to, a described next. The *variable regions* of each chain have a particular shape and charge that enable the molecule to bind a particular antigen. Each of the millions of different immunoglobulins has its own unique pair of identical antigen-binding sites formed from the variable regions at the ends of the L and H chains. These binding sites are identical to the receptors in the membrane of the parent B cell. In fact, the first immunoglobulins made by B cells are inserted into their membranes to form the receptors. When the B cells form plasma cells, they continue to make the same immunoglobulins. When an antibody is cleaved with the enzyme papain at the hinge region, two *Fab* (antibody binding fragment) pieces and one *Fc* (crystallizable fragment) piece result. The Fab fragment binds to the epitope. The Fc region formed by parts of the H chains in the tail of the Y has a site that can bind to and activate complement, participate in allergic reactions, and combine with phagocytes in opsonization.

Classes of Immunoglobulins

Five classes of immunoglobulins have been identified in humans and other higher vertebrates (Table 17.4). Each class has a particular kind of constant region, which gives that class its distinguishing properties. The five classes are identified as IgG, IgA, IgM, IgE, and IgD (➢ Figure 17.8).

IgG, the main class of antibodies found in the blood accounts for as much as 20 percent of all plasma proteins. IgG is produced in largest quantities during a secondary response. The antigen-binding sites of IgG attach to antigens on microorganisms, and their tissue-binding sites attach to receptors on phagocytic cells. Thus, as a microorganism is surrounded by IgG, a phagocytic cell is brought into position to engulf the organism. The tail section of the H chains also activates complement. Complement, as explained in Chapter 16, consists of proteins that lyse microorganisms and attract and stimulate phagocytes.

IgG is the only immunoglobulin that can cross the placenta from mother to fetus and provide antibody protection for it.

IgA occurs in small amounts in blood, in larger amounts in body secretions such as tears, milk, saliva, and mucus and attached to the linings of the digestive, respiratory, and genitourinary systems. IgA is secreted into the blood, transported through epithelial cells that line these tracts, and either released in secretions or attached to linings by tissue-binding sites. Blood IgA consists of a single unit of two H and two L chains. Secretory IgA, which consists of two

There are different subclasses of IgG molecules, distinguished from each other by subtle amino acid differences, affecting their biological activities.

Each day, humans secrete 5–15 g of secretory IgA into their mucous secretions.

➢ **TABLE 17.4** Properties of antibodies

Property	Class of Immunoglobulin				
	IgG	IgM	IgA	IgE	IgD
Number of units	1	5	1 or 2	1	1
Activation of complement	Yes	Yes strongly	Yes, by alternative pathway	No	No
Crosses placenta	Yes	No	No	No	No
Binds to phagocytes	Yes	No	No	No	No
Binds to lymphocytes	Yes	Yes	Yes	Yes	No
Binds to mast cells and basophils	No	No	No	Yes	No
Half-life (days) in serum	21	5–10	6	2	3
Percentage of total blood antibodies in serum	75–85	5–10	10	0.005	0.2
Location	Serum, extravascular, and across placenta	Serum and B cell membrane	Transport across epithelium	Serum and extravascular	B cell membrane

Figure labels: J chain, Heavy, Light, Secretory piece, Nonsecretory, Secretory

IgG

Nonsecretory IgA

JChain

Secretory piece (confers solubility insecretions)

Secretory IgA

Heavy chain

Disulfide bonds

Light chains

IgD

IgE

Jchain

IgM

> Figure 17.8   **The structures of the different classes of antibodies.**

A primary antibody response peaks at about 10–17 days. A secondary response peaks in 2–7 days.

units held together by a J chain (joining chain), has an attached **secretory component,** which protects the IgA from proteolytic (protein-splitting) enzymes and facilitates its transport. The main function of IgA is to bind antigens on microorganisms before they invade tissues. It also activates complement, which helps to kill the microorganisms. IgA does not cross the placenta, but it is abundant in colostrum.

IgM, the first antibody secreted into the blood during the early stages of a primary immune response, can be made by both B cells and plasma cells. IgM consists of five units connected by their tails to a J chain and so has 10 peripheral antigen-binding sites. IgM is found mainly on B cell membranes and is rarely secreted. As IgM binds to antigens it also activates complement and causes microorganisms to clump together. These actions probably account for the initial effects the immune system has on infectious agents. It is also the first antibody formed in life, being synthesized by the fetus. In addition it is the antibody of the inherited ABO blood types. Because of its size, IgM is unable to cross the placenta.

IgE (also called *reagin*) has a special affinity for receptors on the plasma membranes of basophils in the blood or mast cells in the tissues. It binds to these cells by tissue-binding sites, leaving antigen-binding sites free to bind antigens to which humans can develop allergies, such as drugs, pollens, and certain foods. When IgE binds antigens, the associated basophils or mast cells secrete various substances, such as histamine, which produces allergy symptoms. IgE plays a damaging role in the development of allergies to such agents as drugs, pollens, and certain foods. Asthma and hay fever are common allergic diseases discussed in Chapter 18. Levels of IgE are elevated in patients with allergies and in those harboring worm parasites. IgE is found mainly in body fluids and skin and is rare in blood. It has an extremely low concentration in serum.

Like IgM, **IgD** is found mainly on B-cell membranes and is rarely secreted. Although it can bind to antigens, its function is unknown. It may help initiate immune responses and some allergic reactions.

In discussing concentrations of antigens and antibodies, immunologists often refer to titers. A **titer** (ti′ter) is the quantity (concentration) of a substance present in a specific volume of body fluid. For example, during an infection, an individual's antibody titer (the concentration of antibody in the serum) normally increases. An increasing antibody titer serves as an indication of an immune response by the body.

Primary and Secondary Responses

In humoral immunity the **primary response** to an antigen occurs when the antigen is first recognized by host B cells. After recognizing the antigen, B cells divide to form plasma cells, which begin to synthesize antibodies. In a few day antibodies begin to appear in the blood plasma, and they increase in concentration over a period of 1 to 10 weeks. The first antibodies are IgM, which can attack foreign substances directly. As IgM production wanes, IgG production accelerates. Eventually, it too wanes. The concentrations of both IgM and IgG can become so low as to be undetectable in plasma samples. However, memory cells persist in lymphoid tissues. They do not participate in the initial response, but they retain their ability to recognize a particular antigen. They can survive without dividing for many months to many years.

When an antigen recognized by memory cells enters the blood, a **secondary response** occurs. The presence of memory cells makes the secondary response much faster than the primary response. Some memory cells divide rapidly, producing plasma cells, and others remain as memory cells. Plasma cells quickly synthesize and release large quantities of antibodies. In the secondary response, as in the primary response, IgM is produced before IgG. However, IgM is produced in smaller quantities over a shorter period, and IgG is produced sooner and in much larger quantities than in the primary response. Thus, the secondary response is characterized by a rapid increase in antibodies, most of which are IgG. The primary and secondary responses are compared in > Figure 17.9.

Kinds of Antigen-Antibody Reactions

The antigen-antibody reactions of humoral immunity are most useful in defending the body against bacterial infections, but they also neutralize toxins and viruses that have not yet invaded cells. The defensive capability of humoral immunity depends on recognizing antigens associated with pathogens.

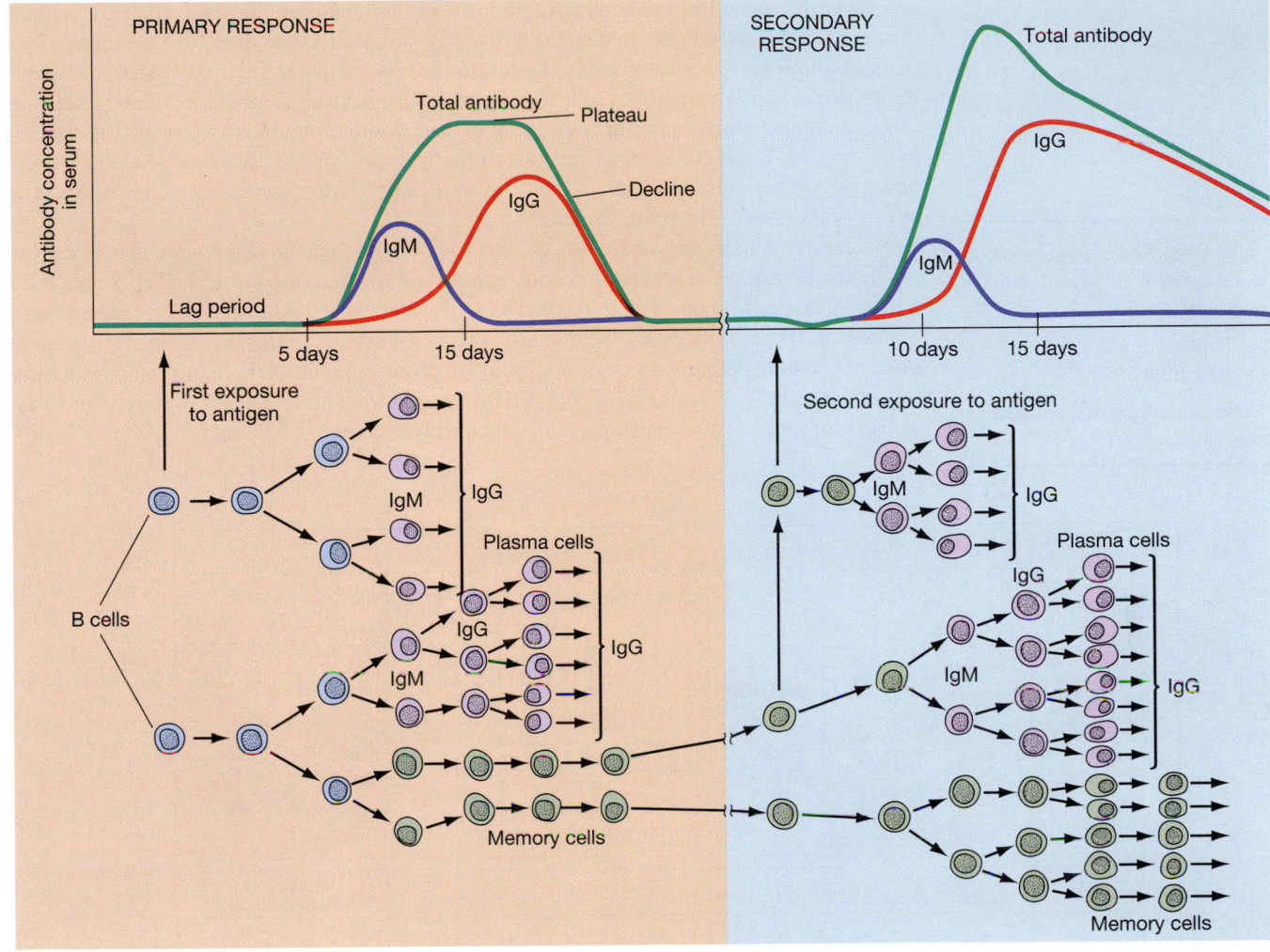

Figure 17.9   **Primary and secondary responses to an antigen.** This shows the correlation of antibody concentrations with the activities of B cells. Cytokines trigger the class switching from IgM to IgG.

For bacteria to colonize surfaces or for viruses to infect cells, these agents first must adhere to surfaces. IgA antibodies in tears, nasal secretions, saliva, and other fluids react with antigens on the microbes. They coat bacteria and viruses and prevent them from adhering to mucosal surfaces.

Microbes that escape IgA invade tissues and encounter IgE in lymph nodes and mucosal tissues. Gut-associated lymphoid tissue releases large quantities of IgE, which bind to mast cells; these cells then release histamine and other substances that initiate and accelerate the inflammatory process. Included in this process is the delivery of IgG and complement to the injured tissue.

Microbes that have reached lymphoid tissue without being recognized by B cells are acted on by macrophages and presented to B cells. B cells then bind the antigens and produce antibodies, usually with the aid of helper T cells. Antibodies binding with antigens on the surfaces of microbes form antigen-antibody complexes.

The formation of antigen-antibody complexes is an important component of the inactivation of infectious agents because it is the first step in removing such agents from the body. However, the means of inactivation varies according to the nature of the antigen and the kind of antibody with which it reacts. Inactivation can be accomplished by such processes as agglutination, opsonization, activation of complement, cell lysis, and neutralization. These reactions occur naturally in the body and can be made to occur in the laboratory. Here we will describe reactions chiefly as they relate to destruction of pathogens. We will discuss their laboratory applications more fully in Chapter 18.

Because bacterial cells are relatively large particles, the particles that result from antigen-antibody reactions also are large. Such reactions result in **agglutination** (ag-lu-tin-a′shun), or sticking together of microbes. IgM produces strong, and IgG produces weak, agglutination re-

> Several bacteria release antigenic proteins that function as superantigens, or proteins that stimulate the immune system much more extensively than do normal antigens.

actions with certain bacterial cells. Agglutination reactions produce results that are visible to the unaided eye and can be used as the basis of laboratory tests to detect the presence of antibodies or antigens. Some antibodies act as opsonins. ➡ (Chapter 16, p. 452) That is, they neutralize toxins and coat microbes so that they can be phagocytized, a process called opsonization.

Complement is an important component in inactivating infectious agents, as was discussed in Chapter 16. Both IgG and IgM are powerful activators of the complement system; IgA is less powerful. Sometimes, antibodies, especially IgM, directly lyse cell membranes of infectious agents without the aid of complement.

Bacterial toxins, being small molecules secreted from the cell, usually are inactivated simply by the formation of antigen-antibody complexes, or **neutralization.** IgG is the main neutralizer of bacterial toxins. Neutralization effectively stops the toxin from doing further damage to the host. It does not destroy the organisms that produce the toxin—antibiotics are needed to prevent persisting organisms from continuing to produce toxin. Viruses too can be inactivated by neutralization (➢ Figure 17.10a); IgM, IgG, and IgA are all effective neutralizers of viruses. Those viruses that have an envelope may then be lysed by complement (➢ Figure 17.10b).

✓ List the five types of immunoglobulins. Compare their structures and properties.

✓ How do primary and secondary responses differ?

✓ What is agglutination? Neutralization?

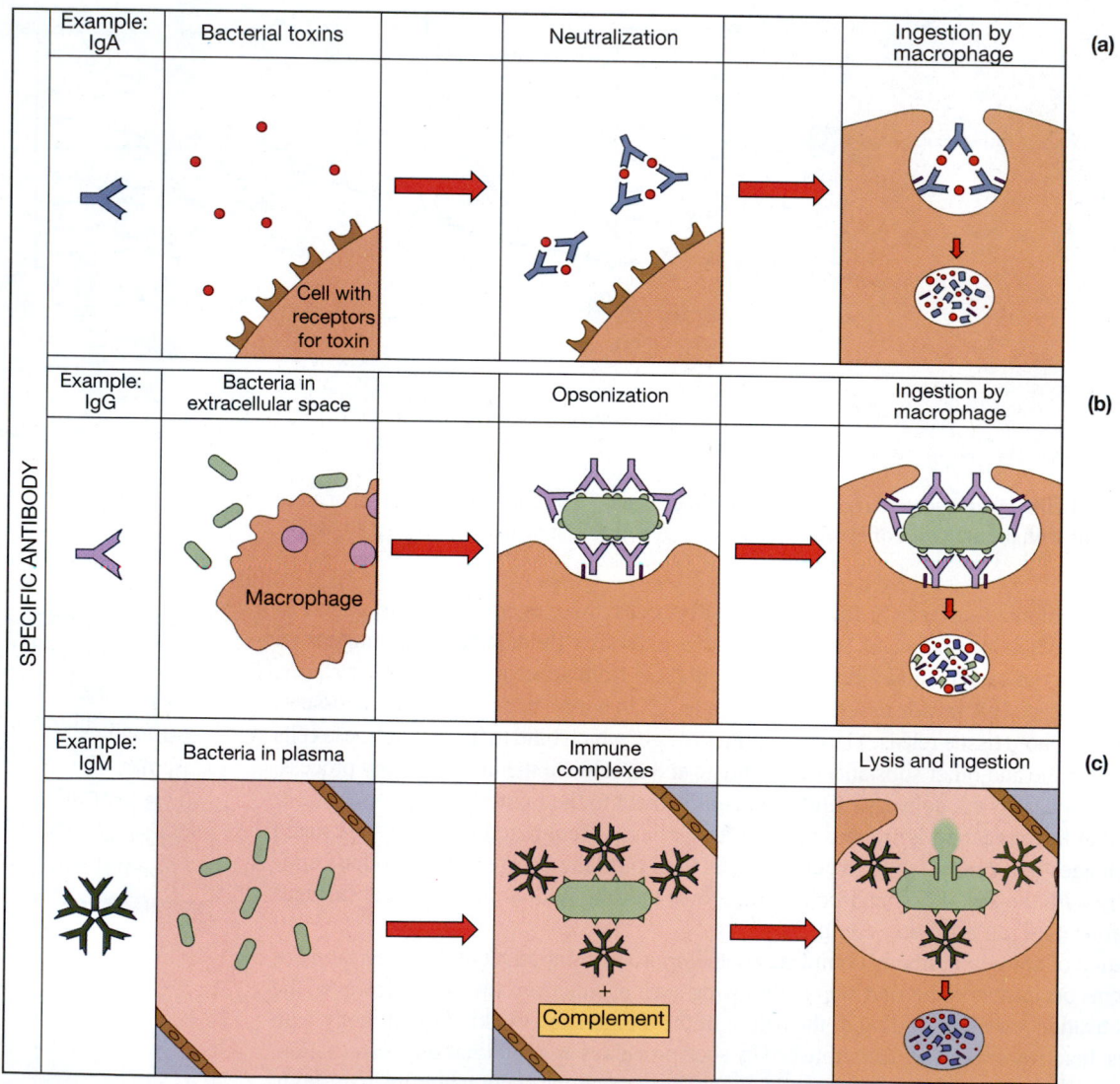

➢ Figure 17.10   **Antibodies produced by humoral immune responses eliminate foreign agents in three ways.** (a) Neutralization of pathogens and toxins by IgA or IgG, (b) opsonization of bacteria by IgG, and (c) cell lysis initiated by IgM or IgG immune complexes allows for the formation of membrane attack complexes involving complement proteins.

➤ **Figure 17.11** **Summary of humoral immunity.**

We have now considered the major characteristics of humoral immunity—how B cells are activated, how antibodies are produced, and how they function. These processes are summarized in ➤ Figure 17.11.

Monoclonal Antibodies

Monoclonal antibodies are antibodies produced in the laboratory by a clone of cultured cells that make one specific antibody. In one method of making monoclonal antibodies, myeloma cells (malignant cells of the immune system) are mixed with sensitized lymphocytes. The malignant cells are used because they will keep dividing indefinitely. The lymphocytes are used because each makes a particular antibody. When the two cell types are mixed in cultures, they can be made to fuse with one another to make a cell called a *hybridoma* (➤ Figure 17.12). ➡ (Chapter 8, p. 219) Hybridomas, which contain genetic information from each original cell, divide indefinitely, all the while producing large quantities of antibody. Which antibody a given hybridoma produces is determined by the antigen to which the lymphocytes were sensitized before their progeny were mixed with myeloma cells.

Generally, when a population of lymphocytes is exposed to an antigen, many different clones of B cells will proliferate, each making a different antibody. Many different hybridomas will therefore be produced by this technique. If one specific antibody is wanted, tests must be used to find which hybridomas are synthesizing that antibody, and those cells are then cloned.

Although monoclonal antibodies were first produced in 1975 as research tools, scientists were quick to recognize their practical uses. With experience, techniques for making monoclonal antibodies have improved. Culture media in which hybridomas thrive and produce large quantities of antibodies have been developed, and methods to grow hybridomas in large vat cultures in commercial laboratories are now available.

Theoretically, a monoclonal antibody can be produced for any antigen, provided lymphocytes sensitized to it can be obtained. Large numbers of hybridomas are now available, each producing a specific antibody. In addition to being used in research, many are produced commercially for use in diagnostic tests and in therapy.

Several diagnostic procedures that use monoclonal antibodies are now available. Generally, these procedures are quicker and more accurate than previously used procedures. For example,

Deodorant Monoclonal Antibodies

A unique commercial use of monoclonal antibodies has been patented by the Gillette Company. The company has devised a deodorant that contains monoclonal antibodies that attack *Staphylococcus haemolyticus,* the organism that causes perspiration-associated body odor. These antibodies block a bacterial enzyme necessary for the production of the unpleasant-smelling compounds.

➤ Figure 17.12   **Production of monoclonal antibodies.** Only the hybridoma cells grown in culture will survive, because any unfused spleen cells cannot divide, and any unfused mouse myeloma cells cannot get the nutrients they need to grow.

a monoclonal antibody can be used to detect pregnancy only 10 days after conception. Other monoclonal antibodies allow rapid diagnosis of hepatitis, influenza, and herpes virus and chlamydial infections. Diagnostic tests for other infectious diseases and allergies are being developed at a rapid rate, and progress is being made in using monoclonal antibodies to diagnose various kinds of cancer. Monoclonal antibodies can detect malignant cells before they have multiplied to produce a tumor.

Methods of using monoclonal antibodies in therapy are under development. These methods first require preparing antibodies to infectious agents or malignant cells. Then an appropriate drug or radioactive substance must be attached to the antibodies. If such antibodies are given to a patient, they carry the toxic substance directly to the cells bearing the appropriate antigen. The great advantage of therapeutic monoclonal antibodies is that they selectively damage infected or malignant cells without damaging normal cells.

Monoclonal antibodies against tumor antigens have been tried in a few cancer patients. Unfortunately, the patients often displayed allergic reactions to myeloma proteins that accompany the antibodies. Researchers are now trying to produce monoclonal antibodies that will kill malignant cells without causing allergic reactions in the patients that receive them. Diphtheria exotoxin delivered to cancer cells by monoclonal antibodies is being tried as a therapy for cancer.

Cell-Mediated Immunity

In contrast to humoral immunity, which involves B cells and immunoglobulins, cell-mediated immunity involves the direct actions of T cells. In cell-mediated immunity, T cells interact directly with other cells that display foreign antigens. These interactions clear the body of viruses and other pathogens that have invaded host cells. They also account for rejection of tumor cells, some allergic reactions, and immunological responses to transplanted tissues.

The cell-mediated immune response involves the differentiation and actions of different types of T cells and the production of chemical mediators call **lymphokines.** Much recent research has been devoted to determining the characteristics, orgins, and functions of T cells, including the functions of secreted lymphokines. Much more research is needed to understand fully cell-mediated immunity. What follows is a brief discussion of our current knowledge.

T cells, as noted earlier, either are processed by the thymus or develop under the influence of thymic hormones. T cells differ from B cells in that they do not make antibodies. However, they do have a particular cell membrane receptor protein that corresponds to the antibodies of B cells and other receptor proteins as well.

The Cell-Mediated Immune Reaction

The cell-mediated immune reaction typically begins with the processing of an antigen—usually one associated with a pathogenic organism—by macrophages. When macrophages phagocytize pathogens, they ingest and degrade the pathogen; then they insert some of the pathogen's antigen molecules into their own cell membranes. This constitutes processing the antigen. When a macrophage presents the antigen to T cells having the proper antigen receptor, the antigen and receptor bind. Macrophages also have on their surfaces genetically determined histocompatibility proteins (Chapter 18) that bind to other T cell receptors. These reactions are summarized in ➤ Figure 17.13.

Binding with macrophages causes T cells to divide and differentiate into different types of T cells, including memory cells (➤ Figure 17.14). Each cell is sensitized to the antigen that initiated the process, and each type has a different function in cell-mediated immune reactions. Some cells act directly and others release lymphokines, which are chemical substances that trigger certain immunologic reactions. The reactions of cell-mediated immunity are summarized in ➤ Figure 17.15. Refer to the figure as you read about the functions of different kinds of T cells.

Macrophages that have processed an antigen secrete the lymphokine interleukin-1 (IL-1), which activates **helper T (T_H) cells.** T_H cells, in turn, secrete lymphokines such as interleukin-2 (IL-2) and gamma interferon. IL-1 from macrophages and IL-2 from T_H cells activate other T cells—**suppressor T (T_S) cells, delayed hypersensitivity T (T_D) cells,** and **cytotoxic (killer) T (T_C) cells.** Also, IL-1, IL-2, and gamma interferon together cause null cells to become **natural killer (NK) cells.**

They're Still Out There

Did you know that scientists are still discovering new diseases? That's sometimes hard to believe, since it seems like there are more than enough diseases that are already known. But it's a scary truth. For example, in May 1998, a new genetic immunodeficiency disease was reported. Patients with this disease are particularly susceptible to Salmonella and Mycobacteria infections because they are deficient in a T_H1 receptor protein that is responsible for initiating the cell-mediated immune reaction. But don't worry too much—people who are born with this disease still seem to handle childhood viral and bacterial infections normally.

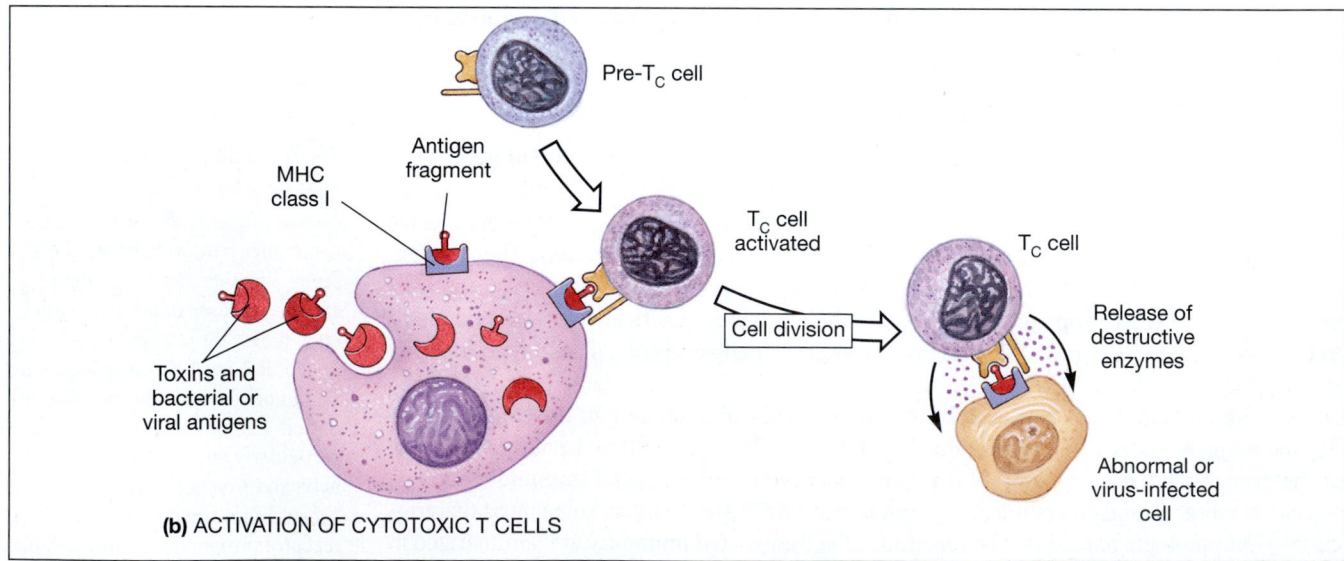

➤ Figure 17.13   **The reactions in cell-mediated immunity.** (a) The macrophage has processed an
antigen and inserted an antigen (peptide) fragment into its plasma membrane as an MHC class II molecule.
T_H cells have receptors that recognize the peptide fragment on MHC class II. Binding causes the T_H cells to
become activated. The activated T cells then differentiate into either T_H1 cells or T_H2 cells. T_H1 cells activate
infected macrophages to destroy internal bacterial infections. T_H2 cells activate B cells (humoral immune responses) by binding to MHC class II:peptide presented by the B cells. (b) Presenting the same peptide fragment on MHC class I to T_C cells activates these cells to attack infected cells, especially abnormal or
virus-infected cells.

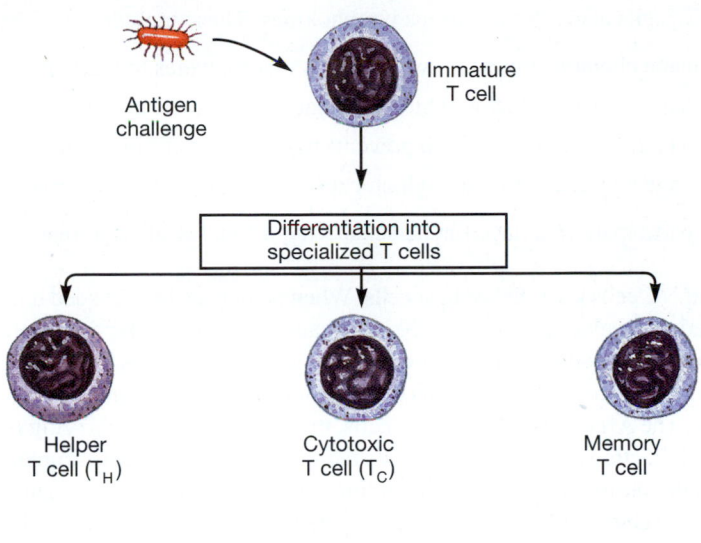

➤ **Figure 17.14** **Types of T cells.** After T cells are challenged by antigens, the cells differentiate into one of several types of functioning T cells.

➤ **Figure 17.15** **Summary of cell-mediated immunity.** (CD stands for "cluster of differentiation.")

At the same time that these cells are differentiating, some T memory cells also are formed. As in humoral immunity, the persistence of memory cells in cell-mediated immunity allows the body to recognize antigens to which T cells have previously reacted and to mount more rapid subsequent responses.

As we noted in the discussion of humoral immunity, T_H cells stimulate the growth and differentiation of B cells, and T_S cells suppress T_H cells. Such suppression apparently helps to prevent both humoral and cell-mediated immune processes from getting out of hand.

Immune Cell Infections

An agent that infects T cells is devastating because the agent destroys the very cells that might have combated the infection. HIV is just such an agent. HIV invades T_H cells, dendritic cells, and macrophages (APCs), the very heart of the immune response. Infection and destruction of these cells prevent humoral and cell-mediated responses from occurring as they should. A lack of T_H1 cells and T_H2 cells impairs both humoral and cell-mediated immune responses, including the destruction of cancer (malignant) cells. Thus, because of extensive destruction of APCs and T_H cells by HIV, persons with AIDS are susceptible to a host of opportunistic microoganisms and to various malignancies.

Activated T_D cells also release various lymphokines. These include:

1. Macrophage chemotactic factor, which helps macrophages to find microbes.
2. Macrophage activating factor, which stimulates phagocytic activity.
3. Migration inhibiting factor, which prevents macrophages from leaving sites of infection.
4. Macrophage aggregation factor, which causes macrophages to congregate at such sites.

T_D cells also participate in delayed hypersensitivity, a kind of allergic reaction explained in Chapter 18.

T_C cells and NK cells kill infected host cells. When pathogens have evaded humoral immunity and established themselves inside cells, they can cause long-term infections unless the infected cells are destroyed by cell-mediated immunity. An agent that infects T cells is especially devastating because it destroys the very cells that might have combated the infection. AIDS is just such a disease. The AIDS virus invades T_H cells, prevents them from carrying out their normal immunological functions, and eventually kills them. The lack of T_H cells impairs both humoral and cell-mediated immune responses, including the destruction of malignant cells. Thus, because of extensive destruction of T_H cells, AIDS patients are susceptible to a host of opportunistic infections and to various malignancies.

How Killer Cells Kill

Recent research shows that T_C cells and NK cells kill other cells by making a lethal protein and firing it at target cells. Eosinophils have a similar protein, which they may use to kill certain helminths and other parasites. But lethal proteins are not the sole property of hosts' defensive cells—the amoebae that cause amoebic dysentery and some other parasites and fungi also have them. Learning more about these lethal proteins may one day enable us to treat amoebic dysentery and other parasitic diseases by blocking the action of these proteins or to treat AIDS and malignant diseases by enhancing their actions.

Cytotoxic T cells act mainly on virally infected cells, whereas NK cells act mainly on tumor cells, cells of transplanted tissues, and possibly on cells infected with intracellular agents such as rickettsias and chlamydias. Each kind of killer cell acts by a different mechanism. Cytotoxic T cells bind to antigens presented by macrophages and then attack virus-infected cells. In contrast, NK cells bind directly to malignant or other target cells without the help of macrophages.

Both kinds of killer cells contain granules of a lethal protein, **perforin,** which is released when they bind to a target cell. Perforin bores holes in the target cell membranes so that essential molecules leak out and the cells die. This process is similar to the action of complement. By killing infected cells while they are few in number and before new virus particles are released from them, cytotoxic T cells prevent the spread of infection—but at the expense of destroying host cells. Similarly, NK cells destroy malignant cells before they have a chance to multiply. Both kinds of killer cells can withdraw from cells they have damaged and move on to other target cells.

The discovery of such an efficient mechanism for killing cells raises two important questions: What prevents perforin from killing adjacent uninfected cells, and what prevents it from attacking the membranes of the killer cells themselves? Perforin doesn't kill adjacent cells because it is effective only when secreted at the binding site between the killer and target cells. Why perforin doesn't attack killer-cell membranes is not known, but it has been suggested that killer cells produce a protein called protectin that inactivates perforin.

The Role of Activated Macrophages

Some bacteria, such as those that cause tuberculosis, Hansen's disease (leprosy), listeriosis, and brucellosis, can continue to grow even after they have been engulfed by macrophages. T_D cells combat such infections by releasing the lymphokine macrophage activating factor. This factor causes macrophages to increase production of toxic hydrogen peroxide, along with enzymes that attack the phagocytized organisms and accelerate the inflammatory response. Organisms that survive these defenses are walled off in granulomas.

We have now completed the discussion of cell-mediated immunity—how it is initiated and how its effects are produced. These processes are summarized in ➤ Figure 17.15. The various

> ## Table 17.5 Characteristics of B cells, T cells, and macrophages

Characteristic	B Cells	T Cells	Macrophages
Site of production	Bursal-equivalent tissues	Thymus or under thymic hormones	
Type of immunity	Humoral	Cell-mediated and assist humoral	Humoral and cell-mediated
Subpopulations	Plasma cells and memory cells	Cytotoxic, helper, suppressor, delayed hypersensitivity, and memory cells	Fixed and wandering
Presence of surface antibodies	Yes	No	No
Presence of foreign surface antigens	No	No	Yes
Presence of receptors for antigens	Yes	Yes	No
Life span	Some long, most short	Long and short	Long
Secretory product	Antibodies	Lymphokines	Interleukin-1
Distribution (% leukocytes)			
Peripheral blood	15–30	55–75	2–12
Lymph nodes	20	75	5
Bone marrow	75	10	10–15
Thymus	10	75	10

functions of B and T cells are summarized in Table 17.5. Comparing Figures 17.11 and 17.15 and studying Table 17.5 will serve to highlight similarities and differences in humoral and cell-mediated immunity and to provide an overview of specific immunity.

Factors That Modify Immune Responses

The host defenses of young healthy human adults living in an unpolluted environment are capable of preventing nearly all infectious diseases. However, a variety of disorders, injuries, medical treatments, environmental factos, and even age can affect resistance to infectious diseases. An individual with reduced resistance is called a **compromised host.**

In the beginning of this chapter we noted that humans are genetically immune to some diseases. It has also been found that different races have different degrees of resistance and susceptibility to various diseases. When black and white military personnel live under the same conditions (same barracks, food, exercise regime, and such), blacks still develop TB at a higher rate.

Age also affects immune responses. In general, the very young and the elderly are most susceptible to infections, and young adults are least susceptible. The young are susceptible because the immune system is not fully developed until age 2 or 3. Infants can, however, produce some IgM shortly after birth, and they receive maternal IgG passively. The elderly are susceptible to infections and malignancies because the immune system, and especially cell-mediated response, is one of the first to decline in function during the aging process. Thus, it makes sense to take special precautions against unnecessarily exposing infants and elderly people to infectious agents. And it also makes sense to obtain recommended immunizations during infancy and early childhood.

Even seasonal patterns affect the immune system. T cells have a yearly cycle, falling to their lowest level in June. People with Hodgkin's disease are most often diagnosed in spring, leading some researchers to feel there is a link between the two cycles.

Genetic and age factors that modify immunity are beyond our control, but we have some control over diet and environment. Let's see how these factors contribute to resistance—or the lack of it.

An adequate diet, especially adequate protein and vitamin intake, is essential for maintaining healthy intact skin and mucous membranes and phagocytic activity. It is likewise important for lymphocyte production and antibody synthesis. Poor nutrition and poor inflammatory response of alcoholics and drug addicts greatly lower their resistance to infection. In the elderly an inadequate diet can further weaken a declining immunological response.

Regular moderate exercise such as 45 minutes of brisk walking, 5 days per week can produce a 20 percent increase in antibody level, which occurs during the exercise and for about

✓ What are monoclonal antibodies, and how are they produced? What are their uses?

✓ Distinguish among the various kinds of T cells and their functions.

✓ What are NK cells? How do they function?

1 hour afterwards. Natural killer-cell activity is also increased. However, excessive exercise such as running more than 20 miles per week depresses the immune system. Marathoners who ran at their fastest pace for 3 hours experienced a drop in natural killer cell activity of more than 30 percent for about 6 hours. Long-distance runners are more vulnerable to infection, especially of the upper respiratory tract, for about 12 to 24 hours after a race, experiencing six times the rate of illness after a race compared with trained runners who did not race.

Traumatic injuries lower resistance at the same time that they provide easier access to tissue for microbes. Tissue repair competes with immune processes because both require extensive protein synthesis. When normal systems that flush away microbes, such as tears, urinary excretions, and mucous secretions, are impaired by injuries, pathogens have easier access to tissues. Antibiotics destroy commensals that sometimes compete with pathogens. Impaired defenses and use of antibiotics allow opportunistic infections to become established.

Environmental factors such as pollution and exposure to radiation also lower resistance to infection. Air pollutants, including those in tobacco smoke, damage respiratory membranes and reduce their ability to remove foreign substances. They also depress the activities of phagocytes. Excessive exposure to radioactive substances damages cells, including cells of the immune system. These factors can be compounded by induced and inherited immunological disorders. Immunosuppressant drugs used to prevent rejection of transplanted tissue impair the functions of lymphocytes and some phagocytes. Diseases such as AIDS destroy T cells. Finally, genetic defects in the immune system itself can result in the absence of B cells, T cells, or both. How these disorders impair immunity is discussed in Chapter 18.

Immunization

Throughout the world each year nearly 3.5 million children, most of them under 5 years old, die of three infectious diseases for which immunization is available. About 2 million die of measles, 800,000 die of tetanus, and 600,000 die of whooping cough. Another 4 million die of various kinds of diarrhea, against which some immunization is possible. Most of these deaths occur in underdeveloped countries.

These statistics dramatize three important facts about immunization. First, immunization can prevent significant numbers of deaths. Second, methods of immunization are not yet available for some infectious diseases, such as certain diarrheas. Most organisms that cause diarrhea exert their effects in the digestive tract, where antibodies and other immune defenses cannot reach them. Finallly, much greater effort is needed to make immunizations available in underdeveloped countries.

Active Immunization

To develop active immunity, as noted earlier, the immune system must be induced to recognize and destroy infectious agents whenever they are encountered. **Active immunization** is the process of inducing active immunity. It can be conferred by administering vaccines or toxoids. A **vaccine** is a substance that contains an antigen to which the immune system responds. Antigens can be derived from living but attenuated (weakened) organisms, dead organisms, or parts of organisms. A **toxoid** is an inactivated toxin that is no longer harmful, but retains its antigenic properties.

Principles of Active Immunization

Regardless of the nature of the immunizing substance, the mechanism of active immunization is eventually the same. When the vaccine or toxoid is administered, the immune system recognizes it as foreign and produces antibodies, or sometimes cytotoxic T cells, and memory cells. This immune response is the same as the one that occurs during the course of a disease. The disease itself does not occur either because whole organisms are not used or because they have been sufficiently weakened to have lost their virulence. In other words, vaccines retain important antigenic properties but lack the ability to cause disease. In fact, organisms in vaccines sometimes do multiply in the host, but without producing disease symptoms. Similarly, toxoids retain antigenic properties but cannot exert their toxic effects.

An important factor in the longevity of immunity from active immunization is the nature of the immunizing substance. In most instances, vaccines made with live organisms confer longer-lasting immunity than those made with dead organisms, parts of organisms, or toxoids. For example, measles (both rubella and rubeola) vaccines and oral poliomyelitis vaccine, which contain live viruses, usually confer lifelong immunity. Intramuscular polio vaccine, which contains killed viruses, and typhoid fever vaccine, which contains dead bacteria, confer immunity that lasts 3 to 5 years. Tetanus and diphtheria toxoids confer immunity of about 10 years' duration.

Because immunity is not always lifelong, "booster shots" are often needed to maintain immunity. As we have noted, the first dose of a vaccine or toxoid stimulates a primary immune response analogous to that during the course of a disease. Subsequent doses stimulate a secondary immune response analogous to that following exposure to an organism to which immunity has already developed. Thus, booster shots boost immunity by greatly increasing the number of antibodies. This increases the length of time sufficient antibodies are available to prevent disease.

The route of administration of a vaccine can affect the quality of immunity. Compared with injecting vaccines into muscle, immunity is more durable when oral vaccines are used against gastrointestinal infections and nasal aerosols are used against respiratory infections.

Vaccines and toxoids, especially those containing live organisms, must be properly stored to retain their effectiveness. Some require refrigeration, and serious failures of measles immunization have resulted from inadequate refrigeration. Others must be used within a certain number of hours or days after a vial of vaccine is opened. Thus, clinics offer some immunizations only on selected days. Immunizations that require expensive vaccines and that are in low demand may be given one day per week or even less frequently.

In general, active immunization cannot be used to prevent a disease after a person has been exposed. This is because the time required for immunity to develop is greater than the incubation period of the disease. Rabies immunization is an exception to this rule. Because rabies typically has a long incubation period, active immunization can be used with some hope that immunity will develop before the rabies-virus reaches the brain. The farther the virus must travel to reach the brain, the greater the chance of effective immunization. So, a bite on the ankle received while kicking off a rabid animal may be less hazardous than one on the trunk or neck. As we shall see later, passive immunization sometimes is used to prevent or lessen the severity of diseases after exposure to them. Several vaccines and toxoids have been licensed for general use in the United States; their properties are summarized in Table 17.6. Many more are available for people with special needs, such as foreign travel, or for experimental purposes. Properties of some special use vaccines are given in Table 17.7.

Recommended Immunizations

Three vaccines that immunize against seven diseases are currently recommended in the United States for routine immunization of normal infants and children. The **DPT vaccine** contains diphtheria toxoid, killed pertussis (whooping cough) bacteria, and tetanus toxoid. Although most children tolerate DPT vaccine, a few suffer severe complications, as described shortly in connection with hazards of vaccines. The **poliomyelitis vaccine** generally used in this country contains three different types of live polioviruses. Vaccines with similar antigens are available for administration orally or intramuscularly. **MMR vaccine** contains live rubella, rubeola, and mumps viruses. This vaccine can be used to immunize against all three diseases simultaneously, or separate vaccines can be given for each disease.

The recommended age for administering vaccines varies. DPT and polio vaccines can be administered effectively as early as 2 months of age. However, MMR vaccine is not recommended before 15 months of age. When it is given to younger infants, the quality of immunity that results usually is not sufficient to protect against infection, probably because of immaturity of the immune system. Vaccines against *Haemophilus influenzae,* type b (Hib), first became available in 1985 but didn't work well in children under 2 years old. The FDA has now approved several new **Hib vaccines** for younger children that could prevent about 10,000 cases per year of meningitis, which kills about 500 children and leaves thousands of survivors mentally retarded, deaf, or otherwise neurologically damaged. Immunizations against Hib are scheduled at 2, 4, 6, and 15 months, or at 2, 4, and 12 months. Children between 15 months and 5 years need only one shot. Immunization is not recommended for those over 5 years old because nearly all children have contracted a Hib infection by then and have thus developed natural active immunity. When given early, the Hib vaccine can be combined with DPT. The recommended immunization schedule for normal infants and children in the United States is given in Table 17.8.

BCG vaccine is given, usually right after birth, to 75 percent of all children worldwide—but not in the United States.

➤ Table 17.6 **Properties of materials available for active immunization**

Disease	Nature of material	Route of administration[a]	Use and comments	Duration of effectiveness
Bacterial meningitis	Polysaccharide– protein conjugate	IM	2, 4, 6, and 15 months; 75% effective	14–34 years
Cholera	Killed bacteria	SC, IM, ID	2 doses a week or more apart, 50% effective, may be required for travel	6 months
Diphtheria	Toxoid	IM	3 doses 4 weeks apart plus boosters, 90% effective	10 years
Hepatitis B	Viral antigen	IM	2 doses 4 weeks apart, booster in 6 months	About 5 years
Influenza (viral)	Inactivated virus	IM	1 or 2 doses, depending on type of virus, recommended for high-risk patients and medical personnel; 75% effective	1–3 years
Lobar pneumonia	Polysaccharide	SC, IM	1 dose before chemotherapy	5 years or more
Measles (rubeola)	Live virus	SC	1 dose at 15 months, revaccination around age 12; 95% effective, may prevent disease if given within 48 hours of exposure	Lifelong
Meningococcal meningitis	Polysaccharide	SC	1 dose, recommended during epidemics and for high-risk patients	Lifelong if given after age 2
Mumps	Live virus	SC	1 dose given after age 1, 95% effective	Lifelong
Pertussis (whooping cough)	Killed bacteria	IM	Same as diphtheria; not given to children subject to seizures	10 years
Plague	Killed bacteria	IM	3 doses 4 weeks apart, for travel to some parts of the world	6 months
Poliomyelitis	Live virus	O	2 doses 6 weeks apart plus booster, 95% effective	Lifelong
Rabies	Killed virus	IM	2 doses 1 week apart with third dose in 2 weeks; 80% effective; used after probable exposure	2 years
Rubella	Live virus	SC	1 dose at 15 months, some recommend second dose at 12 years	Lifelong
Smallpox	Live vaccinia virus	ID	1 dose; 90% effective; used only by laboratory workers exposed to poxviruses	3 years
Tetanus	Toxoid	IM	3 doses 4 weeks apart plus boosters	10 years
Tuberculosis	Attenuated bacteria	ID, SC	1 dose for inadequately treated patients and high-risk groups	Lifelong?
Typhoid fever	Live virus	O	2 doses 4 weeks apart; 70% effective; recommended for travel, epidemics, and carriers	3 years
Yellow fever	Live virus	SC	1 dose, recommended for travel to endemic areas	10 years

[a]SC—subcutaneous, IM—intramuscular, ID—intradermal, O—oral

➢ Table 17.7 **Selected examples of materials for special immunization and experimentation**

Infectious agent	Nature of material	Uses
Adenovirus	Live virus	Military recruits
Bacillus anthracis	Antigen extract	Handlers of animals and hides
Campylobacter	Attenuated bacteria	Experimentation
Vibrio cholerae	Toxoids of *Escherichia coli* and *Vibrio cholerae*	Experimental oral administration to obtain more effective immunization
Cytomegalovirus	Live virus	Experimentation, may produce latent infections
Equine encephalitis virus	Live/inactivated viruses	Laboratory workers and experimentation

➢ Table 17.8 **Recommended immunizations for normal infants and children in the United States**

Disease	Vaccine	Dosage schedule
Diphtheria — DTP	Toxoid	Ages 2, 4, and 6 months, $1\frac{1}{2}$ and 4 to 6 years
Tetanus — DTP	Toxoid	Same as diphtheria; administered in DTP vaccine
Pertussis — DTP	Killed bacteria	Same as diphtheria; administered in DTP vaccine
Tetanus — Td	Toxoid	Age 14–16 years and every 10 years thereafter
Diphtheria — Td	Toxoid	Age 14–16 years and every 10 years thereafter
Poliomyelitis	Live viruses Types I, II, and III	Ages 2 and 4 months, $1\frac{1}{2}$ and 4 to 6 years
Haemophilus influenzae type b (Hib) infection	Polysaccharide–protein conjugate	Ages 2, 4, 6, and 15 months
Measles — MMR	Live virus	Ages 15 months and booster at school age, 11–12 years
Mumps — MMR	Live virus	Ages 15 months, 11–12 years
Rubella — MMR	Live virus	Ages 15 months, 11–12 years
Hepatitis B	Viral antigen fragments	Birth, ages 2 and 6 months

Immunization recommendations vary between developed and underdeveloped countries. Most developed countries use approximately the same immunizations as are recommended in the United States. However, several countries use BCG (Bacille Calmette-Guerin) (➢ Figure 17.16) vaccine to protect against tuberculosis. The World Health Organization (WHO) has administered it to over 150,000 people in various countries in the past several decades. It is not used in the United States because serious controversy existed about its safety and efficacy when it was first developed. Now that it has been proven safe and effective, the incidence of tuberculosis is not sufficient to warrant widespread use. WHO recommends immunization at earlier ages in underdeveloped countries to combat serious contagious diseases: BCG and oral polio vaccines at birth, DPT and polio vaccines at 6, 10, and 14 weeks, and measles vaccine at 9 months.

Hazards of Vaccines

There are overwhelming benefits of using vaccines to prevent serious infectious diseases in populations. However, vaccines also pose hazards that must be weighed when deciding whether they should be administered to entire populations, to certain individuals, or not at all. And, of course, the prevalence and severity of diseases also must be considered in such decisions.

Active immunization often causes fever, malaise, and soreness at the site of injection. Thus, patients already suffering from fever and malaise should not receive immunization because a worsening of their condition might be erroneously attributed to the vaccine. More importantly, the patient's immune system, overburdened by the existing infection, may be unable to mount an adequate response to the antigen in the vaccine. Particular reactions are associated with certain immunizations. For example, joint pain can be caused by rubella vaccine, and convulsions by pertussis vaccine. Allergic reactions sometimes follow the use of influenza and other vaccines that contain egg protein or vaccines that contain antibiotics as preservatives. However, reactions

➢ Figure 17.16 Vaccination mark from inoculation with BCG vaccine. This vaccine used in some countries to immunize against tuberculosis. It is introduced under the skin by a cluster of tines that leave a permanent raised mark.

occur in only a small proportion of vaccine recipients and are generally less severe than the disease. An exceedingly small number of vaccine recipients die or suffer permanent damage from vaccines (see the Polio Vaccine Controversy box in Chapter 24).

Live vaccines pose particular hazards to pregnant women, patients with immunological deficiencies, and patients receiving immunosuppressants such as radiation or corticosteroid drugs. In the case of pregnant women, live viruses sometimes cross the placenta and infect the fetus, whose immune system is immature. They also can cause birth defects. In immunodeficient or immunosuppressed patients, the attenuated virus sometimes has sufficient virulence to cause disease. Therefore, patients who test positive for the AIDS virus should not receive live-virus vaccines. This presents a problem for routine immunization of infants, some of whom, unknown to health care personnel, could have become infected with AIDS before birth. It also means that U.S. military and State Department employees and their accompanying families who are being posted abroad must be tested for AIDS to determine whether they can receive the required live-virus vaccines.

Passive Immunization

To induce passive immunity, ready-made antibodies are introduced into an unprotected individual. Because antibodies are found in the serum portion of the blood, these products are often called **antisera.** Although passive immunity is produced quickly, it is only temporary. It lasts only as long as there is a sufficiently high titer of circulating antibodies in the body. **Passive immunization** is established by administering a preparation such as gamma globulin, hyperimmune serum, or an antitoxin that contains large numbers of ready-made antibodies. However, the specificity and degree of this form of immunization depend on the antibody type and concentration used.

Immune serum globulin, formerly called **gamma globulin,** consists of pooled gamma globulin fractions (the portion of serum containing antibodies) from many individuals. This kind of gamma globulin typically contains sufficient antibodies to provide passive immunity to a number of common diseases, such as mumps, measles, and hepatitis A.

If the donors are specially selected, gamma globulins that have high titers of specific kinds of antibodies can be prepared. Such preparations are often called **hyperimmune sera,** or *convalescent sera.* For example, gamma globulin from persons recovering from mumps or from recent recipients of mumps vaccine contains especially high titers of antimumps antibodies. Similar sera can be collected from donors with high titers of antibodies to other diseases. Hyperimmune sera can also be manufactured by introducing particular antigens—for instance, tetanus toxin—into another animal, such as a horse, and subsequently collecting the antibodies from the animal's serum.

Antitoxins are antibodies against specific toxins, such as those that cause botulism, diphtheria, or tetanus. Passive immunization against tetanus toxins can also be achieved by using tetanus immune globulin, a gamma globulin that contains antibodies against tetanus toxin. The properties of currently available materials used to produce passive immunity are summarized in Table 17.9.

Passive immunization gives immediate immunity to a nonimmune person who is exposed to a disease, or it at least lessens the severity of the disease process. Usually, a vaccine is given after the passive immunizing preparation to provide active immunity. Before the advent of antibiotics, passive immunization was frequently used to prevent or lessen the severity of several kinds of pneumonia and a variety of other infectious diseases. With respect to infectious diseases, the most common current use of passive immunity is to protect people with contaminated wounds against tetanus toxin. Although the incidence of exposure to diphtheria and botulism is lower than that of tetanus, passive immunization can also be used against these diseases.

Passive immunization is also used to counteract the effects of snake and spider bites and to prevent damage to fetuses from certain immunological reactions. Antivenins, or antibodies to the venom of certain poisonous snakes and the black widow spider, are given as emergency treatments. They react with venom molecules that have not already bound to tissues. Thus, the sooner after a bite the antivenin is administered, the more effective it is in counteracting the effects of the venom.

A fetal immunological reaction can occur when a mother with Rh-negative blood carries her second Rh-positive fetus. As is explained in more detail in Chapter 18, the mother becomes sen-

Immunoglobulin in cows' colostrum has been used to successfully treat myasthenia gravis, multiple sclerosis, systemic lupus, and rheumatoid arthritis, among others.

> **Table 17.9** | **Properties of materials available for passive immunization**

Material	Uses
Human gamma globulin	To prevent recurrent infections in patients with deficiencies in humoral immunity and to prevent or lessen disease symptoms after exposure of nonimmune persons to measles or hepatitis A
Specific gamma globulins	
Varicella-zoster immune globulin	To prevent chickenpox in high-risk children; must be given within 4 days of exposure
Hepatitis-B immune globulin	To prevent hepatitis B after exposure (via blood or needles) and to prevent spread of the disease from mothers to newborns
Mumps immune globulin	May prevent orchitis (inflammation of the testes) in adult males exposed to mumps
Pertussis immune globulin	To reduce severity of disease and mortality in children under 3 or in debilitated children
Rabies immune globulin	To prevent rabies after a bite from a possibly rabid animal; applied to the bite if possible and administered intramuscularly
Tetanus immune globulin	To prevent tetanus after injury in nonimmune patients
Vaccinia immune globulin	To halt progress of disease in immunodeficient patients who develop a progressive vaccinia (cowpox) infection

sitized to the Rh-positive red blood cells when her first Rh-positive child is born. The mother's immune system would subsequently make anti-Rh antibodies, which will harm the second Rh-positive child. To prevent this from happening, anti-Rh antibodies are given to the mother within 72 hours after the birth of the first child and after subsequent births, miscarriages, or abortions. Like the antivenins, the anti-Rh antibodies bind to Rh-positive red blood cells, so the cells are destroyed before the mother's immune system can make antibodies to them.

As with active immunization, passive immunization poses some hazards. The most common hazards are allergic reactions. Some antitoxins contain proteins from other animals as a result of their manufacture in eggs or horses. They are particularly likely to cause allergic reactions, especially when the patient receives them for the second time. Thus, vaccines of human origin are safer, at least with respect to the risk of allergic reactions. Allergic reactions to large IgG molecules also can occur if gamma globulins or hyperimmune sera are accidentally given intravenously instead of by their normal intramuscular route. A new immune globulin IV, containing smaller molecules, can be given safely by the intravenous route.

Another hazard, or at least detriment, of passive immunity is that giving ready-made antibodies can interfere with a host's ability to produce its own antibodies. One way this might occur is by ready-made antibodies binding to antigens and preventing them from stimulating the host's immune system. Thus, maternal antibodies, while they protect infants from some infections, may prevent the infant's immune system from making its own antibodies.

Future of Immunization

Immunologists continue to search for new vaccines. Effective vaccines should meet five criteria. (1) Vaccines developed should be protective against the disease for which they were designed. (2) The vaccines must be safe and not have adverse side effects. (3) Protection should be sustained, providing long-term protection from infection. (4) Vaccines should generate neutralizing antibodies or protective T cells to vaccine antigens. (5) Vaccines should be practical in terms of stability and use.

Whole-cell killed vaccines (first-generation vaccines) sometimes produce unwanted side effects due to extraneous cellular materials. Therefore, efforts are made to identify and obtain cellular subunits that contain only the purified immunogenic portion of a microorganism that will produce immunity. *Subunit vaccines* (second-generation vaccines) are safer than *attenuated vaccines,* which use live organisms that are treated to eliminate their virulence. There is always the possibility that the organisms may revert to a virulent state. Many researchers consider attenuated vaccines too risky in the hunt for an AIDS vaccine, but they are functional for some other diseases, as is the BCG vaccine for tuberculosis. In general, live organisms produce higher and longer-lasting immunity than do nonliving organisms. *Recombinant DNA vaccines* (third-generation vaccines) are being produced by inserting the genes for specific antigens into the genomes of nonvirulent organisms. The hepatitis B viral antigens have been cloned in yeast cells, extracted and purified before use, to form a safe and very effective vaccine. Rabies virus antigen has been inserted into vaccinia (cowpox) virus and is being tested as a means for controlling rabies in wild animal populations such as raccoons.

Many vaccine researchers believe that by the year 2025, most Americans will be immunized routinely against some 30 diseases, including AIDS; genital herpes; influenza; hepatitis A, B, C, and E; and chickenpox and shingles. These immunizations will be administered in three stages: infancy and early childhood, for childhood diseases; prior to puberty, for certain sexually transmitted diseases; and adulthood, for influenza.

Immunity to Various Kinds of Pathogens

This chapter has focused on the basic principles of immunity that apply to all pathogens. However, you may find it helpful to note the ways in which the immune system responds to various kinds of pathogens.

Bacteria

As we saw in Chapter 16 ➡ (p. 434), nonspecific defenses such as skin, mucous membranes, and gastric secretions prevent many bacteria from entering host tissues. When bacteria do infect a host, certain immune responses alter the invading organisms so that they can be phagocytized. Once plasma cells produce specific antibodies, the antibodies can interfere with any of several steps in bacterial invasion. They can attach to pili and capsules, preventing bacterial attachment to cell surfaces. Antibodies can work with complement to opsonize bacteria for later phagocytosis or lysis by other cells of the immune system. Or they can neutralize bacterial toxins or inactivate bacterial enzymes.

Viruses

Viruses infect by invading cells—usually first attacking cells that line body passages. Then they directly invade target organs such as the lungs or travel in the blood (viremia) to target organs or organ systems such as the liver or nervous system. Polioviruses invade cells that line the digestive tract, but they also can enter nerve endings.

Immune responses can combat viral infections at any of these locations. Interferons, secretory IgA, and some IgG antibodies act at the surfaces lining cells and prevent or mini-

✓ What is the BCG vaccine? Why is it not used in the United States?

✓ What are some hazards of vaccines?

✓ What is gamma globulin? What is immune globulin? How are they prepared?

✓ How are subunit vaccines prepared? Why are they better than whole-cell killed vaccines?

Cancer and Immunology

Cancer is one of the leading causes of death in developed countries, and its incidence as a cause of death in developing nations is on the rise. Cancer cells can arise by mutation of a normal cell in response to chemical carcinogens, radiation, or viruses; by expression of previously repressed human oncogenes (tumor-producing genes) within cells ➡ (Chapter 10, p. 283) or by a combination of these factors.

Regardless of the means by which they are produced, many cancer cells have certain plasma membrane antigens not found in normal cells. These antigens are an ideal target for destruction by the immune system. According to the theory of *immune surveillance,* T_C cells and NK cells recognize and destroy these abnormal cells before they develop into cancers. If this theory is correct, each of us has within our body many different potentially malignant cells, but we develop cancer only if the T_C cells or NK cells fail to identify and destroy the mutated cells. Unfortunately, many tumors and cancers show little regard for control by the immune system.

One way T_C cells might fail to recognize malignant cells is by the actions of antigens on the malignant cells themselves. The antigens stimulate the formation of antibodies, which bind to the antigens without damaging the malignant cells. Although this binding does not occur before the antigens sensitize T_C cells, it does block T_C cells from attacking the malignant cells. Other tumors can avoid immune system attacks by lacking abnormal antigens or by losing MHC class I molecules through mutation. Such cells escape immune detection by T_C cells but, as described in this chapter, are still attacked by NK cells.

However, the notion that cancer cells can be destroyed by immune reactions has led researchers to work on developing immunotoxins and cancer vaccines. As discussed earlier in this chapter, an immunotoxin is a monoclonal antibody with an anticancer drug, a microbial toxin such as diphtheria

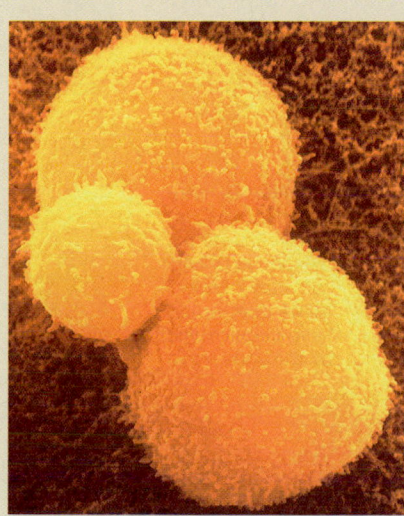

Colorized SEM of a small T lymphocyte attacking two large tumor cells (5700X).

toxin, or a radioactive substance attached to it. The antibody is designed to bind to a specific cancer cell antigen; the attached substance is selected for its ability to destroy the cancer cell. Such immunotoxins are expected to seek and destroy specific cells—the cancer cells that have the appropriate antigen. Cancer vaccines containing one or more cancer cell antigens are also being developed. Some of these vaccines induce immunity to specific cancer cells. Others contain antigens frequently found on certain kinds of cancer cells.

A major difficulty in making immunotoxins and cancer vaccines is that many different kinds of antigens are found on cancer cells from various patients. To be effective against an existing cancer, an immunotoxin or vaccine must be specific for the antigens on the cells of that specific cancer. Similarly, to immunize against common forms of cancer, vaccines need to generate antibodies or T_C cells that will destroy cancer cells should they develop. Another difficulty is that some malignant cells produce immunosuppressive cytokines, such as transforming growth factor-α (TGF-α), that stimulate tumor growth and may interfere with T_H1 cell activity. Immunotoxins and vaccines need to be powerful enough to overcome this inhibition. Finally, some antigens are found on both malignant and normal cells. Great care must be taken to develop immunotoxins and vaccines that will react only with malignant cells.

In October 1991 the first attempt was made to eliminate cancer by immunizing patients against their own tumors. Cells from a malignant melanoma tumor were removed, genes for tumor necrosis (killing) factor were added to the cancer cells' genomes, and the cells were reinjected (infused) into the patient. There, it is hoped, the genetically altered cells will secrete enough of this immune system factor to change the regulation of the immune system so as to overcome its tolerance of the tumor and cause it to begin attacking the cancerous cells. To date, there have been few successes.

Meanwhile, immunization already offers us the ability to prevent about 80 percent of liver cancer cases. Liver cancer is among the most prevalent kinds of cancer. One form of liver cancer, primary hepatocellular carcinoma, accounts for 80 to 90 percent of all cases. It now appears that this form is associated with infection by hepatitis B virus early in life and especially with being a carrier of the virus. Not only has hepatitis B viral DNA been found in chromosomes of cancerous liver cells but also the intact virus has been isolated from cancerous liver cells. The incidence of liver cancer is highest in regions of Africa and Asia where the incidence of hepatitis B infections also is high. Immunization against the hepatitis B virus would protect recipients of the vaccine against infection, prevent them from becoming carriers, and thus protect them against liver cancer.

mize entry of viruses. IgG and IgM act in the blood to neutralize viruses directly or to promote their destruction by complement. Finally, cytotoxins and cellular immunity via T_C cells and NK cells are especially important in clearing the body of cells infected with viruses. The mechanisms by which the immune system combats viral infections are summarized in ➤ Figure 17.17.

Besides specific immune responses to a viral infection, many nonspecific responses can limit infection. Fever is an important defense against viruses. Several viruses, such as influenza, parainfluenza, and rhinoviruses, are temperature-sensitive. They replicate in the lining cells of the respiratory tract, which normally has a temperature between 33 and 35°C—lower than the normal body temperature of 37°C because the cells are cooled as atmospheric air moves over their moist surfaces. When a person has a fever of even 1° to 2°C, the ability of the virus to replicate is reduced. Another benefit of fever in resisting viral infection is that temperature increases cause an increase in interferon production. ➡ (Chapter 16, p. 447)

Fungi

Certain fungal infections progress through a tissue as fungal cells invade and destroy one cell after another. Immunity to fungi is poorly understood, but it appears to be mainly cell-mediated. Fungal skin infections probably are combated by IgA antibodies and T_H cells, which release certain cytokines that activate macrophages. The macrophages, in turn, engulf and digest fungi. Commensal fungi apparently are kept in their place by cell-mediated responses. Supportive evidence comes from studies of individuals with impaired T cell functions. Such

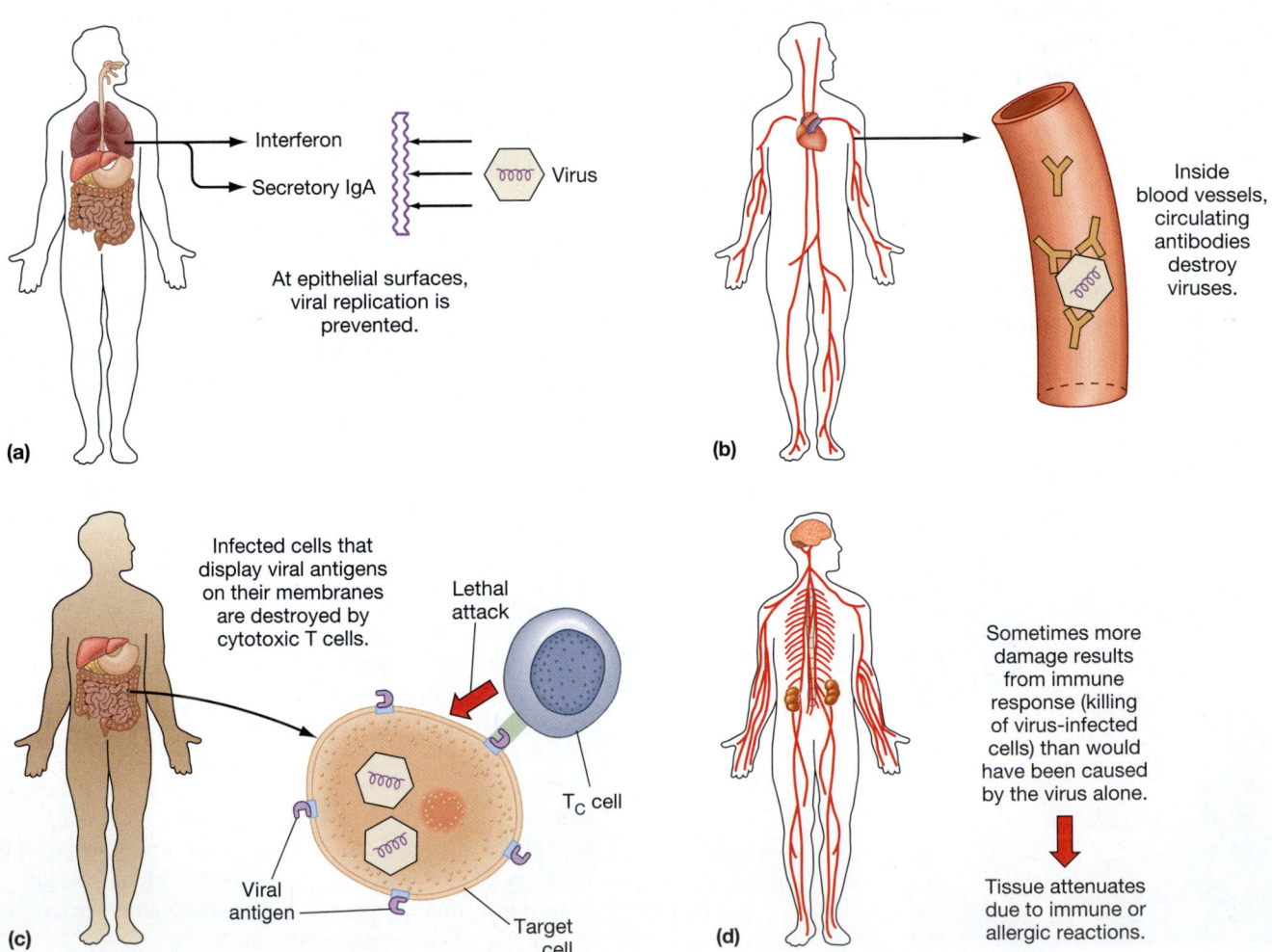

➤ Figure 17.17   **How the immune system combats viruses.**

persons are extremely likely to become infected with opportunistic fungi, such as *Candida albicans*.

Protozoa and Helminths

Protozoa and helminths are quite dissimilar in size and complexity, but many use similar methods of invading the body. Host defenses against them also are similar, except that allergic reactions to helminths can be severe enough to cause more damage to host cells than to the disease agent. Antigens on the surface of the roundworm *Ascaris* are potent inducers of allergic reactions. Individuals with an *Ascaris* allergy can absorb enough antigens through the skin to cause a severe allergic reaction, even just by coming into contact with fluids in which the worms have been preserved.

Parasitic protozoa and helminths interact with their hosts in ways that do not endanger host survival and thereby ensure parasite survival. These pathogens cause chronic, debilitating diseases that usually are not immediately life-threatening. Most of these parasites are relatively large and difficult to phagocytize. Large helminths, such as heartworms in dogs, can block blood vessels and cause sudden death. When attacked by phagocytes, some helminths release toxins that significantly damage the host. Medical intervention also can be hazardous. Giving drugs to kill some helminths causes them to release large quantities of toxic decay products. Thus, once some worm infections have been acquired, coexistence with the parasites may be the best course of action.

Many parasites have complex life cycles with more than one host, and some infect animals that serve as reservoirs for human infection. ➡ (Chapter 11, p. 289) Thus, they stand ready to take advantage of appropriate conditions in various hosts. Each life cycle stage of some parasites can have several surface antigens. Although hosts may produce antibodies against such antigens, the ability of certain parasites to change them provides a way to thwart host defenses. For example, the malaria parasite induces host antibody formation before it invades host cells. While the protozoan multiplies inside cells, it produces different antigens, so the original antibodies formed by the host are not effective against the new parasite progeny when they are released.

> Figure 17.18   **How antigens of parasitic protozoans thwart the immune system.**

The host's immune system also combats parasitic protozoa and helminths by cell-mediated processes. Although T_c cells usually are not effective against such parasites, some T cells release cytokines, such as IL-3, that activate macrophages. These macrophages can attack malarial parasites and several kinds of worms, including blood flukes. Other cytokines, such as IL-5, enhance the ability of eosinophils to combat worm infections.

Parasitic protozoa and helminths have available to them an assortment of mechanisms that thwart immune responses. These mechanisms include the following:

1. Some protozoa protect themselves by invading cells, forming protective cases called cysts at certain life cycle stages, or otherwise becoming inaccessible to host defenses.

2. Some protozoa avoid immune recognition by changing their surface antigens (antigenic variation) regularly or with each reproductive cycle.

3. Some parasites suppress the host's immune responses by releasing toxins that damage lymphocytes, enzymes that inactivate IgG, or soluble antigens that thwart the immune system in a variety of ways (➤ Figure 17.18).

4. Some intracellular protozoa suppress the action of phagocytes by inhibiting fusion of lysosomes with vacuoles, by resisting digestion by lysosomal enzymes, or by impairing oxidative metabolism.

Given the various methods protozoa and helminths have to evade and thwart immune responses of hosts, it is not surprising that they cause chronic debilitating infections.

Retracing Our Steps

Immunology and Immunity

- **Immunology** is the study of **immunity,** which refers to the capacity to recognize and defend against infectious agents and other foreign substances.

- **Susceptibility** is vulnerability to infectious agents. Immunity is **nonspecific** when it acts against any infectious agent and **specific** when it acts against a particular infectious agent. **Immunology** is the study of specific immunity. The **immune system** is the body system that provides the host with specific immunity to particular infectious agents.

Types of Immunity

Innate Immunity

- **Innate immunity** provides a nonspecific, hereditary defense.

Acquired Immunity

- **Acquired immunity** provides a specific, nonhereditary defense.

Active and Passive Immunity

- In **active immunity,** an individual's own immune system makes antibodies.

- In **passive immunity,** ready-made antibodies are introduced into the body.

- Types of immunity are illustrated in ➤ Figure 17.1.

Characteristics of the Immune System

Antigens and Antibodies

- An **antigen,** or **immunogen,** is a foreign substance that can elicit a specific immune response. Most antigens are proteins, but some are polysaccharides, nucleoproteins, or glycoproteins.

- Each antigen has several **epitopes,** or **antigenic determinants.**

Properties of Antibodies (Immunoglobulins)

- An **antibody,** or **immunoglobulin,** is a protein produced in response to the presence of an antigen. Antibodies bind to epitopes on the antigen.

Cells and Tissues of the Immune System

- Lymphocytes differentiate into **B cells** in bursal-equivalent tissues or into **T cells** in the thymus or under the influence of thymic hormones. Lymphocytes develop from **lymphoid stem cells** in the bone marrow.

Dual Nature of the Immune System

- The dual roles of the immune system consist of **humoral immunity,** which is carried out mainly by B cells and plasma cells, and **cell-mediated immunity,** which is carried out mainly by certain T cells.

General Properties of Immune Responses

- Immune responses distinguish **self** from **nonself.**

- According to the **clonal selection theory,** B cells recognize specific epitopes on antigens according to the particular antibody present on the B cell plasma membrane. When a B cell detects an antigen with which it can react, it binds to the antigen, engulfs it, processes it, displays a peptide fragment as MHC class II to T_H2 cells, and divides many times. A **clone** of genetically identical B cells, which differentiate into many plasma cells and some **memory cells,** is produced.

- **Specificity** refers to the ability of immune responses to respond to and distinguish among different antigens and epitopes.

- **Heterogeneity** refers to the ability of immune responses to produce many different antibodies and cell substances on the basis of the different antigens they encounter.

- **Immunological memory** refers to the ability of T and B cells to recognize substances to which the immune system has previously responded.

Humoral Immunity

- B cells are selected to respond to specific antigens in accordance with the particular antibody present on the B cell membrane prior to encountering an antigen.
- When a B cell detects an antigen with whch it can react, it binds to the antigen, divides many times to produce a clone of many plasma cells and some memory cells.
- Many B cells require helper T cells to assist them in binding to antigens; suppressor T cells probably limit the duration of antibody production.
- **Plasma cells** synthesize and release large numbers of antibodies.
- **Memory cells** remain in lymphoid tissue ready to respond to subsequent exposure to the same antigen.

Properties of Antibodies (Immunoglobulins)

- Structurally, antibodies consist of two **heavy** and two **light** polypeptide **chains,** each of which has a variable region capable of reacting with a specific antigen.
- Properties of particular kinds of antibodies **(immunoglobulins)** are summarized in Table 17.4.

Primary and Secondary Responses

- **Primary responses** are the immune system's first encounter with foreign antigens.
- Memory cells remain in lymphoid tissue, ready to respond to subsequent exposure to the same antigen.
- **Secondary responses** bring fast and efficient destruction antigens recognized by B and T memory cells. Primary and secondary responses are summarized in Table 17.5.

Kinds of Antigen-Antibody Reactions

- Humoral immunity is most effective against bacteria, which are destroyed by **agglutination** (clumping) or lysed by complement after opsonization or directly by IgMs or neutralized. Toxins and some viruses can be inactivated by antibody **neutralization.**

Monoclonal Antibodies

- **Monoclonal antibodies** are antibodies produced in the laboratory from a clone of cultured cells that make one specific antibody to one specific epitope.
- Specific monoclonal antibodies can be used in some diagnostic tests, and methods to use them in treating infectious diseases and cancer are being developed.

Cell-Mediated Immunity

- Cell-mediated immunity concerns the direct actions of certain T cells that defend the body against viral infections and reject tumors and transplanted tissues.
- T cells do not make antibodies but instead have membrane receptors for antigens; these receptors bind to MHC molecules displaying foreign peptide fragments.
- Cell-mediated immune responses involve differentiation and activation of several kinds of T cells and the secretion of cytokines.

The Cell-Mediated Immune Reaction

- Processed antigens on MHC class II molecules bind with T cell receptors. Next, IL-1 secreted from macrophages and IL-2 secreted from T cells activate the T cells, which can then differentiate into T_H1 cells and T_H2 cells.
- Certain pathogenic bacteria can grow in macrophages after they have been phagocytized. T_H1 cells can release γ-interferon, a cytokine that causes such infected macrophages to become resensitized to other cytokines.
- AIDS destroys T_H cells, thereby impairing both humoral and cell-mediated immunity.

How Killer Cells Kill

- T_c and NK cells destroy target cells by releasing the lethal protein **perforin.**

The Role of Activated Macrophages

- Certain pathogenic bacteria can grow in macrophages after phagocytosis. The lymphokine macrophage activating factor helps stimulate antimicrobial processes so that macrophages can kill the pathogens.
- When macrophages fail to kill pathogens, the pathogens are walled off in granulomas.

Factors That Modify Immune Responses

- Host defenses in healthy adults in an unpolluted environment prevent nearly all infectious diseases. Individuals with reduced resistance are called **compromised hosts.**
- Factors that reduce host resistance include very young or old age, stress, seasonal patterns, poor nutrition, traumatic injury, pollution, and radiation. Complement deficiencies, immunosuppressants, infections such as HIV, and genetic defects impair immune system function.

Immunization

Active Immunization

- **Active immunization** induces the same response as the one that occurs during a disease. Immunization challenges the immune system to develop specific defenses and memory cells.
- Active immunization is conferred by **vaccines** and **toxoids.** Vaccines can be made from live, attenuated organisms, dead organisms, or parts of organisms. Toxoids are made by inactivating toxins.
- The recommended immunizations for normal infants and children in the United States are summarized in Table 17.8.
- The benefits of active immunization against life-threatening diseases nearly always outweigh the hazards. Reactions to vaccines can cause serious side effects, but their incidence is lower than the incidences of the diseases themselves.

Passive Immunization

- **Passive immunization** occurs by the same mechanism as natural passive transfer of antibodies.
- Passive immunity is conferred by **antisera** such as **immune serum globulin (gamma globulin), hyperimmune** or **convalescent sera,** and **antitoxins.**
- The benefits of passive immunization are limited to providing only temporary protection; the side effects are mainly allergic in nature.

Future of Immunization

- **Subunit vaccines** produce fewer side effects than **whole-cell killed vaccines** and offer greater safety than do **attenuated vaccines.**
- **Recombinant DNA vaccines** contain genes for antigens of pathogens inserted into nonpathogenic organisms' genomes and are very safe.

Immunity to Various Kinds of Pathogens

Bacteria

- Antibodies produced by plasma cells are the chief immunological defense against bacterial antigens. Most immune responses to bacteria serve to promote phagocytosis of the invading cells.

Viruses

- Viral infection is combated by nonspecific defenses, interferon, and antibodies. In addition, the T_c cells of cell-mediated responses and NK cells are important in destroying virus-infected cells.

Fungi

- Immune responses to fungi involve IgA antibodies and are primarily cell-mediated.

Protozoa and Helminths

- Immune responses to parasitic protozoa and helminths are largely cell-mediated. T cells release cytokines that activate macrophages and attract other leukocytes. Allergic reactions to helminths can be damaging to the host than to the parasite.

Terminology Check

acquired immunity   *(p. 459)*
active immunity   *(p. 459)*
active immunization   *(p. 480)*
agglutination   *(p. 471)*
anamnestic response   *(p. 466)*
antibody   *(p. 460)*
antigen   *(p. 459)*
antigenic determinant   *(p. 460)*
antiserum   *(p. 484)*
antitoxin   *(p. 484)*
artificially acquired active immunity
   (p. 459)
artificially acquired immunity   *(p. 459)*
artificially acquired passive immunity
   (p. 459)
B cell   *(p. 462)*
B lymphocyte   *(p. 462)*
cell-mediated immunity *(p. 464)*
clonal selection hypothesis   *(p. 464)*
colostrum   *(p. 459)*
compromised host   *(p. 479)*
cross-reaction   *(p. 464)*
cytotoxic T cell   *(p. 475)*
DPT vaccine   *(p. 481)*
delayed hypersensibility T cells   *(p. 475)*
epitope   *(p. 460)*
gamma globulin   *(p. 484)*

genetic immunity   *(p. 458)*
hapten   *(p. 460)*
heavy chain   *(p. 468)*
helper T cell   *(p. 475)*
heterogeneity   *(p. 464)*
Hib vaccines   *(p. 481)*
humoral immunity   *(p. 463)*
hyperimmune serum   *(p. 484)*
IgA   *(p. 469)*
IgD   *(p. 470)*
IgE   *(p. 470)*
IgG   *(p. 469)*
IgM   *(p. 470)*
immune serum globulin   *(p. 484)*
immune system   *(p. 458)*
immunity   *(p. 458)*
immunoglobulin (Ig)   *(p. 468)*
immunology   *(p. 458)*
innate immunity   *(p. 458)*
light chain   *(p. 468)*
lymphokine   *(p. 475)*
memory   *(p. 466)*
memory cell   *(p. 466)*
MMR vaccine   *(p. 481)*
monoclonal antibody   *(p. 473)*
natural killer (NK) cells   *(p. 475)*
naturally acquired active immunity
   (p. 459)

naturally acquired immunity   *(p. 459)*
naturally acquired passive immunity
   (p. 459)
neutralization   *(p. 472)*
nonself   *(p. 464)*
nonspecific immunity   *(p. 458)*
null cells   *(p. 463)*
passive immunity   *(p. 459)*
passive immunization   *(p. 484)*
perforin   *(p. 478)*
plasma cell   *(p. 466)*
poliomyelitis vaccine   *(p. 481)*
primary response   *(p. 470)*
secondary response   *(p. 470)*
secretory component   *(p. 470)*
self   *(p. 464)*
sensitize   *(p. 466)*
species immunity   *(p. 458)*
specific immunity   *(p. 458)*
specificity   *(p. 464)*
suppressor T cell   *(p. 475)*
T cell   *(p. 463)*
titer   *(p. 461)*
T lymphocyte   *(p. 463)*
tolerance   *(p. 464)*
toxoid   *(p. 480)*
vaccine   *(p. 480)*

See You on the Web

Don't forget I am expecting you on the web. The following is a preparedness checklist for your web explorations. If you have questions on any of these items, you can return to the end of Chapter 1 for more details.

- *Questions We'll Explore*
- *Concept Check questions*
- *Flashcards*
- *Study Guide*

If you are ready, come visit me at

http://www.prenhall.com/black

It will be a journey you'll never forget.

Immunology II: Immunological Disorders and Tests

She has her father's eyes and her mother's AIDS.
Before you get pregnant, find out if you need to be tested.

What do you know about the current state of HIV/AIDS? Add these facts to your information arsenal:

- The rate of HIV infection on college campuses is about 10 times higher than in the general population.

- The vast majority of African AIDS cases were among heterosexuals who did not use injection drugs.

- AIDS is now the 2nd leading cause of death among 25-44 year olds; the 6th leading cause of death among 15-25 year olds; the 7th leading cause of death among 1-4 year olds.

- AIDS now kills about 1 million people per year, worldwide. TB now kills over 3 million people per year, worldwide. Malaria now kills over 2 million people per year, worldwide. Heart disease now kills 12-15 million people per year, worldwide.

Background Basics

In Chapter 17 we emphasized how specific immune responses defend the body against harmful substances. However, such responses are not always beneficial. Sometimes the humoral or cell-mediated responses react in ways that are physiologically unpleasant or even life-threatening. Perhaps you or someone you know gets a runny nose and watery eyes every time hay fever season rolls around. Maybe you know of people who have other allergies, who have had adverse reactions to a blood transfusion, or who suffer from more severe immunological disorders—even AIDS.

In this chapter we will learn more about the immune system by examining the ways the immune system goes awry and reacts inappropriately or inadequately. We also will look at some of the laboratory and clinical methods used to detect and measure immune reactions.

Overview of Immunological Disorders

An **immunological disorder** is a condition that results from an inappropriate or inadequate immune response. Most inappropriate responses involve some type of hypersensitivity, whereas inadequate responses are due to an immunodeficiency.

Hypersensitivity

In **hypersensitivity,** or **allergy,** the immune system reacts in an exaggerated or inappropriate way to a foreign substance. Such responses can be thought of as "too much of a good thing"—the immune system responds to a harmless foreign agent by doing harm instead of protecting the body. Although allergy is another name for hypersensitivity, many disorders that people call "allergies" are not due to immunological reactions. These disorders include toxic responses to drugs, digestive upsets from nonallergic responses to foods, and emotional disturbances.

There are four types of hypersensitivity: (1) immediate hypersensitivity (Type I), (2) cytotoxic hypersensitivity (Type II), (3) immune complex hypersensitivity (Type III), (4) and cell-mediated, or delayed, hypersensitivity (Type IV). The type that develops depends on which components of the immune response are involved and on how

Questions We'll Explore

A What are the different types of hypersensitivities?

B What are the causes, mechanisms, and effects of immediate hypersensitivities?

C What are the causes, mechanisms, and effects of cytotoxic hypersensitivities?

D What are the causes, mechanisms, and effects of immune complex disorders?

E What are the causes, mechanisms, and effects of cell-mediated hypersensitivities?

F How do autoimmune disorders arise, and what are their hypersensitive effects in the body?

G Why are organ transplants sometimes rejected, and how can rejection be prevented?

H How is hypersensitivity involved in some drug reactions?

I What are the causes, mechanisms, and effects of immunodeficiency diseases?

J What tests detect and measure antigens and antibodies in fluids, cells, and tissues?

K What are the epidemiology, origins, treatment, and social perspectives of AIDS?

In 1963, Peter Gell and Robert Coombs developed the classification system that divides hypersensitivities into four types.

quickly the reaction develops. **Immediate (Type I) hypersensitivity,** or *anaphylaxis,* results from a prior exposure to a foreign substance called an *allergen,* an antigen that evokes a hypersensitivity response. Allergies to pollen, foods, and insect stings are examples of immediate hypersensitivity. **Cytotoxic (Type II) hypersensitivity** is elicited by antigens on cells, especially red blood cells, that the immune system recognizes as foreign. This reaction occurs when a patient receives the wrong blood type during a transfusion. **Immune complex (Type III) hypersensitivity** is elicited by antigens in vaccines, on microorganisms, or on a person's own cells. Large antigen–antibody complexes form, precipitate on blood vessel walls, and cause tissue injury within hours. **Cell-mediated (Type IV), or delayed, hypersensitivity** is triggered by exposure to foreign substances from the environment (such as poison ivy), infectious disease agents, transplanted tissues, and the body's own tissues and cells. *Delayed hypersensitivity T cells* react with the foreign cells or substances, causing in some cases extensive tissue destruction.

Autoimmune disorders represent a form of hypersensitivity. In such disorders, the body's immune system responds to its own tissues as if they were foreign. Antibodies or T cells attack self-antigens.

Immunodeficiency

In **immunodeficiency** the immune system responds inadequately to an antigen, either because of inborn or acquired defects in B cells or T cells. The weak responses in immunodeficiency disorders are "too little of a good thing." They do no direct harm, but they leave the individual susceptible to infections, which can be severe and even life-threatening.

Immunodeficiencies can be primary or secondary. **Primary immunodeficiencies** are genetic or developmental defects in which the person lacks T cells or B cells or has defective ones. **Secondary immunodeficiencies** result from damage to T cells or B cells after they have developed normally. These disorders can be caused by malignancies, malnutrition, infections such as AIDS, or drugs that suppress the immune system.

Let us now look more closely at the immunological disorders of hypersensitivity and immunodeficiency.

Immediate (Type I) Hypersensitivity

Immediate (type I) hypersensitivity, or *anaphylactic hypersensitivity,* typically produces an immediate response upon exposure to an allergy-inducing antigen (also known as an *allergen*). Nonallergic persons do not respond to such antigens. **Anaphylaxis** (an″a-fi-lak′sis) is an immediate, exaggerated allergic reaction to antigens. The term *anaphylaxis* refers to detrimental effects to the host caused by an inappropriate immune response. These effects are the opposite of *prophylaxis,* the preventive effects generated by an immune response.

Early investigators discovered that a substance they called **reagin** (re-a′jin) was responsible for this type of hypersensitivity. We now know that reagin consists of IgE antibodies. However, the term reagin is still used in some allergy literature.

Anaphylaxis is the harmful result of IgE antibodies made in response to allergens. It can be local or generalized (systemic). **Localized anaphylaxis** appears as reddening of the skin, watery eyes, hives, asthma, and digestive disturbances. **Generalized anaphylaxis** appears as a systemic life-threatening reaction such as airway constriction or **anaphylactic shock,** a generalized condition resulting from a sudden extreme drop in blood pressure.

The word anaphylaxis comes from the Greek words ana, meaning against; and phylaxis, meaning protection.

Allergen

Immediate hypersensitivity results from two or more exposures to an allergen. An **allergen** is an ordinarily harmless foreign substance (typically a protein or a chemical bound to a protein) that can cause an exaggerated immunological response. The first exposure to the allergen produces no visible signs or symptoms. Allergens include airborne substances such as pollen, household dust, molds, and dander—tiny particles from hair, feathers, or skin. Household dust commonly contains nearly microscopic mites and their fecal pellets. Other allergens include venoms from insect stings, antibiotics, certain foods, sulfites, and foreign substances found in vaccines and in diagnostic or therapeutic materials. Allergens can be introduced into the body by inhalation, ingestion, or injection (Table 18.1).

➤ Table 18.1	Common allergens	
Ingested	**Inhaled**	**Injected**
Animal proteins, especially from milk and eggs	Cocaine	Antibiotics, especially cephalosporins and penicillins
Aspirin	Dander	Heroin
Fruits	Dust (household)	Hormones (adrenocorticotropic hormone and animal insulin)
Grains	Face powder	
Hormone preparations	Insecticides	Insect venoms (from bees, hornets, wasps, and yellow jackets)
Nuts	Mites and their feces	
Penicillin	Pollen (from grass, trees, and weeds)	Snake venoms (from vipers and cobras)
Seafood	Spores (fungal and bacterial)	Spider venoms, especially from black widow and brown recluse

Mechanism of Immediate Hypersensitivity

The typical sequence of events involved with the mechanism of type I hypersensitivity is *sensitization,* which involves the *production of IgE antibodies (antiallergens), allergen-IgE reactions,* and *local and systemic effects* of those reactions. Such reactions occur only in individuals who have previously been exposed to an allergen. In **sensitization,** the initial exposure to an allergen, B cells are activated (➤ Figure 18.1a). The B cells differentiate into plasma cells, which produce IgE antibodies against the specific allergen (➤ Figure 18.1b). The IgE antibodies attach by their Fc tails to the surface of mast cells in the respiratory and gastrointestinal tracts and to basophils in the blood (➤ Figure 18.1c). Attachment leaves the antigen (allergen) binding sites of the IgE antibodies free to react with the same allergen upon future exposure. This sequence of sensitization steps does not occur in all people. Why some people become sensitized to normally innocuous substances whereas others do not is poorly understood.

The sensitized mast cells and basophils now are primed to produce a massive chemical response to a second exposure from the same allergen. Although the *sensitizing* (first) *dose* of an allergen can be fairly high, the *triggering* or *eliciting* (subsequent) *dose* that causes the hypersensitive symptoms can be quite small. When a second or subsequent encounter occurs with the same allergen, the allergen attaches to sensitized mast cells and basophils, cross-linking the IgE antibodies (➤ Figure 18.1d). Cross-linking causes **degranulation,** the rapid release of *preformed mediators* (chemical substances that induce allergic responses) from cytoplasmic granules in mast cells and basophils (➤ Figure 18.1e). **Histamine** is the main preformed mediator in humans. It dilates capillaries, thereby making them more permeable. It also causes bronchial smooth muscle to contract, increases mucus secretion, and stimulates nerve endings that cause pain and itching.

Prostaglandins and **leukotrienes** are *reaction mediators* (chemical substances that control responses) that are also synthesized and released from mast cells after degranulation has occurred. Prostaglandin D_2 is a cellular messenger molecule produced in mast cells and basophils that also causes constriction of bronchial smooth muscle. *Slow-reacting substance of anaphylaxis* (SRS-A) is another mediator that causes a slow, long-lasting airway constriction in animals. SRS-A consists of three leukotriene mediators. These leukotrienes are 100 to 1000 times as potent as histamines and prostaglandin D_2 in causing prolonged airway constriction. Like histamine, leukotrienes and prostaglandin D_2 dilate and increase the permeability of capillaries, increase thick mucus secretion, and stimulate nerve endings that cause pain and itching. Preformed and reaction mediators and their effects are summarized in Table 18.2.

It is a second or later allergen exposure of a sensitized person that produces allergic signs and symptoms. Such hypersensitive responses by mast cells can be triggered by nonallergic factors, too. Emotional stress and temperature extremes often cause mediator release without the involvement of any IgE or allergen.

> Theophylline, which is commonly administered orally or through inhalers to asthmatics, blocks an enzyme that would otherwise lead to degranulation.

Localized Anaphylaxis

Atopy (at′o-pe), which literally means "out of place," refers to localized allergic reactions. Atopic immune reactions occur first at the site where the allergen enters the body. If the allergen enters the skin, it causes a *wheal and flare reaction,* characterized by redness, swelling, and

➤ Figure 18.1   **The mechanism of immediate (Type I) hypersensitivity, or anaphylactic hypersensitivity.** In a first exposure, (a) allergen binds to B cells and is presented as allergen fragments on
the surface of macrophages. Allergen fragment presentation activates T_H cells, which activate B cells.
(b) B cells develop into the plasma cells that secrete IgE antibody. (c) IgE binds by its Fc tail to
basophils and mast cells. In a second or later exposure, (d) allergen binds to sensitized mast cells and
basophils, cross-linking IgE molecules. (e) This cross-linking stimulates degranulation of histamine
and other mediators that cause the symptoms of allergies.

➤ Table 18.2 Mediators of immediate hypersensitivity and their effects	
Mediator	**Effects**
Preformed mediators	
Histamine	Vascular dilation and increased capillary permeability, bronchial smooth muscle contraction, edema of mucosal tissues, secretion of mucus, and itching
Neutrophil and eosino-phil chemotactic factors	Attraction of neutrophils, eosinophils, and other leukocytes to the site of an allergic reaction
Reaction mediators	
Leukotrienes (SRS-A)	Prolonged bronchial smooth muscle contraction, increased capillary permeability, edema of mucosal tissues, and secretion of mucus
Prostaglandin D_2	Formation of minute blood clots, bronchial smooth muscle contraction, and capillary dilation

itching (➤ Figure 18.2). If the allergen is inhaled, mucous membranes of the respiratory tract become inflamed, and the patient has a runny nose and watery eyes. If the allergen is ingested, mucous membranes of the digestive tract become inflamed, and the patient may have abdominal pain and diarrhea. Some ingested allergens, such as foods and drugs, also cause skin rashes.

Hay fever, or *seasonal allergic rhinitis,* is a common kind of atopy. More than 20 million Americans suffer from the typical signs and symptoms of watery eyes, sneezing, nasal congestion, and sometimes shortness of breath. First described in 1819 as resulting from exposure to newly mown hay, hay fever is now known to result from exposure to airborne pollen—tree pollens in spring, grass pollens in summer, and ragweed pollen (➤ Figure 18.3) in fall. Some plants, such as goldenrod and roses, have long been blamed for hay fever but are innocent because their pollens are too heavy to be airborne for any great distance. It is the far less conspicuous green ragweed flowers that cause much of the misery for hay fever sufferers. On occasion, severe allergic rhinitis (inflammation of the nasal surfaces) can progress to sinus infections, middle ear problems, and temporary hearing loss. Although they share many symptoms, hay fever can be distinguished from the common cold by the increased numbers of eosinophils in nasal secretions. Finding elevated numbers of eosinophils in blood also suggests allergy (or infection with helminths).

Generalized Anaphylaxis

Some anaphylactic reactions are generalized, severe, and immediately life-threatening. In a sensitized person, a generalized reaction begins with sudden reddening of skin, intense itching, and hives, especially over the face, chest, and palms of hands. The disorder can then progress to respiratory anaphylaxis or anaphylactic shock.

(a)

(b)

➤ **Figure 18.2 Allergy testing.** (a) Possible allergens are placed on prongs and introduced under the patient's skin. (b) If an individual is hypersensitive, a wheal (white raised area) and flare (reddened area) soon become visible on the skin. Such testing is usually done on the extremities to keep any hypersensitive reaction away from major organs.

(a)

➤ **Figure 18.3**   **False-color SEM of ragweed (*Ambrosia*) pollen.** One of several pollen causes of hay fever. In early spring the culprits are primarily tree pollens such as oak, elm, birch (especially in Europe), and box elder. In late spring and early summer, grass pollens plus those of some broad-leaved plants are most likely to be involved. In late summer and early autumn, the chief allergens are ragweeds, saltbush, and Russian thistle pollens.

(b)

➤ **Figure 18.4**   **Honeybee sting allergy.** An eye has swollen shut but the bee sting has not caused the airway to close, as happens in more severe reactions. (b) Anaphylactic kit for emergency use, showing syringe loaded with epinephrine. Many individuals with severe insect-sting allergies carry such kits with them at all times. They are available only by prescription.

In **respiratory anaphylaxis** the airways become severely constricted and filled with mucus secretions, and the allergic individual may die of suffocation. More than 4000 Americans die each year from respiratory anaphylaxis. Another 15 million Americans suffer from **asthma,** which is often caused by inhaled or ingested allergens, emotional stress, aspirin, or cold, dry air. Asthma can also be caused by hypersensitivity to endogenous microorganisms. For example, some patients become sensitized to *Moraxella catarrhalis,* a normal bacterial resident of respiratory mucous membranes.

In **anaphylactic shock** blood vessels suddenly dilate and become more permeable, causing an abrupt and life-threatening drop in blood pressure. Insect bites and stings are a common cause of anaphylactic shock in people sensitized to insect venoms (➤ Figure 18.4a).

Generalized anaphylaxis must be treated immediately. Unless epinephrine (adrenaline) is administered immediately, death can occur. Epinephrine acts by relaxing smooth muscle of respiratory passageways and constricting blood vessels. People who are sensitized to insect venom often carry an emergency anaphylactic kit (➤ Figure 18.4b). The kit contains a tourniquet, benadryl (antihistamine) tablets, and a syringe containing two doses of epinephrine. Having such a kit on hand could easily mean the difference between life and death because of the rapid onset of life-threatening symptoms in patients who already have had anaphylactic reactions.

Genetic Factors in Allergy

In the United States, 40 million people have some kind of allergy. In many cases, genetic factors are thought to contribute to the development of allergies. Although different family members typically have different allergies (one person may suffer from asthma and another from dust allergies), all will have high levels of IgE antibodies. At least 60 percent of children with atopy

The Origins of Allergies

Allergies account for almost 10 percent of all visits to physicians' offices in the United States. Why does the immune system often react violently to nonharmful substances? And why are some people allergic, whereas others are not? Research has shown that some people who lack the ability to produce IgE antibodies are prone to lung and sinus infections. Also, people who lack the ability to produce IgG or IgM antibodies often produce IgE antibodies to bacterial infections. These observations suggest that the IgE antibody may play a necessary role in immunity, besides causing allergies.

IgE is known to help fight infections by parasitic helminths. ➡ (Chapter 17, p. 470) IgE antibodies also may protect against ectoparasites (ticks, chiggers, fleas). The American biologist Margie Profet believes that IgE antibodies also are a backup system to protect against the ingestion of toxins.

In the book *Why we get sick: The new science of Darwinian medicine,* the authors, physician Randolph Nesse and evolutionist George Williams, suggest that many allergies that exist today were not common 150 years ago. They say that hay fever was almost nonexistent in England in the early 1800s and rare in Japan even as recently as 1950. Yet today, about 10 percent of the Japanese have hay fever. Nesse and Williams suggest that modern comforts may be to blame. Well-insulated houses with thick carpeting are excellent breeding grounds for dust mites and catchalls for allergens such as pollen and mold spores. In fact, studies show that infants raised in homes relatively clean of allergens develop far fewer allergies than do infants raised in the average home. However, these ideas still do not explain fully why some people have allergies whereas others in the same household or family do not.

Because about 15 percent of Americans suffer from allergies, it stands to reason that any one person would have about a 15 percent chance of being allergic to some allergen. In atopic families, if one parent is atopic, a child has a 25 percent chance of having an allergy. If both parents are atopic, the chance jumps to about 50 percent. Thus, some allergies can run in families and may, in part, have a genetic basis. Profet, Nesse, and Williams also suggest that if someone is simultaneously exposed, for example, to a plant toxin and an allergen, the immune system may respond to the toxin by producing IgE antibodies. In this attack, the immune system sees the allergen as "part of the toxin" and so reacts to it as well. Immune cells remain sensitized, and future exposures of the allergen alone will trigger an IgE response even though the toxin is not present.

Unfortunately, no matter what the causes or origins of allergies, we must suffer through them with only our allergy medications to help us along.

have a family history of asthma or hay fever, and half these children later develop other allergies. Thus, allergy probably has a genetic basis, possibly in properties of membranes or the performance of various cells involved in immune responses, such as phagocytes. Normal membranes screen out all but the tiniest microorganisms and virtually all potential allergens. Membranes of allergic individuals, however, are more permeable to larger particles such as pollen grains. Even when allergens pass through membranes, phagocytic cells usually engulf them in normal individuals but sometimes fail to do so completely in allergic individuals.

Treatment of Allergies

One approach to dealing with allergies is to avoid contact with the specific allergen. People with food allergies should not eat a food to which they have had a hypersensitivity reaction. Whatever the allergen, a subsequent exposure generally will trigger degranulation by mast cells (➤ Figure 18.5a). **Desensitization** (hyposensitization) is the only currently available treatment intended to cure an allergy. If denatured allergen is injected subcutaneously ("allergy shots"), it may induce a state of tolerance, preventing the activation of those B cells that mature into IgE-secreting plasma cells (➤ Figure 18.5b). Also, by receiving such injections with gradually increasing doses of the allergen, the patient may produce IgG antibodies, called **blocking antibodies,** against the allergen. Upon reexposure to an allergen, the blocking antibodies combine with the allergen before the allergen has a chance to react with IgE, so mast cells do not release mediators. The number of suppressor T cells sensitized to the allergen also increase significantly during desensitization. Thus, increases in IgG and decreases in IgE may work together to make the patient less sensitive to the allergen.

Desensitization has been very successful against insect venoms and drug allergies such as allergies to penicillin. Unfortunately, desensitization does not alleviate the signs and symptoms in many allergies, such as hay fever. In addition, the treatment itself can cause anaphylactic shock because the injections contain the very substance to which the patient is allergic.

Allergy injections are about 65 to 75 percent effective in individuals whose allergies are caused by inhaled allergens.

➤ Figure 18.5   **A proposed mechanism of action for desensitization allergy shots (hyposensitization).** (a) In a normal allergic response, natural exposure to an allergen causes helper T cells to stimulate those B cells that mature into plasma cells to make IgE antibodies. After binding to mast cells, a second allergen exposure causes degranulation. (b) Desensitization involves the injection of denatured allergen. Such shots may lead to tolerance, preventing B cells from maturing into plasma cells to make IgE antibodies. Exposure to the allergen also may activate those B cells that mature into the plasma cells that make IgG (blocking) antibody. Such IgG antibodies can bind to incoming allergen before it reaches the IgE molecules attached to mast cells. Complexing of allergen with these attached IgE molecules would cause the mast cells to degranulate and release histamine, so blocking this step is the key to preventing allergic responses.

Patients must remain in the physician's office for 20 to 30 minutes after the injection so that emergency treatment will be available if a generalized anaphylactic reaction occurs.

Other allergy treatments alleviate allergy symptoms but do not cure the disorder. Antihistamines counteract the swelling and redness due to histamine but are not effective against SRS-A of asthmatic conditions, which involve constriction of the airways, whereas antiinflammatory agents such as corticosteroids suppress the inflammatory response. Better therapeutic methods of treating allergies are greatly needed. As we learn more about the properties of leukotrienes and IgE antibodies, perhaps this need can be met.

Cytotoxic (Type II) Hypersensitivity

In cytotoxic (Type II) hypersensitivity, specific antibodies react with cell surface antigens interpreted as foreign by the immune system, leading to phagocytosis, killer cell activity, or complement-mediated lysis. The cells to which the antibodies are attached, as well as surrounding tissues, are damaged because of the resulting inflammatory response. Antigens that initiate cytotoxic hypersensitivity typically enter the body in mismatched blood transfusions or during delivery of an Rh-positive infant to an Rh-negative mother.

Mechanism of Cytotoxic Reactions

When an antigen on a plasma membrane is first recognized as foreign, B cells become sensitized and stand ready for antibody production upon a subsequent antigen exposure. During subsequent exposures with the surface antigen, antibodies bind to the antigen and activate complement. Phagocytic cells, such as macrophages and neutrophils, are attracted to the site. The mechanisms of type II hypersensitivity appear to be responsible for the tissue damage in cases of rheumatic fever following a streptococcal infection, in certain viral diseases, in transfusion reactions, and in hemolytic disease of the newborn (mother-infant Rh incompatibility).

Examples of Cytotoxic Reactions

Cytotoxic reactions typical of type II hypersensitivities are exemplified by mismatched blood transfusions and by hemolytic disease of the newborn.

Transfusion Reactions

Normal human red blood cells have genetically determined surface antigens (blood group systems) that form the basis for the different blood types. A **transfusion reaction** can occur when matching antigens and antibodies are present in the patient's blood at the same time. Such reactions can be triggered by any of the blood group antigens. We will focus on antigens A and B, which determine the **ABO blood group system.** As Table 18.3 shows, four blood types—A, B, AB, and O—are named according to whether red blood cells have antigen A, antigen B, both A and B antigens, or neither antigen. Normally, a person's serum has no IgM antibodies against the antigens present on his or her own red blood cells. However, if a sensitized patient receives red blood cells with a different blood cell antigen during a blood transfusion, IgM antibodies cause a Type II hypersensitivity reaction against the foreign antigen. The foreign red blood cells are agglutinated (clumped), complement is activated, and hemolysis (rupture of blood cells) occurs within the blood vessels (➤ Figure 18.6). Symptoms of a transfusion reaction include fever, low blood pressure, back and chest pain, nausea, and vomiting. Transfusion reactions can usually be prevented by careful cross-matching of donor and recipient blood group antigens so that the correct blood type can be selected for transfusion (➤ Figure 18.7).

Transfusion reactions to other erythrocyte antigens, such as Rh (Rhesus), also occur. However they usually are less serious than reactions to foreign A or B antigens because the antigen molecules are less numerous.

Hemolytic Disease of the Newborn

Another example of a cytotoxic reaction is **hemolytic disease of the newborn,** or *erythroblastosis fetalis.* In addition to the ABO blood group, red blood cells can have **Rh antigens,** so named because the antigens were discovered first in Rhesus monkeys. Blood with Rh antigens on red blood cells is designated Rh-positive; red blood cells lacking Rh antigens are designated Rh-negative. Anti-Rh antibodies normally are not present in the serum of either

> In the United States alone, a transfusion takes place about every three seconds—some 12 million units of blood are used every year.

➤ Table 18.3	Properties of the ABO blood group system	
Blood type	Antigens on erythrocytes	Antibodies in serum
A	A	anti-B
B	B	anti-A
AB	A and B	neither anti-A nor anti-B
O	neither A nor B	anti-A and anti-B

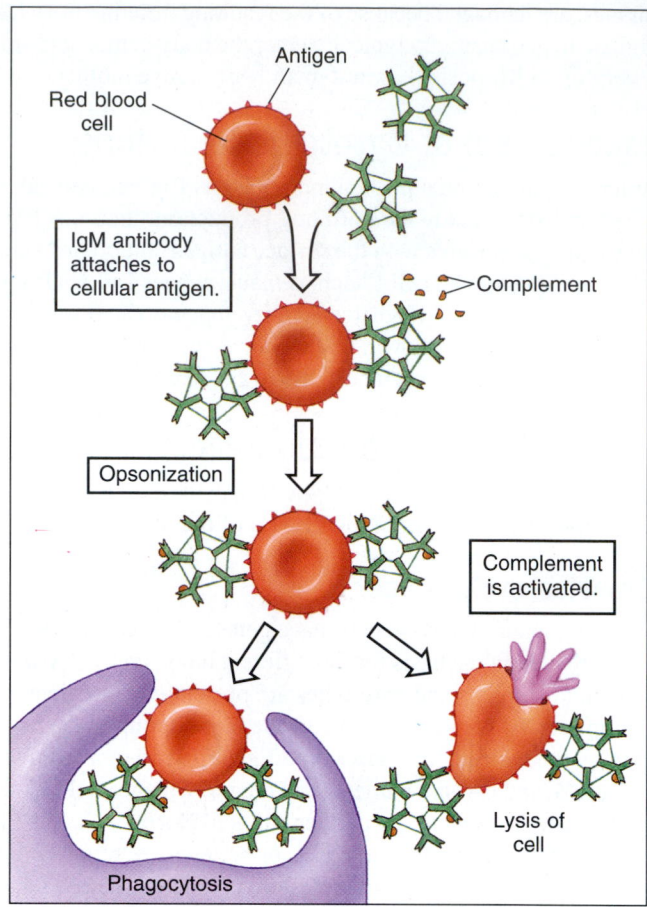

➤ Figure 18.6 **The mechanisms of cytotoxic (Type II) hypersensitivity.** Mismatched red blood cell antigen usually is bound to IgM. Complement is activated and results in either subsequent phagocytosis or lysis of the red blood cells.

➤ Figure 18.7 **Blood typing and transfusions.** Careful blood typing and matching of donor and recipient blood prevents most transfusion reactions. Persons with type AB blood can safely receive a transfusion of any of the four major blood types. Persons with type O blood can safely donate blood to recipients of any blood type. (Can you explain why? Refer to Table 18.3.)

Rh-positive or Rh-negative blood. Consequently, sensitization is necessary for an Rh antigen–antibody reaction.

Sensitization typically occurs when an Rh-negative woman carries an Rh-positive fetus, which inherited this blood type from its father. The fetal Rh antigen rarely enters the mother's circulation during pregnancy but can leak across the placenta during delivery, miscarriage, or abortion (➤ Figure 18.8a). The Rh-negative mother's immune system then becomes sensitized to the Rh antigen and can produce anti-Rh antibodies if it again encounters the Rh antigen.

Because sensitization usually occurs at delivery, the first Rh-positive child of an Rh-negative mother rarely suffers from hemolytic disease. But when a sensitized Rh-negative mother carries a second or subsequent Rh-positive fetus, the mother's anti-Rh antibodies cross the placenta and cause a Type II hypersensitivity reaction in the fetus (➤ Figure 18.8b). If this occurs, fetal red blood cells agglutinate, complement is activated, and the red blood cells are destroyed. The result is hemolytic disease of the newborn. The baby is born with an enlarged liver and spleen caused by efforts of these organs to eliminate damaged red blood cells (➤ Figure 18.8c,d). Such babies exhibit the yellow skin color of jaundice due to excessive bilirubin—a product of the breakdown of red blood cells—in their blood.

Hemolytic disease of the newborn can be prevented by giving Rh-negative mothers intramuscular injections of anti-Rh IgG antibodies (Rhogam) within 72 hours after delivery. The antibodies presumably bind to Rh antigens on the fetal red blood cells that have leaked into the mother's blood. These anti-Rh antibodies destroy the fetal red blood cells before they can act to sensitize her immune system. It is essential to treat all Rh-negative women after delivery, miscarriage, or abortion in case the fetus may have been Rh-positive. Today, anti-Rh antibodies are often administered to Rh-negative women during pregnancy as well. Such treatment at 3 and 5 months prevents sensitization to the fetus in case fetal antigens leak into the mother's circulation, which can result from hard coughing or sneezing. Before anti-Rh antibody (Rhogam) was given preventively, hemolytic disease of the newborn occurred in about 0.5 percent of all pregnancies; 12 percent of these terminated in stillbirths.

Rh–negative mother

Anti–Rh antibodies

Anti–Rh antibodies enter fetal circulation.

Rh–positive fetus

Rh–positive fetus

Fetal red blood cells destroyed

Maternal circulation

First pregnancy

Subsequent pregnancy

(a)

(b)

(c)

(d)

➤ Figure 18.8   **Cause and effect of hemolytic disease of the newborn.** (a) The stage is set for an Rh-incompatibility pregnancy when the mother is Rh$^-$ and the fetus is Rh$^+$ (which is usually the case if the father is Rh$^+$). (b) Rh antigens may cross the placenta and enter the mother's bloodstream before or at delivery. She responds by making anti-Rh antibodies, which also can cross the placenta. Even if antibody production is not stimulated until delivery, the resulting antibodies will persist in the mother's circulation and attack the red blood cells of any subsequent Rh$^+$ fetus. To prevent this situation, Rhogam (anti-Rh antibody) is injected into the mother early in the pregnancy, immediately after delivery, and in cases of miscarriage or abortion. Rhogam reduces exposure to the antigen and thus lessens anti-Rh antibody production. (c) Child affected by hemolytic disease caused by Rh incompatibility. (d) The liver is greatly enlarged.

Immune Complex (Type III) Hypersensitivity

Immune complex (Type III) hypersensitivity results from the formation of antigen–antibody complexes. Under normal circumstances these large immune complexes are engulfed and destroyed by phagocytic cells. Hypersensitivity occurs when antigen–antibody complexes persist or are continuously formed.

Mechanism of Immune Complex Disorders

Like anaphylactic and cytotoxic reactions, immune complex disorders also are initiated after sensitization. Upon subsequent exposure to the sensitizing antigen, specific IgG antibodies combine with the antigen in the blood to form an *immune complex* and activate complement (➤ Figure 18.9). Antibodies bind to live cells or parts of damaged cells in blood vessel walls and other tissues. Normally, large immune complexes are removed by phagocytosis in the liver and spleen. However, immune complexes are often quite small and fail to bind tightly to Kupffer cells in the liver, thereby escaping elimination from the blood, and are deposited in organs, tissues, or joints. Such antigen-antibody complexes and complement, in turn, cause basophils and mast cells to release histamine and other mediators of allergic reactions, with the effects described earlier. Phagocytes attracted chemotactically to these sites of activity release hydrolytic enzymes, causing tissue damage that is acute but can become chronic if the antigen remains for long periods of time.

➤ Figure 18.9 **The mechanism of immune complex (Type III) hypersensitivity.** Immune complexes are formed when antigen is introduced into a previously sensitized individual. When the resulting immune complex is deposited, it activates complement, producing fever, itching, rash or hemorrhagic areas, joint pain, and acute inflammation. On a systemic basis, this can cause serum sickness.

Examples of Immune Complex Disorders

We will illustrate immune complex disorders with two phenomena, the systemic serum sickness and the localized Arthus reaction. Other disorders that involve immune complexes, such as rheumatoid arthritis and systemic lupus erythematosus, are discussed later in connection with autoimmune diseases because the antibodies involved in these disorders react with the person's own tissues. Acute glomerulonephritis following certain streptococcal infections is another immune complex disease, in which the glomeruli of the kidneys can be severely damaged (Chapter 20).

Serum sickness was frequently seen in the pre-antibiotic era when large doses of antitoxin sera were used to immunize people passively against infectious diseases such as diphtheria. Diphtheria toxin given to horses caused them to make antibodies against the toxin. A patient then received the horse serum, which contained not only antidiphtheria toxin antibody but also horse proteins. A sensitized patient's immune system would make sufficient antibodies against these horse proteins to form immune complexes consisting of human antibody reacting against horse serum protein, upon second exposure. These immune complexes, which are removed slowly by phagocytic cells, would attach to the glomeruli of the kidneys. The filtration capacity of the glomeruli was thereby impaired, causing proteins and blood cells to be excreted in the urine. Immune complexes are also deposited in joints and in skin blood vessels.

People with serum sickness usually have fever, enlarged lymph nodes, decreased numbers of circulating leukocytes, and swelling at the injection site. Most people recover from serum sickness as the complexes eventually are cleared from the blood and tissue repair occurs in the glomeruli. However, the disorder became chronic in many diphtheria patients because they received horse serum daily over the course of the disease.

Today, serum sickness is rare and usually is due to second exposure to a foreign substance in a biological product, such as horse serum in a vaccine preparation. An advantage of vaccines made by genetic engineering is that they do not contain such foreign substances. When the use of any biological product is contemplated, the patient should be tested for sensitivity first. A small quantity of the product should be given intradermally (within the skin) or intravenously. A wheal and flare reaction or a drop of 20 points or more in blood pressure following intravenous injection indicate hypersensitivity, and the product should not be given.

The **Arthus reaction** is a local reaction seen in the skin after subcutaneous (under the skin) or intradermal injection of an antigenic substance. It occurs in people who already have large quantities of antibodies (mainly IgG) to the antigen. In 4 to 10 hours, edema and hemorrhage develop around the injection site as immune complexes and complement trigger cell damage and platelet aggregation (➤ Figure 18.10). In severe reactions, tiny clots obstruct blood vessels, and cells normally nourished by the blocked vessels die (➤ Figure 18.11).

Cell-Mediated (Type IV) Hypersensitivity

Cell-mediated (Type IV) hypersensitivity is also called **delayed hypersensitivity** because reactions take more than 12 hours to develop. These reactions are mediated by T cells—specifically, a type of T_H1 cell (sometimes called a **delayed hypersensitivity T [T_{DH}] cell**)—not by antibodies.

Mechanism of Cell-Mediated Reactions

Cell-mediated hypersensitivity occurs as follows. On first exposure, antigen molecules bind to antigen-presenting cells that present antigen fragments to T_H1 (inflammatory T) cells. ➥ (Chapter 17, p. 475) When APCs again present the same antigen during a second, later exposure, the sensitized T_H1 cells release various cytokines, including γ-interferon and migration inhibiting factor (MIF). Gamma-interferon stimulates macrophages to ingest the antigens. If the antigens are on microorganisms, the macrophages usually, but not always, kill the microorganisms. MIF prevents migration of macrophages, so they remain localized at the site of the hypersensitivity reaction. Other cytokines are presumed to cause the hypersensitivity reaction itself. Such re-

➤ Figure 18.10   **The mechanism that produces an Arthus reaction and hemorrhagic areas.**
(a) In severe cases, injection of horse protein antigens leads to (b) immune complex formation. (c) In association with complement, (d) neutrophils release lysosomal enzymes that damage the blood vessel wall. (e) Immune complexes also trigger platelet aggregation that can obstruct blood flow. (f) Complement also attracts more neutrophils to the site and (g) causes mast cell degranulation. (h) Finally, platelet and complement trigger endothelial retraction. Tissue death can occur if cells are cut off from blood vessel flow.

actions account for patches of raw, reddened skin in eczema, swelling, and granulomatous lesions. These processes are summarized in ➤ Figure 18.12.

Examples of Cell-Mediated Disorders

Three common examples of delayed hypersensitivity—contact dermatitis, tuberculin hypersensitivity, and granulomatous hypersensitivity—illustrate the diversity of cell-mediated reactions.

Contact dermatitis occurs in sensitized individuals on second or subsequent exposure to allergens such as oils from poison ivy, rubber, certain metals, dyes, soaps, cosmetics, some

➤ Figure 18.11   **Arthus reaction.** The patient shows an extensive area of hemorrhagic damage that will result in tissue necrosis and sloughing.

> **Figure 18.12** **The mechanism of cell-mediated, or delayed (Type IV), hypersensitivity.** This type of reaction is mediated by T cells rather than by B cells as in types I, II, and III. T cells that have become sensitized to a particular antigen release cytokines on subsequent contact with the same antigen fragment. These cytokines cause inflammatory reactions that attract macrophages to the site. By degranulation, APCs release mediators that add to the inflammatory response. Contact dermatitis and poison ivy rash are examples of cell-mediated hypersensitivity.

plastics, topical medications, and other substances (Table 18.4). Unlike Type I hypersensitivities, Type IV hypersensitivities do not appear to run in families. Molecules too small to cause immune reactions pass through the skin, where they become antigenic by binding to normal proteins on Langerhans cells of the epidermis. These cells, which carry MHC class II antigens, migrate to the lymph nodes, where they act as antigens presenting cells to T_H1 cells. Within 4 to 8 hours after the next exposure, a hypersensitivity reaction begins, and eczema occurs within 48 hours.

Urushiol (u′ru-she-ol), an oil from the poison ivy plant, is a major cause of contact dermatitis in the United States (➤ Figure 18.13). Most people get poison ivy from direct contact with leaves or other plant parts, but some get it by inhaling smoke from burning brush that contains poison ivy plants. Poison ivy is particularly severe when oil droplets come in contact with respiratory membranes. Sensitivity to poison ivy can develop at any age, even among people who have come in contact with it for years without reacting to it. One way to minimize a reaction to poison ivy is to wash exposed areas thoroughly with strong soap or detergent within minutes of contact, before much oil has penetrated the skin and bound chemically to skin cells.

➤ Table 18.4	Selected contact allergens
Allergen	**Common sources of contact**
Benzocaine	Topical anesthetic
Chromium	Jewelry, watches, chrome-tanned leather, cement
Formaldehyde	Facial tissues, nail hardeners, synthetic fabrics
Latex	Surgical or examination gloves
Nickel	Jewelry, watches, objects made of stainless steel and white gold
Mercaptobenzothiazole	Rubber goods
Methapyrilene	Topical antihistamine
Merthiolate	Topical antiseptic
Neomycin	Topical antibiotic
Oleoresin	Oil from poison ivy and similar plants

Poison Ivy? But It Doesn't Grow Here!

Contact dermatitis similar to poison ivy was seen in American military personnel in Japan after World War II. Lesions appeared on elbows, forearms, and in a horseshoe shape on the buttocks and thighs. Knowing that poison ivy did not grow in Japan, the medical staff was puzzled. The puzzle was solved when scientists discovered that oils from a Japanese plant containing a small quantity of urushiol were used to manufacture lacquer. Although the quantity of urushiol in lacquer was not sufficient to sensitize Japanese people, it did elicit an allergic response in Americans previously sensitized to poison ivy. Resting the arms on countertops explained lesions on elbows and forearms; contact with toilet seats explained the horseshoe-shaped lesions.

(a)   (b)

➤ Figure 18.13   **Poison ivy (Type IV) hypersensitivity.** (a) Poison ivy (*Toxicodendron radicans*), showing the leaves with their characteristic three leaflets. Poison ivy vines also contain the irritating oil urushiol, so it is important to be able to recognize them in winter, when leaves may not be present. (b) Poison ivy dermatitis, showing fluid-filled vesicles.

Once a person is sensitized, an exceedingly small quantity of oil will elicit a reaction upon the next exposure. Scratching lesions does not spread the oil, but it can lead to infections. Cashews and mangoes contain substances chemically similar to urushiol, and some people display delayed hypersensitivity (including digestive disturbances) to these substances.

Tuberculin hypersensitivity occurs in sensitized individuals when they are exposed to *tuberculin,* an antigenic lipoprotein from the tubercle bacillus *Mycobacterium tuberculosis.* Similar antigens from the bacterium that causes leprosy (*Mycobacterium leprae*) and the protozoan that causes leishmaniasis (*Leishmania tropica*) produce similar reactions in sensitized individuals. The antigen activates T_H1 cells, which in turn release cytokines that cause large numbers of lymphocytes, monocytes, and macrophages to infiltrate the dermis. The normally soft tissues of the skin then form a raised, hard, sometimes red region called an **induration** (➤ Figure 18.14). In a **tuberculin skin test,** a purified protein derivative (PPD) from *Mycobacterium tuberculosis* is injected subcutaneously. If a person has been exposed to the bacterium or had received the BCG vaccine, an induration will form within 48 hours. The diameter and elevation of the induration, not its redness, indicate whether further tests are needed.

Granulomatous hypersensitivity, the most serious of the cell-mediated hypersensitivities, usually occurs when macrophages have engulfed pathogens but have failed to kill them. Inside the macrophages the protected pathogens survive and sometimes continue to divide. T_H1 cells sensitized to an antigen of the pathogen elicit the hypersensitivity reaction, attracting several cell types to the skin or lung. A granuloma in the skin (leproma) or lung (tubercle) develops. This kind of hypersensitivity is the most delayed of all, appearing 4 weeks or more after exposure to the antigen. Such persistent and chronic antigenic stimuli are also typical of the bacterial disease listeriosis as well as many fungal and helminthic infections.

Characteristics of the four types of hypersensitivity are summarized in Table 18.5.

Autoimmune Disorders

Autoimmune disorders occur when individuals become hypersensitive to specific antigens on cells or tissues of their own bodies, despite mechanisms that ordinarily create tolerance to those self antigens. The antigens elicit an immune response in which **autoantibodies,** antibodies

➤ Figure 18.14   **A positive tuberculin skin test reaction.** The raised area of induration should be observed and measured after 48 to 72 hours. A positive reaction will measure 5 mm or more; 2 mm or smaller is negative; 3 and 4 mm are considered doubtful.

✓ List the four types of hypersensitivities and their names. What are the main mediators for each?

✓ What are the signs and symptoms of anaphylaxis?

✓ How does hemolytic disease of the newborn arise? How can it be prevented?

> ## Table 18.5   Characteristics of the types of hypersensitivity

	Type I	Type II	Type III	Type IV
Characteristic	Immediate	Cytotoxic	Immune Complex	Cell-Mediated
Main mediators	IgE	IgG, IgM	IgG, IgM	T cells
Other mediators	Mast cells, basophils, histamine, prostaglandins, leukotrienes	Complement	Complement, inflammatory factors, eosinophils, neutrophils	Lymphokines, macrophages
Antigen	Soluble or particulate	On cell surfaces	Soluble or particulate	On cell surfaces
Reaction time	Seconds to 30 minutes	Variable, usually hours	3 to 8 hours	24 hours to 4 or more weeks
Nature of reaction	Local wheal and flare, airway restriction, anaphylactic shock	Clumping of erythrocytes, cell destruction	Acute inflammation effects	Cell-mediated cell destruction
Therapy	Desensitization, antihistamines, steroids	Steroids	Steroids	Steroids

against one's own tissues, are produced. An autoimmune response can be T-cell-mediated as well. These disorders are characterized by cell destruction in various types of hypersensitivity reactions. Although autoimmune disorders arise from a response to a self antigen, they range over a wide spectrum—from those that affect a single organ or tissue (organ-specific) to those that are systemic, affecting many organs and tissues (Table 18.6).

Autoimmunization

Autoimmunization is the process by which hypersensitivity to "self" develops. Such a response is usually sustained and long-lasting and can cause long-term tissue damage. Immunologists are beginning to understand this process better. Several different mechanisms of autoimmunity probably exist:

1. *Genetic factors* may predispose a person toward autoimmune disorders. For example, the children of a parent who has autoantibodies to a single organ are likely to develop autoantibodies to the same or to a different single organ. As we shall see later, individuals who have genes for certain histocompatibility antigens are at greater than normal risk of developing particular autoimmune disorders.

2. In addition to predisposing genetic factors, **antigenic,** or *molecular,* **mimicry** can occur. T_H cells might attack tissue antigens that are similar to antigens of some pathogens. Some children who suffer rheumatic fever (caused by *Streptococcus pyogenes*) develop rheumatic heart disease later in life. For some reason the immune system of such individuals "sees" the heart valve tissue as similar to certain streptococcal antigens and attacks the heart valves.

3. The thymus is critical to the normal development of T cells. Aside from the T_H cells that recognize nonself antigens, T_H cells that recognize self antigens can exist if *clonal deletion* fails to remove these self-reactive T cells. ➡ (Chapter 17, p. 466) If they survive and proliferate, they can attack self antigens and trigger B cell activity with antibody production. Antigens hidden in tissues and lacking contact with B or T cells during immune system development or clonal deletion could be released through physical injury. These antigens then will be perceived as foreign by the immune system (see the box "Sympathetic Blindness").

4. Mutations might give rise to aberrant proteins to which B cells react, producing plasma cells that make autoantibodies.

Microbial or viral infections as common as the common cold may initiate or influence autoimmune disorders. The mechanism may be molecular mimicry.

Sympathetic Blindness

Lens proteins are normally confined within the lens capsule of the eye. Because they are never exposed to lymphocytes during development, the immune system never acquires tolerance for them. Sometimes an eye injury allows these proteins to leak into the bloodstream, where they elicit an immune response. The antibodies formed in this way then attack proteins in the undamaged eye. Because circulation (which transports antibodies) tends to be better in that eye, the immune response in the healthy eye can be more intense than that in the injured eye. This phenomenon sometimes leads to sympathetic blindness, or loss of sight in the uninjured eye.

> **Table 18.6** | **The spectrum of autoimmune disorders**

Disorder	Organ(s) or tissues affected	Autoantibody target
Organ-specific disorders		
Addison's disease	Adrenal glands	Adrenal gland proteins
Autoimmune hemolytic anemia	Erythrocytes	Red blood cell membrane proteins
Glomerulonephritis	Kidneys	Streptococcal cross-reactivity with kidney
Graves' disease	Thyroid gland	TSH receptor
Hashimoto's thyroiditis	Thyroid gland	Thyroglobulin
Idiopathic thrombocytopenic purpura	Blood platelets	Platelet glycoproteins
Juvenile diabetes	Pancreas	Beta cells and insulin
Myasthenia gravis	Skeletal muscles	Acetylcholine receptor
Pernicious anemia	Stomach	Vitamin B12 binding site
Postvaccine/postinfection encephalomyelitis	Myelin	Measles cross-reactivity with myelin
Premature menopause	Ovaries	Corpus luteum
Rheumatic fever	Heart	Streptococcal cross-reactivity with heart
Spontaneous male infertility	Testes	Spermatozoa
Ulcerative colitis	Colon	Colon cells
Systemic (disseminated) disorders		
Goodpasture's syndrome	Basement membranes	Basement membrane
Polymyositis/dermatomyositis	Muscles and skin	Cell nuclei
Rheumatoid arthritis	Joints	Cell nuclei, gamma globulins
Scleroderma	Connective tissues	Nucleoli
Sjögren's syndrome	Lacrimal and salivary glands	Cell nuclei
Systemic lupus erythematosus	Many tissues	Cell nuclei, histones

5. Viral components inserted into host cell membranes might act as antigens, or virus–antibody complexes might be deposited in tissues.

6. The sympathetic nervous system, which along with the parasympathetic system controls internal body functions, helps regulate the immune system. When the sympathetic nervous system is damaged, the number of suppressor T cells decreases.

Examples of Autoimmune Disorders

Autoimmune disorders usually are chronic inflammatory disorders with symptoms that can alternately worsen and lessen. They affect about 6 percent of all humans and can affect one organ or many. We now look at three examples to illustrate the diversity of such disorders.

Myasthenia Gravis

Myasthenia gravis is an autoimmune disease that afflicts approximately 25,000 Americans. It affects primarily women in their 20s and 30s and men in their 40s and 50s. The disease usually affects skeletal muscles of the limbs and those involved in eye movements, speech, and swallowing. The principle symptoms of the disease are progressive weakness and muscle fatigue.

For muscles to contract normally, neurons secrete the neurohormone acetylcholine across the gap (junction) between neuron and muscle (➤ Figure 18.15a). When acetylcholine receptors on the muscle cells bind acetylcholine, contraction of the muscle occurs. Evidence suggests that in people with myasthenia gravis, muscle contraction is prevented by IgG autoantibodies that either block the acetylcholine receptor or cause a reduction in the number of acetylcholine receptors (➤ Figure 18.15b). In fact, most myasthenic patients have only 30 to 50 percent as many receptors as do unaffected persons.

Although myasthenia gravis is one of the best-understood autoimmune diseases, the reason why autoantibodies are formed is not well understood. One possibility is that autoantibodies are triggered by an immune response to an infectious virus or bacterium that has antigens mim-

How do myasthenia gravis patients lose acetylcholine receptors? Their IgG autoantibodies induce complement-mediated degradation of the receptors.

(a)

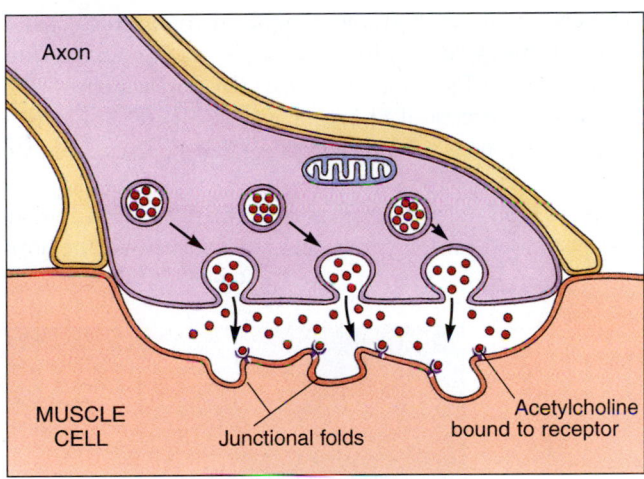

(b)

> Figure 18.15 **Myasthenia gravis.** This disorder involves a loss of acetylcholine receptors from the neuromuscular junction. (a) A normally functioning neuromuscular junction has many acetylcholine receptors that bind acetylcholine. (b) Myasthenic patients have significantly fewer receptors for acetylcholine.

icking a part of the acetylcholine receptor. Myasthenia gravis was once considered a fatal or disabling disease. Today myasthenic patients can be treated with drugs or immunosuppressive steroids so that they can live full lives. However, no means of preventing autoantibodies from forming or of removing them once they have formed is available.

In spite of muscle weakness, many women with myasthenia gravis have children. Their babies have temporary muscular weakness; they are like little rag dolls for the first few weeks of life. Small numbers of autoantibodies from the mother probably cross the placenta, affecting the fetus. Apparently the immune complexes do not permanently damage the fetal neurons because the babies soon have normal muscle function.

Rheumatoid Arthritis

In contrast with myasthenia gravis, which affects single-organs, **rheumatoid arthritis (RA)** affects mainly the joints of the hands and feet, although it can extend to other tissues. Of all forms of arthritis, RA is most likely to lead to crippling disabilities and to develop early in life (between the ages of 30 and 40). It is one of the most common autoimmune diseases, affecting about 2 million Americans. It is two to three times more prevalent in women than in men.

RA is characterized by inflammation and destruction of cartilage in the joints, often causing deformities in the fingers (➤ Figure 18.16). Despite continued research, the cause of RA is unknown. Some researchers believe that an infectious microbe (mycoplasma or virus) is the cause, leading to antigenic mimicry and ultimately to an attack of self antigens. Others believe that a self antigen recognized by the immune system as foreign is involved. Whatever the stimulus, inflammation activates specific cells in the joint and attracts lymphocytes. T_H1 cells rec-

> Figure 18.16 **Rheumatoid arthritis.** Joint inflammation is typical in people suffering from rheumatoid arthritis. In many cases, joint inflammation and destruction are so severe that they result in misshapen digits. In this gamma-ray photograph, swollen joints appear as bright spots.

New Treatment for Rheumatoid Arthritis

Recent research indicates that a protein called tumor necrosis factor-α (TNF-α) can trigger immune responses that attack bone and cartilage in joints, leading to RA. Centocor, a biotechnology company in Pennsylvania, has produced a genetically engineered antibody that binds to TNF-α so that it cannot trigger an immune response. In one study, 80 percent of patients receiving the antibody reported greatly reduced symptoms compared with patients who received a placebo. These antibodies do not survive for long periods in the body, so additional doses of the antibody were needed to prevent a recurrence of the disease.

ognize a self antigen in the joint and trigger the activation of B cells, which differentiate into plasma cells and secrete IgG antibody. These T_H1 cells also release several cytokines that researchers believe trigger resident phagocytic cells to release degrading enzymes from lysosomes. ➡ (Chapter 4, p. 102) These enzymes attack the cartilage, causing inflammation in the joints. People with RA also have circulating IgM antibodies that are specific for their own IgG antibodies. These autoantibodies, called **rheumatoid factors,** are used as a diagnostic test for RA. However, there is no evidence that these factors are involved with the formation of immune complexes. All these enzymes, factors, and cells increase the inflammatory response, leading to swollen, painful joints.

Although no cure exists for RA, treatment can alleviate symptoms. Hydrocortisone lessens inflammation and reduces joint damage, but long-term use weakens bones and causes such undesirable side effects as a reduction of normal immune responses. Aspirin decreases inflammation and reduces pain with fewer side effects. Physical therapy is used to keep joints movable. In severe cases, surgical replacement of damaged joints can restore movement.

Systemic Lupus Erythematosus

About 200,000 Americans suffer from **systemic lupus erythematosus (SLE)**, a systemic autoimmune disease. The name is derived from the reddened skin rash (erythematose) that resembles a wolf's mask (*lupus* is Latin for "wolf"). The butterfly-shaped rash appears over the nose and cheeks of about 30 percent of SLE patients (➢ Figure 18.17). SLE occurs 10 to 20 times as often in women as in men, with 80 percent of cases occurring in women during their reproductive years.

In SLE, autoantibodies (IgG, IgM, IgA) are made primarily against components of DNA but can also be made against blood cells, neurons, and other tissues. As the normal dying process

(a)

➢ Figure 18.17 **Lupus erythematosis.** The characteristic butterfly-shaped rash of systemic lupus erythematosus (a) appears red in fair-skinned people, (b) but white in dark-skinned people.

(b)

of cells (skin, intestinal, kidney) occurs, anti-DNA antibodies attack the remnants of these cells. Immune complexes are deposited between the dermis and epidermis and in blood vessels, joints, glomeruli of the kidneys, and the central nervous system. They cause inflammation and interfere with normal functions at these sites.

Inflammation of blood vessels, heart valves, and joints are common effects of interference. Arthritis is the most common clinical characteristic of SLE; a patchy skin rash on the upper trunk and extremities is a common skin manifestation. This rash often is precipitated by exposure to sunlight. Most SLE patients eventually die from kidney failure as glomeruli fail to remove wastes from the blood. Among individuals with SLE, men tend to have a nonsystemic *discoid* form of the disease. It produces disk-shaped skin lesions and is less serious than the systemic form in its side effects.

SLE cannot be cured. Treatment depends on individual disease characteristics. It can include antipyretics to control fever, corticosteroids to reduce inflammation, and immunosuppressant drugs to prevent or decrease further autoimmune reactions.

Transplantation

Transplantation is the transfer of tissue, called **graft tissue,** from one site to another. An **autograft** involves the grafting of tissue from one part of the body to another—for example, the use of skin from a patient's chest to help repair burn damage on a leg. A graft between genetically identical individuals (identical twins in humans, or members of highly inbred animal strains) is called an **isograft** (*iso,* Greek for "equal"). A graft between two people who are not genetically identical is termed an **allograft** (*allo,* Greek for "different"). Most organ transplants fall into this category. A transplant between individuals of different animal species is known as a **xenograft** (*xeno,* Greek for "foreign").

Early transplantation experiments involved the grafting of skin from one animal to another of the same species. The grafts appeared healthy at first but in a few days to a few weeks became inflamed and fell off. First thought to be due to infection, this reaction, called **transplant rejection,** is now known to be due to the destruction of the grafted tissue by the recipient's (that is, the host's) immune system. This process, which depends on T cells, also accounts for rejection of most organ transplants in humans. Transplants recognized as nonself are rejected.

A much less common transplantation effect is **graft-versus-host (GVH) disease,** in which the transplanted tissue contains immunocompetent cells that launch a cell-mediated response against the recipient's tissues. This response occurs most often when immunodeficient patients receive bone marrow transplants. GVH was more common before immunosuppressive drugs were introduced to block the immunological responses.

Histocompatibility Antigens

All human cells, and those of many other mammals, have a set of self antigens called **histocompatibility antigens** (*histo,* Latin for "tissue"). The genes producing these molecules are called the **major histocompatibility complex (MHC)**. Only identical twins have exactly the same MHC molecules, but all family members have a mixture of similar and different MHC molecules. These antigens are located on the surfaces of cells, including the cells of kidneys, hearts, and other commonly transplanted organs. If donor and recipient histocompatibility antigens are different, as they likely would be when donors and recipients are not related, recipient T cells recognize these cells as foreign and destroy the donor tissue.

To try to prevent allograft rejection, it is necessary to determine if the graft has histocompatibility antigens not found in the recipient. Like red blood cell antigens, histocompatibility antigens can be identified by laboratory tests so that donor and recipient tissues can be as closely matched as possible. Such tests are one of several methods of *tissue typing,* or testing the compatibility of donor and recipient tissues. Because the first studies in humans involved antibody reaction with leukocytes, the MHC molecules on human cells are called **human leukocyte antigens (HLAs)**. Human HLAs are determined by a set of genes located on chromosome 6. They are designated A, B, C, DR, DQ, and DP. The information in each gene specifies a particular antigen. For example, HLA-B is so highly variable that there are alleles for 51 different antigens. Overall, about 120 different antigens are recognized in humans and produced from the HLA genes.

For tissue typing, it would be impossible to do a complete typing of all HLAs present in an individual. However, the HLA-DR antigens are known to generate the strongest rejection reactions. Therefore, the tissues of prospective transplant recipients are typed as to which of the 20 HLA-DR antigens are present. When a donor organ becomes available, it also is typed. It is transplanted into the recipient whose antigens most nearly match. This procedure reduces the chances of rejection. Identical twins are the best match for allografts because all their HLA antigens are the same, but siblings of the same parents can have some matching HLA antigens in common.

Transplant Rejection

Like other immune reactions, transplant rejection displays specificity and memory. ➡ (Chapter 17, pp. 464, 466) Rejection usually is associated with mismatched HLA-DR antigens. Certain cells that present antigens to phagocytes increase the likelihood of rejection. The fact that HLA-DR antigens are found on T cells and macrophages that carry out rejection reactions may explain why these antigens are so important in graft rejection.

T cells are responsible for rejection of grafts of solid tissue, such as a kidney, heart, skin, or other organs. In animal experiments, allografts are retained by animals that lack T cells and are rejected by those lacking B cells. More specifically, T_H2 cells lead to rejection (➤ Figure 18.18). These cells help stimulate cytotoxic T (T_C) cells that reject the transplant through cell-mediated cytotoxicity. The T_H2 cells can also activate B cells to produce plasma cells and antibodies that cause rejection through lytic damage. Macrophages that are activated by T_H1 cells secrete inflammatory mediators and cause cytotoxic damage to the transplant. Natural killer cells can also act in transplant rejection.

The time required for rejection to occur varies from minutes to months. *Hyperacute rejection,* which is a cytotoxic hypersensitivity reaction, occurs when the recipient is already sensitized at the time the graft is done. For example, in kidney transplants, in which the graft is immediately supplied with host blood, extensive tissue destruction occurs within minutes to hours. (Corneal transplants, however, are not rejected because the cornea lacks blood vessels, and antibodies cannot reach them.) *Accelerated rejection* takes several days because it requires cells to reach the graft. *Acute rejection* occurs in days to weeks, requiring T cell sensitization

Disguising Tissues Aids in Transplants

Transplanted tissues are rejected because the recipient's immune system recognizes them as foreign. However, scientists have succeeded in disguising foreign cells by covering HLAs that act as antigens and trigger rejection. Ordinarily the binding of antibodies to antigens on cells starts a process that leads to death of the cells. The researchers modified the antibodies such that although the antibodies fit tightly to foreign cell surface proteins, they did not destroy the cells. Human pancreatic cells with HLA proteins thus "covered" were transplanted into mice. The mouse immune system ignored the human cells, allowing them to live and produce insulin for more than 6 months. Further research may eventually provide transplants to treat diabetics. An advantage of using disguised cells is that immunosuppressants now used with transplants will be unnecessary. These drugs, sometimes taken for the rest of the patients' life, leave patients highly vulnerable to infections and often cause other undesirable side effects, including a predisposition to cancer.

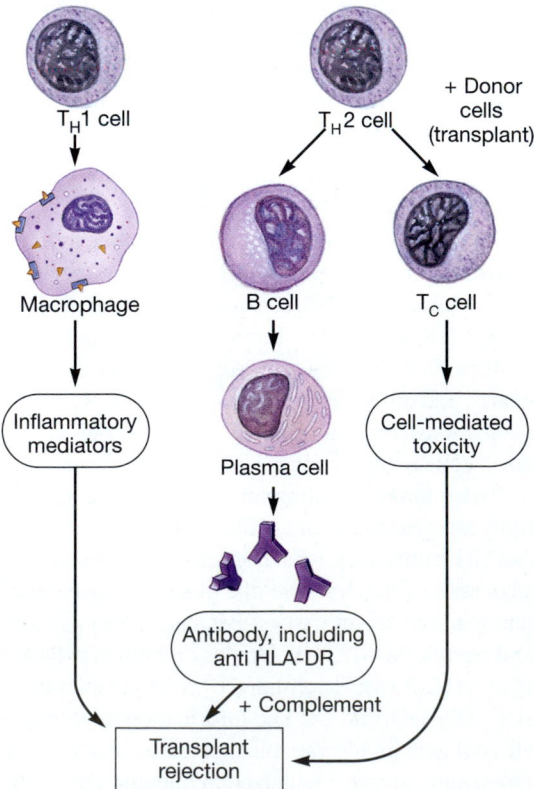

➤ **Figure 18.18   Transplant rejection.** A combination of both cell-mediated and humoral immune reactions is responsible for transplant rejection. T_H1 (inflammatory T) cells activate macrophages, which produce inflammatory mediators. T_H2 cells trigger both T_C and B cell activation. B cell activation leads to the production of plasma cells that synthesize antibodies, including anti–HLA-DR. Inflammatory mediators, T_C cell–mediated toxicity, and antibodies, along with complement, bring about transplant rejection.

after transplantation. Rejection that begins months to years after transplantation represents a *chronic rejection*. This slow process is typical of cardiac and kidney transplants in which an interaction between the immune system and the transplant leads to eventual dysfunction of the transplant.

Immunosuppression

When a patient is facing an organ transplant, the donor's HLA antigens are probably not a complete match. Therefore, it is important to prevent immune reactions that would destroy the organ. The minimizing of immune reactions is called **immunosuppression.** Ideally, immunosuppression should be as specific as possible—it should cause the immune system to tolerate only the antigens in transplanted tissue and allow the immune system to continue to respond to infectious agents.

In practice, radiation or cytotoxic drugs, both of which impair immune responses, are used to minimize rejection reactions. *Radiation* (X-rays) of lymphoid tissues suppresses the immune system, preventing rejection. Radiation also destroys other lymphoid functions, including the ability of the immune system to recognize infectious microbes. **Cytotoxic drugs,** such as azathioprine and methotrexate, damage many kinds of cells. But because they interfere with DNA synthesis, these drugs cause most damage to rapidly dividing cells. Because B cells and T cells divide rapidly after sensitization, the drugs exert a somewhat selective effect on the immune system.

Radiation and cytotoxic drugs impair T cell responses to infections. In contrast, the fungus-derived peptide cyclosporine A (CsA) suppresses, but does not kill, T cells, and it does not affect B cells. It is particularly useful in preventing transplant rejection: It allows T cells to regain function after the drug is stopped, and it does not reduce resistance to infections provided by B cells. The use of immunosuppressive drugs, especially CsA, has greatly increased the success rate of organ transplants. However, CsA may increase the transplant recipient's risk of developing cancer.

> Unfortunately, immunosuppressive drugs, sometimes taken for the rest of the patients' life, leave patients highly vulnerable to infections and cancer.

Drug Reactions

Most drug molecules are too small to act as allergens. If a drug combines with a protein, however, the protein-drug complex sometimes can induce hypersensitivity. All four types of hypersensitivity have been observed in drug reactions.

Type I hypersensitivity can be caused by various drugs. Most reactions are localized, but generalized anaphylactic reactions sometimes occur, especially when drugs are given by injection. Orally administered drugs are less likely to cause hypersensitivity reactions because they are absorbed more slowly. Hypersensitivity reactions require prior sensitization and depend on the production of IgE antibodies. Although penicillin is one of the safest drugs in use, 5 to 10 percent of people receiving it repeatedly become sensitized. Of those sensitized, about 1 percent develop generalized anaphylactic reactions, which account for about 300 deaths per year in the United States.

Type II hypersensitivity (➤ Figure 18.19) can occur when the drug binds to a plasma membrane directly; when it binds to a plasma protein, and the complex binds to a plasma membrane; or when it alters a plasma membrane in such a way that cellular antigens trigger autoantibody production. All such reactions involve IgG or IgM and complement. Their targets—erythrocytes, leukocytes, or platelets—are destroyed by complement-dependent cell lysis. Many antibiotics, sulfonamides, quinidine, and methyldopa elicit Type II reactions.

Type III hypersensitivity appears as serum sickness and can be caused by any drug that participates in the formation of immune complexes. Symptoms appear several days after administration, when sufficient quantities of immune complexes have accumulated to activate the complement system. A few patients sensitized to penicillin develop serum sickness.

Type IV hypersensitivity usually occurs as contact dermatitis after topical application of drugs. Antibiotics, antihistamines, local anesthetics, and additives such as lanolin are frequent agents of Type IV reactions. Medical personnel who handle drugs sometimes develop Type IV hypersensitivities.

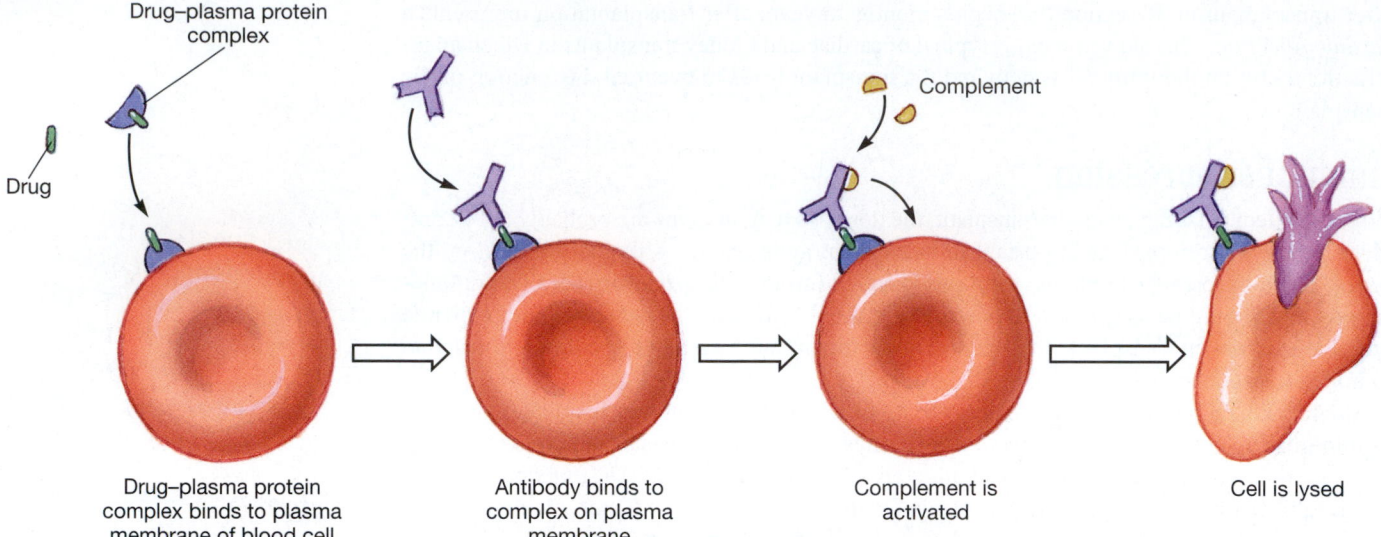

Drug–plasma protein complex

Drug

Complement

Drug–plasma protein complex binds to plasma membrane of blood cell

Antibody binds to complex on plasma membrane

Complement is activated

Cell is lysed

➤ Figure 18.19   **Drug reactions based on Type II hypersensitivity.** A drug (or one of the products of its metabolism in the body) may bind to the plasma membrane of a blood cell, bind to a blood (plasma) protein to form a complex that binds to a plasma membrane, or alter a plasma membrane protein. Autoantibodies— IgG or IgM—are produced and then bind to the complex and activate complement to lyse the cell.

Immunodeficiency Diseases

Immunodeficiency diseases arise from an absence of active lymphocytes or phagocytes, the presence of defective lymphocytes or phagocytes, or the destruction of lymphocytes. Such diseases invariably lead to impaired and inadequate immunity. **Primary immunodeficiency diseases** are caused by genetic defects in embryological development, such as failure of the thymus gland or Peyer's patches to develop normally. The result is a lack of T cells or B cells, or defective T and B cells. **Secondary immunodeficiency diseases** can be caused by (1) infectious agents, such as those responsible for leprosy, tuberculosis, measles, and AIDS; (2) malignancies, such as Hodgkin's disease or multiple myeloma; or (3) immunosuppressants, some chemotherapeutic drugs, certain antibiotics, and radiation. Such agents damage T cells or B cells after the cells have developed normally (➤ Figure 18.20).

Primary Immunodeficiency Diseases

Agammaglobulinemia, the first immunodeficiency disease to be understood, is a B cell deficiency. It occurs primarily in male infants in which B cells, and therefore antibodies, are absent. After maternal antibodies are lost by about 9 months of age, affected infants develop severe infections because they cannot produce IgM, IgA, IgD, and IgE antibodies and produce only small amounts of IgG. Agammaglobulinemia is treated with massive doses of immune serum (gamma) globulin to replace missing antibodies and with antibiotics to prevent infections.

DiGeorge syndrome results from a deficiency of T cells, probably caused by an agent that interferes with embryological development of the thymus gland. Cell-mediated immunity is impaired, so viral diseases pose a greater-than-usual threat. Although B cells are normal, their activation requires T_H cell activation. ➡ (Chapter 17, p. 471) Therefore, humoral immunity also is affected, because there are no functional T_H2 cells. Mice lacking a thymus, known as "nude" mice (➤ Figure 18.21), are bred and raised in germ-free environments for research purposes. They are used in the study of DiGeorge syndrome as well as other areas of immunology and genetics.

Severe combined immunodeficiency (SCID) is particularly debilitating because both B and T cells are absent. SCID can have several genetic origins. For example, stem cells in the bone marrow that normally give rise to lymphocytes fail to develop properly because of a defective

➤ Figure 18.20   **Kinds of immunodeficiencies.**

gene for IL-2, for the enzyme adenosine deaminase (ADA), or for MHC molecules. An infant who inherits this condition is doomed to die within the first few years of life unless he or she is kept in a germ-free environment until satisfactory treatment can be devised. (See the box "The Boy in the Bubble.")

Bone marrow transplants can be effective in SCID patients if a compatible donor (usually a brother or sister) is found. If the transplant is not compatible, the transplanted lymphocytes respond immunologically to antigens in recipient's tissues. This is another example of GVH disease, and it can be lethal.

Gene therapy, which attempts to replace a defective gene with a functional, therapeutic copy of the gene, has been used to treat SCID—with spectacular results. A few children have had

➤ Figure 18.21   **A nude mouse.** These animals lack a thymus gland as well as their fur. They are delivered by cesarean section by sterile technique and must be kept in a germ-free environment all their lives, because they completely lack T cells. They are the equivalent of human cases of DiGeorge syndrome. Researchers use them in many types of studies of the immune system.

The Boy in the Bubble

David, a child with SCID, had to be isolated from all sources of infectious agents because he lacked both B cells and T cells. He lived in a series of specially designed germ-free "bubbles" for most of his life. One such bubble was a self-contained, sterile space suit. At age 12 he received a bone marrow transplant intended to provide him with immune functions. He was watched carefully for signs of GVH disease after the transplant. In about 5 months he developed symptoms similar to those of GVH disease and soon died. On autopsy it was discovered that he died not of GVH disease but of a malignancy caused by the Epstein–Barr virus, which had contaminated the marrow transplant. Researchers concluded that the lack of immune surveillance of tumor cells proved to be the immunodeficiency most responsible for his death.

David, a boy born without an immune system, at age 6 in his self-contained, sterile suit, a mobile isolation system designed for him by NASA. Other equipment included a pushcart with a battery-powered motor and a seat.

their bone marrow cells removed and "infected" with a reproductively deficient retrovirus carrying the missing gene for ADA (which is essential to cell maturation). The cells were then returned to their bodies. Although the patients who received the infected bone marrow transplant must periodically have additional transplants, all are living normal lives!

Secondary (or Acquired) Immunodeficiency Diseases

Immunodeficiency diseases are not always inherited; sometimes they are *acquired* as a result of infections, malignancies, autoimmune diseases, or other conditions. For example, congenital rubella infections can decrease T cell function and antibody production to the extent that infants fail to respond to vaccines. Once patients develop immunodeficiencies, they may suffer from chronic or frequent recurrent infections.

Among malignant diseases that produce immunodeficiencies, those of lymphoid tissues suppress T cell function, and those of bone marrow suppress both T cell function and antibody production. Autoimmune diseases, some kidney disorders, severe burns, malnutrition or starvation, and anesthesia also can cause temporary or permanent immunodeficiencies.

Acquired Immune Deficiency Syndrome (AIDS)

Certainly the most well-known secondary immunodeficiency is **acquired immune deficiency syndrome (AIDS)**, an infectious disease caused by the **human immunodeficiency virus (HIV)**. AIDS can be caused by at least two different types of human immunodeficiency viruses, designated HIV-1 and HIV-2. Most cases of AIDS in the United States, Canada, and Europe are caused by HIV-1. HIV-2, which is most common in certain parts of West Afria, may be less virulent. Both forms are tested for in screening the U.S. blood supply.

Recent studies based on DNA sequencing shows that HIV-2 is very closely related to the simian immunodeficiency virus (SIV) found in African Sooty Mangabey monkeys—so closely that HIV-2 may be the same virus as SIV. However HIV-1 differs significantly enough that it may have separated from the HIV-2/SIV evolutionary tree earlier. Concensus is that HIV-1 has evolved sometime in the last 100 years. The oldest known case of AIDS in Europe was seen in a Danish surgeon who had worked in Zaire. She died in 1976. Cases occurring in Africa earlier this century would probably have been overlooked given the lesser numbers of cases and the state of health care existing then.

Early evidence for the origin of HIV comes from studies of human blood stored in England and Zaire since 1959, in which HIV-1 antibodies were found. The virus may have existed in relatively isolated regions, perhaps in Central Africa, for decades. Migration of rural people to rapidly growing cities, where population density was much higher and sexual contact was more

Cells without the CD4 marker were formerly thought to be immune to the AIDS virus. Now it seems that cells with a CD8 marker begin to make CD4 when stimulated.

casual and more frequent, could have brought about a great increase in the number of infected individuals. The expansion of international travel in recent years could rapidly spread the virus to many other parts of the world. The virus probably made multiple entries into the United States before becoming established.

The virus gradually but relentlessly destroys the immune system. The lack of a functional immune system leaves the body open to a variety of malignancies and opportunistic infections, most of which are rarely seen among people who are not suffering from advanced HIV disease or AIDS. These complications—either alone or in combination—eventually prove fatal (Table 18.7).

HIV specifically targets and damages T_H cells, macrophages, dendritic cells, and Langerhans cells. All these immune cells bear on their surface an antigen called the CD4 marker, to which HIV binds. Infection by HIV is indicated by the presence of anti-HIV antibodies. The infectious cycle for the virus was described in Chapter 10. ➦ (p. 274)

After a person is infected with HIV, an enormous battle ensues between HIV and the immune system. As immune cells destroy virus particles, more virus particles are produced. As virus-infected cells are destroyed by T_C cells, more immune cells replace those killed. Each day as HIV disease progresses, an estimated 1 billion virus particles are produced and destroyed as 2 billion immune cells are replaced! Although these early battles result in a draw, the virus eventually wins the war. As the years pass, it becomes more difficult to replace immune cells such as T_H. In addition, because HIV can mutate quickly, producing virus particles that progressively become more variable in structure, the immune system simply cannot keep up the fight.

Without activated T_H cells and macrophages, the immune system cannot "see" infectious microbes. Because T_H cells are greatly diminished in number, B cells are not stimulated to form plasma cells, which produce antibodies to combat infections. (The anti-HIV antibodies detected early in the course of infection are made before T_H cell populations become too depleted to stimulate B cells.) Similarly, cytokines are produced in amounts insufficient to activate macrophages and T_C cells. A drop in T_H cell count can be used to predict the onset of disease symptoms. A normal T_H count is 800 to 1200 per μl of blood. If the count remains above 400, 8 percent of those infected will develop AIDS within 18 months. With a T_H count of 200, 33 percent progress to AIDS; if the count is below 100, 58 percent will develop AIDS within 18 months.

➤ Table 18.7 Infections frequently found in AIDS patients

Pathogen	Disease
Bacteria	
Mycobacterium tuberculosis	Tuberculosis
Mycobacterium avium-intracellulare	Disseminated tuberculosis
Legionella pneumophila	Pneumonia
Salmonella species	Gastrointestinal disease
Viruses	
Herpes simplex	Skin and mucous membrane lesions, pneumonia
Cytomegalovirus	Encephalitis, pneumonia, gastroenteritis, fevers
Epstein–Barr	Oral hairy leukoplakia, possibly lymphoma
Varicella-zoster	Chickenpox, shingles
Fungi	
Pneumocystis carinii	*Pneumocystis carinii* pneumonia
Candida albicans	Mucous membrane and esophagus infections (thrush)
Cryptococcus neoformans	Meningitis, kidney disease
Histoplasma capsulatum	Pneumonia, disseminated infections, fevers
Other opportunistic fungi	Varies with opportunist
Protozoa	
Toxoplasma gondii	Encephalitis
Cryptosporidium species	Severe diarrhea

Progression of HIV Disease and AIDS

The sequence of events in HIV disease has now been established in some detail. The progression depends heavily on how much of the virus a person is exposed to (the *viral burden*) and how often the exposure is repeated. The CDC classification is based on the absence or presence of certain signs and symptoms and includes laboratory test results. Thus, persons categorized as being in groups 1 to 3 have HIV disease. Persons in group 4 have been diagnosed as having AIDS (➤ Figure 18.22). Although diseases such as hairy leukoplakia (a white lesion appearing on the tongue) are diagnostic of most AIDS patients, other opportunistic diseases or cancers vary among AIDS patients. For example, latent viral infections such as those caused by herpes simplex and cytomegalovirus, normally kept in check by the immune system, flare up and create a variety of disease symptoms. Severe diarrhea can be caused by opportunistic pathogens, including several species of *Cryptosporidium* (Chapter 22); encephalitis can be caused by *Toxoplasma gondii* (Chapter 24); and yeast infections can be caused by *Candida albicans* (Chapter 19). Pneumonia produced by the fungus *Pneumocystis carinii* is common to 80 percent of AIDS patients prior to their death. If patients do not first die of another opportunistic infection, about 50 percent will develop respiratory disease—caused either by *Mycobacterium tuberculosis* or by *Mycobacterium avium-intracellulare*. Overall, 88 percent of AIDS deaths result from an opportunistic infection.

Most AIDS patients develop malignancies not commonly found in the general population. A malignancy called **Kaposi's sarcoma** causes blood vessels to grow into tangled masses that are filled with blood and easily ruptured. In the skin and viscera, this sarcoma shows up as prominent pink or purplish spots (➤ Figure 18.23). It can spread to the digestive tract, lungs,

➤ Figure 18.22   **CDC classification of HIV disease and AIDS.** The progression of an HIV infection to AIDS is shown along with diseases and dysfunctions that can occur.

➤ **Figure 18.23 Kaposi's sarcoma.** This tumor of blood vessels is seen in AIDS patients as dark purple areas.

liver, spleen, and lymph nodes. However, there have been no reported cases of deaths among AIDS patients from Kaposi's sarcoma.

About 30 percent of individuals newly infected with HIV will progress to group 4 within 5 years. Within 15 years, 90 percent of HIV-infected patients will develop AIDS. Some of the drugs (especially protease inhibitors in combinations with other drugs—often called an AIDS cocktail) currently being used or tested against AIDS appear capable of prolonging the life of AIDS patients. However, they do not cure the disease. Furthermore, if discontinued because of side effects or lack of funds, death usually follows swiftly—sometimes within a month!

Epidemiology of AIDS

AIDS has been called the epidemic of the century; certainly, few diseases have had such a dramatic impact. Since 1981 an estimated 1.5 million Americans have been infected with HIV. (The actual number is likely to be higher.) Amazingly, only about 25 percent of the 1.5 million individuals know they are infected! Between 1981 and the end of 1994, about 442,000 Americans had AIDS and 260,000 died from it. By mid-1997 an estimated 600,000 Americans will have AIDS and more than 350,000 will have died from it. In 1994 alone, some 78,000 new AIDS cases were reported, with an estimated 37,000 deaths. AIDS has become the seventh leading cause of death among 1- to 4-year-olds, sixth among 15- to 24-year-olds, and *first* among 25- to 44-year-olds.

The AIDS pandemic is a global phenomenon. WHO officials estimate that by the end of 1998, the worldwide number of people infected with HIV will exceed 35 million (➤ Figure 18.24). But in many developing nations, AIDS cases also are probably undiagnosed or unreported. The region most seriously affected is Sub-Saharan Africa, where over 23.7 million people are infected. The virus is spreading fastest in South and Southeast Asia (India, Myanmar, Thailand), where 6.5 million people are estimated to be infected. Another 2.3 million people in Latin America and the Caribbean are believed to be infected.

Who Gets AIDS, and How

All available evidence suggests that it is virtually impossible to become infected with HIV through casual contact. Rather, a person becomes infected with the AIDS virus only through intimate contact with the body fluids of an infected individual and by transmission from infected mother to fetus. The virus is most commonly transmitted through blood, semen, and vaginal secretions. It seems likely that, however it is transmitted, the virus must make contact with a break or abrasion in the skin or with mucous membranes to cause infection. For this reason, all practices that lead to an exchange of body fluids carry a risk of HIV infection. These include:

1. *Sexual contact with an infected individual.* All forms of sexual intercourse—heterosexual and homosexual, active and passive, vaginal, anal, and oral—carry the risk of HIV infection. Condoms can reduce but not eliminate transmission because they have a significant failure rate (17 to 54 percent in various studies), usually the result of improper

At least a quarter of the 2.5 million units of blood administered in Africa are not screened for the AIDS virus.

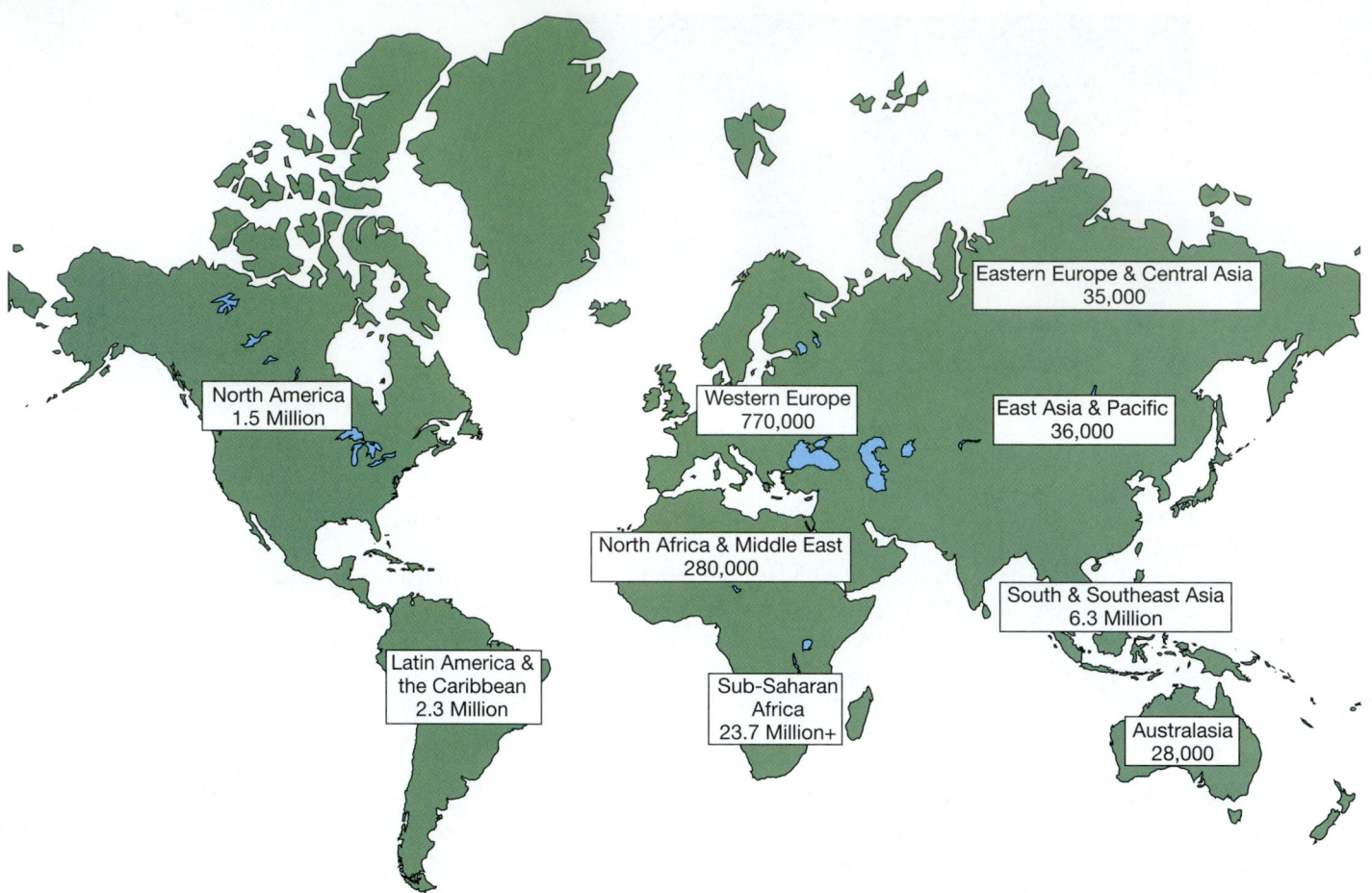

North America
1.5 Million

Eastern Europe & Central Asia
35,000

Western Europe
770,000

East Asia & Pacific
36,000

North Africa & Middle East
280,000

South & Southeast Asia
6.3 Million

Latin America &
the Caribbean
2.3 Million

Sub-Saharan
Africa
23.7 Million+

Australasia
28,000

➤ Figure 18.24   **Estimated number of global HIV infections projected by the World Health Organization through 1998.** Of the 35 million HIV-infected persons, 26 million are living in some state of HIV/AIDS illness. Over 1 million of these are children of 209 countries reporting AIDS cases to the WHO and UNAIDS; 194 have reported AIDS cases. (*Source: Pan American Health Organization Quarterly Report,* September 1996 updated)

use, and not all types of condoms are equally effective in blocking HIV. Natural skin condoms allow passage of the virus; latex condoms are much safer.

2. *The sharing of unsterilized needles by intravenous drug users.*

3. *Receipt of a blood transfusion or blood products contaminated with HIV.* HIV infection by blood transfusion caused many AIDS cases in the early 1980s. Many of the infected persons were hemophiliacs, who received injections of blood products to let their own blood clot properly. Transfusions are much less of a viral threat today, thanks to the testing of donated blood for HIV antibodies and to recombinant DNA technology. And, popular fears to the contrary, it is not possible to acquire the HIV virus by donating blood because new, sterile needles are used.

4. *Passage from an infected mother to an infant.* About 25 percent of infants carried by HIV-positive women become HIV infected. HIV transmission may be possible while the fetus is in the uterus, during delivery, and through breast feeding. Preliminary studies indicate that more newborns are infected during delivery than before birth.

Health care personnel treating AIDS patients or HIV-infected patients are at risk of becoming infected. The CDC has recommended the following precautions to minimize that risk (see also the Universal Precautions in Chapter 15): ➡ (p. 427)

1. *Wear gloves, masks, protective eyewear, and gowns* for touching blood, body fluids, mucous membranes, or skin lesions of patients and for procedures that might release droplets

Treatment with AZT during the last few weeks of pregnancy reduces mother-to-fetus HIV transmission by half.

of body fluids. Discard these items, and wash hands immediately and thoroughly after seeing each patient. Like other medical personnel, dentists and their technicians should consider blood, saliva, and gingival fluids of all patients to be potentially infective and use these procedures to prevent contact with such fluids.

2. *Avoid injury from needles* and other sharp objects, and discard them in puncture-proof containers.

3. *Use mouthpieces, resuscitation bags, or other ventilator devices* for emergency resuscitation.

4. Workers with skin lesions should *avoid direct patient care and handling of contaminated equipment*.

Keep in mind that almost any infectious disease (such as tuberculosis) that an HIV disease patient or AIDS patient may have contracted poses a danger to health care personnel. Of course, this problem is not limited to HIV infection; other types of infections can be a threat to health care workers.

What about an AIDS Vaccine?

The outlook for an AIDS vaccine is not promising. Scientists expect a period of trial and error lasting well beyond the year 2000, because research with HIV presents unusual problems. For one thing, HIV has a high mutation rate due to the imprecise operation of its reverse transcriptase. Thus, even if a successful vaccine against one strain of the virus is produced, another strain might not be affected by the vaccine, and new strains would likely develop. Outbreaks of influenza every few years are caused by similarly high mutation rates in the influenza virus.

Developing a vaccine poses more problems. Attenuated viruses cannot be used in a vaccine because they contain DNA that can be incorporated into the host's genome, possibly later giving rise to AIDS. Whole inactivated viruses, which have been used successfully in polio and influenza vaccines, are inappropriate for an AIDS vaccine. Such vaccines would, for HIV and other lentiviruses, predispose the host to severe infections. Recombinant viruses, such as HIV antigens on a vaccinia virus, are unacceptable because the host might develop disseminated vaccinia. Also, immunocompromised hosts may develop a variety of severe complications. Thus, many problems must be solved before we can have a safe and effective vaccine.

The Social Perspective: Economic, Legal, and Ethical Problems

AIDS and HIV disease will have an increasingly significant economic impact in the coming years. The lifetime cost of medical care for an AIDS patient in the United States can reach $100,000 or more. On the basis of CDC estimates, the total cost for care of all U.S. AIDS patients in 1994 was almost $9 billion. If the care of AIDS patients is a burden in the United States, where average income exceeds $12,000 per person per year, imagine the catastrophic burden in many developing countries, where average annual income is less than $200 per person.

U.S. laws protect the confidentiality of medical information, including AIDS test results. The laws also provide equal opportunity with respect to employment, housing, and education, but many AIDS patients continue to encounter various kinds of discrimination. Protecting the rights of uninfected citizens and those of health professionals who treat AIDS patients also must be considered. Other legal issues concern the responsibility of HIV-infected individuals not to transmit the disease and the liabilities of distributors of blood products.

Many ethical issues are related to the legal issues. A major question is how the epidemic can be curtailed without infringing on individual freedoms. Another question weighs the moral obligation of health professionals to care for all patients against the risk of acquiring a fatal disease. Still other questions relate to allocation of scarce medical resources.

✓ What is an autoimmune disease? What might cause one?

✓ Summarize the causes, signs, and symptoms of myasthenia gravis, rheumatoid arthritis, and systemic lupus erythematosus.

✓ How and why does transplant rejection occur?

✓ Distinguish between primary and secondary immunodeficiency diseases. Give examples of each.

Immunological Tests

In Chapter 17 we considered how certain immunological reactions—agglutination, cell lysis by complement and by IgM antibodies, and neutralization of viruses and toxins—kill pathogens. Now we will consider how those and other reactions are used as laboratory tests to detect and quantify antigens and antibodies. Such laboratory tests make up the branch of immunology called **serology,** so named because many of the tests are performed on serum samples. Today, some laboratory tests also use monoclonal antibodies derived from the culture fluid of animal tissue grown in culture. ➡ (Chapter 17, p. 473) The tests and reactions described here represent a broad but incomplete sampling of laboratory and clinical tests. The selection of the "right" test depends on the nature of the pathogen or disease being analyzed.

The Precipitin Test

Historically, one of the first serologic tests to be developed was the **precipitin test** (➤ Figure 18.25), which can be used to detect antibodies or antigens. This test is based on a **precipitation reaction** in which antibodies called *precipitins* react with antigens, diffuse toward each other, and form a visible precipitate. During such reactions antigen–antibody complexes form within seconds. Latticelike networks of these complexes, which are visually opaque, form minutes to hours later.

Many modifications have been made to the basic precipitin test to increase its sensitivity to detect specific antigen–antibody complexes. **Immunodiffusion tests** are based on the same principle as the precipitin test, but they are carried out in a thin layer of agar that has solidified on a glass slide. Immunodiffusion tests are used to determine if more than one antigen—and therefore more than one antibody—is present in a serum specimen. Small wells are made in the solidified agar, and the antigens and antibodies are placed in separate wells. Antigen–antibody complexes appear as detectable precipitation lines (after staining) in the agar between the wells. After diffusion occurs, one or more bands of precipitation can be detected, each representing a different antigen–antibody complex (➤ Figure 18.26). The bands can be made more visible by washing the agar surface and applying a stain that colors the antigen–antibody complexes. An advantage of immunodiffusion tests is that in a single test medium several antigens can be reacted with one kind of antibody or several kinds of antibodies with one antigen.

When serum samples contain several antigens, **immunoelectrophoresis** can be used to detect separate antigen–antibody complexes. The antigens are placed in a well on an agar-coated slide. An electric current then is passed through the gel. This process is called **electrophoresis.** During electrophoresis different antigen molecules migrate at different rates, depending on the size and electric charges of the molecules. Following electrophoresis, the antibody is placed in a trough made along one or both sides of the slide, and diffusion is allowed to occur. The results of immunoelectrophoresis are similar to those obtained in other immunodiffusion tests—precipitin bands form wherever matching antigen and antibody precipitate (➤ Figure 18.27).

➤ **Figure 18.25** **The precipitin test for antibodies.** When IgG or IgM antibodies (which are soluble) react with soluble antigens, they quickly form small complexes, which over time combine into larger lattices that precipitate from the solution. Such precipitation occurs, however, only when there is an appropriate ratio of antigen to antibody. In the test, antibodies are placed in the bottom of a narrow tube. Soluble antigen is added, and the two are allowed to diffuse toward each other. Where the necessary concentration ratio is achieved (the zone of equivalence), precipitation takes place, visible as a hazy "precipitin ring" in the tube.

Soluble antigen

Zone of excess antigen

Zone of equivalence

Precipitin ring forms

Zone of excess antibody

Antibody

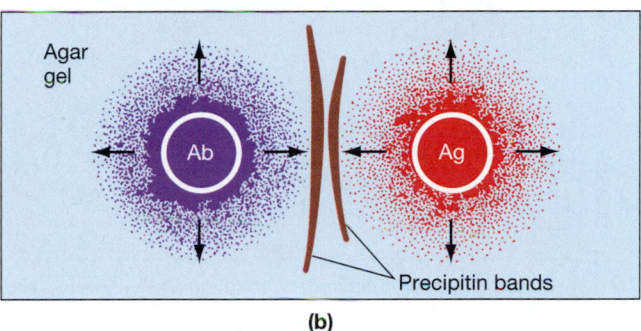

> Figure 18.26 **Immuno-diffusion test.** This test is a modification of the precipitin reaction. Wells are made in an agar gel and filled with test solutions of antigen (Ag) and antibody (Ab). (a) A single antibody and antigen diffuse outward from the wells, meet, react with each other, and precipitate. They form a line called a precipitin band, which is visualized by staining. (b) Two different antigen–antibody complexes diffuse at different rates, producing separate bands.

Thus, the advantage of immunoelectrophoresis is the ability to separate several antigens that might be present in a serum sample.

Radial immunodiffusion provides a quantitative measure of antigen or antibody concentrations. In this test, antibody is added to molten agar, and the solution is allowed to solidify as a thin layer on a glass slide. Antigen samples of different concentrations are placed in wells made in the agar slide. After diffusion the antigen concentration is determined by measuring the diameter of the ring of precipitation around the antigen (➤ Figure 18.28). Similarly, antibody concentrations can be determined by placing antibody samples of different concentrations in wells in a gel containing the antigen.

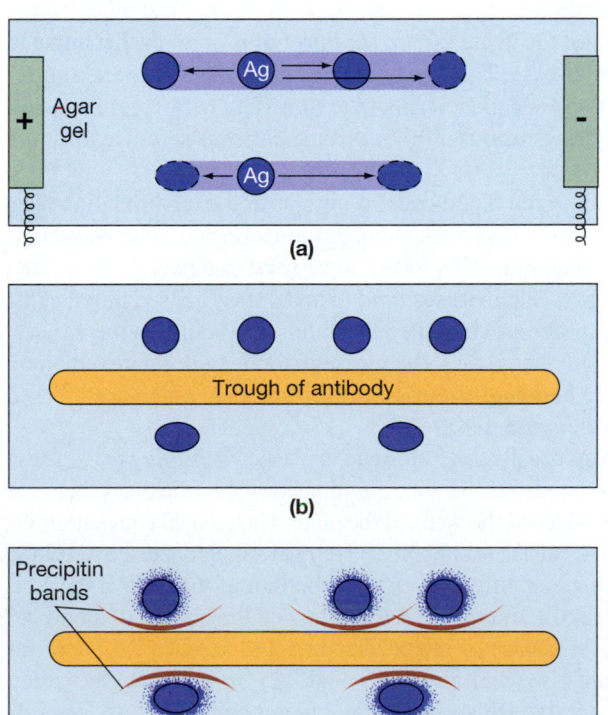

> Figure 18.27 **Immunoelec-trophoresis.** (a) Antigens placed in an agar gel are separated by means of an electric current. Positively charged molecules are drawn toward the negative pole; negatively charged ones move toward the positive pole. (b) A trough is cut into the agar between the wells and filled with antibody. (c) Curved precipitin bands form where antigens and antibodies diffuse to meet and react.

➤ Figure 18.28 **Radial immunodiffusion.** (a) Wells cut into antibody-containing agar sheets are filled with antigen. The antigen diffuses outward, complexing with antibody as it goes. When the ratio of antigen to antibody is optimal, the complexes precipitate in a ring. (b) The diameter of the ring is proportional to the logarithm of the antigen concentration, which can be determined by reference to a standard curve. The concentration of an antibody can also be determined by placing it in a well cut into an antigen-containing agar sheet and comparing the size of the resulting ring with a standard curve for that antibody.

Agglutination Reactions

When antibodies react with antigens on cells, they can cause **agglutination,** or clumping together of the cells. One application of **agglutination reactions** is to determine whether the quantity of antibodies against a particular infectious agent in a patient's blood is increasing. The quantity of antibodies is called the **antibody titer.** It is reported as the reciprocal of the greatest serum dilution in which agglutination occurs. For example, an antiserum that agglutinated an agent's antigen at a dilution of 1:256 but not at 1:512 would be reported as an antibody titer of 256. An increase in the antibody titer over time indicates that the patient's immune system is attacking the agent. Diagnosis of the disease agent is possible when it can be shown that the patient's serum had no antibodies against the agent before the onset of disease or that a rise in titer occurred during the course of the disease. This production of antibodies in the serum resulting from infection (or immunization) is called **seroconversion.** The **tube agglutination test** measures antibody titers by comparing various dilutions of the patient's serum against the same known quantity of the antigen (cells).

Hemagglutination, or agglutination of red blood cells, is similar to the agglutination tests except the antigens are on the surface of red blood cells. Hemagglutination is used in blood typing (➤ Figure 18.29). In addition, hemagglutination tests can be used to detect viruses, such as those that cause measles and influenza. Such viruses bind to and cross-link red blood cells, causing **viral hemagglutination**. This process is inhibited by adding antibodies to the viruses. Because these antiviral antibodies bind to the viruses, the viruses cannot agglutinate red blood cells. Such inhibition is the basis of the **hemagglutination inhibition test,** which can be used to diagnose measles, influenza, and other viral diseases.

Today, these types of agglutination tests usually are performed in plastic *microtiter plates*. These plates contain 96 separate wells, so many tests can be done simultaneously. In the plate shown in ➤ Figure 18.30, antibody dilutions are added to the wells of the plate. Then, equal concentrations of red blood cells are added to each well. If sufficient antibody is present to agglutinate the red blood cells, the antibody–cell complexes sink into a diffuse layer at the bottom of the well. If the antibody titer is too low, the red blood cells settle and form a red "button" at the bottom of the well.

Earlier we saw that severe hemolytic disease of the newborn results from an Rh factor incompatibility between mother (Rh negative) and fetus (Rh positive). In a hemagglutination test, although anti-Rh antibodies will bind to Rh antigen on red blood cells, there are not sufficient Rh antigens to cause clumping with anti-Rh antibodies (➤ Figure 18.31a). The **Coomb's antiglobulin test** is designed to detect such antibodies. If red blood cells coated with anti-Rh

Pink Means You're Pregnant

Did you know that some home pregnancy tests are modified agglutination assays? The tests are based on agglutination inhibition, a highly sensitive assay that can detect small quantities of an antigen. Latex particles are coated with human chorionic gonadotropin (HCG) and antibody to HCG. When a pregnant woman's HCG-containing urine comes in contact with these latex particles, the latex particles can't agglutinate, and so the absence of agglutination indicates pregnancy. And even if you never have to take a home pregnancy test, you still might be subject to an assay involving agglutination inhibition. Many jobs require prospective employees to take drug tests, and agglutination inhibition assays can be used to determine if an individual is using certain types of illegal drugs such as cocaine or heroin.

➤ Figure 18.29 **Hemagglutination.** This test, used for matching blood types, is based on the agglutination reaction. In A, red blood cells were mixed with serum having antibodies to the antigens on the cells. The complex of cells and antibodies clumped together. In B, the serum added did not contain antibodies that recognize the blood type antigens on the cells, so no clumping occurred.

Well

Antibody

Agglutinated
red blood cells
form a diffuse layer.

Nonagglutinated
red blood cells
("red button")

➤ Figure 18.30 **Microtiter plates.** Hemagglutination tests often are carried out in microtiter plates. These plates contain 96 plastic wells, so many tests can be done simultaneously. When hemagglutination occurs, antibody reacts with antigen. All wells contain red blood cells. Positive hemagglutination results in a diffuse clumping of red blood cells (as in row F, column 1). Negative hemagglutination is seen as a clump of red blood cells ("red button") at the bottom of the wells (as in rows G and H).

(a)

Red blood
cells

Anti-Rh
antibodies

Cell-antibody
complexes

Rabbit
anti-human
immunoglobulins

Agglutinated
red blood cells

(b)

➤ Figure 18.31 **The Coomb's antiglobulin test.** (a) Anti-Rh antibodies are allowed to react with red blood cells. If Rh antigens are present on the blood cells, there are not enough of them to produce a hemagglutination reaction. (b) Therefore, anti-human antibodies prepared in rabbits are reacted with the red blood cell–antibody complexes. If Rh antigens are present on the red blood cells, hemagglutination will occur. A person with these red blood cells would be Rh-positive.

antibody are treated with an antibody that recognizes the anti-Rh antibody, the antibody–antibody–cell complexes will agglutinate (➤ Figure 18.31b). Thus, if a patient's serum contains Rh antibodies or if the red blood cells are Rh positive, agglutination will occur.

The body's natural defenses use complement to bind to antigen–antibody complexes, helping to destroy pathogens. This same ability is used in the laboratory or clinic to detect very small quantities of antibodies. The **complement fixation test** is a multistep procedure that begins with the inactivation of complement from a patient's serum by heating. The serum is then diluted, and known quantities of nonhuman complement and the test antigen are added separately (➤ Figure 18.32a). The antigen is specific to the antibody being sought. This mixture is incubated to allow the antigen to react with any antibody present. Next, an indicator system typically consisting of sheep red blood cells and antibody against those cells are added. If the antibody to the test antigen was present in the patient's serum, the antigen–antibody reaction

➤ **Figure 18.32   The complement fixation test for antibodies.** (a) In the first step, the serum to be tested is diluted, and antigen to the antibody being sought is added. If the antibody is present, it reacts with the antigen and forms immune complexes. In the second step, free complement is added. If immune complexes have formed, the complement will interact with them and be fixed; if no immune complexes have formed, the complement will remain free. In the third step, sheep red blood cells with bound sheep red blood cell antibody molecules are added. If free complement is present, it will lyse the red blood cells. This is a negative test result—which indicates that there was no antibody in the original serum. If all the complement has already been fixed by the earlier immune complex, the red blood cells will not be lysed. This is a positive test result—it indicates that antibody was present in the original serum specimen. (b) In a positive test, complement has been fixed, and the red blood cells will not be lysed. Instead, they form a characteristic red button in the bottom of the well.

will have fixed (combined with) the complement. Hence the blood cells will not be lysed, and the test will be positive, forming a red button of undamaged cells (➤ Figure 18.32b). But if the antibody was not present, free complement remaining in the mixture will be fixed by the indicator system, resulting in the lysis of the cells. The test will be negative. The complement fixation test is used to diagnose such bacterial diseases as pertussis and gonorrhea and fungal diseases, including histoplasmosis. The Wassermann test for syphilis uses complement fixation.

Neutralization reactions can be used to detect bacterial toxins and antibodies to viruses. Immunity to diphtheria, which depends on the presence of diphtheria antitoxins (antibodies to diphtheria toxin) can be detected by the **Schick test.** In this test a person is inoculated with a small quantity of diphtheria toxin. If the person is immune to the disease, diphtheria antitoxins (circulating in the blood) will neutralize the toxin, and no adverse reaction will occur. If the person is not immune and the antitoxin is not present, the toxin will cause tissue damage, detected as a swollen reddened area at the injection site after 48 hours.

Viral neutralization occurs when antibodies bind to viruses and neutralize them, or prevent them from infecting cells. In the laboratory or clinic, a patient's serum and a test virus are added to a cell culture or a chick embryo. If the serum contains antibodies to the virus, these antibodies will neutralize the virus and prevent the cells of the culture or the embryo from becoming infected.

Tagged Antibody Tests

The most sensitive immunological tests used to detect antibodies or antigens use antibodies that have a "molecular tag" that is easy to detect even at very low concentrations. In fact, the concentrations are so low that precipitation or agglutination does not occur.

Immunofluorescence makes use of antibodies (usually IgG) to which fluorescent dye molecules are bound (tagged) at the tail (Fc) ends of the antibodies. For example, IgG antibodies tagged with fluorescein isothiocyanate glow a bright yellow green when exposed to ultraviolet (UV) light. Fluorescent tagged antibodies can be used to detect antigens, other antibodies, or complement at their locations on cells or within tissues (➤ Figure 18.33). Because such cells or tissue samples can be examined with a fluorescence microscope, this technique is particularly useful in the research laboratory to locate cellular antigens and autoantibodies. A fluorescent tagged antibody that detects another antibody is known as an *anti-antibody;* one that detects complement is an *anti-complement antibody*. Immunofluorescence is helpful in diagnosing syphilis, gonorrhea, HIV infection, and Legionnaires' disease, to name a few.

Radioimmunoassay (RIA) also can be used to detect very small quantities (nanograms) of antigens and antibodies. To measure an antibody in a test specimen by RIA, a known antigen is placed in a saline (salt) solution and incubated in plastic well plates (➤ Figure 18.34a). Some antigen molecules attach to the plastic; those that do not attach are washed away. The antibody being measured is added and allowed to bind with the antigen (➤ Figure 18.34b). Then a radioactively tagged anti-antibody is applied (➤ Figure 18.34c). After an incubation period, the excess unbound antibody is removed by washing. Radioactive material remaining in the well is measured with a radiation counter to determine the concentration of antibody in the test specimen.

Enzyme-linked immunosorbent assay (ELISA) is a modification of RIA in which the anti-antibody, instead of being radioactive, has an enzyme tag attached to it (➤ Figure 18.35a). After the antibody being measured has reacted with the antigen, the anti-antibody–enzyme complex is added (➤ Figure 18.35b). Finally, a substrate that the enzyme converts to a colored product is applied (➤ Figure 18.35c). The amount of colored product is proportional to the concentration of the antibody. RIA and ELISA are among the most widely used tests for antibodies or for antigens.

An important application of ELISA is the detection of HIV antibodies. The test was developed to screen the nation's blood supply and protect recipients of blood products from infection. ELISA is a sensitive test—so sensitive that the American Red Cross Blood Services reports that false-positive results (color product detected when HIV antibodies are not present) occur at a rate of about 0.2 percent.

Generally, serum antibodies to HIV can be detected by indirect ELISA within 6 weeks of infection.

➤ Figure 18.33   **Immunofluorescence.** Fluorescein is a fluorescent dye molecule that can be complexed with other molecules. When viewed with ultraviolet light (UV) under a fluorescence microscope, it will fluoresce, revealing the presence of the "tagged" molecule. To detect directly the presence of a specific antigen in a tissue, (a) a solution of fluorescein-tagged antibody to that antigen is prepared, added to cells or to a thin section of tissue, incubated, and then washed. Any dye-tagged antibody that has complexed with antigen in tissue will fluoresce when viewed by fluorescence microscopy. (b) In indirect testing the antibody to the antigen being sought is not itself tagged. Instead, its presence is detected by means of a fluorescein-tagged antibody (anti-antibody) to the original antibody. (c) Complement (protein C3) can be added to the tissue section along with the antibody, and a fluorescein-tagged antibody to one of the complement proteins can then be used to detect the presence of antigen–antibody complexes or complement attached to cells. (d) Immunofluorescent staining of influenza virus–infected lung cells. The cells were treated with fluorescent dye–tagged antibodies against the specific virus. The nucleus of infected cells is bright yellow green, indicating the location of the viral antigen.

➤ Figure 18.34   **Radioimmunoassay (RIA) is used to detect very small quantities of antibody.** (a) Antigen first is bound to a well of a plastic plate. After excess unbound antigen is washed out, the solution being tested for antibody is added and allowed to react. (b) If antibody is present, it reacts with the antigen. (c) After any unbound antibody is washed out, a radioactively labeled second antibody (anti-antibody) specific to the first antibody is added. The amount of bound radioactive anti-antibody present is measured; it is proportional to the concentration of the antibody in the original solution.

➤ **Figure 18.35** **Enzyme-linked immunosorbent assay (ELISA) is a modification of RIA.** (a) As in RIA, antigen bound to a well of a plastic plate reacts with the antibody being detected. (b) In one form of the ELISA, an anti-antibody is then added. However, rather than being radioactively labeled as in RIA, this antibody has a covalently attached enzyme. (c) A substrate specific to the enzyme is then added. If this enzyme has bound to the original antibody, it can catalyze a reaction, converting the colorless substrate into a color product. Such ELISA tests are done routinely as an initial test to detect HIV in blood samples.

To confirm the presence of an HIV infection, a more expensive test, called **Western blotting,** can be performed on samples from an individual who has a positive ELISA test. In Western blotting, HIV proteins are isolated from the individual and are first separated in a gel by an electric current, similar to the procedure used in immunoelectrophoresis. The separated proteins are then transferred ("blotted") to cellulose filter paper. Next, serum from the individual is added to the blot. If HIV antibodies are present, they will react with the separated HIV proteins. Such antigen–antibody complexes can be visualized by the addition of an enzyme-labeled anti-human antibody. When an enzyme substrate is added, colored bands appear on the paper (➤ Figure 18.36). Thus, Western blotting can determine the exact viral antigens to which the HIV antibodies are specific.

✓ How do precipitin reactions differ from agglutination reactions?

✓ What is meant by antibody titer?

✓ Compare RIA, ELISA, and Western blotting tests.

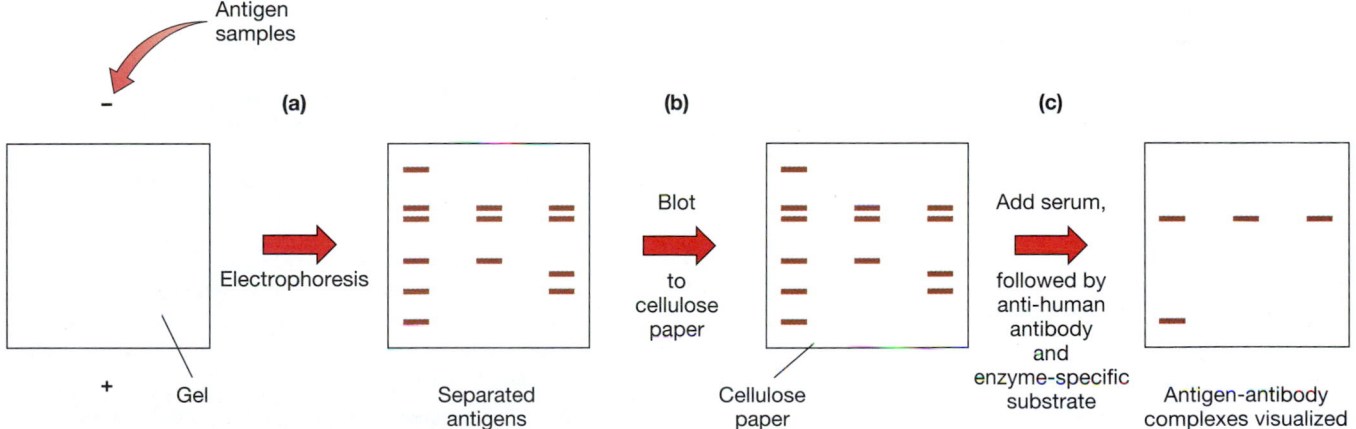

➤ **Figure 18.36**   **Western blotting test for HIV antigens in blood.** (a) HIV antigens in a gel are separated by an electric current, forming bands of separate antigens. (b) The antigen bands are transferred (blotted) to cellulose paper. (c) HIV antibodies tagged with dye are added. Any antibodies recognizing specific HIV antigens attach to that antigen and form a visible band.

Retracing Our Steps

Overview of Immunological Disorders

- An **immunological disorder** results from an exaggerated or an inadequate immune response.

Hypersensitivity

- **Hypersensitivity,** or **allergy,** is due to an exaggerated reaction to an antigen.

Immunodeficiency

- **Immunodeficiency** is due to an inadequate immune response. **Primary immunodeficiencies** are genetic or developmental defects resulting in lack of, or defective B or T cells, or both. **Secondary immunodeficiencies** result from damage to B or T cells after they have developed normally.

Immediate (Type I) Hypersensitivity

- **Anaphylaxis** is due to harmful effects caused by an immediate exaggerated immune response.

Allergens

- An **allergen** is an ordinarily innocuous substance that can trigger a harmful immunological response in an allergen-sensitized person; common allergens are listed in Table 18.1.

Mechanism of Immediate Hypersensitivity

- The mechanism of **immediate hypersensitivity** is summarized in ➤ Figure 18.1; mediators of immediate hypersensitivity are summarized in Table 18.2.
- In the **sensitization** process the affected individual produces IgE antibodies that attach to mast cells and basophils. A second or later exposure to the same allergen will trigger cellular **degranulation** with the release of **histamine** and other preformed mediators. The synthesis of reaction mediators also is responsible for symptoms of immediate hypersensitivity.

Localized Anaphylaxis

- **Atopy** is a localized reaction to an allergen in which histamine and other mediators elicit wheal and flare reactions and other signs and symptoms of allergy.

Generalized Anaphylaxis

- **Generalized anaphylaxis** is a systemic reaction in which the airways are constricted (**respiratory anaphylaxis**) or blood pressure is greatly decreased (**anaphylactic shock**).

Genetic Factors in Allergy

- Genetic factors are thought to contribute to the development of allergies.

Treatment of Allergies

- Allergy is treated by **desensitization** (hyposensitization), as shown in ➤ Figure 18.5; symptoms are alleviated with antihistamines.

Cytotoxic (Type II) Hypersensitivity

Mechanism of Cytotoxic Reactions

- The mechanism of **cytotoxic hypersensitivity** is summarized in ➤ Figure 18.6.

Examples of Cytotoxic Reactions

- **Transfusion reactions** result in red blood cell **hemolysis** due to the production of antibodies to antigens on transfused red blood cells.
- **Hemolytic disease of the newborn** results when anti-Rh antibodies from a sensitized mother react with Rh antigens in an Rh-positive fetus.

Immune Complex (Type III) Hypersensitivity

Mechanism of Immune Complex Disorders

- The mechanism of **immune complex hypersensitivity** is summarized in ➤ Figure 18.9.

Examples of Immune Complex Disorders

- **Serum sickness** occurs when foreign (sometimes animal) antigens in sera combine with antibody to form immune complexes, which are deposited in various tissues.
- The **Arthus reaction** is a local immune response to an antigenic substance (usually an injected substance) that causes edema and hemorrhage.

Cell-Mediated (Type IV) Hypersensitivity

Mechanism of Cell-Mediated Reactions

- The mechanism of **cell-mediated (delayed) hypersensitivity** is summarized in ➤ Figure 18.12.

Examples of Cell-Mediated Disorders

- **Contact dermatitis** occurs after second contact with poison ivy, metals, or another substance and usually appears as eczema.
- Other kinds of cell-mediated hypersensitivity include **tuberculin hypersensitivity** and **granulomatous hypersensitivity.**
- Characteristics of the four types of hypersensitivities are summarized in Table 18.5.

Autoimmune Disorders

- **Autoimmune disorders** arise from a hypersensitivity to self antigens on cells and in tissues. Such diseases often produce **autoantibodies** to these self antigens. Autoimmune diseases can also be cell-mediated.

Autoimmunization

- **Autoimmunization** occurs when the immune system responds to a body component as if it were foreign. Genetic factors and **antigenic mimicry** are among the mechanisms that can cause autoimmune disease.
- Tissue damage from autoimmune disorders can be caused by cytotoxic, immune complex, and cell-mediated hypersensitivity reactions.

Examples of Autoimmune Disorders

- Examples of autoimmune diseases include **myasthenia gravis, rheumatoid arthritis,** and **systemic lupus erythematosus.**

Transplantation

- **Transplantation** involves the transfer of **graft tissue** from one site to another on the same individual (**autograft**), between genetically identical individuals (**isograft**), from one individual to another non-identical individual (**allograft**), or between different species (**xenograft**).

Histocompatibility Antigens

- Genetically determined **histocompatibility antigens** are found on the surface membranes of all cells. Some are correlated with increased risk of certain diseases.
- **Human leukocyte antigens** (HLAs) in graft tissue are a main cause of **transplant rejection,** but immunocompetent cells in bone marrow or other types of grafts sometimes destroy host tissue, as in **graft-versus-host (GVH) disease.**

Transplant Rejection

- Transplant rejection is due to the presence of foreign HLAs. **Hyperacute rejection** is a Type II hypersensitivity, occurring in an already sensitized recipient. **Accelerated rejection** reactions are mainly cell-mediated (Type IV) hypersensitivities. **Chronic rejection** occurs over several months to years.

Immunosuppression

- **Immunosuppression** is a lowering of the responsiveness of the immune system to materials it recognizes as foreign. Immunosuppression is produced by **radiation** and by **cytotoxic drugs.** It minimizes transplant rejection but also can reduce the host's immune response to infectious agents.

Drug Reactions

- All four types of hypersensitivity reactions have been observed in immunological drug reactions.

Immunodeficiency Diseases

- **Immunodeficiency diseases** arise from an absence of formed lymphocytes and other components of the immune system (**primary immunodeficiency**) or from the destruction of already formed lymphocytes (**secondary immunodeficiency**).

Primary Immunodeficiency Diseases

- B cell deficiency, or **agammaglobulinemia,** leads to a lack of humoral immunity.
- T cell deficiency, such as **DiGeorge syndrome,** leads to a lack of cell-mediated and humoral immunity.
- Deficiencies of both B and T cells, or **severe combined immunodeficiency disease (SCID)**, leads to a lack of both humoral and cell-mediated immunity.

Secondary (or Acquired) Immunodeficiency Diseases

- Secondary immunodeficiencies are acquired through infections, malignancies, or autoimmune disorders.
- **Acquired immune deficiency syndrome (AIDS)** is caused by the **human immunodeficiency virus (HIV).** HIV destroys T_H cells and eventually impairs all immune functions.

- HIV-infected individuals progress through a series of stages that lead to AIDS. Group 4 individuals suffer from opportunistic infections and malignancies such as **Kaposi's sarcoma**.
- The AIDS pandemic will have over 35 million people infected with HIV by the end of 1998.
- HIV infection can be acquired by: sexual contact, needle sharing, blood transfusions or products, and transfer across the placenta from mother to fetus.
- Healthcare personnel must practice Universal Precautions at *all* times.
- Vaccine development remains unpromising.
- Social, economic, legal, and ethical problems are becoming increasingly significant.

Immunological Tests

- **Serology** is the use of laboratory tests to detect antigens and antibodies.

The Precipitin Test

- The **precipitin test** can be used to detect antibodies. Modifications of this test include **immunodiffusion, immunoelectrophoresis,** and **radial immunodiffusion.** All these tests rely on the formation of antigen–antibody complexes that precipitate from solutions or in agar gels.

Agglutination Reactions

- **Agglutination reactions** depend on the clumping of antigen and antibody combinations. Such tests can be used to determine one's **antibody titer** or if **seroconversion** has occurred.
- **Hemagglutination** includes the clumping of red blood cells by viruses and the binding of antibodies to specific antigen-bearing red blood cells. The **hemagglutination inhibition test** can be used to diagnose for measles or diseases caused by other viruses. **Coomb's antiglobulin test** is used to detect Rh antibodies.
- The **complement fixation test** indirectly detects antibodies in serum to antigens by determining whether complement combines (is fixed) with antigen-antibody complexes.
- **Neutralization reactions** can detect bacterial toxins and antibodies to certain viruses.

Tagged Antibody Tests

- **Immunofluorescence** allows detection of products of immune reactions on cells or within tissues.
- Assays for antigens and antibodies can be performed with radioactive antibodies (**radioimmunoassay**) or antibodies containing enzymes coupled to them (**enzyme-linked immunosorbent assay, ELISA**). **Western blotting** detects specific antibodies to specific antigens.

Terminology Check

cell-mediated (Type IV) hypersensitivity (p. 494)
complement fixation test   (p. 528)
contact dermatitis   (p. 506)
Coomb's antiglobulin test   (p. 526)
cytotoxic drug   (p. 515)
cytotoxic (Type II) hypersensitivity (p. 494)
degranulation   (p. 495)
delayed (Type IV) hypersensitivity (p. 505)
delayed hypersensitivity (T_{DH}) cell (p. 505)
desensitization   (p. 499)
DiGeorge syndrome   (p. 516)
electrophoresis   (p. 524)
enzyme-linked immunoabsorbent assay (ELISA)   (p. 529)
generalized anaphylaxis   (p. 494)
graft tissue   (p. 513)
graft-versus-host (GVH) disease (p. 513)
granulomatous hypersensitivity   (p. 508)
hemagglutination   (p. 526)
hemagglutination inhibition test (p. 526)
hemolytic disease of the newborn (p. 501)
histamine   (p. 495)
histocompatibility antigen   (p. 513)

human immunodeficiency virus (HIV) (p. 518)
human leukocyte antigen (HLA)   (p. 513)
hypersensitivity   (p. 493)
immediate (Type I) hypersensitivity (p. 494)
immune complex (Type III) hypersensitivity (p. 494)
immunodeficiency   (p. 494)
immunodeficiency disease   (p. 516)
immunodiffusion test   (p. 524)
immunoelectrophoresis   (p. 524)
immunofluorescence   (p. 529)
immunological disorder   (p. 493)
immunosuppression   (p. 515)
induration   (p. 508)
isograft   (p. 513)
Kaposi's sarcoma   (p. 520)
leukotriene   (p. 495)
localized anaphylaxis   (p. 494)
major histocompatibility complex (MHC) (p. 513)
myasthenia gravis   (p. 510)
neutralization reaction   (p. 529)
precipitation reaction   (p. 524)
precipitin test   (p. 524)
primary immunodeficiencies   (p. 494)
primary immunodeficiency disease (p. 516)
prostaglandin   (p. 495)

radial immunodiffusion   (p. 525)
radioimmunoassay (RIA)   (p. 529)
reagin   (p. 494)
respiratory anaphylaxis   (p. 498)
Rh antigen   (p. 501)
rheumatoid arthritis (RA)   (p. 511)
rheumatoid factor   (p. 512)
Schick test   (p. 529)
secondary immunodeficiencies   (p. 494)
secondary immunodeficiency disease (p. 516)
sensitization   (p. 495)
seroconversion   (p. 526)
serology   (p. 524)
serum sickness   (p. 505)
severe combined immunodeficiency (SCID)   (p. 516)
systemic lupus erythematosus (SLE) (p. 512)
transfusion reaction   (p. 501)
transplantation   (p. 513)
transplant rejection   (p. 513)
tube agglutination test   (p. 526)
tuberculin hypersensitivity   (p. 508)
tuberculin skin test   (p. 508)
viral hemagglutination   (p. 526)
viral neutralization   (p. 529)
Western blotting   (p. 531)
xenograft   (p. 513)

See You on the Web

Don't forget I am expecting you on the web. The following is a preparedness checklist for your web explorations. If you have questions on any of these items, you can return to the end of Chapter 1 for more details.

- Questions We'll Explore
- Concept Check questions
- Flashcards
- Study Guide

If you are ready, come visit me at

http://www.prenhall.com/black

It will be a journey you'll never forget.

Diseases of the Skin and Eyes; Wounds and Bites

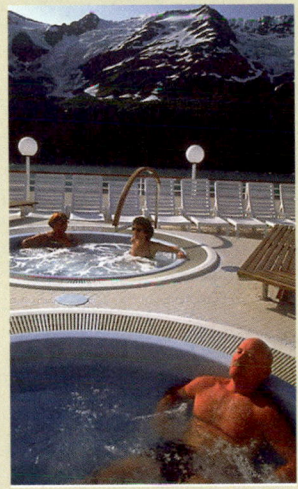

It is really irritating! Despite the fact that you have worked hard to earn this soothing soak in the hot tub, you just can't relax, and for the stupidest reason. The persistent, familiar dum-dum, dum-dum dum-dum warning refrain from JAWS keeps running through your mind. Shaking your head in disgust, you give up and get out to dry off. Unbeknownst to you, you may have just saved yourself from the "jaws" of another pernicious predator, *Pseudomonas aeruginosa.*

Warm water, skin flakes, and body oils make the perfect home for *Pseudomonas,* a bacterium very resistant to disinfection. They live in biofilms in water pipes and can reappear in force in your hot tub's water within hours. Rubbing your eyes and inadvertently scratching your cornea, while poaching in a broth of *Pseudomonas,* may ultimately cost you an eye. Breathing the bacteria from the mist over the tub invites a deadly form of pneumonia. You may count yourself lucky, if you only develop "hot tub rash", a hypersensitivity to *Pseudomonas*-contaminated water.

Background Basics

nonspecific immunity . . . . *(Chapter 16)*
exotoxins . . . . *(Chapter 14)*
hyaluronidase . . . . *(Chapter 14)*
immunization . . . . *(Chapter 17)*
herd immunity . . . . *(Chapter 15)*
Edward Jenner . . . . *(Chapter 1)*
effects of prolonged antibiotic therapy . . . . *(Chapter 13)*
life cycle of schistosomes . . . . *(Chapter 11)*
microfilariae . . . . *(Chapter 11)*

Having considered the general principles of microbiology in the first five units of this text, we are now ready to apply those principles to understanding infectious diseases in humans. As you study diseases, keep in mind the information on normal microflora and disease processes in Chapter 14, and on epidemiology in Chapter 15.

The Skin, Mucous Membranes, and Eyes

The Skin

The skin is the largest single organ of the body. The physical barrier it presents to most microbes is an important nonspecific defense. The surface consists of a thin **epidermis** and a thicker, underlying layer, the **dermis** (➤ Figure 19.1). Lacking blood vessels, the epidermis is nourished by nutrients that diffuse from blood vessels in the dermis. The epidermis has several layers of dead *epithelial* (ep-i-the′le-al) *cells* that function as an excellent barrier against injury and infection of deeper layers of the body. Cells next to the dermis divide throughout the life of an individual and migrate toward the surface, where old cells are sloughed off. Thus the epidermis is renewed every 15 to 30 days. These cells contain a waterproofing protein called **keratin,** which prevents water-soluble substances from entering the body. In addition, thick skin on the palms of the hands and soles of the feet reduces the likelihood that the skin barrier will be broken. *Calluses* represent thickened areas subject to constant wear. Thus, the intact skin surface prevents pathogenic organisms and other foreign substances from entering the body.

Questions We'll Explore

A What are the structures of the skin and eyes?

B What are the normal microflora of the skin and eyes?

C What kinds of pathogens cause skin diseases?

D What are the important epidemiologic and clinical aspects of skin diseases?

E What kinds of pathogens cause eye diseases?

F What are the important epidemiologic and clinical aspects of eye diseases?

G What kinds of pathogens infect wounds and bites?

H What are the important epidemiologic and clinical aspects of wound and bite infections?

➤ **Figure 19.1   The skin.** Consisting of the epidermis and underlying dermis, the skin forms a mechanical barrier to infection. Chemicals secreted onto the skin also retard pathogen attack and infection.

Not only is the skin a physical barrier to infection, its surface is inhabited by a variety of normal microflora. They occupy so much of the surface that it is difficult for pathogens to find a site to attach and colonize ➡ (Chapter 14, p. 378).

The skin generates antimicrobial substances that make it an even more inhospitable environment. Embedded in the dermis, the durable part of the skin, are sebaceous (se-ba'shus) glands and sweat glands. Most **sebaceous glands** produce an oily secretion called **sebum** (se'bum), which consists mainly of organic acids and other lipids. Although these secretions provide some nutrients for the normal microflora, the acidic nature of the sebum helps maintain an acidic skin pH that discourages pathogen growth. The quantity of sebum secretions increases at puberty, contributing to the development of acne in some persons. **Sweat** (sudoriferous) **glands** are distributed over the body and empty a watery secretion through pores in the skin. In the armpits and groin, these glands secrete organic substances in the sweat that lower the skin pH, again inhibiting the growth of most pathogens. The high salt concentration in sweat also inhibits many microorganisms.

Mucous Membranes

Mucous membranes line those tissues and organs that open to the exterior of the body, especially those of the respiratory, digestive, and urogenital systems. Like the skin, mucous membranes have a thin epidermal layer and a deeper connective tissue layer (➤ Figure 19.2). The cells of the epithelial layer block the penetration of microbes and secrete **mucus,** a thick but watery secretion of glycoproteins and electrolytes. Mucus, produced by *goblet cells* and certain glands, forms a protective layer over the epidermal cells, preventing drying and cracking of the mucous membrane. Mucus also traps pathogens before they can establish an infection.

➤ **Figure 19.2   Mucous membranes.** Much like the skin, mucous membranes consist of epithelial cells that line tissues. The secretion of mucus by goblet cells (special epithelial cells) and glands traps and clears away many potential pathogens and other foreign material.

The Eyes

Being exposed to the atmosphere, the eyes contact millions of microbes every day, but have no normal microflora. The eyes have several external protective structures to lessen the chances of an infection. In fact, these structures work so well that deep eye infections are exceedingly rare. Each eye is protected physically by eyelids, eyelashes, the **conjunctiva** (a mucous membrane covering the inner surface of each eyelid and the anterior region of each eye), and the tough **cornea** (the transparent part of the eyeball exposed to the environment) (➤ Figure 19.3). Eyelashes and eyelids help prevent foreign objects from reaching the cornea.

The eyes also produce antimicrobial substances. Each eye has a **lacrimal gland,** which secretes a lacrimal fluid (tears) that continuously flushes the cornea, keeping it moist as the liquid carries away any microorganisms present. *Lacrimal canals* drain tears from the surface of the eyeball through the *nasolacrimal duct* and into the *nasal cavity*. Tears contain *lysozyme,* an enzyme that breaks down bacterial cell walls. It is especially effective in killing Gram-positive organisms, which have a thick peptidoglycan layer (➤ Figure 19.4) ➡ (Chapter 4, p. 78) Lysozyme has no effect on viruses. Tears, like other body secretions, also contain specific defensive chemicals. Glandular cells in the conjunctiva add a mucous substance to the tears. This substance may help trap microorganisms and eliminate them from the eye.

Normal Microflora of the Skin

A huge population of normal microflora colonizes the nearly 2 square meters of skin covering the average human adult. Conditions differ widely at various locations—for example, armpits versus forehead—leading to populations of different compositions in different locations. (See the box "But I Keep Myself So Clean.") Moist areas will support larger populations; oils and sweat provide nutrients. Oily compounds are metabolized by some microbes, yielding fatty acid by-products that accumulate and give an acidic pH (5.5) to certain parts of the skin. This is a first-line, nonspecific defense against infection, since most other microbes are unable to live or grow well at this pH. However, many of these fatty acid compounds have an unpleasant smell and are responsible for body and foot odors. ➡ (Chapter 14, p. 378, Bloodhounds)

Sweat is rich in salt (sodium chloride). The accumulation of salt on the skin is another first-line, nonspecific defense. Relatively few microbes are salt-tolerant. Staphylococci, however, grow well on salty media and are thus one of the predominant organisms on skin. At least a dozen different species of *Staphylococcus* live on our skin.

A third nonspecific line of defense involves the rapid and continuous shedding of skin cells. As younger cells rise up toward the skin surface, they fill with keratin. By the time they reach the top, the older cells are completely filled with keratin and have died. As they slough off, they carry away the microbes that have colonized them. Keratin-degrading fungi living on the skin sometimes, especially under moist conditions, will multiply too rapidly and cause infections such as athlete's foot.

The majority of organisms living on the skin are Gram-positive organisms such as *Staphylococcus* and *Micrococcus,* and coryneform bacteria (e.g., *Corynebacterium* and *Proprionibacterium acnes)*.

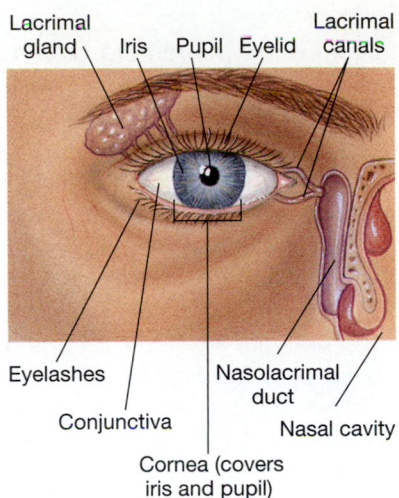

➤ **Figure 19.3   Structures of the eye.** Tears produced over the eye surface contain antibacterial molecules that inhibit potential infections and drain into the nasal cavity. Most eye infections involve the eyelids, conjunctiva, or cornea.

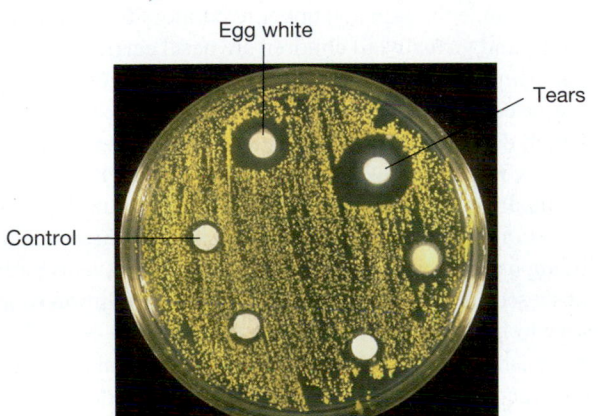

➤ **Figure 19.4   Lysozyme in tears.** Lysis of *Micrococcus luteus* bacteria by lysozyme enzymes obtained from tears and from egg white. The control disks have been dipped in sterile distilled water.

But I Keep Myself So Clean!

Think of your skin as a vast continent on which different habitats exist: forests, deserts, swamps, and so on. Not all places are equally hospitable to microbial life. Why not try an unofficial "census" of your skin? Which sites are the deserts, and which are teeming with microbial inhabitants?

Using sterile technique, press a standard-sized piece of sterile velvet or Velcro against a portion of your skin. Then press it against the surface of a Petri dish containing nutrient agar or tryptic soy agar (much as you would when using a stamp pad). Incubate at 37°C for 48 hours, and count the number of colonies growing on the agar. Comparing colony counts grown from standard-sized samples taken at different locations on the skin should reveal your "hot spots" for microbial growth.

Here are some ideas for sampling sites:

- Forehead, nose (inner versus outer surfaces).
- Cheek, armpit, middle of back, forearm.
- Between toes, upper versus lower surfaces of foot.

Other comparisons:

- Male versus female subjects: a shaved cheek compared to an unbearded cheek.
- Armpits: one side with deodorant, other without.
- Areas of skin with and without makeup (use different sides of one person's face).
- Hands before and after washing. (You should see more colonies after washing; can you think of why?)
- Creases or folds of skin versus adjacent smooth areas.

One unsuspected inhabitant is the mite *Demodex folliculorum,* which dwells inside the openings of oil glands and down in hair follicles. You might raise your eyebrows to discover this population living in your eyebrows and eyelashes.

Diseases of the Skin

Your skin covers your body and accounts for 15 percent of your body weight. It provides an effective barrier to invasion by most microbes except when it is damaged. ➡ (Chapter 16, p. 435) Very few microbes can penetrate unbroken skin; however, mucous membranes are more easily entered. Here we will consider the kinds of skin diseases that occur when the skin surface fails to prevent microbial invasion.

Bacterial Skin Diseases

Many bacteria are found among the normal microflora of the skin. Ordinarily they are kept from invading tissues by an intact skin surface and nonspecific defense mechanisms of the skin. Bacterial and other skin infections usually arise from a failure of these defenses. They are generally diagnosed by appearance and by clinical history.

Staphylococcal Infections

Folliculitis and Other Skin Lesions   Everyone has had a pimple at some time; most likely it was caused by *Staphylococcus aureus,* the most pathogenic of the staphylococci. Staphylococcal skin infections are exceedingly common because the organisms are nearly always present on the skin. Strains of staphylococci colonize the skin and upper respiratory tract of infants within 24 hours of birth. (Half of all adults and virtually all children are nasal carriers of *S. aureus.*) Infection occurs when these organisms invade the skin through a hair follicle, producing **folliculitis** (fol-lik″u-li′tis), also referred to as **pimples** or **pustules.** An infection at the base of an eyelash is called a **sty.** A larger, deeper, pus-filled infection is an **abscess;** an exterior abscess is known as a **furuncle** (fu′rung-kl), or **boil** (➤ Figure 19.5). An estimated 1.5 million Americans have such infections annually. Further spread of infection, particularly on the neck and upper back, creates a massive lesion called a **carbuncle** (kar′bung-kl). Encapsulation of abscesses prevents them from shedding organisms into the blood, but it also prevents circulating antibiotics from reaching the abscesses in effective quantities. Thus, in addition to antibiotic treatment, it is usually necessary to lance and drain abscesses surgically.

Staphylococcal infections are easily transmitted. Asymptomatic carriers, hospital personnel, and hospital visitors often spread staphylococci via the skin as well as by nasal droplets and fomites. Staphylococci usually cause infection in older patients only when foreign bodies such

Boils can become larger than ping-pong balls and are extremely painful.

➤ Figure 19.5   **A furuncle.** This deep, pus-filled lesion is caused by *Staphylococcus aureus.*

as catheters or splinters are present. Although 5 million organisms must be injected into the skin to cause infection, only 100 are needed if they are soaked into a suture and tied into the skin.

Scalded Skin Syndrome   **Scalded skin syndrome** is caused by certain exotoxin-producing strains of *S. aureus.* Two different exotoxins, both called *exfoliatins,* are known. The genes for one are carried on the bacterial chromosome, whereas the genes for the other are located on a plasmid. An individual staphylococcal strain may carry genes for one, both, or neither of these exotoxins. The exotoxins are called exfoliatins because they travel through the bloodstream to sites far from the initial infection, causing the upper skin layers to separate and peel off in leaflike (foliar) sheets.

Scalded skin syndrome is most common in infants but can occur in adults as well. It begins with a slight reddened area, often around the mouth. Within 24 to 48 hours, it spreads to form large, soft, easily ruptured vesicles over the whole body. The skin over vesicles and adjacent reddened areas peels, leaving large, wet, scalded-looking areas (➤ Figure 19.6). The lesions dry and scale, and the skin returns to normal in 7 to 10 days. High fever is generally present. Bacteremia is common and can lead to septicemia and death within 36 hours. Exotoxins are highly antigenic. ➡ (Chapter 14, Table 14.4) They stimulate the production of antibodies that prevent recurrence of this syndrome—but only if reinfection occurs by the same *S. aureus* strain.

Streptococcal Infections

Scarlet Fever   **Scarlet fever,** sometimes called scarlatina, is caused by *Streptococcus pyogenes,* which also causes the familiar strep throat. Strains of the organism that cause scarlet fever have been infected by a temperate phage that enables them to produce an erythrogenic ("red-producing") toxin that causes the scarlet fever rash. Patients who have previously been exposed to the toxin, and so have antibodies that can neutralize it, can develop strep throat without the scarlet fever rash. However, such people can still spread scarlet fever to others. Three different erythrogenic toxins have been identified. A person can develop scarlet fever once from each toxin.

➤ Figure 19.6   **Scalded skin syndrome in an infant.** This infection is caused by *Staphylococcus aureus.* Reddened areas of skin peel off, leaving wet, scalded-looking areas.

Flesh-Eating Bacteria

It was like a real-life *X-Files* episode in June 1994 in England and then again in 1998 in Texas. Newspapers and television sounded alerts after people began to die as a strain of *Streptococcus* ate away at their flesh at the fearful rate of several inches per hour. After killing the initially infected tissue, the bacteria lived on the remains of the dead flesh, churning out toxins that diffused out of the dead area to kill more tissue. Antibiotics were useless against the bacteria because the circulatory system could not function in the dead tissue. The only hope was for the infected areas to be surgically removed, even if it meant amputating a limb. Fortunately, approximately every 10 years, a different strain of Streptococcus gains dominance; perhaps the flesh-eating strain is nearing the end of its 10-year cycle.

In the United States most scarlet fever organisms currently have low virulence. In past decades, when strains were more virulent, it was a much-feared killer. However, even low-virulence strains can cause serious complications, such as glomerulonephritis or rheumatic fever. The use of penicillin has appreciably lowered the mortality rate. Convalescent carriers can shed infective organisms from the nasopharynx for weeks or months after recovery. Fomites also are an important source of streptococcal infections.

Erysipelas   **Erysipelas** (er"i-sip′e-las; from the Greek *erythros,* for red, and *pella,* for skin), also called **St. Anthony's fire,** has been known for over 2000 years and is caused by hemolytic streptococci. Before antibiotics became available, erysipelas often occurred after wounds and surgery and sometimes after very minor abrasions. Mortality was high. Today, it rarely occurs, and mortality is low. The disease begins as a small, bright, raised, rubbery lesion at the site of entry. Lesions spread as streptococci grow at lesion margins, producing toxic products and enzymes such as hyaluronidase. ➥ (Chapter 14, p. 386) Lesions are so sharply defined that they appear to be painted on. The organisms spread through lymphatics and can cause septicemia, abscesses, pneumonia, endocarditis, arthritis, and death if untreated. Curiously, erysipelas tends to recur at old sites. Instead of developing immunity, patients acquire greater susceptibility to future attacks.

Pyoderma and Impetigo

Pyoderma, a pus-producing skin infection, is caused by staphylococci, streptococci, and corynebacteria, singly or in combination. **Impetigo** (im-pe-ti′go), a highly contagious pyoderma, is caused by staphylococci, streptococci, or both (➤ Figure 19.7). Fluid from early pustules usually contains streptococci, whereas fluid from later lesions contains both. Streptococcal strains that cause skin infections usually differ from those that cause strep throat. Impetigo occurs almost exclusively in children; why adults are not susceptible is unknown. Easily transmitted on hands, toys, and furniture, impetigo can rapidly spread through a day-care center. Impetigo rarely produces fever, and it is easily treated with penicillin. Lesions usually heal without scarring, but skin can be discolored for several weeks, and pigment can be permanently lost.

Acne

Acne (acne vulgaris) affects more than 80 percent of teenagers and many adults, too. It is most often the result of male sex hormones that stimulate sebaceous glands to increase in size and secrete more sebum. Acne occurs in both males and females because the hormones are produced by the adrenal glands as well as by the testes. Microorganisms feed on sebum, and ducts of the glands and surrounding tissues become inflamed. "Blackheads" are a mild form of acne in which hair follicles and sebaceous glands become plugged with sebum and keratin. In more severe cases (cystic acne) the plugged ducts become inflamed, rupture, and release secretions. Bac-

➤ **Figure 19.7   Impetigo.** This highly contagious infection is caused by staphylococci, streptococci, or both together.

teria, especially *Propionibacterium acnes,* infect the area and cause more inflammation, more tissue destruction, and scarring. Such lesions can be widely distributed over the body, and some become encysted in connective tissue.

Acne is treated with frequent cleansing of the skin and topical ointments to reduce the risk of infection. Sometimes acne sufferers are advised to avoid fatty foods, but a connection between diet and acne is not well established. Dermatologists often prescribe oral antibiotics such as tetracycline in low doses to control bacterial infections in the lesions. However, continuous use of antibiotics depletes natural intestinal flora and can contribute to the development of antibiotic-resistant strains of bacteria. ➡ (Chapter 13, p. 346) The drug Accutane, derived from a molecule related to vitamin A, is now used to treat severe and persistent acne. It seems to inhibit sebum production for several months after treatment is stopped, but it can cause serious side effects, such as intestinal bleeding. In most cases acne disappears or decreases in severity as the body adjusts to the hormonal changes of puberty and as the functioning of the sebaceous glands stabilizes.

Burn Infections

Severe burns destroy much of the body's protective covering and provide ideal conditions for infection. Burn infections, which are usually nosocomial, account for 80 percent of deaths among burn patients. *Pseudomonas aeruginosa* is the prime cause of life-threatening burn infections, but *Serratia marcescens* and species of *Providencia* also often infect burns. Many strains of these Gram-negative bacilli are antibiotic-resistant.

The thick crust or scab that forms over a severe burn is called **eschar** (es′kar). Bacteria growing in or on eschar pose no great threat, but those growing beneath it cause severe local infections and can move into the blood. Delivering antibiotics to infections under eschar is difficult because the eschar lacks blood vessels. Topical antimicrobials can seep through the eschar; removal of some of the eschar by a surgical scraping technique called **débridement** (da-bred-maw′ or de-bred′ment) helps them to reach infection sites.

Prevention of burn infections is difficult even when patients are isolated within hospital burn units. Having lost skin, patients have lost the benefit of the movement of white blood cells to sites of infection in the skin. Such patients also have fluid and electrolyte deficiencies because of seepage from burned tissue. Finally, their appetites are depressed at a time when extensive tissue repair increases metabolic needs.

Burn infections are as difficult to diagnose as they are to treat. Early signs of infection can consist of only a mild loss of appetite or increased fatigue. Definitive diagnosis is made by finding more than 10,000 bacteria per gram of eschar. *Pseudomonas aeruginosa* infection should be suspected when a greenish discoloration appears at the burn site and cultures have a grapelike odor. This bacterium produces tissue-killing toxins that erode skin. It is extremely resistant to antimicrobial drugs and has been found growing in surgical scrub solutions.

✓ What are some nonspecific lines of defense provided by the skin?

✓ Why are *Staphylococcus* species so numerous on skin?

✓ What are some skin diseases caused by *Staphylococcus*?

✓ What is debridement? Why is it necessary in burn cases?

Viral Skin Diseases

Rubella

The Disease **Rubella,** or **German measles,** is the mildest of several human viral diseases that cause **exanthema** (ex-an-the′mah), or skin rash. A rash, the main symptom of rubella, appears first on the trunk 16 to 21 days after infection, but the virus (a togavirus) spreads in the blood and other tissues before the rash appears. Infected adult women often suffer from temporary arthritis and arthralgia (joint pain) from dissemination of the virus to joint membranes. These complications are seen less frequently in adult men.

Congenital rubella syndrome results from infection of a developing embryo across the placenta. When a woman becomes infected with rubella during the first 8 weeks of pregnancy, severe damage to the embryo's organ systems is likely because they are developing. After the eighteenth week of pregnancy, damage is rare. The spread of rubella viruses in the infant kills many cells, persistently infects other cells, reduces the rate of cell division, and causes chromosomal abnormalities. Many infants are stillborn, and those that survive may suffer from deafness, heart abnormalities, liver disorders, and low birth weight.

To prevent congenital rubella syndrome, women should have their rubella immunity checked before trying to get pregnant.

What Goes Round in Whirlpool Baths?

Whirlpool baths can be quite dangerous if not properly cleaned and disinfected daily. Warm water, plus skin flakes, oils, and secretions makes a perfect breeding ground for microbes. The aerosol from a bubbling whirlpool can contain millions of bacteria per cubic meter of air. *Pseudomonas aeruginosa,* a frequent colonizer, is very difficult to eliminate—it can be back in the water just hours after disinfection. Often it grows attached to the sides of the whirlpool tub or inside the pipes connected to it. Disinfection may kill all the organisms in the water, but when the same bacteria attach to a surface and form a biofilm, they are much more resistant to the disinfecting chemicals.

A person in a whirlpool bath who vigorously rubs his or her eyes could cause a corneal abrasion and thereby become susceptible to bacterial eye infection. *Pseudomonas* infection can cause severe scarring of the cornea and distorted vision, or even blindness. Once such an infection is started, you can lose an eye almost overnight. Also, inhaling aerosols can cause *Pseudomonas* pneumonia, which has a very poor prognosis.

Many people break out in a body rash in whirlpool baths because of a hypersensitivity reaction to being immersed in water filled with *Pseudomonas*. And *Pseudomonas* isn't the only microbe to worry about in or on whirlpools, as two elderly widows discovered. They developed herpes lesions on the backs of their thighs after sitting on the moist rim of a hot tub where a previous occupant with a herpes outbreak had been seated.

Incidence and Transmission   Prior to the development of a rubella vaccine, nearly all humans caught rubella at some time, but many cases were not detected. Half the cases in young children and up to 90 percent of those in young adults are not recognized. In the United States the incidence of rubella was nearly 30 cases per 100,000 people in 1969, when a vaccine was licensed. Rubella, including congenital rubella, has now been nearly eliminated (➤ Figure 19.8).

Transmission is mainly by nasal secretions shortly before, and for about a week following, the appearance of a rash. Many infected individuals do not have a rash and transmit the virus without knowing it. Rubella is highly contagious, especially by direct contact among children aged 5 to 14. Infants infected before birth are rubella carriers; they excrete viruses and expose the hospital staff and visitors, including pregnant women, to the disease.

Diagnosis   Rubella can be positively diagnosed by a variety of laboratory tests. Determination of a four-fold rise in rubella-specific IgM antibody levels is particularly useful in identifying newborn carriers and assessing immunity of pregnant women exposed to rubella.

Immunity and Prevention   The currently available rubella vaccine, part of a combined attenuated viral vaccine preparation (MMR), is the only means of preventing rubella. Children

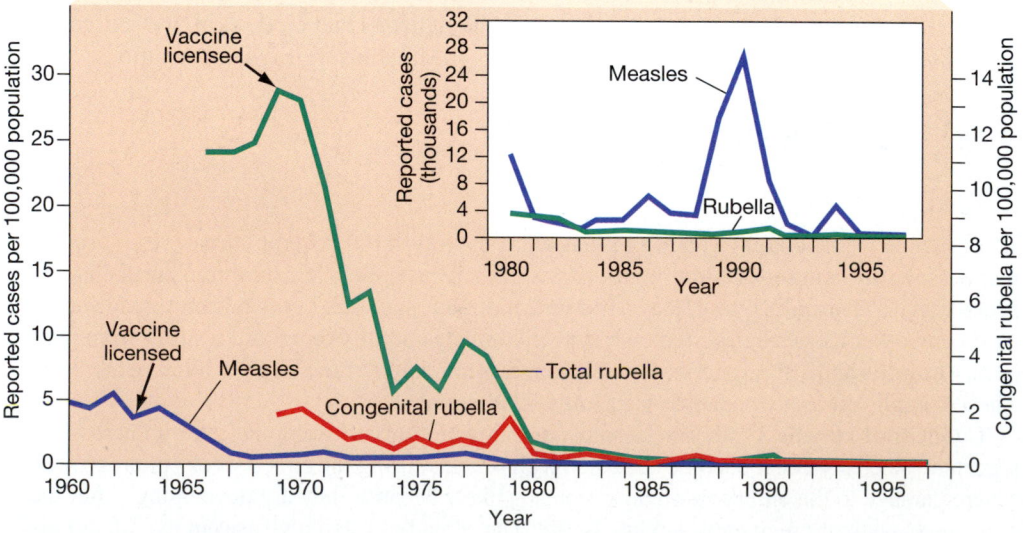

➤ Figure 19.8   **The U.S. incidence of rubella (German measles) and measles (rubeola).** The number of cases dropped rapidly after vaccines were licensed during the 1960s. Failure to vaccinate, however, has led to a recent increase in measles cases. (*Source:* CDC)

should receive it both to protect themselves from measles and to protect infants, too young to receive the vaccine, from being exposed to a source of infection. High levels of immunity (Chapter 15) must be maintained, or outbreaks will occur. ➥ (p. 416) The vaccine produces lower antibody levels than does infection, and immunity probably is not as long-lasting as that produced by infection. Also, viruses appear in the nasopharynx a few weeks after immunization, and some individuals experience mild symptoms of the disease. To prevent infection of fetuses, a second immunization is recommended for females before they become sexually active. If a woman is pregnant when immunized, viruses from the vaccine may be able to infect the fetus. To prevent transmission of the virus from child to mother to fetus, caution must be exercised in immunizing young children whose mothers are pregnant.

Measles

The Disease   **Measles,** or **rubeola,** is a febrile (accompanied by fever) disease with a rash caused by the rubeola virus. This paramyxovirus invades lymphatic tissue and blood. After the virus enters the body through the nose, mouth, or conjunctiva, symptoms appear in 9 to 11 days in children and in 21 days in adults. **Koplik's spots,** white spots with central bluish specks (➤ Figure 19.9), appear on the upper lip and cheek mucosa 2 or 3 days before other symptoms such as fever, conjunctivitis, and cough. These symptoms persist for 3 or 4 days and sometimes progressively worsen. They are followed by a rash, which spreads during a 3- to 4-day period from the forehead to the upper extremities, trunk, and lower extremities and disappears in the same order several days later. The rash is caused by the reaction of T cells with virus-infected cells in small blood vessels. Without such reactions, as in patients with defective cell-mediated immunity, no rash appears, but the virus is free to invade other organs. ➥ (Chapter 18, p. 516) If the virus invades the lungs, kidneys, or brain, the common childhood disease often is fatal.

The most common complications of measles are upper respiratory and middle ear infections. **Measles encephalitis,** a more serious complication, occurs in only 1 or 2 patients per 1000 but has a 30 percent mortality rate and leaves one-third of survivors with permanent brain damage. Another complication, **subacute sclerosing panencephalitis (SSPE),** occurs in only 1 in 200,000 cases and is nearly always fatal. It is due to the persistence of measles viruses in brain tissue and causes death of nerve cells, with progressive mental deterioration and muscle rigidity. SSPE manifests itself 6 to 8 years after measles, usually in children who had measles before age 3. In poorly nourished children, measles causes intestinal inflammation with extensive protein loss and shedding of viruses in stools. More than 15 percent of children infected with measles in developing countries die of the disease or its complications.

Incidence and Transmission   The measles virus is highly contagious; its portal of entry is the respiratory tract. A susceptible person has a 99 percent chance of infection if exposed directly to someone who is releasing the virus while coughing and sneezing. Before widespread immunization, most children had measles before they were 10 years old. In populations such as that of the United States, where periodic epidemics have occurred and most children are well

Measles and More

Did you ever think that you could catch measles from your dog? You can't, since human are the only known hosts of measles, but distemper in dogs is caused by a virus with antigens similar to those of the measles virus. Did you know that measles probably couldn't have existed until 2500 B.C.? Measles can spread only if there is a continuous chain between infected and susceptible people, since the virus doesn't display latency, cannot survive for long outside the body, no other organism survives as a reservoir, and immunity from infection is lifelong.

A world population of at least 300,000 was therefore required to perpetuate the virus, and this population size was not reached until about 2500 B.C. Measles probably appeared as a human pathogen after that time—possibly as a mutant of a virus like the one that causes distemper.

Children exposed to the measles virus, prenatally or shortly after birth, may have increased risk of developing Crohn's disease, a painful intestinal disorder.

➤ **Figure 19.9   Measles (rubeola).** This shows the typical skin rash, plus white Koplik's spots in the mouth on the inner surface of the cheek.

nourished, the disease is serious but rarely fatal. In populations lacking immunity from periodic epidemics or in malnourished children, measles is a killer. In 1875, when measles was first introduced into Fiji, 30 percent of the population died of the disease.

Diagnosis and Treatment   Measles is diagnosed by its symptoms. Treatment is limited to alleviating symptoms and dealing with complications. Secondary bacterial infections can be effectively treated with antibiotics.

Immunity and Prevention   In a nonimmunized population, epidemics of numerous cases over a 3- or 4-week period generally occur every 2 to 5 years over wide regions. Measles vaccine now prevents such epidemics in many developed countries. In the United States, mandatory measles vaccination has greatly reduced the incidence of the disease (see ➤ Figure 19.8). Because the MMR vaccine containing attenuated viruses of measles, mumps, and rubella is generally used, the incidence of all three diseases has decreased simultaneously. ➡ (Chapter 17, p. 481) Eradication of a disease requires that about 90 percent of the population be immune. Religious groups that reject immunization, immigrants who don't understand its importance, and apathetic parents who don't have their children immunized make eradication difficult. Compounding these problems have been federal government cutbacks in funding used to purchase vaccine for clinics. Hence some poor children in the United States are going without vaccine.

Immunity acquired from having measles is lifelong. Recent epidemics among college students immunized as children suggests that immunity from the vaccine may not be lifelong or that the children were immunized before 15 months of age. A few cases of SSPE have been attributed to the vaccine, but the incidence of SSPE is much lower than it was prior to the use of the vaccine.

Chickenpox and Shingles

One Virus, Two Diseases   The **varicella-zoster virus (VZV)**, a herpesvirus, causes both **chickenpox** (varicella) and **shingles** (zoster). The same virus can be isolated from lesions of either disease. Chickenpox is a highly contagious disease that causes skin lesions and usually occurs in children. There are probably over 3 million cases per year in the United States alone. Shingles is a sporadic disease that appears most frequently in older and immunocompromised individuals. A congenital varicella syndrome also is known to occur in about 5 percent of the U.S. population.

In chickenpox the virus enters the upper respiratory tract and conjunctiva and replicates at the site of entry. New viruses are carried in blood to various tissues, where they replicate several more times. Release of these viruses causes fever and malaise. In 14 to 16 days after exposure, small, irregular, rose-colored skin lesions appear. The fluid in them becomes cloudy, and they dry and crust over in a few days. The lesions appear in cyclic crops over 2 to 4 days as the viruses go through cycles of replication. They start on the scalp and trunk and spread to the face and limbs, sometimes to the mouth, throat, and vagina, and occasionally to the respiratory and gastrointestinal tracts.

Chickenpox, although sometimes thought of as a mild childhood disease, can be fatal. The viruses invade and damage cells that line small blood vessels and lymphatics. Circulating blood clots and hemorrhages from damaged blood vessels are common. Death from varicella pneumonia is due to extensive blood vessel damage in the lungs and the accumulation of erythrocytes and leukocytes in alveoli. Cells in the liver, spleen, and other organs also die because of damage to blood vessels within them.

In shingles, painful lesions like those of chickenpox usually are confined to a single region supplied by a particular nerve (➤ Figure 19.10a). Such eruptions arise from latent viruses acquired during a prior case of chickenpox. During the latent period, these viruses reside in ganglia in the cranium and near the spine. When reactivated, the viruses spread from a ganglion along the pathway of its associated nerve(s). Pain and burning and prickling of the skin occur before lesions appear. The viruses damage nerve endings, cause intense inflammation, and produce clusters of skin lesions indistinguishable from chickenpox lesions (➤ Figure 19.10b). Symptoms range from mild itching to continuous, severe pain and can include headache, fever, and malaise. Lesions often appear on the trunk in a girdlelike pattern (*zoster,* girdle) but can

(a)   (b)

➤ **Figure 19.10   Shingles.** Shingles lesions usually result from varicella-zoster virus infections acquired during exposure to childhood chickenpox. The virus can remain latent within the body for many years before being reactivated in adulthood. (a) Vesicles commonly form a belt around the chest or hips, following the pathway of a nerve. (b) The small, yellow vesicles dry up and heal by scabbing, but they can be excruciatingly painful and itchy.

infect the face and eyes. Shingles is most severe in individuals with malignancies or immune disorders. In such patients, lesions may cover wide areas of the skin and sometimes spread to internal organs, where they can be fatal.

The latent virus is activated when cell-mediated immunity drops below a critical minimal level, as can occur in lymphatic cancers, spinal cord trauma, heavy-metal poisoning, or immunosuppression. In other cases, no cause of viral reactivation can be identified. Release of newly replicated viruses increases antibody production, but the antibodies may fail to stop viral replication. Recovery from shingles usually is complete. Second and third cases do occur, and they depend on the degree of development of cell-mediated immunity and local interferon production. Chronic shingles is found in the immunocompromised, including AIDS patients. New vesicles constantly erupt, whereas old ones fail to heal, which can be very debilitating.

Incidence and Transmission   Chickenpox is endemic in industrialized societies in the temperate zone, and its incidence is highest in March and April. The primary infection usually occurs between the ages of 5 and 9. In general, chickenpox in adults who have not had it as children is more severe than in children. Shingles also is age-related, with most cases appearing in people over 45 years of age.

Infection can be spread by respiratory secretions and contact with moist lesions but not from crusted lesions. Children experiencing a mild case, with only a few lesions and no other symptoms, often spread the disease. In rare cases, adults with partial immunity can contract shingles from exposure to children with chickenpox. Susceptible children, however, can easily contract chickenpox from exposure to adults with shingles.

Diagnosis and Treatment   Chickenpox is usually diagnosed by a history of exposure and the nature of lesions, although a rapid laboratory test has recently become available. Shingles may be impossible to distinguish from other herpes lesions without laboratory tests. Treatment is limited to relieving symptoms, but aspirin should not be given to children because of the risk of Reye's syndrome. Acyclovir, once hoped to be useful in early stages of the disease to reduce its severity, has not proven successful. Antiviral agents are being tested on infections in immunosuppressed patients and those with disseminated disease.

Immunity and Prevention   Having chickenpox as a child confers lifelong immunity in most cases. Recurrences (in the form of shingles) are seen only in individuals with low concentrations of VZV antibodies and waning cell-mediated immunity. Chickenpox was the last of the common communicable childhood diseases for which no vaccine was widely available—until 1995, when a vaccine was finally licensed for use in the United States. An attenuated varicella vaccine, developed in Japan, prevents chickenpox when administered as late as 72 hours after exposure. It was used overseas for all children. The U.S. Food and Drug Administration approved

Pregnant women who have never had chickenpox and have not been vaccinated should avoid exposure to children who have chickenpox and to people who have shingles.

the vaccine for general use in 1995. Because the mechanism of latency and reactivation is not well understood, however, immunologists had some reservations about administering vaccines containing viruses that can become latent. Cases have been found in which shingles has been caused by reactivation of the virus in the vaccine. Nevertheless, the vaccine is considered relatively safe and effective.

Other Pox Diseases

Smallpox   In 1980, the World Health Organization (WHO) officially proclaimed that **smallpox** had been eradicated worldwide. This announcement marked the end of centuries of sickness and death from this disease.

The disease first appeared sometime after 10,000 B.C. in a small agricultural settlement in Asia or Africa. The mummy of Ramses V, who died in Egypt in 1160 B.C., has smallpox scars on the face, neck, shoulders, and arms. Smallpox ravaged villages in India and China for centuries, and an epidemic occurred in Syria in A.D. 302. The Persian physician Rhazez clearly described the disease in A.D. 900.

Crusaders brought smallpox to Europe in the twelfth century. The Spaniards carried the virus to the West Indies in 1507, and Cortez's army introduced it into Mexico in 1520. In each instance, the disease was introduced into a population that lacked immunity, and its spread was rampant. As many as 3.5 million Native Americans may have died from smallpox, and by the eighteenth century more than half the inhabitants of Boston had become infected. Slave traders introduced smallpox into Central Africa in the sixteenth and seventeenth centuries, and immigrants from India brought it to South Africa in 1713. The disease reached Australia in 1789.

The idea of immunization itself originated in Asia with the technique of variolation to prevent smallpox. In variolation, threads saturated with fluid from smallpox lesions were introduced into a scratch or dangled in the sleeve of a nonimmune person. This practice immunized some people, but it also started epidemics because live, virulent virus was used. Lady Mary Wortley Montague, wife of the British Ambassador to Turkey, introduced the practice to England in 1717. Despite high morbidity (illness) and even mortality (death) from this technique, General George Washington ordered all his troops variolated in 1777. By 1792, 97 percent of the population of Boston had been variolated.

Smallpox immunization was greatly improved by the English physician Edward Jenner, who noticed that milkmaids with cowpox scars never became infected with smallpox. It is now known that immunity to the less severe cowpox confers immunity to smallpox as well. In 1796, Jenner inoculated an 8-year-old boy with cowpox and 6 weeks later inoculated him with smallpox. As Jenner had expected, the boy remained healthy. In the United States in 1799, the physician Benjamin Waterhouse introduced the cowpox vaccine by vaccinating his own children and then exposing them to smallpox.

Although it was well known by then that smallpox could be prevented by vaccination with cowpoxvirus, in 1947 an unimmunized traveler from Mexico infected 12 people in New York City, and 2 of them died. The last 8 cases of smallpox in the United States occurred in the Rio Grande Valley in 1949.

Smallpox was still endemic in 33 countries in 1967 when the WHO established its immunization campaign; in 1977 only a single natural case of smallpox occurred worldwide (➤ Figure 19.11). When smallpox viruses escaped via air ducts from a laboratory in Birmingham, England, in 1978, an unimmunized medical photographer became infected and died. Her mother suffered a mild case of the disease and recovered. The director of the laboratory from which the virus escaped committed suicide. After this disaster, smallpox stock cultures have been maintained in only two maximum containment laboratories—one in the Atlanta division of the CDC, and the other in the Research Institute for Viral Preparations Moscow. However, smallpox surveillance continues.

The smallpox virus enters the throat and respiratory tract. During a 12-day incubation period, it infects phagocytic cells and later blood cells. The infection spreads to skin cells, causing pus-filled vesicles. Acute systemic symptoms begin with fever, backache, and headache. Vesicles appear first in the mouth and throat. They then rapidly spread to the face, forearms, hands, and finally the trunk and legs. Vesicles become opaque and pustular and encrust within about 2 weeks. Death is most likely to occur 10 to 16 days after onset of the first symptoms.

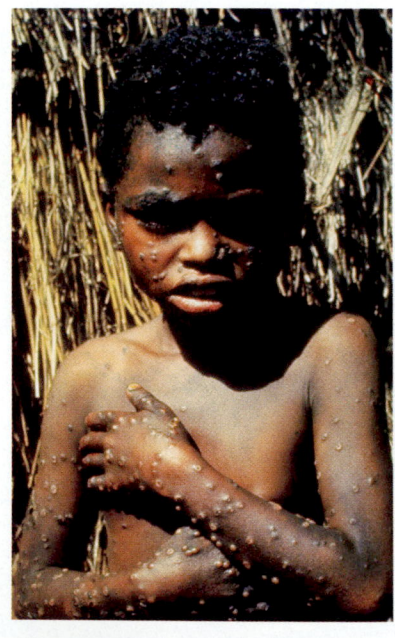

➤ **Figure 19.11   One of the last cases of smallpox.** Lesions are most numerous on the face and arms. As the disease progresses, the blisterlike lesions become opaque and eventually form crusts that fall off. Scarring is common.

Scrapings from lesions are used to differentiate smallpox from leukemia, chickenpox, and syphilis. In a nonimmune population, smallpox is highly contagious; in a largely immunized population, it spreads very slowly. But now that smallpox has been eradicated worldwide, complications of vaccination are far more life-threatening than the disease itself.

Cowpox   **Cowpox,** caused by the vaccinia virus, causes lesions (similar to a smallpox vaccination) at abrasion sites, inflammation of lymph nodes, and fever. Vaccinia viruses can also cause a progressive disease, with numerous lesions and symptoms more like those of smallpox, that can be treated with the drug methisazone. Cattle appear to transmit the disease to humans.

Not only was the vaccinia virus used by Jenner to immunize against smallpox but also it was the first animal virus to be obtained in sufficient quantity for chemical and physical analysis. ➡ (Chapter 1, p. 15) The modern vaccinia virus may have become attenuated by several centuries of passage on calves' skin, a procedure used in making smallpox vaccine.

Molluscum Contagiosum   The **molluscum contagiosum** virus is unusual for several reasons. First, it differs immunologically from both orthopoxviruses and parapoxviruses, the two major groups of poxviruses. Second, it elicits only a slight immune response. Third, although infected cells cease to synthesize DNA, the virus induces neighboring uninfected cells to divide rapidly. Thus, this virus may be intermediate between viruses that cause specific diseases and those that induce tumors. Molluscum contagiosum affects only humans and is distributed worldwide. It causes pearly white to light pink painless tumorlike growths scattered over the skin (➤ Figure 19.12). The disease usually affects children and young adults, and it can persist for years. It is acquired by personal contact or from items such as gym equipment and swimming pools. Treatment generally involves the removal of growths by chemicals or by localized freezing.

Warts

Human **warts,** or **papillomas,** are caused by **human papillomaviruses (HPV)**. HPV specifically attack skin and mucous membranes. Warts grow freely in many sites in the body—the skin, the genital and respiratory tracts, and the oral cavity. Viral infection lasts a lifetime. Even when warts disappear or are removed, the virus remains in surrounding tissue, and warts may recur or form malignant tumors.

The Nature of Warts   Warts vary in appearance, area of occurrence, and pathogenicity. Some are barely visible and self-limiting—that is, they do not grow or spread—and others, such as laryngeal warts, are larger but benign. A few warts are malignant. **Genital warts,** also known

➤ Figure 19.12   **Molluscum contagiosum.** Tumorlike growths develop in cases of this viral infection.

as condylomata acuminata (kon″di-lo-ma′ta a-kum″i-na′ta) (➤ Figure 19.13a), for example, sometimes become malignant, and some of the viral strains that produce them are strongly associated with cervical cancer. (Women who have had genital warts removed are advised to have Pap tests twice a year.) Warts of all types grow larger than normal in people with AIDS or other immunodeficiencies.

Transmission   Papillomaviruses are transmitted by direct contact, usually between humans, or by fomites. **Dermal warts** (➤ Figure 19.13b) form when the virus enters the skin or mucous membranes through abrasions. Genital warts are sexually transmitted, and *juvenile onset laryngeal warts* are acquired during passage through an infected birth canal. The incubation period varies from 1 week to 1 month for dermal warts and 8 to 20 months for genital warts. Genital warts and warts acquired at birth will be discussed in Chapter 20.

Dermal Warts   Epithelial cells become infected and proliferate to form dermal warts, which have distinct boundaries and remain above the basement membrane between the epidermis and the dermis. Children and young adults are more likely than older people to have dermal warts. Only a few warts are present at any one time, and most regress in less than 2 years. Removing one wart often causes regression of all others. In spontaneous regression, all warts usually disappear at the same time. Regression probably is an immunologic phenomenon.

Diagnosis and Treatment   Warts can be distinguished by immunological tests and microscopic examination of tissues. Enzyme immunoassay and immunofluorescent antibody tests can detect about three-fourths of the cases in which viruses are found microscopically. These tests sometimes fail because certain papillomas, especially genital and laryngeal papillomas and those progressing toward malignancy, produce only small quantities of antigens.

Available treatments for various kinds of warts are not entirely satisfactory. The most widely used treatment, cryotherapy, involves the freezing of tissue with liquid carbon dioxide or liq-

✓ What damages occur in congenital rubella syndrome?

✓ What precautions must be observed in rubella immunization?

✓ Distinguish between rubella and rubeola.

✓ How can one virus cause both chickenpox and shingles?

Mind Over Body

Imagine going to the doctor to get rid of a wart and being told that you need to see a hypnotist for treatment. According to Lewis Thomas, warts "can be ordered off the skin by hypnotic suggestion." In a controlled study, 14 patients with multiple warts were hypnotized and the suggestion made that all warts on one side of the body go away. Within a few weeks, this happened to 9 patients, including someone who "got mixed up and destroyed the warts on the wrong side." But if hypnosis doesn't work for you, you could just try waiting your wart out. Thomas also said that warts "inexplicably and often very abruptly . . . come to the end of their lives and vanish without a trace."

(a)

(b)

➤ Figure 19.13 **Warts.** Warts are caused by human papillomaviruses. (a) Genital warts around the vagina, a probable cause of cervical cancer, and (b) dermal warts.

Dermatomycosis	Epidermophyton floccosum	Microsporum canis	Trichophyton mentagrophytes	Trichophyton rubrum	Trichophyton tonsurans
Body ringworm		X	X	X	X
Nail ringworm	X			X	
Groin ringworm	X		X	X	
Scalp ringworm		X	X		X
Beard ringworm			X	X	
Athlete's foot	X		X	X	

➤ Table 19.1 **Dermatomycoses and common organisms found in lesions**

uid nitrogen and excision of the infected tissue. Caustic chemical agents, such as podophylin, salicylic acid, and glutaraldehyde; surgery; antimetabolites such as 5-fluorouracil; and interferon to block viruses also are used to get rid of warts. Recurrences are still common.

Fungal Skin Diseases

The fungi that invade keratinized tissue are called **dermatophytes** (dur'ma'to-fitz), and fungal skin diseases are called **dermatomycoses** (dur″ma-to-mi-ko'ses). These diseases can be caused by any one of several organisms, mainly from three genera: *Epidermophyton, Microsporum,* and *Trichophyton* (Table 19.1). They cause various types of ringworm and attack the skin, nails, and hair.

Fungi that invade subcutaneous tissues live freely in soil or on decaying vegetation and can be found in bird droppings and as airborne spores. These fungi enter tissue through a wound and sometimes spread to lymph vessels. Subcutaneous fungal infections usually spread slowly and insidiously; response to treatment is likewise slow.

Ringworm

Ringworm occurs in several forms (including athlete's foot, discussed next) and is highly contagious. For example, ringworm of the scalp is easily acquired at hairstyling establishments if strict sanitary practices are not followed. Ringworm involves the skin, hair, and nails, and most forms are named according to where they are found. **Tinea corporis** (body ringworm) causes ringlike lesions with a central scaly area. (The shape of these lesions is what originally gave rise to the misleading name "ringworm.") **Tinea cruris** (groin ringworm, or "jock itch") occurs in skin folds in the pubic region (➤ Figure 19.14). **Tinea unguium** (ringworm of the nails) causes hardening and discoloration of fingernails and toenails. In **tinea capitis** (scalp ringworm), hyphae grow down into hair follicles and often leave circular patterns of baldness. **Tinea barbae** (barber's itch) causes similar lesions in the beard.

None of the dermatomycoses result in severe diseases, and lesions usually do not invade other tissues; but they are unsightly, itchy, and persistent. The causative agents grow well at skin temperature, which is slightly below body temperature. Tissue damage caused by dermatophytes can allow secondary bacterial infections to develop.

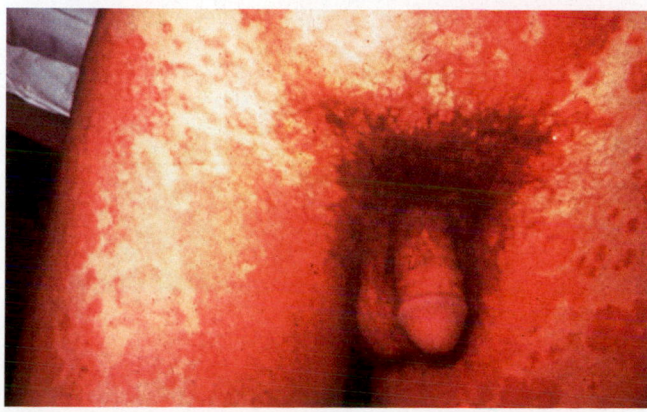

➤ Figure 19.14 **Tinea cruris, ringworm of the groin ("jock itch").** This infection is caused by the fungus *Trichophyton.*

Diagnosis and Treatment   Diagnosis of ringworm can be made by microscopic examination of scrapings from lesions, but observation of the skin itself is often sufficient. Although fungi in tissues generally do not form spores, those in laboratory cultures often do, and workers must be especially careful not to become infected from escaping spores. Large numbers of people were infected when spores from a mistakenly opened Petri plate in a university laboratory were dispersed by the building's ventilation system.

Treatment generally consists of removing all dead epithelial tissues and applying a topical antifungal ointment. If lesions are widespread or difficult to treat topically, as when they infect nailbeds, griseofulvin is administered orally. Prevention of ringworm requires avoiding contaminated objects and spores.

Athlete's Foot   In **athlete's foot,** or **tinea pedis,** hyphae invade the skin between the toes and cause dry, scaly lesions. Fluid-filled lesions develop on moist, sweaty feet. Subsequently, the skin cracks and peels, and a secondary bacterial infection leads to itchy, soggy, white areas between the toes. Athlete's foot, a form of ringworm, results from an ecological imbalance between normal flora and host defenses. The fungi that cause athlete's foot are widespread in the environment; they infect tissues when body defenses fail to repel them. Prevention of athlete's foot depends on maintaining healthy, clean, dry feet that can resist the opportunism of fungi. Antifungal agents, such as miconazole, also are effective in most cases.

Subcutaneous Fungal Infections

Sporotrichosis   **Sporotrichosis** is caused by *Sporothrix schenckii,* which usually enters the body from plants, especially sphagnum moss and rose and barberry thorns. The disease can also be acquired from other humans, dogs, cats, horses, and rodents. It is most common in the midwestern United States, especially in the Mississippi valley. A lesion appears first as a nodular mass at the site of a minor wound. The mass ulcerates, becomes chronic, granulomatous, and pus-filled, and can spread easily to lymphatic vessels. In rare instances it disseminates to internal organs, especially the lungs. Diagnosis is made by culturing specimens of pus or tissue taken from lesions. The cutaneous and lymphatic forms can be treated with potassium iodide; disseminated infections require amphotericin B. People who work with plants or soil should cover injured skin areas to protect themselves from exposure to contaminated materials.

Blastomycosis   North American **blastomycosis** (➤ Figure 19.15), caused by the fungus *Blastomyces dermatitidis,* which is most common in soil of the central and southeastern United States. The fungus enters the body through the lungs or wounds, where it causes disfiguring granulomatous, pus-producing lesions and multiple abscesses in skin and subcutaneous tissue. This condition, called **blastomycetic dermatitis,** for unknown reasons affects chiefly males in their thirties and forties. In some cases, the fungus travels in the blood and invades internal organs, causing **systemic blastomycosis.** The lungs can be infected directly by the in-

(a)   (b)

➤ Figure 19.15   **Blastomycosis lesions.** (a) Before treatment; (b) after treatment with antifungal drugs.

halation of spores, and organisms can travel from the lungs to infect other tissues. The disease usually causes relatively mild respiratory symptoms, fever, and general malaise. It can be diagnosed by finding budding yeast cells in sputum or pus, and is treated with amphotericin B or the less toxic hydroxystilbamidine.

Opportunistic Fungal Infections

Certain fungi, such as some yeasts and black molds, can invade the tissues of humans with impaired resistance. Two yeasts, *Candida albicans* and *Cryptococcus neoformans,* and certain molds, such as *Aspergillus fumigatus, A. niger,* and the zygomycetes *Mucor* and *Rhizopus,* account for many opportunistic fungal diseases.

Candidiasis   *Candida albicans,* an oval, budding yeast, is present among the normal flora of the digestive and urogenital tracts of humans. In debilitated individuals it can cause **candidiasis** (kan"di-di′uh-sis), or **moniliasis,** in one or several tissues. Superficial candidiasis appears as **thrush** (➤ Figure 19.16a), milky patches of inflammation on oral mucous membranes, especially in infants, diabetics, debilitated patients, and those receiving prolonged antibiotic therapy. ➡ (Chapter 13, p. 346) The disease appears as **vaginitis** when vaginal secretions contain large amounts of sugar, as occurs during pregnancy, when oral contraceptives are used, when diabetes is poorly controlled, or when women wear tight synthetic undergarments, which promote warmth and moisture and thus favor yeast growth. Some strains of *Candida* may be sexually transmitted. Cannery workers whose hands are in water for long periods sometimes develop skin and nail lesions (➤ Figure 19.16b). *Candida* can invade the lungs, kidneys, and heart or be carried in the blood, where it causes a severe toxic reaction. Candidiasis, the most common nosocomial fungal infection, is seen in patients with diseases such as tuberculosis, leukemia, and AIDS.

Finding budding cells in lesions, sputum, or exudates confirms the diagnosis of candidiasis. Various antifungal drugs are used to treat it. *Candida* is ubiquitous; infections can be prevented in large part by preventing debilitating conditions.

Aspergillosis   **Aspergillosis** in humans can be caused by various species of *Aspergillus,* but especially *A. fumigatus.* These fungi grow on decaying vegetation. *Aspergillus* initially invades wounds, burns, the cornea, or the external ear, where it thrives in earwax and can ulcerate the eardrum. In immunosuppressed patients, it can cause severe pneumonia. Diagnosis is made by finding characteristic hyphal fragments in tissue biopsies. Antifungal agents are only modestly successful in treatment. Because the mold is so ubiquitous that exposure is inevitable, prevention depends mainly on host defenses. Inhalation of spores by sensitized individuals can give rise to severe allergy symptoms, as well.

Humans aren't the only ones affected by Aspergillus. Rusty splotches on the pages of old books are left by paper-digesting organisms from this genus.

(a)   (b)

➤ Figure 19.16 **Candidiasis.** (a) *Candida* infections of the oral cavity (thrush), seen as white patches, are a common complication of AIDS, diabetes, and prolonged antibiotic therapy. (b) *Candida* infections of the nails are very difficult to eradicate.

(a)   **(b)**

➤ **Figure 19.17   Madura foot.** This disease can be caused by true fungi or by actinomycetes, funguslike bacteria. Lesions filled with pus and pathogens may distort the foot to the point at which amputation is necessary. (a) Bottom view; (b) top view.

Recently, *Aspergillus* has been found to cause a wasting disease of sea fans growing on Caribbean coral reefs. Thus, *Aspergillus* is versatile enough to even invade marine environments.

Zygomycoses   Certain zygomycetes of the genera *Mucor* and *Rhizopus* can infect susceptible humans, such as untreated diabetics, and cause **zygomycoses.** Once established, the fungus invades the lungs, central nervous system, and tissues of the eye orbit and can be rapidly fatal, presumably because it evades body defenses. It can be diagnosed by finding broad hyphae in the lumens and walls of blood vessels. Antifungal drugs may or may not be effective in treating zygomycoses.

Other Skin Diseases

Madura foot, or **maduromycosis,** occurs mainly in the tropics. It is caused by a variety of soil organisms, including fungi of the genus *Madurella* and filamentous actinomycetes, such as *Actinomadura, Nocardia, Streptomyces,* and *Actinomyces.* Organisms enter the body through breaks in the skin, especially in people who do not wear shoes. Initial pus-filled lesions spread and form connected lesions that eventually become chronic and granulomatous. Untreated, the organisms invade muscle and bone, and the foot becomes massively enlarged (➤ Figure 19.17). Madura foot is diagnosed by finding white, yellow, red, or black granules of intertwined hyphae in pus. So-called sulfur granules in pus are yellow hyphae and not sulfur. Unless antibiotic therapy begins early in the infection and is sufficiently prolonged, amputation may be necessary. Keeping soil particles out of wounds prevents this disease.

Skin reactions to cercariae (larvae) of several helminth species of schistosomes cause **swimmer's itch.** These cercariae parasitize birds, domestic animals, and primates and can burrow into human skin. Immune reactions cause reddening and itching and usually prevent cercariae from reaching the blood and causing schistosomiasis. ➡ (Chapter 11, p. 306) Swimmer's itch occurs throughout the United States but is especially common in the Great Lakes area.

Dracunculiasis (drah-kun″kew-li′a-sis) is caused by a parasitic helminth called a guinea worm. This disease is discussed further in Chapter 11.

Diseases of the skin are summarized in Table 19.2.

Diseases of the Eyes

Like skin diseases, eye diseases frequently result from pathogens that enter from the environment, and they cause damage to an exterior portion of our bodies. For that reason we discuss them here.

Algal Infections

Most algae manufacture their own food and are not parasitic, but some strains of the alga *Prototheca* have lost their chlorophyll and survive by parasitizing other organisms. Found in water and moist soil, they enter the body through skin wounds. By 1987, 45 cases of protothecosis had been reported, 2 from cleaning home aquariums. Protothecosis was first observed on the foot of a rice farmer, and most subsequent cases have occurred on legs or hands. In immunodeficient patients the parasite can invade the digestive tract or peritoneal cavity. A few skin infections have responded to oral potassium iodide or intravenous amphotericin B and tetracycline therapy, but no satisfactory treatment has been found for others.

> Table 19.2 | **A summary of diseases of the skin**

Disease	Agent	Characteristics
Bacterial skin diseases		
Folliculitis	*Staphylococcus aureus*	Skin abscess; encapsulated, so not reached by antibiotics
Scalded skin syndrome	*S. aureus*	Vesicular lesions over entire skin surface, fever; most common in infants
Scarlet fever	*Streptococcus pyogenes*	Sore throat, fever, rash caused by toxin; can lead to rheumatic fever and other complications
Erysipelas	*S. pyogenes*	Skin lesions spread to systemic infection; rare today, but common and fatal before antibiotics were available
Pyoderma and impetigo	Staphylococci, streptococci	Skin lesions, usually in children; easily spread by hands and fomites
Acne	*Propionibacterium acnes*	Skin lesions caused by excess of male sex hormones; infection is secoondary; common in teenagers
Burn infections	*Pseudomonas aeruginosa* and other bacteria	Growth of bacteria under eschar, often a nosocomial infection; difficult to diagnose and treat; causative agents typically antibiotic-resistant
Viral skin diseases		
Rubella	Rubella virus	Mild disease with maculopapular exanthema (discolored, pimply rash); infection early in pregnancy can lead to congenital rubella; vaccine has greatly reduced incidence
Measles	Rubeola virus	Severe disease with fever, conjunctivitis, cough, and rash; encephalitis is a complication; disease occurs mainly in children; vaccine has greatly reduced incidence
Chickenpox	Varicella-zoster virus	Generalized macular (discolored) skin lesions
Shingles	Varicella-zoster virus	Pain and skin lesions, usually on trunk; occurs in adults with diminished immunity; susceptible children exposed to cases of shingles can develop chickenpox
Smallpox	Smallpox virus	Eradicated by immunization as a human disease
Other pox diseases	Other poxviruses	Clear or bluish vesicles on skin surfaces; human infections are rare
Warts	Human papillomaviruses	Dermal warts are self-limiting; malignant warts occur in immunologic deficiencies
Fungal skin diseases		
Dermatomycoses	Dermatophytes	Dry, scaly lesions on various parts of the skin; difficult to treat
Sporotrichosis	*Sporothrix schenckii*	Granulomatous, pus-filled lesions; sometimes disseminates to lungs and other organs
Blastomycosis	*Blastomyces dermatitidis*	Granulomatous, pus-filled lesions that develop in lungs and wounds; sometimes disseminates to other organs
Candidiasis	*Candida albicans*	Patchy inflammation of mucous membranes of the mouth (thrush) or vagina (vaginitis); disseminated nosocomial infections occur in immunodeficient patients
Aspergillosis	*Aspergillus* species	Wound infection in immunodeficient patients; also infects burns, cornea, and external ear
Zygomycosis	*Mucor* and *Rhizopus* species	Occurs mainly with untreated diabetes; begins in blood vessels and can rapidly disseminate
Other skin diseases		
Madura foot	Various soil fungi and actinomycetes	Initial lesions spread and become chronic and granulomatous; can require amputation
Swimmer's itch	Cercariae of schistosomes	Itching due to cercariae burrowing into skin; immunological reaction prevents their spread
Dracunculiasis	*Dracunculus medinensis*	Larvae ingested in crustaceans in contaminated water migrate to skin and emerge through lesion; juveniles cause severe allergic reactions

➤ Figure 19.18 **Opthala-mia neonatorum.** This gono-coccal infection of the eye is most often seen in newborns, who acquire it as they pass through an infected birth canal. Adults, however, can also trans-fer bacteria to the eye from the genitals.

Bacterial Eye Diseases

Ophthalmia Neonatorum

Ophthalmia neonatorum, or conjunctivitis of the newborn, is a pyogenic (pus-forming) infection of the eyes caused by organisms such as *Neisseria gonorrhoeae* and *Chlamydia trachomatis*. Organisms present in the birth canal enter the eyes as a baby is born. The resulting infection can cause **keratitis,** an inflammation of the cornea, which can progress to perforation and destruction of the cornea and blindness. Early in the 1900s, 20 to 40 percent of children in European institutions for the blind had suffered from this disease. In developing countries the disease is still prevalent, but the true incidence is unknown because high infant mortality makes it difficult to estimate the number of original infections. Although infections are most common in newborns (➤ Figure 19.18), adults can transfer organisms by hands or by fomites from genitals to their eyes.

Penicillin was once the treatment of choice, but resistant strains have developed, and tetracycline is now used. Tetracycline has the advantage of also being effective against chlamydias and other organisms, with which mothers may also be infected. Preventive measures have almost eradicated ophthalmia neonatorum in developed countries. A few drops of 1 percent silver nitrate solution in the eyes immediately after birth kills gonococci, but it is not effective against chlamydias and can irritate the eyes. Antibiotics such as penicillin, tetracycline, and erythromycin are effective against most, but not all, known bacterial causes of ophthalmia neonatorum.

Bacterial Conjunctivitis

Bacterial conjunctivitis, or pinkeye, is an inflammation of the conjunctiva caused by organisms such as *Staphylococcus aureus, Streptococcus pneumoniae, Neisseria gonorrhoeae, Pseudomonas* species, and *Haemophilus influenzae* biogroup *aegyptius*. The BPF clone of the latter causes Brazilian purpuric fever, in which young children initially have conjunctivitus and then develop potentially fatal septicemia with extensive hemorrhaging and symptoms of meningitis. Bacterial conjunctivitis is extremely contagious, especially among children, and can spread rapidly through schools and day-care centers. Children rub itchy, runny eyes and transfer organisms to their playmates. In warm weather, gnats attracted to the moisture of tears may pick up organisms on their feet and transfer them to other persons' eyes. Topically applied sulfonamide ointment is an effective treatment. Children should not return to school until their infection is completely eliminated.

Trachoma

Trachoma (➤ Figure 19.19), from the Greek word meaning "pebbled," or "rough," is caused by specific strains of *Chlamydia trachomatis*. The disease is marked by severly swollen conjunctiva with a pebbled appearance. Scarring of the eyelids causes the eyelashes to point inward, which leads to scarring and destruction of the cornea and eventually to blindness. Trachoma is the leading cause of preventable blindness worldwide; 500 million people have trachoma, and over 20 million are already blind from it. Recent research has shown that many, if not most, of

Save Your Eyes

When soft contact lenses became popular in the 1970s, fungal and bacterial eye infections became common among wearers. Fungi are not part of the normal microflora of the eye. Some contact lens users developed fungal keratitis (corneal ulcers) because the fungus grew in their hydrogel lens and then attacked the eye surface. Fungi of the genus *Fusarium* are usually the cause. People who complain that they can't get the spots off their contact lenses may not realize that those little spots are fungal colonies growing in the lens material itself. This problem typically results from poor hygiene and improper disinfection.

Fungal eye infections are a special problem for two reasons. First, we don't have an extensive arsenal of effective drugs against fungi, as we do against bacteria. Second, fungal infections are slow to develop, so people may not pay much attention to them at first.

Conversely, bacterial eye infections, such as those caused by *Pseudomonas,* can develop so rapidly that a person can suffer serious eye damage within 24 hours. The careless use of mascara is one common way of getting a corneal ulcer, usually caused by *Pseudomonas*. Within 24 to 48 hours after the initial use of health and beauty aids such as mascara, the material in the container is well colonized by *Pseudomonas* and other organisms. Scratching an eye with a contaminated mascara applicator introduces microbes into the cornea, allowing an infection to start. The person often suffers severe pain and redness within 3 to 4 hours after the injury and seeks treatment, unlike the victim of the slow, insidious attack of fungi.

Cosmetics, especially mascara, should be discarded after 3 to 6 months of use. Preservatives in them do not last forever, and bacteria multiply rapidly. The big "one-year-supply" jar of a cosmetic may not be a bargain after all.

these cases of blindness are actually due to secondary bacterial infections developing in the trachoma-infected tissues, rather than due to the trachoma alone. Although uncommon in the United States, except among Native Americans in the Southwest, the disease is widespread in parts of Asia, Africa, and South America, sometimes affecting 90 percent of the population. Flies are important mechanical vectors, and close mother-child contact facilitates human transfer.

Trachoma was described on Egyptian papyri in 1500 B.C.

Viral Eye Diseases

Epidemic Keratoconjunctivitis

Epidemic keratoconjunctivitis (EKC), caused by an adenovirus, is sometimes called *shipyard eye;* workers are frequently infected from dust particles in the environment. After 8 to 10 days' incubation, the conjunctiva become inflamed, and eyelid edema, pain, tearing, and sensitivity to light follow. Within 2 days the infection spreads to the corneal epithelium and sometimes to deeper corneal tissue. Clouding of the cornea can last up to 2 years. EKC can be nosocomially acquired in eye clinics and ophthalmologists' offices.

Acute Hemorrhagic Conjunctivitis

Another viral eye disease, **acute hemorrhagic conjunctivitis (AHC)**, caused by an enterovirus, appeared in 1969 in Ghana. Serological studies showed it was not prevalent anywhere in the world before then. First seen in the United States in 1981, the disease occurs chiefly in warm,

> Figure 19.19   **Trachoma.** This bacterial disease, caused by *Chlamydia trachomatis,* is the leading cause of preventable blindness in the world, affecting some 500 million people. Notice the pebbled appearance of the vastly swollen conjunctiva.

humid climates under conditions of crowding and poor hygiene. AHC causes severe eye pain, abnormal sensitivity to light, blurred vision, hemorrhage under conjunctival membranes, and sometimes transient inflammation of the cornea. Onset is sudden, and recovery usually is complete in 10 days. A rare complication is a paralysis resembling poliomyelitis.

Parasitic Eye Diseases

Onchocerciasis (River Blindness)

The filarial (threadlike) larvae of the roundworm *Onchocerca volvulus* cause **onchocerciasis** (ong″ ko-ser-si′a-sis), or **river blindness,** in many parts of Africa and Central America (➤ Figure 19.20). Adult worms and microfilariae (small larvae) accumulate in the skin in nodules. ➡ (Chapter 11, p. 313) When a blackfly bites an infected individual, it ingests microfilariae, which mature in the fly and move to its mouthparts. When the fly bites again, infectious microfilariae enter the new host's skin and invade various tissues, including the eyes. In many small villages where the people depend on water from blackfly-infested rivers, nearly all inhabitants who live past middle age are blind.

Adult worms cause skin nodules that can abscess. Microfilariae cause skin depigmentation and severe dermatitis via immunological responses to live filariae or toxins from dead ones. The worst tissue damage occurs as the worms invade the cornea and other parts of the eye. Over several years they cause eye blood vessels to become fibrous, and total blindness ensues by about age 40.

Onchocerciasis can be diagnosed by finding microfilariae in thin skin samples or adult worms visible through the skin. The drug ivermectin kills adult worms quickly and microfilariae over several weeks. Drugs are available to kill microfilariae rapidly, but toxins from very many dying microfilariae can cause anaphylactic shock. Onchocerciasis could be prevented by eliminating blackflies, which congregate along rivers. DDT destroys some flies, but it allows resistant ones to survive, and it accumulates in the environment.

The small size of the pygmies of Uganda is due to onchocerciasis infections. When mothers are infected by *O. volvulus* (and most of them are), the parasite damages the pituitary gland of their fetuses. The result is dwarfism from a deficiency of growth hormone. A forest-dwelling strain of *O. volvulus* does not cause blindness but leads to severe itching as well as psychological distress in more than 9 million infected Africans. Treatment with ivermectin began in 1996.

Loaiasis

The eye worm *Loa loa,* a filarial worm endemic to African rain forests, is transmitted to humans by deer flies. Adult worms live in subcutaneous tissues and eyes (➤ Figure 19.21); microfilariae appear in peripheral blood in the daytime and concentrate in the lungs at night. Deer flies feed during the day and acquire microfilariae from infected humans. The worms develop in the flies, migrate to their mouthparts, and are transferred to other humans when the flies

---

To Bathe or Not To Bathe

The historic baths in Bath, England, have been closed. Why? Because contamination of water sources by an amoeba, Acanthamoeba, led to a fatal amoebic brain infection in one visitor. But Bath is not the only place where Acanthamoeba is found. In the mid-1980s, 24 cases of *Acanthamoeba keratitis* were reported to the CDC, 20 of which were contact lens wearers. Since then, over 100 cases have been reported. The amoeba that causes this condition is found in brackish water, seawater, and hot tubs, as well as contaminated contact lens cleaning solution. It can also travel via airborne dust. The protozoan infection causes severe eye pain and destruction of corneal epithelium. Some patients have been successfully treated with ketoconazole or miconazole, but others have required corneal transplants.

➤ **Figure 19.20 Onchocerciasis, or river blindness.** This disease is caused by the roundworm *Onchocerca volvulus,* whose microfilarial stages are transmitted by bite of the blackfly. In some African villages, nearly all adults are blind and must be led by children, the only ones in the village who can still see, although they are already infected with the worms that will eventually blind them also.

➤ **Figure 19.21** **The removal of a *loa loa* worm from a patient's eye.**

feed again. Microfilariae migrate through subcutaneous tissue, leaving a trail of inflammation, and often settle in the cornea and conjunctiva. Although they usually do not cause blindness, the shock of finding a worm over an inch long in one's eye is undoubtedly a traumatic experience! In fact, the name *Loa loa* is believed to be derived from a voodoo term meaning the rising of a devil within you.

Loaiasis (lo′ah-i″a-sis) is diagnosed by finding microfilariae in blood or by finding worms in the skin or eyes. It is treated by excising adult worms or by the use of suramin or other drugs to eradicate microfilariae. Control could be achieved by eradicating deer flies, but this is an exceedingly difficult task.

Diseases of the eyes are summarized in Table 19.3.

✓ Why is it incorrect to say that you "catch" an athlete's foot infection?

✓ How is river blindness transmitted?

✓ What is thrush, and why is it especially common in AIDS patients?

➤ **Table 19.3** **A summary of diseases of the eyes**

Disease	Agent	Characteristics
Bacterial eye diseases		
Ophthalmia neonatorum corneal lesions and can lead to bacterial conjunctivitus	*Neisseria gonorrhoeae*	Infection acquired during passage through birth canal causes corneal lesions and can lead to blindness; silver nitrate or antibiotics have nearly eradicated it in developed countries
Bacterial conjunctivitis	*Haemophilus influenzae* biogroup *aegyptius,* *Staphylococcus aureus,* *Streptococcus pneumoniae,* *Neisseria gonorrhoeae,* *Pseudomonas* species	Highly contagious inflammation of the conjunctiva in young children
Trachoma	*Chlamydia trachomatis*	Infection and destruction of cornea and conjunctiva; cause of preventable blindness
Keratitis	Bacteria, viruses, and fungi	Ulceration of the cornea; occurs mainly in immunodeficient and debilitated patients
Viral eye diseases		
Epidemic keratoconjunctivitis	Adenovirus	Inflammation of the conjunctiva that spreads to the cornea; transmitted in dust particles; also nosocomial
Acute hemorrhagic conjunctivitis	Enterovirus	Severe pain and hemorrhage under conjunctiva; highly contagious under crowded, unsanitary conditions
Parasitic eye diseases		
Onchocerciasis	*Onchocerca volvulus*	Microfilariae enter skin through blackfly bite and invade eyes and other tissues; causes dermatitis and blindness; occurs in tropics
Loaiasis	*Loa loa*	Microfilariae enter skin through deer fly bite; cause inflammation of conjunctiva and cornea

Wounds and Bites

Intact skin protects against most infectious disease agents, but we have seen that some are able to penetrate unbroken skin and that mucous membranes are not always effective barriers. In other cases, wounds and bites break the protective barrier of skin, allowing organisms to cause disease. (We discuss rabies, probably the most well-known, bite-related disease, in Chapter 23 because it is so closely associated with the nervous system.)

Wound Infections

Gas Gangrene

Associated with deep wounds, **gas gangrene** often is a mixed infection caused by two or more species of *Clostridium,* especially *C. perfringens* (found in 80 to 90 percent of cases), *C. novyi,* and *C. septicum.* Spores of these obligately anaerobic organisms are introduced by injuries or surgery into tissues where circulation is impaired and dead and anaerobic tissue is present. In body regions where oxygen concentration is low, spores germinate, multiply, and produce toxins and enzymes such as collagenases, proteases, and lipases that kill other host cells and extend the anaerobic environment.

The onset of gas gangrene occurs suddenly 12 to 48 hours after injury. As the organisms grow and ferment muscle carbohydrates, they produce gas, mainly hydrogen, and gas bubbles distort and destroy tissue (➤ Figure 19.22). Such tissue is called **crepitant** (rattling) **tissue.** The bubbles audibly "snap, crackle, and pop" when the patient is repositioned. Foul odor is such a prominent feature of gas gangrene that medical personnel can diagnose it without even entering the patient's room. High fever, shock, massive tissue destruction, and blackening of skin accompany this rapidly spreading disease. If left untreated, death occurs swiftly. Diagnosis usually is made on the basis of clinical findings, and treatment is begun before laboratory test results are available. Penicillin is administered, and dead tissue is removed or limbs are amputated. Gas gangrene often follows illegal abortions performed under unsanitary conditions; it usually necessitates a hysterectomy. It is also more common in diabetics with high blood sugar. The increased sugar levels provide an adequate carbohydrate supply in muscles for clostridial organisms to ferment. *C. perfringens* is sometimes present in bile and can occasionally cause gas gangrene of abdominal muscles following gall bladder or bile duct surgery, especially if bile is spilled.

The use of *hyperbaric* (high-pressure) oxygen chambers to treat gas gangrene is somewhat controversial. Patients are placed in chambers containing 100 percent oxygen at 3 atmospheres pressure for 90 minutes two or three times a day. The actual mechanism by which high-pressure oxygen aids recovery is not known, but presumably it kills or inhibits obligate anaerobes. Gas gangrene can be prevented by adequate cleansing of wounds, delaying closing of wounds, and providing drainage when they are closed. No vaccine is available.

> Dogs bite approximately 2 million Americans each year.

> The mouth of an infant contains more bacteria than the mouth of a dog. Human bites are therefore often more dangerous than animal bites.

➤ **Figure 19.22   Gas gangrene.** In this photograph, the disease has blackened the toes of an infected foot.

Other Anaerobic Infections

In addition to clostridia, certain non-spore-forming anaerobes are associated with some infections. *Bacteroides* and *Fusobacterium* species are normally present in the digestive tract, and *Bacteroides* accounts for nearly half of human fecal mass. *Fusobacterium* sometimes causes oral infections. If introduced into the abdominal cavity, genital region, or deep wounds by surgery or human bites, it causes infections there, too. Such infections are difficult to treat because the organisms are resistant to many antibiotics. Abscesses caused by anaerobic organisms must be drained surgically, and appropriate antibiotics used as supportive therapy.

Cat Scratch Fever

Cats are mechanical vectors of **cat scratch fever,** a disease caused by two different organisms: *Afipia felis,* a Gram-negative bacillus with a single flagellum, and, more commonly, the rickettsia *Bartonella (Rochalimaea) henselae.* The organisms are found mainly in capillary walls or in microabscesses. A cat scratch lesion resembling Kaposi's sarcoma has been found in AIDS patients. Another cat scratch lesion found in AIDS patients, as well as in other immunocompromised patients, is bacillary peliosis, in which blood-filled cavities form in bone marrow, liver, and spleen. Presumably cats acquire the organisms from the environment and carry them on claws and in the mouth. When cats scratch, bite, or lick, they transmit the organisms to humans. After 3 to 10 days, a pustule appears at the site of entry, and the patient has a mild fever, headache, sore throat, swollen glands, and conjunctivitis for a few weeks. Diagnosis is based on clinical findings and a history of contact with cats. Treatment with the antibiotics tetracycline or doxycycline is effective against *Bartonella* (*Rochalimaea*)-caused cases, whereas *Afipia*-caused cases generally do not respond, making only symptomatic relief possible. No vaccine is available, and the only means of prevention is to avoid contact with cats.

Rat Bite Fever

One type of **rat bite fever** is caused by *Streptobacillus moniliformis,* which is present in the nose and throat of about half of all wild and laboratory rats. However, only about 10 percent of people bitten by rats develop rat bite fever. Most cases result from bites of wild rats; half are reported in children under 12 living in overcrowded, unsanitary conditions. It can also result from bites and scratches of mice, squirrels, dogs, and cats.

Rat bite fever begins as a localized inflammation at a bite site that heals promptly. In 1 to 3 days, headache begins and new lesions appear elsewhere, especially on palms and soles. Fever is intermittent. By the distribution and appearance of the rash, the disease sometimes is mistaken for Rocky Mountain spotted fever. An arthritis that develops can permanently damage joints.

Another form of rat bite fever, **spirillar fever,** is caused by *Spirillum minor*. First described in Japan as *sodoku* and now known to exist worldwide, this disease is still poorly understood. The initial bite heals easily, but 7 to 21 days later it flares up and occasionally forms an open ulcer. Chills, fever, and inflamed lymph nodes accompany a red or dark purple rash that spreads out from the wound site. After 3 to 5 days symptoms subside, but they can return after a few days, weeks, months, or even years.

Diagnosis of both forms of rat bite fever is made by dark-field examination of *exudates* (oozing fluids). The disease is treated with streptomycin or penicillin; without treatment the mortality rate is about 10 percent. Technicians bitten by rodents should disinfect the bite site, seek medical treatment, and be alert for rat bite fever symptoms.

Arthropod Bites and Diseases

A variety of arthropods, including ticks, mites, and insects, cause human disease either directly or as vectors of pathogens. Many arthropods are ectoparasites that feed on blood. Their bites can be painful, and their toxins can lead to the development of anaphylactic shock. Here we consider the direct effects of bite injuries caused by arthropods; the diseases they transmit are discussed in other chapters.

Tick Paralysis

As ectoparasites, ticks attach themselves to the skin of a host, where they cause both local and systemic effects. The local effect is a mild inflammation at the bite site. Systemic effects usually occur when any of several species of hard-bodied ticks attach to the back of the neck, near the base of the skull, and feed for several days. This allows deep diffusion of anticoagulants and toxins secreted into the bite via the tick's saliva. The anticoagulant prevents the host's blood from clotting while the tick feeds on it. The toxins can cause **tick paralysis,** especially in children. Although the exact chemical composition of these toxins is not known, they appear to be produced by the ovaries of the ticks. They cause fever and paralysis, which affects first the limbs and eventually respiration, speech, and swallowing. Removal of the ticks when symptoms first appear prevents permanent damage. Failure to remove ticks can lead to death by cardiac or respiratory arrest. Both humans and livestock can be affected.

> Though many other methods have been popularized over the years, the best way to extract a tick is to pull it straight out of the skin with tweezers.

Chigger Dermatitis

The term *mites* refers not to a particular species but to an assortment of species. As ectoparasites, adult mites attach to a host long enough to obtain a blood meal and then usually drop off. Chiggers, the larvae of certain species of *Trombicula* mites, burrow into the skin and release proteolytic enzymes that cause host tissue at the bite site to harden into a tube. The chiggers then insert mouthparts into the tube and feed on blood. They cause itching and inflammation in most people and can cause a violent allergic reaction called **chigger dermatitis** in sensitive individuals. Attempts to "smother" a chigger in its tube by painting nail polish, for example, over the itchy bite are misguided. Chiggers are especially prevalent along the southeastern coast of the United States. Unlike some South American chiggers, those found in the United States cease feeding and drop to the ground hours before the itching begins.

Scabies and House Dust Allergy

Scabies, or **sarcoptic mange,** is caused by the itch mite *Sarcoptes scabiei*. By the time intense itching appears, the lesions usually are quite widespread. Scratching lesions and causing them to bleed provides an opportunity for secondary bacterial infections. Scabies is spread by close human contact and can be transmitted by sexual activities. Outbreaks are especially problematic in hospitals and nursing homes. Disinfection of linens and strict isolation are necessary to prevent spread of the infestation.

House dust mites are ubiquitous and present another problem. We all inhale airborne mites or their excrement. Although this phenomenon is unappealing, it causes disease only in individuals with an allergy to house dust. Two other mites are human commensals: *Demodex folliculorum* lives in hair follicles, and *D. brevis* in sebaceous glands. The incidence of these mites in humans increases with age from 20 percent in young adults to nearly 100 percent in the elderly.

Flea Bites

The sand flea, *Tunga penetrans,* is also sometimes called a chigger because it burrows into the skin and lays its eggs. Sand fleas cause extreme itching, inflammation, and pain and provide sites for secondary infections, including infection with tetanus spores. Surgical removal and sterilization of the wound is used to treat sand flea infections. Fleas in homes and on pets are controlled by insecticides with residual effects for days to weeks after application. Wearing shoes and avoiding sandy beaches helps protect against sand fleas. Although all fleas encountered by humans are a nuisance, only those that carry infectious agents are a public health hazard.

> Pubic lice, also called crabs, live in the groin (like their name suggests), but can also be found in the under-arms and eyelashes.

Pediculosis

Common expressions such as "nit-picking," "going over with a fine-toothed comb," "lousy," and "nitty-gritty" attest to association of lice with humans over the ages. To stay alive, lice must remain on their hosts for all but very short periods of their life. They glue their eggs (nits) to fibers of clothing and hair. Two varieties of the louse *Pediculus humanus* parasitize humans. One lives mainly on the body and on clothing in temperate climates where clothing usually is worn; the other lives on hair in any climate. **Pediculosis,** or lice infestation, results in reddened areas at bites, dermatitis, and itching. A lymph exudate from bites provides an ideal medium for secondary fungal infections, especially in the hair. The crab louse, *Phthirus pubis* (➤ Figure 11.19b), clings to skin more tightly than does the body louse and causes intense itching at bites, especially in the pubic area. It is transmitted between humans by close physical contact, but the louse itself is not known to transmit other diseases. Insecticides can be used to eradicate lice, but sanitary conditions and good personal hygiene must be maintained to prevent their return.

Other Insect Bites

Blackflies inflict vicious wounds, and sensitive individuals who are bitten get **blackfly fever,** characterized by an inflammatory reaction, nausea, and headache. Bloodsucking flies, such as the tsetse fly and the deer fly, are related to horse flies. All inflict painful bites and sometimes cause anemia in domestic animals.

Myiasis (mi-i′a-sis) is an infection caused by maggots (fly larvae). Wild and domestic animals are susceptible to myiasis in wounds. Human myiasis can occur when larvae of the botfly, a greenish, metallic-looking fly, penetrate mucous membranes and small wounds. The larvae of more than 20 species of flies, including the common housefly, can inhabit the human intestine. One such species is the botfly, aptly named *Gastrophilus intestinalis*. The Congo floor maggot, the only known bloodsucking larva, sucks human blood. In the daytime it hides in soil and debris or beneath dirt floors; at night it seeks blood meals from hosts that are usually asleep.

Screwworm fly larvae enter cattle via wounds or openings such as the ears (➤ Figure 19.23). The maggots tunnel beneath the animals' skin, causing great pain and opening avenues for infection. One steer may have over 100 maggots tunneling in its head, and the poor beast will run frantically, barely stopping to eat or drink. Ranchers lost billions of dollars before the screwworm was eradicated in North America (see the box "Screwworm Eradication"). It continues to plague other parts of the world today.

Many mosquitoes, bedbugs, and bloodsucking insects feed on humans and leave painful, itchy bites. Bedbugs hide in crevices during the day and come out at night to feed on their sleeping victims. Inflammation at bites results from an allergic reaction to the bug's saliva. Large numbers of bites can lead to anemia, especially in children. Mosquitoes and many other insects can be eradicated by the application of insecticides that have long-term residual effects. They can be kept out of dwellings by the use of tight housing construction and solid roofs instead of thatch, and by household cleanliness.

Wound and bite infections are summarized in Table 19.4.

Screwworm Eradication

An ingenious method of controlling screwworm maggots, which do great harm to cattle, was developed in the 1930s by Edward Knipling of the U.S. Department of Agriculture. He captured and irradiated male screwworm flies with radioactive cobalt, making them incapable of fertilizing eggs. If females, which mate only once, mate with sterile males, they lay infertile eggs, so far fewer new flies are produced. Irradiated males must be released frequently to maintain a high percentage of sterile males in the population. Another control method uses insects, especially wasp larvae, that are parasitic on other insects. Farmers sometimes purposely release parasitic wasps to infect crop-destroying insects and thereby reduce crop damage. Both the use of sterilized males and release of parasitic insects utilize natural habits of insects; neither contributes to environmental pollution, as most pesticides do.

✓ What is crepitant tissue?

✓ What is a "nit"?

✓ What economic impact does the screwworm have in North America today?

➤ Figure 19.23 **Screwworm fly larva (*Cochliomyia hominivorax*) on a steer.** The eggs laid near the eyes have hatched into maggots that are now eating the eye and tunneling into the head.

> **Table 19.4** **A summary of wound and bite infections**

Disease	Agent	Characteristics
Wound infections		
Gas gangrene	*Clostridium perfringens* and other species	Deep wound infections with gas production in anaerobic tissue; tissue necrosis and death can result if not treated promptly
Cat scratch fever	*Afipia felis* and *Bartonella* (*Rochalimaea*) *henselae*	Pustules at scratch site, fever, and conjunctivitis
Rat bite fever	*Streptobacillus moniliformis*	Inflammation at bite site, dissemination of lesions, intermittent fever
Spirillar fever	*Spirillum minor*	Inflammation at rat bite site heals; later reinflammation, fever, and rash
Athropod bites and infections		
Tick paralysis	Various ticks	Toxins introduced with tick bite cause fever and ascending motor paralysis
Chigger dermatitis	*Trombicula* mites	Larvae burrow into skin and cause itching and inflammation; can cause violent allergic reactions
Scabies	*Sarcoptes scabiei*	Widespread lesions with intense itching; other mites cause house dust allergy when their feces are inhaled by allergic individuals
Flea bites	*Tunga penetrans*	Itching and inflammation from adult females in skin
Pediculosis	*Pediculus humanus*	Inflammation at louse bite sites and itching; *Phthirus pubis* (pubic louse) found in pubic areas
Other insect bites		
Blackfly fever	Blackfly	Bites cause severe inflammatory reaction in sensitive individuals
Myiasis	Fly larvae	Maggots infect wounds in animals; congo floor maggot sucks human blood; screwworms injure cattle
Mosquito and other bites	Mosquitoes, some flies, bedbugs	Painful, itchy bites; several insects serve as disease vectors

Retracing Our Steps

- The agents and characteristics of the diseases discussed in this chapter are summarized in Tables 19.2, 19.3, and 19.4. Information in those tables is not repeated in this summary.

The Skin, Mucous Membranes, and Eyes

The Skin

- Skin consists of an outer **epidermis** and an inner **dermis.**
- Skin, a nonspecific defense, is a physical barrier that secretes antimicrobial chemicals, including acidic and salty secretions, waterproofing **keratin,** and **sebum.**
- Skin infections occur anywhere the skin is broken, in ducts of **sebaceous** and **sweat** (sudoriferous) **glands,** in hair follicles, and sometimes on unbroken skin.

Mucous Membranes

- **Mucous membranes** consist of a thin layer of cells that secrete **mucus.** Pathogens can become trapped in mucus.

The Eyes

- Protective structures associated with the eyes include the eyelids, eyelashes, **conjunctiva, cornea,** and **lacrimal glands.**

- **Lysozyme** in tears is microbicidal. Eyes do not have a normal microflora.

Normal Microflora of the Skin

- Large populations of normal microflora colonize the skin, differing widely at various locations. The majority are Gram-positive organisms such as *Staphylococcus* and *Micrococcus,* and coryneform bacteria.
- Metabolic by-products of normal microflora produce an acid pH that is inhospitable to most pathogens. Salt in sweat also inhibits growth of most microbes.
- Continuous shedding of skin cells carries away microbes that have colonized them.

Diseases of the Skin

Bacterial Skin Diseases

- Bacterial skin diseases are usually transmitted by direct contact, droplets, or fomites.
- Most such diseases (including **scalded skin syndrome, scarlet fever, erysipelas,** and **pyoderma**) can be treated with penicillin or other antibiotics. Treating **acne** with antibiotics may contribute

to the development of antibiotic-resistant organisms. Burn infections often are nosocomial and caused by antibiotic-resistant organisms.

Viral Skin Diseases
- **Rubella (German measles), measles (rubeola),** and **chickenpox** viruses are usually transmitted via nasal secretions. Pox diseases (including **smallpox, cowpox,** and **molluscum contagiosum**) and **warts (papillomas)** are usually transmitted by direct contact.
- Treatment of viral skin diseases can usually relieve some symptoms but not cure the disease; warts can be excised or treated chemically.

Fungal Skin Diseases
- Fungal skin diseases are called **dermatomycoses.** They include various types of **ringworm,** including **athlete's foot (tinea pedis); sporotrichosis; blastomycosis; candidiasis (moniliasis); aspergillosis;** and **zygomycoses.**
- Subcutaneous fungal infections often persist despite treatment with topical fungicides. Systemic infections are even more difficult to eradicate.
- Because of the difficulty in treating fungal skin infections, prevention by maintaining healthy skin and avoiding contamination of wounds is especially important.

Other Skin Diseases
- Other skin diseases include those caused by both fungi and bacteria (such as **Madura foot**) and by parasitic helminths (**swimmer's itch** and **dracunculiasis**).

Diseases of the Eyes
Bacterial Eye Diseases
- Bacterial eye diseases include **ophthalmia neonatorum, bacterial conjunctivitis** (pinkeye), and **trachoma.** They are transmitted by direct contact, by fomites, by insect vectors, and (in certain infections) by contamination of infants during delivery.
- Most bacterial eye diseases are treated with antibiotics and prevented by good sanitation.

Viral Eye Diseases
- Viral eye diseases, such as **epidemic keratoconjunctivitis** and **acute hemorrhagic conjunctivitis,** are transmitted by dust particles or direct contact. No effective treatments are available, but good sanitation can help prevent such diseases.

Parasitic Eye Diseases
- Controlling insect vectors reduces transmission of parasitic helminthic eye diseases, which include **onchocerciasis (river blindness)** and **loaiasis.**

Wounds and Bites
Wound Infections
- **Gas gangrene** and other anaerobic infections can be prevented by careful cleaning and draining of deep wounds; they are treated with penicillin, antitoxins, and sometimes with hyperbaric oxygen.

Other Anaerobic Infections
- **Cat scratch** and **rat bite fevers** can be prevented by avoiding injuries from cats and rats, respectively. Laboratory workers should be aware of dangers of infections from rat bites.

Arthropod Bites and Diseases
- Arthropod bites cause diseases such as **tick paralysis, chigger dermatitis, scabies (sarcoptic mange), pediculosis,** and **myiasis.** Such diseases can be prevented by good sanitation and hygiene and protecting the skin from bites.

Terminology Check

See You on the Web

Don't forget I am expecting you on the web. The following is a preparedness checklist for your web explorations. If you have questions on any of these items, you can return to the end of Chapter 1 for more details.

- Questions We'll Explore
- Concept Check questions
- Flashcards
- Study Guide

If you are ready, come visit me at

http://www.prenhall.com/black

It will be a journey you'll never forget.

Urogenital and Sexually Transmitted Diseases

When nature calls, we must answer—but where? We've all surely had to hold it at one time or another until we could get to a socially acceptable place to urinate. However, young children, pets, and wild animals are not too particular about where they urinate. Remember, though, that urine is a portal of exit for disease organisms.

Animals may carry *Leptospira* bacteria in their urinary tracts. Deer, mice, rabbits, cattle, everyone urinates on the ground, and the rain washes residual *Leptospira* bacteria into rivers, streams, lakes, and ponds. Swim in these waters with your mouth open and you could acquire leptospirosis—a debilitating, difficult to diagnose, sometimes even fatal, disease of the kidneys. In chlorinated pools and larger bodies of water you're most likely safe, but think of leptospirosis the next time you let Fido jump in and out of the wading pool.

Background Basics

In this chapter we discuss diseases of the urogenital system, including sexually transmitted diseases. As with the other chapters on diseases, keep in mind what you have learned about normal microflora ➡ (Chapter 14, p. 378), disease processes ➡ (Chapter 14, p. 384), and body defenses (Chapters 16, 17, and 18). In studying urogenital diseases, it is important to recall that the urinary and reproductive (genital) systems are closely associated. Infections in one system easily spread to the other.

Components of the Urogenital System

The **urogenital system** includes the organs of both the urinary system and the reproductive system. Thus, in considering the urogenital system we will look at the structures and sites of infection in the urinary system, the female reproductive system, and the male reproductive system.

The Urinary System

The **urinary system** consists of paired kidneys and ureters, the urinary bladder, and the urethra (➤ Figure 20.1a). The **kidneys** regulate the composition of body fluids and remove nitrogenous and certain other wastes from the body. To accomplish this, each

Questions We'll Explore

A What are the components of the urogenital system? How do they function to prevent infection?

B What are the normal microflora of the urogenital system?

C What bacteria cause urogenital diseases that are not usually sexually transmitted, and what are the important epidemiologic and clinical aspects of these diseases?

D What parasites cause urogenital diseases that are not usually sexually transmitted, and what are the important epidemiologic and clinical aspects of these diseases?

E What bacteria cause sexually transmitted urogenital diseases, and what are the important epidemiologic and clinical aspects of these diseases?

F What viruses cause sexually transmitted urogenital diseases, and what are the important epidemiologic and clinical aspects of these diseases?

kidney contains about 1 million functional units called **nephrons** (➤ Figure 20.1b). In a nephron the fluid part of the blood is filtered from the **glomerulus,** a coiled cluster of capillaries, to the kidney tubules. Starting in the *renal cortex,* nephrons remove solutes, including wastes, and water from the blood. As these materials pass through the tubules known as the *renal medulla,* essential materials such as water, salts, and sugars pass back into the blood. **Urine,** the wastes remaining in the kidney tubules, pass through collecting ducts to the **ureter** of each kidney. The ureters carry urine, which is normally free of microbes, to the **urinary bladder,**

(a)

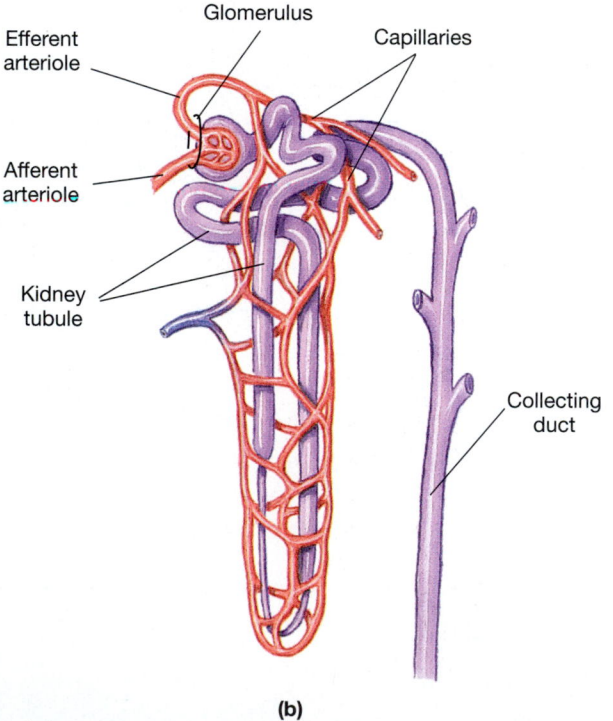

(b)

➤ Figure 20.1   **Structures of the urinary system.** (a) Rear view, including a blowup of the kidney. Infections of the urinary system often occur near the opening of the urethra, which serves as a portal of entry. The flow of urine out of the system tends to prevent many potential infections. (b) The structure of a nephron.

Ovary
(with follicles)
Uterine
tube
Urinary bladder
Urethra
Bartholin gland
Labia minora
Labia majora

Uterus
Endometrium
Colon
Cervix
Vagina
Anus

> Figure 20.2 **Structures of the female reproductive system.** The portal of entry for the reproductive tract (vagina) is separate from the urinary tract (urethra). Chemical defenses, including cervical mucus, have antibacterial actions. In addition, secretions moving from the uterine (fallopian) tubes through the uterus and cervix to the vagina make microbial infections less likely.

where it is stored until released through the **urethra** during micturition (urination). **Urinalysis,** the laboratory analysis of urine specimens, can reveal imbalances in pH or water concentration, the presence of substances such as glucose or proteins, and other conditions associated with infections, metabolic disorders, and other diseases.

The Female Reproductive System

The **female reproductive system** consists of the ovaries, uterine tubes, uterus, vagina, and external genitalia (➤ Figure 20.2). The paired **ovaries** contain cellular aggregations called **ovarian follicles,** each containing an *ovum* (plural: *ova*), or egg, and surrounding epithelial tissue. During a woman's reproductive years, an ovum capable of being fertilized is released once each month. The **uterine tubes** receive ova and convey them to the uterus. Fertilization usually occurs in the uterine tubes. The **uterus** is a pear-shaped organ in which a fertilized ovum develops. It is lined with a mucous membrane called the **endometrium,** the outer portion of which is sloughed during menstruation. The **vagina,** also lined with mucous membrane, extends from the **cervix** (an opening at the narrow lower portion of the uterus) to the outside of the body. It allows passage of menstrual flow, receives sperm during intercourse, and forms part of the birth canal. The female external genitalia include the sexually sensitive *clitoris,* two pairs of *labia* (skin folds), and the mucous-secreting **Bartholin glands.** Because they nourish offspring, **mammary glands** (breasts) are considered part of the female reproductive system. These modified sweat glands develop at puberty and contain gland cells embedded in fat that produce milk and ducts that carry the milk to the nipple.

The Male Reproductive System

The **male reproductive system** consists of the testes, ducts, specific glands, and the penis (➤ Figure 20.3). The **testes** (singular: *testis*) secrete the hormone testosterone into the bloodstream and produce sperm, which are conveyed through a series of ducts to the urethra. Secretions from **seminal vesicles** and the **prostate gland** mix with sperm to form **semen.** Other glands secrete mucus that lubricates the urethra. The **penis** is used to deliver semen into the female reproductive tract during sexual intercourse.

Sites of normal microflora in the urogenital system also are common sites of infections, partly because of their proximity to the rectum and partly because they provide warm, moist conditions suitable for microbial growth. The normal microflora of the urogenital system are found mainly at or near the external opening of the urethra of both sexes and the vagina of

Blue Breasts

The female breast, a part of the reproductive system, sometimes becomes infected. Mastitis (breast infection), usually caused by *Staphylococcus aureus,* is most likely to occur in nursing mothers. Appropriate antibiotic treatment generally gives prompt results with no sequelae. The disease is not limited to humans. In rabbits, for example, mastitis (known as "blue breasts") is common in nursing does and may be treated with penicillin.

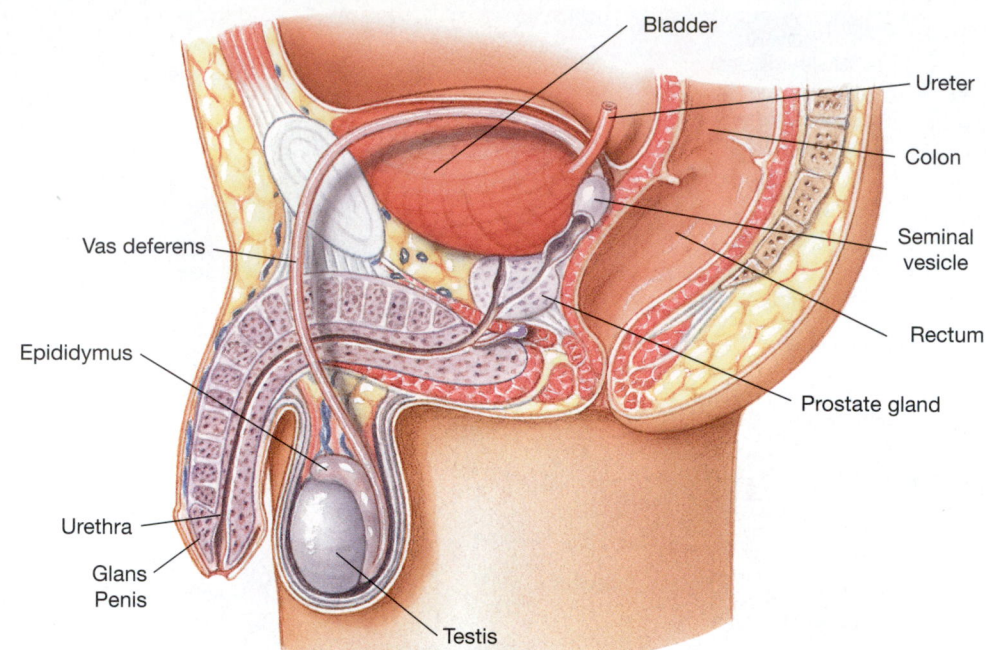

> Figure 20.3 **Structures of the male reproductive system.** Because males have a single portal of entry for the reproductive and urinary systems (the urethra), infections of either system can occur. But chemical secretions containing lysozyme and spermine provide a flow that makes colonization by pathogens more difficult.

Tiny nanobacteria, the size of viruses, live in urine, where by precipitating calcium and other minerals around themselves, they induce formation of kidney stones.

Strains of *Candida albicans* that change from a single cell to a filamentous form seem to be more likely to cause infection than strains that remain in one form.

females. Many infectious diseases of the urogenital system are sexually transmitted, including genital herpes, gonorrhea, syphilis, and nongonoccocal urethritis. Other parts of the urogenital system generally remain sterile, but a breakdown in defense mechanisms can lead to infection.

Defense mechanisms in the urogenital system are numerous. Normal microflora compete with opportunists and pathogens for nutrients and space and prevent them from causing disease. Urinary *sphincters* (muscles that close openings) act as mechanical barriers to microbes and also help prevent the backflow of urine. The flow of urine through the urethra and of mucus through both the urethra and vagina help wash away microbes. The low pH within both the urethra and, during reproductive years, the vagina prevents invasion by pathogens. Semen contains lysozyme and spermine, which help destroy invading pathogens.

Despite its defenses, the urogenital tract is poorly protected against sexually transmitted diseases. The organisms that cause gonorrhea, syphilis, and nongonoccocal urethritis tolerate acidic conditions and successfully compete with natural microflora of the urogenital tract.

Normal Microflora of the Urogenital System

In healthy individuals, all parts of the urinary tract except the portion of the urethra closest to the urethral opening, are sterile. This colonization of the urethral end ensures that even "clean catch" urine specimens will contain bacteria flushed out of the urethra during voiding. *E. coli* is the most common organism, and can reach 100,000 bacteria/ml of urine. However, if urine is collected by direct puncture (suprapubic) of the bladder with sterile needle and syringe, the urine is typically sterile in a healthy person.

Acid pH and high salt and urea concentrations retard the growth of bacteria in urine. However, urine specimens should be delivered to the lab promptly and refrigerated during any delays, as microbes will multiply rapidly in urine left sitting out at room temperature. When this happens, you typically see large numbers of several species of bacteria. A true urinary tract infection most often has large numbers of just a single organism.

Mycobacterium smegmatis (an acid-fast staining bacillus) lives on the external genitalia of both males and females. It is especially common under the foreskin of the penis of uncircumsized males, where it grows in collected secretions called *smegma*. If cleansing beneath the foreskin is inadequate when a voided urine specimen is collected, large numbers of *Mycobacterium smegmatis* bacteria will enter the specimen, where they might be confused with *Mycobacterium tuberculosis* bacilli found in the urine of people with tuberculosis of the kidneys.

In males, other than the last third of the urethra, the genital tract does not have a normal microflora and is sterile. The situation in the female reproductive tract is considerably more complex. There, hormones play a large role. During the child-bearing years, lactobacilli are the most numerous bacteria in the vagina, feeding on glycogen present in vaginal cells. Fermentation of glycogen produces lactic acid and a vaginal pH of about 4.7. This constitutes a nonspecific line of defense, as most microbes other than lactobacilli cannot survive in this acidic environment. During childhood and after menopause, glycogen is absent from vaginal wall cells, and streptococci and staphylococci rather than lactobacilli are the dominant organisms in those alkaline conditions.

Urogenital Diseases Usually Not Transmitted Sexually

Bacterial Urogenital Diseases

Urinary Tract Infections

Urinary tract infections (UTIs) are among the most common of all infections seen in clinical practice. Second only to respiratory infections, they account for 3 million office visits per year in the United States. UTIs cause **urethritis** (u-re-thri′tis), or inflammation of the urethra, and **cystitis** (sis-ti′tis), or inflammation of the bladder. Because infection easily spreads from the urethra to the bladder, most infections are properly named **urethrocystitis** (u-re″thro-sis-ti′tis). Infectious agents reach the bladder more easily through the short (4 cm) female urethra than through the longer (20 cm) male urethra. Thus, women are affected some 40 to 50 times as often as men; by the age of 30, 20 percent of all women have acquired a UTI. In males the prostate gland is closely associated with the urethra and bladder, so **prostatitis** (pros-ta-ti′tis), or inflammation of the prostate gland, often accompanies UTIs.

Each year one out of five women develops **dysuria** (dis-yur′-e-ah), or pain and burning on urination indicative of urethral infection. One-fourth of these infections will develop into chronic cystitis, which will plague the unlucky victims intermittently for years. Symptoms of cystitis include continued dysuria, frequent and urgent urination, and sometimes pus in the urine. Elderly women are prone to UTIs, and up to 12 percent of some groups suffer chronically (Table 20.1).

A major cause of UTIs is incomplete emptying of the bladder during urination. Retained urine serves as a reservoir for microbial growth, thereby fostering infection. Any factor that interferes with the flow of urine and the complete emptying of the bladder can therefore predispose an individual to UTIs. The bladder is sometimes compressed by a "sagging" uterus or by the expansion of the uterus in pregnancy. Pregnancy can also cause decreased flow of urine through the ureters. Even the ring of a diaphragm can exert enough pressure on the bladder or ureters to interfere with voiding. In men the prostate tends to enlarge with age and constrict the urethra. Finally, the problem may be behavorial rather than mechanical: Some people simply do not visit the bathroom often enough. It is important for both females and males to empty the bladder frequently and completely. People with various kinds of paralysis who cannot void completely tend to have frequent UTIs.

UTIs originating in one area often spread throughout the urinary tract by "ascending" or "descending." Infections usually begin in the lower urethra and can ascend to cause inflammation

A newly discovered adhesin, (a protein that sticks to a surface molecule) FimH, helps E. coli cells stick to the walls of the bladder, producing bladder infections.

It is common for a person with a urinary infection to complain that, despite the urge to urinate, only a small amount of urine is passed.

➤ Table 20.1	Urinary tract infections by age and sex	
Age	**Female**	**Male**
First 4 months of life	0.7%	1.3%
Four months to 5 years	4.5%	0.5%
Five years to 60 years	1–4%	< 0.1%
Over 60 years	12%	1–4%
Very elderly	up to 30%	up to 30%

> Figure 20.4 **Analysis of a urine sample.** In this test, the color produced by a diagnostic reagent determines the presence of substances such as protein and nitrate, which can indicate bacterial growth as in a urinary tract infection. Bacteria often produce nitrates while incubating in the bladder. Thus, a first morning urine specimen would contain the highest concentration of nitrates. A positive nitrate test indicates a urinary tract infection. However some bacteria further metabolize nitrates, so a negative test for nitrates does not preclude the presence of bacterial infection.

Are YOU Prone to UTIs?

A simple blood test may soon be able to tell you if you're more prone to UTIs than your classmates. Researchers at the Memorial Sloan-Kettering Cancer Center in New York City have found that women with either one of two specific types of Lewis blood groups are nearly four times as likely to develop UTIs. They aren't sure why this is true, but they think it may be because the cells lining the urinary tracts of people with these two blood types are more receptive to bacterial attachment, which leads to infection. If you are one of the women with susceptible blood types, you might want to think about preventive treatments such as taking antibiotics.

of the kidneys, or **pyelonephritis** (pi″e-lo-ne-fri′tis). Less often, infections begin in the kidneys and descend to the urethra. Although UTIs do not occur more frequently during pregnancy, they usually are more serious because they are likely to ascend. Among pregnant women who have bacteria present in the urine, 40 percent develop pyelonephritis if the infection is not treated promptly. Descending UTIs originate outside the urinary tract. Organisms that enter the bloodstream from a focal infection, such as an abscessed tooth, can be filtered by the kidneys and cause a single infection or a chronic infection. For this reason it is often suggested that persons with chronic or frequent UTIs visit their dentist to see if some undiagnosed gum infection might be the source.

Escherichia coli is the causative agent in 80 percent of UTIs, but other enteric bacteria from feces can also cause such infections. Poor hygiene, such as wiping from back to front with toilet tissue, especially in females, can introduce fecal organisms into the urethra. It is important to teach children good toilet habits and the reasons for them. When *Chlamydia* or *Ureaplasma* are responsible for infection, they usually are sexually transmitted and cause nongonococcal urethritis, which is discussed later.

Between 35 and 40 percent of all nosocomial infections are UTIs. Outpatients have a 1 percent chance of developing a UTI following a single catheterization, whereas hospitalized patients have a 10 percent chance. A great many patients with an indwelling catheter develop a UTI within the first week of use as organisms from skin, the lower urethra, or the catheter colonize the urethra. People with permanent catheters because of paralysis fight a never-ending battle against UTIs. *Staphylococcus epidermidis* often causes infections in patients with indwelling catheters. *Pseudomonas aeruginosa* commonly causes infections that follow the use of instruments to examine the urinary tract. However, *E. coli* causes nearly half of nosocomial UTIs, and *Proteus mirabilis* causes about 13 percent of them.

UTIs are diagnosed by identifying organisms in urine cultures (➤ Figure 20.4). Normal urine in the bladder is sterile, but urine is inevitably contaminated by bacteria as it passes through the lower part of the urethra. Even a clean-catch, midstream urine specimen will contain 10,000 to 100,000 organisms per milliliter. Low numbers of organisms do not necessarily rule out infection; in pyelonephritis and acute prostatitis, organisms sometimes enter the urine only in small numbers. Generally, however, only a single type of pathogen will be present at any one time. Finding several species in urine almost always means that the specimen was contaminated and the test should be repeated.

UTIs are treated with antibiotics, such as amoxicillin, trimethoprim, and quinolones, or with sulfonamides, according to susceptibilities of their causative agents. Prompt treatment helps prevent the spread of infection. UTIs can be prevented by good personal hygiene and frequent, complete emptying of the bladder.

Prostatitis

The symptoms of prostatitis are urgent and frequent urination, low fever, back pain, and sometimes muscle and joint pain. Most men have had at least one prostate infection by age 40. *Escherichia coli* is the cause of 80 percent of the cases, but it is still uncertain how the bacteria reach the prostate. Four routes of infection are possible: (1) by ascent through the urethra, (2) by backflow of contaminated urine, (3) by passage of fecal organisms from the rectum through lymphatics and to the prostate, and (4) by descent of bloodborne organisms. Although uncommon, chronic prostatitis is a major cause of persistent UTIs in males, and it can cause infertility. Acute prostatitis usually responds well to appropriate antibiotic therapy without leaving sequelae.

Pyelonephritis

Pyelonephritis, an inflammation of the kidney, is usually caused by the backup of urine and consequent ascent of microorganisms. Urine backup can be caused by a number of factors, including lower urinary tract blockage or anatomical defects. Young children in particular often have imperfectly formed urinary tract valves that do not prevent urine from backing up. *Escherichia coli* causes 90 percent of outpatient cases and 36 percent of those in hospitalized patients, but yeasts such as *Candida* are occasionally responsible for the infection.

Pyelonephritis—and any other UTI—can be asymptomatic. When present, the symptoms are indistinguishable from those of cystitis, except that sometimes chills and fever occur. Dilute urine is another common finding, leading to frequent urination and **nocturia** (nok-tu′re-ah), or nighttime urination. Patients must be carefully evaluated to identify underlying, predisposing conditions such as kidney stones or other blockages, which need to be relieved. Pyelonephritis is more difficult to treat than lower UTIs, but nitrofurantoin, sulfonamides, trimethoprim, ampicillin, gentamycin, Cipro, and quinolones usually are effective. Often these are started intravenously. During kidney failure or impaired kidney function, drugs must be used with care to prevent toxic accumulations.

Glomerulonephritis

Glomerulonephritis (glom-er″u-lo-nef-ri′tis), or *Bright's disease,* causes inflammation and damage to the glomeruli of the kidneys. It is an immune complex disease that sometimes follows a streptococcal or viral infection. As with rheumatic fever (rheumatogenic strains of *Streptococcus pyogenes* ➡ Chapter 19, p. 600), nephrogenic strains of *S. pyogenes* contain components in their cell walls that, when processed by the immune system, resemble tissue components present in glomerular tissue. This initiates the production of antibodies that cannot distinguish between bacterial cell wall and human glomerular components, resulting in antigen–antibody complexes. Antigen–antibody complexes are filtered out in the kidney, complement is activated, and there is an inflammatory response of glomerular capillaries (➢ Figure 20.5). The inflamed

There is no specific treatment for acute glomerulonephritis, since the illness will usually heal completely within 3 to 12 months after onset.

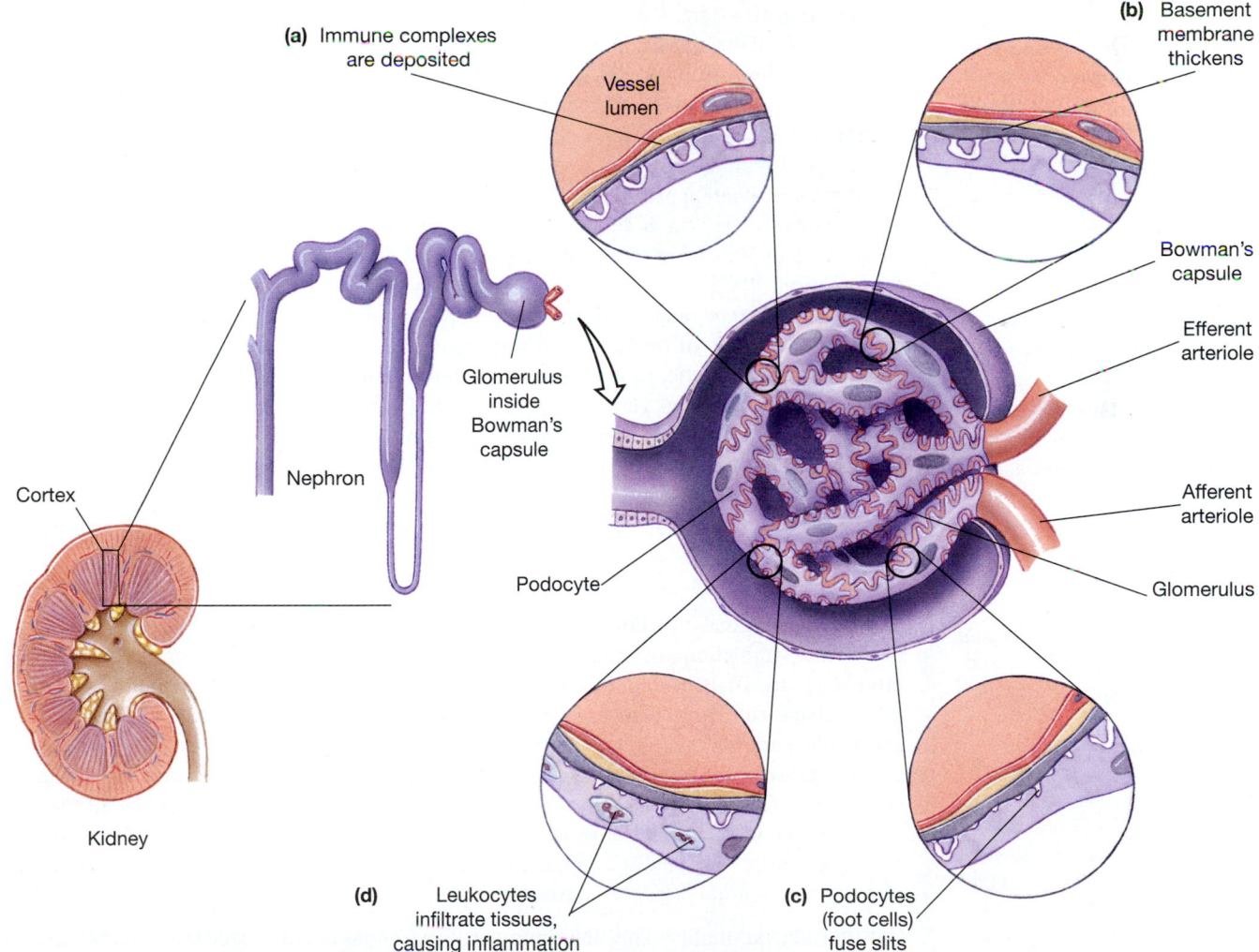

➢ **Figure 20.5   Glomerulonephritis.** This disease occurs when (a) immune complexes are deposited on the filtering surfaces of the glomerulus, which is located inside the Bowman's capsule in the cortex of the kidney. (b) Basement membranes of cells thicken. (c) Podocytes (foot cells) in the glomerulus, which normally act as filters, fuse their slits, and (d) leukocytes are attracted into the tissues. Kidney function is diminished, sometimes to the point of kidney failure.

> Figure 20.6   **A colorized TEM of *Leptospira interrogans.*** Normally found in animals, this spirochete sometimes causes liver infections and jaundice in humans (11,000X).

vessels leak blood and protein into the urine. Attraction and movement of phagocytic cells out of blood vessels, plus the release of hydrolytic enzymes in an attempt to engulf the newly formed complexes, further adds to the damage of the glomeruli of the kidneys.

Because of the risk of glomerulonephritis, organisms from throat and other infections that might be streptococcal (Chapter 21) should be cultured and appropriate antibiotics given. ➡ (p. 600) Although most people recover from glomerulonephritis, some have permanent residual kidney damage, and a few die.

Leptospirosis

Leptospirosis is caused by the spirochete *Leptospira interrogans* (➤ Figure 20.6). It is a zoonosis usually acquired by humans through contact with contaminated urine, directly or in water or soil. Dogs, cats, and many wild mammals carry these spirochetes (recall the "Dog Germs" box in Chapter 15). ➡ (p. 415) In some parts of the world, more than 50 percent of the rats are carriers of *Leptospira*. The bacteria live within the convoluted tubules of the kidneys and are shed into the urine. Often rain washes them from streets and the soil into natural bodies of water. They die quickly in brackish or acidic water but can survive for 3 months or longer in neutral or slightly basic water. Ponds or rivers bordering pastures where animals graze are especially likely to be contaminated.

The organisms enter the body through mucous membranes of the eyes, nose, or mouth or through skin abrasions. These organisms have a hook at the end, and can also burrow into the soft portions of the skin of the patient—usually soles of the feet or palms of the hands. Parents of small children often become infected from contact with pets. The adults frequently clean up after the pets and become infected with leptospires. Dogs, being ignorant of the niceties of toilet training, often account for a cluster infection that affects all who shared a swimming or wading pool with the dogs.

After an incubation period of 10 to 12 days, leptospirosis usually occurs as a febrile and otherwise nonspecific illness. In most cases, an uneventful recovery takes place in 2 to 3 weeks. However, 5 to 30 percent of the untreated cases result in death. A particularly virulent form of the infection, *Weil's syndrome,* is characterized by jaundice and significant liver damage.

Diagnosis can be made by direct microscopic examination of blood, but it is often not even considered because of the nonspecific symptoms and low incidence of leptospirosis. Many cases are asymptomatic, so it is difficult to know the true incidence of the disease. Generally, fewer than 100 cases per year are reported in the United States.

Leptospires are susceptible to almost any antibiotic if it is given in the first 2 or 3 days of the disease, but not after the fourth day. Patients have long-term immunity to the particular strain of *Leptospira* that infected them but not to any other strains. Leptospirosis can be prevented by vaccinating pets and avoiding swimming in or contact with contaminated water.

Bacterial Vaginitis

Vaginitis, or vaginal infection, usually is caused by opportunistic organisms that multiply when the normal vaginal microflora are disturbed by antibiotics or other factors. Predisposing conditions include diabetes mellitus, pregnancy, use of contraceptive pills, menopause, and conditions that result in imbalances of estrogen and progesterone, all of which change the pH and sugar concentration in the vagina. Several organisms account for a share each of vaginitis cases. The bacterium *Gardnerella vaginalis,* in combination with anaerobic bacteria, accounts for about one-third of the cases. *Mobiluncus* is seen in a few infections; it may be a separate organism or a clinical variant of *G. vaginalis.* The protozoan *Trichomonas vaginalis,* which can be opportunistic but usually is transmitted sexually, accounts for about one-fifth of the cases. The fungus *Candida albicans* causes most other cases.

Gardnerella vaginalis   This tiny Gram-negative bacillus or coccobacillus is present in the normal urogenital tract in 20 to 40 percent of healthy women. The normal vaginal pH is 3.8 to 4.4 in women of reproductive age and near neutral in young girls and elderly women. When the vaginal pH reaches 5 to 6, *Gardnerella vaginalis* interacts with anaerobic bacteria such as *Bacteroides* and *Peptostreptococcus* to cause vaginitis. None of these organisms alone produces disease. Because different anaerobes interact with *Gardnerella,* this kind of vaginitis sometimes is called *nonspecific vaginitis. Gardnerella* vaginitis produces a frothy, fishy-smelling vaginal discharge. The discharge, although usually small in volume, contains millions of organisms. Males occasionally get **balantitis** (bal-an-ti´-tis), an infection of the penis that corresponds to female vaginitis. Lesions appear on the penis after sexual contact with a woman who has vaginitis.

Diagnosis can be made when wet mounts of the discharge exhibit "clue cells," vaginal epithelial cells covered with many tiny rods or coccobacilli (➢ Figure 20.7). Metronidazole (Flagyl) suppresses vaginitis by eradicating the anaerobes necessary for continuance of the disease but allows normal lactobacilli to repopulate the vagina. This effect supports the notion that *Gardnerella* vaginitis requires an association with anaerobes. Ampicillin and tetracycline are also sometimes used in treatment. Unflavored (plain) live-culture yogurt used as a douche effectively replaces lactobacilli of normal microflora killed by antibiotic treatment.

> *Predisposing conditions to vaginitis include douching too often, tampons irritating the vaginal walls, tight slacks, and panties or panty hose without a cotton crotch.*

Toxic Shock Syndrome

Infection with certain toxigenic strains of *Staphylococcus aureus* can produce **toxic shock syndrome (TSS)**. Prior to 1977 only 2 to 5 cases were reported each year, but then the incidence rose suddenly, reaching a peak of about 320 cases in September 1980. Since 1986 fewer than 25 cases per year have been reported. The sudden rise was associated with new superabsorbent, but abrasive, tampons, which were left in the vagina for longer than usual periods of time. The tampons caused small tears in the vaginal wall and provided appropriate conditions for bacteria to multiply.

Between 5 and 15 percent of women have *S. aureus* among their vaginal microflora, but only a small fraction of these strains cause TSS. Most cases occur in menstruating women, but men, children, and postmenopausal women with focal infections of *S. aureus* occasionally develop TSS. Organisms enter the blood or grow in accumulated menstrual flow in tampons. They produce *exotoxin C,* which enhances the effects of an endotoxin; how these toxins exert their effects is not clearly understood. Clinical manifestations include fever, low blood pressure (shock), and a red rash, particularly on the trunk, that later peels. Immediate treatment with nafcillin has held the mortality rate to 3 percent. Deaths, when they occur, usually are due to shock. Recurrence is a frequent possibility, especially during subsequent menstrual cycles. Antibiotics can be given prophylactically to prevent recurrences, but women who have had TSS can reduce their risk of recurrence by ceasing to use tampons.

Although infrequent changing of superabsorbent tampons accounts for most cases of TSS, the contraceptive sponge now causes some. Males with boils, furuncles, or other staphylococcal infections also sometimes get TSS.

Parasitic Urogenital Diseases

Trichomoniasis

Although **trichomoniasis** is transmitted primarily by sexual intercourse, it is discussed here because of its similarity to other kinds of vaginitis and because children (and rarely adults) can be infected from contaminated linens and toilet seats. At least three species of protozoa of the

WARNING: Important Information About Toxic Shock Syndrome (TSS)

If you use tampons frequently, you are probably used to throwing out the insert that is found in practically every box. But have you ever read the warning? If you did, you'd know that the warning signs for TSS are sudden fever (usually 102°F or more), vomiting, diarrhea, fainting or near fainting when standing up, dizziness, or a rash that looks like a sunburn. You'd also know that the reported risks are higher to women under 30 years of age and that the incidence of TSS is estimated to be between 1 and 17 cases of TSS per 100,000 menstruating women and girls per year. You can reduce your risk of getting TSS during your period by alternating tampon use with feminine pads, by selecting the minimum absorbency needed to control your menstrual flow, and by asking your doctor about any other questions you might have.

➢ Figure 20.7   **A clue cell of Gardnerella infection in a vaginal smear.** Notice the many bacilli adhering to the surface of the clue cell.

genus *Trichomonas* can parasitize humans, but only *T. vaginalis* causes trichomoniasis. The others are commensals; *T. hominis* is found in the intestine, and *T. tenax* in the mouth. *T. vaginalis* is a large flagellate with four anterior flagella and an undulating membrane (➤ Figure 20.8). It infects urogenital tract surfaces in both males and females and feeds on bacteria and cell secretions. Because the optimum pH for the organism is 5.5 to 6.0, it infects the vagina only when vaginal secretions have an abnormal pH. The symptoms of trichomoniasis are intense itching and a copious white discharge, especially in females.

Trichomoniasis is diagnosed by microscopic examination of smears of vaginal or urethral secretions and treated with metronidazole (Flagyl) and restoration of normal vaginal pH in women. Flagyl cannot be used during pregnancy because it causes abortions, but it is important to get rid of the infection before delivery to prevent infecting the infant. A vinegar douche usually is effective.

The effects of different species of *Trichomonas* illustrate the extreme variation in the degree of damage parasites can do. *T. hominis* and *T. tenax* are considered commensals. However, *T. foetus* causes serious genital infections in cattle. It is the leading cause of spontaneous abortion in cows and, in fact, is responsible for losses to U.S. cattle breeders of nearly $1 million per year.

Sexually Transmitted Diseases

Sexually transmitted diseases (STDs) have become an increasingly serious public health problem in recent years, in part because of changing sexual behaviors. In addition, some causative agents are becoming resistant to antibiotics, and no vaccines have been developed to control any STDs. Consequently, the only means of preventing sexually transmitted diseases is to avoid exposure to them.

Acquired Immune Deficiency Syndrome (AIDS)

Although AIDS is in many cases an STD, it is not transmitted exclusively by sexual contact. We discussed it in Chapter 18 with disorders of the immune system. ➡ (p. 518)

Bacterial Sexually Transmitted Diseases

Gonorrhea

The term **gonorrhea** means "flow of seed" and was coined in A.D. 130 by the Greek physician Galen, who mistook pus for semen. By the thirteenth century, the venereal (from Venus, goddess of love in Roman mythology) transfer of this disease was known. But it was not until the mid-nineteenth century that gonorrhea was recognized as a specific disease; until then it was thought to be an early symptom of syphilis. The causative organism, *Neisseria gonorrhoeae,* was first described by the German physician Albert Neisser in 1879. It is a Gram-negative, spherical or oval diplococcus with flattened adjacent sides and resembles a pair of coffee beans facing each other (➤ Figure 20.9).

Drying kills the organisms in 1 to 2 hours, but they can survive for several hours on fomites. Cases in which *Neisseria* survived improper laundering in the hospital have been documented; in dried masses of pus, the bacteria can survive for 6 to 7 weeks! So although these organisms are ordinarily thought of as being very fragile, some are quite robust.

The infectivity of *Neisseria* is related in several ways to its pili. Attachment pili (fimbriae) enable gonococci to attach to epithelial cells that line the urinary tract so that they are not swept out with the passage of urine. (In fact, sloughing of epithelial cells is one of the body's defenses against such infections.) These bacteria also use their pili to attach to sperm; conceivably, swimming sperm could carry gonococci into the upper part of the reproductive tract. Strains lacking pili usually are nonvirulent. The search for a vaccine against gonococcal pili is under way, but success has not yet been achieved.

Gonococci produce an endotoxin that damages the mucosa in fallopian tubes and releases enzymes such as proteases and phospholipases that may be important in pathogenesis. They also produce an extracellular protease that cleaves IgA, the immunoglobulin present in secretions. ➡ (Chapter 17, p. 469) Gonococci adhere to neutrophils and are phagocytized by them. ➡ (Chapter 16, p. 437) Phagocytosis kills some of the bacteria, but survivors multiply inside

➤ Figure 20.8   *Trichomonas vaginalis.* Notice the characteristic undulating membranes and flagella as seen with the scanning electron microscope.

✓ Why are urinary tract infections more common in women than in men?

✓ Distinguish between the four routes of infection in prostatitis.

✓ Why is kidney tissue damaged in glomerulonephritis, since bacteria are not directly infecting the kidneys?

✓ What are "clue cells"?

(a)

(b)

➤ Figure 20.9 **Diplococci of *Neisseria gonorrhoeae.*** (a) As small, dark purple dots inside the cytoplasm of leukocytes in a urethral smear (1900X), and (b) magnified 40,000X by TEM, showing the internal structure of the pair.

PMNLs. A typical wet mount of a urethral discharge will show PMNLs with diplococci in their cytoplasm (➤ Figure 20.9). Gonococci also obtain iron for their own metabolic needs from the iron transport protein transferrin.

The Disease Humans are the only natural hosts for gonococci. Gonorrhea is transmitted by carriers who either have no symptoms or have ignored them. As many as 40 percent of males and 60 to 80 percent of females remain asymptomatic after infection and can act as carriers for 5 to 15 years. Very few organisms are required to establish an infection; half of a group of males given a urethral inoculum of only 1000 organisms developed gonorrhea. Following a single sexual exposure to an infected individual, about one-third of males become infected. After repeated exposures to the same infected person, three-quarters of males eventually become infected. Of males who develop symptoms, 95 percent have pus dripping from the urethra within 14 days, and many develop symptoms sooner (➤ Figure 20.10a). The typical incubation period is 2 to 7 days.

Both the contraceptive pill and IUDs have contributed to the epidemic of gonorrhea now seen in the Western world (➤ Figure 20.11). Their introduction in the 1960s led to increased sexual freedom and decreased use of condoms and spermicides. Women using the contraceptive pill have a 98 percent chance of being infected upon exposure because the pill alters vaginal conditions in favor of gonococcal growth. Although condoms and spermicides are less reliable contraceptives, they do offer some protection from gonorrhea. Gonorrhea spreads into the endometrial cavity and fallopian tubes two to nine times faster in women who use IUDs than in those who do not use them. The exposure of blood vessels during menstruation allows bacteria to enter the circulatory system and thus facilitates the development of bacteremia.

Although gonorrhea is thought of as a venereal disease (VD), it also affects other parts of the body (➤ Figure 20.10b). Pharyngeal infections, which develop in 5 percent of those exposed by oral sex, are most common in women and homosexual men. The patient may develop a sore throat, but most pharyngeal cases are asymptomatic. The infected tissues act as a focal source for bacteremia. Anorectal infection, especially common in homosexual men, occurs in the last 5 to 10 cm of the rectum. It can be either asymptomatic or painful, with constipation, pus, and rectal bleeding. Women with vaginal gonorrhea also often have anorectal infection. Such infection can occur in the absence of anorectal intercourse due to contamination of the anal opening by organism-laden vaginal discharges. Medical personnel doing internal pelvic examinations can avoid spreading the infection by using a different gloved finger for anal examination than was inserted into the vagina.

The urethra is the most common site of gonorrhea infections in males. The most common site of infection in females is the cervix, followed by the urethra, anal canal, and pharynx. Skene's glands in the urethra and Bartholin's glands near the vaginal outlet also can be infected in females. As many as half of infected females develop **pelvic inflammatory disease (PID)**, in which the infection spreads throughout the pelvic cavity. Studies done in Sweden showed that sterility often follows PID because of tubal occlusion by scarring. The rate of sterility increases

Gonorrhea ranks first among the reported communicable diseases in the United States, with 1 million new cases reported each year.

(a)

➤ Figure 20.10   **Symptoms of gonorrhea.** Symptoms sometimes include (a) in males, a urethral "drip" of pus and (b) an infrequent complication, gonococcally caused arthritis.

(b)

with the number of infections—13 percent after one infection, 35 percent after two infections, and 75 percent after three infections.

Disseminated infections, which occur in 1 to 3 percent of cases, produce bacteremia, fever, joint pain, endocarditis and skin lesions that can be pustular, hemorrhagic, or necrotic. When organisms reach the joints, they can cause arthritis. Gonococcal arthritis is now the most common joint infection in people 16 to 50 years old. Another complication of gonorrhea is infection of the lymphatics that drain the pelvis. Scarring produces tight, inflexible tissue that immobilizes pelvic organs into the condition known as "frozen pelvis."

Transfer of organisms by contaminated hands, or fomites such as towels, can result in eye infections. If untreated, severe scarring of the cornea and blindness can result. Newborns can acquire ophthalmia neonatorum during passage through the birth canal of an infected mother. ➡ (Chapter 19, p. 554) Pus that accumulates behind swollen eyelids can spurt forth with great pressure and infect health care workers. Such infections have occurred despite immediate treatment.

Age affects the site and type of gonococcal infection. During the first year of life, infection usually results from accidental contamination of the eye or vagina by an adult. Between 1 year and puberty, gonorrhea usually occurs as vulvovaginitis in girls who have been sexually molested. A girl's vaginal epithelium has less keratin and is more susceptible to infection than a woman's. The girls exhibit pain on urination, vulvar and perianal soreness, discomfort on defecation, and a yellowish-green discharge from vaginal and urethral openings. Both boys and girls can develop a pus-filled anal discharge indicative of anorectal gonorrhea as a result of sexual abuse. Failure of a hospital to launder sheets properly led to an outbreak of vulvovaginitis in female pediatric patients who acquired it from sitting on the sheets.

(a)

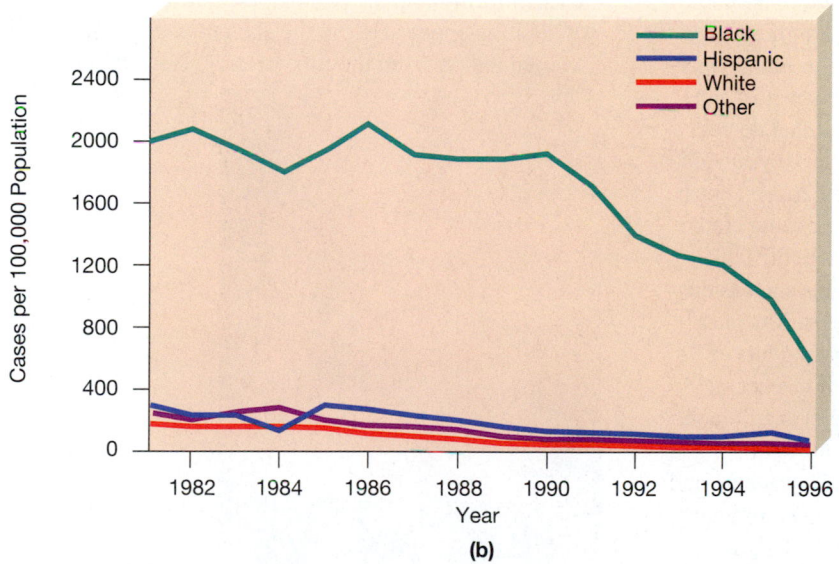

(b)

➤ **Figure 20.11   The U.S. incidence of gonorrhea.** (a) By state (b) by ethnicity. (*Source*: CDC.)

Diagnosis, Treatment, and Prevention   Diagnosis is made by identifying *N. gonorrhoeae* in laboratory cultures. Being a rather fastidious organism, it requires high humidity and ambient carbon dioxide for growth. Temperatures of 35° to 37°C and pH of 7.2 to 7.6 are optimal. Many species of *Neisseria* exist. A positive result on a screening test may not be due to *N. gonorrhoeae;* confirmatory tests should always be done. Samples from patients suspected of having *Neisseria* infections are inoculated from a swab directly into a special medium for transport and incubation in a laboratory. ➡ (Chapter 6, p. 158)

Sulfonamides were the first agents found to treat gonococcal infections. As some strains became resistant to sulfonamides, penicillin became available. At first, penicillin G was effective in very small amounts, but now much larger doses must be used. Some strains of gonococci are entirely resistant to penicillin because they produce the enzyme β-lactamase, which enables them to break down this antibiotic. CDC now recommends a single intramuscular (250 mg) dose of the cephalosporin ceftriaxone. Increases in resistance have made it very difficult to combat gonorrhea. Organisms typically acquire antibiotic resistance by means of their sex pili. One wonders if it will always be possible to find another antibiotic to combat resistant strains.

Homo sapiens and *Treponema pallidum*: An Evolutionary Partnership

Historically, four different diseases—syphilis, pinta, yaws, and bejel—have been caused by the same pathogen, *Treponema pallidum*. DNA studies show the causative organisms to be extremely similar for all four diseases. Only recently has the causative agent of pinta evolved significantly from *T. pallidum;* it has been given a separate species name, *T. carateum*. Researchers C. J. Hackett and Ellis Hudson suggest that the treponemes have evolved in accordance with changing human lifestyles.

The first pathogen infected a wild animal and occasionally infected people as a zoonosis. As such, it caused the set of symptoms known as *pinta,* a mild skin disease that does not reach deeper tissues. Its red, slate blue, and white lesions are disfiguring but not life-threatening. Found in the tropics, it is passed from person to person through direct skin contact. Warm skin, moist with perspiration, provides an ideal environment for the treponeme.

Around 10,000 B.C., probably in Africa, the pathogen mutated and gave rise to *yaws,* a disease that thrived in the somewhat more densely settled human villages. The highest incidence of yaws is still in tropical Africa and in Latin America. Typical symptoms are deep oozing skin lesions that can penetrate underlying tissues and erode bone but that do not reach the viscera. Most victims are children 4 to 10 years old.

By 7000 B.C., people carrying the treponeme moved northward. The climate was cooler and drier, required more clothing, and made transmission via skin contact more difficult. The pathogen survived best in warm, moist areas such as the mouth, groin, and armpits. Transfer to a new host was most likely via oral contact such as kissing or by sharing bits of food or dishes. Today the disease is known as nonvenereal syphilis, or *bejel*. The strain responsible for bejel is still more invasive; it causes granulomatous lesions like those of syphilis and attacks tissues of the skeletal, cardiovascular, and nervous systems. It is often transferred by young children, because older ones are generally past the infectious stage.

Bejel is almost always acquired prior to puberty and is not sexually transmitted. It is never transmitted congenitally, presumably because the mother's infection becomes nontransmissible before she is old enough to become pregnant. Bejel epidemics occurred during the fifteenth century in Africa and in the seventeenth and eighteenth centuries in Scotland and Norway. Today the worldwide incidence of bejel is higher than that of venereal syphilis.

As standards of living improved and people covered more of their bodies, the surest means of transmitting the treponeme was via sexual activity and warm, moist reproductive tracts. The pathogen, again evolving in accordance with changing human lifestyles, came to cause venereal syphilis, in which the primary chancre appears to be analogous to yaws lesions. The disease is more severe than bejel because it penetrates organs to a greater and more devastating extent.

There is further evidence for the evolution of treponemes with human culture. First, none of the four diseases can be found in the same geographic area as any other. Second, infection with yaws prevents concurrent syphilis. Third, people with yaws who move to cooler climates lose their yaws symptoms and develop those of bejel instead. Evidently, the treponeme is extremely adaptive to its environment. Indeed, syphilis itself seems to have evolved considerably over the past five centuries. Accounts that date from the 1490s describe the disease as killing its victims in a matter of weeks. Today it is far less virulent; the incubation period alone is longer than that, and people with syphilis may survive for decades.

If the theory of parallel evolution of pathogen and humans is correct, historians who blamed Columbus's sailors for bringing the organism back from the New World were mistaken. The fact that the disease appeared in Europe at about the same time as Columbus's voyages is probably just a coincidence. The treponeme most likely arrived in Europe with slaves from Africa before the discovery of America, eventually evolving to the point at which it could cause venereal syphilis. Evidence for this hypothesis has been found in Native American burial grounds in New England. The remains of individuals who died before the arrival of settlers from the Old World show no trace of syphilis. Signs of syphilis appear first in the skeletons of young native women dating from the time of early European settlement.

(a)

(b)

(a) The rash of pina. (b) Lesions and bone damage caused by yaws. The bone damage can be similar to that of saber shin produced by syphilis (Figure 20.15).

Gonorrhea patients often have other STDs. A 7-day course of treatment with oral doxycycline has the advantage of killing *Chlamydia,* which may concurrently be present. Forty-five percent of patients having gonorrhea also test positive for *Chlamydia.* Follow-up cultures should be done 7 to 15 days after completion of treatment to be sure the infection is cured. Additional follow-up cultures are also advised at 6 weeks, as 15 percent of women who are negative at 7 to 14 days are positive again at 6 weeks—possibly because of reinfection. All sexual partners must be treated.

The best means of preventing gonorrhea is to avoid sexual contact with infected individuals. No vaccine is available.

Syphilis

Syphilis is caused by the spirochete *Treponema pallidum,* an active motile organism with fastidious growth requirements. ➡ (➤ Figure 3.27b, p. 64) The evolution of this organism has paralleled human evolution. *Treponema pallidum* eluded discovery until it was finally stained in 1905. Today in the United States, syphilis is much less common than gonorrhea. However, the incidence of syphilis has, on average, been on the rise since 1960, although it has been declining steadily since 1990 (➤ Figure 20.12).

The disease is ordinarily transmitted by sexual means, but it can be passed in body fluids such as saliva. It thereby creates a hazard for dentists, dental hygienists, and those who kiss. It is not transmitted in food, water, or air or by arthropod vectors. Humans are its only reservoir. Donated blood need not be screened for syphilis because any spirochetes present are killed when the blood is refrigerated.

The Disease   A typical case of syphilis progresses as follows:

1. *Incubation stage:* Over a period of 2 to 6 weeks after entering the body, the organisms multiply and spread throughout the body.

2. *Primary stage:* An inflammatory response at the original entry site causes formation of a **chancre** (shang′ker), a hard, painless, nondischarging lesion about $1/2$ inch in diameter. One or more primary chancres usually develop on the genitals but can develop on lips or hands (➤ Figure 20.13). In females, chancres on the cervix or another internal location sometimes escape detection. The patient often is embarrassed to seek medical attention for a lesion in a genital location and hopes that it will just "go away." And go away it does after about 4 to 6 weeks, without leaving any scarring. The patient thinks all is well, but the disease has merely entered the next stage.

3. *Primary latent period:* All external signs of the disease disappear, but blood tests diagnostic for syphilis are positive.

4. *Secondary stage:* Symptoms appear, disappear, and reappear over a period of about 5 years, during which the patient is contagious. These symptoms include a copper-colored rash, particularly on the palms of the hands and the soles of the feet (➤ Figure 20.14a), and various pustular rashes and skin eruptions. Painful, whitish mucous patches swarming with spirochetes appear on the tongue, cheeks, and gums. Kissing spreads the spirochetes to others. These lesions heal uneventfully, and the patient again thinks all is well. But the disease now has entered the next stage.

5. *Secondary latent stage:* Again all symptoms disappear, and blood tests can be negative. This stage can persist for life or for a highly variable period, or may never occur. Symptoms can reoccur at any time during latency. In some patients syphilis does not progress beyond this stage, but in many patients it progresses to the tertiary stage.

6. *Tertiary stage:* Permanent damage occurs throughout various systems of the body. A wide assortment of symptoms can appear; syphilis has been called the "great imitator" because its symptoms can mimic those of so many other diseases. Most involve the cardiovascular and nervous systems. Blood vessels and heart valves are damaged. In long-standing cases, calcium deposits in heart valves can be so extensive as to be visible on a chest X-ray. Neurological damage, called **neurosyphilis,** can include thickening of the meninges; ataxia (a-tax′e-ah), or an unsteady gait or inability to walk; and paresis (par′e-sis),

Finished Before It Started

Here in the United States, we rarely use the relatively new quinolone antibiotics to treat gonorrhea. We don't want to selectively breed resistant strains, so we have restricted their use to special cases. Unfortunately, the quinolones have been so used in the Philippines that quinolone-resistant strains of *Neisseria gonorrhoeae* have already emerged there. If these strains make their way to the United States, the quinolones will be obsolete for treating gonorrhea before they can even be used generally.

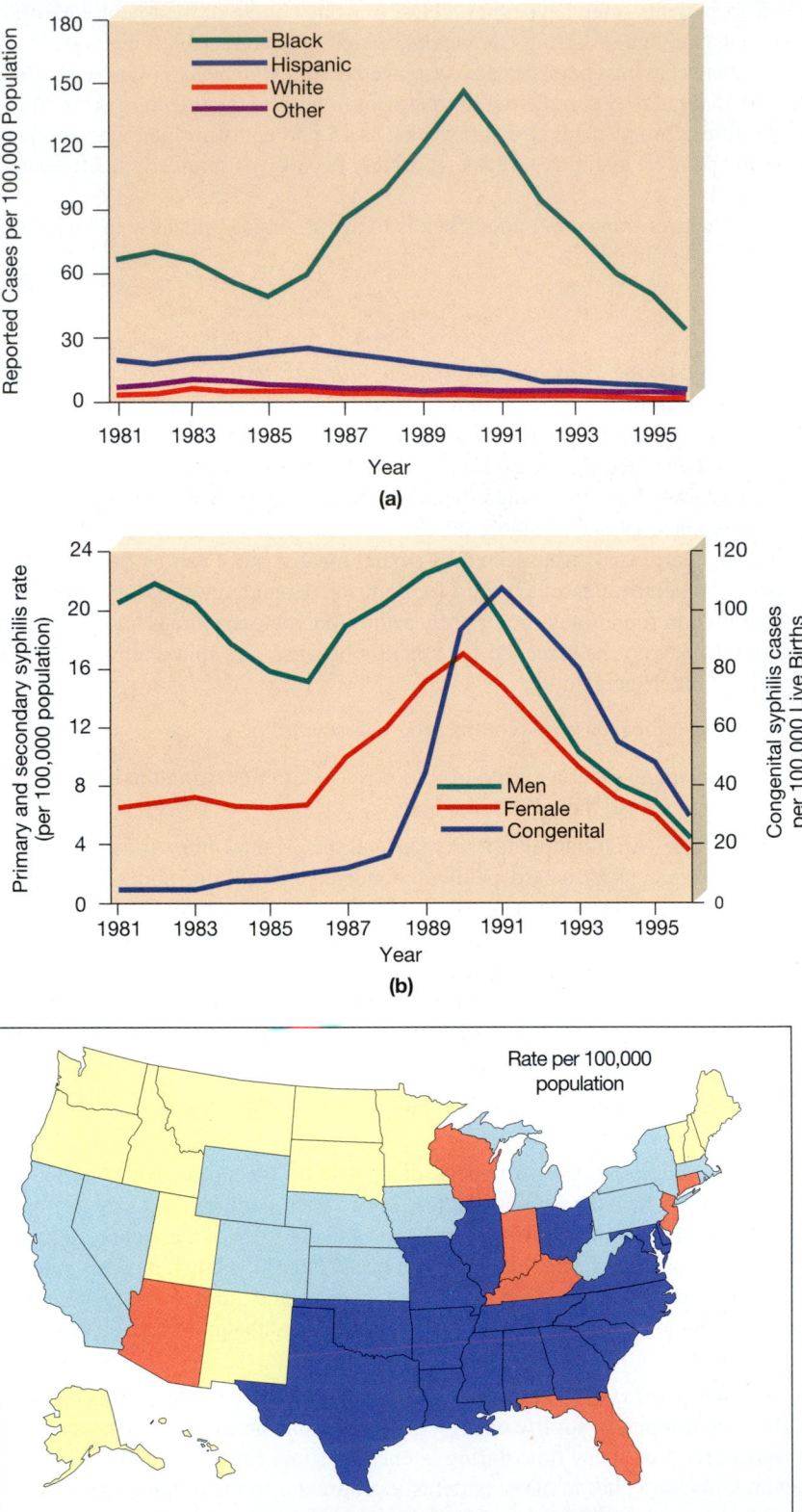

➤ **Figure 20.12   The U.S. incidence of syphilis.** (a) By ethnicity. (b) Among women and congenital syphilis rates. As more women contract syphilis, transmission to the fetus becomes more common, leading to a parallel increase in incidence of congenital syphilis. (c) Geographic distribution, 1996. (*Source*: CDC).

(a)

(b)

➢ Figure 20.13 **Primary chancres of syphilis.** (a) A genital site (penis); (b) an extragenital site (the face).

(a)

(b)

➢ Figure 20.14 **Signs of secondary and tertiary syphilis.** (a) A typical papular rash (secondary stage); (b) a gumma (tertiary stage).

or paralysis and insanity. These symptoms often are due to formation of granulomatous inflammations called **gummas** (gum'ahz). ➡ (Chapter 16, p. 446) Internal gummas typically destroy neural tissue, whereas external gummas destroy skin tissue (➢ Figure 20.14b). Mental illness accompanies neural damage. In the preantibiotic era, as many as half the beds in mental hospitals were occupied by patients with tertiary syphilis.

Diagnosis, Treatment, and Prevention   Diagnostic tests include fluorescent antibody and treponemal immobilization tests. Actively motile organisms can be observed under the dark-field microscope while specific antibody against *Treponema pallidum* is added. Immobilization of organisms by the antibody is 98 percent confirmatory for syphilis. Other blood tests such as VDRL (Venereal Disease Research Laboratory) and Wasserman tests have a high frequency of false positives, as they are based on the detection of tissue damage. A severe case of influenza, a myocardial infarction, or an autoimmune disease can cause sufficient damage to elicit a false positive reaction. Therefore, screening tests such as the VDRL must always be followed up by a confirmatory test, such as the fluorescent antibody test.

Syphilis usually is treated with penicillin, but tetracycline and erythromycin also are effective. The longer the patient has had syphilis, the more important are continued treatment and testing to ensure that organisms have been eradicated. No vaccine is available, and recovery does not confer immunity.

Congenital Syphilis   **Congenital syphilis** occurs when treponemes cross the placenta from mother to baby. At birth or shortly thereafter, the infant may show such signs as notched incisors, or Hutchinson's teeth (➢ Figure 20.15a), a perforated palate, saber shins (in which the shin bone projects sharply on the front of the leg) (➢ Figure 20.15b), an aged-looking face with saddle-

(a)

(b)

(c)

➢ Figure 20.15   **Signs of congenital syphilis.** (a) Hutchinson's teeth, showing notched central incisors. (b) Saber shin, displaying arching of the anterior edge of the tibia. (c) Saddle-nose, which causes a snuffling type of breathing.

nose (a flat, saddle-shaped nose) (➤ Figure 20.15c), and a nasal discharge. Women should be tested for syphilis prior to becoming pregnant, and as part of their prenatal testing if they are already pregnant, to prevent such congenital infections.

Chancroid

Chancroid (shang´kroid), called soft chancre to distinguish it from the hard, painless chancre of syphilis, is caused by *Haemophilus ducreyi*. Named for Augusto Ducrey, the Italian dermatologist who first observed it in skin lesions in 1889, the organism is a small Gram-negative rod that occurs in strands.

Relatively rare in the United States, with approximately 800 cases reported per year, chancroid is seen most frequently in developing countries of Africa, the Caribbean, and Southeast Asia. Its worldwide incidence is believed to be greater than that of either gonorrhea or syphilis. Most cases in the United States occur in immigrants and a few in military personnel. In 1982 CDC reported a sudden increase in the number of cases in southern California—from an average of 29 per year to over 400 per year. Of these, 90 percent were in Hispanic men, most of whom had recently immigrated from Mexico.

The Disease   Chancroid begins with the appearance of soft, painful lesions called chancres, which bleed easily, on the genitals 3 to 5 days after sexual exposure. They often occur on the labia and clitoris in females and on the penis in males. However, the infection can be present without apparent lesions, the only symptom being a burning sensation after urination. Chancres also can occur on the tongue and lips. Regardless of their location, chancres are extremely infective. Medical personnel sometimes acquire lesions on the hands merely from contact with chancres. In about one-third of the patients, chancroid spreads to the groin, where it forms enlarged masses of lymphatic tissue called *buboes*. These appear about 1 week after infection, swell to great size, and can break through the skin, discharging pus to the surface (➤ Figure 20.16).

Diagnosis and Treatment   Chancroid is diagnosed by identifying the organism in scrapings from a lesion or in fluids from a bubo. Patients with chancroid also often have syphilis and other STDs. Thus, a patient with a positive diagnosis for one STD should be tested for other STDs. Untreated lesions can persist for months. Often the disease resolves on its own. Infection does not confer permanent immunity, and the disease can be acquired again and again. It is treated with antibiotics such as tetracycline, erythromycin, sulfanilamide, or a combination of trimethoprim and sulfamethoxazole. With treatment lesions heal rapidly, but they often leave deep scars with much tissue destruction.

Nongonococcal Urethritis

As the name suggests, **nongonococcal urethritis (NGU)** is a gonorrhea-like STD caused by organisms other than gonococci. Most cases are caused by *Chlamydia trachomatis,* but some are caused by mycoplasmas. The prevalence of chlamydial infections is greater than that of any other STDs and is increasing dramatically.

➤ Figure 20.16   **Lesions of chancroid on the penis.** A draining buboe is in the adjacent groin area. Chancroid is caused by *Haemophilus ducreyi.*

It may be difficult to detect chlamydia; about 80 percent of women and 10 percent of men with the disease have no symptoms.

Chlamydial Infections   *Chlamydia trachomatis* is a tiny spherical bacterium with a complex intracellular life cycle. ➡ (Chapter 9, p. 249) In addition to causing NGU, subspecies strains of *C. trachomatis* cause a wide range of disorders, including conjunctivitis and lymphogranuloma venereum (discussed later). The CDC has estimated that 3 to 5 million Americans contract *Chlamydia* NGU each year. The subspecies that causes inclusion conjunctivitis in newborns also causes 30 to 50 percent of NGU cases in males as well as 30 to 50 percent of vulvovaginitis cases and half of cervicitis (inflammation of the cervix) cases in women. The large numbers of humans who are infected and who have organisms in their body secretions makes transmission of chlamydial infections especially easy.

Following an incubation period of 1 to 3 weeks, symptoms of NGU similar to but milder than gonorrhea begin. A scanty watery urethral discharge is observed, especially after passage of first morning urine. Sometimes this is accompanied by tingling sensations in the penis. Many chlamydial NGU infections are asymptomatic, and fortunately most chlamydial STDs produce no significant complications or aftereffects. However, inflammation of the epididymis (the tube through which sperm pass from the testis) can lead to sterility.

Pelvic inflammatory disease, which can be caused by more than 20 different infectious agents, is a common complication of NGU as well as of gonorrhea. Chlamydial PID increases the risk of sterility and ectopic pregnancy, a pregnancy in which the embryo begins development outside the uterus (for example, in a fallopian tube or the peritoneal cavity). Surveys of pregnant women reveal that 11 percent have *Chlamydia* in the cervix; postpartum fever is common among infected women. Their infants can develop neonatal chlamydial pneumonia, a rarely fatal disease that accounts for 30 percent of all pneumonias in children under 6 months of age. Because the condition typically does not develop until 2 or 3 months after delivery, its connection with cervical chlamydial infection is often overlooked.

Chlamydial infections are difficult to control. Infants can become infected while passing through the birth canal of an infected mother. Silver nitrate used to prevent gonorrheal ophthalmia neonatorum does not protect against *Chlamydia,* but erythromycin protects against both. In venereal disease clinics, penicillin used to treat syphilis and gonorrhea fails to eliminate chlamydial infections. The recent rise in chlamydial infections could be stopped with tetracycline and sulfa drugs if all sexual partners were treated.

Adult **inclusion conjunctivitis** can result from self-inoculation with *C. trachomatis* from genitals via fingers or towels; it is especially common in sexually active young adults. It closely resembles trachoma. ➡ (Chapter 19, p. 555) Prior to the widespread use of chlorine in pools, it was called "swimming pool conjunctivitis." Reinfection can occur, but whether by inadequate immune response or by infection with another of eight different infectious strains is unknown.

Each year about 75,000 infants acquire chlamydial **inclusion blennorrhea** (blen-or-e′ah), a name derived from the Greek for "mucus flow." This usually benign conjunctivitis begins as a pus-filled mucous discharge 7 to 12 days after delivery and subsides with erythromycin treatment or spontaneously after a few weeks or months. If it persists, it becomes indistinguishable from childhood trachoma and can lead to blindness.

Mycoplasmal Infections   NGU also can be caused by *Mycoplasma hominis,* which is frequently among the normal urogenital microflora especially in women. Mycoplasmas, which have no cell wall, apparently infect by fusing their cell membranes with the host cell membranes. Such infections are very common; more than half of normal adults have antibodies to *M. hominis.* Although most often associated with NGU, *M. hominis* sometimes causes PID in women and opportunistic urethritis in men. Mycoplasmas on the cervix during pregnancy probably colonize the placenta and cause spontaneous abortions, premature births, and low birth rate. They also foster ectopic pregnancies.

Yet another causative agent of NGU is *Ureaplasma urealyticum,* formerly called T-strain (T for tiny) mycoplasma. Between 1 and 2.5 million people in the United States are infected. One of the smallest bacteria known to cause human disease, its name is based on the fact that it requires a 10 percent urea medium to grow. Of patients seen in venereal disease clinics, 50 to 80 percent carry *U. urealyticum* in addition to other sexually transmitted pathogens. The organism accounts for more than half of all infections that make couples infertile. Low sperm counts and poor sperm mobility have been observed in males; the organisms bind tightly to sperm and can be transmitted by them to sexual partners. They are a major cause of fetal death, recurrent miscarriage, prematurity, and low birth weight—itself a leading cause of neonatal death.

Diagnosis is made by culture from urethral and vaginal discharges and from placental surfaces. When both members of an infected couple are treated, pregnancy is achieved in 60 percent of cases, compared with a success rate of only 5 percent in untreated couples infected with this organism. Because mycoplasmas lack a cell wall, penicillin is not effective against them. Therefore, people being treated with penicillin for other STDs will not be cured of NGU. Tetracycline is most often used because it also controls *Chlamydia*. The 15 percent of *Mycoplasma* strains resistant to tetracycline can be treated with erythromycin and spectinomycin.

Lymphogranuloma Venereum

Another STD, **lymphogranuloma venereum (LGV)**, is common in tropical and subtropical regions. Although only about 250 cases per year occur in the United States, mostly in southeastern states, as many as 10,000 cases are treated annually in a single Ethiopian clinic. This nonreportable disease is about 20 times more common in males than in females.

The Disease   The causative agent of LGV was identified in 1940 as a highly invasive strain of *Chlamydia trachomatis*. Within 7 to 12 days after contact with the organism, lesions appear at the site of infection—usually the genitals, but sometimes the oral cavity. In most cases the lesions rupture and heal without scarring. Other symptoms include fever, malaise, headache, nausea and vomiting, and a skin rash. One week to 2 months later, the organisms invade the lymphatic system and cause regional lymph nodes to become enlarged, painful, and pus-filled buboes (➤ Figure 20.17a). Victims have been known to use a razor blade to open the buboes to gain relief (➤ Figure 20.17b), but aspiration with a sterile needle is a safer treatment. Lymph node inflammation occasionally obstructs and scars lymph vessels, causing edema of genital skin and elephantiasis (massive enlargement) of the external genitalia in both males and females.

(a)

(b)

➤ Figure 20.17   **Bilateral buboes of lymphogranuloma venereum, caused by *Chlamydia trachomatis.*** (a) Early in development; (b) after reaching such size that the lesion was opened for drainage.

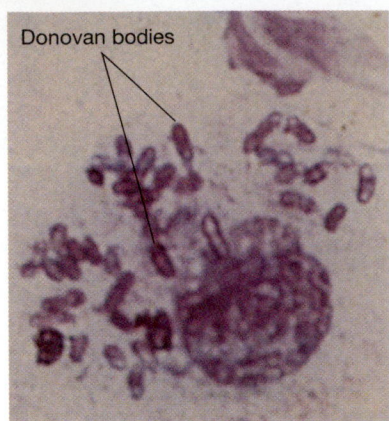

Donovan bodies

➤ **Figure 20.18** **Scrapings of lesions of granuloma inguinale.** Caused by *Calymmatobacterium granulomatis,* these lesions reveal encapsulated bacterial cells inside the much larger macrophages. The bacterial cells resemble closed safety pins and are called Donovan bodies.

✓ Which body sites can be affected by gonorrhea?

✓ Does one develop immunity following a case of gonorrhea? After a case of syphilis?

✓ What is a buboe?

✓ What is PID? What can cause it?

There are an estimated 20 million persons with genital herpes in the United States alone, with 200,000 to 500,000 cases reported annually.

Hold the Cortisone

If you are used to using cortisone to treat your inflammatory lesions, make sure you don't use it on herpes lesions, especially ocular lesions. Because cortisone suppresses the immune system, it allows viral multiplication to increase and to cause more extensive cellular injury. The damage can cause perforation of the cornea.

Rectal infections often occur in homosexual men. In women, lymph from the vagina drains toward the rectum, and lymph nodes in the walls of the rectum become chronically enlarged in 25 percent of the cases. This causes rectal blockage that usually requires surgery. Untreated cases produce a bloody, pus-filled anal discharge and can eventually lead to rectal perforation. Organisms transferred from the hands to the eyes can cause conjunctivitis. In rare instances the disease progresses to meningitis, arthritis, and pericarditis.

Spontaneous "cures" sometimes represent latent infections. Latency can be prolonged, as is evidenced by males who infect sex partners many years after their own initial infection. The genital tracts and rectums of chronically infected, but at times asymptomatic, persons serve as reservoirs of infection.

Diagnosis and Treatment   LGV is diagnosed by finding chlamydias as inclusions stained with iodine in pus from lymph nodes. Serological tests are available, but they frequently give false-positive results. Tetracyline is the drug of choice for treating LGV, but erythromycin and sulfamethoxasole also are effective. Enlarged lymph nodes can take 4 to 6 weeks to subside even after successful antibiotic therapy.

Granuloma Inguinale

Granuloma inguinale, or donovanosis, is caused by the small Gram-negative encapsulated rod *Calymmatobacterium granulomatis,* which is related to *Klebsiella.* The disease is uncommon in the United States, with about 50 cases reported per year, mostly among male homosexuals. It is common in India, western coastal Africa, South Pacific islands, and some South American countries, from which it is occasionally brought to the United States. The epidemiology of granuloma inguinale is not completely understood. Some cases appear to be sexually transmitted and others not. Even in genital cases, infectivity is low and many partners of infected individuals do not get it.

Granuloma inguinale appears as irregularly shaped, painless ulcers on or around the genitals 9 to 50 days after intercourse. Fever is absent. Ulcers can be spread to other body regions by contaminated fingers. As ulcers heal, skin pigmentation is lost. Without treatment, tissue damage can be extensive. Diagnosis from scrapings of lesions is confirmed by finding large mononuclear cells called **Donovan bodies** (➤ Figure 20.18). Antibiotics such as ampicillin, tetracycline, erythromycin, and gentamicin offer effective treatment.

Viral Sexually Transmitted Diseases

Herpesvirus Infections

Two closely related herpesviruses cause disease in humans (Chapter 10). ➡ (p. 265) **Herpes simplex virus type 1 (HSV-1)** typically causes fever blisters (cold sores), and **herpes simplex virus type 2 (HSV-2),** sometimes called *herpes hominis virus,* typically causes **genital herpes.** Both HSV-1 and HSV-2 have an incubation period of 4 to 10 days, cause the same kind of lesions, and have been isolated from skin and mucous membranes of oral and genital lesions. Among oral infections, 90 percent are caused by HSV-1 and 10 percent by HSV-2. Among genital infections, 85 percent are caused by HSV-2 and 15 percent by HSV-1. The presence of HSV-1 in the genital region and HSV-2 in the oral region is most commonly due to the practice of oral sex. Genital herpes is by far the most common and most severe of the herpes simplex viral infections.

Initial HSV-1 or HSV-2 infection can be asymptomatic, especially in children, or cause localized lesions with or without symptoms of acute infection. Most adults have antibodies to herpesviruses, but only 10 to 15 percent have experienced symptoms. In both HSV-1 and HSV-2 infections, vesicles form under keratinized cells and fill with fluid from virus-damaged cells, particles of cell debris, and inflammatory cells. Vesicles are painful, but they heal completely in 2 to 3 weeks without scarring unless there is a secondary bacterial infection. Adjacent lymph nodes enlarge and are sometimes tender.

Latency ➡ (Chapter 10, p. 264) is a hallmark of herpes infections. More than 80 percent of the adult population worldwide harbors these viruses, but only a small proportion experience recurrent infections. Within 2 weeks of an active infection, the viruses travel via sensory neurons to ganglia (➤ Figure 20.19). Within the ganglia they replicate slowly or not at all. They can reactivate spontaneously or be activated by fever, ultraviolet radiation, stress, hormone imbalance, menstrual bleeding, a change in the immune system, or trauma. The infection can spread to and kill cells in the adrenal glands, liver, spleen, and lungs. In fatal herpes encephalitis, soft, discolored lesions appear in both gray and white matter of the brain.

After reactivation, the virus moves along the nerve axon to the epithelial cells. There it replicates, causing recurrent lesions. These lesions, which always recur in exactly the same place as the original infection, are smaller, shed fewer viruses, contain more inflammatory cells, and heal more rapidly than primary lesions. Successive recurrences usually become milder until they finally cease. While the virus is in a neuron, neither humoral nor cellular immunity can combat it. Once the virus reaches target epithelial cells and starts to replicate, antibodies can neutralize the viruses, and T cells can eliminate virus-infected cells. These immunological processes that make it difficult to isolate viruses from vesicular fluid in recurrent lesions also reduce their severity and duration. Recurrences can be limited to one or two episodes or can appear periodically for the life of the patient but typically occur five to seven times. Even if recurrences

Trigeminal ganglion

Brachial plexus ganglia

Sacral ganglia

Herpes of the eye

Fever blister (labial herpes)

Whitlow

Genital herpes

➤ **Figure 20.19   Herpes simplex viruses are permanent "houseguests" once acquired.** After an occurrence of a lesion, they migrate to ganglia, from which they travel back to the site of the original lesion at the time of the next outbreak. An infrequent exception to this pattern sometimes occurs at the trigeminal ganglion, at which three branches of the trigeminal (fifth cranial) nerve join. Virus from a lesion on the lower lip could migrate out along the ophthalmic branch on a subsequent episode, causing the eye to become affected, or even migrate backward into the brain, causing meningitis and brain damage. Why this sometimes happens is not known; fortunately, it is very rare.

cease, the viruses remain latent in ganglia, and long-dormant viruses can be reactivated by severe stress, trauma, or impaired immune function (as for example in AIDS patients).

"HSV shedders," people who shed viruses while remaining asymptomatic, pose a significant problem. As many as 1 in 200 women have been shown to shed viruses even when they have no observable lesions and in some cases no knowledge of ever having had a herpes infection. Such women pose a serious threat to any infant they might bear.

Genital Herpes   Genital herpes infections usually are acquired after the onset of sexual activity because the virus is transmitted mainly by sexual contact. However, the virus can survive for short periods of time in moist areas such as hot tubs. Changes in sexual practices, especially an increase in oral sex, have increased the incidence of HSV-1 in genital lesions and HSV-2 in oral lesions. More than 20 million Americans now have genital herpes, and half a million new cases are seen each year.

In females the vesicles appear on the mucous membranes of the labia, vagina, and cervix. Ulcerations sometimes spread over the vulva and can even appear on the thighs. In males tiny vesicles appear on the penis and foreskin and are accompanied by urethritis and a watery discharge (➤ Figure 20.20). The prostate gland and seminal vesicles also can be affected. Both sexes experience intense pain and itching at the sites of lesions and swelling of lymph nodes in the groin.

A person infected with herpesvirus is contagious any time viruses are being shed. Shedding always occurs when active lesions are present and usually starts a few days before lesions appear. It can occur continuously even when lesions are not present. Therefore, abstention from sexual contacts when lesions are present does not always prevent spread of the disease. In recent years promiscuous sexual practices and ignorance of, or lack of concern about, transmitting the disease have greatly increased the number of cases of genital herpes. This incurable disease has now become one of the most common STDs.

Women infected with genital herpes may be subject to three other serious problems. First, the incidence of miscarriages among women with genital herpes is higher than that for uninfected women. Second, when infected women become pregnant, the infant must be delivered by Caesarean section. Finally, infected women have an increased risk of developing a kind of cancer called cervical carcinoma. Among women with cervical carcinoma, 80 percent have antibodies to HSV-2; they have been infected and probably still harbor the viruses. Most have also had multiple sexual partners. The exact roles of HSV-2, wart viruses (discussed later), or other factors associated with sexual activity in the development of the malignancy are not yet known. Until more information is available, avoiding sexual contact with virus-infected individuals and with multiple partners seems prudent.

Neonatal Herpes   **Neonatal herpes** (➤ Figure 20.21) can appear at birth or up to 3 weeks after birth. Babies most often become infected by delivery through a birth canal contaminated with HSV-2, but they can also become infected through contaminated equipment and hospital procedures. In rare instances, infants are infected *in utero*. Because neonates are highly susceptible to HSV infections, they should not be cared for by individuals with such infections. Infected mothers who must care for their infants must scrupulously follow sanitary procedures.

At diagnosis, two-thirds of infected infants have skin vesicles; the others already have disseminated infection with neural or visceral lesions. Skin infections disseminate in 70 percent

➤ Figure 20.20   **Penile herpes lesions.**

Herpes Singles Club

Are you single? Do you have herpes? Then you might want to join a "herpes singles club." Some herpes victims join these clubs in hopes of being able to enjoy sexual activity with other members without inflicting the disease on uninfected people. Although this may prevent the spread of herpes infections to "outsiders," it can also give herpes sufferers a false sense of security. Because several different strains of both HSV-1 and HSV-2 exist, people can infect each other with different strains of HSV. Each new strain can cause a new primary infection and very painful lesions.

➤ Figure 20.21   **Neonatal herpes.** This type of herpes can be acquired when the baby passes through the birth canal of an infected mother. This could be avoided by Caesarean section delivery. Sometimes lesions are so extensive that they cover almost the entire skin surface. Such children either suffer profound brain damage or do not survive.

of infected infants. Infants with disseminated infections display poor appetite, vomiting, diarrhea, respiratory difficulties, and hypoactivity. Some also have neurological symptoms, jaundice, and eye disorders. Neonates with disseminated infections deteriorate rapidly and usually die within 10 days. The few that survive usually have central nervous system and eye damage. Occasionally infants have only a few vesicles, but latent viruses may later cause significant damage. Early diagnosis and treatment of neonatal herpesvirus infections are essential for survival and to reduce the likelihood of neurological damage.

Other Herpes Simplex Infections   A variety of manifestations of herpesvirus infections have been seen. Most are thought to be caused by HSV-1, but HSV-2 may be responsible for some. They include gingivostomatitis, herpes labialis, keratoconjunctivitis, herpes meningoencephalitis, herpes pneumonia, eczema herpeticum, traumatic herpes, herpes gladiatorium, and whitlows. HSV-1 is spread by contaminated secretions and through contact with lesions and usually is acquired in childhood from relatives, nurses, or other children. The incidence of the virus is especially high within families, hospitals, and other institutions. Most humans are infected with HSV-1 in the first 18 months of life. Many of these infections are inapparent, and the first apparent lesions result from reactivations.

Gingivostomatitis (jin″gi-vo-sto″mah-ti′tis), lesions of the mucous membranes of the mouth, is most common in children 1 to 3 years old. After an incubation period of 2 to 20 days, small vesicles appear around the mouth over a 7-day period. Each vesicle accumulates fluid, scabs over, and heals in 2 to 3 weeks. Recurrent lesions typically occur in the form of **herpes labialis** (la′be-ah-lis), or fever blisters on the lips. If the eyes instead of the mouth are the site of initial infection, vesicles appear on the cornea and eyelids, causing **keratoconjunctivitis** (ker″a-to-kon-junk-ti-vi′tis).

The most serious manifestation of HSV infection is *herpes meningoencephalitis,* which can follow a generalized herpes infection in a newborn, child, or adult. How the virus crosses the blood–brain barrier to enter the central nervous system has recently been discovered. The virus remains latent in the ganglion of the trigeminal nerve, which supplies parts of the face and mouth, until some unknown factor reactivates it. Then, instead of moving outward to epithelial sites such as the tongue or lip, it ascends along the nerve into the brain. The disease has a rapid onset, with fever, headache, meningeal irritation, convulsions, and altered reflexes. In the middle-aged and elderly, meningoencephalitis can appear without any preceding symptoms. The patient experiences increasing confusion, hallucinations, and sometimes seizures. Most patients die in 8 to 10 days; survivors usually have permanent neurological damage.

Herpes pneumonia is rare. It usually is seen only in burn patients, alcoholics, and patients with AIDS or other immunodeficiencies. **Eczema herpeticum** is a generalized eruption caused by entry of the virus through the skin. **Traumatic herpes** occurs when the virus enters traumatized skin in the area of a burn or other injury. **Herpes gladiatorium** occurs in skin injuries of wrestlers. A **whitlow** (➤ Figure 20.22) is a herpetic lesion on a finger that can result from exposure to oral, ocular, and probably genital herpes lesions. Dental technicians, nurses, and other medical personnel should therefore use rubber gloves when treating patients with herpes lesions. A person with a whitlow can spread the infection to the mouth, eyes, and genital areas or to other persons.

Diagnosis, Treatment, and Prognosis   Herpesviruses are most easily isolated from vesicular fluid and cells from the base of a lesion and can be grown in the laboratory in a variety of cell types. The cytopathic effect of the viruses appears in the culture quickly. The time to obtain a diagnosis has been greatly decreased by rapid immunological tests. Speedy diagnosis is especially important for women in labor or near delivery, as a Caesarean delivery can prevent exposure of the infant to the virus.

In recent years several drugs have been used with varying degrees of success in treating HSV-1 and HSV-2 infections. Trifluorothymidine is effective in treating ocular herpes. Although acyclovir fails to prevent recurrences reliably, it does prevent the spread of lesions, decrease viral shedding, and shorten healing time. Best results are obtained with acyclovir when it is used at the beginning of a primary infection. The use of acyclovir, Ara-A, and vidarabine in various combinations early in the infection has increased survival rates in herpes meningoencephalitis and neonatal herpes. The prognosis for alleviating genital herpes symptoms is

➤ **Figure 20.22   Herpetic whitlow.** This type of herpes virus infection is a very painful infection of the finger, which can be spread to other areas of the body, for example, by rubbing one's eyes. Health care workers should protect themselves by wearing latex gloves.

reasonably good. Most patients have few if any recurrences if they take acyclovir daily but may have a recurrence if they stop taking it. No treatment can eradicate latent viruses.

Immunity and Prevention   Although immunological knowledge of herpesviruses is sufficient to make a vaccine, such a vaccine is not yet available. Even if a vaccine were available and administered to all who lack HSV antibodies, it would take years to eradicate the disease. Large numbers of people already harbor latent viruses, so a vaccine to inactivate the viral genetic information responsible for *latency* is needed. A vaccine that elicits antibodies like those from natural infections would probably not be effective; such antibodies apparently have no effect on latent viruses and cannot alone prevent recurrences.

The best available means of preventing HSV infections is to avoid contact with individuals with HSV-1 or HSV-2 lesions. If all infected individuals were to refrain from sexual activity, especially when they have active lesions, some new cases of genital herpes could be prevented. If pregnant women would advise their obstetricians of herpes infections, their infants could be protected from exposure during delivery. Even these precautions will not prevent the spread of genital herpes. Most infected individuals shed viruses a few days before lesions appear, and some shed viruses continuously without having any lesions.

Genital Warts

Condylomas (kon-dil-o'mahz), or **genital warts,** caused by the human papillomavirus, most often occur in the sexually promiscuous young adult population. The incidence of genital warts has increased rapidly in recent years to the extent that such warts are now among the most prevalent STDs. Two-thirds of the sexual partners of infected individuals also develop warts. The warts can be papillary or flat. In males the warts appear on the penis, anus, and perineum; in females they are on the vagina, cervix, perineum, and anus.

Like dermal warts (Chapter 19), genital warts cause irritation and sometimes intense itching. They can persist or regress spontaneously. ➥ (p. 548) Genital warts often become infected with bacteria, and those that persist for many years can transform into malignant growths. Some victims suffer psychological damage from the presence of warts. Warts temporarily increase in number and size during pregnancy but decrease after delivery. Infants can be infected during delivery. The number of warts increases in immunosuppressed patients and in those with AIDS and other immunological deficiencies. (Treatment of warts was discussed in Chapter 19. ➥ (p. 548)

A special hazard of genital warts is their epidemiologic association with cervical carcinoma. Lesions previously diagnosed as cervical dysplasia (dis-pla'se-ah), or abnormal cell growth, are now more often properly recognized as condylomas even though both can be found side by side on the cervix. Distinguishing these conditions from cervical carcinoma is even more difficult. Examining lesions for viral capsid antigens and viral nucleic acids can be helpful. Such examinations suggest that two strains of human papillomaviruses, HPV-8 and HPV-16, are most likely to be found in cervical carcinomas. The extent to which they contribute to such malignancies is not clear, but the virus has been found in 90 percent of patients with cervical carcinoma.

Laryngeal Papillomas

Laryngeal papillomas (pap-il-lo'maz) are benign growths that can be dangerous if they block the airway. Hoarseness, voice changes, and respiratory distress occur when the airway becomes obstructed. Children are more likely to have laryngeal papillomas than adults. Surgical removal, sometimes every 2 to 4 weeks, is the only treatment for these obstructive growths. Also, there is danger of spreading the virus to the lungs during surgery. Laryngeal papillomas are usually caused by HPV-6 and HPV-11, which are thought to infect infants during birth to women with active genital warts.

Cytomegalovirus Infections

The **cytomegaloviruses (CMVs)** constitute a widespread and diverse group of herpesviruses. In general, each strain of CMV is capable of infecting a single species. The majority of human

CMV infections occur in older children and adults and go unnoticed because they do not produce clinically apparent symptoms. An estimated 80 percent of U.S. adults carry the virus. When there are symptoms, they include malaise, myalgia, protracted fever, abnormal liver function, and lymph node inflammation without swelling. Symptoms are more severe in patients with AIDS and other immunodeficiencies.

Initially the virus can be recovered from the oropharynx; viremia with many virus-infected neutrophils can last for months. The viruses replicate and have low pathogenicity but are excreted intermittently over many months, during which time they can infect others. Viruses are shed in all body fluids—saliva, blood, semen, breast milk—but they are found most often and in the largest quantities in urine even a year or more after infection. In a symptomatic primary attack of CMV, cell-mediated immunity is depressed, and the ratio of T helper to T suppressor cells is reversed. However, the immune system returns to normal during convalescence. In subclinical cases cell-mediated immunity remains active. Large quantities of long-lasting antibodies are produced in response to CMV infection, but they do not prevent viral shedding. The virus is usually spread through long, close contact with children who are shedding the virus, but it can be spread by blood transfusions, organ transplants, and sexual intercourse.

Fetal and Infant CMV Infections   In fetuses and infants, CMV infections can be life threatening because the virus disseminates widely to various organs. Fetuses become infected by viruses that cross the placenta from infected mothers. Unlike rubella infection, maternal CMV infections are rarely detected, so the risk to the fetus is unknown. Maternal antibodies also can cross the placenta and inactivate small quantities of virus. Maternal hormones suppress CMV, but their effect diminishes as the pregnancy progresses. Both fetuses and neonates are dependent on maternal defenses against CMV because their own immune systems are too immature to mount a successful defense. Tests for maternal antibody levels are available. Young children are the most likely source of infection to nonimmune women.

In severe CMV infections in which the fetus has been infected with large numbers of viruses (about 4000 cases per year in the United States), intrauterine growth retardation and severe brain damage can occur (➤ Figure 20.23). Many babies have CMV-infected cells in the inner ear and hearing loss due to nerve damage. Some have jaundice with liver damage, and others have impaired vision. Less severe infections (another 4500–6000 cases per year) cause damage to certain brain areas and mild central nervous system disorders with or without damage to hearing or sight. Mortality can be as high as 30 percent.

CMV infections contracted after birth generally cause fewer permanent defects than those contracted before birth. However, these infections can cause severe illness. Among babies infected with CMV, 5 percent have typical generalized *cytomegalic inclusion disease (CID)*; another 5 percent have atypical, less generalized infections; the rest have subclinical but chronic infections. Many infants with CID infections have significant mental or sensory disorders. These include an abnormally small brain accompanied by intellectual deficits and inflammation of the eyes with impaired vision. In subclinical cases the prognosis is better, but 10 percent will have deafness or other sensory problems. Some will experience intellectual and

Studies show that CMV spreads to 1 in 5 nonimmune parents who have infected children.

➤ **Figure 20.23   A baby with birth defects due to congenitally acquired cytomegalovirus.** The virus, sometimes called salivary gland virus, can cause grave damage to people who have impaired immune systems.

behavioral disorders later. Babies who acquire CMV infections from blood transfusions have a gray pallor and symptoms like those of CID. Pneumonia, respiratory deterioration, and death may follow because the infants receiving transfusions usually are premature and debilitated before transfusions are given. Moreover, CMV infections sometimes are not diagnosed in the presence of many other disorders. Finally, CMV sometimes causes pneumonia in infants 1 to 6 months of age in the presence of other infections, such as *Chlamydia trachomatis* and *Ureaplasma urealyticum*.

Disseminated CMV In severe disseminated disease, the virus can be present in many organs. When the kidneys become infected, immune complexes are deposited in the glomeruli, but renal dysfunction is rare except in renal transplants. Subclinical hepatitis occurs in both children and adults. Lung infections are common in infants and immunosuppressed adults. Brain involvement is rare except in fetuses.

CMV is most virulent when present as a primary infection in immunosuppressed patients. Between 1 and 4 months after an organ transplant, the patient develops symptoms like those of infectious mononucleosis, with prolonged fever and enlargement of the liver and spleen. Pneumonia is frequent and severe in bone marrow transplant patients, and mortality can be as high as 40 percent. Hepatitis is usually reversible, but damage to the eyes is likely to be progressive and irreversible.

Diagnosis, Treatment, and Prevention Definitive diagnosis of CMV infections requires identification of the virus from clinical specimens, a procedure that can take up to 6 weeks. New, faster techniques use monoclonal antibodies to detect viral antigens. It is anticipated that nucleic acid probes, which make use of the base sequences unique to CMV nucleic acids, will allow identification of the virus in clinical specimens in less than 24 hours.

No effective treatment for CMV infections in infants is available. The prognosis is poor when infections occur in fetuses because the damage is done before a diagnosis is made in the neonate. Interferon and hyperimmune gamma globulin given before and after organ transplantation reduce the incidence and severity of CMV infections in transplant patients. When blood transfusions are required, donor and recipient should be matched for CMV antigens to minimize introduction of CMV to which the recipient is not immune. Also, donors serologically negative for CMV should be used with infants, especially premature infants, to avoid the possibility of transmitting CMV through a transfusion. No effective vaccine exists.

The agents and characteristics of the diseases discussed in this chapter are summarized in Table 20.2.

✓ How can gential herpes viral infections be acquired?

✓ How can neonatal herpes infections be avoided?

✓ Why is cervical cancer now listed as a sexually transmitted disease?

➤ Table 20.2 **A summary of urogential and sexually transmitted diseases**

Disease	Agent(s)	Characteristics
Bacterial urogenital diseases		
Urinary tract infections	*Escherichia coli, Proteus mirabilis,* and other bacterial species	Dysuria; sometimes leads to chronic cystitis; often ascend or descend in urinary tract
Prostatitis	*E. coli* and other bacterial species	Dysuria, urgent and frequent urination, low fever, back pain; can cause infertility
Pyelonephritis	E. coli and other bacterial species, sometimes the yeast *Candida*	Inflammation of pelvis of kidney, often caused by urinary tract blockage; dysuria, nocturia, sometimes fever
Glomerulonephritis	Streptococcal or viral infections from other sites	Deposition of immune complexes causes inflammation of glomeruli; can cause permanent kidney damage
Leptospirosis	*Leptospira interrogans*	Fever, nonspecific symptoms; can lead to Weil's syndrome with jaundice and liver damage
Bacterial vaginitis	*Gardnerella vaginalis* with anaerobes	Frothy, fishy-smelling discharge, pain and inflammation
Toxic shock syndrome	*Staphylococcus aureus*	Toxins reach blood and cause fever, rash, and shock that can lead to death
Parasitic urogenital disease		
Trichomoniasis	*Trichomonas vaginalis*	Intense itching, copious white discharge
Bacterial sexually transmitted diseases		
Gonorrhea	*Neisseria gonorrhoeae*	Infectious organisms release endotoxin that damages mucosa; pus-filled discharge; can cause PID and infect other systems
Syphilis	*Treponema pallidum*	Chancre develops in primary stage; mucous membrane lesions and rash occur in secondary stage; permanent cardiovascular and neurological damage often occur in tertiary stage
Chancroid	*Haemophilus ducreyi*	Painful, bleeding lesions on genitals; often enlarged lymphatic buboes
Nongonococcal urethritis	*Chlamydia trachomatis* and mycoplasmas	Scanty watery urethral discharge, inflammation, sometimes sterility; can cause neonatal infections and fetal death
Lymphogranuloma venereum	*Chlamydia trachomatis*	Genital lesions, fever, malaise, headache, nausea, vomiting, skin rash; lymph nodes become pus-filled buboes
Granuloma inguinale	*Calymmatobacterium granulomatis*	Painful ulcers on genitals and other sites; loss of skin pigmentation as ulcers heal
Viral sexually transmitted diseases		
Herpes simplex infections	Herpes simplex viruses	Fever blisters usually caused by HSV-1, genital herpes usually caused by HSV-2 (both are latent viruses); recurrent painful vesicular lesions, neonatal herpes and a variety of other manifestations
Genital warts	Human papillomaviruses	Warts on genitals, vagina, and cervix; irritation and sometimes intense itching; can contribute to cervical carcinoma
Cytomegalovirus infections	Cytomegaloviruses	Often asymptomatic but severe in fetuses, neonates, and immunodeficient patients; malaise, myalgia, fever, inflamed lymph nodes, neural damage and death in fetuses and neonates

Retracing Our Steps

- The agents and characteristics of the **sexually transmitted diseases (STDs)** discussed in this chapter are summarized in Table 20.2. Information in that table is not repeated in this summary.

Components of the Urogenital System

- The **urogenital system** includes the **urinary system** and the reproductive system. The urinary system consists of the **kidneys, ureters, urinary bladder,** and **urethra.** The **female reproductive system** consists of the **ovaries, uterine tubes, uterus, vagina,** external genitalia, and **mammary glands.** The **male reproductive system** consists of the **testes,** a system of ducts, glands, and the **penis.**
- Urogenital infections are most common near the openings of the vagina and urethra.
- Nonspecific defenses include urinary sphincters, cleansing action of the outflow of urine, acidity of mucous membranes, and competition from normal microflora.

Normal Microflora of the Urogenital System

- Normally all parts of the urinary tract, except for the portion of the urethra nearest its opening, are sterile.
- In males, other than the last third of the urethra, the genital tract lacks a normal microflora and is sterile.
- In females, during childbearing years, lactobacilli predominate in the vagina producing an acid pH inhibitory to growth of most other types of microbes. During childhood and after menopause, streptococci and staphylococci predominate in the alkaline conditions of the vagina.

Urogential Diseases Usually Not Transmitted Sexually

Bacterial Urogenital Diseases

- **Urinary tract infections (UTIs),** which are exceedingly common, can ascend or descend to spread through the urogenital system. They are diagnosed from urine cultures, treated with various antibiotics, and can be prevented by good personal hygiene and complete emptying of the bladder.
- **Prostatitis** can result from the spread of a urinary tract infection and usually can be treated with antibiotics.
- **Pyelonephritis** also can result from the spread of a urinary tract infection; it can be difficult to treat because renal failure allows drugs to accumulate to toxic levels.
- **Glomerulonephritis** usually results from throat and other infections in which immune complexes deposit in glomeruli. Prompt treatment of other infections minimizes the risk of glomerulonephritis.
- **Leptospirosis** is acquired through contact with contaminated urine and is diagnosed by microscopic examination of blood. It can be

treated with antibiotics and prevented by vaccinating pets, which are usually responsible for transmitting the disease to humans.
- Bacterial and other forms of **vaginitis** occur when normal vaginal microflora are disturbed. *Gardnerella vaginitis* is diagnosed by the presence of clue cells. It is most often treated with metronidazole to eradicate anaerobes and allow the restoration of normal microflora.
- **Toxic shock syndrome (TSS)** most often arises from the use of superabsorbent tampons. It should be treated promptly with nafcillin and can be prevented by avoiding such tampons.

Parasitic Urogenital Diseases

- **Trichomoniasis** usually is transmitted sexually, is diagnosed from smears of secretions, and is treated with metronidazole.

Sexually Transmitted Diseases

Bacterial Sexually Transmitted Diseases

- **Gonorrhea** is diagnosed by finding organisms in cultures and is treated with penicillin. With penicillin-resistant organisms, another antibiotic is used.
- **Syphilis** is diagnosed by immunological tests and is treated with tetracycline or erythromycin.
- **Chancroid** occurs mainly in developing countries. It is diagnosed by finding organisms from lesions or buboes and is treated with tetracycline and other antibiotics.
- The incidence of **nongonococcal urethritis (NGU)** is increasing dramatically. It can be diagnosed by culturing samples from discharges and is treated with erythromycin, tetracycline, or other antibiotics, depending on the susceptibility of the causative agent.
- **Lymphogranuloma venereum (LGV)** occurs mainly in tropical and subtropical regions. It is diagnosed by finding chlamydias as inclusions in pus and is treated with tetracycline or another antibiotic.
- **Granuloma inguinale** is common in India and several other countries. It is diagnosed by finding **Donovan bodies** in scrapings from lesions and can be treated with a variety of antibiotics.
- All the preceding diseases are usually transmitted by sexual contact and can be prevented by avoiding such contacts. No vaccines are available.

Viral Sexually Transmitted Diseases

- **Genital herpes** is the most severe of herpes simplex virus infections. It is diagnosed by finding viruses in vesicular fluids or by immunological tests and can be treated, but not cured, with acyclovir and other antiviral agents.
- **Genital warts** must be distinguished from cervical dysplasia and carcinoma; treatment of warts was discussed in Chapter 19.
- **Cytomegalovirus (CMG)** infections are of greatest significance in fetuses, newborns, and the immunosuppressed. Rapid diagnosis can be made by monoclonal antibody tests, but no effective treatment is available.

Terminology Check

balantitis   (p. 573)
Bartholin gland   (p. 567)
cervix   (p. 567)
chancre   (p. 579)
chancroid   (p. 583)
condyloma   (p. 590)
congenital syphilis   (p. 582)
cystitis   (p. 569)
cytomegalovirus (CMV)   (p. 590)
Donovan body   (p. 586)
dysuria   (p. 569)
eczema herpeticum   (p. 589)
endometrium   (p. 567)
female reproductive system
   (p. 567)
genital herpes   (p. 586)
genital wart   (p. 590)
gingivostomatitis   (p. 589)
glomerulonephritis   (p. 571)
glomerulus   (p. 566)
gonorrhea   (p. 574)
granuloma inguinale   (p. 586)
gumma   (p. 582)
herpes gladiatorium   (p. 589)
herpes labialis   (p. 589)
herpes pneumonia   (p. 589)

herpes simplex virus type 1 (HSV-1)
   (p. 586)
herpes simplex virus type 2 (HSV-2)
   (p. 586)
inclusion blennorrhea   (p. 584)
inclusion conjunctivitis   (p. 584)
keratoconjunctivitis   (p. 589)
kidney   (p. 565)
laryngeal papilloma   (p. 590)
leptospirosis   (p. 572)
lymphogranuloma venereum (LGV)
   (p. 585)
male reproductive system   (p. 567)
mammary gland   (p. 567)
neonatal herpes   (p. 588)
nephron   (p. 566)
neurosyphilis   (p. 579)
nocturia   (p. 571)
nongonococcal urethritis (NGU)
   (p. 583)
ovarian follicle   (p. 567)
ovary   (p. 567)
pelvic inflammatory disease (PID)
   (p. 575)
penis   (p. 567)
prostate gland   (p. 567)

prostatitis   (p. 569)
pyelonephritis   (p. 570)
semen   (p. 567)
seminal vesicle   (p. 567)
sexually transmitted disease (STD)
   (p. 574)
syphilis   (p. 579)
testis   (p. 567)
toxic shock syndrome (TSS)   (p. 573)
traumatic herpes   (p. 589)
trichomoniasis   (p. 573)
ureter   (p. 566)
urethra   (p. 567)
urethritis   (p. 569)
urethrocystitis   (p. 569)
urinalysis   (p. 567)
urinary bladder   (p. 566)
urinary system   (p. 565)
urinary tract infection (UTI)   (p. 569)
urine   (p. 566)
urogenital system   (p. 565)
uterine tube   (p. 567)
uterus   (p. 567)
vagina   (p. 567)
vaginitis   (p. 572)
whitlow   (p. 589)

See You on the Web

Don't forget I am expecting you on the web. The following is a preparedness checklist for your web explorations. If you have questions on any of these items, you can return to the end of Chapter 1 for more details.

- Questions We'll Explore
- Concept Check questions
- Flashcards
- Study Guide

If you are ready, come visit me at

http://www.prenhall.com/black

It will be a journey you'll never forget.

Diseases of the Respiratory System

Flashbulbs, music, wisps of fabric as the top designers display their new season of clothing on toothpick-thin supermodels. What does this have to do with the painting you see here? This painting, *La Vie De Boheme*, reflects how fashionable it was to be languorous and pale in the latter part of the nineteenth century. Much of the art of this period shows the same: pale, lethargic women reclining across sofas with the bright pink cheeks that marked tuberculosis.

But why was the tuberculosis look admired? It was believed then that highly intellectual and creative individuals were especially susceptible to tuberculosis. The merits of poets, artists, and musicians were validated by their deaths in unheated attics. Death from tuberculosis was also a prominent theme in the opera. Indeed, tuberculosis became a fashionable disease and set the standards of beauty. Could that still be echoing in the emaciated look of many of today's high fashion models?

Background Basics

The human respiratory system can be infected by various bacteria, viruses, fungi, and at least one helminth. Whether respiratory infections become established depends on host–microbe relationships (Chapter 14) and the condition of the respiratory system and its nonspecific defenses. (Chapter 16) Respiratory infections are divided into upper respiratory infections, including ear infections, and lower respiratory infections.

Components of the Respiratory System

The **respiratory system** consists of the **upper respiratory tract**—consisting of the nasal cavity, pharynx (far'ingks), larynx (lar'ingks), trachea (tra'ke-a), and bronchi (brong'ki)—and the **lower respiratory tract**—composed of the lungs (➤ Figure 21.1). This entire system is lined with moist epithelium. However, in the upper respiratory tract this epithelium contains mucus-secreting cells and is covered with cilia.

The Upper Respiratory Tract

Because the respiratory system moves oxygen from the atmosphere to the blood and removes carbon dioxide from the blood to the atmosphere, every breath of air contains millions of microorganisms and other suspended particles. Some microorganisms and

Questions We'll Explore

A What are the components of the respiratory system? How do they function to prevent infection?

B What are the normal microflora of the respiratory system?

C Which bacteria cause upper respiratory infections, and what are the important epidemiologic and clinical aspects of these diseases?

D Which viruses cause upper respiratory infections, and what are the important epidemiologic and clinical aspects of these diseases?

E Which bacteria cause lower respiratory infections, and what are the important epidemiologic and clinical aspects of these diseases?

F Which viruses cause lower respiratory infections, and what are the important epidemiologic and clinical aspects of these diseases?

G Which fungi cause lower respiratory infections, and what are the important epidemiologic and clinical aspects of these diseases?

H Which parasites cause respiratory infections, and what are the important epidemiologic and clinical aspects of these diseases?

➤ **Figure 21.1   Structures of the respiratory system.** (a) Much of the upper respiratory tract contains resident microbiota. Parts of the tract are covered with mucus and cilia that trap and move airborne material out. The lower respiratory tract lacks host microflora and is subject to more severe and dangerous infections. (b) Bronchioles terminate in alveoli. Any microbes entering an alveolus can be destroyed by alveolar macrophages (not shown).

particles breathed in are removed by hairs and mucus as the air passes through the **nasal cavity.** If microbes enter the **nasal sinuses,** which are hollow cavities lined with mucous membrane within the skull, sinus infections can occur. Air with any remaining microbes and particles then passes through the **pharynx** (throat), a common passageway for the respiratory and digestive systems. The auditory (Eustachian) tube connects the pharynx with the middle ear.

From the pharynx, air and any remaining microbes pass through a series of rigid-walled tubes, the larynx and the trachea. The **larynx** (voice box) contains the vocal cords, which produce sound when they vibrate. The *epiglottis* is a flap of tissue that prevents food and fluids from entering the larynx. The **trachea** (windpipe) branches into primary **bronchi** (singular: *bronchus*), all of which are lined with cilia.

The upper respiratory tract contains a variety of normal microflora that help prevent infection by pathogens that may be inhaled. In addition, mucus from the membranes that line the nasal cavity and pharynx traps microorganisms and most particles of debris, preventing them from passing beyond the pharynx. Mucus also contains lysozyme. Coughing and sneezing mechanically agitate mucus, increasing exposure of microorganisms to mucus and helping to expel them.

The beating of cilia generally serves to move a cell through its environment. ➡ (Chapter 4, p. 95) In the nasal cavity and bronchi, however, cilia extend from the epithelial cells. Because these cells are anchored in place, they function instead to trap and move microbes and particles near the cell surfaces. There mucus with debris trapped in it is moved up into the pharynx. This mechanism, the **mucociliary escalator,** allows materials in the bronchi to be lifted to the pharynx and to be spit out or swallowed. Nevertheless, the mucous membranes of the upper respiratory system are common sites of infection. Such infections often spread to the sinuses, the middle ear, and even to the lower respiratory tract.

The Lower Respiratory Tract

From the bronchi, air passes into the lungs. Secondary bronchi divide into smaller **bronchioles,** forming a branching structure known as the *bronchial tree*. This complex branching arrangement greatly increases the surface area exposed to oxygen flowing into, and carbon dioxide flowing out of, the lungs. Air passing through the terminal bronchioles then enters the **respiratory bronchioles** (➤ Figure 21.1b). These microscopic channels end in a series of saclike **alveoli,** (singular: *alveolus*) which form clusters (nodules). It is in the alveoli that gas exchange occurs. There oxygen diffuses into the blood, and carbon dioxide from the blood diffuses into the alveoli. The bronchial tree, alveoli, blood vessels, and lymphatic vessels form the bulk of the lungs. The surfaces of the lungs and the cavities they occupy are covered by a membrane called the **pleura,** which secretes *serous fluid,* a watery fluid that lubricates some tissues.

Mucus from the lining of the bronchial tree also traps foreign materials that have passed beyond the pharynx. If the upper respiratory tract defense mechanisms fail and microorganisms get into bronchi and bronchioles, help remove them. Only when the numbers of organisms exceed the capacity of the phagocytes to destroy them, or when an infection reaches the lungs by way of blood or lymph vessels, does a lower respiratory infection occur.

The Ears

Like the eyes, the ears are exposed to the environment and thus are subject to microbial attack. The ears contain physical protective structures to prevent infection. Each ear is divided into the outer ear, the middle ear, and the inner ear (➤ Figure 21.2). The *outer ear* has a flaplike **pinna,** or auricle (commonly called the ear), covered with skin, and an **auditory canal,** which is is lined with skin that has many small hairs and many ceruminous glands. The **ceruminous** (se-ru′mi-nus) **glands** are modified sebaceous glands that secrete **cerumen** (earwax). Both hairs and wax help trap microorganisms and other foreign objects and prevent them from entering the auditory canal. Nevertheless, this canal can become infected with fungi.

The **tympanic membrane,** or *eardrum,* separates the outer and middle ears. The *middle ear* is a small, air-filled cavity containing the small bones called *ossicles* that transmit sound waves from the tympanic membrane to the inner ear. If the tympanic membrane is intact, most middle ear infections arise from microorganisms that move up the *auditory (Eustachian) tube* from the nasopharynx. The *inner ear* converts sound waves into nerve impulses carried by the vestibulocochlear cranial nerve (VIII).

Outer and middle ear infections are relatively common—especially in children, because their auditory tubes are shorter and wider, facilitating the transmission of microbes. Because of the relative inaccessibility of the inner ear to pathogens, inner ear infections are rare. How-

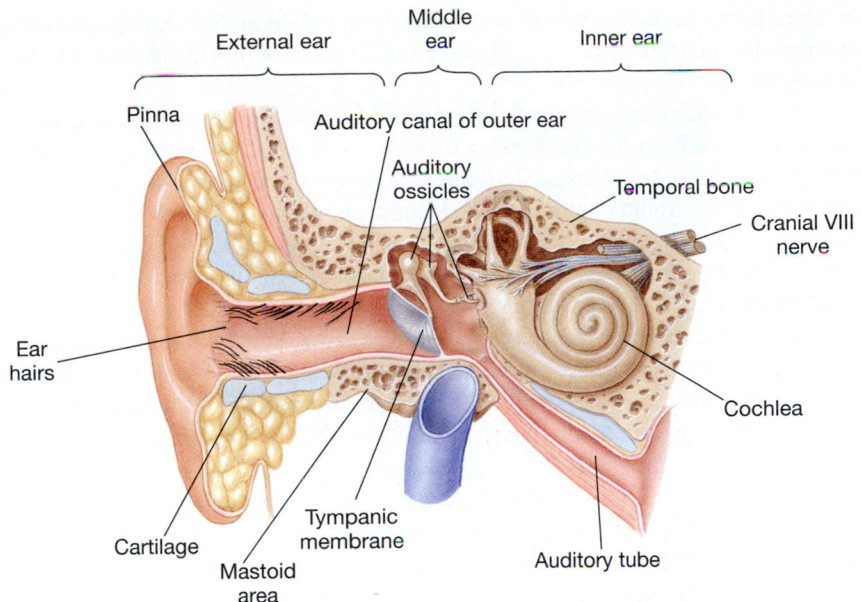

External ear | Middle ear | Inner ear

Pinna
Auditory canal of outer ear
Auditory ossicles
Temporal bone
Cranial VIII nerve
Ear hairs
Cochlea
Cartilage
Tympanic membrane
Mastoid area
Auditory tube

➤ **Figure 21.2 Structures of the ear.** The outer ear is protected from foreign material by ear hairs and earwax. Nevertheless, some ear infections involve the outer ear. Infections of the middle ear usually arise from infections moving up the auditory (Eustachian) tube.

ever, if such an infection does occur, it can easily spread to the **mastoid area,** a bony projection of the skull located behind and below the auditory canal. There is only a thin bony separation between mastoid and brain. Should the mastoid area become infected, therefore—a condition known as *mastoiditis*—the results can be quite serious, as there is danger that the infection will reach the brain.

Normal Microflora of the Respiratory System

Because of the functions of the respiratory system, its surfaces cannot be lined with a tough, keratinized, thick, multiple-layered type of tissue such as we have on our skin. Gas exchange, along with warming and humidifying the air we breathe, require a very thin, moist, delicate layer of cells in close contact with the circulatory system. In some cases, such as alveoli, only a single flat epithelial cell lies pressed against a capillary. Microbial colonization of such surfaces would hinder gas exchange and very likely damage the delicate cells. So it is no surprise that the upper respiratory tract has many protective mechanisms built into it to prevent colonization of the lower respiratory tract.

Healthy lungs are usually sterile. In the trachea (below the level of the larynx), bronchi, and larger bronchioles, there likewise is no normal microflora. Organisms found there are merely transient microbes that have been inhaled. They are removed by the mucociliary escalator. Cigarette smoking and inhalation of toxic fumes decrease the activity of the mucociliary escalator. Failure to clear organisms can then lead to infection.

Within the alveoli, macrophages ➡ (Chapter 16, p. 438) engulf particles and microbes. These macrophages can also move up the mucociliary escalator to the throat. However, not all microbes are trapped in the mucus of the middle portion of the respiratory tract. Some, such as *Mycobacterium tuberculosis* and *Coccidioides immitis,* are able to remain airborne through the trachea and bronchi, and so reach the alveoli. There, macrophages may be unable to destroy them. In fact, *M. tuberculosis* can survive and multiply inside of macrophages, eventually causing infection within the alveoli. Conditions that impair the coughing reflex, or that prevent the epiglottis from fitting tightly over the glottis, also predispose to lower respiratory infection. Most infections of the larynx, trachea, and bronchi are viral in nature.

The pharynx has a normal microflora similar to that of the mouth. Aspiration of secretions from this area can cause infections such as pneumonia lower in the respiratory tract. The pharynx has its own set of defenses, a ring of lymphoid tissues, such as the tonsils, which are an important part of the immune system ➡ (Chapter 16). Still, potential pathogens, such as *Klebsiella pneumoniae* and species of *Haemophilus,* are able to survive in and colonize the pharynx. Their presence does not necessarily indicate an infection. For example, you do not develop pneumonia of the pharynx—the organisms must successfully invade the bronchi and/or lungs in order to cause pneumonia.

The upper respiratory tract forms the first line of defense against infection. These areas have a normal microflora similar to that of skin, with *Staphylococcus epidermidis* and corynebacteria being most numerous. *Staphylococcus aureus* is also often found here, especially in the anterior nostrils. About one-third of the healthy population carries *S. aureus* at any given time—half permanently, and half transiently. Nasal carriers of *S. aureus* can easily spread it to other individuals, especially newborns who are not yet fully colonized. A nasal carrier of a highly virulent strain of *S. aureus* working in a neonatal unit presents a very real hazard, as most newborns are colonized by a single strain of *S. aureus* from their environment within the first 24 hours after delivery. However, it has been shown that deliberate colonization of the infant's nasal mucosa with a fairly harmless strain of *S. aureus* during this first 24-hour period prevents subsequent colonization with a different (potentially more virulent) strain of *S. aureus*.

Hairs lining the anterior nostrils prevent the passage of larger particles. But they are also important because they cause air to swirl and eddy, thereby slowing it down and forcing particulate matter to settle out onto mucus. The mucus is propelled toward the pharynx at a rate of 0.5 to 1.0 cm per minute, where it forms a postnasal drip. Swallowed into the stomach, trapped microbes are destroyed. The nasal passages have projections and plates of bone that, like the hairs, cause turbulence and slowing of air flow.

All in all, the defenses of the respiratory system are quite efficient, considering that the average adult male inhales an estimated average of eight microbes per minute at rest, and more while breathing harder, yet develops only a few respiratory infections each year.

Diseases of the Upper Respiratory Tract

Bacterial Upper Respiratory Diseases

Bacterial infections of the upper respiratory tract are exceedingly common. They are easily acquired through the inhalation of droplet nuclei from infected persons or carriers, especially in winter, when people are crowded indoors in poorly ventilated areas.

Pharyngitis and Related Infections

Pharyngitis, or sore throat, is an infection of the pharynx. It is frequently caused by a virus but is sometimes bacterial in origin. **Laryngitis** is an infection of the larynx, often with loss of voice. **Epiglottitis,** an infection of the epiglottis, can close the airway and cause suffocation. *Croup,* a viral infection found in children, also involves the larynx and the epiglottis. Difficult breathing, spasms of the larynx, and membrane formation may occur. When an infection reaches the sinus cavities, bronchi, or tonsils, the conditions are referred to as **sinusitis, bronchitis,** and **tonsilitis,** respectively. When an infection spreads to the lungs, it is no longer an upper respiratory disease and is called *pneumonia.*

Streptococcal Pharyngitis   Less than 10 percent of cases of pharyngitis are caused by the group A β-hemolytic *Streptococcus pyogenes*. This infection, familiarly known as *strep throat,* is most common in children 5 to 15 years old. It is acquired by inhaling droplet nuclei from active cases or healthy carriers. Dogs and other family pets also can be carriers. The detection and elimination of carrier states is often difficult in recurrent or cluster outbreaks. Contaminated food, milk, and water also can spread the disease, so it is important that infected persons and carriers not handle food. In strep throat, the throat typically becomes inflamed, and the adenoids and lymph nodes in the neck swell. The tonsils become quite tender and develop white, pus-filled lesions (➤ Figure 21.3). Onset is usually abrupt, with chills, headache, acute throat soreness, especially upon swallowing, and often nausea and vomiting. Fever is generally high. The absence of cough and nasal discharge helps distinguish strep throat from the common cold. Strains of *S. pyogenes* infected with a specific toxin-producing temperate phage cause scarlet fever along with pharyngitis. ➡ (Chapter 19, p. 539)

Diagnosis is made by positive throat culture. A rapid enzyme-labeled antibody screening test using a throat swab can be completed in a physician's office within minutes. Immediate treat-

It's more likely that you will catch a cold when you are around more infected people, but the more people you associate with, the greater your resistance will be.

With a strep test now available, it takes a physician about 15 minutes to detect a streptococcal infection, instead of the 24 hours or more required for a culture.

(a)

(b)

➤ Figure 21.3   **Strep throat.** This common form of pharyngitis is characterized by (a) enlarged and reddened adenoids at the sides of the throat and (b) white, pus-filled lesions on tonsils.

ment is important. If treatment is delayed, *S. pyogenes* can interact with the immune system and give rise to rheumatic fever (Chapter 23), which occurs in 3 percent of untreated cases. For this reason treatment with penicillin or one of its derivatives is often begun even before culture results are available.

Laryngitis and Epiglottitis   Laryngitis can be caused by bacteria such as *Haemophilus influenzae* and *Streptococcus pneumoniae,* by viruses alone, or by a combination of bacteria and viruses. Acute epiglottitis is almost invariably caused by *H. influenzae.* Inflammation of the tissues rapidly closes the airway, causing difficulty in breathing or even death.

Sinusitis   More than half the cases of sinusitis are caused by *Streptococcus pneumoniae* or by *Haemophilus influenzae,* but some cases are caused by *Staphylococcus aureus* or *Streptococcus pyogenes.* Swelling of sinus cavity linings slows or prevents drainage and leads to pressure and severe pain. When drainage remains impeded, mucus accumulates and fosters bacterial growth. Secretions consisting of mucus, bacteria, and phagocytic cells then collect in the sinuses. Chronic sinusitis can permanently damage sinus linings and cause *polyps* (smooth, pendulous growths) to form. Applying moist heat over the sinuses, instilling drops of vasoconstrictors, such as ephedrine, into the nasal passages, humidifying the air, and holding the head in a position to promote drainage help alleviate symptoms. Treatment with antibiotics such as penicillin and medications to lessen discomfort may be necessary. Swimmers and divers often suffer from sinusitis if water is forced into the sinuses. Such people can prevent water from entering their sinuses by using nose clips or by exhaling as they submerge and continuing to exhale while their head is underwater.

Bronchitis   Bronchitis involves the bronchi and bronchioles but does not extend into the alveoli. About 15 percent of the general population has chronic bronchitis. It is most common in older people and is linked to smoking, air pollution, inhalation of coal dust, cotton lint, and other particles, and heredity. Patients cough up sputum containing mucus, organisms, and phagocytic cells. Common causative agents include *Streptococcus pneumoniae, Mycoplasma pneumoniae,* and various species of *Haemophilus, Streptococcus,* and *Staphylococcus.* Infections can spread to the alveoli of the lungs and cause pneumonia. Diagnosis is made from sputum cultures. By the time an infected individual seeks medical attention, respiratory membranes may have been permanently damaged. Antibiotic treatment can halt further deterioration but cannot reverse the damage already done. Eventually severe shortness of breath develops.

Diphtheria

Although fewer than 10 cases per year are now seen in the United States, **diphtheria** was once a feared killer. A century ago, 30 to 50 percent of patients died; most deaths occurred by suffocation in children under age 4. Diphtheria is still a problem today in the former Soviet Union, where the disease exists in epidemic proportions.

You should suspect epiglottitis when swallowing is extremely painful (causing drooling), when speech is muffled, and when breathing becomes difficult.

Every year, over $1.5 billion worth of "sinus" medicine is purchased in America for the symptoms of sinus disease.

It is thought that George Washington may have died from diphtheria.

➤ **Figure 21.4** *Corynebacterium diphtheriae,* **the cause of diphtheria.** This stained photomicrograph shows metachromatic granules of phosphates that have stained a deep reddish blue.

Diphtheria and the "Desert Fox"

German general Erwin Rommel, the "Desert Fox," led the North African tank warefare during World War II. He was always seen in newsreels with a handkerchief pressed to his nose. This was not because of desert dust, but because of nasal diphtheria. He periodically returned to Berlin for treatment, leaving his troops without his brilliant and charismatic leadership. In fact, the Allies may have defeated the Nazis in North Afria because of the battles lost due to troops who were disheartened in his absence. In Berlin, Rommel joined other generals in a plot to assassinate Hitler. The plot failed; Rommel's role was discovered. As Gestapo agents were taking him to headquarters, he died of a heart attack. This could have been the result of a heart weakened by years of exposure to diphtheria toxin. Had Rommel never contracted diphtheria, would the outcome of World War II have changed?

Sequelae, or adverse signs that follow a disease, are common in diphtheria. Myocarditis, an inflammation of the heart muscle, and polyneuritis, an inflammation of several nerves, account for deaths even after apparent recovery. Significant cardiac abnormalities occur in 20 percent of patients. Neurologic problems, including paralysis, can occur after particularly severe cases.

Causative Agent   Diphtheria is caused by strains of *Corynebacterium diphtheriae* infected with a prophage that carries an exotoxin-producing gene. ➡ (Chapter 10, p. 272) Club-shaped cells of this Gram-positive rod grow side by side in palisades (like the upright logs used to fortify old forts). These cells contain metachromatic granules of phosphates, which turn reddish when stained with methylene blue (➤ Figure 21.4). **Diphtheroids,** are corynebacteria that are found in or on such body sites as the nose, throat, nasopharynx, urinary tract, and skin. These organisms differ from *C. diphtheria* by not being toxin-producers. On inoculation into a guinea pig, toxin-producing *C. diphtheriae* strains cause disease, but diphtheroids do not. Some laboratories also use a gel diffusion test to detect toxin-producing strains.

To produce toxin, the bacterium must be infected by an appropriate strain of bacteriophage in the lysogenic prophage state. ➡ (Chapter 10, p. 271) In other words, the bacteriophage DNA must be integrated into the bacterial chromosome, where its toxin-producing genes are expressed. "Curing" the bacterium of its phage infection eliminates its ability to produce toxin. The phage-infected cell also fails to produce toxin unless the patient's blood iron concentration drops to a critically low level.

The Disease   Humans are the only natural hosts for *C. diphtheriae.* Diphtheria usually is spread by droplets of respiratory secretions. Infection usually begins in the pharynx 2 to 4 days after exposure. The organism, damaged epithelial cells, fibrin, and blood cells combine to form a **pseudomembrane** (➤ Figure 21.5). Although it is not a true membrane, removing it leaves a raw, bleeding surface that is soon covered by another pseudomembrane. The pseudomembrane can block the airway and cause suffocation. The organisms very rarely invade deeper tissues or spread to other sites, but the extremely potent toxin spreads throughout the body and kills cells by interfering with protein synthesis. The heart, kidneys, and nervous system are most susceptible.

Diphtheria organisms are sometimes found in unusual sites. They can invade the nasal cavities, which have relatively few blood vessels. There they cause a milder disease because less toxin enters the blood. They also can invade the skin in *cutaneous diphtheria* and cause tissue injury. Only a small amount of toxin is absorbed by the blood. Cutaneous diphtheria is a tropical disease associated with poor hygiene; it is rare in the United States.

Treatment and Prevention   Diphtheria is treated by administering both antitoxin to counteract toxin and antibiotics such as penicillin or erythromycin to kill the organisms. The disease can be prevented with DTP (*diphtheria, tetanus, pertussis*) vaccine, which contains diphtheria toxoid, tetanus toxoid, and killed *Bordetella pertussis* and is administered in a series of shots beginning at age 2 months. Boosters are needed throughout childhood and adulthood. In the past, when diphtheria was endemic in the United States, most of the adult population maintained, or even acquired, immunity by continual exposure to small quantities of the bacterium.

➤ **Figure 21.5** **A pseudomembrane characteristic of diphtheria.** This is not a true membrane; rather, it adheres to the underlying tissue. If torn off, it will leave a raw, bloody surface and will eventually reform. However, it must be removed when it blocks the airway.

Such exposures acted as natural booster shots. Now that diphtheria is no longer endemic, immunity depends on the vaccine. Because diphtheria is not well controlled elsewhere in the world, especially in the former Soviet Union, vaccination must be continued in the United States to prevent epidemics should the disease spread from other places.

Ear Infections

Ear infections commonly occur as **otitis media** in the middle ear and as **otitis externa** in the external auditory canal. Because otitis media usually produces a puslike exudate, it often is referred to as otitis media with effusion (OME). *Streptococcus pneumoniae, S. pyogenes,* and *Haemophilus influenzae* account for about half of acute cases. Organisms such as species of *Proteus, Klebsiella,* and *Pseudomonas* are often responsible for chronic cases. Otitis externa usually is caused by *Staphylococcus aureus* or *Pseudomonas aeruginosa.* Pseudomonad infections are common in swimmers because the organisms are highly resistant to chlorine.

OME infections arise from the passage of pharyngeal organisms through the Eustachian tube. Fever and earache, which arise from pressure the pus creates in the middle ear, usually are present, but some cases are asymptomatic. The disease is treated with antibiotics, usually penicillin. It is important to continue treatment until all organisms are eradicated to prevent complications. Even after successful therapy, sterile fluid can remain in the middle ear, impairing vibration of the tiny bones of hearing there and decreasing sound transmission. Tubes are sometimes inserted to prevent fluid accumulation and repeated middle ear infections (➤ Figure 21.6). If the impairment occurs during speech development, as it often does, speech can be adversely affected. Some children thought to be inattentive or defiant really cannot hear. School can be a nightmare for such children. As children grow older, the Eustachian tube changes shape and develops an angle that prevents most organisms from reaching the middle ear—much to the relief of the children and their parents.

Viral Upper Respiratory Diseases

The Common Cold

The common cold, or **coryza,** probably causes more misery and loss of work hours than any other infectious disease, but it is not life-threatening. Although exact statistics are not available on the number of infections per year, Americans lose more than 200 million days of work and

- Auditory ossicle
- Middle ear space
- Tube
- Fluid trapped in middle ear by infection
- Tympanic membrane
- External ear canal
- Eustachian tube
- Pharynx

➤ Figure 21.6   **Treating middle ear infections.** Tubes are placed through the tympanic membrane (eardrum) to promote drainage. This can be done in the physician's office. The phlange on the rear end of the tube is designed to hold it in place.

school per year because of colds. Economically, colds are a bane to employers, who must deal with lost work time, and a boon to manufacturers and retailers of cold remedies. Cold viruses are ubiquitous and present year round, but most infections occur in early fall or early spring. After an incubation period of 2 to 4 days, signs and symptoms such as sneezing, inflammation of mucous membranes, excessive mucus secretion, and airway obstruction appear. Sore throat, malaise, headache, cough, and occasionally tracheobronchitis occur. Pus and blood may be present in nasal secretions. The illness lasts about 1 week. The severity of symptoms is directly correlated with the quantity of viruses released from infected epithelial cells of the upper respiratory tract. Viral cold infections can also predispose an individual to secondary bacterial infections such as pneumonia, sinusitis, bronchitis, and otitis media. These infections add time, misery, and additional cost to the original cold. More serious diseases such as whooping cough and respiratory syncytial virus infections can be mistaken early on for colds, so correct diagnosis and subsequent treatment are delayed.

Causative Agents   Different cold viruses predominate during different seasons. In fall and spring, rhinoviruses are in the majority. Parainfluenza virus is present all year but peaks in late summer. In mid-December the coronaviruses appear. Adenoviruses are present at a low level all year. About 200 different viruses can cause colds.

Rhinoviruses are the most common cause of colds, accounting for about half of all cases. They are resistant to antibiotics, chemotherapeutics, and disinfectants but are quickly inactivated by acidic conditions. They grow best at 33° to 34°C and thus replicate in the epithelium of the upper respiratory tract, where air movements lower the tissue temperature. Rhinoviruses can be isolated from nasal secretions and throat washings and identified by their sensitivity to low pH and resistance to ether and chloroform, a unique combination of traits among viruses. At least 113 different rhinoviruses have been identified, all with different antigens. Natural immunity is short-lived, and no effective vaccine has been developed. Even if a person becomes immune to some rhinoviruses, there are always others to cause another cold.

The second most common cause of colds is another group of ubiquitous viruses, the **coronaviruses.** Coronaviruses have clublike projections that give them a halo (*corona,* Latin for "halo" or "crown"). The projections are responsible for attaching the virus to host cells as well as stimulating the immune system to produce antibodies against the virus. In addition to causing colds, these viruses also cause acute respiratory distress, in some cases mild pneumonia, and acute gastroenteritis. They infect the epithelium of both the respiratory and digestive tracts. In the digestive tract they reduce absorptive capacity and cause diarrhea, dehydration, and electrolyte imbalances.

Transmission   Cold viruses are more often spread by fomites than by close contact with infected persons. Blowing the nose and handling used tissues contaminates the fingers, so that anything touched becomes contaminated. Conscientiously using tissues once, discarding them immediately in covered containers, and thoroughly washing hands after each nose wipe can significantly reduce the spread of rhinoviruses.

Diagnosis and Treatment   Most people diagnose and treat their own colds with remedies that alleviate some symptoms. Over-the-counter antihistamines are reasonably effective in counteracting inflammatory reactions as the body attempts to defend itself against the viruses. Experiments with certain kinds of human interferon have shown potential in blocking or limiting rhinovirus infections, but they must be given with another agent that helps the interferon reach epithelial cells beneath a thick blanket of mucus. A quantity of alpha interferon sufficient to block infection causes irritation of membranes and bleeding. Different combinations of interferons and other factors are being explored to control rhinovirus infections.

Parainfluenza

Parainfluenza is characterized by rhinitis (nasal inflammation), pharyngitis, bronchitis, and sometimes pneumonia, mainly in children. **Parainfluenza viruses** (paramyxoviruses; Chapter 10) initially attack the mucous membranes of the nose and throat. ➡ (p. 263) In very mild cases, symptoms may be inapparent. When symptoms do appear, the first symptoms are cough and hoarseness for 2 or 3 days, harsh breathing sounds, and a red throat. Symptoms can progress

---

A Cure for the Common Cold?

Why is there a vaccine against polio but not against the common cold? There are only three distinct forms (serotypes) of wild poliovirus. Once a separate vaccine was made against each serotype, the job was done. However, there are at least 113 different rhinoviruses; we would need 113 vaccines to attack the rhinoviruses, and then more for the coronaviruses and adenoviruses that also cause colds. Why are there only three types of poliovirus? Polioviruses must survive the stomach's acidic conditions in order to establish an infection. Mutations in their capsid structure almost always leave them vulnerable to low (acidic) pH, so mutated forms do not succeed. Rhinoviruses are spared this harrowing route. Capsid changes due to mutations enable them to evade the upper respiratory tract's defenses easily, so some mutated forms succeed. The result? There are too many versions of cold viruses for vaccine researchers to keep up with.

> **Table 21.1** **A summary of diseases of the upper respiratory tract**

Disease	Agent(s)	Characteristics
Bacterial upper respiratory diseases		
Pharyngitis	*Streptococcus pyogenes*	Inflammation of the throat; fever without cough or nasal discharge
Laryngitis and epiglottitis	*Haemophilus influenzae, Streptococcus pneumoniae*	Inflammation of the larynx and epiglottis, often with loss of voice
Sinusitis	*H. influenzae, S. pneumoniae, S. pyogenes, Staphylococcus aureus*	Inflammation of sinus cavities, sometimes with severe pain
Bronchitis	*Streptococcus pneumoniae, Mycoplasma pneumoniae,* and others	Inflammation of bronchi and bronchioles with a mucopurulent (mucus- and pus-filled) cough; shortness of breath in chronic cases
Diphtheria	*Corynebacterium diphtheriae*	Inflammation of the pharynx with pseudomembrane and systemic effects of toxin
Otitis externa	*Staphylococcus aureus, Pseudomonas aeruginosa*	Inflammation of external ear canal; common in swimmers
Otitis media	*Streptococcus pneumoniae, S. pyogenes, Haemophilus influenzae*	Pus-filled infection of the middle ear with pressure and pain
Viral upper respiratory diseases		
Common cold	Rhinoviruses, coronaviruses	Sore throat, malaise, headache, and cough
Parainfluenza	Parainfluenza viruses	Nasal inflammation, pharyngitis, bronchitis, croup, sometimes pneumonia

to a barking cough and a high-pitched, noisy respiration called *stridor* (strī′dor). Recovery is usually quick—within a few days.

Of the four parainfluenza viruses capable of infecting humans, two can cause croup. **Croup** is defined as any acute obstruction of the larynx and can be caused by a variety of infectious agents, including parainfluenza viruses. Both the larynx and the epiglottis become swollen and inflamed; the high-pitched barking cough of croup results from partial closure of these structures. Increased humidity from a cool-mist vaporizer or from a hot shower helps to relieve croup symptoms.

By age 10, most children, whether or not they have had recognizable illness, have antibodies to all four parainfluenza viruses. Thus, the incidence of infection is very high, although the incidence of clinically apparent disease is much lower. Epidemics and smaller outbreaks of parainfluenza infections occur primarily in fall but also sometimes in early spring after the "flu" season. The viruses are spread by direct contact or by large droplets. The causative viruses can be inactivated by drying, increased temperature, and most disinfectants, so they do not remain long on surfaces or in the environment. Resistance to infection comes from secretory IgA that defends mucous membranes against infection, and not from blood-borne IgG. (Chapter 17, p. 469) Reinfection with parainfluenza viruses is rare, so secretory immunoglobulins must create effective immunity. However, efforts to make a vaccine for parainfluenza viruses have not been successful.

Diseases of the upper respiratory tract are summarized in Table 21.1.

Diseases of the Lower Respiratory Tract

Bacterial Lower Respiratory Diseases

Among the bacterial diseases of the lower respiratory tract are two of the great killer infections of history: pneumonia and tuberculosis. The advent of antibiotic therapy brought these diseases under control to a considerable extent. Both are making comebacks today as a result of the spread of AIDS and of treatment with immunosuppressive drugs for transplant patients and with anti-inflammatory agents for autoimmune disorders, such as rheumatoid arthritis and multiple sclerosis. Lowered resistance, overcrowding, chronic diseases, aging, and other immunosuppressive factors also contribute to the severity of the problem.

Croup's main symptom is a harsh cough that sounds like a seal's bark.

✓ Of what importance is bacteriophage infection in *Corynebacterium diphtheriae*?

✓ How do tubes in the tympanic membrane help to prevent frequent middle ear infections?

✓ How is the common cold usually transmitted?

Whooping Cough

Whooping cough, also called **pertussis,** is a highly contagious disease known only in humans. The word pertussis means "violent cough," and the Chinese call it the "cough of 100 days." Although distributed worldwide, strains found in the United States are less virulent than most. However, jet travel could bring virulent strains from North Africa or other parts of the world at any time. Whooping cough is a major health problem in developing nations, where lack of immunization allows 80 percent of those exposed to contract the disease. In the United States, concern and negative publicity regarding vaccine safety discouraged some parents from having very young children immunized. As a result, incidence of the disease more than doubled during the 1980s and reached nearly 6600 in 1993 (➤ Figure 21.7a). The disease tends to occur sporadically, especially in infants or young children; 50 percent of cases occur in the first year of life (➤ Figure 21.7b).

Before the development of the vaccine, nearly every child got whooping cough. Adults who contract it today either were not vaccinated, or their immunity has declined. Partial immunity lessens the severity of the disease. Many adults may fail to display the characteristic "whooping" sound and so are misdiagnosed. Such people serve as a reservoir to spread the infection.

➤ Figure 21.7   **The U.S. incidence of whooping cough,** (a) By year; (b) by age group. (c) The technique of culturing for whooping cough. A swab attached to a thin, flexible wire is inserted through the nostrils, and the patient is asked to cough several times. (*Source:* CDC)

Causative Agent   *Bordetella pertussis,* a small, aerobic, encapsulated Gram-negative coccobacillus first isolated in 1906, is the usual causative agent of whooping cough. Only about 5 percent of the cases are due to *Bordetella parapertussis* and *B. bronchiseptica,* which usually produce a milder disease. *B. bronchiseptica* is a normal resident of canine respiratory tracts, where it sometimes causes "kennel cough."

Susceptible people become infected by inhaling respiratory droplets. The organisms colonize cilia lining the respiratory tract. Only active cases of whooping cough are known to shed organisms; carriers of the disease are unknown. *Bordetella pertussis* does not invade tissues or enter the blood, but it does produce several substances that contribute to its virulence. It produces an endotoxin, an exotoxin, and *hemagglutinins* (hem"ah-gloo'ti-ninz), surface antigens that help it attach to the cilia of epithelial cells in the upper respiratory tract. There, the toxins destroy the ciliated epithelial cells. Those cells are then sloughed, leaving a surface of nonciliated cells, which in turn allows mucus to accumulate within the respiratory airway.

The Disease   After an incubation period of 7 to 10 days, the disease progresses through three stages: *catarrhal, paroxysmal,* and *convalescent.* The **catarrhal stage** is characterized by fever, sneezing, vomiting, and a mild, dry, persistent cough. A week or two later the **paroxysmal** (paroks-iz'mal) (or *intensifying stage*) begins as mucus and masses of bacteria fill the airway and immobilize the cilia. Strong, sticky, ropelike strings of mucus in the airway elicit violent, paroxysmal coughing. Failure to keep the airway open leads to **cyanosis** (si-an-o'sis), or a bluing of the skin, because too little oxygen gets to the blood. Keeping the airway open is especially difficult in infants; airway blockage accounts for the high death rate in patients under 1 year of age. Straining to draw in air gives the characteristic loud "whooping" sound. Coughing sieges occur several times a day and cause exhaustion. Sometimes coughing is so severe that it causes hemorrhage, convulsions, and rib fractures. Vomiting that usually follows a coughing siege leads to dehydration, nutrient deficiency, and electrolyte imbalance—all especially dangerous in infants.

After a paroxysmal stage, usually of 1 to 6 weeks duration but sometimes longer, the patient enters the **convalescent stage.** Milder coughing can continue for several months before it eventually subsides. Secondary infections with other organisms are common at this stage.

Diagnosis and Treatment   Whooping cough is diagnosed by obtaining organisms from the posterior nasal passages (➤ Figure 21.7c) and culturing them on freshly made charcoal–blood agar medium, which has largely replaced the classic potato–blood–glycerol agar (Bordet-Gengou) medium. Because penicillin does not kill *B. pertussis,* the antibiotic is often added to the medium to suppress the growth of other organisms. Once typical colonies appear, fluorescent antibody stain can be used for identification. Whooping cough treatment includes antitoxin given early in the disease to combat toxin, and erythromycin to shorten the paroxysmal, or second, stage. ➥ (Chapter 17, p. 484) The antibiotic cannot entirely eliminate that stage, but it does reduce the number of viable organisms being shed. Supportive measures such as suctioning, rehydration, oxygen therapy, and attention to nutrition and electrolyte balance are very important, as is appropriate treatment of secondary infections.

Prevention   Pertussis vaccine has saved many lives, but it is not entirely safe. In the United States it lowered the number of cases from 227,319 in 1938 to 3590 in 1994, but each year vaccine reactions cause 5 to 20 children to die and another 50 to suffer permanent brain damage. Yet the number of children hurt by the vaccine is far smaller than the number who would die without it. Vaccination begins with the DTP series at 2 months of age, as soon as the immune system is capable of responding to antigens in the vaccine. Early immunization is important because pertussis antibodies do not cross the placenta, so infants have no passive immunity to whooping cough. Immunity gradually decreases after vaccination. However, immunization of unvaccinated children over 7 years of age or adults is not recommended, due to the risks associated with the vaccine. This means that a large part of the teenage and adult population is actually susceptible to disease, but mistakenly feel they are protected by their early vaccinations. Recombinant DNA technology is being used for the development of acellular vaccines. Because they contain only bacterial proteins, such preparations could reduce or even eliminate the side effects associated with current preparations. Acellular pertussis vaccines are approved for use as fourth or fifth doses, but not as the first three primary doses for routine immunization.

Recovery from the disease confers immunity, but not for a lifetime. Second cases are known, especially in adulthood, but they are usually milder than the first. The United States is fortunate right now in having low-virulence strains of *B. pertussis*. African strains are far more virulent and would probably cause massive outbreaks if they were to become established here.

Classic Pneumonia

In the United States, more deaths result from pneumonia than from any other infectious disease. **Pneumonia,** an inflammation of lung tissue, can be caused by bacteria, viruses, fungi, certain helminths, chemicals, radiation, and some allergies. Infectious forms of the disease develop when pathogens that are able to evade upper respiratory defenses are inhaled. Such organisms initiate the disease process by colonizing the upper respiratory tract and then entering the lower respiratory tract accidentally during a deep breath or suppressed cough, or by means of a large amount of mucus. Several bacteria are known causes of pneumonia, including *Streptococcus pneumoniae,* also known as the pneumococcus; *Staphylococcus aureus; Klebsiella pneumoniae;* and *Mycoplasma pneumoniae,* a recently recognized cause of human respiratory infections that is responsible for about 10 percent of community-acquired cases of pneumonia. *Pseudomonas aeruginosa* is another respiratory pathogen. It causes pneumonia in immunocompromised persons and hospitalized patients. Once the organism gains access to damaged respiratory epithelium, it attaches by means of its pili and multiplies (➤ Figure 21.8).

Classification of Pneumonias   Pneumonias are classified by site of infection as lobar or bronchial. **Lobar pneumonia** affects one or more of the five major lobes of the lungs. It is a serious primary disease that affects about 500,000 people in the United States annually. *Streptococcus pneumoniae* causes 95 percent of all cases. However, back in 1881 when *Klebsiella pneumoniae* was first discovered, it was thought to be the primary cause. Fibrin deposits are characteristic of lobar pneumonia. When they solidify, they cause **consolidation,** or blockage of air spaces. **Pleurisy,** inflammation of the pleural membranes that causes painful breathing, often accompanies lobar pneumonia.

Bronchial pneumonia begins in the bronchi and can spread into the surrounding tissues toward the alveoli in a patchy manner. Although also most likely to be caused by pneumococci, bronchial pneumonia differs from lobar pneumonia in two ways: (1) It often appears as a secondary infection after a primary infection such as a viral influenza or heart disease, or another lung disease, and (2) it lacks the plentiful fibrin deposits of lobar pneumonia. Bronchial pneumonia can follow exposure to chemicals or aspiration of vomitus or other fluids. Infants sometimes aspirate amniotic fluid during birth, especially during delivery by Caesarean section. Bronchial pneumonia is common in elderly and debilitated patients. In fact, it is sometimes called the "old man's friend" because it spares an elderly person a possibly longer, more painful death from some other cause.

Transmission   Both lobar and bronchial pneumonia are transmitted by respiratory droplets and, in the winter, by carriers—including health care workers—who have had contact with pneumonia patients. As much as 60 percent of a population can be carriers in closely associated groups, such as personnel on military bases or children in preschools.

The Disease   After a few days of mild upper respiratory symptoms, the onset of pneumococcal pneumonia is sudden. The infected person suffers violent chills and high fever (up to 106°F). Chest pain, cough, and sputum containing blood, mucus, and pus follow. Fever may end 5 to 10 days after onset when untreated or within 24 hours after antibiotics are given.

Pneumococcal pneumonia is the only infectious disease in the top 10 causes of death in the United States. With immediate and adequate antibiotic therapy, the mortality rate is 5 percent; without treatment, it is 30 percent. This compares favorably with *Klebsiella* pneumonia, which despite best treatment still has a mortality rate of 50 percent. *Klebsiella* causes an extremely severe pneumonia that can lead to chronic ulcerative lesions in the lungs and extensive destruction of lung tissue. Failure to seek medical attention promptly keeps the overall mortality rate for pneumonia in the United States at 25 percent. Conditions that predispose toward pneumonia include old age, chilling, drugs, anesthesia, alcoholism, and a variety of disease states.

➤ Figure 21.8 **Pneumonia-causing bacteria.** This SEM shows *Pseudomonas aeruginosa* (see arrow) attaching to human respiratory epithelium.

Diagnosis, Treatment, and Prevention   Diagnosis of pneumonia is based on clinical observations, X-rays, or sputum culture. *Klebsiella* pneumonia is usually treated with cephalosporins. Penicillin is the drug of choice for treatment of pneumococcal pneumonia. Following recovery, immunity exists for a few months only against the particular serotype that caused the infection. Thus, a patient can develop case after case of pneumonia from infection with other serotypes or other species. More than 85 serotypes are recognized. Of these, only 23 are responsible for 85 percent of pneumococcal disease in the United States. Artificial immunity can be induced with the polyvalent vaccine Pneumovax. This vaccine contains 23 serotype antigens, so it protects against 80 percent of pneumococcal pneumonias in all age groups except children under 2 years of age. Immunization is especially recommended for the elderly and at-risk populations.

Mycoplasma Pneumonia

One of the tiniest bacterial pathogens known, *Mycoplasma pneumoniae* ordinarily causes mild, and sometimes inapparent, upper respiratory tract infections. In 3 to 10 percent of infections it causes **primary atypical pneumonia**, or *Mycoplasma* pneumonia, usually a mild pneumonia with an insidious onset. The disease is said to be atypical because the symptoms are different from those of classic pneumonia. Patients often remain ambulatory, so the disease is sometimes called **walking pneumonia.** The mortality rate is less than 0.1 percent. It is unusual in preschool children and is most common among young people 5 to 19 years old. *Mycoplasma pneumoniae* may cause up to 20 percent of all noninfluenza community-acquired pneumonias.

Transmission is by respiratory secretions in droplet form, and onset of symptoms follows an incubation period of 12 to 14 days. Fever lasts 8 to 10 days and gradually declines along with the cough and chest pain. Unusual features of M*ycoplasma* pneumonia are that alveoli decrease in size by inward swelling of the alveolar walls and that the alveoli do not fill with fluid.

Diagnosis can be made by isolating *M. pneumoniae* from sputum or from a nasopharyngeal swab. This process takes 2 to 3 weeks due to the slow growth rate of the organisms (➤ Figure 21.9). Serologic tests such as indirect immunofluorescence, latex agglutination, and ELISA are also of value during the early stages of the disease. Commercially available DNA probes are also used and produce results similar to those of culture. However, treatment is usually based on clinical symptoms.

Erythromycin and tetracycline are the drugs of choice. Penicillin has no effect because *Mycoplasma* lack cell walls (the site of penicillin's antibacterial action). Even untreated cases have a favorable prognosis. No vaccine is currently available, so prevention requires avoidance of contact with infected persons and their secretions.

Legionnaires' Disease

In 1976 many war veterans attending a convention in Philadelphia became victims of a mysterious ailment that became known as **Legionnaires' disease.** After 29 deaths and much frantic investigation, the previously unidentified causative organism, *Legionella pneumophila,* was finally isolated (➤ Figure 21.10). Researchers at CDC later found antibodies to *L. pneumophila* in frozen blood samples from unidentified outbreaks that occurred decades ago. One wonders how the organism was overlooked for such a long time. However, it is sufficiently different from previously classified bacteria that a new genus had to be created for it.

Legionella pneumophila is a weakly Gram-negative, strictly aerobic bacillus with fastidious nutritional requirements. It does not ferment sugars and has an obscure life cycle. More than 20 *Legionella* species have been identified. Most are free-living in soil or water and do not ordinarily cause disease. However, some strains live as intracellular parasites of amoebas of many species such as *Acanthamoeba, Naegleria, Hartmanella,* and *Echinamoeba*. Some of these amoebas colonize cooling towers, shower heads, and other wet areas. Legionellosis is transmitted when organisms growing in soil or water become airborne and enter the patient's lungs as an aerosol. Person-to-person transmission has never been documented. Air conditioners, ornamental fountains, produce sprayers in grocery stores, humidifiers, and vaporizers in patient rooms have been implicated in the spread of the disease. Such devices should be regularly disinfected. Once inhaled, the *Legionella* organisms are taken up into ameoboid phagocytes by a

The Prowler Who's Hard to Stop
Our cells are pretty impressive at killing intruders. For example, most of our cells have little sacs of digestive enzymes called lysosomes. These enzymes can break down carbohydrates, proteins, nucleic acids, some lipids, making them pretty effective at killing just about anything foreign that enters the cell. But some bacteria can get into our cells and survive, anyway. How do they do it? Legionella pneumophila, which causes Legionnaires' pneumonia, invades macrophages. Somehow, the bacterium alters its host so that the host's lysosome can't release its destructive enzymes. That's why antibiotics are necessary—our cells can't always kill the bacterium on their own.

(a)

➤ Figure 21.9 **Mycoplasma.** (a) These cells lack a cell wall and assume irregular shapes (151,000X). (b) A cell infected by numerous *Mycoplasma* cells.

(b)

➤ Figure 21.10 *Legionella pneumophila,* **the cause of Legionnaires' disease.** (Colorized TEM 25,000×.)

A German team has recently found molecular evidence of Mycobacterium tuberculosis in a 3000-year-old mummy, the oldest evidence of TB in any population.

spirally progressing form of phagocytosis. They thrive inside the acidic conditions of the phagolysosome, multiplying and eventually rupturing the cell. ➥ (Chapter 16, p. 439) They adapt well to living inside amoeboid white blood cells.

After an incubation period of 2 to 10 days, Legionnaires' disease appears with fever, chills, headache, diarrhea, vomiting, fluid in the lungs, pain in the chest and abdomen, and, less frequently, profuse sweating and mental disorders. When death occurs, it is usually due to shock and kidney failure. In nonpneumonic legionellosis, after about 48 hours' incubation, the patient suffers 2 to 5 days of flulike symptoms without infiltration of the lungs.

Pontiac fever is a mild legionellosis named for an outbreak in 1968 affecting 144 people—95 percent of the employees of the Pontiac, Michigan, Health Department. None died; all recovered within 3 to 4 days.

Direct fluorescent antibody tests, ELISA, and commercially available genetic probes are used to diagnose *Legionella* infections. Erythromycin is used to treat them; most other antibiotics have no effect. Because it is difficult to distinguish Legionnaires' disease from other pneumonias, erythromycin is typically given along with a second antibiotic such as penicillin to treat any pneumonia-like disease.

Control of *Legionella*-associated infections depends on maintaining adequate chlorine levels in all potable water sources, cooling towers, and other reservoirs of potable water when not in use. Periodic cleaning of surfaces in air conditioners, humidifiers, and similar equipment is valuable in reducing the incidence of outbreaks, especially in hospitals.

Tuberculosis

Tuberculosis, or TB, (formerly called *consumption*) has plagued humankind since ancient times, as indicated by skeletal damage in Egyptian mummies and earlier human remains. It remains a massive global health problem today. One and a half billion people (six times the U.S. population) have tuberculosis. Each year 600,000 die, and 3 to 5 million new cases arise.

Incidence in the United States   The incidence of tuberculosis in the general U.S. population decreased by nearly 6 percent per year from 1953, when uniform national reporting began, to 1986. Temporary increases in the 1980s were caused by Asian and Haitian refugees, many of whom became infected in crowded, unsanitary camps and escape boats. Disproportionately large numbers of resident nonwhite groups—blacks, Eskimos, Native Americans, and Hispanics—suffer from tuberculosis, sometimes in epidemic proportions (➤ Figure 21.11). In 1986 the incidence increased by 2.6 percent. Many new cases occurred among AIDS patients; others appeared to represent reactivations of old infections triggered by immunodeficiency, crowding, stress, and use of anti-inflammatory drugs. In 1993 the number of new cases reported per month rose from 778 in January to 5130 in December. The increase continues and people are afraid that tuberculosis may return and grow into an even bigger and more frightening plague than AIDS. Tuberculosis is an airborne disease to which anyone could be exposed, whereas behavior modification can lessen much of one's chances of exposure to the HIV virus.

Before 1900, approximately one-third of adults died of TB before reaching old age.

(a)

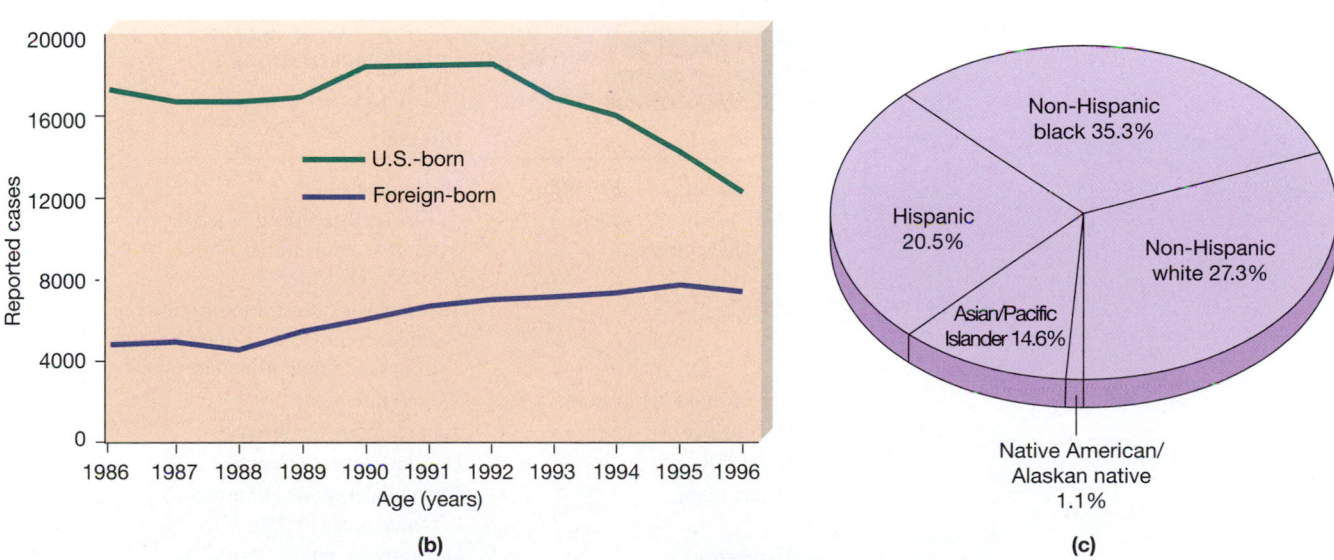

(b)

(c)

➤ **Figure 21.11   The U.S. incidence of TB, 1996.** (a) By state; (b) by place of birth; (c) by ethnicity. (*Source*: CDC)

In June 1998 microbiologists from the Pasteur Institute unveiled the complete DNA sequence of Mycobacterium tuberculosis, revealing more than 4000 genes.

Causative Agents   The causative agents of tuberculosis are members of the genus *Mycobacterium,* with *M. tuberculosis* causing the vast majority of cases (Table 21.2). *Mycobacterium tuberculosis* was discovered by Robert Koch in 1882, when the disease was called the "White Plague" of Europe. Certain other agents, referred to as atypical mycobacteria, also cause tuberculosis, especially *M. avium-intracellulare* complex (MAC) in AIDS patients. All mycobacteria are straight or slightly curved rods that stain acid-fast, as illustrated in ➤ Figure 3.29. ➡ (p. 66)

Certain properties of mycobacteria are closely associated with their role in tuberculosis. Waxes and long-chain mycolic acids in mycobacterial cell walls make mycobacteria difficult to Gram stain, contribute to environmental survival of these organisms, and protect them from some host defenses. Being obligate aerobes sensitive to slight decreases in oxygen concentration, mycobacteria grow best in the apical, or upper portions of the lungs, which are the most highly oxygenated. Pathogenic mycobacteria have an exceedingly long generation time (12 to 18 hours, compared with 20 to 30 minutes for most bacteria), which accounts for the long time (up to 8 weeks) it takes to produce a visible colony on laboratory media. Mycobacteria are highly resistant to drying and can remain viable for 6 to 8 months in dried sputum, a property that contributes to public health problems. They are, however, quite sensitive to direct sunlight.

The Disease   Tuberculosis is acquired by the inhalation of droplet nuclei of respiratory secretions or particles of dry sputum containing tubercle bacilli. Young children and elderly people are particularly at risk, so screening school, day-care, and nursing home workers for tuberculosis is important. After organisms are inhaled, they multiply very slowly *inside* white blood cells that have phagocytized them. They elicit a host response that includes neutrophil infiltration and fluid accumulation within the alveoli of the lungs. The organisms eventually rupture and destroy the neutrophils. Later, macrophages and lymphocytes move into the area. Alveolar macrophages also phagocytize the living tubercle bacilli, which are again able to multiply within and destroy their new hosts. Rupture of the dead phagocytes releases infective organisms. No toxins are produced. As additional cells are infected, an acute inflammatory response occurs. A large quantity of fluid is released, especially in lung tissue, where it produces pneumonia-like symptoms. Lesions sometimes heal, but more often they produce massive tissue necrosis or solidify to become chronic granulomas, or **tubercles** (➤ Figure 21.12). Tubercles consist of central accumulations of enlarged macrophages, multinucleate Langerhans giant cells containing tubercle bacilli, peripheral lymphocytes, macrophages, and newly formed connective tissue. The central portion of the tubercle undergoes destruction, typically giving it a cheesy, or **caseous,** appearance. Some organisms gain access to the lymphatic and circulatory systems. About 3 to 4 weeks after exposure, delayed hypersensitivity and cell-mediated immunity develop. On occasion, however, the host fails to respond immunologically. Uncon-

➤ Table 21.2	Mycobacteria that cause human disease
Species	**Disease**
Mycobacterium tuberculosis	Tuberculosis
M. avium-intracellulare (MAC complex)	Tuberculosis-like disease in humans, transmitted from birds and swine
M. bovis	Tuberculosis, transmitted from cattle; could be transmitted from nonhuman primates
M. fortuitum complex	Wound infections
M. kansasii	Tuberculosis-like disease
M. leprae	Hansen's disease (leprosy)
M. marinum	Cutaneous lesions in humans, tuberculosis in fish
M. ulcerans	Ulcerative lesions

(a)   (b)

➤ **Figure 21.12**   **TB X-ray photos.** (a) In this normal chest X-ray, the faint white lines are arteries and other blood vessels. The heart is visible as a white bulge in the lower right quadrant. (b) An advanced case of pulmonary tuberculosis; white patches (arrows) indicate areas of disease. These patches, or tubercles, may contain live *Mycobacterium tuberculosis*. Lung tissue and function in these lesions is permanently destroyed.

trolled multiplication of tubercle bacilli then takes place in the lungs, resulting in the formation of numerous tubercles. Organisms are subsequently spread by the circulatory system to other body tissues and organisms. This condition is called **miliary tuberculosis,** named after the small lesions that form that resemble millet seeds. Lesions can be walled off from the rest of the lung by encapsulation when the host has sufficient resistance. Lesions near blood vessels can perforate the vessels and cause hemorrhage, which leads to bloody sputum, a major symptom of tuberculosis.

Granulomas can keep viable organisms walled off for decades. When the immune system becomes impaired by age or other infections, granulomas can open and the disease can be reactivated. People with AIDS often have tuberculosis, either from new infections or reactivation of old ones. Except for initial infections among immigrants, AIDS patients, and residents of certain large cities such as New York and Washington, D.C., most tuberculosis cases in the United States are reactivations rather than first-time infections.

Tuberculosis of the bones can cause extensive erosion, especially in the spine (➤ Figure 21.13). The urogenital tract, meninges, lymphatic system, and peritoneum also are prone to develop extrapulmonary tuberculosis. In **disseminated tuberculosis,** now frequently seen in AIDS patients, infected cells become casts: As the tubercle bacillus multiplies in a cell, the cell's organelles and membranes are destroyed, and a clump of microorganisms in the shape, or cast, of the former cell remains. When this process occurs in the intestine, a replica of the intestine consisting of live *Mycobacterium* can be seen.

Although primary tuberculosis is nowadays most likely to affect the lungs, it can also affect the digestive tract. Before milk pasteurization became widespread, primary digestive tract infections were often seen, especially in children. Most were caused by *Mycobacterium bovis* transmitted in raw milk or other dairy products. By law, dairy cattle are now regularly screened for tuberculosis, and infected animals are destroyed. However, some cows escape testing, so drinking raw, unpasteurized milk is a health risk. Today tuberculosis of the digestive tract usu-

Relief Could Cause Disease
If you were suffering from inflamed arthritic joints, you might not be happy to hear that your physician wouldn't let you have cortisone. But there would be a reason for caution. When cortisone is used to treat the arthritis of older patients, the drug also affects the granulomatous lesions of tuberculosis; it actually reactivates tuberculosis organisms as it counteracts the factors that keep the lesions walled off. Bringing relief to the aching joints of arthritic patients can activate hidden tuberculosis lesions, and then these patients can then go on to infect others, particularly children. So listen to your doctor when he says that cortisone may not be the best option for arthritis relief!

➤ Figure 21.13 **Tuberculosis of the spinal column.** Notice that several of the lower vertebrae have fused as a result of infection damage.

ally occurs as a secondary infection site, as when pathogen-laden sputum coughed up from a primary lung infection is swallowed.

Diagnosis, Treatment, and Prevention Tuberculosis can be diagnosed by sputum culture, but because the organisms grow very slowly, cultures must be kept for at least 8 weeks before they are declared negative. Genetic probes are also of value in identification of mycobacteria. Chest X-rays fail to reveal lesions outside the lungs and detect only relatively large ones within them. Therefore, screening is now done by skin tests rather than by X-ray examinations. In a skin test, a small quantity of *purified protein derivative (PPD)*, a protein from tuberculosis bacilli, is injected intracutaneously and examined 48 to 72 hours later for an *induration,* or a raised, not necessarily reddened bump. Induration is a delayed hypersensitivity reaction to the PPD. A positive skin test indicates previous exposure and some degree of immune response. It does not indicate that the person now has or has ever had tuberculosis—the immune response may have prevented infection or eliminated or walled off organisms. In fact, health care workers, teachers, and others who have frequent skin tests may eventually develop a positive response due to sensitization to the test material itself.

Treatment is with isoniazid and rifampin for at least 1 year. Many strains of *Mycobacterium,* however (particularly atypical species that are often found in AIDS patients but are rare in healthy individuals), are now resistant to isoniazid. Such strains must be treated with a "second-line" or sometimes even a "third-line" drug. If treatment is successful, sputum cultures become negative for mycobacteria within 3 weeks. Before the development of isoniazid, patients were often sent to tuberculosis sanatoriums, where cold, fresh air was believed to have curative effects. Patients were bundled up and left outdoors for most of the day. In reality, the rest and proper nutrition were what really helped some patients survive.

Tuberculosis can be prevented by vaccination with attenuated organisms in the vaccine BCG, or bacillus of Calmette and Guérin. The French bacteriologists Albert Calmette and Camille Guérin developed the vaccine at the Pasteur Institute in Paris in the early 1900s. BCG immunization is widely practiced in other parts of the world where the prevalence of tuberculosis is high. ➡ (Chapter 17, p. 483) The vaccine's use in the United States is limited to high-risk individuals, such as health care workers regularly exposed to tuberculosis patients, or to personnel specializing in pulmonary and chest diseases.

Psittacosis and Ornithosis

In the early 1900s, **psittacosis** (sit″ah-ko′sis), or *parrot fever,* a respiratory disease associated with psittacine birds, such as parrots and parakeets, was discovered. Today at least 130 different species of birds, including ducks, chickens, and turkeys, are found to carry a form of this disease. Because most of these birds are not psittacines, their disease is referred to as **ornithosis** (from *ornis,* Greek for "bird"). Wild and domestic birds can be infected, often without showing symptoms. However, stresses such as overcrowding, chilling, or shipping to pet shops can activate the disease. Both forms of the disease are caused by *Chlamydia psittaci* and are spread by direct contact, infectious nasal droplets, and feces. Organisms can be found in every organ of an infected bird. The birds have diarrhea and a mucopurulent discharge from the nose and mouth.

Humans usually acquire the disease from birds. Poultry-plant workers are especially susceptible. Human-to-human transmission also has been documented, and medical personnel caring for patients can contract the disease. Organisms are inhaled and spread systemically to the lungs and reticuloendothelial system, especially Küpffer's cells. Most cases of the disease are mild and self-limiting, but some patients develop a serious pneumonia. After an incubation time of 1 to 2 weeks, onset of symptoms is sudden, with sore throat, coughing, difficulty in breathing, headache, fever, and chills. This disease is difficult to distinguish from other pneumonias on a clinical basis. Definitive diagnosis is made by inoculation into tissue culture. However, the organism is so highly infective that this should be undertaken only by very experienced technicians, in specially equipped labs.

With tetracycline treatment, the mortality rate from ornithosis pneumonia is about 5 percent; in untreated cases, it is around 20 percent. No vaccine is available, so prevention involves strict quarantine regulations for birds entering the United States.

Q Fever

Q fever was first described in Queensland, Australia. The *Q* stands not for Queensland but for "query," because determining which organism caused it remained a question for a long time. Q fever is now known to be caused by *Coxiella burnetii,* an organism included among the rickettsias. Unlike other rickettsias, this organism survives long periods outside cells and can be transmitted aerially as well as by ticks. For a long time it was a mystery how Q fever organisms can survive 7 to 9 months in wool, 6 months in dried blood, and over 2 years in water or skim milk. The mystery now appears to have been solved. *C. burnetii* has two forms, called large- and small-cell variants (➤ Figure 21.14). Electron microscope studies suggest that a type of endospore may be formed at one end of the large-cell variant, thereby giving resistance uncharacteristic of other rickettsias.

Once *C. burnetii* cells with endospore-like bodies inside them have been inhaled, they are phagocytized by host cells. *Coxiella* grow in phagolysosomes of host cells. There they display the unusual property of responding to the acidic conditions by increasing their metabolism and multiplying rapidly until they almost fill infected cells. Eventually the cells rupture, releasing new groups of pathogenic bacteria. These pathogens, when spread by the bloodstream, then infect other body cells. There appears to be no exit from the human body.

Coxiella burnetii exist all over the world, especially in cattle- and sheep-raising areas. Wild animals and domestic sheep and cattle are the normal hosts of *C. burnetii.* The bacteria are transmitted via tick bites, feces, and genital secretions of infected animals. Humans become in-

Small cell forms   Large cell forms

➤ **Figure 21.14** *Coxiella burnetii.* (a) Small and large cell forms. The cell walls of the large forms have less peptidoglycan with no cross-links—hence the diversity of cell shapes. (b) This cell has just divided; notice the endospore-like body (E), which is probably responsible for the relative resistance of the organism.

fected by inhaling aerosol droplets from infected domestic animals, which usually do not appear ill. Farmers become infected while attending a cow giving birth or miscarrying if the placenta is laden with organisms. (In Canada recently a dozen card players all contracted Q fever after an infected pet cat had a litter of kittens in the same room.) Slaughterhouse and tannery workers become infected by inhaling dried tick feces from the hides of animals. Because of its resistance to drying, *Coxiella* can remain viable in the environment for long periods of time. Transmission to humans also occurs by the ingestion of milk from contaminated animals. In areas such as Los Angeles, an estimated 10 percent of the cows shed organisms into their milk. Rates are lower in some other areas. Flash pasteurization eliminates this hazard. ➡ (Chapter 12, p. 329) In fact, flash pasteurization was designed to kill Q fever organisms without giving the milk a "cooked" flavor.

Symptoms of Q fever—chills, fever, headache, malaise, and severe sweats—are very similar to those of primary atypical pneumonia. The incubation period is 18 to 20 days. Diagnosis is made by serologic testing or by direct immunofluorescent antibody staining. Treatment is with antibiotics such as tetracycline or fluoroquinolone. Lifelong immunity generally follows an attack of Q fever; a vaccine is available for workers with occupational exposure.

Untreated or inadequately treated cases of Q fever can go into long periods of remission, sometimes lasting years. Cortisone treatment can reactivate it. In chronic cases, endocarditis and heart valve infections are sometimes seen. Endocarditis is invariably fatal, as the organisms do not respond to antibiotic therapy. Special strains appear to be genetically adapted to infect the heart and are very resistant, causing the most serious disease. Other genetic strains cause milder forms of disease.

Nocardiosis

Nocardiosis is characterized by tissue lesions and abscesses. It is caused by the aerobic, acid-fast staining filamentous bacterium *Nocardia asteroides* (➤ Figure 21.15). The organisms are found in soil and water. Disease is usually initiated by inhalation of *N. asteroides*. The bacterium was first identified in 1888 as the causative agent of a disease in cattle known as *farcy*. In humans the primary infection site is the lungs in about three-fourths of all cases, but the disease can also originate in the skin and in other organs. The disease usually occurs in immunosuppressed patients. Human-to-human transmission is rare. Mortality rates as high as 85 percent have been reported. However, with early diagnosis and aggressive treatment with sulfonamides in combination with trimethoprim, mortality can be kept below 50 percent. Diagnosis is based on finding the organism in sputum or other specimens and on culture results.

Viral Lower Respiratory Diseases

Influenza

Influenza is the last remaining great plague from the past. Since Hippocrates described an influenza-like outbreak in 412 B.C., repeated influenza epidemics and pandemics have been recorded, even in recent times (➤ Figure 21.16). One of the greatest killers of all time was the

The genus *Nocardia* was named after Edmond Nocard (1850–1908), a French bacteriologist and veterinary pathologist.

✓ What dangers are associated with pertussis vaccine?

✓ How does bronchial pneumonia differ from lobar pneumonia?

✓ What sanitary measures can be taken to prevent Legionnaires' disease?

✓ How is AIDS related to TB?

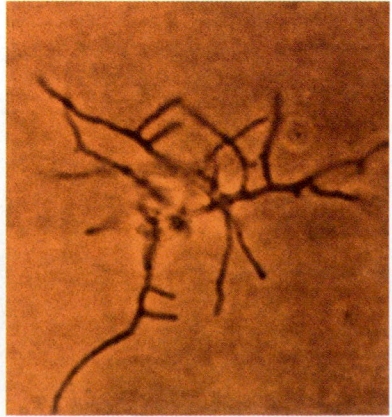

➤ Figure 21.15 *Nocardia asteroides.* This clearly shows the branching, filamentous shapes of the bacteria 1400X).

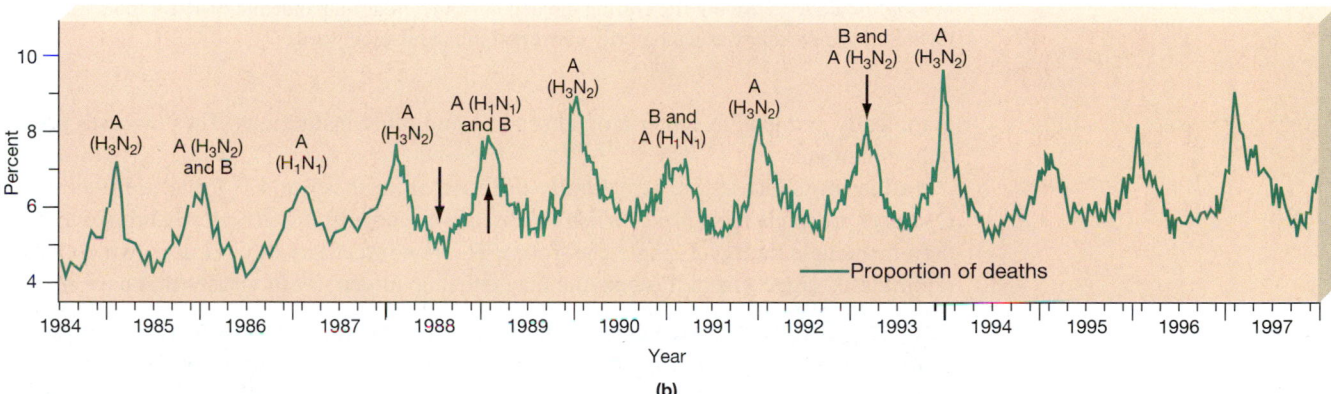

> **Figure 21.16**   **The U.S. incidence of influenza.** (a) Since 1900; (b) recently (as a proportion of all deaths in 121 cities). Influenza varieties are labeled by type and by hemagglutinin (H) and neuraminidase (N) numbers. (*Source*: CDC)

pandemic of swine flu (also known as Spanish flu) of 1918–1919 when 20 to 40 million people died (➤ Figure 21.17). The causative virus was unusually virulent, and the crowded, unsanitary conditions created by World War I increased the spread of the virus.

Causative Agents   Influenza is caused by **orthomyxoviruses.** These RNA viruses have an envelope surface antigen hemagglutinin that is responsible for their infectivity. It attaches specifically to a receptor on erythrocytes and other host cells. Some influenza viruses have an enzyme called *neuraminidase* (nur″am-in′i-das), which helps the virus penetrate the mucus layer protecting the respiratory epithelium (➤ Figure 21.18). The enzyme also plays a role in the budding of new virus particles from infected cells. These properties allow identification of influenza viruses in the laboratory.

On the basis of their nucleoprotein antigens, three major influenza virus serotypes are recognized: types A, B, and C. Influenza viruses have a tendency to undergo **antigenic variations** (*changeability*), or mutations that affect viral antigens. Thus, immunity developed through infection with one influenza virus is often insufficient to prevent infection by a variant.

Influenza virus type A, first isolated in 1933, causes epidemics and pandemics at irregular intervals because it periodically undergoes antigenic variation. Secondary bacterial pneumonias are the most important and serious complications of influenza virus A infections. Various species of birds and mammals are infected with strains of influenza A and may serve as reservoirs of infection.

Influenza B viruses also undergo antigenic changes, but less extensively and at a slower rate than do influenza A viruses. Epidemics caused by influenza B viruses are limited geo-

➤ **Figure 21.17**   **Swine flu.** During the height of the great flu pandemic of 1918, the gymnasium of Iowa State University was temporarily converted into a hospital ward.

graphically and tend to center around schools and other institutions. These viruses are found only in humans.

Influenza C viruses are structurally different from the A and B viruses. Infections with C viruses are rarely recognized. When the disease is recognized, it is typically limited to the children in a single family or single classroom. The low infectivity of C viruses, which lack neuraminidase, suggests that this enzyme may enhance infectivity in viruses that have it.

Antigenic Variation   Antigenic variation in influenza viruses occurs by two processes, *antigenic drift* and *antigenic shift*. **Antigenic drift** results from mutations in genes that code for hemagglutinin and neuraminidase. Such mutations change the configuration of the part of the

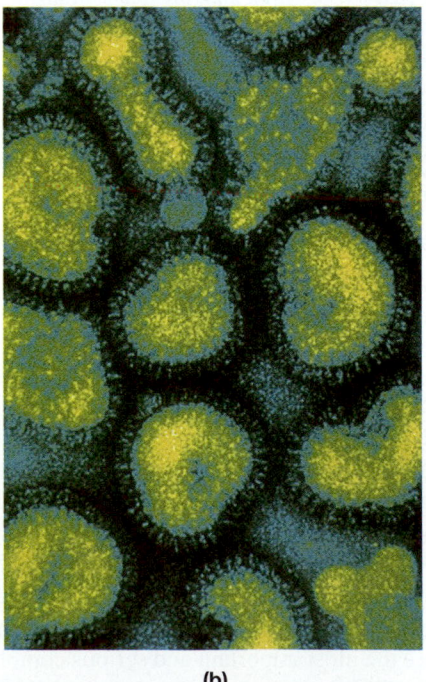

(a)   (b)

➤ **Figure 21.18**   **The influenza virus.** (a) The virus shows tinin and neuraminidase spikes on its outer surface and an RNA core. (b) A colorized TEM of influenza virus type A.

The Great Flu Pandemic of 1918

Where could he be? He should have been home an hour ago! In normal times, this wouldn't have been so alarming—but those weren't normal times. Was his body one of those piled up like cordwood on a street corner downtown?

Such fears raced through the minds of frantic families during the great Spanish flu pandemic of 1918. Seemingly healthy people dropped dead without warning. Undertakers couldn't meet the demands of a pandemic that killed half a million Americans in only 10 months. (The U.S. Civil War claimed about that many lives over 4 years.) Philadelphia alone recorded 528 bodies piled up, awaiting burial, in just one day. Numerous others lay undiscovered. Rescue teams went door to door, seeking victims who were too sick to call out for aid. Wagons made regular rounds, stopping at corners to pick up bodies that families, neighbors, storekeepers, or passers-by had carried there. No wonder families worried when someone was late getting home. Who knew which of the 25 million infected Americans would die?

Unlike many diseases, influenza took its heaviest toll among young, healthy individuals. Some died of high fever; others succumbed to secondary bacterial infections, chiefly pneumonia. The immediate cause of death for many patients was lung damage. Tremendous pressure built up until the lungs degenerated completely (on autopsy, they typically had the consistency of pudding). If the crushing pressure was relieved, the patient had a chance of survival. One physician called to the bedside of a teenage girl could do little but comfort the parents. Suddenly a stream of blood shot from her nose and soaked the trio. Projectile hemorrhaging had reduced the pressure in her lungs, and she lived. But people who survived cases with high fever often developed Parkinson's disease. We now associate Parkinson's disease with the elderly. But after the 1918 pandemic, it affected all ages, including many small children.

The cause of this grim experience was something too tiny for the researchers of 1918 to see—the swine flu virus. What they did see with light microscopes were secondary bacterial invaders. (One bacillus was named *Haemophilus influenzae* in the mistaken belief that it was the cause of the pandemic.) Indeed, in one New York City family, all six members were found to be infected with different species of bacteria. No wonder the medical world was confused. Attempts at vaccine production proved futile. Few people even guessed at the existence of the virus.

Where had this virus come from? A new type of influenza had appeared in France in April 1918 as American and European troops fought during the final months of World War I. From France the virus traveled to Spain. That summer it spread throughout Europe, China, and West Africa. By then, the English-speaking countries had dubbed it the "Spanish lady." The Lady entered Boston in August and crossed the United States within a month. Around the world she went, causing a pandemic. By the time it had subsided in 1920, more than 25 million had died. The country hardest hit was India, where death wiped out the entire population increase of a decade. In some parts of the world, nearly half the population died. There is some evidence that the virus took nonhuman victims as well—baboons in South Africa, for example.

It is not certain what brought on this deadly kind of flu. The influenza virus, with a long history as a human pathogen, is one of the most rapidly mutating viruses known. But the natural mutation process may have had human help. Mustard gas, which was widely used in trench warfare in France, is a powerful alkylating agent—a known mutagen. Could it have interacted with the virus to produce the new, virulent strain? No one can be sure.

The term *swine flu* was not used until years after the pandemic was over. Influenza was unknown in pigs before the Spanish flu reached the United States in 1918. Thereafter, swine were found to suffer from a complex disease involving both an influenza virus and a bacterium. The virus was believed to be the one that caused the Spanish flu. Today, the virus that causes influenza in swine is not the one that causes "swine flu" in humans. The name swine flu may not have been a mistake, however, for the rapidly evolving virus strains are likely to have mutated quite a bit since 1918.

Is the virus that caused the great pandemic still a threat? In 1976, the U.S. government mounted a massive immunization program against swine flu. Flu viruses almost always march westward around the globe. By looking at the viruses that have caused problems to the east, U.S. public health officials decide which viral strains to incorporate into next year's vaccine. So when swine flu was detected in Asia in 1975, it sent chills up and down medical spines. Not only could millions of deaths occur—there were also the sequelae. How many Parkinson's cases would be survivors of a 1976 pandemic?

Nervously the government exhorted all to be immunized. Then came a disturbing development. Some recipients of swine flu vaccine had subsequently developed *Guillain-Barré syndrome,* resulting in complete paralysis. (Although most persons recover completely, a small percentage die of the paralysis.) Which would be worse: swine flu or paralysis? Lines at vaccine clinics shortened. With fewer people immunized, what would happen when the swine flu arrived? The medical community waited. But swine flu never came. Was it a reprieve or just a postponement? No one can say.

antigen molecule that stimulates the production of, and combines with, specific antibodies. Thus, antibodies formed against the parental hemagglutinin or neuraminidase are less effective in inhibiting the mutated forms of these viral components of viral offspring. If sufficient antigenic drift occurs naturally, an influenza epidemic will follow. **Antigenic shift** results from gene reassortment, possibly after two different viruses infect the same cell and exchange large segments of their genome. It represents more dramatic changes; the viral strains that emerge are significantly different antigenically from previously known strains. Thus, antibodies formed against one type of hemagglutinin are not protective against another type. Antigenic shift generally precedes a major pandemic. It is fortunate that antigenic shift is rare because the strains it produces cause severe pandemics. Six have occurred in the last century—in 1890, 1900, 1918, 1957, 1968, and 1977.

Immune responses vary according to antigenic changes. It is as if a contest occurs between the virus's ability to elude the immune system and the immune system's ability to recognize and inactivate the virus. The antigenic change in most viruses is small, and existing antibodies prevent the viruses from causing infections. The viruses cause severe disease in a few people who lack antibodies, mild disease in others who have some effective antibodies, and no disease in others. The diverse assortment of antibodies that hosts acquire from exposure to several different viral strains tends to protect them from new strains. For a virus to cause an epidemic, it must have undergone sufficient antigenic change to elude most available antibodies.

Influenza viruses replicate in the cytoplasm of nucleated host cells but not in erythrocytes, which lack a nucleus. The effects of influenza viruses on cells they infect are noteworthy. The viruses acquire their envelopes by budding through host cell membranes and can do so without immediately killing the host cell. ➡ (Chapter 10, p. 277) A host cell typically produces thousands of viruses per minute over many hours before the cell's macromolecules are exhausted and the cell dies.

The Disease   Influenza, like many other viral respiratory infections, is a relatively superficial infection. The viruses enter the body through inhalation of virus-containing droplets or indirect contact with infectious respiratory secretions. Immediate invasion of the oropharyngeal epithelial lining results. The invading viruses multiply and spread quickly to other portions of the respiratory tract, including mucus-secreting and ciliated epithelial cells. First the cilia are destroyed; then the cells are damaged. Severely damaged cells die and are sloughed. Although host cells begin regeneration immediately, it takes 10 days or more to restore an intact ciliated epithelium. Loss of the mucociliary escalator, ordinarily a major host defense, allows bacterial invasion and enhanced adherence of bacteria to virus-infected cells. ➡ (p. 600) Impaired phagocytosis and accumulation of fluid in the lungs add to the risk of secondary bacterial infections, especially pneumonia. Death can result from influenza alone, secondary bacterial infection alone, or a combination. Although antibiotics reduce the risk of death from bacterial infections, they have no effect on viral infections. ➡ (Chapter 13, p. 366)

In a nonimmune person, disease signs and symptoms begin to appear 36 to 48 hours after infection. Fever, malaise, and muscle soreness are the most common symptoms; cough, nasal discharge, sore throat, and gastroenteritis (usually only in children) also occur frequently. The fever usually lasts about 3 days. As systemic symptoms decrease, respiratory symptoms increase. The severity of the disease is directly proportional to the quantity of viruses released from cells. Release of viruses starts as the first malaise appears and peaks about a day before maximum fever and interferon production occur. Viral shedding usually ceases within 8 days of initial exposure. Although the acute phase of the illness is over in about a week, fatigue, cough, and weakness may persist over several weeks.

Incidence and Transmission   In northern temperate zones, influenza appears in late November or early December and disappears in April. In most years the greatest number of cases are seen between January and mid-March. In any one region influenza has a 5- to 7-week period of high prevalence. Indoor crowding, poor air circulation, and dry air expedite the spread of the virus. Schools furnish almost ideal conditions for influenza transmission and usually are the focal points of outbreaks. People often claim to have the "flu," however, when they are actually suffering from another condition such as an allergy or a bad cold (Table 21.3). Gas-

Grain dust harbors bacterial endotoxins that can cause lung inflammation in both farm animals and agricultural workers.

Some cases of asthma, as well as some cases of chronic obstructive lung disease, may actually be caused by the adenovirus.

> **Table 21.3** **Differences between a common cold, influenza, and pneumonia**

Symptoms	Cold	Influenza	Pneumonia
Fever	Rare	Characteristic high (100.4–104°F) sudden onset, lasts 3 to 4 days	May or may not be high
Headache	Occasional	Prominent	Occasional
General aches and pains	Slight	Usual; often quite severe	Occasionally quite severe
Fatigue and weakness	Quite mild	Extreme; can last up to a month	May occur depending on type
Exhaustion	Never	May occur early and prominent	May occur depending on type
Runny, stuffy nose	Common	Sometimes	Not characteristic
Sneezing	Usual	Sometimes	Not characteristic
Sore throat	Common	Sometimes	Not characteristic
Chest discomfort, cough	Mild to moderate; hacking cough	Can become severe	Frequent and may be severe
Complications	Sinus and ear infections	Bronchitis, pneumonia; can be life-threatening	Widespread infections of other organs; can be life-threatening, especially in elderly and debilitated persons

trointestinal problems, often called "stomach flu," are probably not flu at all but are most likely due to another viral infection.

Diagnosis and Treatment   The best specimens for isolation of viruses are throat swabs taken as early in the illness as possible. The viruses can be cultured in embryonated chick eggs and various cell lines and identified by hemagglutination inhibition and immunofluorescent antibody tests.

Moderately effective treatment is now becoming available for influenza. The drug amantadine blocks influenza A virus replication, probably by interfering with uncoating. ➡ (Chapter 13, p. 367) If given soon after the onset of symptoms, amantadine is effective in shortening the disease's course and reducing the severity. It is useful for short-term protection of selected individuals but impractical to use for the duration of an influenza epidemic unless patients are especially compromised. Furthermore, it prevents users from developing antibodies to the virus and has some unpleasant side effects. Rimantidine is more effective and less toxic. ➡ (Chapter 13, p. 367) Ribavirin has been demonstrated to inactivate both A and B viruses.

Immunity and Prevention   Because of influenza viruses' ability to mutate, annual immunization is recommended, especially for high-risk persons, such as those with chronic conditions. The killed virus vaccine provides effective protection against influenza, and annual immunization increases the diversity of the recipient's antibodies—which may be of use to him in some future flu outbreak. Even less frequent immunization can provide some degree of protection. Although immunized individuals sometimes become infected with flu viruses, they generally have a milder disease of shorter duration than do nonimmunized individuals. Furthermore, they shed fewer viruses over a shorter duration of time. Thus, immunization of some people tends to reduce the spread of the disease to other people.

A human gene that confers resistance to influenza was identified at the University of California in Santa Barbara. This gene is turned on by interferon. It produces a specific protein, called the Mx protein, which prevents the virus from making viral RNA and proteins.

Respiratory Syncytial Virus Infection

The **respiratory syncytial virus (RSV)** is the most common and most important and costly cause of lower respiratory tract infections in children under 1 year and especially in male infants 1 to 6 months old. The virus derives its name from the fact that it causes cells in cultures to fuse their plasma membranes, lose independent identity, and become multinucleate masses, or **syncytia** (sin-sish′ya; singular: *syncytium*). The disease, a form of **viral pneumonia,** be-

What's in a Flu Vaccine?

There are so many different strains of the flu virus, due to antigenic variation. How does the CDC decide which to include in each year's vaccine? Each year around June, the CDC looks at which strains are most prevalent and what alterations they have undergone in order to select the strains to be used in the coming flu season's vaccine. For example, the vaccine for the 1998–1999 season contained three strains: influenza A/Sydney/5/97-like (H3N2), A/Beijing/262/95-like (H1N1), and B/Beijing/184/93-like antigens. These strains were named for the geographic location where they were first isolated, a number assigned by the laboratory that isolated them, the year of isolation, and—for type A viruses—the numbers of their envelope antigens, hemagglutinin (H) and neuraminidase (N). The vaccines were thoroughly evaluated for safety and effectiveness before they are given to humans. If you want to avoid the flu each year, the CDC will help you.

gins with a 3- or 4-day period of fever as the virus infects the respiratory tract. Fever is followed by hyperventilation, hyperinflation, and infiltration of the lungs with fluid. If an infant with RSV is placed on a rapidly pulsing respirator, the infant can shed enough virus to infect an entire intensive care nursery. Major outbreaks of this nature generally result in fatalities. Viruses are released for weeks to months after infection, and reinfection is common. In older children and adults, RSV affects mainly the upper respiratory tract, and adults who carry the virus in their nose can be the source of nursery infections.

The incidence and transmission of RSV is quite similar to that for parainfluenza viruses, and simultaneous infection by both kinds of viruses is not uncommon. The virus can be identified by enzyme immunoassay performed on nasal secretions or by a direct immunofluorescent test on cells from such secretions. Ribavirin shortens the duration of the disease. There is no vaccine. The best means of prevention is to reduce an infant's exposure to crowds and other sources of infection.

Hantavirus Pulmonary Syndrome

In May and June 1993, 24 cases of severe respiratory illness among residents of the Four Corners area of the southwestern United States were reported. Healthy adults suddenly became ill and died within a few hours. Fear raced through Arizona, Colorado, New Mexico, and parts of the Navajo reservations where deaths had occurred. Investigation identified a previously unreported hantavirus as the cause of the disease, later named **hantavirus pulmonary syndrome (HPS)**. The virus was identified by PCR techniques and RNA sequencing (Chapter 7) and was shown to be distinct from all other known hantaviruses. ➡ (p. 192) Hantavirus illness had been known elsewhere in the world for years (Chapter 23), but with renal and hemorrhagic symptoms—not pulmonary illness. The new virus was tentatively named Muerto Canyon virus, after the area from which it was isolated. After complaints from residents of the area, however, the name Sin Nombre ("No-Name") virus was proposed and has been adopted.

The fatality rate (60 percent) has been more than 10 times higher than for other hantaviruses. During 1994, other cases of HPS were found throughout the United States—in Florida, Louisiana, Indiana, and Rhode Island. In each case, a different, new hantavirus was found to be the cause. Each virus was carried by rodents that appeared healthy but shed virus in urine, feces, and saliva. Muerto Canyon virus is carried by the deer mouse, *Peromyscus maniculatus.* The Florida variety of the virus is carried by the cotton rat, *Sigmodon hispidus;* the vector for the Louisiana strain has not yet been identified. Virus from dried excreta becomes airborne and is inhaled. Exposure to rodent-contaminated areas, both indoors and outdoors, should be avoided. Ribavirin has been somewhat useful in treatment, but further studies are needed.

Acute Respiratory Disease

Acute respiratory disease (ARD) is a lower respiratory tract viral disease that ranges from mild to severe. Respiratory symptoms such as sore throat, cough, and other cold symptoms, fever, headache, and malaise are frequently seen. Severe viral pneumonia that lasts about 10 days sometimes occurs in ARD.

Cases of viral ARD have been seen in military training facilities in the United States and in Europe. Epidemics usually start 3 to 6 weeks after the start of training. The epidemic nature of the disease has been attributed to crowding of people from different geographic areas under stressful conditions. However, epidemics have not been observed in colleges and other institutions where conditions are similar, so other factors may be involved in such epidemics.

Adenoviruses cause about 5 percent of ARD cases in children under 5 years old. Usually the symptoms are mild and nonspecific—stuffy nose, cough, and nasal discharge. More severe symptoms such as tonsillitis, pharyngitis, bronchitis, bronchiolitis, croup, and often conjunctivitis and abdominal pain may also appear. Adenovirus pneumonia accounts for about 10 percent of all childhood pneumonia and is occasionally fatal.

Fungal Respiratory Diseases

Compared with bacteria and viruses, fungi are much less frequent causes of respiratory diseases. Most fungal infections are seen in immunodeficient and debilitated patients. Blastomycosis, usually a skin disease, can also cause mild respiratory symptoms. ➡ (Chapter 19, p. 550)

The Medicine Men Knew

In May 1993, public health officials were stumped. What caused the mysterious outbreak of respiratory illness in the Four Corners area, they wondered? The answer came from the Navajo medicine men, who have always been careful observers of nature. These medicine men pointed out that there was an exceptional crop of pinyon pine nuts that year. Almost every tree on the Navajo reservation had borne nuts for the entire year, whereas normally just a few trees do so for a few weeks each year. Rodents buried the biggest nuts from this bumper crop in burrows. People hunting pine nuts dug up these storehouses and exposed themselves to rodent feces and urine containing hantavirus. CDC investigators were able to tie the changes in the nut crop to the outbreak, thanks to the medicine men.

(a)

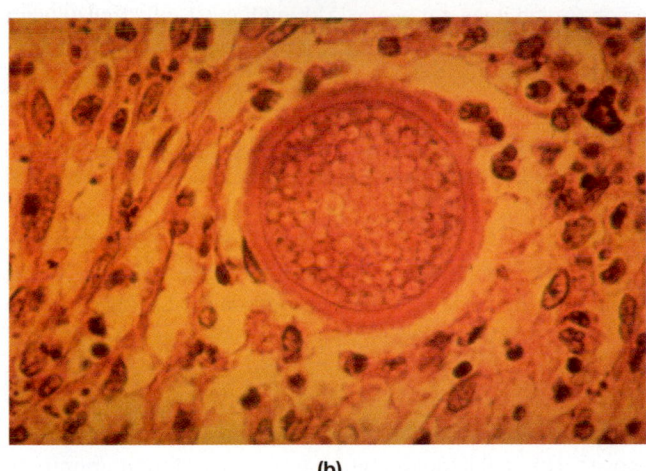

(b)

> Figure 21.19   **Coc-cidioidomycosis.** (a) The skin rash of coccidioidomycosis on the legs. (b) A tissue section showing a spherule containing many endospores.

Coccidioidomycosis

The soil fungus *Coccidioides immitis* causes **coccidioidomycosis** (kok″sid-i-oy′do-mi-ko″sis), or San Joaquin valley fever (➤ Figure 21.19). This organism is found mainly in warm, arid regions of the southwestern United States and Mexico. Infection usually occurs as a result of inhaling dust particles laden with fungal spores. Dogs, cattle, sheep, and wild rodents deposit spores in feces, and the spores easily become airborne. Susceptible individuals have become infected merely from standing on the platform at a train station for a short period of time to get a breath of "fresh air." Travelers passing through dust storms in the Southwest often become infected and return to states where the disease is not expected. Recent increases in disease incidence in California are due in part to earthquake activity, which has produced excess dust. When spores are inhaled by susceptible individuals, an influenza-like illness results. Coccidioidomycosis is always highly infectious, and it can be self-limiting or progressive. In fewer than 1 percent of these individuals, dissemination to the meninges or the bones occurs within 1 year of the initial infection. Dissemination is much more common in blacks than in whites.

Large spherules containing numerous small endospores (reproductive spores that are not ordinarily infectious—not to be confused with bacterial endospores) can be found in sputum, pus, spinal fluid, or biopsied tissue (➤ Figure 21.19b). The fungus forms cottony, white mycelia (colonies) in culture. Various immunological tests are available to aid in diagnosis, but some give false positive results in individuals who have antibodies against other fungal diseases. A skin test is also available to determine previous exposure to the fungus. The test is performed in the same manner as the tuberculosis skin test and measures the delayed hypersensitivity to antigens of *C. immitis*. Individuals exhibiting a positive skin test are immune to a second attack of the disease.

Coccidioidomycosis in its common form is an acute, self-limiting infection that generally does not require treatment. However, in cases of disseminated disease, therapy is needed. Amphotericin B is the most effective agent available. Newer polyenes, such as the azoles, show promise. If treatment fails, the disease is often fatal. Prevention is difficult, but reducing dust in endemic regions may be helpful. A vaccine is being developed for coccidioidomycosis.

Histoplasmosis

The soil fungus *Histoplasma capsulatum* causes **histoplasmosis,** or *Darling's disease*. The disease is endemic to the central and eastern United States but is found globally in major river valleys. The highest incidence is seen in the Mississippi and Ohio valleys, where 80 percent of the population shows immunological evidence of exposure to the fungus. *H. capsulatum* thrives in soil mixed with feces and especially in chicken houses and in caves containing bat guano (feces). Cave dust is responsible for so many cases that the disease is sometimes referred to as cave sickness.

The fungus enters the body by the inhalation of conidia (fungal spores). ➥ (Chapter 11, p. 301) These conidia are engulfed, but not killed, by macrophages and thus travel in macrophages throughout the body. Person-to-person spread does not occur. Although most infections do not produce disease symptoms, granulomatous lesions occur in the lungs and spleen of susceptible individuals. Inhaling large numbers of conidia can cause pulmonary infection resembling pneumonia. In a few individuals, especially the very young, very old, or those receiving immunosuppressants, *H. capsulatum* can spread to the spleen, liver, and lymph nodes. Anemia, high fever, and an enlarged spleen and liver are common in disseminated histoplasmosis, and death often results.

Diagnosis is made by microscopic identification of small ovoid cells of the organism inside infected human cells. When cultured at body temperature, the fungi look like budding yeasts; at room temperature, they form mycelia with hyphae and spores. An intradermal skin test is also available to determine previous exposure to *H. capsulatum*.

Supportive therapy is used for pulmonary histoplasmosis, and amphotericin B is sometimes effective in treating the disseminated disease. Humans can very easily become infected in environments such as chicken houses and caves, where the air is laden with spores. Some employers hire only individuals with positive histoplasmosis skin tests, and therefore immunity to histoplasmosis, to work in high-risk environments. Spraying infected soil and fecal deposits with 3 percent formaldehyde destroys some of the spores.

Cryptococcosis

Cryptococcosis is caused by *Filobasidiella* (formerly *Cryptococcus*) *neoformans,* a budding, encapsulated yeast. Organisms typically enter the body through the skin, nose, or mouth. Birds carry the fungi on their feet and beak. Although birds do not suffer from cryptococcosis, they do disseminate the opportunistic yeast. These organisms thrive on the nitrogenous waste creatinine, which is present in high concentration in bird feces.

Cryptococcosis is usually characterized by mild symptoms of respiratory infection, but it can become systemic if large quantities of the spores are inhaled by debilitated patients. *Filobasidiella neoformans* often spreads to the meninges, which become thickened and matted, and the organism can invade brain tissue. As with other fungal opportunists, cryptococcosis is increasing in incidence among AIDS patients.

Observing the organism in body fluids confirms the diagnosis. A latex agglutination test is also used to detect the presence of capsular material in body fluids. Flucytosine and amphotericin B can be used in combination to treat systemic disease. Because *F. neoformans* thrives in pigeon droppings, some degree of prevention can be achieved by reducing pigeon populations and decontaminating droppings with alkali.

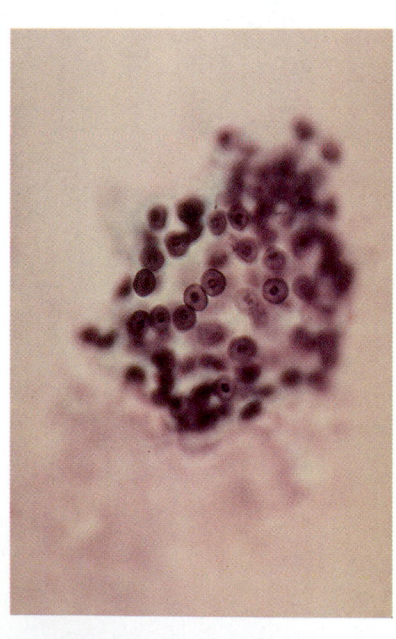

➤ **Figure 21.20** *Pneumocystis carinii* **in sputum.** The organism is a frequent cause of pneumonia in AIDS patients (353X).

Pneumocystis Pneumonia

Pneumocystis carinii (➤ Figure 21.20) was long thought to be a protozoan of the sporozoan group. It is now thought to be an opportunistic fungus. It invades cells of the lungs and causes alveolar septa to thicken and the epithelium to rupture. Then the parasites and a foamy exudate from cells collect in the alveoli. The disease, called ***Pneumocystis* pneumonia,** occurs in infants, the elderly, and the immunocompromised. The marked increase in the incidence of this disease in recent years is due mainly to its ability to infect persons with AIDS. The fungus can spread to other organs and cause extrapulmonary infections.

Diagnosis is made by finding organisms in biopsied lung tissue or bronchial lavage (washings from the bronchial tubes). Treatment of *Pneumocystis* infection generally involves the administration of a combination of trimethoprim and sulfamethoxazole or pentamidine. HIV-infected persons and AIDS patients are treated with these drugs as a preventive measure.

Aspergillosis

Aspergillosis (Chapter 19), called *farmer's lung disease* when it occurs in the lungs, is usually caused by *Aspergillus fumigatus* or *A. flavus*. ➥ (p. 551) Other species of *Aspergillus* and even other genera of fungi also can cause farmer's lung disease. Fungal spores inhaled from piles of rotting vegetation or compost may cause clinical allergy, such as asthma, or may produce invasive infection in the lower respiratory tract. Masses of fungal mycelium may grow large enough to be visible on X-rays as a *fungus ball,* or *aspergilloma.* These masses can obstruct gas exchange and cause death by asphyxiation. *Aspergillus* growing in the lungs may also serve as antigens that trigger chronic asthma. Amphotericin B is the drug of choice for invasive infections. Immunosuppressed, immunodeficient (AIDS), and diabetic patients are at higher than normal risk. Fungus balls are very difficult to treat, and surgery is sometimes needed to remove them.

➤ Figure 21.21   **The lung fluke** ***Paragonimus westermani.*** This is an adult worm stained to show internal structures (13X).

Parasitic Respiratory Diseases

The lung fluke *Paragonimus westermani* is found in many parts of Asia and the South Pacific (➤ Figure 21.21). The fluke's life cycle starts when egg-laden feces are released into water; the eggs hatch and invade first a snail and then a crab or crayfish, in which the last larval form, or metacercariae, develop. ➥ (Chapter 11, p. 306) When a human eats an infected shellfish, the metacercariae leave the crab or crayfish as it is digested in the human's small intestine. The larvae then bore through the intestine and embed in the abdominal wall temporarily. They soon leave it and penetrate the diaphragm and membranes around the lungs to reach the bronchioles. The larvae mature into adults and lay eggs in the bronchioles. When the host coughs, the eggs move into the pharynx, are swallowed, and exit the body in the feces. Infected humans have a chronic cough, bloody sputum, and difficulty breathing. Diagnosis can be made by finding eggs in sputum or by any one of several immunologic tests. The drug praziquantel is effective in treating lung fluke infections. Infections can be avoided by cooking shellfish before eating them.

Diseases of the lower respiratory tract are summarized in Table 21.4.

✓ A patient has a sore throat, stuffy nose, and is sneezing. What is the probability that this is a cold? The flu? Pneumonia?

✓ Give an example of a fungal respiratory disease that has a distinct geographic pattern of occurrence.

✓ What types of different symptoms can hantaviruses cause?

✓ How does antigenic shift relate to influenza epidemics?

➤ Table 21.4 | A summary of diseases of the lower respiratory tract

Disease	Agent(s)	Characteristics
Bacterial lower respiratory diseases		
Whooping cough	*Bordetella pertussis*	Catarrhal stage with fever, sneezing, vomiting, and mild cough; paroxysmal stage with ropy mucus and violent cough; convalescent stage with mild cough
Classic pneumonia	*Streptococcus pneumoniae, Staphylococcus aureus, Klebsiella pneumoniae*	Inflammation of bronchi or alveoli of lungs with fluid accumulation and fever
Mycoplasma pneumonia	*Mycoplasma pneumoniae*	Mild inflammation of bronchi or alveoli
Legionnaires' disease	*Legionella pneumophila*	Inflammation of the lungs, fever, chills, headache, diarrhea, vomiting, and fluid in lungs
Tuberculosis	*Mycobacterium tuberculosis*	Tubercles in lungs and sometimes in other tissues; organisms can persist in walled-off lesions and be reactivated
Ornithosis	*Chlamydia psittaci*	Pneumonia-like disease transmitted to humans by birds
Q fever	*Coxiella burnetii*	Disease similar to Mycoplasma pneumonia but transmitted by ticks, aerosols, and fomites
Nocardiosis	*Nocardia asteroides*	Pneumonia-like disease seen in immunodeficient patients
Viral lower respiratory diseases		
Influenza	Influenza viruses	Viruses subject to antigenic variation, with new strains causing epidemics; inflammation of oropharyngeal membranes, fever, malaise, muscle pain, cough, nasal discharge, and gastroenteritis
Respiratory syncytial virus infection	Respiratory syncytial virus	Febrile disease of the respiratory tract; can cause viral pneumonia
Hantavirus pulmonary syndrome	Hantaviruses	Fever, kidney abnormalities; in severe cases shock, bleeding, and pulmonary edema occur
Acute respiratory disease	Adenoviruses	Mild cough and nasal discharge; can cause viral pneumonia
Fungal respiratory diseases		
Coccidioidomycosis	*Coccidioides immitis*	Influenza-like illness; dissemination to meninges and bones can occur
Histoplasmosis	*Histoplasma capsulatum*	Granulomatous lesions in lungs and spleen in susceptible individuals; can cause pneumonia
Cryptococcosis	*Filobasidiella (Cryptococcus) neoformans*	Usually a mild pulmonary disease; pneumonia and dissemination to meninges can occur
Blastomycosis	*Blastomyces dermatitidis*	Usually a skin disease (see Chapter 19); sometimes disseminated to lungs or acquired directly by inhalation; mild respiratory symptoms
Pneumocystis pneumonia	*Pneumocystis carinii*	Rupture of alveolar septa, foamy sputum; occurs mainly in immunodeficient patients
Aspergillosis	*Aspergillus* species	Allergic asthmatic response to inhalation of spores or invasive infection of lung; fungal balls can cause asphyxiation
Parasitic respiratory diseases		
Lung fluke infection	*Paragonimus westermani*	Larvae mature in bronchioles and cause chronic cough, bloody sputum, and difficulty breathing

The agents and characteristics of the diseases discussed in this chapter are summarized in Tables 21.1 and 21.4. Information in these tables is not repeated in this summary.

Components of the Respiratory System

The Upper Respiratory Tract

- The **upper respiratory tract** consists of the nasal cavity, pharynx, larynx, trachea, and bronchi. It is lined with mucus-secreting epithelium and is covered with cilia.

- The **mucociliary escalator** is a nonspecific defense whereby microbes trapped in mucus are moved up to the pharynx.

The Lower Respiratory Tract

- The **lower respiratory tract** consists of the **bronchioles, respiratory bronchioles,** and **alveoli,** forming the lungs. The surfaces of the lungs and the cavities they occupy are covered by the **pleura,** which secretes **serous fluid.**

The Ears

- The outer ear consists of a **pinna** and an **auditory canal,** lined with **ceruminous glands** that secrete cerumen (earwax).

- The **tympanic membrane** separates the outer and middle ears. The middle ear contains three ossicles that transmit sound waves to the inner ear where the vestibulocochlear cranial nerve carries impulses to the brain.

Normal Microflora of the Respiratory System

- Healthy lungs are usually sterile. Below the level of the larynx, the trachea, bronchi, and larger bronchioles have a transient rather than a normal resident microflora.

- The pharynx has a normal microflora similar to that of the mouth.

- The upper respiratory tract above the pharynx has a normal microflora similar to that of skin.

Diseases of the Upper Respiratory Tract

Bacterial Upper Respiratory Diseases

- Infections related to **pharyngitis,** or sore throat, include **laryngitis, epiglottitis, sinusitis,** and **bronchitis.** These diseases usually are transmitted by respiratory droplets. If severe enough, they can be treated with penicillin or other antibiotics.

- **Diphtheria,** no longer common in the United States, occurs only in humans. Both the organism and its toxin, produced by genes from a lysogenic prophage, contribute to disease signs and symptoms. A major disease sign is the formation of a **pseudomembrane** that can block the airway. Diphtheria is spread by respiratory droplets, treated with antitoxin and antibiotics such as penicillin, and prevented by DTP vaccine.

- Ear infections occur in the middle (**otitis media**) and outer (**otitis externa**) ear. Organisms reach the middle ear via the Eustachian tube and usually can be eradicated by penicillin.

Viral Upper Respiratory Diseases

- The common cold, **coryza,** is transmitted by fomites and aerosols. Treatment is limited to alleviating symptoms; no vaccine is available.

- **Parainfluenza** infections range from inapparent disease to severe **croup.** Most children develop antibodies to **parainfluenza viruses** by age 10.

Diseases of the Lower Respiratory Tract

Bacterial Lower Respiratory Diseases

- **Whooping cough,** or **pertussis,** is distributed worldwide, but immunization has decreased its incidence. It is transmitted by respiratory droplets and treated with antitoxin and erythromycin. Vaccine prevents the disease but may also cause some complications and deaths.

- Classic **pneumonia** can be **lobar** or **bronchial;** it is transmitted by respiratory droplets and carriers. *Klebsiella* pneumonia is more severe than pneumococcal pneumonia. Penicillin is the drug of choice for pneumococcal pneumonia, and vaccine is recommended for high-risk populations. *Mycoplasma* pneumonia is transmitted by respiratory droplets and is treated with erythromycin or tetracycline.

- **Legionnaires' disease** is transmitted via aerosols from contaminated water and treated with erythromycin.

- **Tuberculosis (TB)** has been a major health problem worldwide for centuries, and its incidence in the United States is increasing. It is transmitted by respiratory droplets. Inactive but viable organisms can persist for years walled off in **tubercles.** Treatment with isoniazid is effective except for resistant organisms, which must be treated with "second- or third-line" drugs. A vaccine is available worldwide, but its use in the United States is limited to high-risk individuals.

- **Ornithosis** is transmitted to humans from infected birds. It is usually mild but can cause a serious pneumonia. The causative organism is dangerous to handle in the lab.

- **Q fever** is transmitted by ticks, aerosol droplets, and fomites. It is treated with tetracycline, and a vaccine is available for workers with occupational exposure.

- **Nocardiosis,** marked by lesions and abscesses, infects primarily the lungs but can infect the skin or other organs.

Viral Lower Respiratory Diseases

- The viruses that cause **influenza** display **antigenic variation** (mutation that affects antigens). The disease occurs mainly from December through April. It is transmitted in crowded, poorly ventilated conditions and diagnosed by immunological tests. Vaccine can prevent the disease, but its effectiveness can be lessened by antigenic variation.

- **Respiratory syncytial virus (RSV)** infections and **acute respiratory disease (ARD)** are important acute respiratory tract illnesses. RSV infections tend to be severe, especially in young children.

- **Hantavirus pulmonary syndrome (HPS),** a severe infection, has a high mortality. The hantavirus infects primarily rodents and is found in their excreta.

Fungal Respiratory Diseases

- Opportunistic fungal infections occur mainly in immunocompromised and debilitated patients. These infections usually are transmitted by spores, and some can be treated with amphotericin B.

- In the United States, **coccidioidomycosis** occurs in warm, arid regions; **histoplasmosis** is endemic in eastern states, and **cryptococcosis** occurs wherever there are infected birds, especially pigeons.
- *Pneumocystis* **pneumonia,** an opportunistic fungal infection, is a common cause of death among AIDS patients.

- **Aspergillosis** is contracted by the inhalation of fungal spores and can involve the growth of large masses of fungal mycelia (fungus balls) that are visible on X-rays.

Parasitic Respiratory Diseases

- Lung fluke infections occur mainly in Asia and the South Pacific where infected shellfish are eaten. The disease can be treated with drugs and prevented by cooking shellfish adequately.

Terminology Check

acute respiratory disease (ARD)   (p. 622)
alveolus   (p. 598)
antigenic drift   (p. 618)
antigenic shift   (p. 620)
antigenic variation   (p. 617)
aspergillosis   (p. 625)
auditory canal   (p. 598)
bronchial pneumonia   (p. 608)
bronchiole   (p. 598)
bronchitis   (p. 600)
bronchus   (p. 598)
caseous   (p. 612)
catarrhal stage   (p. 607)
cerumen   (p. 598)
ceruminous gland   (p. 598)
coccidioidomycosis   (p. 623)
consolidation   (p. 608)
convalescent stage   (p. 607)
coronavirus   (p. 604)
coryza   (p. 603)
croup   (p. 605)
cryptococcosis   (p. 624)
cyanosis   (p. 607)
diphtheria   (p. 601)
diphtheroid   (p. 602)
disseminated tuberculosis   (p. 613)

epiglottitis   (p. 600)
hantavirus pulmonary syndrome (HPS)   (p. 622)
histoplasmosis   (p. 624)
influenza   (p. 616)
laryngitis   (p. 600)
larynx   (p. 598)
Legionnaires' disease   (p. 609)
lower respiratory tract   (p. 596)
lobar pneumonia   (p. 608)
mastoid area   (p. 599)
miliary tuberculosis   (p. 613)
mucociliary escalator   (p. 598)
nasal cavity   (p. 597)
nasal sinus   (p. 597)
nocardiosis   (p. 616)
ornithosis   (p. 615)
orthomyxovirus   (p. 617)
otitis externa   (p. 603)
otitis media   (p. 603)
parainfluenza   (p. 604)
parainfluenza virus   (p. 604)
paroxysmal stage   (p. 607)
pertussis   (p. 606)
pharyngitis   (p. 600)
pharynx   (p. 597)

pinna   (p. 598)
pleura   (p. 598)
pleurisy   (p. 608)
Pneumocystis pneumonia   (p. 625)
pneumonia   (p. 608)
Pontiac fever   (p. 610)
primary atypical pneumonia   (p. 609)
pseudomembrane   (p. 602)
psittacosis   (p. 615)
Q fever   (p. 615)
respiratory bronchiole   (p. 598)
respiratory syncytial virus (RSV)   (p. 621)
respiratory system   (p. 596)
rhinovirus   (p. 604)
sinusitis   (p. 600)
syncytium   (p. 621)
tonsilitis   (p. 600)
trachea   (p. 598)
tubercle   (p. 612)
tuberculosis (TB)   (p. 610)
tympanic membrane   (p. 598)
upper respiratory tract   (p. 596)
viral pneumonia   (p. 621)
walking pneumonia   (p. 609)
whooping cough   (p. 606)

See You on the Web

Don't forget I am expecting you on the web. The following is a preparedness checklist for your web explorations. If you have questions on any of these items, you can return to the end of Chapter 1 for more details.

- Questions We'll Explore
- Concept Check questions
- Flashcards
- Study Guide

If you are ready, come visit me at

http://www.prenhall.com/black

It will be a journey you'll never forget.

Oral and Gastrointestinal Diseases

Just make it stop—that burning pain in the stomach! Quick! Something to ease the pain of that miserable ulcer. If only there were a cure . . .

Ah, but there is a cure now that we know the vast majority (70–95%) of stomach and duodenal ulcers are caused by the bacterium, *Helicobacter pylori*. Vigorous antibiotic treatment can exterminate the microbes and cure you of your ulcers—unless, of course, you're reinfected. Scientists are hard at work searching for a vaccine.

And where does one acquire this infection? Later in this chapter we will discuss this problem as well as examine its relationship to stomach cancer. For those of you who have had first hand experience with ulcers or stomach cancer, it must feel great to know that the days for these afflictions are ending.

Background Basics

We all have a wealth of firsthand experience with diseases of the oral cavity and gastrointestinal tract. We've had plaque removed from our teeth and cavities filled. We've had episodes of nausea, vomiting, and diarrhea—usually of short duration—after we have eaten contaminated food or drank impure water. Fortunately, for most of us these problems have been minor inconveniences rather than serious illnesses. In other parts of the world, however, especially where water and sewage treatment are inadequate, gastrointestinal infections are major problems. In this chapter we will consider diseases of the oral cavity and intestinal tract, both minor and serious.

Like the preceding chapters dealing with diseases of body systems, this chapter presumes a knowledge of disease processes, normal and opportunistic microflora (Chapter 14), epidemiology (Chapter 15), host systems and host defenses (Chapter 16), and immunity (Chapters 17 and 18). It also assumes familiarity with the characteristics of bacteria, viruses, and eukaryotic microorganisms and helminths (Chapters 9 through 11).

Components of the Digestive System

The **digestive system** consists of an elongated tube, or tract, that extends from the mouth to the rectum and includes accessory organs that help in digestion of food materials (➤ Figure 22.1). The digestive tract consists of various organs—mouth, pharynx, esophagus, stomach, and intestines—whereas the accessory organs include the teeth, salivary glands, liver, gallbladder, and pancreas. The digestive system has five major functions:

Questions We'll Explore

A What are the components of the digestive system? How do they function to prevent infection?

B What are the normal microflora of the mouth and digestive system?

C What kinds of pathogens cause diseases of the oral cavity, and what are the important epidemiologic and clinical aspects of these diseases?

D Which bacteria cause gastrointestinal diseases, and what are the important epidemiologic and clinical aspects of these diseases?

E Which viruses cause gastrointestinal diseases, and what are the important epidemiologic and clinical aspects of these diseases?

F Which protozoa cause gastrointestinal diseases, and what are the important epidemiologic and clinical aspects of these diseases?

G Which fungi cause gastrointestinal diseases, and what are the important epidemiologic and clinical aspects of these diseases?

H Which helminths cause gastrointestinal diseases, and what are the important epidemiologic and clinical aspects of these diseases?

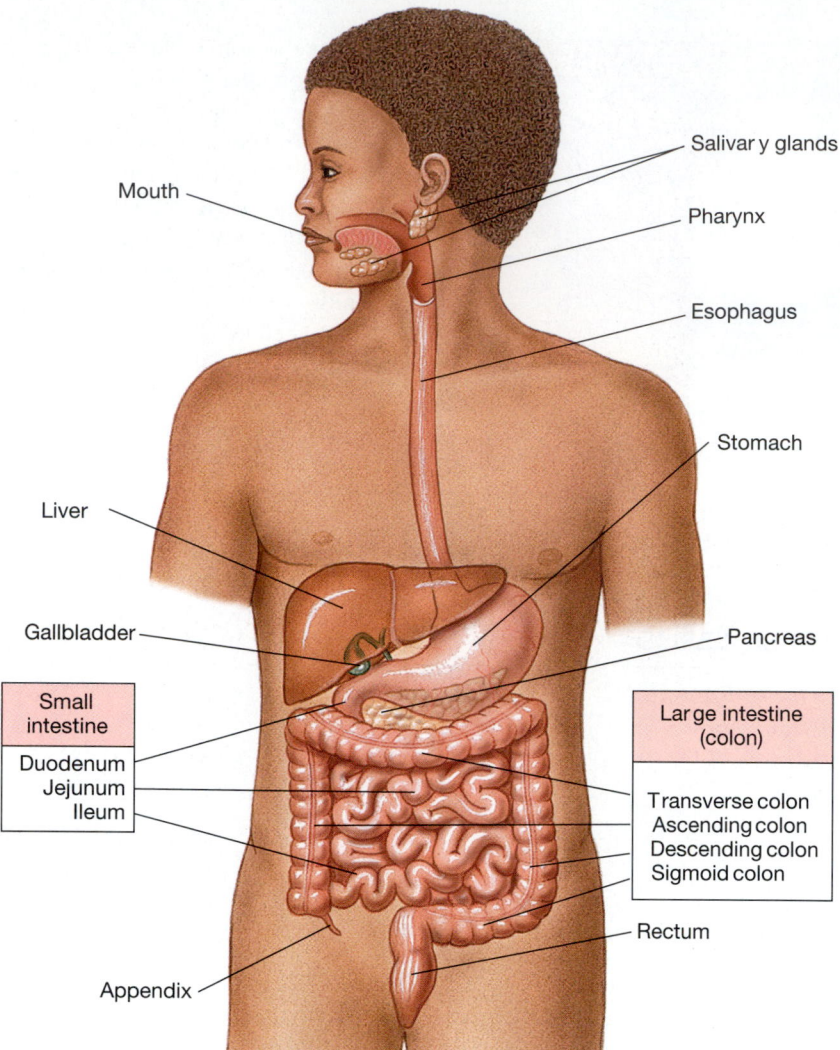

➤ Figure 22.1 **The structure of the digestive system.** Normal microflora of the digestive system prevent the easy growth of pathogens, and stomach acids and enzymes destroy most disease agents. Vomiting and diarrhea are two expulsion mechanisms used to rid the system of toxic materials, including bacterial toxins.

1. *Movement* of food materials through the tract.
2. *Secretion* of digestive juices and mucus to break down food materials.
3. *Digestion,* the process of breaking large food molecules into small molecules that can be passed from the digestive tract into the bloodstream.
4. *Absorption* of digested food substances into the blood or lymph.
5. *Elimination* of indigestible food materials and intestinal microflora.

Although food contains many microorganisms, most are killed by various defense mechanisms in the digestive tract. Throughout the digestive tract, **mucin** (myu'sin), a glycoprotein in mucus, coats bacteria and prevents their attaching to surfaces.

The Mouth

The mouth, or *oral cavity,* is lined with a mucous membrane and contains the tongue, teeth, and salivary glands. The mouth is a portal of entry for microorganisms; a normal mouth contains more resident microbes than there are people on earth.

Each tooth has a *crown* covered with **enamel** above the gum and a *root* covered with **cementum** below the gum (➤ Figure 22.2). Under these coverings is a porous substance called *dentin,* a central *pulp cavity,* and the *root canals,* where blood vessels and nerves are located. Each tooth is held in a tooth socket by fibers running from the cementum to the bone of the socket. Although enamel is the hardest substance in the body, it can be attacked by microbial-

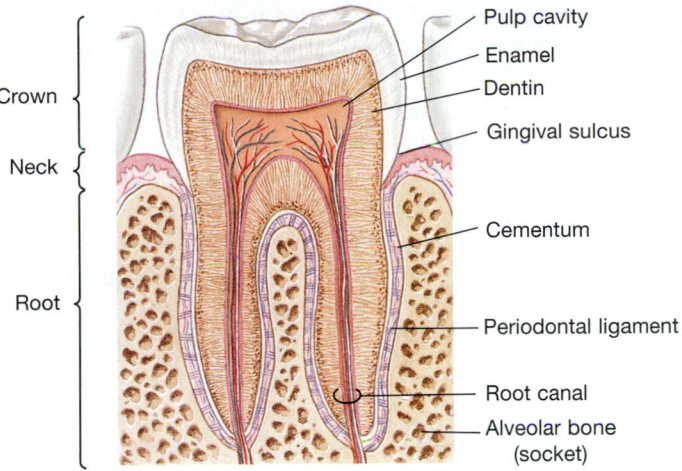

Crown

Neck

Root

Pulp cavity
Enamel
Dentin
Gingival sulcus

Cementum

Periodontal ligament

Root canal
Alveolar bone
(socket)

➤ **Figure 22.2   A section through a typical tooth and gum.**

produced acids and enzymes. Microbes also can infect the gums, form pockets of infection between teeth and gums, and spread to the underlying bone.

The salivary glands secrete saliva, containing both antibodies that can coat bacteria and lysozyme that kills some bacteria. The salivary glands themselves, however, are subject to infection.

The Stomach

After food is chewed and mixed with saliva in the mouth, it passes through the pharynx and esophagus to the stomach. There the food mixes with hydrochloric acid and the enzyme pepsin, which together begin the digestion of proteins. The lining of the stomach is protected from the acid by viscous mucus. Only alcohol, aspirin, and some lipid-soluble drugs can cross the mucous barrier and be absorbed in the stomach.

The Small Intestine

Partially digested food leaving the stomach enters the **small intestine** and is mixed immediately with secretions from the liver and pancreas. Liver cells secrete bile, a mixture of bile salts, cholesterol, and other lipids that aids in the digestion of fats.

Materials digested in the small intestine enter the bloodstream and are carried directly to the liver via the portal vein. There is a variety of toxins in the foods we eat; the liver has the job of detoxifying these materials. In tissues such as the liver, capillaries are enlarged to form networks of vessels called **sinusoids.** These sinusoids are lined with phagocytic *Kupffer cells,* which remove any dead blood cells, bacteria, and toxins from the blood as it passes through the sinusoids. The pancreas releases hormones into the blood and digestive juices into ducts that empty into the small intestine. These digestive juices include enzymes that digest starch, proteins, lipids, and nucleic acids, along with bicarbonate that neutralizes the acidic materials from the stomach.

The small intestine has an extensive internal surface area (about one-tenth the area of a football field) due to folds in the wall—fingerlike projections called **villi** (singular: *villus*)—and folds called **microvilli** in membranes of mucosal cells. Villi contain blood and lymph vessels. Digestion is completed in the small intestine; simple sugars and amino acids are absorbed into blood vessels, and fats into lymph vessels. Enzymes in the small intestine kill most microorganisms and inactivate most viruses in food. Also, the lower end of the intestine, called the *ileum,* contains lymphoid tissue clustered into *Peyer's patches.* These patches contain lymphocytes and macrophages that protect the small intestine from microbial invasion from the large intestine.

The Large Intestine

The **large intestine,** or *colon,* joins the small intestine near the appendix and ends at the rectum. What digestion occurs in the large intestine is done by the bacteria of the normal microflora. By-products of bacterial metabolism, such as amino acids, thiamine, riboflavin, and vitamins K and B12, are absorbed but in quantities too small to satisfy nutritional needs. Much water is absorbed, and undigested food is converted to feces. **Feces,** which consist of about three-fourths water and one-fourth solids, are stored in the rectum until they are eliminated from the body.

The large intestine also contains many resident microbes that ultimately come from the mouth or from the anus, another portal of entry. However, the normal microflora of the large intestine compete with opportunists and pathogens and prevent their growth and reproduction under normal conditions. Gastrointestinal infections generally occur only when defense mechanisms have been overwhelmed by large numbers of a pathogenic organism. Even when an infection occurs, vomiting and diarrhea help rid the body of the pathogens and their toxins.

Normal Microflora of the Mouth and Digestive System

Unless a microbe entering the mouth can quickly attach to a surface, it will be washed down into the digestive tract by the flow of saliva. Patterns of salivary flow vary from person to person; some are highly efficient at rinsing food and microbes away, while others are inefficient. The development of tooth decay is greatly affected by the degree of efficiency.

The ecology of the oral microflora is poorly understood. Although over 400 species of microbes have been identified as living there, probably even more than that number of species have not yet been identified. Research has focused on those organisms associated with tooth decay and gum disease, but there are many other microbes living on the cells lining the cheek, on the tongue, palate, and floor of the mouth. These may contribute to bad breath.

The human esophagus does not appear to have a permanent normal microflora. The stomach's acid pH usually prevents colonization by microbes. It and the first two-thirds of the small intestine contain very few microbes, mainly lactobacilli and streptococci that are merely passing through. In the last third of the small intestine, motility of contents is slower, and some microbes are able to colonize its surfaces. These are mainly Gram-negative, facultatively anaerobic bacteria, especially member of the family Enterobacteriaceae (e.g., *E. coli*), plus obligate anaerobes (e.g., *Bacteroides* and *Clostridium*). In the large intestine, food may remain as long as 60 hours, allowing ample time for microbial colonization and replication. Feces are composed about 50 percent by weight and volume of bacteria. Most of these are species of *Bacteroides*. The microbes of the lower digestive tract utilize foodstuffs that passed through the upper digestive tract without being completely digested. One important by-product of their activities is the production of vitamin K, necessary for the proper clotting of blood.

Diseases of the Oral Cavity

Although we would like to believe otherwise, our mouths are microbial breeding grounds. About 2 billion organisms live in a healthy human mouth. We pick up microbes in many ways—from food, kissing, and dirty fingers, to name a few—and the microbes thrive.

Bacterial Diseases of the Oral Cavity

Dental Plaque

Dental plaque is a continuously formed coating of microorganisms and organic matter on tooth surfaces. Plaque formation, although not a disease itself, is the first step in tooth decay and gum disease. Scrupulous and frequent cleaning of teeth minimizes but does not entirely prevent plaque formation. Plaque begins to form within 24 hours after cleaning. Unless it is removed regularly, plaque can become so securely attached to teeth that it can no longer be removed by home methods. Professional cleaning removes plaque, but it begins to form again even before you get home from the dentist's office.

Plaque formation begins as positively charged proteins in saliva adhere to negatively charged enamel surfaces and form a *pellicle* (film) over the tooth surface. Cocci such as *Streptococcus mutans* and some filamentous bacteria among the normal oral flora microbiota attach to the newly formed pellicle. These organisms may hydrolyze sucrose (table sugar) to glucose and fructose, both of which can be metabolized for energy production and growth. Plaque-forming bacteria such as *S. mutans* can also convert sucrose to polymers of glucose units (such as the polysaccharide dextran) that serve as bridges holding together cells in plaque. Plaque consists of up to 30 different genera of bacteria and their products, such as dextran, saliva proteins, and minerals (➤ Figure 22.3). If plaque is not removed thoroughly and regularly, streptococci, lactobacilli, and other acid-producing bacteria may accumulate within it in layers 300 to 500 cells thick. These organisms metabolize fructose and other sugars that diffuse into the plaque and set the stage for tooth decay. Plaque that accumulates near the gumline also offers protection to bacteria in the gingival (gum) crevices between the teeth and the gums. These bacteria include species of *Actinomyces, Veillonella, Fusobacterium,* and sometimes spirochetes as well as the streptococci just noted. If plaque is allowed to accumulate, some crevices change into anaerobic pockets full of bacteria that can irritate the gums or destroy the bone in which the teeth are set. In some people, this gumline plaque mineralizes into *calculus* (*tartar*) that can also irritate the gums and contributes to inflammation and bleeding.

Dental Caries

The Disease   **Dental caries,** or tooth decay, is the chemical dissolution of enamel and deeper parts of teeth. Taking its name from the Latin *cariosus,* which means rotten, caries is the most common infectious disease in developed countries where the diet contains relatively large amounts of refined sugar. Unchecked, the decay can proceed through enamel, into dentin, into the pulp cavity, and eventually can cause an abscess in the bone that supports the tooth. Sugars easily diffuse through plaque to bacteria embedded in it, but acids produced by bacterial fermentation fail to diffuse out. The acids gradually dissolve enamel, after which protein-digesting enzymes break down any remaining material.

The combination of sucrose and the action of *S. mutans* on it account for much tooth decay. Consequently, the more sucrose you eat and the more frequently you eat it, the greater your risk

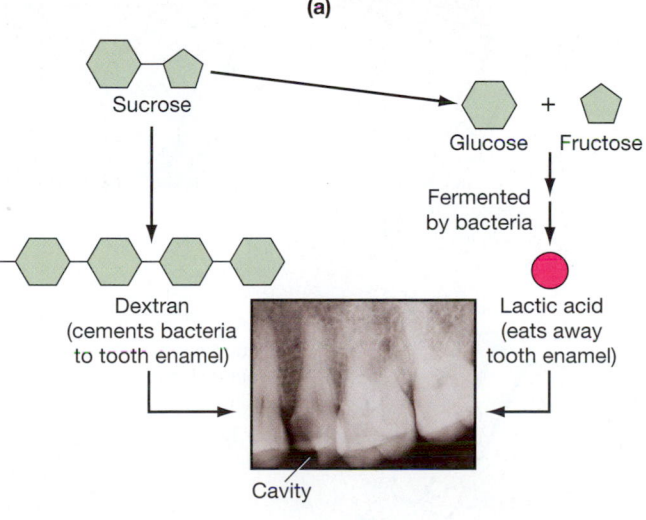

(a)

(b)

Sucrose → **Glucose** + **Fructose**

Fermented by bacteria → **Lactic acid** (eats away tooth enamel)

Dextran (cements bacteria to tooth enamel)

Cavity

➤ **Figure 22.3   Dental plaque.** (a) A variety of organisms that accumulate in plaque deposits. (b) Production of sticky dextran from sucrose enables bacteria to adhere to tooth surfaces, where the lactic acid produced in this process eats away tooth enamel and forms cavities.

of dental caries. Saliva helps rinse sugars from the mouth, but its rinsing efficiency varies with flow rate and mouth shape. Sugar accumulates in poorly rinsed areas. Although starchy foods are only partially digested in the mouth, those that are sticky can adhere to tooth surfaces and remain on the teeth long enough for bacterial action to contribute to tooth decay. Sugar alcohols, such as sorbitol and xylitol used in "sugar-free" chewing gums, do not contribute to tooth decay because many bacteria cannot metabolize them. Bacteria that can utilize sorbitol, such as *S. mutans,* metabolize it slowly. They release acid at a rate that allows buffers in saliva to neutralize it, preventing a drop in pH in areas of plaque.

Treatment and Prevention   Dental caries are treated by removing decay and filling the cavity with *resins* (plastic materials) or with *amalgam* (a mixture of silver and other metals). Dental caries can be prevented or their incidence greatly reduced by limiting the intake of sugary and sticky foods, brushing regularly with plaque-removing toothpaste, and flossing between teeth. Vaccines to prevent dental caries are being developed against strains of *S. mutans,* which are most prevalent at decay sites. An injectable vaccine that elicits circulating IgG production has been successful in monkeys. An oral vaccine that causes secretory IgA production has been successful in rats. Neither vaccine has yet been adequately tested in humans. Stress may be a factor in the effectiveness of a vaccine because it appears to depress the immune system. This was noted when dental students were found to secrete less salivary IgA while taking examinations than while on summer vacation. ➥ (Chapter 17, p. 471)

The use of **fluoride** has been the most significant factor in reducing tooth decay. It works by hardening the surface enamel of teeth. Fluoride reduces the solubility of tooth enamel by inhibiting demineralization and also enhances remineralization. To understand how fluoride affects enamel, imagine the tooth surface as consisting of the ends of rods packed together like a fistful of pencils (➤ Figure 22.4). No matter how tightly the rods are packed, channels exist

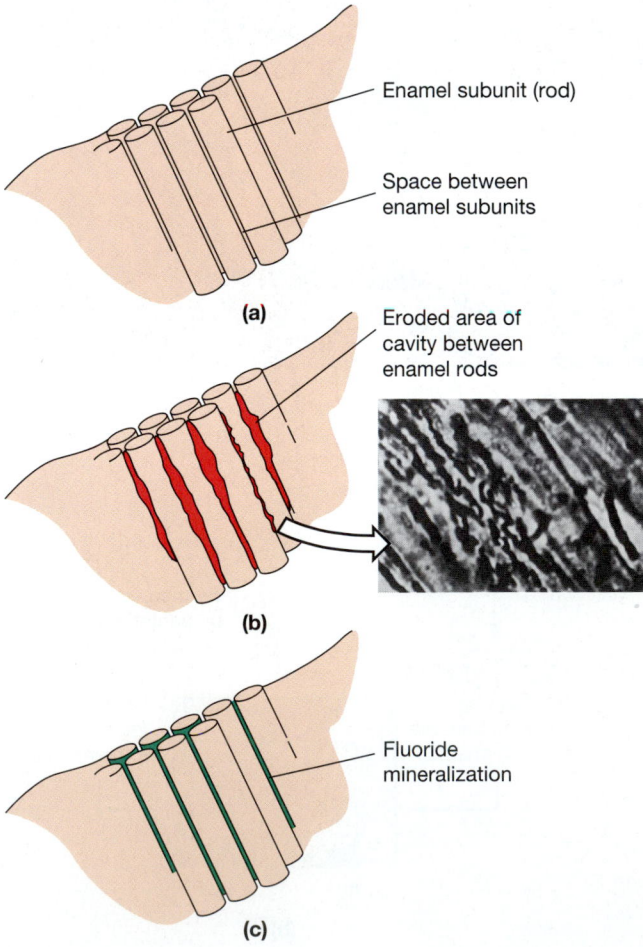

Enamel subunit (rod)

Space between enamel subunits

(a)

Eroded area of cavity between enamel rods

(b)

Fluoride mineralization

(c)

➤ Figure 22.4   **The effect of fluoride on tooth decay.** (a) Normal tooth structure; (b) dental caries forming as a result of bacterially produced acid seeping down between rods of enamel; (c) mineralization by application of fluoride, which fills in the spaces between enamel rods, thereby preventing acid from seeping in.

between them. Acids produced by bacteria in plaque seep through channels and dissolve the enamel rods. As channels are enlarged, more acid enters and dissolves more enamel. Eventually, sufficient enamel is eroded to form cavities, in which dental caries can form. Fluoride fills the spaces between rods with a hard mineralized material that strengthens the tooth surface and prevents acid penetration. Fluoride may also inhibit certain enzymes that produce phosphates needed by bacteria to capture energy from nutrients. Without the phosphates, the bacteria die.

Numerous studies have shown fluoride to be safe and effective in the prevention of tooth decay. Added to city water supplies in a 1-part-per-million concentration, fluoride reduces tooth decay in children by as much as 60 percent. Because caries occur mainly in childhood and adolescence, a successful program to prevent childhood tooth decay is of major importance. Yet nearly half the U.S. population obtains water from wells or from nonfluoridated community supplies. In some communities that lack fluoridated water, children receive fluoride tablets or gels at school. Ingestion of fluoride by children during tooth formation strengthens the entire tooth, whereas topical application affects only the surface of the tooth.

Although fluoride is most beneficial before age 20, it provides some benefits at any age, even when water is fluoridated and fluoride toothpastes and mouthwashes are used. Fluoride gel treatments should be given every 6 months from about age 4 to adulthood. Each time teeth are cleaned, a little surface is removed, and remineralization with topical fluoride helps restore the surface (➤ Figure 22.5).

Another means of preventing dental caries is sealing of the teeth with a type of resin that mechanically bonds to the tooth structure. Because the grinding surfaces of teeth have tiny pits and crevices that fill with plaque that cannot be removed with a toothbrush, these surfaces are more susceptible to decay than are smooth sides. Applying sealant to teeth provides nearly complete protection against decay as long as the sealant remains, which may be 10 years or longer. Applied at about age 6 or 7 after secondary grinding teeth have erupted, sealants will last through most of the decay-prone years. In the sealing process, teeth are first thoroughly cleaned and etched with acid; then the sealant is applied. Tiny cavities cease to grow in sealed teeth and eventually become sterile as the decay organisms die. Sealant can be applied only to teeth without fillings, but it can be used on adults as well as children.

Periodontal Disease

The Disease   When bacteria become trapped in gingival crevices, they cause both tooth decay and gum inflammation. **Periodontal disease** is a combination of gum inflammation and erosion of periodontal ligaments and the alveolar bone that supports teeth (➤ Figure 22.6). The disease is a chronic, generally painless infection that affects over 80 percent of teenagers and adults and is the major cause of tooth loss.

Plaque formation is the initiating event in periodontal disease. Organisms in gingival crevices are believed to produce endotoxins and acids that in turn produce an inflammatory response. This response breaks down the epithelial cells of the gums, and new groups of organisms re-

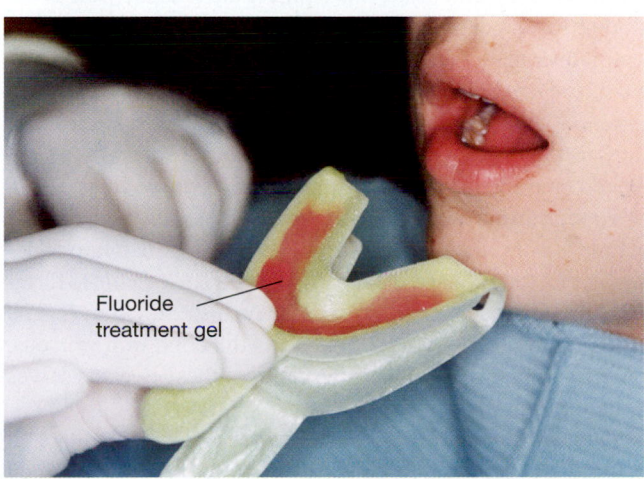

Fluoride treatment gel

➤ Figure 22.5   **Fluoride treatment of teeth.** Visits to the dentist can be minimized by proper oral hygiene and the use of fluoride-containing toothpaste or oral rinse. Fluoride can also be applied in the dentist's office.

Smile

If one of your teeth decayed and you had to get an artificial tooth, you probably wouldn't worry too much; your dentist could probably restore your smile to normal. But in addition to the dentist, you might owe microbiologists a "thanks." Medical implants, from artificial teeth to replacement joints, often become contaminated with such bacteria as *Staphylococcus aureus* and antibiotics are usually useless against them. Now microbiologists have synthesized a protein that binds to the *Staphylococcus aureus* receptors. This protein can be loaded right into the medical implant's plastic, keeping bacteria from adhering and greatly reducing the infection rate.

(a)

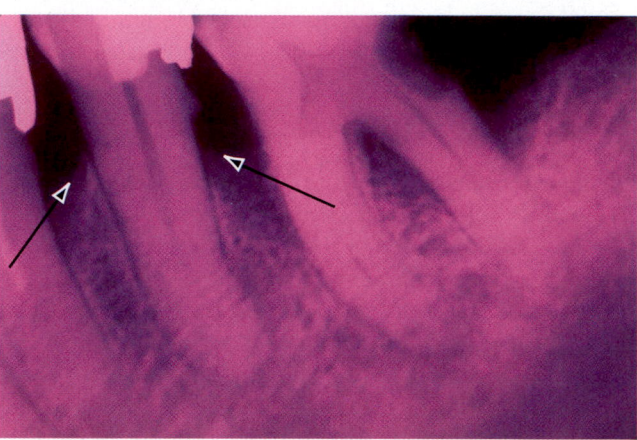

> Figure 22.6   **Advanced periodontal disease.** (a) Severe gum inflammation, which can lead to (b) loss of bone (see arrows) surrounding roots of teeth, causing loosening and eventual loss of teeth.

(b)

Keep On Brushing

Your dentist probably never told you that you could suffer from autoimmune diseases and heart abnormalities if you didn't brush your teeth every day, but it's true. Chronic periodontal disease can affect systems throughout the entire body. The cardiovascular system seems to be especially susceptible. How? Chronic infections activate the immune system and sometimes cause the body to turn against itself, leading to cardiovascular autoimmune diseases. Periodontal bacteria can also invade the cardiovascular system from open wounds in the gums themselves. At least one strain of *Streptococcus sanguis,* a bacterium found in plaque, is believed to cause heart abnormalities in animals and platelet clumping in humans. So brush those pearly whites faithfully—you'll be protecting more than just your mouth.

place previous crevice inhabitants as the disease progresses. If the process is not arrested, the gums recede and can become necrotic, and the teeth loosen as surrounding bone and ligaments are eroded and weakened.

In its mildest form, periodontal disease is called **gingivitis** (jin-jiv-i′tis), which affects only the gums. In its most severe form, gingivitis is called **acute necrotizing ulcerative gingivitis (ANUG)**, or trench mouth. The disease got its name because it was common among soldiers in World War I under stress "in the trenches." It is now common in young people, and stress seems to be an important factor in its development. ANUG responds to antibiotics, but more advanced forms of periodontal disease usually do not. Unchecked, periodontal disease can lead to chronic **periodontitis,** which affects the bone and tissue that support the teeth as well as the gums. The loss of bone due to periodontitis is not reversible.

Periodontal disease occurs when plaque is allowed to accumulate, resulting in overgrowth of potentially virulent bacteria such as *Porphyromonas gingivalis, Actinobacillus actinomycetemcomitans, Prevotella intermedia, Bacteroides forsythus, Fusobacterium nucleatum,* and other Gram-negative rods. Many of the organisms living in the mouth, especially, spirochetes, have yet to be isolated, cultured, and identified. Certain other bacteria, such as *Streptococcus mitis, S. sanguis, Veillonella* species, and some *Actinomyces* species, may help control virulent bacterial populations. If plaque is not removed by brushing or flossing, calcium is deposited on plaque surfaces, forming a very rough, hard crust called **tartar,** or *calculus.* Tartar strongly binds to teeth surfaces. As more plaque accumulates, tartar forms a thick layer. The calculus contributes to the conditions leading to bleeding of gums. As the condition worsens, pockets of inflammation form in the gum crevices. Measuring the depth of these pockets serves as an indication of the extent of periodontal disease.

A 1982 study of organisms associated with periodontal disease, which was done by Paul H. Keyes, illustrates a method of studying a complex microbial environment. Keyes obtained plaque from individuals with healthy gums, mildly inflamed gums, and severely diseased gums. When he examined them by phase-contrast microscopy, he found distinctly different populations of microorganisms. Plaque from healthy gums contains nonmotile filamentous bacteria, a few colonies of cocci, fewer than five leukocytes per microscopic field, and no amoebas or spirochetes. Plaque from mildly inflamed gums shows a greater variety of microbes. Nonmotile organisms form dense masses surrounded by clusters of motile spirochetes and rapidly spinning bacilli too numerous to count. Some spirochetes and leukocytes are present, but amoebas and trichomonads are absent. Plaque from severely diseased gums contains dense mats of nonmotile filaments, cocci, and numerous motile organisms. Extending from the dense mats are brushlike aggregations of spirochetes and flexible rods that move in rippling waves and migrate from one surface to another. Amoebas always are present, trichomonads sometimes are seen, and leukocytes are numerous.

More recent studies suggest that of 300 species of bacteria in the mouth, *Porphyromonas gingivalis* may be a specific cause of some cases of periodontal disease. Researchers at the University of Texas have succeeded in causing a burst of periodontal disease in monkeys given doses of the bacterium. The researchers have also had some success in treating the monkeys with rifampin.

Plaque organisms do not appear to penetrate gum tissue. Their effects are caused largely by the secretion of destructive enzymes, inflammation, and allergic reactions to bacterial products spreading from the plaque.

Treatment and Prevention   The treatment of chronic periodontal disease is a somewhat controversial subject in dentistry today, and patients vary in their responses to different treatments. Treatments include antimicrobial mouth rinses, brushing with a mixture of bicarbonate of soda and hydrogen peroxide, surgery to eliminate pockets, and antibiotic therapy in unresponsive or rapidly progressing cases. Chronic periodontal disease can be prevented or its onset delayed by daily thorough cleaning of teeth and, most important, frequent professional removal of plaque from the pockets. Once organisms erode gums and form pockets between the teeth and gums, infections must be kept under control.

Viral Diseases of the Oral Cavity

Mumps

Mumps is caused by a paramyxovirus somewhat similar to the measles (rubeola) virus. The virus is transmitted by saliva or as aerosol droplets, enters via the oral cavity or via the respiratory tract, and invades cells of the oropharynx. After initially replicating in the upper respiratory tract, the virus travels in the blood to the salivary glands and sometimes to other glands and organs, such as the testes and the meninges. Swelling of the parotid glands appears 14 to 21 days after initial infection and can persist as long as 7 days. Viruses are released and the infected person is contagious for 7 days before the glands swell and for up to 9 days after swelling subsides. Mumps viruses are excreted in urine for up to 2 weeks after the onset of symptoms.

Humans are the only known hosts for the mumps virus. The virus is found worldwide, occurring especially in the spring. Infections are most common in children 6 to 10 years of age. Although up to 85 percent of an exposed susceptible population will be infected by mumps, 20 to 40 percent will have no symptoms. When the disease appears in postpubertal males, 20 to 30 percent develop **orchitis,** inflammation of the testes. Such infections are capable of causing sterility but rarely do so. Other complications of mumps, regardless of the infected person's age or sex, include meningoencephalitis, eye and ear infections, and inflammation of glands other than the parotid, such as the ovaries and pancreas (where it may be a cause of juvenile onset diabetes). An effective vaccine for mumps contains viruses weakened by passage through a sequence of embryonated egg inoculations. The vaccine usually is administered in combination with measles and rubella vaccines—collectively known as the MMR vaccine. The incidence of mumps has decreased dramatically since the vaccine was licensed and put to use in 1967 (➤ Figure 22.7). The vaccine is recommended for all persons who were born after 1957 and have no immunity to the disease. Infants are immunized after 15 months of age.

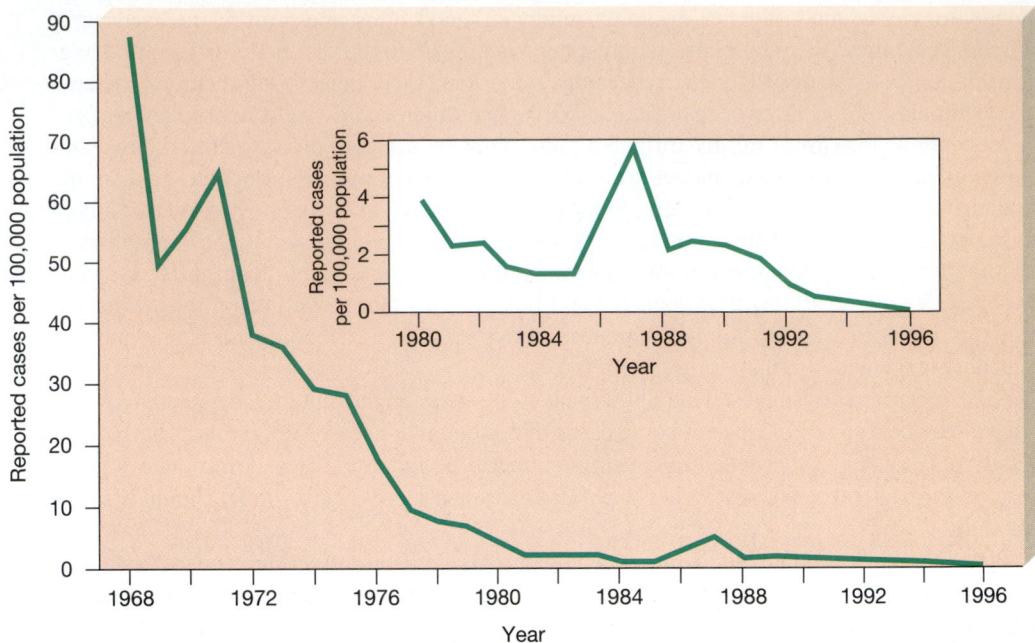

> ➤ Figure 22.7   **The U.S. incidence of mumps.** Since a vaccine became available in 1967, the number of cases has sharply declined. (*Source:* CDC)

Other Diseases

Thrush, an oral infection by the yeast *Candida albicans,* was discussed in Chapter 19 and illustrated in ➤ Figure 19.16a. ➡ (p. 551) Herpes simplex virus infections, a cause of fever blisters and cold sores on the lips and in the mouth, are discussed in Chapter 20.

Diseases of the oral cavity are summarized in Table 22.1.

Gastrointestinal Diseases Caused by Bacteria

Bacterial Food Poisoning

Food poisoning is caused by ingesting food contaminated with preformed toxins. It also can be caused by the ingestion of foods contaminated with pesticides, heavy metals, or other toxic substances. In food poisoning caused by microbial toxins, organisms that can continue to produce toxin may also be ingested with the toxins. However, tissue damage is due to action of the toxin, so most cases of microbial food poisoning are intoxications rather than infections. Be-

✓ How does fluoride help reduce dental decay?

✓ Distinguish between plaque and tartar.

Foodborne microbes trigger some 81 million cases of disease in the U.S. each year, according to the Government Accounting Office.

➤ Table 22.1	A summary of oral diseases	
Disease	**Agent(s)**	**Characteristics**
Bacterial diseases of the oral cavity		
Dental caries	*Streptococcus mutans* and other species	Erosion of tooth enamel and other structures by acids from microbial metabolism
Periodontal disease	Various bacteria, *Porphyromonas gingivalis*	Inflammation and destruction of gums, loosening of teeth, erosion of bone
Viral disease of the oral cavity		
Mumps	Paramyxovirus	Inflammation and swelling of salivary glands and sometimes testes, epididymis, and other tissues

cause the toxin is preformed, the onset of symptoms in intoxication is more rapid than in infection. Food poisoning can be prevented by following proper food-handling procedures, described in Chapter 26.

Bacteria that produce toxins responsible for food poisoning include *Staphylococcus aureus, Clostridium perfringens, C. botulinum,* and *Bacillus cereus.* Staphylococci usually enter food by way of infected food handlers. Other organisms that cause food poisoning are ubiquitous soil organisms found in water, feces, sewage, and nearly all foods. We will consider the nature of the food poisoning intoxication caused by each of these organisms.

Staphylococcal Enterotoxicosis

Certain strains of *Staphylococcus aureus* cause food poisoning, or **enterotoxicosis** (en″ter-o-tox-i-ko′sis), by releasing certain **enterotoxins**—enterotoxins A or D—which are exotoxins that inflame the intestinal lining and inhibit water adsorption from the intestine. These enterotoxins also cause neural stimulation of the vomiting center of the brain. Because the organisms are relatively resistant to heat and drying, foods easily become contaminated with them from food handlers or from the environment. The organisms multiply and release toxin in uncooked or inadequately cooked foods, especially if the foods are unrefrigerated. Nearly any food can be contaminated with *S. aureus,* but those with a starch or cream base are most likely candidates. Cream pies, dairy products, poultry products, and picnic foods such as potato salad are common culprits. Contamination is difficult to detect because it produces no change in food's appearance, taste, or odor. Unlike most exotoxins, the toxin is heat-stable and withstands boiling for 30 minutes. Hence cooking foods can kill the organisms but does not destroy the toxin. ➡ (Chapter 14, p. 388)

When food contaminated with *S. aureus* enterotoxin enters the intestine, the toxin acts directly. The organisms usually continue to produce toxin (but do not multiply). When it comes in contact with the mucosa, the toxin causes tissue damage only after it has entered the blood and has circulated back to the intestine. Symptoms such as abdominal pain, nausea, vomiting, and **diarrhea** (excessive frequency and looseness of bowel movements), but usually not fever, appear 1 to 6 hours after ingestion of contaminated food. The time required for symptoms to appear depends on how long it takes for absorption of a sufficient quantity of toxin to produce them. A food loaded with toxin elicits symptoms quickly; one with only a little toxin takes longer. Once elicited, symptoms usually last about 8 hours. For otherwise healthy adults, no treatment is required because the disease is self-limiting. It can be severe in infants, the elderly, and debilitated patients. Recovery from food poisoning does not confer immunity, as sufficient antibody is not produced. The best way of preventing food poisoning by *S. aureus* is to use sanitary food-handling procedures.

Other Kinds of Food Poisoning

An enterotoxin from *Clostridium perfringens* also causes food poisoning. ➡ (Chapter 14, p. 388) The enterotoxin, which is released only during sporulation, is produced under anaerobic conditions, as when undercooked meats and gravies are kept warm for a while. The main symptom is diarrhea. Compared with *S. aureus* food poisoning, *C. perfringens* food poisoning takes longer to appear—8 to 24 hours after ingestion—and lasts longer (about 24 hours). It too is self-limiting and can be prevented by sanitary food handling.

Although *C. perfringens* enterotoxin causes food poisoning, the organism itself can infect tissues. Upon introduction of spores into a wound, germination to vegetative cells can result in gas gangrene (Chapter 19) and anaerobic cellulitis (inflammation of connective tissue). ➡ (p. 558) Vegetative cells multiply and release toxins and enzymes that diffuse into surrounding healthy tissue, causing cellular destruction and death.

Botulism, which is caused by a neurotoxin produced by *Clostridium botulinum,* is acquired when toxin-contaminated food is eaten. Although it is a kind of food poisoning, it has little effect on the digestive system. Its effects on the nervous system are discussed in Chapter 24.

Bacillus cereus secretes a toxin that acts as an emetic—that is, it induces vomiting. The incubation period and signs and symptoms resemble those of staphylococcal food poisoning. Symptoms occur less than 12 hours after ingestion and last only a short time. Such toxins are

---

Mom Knows Best, But She Doesn't Always Know Everything

Mom probably said to wash fresh fruits and vegtables to rid them of germs. That's a great habit, but it takes more than simple washing to eliminate all of the bacteria. In May 1998, researcher Elizabeth Ehrenfeld deposited up to 100 million *E. coli* or *Salmonella* on bacteria-free bean sprouts, then washed them three times in clean water. While this reduced the bacterial population, it still left up to 1 million microbes per gram of food—many thousands of times the number needed to sicken someone with a weakened immune system. What can you do to protect yourself? Researchers are looking into nontoxic mixes of vinegar and off-the-shelf hydrogen peroxide that will eliminate all *E. coli* organisms from foods. Others are considering other chlorine-free, antibacterial products that you dissolve in tap water and spray on foods to kill all germs. Some of the disinfectants are already being marketed.

The Price You Probably Don't Mind Paying

Ever wonder why seafood is so expensive? When it comes to mussels, it's probably because they are so tough to raise. Mussels require a current of running water to feed, breathe, mate, and spawn. And though they may be easy to protect from muskrats, raccoons, and crows when they are being raised in captivity, they are much harder to protect against other enemies. Researcher Francis O'Beirn says, "They have to be scrubbed down individually to rid them of fungi and bacteria. . . . Mussels accomplish this toiletry themselves in the wild by digging themselves in the mud." So don't complain about the high price of mussels. Think about it this way: Your money is keeping you from getting food poisoning.

often found in contaminated rice or meat dishes. *Bacillus cereus* exists as a saprophyte in water and soil.

Food poisoning by *Pseudomonas cocovenenans,* which occurs in Polynesia, is named **bongkrek disease,** after its association with the native coconut delicacy bongkrek. The bacterium produces a potent, and often fatal, toxin that is frequently found in dishes prepared with coconut.

Bacterial Enteritis and Enteric Fevers

For most of us diarrhea is an unpleasant inconvenience, but it can be deadly. In 1900 in New York City the death rate among infants from diseases grouped together as diarrhea was 5603 per 100,000. Although today the death rate is less than 60 per 100,000, the disease is still life-threatening, and death from diarrhea can occur in hours in infants.

Enteritis is an inflammation of the intestine. **Bacterial enteritis** is an intestinal infection, not an intoxication, as is food poisoning. The causative bacteria actually invade and damage the intestinal mucosa or deeper tissues. Enteritis that affects chiefly the small intestine usually causes diarrhea. When the large intestine is affected, the result is often called **dysentery,** a severe diarrhea that often contains large quantities of mucus and sometimes blood or even pus. Some pathogens spread through the body from the intestinal mucosa and cause systemic infections, as is the case with typhoid fever. Such infections are called **enteric fevers.**

Salmonellosis

Salmonellosis (sal″mo-nel-o′sis) is a common enteritis caused by some members of the genus *Salmonella*. The annual reported incidence of salmonellosis (excluding typhoid fever) in the United States climbed from 20,000 cases in the early 1970s to more than 65,000 cases in 1986 but has dropped to about 42,000 cases in 1993. Many more cases go unreported: Some health experts estimate the true prevalence to exceed 2 million. The classification of the salmonellas has recently been revised. The number of species within the genus *Salmonella* has been reduced to three: *Salmonella typhi, S. choleraesuis* (usually a swine pathogen), and *S. enteritidis*. About 2000 strains of salmonellas have been identified by their surface antigens and grouped into **serovars** (se′ro-varz). Prior to 1972 each was named as a separate species. Most of the 2000 strains have been consolidated into the species *S. enteritidis*. The former *S. typhimurium* is now considered to be a serovar (strain) of *S. enteritidis*. However, most investigators still refer to *S. typhimurium* rather than more properly to *S. enteritidis* serovar *typhimurium*. Identification of serovars is sometimes useful in tracing the source of disease outbreaks.

Salmonella species other than *S. typhi,* the cause of typhoid fever, can be found in the gastrointestinal tracts of many animals, including poultry, wild birds, and rodents. In these hosts the bacteria either cause obvious disease or are carried without producing harmful effects. It is now known that chicken eggs can become infected if the hens laying them are infected. Salmonellas have also been traced to contaminated water and to food contaminated by carriers. The practice of confining poultry and pigs to small spaces while being reared for market makes feeding and caring for the animals more efficient, but it also facilitates the transmission of *Salmonella* and other pathogens among them.

Salmonella infection is generally associated with the ingestion of improperly prepared, previously contaminated food. Meat and dairy products are most likely candidates. Foods containing uncooked eggs can also be a source.

Signs and symptoms of salmonellosis include abdominal pain, fever, and diarrhea with blood and mucus. They appear 8 to 48 hours after ingestion of organisms and are associated with the organisms' invading the mucosa of both the small and large intestines. Fever probably is caused by endotoxins, toxins that are released from a cell only when it is lysed. In otherwise healthy adults, salmonellosis lasts 1 to 4 days and is self-limiting. Antibiotics usually are not given because they tend to induce carrier states and to contribute to the development of antibiotic-resistant strains. Infants and elderly or debilitated patients often have more severe and prolonged symptoms. In such cases, antibiotics may be prescribed.

Other serovars of *Salmonella* also cause disease. Because of their ability to invade intestinal tissue and enter the blood, *S. typhimurium* and *S. paratyphi* cause a somewhat more seri-

---

Clostridium perfringens wreaked havoc in the trenches during World War I, as soil-contaminated wounds frequently became infected, leading to gas gangrene.

In the American Civil War, 81,360 Union Army soldiers died of dysentery and typhoid; 93,443 were killed in combat.

Homeless Bacteria Looking for a New Host

Would you believe that chicken farmers are spraying newly hatched chicks with bacteria? It's true. But the bacteria that they are spraying, known collectively as Preempt, consists of 29 types of living, nontoxic bacteria isolated from the gut of mature chickens. As the hatchlings preen, the chicks ingest the bacteria, seeding their intestines with beneficial microbes. These microbes cover niches in the intestines so quickly and heavily that intestinal pathogens find it all but impossible to move in. *Salmonella, Campylobacter, Listeria,* and the deadly *Escherichia coli* 0157:H7 are going to have to find new animals to infect.

ous condition called **enterocolitis,** or enteric fever. Symptoms and bacteremia appear after an incubation period of 1 to 10 days. Enteric symptoms such as fever and chills can last 1 to 3 weeks. Chronic infections of the gallbladder and other tissues are not uncommon. A carrier state becomes established when, after the patient's recovery, organisms from chronically infected tissues continue to be excreted with feces. Broad-spectrum antibiotics rid carriers of organisms but can activate the disease in some carriers by upsetting the balance of intestinal microflora.

Prevention of salmonellosis and enterocolitis depends on the maintenance of sanitary water and food supplies and eradication of organisms from carriers. The organisms cannot be entirely eradicated because poultry and other animals serve as reservoirs, and no effective vaccine is available.

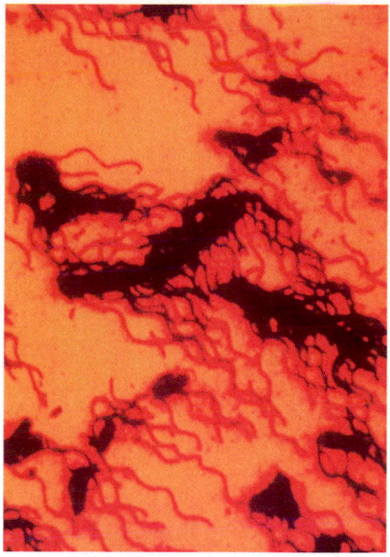

> Figure 22.8   **Salmonella typhi, the cause of typhoid fever.** Notice the flagella, which have been made visible by the use of stain. They would not otherwise be visible when using a light microscope (2000X).

Typhoid Fever

Typhoid fever, one of the most serious of the epidemic enteric infections, is caused by *Salmonella typhi* (➤ Figure 22.8). The disease is rare in places where good sanitation is practiced but is more common where faulty water and sewage systems exist. Uncooked shellfish, raw fruit, or raw vegetables have often served as sources of the pathogen in outbreaks in the United States. In 1942 the total number of U.S. cases was 5000. For most of the last 30 years, however, fewer than 500 cases per year have been reported in the United States. The organisms enter the body in food or water and invade the mucosa of the upper small intestine. From there they invade lymphoid tissues and are phagocytized and disseminated. The organisms multiply in the phagocytes, emerge, and continue to multiply intracellularly.

Bacteremia and septicemia occur at the same time as symptoms appear. During the first week, the patient suffers from headache, malaise, and fever, probably due to an endotoxin. During the second week, the patient's condition worsens. The organisms invade many tissues, including the intestinal mucosa, and are excreted in the stools. *Salmonella typhi* thrive and multiply in bile; organisms from the gallbladder reinfect the intestinal mucosa and lymphoid tissue such as Peyer's patches. Characteristic "rose spots" often appear on the trunk and abdomen for a few days. Abdominal distention and tenderness and enlargement of the spleen are common complaints, but diarrhea usually is absent. Unlike most other infections, in which leukocytes increase in number, leukocytes decrease in number in typhoid fever. Some patients become delirious and can suffer complications such as internal hemorrhage, perforation of the bowel, and pneumonia. Chloramphenicol is the antibiotic of choice in treating typhoid fever, but some strains of *S. typhi* are resistant to it.

By the fourth week, symptoms subside, convalescence begins, and immunity develops. Cell-mediated immunity provides protection against future infection. Antibodies to surface antigens of the bacterium also are produced, but these are of more use in laboratory diagnosis than in protecting the patient against infection. The Widal test detects antibodies and is used to confirm the diagnosis of typhoid fever. A live, attenuated oral typhoid vaccine is now available. It stimulates cell-mediated as well as humoral immunity, including the production of secretory antibody (IgA), which aids in prevention of mucosal invasion. ➥ (Chapter 17, p. 469) After the initial series of doses, a booster dose is needed every 3 years. Means of protection against typhoid fever include good sanitation and sewage disposal and proper hygiene.

Shigellosis

Shigellosis (shig″el-o′sis), or **bacillary dysentery,** can be caused by several serovars (se′-ro-var), of *Shigella* (➤ Figure 22.9). They include *S. dysenteriae* (serovar A), *S. flexneri* (serovar B), *S. boydii* (serovar C), and *S. sonnei* (serovar D). The latter causes 80 percent of the cases in the United States. Shigellosis was first described in the fourth century B.C.. Although less invasive than salmonellosis, it spreads rapidly in overcrowded conditions with poor sanitation. Humans and other higher primates such as chimpanzees and gorillas are the only reservoirs of infection, but the organisms can persist in foods for up to a month. ➥ (Chapter 13, p. 348)

The pathogens are spread by contaminated food, fingers, flies, feces, and fomites. Playing, bathing, and washing clothes in contaminated water play significant roles in transmitting *Shigella*. In areas of good sanitation, infection is usually acquired by poor handwashing.

> Figure 22.9   **A colorized TEM of *Shigella*, the cause of shigellosis (bacillary dysentery).** (41,250X)

Children aged 1 to 10 are most susceptible to *Shigella*, which accounts for 15 percent of the infant diarrhea cases in the United States. For the past several decades, the total number of annual cases of shigellosis in the United States has ranged between 15,000 and 20,000. Many recent outbreaks have been recorded in day-care centers. The ingestion of just 10 organisms can be sufficient to cause infection. Thus, even slight lapses in hygiene can allow easy spread of the disease. In developing nations it is a major cause of infant mortality, with one-third of all infant deaths due to dehydration from shigellosis and other enteric pathogens. Essential fluid replacement treatment often is not available in developing countries.

Once contaminated food or water is ingested, *Shigella* survive the acidity of the stomach, pass through the small intestine, and attach to portions of it and the large intestine. The pathogens invade host cells and induce the cells to create special filaments for invading adjacent host cells. After an incubation period of 1 to 4 days, abdominal cramps, fever, and profuse diarrhea with blood and mucus suddenly appear. The severity of signs and symptoms ranges from most to least serious in the same order as the serovar designations A to D. In the most serious cases, diarrhea can cause dangerous protein deficiencies—called *kwashiorkor* (kwash-e-or'kor)—and vitamin B_{12} deficiencies, which, together with loss of electrolytes, can result in neurological damage. Along with the fever-eliciting endotoxin found in all serovars, *S. dysenteriae* also produces Shiga toxin, which acts as a neurotoxin. The neurotoxin activity is believed to be responsible for the severity and relatively high fatality rate of *S. dysenteriae* disease due to convulsions and coma. All serovars cause ulceration and bleeding of the intestinal lining and sometimes deeper intestinal layers. Symptoms persist for 2 to 7 days and usually are self-limiting, but they can cause severe dehydration and fluid and electrolyte imbalances.

Specific diagnosis of shigellosis can be difficult because the organisms are very sensitive to acids in feces. The organisms die if fecal specimens are not maintained in a buffered transport medium. Viable specimens can be obtained by swabbing a bowel lesion during internal examination of the bowel. Stool specimens are inoculated into selective media. Distinctive colonies usually appear after 24 hours of incubation.

Treatment is necessary in children and debilitated patients. Restoring fluid and electrolytes is essential to recovery. Usually hydration fluids, with or without antibiotics, are used for this purpose. A combination of antibiotics—ampicillin, tetracycline, and nalidixic acid—is used. Nalidixic acid is a synthetic agent that inhibits bacterial DNA synthesis, especially in Gram-negative intestinal pathogens. Prevention is difficult because many people have inapparent infections. A carrier state, usually lasting less than 1 month, exists and accounts for many new cases by fecal–oral transmission. Any breakdown in sanitation can lead to transmission by way of feces, fingers, and flies. Immunity following recovery from shigellosis is transient. From limited experience in recent years, oral vaccines containing certain strains of *S. flexneri* and *S. sonnei* seem to be safe and effective.

Asiatic Cholera

Asiatic cholera, so named because of its high incidence in Asia, can affect people anywhere sanitation is poor and fecal contamination of water occurs. Worldwide, more than 100,000 cases are reported annually. In the United States, fewer than 10 cases are reported per year. Some of those are transmitted to humans from contaminated shellfish in Gulf Coast states, where mild winters do not kill off the causative vibrios, leaving waters contaminated year-round. Occasionally some immunesuppressed individuals die from *Vibrio vulnificus* and *V. parahemolyticus* infections, which are ordinarily rather mild and do not cause true cholera. In endemic regions, such as parts of Asia, 5 to 15 percent of patients die; when a seasonal epidemic occurs, as many as 75 percent of patients die.

The causative organism, *Vibrio cholerae* (➤ Figure 22.10), can survive outside the body in cool alkaline water, especially if organic and/or fecal matter is present. When ingested, it invades the intestinal mucosa, multiplies, and releases a potent enterotoxin. The enterotoxin, known as *choleragen,* binds to epithelial cells of the small intestine and makes plasma membranes highly permeable to water. This action results in a significant secretion of fluids and chloride ions and the inhibition of sodium absorption. At this time the intestinal lining becomes shredded, causing numerous small, white flecks resembling rice grains to be passed in the feces. Infected

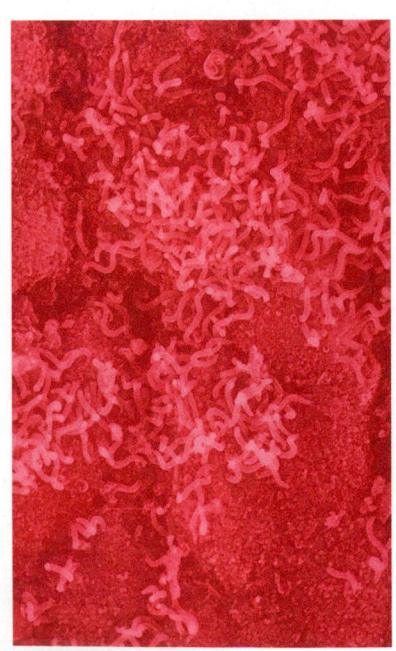

➤ Figure 22.10 **A colorized SEM of *Vibrio cholerae,* the cause of asiatic cholera.** The bacterium is slightly curved and has a single polar flagellum. It is seen as paler pink curved rods lying on darker red intestinal lining cells (3000X).

individuals experience severe nausea, vomiting, abdominal pain, and diarrhea. Stools rapidly become clear, containing numerous mucus plugs, thus giving rise to the term "rice-water stools." As many as 22 liters of fluids and electrolytes can be lost per day, so all patients, regardless of age, are subject to severe dehydration. Special "cholera cots" made of canvas with a hole cut beneath the buttocks have been used. A bucket marked in liters is placed below the hole to measure the fluid lost so that it can be replaced. Most deaths are attributable to shock due to greatly reduced blood volume.

Fluid and electrolyte replacement is the most effective treatment for cholera. During the 1971 epidemic in India and Pakistan, medical personnel were able to save large numbers of victims of the disease largely because of the availability of replacement therapy. Treatment with tetracycline reduces the duration of the symptoms but does not eliminate the organism or the toxin. Recovery from the disease confers only temporary immunity. Many recovered patients remain in a carrier state and can infect others and reinfect themselves. The only available vaccine is not very effective and is not widely used. However, a toxoid-type vaccine is being tested. It also may be possible to develop a vaccine that makes use of secreted IgA against the cholera organism.

A strain of *V. cholerae* known as the El Tor biotype (named after the quarantine camp where it was first isolated) causes a form of cholera that is slower and more insidious in its onset than the classical form of the disease. Cholera outbreaks have often resulted in the imposition of quarantine and the halting of shipping from port cities. In the past, therefore, countries in South and Central America that were economically dependent on their exports would sometimes report that cholera was not present, only "El Tor." The two forms of the disease, however, are essentially the same.

Vibriosis

An enteritis called **vibriosis** (vib-re-o′sis) is caused largely by *Vibrio parahaemolyticus*. Although the disease is most common in Japan, where raw fish is considered a delicacy, the organism is widely distributed in marine environments. In the United States, infections usually are acquired from contaminated fish and shellfish that have not been thoroughly cooked. Most outbreaks in the United States occur at outdoor festivities where crab and shrimp are served without thorough cooking and proper refrigeration. Some outbreaks have been traced to the eating of contaminated raw oysters. *Vibrio parahaemolyticus* also can infect skin wounds of people exposed to contaminated water. Once inside the intestine, the organisms colonize the mucosa and release an enterotoxin. Symptoms of nausea, vomiting, diarrhea, and abdominal pain appear about 12 hours after contaminated food or water has been ingested, and they last 2 to 5 days. The disease usually is not treated, and no vaccine is available.

Traveler's Diarrhea

Among the 250 million people who travel internationally each year, it has been estimated that over 100 million suffer from a self-limiting mild to severe diarrhea. Of these travelers, 30 percent are confined to bed, and another 40 percent are forced to curtail their activities. This disorder, officially called **traveler's diarrhea,** also has been called "Delhi belly," "Montezuma's revenge," and some even less attractive names.

The most common causes of traveler's diarrhea are pathogenic strains of *Escherichia coli,* which account for 40 to 70 percent of all cases. Strains differ with geographic location, so travelers are likely to be exposed to new strains. Some strains of *E. coli* are normal inhabitants of the human digestive tract, and only certain strains are capable of causing enteritis. **Enteroinvasive strains** have a plasmid with a gene coding for a particular surface antigen called K antigen, which enables the strains to attach to and invade mucosal cells. **Enterotoxigenic strains** contain a plasmid that enables them to make an enterotoxin. They attach to the mucosa by attachment pili or fimbriae. These organisms also cause numerous cases of infant diarrhea. Other causes of traveler's diarrhea include bacteria such as *Shigella, Salmonella, Campylobacter,* rotaviruses, and protozoa such as *Giardia* and *Entamoeba.* Jet lag and other stresses of travel do not cause diarrhea, but they can lower resistance to infection. Travelers can experience symptoms of diarrhea even when no pathogens are present. Such cases are due to unusual kinds and amounts of substances dissolved in water, which lead to gastrointestinal upsets.

Cholera epidemics may be decreased using satellites to detect algal blooms, the food of cholera-carrying copepods.

In 1994, 10,000 cases of Gatorade® were shipped to Rwandans suffering a cholera outbreak to help restore the electrolyte balance in those afflicted.

Beans, Anyone?

Troubled by intestinal gas? Well, everyone has it. The average person releases about a liter per day. This gas, called *flatus,* is produced by colon bacteria as they metabolize compounds that you have been unable to digest. Complex sugars from foods such as beans, peas, cabbage, brussels sprouts, oat bran, bananas, apples, and high-fiber cereals are broken down by some bacteria, releasing hydrogen gas plus other gases and compounds as by-products. Other intestinal bacteria consume the hydrogen. The balance between these two groups of bacteria partially explains why some people produce more gas than others.

The commercial product Beano, produced by and collected from the mold *Aspergillus niger,* contains the sugar-digesting enzyme that people lack. The enzyme, an α-galactosidase, breaks the alpha linkages in the complex sugars, digesting them before they reach the bacteria in the colon, thus reducing the amount of flatulence.

Cholera Around the World—and on Our Doorstep

In 1892 the great German hygienist Max von Pettenkofer, in the presence of witnesses, raised a broth culture of *Vibrio cholerae* to his lips and swallowed approximately 1 billion organisms. His aim was to disprove Robert Koch's studies, done in India, showing that cholera is a contagious disease spread by polluted water. Von Pettenkofer firmly believed that cholera was neither contagious nor related to drinking water. He thought the disease was due instead to interactions of microbes and soil, with "soil factors" playing the greatest role. Control of the disease would depend on the removal of these "soil factors." Within a few days, von Pettenkofer developed a mild but genuine case of cholera, and *V. cholerae* organisms were recovered from his stool. However, he insisted that his was *not* a real case of cholera. So, a few days later, his assistant repeated the experiment, fell severely ill with cholera, but recovered. The culture used in these experiments was a weakly virulent strain supplied by the bacteriologist Georg Gaffky, who had guessed at what von Pettenkofer intended to do and did not want the 74-year-old man to die.

In February 1991 Peruvians saw a similar scene broadcast on their televisions. The president of the country, Alberto Fujimori, and his wife were filmed eating raw fish and proclaiming that this popular dish was safe to eat. At that time 45,000 Peruvians had become cholera victims, and at least 193 had died from the disease. To slow the epidemic, the Peruvian Health Ministry declared that raw fish was likely to be contaminated and therefore unsafe to eat. Physicians worldwide criticized President Fujimori for his foolhardy defense of the Peruvian fisheries industry. Fujimori neglected to make it clear to his viewers that the fish he and his wife ate had been caught far out to sea, away from the sewage-contaminated coastal and river waters where poorer Peruvians caught their fish. Ironically, the Peruvian minister of fisheries, Felix Canal, attempted the same demonstration and promptly came down with cholera.

In April 1991 eight people in New Jersey contracted cholera by eating crab meat illegally imported from Ecuador. And in September 1991, Alabama state health officials closed an oyster reef near Dauphin Island, in the Gulf of Mexico, that they said was teeming with the deadly strain of *V. cholerae* that was causing the outbreak in South America. Thankfully, health officials swiftly caught and managed these two outbreaks. But cholera can quickly veer out of control, causing untold suffering and death.

Cholera had existed on the Indian subcontinent for centuries before the first Europeans arrived. Portuguese explorers described it early in the sixteenth century, but it did not invade other areas until 1817. The world is now gripped by the seventh cholera pandemic to have swept the globe. The United States experienced the second, third, and fourth of these, beginning in 1832, 1849, and 1866, respectively. The fourth pandemic left more than 50,000 Americans dead, with outbreaks continuing until about 1878. In 1887 and 1892 boatloads of infected passengers arrived in New York City as part of the fifth pandemic. However, a drastic shift in knowledge and attitudes in the United States prevented its spread. During the 1832 outbreak, cholera was viewed as a punishment from God. Indeed, health authorities resisted pleas to clean up streets and water supplies; to do so, they felt, would be to oppose God's will. Pigs were allowed to wander the New York streets, and they left feces and garbage in their wake. In contrast, by 1866 most intelligent physicians realized that cholera was a contagious disease spread by filth and ignorance. During that pandemic, physicians and city officials learned a great deal about control measures. Implementing these measures prevented future pandemics from spreading extensively into the United States.

The current pandemic, caused by the El Tor strain of *V. cholerae,* began in Indonesia in 1958. From there it spread to Macao, Hong Kong, and the Philippines and reached eastern Europe by 1965. It is still rampant in parts of Africa and

---

Symptoms of traveler's diarrhea vary from mild to severe and include nausea, vomiting, diarrhea, bloating, malaise, and abdominal pain. A typical case causes 4 to 5 loose stools per day for 3 or 4 days. Fluid loss is greater with invasive strains than with toxigenic strains. The disease is especially hazardous in infants, who are subject to severe dehydration. Bottle-fed infants are much more likely to become infected than breast-fed infants. Before their normal microflora are sufficiently established to compete with the pathogens, newborns are especially at risk of acquiring pathogenic strains from hospital workers. After infancy, children most often acquire *E. coli* infections during travel in foreign countries, especially those with poor sanitation.

Travelers sometimes medicate themselves with antibiotics before and during a stay in a foreign country. This treatment is not recommended because antibiotics usually are not effective and because such use contributes to the development of antibiotic-resistant strains. A better practice is to keep antidiarrhea medicine available and to use it only after symptoms appear.

Traveler's diarrhea can persist for months or years as a postinfectious irritable bowel syndrome. It can also cause lactose intolerance by damaging intestinal lining cells that normally produce the enzyme lactase, which digests lactose (milk sugar). During bouts of diarrhea, patients should eliminate from their diet dairy products and other foods containing lactose. After

In December 1997, the Food and Drug Administration at last approved the use of radiation for eliminating harmful microorganisms such as *E. coli* from red meat.

Asia. In 1989 sub-Saharan Africa accounted for three-quarters of all cases in the world. In the summer of 1994, tribal conflicts in the central African nation of Rwanda claimed more than 500,000 lives, and 3 million Rwandans fled to surrounding nations, including Zaire, Burundi, and Tanzania. There, the refugees set up primitive camps while they waited for the slaughter to subside. With no running water, no sanitation facilities, and little food, the refugees succumbed to cholera, dysentery, and malaria before worldwide relief efforts came to their aid. About 50,000 Rwandan refugees contracted cholera in the camps, and thousands died due to lack of treatment. With proper treatment, cholera victims can recover, but treatment requires clean water for victims to drink to replace lost fluids. In late July 1994 the United States and Canada joined the relief efforts by sending physicians, food, clean water, and portable latrines to the region.

Curbing the disease is a different matter where sanitation remains primitive over the long term. Health officials fear that cholera will establish itself as an endemic disease in South America and will bring death for decades to poor areas that lack proper sewage and water treatment systems. Carried by ship from Africa, cholera reached a port city north of Lima, Peru, in January 1991. It spread through the untreated drinking waters and via infected people and contaminated food to Lima. Within 3 months, some 175,000 people had sickened, and 1258 had died. Because only about one-fourth of infected people show symptoms, vast numbers of Peruvians must have been infected. Meanwhile, cholera spread into nearby countries, including Colombia, Ecuador, Chile, Guatemala, Brazil, Mexico—and even the United States. Because the current vaccine protects only half of those who receive it, and then only for 6 months, the World Health Organization is not recommending mass vaccination campaigns.

Many South American countries have instituted educational programs about cholera transmission and offer practical solutions to the sanitation problems. For instance, the Peruvian government now uses television and radio to broadcast information about boiling water, proper handling of food, and the necessity of disposing of garbage and sewage away from drinking water supplies. The government has distributed millions of chlorination tablets to its citizens. In addition, physicians and nurses take special courses about the treatment of cholera, and food handlers are trained in workshops about how to handle food safely. As a result of this massive educational effort, many areas of Peru have witnessed a decrease in the number of cholera cases.

Overcoming ingrained habits, however, is not easy. One man in Peru said he would not boil his water because boiled water doesn't quench his thirst. Children continue to play in sewage outfalls, and the livestock that roam the streets in many Peruvian cities are an ongoing threat to the health of its citizens. The fight against cholera is also a battle against ignorance, but with proper education, perhaps this fight can be won.

A Rwandan refugee fills his water jug and women wash clothes in the same lake in which the body of a cholera victim decomposes in 1994.

a few weeks, small quantities of such foods can be reintroduced and gradually increased as long as they are well tolerated. In some cases normal lactose tolerance is lost forever. *Escherichia coli* and other bacteria, the protozoan *Giardia,* and certain helminths frequently cause lactose intolerance.

Escherichia coli has significance far beyond its ability to cause diarrhea. It is an important indicator organism because it is always present in water contaminated with fecal material. *Escherichia coli* is usually more numerous than other organisms and is easier to isolate. Finding *E. coli* in water indicates that any pathogens found in feces might also be present. Deadly outbreaks of toxin-producing *E. coli* strain 0157:H7 in 1992 and since have been attributed to undercooked hamburgers served at fast-food restaurants. Several cases involved the kidneys and resulted in the condition known as *hemolytic-uremic syndrome.*

Escherichia coli is also an extremely versatile opportunistic pathogen—it can infect any part of the body subject to fecal contamination, including the urinary and reproductive tracts and the abdominal cavity after perforation of the bowel. It is present in many bacteremias, causes septicemias, and can infect the gallbladder, meninges, surgical wounds, skin lesions, and lungs, especially in debilitated and immunodeficient patients.

E. coli can survive easily for 2 months on a dry stainless-steel work surface, and 2 to 5 months in natural bodies of water.

Other Kinds of Bacterial Enteritis

Certain strains of *Campylobacter jejuni* and *C. fetus* can be found in food and water and are becoming increasingly associated with human gastroenteritis, especially in infants and in elderly or debilitated patients. Some strains of *C. fetus* cause infectious abortions in several kinds of domestic animals. Although these organisms apparently do not multiply in foods, they are transmitted passively in undercooked chicken, unpasteurized milk, and poultry held in unchlorinated water during processing. Improper cooking of contaminated poultry can lead to enteritis. In some areas *Campylobacter* surpasses *Salmonella* as the major cause of foodborne enteric disease. Many health departments have begun testing for *Campylobacter* in food-handling areas and on equipment.

Campylobacter infections cause copious diarrhea, foul-smelling feces, fever, and abdominal pain. They also cause arthritis in 2 to 10 percent of infected children but rarely in adults. Because large quantities of fluid can be lost, dehydration and fluid and electrolyte imbalances are common among the populations most affected. The disease is treated with fluid and electrolyte replacement and sometimes with tetracycline and erythromycin.

Yersiniosis (yer-sin″e-o′sis), a severe enteritis, is caused by *Yersinia enterocolitica*. The disease is most common in western Europe, but some cases are seen in the United States. This free-living organism is found mainly in marine environments, but it can survive in many places.

(a) Normal gastric mucosa

(b) Invasion by *Helicobacter pylori*

Mucus layer
Helicobacter pylori
Epithelium
Blood vessel

(c) Gastritis

Gastric acid
Dissolving epithelium
Lymphocyte
Neutrophil

Cofactor?

(d) Peptic (gastric) ulcer

(e) Stomach (gastric) ulcer

➤ **Figure 22.11 Stomach (gastric) ulcers, caused by the bacterium *Helicobacter pylori*.** (a) Healthy stomach lining is (b) invaded by *Helicobacter pylori,* a spiral-shaped bacterium. Recent research indicates that the organism is the cause of (c) chronic gastritis and (d) peptic ulcers and (e) may be involved in stomach cancer. Antibiotic treatment can lead to permanent cure of ulcers if reinfection does not occur.

Infection can be acquired from water, milk, seafoods, fruits, and vegetables, even when they are refrigerated, because the organism grows more rapidly at refrigerator temperatures than at body temperature. The organism is most easily identified when cultured at 25°C, a temperature at which the organisms are motile and easily distinguished from other bacteria. Yersiniosis symptoms, which are related to release of an enterotoxin, are similar to other kinds of enteritis, but abdominal pain usually is more severe, and white blood cells increase in number. Yersiniosis sometimes is misdiagnosed as appendicitis because of the similarity of symptoms.

A new source of *Yersinia* infection has caused the CDC to issue a warning to preparers of the Southern delicacy chitterlings, or chitlins (pork intestines). Until the preparers have very carefully washed their hands, they should avoid touching children or anything used by children. Children should not be allowed to handle raw chitterlings. Fifteen children in Atlanta recently became ill, mainly from contact with preparers, whereas the adults remained well.

Bacterial Infections of the Stomach, Esophagus, and Intestines

Peptic Ulcer and Chronic Gastritis

Recent studies have revealed a bacterial cause of peptic ulcers and chronic gastritis and a probable cofactor of stomach cancer (➤ Figure 22.11). The organism, *Helicobacter pylori* (formerly called *Campylobacter pylori*) (➤ Figure 22.12), was first cultured in 1982 from gastric biopsy tissues by Barry Marshall and J. Robin Warren. It is able to survive the very acidic conditions of the stomach by generating ammonia from urea. The ammonia is believed to neutralize gastric acidity around the *Helicobacter* cells, thereby allowing the organisms to survive and reproduce. They colonize and multiply in the gastric mucosa directly above the epithelial cell layer of the stomach.

Peptic ulcers are lesions of the mucous membranes lining the esophagus, stomach, or duodenum. The lesions are caused by the sloughing away of dead inflammatory tissue and exposure to acid; they eventually result in an excavation into the surface of the organ. Four million Americans suffer from ulcers each year. Approximately 10 percent of the population will suffer from an ulcer at some point in their lives. Ulcers are responsible for about 46,000 operations and 14,000 deaths per year.

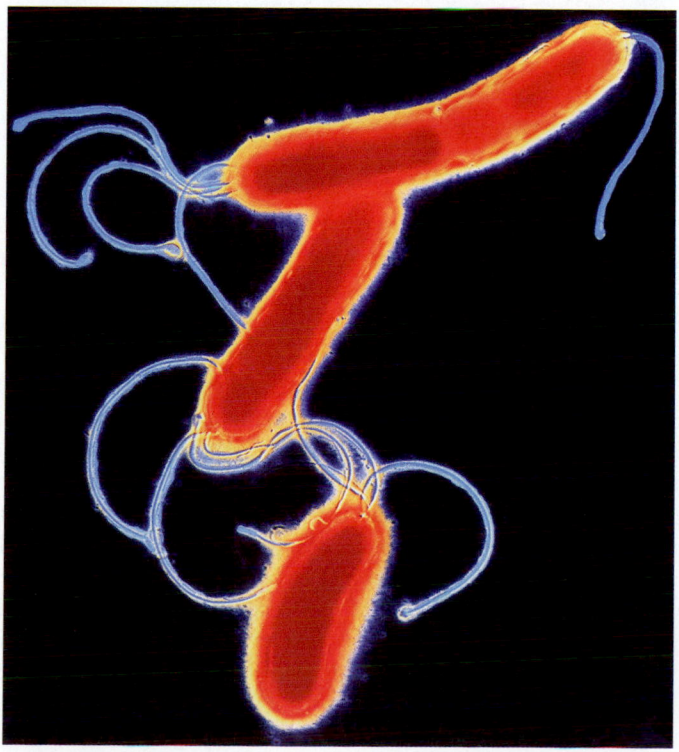

The Bacterium That Just Won't Quit

Helicobacter pylori really is a nasty organism. As if causing ulcers (and maybe even stomach cancer) weren't enough, one study found *H. pylori* in 62 percent of people with heart disease. In comparison, *H. pylori* affected only 40 percent of matched control subjects. That's cause enough for greater concern. Add to that the fact that roughly one-third of people in the United States and Western Europe harbor *H. pylori,* with the proportion even higher in Asia and developing countries, and the fact that *H. pylori* is developing resistance to metronidazole, the potent drug often used to fight this bacteria. Resistance to metronidazole among infected people is estimated at 40 percent in Europe and more than 70 percent in the developing world. Let's hope that science will be able to stay one step ahead of these organisms.

The adhesin used by Helicobacter pylori to bind to the stomach lining has been identified. This may make a vaccine to prevent H. pylori's colonization possible.

➤ **Figure 22.12** *Helicobacter pylori.* Stomach (gastic) ulcers are now known to be caused by this bacterium. The ulcers can be cured by killing the bacteria with antibiotics. Note the knob at the end of each flagellum. These make it possible for *H. pylori* to move through the thick mucus that coats the stomach lining. Other bacteria with normal flagella are unable to move forward and get trapped in the mucus.

The Saga of Typhoid Mary

1901—Mary Mallon is working as a cook for a family. A visitor with typhoid fever arrives, and Mary develops typhoid fever. One month later the family laundress falls ill with typhoid.

1902—Mary changes employers. Two weeks later the laundress develops typhoid and is soon followed by six more members of the household. Mary quickly leaves.

1903—Now working as a cook in a household in Ithaca, New York, Mary is believed to have started a waterborne typhoid outbreak that spreads widely, killing 1300 people. Mary quickly leaves.

1904—Mary moves on to a household on Long Island; within three weeks, four fellow servants develop typhoid.

1906—Moving again, Mary arrives, and six people fall ill with typhoid, all within 1 week. One girl dies. Two weeks later, Mary—having been suspected as the cause—flees to another family. After another 2 weeks, the family's laundress develops typhoid fever.

1907—Using a false name, Mary starts work as a cook for a family in New York City. Two months later, two household members develop typhoid; one dies. Mary leaves soon after the disease breaks out, but this time the New York City Health Department tracks her down. In March 1907 she refuses to cooperate with Health Department workers who want to test her stool for the presence of typhoid bacilli that would mark this otherwise healthy woman as a carrier of typhoid. Police arrive and forcibly remove her to a hospital on an island in the East River. Tests reveal typhoid bacilli in her stool specimens; presumably they are being shed from her bacilli-colonized gallbladder. No antibiotics are available at this time—nor will they be until the late 1940s. Surgical removal of Mary's gallbladder is the only way to end her carrier state. Not a believer in the germ theory of disease, Mary is certain that she will die under the surgeon's knife, all for foolish nonsense, and refuses to undergo the surgery. She is sentenced to involuntary isolation in the hospital until such time as she is no longer a menace to public health.

1907–1910—Mary's stools are cultured every few days. Some days there are no typhoid bacilli; other days huge numbers appear. Meanwhile, Mary has attained public celebrity status—imprisoned against her will, threatened by the surgeon's knife!

1910—Public sympathy leads to Mary's release. She promises never again to work as a cook.

1915—Twenty-five cases of typhoid occur at the Sloane Hospital for Women in New York City. Eight die—mostly doctors and nurses. The hospital cook steps out "for a few minutes," never to return. She does not use the name Mary Mallon. But when the Health Department catches up with her, it is indeed Mary again, insisting on her right to be employed as a cook—the only job she likes.

It is back to the island hospital again for Mary. On arrival at the quarantine facilities, she offers to work in the kitchen—an offer authorities refuse. Mary steadfastly refuses a gallbladder operation and forcefully insists that she will work as a cook if she gets out. What to do? Mary remains quarantined there for the rest of her life.

1938—Mary Mallon dies at age 70 from a stroke.

Had you been a Health Department official, a judge, a lawyer, or someone else associated with Mary, what would you have done? Before antibiotics and vaccines were available, quarantine was a common practice.

Data linking chronic H. pylori infection to stomach cancer have led the International Agency for Research on Cancer to classify it as a Group I carcinogen.

Chronic gastritis (stomach inflammation) may be so mild as to cause no noticeable signs or symptoms, or it can produce pain and indigestion. It is observed in 70 to 95 percent of people who have peptic ulcers. Severe gastritis may lead to ulceration.

Helicobacter pylori is present in 95 percent of patients with duodenal ulcers and in 70 percent of those with gastric ulcers. In developed countries, it is uncommon for children to be infected, but the rate of infection increases by about 1 percent per year of age above 20, eventually reaching an average of 20 to 30 percent among U.S. adults. In developing countries, it is commonly found in children and reaches levels of 80 percent of the adult population. In Latin America, stomach cancer rates are the highest in the world, and so is the rate of *H. pylori* infection. Interestingly, U.S. Hispanics also have a 75 percent infection rate. In Japan, rates of stomach cancer and of *H. pylori* infection are dropping together. It may be that just having chronic inflammation of the stomach for a long period of time—which occurs when a person suffers *H. pylori* infection early in life—predisposes toward stomach cancer. Thus some scientists are unwilling to call *H. pylori* a definite cause of stomach cancer but feel it may be only a cofactor because most people infected with *H. pylori* never develop stomach cancer. However, among the various kinds of stomach cancer, the most common type (gastric intestinal) has an 89 percent correlation with *H. pylori* infection.

No one is certain yet of the route of infection or the portal of exit. It is of great importance, however, if we find that we can prevent or cure ulcers by eliminating *H. pylori* infection. Currently available drugs such as Tagament only control, but cannot cure, ulcers. Experimental treatments with antibiotics have reported cure rates as high as 80 to 90 percent, but these drugs must be used very carefully. Patients treated only with drugs that suppress stomach acid had a relapse rate of 75–95 percent within 2 years. Researchers expect to have better tests, drugs, and treatment plans worked out in a few years.

Approaches to *H. pylori* identification include direct detection of the organism in gastric biopsies, testing for the presence of the enzyme urease in biopsy tissue, and culturing specimens on special media. Serological tests for specific antibody against *H. pylori* have been developed and are undergoing evaluation.

Pseudomembranous Colitis

Pseudomembranous colitis, a condition characterized by the formation of a membrane covering on the mucosal surface of the colon, is caused by *Clostridium difficile*. The signs and symptoms, diagnosis, and treatment for *C. difficile* disease were discussed in Chapter 13. ➡ (p. 359)

Bacterial Infections of the Gallbladder and Biliary Tract

The gallbladder and its associated ducts can also be sites of bacterial infection. The liver produces bile, which is stored in the gallbladder and slowly released into the small intestine. Organisms with lipid-containing envelopes are destroyed by the action of bile, which breaks down lipids. Most enteric viruses, poliovirus, and hepatitis A virus (discussed in the next section) lack lipid envelopes and are protected from bile activity. Organisms such as the typhoid bacillus are so resistant to the action of bile that they can actually grow in the gallbladder itself. They are shed from the gallbladder into the intestine and are eliminated in feces. People carrying these organisms in the gallbladder are asymptomatic.

Gallstones, formed from crystals of cholesterol and calcium salts, can block the bile ducts, decreasing the flow of bile and predisposing the individual to inflammation of the gallbladder (*cholecystitis*) or biliary ducts (*cholangitis*). Distension of the gallbladder due to the accumulation of bile fosters the entry of microbes—most commonly *E. coli*—into the bloodstream, probably through minute tears in the gallbladder wall. Infection of the gallbladder can also ascend into the liver.

Diagnosis of gallbladder infections is based on clinical findings of recurring pains (*biliary colic*), nausea, vomiting, chills, fever, and often jaundice due to the absorption of blocked bile into the bloodstream. Despite prompt treatment with antibiotics, gallbladder infections cannot be cured unless the causative obstruction is removed, either by surgery or by the spontaneous passage of the gallstone.

Gastrointestinal diseases caused by bacteria are summarized in Table 22.2.

Gastrointestinal Diseases Caused by Other Pathogens

Viral Gastrointestinal Diseases

Viral Enteritis

Rotavirus infection is a major cause of **viral enteritis** among infants and young children. Rotaviruses are transmitted by the fecal–oral route, replicate in the intestine, damage the intestinal epithelium, and cause a watery diarrhea within 48 hours. Rotavirus infection is a major cause of infant morbidity and mortality in developing countries. There 3 to 5 billion cases occur annually, with 5 to 10 million deaths in children under age 5. Rotavirus infections account for a third of childhood deaths in some countries. Rotavirus infections also occur in infants and young children in developed countries, where the infections are often nosocomial. Special care should be used with hospitalized children to prevent such infections. The number of cases rises

---

✓ Distinguish between diarrhea and dysentery.

✓ Is cholera a problem in the Western hemisphere? Explain.

✓ Why are travelers advised not to take antibiotics before and during a trip, in order to avoid traveler's diarrhea?

✓ What causes stomach (gastric) ulcers? How can they be cured?

Intestinal bacteria have been linked to chronic gallstone formation. Antibiotics halt this until discontinued. Then stone formation resumes.

Another Plug for Breast Feeding

If you're a female, and even if you're a male, you've probably heard more than you ever wanted to know about the benefits of breast feeding. You probably know that breast milk allows antibodies to pass from the mother to her infant. But you might be interested to know that one of these antibodies acts specifically against rotavirus, the most common cause of diarrhea in infants, and that a complex carbohydrate in breast milk affords babies even more protection. How? The compound, called lactadherin, doesn't break down in the baby's stomach. Instead, it acts like it's one of the natural carbohydrates found in the child's intestines. The rotavirus attaches to the lactadherin, believing that it's anchoring itself in the gut, and then gets flushed out of the child's system as the lactadherin passes through.

> Table 22.2   **A summary of gastrointestinal diseases caused by bacteria**

Disease	Agent(s)	Characteristics
Bacterial food poisoning		
Staphylococcal enterotoxicosis	*Staphylococcus aureus*	Heat-stable enterotoxin causes tissue damage, abdominal pain, nausea, vomiting
Other kinds of food poisoning	*Clostridium perfringens, C. botulinum, Bacillus cereus*	Diarrhea and sometimes intestinal infection and gas gangrene
Bacterial enteritis and enteric fevers		
Salmonellosis	*Salmonella typhimurium, S. enteritidis*	Abdominal pain, fever, diarrhea with blood and mucus from toxin; enterocolitis from invasion of organisms; chronic infections and carrier states occur
Typhoid fever	*Salmonella typhi*	Organisms invade mucosa and lymphatics, multiply in phagocytes and other tissues; high fever and "rose spots"; carrier state and life-threatening complications can occur
Shigellosis	*Shigella* strains	Organisms cause intestinal lesions and release toxins; symptoms include cramps, fever, profuse diarrhea with blood and mucus
Asiatic cholera	*Vibrio cholerae*	Organisms invade intestinal lining, release potent toxin that increases lining permeability; symptoms include nausea, vomiting, copious diarrhea, fluid imbalances
Vibriosis	*Vibrio parahaemolyticus*	Organisms colonize mucosa and release toxin; nausea, vomiting, diarrhea are self-limiting
Traveler's diarrhea	Pathogenic strains of *Escherichia coli,* other bacteria (also viruses and protozoa)	Organisms can invade mucosa and/or produce toxin, cause nausea, vomiting, diarrhea, bloating, malaise, abdominal pain; self-limiting except for postinfection complications; dehydration and death in infants
Other kinds of bacterial enteritis	*Campylobacter jejuni, C. fetus, Yersinia enterocolitica*	*Campylobacter* species cause enteritis in infants and debilitated patients; *Yersinia* releases a toxin that causes enteritis with pain resembling appendicitis
Bacterial infections of the upper gastrointestinal tract		
Ulcers, stomach cancer	*Helicobacter pylori*	Definite association with ulcers; probable co-factor in stomach cancer
Cholecystitis, cholangitis	Usually *E. coli*	Blockage of bile ducts by gallstones causes inflammation of gallbladder and bile ducts; accumulation of bile can cause infection to spread to bloodstream or liver
Pseudomembranous colitis	*Clostridium difficile*	Formation of a pseudomembrane on the mucosal surface of the colon; occurs in patients whose gastrointestinal microflora have been altered by the use of antibiotics

dramatically during the winter in the United States. This timing helps distinguish it from bacterial diarrheas.

Rotaviruses are reoviruses that replicate their double-stranded RNA in such great numbers that they can easily be found by electron microscopy in fecal suspensions (> Figure 22.13); no effort to concentrate the viruses is required. Immunoelectron microscopy (immunologic techniques combined with electron microscopy) can be used to identify rotaviruses when antibodies from a patient's serum are reacted with virions in diagnostic specimens. In addition, many hospitals now use ELISA tests that detect rotaviruses in stool specimens. Although there is no specific treatment, restoring fluid and electrolyte balance is crucial and should be prompt.

Rotavirus infection in humans is important in medical practice in the United States. Antibodies to the virus have been identified in as many as 90 percent of groups of children tested, even though the infection may not have been identified at the time it occurred. Much more research is needed to determine how immunity is produced and how the disease might be controlled.

Enteritis can also be caused by viruses other than rotaviruses. Species of *Enterovirus* such as the echoviruses (*e*nteric *c*ytopathic *h*uman *o*rphan *v*iruses) can cause mild gastrointestinal symptoms and damage intestinal cells. They also sometimes infect other tissues, and certain species can cause meningoencephalitis (inflammation of the brain and meninges).

Patients who have received bone marrow transplants are especially susceptible to infection with rotaviruses, enteroviruses, and the bacterium *Clostridium difficile*. As many as 55 percent of such patients succumb to these infections.

The *Norwalk virus* (named for a 1968 outbreak in Norwalk, Ohio) is responsible for nearly half of all U.S. outbreaks of acute infectious nonbacterial enteritis. Norwalk virus infection affects older children and adults more often than it affects preschoolers or infants. Outbreaks occur throughout the year and are common at schools, camps, and nursing homes and on cruise ships. The infection is the second most common cause of illness (after respiratory disease) among U.S. families and occurs worldwide. It is characterized by 1 to 2 days of diarrhea, vomiting, or both. Immunity does not follow an attack, which makes development of a vaccine unlikely. Careful sanitary practices are the best means of prevention.

Hepatitis

Hepatitis, an inflammation of the liver, usually is caused by viruses (➤ Figure 22.14 and Table 22.3). It also can be caused by an amoeba and various toxic chemicals. The most common viral hepatitis is **hepatitis A,** formerly called **infectious hepatitis.** It is caused by the hepatitis A virus (HAV), a single-stranded RNA virus usually transmitted by the fecal–oral route. **Hepatitis B,** formerly called **serum hepatitis,** is caused by the hepatitis B virus (HBV), a double-stranded DNA virus usually transmitted via blood. A third type of hepatitis is transmitted parenterally (by blood) and is probably caused by at least two viral agents. This type is diagnosed in the absence of HAV and HBV as **hepatitis C (HCV),** formerly called *non-A, non-B (NANB) hepatitis.* A fourth type of hepatitis, transmitted by the fecal–oral route and formerly called *non-A, non-B, non-C hepatitis,* has been separated out as **hepatitis E (HEV).** An especially severe form of the disease **hepatitis D,** or **delta hepatitis,** is caused by the presence of both hepatitis D virus (HDV) and HBV. However, HDV alone does not cause disease; it cannot infect without HBV.

Hepatitis A

Hepatitis A occurs most often in children and young adults, especially in autumn and winter. It can occur in epidemics if a population is subjected to water or food, especially shellfish, contaminated with HAV. Outbreaks due to contaminated food in fast-food restaurants have been on the rise.

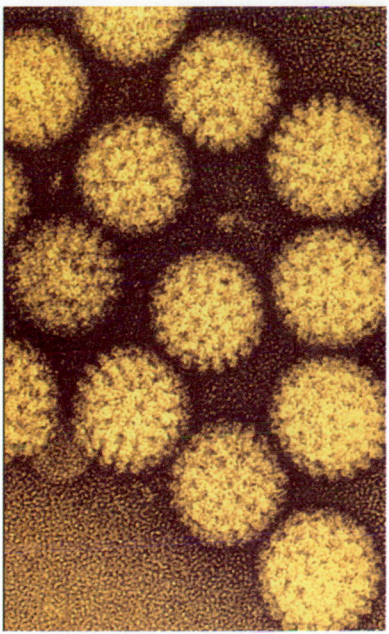

➤ Figure 22.13   **Rotaviruses.** Seen here as large yellow spheres in fecal suspension (625,000X), they resemble little wheels and are a cause of diarrhea.

➤ Figure 22.14   **Hepatitis A virus.**

> ➤ Table 22.3   **Comparison of types of viral hepatitis**

Characteristic	Hepatitis A	Hepatitis B	Hepatitis C	Hepatitis D	Hepatitis E
Alternate names	Infectious hepatitis; epidemic hepatitis; short-term hepatitis	Serum hepatitis	Parenterally transmitted non-A, non-B, post-transfusion hepatitis	Delta hepatitis	Enterically transmitted non-A, non-B, non-C hepatitis
Agent	HAV RNA virus, Picornaviridae	HBV DNA virus, Hepadnaviridae	HCV At least two unclassified RNA viruses; ? Flavivirus, ? Togavirus	HDV Defective RNA virus; has hepatitis B capsid	HEV Unclassified RNA virus; ? Calcivirus
Transmission	Fecal–oral	Blood and other body fluids; crosses placenta with high frequency	Blood and blood products; occasionally crosses placenta	Blood, must coinfect or superinfect with hepatitis B; can cross the placenta	Fecal–oral; more common in adults than in children
Incubation period	15–40 days; average; 28 days	45–180 days; average, 90 days	Short, 2–4 weeks; long, 8–12 weeks	2–12 weeks	2–6 weeks
Severity of disease	Self-limiting; usually mild, rarely severe	Subclinical to severe; most recover completely	Subclinical to severe; most resolve spontaneously	Severe; high mortality rate	Moderate but high mortality in pregnant women
Carrier state	No	Yes, is associated with 80% of liver cancer	Yes, possible association with liver cancer	Yes	No
Chronic liver disease	No	Yes	Yes	Yes	No
Vaccines	Yes	Yes	No	No	No

Hepatitis A has an incubation period of 15 to 40 days and begins as an acute febrile illness. After entering the body through the mouth, the viruses (an RNA picornavirus) replicate in the gastrointestinal tract and spread through the blood to the liver, spleen, and kidneys. Jaundice, a yellowing of the skin common in hepatitis, is caused by impaired liver function: The liver fails to rid the body of a yellow substance called **bilirubin,** which is a product of the breakdown of hemoglobin from red blood cells. Other symptoms of hepatitis are malaise, nausea, diarrhea, abdominal pain, and lack of appetite for a period of 2 days to 3 weeks. Chronic infections are rare, and recovery usually is complete. Immunological tests are available to detect hepatitis A viruses and host antibodies against them. There is no treatment for hepatitis other than alleviating symptoms. A vaccine for hepatitis A is now available. Gamma globulin injections are also used to provide temporary immunity.

Hepatitis B

Hepatitis B occurs in people of all ages with about the same incidence throughout the year. It can be transmitted by intravenous or percutaneous (into the skin) injections, by anal/oral sexual practices (common among homosexual males), contact with other virus-containing body secretions (including semen and breast milk), and by contaminated needles among intravenous drug users. Health care workers who have routine contact with patients' body fluids (especially blood) have a higher incidence of the disease than does the general community. Transmission via contaminated semen in artificial insemination has been documented.

Hepatitis B has an incubation period of 45 to 180 days, with an average of 90 days. The virus replicates in cells of the liver, lymphoid tissues, and blood-forming tissues. It can persist in the blood for years, thus creating a carrier state. The onset of symptoms is insidious, and fever is uncommon. Otherwise the symptoms are similar to those of hepatitis A, except that chronic active hepatitis B frequently destroys liver cells.

Immunological methods are available to detect hepatitis B virus and the host's antibodies. Treatment relieves some symptoms but does not cure the disease. An effective vaccine is available, and government regulations require that it be provided by employers for health care workers who may have contact with blood or body fluids that might contain HBV. Sexually active teenagers and adults should also be vaccinated. Pediatricians recommend vaccination of all preteens, as parents can not be sure of when their teenagers may become sexually active. Babies in the United States are now routinely immunized. However, most of the older population remains unprotected. The vaccine is safe. It is produced by recombinant DNA technology—the insertion of the appropriate hepatitis B virus genes into a plasmid. The plasmid is then inserted into yeast cells, thereby avoiding all contact with human cells. This yeast-produced vaccine has been given to over 2 million Americans and is 95 percent effective. When given to hepatitis B-infected pregnant women, it reduces from 90 to 23 percent the number of infants who become carriers. This number can be reduced to 5 percent when gamma globulin is given along with the vaccine. Some 40 percent of carriers die of liver disease, and in some parts of the world nearly 90 percent of mothers are infected. Health officials in the United States currently recommend that all infants be vaccinated at birth. But such vaccination is expensive and is probably beyond the means of many developing countries, where it is most important.

The hepatitis B virus, a member of the Hepadnaviridae family, is unusually stable and resists drying and irradiation. Its double-stranded, circular DNA has a gap in one strand that may help it insert into liver cell DNA. Insertion of viral DNA into liver cell DNA, in turn, may contribute to liver cell carcinoma, a kind of cancer that occurs much more frequently in people who have had hepatitis B than in the general population.

Hepatitis C

Hepatitis C virus has been recovered from parenterally transmitted non-A, non-B hepatitis cases. Because hepatitis C disease has two different incubation periods—2 to 4 weeks and 8 to 12 weeks—some researchers believe that there may be two different causative agents, both RNA viruses: one a flavivirus, the other a togavirus. Hepatitis C can be distinguished from other kinds of hepatitis by the high blood concentration of a liver enzyme, alanine transferase. Various enzymes are released into the blood from damaged liver cells in all types of hepatitis, but this particular enzyme is unusually elevated in HCV hepatitis. Although the disease usually is mild or even inapparent, the infection can be severe in compromised individuals and becomes chronic in about half those infected. No vaccine is available, and immunity does not follow infection.

Hepatitis E

Transmitted through fecally contaminated water supplies, hepatitis E has caused large outbreaks in Asia and Africa. It is more common in adults than in children. The mortality rate is low (1 percent), except in pregnant women, for whom it is about 20 percent. No vaccine is available, and immunity does not follow infection. It is caused by an RNA virus, probably a calcivirus.

Hepatitis D

Hepatitis D has an incubation period of 2 to 12 weeks. That period is shorter when HBV carriers are superinfected with HDV than when individuals are infected with both viruses at the same time. HDV alone fails to cause disease because it requires HBV antigens for replication. HDV and HBV together can result in death.

The hepatitis virus story has not ended. Recently three more viruses have been discovered that are associated with hepatitis-like disease. If these viruses turn out to be causative agents for true hepatitis cases, it has been proposed that they be named hepatitis F, G, and H viruses. Hepatitis G, curiously enough, seems to spread by both the fecal–oral route and the blood route. Some investigators believe that there may be dozens more viruses that can cause little-known forms of hepatitis.

Best Stunt Yet

Daredevil Evel Knievel's stunts didn't always go as planned. Multiple accidents led to many trips to the hospital, often involving blood transfusions (over 200 units of blood in the course of his lifetime). Now 59 years old, Knievel waits for a liver transplant. He suffers from hepatitis.

One or more of those units of blood probably transmitted the hepatitis C virus to Knievel. In his own words, he describes the virus: "It attacks the liver like a rattlesnake." "They put it in me unknowingly. That's just the breaks of life. I blame no one for it." He spends most of his time publicizing the virus, emphasizing the need to find a cure. But finding that might be a bigger stunt than any of those Knievel ever performed.

Protozoan Gastrointestinal Diseases

Giardiasis

The flagellated protozoan *Giardia intestinalis* (➤ Figure 22.15a) was first observed by Leeuwenhoek in 1681 when he was studying organisms in his own stools. It is, however, a far older organism. Examination of *Giardia's* DNA has shown it to be the most primitive DNA of any eukaryote, very similar to that of older prokaryotes.

Giardia infects the small intestine of humans, especially children, and causes a disorder called **giardiasis** (je´ar-di´a-sis). After cysts ingested from fecal material pass through the stomach and small intestine, motile trophozoites are released in the colon. ➡ (Chapter 11, p. 294) The parasite has an adhesive disk by which it attaches to the bowel wall (➤ Figure 22.15b). It feeds mainly on mucus, forming cysts that are deposited in the mucus and passed intermittently in mucous stools. Symptoms of giardiasis include inflammation of the bowel, diarrhea, dehydration, and weight loss. Nutritional deficiencies are common in infected children because the parasites can occupy much of the intestinal absorptive area. Fat absorption is greatly reduced, and deficiencies of fat-soluble vitamins are common. Diarrhea is copious and frothy from bacterial action on unabsorbed fats, but it is not bloody because the parasitic protozoa usually do not invade cells. Some infected persons experience severe joint inflammation and an itchy rash even before they have diarrhea. This *reactive arthritis of Giardia* fails to respond to anti-inflammatory drugs ordinarily used to treat arthritis. The arthritis disappears along with the diarrhea when antiprotozoan drugs are given.

Giardiasis is transmitted through food, water, and hands contaminated with fecal matter. It is occasionally transmitted from wild animals and is thus called "beaver fever" by backpackers and hunters in the western United States. Contaminated water supplies in Aspen, Colorado, Leningrad, Russia, and probably many other places have caused large numbers of cases. In some child-care centers, up to 70 percent of the children are infected with the pathogen. *Giardia* cysts are not killed by ordinary sewage treatment and chlorination. Some localities in Pennsylvania are being forced to drink bottled water until they can afford to add sand filters to their treatment plants to trap *Giardia* cysts.

Diagnosis is made by microscopic examination and finding cysts of the protozoan in stools. *Giardia* is found in the mucous sheets that line the intestinal surface. These sheets are released

> A little-known reservoir of the protozoan *Giardia intestinalis* is the garden, especially if the garden is located somewhere near a major river.

(a)

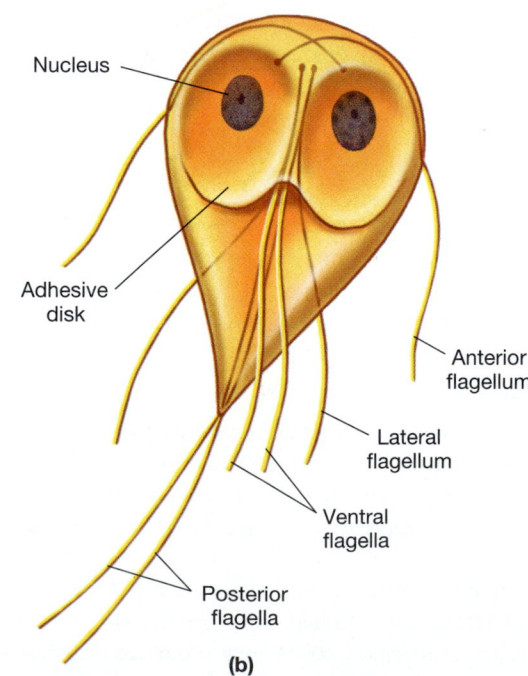

Nucleus

Adhesive disk

Anterior flagellum

Lateral flagellum

Ventral flagella

Posterior flagella

(b)

➤ Figure 22.15   *Giardia intestinalis*, **a protozoan parasite that causes diarrhea.** (a) SEM (35,600X). (b) The structure of *G. intestinalis*.

every few days. Samples such as those taken from a bedpan should include some mucus. Because of the intermittent passing of cyst-containing mucus, daily stool samples for several days are needed to increase the likelihood of a positive diagnosis. Trophozoites are sometimes found in watery stools. Immunofluorescent antibody techniques and ELISA procedures are also used for diagnosis.

Quinacrine (Atabrine), furazolidine, and metronidazole (Flagyl) are used to treat giardiasis. The disease can be prevented by maintaining pure water supplies uncontaminated by human or animal wastes.

Amoebic Dysentery and Chronic Amebiasis

Amebiasis (am-e-bi'a-sis) is caused by *Entamoeba histolytica* (➤ Figure 22.16), a major pathogenic amoeba. It was originally isolated from intestinal ulcers of a patient who had succumbed to severe diarrhea, used to infect a dog in which the same disease appeared, and recovered from the dog, thereby demonstrating Koch's postulates. ➥ (Chapter 14, p. 381) Amebiasis can appear as a severe acute disease called **amoebic dysentery** or as **chronic amebiasis,** which can suddenly revert to the acute stage. Approximately 400 million people are infected worldwide, most with chronic amebiasis. The proportion of the infected population varies from 1 percent in Canada to 5 percent in the United States to 40 percent in tropical areas.

➤ **Figure 22.16** *Entamoeba histolytica* **trophozoite.** Active amoebic form; 1550X.

Humans become infected with the parasite by ingesting cysts in food or water contaminated with fecal matter. After ingestion and passage through the stomach and small intestine, cysts rupture and release amoeboid trophozoites in the colon. Trophozoites reproduce asexually within the colon. There they feed on bacteria that make up the normal microflora of the large intestine. They may cause very little trouble to the host, or they may invade the intestinal mucosa, where they can live indefinitely. Once they have invaded the intestinal mucosa, the parasites multiply and cause significant ulceration. Sometimes their proteolytic enzymes digest deep into, or even through, the bowel wall. Thus, the protozoa sometimes enter blood vessels and travel to other tissues, or they allow bacteria in fecal material to enter the body cavity and cause peritonitis. Patients with amebiasis have abdominal tenderness, 30 or more bowel movements per day, and dehydration from excessive fluid loss. If the parasites invade liver and lung tissue, they can cause abscesses. Bacterial infection of lesions in any tissues can occur.

Because fecal material becomes dehydrated as it passes through the colon, *E. histolytica* trophozoites tend to encyst there. The cysts are passed with the feces. Cysts can survive up to 30 days in a cool, moist environment and are not killed by normal chlorine concentrations in water. The most common means of transmission is through the fecal–oral route. Flies and cockroaches can also be mechanical vectors. Infection can be acquired through sexual practices in which there is ingestion of fecal material.

Amoebic infections can be diagnosed by finding trophozoites or cysts in stools, but several stool samples on consecutive days may be needed to find them. Immunofluorescent antibody and ELISA procedures are also available for diagnosis. Metronidazole is widely used in treatment, even though it has been found to be mutagenic in bacteria and carcinogenic in rats. Antibiotics also are used to prevent or cure secondary bacterial infections. Such infections can be prevented by sanitary handling of water and food.

Balantidiasis

Balantidium coli (➤ Figure 22.17) is the only ciliated protozoan that causes human disease. It is distributed worldwide, particularly in the tropics, but human infection is rare except in the Philippines. *Balantidium coli* is transmitted by cysts in fecal matter. After being ingested, cysts rupture and release trophozoites that invade the walls of the large intestine, causing a dysentery known as **balantidiasis** (bal"an-tid-i'a-sis). Symptoms of the disease are similar to those of amoebic dysentery; as with amoebic dysentery, perforation of the intestine from balantidiasis can lead to fatal peritonitis.

Diagnosis is made by finding trophozoites or cysts in fecal specimens. Tetracycline or metronidazole are used to treat the disease, but some people remain carriers even after treatment. As with other organisms transmitted through fecal matter, infection can be prevented by good sanitation. Pigs serve as a reservoir of infection, so contact with their feces must be avoided.

(a)

Food vacuole

Cytostome (mouth)

Macronucleus

Micronucleus

Cilia

Contractile vacuole

(b)

> Figure 22.17 *Balantidium coli.* This very large ciliated protozoan causes diarrhea. It is the only ciliate known to infect humans. (a) It is photographed here in a fecal smear. (b) The structure of *B. coli.*

Diarrhea hit 403,000 residents of Milwaukee, Wisconsin—4,400 badly enough to be hospitalized—in 1993 when *Cryptosporidium* contaminated the city water supply.

When Napoleon was in the Ukraine, he received bad grain and his horses became afflicted with ergot poisoning.

Hold the Aflatoxin!

That peanut butter and jelly sandwich you packed for lunch might give you more than the sugary protein-filled energy rush you hoped for. Aflatoxin can be present in peanut butter if any moldy peanuts are used to make it. In one study, 7 percent of peanut butter samples tested contained aflatoxin. Similar toxins can be present in jellies. Even when the moldy top layer of jelly is discarded, some toxins may have diffused down into the jelly—in other words, don't scrape the top off and eat what's underneath! If, in addition, moldy bread is used, you may be eating a potentially highly carcinogenic sandwich for lunch.

Cryptosporidiosis

Protozoans of the genus *Cryptosporidium* commonly cause opportunistic infections worldwide, probably by fecal–oral transmission from kittens and puppies. These organisms live in or under the membrane of cells lining the digestive and respiratory systems. After being swallowed, cysts burst in the intestine, releasing protozoa that invade intestinal cells or migrate to other tissues. In 1993 an outbreak of watery diarrhea affected some 403,000 Milwaukee residents; the outbreak was attributed to *Cryptosporidium* infection of one of the city's water treatment plants. In immunocompetent individuals, the disease is self-limiting, but in immunosuppressed patients, it causes severe diarrhea—up to 25 bowel movements per day with a loss of as much as 17 liters of fluid. The majority of severe cases of **cryptosporidiosis** (krip″to-spor-id-e-o′sis) have been seen in AIDS patients. No effective treatment has been found.

Effects of Fungal Toxins

Fungi produce a large number of toxins, and most come from members of the genera *Aspergillus* and *Penicillium*. Their various effects on humans include loss of muscle coordination, tremors, and weight loss. Some are carcinogenic.

Aspergillus flavus and other aspergilli produce poisonous substances called **aflatoxins** (af″lah-tox′inz). Aflatoxins are the most potent carcinogens yet discovered. Although the effects of the toxins on humans are not fully understood, their presence in foodstuffs may cause cancer of the liver. The toxins reach humans in food made from mold-infested grain and peanuts.

Claviceps purpurea is a fungus that grows parasitically on rye and wheat (➤ Figure 22.18). **Ergot** (er′got) is the name of the vegetative structure *C. purpurea* forms as well as the name

(a)

Ergotamine

(b)

(a)

➤ **Figure 22.18   Ergot producing fungus.** (a) Micrograph of a cross-section through the parasitic fungus *Claviceps purpurea* (65X), which produces ergot. (b) The structure of ergotamine, a toxic alkaloid produced by the ergot-producing mold.

of the toxin and the disease it causes in grains. Although most strains of these grains grown in the United States are genetically resistant to the ergot fungus, many strains grown in other parts of the world are not. Ergot causes a variety of effects on humans. When the fungus is harvested with rye and incorporated into foodstuffs, it can cause **ergot poisoning,** or *ergotism*—hallucinations, high fever, convulsions, gangrene of the limbs, and ultimately death. The same substance that causes ergot poisoning in large quantities can be used therapeutically in small quantities. Drugs derived from ergot are used in carefully measured doses to control bleeding in childbirth, induce abortions, treat migraine headaches, and lower high blood pressure.

Mushroom toxins are found mainly in various species of *Amanita,* which are widely distributed around the world (➤ Figure 22.19). The mushroom toxins phallotoxin and amatoxin act on liver cells. They cause vomiting, diarrhea, and jaundice. Ingestion of a sufficient quantity of the toxin can be lethal or can so damage the liver that an organ transplant is necessary.

Helminth Gastrointestinal Diseases

A wide variety of helminths can parasitize the human intestinal tract, and some also invade other tissues. Although most are prevalent only in tropical regions, several are endemic to the United States. Health care workers in the United States should be alert to the possibility that patients may have acquired such parasites while traveling in tropical areas.

Fluke Infections

Foodborne fluke infections affect 40 million people worldwide. The sheep liver fluke *Fasciola hepatica* (➤ Figure 22.20) is the most thoroughly studied of all flukes. It is found in humans in South America, Cuba, northern Africa, and some parts of Europe. Its intermediate host is a snail. Cercaria (larval forms) develop in the snail and upon their release mature in water and encyst as metacercaria on water vegetation. When humans eat such vegetation, especially watercress, the metacercaria are released in the large intestine, bore through the intestinal wall, and migrate to the liver. There they feed on blood, block bile ducts, and cause inflammation.

(b)

➤ **Figure 22.19   Deadly toxin-producing mushrooms.** *Amanita* mushrooms kill by means of a toxin that inhibits RNA polymerase. (a) *Amanita muscaria,* which was formerly used as an insecticide. It was sprinkled with sugar and set out to attract flies, which died after nibbling on the insecticide-laced sugar. (b) *Amanita virosa,* commonly called the "destroying angel."

✓ Which types of hepatitis can be prevented by vaccines?

✓ Who should receive hepatitis vaccines?

✓ What is aflatoxin? Its source? Its effects?

> Figure 22.20   *Fasciola hepatica,*
the sheep liver fluke. This specimen
has been stained to reveal its internal
structures (2.5X).

Sometimes they migrate to the eyes, brain, or lungs. Adult flukes can be found in the bile ducts and the gallbladder. They live about 10 years. Infections can be diagnosed by finding eggs in stool specimens. Infected humans can be treated with bithionol and other antihelminthic agents. Infection can be prevented by not eating water vegetation unless it has been cooked.

The Chinese liver fluke *Clonorchis sinensis* is widely distributed in Asia. As much as 80 percent of the population in rural areas is infected; some travelers and users of raw imported products also become infected. The life cycle is similar to *Fasciola*'s except that a second intermediate host, typically a fish but sometimes a crustacean, is required. Metacercaria excyst (emerge from a cyst) in the duodenum and migrate to the liver, and adult flukes take up residence in bile ducts. They destroy bile duct epithelium, block ducts, and sometimes perforate and damage the liver. The incidence of liver cancer is unusually high in areas where fluke infections are high, but it is unknown whether the fluke is responsible. Finding eggs in feces confirms the diagnosis, but there is no effective treatment. The parasite could be eradicated by cooking fish and crustaceans. However, the cultural habit of eating raw fish and shellfish and the lack of fuel for cooking maintain infections.

Yet another fluke, *Fasciolopsis buski,* is common in pigs and humans in the Orient. It lives in the small intestine and causes chronic diarrhea and inflammation. If several flukes are present, they can cause obstruction, abscesses, and **verminous intoxication,** an allergic reaction to toxins in the flukes' metabolic wastes. Antihelminthic drugs can be used to rid the body of flukes. Human infection can be prevented by controlling snails, avoiding uncooked vegetation, and stopping the use of human excreta for fertilizer.

Tapeworm Infections

Human tapeworm infections can be caused by several species, most of which have worldwide distribution. Illustrations of the tapeworm life cycle were provided in Chapter 11. ➡ (p. 307) Humans are most often infected by eating uncooked or poorly cooked, contaminated pork or beef. Tapeworm infections can also occur through contact with infected dogs or from eating infected raw fish.

The pork tapeworm *Taenia solium* (➤ Figure 22.21a) reaches a length of 2 to 7 m, and the beef tapeworm *T. saginata* reaches a length of 5 to 25 m. These worms usually enter the body as larvae in raw or poorly cooked meat, especially pork. Viable larvae can develop into adult tapeworms. When adult tapeworms develop in the intestine, they absorb large quantities of nutrients and lead to malnutrition even when the person has an adequate diet. Long, ribbonlike worms may tangle up into a mass that blocks passage of materials through the intestine.

When humans ingest tapeworm eggs (or ova) instead of larvae (➤ Figure 22.21b), the egg coverings disintegrate in the small intestine, and the released larvae penetrate the intestinal wall and enter the blood. In 60 to 70 days, a larva migrates to various tissues and develops into a **cysticercus,** or bladder worm. ➡ (Chapter 11, p. 308) The bladder worm consists of an oval white sac with the tapeworm head invaginated into it. In humans, cysticerci are frequently found in the brain, where they can reach a diameter of 6 cm. When pigs ingest tapeworm eggs, the cysticerci migrate to muscles and encyst. The pigs' defenses cause the encysted larvae to be surrounded by calcium deposits. In humans, calcification fails to stop the growth of cysticerci; if vital organs are involved, patients may suffer paralysis and convulsions. When a cysticercus dies, it releases toxins and usually causes a severe, or even fatal, allergic response. Human tapeworm infections can be prevented by the sanitary disposal of human wastes and by thorough cooking of meats and fish. Even freezing meats at −5°C for at least a week appears to kill the parasites.

Humans ingest eggs of *Echinococcus granulosus* tapeworms through contact with infected dogs, especially when the dogs lick the faces of small children. Eggs of this tapeworm are especially likely to produce cysts, called **hydatid cysts** (➤ Figure 22.21c), in vital tissues such as the liver, lungs, and brain. ➡ (Chapter 11, p. 308) The cysts, which can contain hundreds of tiny immature worm heads and often reach the size of a grapefruit or larger, exert pressure on organs. If a cyst ruptures, as may happen during an attempt at surgical removal, it can release all these infective units. A ruptured cyst can also cause severe allergic reactions such as anaphylactic shock.

Humans can ingest *Hymenolepis nana* tapeworm eggs in cereals or other foods that contain parts of infected insects. Such tapeworms in the intestine, especially in children, can cause diarrhea, abdominal pain, and convulsions.

(a)

(b)

(c)

➤ **Figure 22.21** **Tapeworms.** (a) A colorized SEM of the scolex, or head, of *Taenia solium,* showing suckers and hooks, which aid in attaching to the intestinal lining of its host. (b) Proglottids, or body segments, of *T. pisiformis* (2X). Small new ones grow behind the scolex; they increase in size as they age and move backward away from the scolex. The last row of proglottids is filled with mature eggs. (c) Protoscolices of *Echinococcus granulosus.* Each round structure, the size of a grain of sand, contains the head of a tapeworm and can grow into a complete worm. As many as millions of such structures are found in a fluid-filled hydatid cyst, which can contain up to 15 liters of fluid. When even a more moderately sized cyst forms in the brain, it can do extensive damage.

The broad fish tapeworm, *Diphyllobothrium latum,* is common in fish-eating carnivores. It reaches humans through ingestion of raw or poorly cooked, contaminated fish in sushi and other dishes. Fish tapeworm infections are common in Scandinavia, Russia, and the Baltic—approaching 100 percent infestation of the population in some areas—and have occurred in the Great Lakes area of the United States. The worm requires as intermediate hosts both a small crustacean and a fish to complete its life cycle. When humans ingest infected fish tissue, the worms coiled up in the fish muscle reach and mature in the intestine. Adult worms attach to the intestine and begin producing eggs. While attached, the parasites absorb large quantities of vitamin B_{12} and impair the victim's ability to absorb the vitamin. Vitamin B_{12} deficiency, or pernicious anemia, from tapeworm infections is especially high in Finland.

Tapeworm infections are diagnosed by finding eggs or proglottids in feces. Infections can be treated with niclosamide and other antihelminthic agents. Prompt diagnosis and treatment are important to remove worms before they invade tissues beyond the intestine. Human infections could be completely prevented by avoiding raw meats and fish and infected dogs.

Trichinosis

Trichinosis (trik″in-o′sis) is caused by the small roundworm *Trichinella spiralis,* also sometimes called *trichina worm.* This parasite, unlike most, is more common in temperate than in tropical climates. (Almost all U.S. adults have antibodies to *Trichinella* and therefore carry around a few worms. But a few will not cause symptoms.) The parasite usually enters the digestive tract as encysted larvae (➤ Figure 22.22) in poorly cooked pork, but infections have been traced to venison and meat from other game animals and to horse meat in France. In the intestine, cysts release larvae that develop into adults. The adults mate, the males then die, and the females produce living larvae before they too die. The larvae migrate through blood and lymph vessels to the liver, heart, lungs, and other tissues. When they reach skeletal muscles, especially eye, tongue, diaphragm, and chewing muscles, they form cysts. The formation of cysts in humans represents a dead-end for the worms, as they will not be eaten and pass on to another host. Encysted worms remain alive and infectious for years.

➤ **Figure 22.22** *Trichinella spiralis.* This specimen is curled up as a cyst, embedded in striated muscle fibers (1000X).

These parasites cause tissue damage as adults and as migrating and encysted larvae. The adult females penetrate the intestinal mucosa and release toxic wastes that produce symptoms similar to those of food poisoning. Wandering larvae damage blood vessels and any tissues they enter. Death can result from heart failure, kidney failure, respiratory disorders, or reactions to toxins. Encysted larvae cause muscle pain. Trichinosis is difficult to diagnose, but muscle biopsies and immunological tests are sometimes positive. Treatment is directed toward relieving symptoms because the disease cannot be cured. It can be prevented by eating only thoroughly cooked meat. Freezing does not necessarily kill encysted larvae, and microwave cooking is safe only if the internal temperature of the meat reaches 77°C. Microwave cooking depends on geometry. Because pieces of meat are irregularly shaped, they must be rotated during cooking to heat evenly. Studies of pork experimentally infected with *Trichinella* show survival of some worms whenever microwave cooking is done without rotation. By law, pigs in the United States must not be fed garbage or restaurant scraps unless it has been cooked first. This prevents raw pork scraps, containing live trichina worms, from being recycled back into pigs, which they will then infect. Therefore, it is not very likely that you will get trichinosis from U.S.-raised pork. The average number of cases diagnosed per year in the United States is approximately 100.

Hookworm Infections

Hookworm is most often caused by one of two species of small (11 to 13 mm long) roundworms—*Ancylostoma duodenale* and *Necator americanus*. Although these parasites have a complex life cycle, that cycle can occur in a single host, and the host is often a human. Eggs in feces quickly hatch in moist soil. There the eggs release free-living larvae that feed on bacteria and organic debris, grow, molt, and become mature parasitic larvae. If these larvae reach the skin, typically of the feet or legs, they burrow through it to reach blood vessels that carry them to the heart and lungs. The larvae then penetrate lung tissue, and some are coughed up and swallowed. In the intestine the larvae burrow into villi and mature into adult worms. The adult worms mate and start the cycle over again. Worldwide, an estimated 500 million people are infested with hookworms.

As hookworm larvae burrow through the skin, host inflammatory reactions kill many of them, although bacterial infection of penetration sites causes **ground itch.** In the lungs the parasites cause many tiny hemorrhages, but they cause the greatest damage to the lining of the entire small intestine. They feed on blood and cause abdominal pain, loss of appetite, and protein and iron deficiencies, so people infected with hookworms often appear to be lazy. These effects are especially debilitating in people whose diets are barely adequate without the burden of worm infections.

Diagnosis is made by finding eggs or worms in feces, but samples must be concentrated to find them. Tetrachloroethylene is effective against *Necator* infections. It is inexpensive and easy to administer in mass treatment efforts. Bephenium hydroxynaphthalate (Alcopar), mebendazole, and several other drugs, although more expensive, kill both species of hookworms. Dietary supplements, especially iron, should be provided for all hookworm patients. Hookworm is preventable through sanitary disposal of human wastes, but stopping the use of human excreta as fertilizer and getting uneducated people to use latrines can be difficult. Plantation workers often defecate repeatedly in areas near fields where they work. If infected, such individuals serve as a continuous source of larvae to themselves and to others.

In 1991, cases of human hookworm infection by the dog hookworm *Ancylostoma caninum* (➤ Figure 22.23) were linked with intestinal problems such as diarrhea, abdominal pain, and weight loss. Larvae of other species of hookworm for which humans are not the normal host sometimes penetrate the skin and cause *cutaneous larva migrans,* or *creeping eruption.* Severe skin inflammation results from body defenses that prevent further migration of the parasites. Such infections often are acquired from infected cats and dogs and can be treated with thiabendazole.

Ascariasis

Ascaris lumbricoides (➤ Figure 22.24) is a large roundworm 25 to 35 cm long that causes **ascariasis** (as″kar-i′a-sis). People become infected by ingesting food or water contaminated with *Ascaris* eggs. Once in the intestine, the eggs hatch, and larvae penetrate the intestinal wall and enter lymph vessels and venules. Although the larvae can invade and cause immunological reactions in almost any tissue, most move through the respiratory tract to the pharynx and are swallowed. Larvae move to the small intestine, mature, and begin to produce eggs. A single female

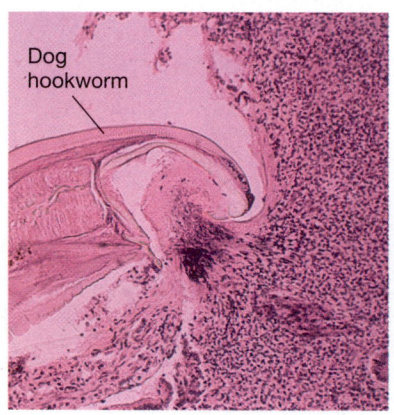

Dog hookworm

➤ Figure 22.23   **Dog hookworm, *Ancylostoma caninum.*** This roundworm is about 1 cm long, attached to intestinal epithelium, where it sucks blood (250X). It has taken a large piece of intestine into its mouth. (Heavy infestations may result in anemia.)

> ➤ Figure 22.24 *Ascaris lumbricoides*, **a large roundworm.** Females may reach a length of 35 cm and can produce several hundred eggs per day. The eggs are passed out in feces and can survive in soil for months or even years. These worms were all removed from one child.

The intestinal tracts of many cats dissected in A&P labs are infested with Ascaris lumbricoides, whose eggs may still be viable.

worm produces an average of 200,000 eggs per day, with a total lifetime output of about 26 million eggs. A female lives 12 to 18 months. Eggs are especially resistant to acids and can develop in 2 percent formalin. They also resist drying, and people can be infected by airborne eggs. In some areas of the southern United States where the soil never freezes, 20 to 60 percent of children are infected, usually with 5 to 10 worms. Worldwide, 25 percent of the population is infested with *Ascaris* worms. Mebendazole and pyrantel are effective in treatment.

Ascaris worms cause three kinds of damage:

1. Larvae burrowing through lung tissue cause *Ascaris* pneumonitis, which involves hemorrhage, edema, and blockage of alveoli from worms, dead leukocytes, and tissue debris. If secondary bacterial pneumonia develops, it can be fatal.

2. Adult worms cause malnutrition, but because they feed mainly on the contents of the intestine, they do little damage to the mucosa. They also release toxic wastes that elicit allergic reactions. If sufficiently numerous, they cause intestinal blockage and sometimes perforation. The peritonitis that follows perforation is nearly always fatal.

3. Wandering worms cause abscesses in the liver and other organs and sometimes traumatize victims or doctors examining them for other disorders by crawling from body openings such as the nose or umbilicus.

Diagnosis is made by finding eggs or worms in feces. The adult worms can be eradicated from the body by piperazine, mebendazole, and several other drugs, but no treatment is available to rid the body of larvae. Piperazine temporarily relaxes the worm's muscles. Failing to move against the current, and to press closely against the wall of the intestine, allows peristalsis to carry the worms out through the anus. Infection is fully preventable by good sanitation and personal hygiene.

Toxocara species, another type of roundworm, ordinarily parasitize cats and dogs. It has been estimated that in the United States, 98 percent of puppies, including those from good kennels, are infected. Infection rates may be almost as high in dogs and cats of any age. *Visceral larva migrans* is the migration of larvae of these parasites in human tissues such as liver, lung, and brain, where they cause tissue damage and allergic reactions. The risk of such human infections can be minimized by worming pets periodically, disposing of pet wastes carefully, and keeping children's sandboxes inaccessible to pets.

Trichuriasis

Trichuriasis (trik″u-ri′a-sis) is caused by the **whipworm,** *Trichuris trichiura,* which is distributed nearly worldwide. It is estimated that nearly 300 million people are infected, some of them in the southeastern United States. For humans to be infected, human feces must be deposited on warm, moist soil in shady areas. Small children, who put dirty hands in their mouth, are particularly susceptible to infection. Eggs deposited with feces contain partially developed embryos. When the eggs are swallowed, they hatch. Juveniles crawl into enzyme-secreting glands of the intestine called crypts of Lieberkühn, where they develop. They return to the intestinal lumen (central space), where they reach full maturity within 3 months of initial infection.

Adult whipworms damage the intestinal mucosa and feed on blood. They cause chronic bleeding, anemia, malnutrition, allergic reactions to toxins, and susceptibility to secondary

Sushi

In Japan, several hundred people each year find out what it is like to ingest a live worm with their tasty snack. Typically they consume sushi at dinner and awake in the small hours of the morning with such agonizing pain that they are rushed to the hospital. There, examination of the stomach with a fiber-optic gastroscope shows *Anisakis* larvae penetrating the gastric or duodenal mucosa. In some cases, the damage by this roundworm is so severe that a portion of the stomach wall must be removed. In more fortunate cases, the worm larvae can be removed by forceps attached to the gastroscope. This infection is called *anisakiasis* (an″is-a-ki′a-sis). There is no treatment for it other than removal of the larvae.

What are the odds that a sushi dinner will give you anisakiasis in the United States? Only a half dozen or so cases have been reported in the United States; curiously, these cases tended to involve problems somewhat higher up in the digestive tract. A California man had eaten raw white sea bass 10 days before he felt a peculiar sensation at the back of his throat. Coughing, he reached into his mouth and pulled out a lively 7.5-cm-long worm. The other U.S. victims have similarly coughed up and removed their own worms—one as quickly as 4 hours after a meal of cod fillet. Still, considering the number of people in the United States who eat raw or undercooked fish, the total number of diagnosed cases of anisakiasis is so low that there is not much to worry about—so far!

bacterial infection. Infections can result in rectal bleeding in children. Infection is diagnosed by finding worm eggs in stools. The drug mebendazole is effective in ridding the intestine of parasites. Sanitary disposal of wastes is essential to prevent reinfection.

Strongyloidiasis

Strongyloidiasis (stron″jil-oi-di′a-sis) is caused by *Strongyloides stercoralis* (➤ Figure 22.25). This parasite is unusual in that females produce eggs by *parthenogenesis*—that is, without fertilization by a male. In fact, no male of this species has ever been reliably identified in the parasitic stage. Adult females, about 2.2 mm long by 0.04 mm wide, attach to the small intestine, burrow into underlying layers, and release eggs containing noninfective larvae. Many eggs hatch in the intestine and are passed with the feces. In soil, larvae can become free-living adults or develop into infective larvae and penetrate the skin of new hosts. Infective larvae, which penetrate the skin, are carried by blood to the lungs. There they bore their way to the trachea, travel to the pharynx, and are swallowed. When they arrive in the small intestine, they develop into adults and restart the life cycle.

The *Strongyloides* larvae cause itching, swelling, and bleeding at penetration sites, which often become infected with bacteria. Migrating larvae cause immunological reactions in the host, but the reactions usually do not stop the parasites. Coughing and burning sensations in the chest occur in lung infections, and burning and ulceration occur in intestinal infections. Secondary bacterial infection in any tissue can lead to septicemia. The diarrhea and fluid loss associated with intestinal parasites is severe and difficult to control even with electrolyte therapy. Consequently, patients often die of complications such as heart failure or paralysis of respiratory muscles. Immunosuppressed patients can die when worms are disseminated beyond the gut, into lungs and meninges.

Strongyloides infection in humans usually occurs when larvae are encountered in contaminated soil or water. Diagnosis is difficult because larvae can be found in fecal smears only in massive infections. Work is in progress to develop a reliable immunological test. The drugs thiabendazole and cambendazole have the greatest effect on the parasites with the fewest undesirable side effects.

Pinworm Infections

Pinworm infections are caused by the small roundworm *Enterobius vermicularis*. Like the hookworm, this parasite can complete its life cycle without an alternate host. In fact, humans are its only known host. It has the greatest geographic distribution of any human worm parasite in the world. While it is more common in lower socioeconomic groups, it can be found even in the seats of the mighty. An estimated 209 million people worldwide are infested with pinworms; 18 million of these in the United States and Canada. Adult pinworms attach to the epithelium of the large intestine and mate, and the females produce eggs. Egg-laden females containing 11,000 to 15,000 eggs migrate toward the anus during the night, release their eggs on the exterior of the anus, and then crawl back in. Lower temperature and aerobic environment are the factors that trigger egg laying. These eggs are easily transmitted to other people by bedclothes, by debris under fingernails of those who scratch the itchy area around the anus, and even by inhalation of airborne eggs. Ingested eggs hatch in the small intestine and release larvae that mature and reproduce in the large intestine. Some few worms may travel upward to the stomach, esophagus, and nose. Others may reach the bladder, the uterus, and via the fallopian tubes the peritoneal cavity.

Although pinworm infection usually is not debilitating, it does cause considerable discomfort and can interfere with adequate rest and nutrition, especially in children. Infection with large numbers of the worms can cause the rectum to protrude from the body. Pinworm infections are diagnosed by finding eggs around the anus; at night or right after the host awakens, the pinworms can be picked up with the sticky side of transparent cellophane tape (translucent will not work) affixed to a wooden tongue depressor. If one member of a family has pinworms, all are presumed infected and are treated with piperazine or another antihelminthic agent. These agents are generally inexpensive and nontoxic. Bed linens, clothing, and towels should be washed and the house thoroughly cleaned at the time of treatment. The treatment and cleaning are repeated in 10 days. Despite these efforts, reinfection in families is very likely. In the absence of reinfection, the infection is self-limiting and will cease without treatment. Parents sometimes need to be reminded that this is a nonfatal infection. (➤ Figure 14.2, p.376).

Gastrointestinal diseases caused by pathogens other than bacteria are summarized in Table 22.4.

➤ Figure 22.25 *Strongyloides stercoralis.* This is a roundworm parasite of the small intestine (250X).

✓ Describe the life cycle of a tapeworm. Where do the stages of cysticercus and hydatid cysts fit in the life cycle?

✓ Can you safely cook pork in a microwave oven?

✓ Describe the technique used to diagnose a case of pinworms.

➤ **Table 22.4** **A summary of gastrointestinal diseases caused by pathogens other than bacteria**

Disease	Agent(s)	Characteristics
Viral gastrointestinal diseases		
Viral enteritis	Rotaviruses	Viral replication destroys intestinal epithelium; causes diarrhea and dehydration that can be fatal under age 5
Hepatitis A	Hepatitis A virus	Viral replication in intestinal and other cells causes malaise, nausea, diarrhea, abdominal pain, lack of appetite, fever, and jaundice; usually self-limiting; vaccine available
Hepatitis B	Hepatitis B virus	Viral replication and symptoms similar to hepatitis A, except onset is insidious and fever usually absent; chronic infections and carrier state occur; transmitted parenterally; vaccine available
Hepatitis C	Unknown, but may involve two agents	Mild symptoms, but disease often chronic
Hepatitis D	Hepatitis D and B viruses	Can result in fatal hepatitis
Hepatitis E	Hepatitis E virus	Moderate disease, but high mortality in pregnant women
Protozoan gastrointestinal diseases		
Giardiasis	*Giardia intestinalis*	Parasite attaches to intestinal wall; feeds on mucus and causes inflammation, diarrhea, dehydration, nutritional deficiencies, and sometimes reactive arthritis
Amoebic dysentery	*Entamoeba histolytica*	Parasites ulcerate mucosa and cause severe acute diarrhea, abdominal tenderness, and dehydration; parasites can live indefinitely in intestine, causing latent amebiasis
Balantidiasis	*Balantidium coli*	Organisms invade walls of intestine; cause dysentery, and sometimes perforation and peritonitis
Cryptosporidiosis	*Cryptosporidium* species	Organisms live in or under mucosal cells and cause severe diarrhea in immunocompromised patients
Helminth gastrointestinal diseases		
Fluke infections	*Fasciola hepatica, Clonorchis sinensis, Fasciolopsis buski*	Organisms excyst in intestine and migrate to liver, block ducts, and damage tissues; *F. buski* causes intestinal obstruction, abscesses, and verminous intoxication
Tapeworm infections	*Taenia solium, T. saginata, Echinococcus granulosis, Hymenolepis nana, Diphyllobothrium latum*	Most infections caused by ingesting encysted larvae, which mature and erode intestinal mucosa; ingestion of eggs allows larval forms to develop in humans as cysticerci or hydatid cysts that can damage the brain and other vital organs
Trichinosis	*Trichinella spiralis*	Larvae excyst in intestine, mature, and produce new larvae, which migrate to various tissues and encyst in muscles, where they cause pain; larvae cause tissue damage, and adults release toxins
Hookworm infections	*Ancylostoma duodenale, Necator americanus*	Larvae penetrate skin and migrate via blood vessels to heart and lungs; coughed-up larvae enter digestive tract and burrow into villi; mature worms feed on blood, cause abdominal pain, loss of appetite, and protein and iron deficiencies
Ascariasis	*Ascaris lumbricoides*	Eggs hatch in the intestine, and larvae cause immunological reactions in many tissues; adults cause malnutrition by feeding on nutrients in gut; wandering worms cause abscesses
Trichuriasis (whipworm)	*Trichuris trichiura*	Eggs hatch in the intestine, and juveniles invade crypts of Lieberkühn and mature; adults damage intestinal mucosa, causing chronic bleeding, anemia, malnutrition, and allergic reactions to toxins
Strongyloidiasis	*Strongyloides stercoralis*	Filariae penetrate the skin, bore to trachea, climb to pharynx and are swallowed; mature worms bore into intestine and release embryos; cause inflammation, bleeding, and immunologic reactions at various sites, severe diarrhea and fluid loss
Pinworm infections	*Enterobius vermicularis*	Ingested eggs give rise to mature worms in the intestine; infection interferes with nutrition, especially in children

Retracing Our Steps

The agents and characteristics of the diseases discussed in this chapter are summarized in Tables 22.1, 22.2, and 22.4. Information in those tables is not repeated here.

Components of the Digestive System

- The **digestive system** is an elongated tube comprised of: the mouth, pharynx, esophagus, stomach, and intestines, along with the accessory organs: teeth, salivary glands, liver, gallbladder, and pancreas.

- The five major functions of the digestive system are:
  1. *movement* of food through the tract
  2. *secretion* of digestive juices and mucus
  3. *digestion* of food
  4. *absorption* of digested food into the bloodstream
  5. *elimination* of undigested food components and intestinal microflora.

- Nonspecific defenses include mucus, stomach acid, Kupffer cells in liver sinusoids, patches of submucosal lymphatic tissue, and competition from normal flora.

Normal Microflora of the Mouth and Digestive System

- Perhaps a thousand species of microbes live within the human oral cavity and over half of these have not yet been identified. Their ecology is poorly understood.

- The esophagus lacks a permanent normal microflora, as does the stomach. The first two-thirds of the small intestine contain mainly transient microbes, but the last third's surface is colonized mainly by Enterobacteriaceae and obligate anaerobes. In the large intestine, feces are approximately 50 percent bacteria, mostly *Bacteroides.*

- Bacteria of the large intestine produce important metabolic by-products, such as vitamin K.

Diseases of the Oral Cavity

Bacterial Diseases of the Oral Cavity

- **Dental plaque** is a continuously formed coating on teeth that consists of microorganisms in organic matter.

- **Dental caries** (tooth decay) is the chemical breakdown of enamel and deeper parts of the teeth. Caries can be prevented by good oral hygiene, use of **fluoride,** and application of sealants.

- **Periodontal disease** involves the gums, periodontal ligaments, and alveolar bone and can be prevented by prevention of plaque buildup. It is treated with plaque removal techniques, mouth rinses, peroxide—sodium bicarbonate, surgery, or antibiotics.

Viral Diseases of the Oral Cavity

- **Mumps** is transmitted by saliva or as an aerosol and occurs worldwide, mainly in children. It can be prevented by a vaccine.

Gastrointestinal Diseases Caused by Bacteria

Bacterial Food Poisoning

- **Food poisoning** is caused by ingesting food containing pre-formed toxins. It can be prevented by sanitary handling and proper cooking and refrigeration of foods. Bacterial toxins in foods cause intoxications.

- Staphylococcal **enterotoxicosis** usually results from eating contaminated, poorly refrigerated foods, especially dairy and poultry products.

- Other kinds of food poisoning usually arise from eating undercooked, contaminated meats and gravies and rice.

Bacterial Enteritis and Enteric Fevers

- **Enteritis** is an inflammation of the intestine. **Enteric fever** is a systemic disease caused by pathogens that invade other tissues. All enteritises and enteric fevers are transmitted via the fecal–oral route and can be prevented by good sanitation.

- **Salmonellosis** is a self-limiting form of enteritis that is treated with antibiotics only in high-risk patients.

- **Typhoid fever** is an enteric infection caused by the bacterium *Salmonella typhi.* The disease is treated with chloramphenicol. Cell-mediated immunity follows infection. The available vaccine is limited in its effectiveness.

- **Shigellosis,** or **bacillary dysentery,** is treated with antibiotics, and recovery does not produce immunity.

- **Asiatic cholera,** common in Asia and other regions with poor sanitation, is treated with fluid and electrolyte replacement and tetracycline. Recovery can produce a long-lasting immunity, but the duration is unknown. Vaccines are not effective in cases of massive exposures.

- **Vibriosis** is a mild disease common where raw seafood is eaten.

- **Traveler's diarrhea** occurs in more than 1 million travelers each year. It is usually self-limiting but can have complications.

- *Escherichia coli* is an indicator of fecal contamination and an opportunistic pathogen.

Bacterial Infections of the Stomach, Esophagus, and Intestines

- *Helicobacter pylori* is considered to be the cause of peptic ulcers and chronic gastritis and a probable cofactor of stomach cancer.

Bacterial Infections of the Gallbladder and Biliary Tract

- Bile destroys most organisms that have lipid envelopes. *Salmonella typhi* is resistant to bile; it can live in the gallbladder and be shed in feces without causing symptoms. Gallstones blocking bile ducts can cause infections of the gallbladder and ducts, usually by *E. coli.* Infection can spread to the bloodstream or ascend to the liver.

Gastrointestinal Diseases Caused by Other Pathogens

Viral Gastrointestinal Diseases

- Most viral gastrointestinal diseases arise from the ingestion of contaminated water or food and are transmitted via the fecal–oral route, but hepatitis B, C, and D are transmitted parenterally from contaminated blood and other body fluids.

- **Rotavirus** infections kill many children in developing countries.

- Treatment for viral **hepatitis** relieves symptoms. Gamma globulin gives temporary immunity to **hepatitis A.** Vaccines are available for both hepatitis A and **hepatitis B.**

Protozoan Gastrointestinal Diseases

- Protozoan gastrointestinal diseases arise from contaminated food and water via the fecal–oral route and can be prevented by good sanitation.

- Diagnosis usually is made by finding protozoan cysts in fecal matter. Most such diseases can be treated with antiprotozoan agents.
- **Giardiasis** is especially common in children. **Amoebic dysentery, chronic amebiasis,** and **balantidiasis** occur worldwide but mainly in tropical areas. **Cryptosporidiosis** occurs mainly in immunocompromised patients, but recent outbreaks have involved persons with normal immune systems.

Effects of Fungal Toxins
- **Aflatoxins** are potent carcinogens produced by fungi of the genus *Aspergillus;* humans ingest them from moldy grain and peanuts.
- **Ergot poisoning** comes from eating grains contaminated with *Claviceps purpurea.* Small quantities of **ergot** toxin can be used medicinally.

- Mushroom toxins, which are associated mainly with species of *Amanita,* cause vomiting, diarrhea, jaundice, and hallucinations. In sufficient quantities, they are fatal.

Helminth Gastrointestinal Diseases
- Helminth gastrointestinal diseases are acquired mainly in tropical regions and include several kinds of fluke, roundworm, and tapeworm infections.
- Helminths that infect humans often have complex life cycles in which some animals, fish, vegetation, snails, and crustaceans may also serve as hosts.
- Most such diseases can be diagnosed by finding eggs in fecal specimens. They can be prevented by good sanitation—avoidance of contaminated water and soil and thoroughly cooking foods that might be contaminated.

Terminology Check

acute necrotizing ulcerative gingivitis
  (ANUG) *(p. 636)*
aflatoxin *(p. 656)*
amoebic dysentery *(p. 655)*
ascariasis *(p. 660)*
Asiatic cholera *(p. 642)*
bacillary dysentery *(p. 641)*
bacterial enteritis *(p. 640)*
balantidiasis *(p. 655)*
bilirubin *(p. 652)*
bongkrek disease *(p. 640)*
cementum *(p. 630)*
chronic amebiasis *(p. 655)*
cryptosporidiosis *(p. 656)*
cysticercus *(p. 658)*
delta hepatitis *(p. 651)*
dental caries *(p. 633)*
dental plaque *(p. 632)*
diarrhea *(p. 639)*
digestive system *(p. 629)*
dysentery *(p. 640)*
enamel *(p. 630)*
enteric fever *(p. 640)*
enteritis *(p. 640)*
enterocolitis *(p. 641)*

enteroinvasive strain *(p. 643)*
enterotoxicosis *(p. 639)*
enterotoxigenic strain *(p. 643)*
enterotoxin *(p. 639)*
ergot *(p. 656)*
ergot poisoning *(p. 657)*
feces *(p. 632)*
fluoride *(p. 634)*
food poisoning *(p. 638)*
giardiasis *(p. 654)*
gingivitis *(p. 636)*
ground itch *(p. 660)*
hepatitis *(p. 651)*
hepatitis A *(p. 651)*
hepatitis B *(p. 651)*
hepatitis C (HCV) *(p. 651)*
hepatitis D *(p. 651)*
hepatitis E (HEV) *(p. 651)*
hookworm *(p. 660)*
hydatid cyst *(p. 658)*
infectious hepatitis *(p. 651)*
large intestine *(p. 631)*
microvillus *(p. 631)*
mucin *(p. 630)*
mumps *(p. 637)*

orchitis *(p. 637)*
periodontal disease *(p. 635)*
periodontitis *(p. 636)*
pinworm *(p. 662)*
rotavirus *(p. 649)*
salmonellosis *(p. 640)*
serovar *(p. 640)*
serum hepatitis *(p. 651)*
shigellosis *(p. 641)*
sinusoid *(p. 631)*
small intestine *(p. 631)*
strongyloidiasis *(p. 662)*
tartar *(p. 636)*
traveler's diarrhea *(p. 643)*
trichinosis *(p. 659)*
trichuriasis *(p. 661)*
typhoid fever *(p. 641)*
verminous intoxication *(p. 658)*
vibriosis *(p. 643)*
villus *(p. 631)*
viral enteritis *(p. 649)*
whipworm *(p. 661)*
yersiniosis *(p. 646)*

See You on the Web

Don't forget I am expecting you on the web. The following is a preparedness checklist for your web explorations. If you have questions on any of these items, you can return to the end of Chapter 1 for more details.

- *Questions We'll Explore*
- *Concept Check questions*
- *Flashcards*
- *Study Guide*

If you are ready, come visit me at

http://www.prenhall.com/black

It will be a journey you'll never forget.

Cardiovascular, Lymphatic, and Systemic Diseases

"Home, home on the range . . ." More than deer and antelope make their home on the range at Yellowstone. You're also likely to find brucellosis roaming free among herds of bison. Brucellosis is a contagious bacterial disease that affects the circulatory system. It causes the reduction of meat and milk production, as well as abortion. Surveys show that half of the Yellowstone animals are infected with brucellosis.

This disease costs U.S. cattle producers over $30 million each year. When infected animals are found on cattle ranches, they are slaughtered. An improved vaccine is now available, and the Park Service hopes to eliminate the disease from its bison by the year 2010. Until then, ranchers and Park Rangers will have to rely on all possible preventive measures. To discourage contact with ranch herds, some have established separate feeding sites for Park animals. Others simply keep their shotguns handy.

Background Basics

bacteremia . . . . *(Chapter 14)*
septicemia . . . . *(Chapter 14)*
broad-spectrum antibiotics . . . . *(Chapter 13)*
endotoxin, exotoxin . . . . *(Chapter 14)*
cross-reactive antibodies . . . . *(Chapter 17)*

D iseases of the cardiovascular and lymphatic systems frequently affect several other systems because the infectious agents are easily disseminated through blood and lymph. Therefore, diseases that usually affect multiple systems are included in this chapter.

The Cardiovascular System

The **cardiovascular system** consists of the heart, blood vessels, and blood (➤ Figure 23.1). This system supplies oxygen and nutrients to all parts of the body and removes carbon dioxide and other wastes from them. It also carries hormones from endocrine glands to the appropriate tissues and cells and regulates the pH of the blood via buffers.

The Heart and Blood Vessels

The heart lies between the lungs in a tough, membranous *pericardial sac* lubricated by serous fluid. Its wall consists of a thin internal *endocardium*, a thick muscular *myocardium*, and an outer *epicardium*. The heart has four chambers—two *atria* and two *ventricles*, with valves that direct blood flow at the exits of each chamber. The right side of the heart pumps blood to the respiratory portion of the lungs, the left side to all other organs and tissues.

Blood leaving the heart circulates through a closed system of blood vessels and subsequently returns to the heart. Its flow is regulated so that all cells receive nutrients and get rid of wastes according to their needs. The blood vessels include *arteries* that receive blood from the heart, *arterioles* that branch from arteries, *capillaries* that branch from arterioles, *venules* that receive blood from capillaries, and *veins* that receive blood from venules and return it to the heart. Capillary walls are composed of a single layer of cells that allows exchange of materials between the blood and the tissues. Leukocytes, such as neutrophils and macrophages, sometimes force their way between cells of the capillary walls (diapedesis) during infections.

Questions We'll Explore

A What are the components of the cardiovascular and lymphatic systems? How do they function to prevent infection?

B Which pathogens cause bacterial septicemias and related diseases, and what are the important epidemiologic and clinical aspects of these diseases?

C Which helminths cause diseases of the blood and lymph, and what are the important epidemiologic and clinical aspects of these diseases?

D Which bacterial pathogens cause systemic diseases, and what are the important epidemiologic and clinical aspects of these diseases?

E Which rickettsia cause systemic diseases, and what are the important epidemiologic and clinical aspects of these diseases?

F Which viruses cause systemic diseases, and what are the important epidemiologic and clinical aspects of these diseases?

G Which protozoa cause systemic diseases, and what are the important epidemiologic and clinical aspects of these diseases?

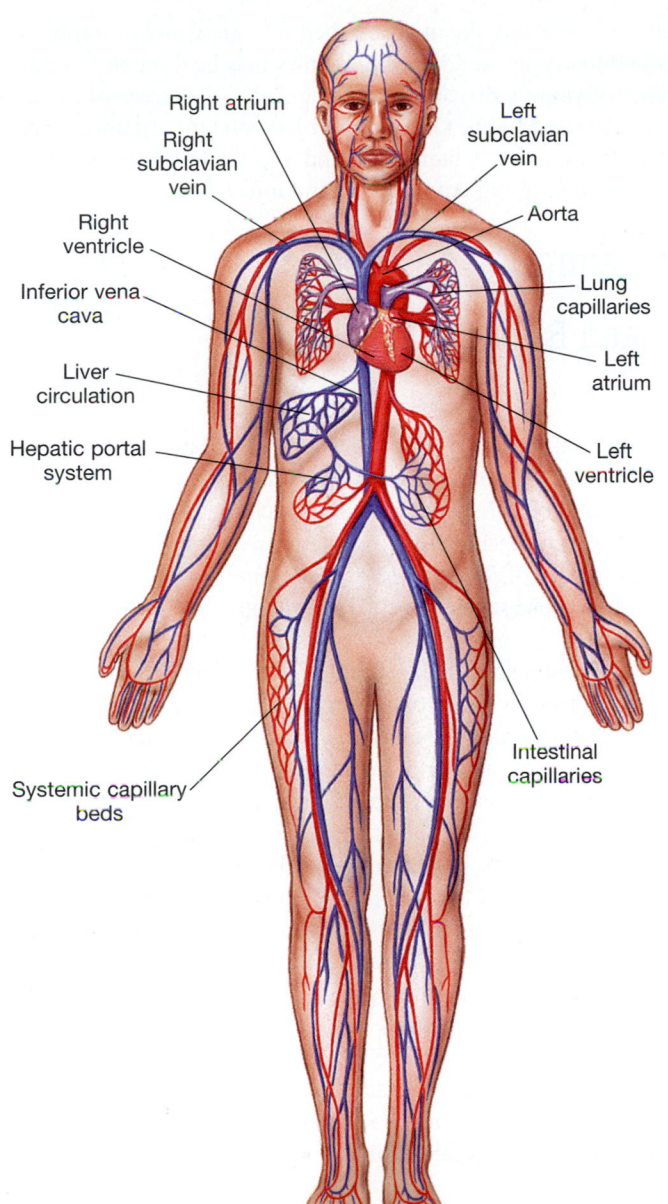

Right atrium

Right subclavian vein

Right ventricle

Inferior vena cava

Liver circulation

Hepatic portal system

Systemic capillary beds

Left subclavian vein

Aorta

Lung capillaries

Left atrium

Left ventricle

Intestinal capillaries

➤ Figure 23.1 **The general structure of the cardiovascular system.** Oxygenated blood is shown in red, unoxygenated blood in blue. The cardiovascular system is normally sterile and contains no resident microbes.

The Blood

The blood consists of plasma and formed elements (cells and cell fragments). Plasma is more than 90 percent water and contains proteins such as albumins, globulins, and fibrinogen. Certain globulins, the antibodies, are important in defending the body against infection, and fibrinogen is important in blood clotting. Plasma also contains **electrolytes** (ions such as Na^+, K^+, and Cl^-), gases (such as oxygen and carbon dioxide), nutrients, and waste products. In contrast with plasma, serum is the fluid that remains after both formed elements and clotting factors have been removed.

Besides leukocytes and platelets, formed elements of the blood include erythrocytes. Erythrocytes are the most abundant of the formed elements, accounting for 40 to 45 percent of the total blood volume. Erythrocyte volume is an important indicator of the oxygen-carrying capacity of the blood because erythrocytes contain the oxygen-binding molecule hemoglobin.

Normal Microflora of the Cardiovascular System

The cardiovascular system does not have any normal resident microflora. Because it is a closed system, blood, blood vessels, and the heart should ordinarily be sterile. However, even in healthy individuals, microorganisms will occasionally enter the bloodstream, causing a transient **bacteremia**

➥ (Chapter 14, p. 393). If not killed or removed, the microbes begin to grow and multiply in the bloodstream, causing **septicemia** (blood poisoning). Points of entry may be through wounds (some as minor as vigorous brushing of your teeth), or by shedding from other areas of infection such as a boil, a pimple, or an abcessed tooth. Once in the bloodstream, organisms are eliminated by antibodies and/or phagocytosis ➥ (Chapters 16 and 17). Certain sites, such as heart valves, are especially susceptible to bacterial colonization and infection.

Cardiovascular and Lymphatic Diseases

Bacterial Septicemias and Related Diseases

Septicemias

Before antibiotics, septicemia was often fatal; even with antibiotics, it is still not easy to treat. Gram-positive organisms such as *Staphylococcus aureus* and *Streptococcus pneumoniae* once commonly caused most septicemias. Today, broad-spectrum antibiotics have made septicemias from these organisms less frequent. However, other bacterial species—including *Pseudomonas aeruginosa, Bacteroides fragilis,* and species of *Klebsiella, Proteus, Enterobacter,* and *Serratia* have come into play. These organisms cause **septic shock,** a life-threatening septicemia accompanied by low blood pressure and the collapse of blood vessels. Probably one-third of all septicemias are now Gram-negative septic shock, and 10 percent of all septicemias are caused by multiple organisms. Endotoxins produced by these organisms are directly responsible for shock. Antibiotics often worsen the situation; when they kill organisms, the disintegrating organisms release larger quantities of endotoxin, causing more damage to the host's blood vessels, and blood pressure drops even further. A few cases of septic shock are caused by organisms that produce exotoxins.

Symptoms of septicemia include fever, shock, and **lymphangitis** (lim-fan-ji'tis), or red streaks due to inflamed lymphatic vessels beneath the skin (➤ Figure 23.2). One-third of all septicemia cases are nosocomial and appear within 24 hours after an invasive medical procedure has been performed. The transition from bacteremia to septicemia can be sudden or gradual. Therefore, hospital patients who have undergone invasive procedures should be carefully watched for signs of septicemia. Septicemia has a mortality of 50 to 70 percent and accounts for about 100,000 deaths per year in the United States alone.

Diagnosis of septicemia is made by culturing of blood, catheter tips, urine, or other sources of infection. In treating septicemias, blood pressure must be elevated and stabilized; then the infectious organisms must be eliminated by appropriate antibiotic therapy.

Puerperal Fever

Puerperal (pu-er'per-al) **fever,** also called *puerperal sepsis* or **childbed fever,** was a common cause of death before antibiotics became available. ➥ (Chapter 1, p. 14) It is caused by β-hemolytic streptococci (*Streptococcus pyogenes*), which are normal vaginal and respiratory microflora. They can also be introduced during delivery by medical personnel. Streptococci pass through irritated uterine surfaces and invade the blood, giving rise to septicemia. Signs and symptoms of the disease are chills, fever, pelvic distention and tenderness, and a bloody

➤ **Figure 23.2 Lymphangitis from an infected burn.** The reddened streaks in the arm indicate the spread of organisms through lymph vessels, a symptom of septicemia. The old-fashioned name for such an infection was blood poisoning.

vaginal discharge. Streptococci can be isolated from blood cultures to diagnose puerperal fever. Penicillin is effective except against resistant organisms, and mortality is low with prompt treatment, but recovery usually takes many weeks, and relapses are common.

Rheumatic Fever

Rheumatic fever is a multisystem disorder following infection by β-hemolytic *Streptococcus pyogenes*. That rheumatic fever can follow such infections has been known for decades, but the mechanisms by which it occurs are not yet completely understood. Some form of genetic predisposition is suspected because a certain HLA antigen is present in 75 percent of rheumatic fever patients but in only 12 percent of the general population.

Most rheumatic fever patients are between the ages of 5 and 15. Onset of the disease usually occurs 2 to 3 weeks after a strep throat, but it can occur within 1 week or as late as 5 weeks after the initial infection. Strep throat symptoms have disappeared by the time the rheumatic fever symptoms begin. Classic signs and symptoms include fever, arthritis, and a rash. Evidence of damage to the mitral valve of the heart confirms a specific diagnosis of rheumatic fever. Weeks or months later subcutaneous nodules appear, especially near the elbows. Approximately 3 percent of untreated strep throat cases progress to rheumatic fever. Culturing of streptococci and serological tests are helpful in diagnosis, as is a previous history of streptococcal infection.

Heart damage in rheumatic fever results from immunological events. Certain streptococcal strains have an antigen that is very similar to heart cell antigens. Antibodies that bind to one antigen will bind to the other; that is, the antibodies are *cross-reactive*. In the immune reaction, lymphocytes probably become sensitized to the antigen and attack heart tissue as well as the streptococci. The resulting heart damage can be fatal. Antibiotic therapy will not reverse existing damage, but it can prevent further damage and can be used to prevent recurrences. Once rheumatic fever occurs, mitral valve deformities contribute to eddies in blood flow that predispose toward bacterial colonization of heart valve surfaces. (This condition, *bacterial endocarditis,* is discussed next.) Individuals at risk of rheumatic fever should receive a prophylactic antibiotic—usually penicillin—before dental work or other invasive procedures to prevent possible streptococcal infection.

Prompt treatment of β-hemolytic *S. pyogenes* infections with antibiotics before cross-reactive antibodies can form is the only practical way to prevent rheumatic fever. No effective vaccine exists; vaccines produced so far elicit damaging antibodies, but this problem might someday be solved. Anti-inflammatory drugs such as steroids and aspirin can lessen scarring of heart tissue.

Bacterial Endocarditis

Bacterial endocarditis (en′-do-kar-di″tis), or *infective endocarditis,* is a life-threatening infection and inflammation of the lining and valves of the heart. It can be subacute or acute. Two out of three patients have the subacute type, which manifests itself as fever, malaise, bacteremia, and regurgitating heart murmur usually lasting 2 weeks or more. It occurs primarily in people over age 45 who have a history of valvular disease from rheumatic fever or congenital defects. Many microbes, including fungi, can cause endocarditis, but most cases are due to bacteria, especially strains of *Streptococcus* or *Staphylococcus,* many of them normal residents of the mouth or throat. Acute endocarditis is a rapidly progressive disease that destroys heart valves and causes death in a few days.

In bacterial endocarditis, organisms from another body site of infection are transported to the heart. A **vegetation** develops, in which exposed collagen fibers on damaged valvular surfaces trigger fibrin deposition (➤ Figure 23.3). Transient bacteria attach to fibrin and form a bacteria–fibrin mass. Vegetations deform heart valves, decrease their flexibility, and prevent them from closing completely. Blood flows backward from ventricles into atria when the ventricles contract, decreasing the pumping efficiency of the heart. Congestive heart failure, an accumulation of fluids around the heart, is the most common complication and direct cause of death from bacterial endocarditis.

Bacterial endocarditis is diagnosed from blood cultures and is treated with penicillin or other antibiotics, depending on the susceptibilities of causative organisms. In some cases surgical valve replacement is necessary. Untreated endocarditis results in death. Antibiotics cure

---

> Rheumatic fever can cause aortic stenosis, a narrowing of the aortic valve opening, which causes the left ventricle of the heart to work harder.

Man's Best Friend's Worst Enemy

Keep the mosquitoes away from Fido, or better yet give him preventative medication. The heartworm, *Dirofilia immitis,* is transmitted by mosquitoes. Once in the dog's bloodstream, the larvae migrate to the skin, mature and move to the heart. There they mate and release microfilariae that become infective larvae only when ingested by a mosquito. Adult worms 15 to 30 cm long accumulate in the right side of the heart and are also found in the lungs and liver. The unfortunate dog dies an early death from heart damage and circulatory damage.

The drug Ivermectin, when given in monthly doses, prevents heartworm infection. However, it can be administered only to dogs who are not already infected. If adult worms are present, the drug kills them. But disintegration of worm remains releases toxic substances and debris that can clog blood vessels while further weakening the wall of the heart where they had resided.

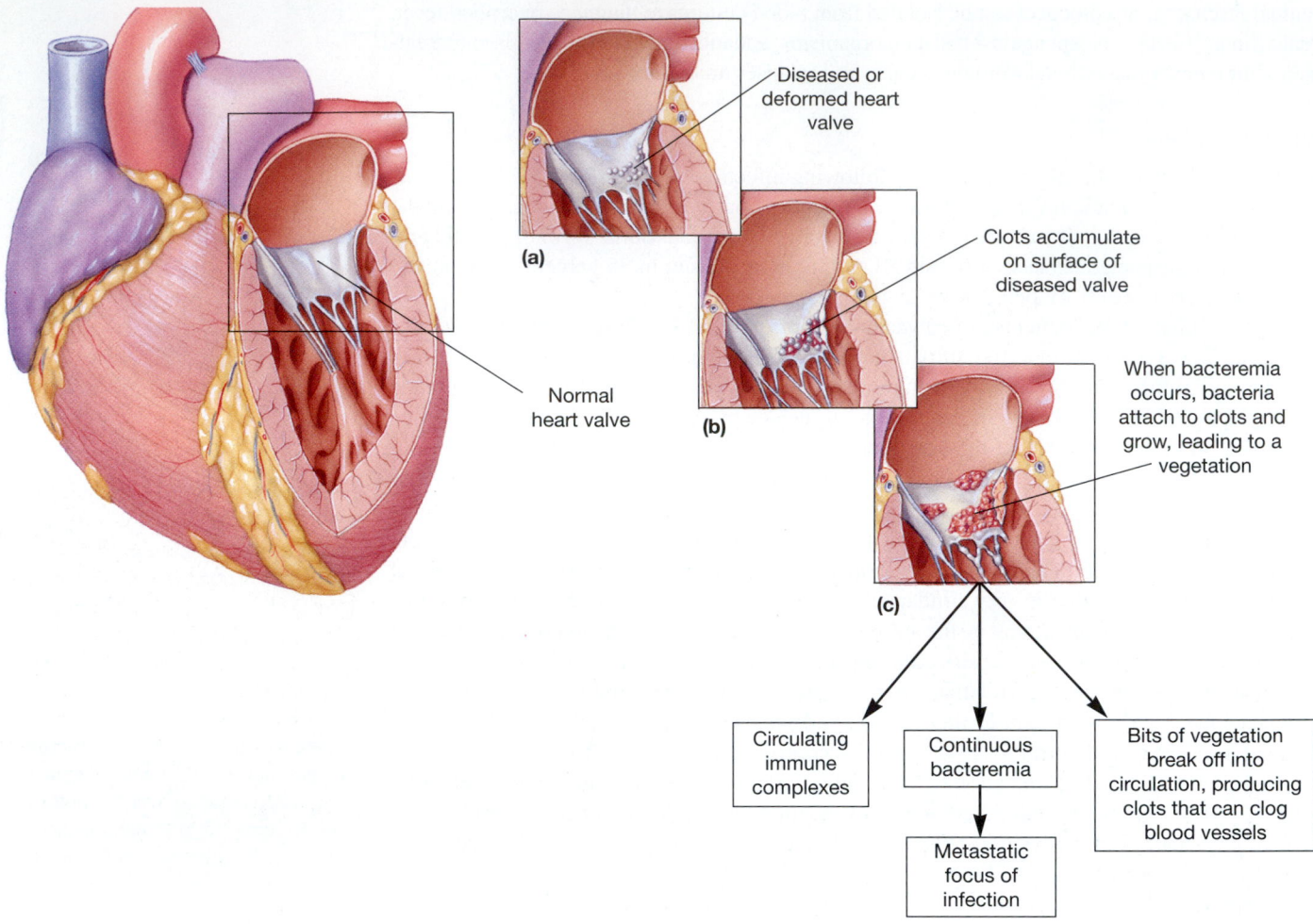

Diseased or deformed heart valve

(a)

Normal heart valve

Clots accumulate on surface of diseased valve

(b)

When bacteremia occurs, bacteria attach to clots and grow, leading to a vegetation

(c)

Circulating immune complexes

Continuous bacteremia

Bits of vegetation break off into circulation, producing clots that can clog blood vessels

Metastatic focus of infection

➤ Figure 23.3   **Bacterial endocarditis and vegetation.** (a) A heart valve impaired by bacterial (infective) endocarditis. (b) and (c) The process by which vegetation develops on heart valves.

✓ Describe the normal flora of the cardiovascular system.

✓ Why could treatment with antibiotics bring on cases of septic shock?

✓ What does a vegetation found in the heart consist of? How and why does it form? What effects does it have?

about half of all patients; surgery cures another quarter, and one-quarter die. Deaths are most frequent among intravenous drug abusers and others with compromising conditions.

Myocarditis (mi′o-kar-di″tis), an inflammation of the heart muscle (myocardium), and **pericarditis** (per′e-kar-di″tis), an inflammation of the protective membrane around the heart (pericardial sac), also can be caused by microbial infections. Although most such infections are viral, *Staphylococcus aureus* causes 40 percent of all pericarditis cases. Untreated cases have a mortality rate of nearly 100 percent, whereas cases receiving appropriate treatment have a mortality rate between 20 and 40 percent.

Helminthic Diseases of the Blood and Lymph

Schistosomiasis

Three species of blood flukes of the genus *Schistosoma* cause **schistosomiasis** (shis″to-so-mi′a-sis), and each requires a particular snail intermediate host to complete its life cycle (see ➤ Figure 11.15). ➡ (p. 306) The first of these helminths was identified by the German parasitologist Theodor Bilharz in the 1850s, so the disease is also called *bilharzia* (bil-har′ze-a). Bilharzia has been known since biblical times; some think Joshua's curse on Jericho was the placing of blood flukes in communal wells. In fact, the Egypt of the pharaohs was called "land of the menstruating men" by ancient writers because the prevalence of flukes made bloody urine so common. Eggs of these parasites have been found in the bladder walls of Egyptian mummies.

WHO reports some 200 million cases of schistosomiasis worldwide. Incidence of the disease has increased significantly in Egypt since the building of the Aswan Dam in 1960 because collection of water behind the dam created exceptionally favorable conditions for snail hosts. *Schistosoma japonicum* is found in Asia, *S. haematobium* in Africa, and *S. mansoni* (➤ Figure 23.4) in Africa, South America, and the Caribbean. The latter two species probably reached South America during slave trading days, but only *S. mansoni* found an appropriate snail host there. New species have been reported in Vietnam (for instance, *S. mekongi*) and in other geographic areas.

Humans become infected by free-swimming cercariae that have emerged from their snail hosts (➤ Figure 23.4b). Cercariae penetrate the skin when humans wade in snail-infested waters, migrate to blood vessels, and are carried to the lungs and liver. The flukes mature and migrate to veins between the intestine and liver or sometimes the urinary bladder, where they mate and produce eggs. Adults have a special ability to coat themselves with host antigens, thereby evading the host's immune system. Some eggs become trapped in the tissues and cause inflammation; others penetrate the intestinal wall and are excreted in the feces. Cercariae cause dermatitis at penetration sites and tissue damage during migration. Metacercariae and adults migrate to and invade the liver (➤ Figure 23.4c), where they cause cirrhosis. These flukes also can invade and damage other organs.

(a)

(b)

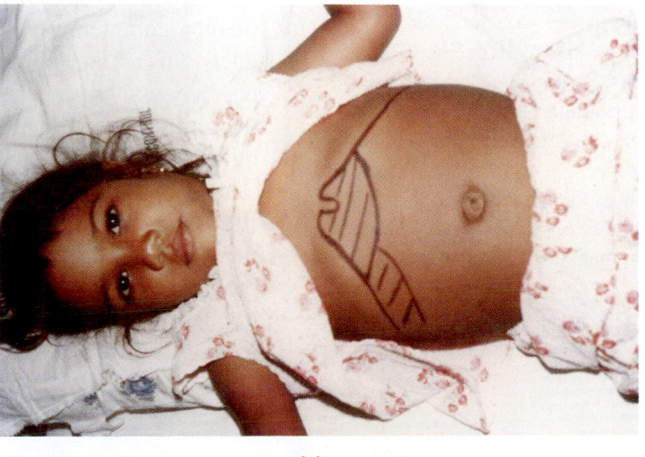

(c)

➤ **Figure 23.4   Schistosomiasis.** Schistosomes, or blood flukes, are trematodes that cause schistosomiasis in Africa, South America, and the Caribbean. (a) A mating pair of schistosomes. The larger, male worm holds the smaller, female worm in a groove of his body. The male attaches itself to the wall of a human blood vessel by means of the suction cup just below its head. The worms, thinner than a cotton fiber, are barely visible to the unaided eye. A mating pair can live in the human body for up to 10 years. (b) The cercarial stage (160X) leaves snails and penetrates the skin of humans wading or swimming in water. (c) A child suffering from schistosomiasis. The tracing delineates her enlarged liver, typical of this disease.

Schistosome eggs are highly antigenic, and allergic reactions to them are responsible for much of the damage caused by blood flukes. If the eggs are released near the spinal cord, the resulting inflammation can cause neurological disorders. The eggs most often damage blood vessels, but which vessels are damaged depends on the species. *Schistosoma japonicum* severely damages blood vessels in the small intestine; *S. mansoni,* those of the large intestine; and *S. haematobium,* those of the urinary bladder. Symptoms of schistosome bladder infections include pain on urination, bladder inflammation, and bloody urine.

Diagnosis can be made by finding eggs in feces or urine, but eggs may not be present in chronic cases. Intradermal injection of schistosome antigen and measurement of the area of the wheal or a complement fixation test are good immunological methods of diagnosis. ➡ (Chapter 18, p. 528) Until recently, toxic antimony compounds were used to treat the disease. Several new drugs, especially praziquantal, seem to be quite effective and less toxic. Some investigators, however, believe that praziquantal-resistant schistosomes are on the increase.

Schistosomiasis could be prevented entirely if human wastes were not emptied into rivers or if humans never waded in snail-infested waters. The practice of wading in the local river to wash the body after defecation or urination is an important means of transmission wherever the infection occurs. Chemical molluscicides have been used to reduce snail populations, but it is difficult to determine the proper concentrations under varying river conditions. Recent experiments using predatory snails to destroy cercariae-carrying snails have shown great promise. Finally, work toward developing a vaccine is underway. One new vaccine in testing does not prevent infection but protects against damage caused by schistosomes.

> Over 1 billion people in 73 countries are at risk of contracting elephantiasis, and there are over 100 million people already infected.

A Good Deed

Elephantiasis is actually a lymphedema caused by the congestion of the lymphatic vessels by filarial worms. To end the scourge of lymphatic filariasis, the World Bank, the World Health Organization, and the healthcare company, SmithKline Beecham, are cooperating on a program where albendazole, the drug to treat the disease, will be donated free-of-charge to needy countries. Because it takes one yearly dose for 4 to 5 years to treat all people in infected areas, SmithKline Beecham's donation commitment is several billion doses of albendazole. It is estimated that because one-fifth of the world's population is at risk of infection it will take approximately 20 years to eliminate the disease.

Filariasis

Filariasis (fil″ar-i′a-sis) can be caused by several different roundworms. *Wuchereria bancrofti* (see ➤ Figure 11.18) is a common cause of this tropical disease, for which WHO reports over 100 million cases worldwide. ➡ (p. 309) Adult worms are found in the lymph glands and ducts of humans. Females release embryos called *microfilariae.* They are present in peripheral blood vessels during the night and retreat to deep vessels, especially those of the lungs, during the day. Mosquitoes also are essential hosts in the life cycle of this parasite, and several species of night feeders among the genera *Culex, Aedes,* and *Anopheles* serve as hosts. When a mosquito bites an infected person, it ingests microfilariae that develop into larvae and migrate to the mosquito's mouthparts. When the mosquito bites again, the larvae are infected and can infect another person. They enter the blood, develop, and reproduce in the lymph glands and ducts, thereby completing the life cycle. Adult worms are responsible for inflammation in lymph ducts, fever, and eventual blockage of lymph channels in affected areas. Repeated infections over a period of years can lead to **elephantiasis** (el″ef-an-ti′a-sis), gross enlargement of limbs, the scrotum, and sometimes other body parts due to the blockage of lymph vessels and subsequent accumulation of fluid and connective tissue in those vessels (➤ Figure 23.5a).

Filariasis is diagnosed by finding microfilariae in thick blood smears (➤ Figure 23.5b) made from blood samples taken at night or by an intradermal test. The drugs diethylcarbamizine (Hetrazan) and metronidazole are effective in treating the disease. Swollen limbs are wrapped in pressure bandages to force lymph from them; if distortion is not too great, nearly normal size can be regained. To control the disease, it would be necessary to treat all infected individuals and to eradicate the species of mosquitoes that carry the parasite. Limited progress has been made.

Cardiovascular and lymphatic diseases are summarized in Table 23.1.

Systemic Diseases

Bacterial Systemic Diseases

Anthrax

Anthrax is a zoonosis that affects mostly plant-eating animals, especially sheep, goats, and cattle. Meat-eating animals can acquire the disease by eating infected flesh or by inhaling anthrax spores but the disease is not spread from live animal to animal. Each year, many thousands of

> *Bacillus anthracis* produces a toxin that inhibits a protein kinase used in cell signaling. This may mean that protease inhibitors could be used against anthrax.

(a)

(b)

> Figure 23.5 **Elephantiasis.** Elephantiasis of the scrotum, caused by the roundworm *Wuchereria bancrofti.* Swelling results from blockage of the lymphatic system by adult worms. Another common site of elephantiasis is the leg. (b) The microfilaria stage of the life cycle (2900X) is transmitted to humans by the bite of mosquitoes.

> **Table 23.1** | **A summary of cardiovascular and lymphatic diseases**

Disease	Agent(s)	Characteristics
Bacterial septicemias and related diseases		
Septicemia	Various bacterial species	Septic shock due to endotoxins of causative agent(s), fever, lymphangitis
Puerperal fever	*Streptococcus pyogenes*	Organisms from uterus invade blood and cause septicemia, pelvic distention, bloody discharge
Rheumatic fever	*Streptococcus pyogenes*	Fever, arthritis, rash, mitral valve damage due to immunological reaction
Bacterial endocarditis	*Staphylococcus* or *Streptococcus* strains	Inflammation and vegetation of heart valves and lining, fever, malaise, bacteremia, heart murmur; congestive heart failure that can cause death
Parasitic diseases of the blood and lymph		
Schistosomiasis	*Schistosoma haematobium, S. mansoni, S. japonicum*	Dermatitis from cercaria, cirrhosis of liver from eggs, allergic reactions to eggs, tissue damage in intestine and urinary bladder
Filariasis	*Wuchereria bancrofti*	Inflammation and blockage of lymph ducts, leading to elephantiasis, fever

Disarming the Anthrax Bomb

Imagine the United States at war with another country. The opposing leaders, knowing they cannot beat us with their armies, decide to wipe us out in a different way—by sending over the deadly disease anthrax. That's a scary scenario because germ warfare with the anthrax bacillus is so realistic. Maybe it's not quite as scary anymore, thanks to the efforts of a research team led by George Vande Woude at the National Cancer Laboratory Institute (NCI) in Frederick, Maryland. Researchers there may have learned how a toxic protein produced by the anthrax bacillus—the one that is thought to be one of the principal causes of death in infected individuals—works. New treatments will focus on neutralizing this toxin. As Vande Woude puts it, "Conceivably, we could find a drug that would make anthrax as a weapon of destruction as powerful as a water pistol."

animals worldwide have anthrax, but only 20,000 to 100,000 humans develop it, mostly in Africa, Asia, and Haiti. An anthrax outbreak in Russia in 1979 killed at least 64 people. Health authorities in the United States have made vigorous efforts to eradicate anthrax and to prevent its import from other countries. Only five human cases have occurred in the United States since 1980, and no more than six per year have been reported since 1970.

The Disease   The causative agent of anthrax, *Bacillus anthracis,* was discovered in 1877 by Robert Koch. The bacillus is a large, Gram-positive, facultatively anaerobic, endospore-forming rod. The endospores form only under aerobic conditions and are not found in tissues or circulating blood. But if an infected animal's blood is spilled during an autopsy, bacilli exposed to air rapidly form endospores. Veterinarians and farmers should be very careful to avoid contaminating soil or other materials, as endospores can remain viable over 50 years, ruining a pasture for animal usage during those years.

Most human anthrax results from contact with endospores during occupational exposure on farms or in industries that handle wool, hides, meat, or bones. Respiratory anthrax, or "woolsorters' disease," was such a problem in nineteenth-century England that legislation was enacted to protect textile workers from this occupational disease.

Of human anthrax cases, 90 percent are cutaneous, 5 percent are respiratory, and 5 percent are intestinal. Cutaneous anthrax has a mortality rate of 10 to 20 percent when untreated, but only 1 percent with adequate treatment. Respiratory anthrax is always fatal, regardless of treatment. Clots form inside pulmonary capillaries and lymph nodes, causing swelling that obstructs airways. Intestinal anthrax has a fatality rate of 25 to 50 percent. In addition, regardless of the initial site of infection, septicemia leads to meningitis (which is almost always fatal in 1 to 6 days) in about 5 percent of patients.

Cutaneous anthrax develops 2 to 5 days after endospores enter epithelial layers of the skin. Lesions 1 to 3 cm in diameter develop at the site of entry (➤ Figure 23.6). Eventually, the center of the lesion becomes black and necrotic; finally, it heals but leaves a scar. Recovery probably confers some—but not total—immunity.

Symptoms of intestinal anthrax closely mimic those of food poisoning. ➥ (Chapter 22, p. 639) An outbreak in the United States resulted from consumption of imported cheese made from unpasteurized goat's milk. The cheese was served at a wine and cheese party to a group of physicians. Some of them traced the source of the infection to a particular cheese factory in France, which was closed as a result.

Diagnosis, Treatment, and Prevention   Anthrax is diagnosed by culturing blood or examining smears from cutaneous lesions of patients with a history of possible exposure. The disease is treated with penicillin or tetracycline. A vaccine is available and can be given annually to workers with occupational exposure to anthrax, but industries still should maintain dust-free environments and provide respirators to prevent endospore inhalation. Worker education and on-site employee health services for prompt detection of infections also are important. Unimmunized visitors must be kept away from work areas. Clothing worn by workers should be sterilized and cleaned at the facility to prevent family members from becoming infected by handling it.

Animal immunization is an important means of prevention. Farmers must avoid using bone meal contaminated with anthrax spores and must dispose of infected animals by burying them in deep, lime-lined pits. Lime prevents earthworms from bringing anthrax endospores to the surface; incineration is used but must be done properly to avoid the spread of bits of contaminated carcass and spores by the wind. Veterinarians must be especially careful when working with infected animals or giving vaccinations because accidental vaccination of humans with a vaccine intended for animals can cause anthrax.

➤ **Figure 23.6   Cutaneous anthrax.** Spores introduced into the skin through abrasions or cuts germinate. The newly emerged organisms multiply, release exotoxin locally, and invade adjacent tissue, where they produce extensive damage.

Plague

From 1937 through 1974, fewer than 10 cases per year of plague, a zoonosis, were reported in the United States, with no cases reported in some years. Then 20 cases were reported in 1975 and 40 in 1983, mostly in rural settings in the Rocky Mountain states (➤ Figure 23.7). There, the disease is referred to as *sylvatic plague* because it is carried by wild rodents such as ground squirrels, chipmunks, and desert rats. The United States is fortunate in not suffering from *urban plague,* in which the European or city rat is the main host. Plague remains an endemic disease

Prairie Heritage: Deadly Spores

For centuries, vast herds of bison roamed the Great Plains of the United States. The bison periodically suffered outbreaks of anthrax. Their dead carcasses rotted, leaving anthrax spores in the soil. Then came the great cattle drives of the 1800s, from Texas to Canada. Carcasses were left along the trails, adding more anthrax spores to the soil. By the 1930s, ranchers realized that the bodies of anthrax-infected animals must be dealt with in order to prevent soil contamination. Some ranchers still remember the huge bonfires of burning carcasses in the 1930s and 1940s. But by then the soil was contaminated with anthrax spores—which remain viable even today.

In dry years, wind blows away the topsoil, exposing spores from the older, deeper layers. In wet years, or along river and creek beds, erosion washes away topsoil, likewise exposing and spreading spores. From Texas to Canada, yearly anthrax immunization of cattle is largely routine. But after 10 or 15 years without an outbreak, ranchers may get a false sense of security and discontinue using the vaccine. That is what happened in South Dakota during the summer of 1993.

In August 1993, an unvaccinated herd of beef cattle in southeastern South Dakota began to sicken. On August 13, after three animals had died, a veterinarian was summoned. His diagnosis was pulmonary emphysema. On August 15, a second veterinarian arrived and confirmed that diagnosis. The rancher was told that it was safe to send the rest of the herd to slaughter. Soon 19 animals had died, 9 others went to slaughter immediately, and the rest were shipped to the stockyards to be sold.

At the slaughterhouses the animals' livers did not look healthy to inspectors. The rancher was notified of the possible diagnosis of anthrax. Laboratory testing on August 17 confirmed the presence of anthrax. An immediate order of quarantine was issued the following morning.

The slaughtered animals' remains were retrieved, and the meat plants and storage lockers had to be disinfected by steaming. Fortunately, none of the animals at the stockyards had been sold yet. All were shipped back to the ranch from which they had come and were held in quarantine. There they received anthrax vaccine plus antibiotics, and a booster shot 2 weeks later. The manure and bedding from the areas they had occupied in the stockyards had to be collected, hauled away, and sterilized.

The dead animals had been sent to a rendering plant, where carcasses or waste parts such as bones and fat are boiled down. Products from rendering plants may eventually be used in food products for humans or other animals. Thus, these 19 carcasses were a danger. Had the animals been milk cows, their milk could also have been contaminated.

State health inspectors and others who had handled the herd were given tetracycline to prevent the development of anthrax infection. Happily, no human cases occurred. But by then, 32 head of cattle had died. The last 13 carcasses were burned. Thirty days after the last death, the rest of the herd was released from quarantine. Nearby ranchers who had neglected to vaccinate their herds swiftly administered the vaccine.

South Dakota and the other Great Plains states have periodic outbreaks. It is part of life there. Health authorities must be ever vigilant against the prairie heritage of spores from the distant past.

---

in certain other areas of the world, but the number of cases worldwide is much smaller than in the great pandemics mentioned in Chapter 1. ➡ (p. 8)

The Disease   The causative agent of plague, *Yersinia pestis,* is a short Gram-negative rod. As a zoonosis, the disease is spread by infected rodents, especially rats, which transmit the organisms by animal-to-animal contact and occasionally to humans by flea bites. As infected rats die of plague, their body temperature drops, and their blood coagulates; hungry fleas jump to nearby sources of warmth and liquid blood. The new host usually is another rat, but in crowded, rat-infested living quarters or when contact is made with the carcass of a plague-infected animal, the next host can easily be a human.

The flea itself suffers from plague infection. The plague organisms ingested from a sick rat multiply and block the flea's digestive tract until food (blood meals) cannot pass through it. The flea gets hungrier, bites more ferociously, and infects new victims as it dispenses plague organisms with each bite. Eventually the flea dies, but this is little consolation to a new human victim, who has a 50 to 60 percent chance of dying if not treated.

Once inside the host, plague bacilli multiply and travel in lymphatics to lymph nodes, where they cause hemorrhages and immense lymph node enlargements called **buboes** (singular: *bubo*), especially in the groin and armpit (➤ Figure 23.8). Buboes are characteristic of **bubonic plague** and appear after an incubation time of 2 to 7 days. Hemorrhages turn the skin black—hence the name Black Death. Deaths from bubonic plague can be prevented with adequate timely antibiotic treatment. If organisms move from the lymphatics into the circulatory system, **septicemic plague** develops. It is characterized by hemorrhage and necrosis in all parts of the

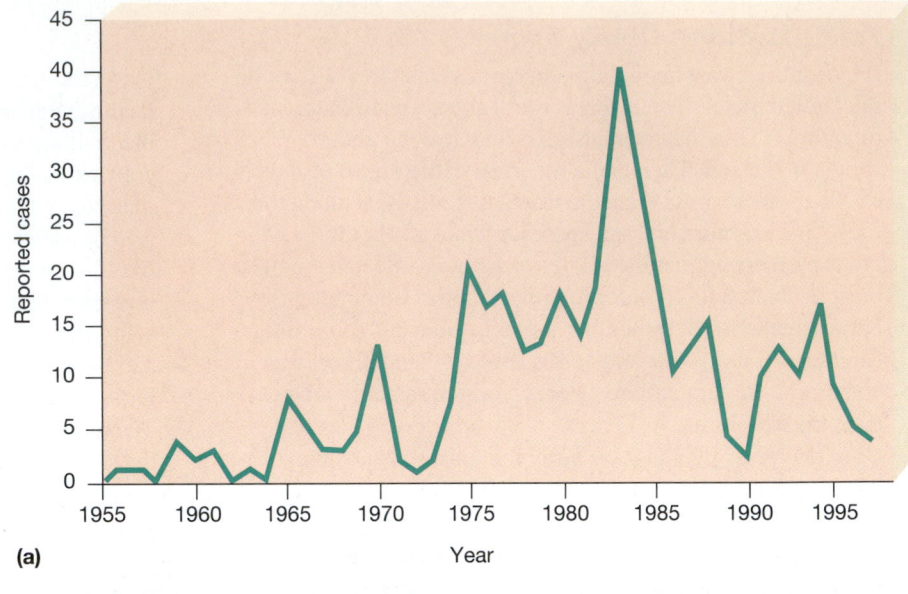

(a)

Plague in the United States?

Here? Today? Yes! In fact, plague is quite prevalent in gophers, chipmunks, pack rats, prairie dogs, and ground squirrels in 15 western states. Indeed, plague is thought to have persisted for years among similar wild rodent populations of the Central Asian steppe before breaking out to decimate Europe as the "Black Death" in 1346. Should our rural (*sylvatic*) plague infect city rats, giving rise to *urban* plague, we may be in for big trouble. So far only sporadic cases of human plague have occurred in the American West, often among hunters and Native Americans on reservations. But we've been lucky. Some years ago, a pet-store owner in Los Angeles trapped some desert rodents with the intention of selling them. Unfortunately for him—but perhaps fortunately for the community—he died of plague before he could sell any of the dangerous animals as pets.

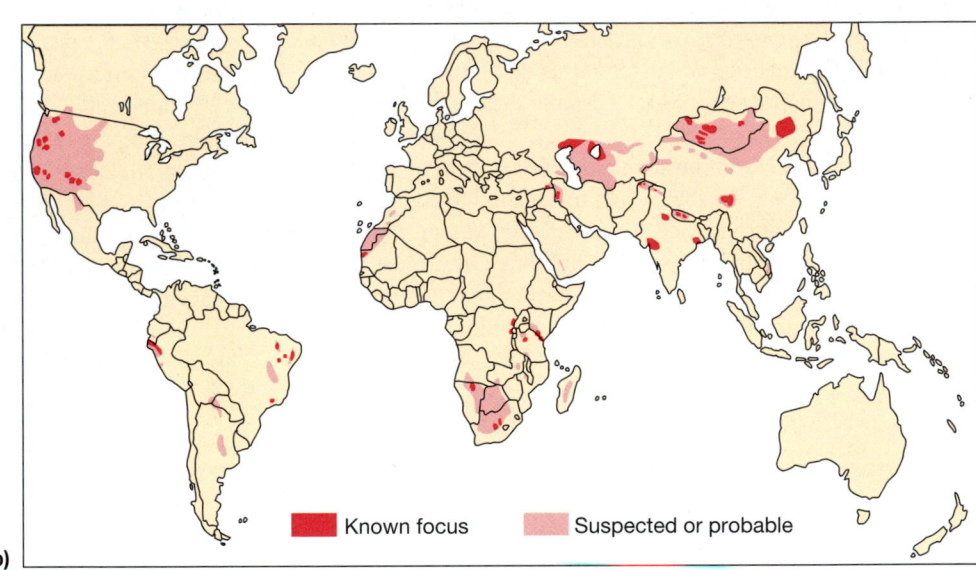

(b)

➤ Figure 23.7   **The incidence of plague.** (a) U.S. incidence, 1955–1997. (b) Worldwide regions of endemic plague. (*Source*: CDC)

➤ Figure 23.8   **Buboes.** In this drawing from a fourteenth-century Flemish manuscript, a physician lances a plague-caused bubo. A second patient awaits the lancing of his underarm bubo.

body, meningitis, and pneumonia. This form of plague is invariably fatal despite the best modern care. **Pneumonic plague** occurs with lung involvement and can be spread when aerosol droplets from a coughing patient are inhaled. It, too, has a mortality rate approaching 100 percent despite excellent care. Medical personnel working with patients are more likely to acquire pneumonic than bubonic plague.

Diagnosis, Treatment, and Prevention   Plague can be diagnosed by fluorescent antibody tests or by the identification of *Y. pestis* from stained smears of sputum or from fluid aspirated from lymph nodes. It is treated with streptomycin, tetracycline, or both. Fortunately no drug-resistant strains have yet appeared.

Recovery from a case of plague confers lifetime immunity. A vaccine is available to protect medical personnel and the families of victims. However, during treatment of plague patients in the Vietnam War, it was discovered that workers, even those protected by immunization, can become pharyngeal carriers of plague organisms for a short period of time. Plague can be prevented by immunizing those who travel to endemic regions, controlling rat populations, and maintaining surveillance for infections in wild rodent populations. CDC surveys have found plague only among wild rural (usually desert) rodents, which are not likely to come in contact with urban populations. However, plague has moved eastward from the California coast, where it first arrived on a ship from China that docked at San Francisco in 1899. If the disease spreads to urban rats, many of which are resistant to rat poisons, the risk to humans will increase.

Tularemia

The Disease   **Tularemia,** caused by *Francisella tularensis,* is a zoonosis found in more than 100 mammals—especially cottontail rabbits, muskrats, and rodents—and arthropod vectors, such as ticks and deer flies. In ticks the pathogen is incorporated in eggs as they leave the ovaries in **transovarian transmission**—infection of eggs before they are fertilized—and thereby passes from one generation to the next. Although in about half of human cases the vector is never identified, tularemia is most often associated with cottontail rabbits; the number of cases reported always rises significantly during rabbit-hunting season.

Francisella tularensis, a small Gram-negative coccobacillus with worldwide distribution, was first isolated in 1911 from Tulare County, California, for which the species was named. The genus was named after Edward Francis, who did much of the early work on this organism. The annual incidence of tularemia in the United States has dropped from more than 2000 cases in 1939 to fewer than 200 cases in recent years (➤ Figure 23.9). The disease is an occupational hazard for taxidermists.

Tularemia can be acquired in three ways. First, organisms usually enter through minor cuts, abrasions, or bites. Second, organisms can be inhaled, especially from aerosols formed during

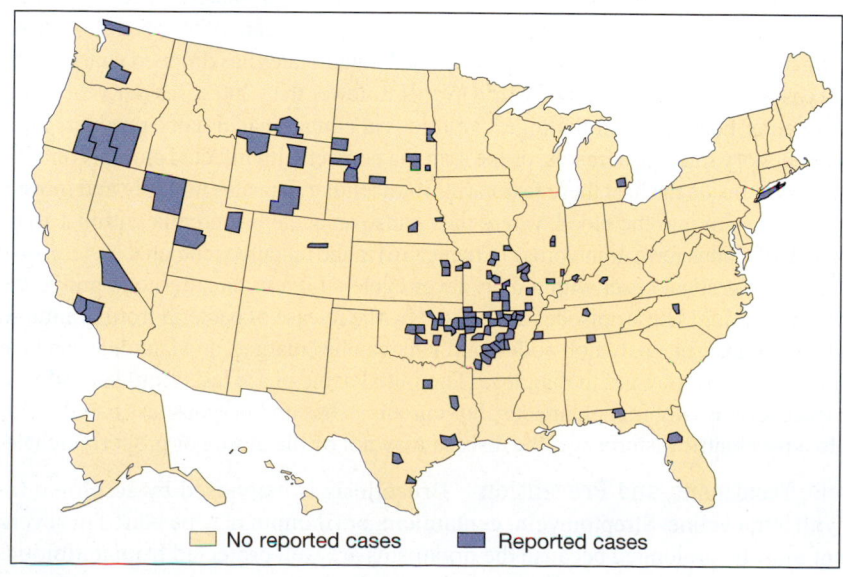

☐ No reported cases   ■ Reported cases

➤ Figure 23.9   **The U.S. incidence of tularemia, 1993.** (*Source:* CDC).

the skinning of infected animals. Third, organisms can be acquired by consumption of contaminated water or meat, resulting in an intestinal form of the disease. Swimming in a river near a colony of infected river rats and eating undercooked infected rabbit have been reported as sources of infection. Freezing, even for years, does not destroy the organisms.

Entry through the skin results in the **ulceroglandular** form of disease. After a 48-hour incubation period, symptoms begin with abrupt high fever of 40° to 41° C (104° to 106°F) with chills and shaking. If untreated, fever, severe headache, and buboes can last a month. An ulcer sometimes forms at the site of entry. The handling of animals and skins is most likely to cause ulcers on hands, whereas a bite by an arthropod vector is most likely to cause buboes (lymph node lesions) in the groin or armpit. The patient is initially disabled for 1 to 2 months and can have frequent relapses. The mortality is about 5 percent if left untreated. In the days of American pioneers, rabbit was a major part of the diet, and tularemia also must have been a major feature of pioneer life. Until the 1960s, when biological warfare was banned, *F. tularensis* was one organism studied by government scientists for possible use in biological warfare.

Bacteremia from lesions can lead to **typhoidal tularemia,** a septicemia that resembles typhoid fever. Touching the eyes with contaminated hands can lead to conjunctivitis, but this happens in very few cases. Inhalation of organisms or their spread from blood produces a patchy bronchial pneumonia that leads to lung tissue necrosis and 30 percent mortality.

Diagnosis, Treatment, and Prevention   Diagnosis from blood cultures is difficult; the highly infectious organisms are hard to grow on ordinary laboratory media. Only 50 organisms are sufficient to produce a human infection, regardless of the route of administration. Laboratory infections are easily acquired, so culturing of *Francisella tularensis* should be attempted only by very experienced personnel in isolation laboratories equipped with special air-flow hoods and other safety devices. Animal inoculations should be avoided. Agglutination tests are the standard methods of diagnosis ➡ (Chapter 18, p. 526) Streptomycin is the drug of choice for all forms of tularemia.

Prevention by eliminating the organism from wild reservoir populations is impractical. Therefore, one should avoid handling sickly animals, wear gloves when handling or skinning wild game, and, in tick-infested areas, wear protective clothing and frequently search clothing and skin for ticks. A vaccine exists, but it is not always protective and must be readministered every 3 to 5 years.

Brucellosis

The Disease   **Brucellosis,** also called **undulant fever** or **Malta fever,** is a zoonosis highly infective for humans. It is caused by several species of *Brucella. Brucella melitensis* was isolated in 1887 on the Mediterranean isle of Malta by Sir David Bruce. *Brucella* are small Gram-negative bacilli, each of which has a preferred host: *B. abortus,* cattle; *B. melitensis,* sheep and goats; *B. suis,* swine; and *B. canis,* dogs. In addition to preferred hosts, each species can infect several other hosts, including humans. The incidence of brucellosis in the United States has dropped sharply from more than 6000 cases per year at the end of World War II to fewer than 200 cases per year since 1978.

Brucella enter hosts through the digestive tract via contaminated dairy products and animal feed, the respiratory tract via aerosols, or the skin via contact with infected animals on farms or in slaughterhouses. Inside the host these facultative intracellular parasites multiply and move through the lymphatic system into the blood, where they cause an acute bacteremia within 1 to 6 weeks. Uncontrolled infections form granulomas (Chapter 16) in the reticuloendothelial system. ➡ (p. 446) Brucellosis has a gradual onset with a daily fever cycle—high in the afternoon and low at night after profuse sweating. These episodes are caused by the release of bacteria from granulomas into the bloodstream. The spleen, lymph nodes, and liver can be enlarged, and jaundice can be present, but symptoms may be too mild to diagnose. The initial acute phase lasts from several weeks to 6 months. Recovery usually is spontaneous, but chronic aches and nervousness can develop. If attributed to a psychiatric disturbance, the disease may not be diagnosed or properly treated.

Diagnosis, Treatment, and Prevention   Brucellosis is diagnosed by serologic tests and treated with tetracycline. Streptomycin, gentamicin, or rifampin may be added in severe cases. Treatment must be prolonged because the organisms are well protected from antibiotics in the

In Alaska, reindeer and caribou are known to be infected with brucellosis and serve as reservoir hosts. Eating their meat can cause human infection.

blood. When death does occur, it is usually from endocarditis. Brucellosis can be prevented by pasteurizing dairy products, immunizing animal herds, and providing education and protective clothing for workers with occupational exposure. No human vaccine is available at present.

Cattle ranchers in the vicinity of Yellowstone, Montana, face a danger from diseased bison who wander outside the park boundaries. Over 50 percent of the Yellowstone Park bison herd are infected with brucellosis. They can spread it to cattle when they mingle with them, especially during hard winters, when food may be more abundant on ranchers' lands. Hunters are allowed to kill bison who leave the park. Nearly 600 are shot in some years. The state of Montana has spent millions of dollars in the last decade to eradicate brucellosis from its herds.

Relapsing Fever

The Disease   **Relapsing fever** is an acute arthropodborne disease characterized by alternating fever and nonfever periods. Nearly a dozen species of the genus *Borrelia* are the causative agents (➤ Figure 23.10). Two different vectors transmit relapsing fever: soft ticks of the genus *Ornithodoros,* and human body and head lice of the genus *Pediculus.* Fifteen species of soft ticks are known to transmit relapsing fever. Louseborne infections are called **epidemic relapsing fever,** whereas tickborne infections are called **endemic relapsing fever.** Relapsing fever appeared in ancient Greece; since then outbreaks have occurred during times of war, poverty, and in situations of extreme overcrowding that occur, for example, after natural disasters. During epidemics, mortality can reach 30 percent of untreated cases.

Borrelia are large, spiral, Gram-negative bacteria; their coarser, more irregular spirals and ease of staining distinguish them from *Treponema* (syphilis) spirochetes.

Transmission of relapsing fever varies with the vector involved. Lice must be crushed and their body contents scratched into the skin to transmit the disease. Each louse acquires the organisms by biting an infected host. Ticks transmit disease organisms in their salivary secretions

(a)

(b)

➤ Figure 23.10   **Relapsing fever.** This disease is caused by various *Borrelia* species, such as (a) *Borrelia recurrentis* (among blood cells; 1250X), carried by ticks and by (b) lice such as the body louse.

Outwitted by the Spirochetes Again

Some microorganisms are really "smart." They'd have to be in order to survive against our modern-day arsenal of medicines and vaccines and outwit our ever-watchful immune systems. If you look at pathogens that way, then the smartest ones are those that have been around the longest. That would make the spirochete that causes relapsing fever quite clever, since it has been around for thousands of years. The *Borrelia* spirochete is transmitted to humans by tick bites. Within the mammalian host, these spirochetes periodically replace one outer surface antigen with another, a process that presumably allows them to evade our immune systems. You can't help but give these microbes some respect.

Borrelia burgdorferi, the causative agent of Lyme disease, penetrates the brains of some people but not the brains of others.

when biting and by transovarian transmission. The ticks can survive for up to 5 years without a meal and still contain live infectious *Borrelia*. Eradication of these vectors is impossible. The ticks feed mainly at night for only about half an hour; victims such as campers or occupants of houses infested with tick-carrying rodents never realize they have been bitten. Lack of this important piece of clinical history makes diagnosis difficult.

Much of what we know about relapsing fever was learned before the antibiotic era, when syphilis was treated by inducing high fever. Patients were purposely infected with relapsing fever or with malaria; the high fever killed the syphilis organisms and left the patient with a more manageable disease. After an incubation period of 3 to 5 days, relapsing fever begins with a sudden onset of chills and high fever. Fever persists for 3 to 7 days and ends by crisis. About 16 percent of patients never experience a relapse, but most, after 7 to 10 days, have fever for 2 or 3 days. Typically, more fevers followed by intervals of relief occur—hence the name relapsing fever. The disease is particularly dangerous in pregnant women because the organisms can cross the placenta and infect the fetus.

Relapses are explained by changes in the organisms' antigens. During a febrile period, the body's immune response destroys most of the organisms. The few that remain have surface antigens the host's immune system fails to recognize. These organisms multiply during a relief period until they are numerous enough to cause a relapse. Each relapse represents a new population of organisms that have evaded the host's defense mechanisms.

Diagnosis, Treatment, and Prevention   Diagnosis is made by identifying the organisms in stained blood smears prepared during a rising fever phase. Tetracycline or chloramphenicol are used to treat relapsing fever. Immunity following recovery is usually short-lived. No vaccine is available, so prevention primarily is directed toward tick and louse control as well as educating the public.

Lyme Disease

The Disease   Alteration of ecosystems can give rise to new human diseases or to increased incidence and recognition of previously unidentified diseases (*emerging diseases*), as has been demonstrated in the case of **Lyme disease.** More Virginia white-tailed deer, which thrive along borders between forests and clearings and are a major reservoir of the disease agent, now inhabit the United States than when the Pilgrims landed. Settlers clearing fields created suitable habitats; as hunting of deer for food has declined, their populations have increased to record levels. With this increase has come Lyme disease, first described in 1974 by Allen Steere and his colleagues at Yale University and named after the Connecticut town where the earliest recognized cases occurred. The disease has now been identified on three continents and in more than 46 states in the United States. It is common along with deer from Cape Cod to Virginia and in Minnesota (➤ Figure 23.11).

➤ **Figure 23.11   The U.S. incidence of Lyme disease, 1994.** What appears to be a dramatic spread of the disease may really be better diagnosis due to heightened awareness. However, the disease is also spreading in some areas. (*Source*: CDC)

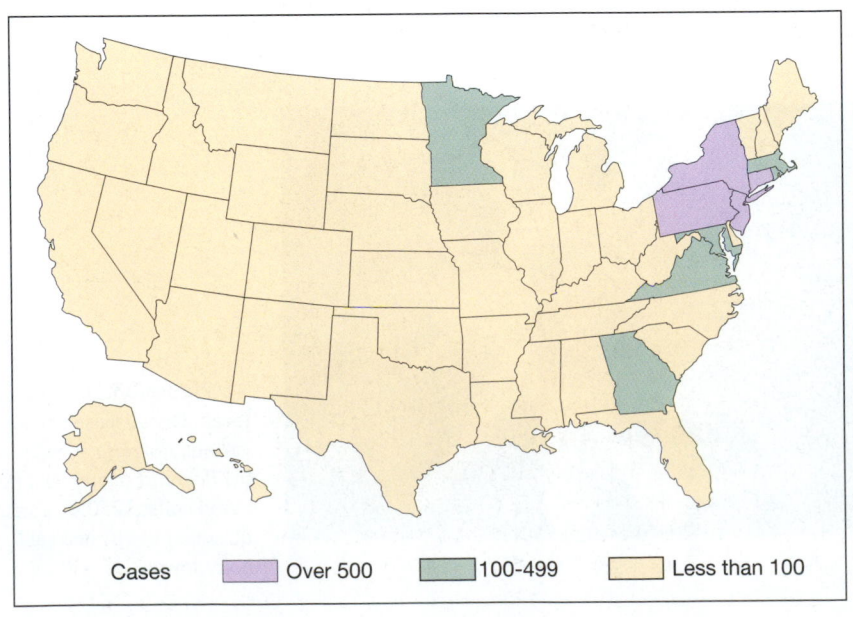

| Cases | ▮ Over 500 | ▮ 100-499 | ▮ Less than 100 |

In 1982 Willy Burgdorfer of the National Institutes of Health laboratory in Montana isolated and described the causative organism, *Borrelia burgdorferi,* a previously unknown spirochete. By 1985 Lyme disease was the most commonly reported tickborne disease in this country. It is carried by *Ixodes dammini,* the deer tick, which feeds on deer and small mammals such as mice. The tick takes three blood meals during its 2-year life cycle ➤ (Figure 23.12a). It ingests contaminated blood in one meal and transmits disease during a subsequent one. Dogs, horses, and cows as well as humans can be infected. As the tick spreads to areas of high human density, the risk of tickborne infections increases.

Diagnosis, Treatment, and Prevention   Symptoms of Lyme disease vary, but most patients develop flulike symptoms shortly after being bitten by an infected tick. A bull's-eye rash at the site of the bite is also seen in about half of all cases (➤ Figure 23.12b). Weeks or months later, other symptoms occur. Arthritis is the most common symptom, but loss of insulating myelin from nerve cells can cause symptoms resembling Alzheimer's disease and multiple sclerosis. Myocarditis occurs in some patients. Because most patients do not seek medical attention early, Lyme disease usually is diagnosed by clinical symptoms after arthritis appears. An antibody test is available. Lyme disease is treated with antibiotics such as doxycycline and amoxicillin, which are more effective the earlier they are administered in the course of the disease. Unfortunately, the earlier stages are often misdiagnosed. Furthermore, undiagnosed subclinical infections are common in endemic areas.

Although no vaccine for humans currently exists, a vaccine to protect dogs from Lyme disease is now available. Dogs are six times more likely than humans to develop the disease. Cats

Doxycyline is a good, inexpensive, broad-spectrum antibiotic that can be used to treat Lyme disease, bronchitis, and even malaria.

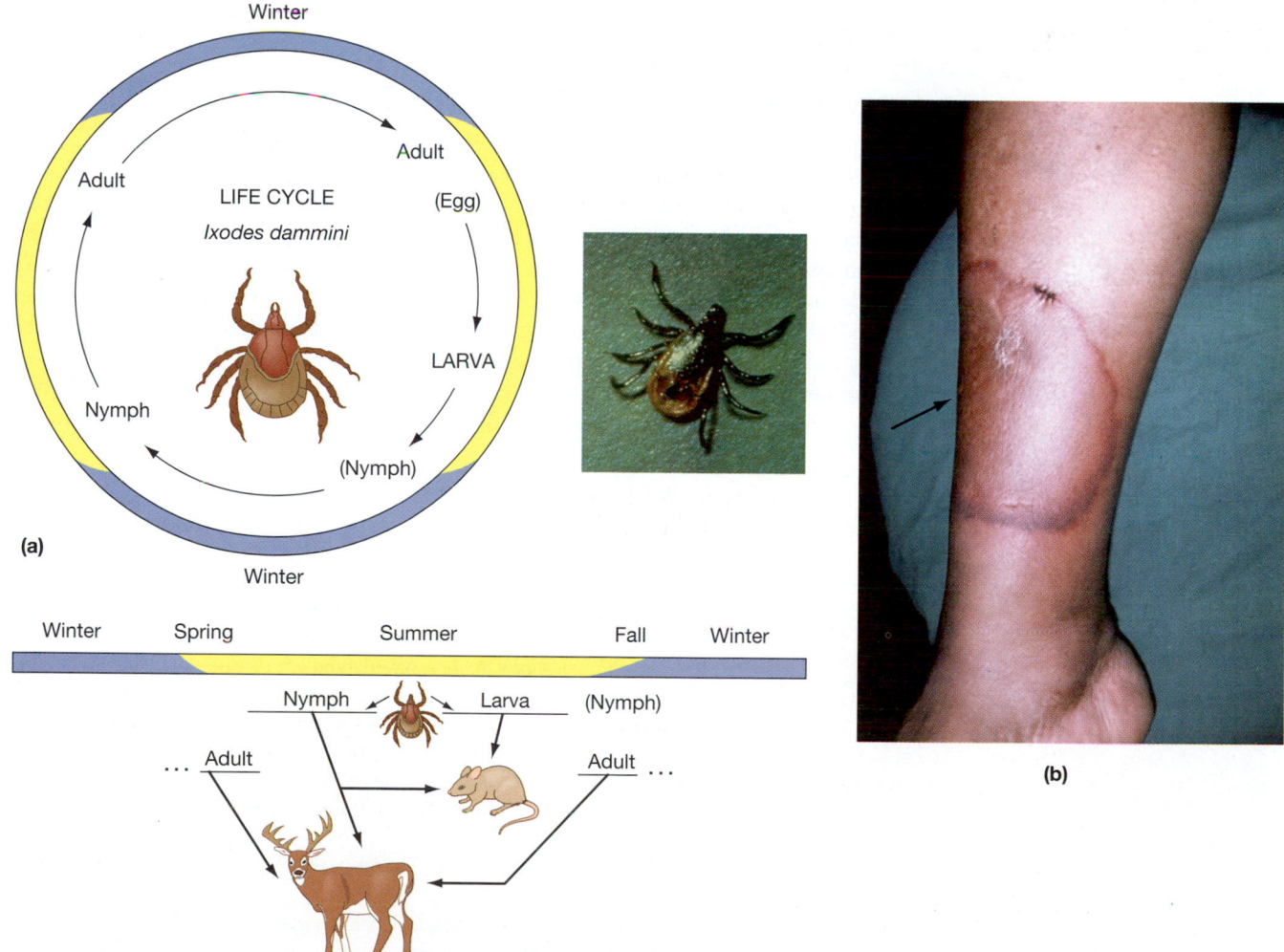

➤ **Figure 23.12   Lyme disease.** (a) The life cycle of the tick *Ixodes dammini.* (b) The typical "bull's-eye" rash of Lyme disease, showing concentric rings around the initial site of the tick bite.

are less likely to develop Lyme disease than dogs, possibly because of their continual grooming. Other animals such as cattle are also susceptible, as a Wisconsin dairy farmer discovered when three-quarters of his 60 dairy cows went lame from Lyme arthritic enlargements of leg joints, some nearly as large as basketballs. Only about 10 percent of infected people develop crippling Lyme arthritis. Researchers have found a genetic predisposition in those who do, with 89 percent having HLA-DR 4 or RLA-DR 2 tissue antigens. ➡ (Chapter 18, p. 513) Antibiotics given in the first 6 weeks after infection generally prevent this chronic arthritis. After a certain point in the course of the infection, however, antibiotics are no longer able to prevent the development of arthritis in susceptible individuals.

Because no vaccine for humans is available, Lyme disease can be controlled only by avoiding tick bites. Bites are difficult to avoid, however, because the tick—especially the immature form, which often carries the spirochete—is very tiny, the size of a poppy seed, and may go unnoticed. In tick-infested areas, a hat, long sleeves, and long pants should be worn, with socks pulled up over the bottom ends of the pants. Tick repellant can also be used. Prevention is a lot easier than cure. Once chronic Lyme disease is established, it can even cross the placenta and infect a fetus.

A drastic attempt to control the disease was tried on Great Island, Massachusetts, where an entire 52-member deer herd was killed. Putting antitick eartags on deer and mice, a difficult job, has not been very effective. Investigators have also tried saturating cotton balls with insecticide and leaving them around for the mice to take back to their nests, thereby killing off many of their ticks. Now, a European wasp that parasitizes the ticks is being released in tick-infected areas. This biological control method has given excellent results, almost exterminating the tick in some areas.

Bacterial systemic diseases are summarized in Table 23.2.

Rickettsial and Related Systemic Diseases

Rickettsias are named after Howard T. Ricketts, who identified them as the causative agents of typhus and Rocky Mountain spotted fevers. Both he and another investigator, Baron von Prowazek, died of laboratory infections from these highly infectious organisms. Rickettsias are small, Gram-negative coccobacilli and are microscopic obligate intracellular parasites. ➡ (Chapter 9, p. 251) Those that cause typhus fever grow in the cytoplasm of infected cells; those that cause spotted fever grow in both the nucleus and the cytoplasm. Rickettsias can be cultured only in cells. Embryonated eggs also are often used, but they should be handled only by very experienced technicians and only in specially equipped isolation laboratories. Despite improvements in laboratory design and techniques and the availability of vaccine, laboratory infections are still common, and ocaisionally fatal.

Symbiotic microbes and nematodes may be used in the future to control deer ticks, the arthropod vector responsible for the spread of Lyme disease.

✓ Why would anthrax made such a good agent for biological warfare?

✓ How does portal of entry affect plague symptoms and mortality rate?

✓ Why is it important to seek early treatment for Lyme disease?

➤ **Table 23.2** | **A summary of bacterial systemic diseases**

Disease	Agent(s)	Characteristics
Anthrax	*Bacillus anthracis*	Cutaneous lesions become necrotic; respiratory infections always fatal; intestinal infections similar to food poisoning
Plague	*Yersinia pestis*	Bubonic plague causes buboes, or enlarged lymph nodes, and hemorrhages turn skin black; septicemic plague occurs when organisms invade blood and cause hemorrhage and necrosis in many tissues; pneumonic plague from inhalation of organisms causes pneumonia
Tularemia	*Francisella tularensis*	Ulceroglandular form causes high fever, headache, buboes; bacteremia leads to typhoidal tularemia; inhalation leads to bronchopneumonia; relapses often occur
Brucellosis	*Brucella* species	Gradual onset of symptoms, cyclic fever, enlarged lymph nodes and liver, jaundice
Relapsing fever	*Borrelia recurrentis*	Several days of high fever, respites, and shorter periods of fever due to changes in organisms' antigens; can cross placenta
Lyme disease	*Borrelia burgdorferi*	Rash and flulike symptoms at onset, later arthritis and nerve and heart damage; can cross placenta

Rickettsial diseases have several properties in common. The organisms invade and damage the cells of blood vessel linings and cause the linings to leak. This leakage causes skin lesions and especially **petechiae** (pe-te′ke-e; singular: *petechia*), pinpoint-size hemorrhages most common in skin folds. It also causes necrosis in organs such as the brain and heart. Although each disease produces a particular kind of skin rash, all rickettsial diseases cause fever, headache, extreme weakness, and liver and spleen enlargement. Headaches can be very painful, causing the patient to appear confused.

Except for Q fever (Chapter 21), rickettsial diseases are transmitted among vertebrate hosts by an arthropod vector. ➥ (p. 615) Humans often are accidental hosts of zoonoses, but they are the sole hosts of epidemic typhus fever and trench fever. Because of the hazards of culturing organisms, rickettsial diseases usually are diagnosed by clinical findings and serologic tests. The diseases are treated with tetracycline or chloramphenicol; even these antibiotics merely inhibit, but do not kill, rickettsias. Therapy must be prolonged until the body's defenses can overcome the infection. Often rickettsias are not totally eliminated; they can remain latent in lymph nodes and have been known to reactivate 20 years after initial infection.

Typhus Fever

Typhus fever occurs in a variety of forms, including *epidemic, endemic (murine),* and *scrub typhus. Brill–Zinsser disease* is a recurrent form of endemic typhus.

Epidemic typhus, also called *classic, European,* or *louseborne typhus,* is caused by *Rickettsia prowazekii.* The disease is most frequently seen during wars and other conditions of overcrowding and poor sanitation. In 1812 epidemic typhus helped drive Napoleon from Russia; more recently, during World War I, it infected over 30 million Russians and killed 3 million. The history of warfare is filled with instances in which typhus was the "commanding general." These and other examples of the effects of epidemic typhus on human affairs are well documented in Hans Zinsser's book *Rats, Lice, and History.* Only after the discovery of the pesticide DDT during World War II were typhus epidemics halted.

Epidemic typhus is transmitted by human body lice. After a louse feeds on an infected person, rickettsias multiply in its digestive tract and are shed in its feces. When a louse bites, it also defecates; infected lice deposit organisms next to a bite and themselves die of typhus in a few weeks. As victims scratch bites, they inoculate organisms into the wound. Lice become infected by biting an infected human. They abandon dead bodies or people with high fevers, moving to and infecting new hosts.

After about 12 days' incubation, onset of fever and headache is abrupt and is followed 6 or 7 days later by a rash on the trunk that spreads to the extremities but rarely affects palms or soles. Antibiotic therapy should be started immediately. Without treatment the disease typically lasts up to 3 weeks, and mortality ranges between 3 and 40 percent. The disease can be prevented by eradicating lice with insecticides and by maintaining hygienic living conditions. A vaccine is available. Recovery generally gives lifetime immunity, except when Brill–Zinsser disease occurs.

Brill–Zinsser disease, or *recrudescent typhus,* is a recurrence of a typhus infection. It is named for Nathan Brill and Hans Zinsser, who studied it in the 1930s among New York City's Eastern European immigrants. Compared with first infections, this disease has milder symptoms, is shorter in duration, and often does not cause a skin rash. It is caused by reactivation of latent organisms harbored in lymph nodes, sometimes for years. Lice feeding on patients with Brill–Zinsser disease can transmit the organism and cause initial typhus infections in susceptible individuals. The disease can be prevented by preventing typhus itself and can be reduced in incidence by adequate antibiotic therapy of those already infected. Despite rigorous treatment, some victims of typhus still develop Brill–Zinsser disease later. It can be distinguished from epidemic typhus by the type of antibodies formed shortly after onset. Epidemic typhus first elicits IgM and then IgG antibodies, whereas Brill–Zinsser disease, being a secondary response, elicits primarily IgG antibodies. ➥ (Chapter 17, p. 471).

Endemic typhus, or **murine typhus** (so named for its association with rats—*murine* refers to rats and mice) is a fleaborne typhus caused by *Rickettsia typhi.* It occurs in isolated pockets around the world, including southeastern and Gulf Coast states in the United States, especially Texas. Its incidence in the United States is fewer than 100 cases per year. Fleas from infected rats defecate while biting, infecting the humans they bite. The host rubs organisms into the

---

Microbes and War

"Soldiers have rarely won wars. They more often mop up after the barrage of epidemics. And typhus, with its brothers and sisters—plague, cholera, typhoid, dysentery—has decided more campaigns than Caesar, Hannibal, Napoleon, and all the . . . generals of history. The epidemics get the blame for defeat, the generals the credit for victory. It ought to be the other way 'round. . . ."

—Hans Zinsser, 1935

bite wound or transfers them to mucous membranes, another portal of entry. After 10 to 14 days' incubation, onset of fever, chills, and a crushing headache is abrupt, followed by a rash in 3 to 5 days. The disease is self-limiting and lasts about 2 weeks if untreated. Mortality is about 2 percent.

Scrub typhus, or *tsutsugamushi* (soot″soo-ga-moosh′e) *disease,* is caused by *Rickettsia tsutsugamushi*. "Tsutsugamushi" is Japanese for "bad little bug." The "bug" that transmits this disease is a mite that feeds on rats in Japan, Australia, and parts of Southeast Asia. Mites drop off rodent hosts and infect humans with their bites. Scrub typhus was a problem during World War II and the Vietnam War when soldiers crawled on their bellies through low scrub vegetation to avoid snipers. After 10 to 12 days' incubation, scrub typhus begins abruptly with fever, chills, and headache. Many patients develop sloughing lesions at the bite sites and later a generalized spotty rash. In untreated cases, the fatality rate can reach 50 percent, but with prompt antibiotic treatment, fatalities are rare. Although no vaccine is available, infections can be prevented by controlling mite populations.

Rocky Mountain Spotted Fever

Rocky Mountain spotted fever was first recognized around 1900 in Rocky Mountain states such as Idaho and Montana. However, its geographic distribution suggests that "Appalachian spotted fever" would be a more appropriate name (➤ Figure 23.13). The incidence has stayed

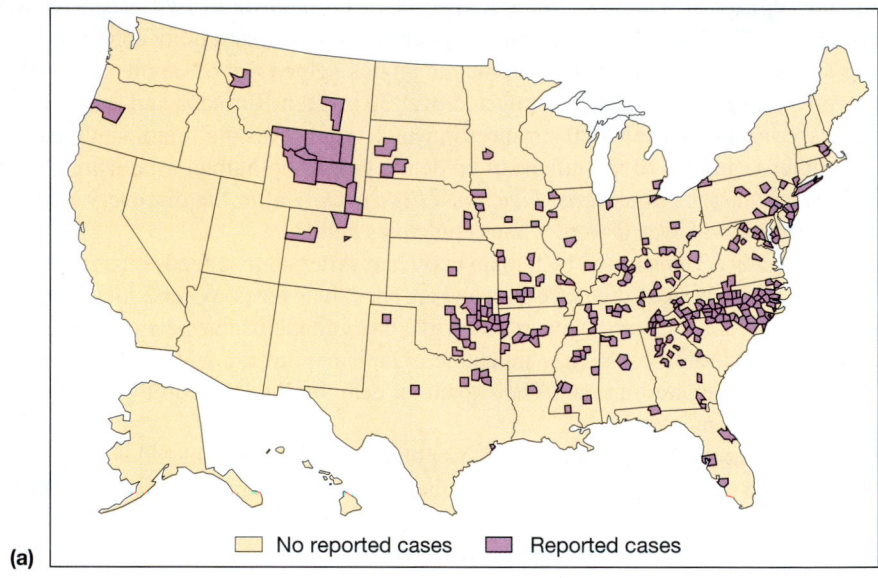

(a)   □ No reported cases   ▨ Reported cases

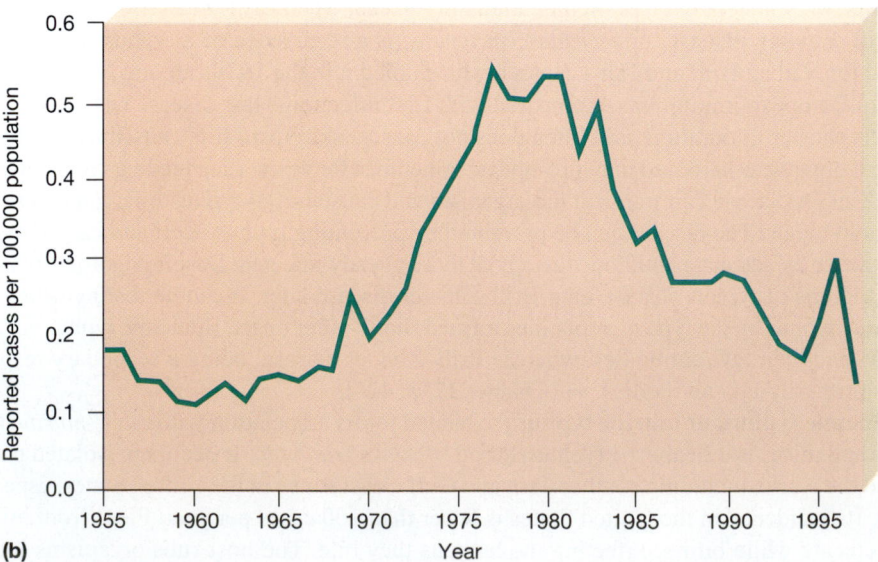

(b)

➤ Figure 23.13 **The U.S. incidence of Rocky Mountain spotted fever.** (a) By county, 1997; (b) by year. (*Source*: CDC)

(a)

(b)

> **Figure 23.14   Rocky Mountain spotted fever.** (a) Vectors include the dog tick *Dermacentor variabilis* (magnified 8X). (b) The characteristic rash of the disease on a patient's feet.

below 1000 cases per year since the 1930s but might increase as more people spend time in tick-infested areas. This disease is caused by *Rickettsia rickettsii* and is transmitted by ticks of the genus *Dermacentor* (➤ Figure 23.14a). After 3 to 4 days' incubation, onset of fever, headache, and weakness is abrupt, followed in 2 to 4 days by a rash (➤ Figure 23.14b). The rash begins on ankles and wrists, is prominent on palms and soles, and progresses toward the trunk—just the reverse of the progression in typhus. Spots are caused by blood leaking out of damaged blood vessels beneath the skin surface; they coalesce as blood leaks from many damage sites. Blood vessels in organs throughout the body are similarly damaged.

Strains vary considerably in virulence; likewise, mortality varies from 5 percent to 80 percent, with an average of 20 percent in untreated cases. Prompt antibiotic treatment keeps the mortality rate to between 5 and 10 percent. Rocky Mountain spotted fever can be prevented by wearing protective clothing and by vigilantly inspecting clothing and skin during visits to tick-infested areas. Inspecting children's hair is especially important. The only vaccine is not completely effective.

Worldwide, there are many other types of spotted fevers. Each is caused by a specific rickettsial species and is typically named for its location, such as "Siberian spotted fever."

Rickettsialpox

Rickettsialpox, caused by *Rickettsia akari,* was first discovered in New York City in 1946 and is now known to occur in Russia and Korea as well. It is carried by mites found on house mice. The disease is relatively mild, and its lesions resemble those of chickenpox. Because of misdiagnosis as chickenpox or other diseases, incidence and mortality data are unreliable, but no fatalities have been reported. It can be prevented by controlling rodents.

Trench Fever

Trench fever, also called *shinbone fever,* resembles epidemic typhus in that it is transmitted among humans by lice and is prevalent during wars and under unsanitary conditions. Stress probably is a predisposing factor. The causative agent is the bacterium *Bartonella (Rochalimaea) quintana,* classified with rickettsias even though it is not an obligate intracellular parasite. It can be cultured in artificial media and has worldwide distribution, but it rarely causes disease.

Trench fever was first seen in World War I among soldiers living in trenches and wearing the same clothing day after day. British "trench coats" were developed to protect troops who were without shelter during bad weather, but they were not entirely effective. Soldiers and their clothing, including the trench coats, became infested with body lice; exhausted soldiers crowded together in filth fell victim to the disease. After World War I the disease disappeared and did not reappear until World War II. Cases of trench fever have recently been found among the urban poor. The symptoms of trench fever are a 5-day fever and severe leg pain, but many soldiers have

➢ Figure 23.15   **Lesions of verruga peruana.**

reported recurrent symptoms, including mental confusion and depression, as long as 19 years after infection. No vaccine is available, and prevention depends on control of lice.

Bartonellosis

Bartonellosis (bar″to-nel-o′sis) is caused by *Bartonella bacilliformis,* named for the Peruvian physician A. L. Barton, who first described it in 1901. The disease occurs in two forms: **Oroya fever,** or *Carrion's disease,* an acute fatal fever with severe anemia, and **verruga peruana** (ver-oo′gah per-oo-a′nah), a chronic, nonfatal skin disease (➢ Figure 23.15). Both are found only on the western slopes of the Andes in Peru, Ecuador, and Colombia—the habitat of the sandfly *Phlebotomus,* which transmits the organism. In 1885 Daniel Carrion, a Peruvian medical student, inoculated himself with material from a wartlike verruga peruana lesion to show a connection between it and Oroya fever. His death 39 days later from Oroya fever clearly demonstrated the connection.

After being transmitted to a human host by the bite of an infected sandfly, *B. bacilliformis* enters the blood and multiplies during an incubation period of a few weeks to 4 months. Little is known of the epidemiology of bartonellosis, but humans appear to be the sole reservoir. Oroya fever is a severe, febrile, hemolytic anemia. Verruga peruana causes only skin lesions that persist for 1 month to 2 years but usually last for about 6 months. Lesions heal spontaneously but can recur. Oroya fever probably develops in people with no immunity, and verruga peruana occurs in those with partial immunity. Penicillin, tetracyline, or streptomycin can cure Oroya fever but not verruga peruana. No vaccine is available, and prevention depends on the control of sandflies.

Ehrlichiosis

Some recently identified human pathogens related to the rickettsias belong to the genera *Ehrlichia* and *Bartonella.* They are Gram-negative coccobacilli and are microscopic obligate intracellular parasites.

Ehrlichiosis, originally recognized as a disease in dogs, has now been found in humans. *Ehrlichia canis* and *E. chaeffeensis,* the causative agents, are spread by dog ticks, as well as by ticks that spread Lyme disease. Clinically, the human form of the disease resembles other rickettsial diseases. Typical symptoms include fever, headache, hepatitis, and muscle pain. Blood abnormalities, such as a reduction of white blood cells, are common. The absence of a rash distinguishes ehrlichiosis from Rocky Mountain spotted fever. Diagnosis requires special serological tests, but finding typical erhlichial intracytoplasmic inclusions in white blood cells may suggest the presence of *Ehrlichia.*

The characteristics of rickettsial diseases are summarized in Table 23.3.

Bacillary Angiomatosis

Bacillary angiomatosis is caused by *Bartonella henselae,* another rickettsial organism. The disease involves small blood vessels of the skin and internal organs. It is seen in persons with AIDS and other immunocompromised patients. Molecular probes and related techniques are used in diagnosis. The organism is also the cause of cat scratch fever. ➥ (Chapter 19, p. 559)

✓ How does Q fever differ from other rickettsial diseases?

✓ How does *Bartonella (Rochalimaea) quintana* differ from other rickettsiae?

> **Table 23.3** **A summary of rickettsial diseases**

Disease	Causative organism	Geographic area of prevalance	Arthropod vector reservoir	Vertebrate reservoir
Typhus group				
Epidemic (classic, European) typhus	*Rickettsia prowazekii*	Worldwide	Louse	Human
Brill–Zinsser disease (recrudescent typhus)	*R. prowazekii*	Worldwide	(Recurring infection)	Human
Endemic (murine) typhus	*R. typhi*	Worldwide, small scattered foci	Flea	Rodents
Scrub typhus group				
Scrub typhus (tsutsugamushi disease)	*R. tsutsugamushi*	Japan, Southeast Asia	Mite	Rat
Spotted fever group				
Rocky Mountain spotted fever	*R. rickettsii*	Western Hemisphere	Tick	Rodents, dogs
Rickettsialpox	*R. akari*	United States, Korea, Russia	Mite	House mouse
Trench fever	*Bartonella (Rochalimaea) quintana*	Worldwide, but disease only during wars	Louse	Human
Bartonellosis	*Bartonella bacilliformis*	Western slopes of the Andes	Sandfly	Humans only known host
Ehrlichiosis	*Ehrlichia canis, E. chaffeensis*	Southeastern and south central United States	Tick	Dogs, humans

Viral Systemic Diseases

Dengue Fever

Dengue (den′ga or den′′ge) **fever,** first characterized in 1780 by the American physician Benjamin Rush in Philadelphia, has also been called *breakbone fever* because of the severe bone and joint pain it causes. Other symptoms include high fever, headache, loss of appetite, nausea, weakness, and, in some cases, a rash. The disease is self-limiting and runs its course in about 10 days. Four distinct immunological types of the dengue virus—an arbovirus of the family Flaviviridae—have been identified, and two of them have been correlated with disease symptoms. A first dengue fever infection produces the symptoms just noted. A second dengue fever infection with a different immunologic type of the virus causes a hemorrhagic form of the disease. The hemorrhage occurs while the virus is replicating in circulating lymphocytes. It involves an immune response made possible by prior infection. Other symptoms are rapid breathing and low blood pressure that may progress to shock. The shock is reversible if treatment is initiated promptly. Serologic tests are available to diagnose dengue fever, and a vaccine against one immunological type of the virus appears to confer immunity.

Dengue fever is distributed worldwide in tropical areas, causing 500,000 to 3 million cases per year, with occasional episodes in the subtropics. It is endemic in 101 countries. Currently South America and the Caribbean are experiencing serious outbreaks of dengue fever—enough to now make health authorities refer to it as an *emerging disease.* Its main vector is *Aedes aegypti,* although in some areas the Asian tiger mosquito *A. albopictus* can be important. Since 1985, health officials have been concerned about the arrival and spread of *A. albopictus* in the United States. The mosquito apparently arrived in used tire casings imported from Asia. In 1980 the first outbreak of dengue fever in the United States in about 40 years occurred. Since then, all four serotypes have arrived in the United States. The rapid spread and aggressive biting habits of *A. albopictus* could bring dengue fever to areas that previously had been safe.

More Than Those Slave Owners Bargained For

The slave trade brought more to the Americas than slaves—it brought the yellow fever mosquito, *Aedes aegypti.* The mosquitoes laid their eggs in the water cisterns of sailing ships and lived almost exclusively on the blood of their human shipmates after hatching. And once the boat docked in the Western Hemisphere, the mosquitoes promptly began spreading diseases: in particular, the devastating epidemics of yellow fever that didn't end until the dawn of the twentieth century, and dengue and dengue hemorrhagic fever in even more ferocious epidemics.

In 1802, Napoleon lost 22,000 of his 25,000 troops to yellow fever in Haiti while trying to stop a slave revolt.

> ➤ **Figure 23.16 The Epstein–Barr virus.** A colorized TEM (200,000X) of the virus that causes infectious mononucleosis and, along with other factors, leads to Burkitt's lymphoma.

Mosquito control is the primary method of preventing the disease, as vaccines are not available for all dengue viral serotypes.

Yellow Fever

Yellow fever was first studied by Carlos Finlay and Walter Reed in the early 1900s when infection of workers threatened to disrupt the construction of the Panama Canal. Although adequate techniques to identify the causative flavivirus (Chapter 10), an arbovirus, were not available, Finley and Reed identified the mosquito vector, *Aedes aegypti,* and instituted control measures that prevented transmission of the disease. ➡ (p. 263) The disease is now limited to tropical areas of Central and South America and Africa. Incidence is greatest in remote jungle areas, where monkeys serve as reservoirs of infection and carrier mosquitoes bite both monkeys and people. In recent decades the number of reported cases each year has varied from 12 to 304, but yellow fever incidence has been vastly underreported. The actual number of cases worldwide is probably 200,000 annually, 30,000 of whom die each year.

Yellow fever causes fever, nausea, and vomiting, which coincide with viremia. Liver damage from viral replication in liver cells causes the jaundice for which the disease is named. The disease is of short duration: In less than a week, the patient has either died or is recovering. In most instances the fatality rate is about 5 percent, but in some epidemics it reaches 30 percent. Two strains of yellow fever viruses are used to produce vaccines. The Dakar strain is scratched into the skin, whereas the 17D strain is administered subcutaneously. Both are effective in establishing immunity.

Infectious Mononucleosis

In 1962 Dennis Burkitt suggested that a virus caused the lymphoid malignancy now called *Burkitt's lymphoma,* found in children in East Africa. This herpesvirus, now called **Epstein–Barr virus (EBV)** or human herpesvirus number 4 (➤ Figure 23.16), is known to cause both Burkitt's lymphoma, most cases of **infectious mononucleosis,** and oral hairy leukoplakia, a disease found among AIDS cases (see Table 10.3). ➡ (Chapter 10, p. 267) EBV infects primarily human B lymphocytes. It replicates like most other herpesviruses and derives its envelope from the inner nuclear membrane of the host cell. The virus has an unusually large number of genes—more than 50 different proteins are produced by complete expression of the DNA in EBV.

EBV enters the body through the oropharynx. The virus first infects epithelial cells and eventually B cells. It establishes a persistent infection in which viruses are shed for months to years. The viruses invade such sites as lungs, bone marrow, and lymphoid organs, where they infect certain kinds of mature B lymphocytes. They penetrate B lymphocytes over a 12-hour period, and EBV replication begins within 6 hours after penetration. Viral DNA replicates much faster than cellular DNA. Viral DNA can exist as circular plasmids or can become integrated into the cellular DNA.

EBV exerts three significant effects on lymphocytes:

1. The virus acts on antibody-producing cells and elicits EBV antibodies.

2. Infection and transformation of viral genes to the cellular DNA (or presence of the viral chromosome as a free plasmid in the host cell) are complex events that occur primarily in B lymphocytes, which have receptors for EBV. The cells produce a variety of antigens, some of which are recognized by T cells. This recognition causes the T cells to proliferate, which accounts for the excess of lymphocytes seen in infectious mononucleosis.

3. Other antigens are induced on the surface of some infected B cells. They appear to play a role in some B cell and T cell interactions and may account for some symptoms of infectious mononucleosis.

The proliferation of EBV-infected lymphocytes is limited by cytotoxic T cells and cells that make humoral antibodies and complement. ➡ (Chapter 17, p. 477) If these defenses fail to limit lymphocyte proliferation, uncontrolled B cell proliferation can lead to B cell cancer or Burkitt's lymphoma.

Infectious mononucleosis is an acute disease that affects many systems. Lymphatic tissues become inflamed, some liver cells become necrotic, and monocytes accumulate in liver sinusoids. In some cases myocarditis and glomerulonephritis are seen. The incubation period for the disease is from 30 to 50 days. Mild symptoms—headache, fatigue, and malaise—occur during the first 3 to 5 days of the disease and worsen as the disease progresses. About 80 percent of patients have a sore throat during the first week. The spleen is enlarged, and cells in lymphoid tissues in the oropharynx multiply. The tonsils are coated with a gray exudate, and the soft palate may be covered with petechiae. Secondary infection with β-hemolytic streptococci frequently occurs. Although the disease causes great discomfort and requires several weeks of recuperation, fatalities are rare and usually result from underlying immunological defects.

Diagnosis of EBV infections is complicated by the fact that the disease resembles cytomegalovirus infections, toxoplasmosis, and acute leukemia. The distinguishing symptoms of EBV infections are the concurrent sore throat, multiplication of lymphocytes, and the presence of antibodies against the antigens on sheep and human erythrocytes. Infectious mononucleosis is treated with bed rest and antibiotics for secondary infections. Ampicillin is not used because it causes a rash in infectious mononucleosis patients. The presence of IgG antibodies to viral capsid protein indicates a past infection, and their numbers furnish an index of immunity. Increasing numbers of IgM antibodies to the protein are evidence of a current infection. No vaccine is available.

In developing countries, the entire population has antibodies to EBV by age 1. Exposure to the virus in infancy produces mild symptoms or no symptoms and confers immunity to later infection. Where living standards are higher, a more severe disease is seen later in life. The incidence of infectious mononucleosis in the United States is highest among relatively affluent teenagers and young adults; 10 to 15 percent become infected. The affected age group and the large inoculum required to transmit the disease may be responsible for its being called the "kissing disease." Actually, infectious mononucleosis is not highly contagious. About 15 percent of people who have recovered from the disease are shedding low levels of virus in their saliva at any one time. Those who have recently been infected will shed the virus continuously for about 18 months. About half of immunosuppressed individuals constantly shed large amounts of the virus. Once infected by EBV, even after symptoms have disappeared, the virus remains in B cells in a latent state as a permanent infection. Immunosuppression allows the virus to resume replication. In a person with a normal immune system, any reactivation of EBV is quickly halted.

Burkitt's Lymphoma   **Burkitt's lymphoma,** a tumor of the jaw and viscera such as liver and spleen, is seen mainly in children (➤ Figure 23.17). It occurs about 6 years after the primary infection with EBV, but genetic and environmental factors play a role in the development of the disease. The tumor frequently arises from a single cell. The immune system of affected individuals appears to be normal, but it is incapable of eliminating the tumor cells. Burkitt's lymphoma is found mainly in regions of Africa where malaria is endemic, and infection with malarial parasites may enhance growth of the virus or interfere with the immune response.

Other Effects   Another tumor associated with EBV, which also shows a distinct geographic localization, is *nasopharyngeal carcinoma,* found most often in China and rarely in the Western Hemisphere. It is the most common tumor in the male population of southern China, comprising about 20 percent of all cancers. Genetic and cultural patterns predispose to this tumor. Chemicals consumed in salted fish, plus plant extracts (from the plant family *Euphorbiaceae*) used in Chinese traditional medicine, can induce replication of EBV.

Individuals with immunological defects are especially susceptible to developing lymphomas, presumably because they lack the necessary immune mechanisms to eliminate the malignant cells. Cyclosporine A, used to depress immunity in organ transplant patients, enhances lymphoid cell growth in donor organs. Organs from donors who have had EBV infections may contain EBV; immunosuppression from any cause, including AIDS, can release EBV and lead to mononucleosis or malignancy. Recipients of EBV-free organs will probably have latent EBV infections of their own, which can then proliferate due to the use of immunosuppresive drugs to prevent organ rejection. Blood transfusions can also transmit EBV.

➤ **Figure 23.17   Burkitt's lymphoma.** This condition is a form of cancer of the jaw that results from infection with the EBV; it is usually seen only in African children.

Chronic Fatigue Syndrome   Since 1985, researchers have been attempting to determine whether a *chronic EBV syndrome* exists. Patients report fever and persistent fatigue along with a variety of other nonspecific symptoms similar to those of infectious mononucleosis. Some,

Ebola Virus Scare

In April and May of 1995, an outbreak of the Ebola virus occurred in Zaire, Africa. The outbreak was traced to a forest worker who had become infected in December 1994. People infected with the virus develop symptoms of hemorrhagic fever, including fever and muscle aches. Most patients experience respiratory and kidney problems, abdominal pain, sore throat, and severe bleeding as well. Because the blood of Ebola patients cannot clot properly, bleeding occurs from needle injection sites, the digestive system, internal organs, and the skin. The virus is spread by direct contact with body fluids. Thus, outbreaks can be limited by isolating infected individuals, sterilizing needles and syringes, disposing of hospital waste and corpses properly, and using masks, gowns, boots, and gloves by hospital personnel—who are at special risk. Thousands of dollars worth of such supplies, as well as medicines and blood plasma, have been donated from around the world to limit the spread of the latest Ebola outbreak.

The reservoir for the Ebola virus is unknown. The first known outbreaks occurred in 1976 near the Ebola River (for which the virus was named) in Zaire and in western Sudan. That first Zaire outbreak resulted from Ebola-contaminated needles and syringes that were reused without sterilization. In fact, most of the outbreaks, including the one in 1995, were the result of poor medical conditions in hospitals.

As of mid-1998 there were four known Ebola viruses: subspecies Zaire, Sudan, Tai, and Reston. Ebola Zaire and Ebola Sudan can infect humans. Ebola Tai is the first Ebola virus known to be passed to humans from other animals—naturally infected chimpanzees of the Tai Forest in the Ivory Coast. Because infections of chimps, like humans, can lead to death, the original source of the virus is not the chimps. Ebola Reston, isolated from infected monkeys delivered to Reston, Virginia, in 1989, appears to be spread only among monkeys, but in this case aerosol transmission was documented.

Ebola victim in Kikwit, Zaire, in 1995.

You're in Luck, As Long As You're a Guinea Pig

Imagine waking up to a real-life scene from a horror film. You are vomiting blood and bleeding profusely, both internally and from the nose, eyes, and gums. You know that there is a 50 to 90 percent chance that you will die within 2 weeks—or maybe even within days—from blood loss and shock. You have Ebola. There is no cure and no vaccine available. Now imagine waking up to that same horror scene, except this time in the body of a guinea pig. You have much less need to fear. An experimental vaccine has already been developed by scientists at the University of Michigan and at the CDC in Atlanta that will give your immune system enough time to muster a response against Ebola. You sigh in relief. Now if you could only change back into a human.

but not all, have EBV antibodies. Others have had measles or herpesvirus infections. Because a direct relationship between the symptoms and previous EBV infection has not been established, the illness has been renamed **chronic fatigue syndrome.** More study is needed to determine whether the syndrome is associated with one or more previous viral illnesses or whether it might be a psychological disorder. Some recent studies point to possible immune system defects. Herpes human virus number 6 (HHV6), the recently discovered cause of the childhood disease roseola, is also a candidate for causative agent of chronic fatigue syndrome.

Other Viral Infections

Filovirus Fevers   **Filoviruses,** or filamentous viruses, display unusual variability in shape. Some are branched, others are fishhook- or U-shaped, and still others are circular. They contain negative-sense RNA in a helical capsid and vary in length from 130 to 4000 nm. ➡ (Chapter 10, p. 263) Two filoviruses have been associated with human disease. The **Ebola virus** caused outbreaks of hemorrhagic fever first in 1976, with a mortality of 88 percent in Zaire and 51 percent in Sudan. Nearly one-fifth of the population of rural areas of Central Africa have antibodies to Ebola. Transmission is person to person. A 1995 outbreak of Ebola virus in Zaire became world news; more than 200 cases were documented, with a mortality rate of about 75 percent. The *Marburg virus* was first recognized in Germany when technicians preparing monkey kidney cell cultures died of a hemorrhagic disease. Nosocomial Marburg virus infections have since been encountered with a mortality of about 25 percent. Hemorrhage into skin, mucous membranes, and internal organs, death of cells of the liver, lymph tissue, kidneys, and gonads, and brain edema also have been observed. The virus has been isolated directly from monkeys and from laboratory inoculation of guinea pigs.

Bunyavirus Fevers   Infections caused by bunyaviruses begin suddenly with fever and chills, headache, and muscle aches. Although usually not fatal and without permanent effects, the infections are temporarily incapacitating. When encephalitis occurs, it progresses slowly, either because the viruses replicate slowly in neural tissues or because certain viruses capable of replicating in neural tissue are selected. Rats, bats, and animals with hooves serve as reservoirs of infection. Tropical and temperate forest mosquitoes are vectors.

The *LaCrosse bunyavirus* has been identified in the northeastern and north-central United States. It causes a mild disease in adults but can cause seizures, convulsions, mental confusion, and paralysis in children. The *California encephalitis virus,* a bunyavirus initially isolated from mosquitoes in the San Joaquin Valley, has similar effects on humans.

Bunyaviruses called **phleboviruses** (because they are carried by the sandfly *Phlebotomus papatasii*) have been recovered from human infections. The **Rift Valley fever** virus causes epidemics and has unpredictable virulence. It causes sudden vomiting, joint pain, and slowing of the heart rate. One epidemic of Rift Valley fever in 1975 in central Africa infected thousands but left only 4 people dead. Only 2 years later the disease appeared in Egypt, where 200,000 cases and 598 deaths were recorded.

The *hantaviruses* are associated with hemorrhagic fevers, including Korean or epidemic hemorrhagic fever, and kidney disease. They are distributed widely over Eurasia and find reservoirs in rodents. They cause capillary leakage, hemorrhage, and cell death in the pituitary gland, heart, and kidneys. Kidney damage can be severe, and low blood pressure can proceed quickly to shock, from which one-third of patients die. Even more die if bleeding occurs in the gastrointestinal tract and the central nervous system or if fluid accumulates in the lungs. The major source of such infections is contact with infected rodents or their excreta. Some rodents shed viruses in feces and saliva for 30 days and into their urine for a year. Humans under 10 or over 60 years are rarely infected, perhaps because they are less likely to come in contact with infected materials.

Arenavirus Fevers   Like bunyaviruses, arenaviruses cause hemorrhagic fevers. Of these **Lassa fever** is perhaps the most widely known. It is an African disease that begins with pharyngeal lesions and proceeds to severe liver damage. The prognosis is poor in 20 to 30 percent of cases in which hemorrhage from mucous membranes occurs. Several other arenavirus infections, including **Bolivian hemorrhagic fever,** have been identified in humans, especially in Africa and South America.

Bolivian and other *South American hemorrhagic fevers* are multisystem diseases with insidious onset and progressive effects. The viruses attack lymphatic tissues and bone marrow and cause vascular damage, bleeding, and shock. However, death, which occurs in about 15 percent of cases, usually results from damage to the central nervous system. How the viruses affect the nervous system is not known.

Colorado Tick Fever   **Colorado tick fever** is caused by an **orbivirus** (a member of the Reoviridae) that is transmitted to humans by dog ticks from reservoir animals such as squirrels and chipmunks. High levels of viremia with infection of immature erythrocytes occur. The patient suffers from headache, backache, and fever, but recovery usually is complete.

Parvovirus Infections   Two parvovirus infections are now known to affect cats and dogs, respectively. **Feline panleukopenia virus (FPV)** causes severe disease with fever, decreased numbers of white blood cells, and enteritis in cats. The virus replicates in blood-forming and lymphoid tissues and secondarily invades the intestinal mucosa. In 1978 a new virus, the **canine parvovirus,** appeared and infected dogs in widespread geographic areas. This virus first surfaced in North America, Europe, and Australia and rapidly spread worldwide. It causes severe vomiting and diarrhea in dogs of all ages and sudden death with myocarditis in puppies under 3 months old. When the canine parvovirus first appeared in an area, the death rate often exceeded 80 percent. Vaccines are now available for both feline panleukopenia and canine parvovirus.

Aplastic Crisis   A member of the genus *Erythrovirus* also called B19 has been identified as the probable cause of **aplastic crisis**—a period in which erythrocyte production ceases—in sickle-cell anemia. The virus appears to replicate in rapidly dividing cells of the bone marrow. Afflicted children soon are in acute distress. Although normal erythrocytes remain functional

No Good Choices

If you had a choice, which would you choose: Bolivian hemorrhagic fever or malaria? Probably neither. But this was a dilemma facing the Bolivians. For years, hemorrhagic fever existed in Bolivia without causing major problems. This situation changed when a campaign was initiated to control malaria by eradicating mosquitoes. Large quantities of pesticides were used, killing many village cats; in the absence of cats, the rodent population multiplied and brought arenavirus into human dwellings. This caused the incidence of hemorrhagic fever to rise to epidemic levels.

in the blood for about 120 days, sickled cells survive only 10 to 15 days. Under such circumstances severe destruction of erythrocytes occurs. A child generally experiences only one such crisis, possibly because immunity to the virus develops. Another bit of evidence that aplastic crisis is infectious is that, although it occurs only in sickle-cell anemia patients, it does occur in 3- to 5-year cycles within particular communities.

Fifth Disease   The erythrovirus B19 destroys the stem cells that give rise to red blood cells. This is not a major problem in healthy adults or children, but it is a serious danger to those who have chronic hemolytic anemias, such as sickle-cell anemia, and who therefore have difficulty maintaining normal levels of red blood cells. It is also a danger to the fetus if a pregnant woman contracts the virus. The virus can be transmitted across the placenta and cause the fetus to develop fatal anemia. The virus does not, however, cause birth defects. Immunodeficient patients cannot control replication of the virus and may develop chronic anemia. After B19 was recognized as the cause of aplastic crisis in sickle-cell anemia, it was also found to cause **fifth disease** (*erythema infectiosum*) in normal children. It is common in children of ages 5 to 14 and often goes unnoticed. Infected children have a bright red rash on the cheeks that may spread to the trunk and extremities. A low-grade fever may accompany the rash. Often the infection is totally asymptomatic. The virus appears to be spread via the respiratory route. The disease is self-limiting and confers lifetime immunity. The name "fifth disease" comes from a nineteenth-century list of childhood rash diseases. The first disease on this list is scarlet fever; the second is rubeola; the third, rubella; the fourth, epidemic pseudoscarlatina (a type of sepsis); and the fifth, erythema infectiosum.

Coxsackie Virus Infections   *Coxsackie viruses* have an affinity for the pericardium and myocardium. They can cause meningoencephalitis, diarrhea, rashes, pharyngitis, and liver disease. Epidemic muscle pain, diabetes, and inflammation of the pancreas, heart muscle, and the pericardium are associated with coxsackie B viruses. Coxsackie viruses are highly infectious and readily spread among family members and in institutions. Most infections probably arise from fecal–oral transmission, but because the viruses can be isolated from nasal secretions, infection also may occur by respiratory fomites. Coxsackie virus infections during pregnancy can cause congenital defects, but their incidence is much lower than that for rubella virus infections, and aborting the fetus is usually not recommended. No effective means are available for treatment, immunization, or prevention.

Protozoan Systemic Diseases

Leishmaniasis

Three species of protozoa of the genus *Leishmania* can cause **leishmaniasis** (lish″man-i′a-sis) in humans (➢ Figure 23.18a). The protozoa are transmitted by sandflies. When an infected sandfly bites, the parasites enter the blood of the host and are phagocytized by macrophages. ➡ (Chapter 16, p. 440) The parasites multiply inside the macrophages, and new parasites are released when the macrophages rupture. Leishmaniasis is endemic to most tropical and subtropical countries, where appropriate species of sandfly vectors are available. Rodents can be reservoir hosts of the disease. Some 12 million cases worldwide are reported by WHO.

The Disease   *Leishmania donovani* causes **kala azar** (kah′lah ah-zar′; Hindi for "black poison"), or *visceral leishmaniasis*. Symptoms include high irregular fever, progressive weakness, wasting, and protrusion of the abdomen due to extensive liver and spleen enlargement. Extensive damage to the immune system results when parasites destroy large numbers of phagocytic cells. If untreated, the disease is usually fatal in 2 to 3 years, and it can be fatal in 6 months in patients with impaired immunity and secondary infections.

Other leishmanias are more localized in their effects and are rarely fatal. *L. tropica* causes a cutaneous lesion, sometimes called an *oriental sore,* at the site of a sandfly bite. *L. braziliensis* causes skin and mucous membrane lesions and sometimes nasal and oral polyps (➢ Figure 23.18b). Some parents, observing that people who had oriental sores rarely got kala azar, have purposely infected children with oriental sores on inconspicuous parts of the body to protect them from the more serious disease.

✓ If you have had infectious mononucleosis, do you still carry the virus? If so, what can it do?

✓ What diseases are caused by B19?

Desert Storm Leishmaniasis

The soldiers who served our country in the Persian Gulf War in 1991 probably didn't expect to die at the hands of a microorganism, but several dozen cases of leishmaniasis have been confirmed in U.S. military personnel involved in Desert Storm. The disease is fatal in 90 percent of untreated cases. Currently, available treatment lowers the mortality rate to 10 percent. Nevertheless, for several years, health officials banned blood transfusions and organ transplants from Gulf War veterans to prevent the spread of the disease.

(a)

(b)

➤ Figure 23.18   **Leishmaniasis.** (a) *Leishmania donovani,* seen as small dots among cells. (b) A patient suffering from leishmaniasis, probably caused by *L. braziliensis.*

Diagnosis, Treatment, and Prevention   Diagnosis is made by identifying the protozoa in blood smears in kala azar and from scrapings of skin and mucous membrane lesions. Antimony compounds are used to treat both kala azar and skin and mucous membrane lesions. However, such drugs are very toxic. Prevention depends mainly on controlling sandfly breeding and elimination of rodent reservoir infections.

Malaria

Several species of the protozoan *Plasmodium* are capable of causing **malaria,** one of the most severe of all parasitic diseases. Malaria is one of the world's greatest public health problems. It is endemic in most tropical areas (➤ Figure 23.19). The number of clinical cases has been estimated at up to 500 million cases worldwide, with 1.5 million to 3 million deaths—of which more than 1 million occur in children under age 5. Nearly all adults in Africa and India have been infected, and annual economic losses due to malaria exceed $1 billion in Africa alone. Drug-resistant strains are increasing rapidly, driving the death rates up. At one time malaria was thought to have been eradicated from the United States, but military personnel, travelers, and immigrants have carried the disease with them to the United States from endemic areas.

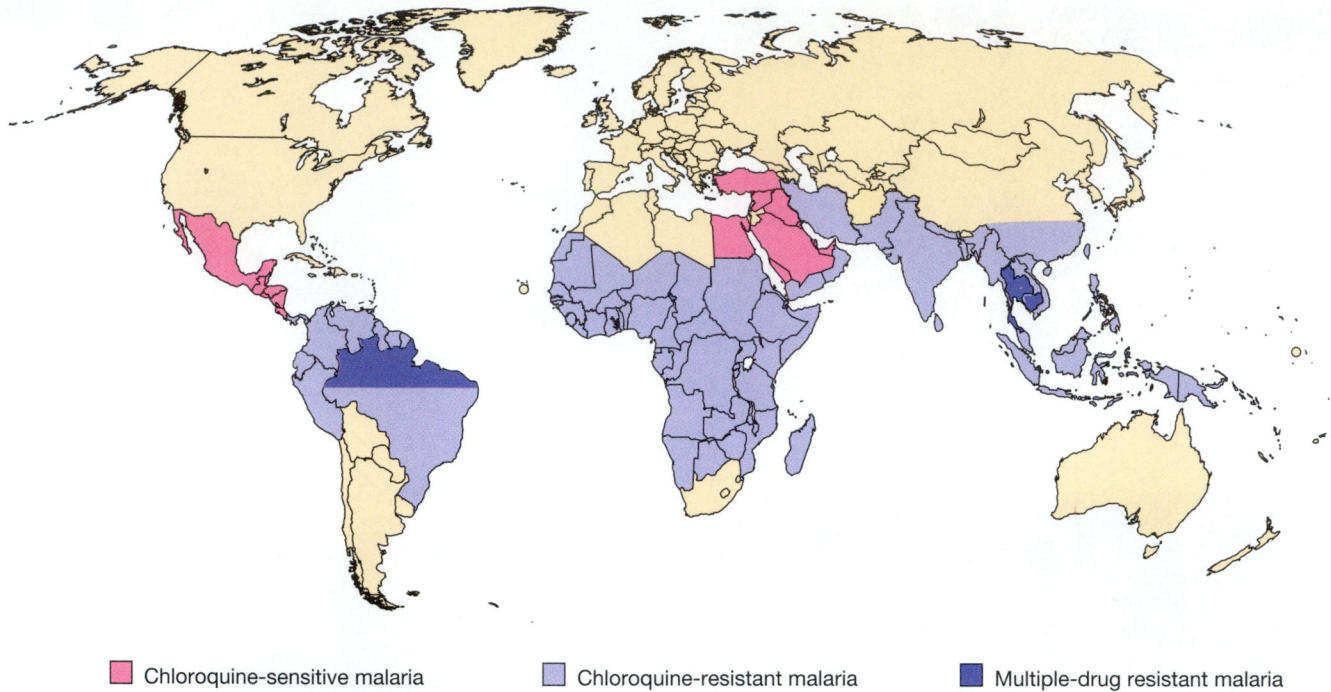

■ Chloroquine-sensitive malaria   ■ Chloroquine-resistant malaria   ■ Multiple-drug resistant malaria

➤ Figure 23.19   **The worldwide incidence of malaria, 1996.** New drugs for malaria are being avidly sought as drug-resistant forms spread in sub-Saharan Africa, Asia, and parts of Latin America. The main drug treatment, chloroquine, has virtually stopped working in parts of Africa and Asia. (*Source*: CDC).

Anemia Protects Against Malaria

If you are suffering from anemia, some scientists might think that you are lucky. Low blood iron levels, which are characteristic of certain types of anemia, seem to boost the immune system's ability to fight malaria. Studies suggest that compounds that remove iron from the blood, called iron chelators, interfere with the DNA replication of the parasite that causes malaria, helping to clear the parasite from the body. These iron chelators also appear to boost levels of NO_3, leading to an immune response. But this immune response also causes fever and inflammation. That's not so lucky, and scientists agree that much more work must be done to develop an effective iron-chelating drug with only mild side effects.

Members of the genus *Plasmodium* are amoeboid, intracellular protozoa that infect erythrocytes and other tissues. They are transmitted to humans through the bite of *Anopheles* mosquitoes. *Plasmodium* species have a complex life cycle (see ➤ Figure 11.4). ➡ (p. 295) At least four species, *P. vivax, P. malariae, P. ovale,* and *P. falciparum,* infect humans. They can be identified by their effects on, and in some cases by their appearance in, red blood cells and by the nature of the disease the parasites cause.

Certain individuals, especially West African blacks, and persons of Mediterranean origin, are protected from malaria by carrying the gene for sickle-cell anemia. The presence of two such genes causes sickle-cell anemia, but the presence of a single gene prevents malarial parasites from growing in erythrocytes. When the parasite enters the cell, it uses up oxygen. Under low-oxygen conditions, the cell sickles (takes on a distorted spiney or sickle shape). The spleen removes abnormally shaped blood cells, such as sickle-cells, usually destroying them before the malaria parasite has time to complete its life cycle, thus reducing the number of infected cells to a level where symptoms are minor or do not occur.

The Disease   In the pathogenesis of malaria, sporozoites (Chapter 11) enter the blood from the bite of an infected female mosquito. ➡ (p. 294) (Male mosquitoes are not equipped to take a blood meal.) The parasites disappear from the blood within an hour and invade cells of the liver and other organs. In about a week they begin releasing merozoites, which invade and reproduce in red blood cells as trophozoites (➤ Figure 23.20). At intervals of 48 to 72 hours, depending on the infecting species of *Plasmodium,* blood cells rupture in a characteristic pattern and release more merozoites, which infect other red blood cells. The release of merozoites soon becomes synchronized and corresponds to intervals of high fever. Some merozoites become gametocytes, which can undergo sexual reproduction in mosquitoes, should they feed on the patient's blood. The average malarial parasites destroy 25 to 75 percent of the hemoglobin in an erythrocyte, degrading it in acidic food vacuoles, in order to satisfy their protein requirements. Even after the initial disease has subsided, patients are subject to relapses as dormant protozoa become activated, emerge from the liver, and initiate a new cycle of disease. Relapses do not occur after infections caused by *P. falciparum* because this species does not remain in the liver.

➤ Figure 23.20 **Plasmodium, the cause of malaria.** The "ring" stage of the malarial parasite *Plasmodium falciparum,* seen as dark circular structures within red blood cells.(575X) At this stage, merozoites have become trophozoites upon invading the host's red blood cells.

Of the four species of malarial parasites, *P. falciparum* causes the most severe disease because it agglutinates red blood cells and obstructs blood vessels. Such obstruction causes tissue **ischemia,** or reduced blood flow with oxygen and nutrient deficiency and waste accumulation. This species can also cause malignant malaria—an especially virulent, rapidly fatal disease—and a condition called **blackwater fever.** In blackwater fever, large numbers of erythrocytes are lysed, probably because of the host's autoimmune reaction to the parasites. Products of hemoglobin breakdown cause jaundice and kidney damage. Pigments from hemoglobin blacken the urine and give the disease its name.

Diagnosis, Treatment, and Prevention   The main means of diagnosing malaria is by identifying the protozoa in red blood cells. The species of *Plasmodium* responsible for a particular infection can be identified by the distinctive appearance of erythrocytes invaded by the parasite. Chloroquine (Aralen) is the drug of choice for all forms of malaria in the acute stage. A serious problem in the treatment of malaria is that some strains, especially strains of *P. falciparum,* have become resistant to chloroquine. Drugs have recently been found that can be administered with chloroquine to overcome such resistance. This strategy has been tested in monkeys but not yet in humans. A traveler who will enter a malarial region can take chloroquine prophylactically for a week before entry, during the stay, and for 6 weeks after leaving the area. The drug suppresses clinical symptoms of malaria, but it does not necessarily prevent infection. Disease caused by *P. vivax* or *P. ovale* can appear months or years after a person leaves a malarial area, even when the suppressant drug has been taken. Primaquine is the drug of choice for eliminating parasites from the liver and other tissues if they have been infected.

Attempts to destroy mosquitoes that carry malaria have been an important component of malaria control efforts. In the early 1960s the pesticide DDT was used successfully to eradicate malaria-carrying mosquitoes from the United States. (But it was soon banned in the United States due to its toxicity.) DDT and other insecticides also have been tried in other areas, especially in Africa. Unfortunately, the region in which such mosquitoes thrive in Africa is so vast that no insecticide spraying program has been effective. Some of the mosquitoes, particularly those that carry *P. falciparum,* have now become resistant to DDT and probably to other pesticides. Thus, the use of insecticides has made malaria more deadly by increasing the proportion of mosquitoes that carry the more virulent parasite.

Researchers at the CDC and their colleagues at other institutions have recently developed a strain of *Anopheles gambiae* that is highly resistant to infection with malaria parasites. The researchers believe that the resistance is due to a relatively simple genetic change and that it might be possible to induce such resistance in natural vector populations. If large numbers of *A. gambiae,* the most important malaria vector in Africa, could be made resistant, transmission of malaria could be greatly reduced.

Another control effort is directed toward developing a malaria vaccine. One problem has been to identify the stage of the parasite responsible for triggering the immune response in humans. An antigen on the sporozoite has now been identified, and the gene for this antigen can be cloned using recombinant DNA technology. Consequently, the antigen can be manufactured and

Using Malaria to Unite the World

Who would have thought that malaria would play a role in world politics? These days, that's exactly what it's doing. The World Health Organization has united the world in and effort to halve world deaths from malaria by 2010. The program, called Roll Back Malaria, involves the Multilateral Initiative on Malaria (MIM), which involves 37 African countries, the United States, Europe, and the African Initiative on Malaria. The coalition brings together governments, the World Bank, and other international agencies. And it's a good thing that we're uniting now; malaria already causes about 500 million cases of illness and up to 3 million deaths a year worldwide as the parasite develops increasing drug resistance and spreads into new regions such as Central Asia and Eastern Europe.

used to make a vaccine. Thus, great strides have been made toward the development of an effective vaccine; the vaccine will hopefully be available in the near future. But even when a vaccine becomes available, administering it to the vast numbers of people living where malaria is endemic will be a monumental task. Mechanisms to distribute the vaccine and ways of gaining the people's cooperation will be needed. The cost of the vaccine in the quantities required will be another major problem.

Toxoplasmosis

Toxoplasma gondii (➤ Figure 23.21) is a widely distributed protozoan that infects many warm-blooded animals—both domestic and wild. It is an intracellular parasite and can invade many tissues. Humans usually become infected through contact with the feces of domestic cats that forage for natural foods, especially infected rodents (➤ Figure 23.22). Cats that are kept inside and are fed on canned or cooked foods are unlikely to acquire the parasites. Another common means of transmission is the consumption of raw or undercooked contaminated meat. The French, who consume large amounts of steak tartare (raw ground beef), have the highest incidence of infection in the world.

The Disease   *Toxoplasma gondii* causes only mild lymph node inflammation in most humans. In the majority of cases, the infection is chronic, asymptomatic, and self-limiting. However, *T. gondii* can cause serious **toxoplasmosis,** especially in developing fetuses, newborn infants, and sometimes in young children. The organism can be transferred across the placenta of an infected mother to the fetus. There it causes serious congenital defects, including the accumulation of cerebrospinal fluid, abnormally small head, blindness, mental retardation, and disorders of movement. It can also be responsible for stillbirths and spontaneous abortions. If infection occurs after birth, the symptoms are similar to, but less severe than, those seen in fetuses. In patients with severe immunosuppression, such as AIDS patients, the disease can appear as encephalitis and can also cause dermatologic problems.

Diagnosis, Treatment, and Prevention   Toxoplasmosis can be diagnosed by finding the parasites in the blood, cerebrospinal fluid, or tissues, by animal inoculation with subsequent isolation of the organisms, or by indirect immunofluorescence tests. ➡ (Chapter 18, p. 531) Pyrimethamine and trisulfapyridine are used in combination to treat toxoplasmosis, but no treatment can reverse permanent damage from prenatal infection. To prevent this disease, pregnant women should avoid contact with raw meat and cat feces, and cats should be kept out of sandboxes where children play, especially if a child might carry the organism to a pregnant woman.

Babesiosis

Several species of the sporozoan *Babesia* can cause **babesiosis** (ba-be″se-o′sis). Cattle are affected with babesiosis caused by the tickborne protozoan *Babesia bigemina,* but *B. microti* is most often associated with human infections. The parasites enter the blood via bites of infected ticks and invade and multiply in red blood cells.

➤ Figure 23.21   **Toxoplasmosis.** Crescent-shaped protozoans, *Toxoplasma gondii,* leaving an infected cell in which they have multiplied (SEM, 3600X). This organism can be a danger to immunocompromised patients and to pregnant women, in whom it can cause serious birth defects and miscarriages.

Infected cell

Toxoplasma gondii

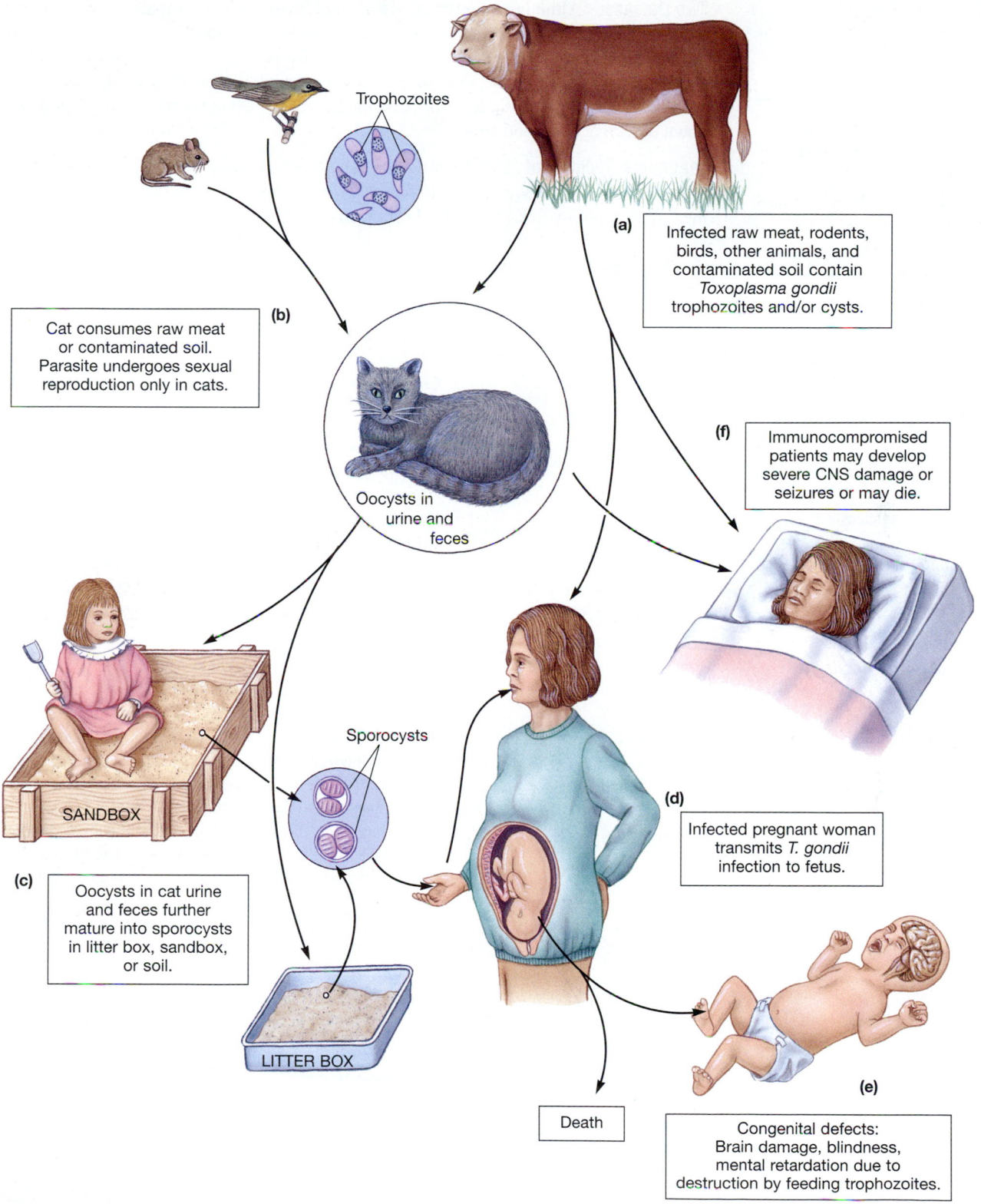

Trophozoites

(a) Infected raw meat, rodents, birds, other animals, and contaminated soil contain *Toxoplasma gondii* trophozoites and/or cysts.

(b) Cat consumes raw meat or contaminated soil. Parasite undergoes sexual reproduction only in cats.

Oocysts in urine and feces

(f) Immunocompromised patients may develop severe CNS damage or seizures or may die.

Sporocysts

(c) Oocysts in cat urine and feces further mature into sporocysts in litter box, sandbox, or soil.

SANDBOX

LITTER BOX

(d) Infected pregnant woman transmits *T. gondii* infection to fetus.

Death

(e) Congenital defects: Brain damage, blindness, mental retardation due to destruction by feeding trophozoites.

➤ **Figure 23.22** *Toxoplasma gondii* **cycle of natural transmission.** (a) Contaminated raw meat or soil containing trophozoites (feeding forms) of *T. gondii* are consumed. (b) In the intestinal tract of cats, the trophozoites undergo sexual reproduction. Oocysts are released in urine and feces and further mature into sporocysts in litter boxes, sandboxes, or soil. (c) If the trophozoites are consumed by older children or adults, a mild flulike syndrome may result. (d) Pregnant women suffer only mild or inapparent infection. (e) But they can transmit this infection to their fetus, which may develop congenital defects or die. (f) Immunocompromised individuals, who may already have the organism in their body, rapidly die of central nervous system destruction.

The Disease   Although many cases are asymptomatic, when symptoms appear, they usually begin with sudden high fever, headache, and muscle pain. Anemia and jaundice can occur as red blood cells are destroyed. The symptoms last for several weeks and are followed by a prolonged carrier state. If babesiosis occurs in a person who has undergone spleen removal, it is usually fatal in 5 to 8 days. This is because the lack of a spleen impairs the body's ability to break down defective red blood cells.

Diagnosis, Treatment, and Prevention   Diagnosis is made from blood smears, but the parasite can be confused with *Plasmodium falciparum*. Chloroquine is the drug of choice for treatment, and avoiding tick bites is the best means of protection.

The properties of nonbacterial systemic diseases are summarized in Table 23.4.

✓ How does carrying a gene for sickle-cell anemia protect against malaria?

✓ Over what period should travelers take antimalarial drugs? Why?

✓ Who is at risk for toxoplasmosis?

➤ Table 23.4   Viral and protozoan systemic diseases

Disease	Agent(s)	Characteristics
Viral systemic diseases		
Dengue fever	Dengue fever virus	Severe bone and joint pain, high fever, headache, loss of appetite, weakness, sometimes a rash
Yellow fever	Yellow fever virus	Fever, anorexia, nausea, vomiting, liver damage, jaundice
Infectious mononucleosis	Epstein–Barr virus	Headache, fatigue, malaise, usually sore throat, secondary streptococcal infections common
Other viral fevers	Filoviruses, bunyaviruses, phleboviruses, arenaviruses, orbivirus, and coxsackie viruses	Some cause hemorrhagic fevers; some cause encephalitis, joint pain, slow heart rate, infection of erythrocytes, diarrhea, rashes, sore throat, liver disease, meningitis, inflammation of the heart and the sac around it
Protozoan systemic diseases		
Kala azar	*Leishmania donovani*	Visceral leishmaniasis with irregular fever, weakness, wasting, enlarged liver and spleen
Localized leishmaniasis	*L. tropica, L. braziliensis*	Oriental sore and skin and mucous membrane lesions
Malaria	*Plasmodium* species	Periods of high fever associated with release of parasites from red blood cells; relapses can occur; one species can cause malignant malaria and blackwater fever
Toxoplasmosis	*Toxoplasma gondii*	Mild lymph node inflammation in adults; can cross placenta and cause serious damage to nervous system of fetus; also causes damage in small children and immunosuppressed patients
Babesiosis	*Babesia microti*	High fever, headache, muscle pain, anemia, and jaundice; fatal in patients whose spleens have been removed

Retracing Our Steps

- The agents and characteristics of the diseases discussed in this chapter are summarized in Tables 23.1 through 23.4. Information in those tables is not repeated in this summary.

The Cardiovascular System

- The **cardiovascular system** consists of the heart, an extensive system of blood vessels, and blood.
- Although the cardiovascular systems normally sterile, pathogens can be transported in blood (*bacteremia*), multiply in blood (*septicemia*), and infect the heart valves and pericardium.

Cardiovascular and Lymphatic Diseases

Bacterial Septicemias and Related Diseases

- **Septicemia,** or blood poisoning, involves multiplication of bacteria in the blood. Septicemias and related diseases are diagnosed by culturing appropriate samples and are treated with antibiotics.
- **Rheumatic fever** and **bacterial endocarditis** occur most often in patients who have had previous streptococcal infections.

Helminthic Diseases of the Blood and Lymph

- **Schistosomiasis** is acquired from schistosome larvae that penetrate the skin and is diagnosed by finding eggs in feces. It is treated with praziquantel and could be prevented by the eradication of infected snails or avoidance of snail-infested water.
- **Filariasis** is transmitted by mosquitoes and is diagnosed by finding microfilariae in blood. It is treated with diethylcarbamizine or metronidazole and could be prevented if infected mosquitoes could be eradicated.

Systemic Diseases

Bacterial Systemic Diseases

- **Anthrax** is acquired through contact with *Bacillus anthracis* spores from infected domestic animals or their hides. It is diagnosed by culturing blood or from smears from lesions and is treated with penicillin or tetracycline. The disease can be prevented by immunizing animals and humans with occupational exposure and by careful burial of infected animals.
- Plague has occurred in periodic epidemics since the Middle Ages, remains endemic in some regions, and is increasing in incidence in the United States. The form that is transmitted by fleas from infected rats is known as **bubonic plague,** which forms enlarged lymph nodes, called **buboes.** If the disease progresses to the circulatory system, it is called **septicemic plague.** Lung involvement results in **pneumonic plague,** which is contagious and transmitted by aerosols. Plague is diagnosed by stained smears and antibody tests and treated with streptomycin or tetracycline. It can be prevented by controlling rat populations and immunizing people who enter endemic areas.
- **Tularemia** is transmitted through the skin, inhaled, or ingested. It can be diagnosed by agglutination tests and treated with streptomycin. It is prevented by avoiding infected mammals and arthropods; the vaccine is short-lasting and not fully protective.
- **Brucellosis** is transmitted to humans through the skin from domestic animals by the ingestion of contaminated dairy products

and by inhalation or ingestion and is diagnosed by serological tests. It is treated with prolonged antibiotic therapy and can be prevented by avoiding infected animals and contaminated fomites.

- **Relapsing fever** is transmitted by lice and ticks and can be diagnosed from blood smears. It is treated with tetracycline or chloramphenicol and can be prevented by avoiding or controlling ticks and lice.
- **Lyme disease** is transmitted by ticks from infected deer and other animals and is diagnosed by clinical signs and serological tests. It is treated with antibiotics and can be prevented by avoiding tick bites.

Rickettsial and Related Systemic Diseases

- **Typhus fever** occurs in several forms. **Epidemic typhus,** transmitted by human body lice, usually occurs in unsanitary, overcrowded conditions. It has a high mortality rate unless treated with antibiotics.
- **Brill–Zinsser disease,** or recrudescent typhus, is a recurrence of a latent typhus infection. **Endemic,** or **murine, typhus** is carried by fleas, and **scrub typhus,** by mites from infected rats.
- **Rocky Mountain spotted fever,** carried by ticks, damages blood vessels. Strains of the causative rickettsia vary in virulence, and the mortality of the untreated infection can be high.
- **Rickettsialpox** is carried by mites that live on house mice. **Trench fever,** transmitted by lice, is prevalent under unsanitary conditions, most often among persons under stress. **Bartonellosis,** transmitted by sandflies, occurs in two forms: **Oroya fever,** an acute fever that causes life-threatening anemia, and **verruga peruana,** a self-limiting skin rash.
- Recently identified human pathogens resembling the rickettsias include *Ehrlichia canis* and *E. chaeffeenis,* which cause **ehrlichiosis,** and *Bartonella henselae,* which causes **bacillary angiomatosis.**

Viral Systemic Diseases

- **Dengue fever,** an arbovirus disease, can be diagnosed by serologic tests; a vaccine is available against one immunological type of dengue fever virus.
- **Yellow fever,** another arbovirus disease, is diagnosed by symptoms and can be prevented with vaccine.
- **Infectious mononucleosis** is caused by the **Epstein-Barr virus,** diagnosed by symptoms, and treated symptomatically and with antibiotics for secondary infections. In developing countries babies have mild symptoms and produce antibodies by age 1, but in developed countries patients are teenagers or young adults, who have a much more serious disease.
- **Chronic fatigue syndrome** has been linked with the Epstein–Barr virus, which is responsible for infectious mononucleosis, **Burkitt's lymphoma,** nasopharyngeal carcinoma, and oral hairy leukoplakia.
- Other viral infections include **filovirus** fevers (such as **Ebola virus** infection), bunyavirus fevers (such as **Rift Valley fever**), arenavirus fevers (such as **Lassa fever** and **Bolivian hemorrhagic fever**), **Colorado tick fever, feline panleukopenia virus** and **canine parvovirus** infections, and **fifth disease.**

Protozoan Systemic Diseases

- **Leishmaniasis** occurs in tropical and subtropical countries where appropriate species of sandfly vectors are available. It is diagnosed from blood smears or scrapings from lesions and is treated with antimony. The disease could be prevented by controlling sandfly breeding and eliminating rodent reservoir infections.

- **Malaria** is one of the world's greatest public health problems; it kills a million people annually, most of them children. Cases in the United States come from people who have been in endemic areas. Malaria is transmitted by female *Anopheles* mosquitoes and is diagnosed by identifying protozoa in blood smears. Active disease is treated with chloroquine (except for resistant strains),

and latent parasites are eliminated with primaquine. Research is in progress to find a way to control mosquitoes and to develop an effective vaccine.

- **Toxoplasmosis** is usually transmitted from feces of cats that have consumed infected rodents and by human ingestion of contaminated raw or undercooked meat. It is diagnosed by finding parasites in body fluids or tissues and is treated with pyrimetamine and trisulfapyridine. The disease can be prevented by avoiding contaminated materials.

- **Babesiosis** is transmitted by ticks and is diagnosed from blood smears. It is treated with chloroquine and can be prevented by avoiding contact with ticks.

Terminology Check

anthrax *(p. 672)*
aplastic crisis *(p. 691)*
babesiosis *(p. 696)*
bacillary angiomatosis *(p. 686)*
bactermia *(p.667)*
bacterial endocarditis *(p. 669)*
bartonellosis *(p. 686)*
blackwater fever *(p. 695)*
Bolivian hemorrhagic fever *(p. 691)*
Brill–Zinsser disease *(p. 683)*
brucellosis *(p. 678)*
bubo *(p. 675)*
bubonic plague *(p. 675)*
Burkitt's lymphoma *(p. 689)*
canine parvovirus *(p. 691)*
cardiovascular system *(p. 666)*
childbed fever *(p. 668)*
chronic fatigue syndrome *(p. 690)*
Colorado tick fever *(p. 691)*
dengue fever *(p. 687)*
Ebola virus *(p. 690)*
ehrlichiosis *(p. 686)*
electrolyte *(p. 667)*
elephantiasis *(p. 672)*

endemic relapsing fever *(p. 679)*
endemic typhus *(p. 683)*
epidemic relapsing fever *(p. 679)*
epidemic typhus *(p. 683)*
Epstein–Barr virus (EBV) *(p. 688)*
feline panleukopenia virus (FPV)
   (p. 691)
fifth disease *(p. 692)*
filariasis *(p. 672)*
filovirus *(p. 690)*
infectious mononucleosis *(p. 688)*
ischemia *(p. 695)*
kala azar *(p. 692)*
Lassa fever *(p. 691)*
leishmaniasis *(p. 692)*
Lyme disease *(p. 680)*
lymphangitis *(p. 668)*
malaria *(p. 693)*
Malta fever *(p. 678)*
murine typhus *(p. 683)*
myocarditis *(p. 670)*
orbivirus *(p. 691)*
Oroya fever *(p. 686)*
pericarditis *(p. 670)*
petechia *(p. 683)*

phlebovirus *(p. 691)*
pneumonic plague *(p. 677)*
puerperal fever *(p. 668)*
relapsing fever *(p. 679)*
rheumatic fever *(p. 669)*
rickettsialpox *(p. 685)*
Rift Valley fever *(p. 691)*
Rocky Mountain spotted fever *(p. 684)*
schistosomiasis *(p. 670)*
scrub typhus *(p. 684)*
septicemia *(p. 668)*
septicemic plague *(p. 675)*
septic shock *(p. 668)*
toxoplasmosis *(p. 696)*
transovarian transmission *(p. 677)*
trench fever *(p. 685)*
tularemia *(p. 677)*
typhoidal tularemia *(p. 678)*
typhus fever *(p. 683)*
ulceroglandular *(p. 678)*
undulant fever *(p. 678)*
vegetation *(p. 669)*
verruga peruana *(p. 686)*
yellow fever *(p. 688)*

See You on the Web

Don't forget I am expecting you on the web. The following is a preparedness checklist for your web explorations. If you have questions on any of these items, you can return to the end of Chapter 1 for more details.

- *Questions We'll Explore*
- *Concept Check questions*
- *Flashcards*
- *Study Guide*

If you are ready, come visit me at

http://www.prenhall.com/black

It will be a journey you'll never forget.

Diseases of the Nervous System

A law, passed following the British "mad cow disease" scare in 1997, prohibits butchers from selling meat on the bone. Britons are still bewailing the loss of their national favorite, standing rib roast. But why was this law passed?

Nerve fibers exiting from the bone into adjacent meat could be carrying large numbers of prions, the infectious agent responsible for mad cow disease and Creutzfeld–Jakob disease. Concern over the consumption of British beef arose after a relatively large number of Britons had died of an unusual form of Creutzfeld–Jakob disease thought to be acquired by eating infected beef.

Could anything be done to protect British beef eaters? The methods of butchering have been radically changed. Cuts are no longer made through the spinal cord, as the infectious prions concentrate in nervous tissue. A metal knife blade passing through this tissue could drag prions out into the meat. (And, blades can remain contaminated for months despite regular normal cleaning.) Today butchers cut up alongside the spine, but not through it or close to it.

Background Basics

As with cardiovascular and lymphatic diseases, the diseases of the nervous system also often affect other systems. This chapter assumes a knowledge of disease processes (Chapter 14) and host systems and host defenses (Chapters 16 and 17).

Components of the Nervous System

As you read this sentence, millions of neural signals allow you to understand words while you maintain an upright posture and carry on various internal processes, such as breathing. The ability of the nervous system to control body functions at the same time depends on many *neurons,* or nerve cells, working together. Structurally, the **nervous system** has two components: the central and peripheral nervous systems (➤ Figure 24.1). The **central nervous system (CNS)**, which consists of the brain and spinal cord, receives signals from and issues commands via the **peripheral nervous system (PNS)**, composed of nerves that supply all parts of the body. **Nerves** of the peripheral nervous system consist of nerve fibers that transmit sensory information and motor responses. Aggregations of these cell bodies in the PNS are called **ganglia** (singular: *ganglion*). The brain and spinal cord are covered and protected by membranes called **meninges** (menin´jez), which are sheets of connective tissues. Hollow chambers in the brain and spinal cord and spaces between meninges are filled with *cerebrospinal fluid*.

Like the cardiovascular system, the nervous system is ordinarily sterile and has no normal microflora. Bacterial and viral pathogens, however, can enter cerebrospinal fluid from the blood. The CNS is well protected by bone and meninges from invasion by pathogens. In addition, phagocytes of the nervous system called *microglial cells* can destroy invaders that reach the brain and spinal cord. The brain itself has special thick-walled capillaries without pores in their walls. These capillaries form the **blood–brain barrier,** which limits entry of substances into brain cells. Although it does protect the cells from microorganisms and toxic substances, the blood–brain barrier prevents the cells from receiving medications that easily reach other cells.

Questions We'll Explore

A What are the components of the nervous system? What defenses against infection does it have?

B Which pathogens cause bacterial diseases of the brain and meninges, and what are the important epidemiologic and clinical aspects of these diseases?

C Which pathogens cause viral diseases of the brain and meninges, and what are the important epidemiologic and clinical aspects of these diseases?

D Which pathogens cause bacterial nerve diseases, and what are the important epidemiologic and clinical aspects of these diseases?

E Which pathogens cause viral nerve diseases, and what are the important epidemiologic and clinical aspects of these diseases?

F Which pathogens cause prion nervous system diseases, and what are the important epidemiologic and clinical aspects of these diseases?

G Which pathogens cause parasitic diseases of the nervous system, and what are the important epidemiologic and clinical aspects of these diseases?

Central nervous system
- Cerebrum
- Cerebellum
- Brain stem
- Spinalcord

Peripheral nervous system
- Peripheral nerves

Ganglion

➤ Figure 24.1   **The structure of the nervous system.** The nervous system normally is free of resident microbes. Infections typically affect the meninges (that cover the brain and spinal cord) or the sensory nerve ganglia.

Normal Microflora of the Nervous System

The nervous system does not have a normal microflora.

Diseases of the Brain and Meninges

Bacterial Diseases of the Brain and Meninges

Bacterial Meningitis

Bacterial meningitis is an inflammation of the *meninges,* the membranes that cover the brain and spinal cord. This life-threatening disease can be caused by several kinds of bacteria, each of which has a prevalence that can be correlated with the age of the host (Table 24.1). Meningitis causes *necrosis* (death of tissues in a given area), clogging of blood vessels, increased pressure within the skull from edema, decreased cerebrospinal fluid flow, and impaired central nervous system function. The early symptoms are headache, fever, and chills. In rare instances, seizures develop. Onset can be insidious or fulminating. Death occurs from shock and other serious complications within hours of the appearance of symptoms.

Most cases of meningitis are acute, but some are chronic. Acute meningitis is acquired from carriers or endogenous organisms. Organisms gain access to the meninges directly during surgery or trauma, or spread to them in blood from other infections such as pneumonia and otitis media. Host defenses in the arachnoid layer, one of the meninges, ordinarily combat bacteremia. If organisms overwhelm the defenses, however, meningitis results. Chronic meningitis occurs with extension of underlying diseases such as syphilis or tuberculosis. The bacteria causing these underlying diseases are slow growers; in such cases, onset of typical meningitis symptoms is insidious, occurring over a period of weeks.

Meningitis is diagnosed by culturing cerebrospinal fluid. The fluid usually is turbid—sometimes so thick with pus that it is difficult to remove with a syringe. Antibiotic treatment varies with the causative organism. If tubercular meningitis is suspected, isoniazid therapy is called for immediately. Treatment for tubercular meningitis lasts a year of more, whereas some very rare cases of fungal meningitis may require several years of treatment.

Meningococcal Meningitis   The bacterium *Neisseria meningitidis* has caused 2000–3000 cases of meningitis per year for the past decade in the United States. Mortality is about 85 percent when the disease is untreated but only 1 percent with the best treatment. The 15 percent mortality rate in the United States probably reflects delay in seeking treatment. This disease was the leading cause of death from infectious disease among U.S. armed forces during World War II.

In meningococcal meningitis the organisms colonize the nasopharynx, spread to the blood, and make their way to the meninges, where they grow rapidly (➤ Figure 24.2). In a complication called Waterhouse–Friderichsen syndrome, the meningococci invade all parts of the body, and death occurs within hours from endotoxin shock. The immediate cause of death is usually clotting of blood followed by massive hemorrhage in the adrenal glands (located above the kidneys), leading to fatal deficiency of essential adrenal hormones. A lesser degree of hemorrhaging is sometimes seen in meningitis patients who develop a petechial skin rash.

Penicillin is the drug of choice for treatment; the prevalence of resistant strains makes sulfonamides no longer useful. Third-generation cephalosporins and ampicillin are also used now. A vaccine is available, but it is not effective against the most common type B meningococci. The risk of

Meningitis can cause what is known as "pseudodementia," a treatable condition that is completely reversible as long as the brain has not suffered severe damage.

➤ Table 24.1	Types of bacterial meningitis	
Age	**Most frequent causative agents**	**Comments**
Newborn (0–2 months)	*Escherichia coli,* other Enterobacteriaceae, *Streptococcus* species	Average mortality about 50%; incidence 40–50/100,000 live births; maternal transmission
Preschool (2 months–5 years)	*Haemophilus influenzae,* type b; *Neisseria meningitidis*	Maximum incidence 6–8 months; overall incidence 180/100,000 children
Youth and young adult (5–40 years)	*Neisseria meningitidis, Streptococcus pneumoniae*	Sporadic or epidemic
Mature adult (over 40 years)	*Streptococcus pneumoniae, Staphylococcus* species	Sporadic

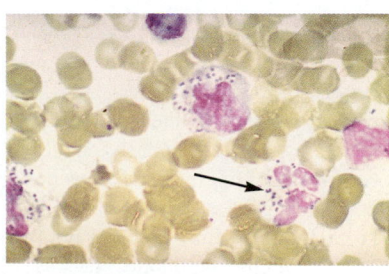

➤ Figure 24.2   **Meningococci.** These organisms are the cause of meningococcal meningitis. They have been phagocytized by white blood cells in a cerebrospinal fluid specimen.

this disease can be decreased by prevention of overtiring and overcrowding. The number of inches required between bunks in military barracks is based on experience with meningococcal outbreaks.

In some closed environments, such as military bases, dormitories, and day-care centers, 90 percent of the population can be carriers of meningococci, yet only 1 per 1000 carriers develops the disease. Among members of patient households, 80 to 90 percent are carriers, compared with only 5 to 30 percent in the general public. Antibiotics can eliminate the carrier state.

Haemophilus Meningitis   About two-thirds of bacterial meningitis cases during the first year of life are caused by *Haemophilus influenzae* type B (Hib). Among children, 30 to 50 percent are carriers of this organism; among adults, the figure is only 3 percent. Humans are exposed to *H. influenzae* early in life and rapidly acquire immunity, so the disease is rare in adults. Only 10 percent of children between ages 3 and 6 lack antibodies, and all those over age 6 have antibodies. Without treatment this disease is nearly always fatal. Even with treatment, one-third die. Of those who recover, 30 to 50 percent have serious mental retardation, and 5 percent are permanently institutionalized because of damage to the central nervous system. *Haemophilus* meningitis is the leading cause of mental retardation in the United States and worldwide. Hib vaccines are available, which all children age 5 and under should receive. ➡ (Chapter 17, p. 481) These vaccines have dramatically reduced incidence of the disease.

Streptococcus Meningitis   Among adults *Streptococcus pneumoniae* is the most common cause of meningitis. Organisms generally spread via the blood from lung, sinus, mastoid, or ear infections. Mortality is 40 percent.

Listeriosis

Another kind of meningitis, **listeriosis,** is caused by *Listeria monocytogenes,* a small, Gram-positive bacillus that is widely distributed in nature. Foodborne transmission by improperly processed milk, cheese, meat, and vegetables is the most common source of infection. It is sometimes acquired as a zoonosis and is particularly threatening to those with impaired immune systems. Although not an especially significant human disease for several decades, listeriosis is now a leading cause of infection in kidney transplant patients. In pregnant women the bacillus can cross the placenta, infect the fetus, and cause abortion, stillbirth, or neonatal death. Listeriosis is responsible for many cases of fetal damage.

Brain Abscesses

Microorganisms that cause **brain abscesses** reach the brain from head wounds or via blood from another site. As would be expected with wounds, multispecies infections are common, and anaerobes are as likely to be responsible as are aerobes. Most such abscesses occur in patients under age 40, but two age periods—birth to 20 years and 50 to 70 years—show peak incidences. The infection gradually grows in mass and compresses the brain. The masses can be detected by CAT scans or X-rays, and causative agents can be identified by serologic tests and by culturing cerebrospinal fluid. In very early stages, antibiotic treatment can be sufficient, but later surgical drainage or removal of abscesses is usually necessary. Abscesses in areas of the brain that control the heart or other vital organs cannot be treated surgically. Without treatment, half the patients die, but with the current best treatment, only 5 to 10 percent die.

Viral Diseases of the Brain and Meninges

Viral Meningitis

Unlike bacterial meningitis, which is always fatal if untreated, **viral meningitis** is usually self-limiting and nonfatal. Enteroviruses account for approximately 40 percent of viral meningitis cases, and mumps virus for 15 percent. The causative virus in 30 percent of cases remains unidentified.

Rabies

Rabies was described by Democritus in the fifth century B.C. and by Aristotle in the fourth century B.C. Pasteur made real progress in understanding the disease when he found evidence of the infectious agent in saliva, the central nervous system, and peripheral nerves. He attenuated the agent and proved that a suspension of it could be used to prevent rabies. In 1903 the

---

Certain nontoxic bacteria that are also found in hot dogs produce bacteriocins that effectively kill the Listeria.

A Road Less Travelled

Why are brain infections hard to treat? Because the blood-brain barrier separates the brain from the rest of the body's cardiovascular system. This barrier allows only selective substances to pass through it. The body's own antibodies and complement proteins have a hard time passing from the blood into brain tissue to fight off infections. Antibiotics such as penicillin cannot easily pass through the blood-brain barrier, and if the meninges are inflamed, passage is easier. In most cases, average concentration antibiotics (administered orally or intravenously) in the cerebrospinal fluid generally reaches only about 15 percent of their concentration in plasma. But all is not lost. Some other antibiotics, such as chloramphenicol and tetracycline, are lipid-soluble and diffuse easily across the blood-brain barrier. Although the road to the brain is less traveled, at least it is not abandoned.

Italian physician Adelchi Negri discovered inclusion bodies ➡ (Chapter 14, p. 390) called *Negri bodies,* or clusters of viruses, in neurons (➤ Figure 24.3). Negri bodies were used to diagnose rabies for more than 50 years until an immunofluorescent antibody test (IFAT) was developed in 1958. Still used today, the IFAT is so sensitive that after an animal suspected of having rabies bites a human, it can be killed immediately and its brain examined for rabies antigens. Prior to the availability of IFAT, such animals had to be held for 30 days, or until symptoms of rabies developed, before a search for Negri bodies could be made. Most biting animals are not rabid, so the IFAT has saved many people both the anguish of wondering if they have been exposed to rabies and the discomfort and risk of treatment.

The virus that causes rabies has a worldwide distribution. It infects all mammals exposed to it, so the possibilities for reservoir infections are almost limitless. Different types of rabies are found in different regions of the world. In almost all of Asia, Africa, Mexico, and Central and South America, rabies is endemic in dogs. In Canada, the United States, and Western Europe, wildlife rabies predominates, and dog rabies is controlled. The WHO lists 60 countries, including England, Australia, Japan, Sweden, and Spain, as rabies-free. This success is due to animal vaccination and quarantine programs.

Identifying rabid animals can be a problem. About half of all rabid dogs release viruses into saliva 3 to 6 days before they show symptoms of rabies. In contrast, 90 percent of rabid cats have viruses in their saliva about 1 day before they become symptomatic. Any change in the behavior of an animal can be a warning that it might be rabid. A "friendly" wild animal approaching people or a gentle family animal snapping without provocation sometimes indicates impending rabies symptoms. Small wild animals and some domestic ones, too, will bite if suddenly grabbed and held. Biting is their only defense in such a stressful situation. Much unnecessary anguish could be avoided if people, especially children, were taught to appreciate unfamiliar animals without touching them.

Animals vary in their susceptibility to rabies, and susceptibility is directly correlated with the role of the animals in maintaining reservoir infections. Foxes, coyotes, skunks, raccoons, and bats are highly susceptible. No infected fox has ever been known to survive, whereas 10 to 20 percent of dogs and 30 to 40 percent of mongooses do survive. Bats are particularly dangerous because they are asymptomatic and shed viruses into their feces, urine, and saliva. Two explorers of bat-infested caves in Texas died of rabies. Dogs, cats, cattle, horses, and sheep are less susceptible. Whether a person becomes infected depends mainly on whether an animal is shedding viruses in its saliva at the time of the bite. Even animals later proved to have rabies may not have been shedding viruses at the time they inflicted bites.

The Disease   Rabies is caused by the **rabies virus,** an RNA-containing rhabdovirus. ➡ (➤ Figure 10.3c, p. 262) After entering the body through an animal bite or other break in the skin, the rabies virus first replicates in injured tissue for 1 to 4 days and then migrates to nerves, where it replicates slowly until it reaches the spinal cord. It progresses rapidly up the spinal cord to the brain by the flow of cytoplasm through axons. The time from infection to the appearance of symptoms varies from 13 days to 2 years but is usually between 20 and 60 days. The length of time required for symptoms to appear is proportional to the distance between the wound and the brain and is affected by the accessibility of nerve fibers. Thus, a bite on the face, which is well supplied with nerves and close to the brain, produces symptoms much more quickly than does a bite on the leg. The rabies virus has a predilection for nervous tissue but also infects salivary glands and the respiratory tract lining. There are even documented cases of transmission of rabies virus via corneal transplants.

It is the typically lengthy incubation period of rabies that allows postexposure immunization to be possible. Ordinarily there is sufficient time for the bitten individual to be vaccinated and to respond by making enough protective antibodies to prevent onset of the disease. Once symptoms have occurred, it is too late to vaccinate, and death usually follows quickly.

In humans the first symptoms are headache, fever, nausea, and partial paralysis near the bite site. These symptoms persist for 2 to 10 days and then worsen until the acute neurological phase of the disease ensues. The patient's gait becomes uncoordinated as paralysis becomes more general. Hydrophobia (fear of water) occurs as throat muscles undergo painful spasms, especially during swallowing. Aerophobia (fear of moving air) occurs because the skin is hypersensitive to any sensations. Confusion, hyperactivity, and hallucinations also occur. Within 10 to 14 days of the onset of

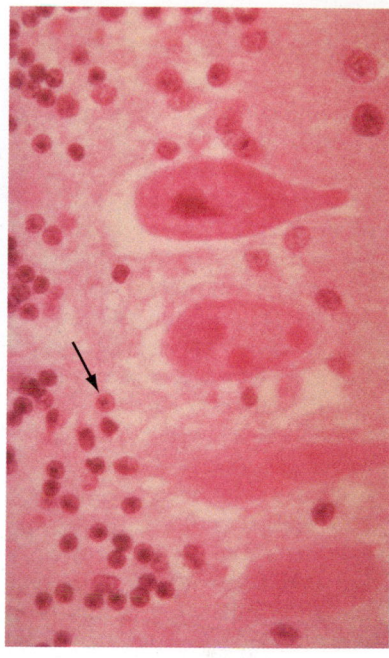

➤ **Figure 24.3**   **Negri bodies.** These characteristic signs of rabies are shown here in the cerebellum of a human brain.

Even a healthy pet that has bitten someone should be confined and watched for 10 days to see if it develops symptoms of rabies.

Watch Out for These Symptoms in Your Dog

Your pet dog probably won't ever contract rabies, thanks to vaccination requirements. But how would you know if your dog did contract rabies? Symptoms would begin with the dog acting as if it had a sore throat or something caught in its throat. As the disease progressed, the dog could become groggy or paralyzed (dumb rabies) or agitated and aggressive, biting anything that disturbs it (furious rabies). Throat muscle spasms, difficulty in swallowing, and drooling would also be indicative of rabies. Eventually, the dog would become apathetic and stuporous and slide into a final coma.

symptoms, the patient typically goes into a coma and dies. Of all human patients who have suffered from clinical rabies, only two have been known to survive and make a complete recovery. Both had some degree of protection from an earlier immunization, and they were known to have been bitten by a rabid animal, so postexposure immunological treatment was started immediately.

Diagnosis, Treatment, and Prevention   A sample from a brain or skin biopsy can be stained by the IFAT to identify rabies virus antigen before the patient dies. Finding the antigen confirms the diagnosis, but failure to find it does not rule out rabies. Sometimes a diagnosis can be made before death by testing cerebrospinal fluid or serum for neutralizing antibodies, which increase 10 to 12 days after the symptoms appear.

The bite of a rabid animal is treated by first thoroughly cleaning it with soap and flushing it with large amounts of water. Hyperimmune rabies serum is introduced into and around the wound in hope of neutralizing viruses before they reach the nervous system, where they are beyond reach of the antibodies. Interferon also can be applied to the wound. A series of injections of vaccine is given to induce the production of neutralizing antibodies. The biting animal should be located and confined for examination by IFAT.

The best means of preventing rabies is to immunize pets, and such immunization is required in many countries. Attempts to reduce rabies in raccoons have been made by using vaccine in small sponges covered with food bait. When the raccoons eat the bait, they also ingest enough vaccine to prevent them from acquiring rabies. Preliminary studies conducted in the United States show promising results. When raccoons from a bait-treated wildlife area were trapped, 15 out of 16 survived a challenge dose of rabies virus. All 16 caught in the nonbaited control area died of the same challenge doses. It is not certain whether the animal that died in the experimental group ever ate any bait. It is estimated that a distribution of one bait pellet per acre, at a cost of $1 per pellet, could prove sufficient to reduce rabies dramatically in wild raccoons.

Rabies immunization is recommended for veterinarians and their staffs, hunters who may have contact with wild animals, and technicians who work with the virus. The first vaccine, and for many years the only one available, was developed by Pasteur. This vaccine contained viruses modified by 50 transfers, involving drying of infected spinal cords, from one rabbit to another. It was used in all cases of suspected rabies because the disease might develop while the patient waited for test results on the biting animal. The vaccine was given subcutaneously (into the skin) in 14 or more daily injections into the abdomen, where the thick tissue slows the rate of absorption. These injections had unpleasant to serious side effects, including severe abdominal pain, fatigue, fever, and sometimes rabies infection. A relatively new vaccine produced from viruses grown in human diploid fibroblast cultures elicits high levels of neutralizing antibodies with only a few injections and minimal side effects. The current vaccine is given intramuscularly on days 0, 3, 7, 14, and 28. In addition, hyperimmune globulin is placed deep in the wound and infiltrated around the wound.

Encephalitis

The Disease   **Encephalitis** is an inflammation of the brain caused by a variety of togaviruses or by a flavivirus. We will consider the following four diseases, each of which is caused by a different virus: **eastern equine encephalitis (EEE)**, seen most often in the eastern United States; **western equine encephalitis (WEE)**, seen in the western United States; ➡ (➢ Figure 15.6, p. 406); **Venezuelan equine encephalitis (VEE)**, seen in Florida, Texas, Mexico, and much of South America; and **St. Louis encephalitis (SLE)**, seen from east to west in the central United States. ➡ (➢ Figure 15.3, p. 404) The equine varieties, caused by togaviruses, are so named because they infect horses more often than humans. The life cycles of these viruses generally involve transmission from a mosquito to a bird, back to a mosquito, and then to a horse, human, or other mammal, and finally back to a mosquito. The St. Louis variety, caused by a flavivirus, is so named because the first epidemic was identified in St. Louis in 1933. That virus appears to be transmitted mostly between English sparrows, mosquitoes, and humans.

The viruses, which are introduced into the body through bites of infected mosquitoes, first multiply in the skin and spread to lymph nodes. Viremia involving especially large numbers of viruses follows. In a few infections, the viruses invade the central nervous system, where they cause shrinkage and lysis of neurons. WEE appears every summer, and about a third of the cases occur in children under 1 year of age. Fever and headache are common symptoms, and convulsions sometimes occur. EEE is a much more serious disease; it causes a severe necrotizing infection of the brain. The disease is fatal in 50 to 80 percent of cases, and survivors often suffer permanent brain damage. Fortunately, because swamp birds are the major reservoir for the virus and swamp mosquitoes the major vectors, very few humans become infected. VEE is mainly a disease of horses; when it occurs in humans, it resembles influenza.

SLE occurs in late summer epidemics about every 10 years and causes the most severe symptoms in elderly patients. The illness starts with malaise, fever, and chills as consequences of viremia. Other common symptoms include anorexia, myalgia (muscle pain), sore throat, and drowsiness. In addition, certain patients have symptoms of a urinary tract infection, neurological disorders, altered states of consciousness, and convulsions. Complications can include secondary bacterial infections, blood clots in the lungs, and gastrointestinal hemorrhage. Most patients escape the complications and recover fully.

Diagnosis, Treatment, and Prevention   Encephalitis sometimes can be diagnosed by isolating the causative agent from cell cultures or mice inoculated with blood or spinal fluid. Cultures can be negative when the disease is present because the viremic phase of the disease usually is over before the patient seeks medical attention. Serological methods can be used to identify antibodies at any time during and following the illness. Treatment only alleviates symptoms. Vaccines are available for immunizing horses, but they are rarely used in humans because of the danger of inducing a virulent form of the disease. Prevention by eradicating mosquito vectors is a more appropriate means of decreasing the already low incidence of encephalitis among humans.

Other Viral Diseases of the Brain and Meninges

Herpes Meningoencephalitis   Herpes simplex virus, which usually is responsible for cold sores, also can cause **herpes meningoencephalitis.** This disease often follows a generalized herpes infection in a newborn infant, child, or adult. The virus reaches the brain by ascending from the trigeminal ganglion. The disease has a rapid onset with fever and chills, headache, convulsions, and altered reflexes. In the middle-aged or elderly, meningoencephalitis causes confusion, loss of speech, hallucinations, and sometimes seizures. Most patients die in 8 to 10 days; survivors usually display neurological damage.

Polyomavirus Infections   Polyomaviruses enter the body through the respiratory or gastrointestinal tract. Initial replication takes place in the cells the virus first enters. The viremia that follows allows the viruses to reach target organs, particularly the kidneys, lungs, and brain. Polyomaviruses, which are papovaviruses, were first recognized in the 1960s as viral particles in the enlarged nuclei of oligodendrocytes. These are the cells that produce myelin, the lipoprotein that coats nerve fibers in the central nervous system. Infected oligodendrocytes are observed to surround areas that lack myelin in the brains of patients dying from **progressive multifocal leukoencephalopathy.**

One cause of progressive multifocal leukoencephalopathy is now known to be the JC virus, a polyomavirus named with the initials of a victim from whom it was isolated. Onset is insidious, with vision and speech impairment being the first signs. Typical symptoms of viral infections, such as fever and headache, are absent. Mental deterioration, limb paralysis, and blindness follow. Diagnosis is difficult because cerebrospinal fluid remains normal, and only nonspecific changes are seen in electroencephalograms. The JC virus infects and kills oligodendrocytes but does not affect neurons. On occasion a young patient develops this disease as a complication of a primary infection, but most cases result from reactivated latent viruses from childhood infections.

Other polyomaviruses have been isolated from various patients. The BK virus was isolated from the urine of a kidney-transplant patient. In one instance a 16-year-old boy with an immunodeficiency had BK viremia and developed kidney inflammation with viruses present in kidney cells. Irreversible kidney failure resulted. The BK virus also has been associated with, but not isolated from, respiratory illness and cystitis, an inflammation of the urinary bladder.

Half the children in the United States have antibodies to the JC virus by age 14; antibodies to the BK virus are found by age 4. Although JC and BK viruses apparently persist in most humans for years without causing disease, they sometimes reappear as complications of chronic diseases, immunodeficiencies, and disorders in which lymphocytes proliferate. Pregnancy, diabetes, organ transplantation, antitumor therapy, and immunodeficiency diseases, including AIDS, are among the

Amoebic Invaders

If you are an avid swimmer or an avid hot tubber, here's something you might want to know: Two different genera of soil amoebas, Naegleria and Acanthamoeba, are opportunistic human pathogens you could encounter. *Naegleria fowleri* is usually seen in swimmers. The amoebas probably enter through the nasal passage and make their way up along nerves to the meninges. There, they cause meningoencephalitis. *Acanthamoeba polyphaga* accumulates on the water surface of contaminated hot tubs when the tubs are covered. They are dispersed after the lid is removed and bathers begin to enter the water. These amoebas cause ulceration of the eyes and the skin; if they invade the central nervous system, death can occur within a few weeks due to meningoencephalitis.

✓ List five organisms that can cause meningitis.

✓ Why are bats a particularly dangerous source of rabies infection?

✓ How does encephalitis differ from meningitis?

conditions that can reactivate polyomaviruses. Many kidney-transplant patients, for example, excrete either BK or JC viruses, but the shedding of these viruses rarely has serious consequences. However, unchecked viral multiplication, which is particularly likely with T cell deficiencies, can sometimes occur and cause clinically apparent disease. Both JC and BK viruses are oncogenic in laboratory animals and possibly also in humans. No diagnostic tests for polyomaviruses are available for routine use; nor is any treatment available for the infections, even if they can be recognized.

Other Diseases of the Nervous System

Bacterial Nerve Diseases

Hansen's Disease

Hansen's disease, the currently preferred name for **leprosy,** has been known since biblical times, when many things, even houses, were said to have "leprosy." Many so-called leprosy cases were other skin diseases such as fungal and viral infections, and the houses probably had fungi growing on their walls.

The disease is vastly underreported, but an estimated 15 million cases exist worldwide today, mainly in Asia, Africa, and South America (➢ Figure 24.4). Hansen's disease also occurs in

(a)

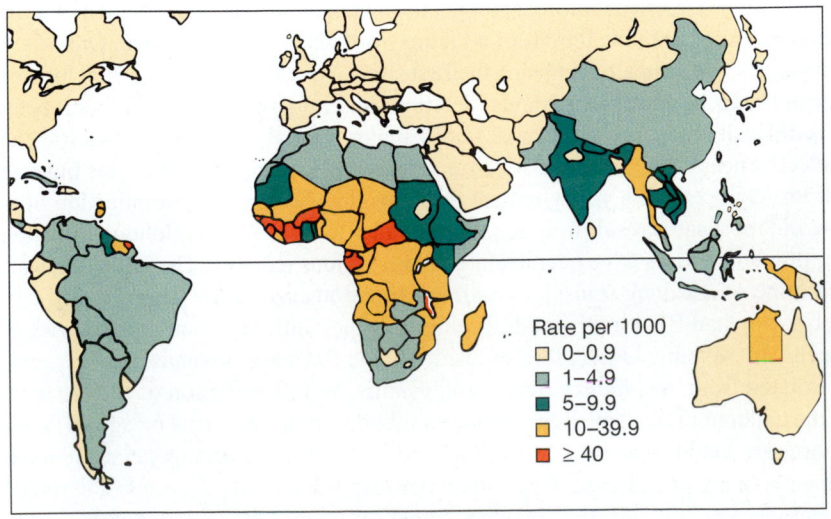

Rate per 1000
- ☐ 0–0.9
- ☐ 1–4.9
- ☐ 5–9.9
- ☐ 10–39.9
- ☐ ≥ 40

(b)

➢ **Figure 24.4   The incidence of Hansen's disease (leprosy).** (a) In the United States; (b) worldwide. (*Source:* CDC) the United States, where a peak number of 361 cases were reported in 1985, mostly among immigrants from countries where Hansen's disease is endemic. Infected people sometimes show no symptoms when they enter the United States. No reliable test is available to disclose all these subclinical cases, although the **lepromin skin test,** similar to the tuberculin skin test for tuberculosis, detects some of them. Health care workers should watch for Hansen's disease among immigrants.

The Disease   The acid-fast bacillus *Mycobacterium leprae* is found in all cases of Hansen's disease. Although *M. leprae* was the first bacterium to be recognized as a human pathogen, demonstrating that it fulfills Koch's postulates has been slow because the organism is difficult to grow in the laboratory. For one thing, it reproduces very slowly, having a 12-day division cycle. Recently developed methods for growing *M. leprae* in such diverse hosts as nine-banded armadillos, chimpanzees, mangabey monkeys, and mice have made the bacterium available for research. It may now be possible to develop a vaccine. PCR techniques allow *M. leprae* to be identified in samples of skin or nasal tissues, thus simplifying diagnosis.

Clinical forms of Hansen's disease vary along a spectrum from tuberculoid to lepromatous. In the **tuberculoid,** or anesthetic, form (➢ Figure 24.5a), areas of skin lose pigment and sensation. In the **lepromatous** (lep-ro′mat-us), or nodular, form (➢ Figure 24.5b), a granulomatous response (Chapter 16) causes enlarged, disfiguring skin lesions called **lepromas.** ➥ (p. 448) Incubation time averages 2 to 5 years for the tuberculoid form and 9 to 12 years for the lepromatous form.

Mycobacterium leprae is the only bacterium known to destroy peripheral nerve tissue; it also destroys skin and mucous membranes. The organism has a predilection for cooler parts of the human body, such as the nose, ears, and fingers, but large numbers of organisms are seen throughout the body, except for the central nervous system. Continuous bacteremia of 1000 organisms per milliliter of blood has been demonstrated in lepromatous cases. Large numbers of bacilli are shed in respiratory secretions and in pus discharged from lesions. Although Hansen's disease is not highly contagious, shedding of organisms probably transmits the disease to those with extensive close contact with patients, such as children of infected parents.

As Hansen's disease progresses, it deforms hands and feet (➢ Figure 24.6). Severe lepromatous disease erodes bone: Fingers and toes become needlelike, pits develop in the skull, nasal bones are destroyed, and teeth fall from the jaw as bone surrounding them is lost. Surgery sometimes can restore the use of extremely crippled hands and feet. The National Hansen's Disease Center, a U.S. Public Health Service facility in Carville, Louisiana, has pioneered development of these special surgical techniques. In the past decade, it was recognized that changes in the feet of diabetic patients can be helped by these same surgical techniques. Thus,

> *M. leprae bacilli occur within cells in a characteristic arrangement, grouped together like a bundle of logs, and are covered by capsules.*

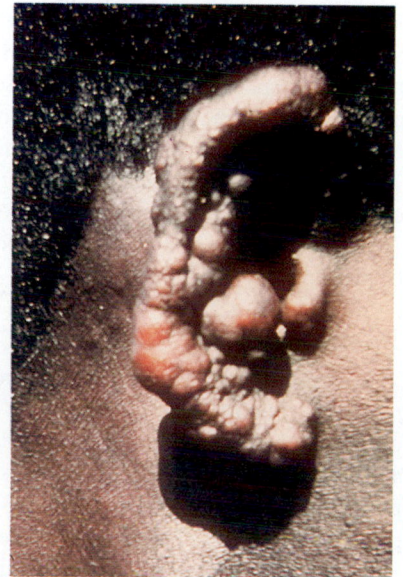

(a)   (b)

➢ **Figure 24.5   Hansen's disease: two extremes with gradations and combinations between.** (a) In the tuberculoid, or anesthetic, form, areas of skin lose pigment and sensitivity. A pin may be stuck into these "anesthetized" areas and not be felt because of destruction of nerves and nerve endings. (b) The nodular form is characterized by disfiguring granulomas called lepromas.

➤ Figure 24.6 **A deformed "claw" hand of Hansen's disease.** This deformity can be treated surgically in its early stages, thereby preventing crippling.

Endospore

(a)

(b)

➤ Figure 24.7 **Tetanus spores.** (a) A TEM of *Clostridium tetani* bacillus with large, dark terminal endospore (105,800X). (b) A freeze-etch TEM preparation, showing the rounded endospore inside a *Clostridium* bacillus (107,250X).

Carville now has an active teaching program for surgeons who will use knowledge gained from one of the most shunned diseases to help thousands of victims of diabetes.

Examination of ancient skeletons has provided insights into the epidemiology of Hansen's disease in past centuries. It is clear that the disease traveled from the Old World to the New World and that in the past, even allowing for misdiagnosis, its incidence in Europe was much higher than it is today. In England, lepers were shunned even in death, being buried in separate cemeteries. Examination of skeletons from these graves reveals that, by far, the majority of the people buried there did in fact have leprosy. Genetic factors may predispose toward resistance to Hansen's disease; as susceptible individuals have died, resistant ones have made up a greater percentage of the population.

Diagnosis, Treatment, and Prevention   Hansen's disease is diagnosed by PCR or by finding the organism in acid-fast-stained smears and in scrapings from lesions or biopsies. The disease is treated with dapsone and rifampin, but dapsone-resistant strains are beginning to appear. Treatment greatly reduces nodules of lepromatous disease, but it cannot restore lost tissue. Until recently, victims of Hansen's disease were isolated in special hospitals called leprosariums. Now the disease can be arrested, and the people can live nearly normal lives without infecting others in their community. (They must still sleep in separate bedrooms and use only their own linens and utensils, however, and they cannot live in a household where children are present.)

Immune responses in Hansen's disease are cell mediated and vary from strong to weak. Strong responses and distinctly positive skin tests are seen in patients with the less serious tuberculoid disease. Weak responses and negative skin tests are seen in patients with rapidly progressing lepromatous disease. However, test results may change over time from positive to negative and vice versa as immune response rises and falls. Lepromatous patients usually have adequate cell-mediated response to other antigens, so their lack of immunity is not due to generalized T cell absence or dysfunction. The absence of T cells in "nude" mice, which are hairless and lack a thymus, makes them suitable organisms for growing large quantities of organisms.

Vaccine is not available for Hansen's disease. Even if a vaccine became available now, its effectiveness would take years to determine because of the disease's long incubation period. Avoiding exposure and receiving prophylactic chemotherapy after exposure are the only means of prevention.

Tetanus

Tetanus is caused by the obligately anaerobic, Gram-positive, spore-forming rod *Clostridium tetani* (➤ Figure 24.7). The organism can be cultured in the laboratory only under strict anaerobic conditions. *Clostridium tetani* endospores are exceedingly resistant to drying, disinfectants, and heat. Boiling for 20 minutes does not kill them, and they can survive for years if not exposed to sunlight. ➡ (Chapter 4, p. 83) Spores are found in all soils but especially in those enriched with manure. The organisms are part of the normal bowel microflora of horses and cattle and about 25 percent of humans. Therefore, handling bedpans, dirty diapers, or other objects contaminated with feces can transmit organisms to persons who have any breaks in their skin.

Since the development of tetanus vaccine in 1933, the incidence of tetanus in the United States has steadily dropped, with the annual number of cases remaining below 100 since 1975. The highest incidence is in older people, especially women. Vaccine was not available during their childhoods, and they did not receive it during military service, as men did. They remain susceptible to tetanus spores as they enjoy gardening in their retirement years. The elderly should be immunized for their protection.

The Disease   To cause tetanus, spores must be deposited deep in tissues, where oxygen is unavailable. This occurs in deep cuts and puncture wounds. Stepping on a rusty nail has a reputation for leading to tetanus, but it is tetanus endospores, not rust, that cause the disease—a shiny new nail can be just as dangerous if the spores are present. Making puncture wounds bleed helps flush tetanus spores and other organisms from them. Once inside the host, the noninvasive tetanus organisms stay at the wound site and release a powerful exotoxin; tetanus is a toxin-mediated disease. After 4 to 10 days' incubation, symptoms begin, with generalized muscle stiffness followed by spasms that affect every muscle. An arched back and clenched fists and jaws (hence the term *lockjaw*), are classic symptoms (➤ Figure 24.8). Spasms can be violent enough to break bones. Eventually, respiratory muscles become paralyzed, heart function is disturbed, and, with rare exceptions, the patient dies. Survivors experience a period of sore muscles but suffer no further sequelae. Before vaccine was available, many soldiers died from tetanus. On battlefields strewn with horses and manure, contamination of wounds with tetanus spores was inevitable. War-related cases were virtually eliminated by vaccinating soldiers; only 12 cases occurred during World War II.

Treatment and Prevention   Tetanus toxoid vaccine given prior to injuries protects against the toxin. Antitoxin and antibiotics are given to nonimmunized patients when injuries are treated. Because antitoxin must be administered to inactivate the toxin before the immune system has time to become sensitized to it, infection treated in this way confers no immunity. Patients should receive toxoid immunization after they recover.

Tetanus neonatorum is acquired through the raw stump of the umbilical cord. In some societies, contaminated knives are used to cut umbilical cords after a baby is delivered, and mud is smeared on the cut end. In parts of some developing nations, 10 percent of deaths within a month of birth are due to neonatal tetanus.

Botulism

Botulism derives its name from the Latin word *botulus,* which means "sausage." It was coined at a time when the disease was often acquired from eating sausages. Botulism is caused by *Clostridium botulinum,* a spore-forming obligate anaerobe that releases a potent exotoxin

Kitasato isolated C. tetani, established it as the causative organism of tetanus, showed that it could not invade the bloodstream, and showed that the disease is an intoxication.

Clostridium difficile, which causes pseudomembraneous colitis, has the distinction of producing the two largest-known bacterial toxins.

To prevent botulism, follow the 2-40-140 rule. Don't eat meats, dressings, or salads that have been kept for more than 2 hours between 40°F and 140°F.

➤ **Figure 24.8   A soldier dying of tetanus.** Tetanus was a common cause of death in the days of cavalry troops. The extreme contraction of all muscles, from those of the face to those of the toes, is a classic symptom of the disease.

Botulism in Waterfowl

We're not the only animals who need to fear botulism food poisoning. Botulism is also an important disease in wetland birds especially during ecological disturbances. In fact, the disease is a leading cause of death among ducks in the western United States. How do the birds become infected? After storms uproot marsh plants, the plants begin to decompose, using up available oxygen and leading to the death of small aquatic invertebrates. Botulism endospores in these animals' digestive tracts or in the mud germinate and produce toxin. Ducks eat the dead invertebrates, die, and fill with toxin. Flies lay eggs on the dead ducks; the eggs hatch into toxin-filled maggots. Healthy ducks eat the maggots, die, and provide a site for new fly eggs. The cycle is repeated until thousands of ducks have died and no more ducks are left at that site.

(neurotoxin). (See the box "Clinical Use of Botulinum Toxin" in Chapter 14.) ➡ (p. 390) The disease occurs in three forms: foodborne, infant, and wound. Foodborne botulism accounts for 90 percent of cases and is caused by ingestion of toxin, usually from improperly home-canned nonacid foods, especially green beans and green peppers (➢ Figure 24.9). Thus, foodborne botulism is an intoxication; the organisms do not infect tissues. Infant botulism and wound botulism involve both infection and intoxication because the organisms grow in tissues and produce toxin.

Endospores of *C. botulinum* are more heat resistant than those of any other anaerobe; they withstand several hours at 100°C and 10 minutes at 120°C. They also are very resistant to freezing (down to −190°C) and irradiation. Found in most soils in the Northern Hemisphere, these endospores remain viable for long periods of time and enable the organism to withstand aerobic conditions. Endospores will germinate only in anaerobic conditions.

The ability of *C. botulinum* to form toxin depends on infection with a bacteriophage. This phage carries the information for botulism toxin production. If infected with an appropriate bacteriophage, *C. botulinum* produces one of eight different toxins, of which only four cause human disease. The other toxins cause disease in various other animals. If a strain of the bacillus is "cured" of its phage infection, it no longer produces toxin. If later infected with a different phage, the bacillus will produce another toxin.

Botulism toxin is the most potent toxin known—even more toxic than *Shigella* and tetanus toxins. As little as 0.000005 μg can kill a mouse! One ounce could kill the entire U.S. population. Originally thought to be an exotoxin, it is now known to be produced inside the cytoplasm and released only upon death and autolysis of the cell. It is activated by proteolytic enzymes, possibly including trypsin in the host's intestine. The toxin is colorless, odorless, and tasteless; people have died from a single taste of affected food. If endospores are not destroyed, they germinate in food during storage under anaerobic conditions and can release large quantities of toxin. Whereas endospores are highly heat resistant, the toxin can be inactivated by only a few minutes' boiling. Boiling home-canned foods vigorously before serving would eliminate most foodborne botulism.

The Disease   Botulism is a neuroparalytic disease with sudden onset and rapidly progressing paralysis. It ends in death from respiratory arrest if not treated promptly. The toxin acts at junctions between neurons and muscle cells and prevents the release of acetylcholine, the chemical that neurons release to cause muscle cells to contract. The toxin thus paralyzes muscles in a relaxed state—starting with small eye muscles, progressing to the larynx and pharynx, and on to the respiratory muscles. This causes double vision, difficulty in speaking and swallowing, and difficulty in breathing. It causes no fever but can cause gastrointestinal disturbances. Although the toxin is an antigen, people who have recovered do not have antibodies, so the amount of antigen required to elicit antibodies must be greater than the lethal dose.

Diagnosis and Treatment   Diagnosis is based on clinical symptoms and history, with confirmation later by demonstrating toxin in serum, feces, or food remains. Although the confirmation test takes 24 to 96 hours, treatment with a polyvalent antitoxin is started immediately.

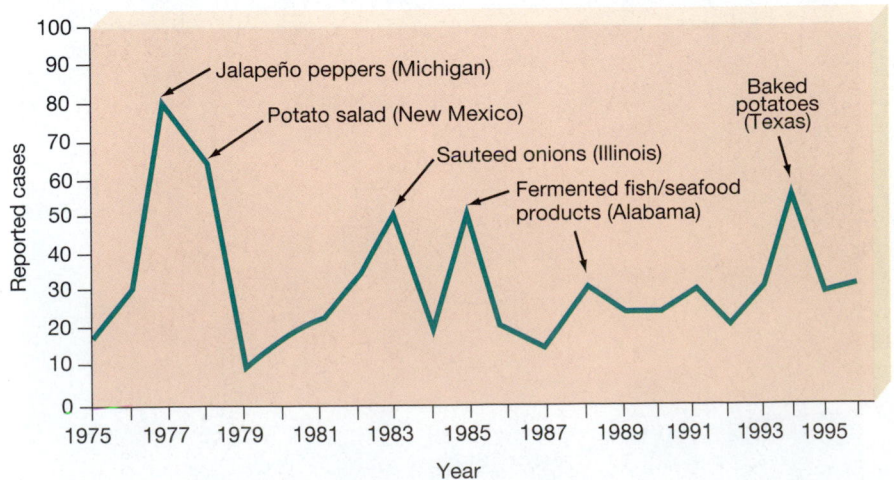

➢ **Figure 24.9   The U.S. incidence of foodborne botulism.** The foods that caused certain outbreaks are included. (*Source*: CDC)

A polyvalent antitoxin is used to ensure effectiveness against all toxins that affect humans. Help in maintaining respiration is important and may be continued for up to 2 months. Antibiotics are of no use because foodborne botulism is due to preformed toxin and not to growth of organisms. With proper treatment, the mortality rate is less than 10 percent.

Infant botulism was first recognized in 1976, and incidence has ranged between 30 and 100 cases per year, mostly in California. The disease is associated with feeding honey to infants. Studies in California have shown that 10 percent of the jars of honey sold there contain botulism endospores. Endospores germinate and grow in the immature digestive tract of infants, probably because they lack appropriate competing microbiota. As toxin is absorbed, the infant becomes lethargic and loses the ability to suck and swallow, so the disease is often called "floppy baby" syndrome. Infant botulism usually occurs in infants under 6 months of age and rarely after 12 months. If parents would not give honey to children under 1 year, most cases could be prevented. The prognosis is excellent, and death is rare, but the child usually must remain hospitalized for several months.

Wound botulism is the least common form of botulism; no more than one case per year has been seen in the United States since 1942. It occurs in deep, crushing wounds. Tissue damage impairs circulation and creates anaerobic conditions, so endospores germinate, multiply, and produce toxin. Toxin enters the blood and is distributed throughout the body. It reaches junctions between neurons and muscle cells about a week after injury and causes progressive paralysis. The mortality rate is about 25 percent.

✓ Distinguish between tuberculoid and lepromatous forms of Hansen's disease.

✓ Why is Hansen's disease no longer a common disease in Europe? More common in South America?

✓ Why must patients who recover from tetanus be immunized against tetanus afterwards? Why hasn't having a case of the disease made them immune?

✓ What symptoms would you experience as you develop a case of botulism food poisoning?

Viral Nerve Diseases

Poliomyelitis

Poliomyelitis is a very ancient disease; its effects are clearly depicted in Egyptian wall paintings thousands of years old. As recently as the early 1950s, it was a dreaded disease in the United States, with nearly 58,000 cases reported in the peak year of 1952. The coming of summer struck terror in the minds of parents, and the diagnosis of a case of paralytic polio in the community was cause for outright panic. In the United States today, those most likely to become infected are members of religious groups who are opposed to immunization, and illegal aliens who are unprotected by a vaccine.

The Disease   Poliomyelitis is caused by three strains of polioviruses (picornaviruses) that have an affinity for motor neurons of the spinal cord and brain. Although most poliovirus infections are inapparent or mild and nonparalytic, the virus reaches the central nervous system in 1 to 2 percent of cases. High fever, back pain, and muscle spasms result. In less than 1 percent of cases, these symptoms are accompanied by partial or complete paralysis of muscles in a relaxed state. The nature and degree of paralysis depend on which neurons in the spinal cord and brain are infected and how severely they are damaged, or if they are lysed. Any paralysis remaining after several months is permanent. The very old and the very young are likely to suffer paralysis as a result of poliovirus infection. Malnutrition, physical exhaustion, corticosteroids, radiation, and pregnancy can increase the severity of the disease.

Poliovirus infections in small children in impoverished areas may go undetected, whereas teenagers and young adults in affluent areas sometimes acquire severe, paralytic poliovirus infections. Good sanitation in the affluent areas reduces exposure and, therefore, natural immunity to the viruses.

Diagnosis, Treatment, and Prevention   Diagnosis of poliomyelitis is made by isolating the virus from pharyngeal swabs or feces, culturing it, and noting its cytopathic effects. Methods also are available for identifying antibodies to the virus in serum. Treatment alleviates symptoms, but patients with paralyzed breathing muscles must forever live in an "iron lung" (➤ Figure 24.10).

Before vaccine became available in 1955, only nonspecific public health measures were available to prevent the spread of poliomyelitis. Schools, swimming pools, and other places where crowds, especially crowds of children, gathered were closed. Large quantities of insecticides were sprayed in the mistaken belief that insect bites somehow played a role in transmission of

Absence of animal reservoirs plus availability of effective vaccines have made health authorities choose polio as the next disease to eradicate totally from our planet.

Scientists have discovered poliovirus receptor-related protein. This is the first receptor found to allow the entrance of the poliovirus and herpes simplex viruses types 1 and 2.

(a)

(b)

➤ Figure 24.10   **Polio.**
During polio epidemics (before
1955) in the United States,
(a) row after row of iron lungs
was filled with patients (b) such
as this 2-year-old girl. Patients
sometimes remained in them for
years, or until their death.

the disease. Transmission is now known to occur both by the fecal–oral route and from pharyngeal secretions, thus explaining the dangers of fecally contaminated swimming pools in
summer. During the first few years that vaccine was available, it could not be made in sufficient quantities to immunize the whole population. Clinics were set up to immunize pregnant
women and young children.

Vaccines   In 1955 the injectable Salk polio vaccine became available. It contained viruses that
were inactivated by formalin at neutral pH but that still retained their antigenic properties. Unfortunately, before the technique was perfected, some batches of vaccine still contained infectious viruses. More than 200 cases of polio and 10 deaths resulted. In 1963 the oral Sabin vaccine,
which contained attenuated live viruses, was introduced. In addition to ease of administration,
this vaccine has the advantages of longer-lasting immunity and prevention of fecal–oral transmission by eliminating viruses in the gastrointestinal tract, where they multiply. Vaccine use has
reduced polio incidence in the United States from about 29,000 cases in 1955 to 20 cases in
1969 in unimmunized and immunosuppressed individuals (➤ Figure 24.11). In October 1995
CDC recommended a combination of polio vaccines to reduce the incidence of vaccine-related
polio. (See the box on p. 715.) Infants would receive shots of the inactivated vaccine at 2 and 4
months, then oral doses of the attenuated vaccine at 6 months and up to 18 months. The oral vaccine is needed because it is effective against "wild" polio virus that, although eradicated from
the United States, could be brought in from other areas by travelers.

The Sabin polio vaccine
is administered orally to
children on a sugar cube
or in sugar liquid.

The Polio Vaccine Controversy

In the United States, the attenuated-virus vaccine developed by Albert Sabin is almost universally used. Some other countries, however, have returned to the inactivated-virus vaccine developed by Jonas Salk. Which is preferable?

Both vaccines have advantages and disadvantages. The inactivated-virus vaccine can be incorporated with other pediatric vaccines, and at the appropriate dose it can be given to immunodeficient patients. However, it does not produce immunity in all recipients, and recipients may require booster shots. Moreover, virulent viruses are used to make the vaccine; a tragedy could occur if they are not completely inactivated.

The attenuated-virus vaccine induces immunity similar to that produced by a natural infection in that the vaccine causes antibodies to be produced in the gut (secretory IgA) as well as in the bloodstream. This reduces the possibility that people who are themselves immune can serve as reservoirs of infection by carrying the virus in their intestines and passing it on to others. Immunity develops quickly and may be lifelong. Oral administration requires less skill and is more acceptable than injection to many people. Finally, the oral vaccine remains potent without refrigeration for a longer time than does the injectable vaccine.

Unfortunately, the live virus used in the oral vaccine does occasionally mutate, and some mutants are virulent. The viruses multiply in gut tissue of the vaccinated individual and are shed in their feces. Thus, changing the diaper of a recently vaccinated child can pose a hazard, especially to family members or day-care workers and immunodeficient patients, depending on the degree of hygiene practiced. About six cases of polio are caused by the vaccine each year in the United States, and about that many are caused by close contact with these immunized children. Since 1980, a total of 143 out of 145 cases of paralytic polio have been associated with oral polio vaccine. The other two cases were classified as indeterminate. Liability costs are now included in the price of the oral vaccine to cover awards that may have to be made to victims of vaccine-induced polio. In the United States, these costs have made the oral vaccine more expensive than the injected vaccine, although the oral vaccine is cheaper to make and administer. In warm countries with endemic poliomyelitis and other viral infections, repeated administration has often failed to induce immunity, probably because immune responses being mounted against other viruses prevent an adequate immune response to the vaccine organisms. Finally, the attenuated-virus vaccine, like all vaccines that use live organisms, cannot be administered to immunosuppressed or immunocompromised persons.

Polio has been virtually eliminated in Israel through the use of a combination of inactivated-virus and attenuated-virus vaccines. This procedure shows promise for controlling polio in developing countries with warm climates, fecally contaminated water, and endemic polio. However, the cost of obtaining and administering both vaccines will delay implementation.

Postpolio syndrome is a condition in which people who survived polio, years before, suffer weakening or paralysis of muscles, which requires them again to use crutches and braces. It is neither infectious nor a recurrence of the disease. It is believed to be due to overuse of compensating muscles that have labored too hard for too many years and now cannot function properly.

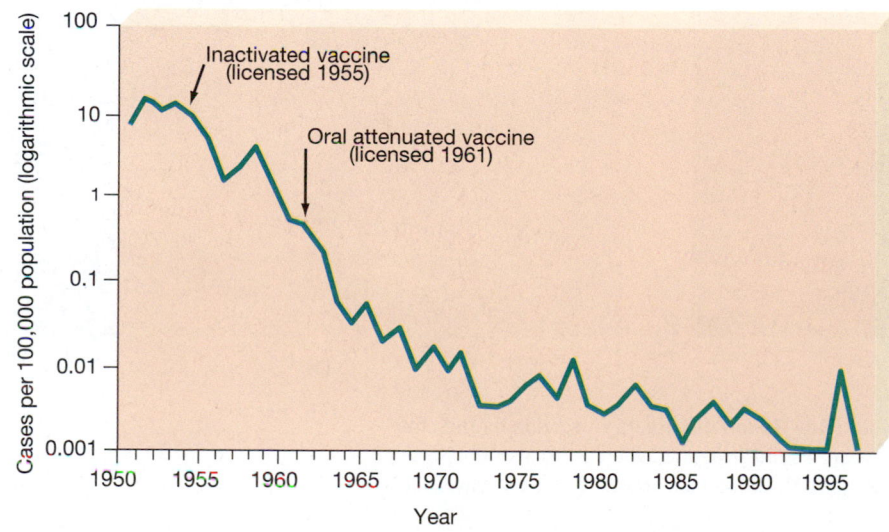

> Figure 24.11 **The U.S. incidence of polio.** The drastic drop since vaccines became available (first Salk, then Sabin) has not happened in South and Central America. Thus, many people still suffer from this preventable disease. In fact, more people die each year from vaccine-preventable diseases than die from AIDS. (*Source*: CDC)

Prion Diseases of the Nervous System

Since the 1920s, when **Creutzfeldt–Jakob disease (CJD)** was first investigated, several degenerative diseases of the nervous system have been identified. Although the existence of **prions** was not universally accepted, many researchers believe that such diseases are associated with prions. ➡ (Chapter 10, p. 283) Stanley Prusiner received the 1997 Nobel Prize for his work on prions and disease.

These diseases are referred to collectively as **transmissible spongiform encephalopathies** because in damaging neurons, they give brain tissue a "spongy" appearance (➢ Figure 24.12). They include *kuru,* CJD, and a special form of CJD called *Gerstmann-Strassler disease* in humans; *scrapie* in sheep and goats; *transmissible mink encephalopathy; chronic wasting disease* of elk and mule deer; and *"mad cow disease,"* or *bovine spongiform encephalopathy,* in British dairy cattle. In addition, in 1991, 29 cases of transmissible spongiform encephalopathies were reported in British cats, and 2 were reported in ostriches in the Berlin Zoo.

A prominent feature of all prion-associated diseases is the lack of any inflammatory response, which is a hallmark of other infectious diseases. There is, however, an increase in the size of *astrocytes*—the cells that regulate the passage of materials from the blood to neurons—throughout the central nervous system. These cells apparently produce large quantities of a filamentous protein called *amyloid,* which is characteristic of a variety of degenerative diseases of the nervous system, including Alzheimer's disease.

Like other infectious agents, prions are transmissible, although disease symptoms may not appear until years after initial infection. **Kuru,** which occurred mainly in New Guinea, is transmitted through small breaks in the skin. Why this disease appeared mainly in women was puzzling until it was discovered that New Guinean women prepared the bodies of the dead for cannibalistic consumption and smeared their own bodies with the raw flesh of the corpses. Prions in diseased tissues entered the blood of the women or the children playing at their feet, traveled to the brain, and eventually caused kuru. Men and adolescent boys live apart from the women and small children and so were spared infection with kuru.

D. Carleton Gajdusek, a U.S. investigator at the National Institutes of Health (NIH), won a Nobel Prize in 1976 for his work on kuru. He found that 1 to 15 years following inoculation with the kuru prion, onset of symptoms began with headaches, minor loss of coordination, and a tendency to giggle at inappropriate times. Three months later, victims needed crutches to walk or stand; a month after that, victims could not move except for spasms. At that point, the swallowing muscles no longer functioned, and malnutrition became a serious problem. Rela-

(a)

(b)

➢ Figure 24.12   **A prion-caused disease, Creutzfeldt–Jakob disease.** (a) A section through the cerebral cortex of a normal human brain (370X) reveals a solid structure, whereas (b) a section through the brain of a patient with Creutzfeldt–Jakob disease (370X) shows many holes. It is clear why CJD is referred to as a subacute *spongiform* encephalopathy.

➤ Figure 24.13 **Kuru.** Kuru victims have reached the point where they must be fed prechewed food. Since cannibalistic rites have been stopped in New Guinea, the disease has disappeared there.

tives would prechew food and massage it down the esophagus of victims (➤ Figure 24.13), but within 1 year, the patients died.

Some cases of kuru have been traced to inadvertent inoculation with prions present in corneal transplant tissue from patients having undiagnosed CJD, to dwarfed children who received human growth hormone injections made from cadaver pituitaries, and to silver electrodes implanted in the brain during surgical procedures following use in a CJD patient. The same electrodes, located 17 months later after numerous supposed sterilizations, were implanted into a chimpanzee's brain, where they caused CJD. These findings have caused some newer methods of instrument sterilization to be abandoned. Autoclaving at 121°C at 15 psi of pressure for 1 hour destroys infectivity, something that years of storage in formaldehyde fails to do.

In one experiment, Gajdusek and Paul Brown (also at NIH) buried scrapie-infected hamster brains in soil and left them undisturbed for 3 years. When the brain material was dug up, it was still infectious. Gajdusek and Brown suspect that infected material may retain its deadliness for more than a decade under such circumstances. Thus, sheep that graze in fields where contaminated carcasses have been placed have become infected with **scrapie** (➤ Figure 24.14). Marking of burial sites and reexamination of the accepted technique of adding corrosive quicklime to the bodies need to be done before eradication can be achieved.

In most cases of CJD, no source of prions has been identified. In the most thoroughly studied instance of Gerstmann–Strassler syndrome, CJD developed in every generation of a family line for over 100 years. In this form of CJD, a genetic mechanism seems to facilitate prion infection, but how this mechanism works is not clear. Perhaps the presence of a particular gene renders patients susceptible to infection by prions from outside the body, or perhaps the gene activates synthesis of prions within the body.

➤ Figure 24.14 **Scrapie.** Sheep infected with scrapie will rub or scrape—sometimes until they are bloodied—against fences, poles, or trees. This fatal disease has no cure.

➤ Figure 24.15   **Mad cow disease.** Carcasses of infected cattle are burned rather than buried.

Mad cow disease (➤ Figure 24.15) reached a peak rate in the early 1990s, with 400–500 deaths of British cattle per week, but it has now diminished. It is thought to have existed in Britain since at least the late 1960s, and the number of cases began to mushroom in early 1987. At that time the method of rendering (boiling down) animal remains for livestock feed was changed so as to omit a solvent extraction step and to increase the number of sheep heads used (including the brains). There had been considerable debate as to whether prions can cross species lines in nature. But prions have clearly been transmitted from one species to another in laboratory trials. The British government has prohibited the addition of beef brains to hamburger, formerly a common practice. The British have also changed their butchering methods so that knife or saw blades do not pass through the spinal cord, thus preventing blades from contaminating edible parts. Many countries, including the United States, have banned the import of British beef, cattle, and beef products.

The possibility that slow-onset prions play a role in other neurodegenerative diseases such as Alzheimer's and Parkinson's diseases is being investigated. Alzheimer's disease, first described by Alois Alzheimer in 1907, now affects more than 2 million people in the United States—more than 5 percent of the population over age 65. Amyloid proteins arranged in structures called *neurofibrillary tangles,* or *plaques,* have been found at autopsy in the brains of Alzheimer's patients, but the holes found in kuru and other spongiform encephalopathies are missing. Whether prions play a role in the deposition of the proteins is not yet known. So far researchers have not been able to transmit Alzheimer's disease to laboratory animals, as has been done with kuru, CJD, and scrapie. Whether this is due to nonsusceptibility of the animals or the absence of an infectious agent also is not yet known. However, bits of β-amyloid protein injected into rat brains have recently produced a disease similar to Alzheimer's in these rats. Genetic factors may be involved, at least in Alzheimer's disease and CJD, because multiple cases of both diseases have been seen in some families. Much remains to be learned about prions, neurodegenerative diseases, and the relationship between them.

Parasitic Diseases of the Nervous System

African Sleeping Sickness

African sleeping sickness, or **trypanosomiasis** (tri-pan″o-so-mi′a-sis), is a disease of equatorial Africa caused by protozoan blood parasites of the genus *Trypanosoma.* Although 100 or more species of this parasite infect various vertebrates and invertebrates, two species, *T. bru-*

Both male and female tsetse flies bite and transmit disease. They bite only during daylight hours.

cei gambiense and *T. brucei rhodesiense,* cause disease in humans. Typically, trypanosomes have an undulating membrane and a flagellum (➤ Figure 24.16a), but in some stages they are shorter and lack a flagellum. For humans to get African sleeping sickness, they must be bitten by an infected tsetse (tset´se) fly (➤ Figure 24.16b). When a tsetse fly bites, it injects infectious trypanosomes, sometimes hundreds in a single bite, into the blood of its victim. The flies serve as vectors and as hosts for part of the life cycle of trypanosomes. Although transmission usually is from one human to another via the fly, game animals serve as natural reservoirs for *T. brucei rhodesiense.*

The Disease   African sleeping sickness is a progressive disease characterized according to the tissue in which the parasites congregate during stages that affect first the blood, then the lymph nodes, and finally the central nervous system. Although the parasites do not actually invade cells, they can damage every tissue and organ in the body.

The bites of tsetse flies cause a local inflammatory reaction. After an incubation period of 2 to 23 days, fever appears initially for about a week while parasites are in the blood and at irregular intervals as the parasites are released from lymph nodes. Patients are able to work through the first and second stages, but they suffer from various symptoms—shortness of breath, cardiac pain, disturbed vision, anemia, and weakness—that become increasingly severe. Invasion of the nervous system by the parasites causes headache, apathy, tremors, and an uncoordinated, shuffling gait. Pain and stiffness in the neck and paralysis occur as the disease progresses. Eventually, the patient cannot be roused to eat, becomes emaciated, has convulsions, sleeps continuously, goes into a profound coma, and dies.

Infection with *T. brucei gambiense* produces a slowly progressive, chronic disease that, if untreated, lasts several years before central nervous system symptoms intensify and death occurs. Infection with *T. brucei rhodesiense* produces a more rapidly progressive disease; it is often fatal within a few months, before central nervous system damage is apparent.

Diagnosis, Treatment, and Prevention   Diagnosis of African sleeping sickness is made by finding the trypanosome parasite in the blood. Until recently, arsenic-containing drugs were used to treat the disease, but they cause eye damage, and the parasites quickly become tolerant to them. Pentamidine, suramin, and melarsoprol are now used, usually in that order. If pentamidine, the least toxic drug, fails to combat the infection, more toxic ones are tried. Results of treatment with any drug are generally more successful if treatment is begun before central nervous system involvement occurs. A combination of Berenil and nitroimidazole is being used to treat the disease after it has progressed to the central nervous system.

Preventing human infection is nearly impossible because of the wide range of tsetse flies (4.5 million square miles) and possibly because of the reservoir of infection in large game animals. Some control has been achieved by clearing brush where flies congregate and by applying aerial pesticides. Another significant means of reducing the tsetse fly population is to release irradiated male flies, which fail to produce viable sperm. The eggs of females that mate with them fail to develop.

Trypanosomes that cause African sleeping sickness have a special means of evading the host's defenses. The intermittent fever of the disease is directly correlated with increasing numbers of parasites in the blood. The unusual thing about these parasites is that each time parasites appear in the blood, they have a glycoprotein coat different from that of previously released parasites. By the time the immune system has developed antibodies to a trypanosome surface antigen, the trypanosome has a different surface antigen. This ability to alter antigens has thwarted efforts to produce a vaccine for African sleeping sickness. When trypanosomes first enter a human, they seem to be able to make only about 15 antigens; later they can make 100 or more. Researchers hope to produce a vaccine that will confer immunity against any antigen the trypanosome might have when it enters the human body. Such a vaccine would allow the body to attack the trypanosome before an infection becomes established.

Should a vaccine be developed, a means to mount a costly, massive vaccination program would be needed. Because available vaccines, such as those for measles and mumps, have not been administered to many of the children in the region endemic to African sleeping sickness, the prospect for mass immunization is not bright. More than 20,000 people per year are likely to continue to die of the disease.

(a)

(b)

➤ **Figure 24.16   African sleeping sickness.** (a) *Trypanosoma brucei gambiense* (1855X), seen in a blood sample, causes African sleeping sickness. (b) Trypanosomes are spread by the bite of the tsetse fly.

Chagas' Disease

Chagas' (Chah′gahs) **disease,** named for the Brazilian physician Carlos Chagas, who first described it in 1909, is caused by *Trypanosoma cruzi.* The disease occurs sporadically in the southern United States and is endemic in Mexico, Central America, and much of South America. It affects more than 18 million people (➤ Figure 24.17). *T. cruzi* looks like the trypanosomes that cause African sleeping sickness. It is transmitted by several kinds of reduviid bugs (Chapter 11) that are hosts for the sexual phase of the trypanosome's life cycle. ➡ (p. 314) Each species of bug occupies a particular region, so the bugs that transmit Chagas' disease in Mexico are different from those that transmit it in South America. Reduviid bugs often bite near the eyes; they defecate as they bite, depositing infectious parasites on the skin. Humans almost automatically rub such a bug bite and thereby transfer parasites to the eyes or the bite wound.

The Disease   Chagas' disease begins with subcutaneous inflammation around the bug bite. After 1 to 2 weeks, the parasites have made their way to lymph nodes, where they repeatedly divide and form aggregates called **pseudocysts.** Wherever the pseudocysts rupture, they cause inflammation and tissue necrosis. These parasites enter cells either by invasion or by phagocytosis and can damage lymphatic tissues, all kinds of muscle, and especially supporting tissues around nerve ganglia. Destruction of nerve ganglia in the heart accounts for nearly three-fourths of deaths from heart disease among young adults in endemic areas.

Chagas' disease appears in an acute form and a chronic form. The acute disease, which is most common in children under 2, is characterized by severe anemia, muscle pain, and nervous disorders. In especially virulent acute disease, death can occur in 3 to 4 weeks, but many patients recover after several months of less virulent disease. The chronic disease, which is seen mainly in adults, probably arises from a childhood infection. It is a mild disease and is sometimes asymptomatic but often causes enlargement of various organs. Insidious damage to nerves can cause several severe effects. In the digestive tract, it slows or stops muscle contractions; in the heart, it can cause irregular heartbeat and accumulation of fluid around the heart; in the

Assassin Bugs

A few cases of Chagas' disease usually occur in the United States, along the Texas border. Fortunately the infective reduviid bugs (often called "kissing bugs" or "assassin bugs") don't live much farther north than that. However, there are other species of reduviids living in most of the United States. Could they also spread Chagas' disease? The trypanosomes live well in these other bugs. But, our northern bugs don't necessarily defecate when they bite. The two activities are linked in Mexican and South American reduviids. A northern bug could defecate at or near a bite site, just by accident. So there is a small likelihood of infection. A greater worry is global warming, which could extend the range of southern species well into the United States. And just think, there is no known cure for Chagas' disease.

Different agents in the body usually activate T cells of the immune system; in the presence of Trypanosoma cruzi, T cells are not activated by any of these agents.

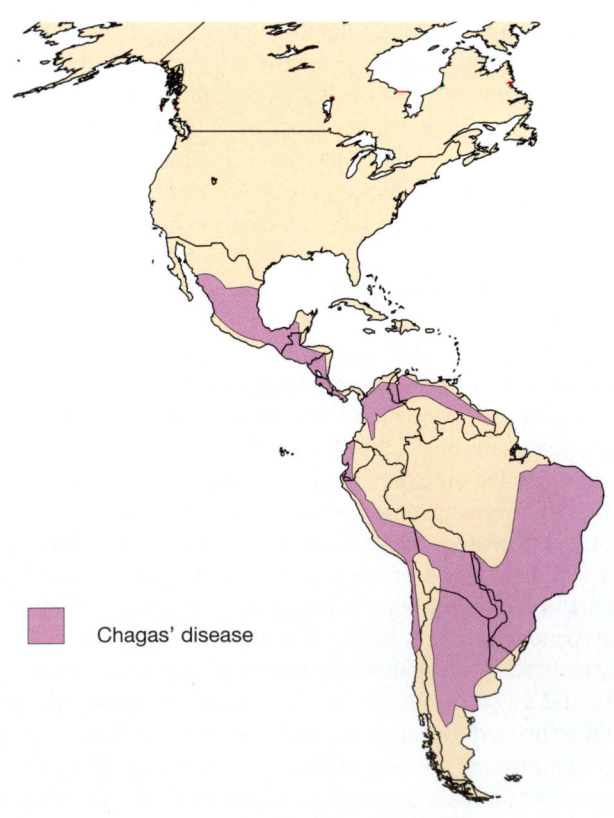

■ Chagas' disease

➤ Figure 24.17   **Worldwide distribution of Chagas' disease, 1994.** (*Source*: CDC)

Nagana and HDL

In up to 25 percent of Africa, it is impossible to raise live-stock because of the tsetse-fly-borne disease *nagana*. The causative agent, the flagellated parasite *Trypanosoma brucei brucei,* probably originated in Africa. It lives in the blood of large native grazing animals without harming them, presumably because a long association has allowed host and parasite to adapt to each other. But when humans migrated into the areas inhabited by tsetse flies, the domesticated livestock the humans brought with them were not able to survive infection with *T. brucei brucei*. When bitten by tsetse flies, humans are also infected by the parasite but do not become ill; the parasites rapidly disintegrate in human blood.

Which component of human blood protects against *T. brucei brucei?* The surprising answer involves a protein never before recognized as providing protection against infectious disease. The high-density lipoprotein (HDL) fraction of blood serum, known for its role in removing cholesterol from the blood, contains a small subfraction that binds to hemoglobin released from dying red blood cells. This binding allows the iron in hemoglobin to be recycled within the body. When a tsetse fly bite releases *T. brucei brucei* into the human bloodstream, the protein–hemoglobin complex is taken in by the trypanosome and eventually enters the parasite's lysosomes. The acidic pH inside the lysosomes causes the protein–hemoglobin to undergo enzymatic changes that result in the release of free radicals. These extremely reactive molecules destroy the membranes of the lysosomes, whose contents—lysozyme enzymes—are expelled into the cytoplasm. There the enzymes digest the parasite from within.

Only humans, some apes, and Old World monkeys have the gene for this *trypanosome lytic factor* (TLF). However, it is hoped that genetic engineering techniques can be used to transfer the gene into livestock, thereby producing trypanosome-resistant breeds of cattle, for example.

---

central nervous system, it can cause paralysis by destroying motor centers. *T. cruzi* also crosses the placenta, so chronically infected mothers often give birth to infants with severe acute disease.

Diagnosis, Treatment, and Prevention   Several diagnostic techniques are available for Chagas' disease. The parasites can be found in the blood during fever in acute cases. Small animals such as guinea pigs and mice can be inoculated with blood from patients and observed for disease symptoms. This technique is known as *xenodiagnosis. Xenos* is Greek for "strange" or "foreign"; in this context it refers to the use of an organism different from a human. Chronic Chagas' disease sometimes can be detected by allowing patients to be bitten by uninfected laboratory-reared reduviid bugs and examining the bugs in 2 to 4 weeks for trypanosomes, which develop in the bug's intestine.

Despite several ways to diagnose the disease, no effective treatment is available. Drugs used to treat other trypanosome infections are of no use because they fail to reach the parasites inside cells. Work is underway to develop new drugs and a vaccine, but until these are available, control of the reduviid vectors is the only means of reducing misery from this disease. Treating homes with insecticides offers some protection, but the bugs crawl into crevices in walls and thatched roofs and are difficult to eradicate.

The agents and characteristics of the diseases discussed in this chapter are summarized in Table 24.2.

✓ Compare the live (oral) versus killed polio vaccines as to advantages and disadvantages.

✓ How is African sleeping sickness transmitted?

✓ What damage occurs in Chagas' disease?

> Table 24.2 | **A summary of diseases of the nervous system**

Disease	Agent(s)	Characteristics
Bacterial diseases of the brain and meninges		
Bacterial meningitis	See Table 24.1	Tissue necrosis, brain edema, headache, fever, occasionally seizures
Listeriosis	*Listeria monocytogenes*	A kind of meningitis seen in fetuses and immunodeficient patients
Brain abscesses	Various anaerobes	Infection that grows in mass and compresses brain
Viral diseases of the brain and meninges		
Rabies	Rabies virus	Invades nerves and brain; headache, fever, nausea, partial paralysis, coma, and death ensue unless patient has immunity
Encephalitis	Several encephalitis viruses	Shrinkage and lysis of neurons of the central nervous system; headache, fever, and sometimes brain necrosis and convulsions
Herpes meningoencephalitis	Herpesvirus	Fever, headache, meningeal irritation, convulsions, altered reflexes
Progressive multifocal leukoencephalopathy	Polyomavirus, JC virus	Infects oligodendrocytes in areas of brain lacking myelin; mental deterioration, limb paralysis, and blindness
Bacterial nerve diseases		
Hansen's disease	*Mycobacterium leprae*	Range of symptoms from loss of skin pigment and sensation sensation to lepromas and erosion of skin and bone
Tetanus	*Clostridium tetani*	Toxin-mediated disease; muscle stiffness, spasms, paralysis of respiratory muscles, heart damage, and usually death
Botulism	*Clostridium botulinum*	Preformed toxin from food prevents release of acetylcholine; paralysis and death result unless treated promptly; in infants and wounds, endospores germinate and produce toxin
Viral nerve diseases		
Poliomyelitis	Several types of polioviruses	Fever, back pain, muscle spasms, partial or complete flaccid paralysis from destruction of motor neurons
Prion diseases of the nervous system		
Transmissible spongiform encephalopathies	Prions	Death of brain cells leave holes creating spongiform brain tissue; amyloid plaques form; long delay before symptoms appear; then spasms rapidly worsening to collapse; no cure
Parasitic diseases of the nervous system		
African sleeping sickness	*Trypanosoma brucei gambiense,* *T. brucei rhodesiense*	Fever, weakness, anemia, tremors, shuffling gait, apathy; as parasites invade nervous system, emaciation, convulsions, and coma ensue
Chagas' disease	*Trypanosoma cruzi*	Subcutaneous inflammation, damage to lymphatic tissues, muscle, and nerve ganglia; muscle pain and paralysis of intestinal, heart, and skeletal muscle

Retracing Our Steps

- The agents and characteristics of the diseases discussed in this chapter are summarized in Table 24.2. Information in that table is not repeated in this summary.

Components of the Nervous System

- The **central nervous system (CNS)** and the **peripheral nervous system (PNS)** are the two main components. The brain and spinal cord form the CNS; **nerves** reaching all parts of the body compose the PNS. Aggregations of nerve cell bodies in the PNS are called **ganglia. Meninges** cover the brain and spinal cord.

- The CNS is protected by special thick-walled capillaries without pores in their walls, which form the **blood–brain barrier.**

- The nervous system has no normal microflora.

Diseases of the Brain and Meninges

Bacterial Diseases of the Brain and Meninges

- **Bacterial meningitis,** acquired from carriers or endogenous organisms, can usually be treated with penicillin. A vaccine is available to protect children from *Haemophilus* meningitis.

- **Listeriosis** can be transmitted by improperly processed dairy products and can cross the placenta. It is a great threat to immunodeficient patients.

- **Brain abscesses** arise from wounds or as secondary infections. Antibiotics are effective if given early in an infection, and surgery can be used later unless the abscess is in a vital area of the brain.

Viral Diseases of the Brain and Meninges

- **Viral meningitis** is usually self-limiting and nonfatal.

- **Rabies** has a worldwide distribution except for a few rabies-free countries. It is difficult to control because of the large number of small mammals that serve as reservoirs.

- Rabies is diagnosed by IFAT and treated by thoroughly cleaning bite wounds and injecting hyperimmune rabies serum and giving vaccine. It can be prevented by immunizing pets and people at risk and by avoidance of contact with wild animals by those who are not immunized.

- **Encephalitis** is transmitted by mosquitoes, often from horses, and can sometimes be diagnosed by culturing blood or spinal fluid in cells or mice. Vaccine is available for horses.

- **Herpes meningoencephalitis** often follows a generalized herpes infection.

Other Diseases of the Nervous System

Bacterial Nerve Diseases

- **Hansen's disease (leprosy)** affects millions of people worldwide. Many patients are asymptomatic for years after infection, and no diagnostic test is available to identify such patients. Dapsone and rifampin are used to treat Hansen's disease, but no vaccine is available.

- **Tetanus** typically follows introduction of spores into deep wounds. It is treated with antitoxin and antibiotics. If all people received the available vaccine, tetanus could be prevented.

- **Botulism** is acquired by ingesting foods that contain a potent preformed neurotoxin and is treated with polyvalent antitoxin. In **infant botulism** and **wound botulism,** spores germinate and produce toxin.

Viral Nerve Diseases

- Prior to the development of vaccines in the 1950s, **poliomyelitis** was a common and dreaded disease. It can be diagnosed from cultures and by immunological methods. Treatment merely alleviates symptoms and includes the use of an iron lung to maintain breathing when respiratory muscles are paralyzed.

- Both an injectable and an oral polio vaccine are available; each has advantages and disadvantages. Worldwide use of vaccine could eradicate poliomyelitis.

Prion Diseases of the Nervous System

- Following a long delay after inoculation, brain cells die causing a spongiform appearance of the tissue. Amyloid plaques accumulate.

- Symptoms begin with spasms, which rapidly worsen to collapse, followed by death.

- Prions cause **transmissible spongiform encephalopathy** in humans and many animals. Prions can be transmitted across species lines.

- **Kuru** and **Creutzfeldt–Jakob disease (CJD)** affect humans, **scrapies** attacks sheep, and **mad cow disease** occurs in cattle.

Parasitic Diseases of the Nervous System

- **African sleeping sickness** occurs in equatorial Africa and is transmitted by the tsetse fly. It is diagnosed by finding parasites in the blood and can be treated with pentamidine and other drugs.

- **Chagas' disease** occurs from the southern United States to all but the southernmost part of South America and is transmitted by various species of bugs. It is diagnosed by finding parasites in the blood and by xenodiagnosis. No effective treatment is available.

Terminology Check

See You on the Web

Don't forget I am expecting you on the web. The following is a preparedness checklist for your web explorations. If you have questions on any of these items, you can return to the end of Chapter 1 for more details.

- Questions We'll Explore
- Concept Check questions
- Flashcards
- Study Guide

If you are ready, come visit me at

http://www.prenhall.com/black

It will be a journey you'll never forget.

Environmental Microbiology

Get your nose up close to the page and see if you can catch a whiff of the tart aroma of this bleu cheese. If you could roll a bit of this wonderful odorous cheese around in your mouth, you would understand the appeal of this cheese to gourmands worldwide. But what you may not realize is that bleu cheese owes its appeal to mold spores of *Penicillium roqueforti* or *Penicillium glaucum.* Blue-veined cheeses are carefully innoculated with these mold spores and then pockmarked with pits and holes to allow the mold to spread and congregate during the aging process. The same mold can turn a melon or a mango into unappetizing mush, but for bleu cheese it is a match made in heaven.

Background Basics

Microorganisms are found in every environment—in the air we breathe, the food we eat, the soil where food is grown, and the water we drink. Many microbes benefit humans, whereas only a few cause human disease. One might consider environmental microbiology to be of little significance in the health sciences, but that is not the case. Humans, being one of many organisms in an environment, affect and are affected by both living and nonliving components of an environment, including its microorganisms. To control disease, health scientists need to know how to control microorganisms in air, food, soil, and water. To do that, they need to understand the roles of microorganisms in the environment.

Fundamentals of Ecology

Ecology is the study of the relationships among organisms and their environment. These relationships include interactions of organisms with physical features—the **abiotic factors**—of the environment and interactions of organisms with one another—the **biotic factors**—of the environment. An **ecosystem** comprises all the organisms in a given area together with the surrounding abiotic and biotic factors.

The Nature of Ecosystems

Ecosystems are organized into various biological levels. The **biosphere** is the region of the earth inhabited by living organisms. It consists of the *hydrosphere* (earth's water supply), the *lithosphere* (the soil and rock that include the earth's crust), and the *atmosphere* (the gaseous envelope surrounding earth). A wide diversity of organisms exists within the biosphere. A terrestrial ecosystem, such as a desert, tundra, grassland,

Questions We'll Explore

A What is ecology, and how does energy flow in ecosystems?

B Why is recycling important, and how are water and carbon recycled?

C What other biogeochemical cycles exist, and what roles do microorganisms play in them?

D What kinds of microorganisms are found in air, and how are they detected and controlled?

E What kinds of microorganisms are found in soil, and what are their roles in biogeochemical cycles and as pathogens?

F How do freshwater and marine environments, and their microorganisms, differ?

G How do water pollution and waterborne pathogens affect humans?

H How is water purified, and how is it tested to determine purity?

I What is sewage, and what processes are involved in primary, secondary, and tertiary sewage treatment?

J What is bioremediation, and what are its advantages and disadvantages?

or tropical rain forest, is characterized by a particular climate, soil type, and organisms. The hydrosphere is divided into freshwater and marine ecosystems.

The organisms within an ecosystem live in communities. An ecological **community** consists of all the kinds of organisms that are present in a given environment. Microorganisms can be categorized as indigenous or nonindigenous to an environment. **Indigenous,** or *native,* **organisms** are always found in a given environment. They generally are able to adapt to normal seasonal changes or changes in the quantity of available nutrients in the environment. For example, *Spirillum volutans* is indigenous to stagnant water, various species of *Streptomycetes* are indigenous to soil, and *Escherichia coli* is indigenous to the human digestive tract. Regardless of variations in the environment (except for cataclysmic changes), an environment will always support the life of its indigenous organisms. **Nonindigenous organisms** are temporary inhabitants of an environment. They become numerous when growth conditions are favorable for them and disappear when conditions become unfavorable.

Communities are made up of *populations,* groups of organisms of the same species. In general, communities composed of many populations of organisms are more stable than those composed of only a few populations—that is, of only a few different species. The various species create a system of "checks and balances" such that the numbers of each species remain relatively constant.

The basic unit of the population is the individual organism. Organisms occupy a particular habitat and niche. The *habitat* is the physical location of the organism. Microorganisms are so small that they often occupy a **microenvironment,** a habitat in which the oxygen, nutrients, and light are stable, including the environment immediately surrounding the microbe. A particle of soil could be the microenvironment of a bacterium. This environment is more important to the bacterium than the more extensive *macroenvironment*. An organism's *niche* is the role it plays in the ecosystem—that is, its use of biotic and abiotic factors in its environment. Microbes may be *producers, consumers,* or *decomposers*. We will discuss each of these groups in the next section.

The Flow of Energy in Ecosystems

Energy is essential to life, and radiant energy from the sun is the ultimate source of energy for nearly all organisms in any ecosystem. (The chemolithotrophic bacteria that extract energy from inorganic compounds are an exception.) ➡ (Chapter 5, p. 130) Organisms called **producers** (autotrophs) capture energy from the sun. They use this energy and various nutrients from soil or water to synthesize the substances they need to grow and to support their other activities. Energy stored in the bodies of producers is transferred through an ecosystem when **consumers** (heterotrophs) obtain nutrients by eating the producers or other consumers. **Decomposers** obtain energy by digesting dead bodies or wastes of producers and consumers. The decomposers release substances that producers can use as nutrients. The flow of energy and nutrients in an ecosystem is summarized in ➤ Figure 25.1.

Microorganisms can be producers, consumers, or decomposers in ecosystems. Producers include photosynthetic organisms among bacteria, cyanobacteria, protists, and eukaryotic algae. Although green plants are the primary producers on land, microorganisms fill this role in the ocean. Consumers include heterotrophic bacteria, protists, and microscopic fungi. (To the extent that viruses divert a cell's energy to the synthesis of new viruses, they too act as consumers.) Many microorganisms act as decomposers. In fact, they play a greater role in the decomposition of organic substances than larger organisms do.

Biogeochemical Cycles

As they carry on essential life processes, living organisms incorporate water molecules and carbon, nitrogen, and other elements from their environment into their bodies. Without decomposers to ensure the flow of nutrients through ecosystems, much matter would soon be incorporated into bodies and wastes, and life would soon become extinct. Although the supply of energy from sunlight is continuously renewed, the supply of water and the chemical elements that serve as nutrients is fixed. These materials must be recycled continuously to make

Bacteria Make the World Go 'Round

"The bacteria particularly . . . are still more important than we. Omnipresent in infinite varieties, . . . they release the carbon and nitrogen held in the dead bodies of plants and animals which would—without bacteria and yeasts—remain locked up forever in useless combinations, removed forever as further sources of energy and synthesis. Incessantly busy in swamp and field, these minute benefactors release the frozen elements and return them to the common stock, so that they may pass through other cycles as parts of other living bodies. . . . Without the bacteria to maintain the continuities of the cycles of carbon and nitrogen between plants and animals, all life would eventually cease. . . . Without them, the physical world would become a storehouse of well-preserved specimens of its past flora and fauna . . . useless for the nourishment of the bodies of posterity. . . ."

—Hans Zinsser, 1935

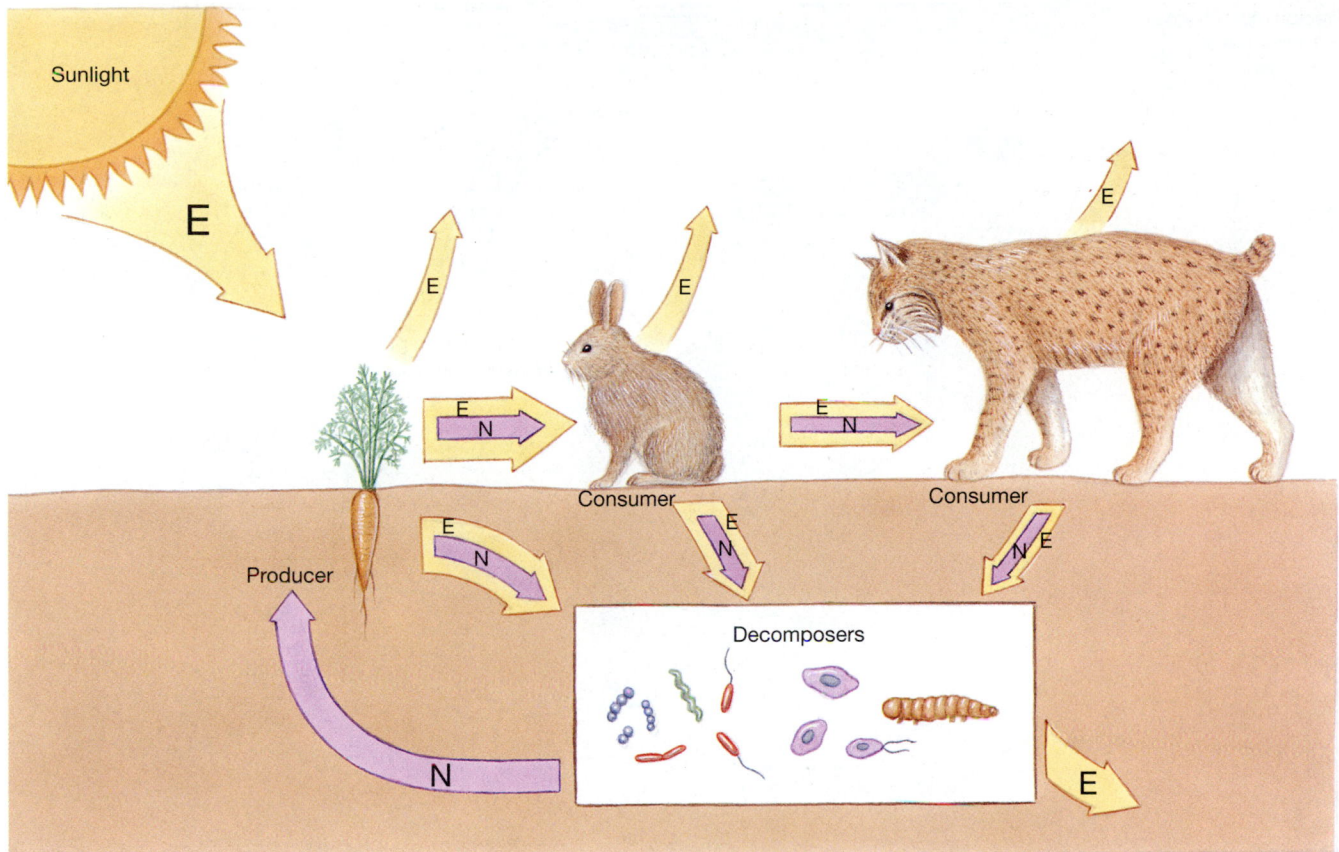

➤ Figure 25.1   **The flow of energy (E) and nutrients (N) in ecosystems.** Energy flows through the system (it is obtained continuously from the sun), whereas nutrients in the environment must be recycled for new life to continue.

them available to living organisms. The mechanisms by which such recycling occurs are referred to collectively as **biogeochemical cycles;** *bio* refers to living things, and *geo* refers to the earth, the environment of the living things.

The Water Cycle

The **water cycle,** or **hydrologic cycle** (➤ Figure 25.2), recycles water. Water reaches the earth's surface as precipitation from the atmosphere. It enters living organisms during photosynthesis and by ingestion. It leaves them as a by-product of respiration, in wastes, and by evaporation from the surfaces of living things, such as by *transpiration* (water loss from pores in plant leaves). Like all other living things, microorganisms use water in metabolism, but they also live in water or very moist environments. Many form spores or cysts that help them survive periods of drought, but vegetative cells must have water.

The Carbon Cycle

In the **carbon cycle** (➤ Figure 25.3), carbon from atmospheric carbon dioxide (CO_2) enters producers during photosynthesis or chemosynthesis. Consumers obtain carbon compounds by eating producers, other consumers, or the remains of either. Carbon dioxide is returned to the atmosphere by respiration and by the actions of decomposers on the dead bodies and wastes of other organisms. Carbon compounds can be deposited in peat, coal, and oil and released from them during burning. A small but significant quantity of carbon dioxide in the atmosphere comes from volcanic activity and from the weathering of rocks, many of which contain the carbonate ion, CO_3^{2-}. The oceans and carbonate rocks are the largest reservoirs of carbon, but recycling of carbon through these reservoirs is very slow.

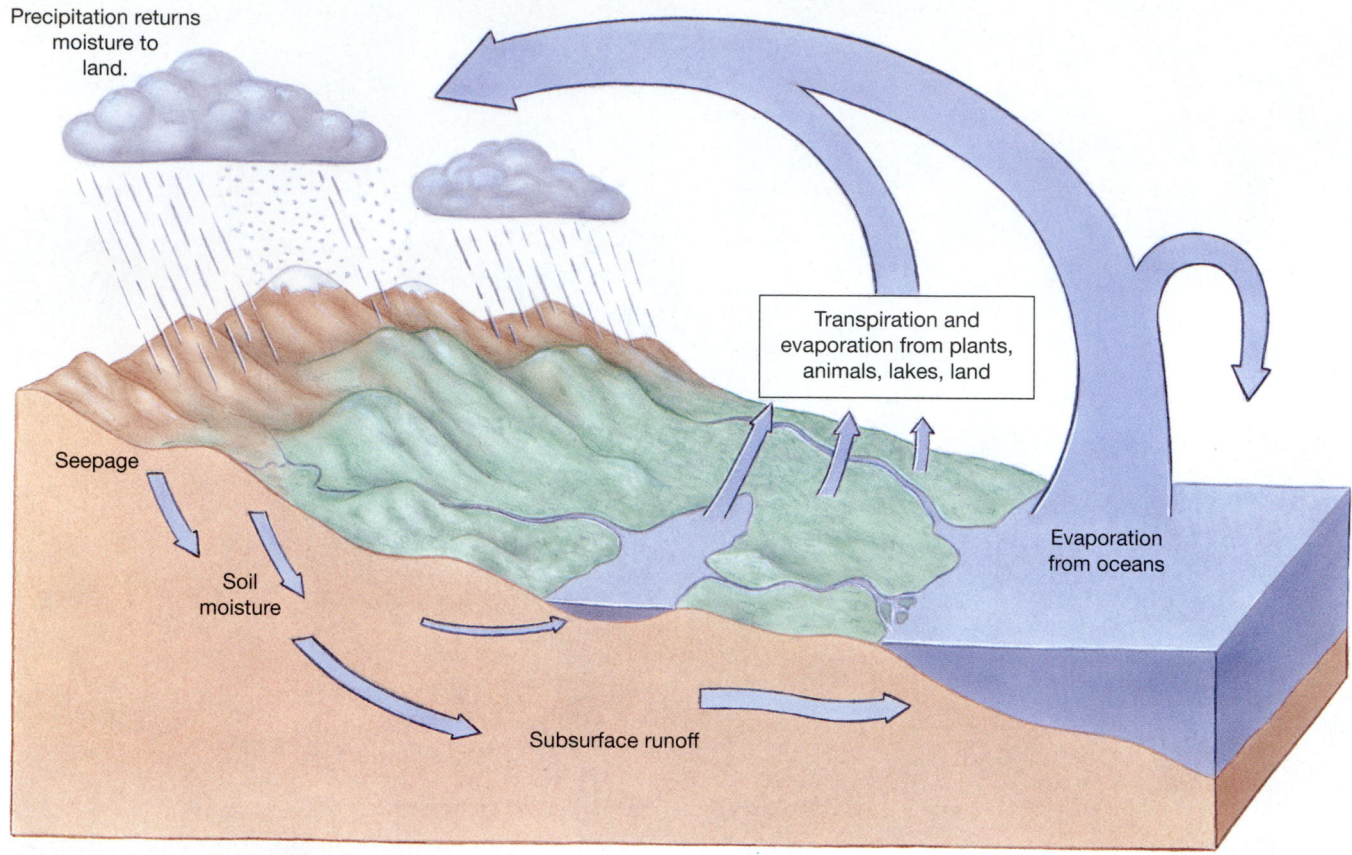

➤ Figure 25.2   **The water cycle.**

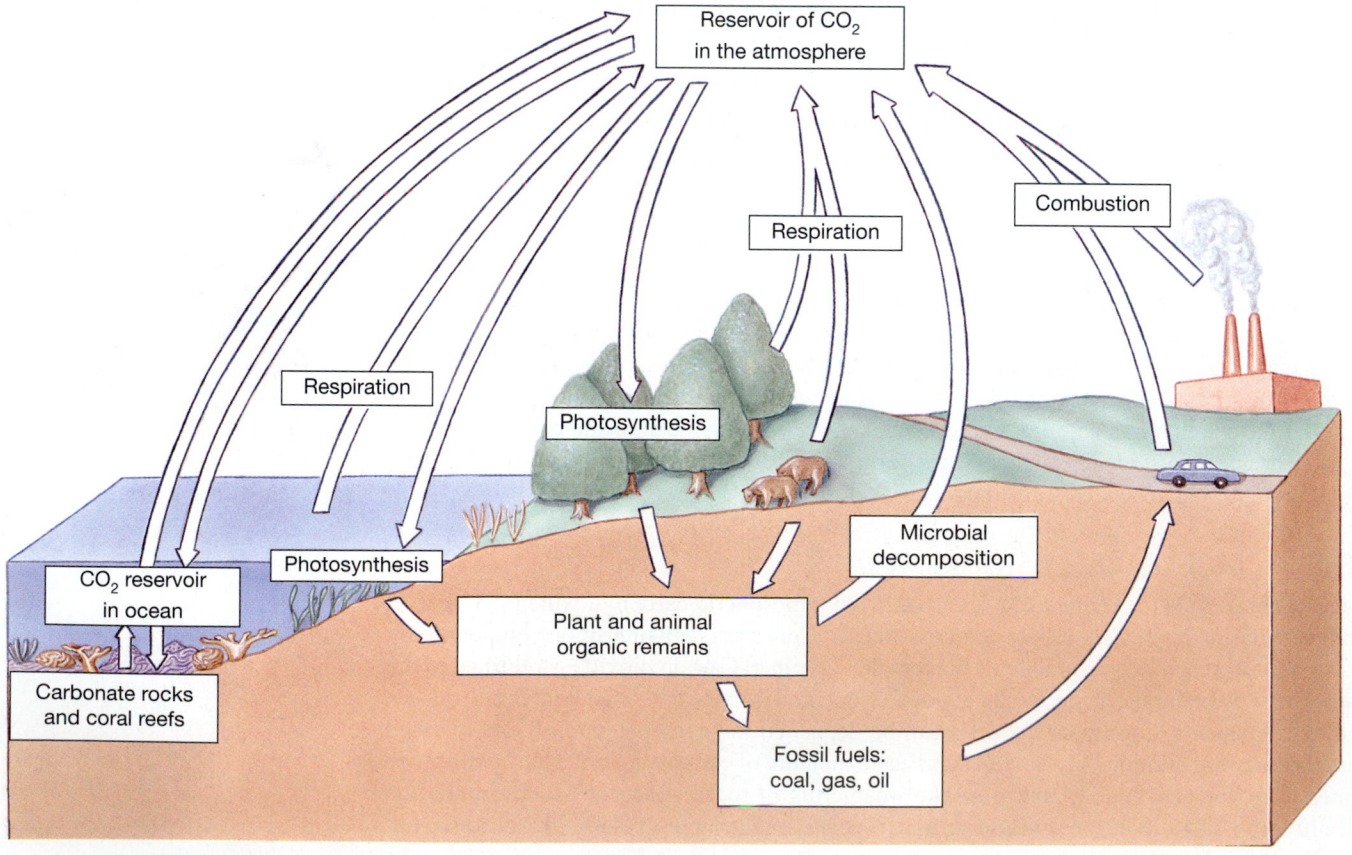

➤ Figure 25.3   **The carbon cycle.**

The Greenhouse Effect

Atmospheric carbon dioxide and water vapor form a blanket over the earth's surface, creating a "greenhouse effect." These gases allow the sun's radiation to penetrate the atmosphere, thereby reaching the earth's surface and warming it and the atmosphere. But they trap much of the infrared (heat) radiation produced by the warm surface, reflecting it back to earth. Solar energy is thus captured within the "greenhouse." The overall effect is to reduce temperature variation and to raise the temperature near the earth's surface.

Over the past century, human activities have released relatively large quantities of carbon dioxide from coal and oil combustion. In fact, the concentration of CO_2 in the atmosphere has increased by more than 10 percent since 1958. Yet forests, which absorb carbon dioxide and replenish the earth's oxygen through photosynthesis, have been cut down and burned at an ever-increasing rate. Some scientists think these processes are causing a general warming trend, increasing average temperatures around the world and possibly changing the balance of organisms in ecosystems. This trend would make now-temperate regions too warm to grow wheat and other food crops, create droughts in other areas, and make new deserts. It might even melt polar ice, raising the level of oceans and flooding coastal cities.

During the first 5 months of 1998, global temperature increased by 0.5°C! Global warming might lead to an increased incidence of infectious diseases as the ranges of the vectors (mosquitos, snails, flies, and such) carrying the diseases expand. Several scenarios have been developed by computer modeling. According to one model, a global temperature rise of just 3°C over the next century could result in 50 million to 80 million new malaria cases per year. Other diseases, including schistosomiasis, African trypanosomiasis, and dengue and yellow fevers, are also likely to spread with global warming.

Marine photosynthetic plankton (*phytoplankton*) utilize CO_2 and are thought to affect the earth's temperature. (See the box And You Thought It Was El Nino, p. 742.) In June 1995, researchers reported that iron added in regular doses to part of the Pacific Ocean promoted extensive phytoplankton growth. The plankton absorbed huge amounts of CO_2 from the air. Scientists are quick to warn, however, against the prospect of seeding the oceans with iron to reduce global CO_2 levels. Iron inputs would alter the food chain and could increase levels of other greenhouse gases.

As we have seen in earlier chapters, all microorganisms require some carbon source to maintain life. Most carbon entering living things comes from carbon dioxide dissolved in bodies of water or in the atmosphere. Even the carbon in sugars and starches ingested by consumers is derived from carbon dioxide. Because the atmosphere contains only a limited quantity of carbon dioxide (0.03 percent), recycling is essential for maintaining a continuous supply of atmospheric carbon dioxide.

The Nitrogen Cycle and Nitrogen Bacteria

In the **nitrogen cycle** ($\succ$ Figure 25.4), nitrogen moves from the atmosphere through various organisms and back into the atmosphere. This cyclic flow depends not only on decomposers but also on various nitrogen bacteria.

Decomposers use several enzymes to break down proteins in dead organisms and their wastes, releasing nitrogen in much the same way they release carbon. Proteinases break large protein molecules into smaller molecules. Peptidases break peptide bonds to release amino acids. Deaminases remove amino groups from amino acids and release ammonia. Eventually, free nitrogen gas finds its way back into the atmosphere. Many soil microorganisms produce one or more of these enzymes. Clostridia, actinomycetes, and many fungi produce extracellular proteinases that initiate protein decomposition.

Nitrogen bacteria fall into one of three categories according to the roles they play in the nitrogen cycle:

- Nitrogen-fixing bacteria
- Nitrifying bacteria
- Denitrifying bacteria.

Nitrogen-Fixing Bacteria

Nitrogen fixation is the reduction of atmospheric nitrogen gas (N_2) to ammonia (NH_3). Organisms that can fix nitrogen are essential for maintaining a supply of physiologically usable

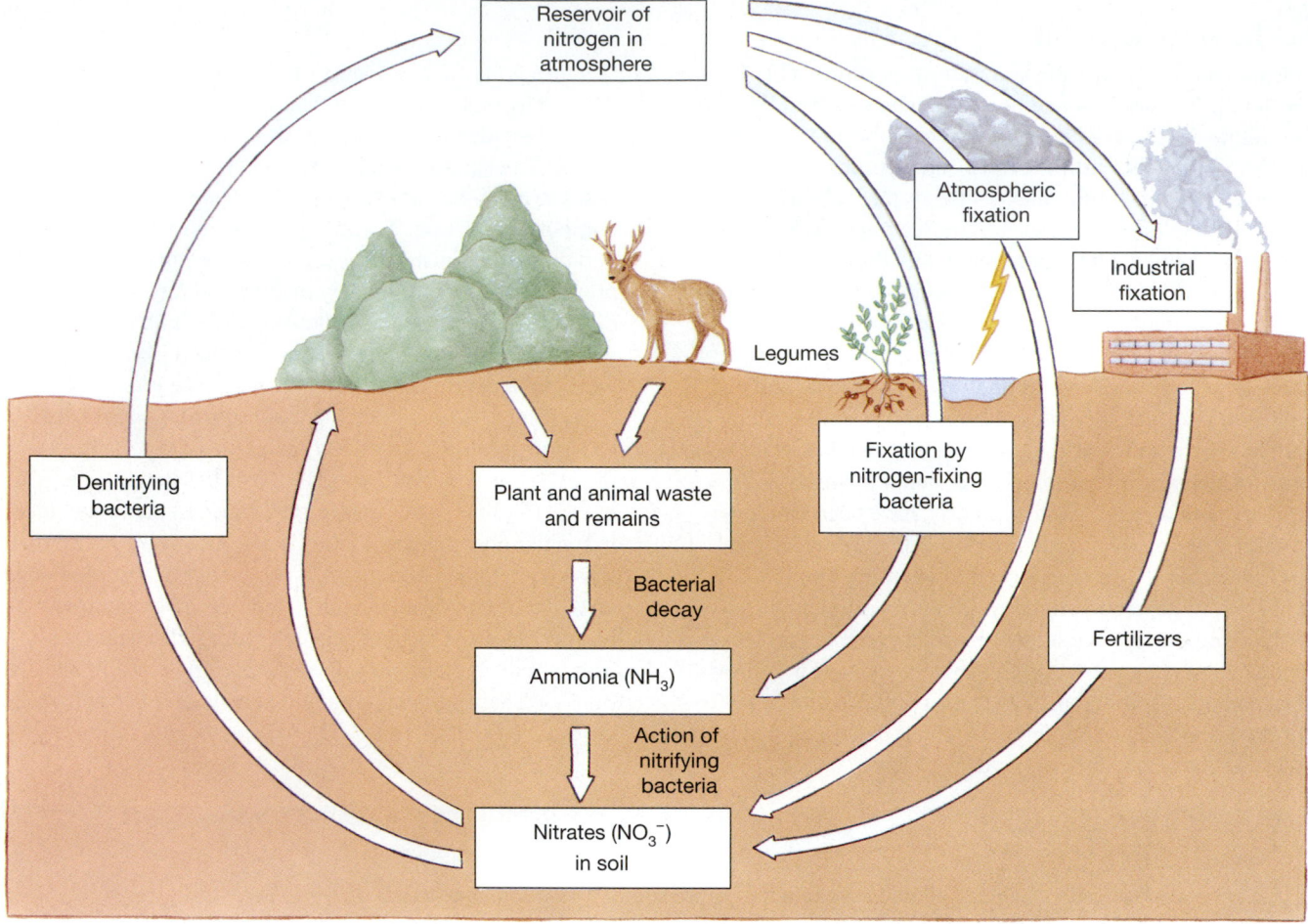

> Figure 25.4   **The nitrogen cycle.**

Nitrogen-Fixing Bacteria in Root Nodules

It's easy to find and visualize nitrogen-fixing bacteria in root nodules. To do this, first dig up the root system from a leguminous plant such as beans or peas. Carefully wash away the dirt to expose the nodules. Crush a nodule on a clean glass slide. Add a small drop of water, and use the nodule pieces or an inoculating loop to spread the fluid out into a thin film. Remove the nodule pieces. Allow the slide to air-dry, and then quickly pass it through a Bunsen burner flame three or four times to heat-fix the organisms. Flood it with methylene blue stain for 1 minute. Rinse the slide with water, dry it, and examine it under the oil immersion lens. The bacilli you see are likely to belong to the genus *Rhizobium*.

nitrogen on earth. About 255 million metric tons of nitrogen is fixed annually—70 percent of it by nitrogen-fixing bacteria. Bacteria and cyanobacteria fix nitrogen in many different environments—from Antarctica to hot springs, acid bogs to salt flats, flooded lands to deserts, in marine and fresh water, and even in the gut of some organisms.

The energy for nitrogen fixation can come from fermentation, aerobic respiration, or photosynthesis. The various organisms that fix nitrogen live independently, in loose associations, or in intimate symbiosis. Regardless of the environment or the associations of the organisms, nitrogen-fixing bacteria must have a functional nitrogen-fixing enzyme called **nitrogenase,** a reducing agent that supplies hydrogen as well as energy from ATP. In aerobic environments, nitrogen fixers must also have a mechanism to protect the oxygen-sensitive nitrogenase from inactivation.

Free-living aerobic nitrogen fixers include several species of the genus *Azotobacter* and some methylotrophic bacteria. Cyanobacteria are also free-living nitrogen fixers. *Azotobacter* is found in soils and is a versatile heterotroph whose growth is limited by the amount of organic carbon available. *Methylotrophic bacteria* can fix nitrogen when provided with methane, methanol, or hydrogen from various substrates. Cyanobacteria can fix nitrogen by using hydrogen from hydrogen sulfide, so they increase the availability of nitrogen in sulfurous environments.

Nitrogen-fixing facultative anaerobes include species of *Klebsiella, Enterobacter, Citrobacter,* and *Bacillus.* In addition, a number of obligate anaerobes, including photosynthetic Rhodospirillaceae and bacteria of the genera *Clostridium, Desulfovibrio,* and *Desulfotomaculum,* also fix nitrogen. Several species of *Klebsiella* capable of fixing nitrogen are found in *rhizomes* (subsurface stems) of legumes such as peas and beans and in the intestines of humans and other animals. Nitrogen fixation has been observed in about 12 percent of *Klebsiella pneumoniae* organisms from patients. Various species of *Clostridium* are found in soils and muds. They use a variety of organic substances for energy and withstand unfavorable conditions as spores. They tolerate a range of pH from 4.5 to 8.5 but fix nitrogen best at pH 5.5 to 6.5. *Desulfovibrio* and *Desulfotomaculum,* anaerobic sulfate reducers that live in mud and soil sediments, fix nitrogen at pH 7 to 8.

Some nitrogen fixers are found in symbiotic association with other organisms, which provide them with organic carbon sources. For example, the cyanobacterium *Anabaena* (➤ Figure 25.5a) is found in pores of the leaves of *Azolla,* a small water fern found in many parts of the world. The nitrogen-fixing *Anabaena* supplies the necessary nitrogen; the fern supplies substrates for energy capture in ATP and reductants for nitrogen fixation. Together, they fix 100 kg of nitrogen per hectare (about two football fields in area) per year and are used as "green manure" in rice cultivation in southeast Asia.

Rhizobium is the primary symbiotic nitrogen fixer. It lives in *nodules,* growths extending from the roots of certain plants, usually legumes (➤ Figure 25.5b and c). In this symbiotic relationship, the plant benefits by receiving nitrogen in a usable form, and the bacteria benefit by receiving nutrients needed for growth. When paired with legumes, *Rhizobium* can fix 150 to 200 kg of nitrogen per hectare of land per year. In the absence of legumes, the bacteria fix only about 3.5 kg of nitrogen per hectare per year—less than 2 percent of that fixed in the symbiotic relationship. Farmers often mix nitrogen-fixing bacteria with seed peas and beans before planting to ensure that nitrogen fixation will be adequate for their crops to thrive.

The mechanism by which *Rhizobium* establishes a symbiotic relationship with a legume has been the subject of many research studies. *Rhizobium* multiplies in the vicinity of legume roots, probably under the influence of root secretions. As the rhizobia increase in numbers, they release enzymes that digest cellulose and the substance that cements cellulose fibers together in the root cell walls. The rhizobia then change from their free-living rod shape and become spherical, flagellated cells called **swarmer cells.** These cells are thought to produce indoleacetic acid, a plant growth hormone that causes curling of root hairs. The swarmer cells then invade the root hairs and form hyphalike networks, killing some root cells and proliferating in others. Swarmer cells become large, irregularly shaped cells called **bacteroids,** which are tightly packed into root cells, probably under the influence of chemical substances in the plant cells. Accumulations of bacteroids in adjacent root cells form nodules on the plant roots.

Bacteroids contain the enzyme nitrogenase, which catalyzes the following reaction:

$$N_2 + 3 H_2 \xrightarrow{\text{\textit{Rhizobium}}} 2 NH_3$$
nitrogen gas   hydrogen gas   ammonia

However, this enzyme is inactivated by oxygen, so nitrogen fixation can occur only when oxygen is prevented from reaching the enzyme. Nitrogenase is protected from oxygen by a kind of hemoglobin, a red pigment that binds oxygen. This particular hemoglobin is synthesized only in root nodules containing bacteroids, because part of the genetic information for its synthesis is in the bacteroids and part is in the plant cells. Synthesis of nitrogenase is repressed in the presence of excess ammonium (NH_4^+) and derepressed in the presence of free nitrogen. Thus, fixation occurs only when "fixed" nitrogen is in short supply and free nitrogen is available.

Rhizobium species vary in their capacity to invade particular legumes and in their capacity to fix nitrogen once they have invaded. Some species cannot invade any legumes, whereas others invade only certain legumes. This invasive specificity is determined genetically (probably by a single gene or a group of closely related genes) and can be altered by genetic transformation. ➡ (Chapter 8, p. 199) Such a transformation might enable a species of *Rhizobium* to invade a group of legumes it previously could not colonize.

Although symbiotic nitrogen fixation occurs mainly in association between rhizobia and legumes, other such associations are known. The alder tree, which grows in soils low in nitrogen, has root nodules similar to those formed by rhizobia. These nodules contain nitrogen-fixing actinomycetes of the genus *Frankia.*

Nitrifying Bacteria

Nitrification is the process by which ammonia or ammonium ions are oxidized to nitrites or nitrates. Carried out by autotrophic bacteria, nitrification is an important part of the nitrogen cycle because it supplies plants with *nitrate* (NO_3^-), the form of nitrogen most usable in plant

(a)

(b)

Root   Rhizobium

(c)

➤ **Figure 25.5   Rhizobium bacteria and root nodules.** (a) SEM of the beadlike, filamentous *Anabaena azollae,* a nitrogen-fixing cyanobacterium that lives in a mutualistic relationship with the water fern *Azolla.* (b) Nodules on the roots of a bean plant, resulting from invasion of the plant by nitrogen-fixing bacteria. (c) A cross section through a root of a leguminous plant shows densely packed *Rhizobium* bacteria inside the nodule.

Bacterial nitrogen-fixing genes added to major crop plants such as corn could make them self-fertilizing, eliminating costs and pollution problems.

metabolism. Nitrification occurs in two steps. In each step, nitrogen is oxidized and energy is captured by the bacteria that carry out the reaction. Various species of *Nitrosomonas* (➤ Figure 25.6) and related genera, Gram-negative rod-shaped bacteria, produce *nitrite* (NO_2^-), the reduced form of nitrates:

$$\underset{\text{ammonium ion}}{NH_4^+} \quad + \quad \underset{\text{oxygen gas}}{1\tfrac{1}{2}O_2} \quad \xrightarrow{\text{\textit{Nitrosomonas}}}$$

$$\underset{\text{nitrite}}{NO_2^-} \quad + \quad \underset{\text{water}}{H_2O} \quad + \quad \underset{\text{hydrogen ions}}{2\,H^+} \quad + \quad \text{energy}$$

Species of *Nitrobacter* and related genera, also Gram-negative rods, produce nitrates:

$$\underset{\text{nitrite}}{NO_2^-} \quad + \quad \underset{\text{oxygen}}{\tfrac{1}{2}O_2} \quad \xrightarrow{\text{\textit{Nitrobacter}}} \quad \underset{\text{nitrate}}{NO_3^-} \quad + \quad \text{energy}$$

These bacteria use energy derived from the preceding reactions to reduce carbon dioxide in autotrophic metabolism. Because oxygen is required for the nitrification reactions, the reactions occur only in oxygenated water and soils. Furthermore, because nitrite is toxic to plants, it is essential that these reactions be carried out in sequence to provide nitrates and to prevent excessive accumulation of nitrites in soil.

Denitrifying Bacteria

Denitrification is the process by which nitrates are reduced to nitrous oxide (N_2O) or nitrogen gas (unbalanced equation):

$$\underset{\text{nitrate}}{NO_3^-} \quad \longrightarrow \quad \underset{\text{nitrite}}{NO_2^-} \quad \longrightarrow \quad \underset{\substack{\text{nitrous}\\\text{oxide}}}{N_2O} \quad \longrightarrow \quad \underset{\substack{\text{nitrogen}\\\text{gas}}}{N_2}$$

Although this process does not occur to any significant degree in well-oxygenated soils, it does occur in oxygen-depleted, waterlogged soils. Most denitrification is performed by *Pseudomonas* species, but it can also be accomplished by *Thiobacillus denitrificans, Micrococcus denitrificans,* and several species of *Serratia* and *Achromobacter*. These bacteria, although usually aerobic, use nitrate instead of oxygen as a hydrogen acceptor under anaerobic conditions. Another process that reduces soil nitrate is the reduction of nitrate to ammonia.

➤ Figure 25.6   **A TEM of a microcolony of the nitrifying bacterium** *Nitrosomonas.* (17,750X).

Several anaerobic bacteria carry out this process (called *dissimilative nitrate reduction*) in a complex reaction that can be summarized as follows (unbalanced equation):

$$NO_3^- \longrightarrow H_2 \longrightarrow NH_3 \longrightarrow N_2O$$

| nitrate | hydrogen gas | ammonia | nitrous oxide |

Although the latter process reduces the quantity of available nitrate, it retains the nitrogen in the soil in other forms.

Denitrification is a wasteful process because it removes nitrates from the soil and interferes with plant growth. It is responsible for significant losses of nitrogen from fertilizer applied to soils. Another unfortunate effect of denitrification is the production of nitrous oxide, which is converted to nitric oxide (NO) in the atmosphere. Nitric oxide, in turn, reacts with ozone (O_3) in the upper atmosphere. Ozone provides a barrier between living things on earth and the sun's ultraviolet radiation. If enough ozone is destroyed, it no longer serves as an effective screen, and living things can be exposed to excessive ultraviolet radiation, which can cause cancer and mutations. ➡ (Chapter 7, p. 189)

The Sulfur Cycle and Sulfur Bacteria

The **sulfur cycle** (➤ Figure 25.7), which involves the movement of sulfur through an ecosystem, resembles the nitrogen cycle in several respects. Sulfhydryl (—SH) groups in proteins of dead organisms are converted to hydrogen sulfide (H_2S) by a variety of microorganisms. This process is analogous to the release of ammonia from proteins in the nitrogen cycle. Hydrogen sulfide is toxic to living things and thus must be oxidized rapidly. Oxidation to elemental sulfur is followed by oxidation to sulfate (SO_4^{2-}), which is the form of sulfur most usable by both microorganisms and plants. This process is analogous to nitrification. The sulfur cycle is of special importance in aquatic environments, where sulfate is a common ion, especially in ocean water.

And It Really Smells Bad, Too.

In 1996 researchers found that common agricultural activities were responsible for approximately 6 percent of total U.S. greenhouse gas emissions. Agricultural activites that contribute the most to the emission of greenhouse gases are enteric fermentation in domestic livestock and management of the manure they produce. Methane and nitrous oxide constitute the major portion of the emitted gases. Methane is produced as part of the normal digestive process when microbes ferment food consumed by the animals. Ruminants—cattle, buffalo, sheep, goats and camels—are the major emitters of methane. The management of livestock manure produces methane and nitrous oxide. Methane is produced by the anaerobic decomposition of the manure and nitrous oxide is produced through the denitrification of the organic nitrogen in the manure. Much of these emissions can be lessened by controlling livestock diet and encouraging aerobic decomposition of manure.

➤ Figure 25.7   **The sulfur cycle.**

> ➤ Figure 25.8   **The sulfate-reducing bacterium *Desulfotomaculum acetoxidans.*** Notice the spherical spores and gas vacuoles (phase contrast, 5000X).

The various sulfur bacteria can be categorized according to their roles in the sulfur cycle. These roles include sulfate reduction, sulfur reduction, and sulfur oxidation.

Sulfate-Reducing Bacteria

Sulfate reduction is the reduction of sulfate (SO_4^{2-}) to hydrogen sulfide (H_2S). The sulfate-reducing bacteria are among the oldest life forms, probably more than 3 billion years old. They include the closely related genera *Desulfovibrio, Desulfomonas,* and *Desulfotomaculum.* In these bacteria, sulfate is the final electron acceptor in anaerobic oxidation, as oxygen is the final electron acceptor in aerobic oxidation. By reducing sulfate, these bacteria produce large amounts of hydrogen sulfide. However, for this process to occur, energy from ATP is required to phosphorylate the sulfate and convert it to ADP—SO_4. ADP—SO_4 then can act as an electron acceptor and compete successfully for substrates.

Sulfate-reducing bacteria (➤ Figure 25.8) are strict anaerobes. They are widely distributed and predominate in nearly all anaerobic environments. As discussed in Chapter 6, sulfate-reducing bacteria can be psychrophilic, mesophilic, thermophilic, or halophilic. ➡ (p. 146) The variety of organic carbon sources they can metabolize is limited. Most use lactate, pyruvate, fumarate, malate, or ethanol, and some can use glucose and citrate. End products of such metabolism are typically acetate and carbon dioxide. Some sulfate reducers use products derived from the anaerobic degradation of plant material by other organisms. *Desulfovibrio, Desulfomonas,* and *Desulfotomaculum* oxidize fatty acids and a variety of other organic acids.

Sulfur-Reducing Bacteria

Sulfur reduction is the reduction of sulfur to hydrogen sulfide. Like sulfate-reducing bacteria, sulfur-reducing bacteria are anaerobes. They can use intracellular or extracellular sulfur as an electron acceptor in fermentation. The sulfur can be in elemental form or in disulfide bonds of organic molecules. This process provides energy for the organisms when sunlight is not available for photosynthesis.

Sulfur-Oxidizing Bacteria

Sulfur oxidation is the oxidation of various forms of sulfur to sulfate. *Thiobacillus* and similar bacteria oxidize hydrogen sulfide, ferrous sulfide, or elemental sulfur to sulfuric acid (H_2SO_4). When this acid ionizes, it greatly decreases the pH of the environment, sometimes lowering the pH to 1 or 2. Sulfur-oxidizing organisms are responsible for oxidizing ferrous sulfide in coal-mining wastes, and the acid they produce is extremely toxic to fish and other organisms in streams fed by such wastes.

Other Biogeochemical Cycles

In addition to water, carbon, nitrogen, and sulfur—whose cycles we have examined—phosphorus and other elements also move through ecosystems cyclically. Any element (including trace elements) that appears in the cells of living organisms must be recycled to extract it from dead organisms and make it available to living ones. We conclude our discussion of biogeochemical cycles with a brief description of the phosphorus cycle.

The **phosphorus cycle** (➤ Figure 25.9) involves the movement of phosphorus among inorganic and organic forms. Soil microorganisms are active in the phosphorus cycle in at least two important ways: (1) They break down organic phosphates from decomposing organisms to inorganic phosphates, and (2) they convert inorganic phosphates to orthophosphate (PO_4^{3-}), a water-soluble nutrient used by both plants and microorganisms. These functions are particularly important because phosphorus is often the limiting nutrient in many environments.

Air

Having discussed briefly some fundamentals of ecology and of biogeochemical cycles, we now explore different kinds of environments and the microorganisms they contain. We will begin with the air around us and then consider soil and water.

✓ Distinguish between autotrophs, heterotrophs, and decomposers.

✓ What is meant by nitrogen fixation? Who does it? What is its importance to us?

✓ What is the importance of denitrification?

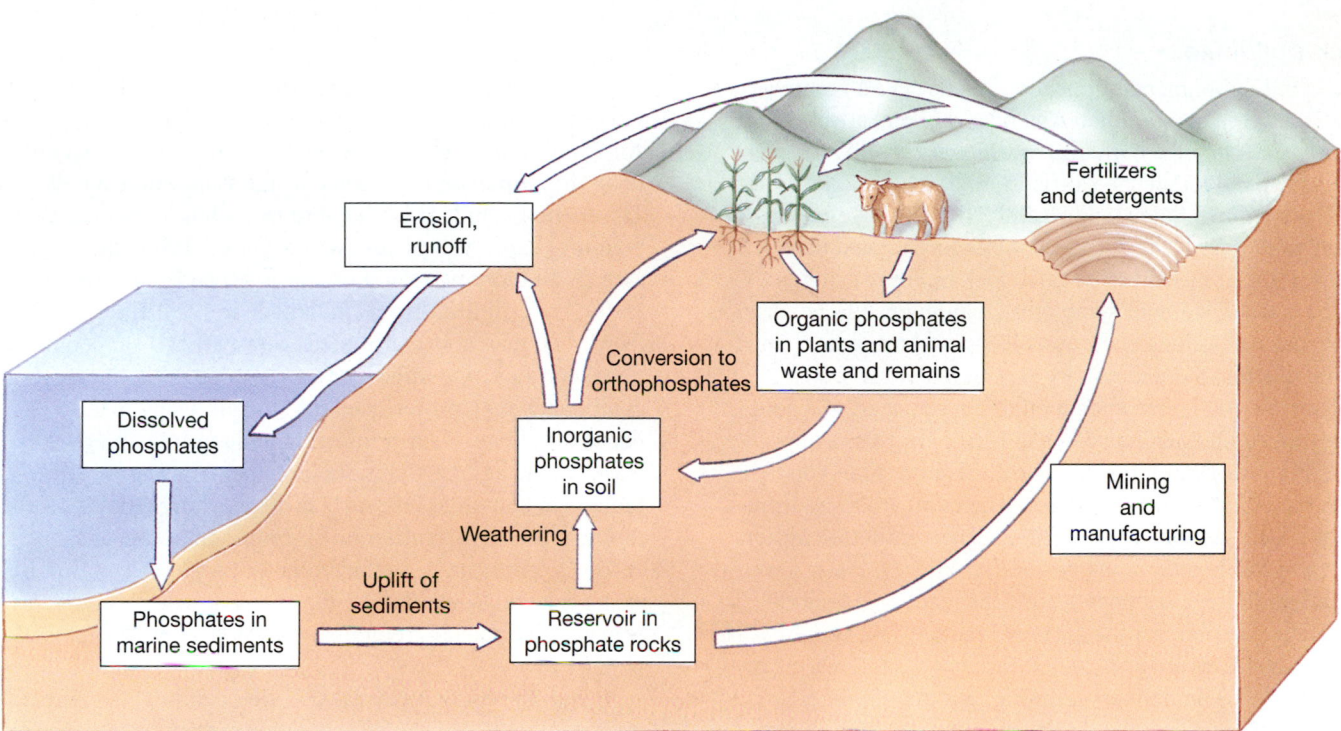

➤ Figure 25.9 **The phosphorus cycle.**

Microorganisms Found in Air

Microorganisms do not grow in air, in part because it lacks the nutrients needed for metabolism and growth. However, spores are carried in air, and vegetative cells can be carried on dust particles and water droplets in air. The kinds and numbers of airborne microorganisms vary tremendously in different environments. Large numbers of many different kinds of microorganisms are present in indoor air where humans are crowded together and building ventilation is poor. Small numbers have been detected at altitudes of 3000 m.

Among the organisms found in air, mold spores are by far the most numerous, and the predominant genus is usually *Cladosporium*. Bacteria commonly found in air include both aerobic spore formers such as *Bacillus subtilis* and non-spore formers such as *Micrococcus* and *Sarcina*. Algae, protozoa, yeasts, and viruses also have been isolated from air. While coughing, sneezing, or even talking, infected humans can expel pathogens along with water droplets. Health care workers should handle wastes from patients carefully to avoid producing aerosols (tiny droplets that remain suspended for some time) of pathogens. ➥ (Chapter 15, p. 415)

Determining the Microbial Content of Air

Airborne microbes can be detected by collecting those that happen to fall onto an agar plate or in a liquid medium. A special centrifuge-like air-sampling instrument provides a better measure of airborne microbes (➤ Figure 25.10a).

Methods for Controlling Microorganisms in Air

Chemical agents, radiation, filtration, and laminar airflow all can be used to control microorganisms in air. Certain chemical agents, such as triethylene glycol, resorcinol, and lactic acid dispersed as aerosols, kill many if not all microorganisms in room air. These agents are highly bactericidal, remain suspended long enough to act at normal room temperature and humidity, are nontoxic to humans, and do not damage or discolor objects in the room.

Ultraviolet radiation has little penetrating power and is bactericidal only when rays make direct contact with microorganisms in air. ➥ (Chapter 12, p. 334) Thus, ultraviolet lamps must

Sick Buildings

The U.S. Environmental Protection Agency (EPA) estimates that 30 million to 75 million American workers are at risk of getting sick from the buildings in which they work. Since the late 1970s, indoor air specialists from the National Institute for Occupational Safety and Health (NIOSH) have investigated more than 1000 instances of building-related illness. Some of the ailments can be fatal. Sick building syndrome (SBS) is a collection of symptoms including headache, dizziness, nausea, burning eyes, nosebleeds, and respiratory and skin problems. Symptoms vary among individuals.

How do you know when a building—instead of a cold, allergies, or another cause—is to blame for SBS symptoms? If your symptoms get worse the more time you spend in a building and improve when you are away from it, take a close look at the building. The ventilation system has been implicated in many cases of SBS. In many new "tight" buildings, occupants are dependent on a central ventilation system for the air they breathe. The system pulls in air from outside, runs it through filters, warms or cools it, and then delivers it throughout the building by a series of ducts. Once the air has circulated, return ducts channel some of it outside. Dirt, dust, insects, mold, mildew, particles of building materials, and other detritus are found on the system's filters. Whatever gets into the ventilation system gets into workers' lungs as well. New filters and proper cleaning of the system's equipment often correct ventilation problems.

Another possible cause of SBS is too little air. In the 1930s the American Society of Heating, Refrigeration, and Air Conditioning Engineers set 15 cubic feet of fresh outdoor air per person per minute as the ventilation standard. The current standard is 20 cubic feet—but it assumes that no more than seven people will occupy a 1000-square foot area. With more people, more air is needed. The "fleece factor," or pollution from "manmade mineral fibers" (MMMF), may also cause SBS. New carpets, padded partitions, curtains, and upholstery are treated with volatile compounds that are released into the air. Copy machines, paper shredders, laser printers, and blueprint copiers release vapors and particles that may be irritating. Finally, biological contaminants such as molds, fungi, and bacteria thrive in building ventilation systems and are spread as aerosols. In recent years, bacteria that collect in ventilation systems have proved fatal—for example, when *Legionella* species have caused Legionnaires' disease. ➥ (Chapter 21, p. 610)

If you suspect that you are in a sick building, document your symptoms, check the ventilation system, inspect the air vents for dirt, check for exhaust vents around copy machines, and so on. Ask the maintenance engineers how much fresh outdoor air is circulating in cubic feet per minute per person. Once you have targeted hazards, convince someone to do something about them.

(a)

(b)

➢ **Figure 25.10   Measuring airborne microbes.** (a) An air-sampling device used to measure airborne bacteria and fungi. (b) Technicians are protected from the airborne spread of organisms by the use of a laminar flow hood, which suctions air away from the opening and filters it before expelling it.

be carefully placed to ensure treatment of all air in a room. They are most useful in maintaining sterile conditions in rooms only sporadically occupied by humans. They can be turned off while technicians are performing sterile techniques and turned on again when the room is not in use. Humans entering a room while ultraviolet lights are on must wear protective clothing and special glasses to protect the eyes from burns. Ultraviolet lamps can also be installed in air ducts to reduce the number of microorganisms entering a room through its ventilation system.

Air filtration involves passing air through fibrous materials, such as cotton or fiberglass. It is useful in industrial processes in which sterile air must be bubbled through large fermentation vats. Cellulose acetate filters can be installed in a *laminar airflow system* (➤ Figure 25.10b) to remove microorganisms that may have escaped into the air underneath the laminar flow hood. The air is suctioned away from the opening, filtered, and then returned to the room.

Soil

We might think of the soil we walk on as an inert substance, but nothing could be further from the truth. Soil, in fact, is teeming with microscopic and small macroscopic organisms, and it receives animal wastes and organic matter from dead organisms. Microorganisms act as decomposers to break down this organic matter into simple nutrients that can be used by plants and by the microbes themselves. (Animals, of course, obtain their nutrients from plants or from other animals.) Soil microorganisms are thus extremely important in recycling substances in ecosystems.

Components of Soil

Soil is divided into a number of layers, or *soil horizons* (➤ Figure 25.11). These horizons include the *topsoil, subsoil,* and *parent material,* which overlie bedrock. Soil contains inorganic

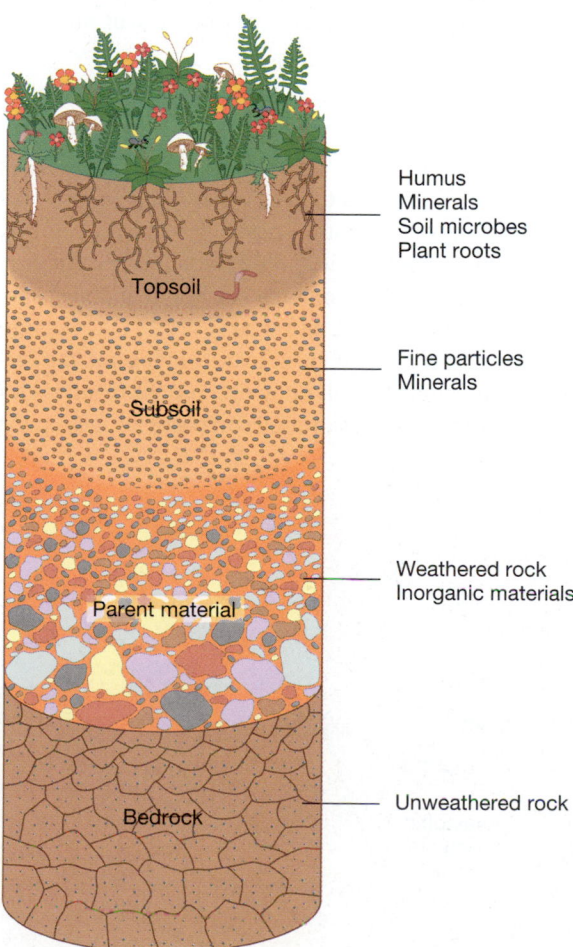

Topsoil

Subsoil

Parent material

Bedrock

Humus
Minerals
Soil microbes
Plant roots

Fine particles
Minerals

Weathered rock
Inorganic materials

Unweathered rock

➤ Figure 25.11   **Soil horizons.**
Shown here is a vertical segment of the main soil layers.

and organic components. The inorganic components are rocks, minerals, water, and gases. The organic components are **humus** (nonliving organic matter) and living organisms. Soils differ greatly in the relative proportions of these components. Topsoil, the surface layer of soil, contains the greatest number of microorganisms because it is well supplied with oxygen and nutrients. Lower layers of soil (subsoil and parent material) and soils depleted of oxygen and nutrients contain fewer organisms.

The most abundant inorganic components of soil are pulverized rocks and minerals. *Weathering* of rocks—mechanical or chemical breakdown—releases minerals, and the elements that make up those minerals, into the soil. The most abundant elements in many soils are silicon, aluminum, and iron. Small quantities of other elements, such as calcium, potassium, magnesium, sodium, phosphorus, nitrogen, and sulfur, also are present in soil.

In addition to rocks and minerals, soil contains water and the gases carbon dioxide, oxygen, and nitrogen. Water molecules adhere to soil particles or are interspersed among them. The amount of water in soil is highly variable and depends on the climate, the quantity of recent rainfall, and the drainage of the soil. Gases are dispersed among soil particles or dissolved in water. The concentration of gases in the soil also varies with the metabolic activity of soil organisms. Compared with atmospheric air, soil is lower in oxygen and higher in carbon dioxide.

Humus is constantly changing as organisms die and as decomposers degrade complex molecules to simpler ones. Soils differ greatly in the amount of humus they contain. Most soils are 2 to 10 percent humus, but a peat bog can be 95 percent humus.

In addition to microorganisms, soil also contains the root systems of many plants, a variety of invertebrate animals (nematodes, earthworms, snails, slugs, insects, centipedes, millipedes, and spiders), and a few burrowing reptiles and mammals. Despite the great diversity of soil organisms, microorganisms are by far the most numerous both in total numbers and in the number of species present. As we have seen, microorganisms convert humus to nutrients usable by other organisms. They also make great demands on soil nutrients. When nutrients are abundant, the number of microbes increases rapidly. As nutrients are depleted, the number of microbes decreases accordingly.

Microorganisms in Soil

All the major groups of microorganisms—bacteria, fungi, algae, and protists, as well as viruses—are present in soil, but bacteria are more numerous than all other kinds of microorganisms (➤ Figure 25.12). Among the bacteria in soil are autotrophs, heterotrophs, aerobes, anaerobes, and, depending on the soil temperature, mesophiles and thermophiles. In addition to nitrogen-fixing, nitrifying, and denitrifying bacteria, soil contains bacteria that digest special substances such as cellulose, protein, pectin, butyric acid, and urea.

Soil fungi are mostly molds. Both mycelia and spores are present mainly in topsoil, the aerobic surface layer of the soil. Fungi serve two functions in soil. They decompose plant tissues such as cellulose and lignin, and their mycelia form networks around soil particles, giving the

---

Fungal Confusion

The rust fungus, *Puccinia recondita,* causes millions of dollars of crop loss in wheat and rye each year. When its spores land on a leaf, they send out tiny "infection tubes" that enter pores (stomata) on the leaf surface. Hairy leaves developed by breeders "confuse" the fungus and it dies before locating a stoma.

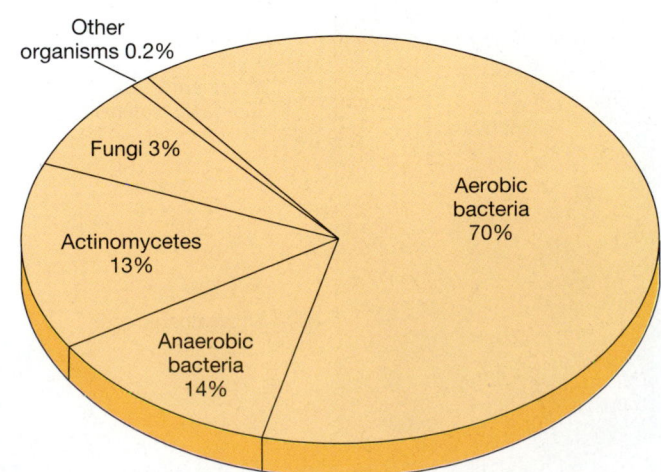

➤ Figure 25.12   **The relative proportions of various kinds of organisms found in soil.** "Other organisms" include such things as algae, protists, and viruses.

Other organisms 0.2%
Fungi 3%
Actinomycetes 13%
Aerobic bacteria 70%
Anaerobic bacteria 14%

soil a crumbly texture. ➡ (Chapter 11, p. 299) In addition to molds, yeasts are abundant in soils in which grapes and other fruits are growing.

Small numbers of cyanobacteria, algae, protists, and viruses are found in most soils. Algae are found only on the soil surface, where they can carry on photosynthesis. In desert and other barren soils, algae contribute significantly to the accumulation of organic matter in the soil. Protists, mostly amoebas and flagellated protozoa, also are found in many soils. They feed on bacteria and may help control bacterial populations. Soil viruses infect mostly bacteria, but a few infect fungi, and quite a few infect plants . Very little is known about the identity or classification of many plant viruses.

Viroids in soil are often spread by mechanical means and can cause serious plant diseases. Viruses that attack insects are common, most of them belonging to the families Baculoviridae and Iridoviridae. Animal viruses are not indigenous to soil but are often added to soil through human activities, such as manure spreading. The survival of viruses in soil varies with environmental conditions and the particular virus. It ranges from hours to years.

It is possible to use viruses for biocontrol of insect pests in soil. Viruses of an appropriate type are inoculated into soil. When all insect pests have been killed, the virus tends to disappear, offering no further risk—an attribute important for government approval. Caterpillar stages of the celery looper butterfly can be controlled by the addition of the *Anagrapha falcifera* multiple nuclear polyhedrosis virus to the soil.

Factors Affecting Soil Microorganisms

Soil microorganisms, like all other organisms, interact with their environment. Their growth is influenced both by abiotic factors and by other organisms. The microorganisms, in turn, affect the physical characteristics of soil and other organisms in the soil.

Abiotic factors in soil, as in any other environment, include moisture, oxygen concentration, pH, and temperature. The moisture and oxygen content of soil are closely related. Spaces among soil particles ordinarily contain both water and oxygen, and aerobic organisms thrive there. However, in waterlogged soils all the spaces are filled with water, so only anaerobic bacteria grow there.

Soil pH, which can vary from 2 to 9, is an important factor in determining which microorganisms will be present. Most soil bacteria have an optimum pH between 6 and 8, but some molds can grow at almost any soil pH. Molds thrive in highly acidic soils partly because of reduced competition from bacteria for available nutrients. Lime neutralizes acidic soil and increases the bacterial population. Fertilizer containing ammonium salts has two effects on soil: (1) It provides a nitrogen source for plants; and (2) when it is metabolized by certain bacteria, the bacteria release nitric acid, which decreases the soil pH and increases the mold population.

Soil temperature varies seasonally from below freezing to as high as 60°C at soil surfaces exposed to intense summer sunlight. Mesophilic and thermophilic bacteria are quite numerous in warm to hot soils, whereas cold-tolerant mesophiles (and not true psychrophiles) are present in cold soils. Most soil molds are mesophilic and are found mainly in soils of moderate temperature.

Exceedingly wide variations exist in physical characteristics of soil and in the numbers and kinds of organisms it contains—even in soil samples taken only a few centimeters apart. Such observations led ecologists to develop the concept of microenvironments. The interactions among organisms and between the organisms and their environment can be quite different in different microenvironments, no matter how close together they are.

The Importance of Decomposers in Soil

Soil microorganisms that are decomposers are important in the carbon cycle because of their ability to decompose organic matter. The decomposition of complex organic substances from dead organisms is a stepwise process that requires the action of several kinds of microorganisms. The organic substances include cellulose, lignins, and pectins in the cell walls of plants, glycogen from animal tissues, and proteins and fats from both plants and animals. Cellulose is degraded by bacteria, especially those of the genus *Cytophaga,* and by various fungi. Lignins and pectins are partially digested by fungi, and the products of fungal action are further digested by bacteria. Protozoa and nematodes also can play a role in the degradation of lignins and pectins. Proteins are degraded to individual amino acids mainly by fungi, actinomycetes, and clostridia.

Under the anaerobic conditions of waterlogged soils in marshes and swamps, methane is the main carbon-containing product. It is produced by three genera of strictly anaerobic bacteria—*Methanococcus, Methanobacterium,* and *Methanosarcina.* In addition to degrading carbon compounds to methane, they also obtain energy by oxidizing hydrogen gas:

$$4\ H_2\ +\ CO_2\ \longrightarrow\ CH_4\ +\ 2\ H_2O$$

hydrogen	carbon	methane	water
gas	dioxide		

In one way or another, organic substances are metabolized to carbon dioxide, water, and other small molecules. In fact, for each naturally occurring organic compound, there is one or more organisms that can decompose it. Thus, carbon is continuously recycled. However, certain organic compounds manufactured by humans resist the actions of microorganisms. Accumulations of these synthetic substances create environmental hazards.

Nitrogen enters the soil through the decomposition of proteins from dead organisms and through the actions of nitrogen-fixing organisms. In addition to protein decomposition, which introduces nitrogen into the soil, gaseous nitrogen is fixed both by free-living microorganisms and by symbiotic microorganisms associated with the roots of legumes, as previously described.

Soil Pathogens

Soil pathogens are primarily plant pathogens, many of which have been discussed in earlier chapters. A few soil pathogens can affect humans and other animals. The main human pathogens found in soil belong to the genus *Clostridium.* ➡ (Chapter 24, p. 710) All are anaerobic spore formers. *Clostridium tetani* causes tetanus and can be introduced easily into a puncture wound. *Clostridium botulinum* causes botulism. Its spores, found on many edible plants, can survive in incompletely processed foods to produce a deadly toxin. *Clostridium perfringens* causes gas gangrene in poorly cleaned wounds. Grazing animals can contract anthrax from spores of *Bacillus anthracis* in the soil. In fact, most soil organisms that infect warmblooded animals exist as spores because soil temperatures are usually too low to maintain vegetative cells of these pathogens.

Water

All aquatic ecosystems—freshwater, ocean water, even rainwater—contain microorganisms as well as inorganic substances. Most of the organisms in such environments have been considered in earlier chapters. Here we will consider properties of the environments, interactions of microorganisms with their environment, transmission of human pathogens in water, and methods for maintaining safe water supplies.

Freshwater Environments

Freshwater environments include surface water, such as lakes, ponds, rivers, and streams, and groundwater that runs through underground strata of rock. Although groundwater contains few microorganisms, surface water contains large numbers of many different kinds of microorganisms.

Ponds and lakes are divided vertically into zones. The shoreline, or *littoral zone,* is an area of shallow water near the shore where light penetrates to the bottom. The *limnetic zone* comprises the sunlit water away from the shore; resident microorganisms include algae and cyanobacteria. Between the limnetic zone and the lake sediment is the *profundal zone.* When organisms in the limnetic zone die, they sink to the profundal zone, where they provide nutrients for other organisms. The sediment, or *benthic zone,* is composed of organic debris and mud (➤ Figure 25.13).

In aquatic environments, water temperatures vary from 0°C to nearly 100°C. Most microorganisms thrive in water at moderate temperatures. However, some thermophilic bacteria

The Nose of the Beholder

Thermal hot springs derive their heat from masses of magma that have pushed their way near to the earth's surface. Natural hot springs, many used as health and bathing resorts, are home to thermophilic and metal-reducing bacteria. The water in thermal hot springs, which may emerge from rocks at temperatures well over 100°C, is rich in dissolved minerals. It is the hot water and high mineral content, that many people feel is the source of benefit to bathe in these natural spas. Thermal springs that are rich in dissolved sulfur oxides may support the growth of a range of sulfur fixing bacteria. These bacteria, which include *Desulfovibrio* and *Desulfomonas,* reduce sulfate and produce hydrogen sulfide. Release of the hydrogen sulfide gas from the hot bubbling water smells like sewer gas or rotten eggs. However, in the context of a natural thermal hot spring, the odor is not unpleasant.

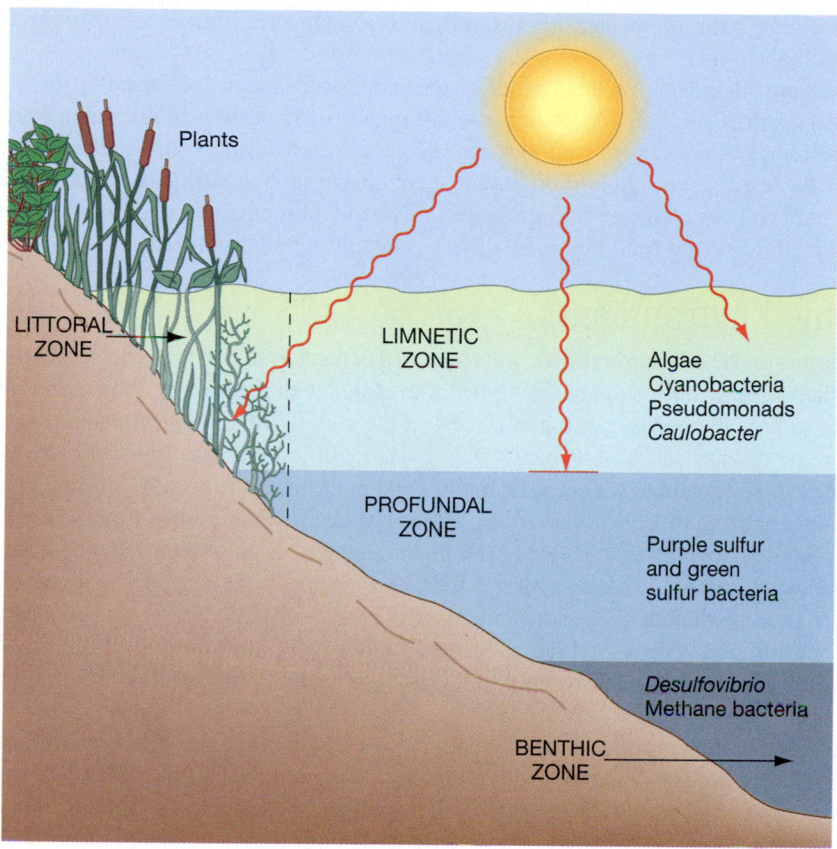

Plants

LITTORAL ZONE

LIMNETIC ZONE

Algae
Cyanobacteria
Pseudomonads
Caulobacter

PROFUNDAL ZONE

Purple sulfur
and green
sulfur bacteria

Desulfovibrio
Methane bacteria

BENTHIC ZONE

➤ Figure 25.13   **Zonation in a typical lake or pond.** Organisms commonly found in each zone are listed at the right.

have been found in geysers at a water temperature above 90°C, and psychrophilic fungi and bacteria have been found in water at 0°C. The pH of fresh water varies from 2 to 9. Although most microorganisms grow best in waters of nearly neutral pH, a few have been found in both extremely acidic and extremely alkaline waters. Most natural waters are rich in nutrients, but the quantities of various nutrients in different bodies of water vary considerably. Sometimes nutrients become so abundant that a "bloom," or sudden proliferation of organisms in a body of water, occurs (➤ Figure 25.14). ➡ (See the box Red Tides in Chapter 11, p. 291.)

In aquatic environments, oxygen can be the limiting factor in the growth of microorganisms. Because of oxygen's low solubility in water, its concentration never exceeds 0.007 g per 100 g of water. When water contains large quantities of organic matter, decomposers rapidly deplete the oxygen supply as they oxidize the organic matter. Oxygen depletion is much more likely in standing water in lakes and ponds than in running water in rivers and streams because the movement of running water causes it to be continuously oxygenated.

Yet another factor affecting microorganisms in aquatic environments is the depth to which the sunlight penetrates the water. Of minor importance in fresh water except for deep lakes, it

Though the earth is 70 percent water, only 1 percent is available for human use. The remainder is unavailable in oceans, seas, and glaciers.

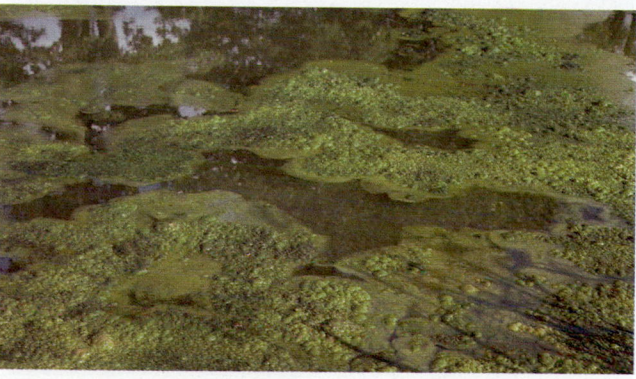

➤ Figure 25.14   **Algal bloom.** An overabundance of nutrients has allowed an algal bloom to develop on this pond. Bubbles of oxygen can be seen collecting underneath some of the algae. However, algal blooms often lead to increased biological oxygen demand (BOD) as they die off and decompose, using up oxygen needed by other pond organisms such as fish.

is very important in the ocean. Photosynthetic organisms are limited to locations with adequate sunlight.

Which organisms are present depends on the temperature and pH of the water, dissolved minerals, the depth of sunlight penetration, and the quantity of nutrients in the water. Aerobic bacteria are found where oxygen supplies are adequate, whereas anaerobic bacteria are found in oxygen-depleted waters. Eukaryotic algae, cyanobacteria, and sulfur bacteria are limited to water that receives adequate sunlight. *Desulfovibrio* and methane bacteria are found in lake sediments. Protists are found in many different freshwater environments.

Marine Environments

The marine, or ocean, environment covers about 70 percent of the earth's surface and is therefore larger than all other environments combined. Compared with fresh water, the ocean is much less variable in both temperature and pH. Except in the vicinity of volcanic vents in the sea floor, where water reaches 250°C ➡ (Chapter 9, p. 231), ocean water ranges from 30°C to 40°C at the surface near the equator to 0°C in polar regions and in the lowest depths. At any one location and depth, the temperature is nearly constant. The pH of ocean water ranges from nearly neutral to slightly alkaline (pH 6.5 to 8.3). This pH range is suitable for growth of many microorganisms, and sufficient carbon dioxide dissolves in the water to support photosynthetic organisms.

Ocean water, which is about seven times as salty as fresh water, displays a remarkably constant concentration of dissolved salts—3.3 to 3.7 g per 100 g of water. Thus, organisms that live in marine environments must be able to tolerate high salinity but need not tolerate variations in salinity.

Marine environments display a far greater range of hydrostatic pressure than do fresh waters. Hydrostatic pressure increases with depth at a rate of approximately 1 atm per 10 m, so the pressure at a depth of 1000 m is 100 times that at the surface. A few microorganisms, including archaeobacteria, have been isolated from Pacific Ocean trenches at depths greater than 1000 m!

Other factors that vary with the depth of ocean water are penetration of sunlight and oxygen concentration. Sunlight of sufficient intensity to support photosynthesis penetrates ocean water to depths of only 50 to 125 m, depending on the season, latitude, and transparency of the water. Oxygen diffuses into surface water and is released by photosynthetic organisms in sunlit water. However, deep water lacks both sunlight and oxygen.

Nutrient concentrations vary in ocean water, depending on depth and proximity to the shore (➤ Figure 25.15). Nutrients are most plentiful near the mouths of rivers and near the shore, where runoff from land enriches the water. However, ocean water is generally lower in phos-

And You Thought It Was El Nino

Oceanic phytoplankton absorb large amounts of carbon dioxide as they perform photosynthesis. Atmospheric carbon dioxide is one of the greenhouse gases that trap heat and raise global temperature. Seasonal fluctuations of phytoplankton populations affect the earth's temperature more than we had previously realized. Recent research has revealed that massive populations of marine viruses may be at the root of these processes. Earlier, the ocean was thought to contain relatively few viruses. Viral infections destroy phytoplankton and thus may help determine the earth's weather.

There are over a million bacteria/ml of open ocean water.

➤ Figure 25.15 **Nutrients flow into the ocean with the influx of water from rivers.** The light-colored area of water is silt-bearing, nutrient-rich river water.

phates and nitrates than fresh water. Nutrients in the waters of the open ocean are relatively dilute. Photosynthetic organisms near the surface serve as food for the heterotrophic organisms at the same or deeper levels. Decomposers are usually found in bottom sediments, where they release nutrients from dead organisms.

Large numbers of many different kinds of microorganisms live in the ocean. Yet, much remains to be learned about the exact numbers and the particular species living in various parts of the ocean. The primary producers of the ocean are photosynthetic microorganisms called *phytoplankton*. They are motile and contain oil droplets or other devices for buoyancy, allowing them to remain in sunlit waters. Phytoplankton include cyanobacteria, diatoms, dinoflagellates, chlamydomonads (kla″mi-do-mo′nadz), and a variety of other protists and eukaryotic algae (➤ Figure 25.16).

Many of the consumers of the ocean are heterotrophic bacteria. Which species inhabit a given region is determined by the temperature and pH of the water and by the available nutrients. Members of the genera *Pseudomonas, Vibrio, Achromobacter,* and *Flavobacterium* are common in ocean water. Protozoa, especially radiolarians and foraminiferans, and a variety of fungi also feed on producers in the open ocean. However, most of these organisms inhabit an aquatic zone beneath the region of intense sunlight.

Between the stratum in which these consumers live and the bottom of the ocean is a relatively uninhabited stratum. In the sediments at the bottom of the ocean, the number of microorganisms again increases. The bottom dwellers are generally strict or facultative anaerobic decomposers. Many of them contribute significantly to the maintenance of biogeochemical cycles and produce such substances as ammonium ion, hydrogen sulfate, and nitrogen gas.

Water Pollution

Water is considered polluted if a substance or condition is present that renders the water useless for a particular purpose. Thus, the concept of water pollution is a relative one, depending on both the nature of the pollutants and the intended uses of the water. For example, human drinking water is considered polluted if it contains pathogens or toxic substances. Water that is too polluted for drinking may be safe for swimming; water that is too polluted for swimming may be acceptable for boating, industrial uses, or generating electrical power. The EPA has established drinking water standards and methods for water testing. Water that is fit for human consumption is termed **potable water.**

Pollutants

The major types of water pollutants are organic wastes such as sewage and animal manures, industrial wastes, oil, radioactive substances, sediments from soil erosion, and heat. Organic wastes suspended in water are usually decomposed by microorganisms provided that the water contains sufficient oxygen for oxidation of the substances. The oxygen required for such decomposition is called **biological oxygen demand (BOD).** When BOD is high, the water can be depleted of oxygen rapidly. Anaerobes increase in number while populations of aerobic decomposers dwindle, leaving behind large quantities of organic wastes. Organic wastes also can contain pathogenic organisms from among the bacteria, viruses, and protozoa.

Industrial wastes contain metals, minerals, inorganic and organic compounds, and some synthetic chemicals. The metals, minerals, and other inorganic substances can alter the pH and osmotic pressure of the water, and some are toxic to humans and other organisms. Synthetic chemicals can persist in water because most decomposers lack enzymes to degrade them. Oil is another important water pollutant. Radioactive substances released into water persist as a hazard to living organisms until they have undergone natural radioactive decay.

Soil particles, sand, and minerals from soil erosion enter water from agricultural, mining, and construction activities. Nitrates, phosphates, and other nutrients enter the water from detergents, fertilizers, and animal manures. Abundant nutrient enrichment of water, called **eutrophication,** leads to excessive growth of algae and other plants. Eventually the plants become so dense that sunlight cannot penetrate the water. Many algae and other plants die, leaving large quantities of dead organic matter in the water. The high BOD of this organic matter leads to oxygen depletion and the persistence of undecomposed matter in the water.

(a)

(b)

➤ Figure 25.16 **Phytoplankton: producers of the ocean.** They include organisms such as (a) The marine diatom *Isthmia nervosa* (625X) (b) diatoms collected in a plankton-tow from Rhode Island Sound (100X).

30,000 people die every day in developing countries due to lack of clean water (WHO).

Even heat can act as a water pollutant when large quantities of heated water are released into rivers, lakes, or oceans. Increasing the temperature of water decreases the solubility of oxygen in it. The altered temperature and decreased oxygen supply significantly change the ecological balance of the aquatic environment.

The effects of water pollution are summarized in Table 25.1.

Pathogens in Water

Human pathogens in water supplies usually come from contamination of the water with human feces. When water is contaminated with fecal material, any pathogens that leave the body through the feces—many bacteria and viruses and some protozoa—can be present. The most common pathogens transmitted in water are listed in Table 25.2. Water usually is tested for

✓ What is the effect of UV radiation on organisms in air? In water?

✓ Where are most organisms found in soil?

✓ List four factors affecting soil microorganisms.

✓ What is BOD? What do microbes have to do with this?

➤ Table 25.1 **The effects of water pollution**

Pollutant	Effects	Comment
Organic wastes (sewage, decaying plants, animal manures, wastes from food-processing plants, oil refineries, and leather, paper, and textile plants)	Increase biological oxygen demand of water	If adequate oxygen is available, these substances can be degraded by microorganisms usually present in water. If oxygen becomes depleted, decomposition is limited to what can be done by anaerobic decomposers. Water plants may be killed, and animals may be killed or caused to migrate.
Pathogenic organisms	Cause disease in humans who drink the water	Most bacteria are well controlled in public drinking water, but certain viruses, especially those that cause hepatitis, still cause human disease. More effective means of removing viruses during purification are needed.
Inorganic chemicals and minerals	Increase the salinity and acidity of water and render it toxic	Such chemicals should be removed during waste treatment. Heavy metals such as mercury, which are toxic to humans, should be prevented from entering water supplies.
Synthetic organic chemicals (herbicides, pesticides, detergents, plastics, wastes from industrial processes)	Can cause birth defects, cancer, neurological damage, and other illness	Because these substances are not biodegradable, chemical or physical means must be used to remove them during waste treatment. Many such substances become magnified (increased in concentration) as they are passed along food chains.
Plant nutrients	Cause excessive and sometimes uncontrolled growth of aquatic plants (eutrophication); impart undesirable odors and tastes to drinking water	Removal of excess phosphates and nitrates from water during waste treatment is costly and difficult.
Sediments from land erosion	Cause silting of waterways and destruction of hydroelectric equipment near dams; reduce light reaching plants in water and oxygen content of water	
Radioactive wastes	Can cause cancer, birth defects, radiation sickness when in large doses	Effects can be magnified through food chains. Because such wastes are difficult to remove from water, preventing them from reaching water is exceedingly important.
Heated water	Reduces oxygen solubility in water; alters habitats and kinds of organisms present; encourages growth of some aquatic life, but can decrease growth of desired organisms such as fish	

> ➤ Table 25.2   **Human pathogens transmitted in water**

Organisms	Diseases caused
Salmonella typhi	Typhoid fever
Other *Salmonella* species	Salmonellosis (gastroenteritis)
Shigella species	Shigellosis (bacillary dysentery)
Vibrio cholerae	Asiatic cholera
Vibrio parahaemolyticus	Gastroenteritis
Escherichia coli	Gastroenteritis
Yersinia enterocolitica	Gastroenteritis
Campylobacter fetus	Gastroenteritis
Legionella pneumophilia	Legionnaire's disease (pneumonia)
Hepatitis A virus	Hepatitis
Poliovirus	Poliomyelitis
Giardia intestinalis	Giardiasis
Balantidium coli	Balantidiasis
Entamoeba histolytica	Amoebic dysentery
Cryptosporidium parvum	Cryptosporidiosis (gastroenteritis)

fecal contamination by isolating *Escherichia coli* from a water sample. *Escherichia coli* is called an **indicator organism** because, as *E. coli* is a natural inhabitant of the human digestive tract, its presence in water indicates that the water is contaminated with fecal material.

Water Purification

Purification Procedures

Purification procedures for human drinking water are determined by the degree of purity of the water at its source. Water from deep wells or from reservoirs fed by clean mountain streams requires very little treatment to make it safe to drink. In contrast, water from rivers that contain industrial and animal wastes and even sewage from upstream towns requires extensive treatment before it is safe to drink. Such water is first allowed to stand in a holding reservoir until some of the particulate matter settles out. Then alum (aluminum potassium sulfate) is added to cause **flocculation,** or precipitation of suspended colloids (Chapter 2) such as clay. ➡ (p. 32) Many microorganisms are also removed from the water by flocculation.

Following flocculation treatment, water is filtered. **Filtration,** the passage of water through beds of sand, removes nearly all the remaining microorganisms. Charcoal can be used instead of sand for filtration; it has the advantage of removing organic chemicals that are not removed by sand. Finally, water is chlorinated. **Chlorination,** the addition of chlorine to water, readily kills bacteria but is less effective in destroying viruses and cysts of pathogenic protozoa. The amount of chlorine required to destroy most of the microorganisms is increased by the presence of organic matter in the water. Chlorine may combine with some organic molecules to form carcinogenic substances. Although current water chlorination procedures have not been proved to increase the risk of cancer in humans, the long-term effects of the interaction of chlorine and organic compounds are difficult to assess.

Tests for Water Purity

Water purity is usually tested by looking for **coliform bacteria.** The coliform bacteria, which include *E. coli,* are Gram-negative, non-spore-forming, aerobic or facultative anaerobic bacteria that ferment lactose and produce acid and gas. Most municipal water supplies are regularly tested for the presence of coliform bacteria. The presence of a significant number of coliforms is evidence that the water may not be safe for drinking. Three methods of testing for coliform bacteria currently in use are the multiple-tube fermentation method, the membrane filter method, and the ONPG and MUG test.

On the basis of CDC estimates, 940,000 Americans become ill each year from contaminated water, and 900 die.

In the United States over 98 percent of the water supply systems that disinfect water, use chlorine-based compounds as the disinfectant.

The **multiple-tube fermentation method** (➤ Figure 25.17) involves three stages of testing: the presumptive test, the confirmed test, and the completed test. In the **presumptive test,** a water sample is used to inoculate lactose broth tubes. Each tube receives a water volume of 10 ml, 1 ml, or 0.1 ml. The tubes are incubated at 35°C and observed after 24 and 48 hours for evidence of gas production. Gas production provides presumptive evidence that coliform bacteria are present. In using the presumptive test, microbiologists analyzing water samples determine approximate numbers of organisms by means of the most probable number method (Chapter 6). ➡ (p. 143)

Because certain noncoliform bacteria also produce gas, additional testing is necessary to confirm the presence of coliforms. In the **confirmed test,** samples from the highest dilution showing gas production are streaked onto eosin–methylene blue (EMB) agar plates. EMB prevents growth of Gram-positive organisms. Coliforms (which are Gram-negative) produce acid; under acidic conditions the eosin and methylene blue dyes are absorbed by the organisms of a colony. Thus, after 24-hour incubation, coliform colonies have dark centers and may also have a metallic greenish sheen. Observing such colonies confirms the presence of coliforms. In the **completed test,** organisms from dark colonies are used to inoculate lactose broth and agar slants. The production of acid and gas in the lactose broth and the microscopic identification of Gram-negative, non-spore-forming rods from slants constitute a positive completed test.

In the **membrane filter method** (➤ Figure 25.18), a 100-ml water sample is drawn through a sterile membrane filter that has pores about 0.45 μm in diameter. This membrane, which traps bacteria on its surface, is then incubated on the surface of a sterile absorbent pad previously saturated with an appropriate growth medium. After incubation, colonies form on the filter where bacteria were trapped during filtration. The presence of more than one colony per

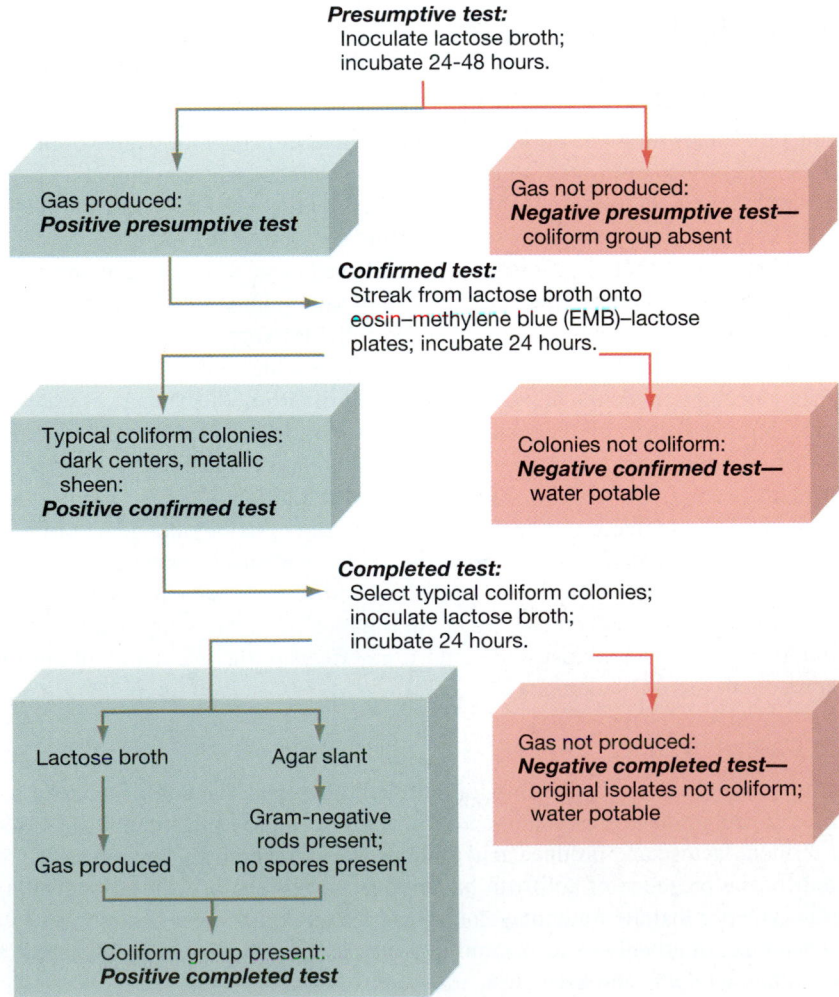

➤ Figure 25.17   The multiple-tube fermentation test for water purity.

(a)   (b)

➤ Figure 25.18   **The membrane filter test for water purity.** (a) Filtering water samples and trapping organisms on filter pads, which are then incubated. (b) After incubation, the coliform colonies that appear are counted.

100 ml of water indicates that the water may be unsafe for human consumption. If colonies are observed, additional tests can be performed to identify them specifically. The membrane filter method is much faster, and permits larger volumes of water to be tested, than the multiple-tube fermentation method.

The **ONPG and MUG test** relies on the ability of coliform bacteria to secrete enzymes that convert a substrate (Chapter 5) into a product that can be detected by a color change. ➥ (p. 111) In this test, a water sample is inoculated into a nutrient-broth tube containing the substrates ONPG (*o*-nitrophenyl-β-D galactopyranoside) and MUG (4-methylumbelliferyl-β-D-glucuronide). After a suitable incubation period, if coliforms are present, they will have secreted the enzymes β-galactosidase and β-glucuronidase and other substances. Beta-galactosidase hydrolyzes OPNG into a yellow-colored product, whereas β-glucuronidase hydrolyzes MUG into a product that fluoresces blue when illuminated with ultraviolet light (➤ Figure 25.19). This test can be used in conjunction with the other tests for water purity.

In addition to contamination with coliforms, drinking water can also contain other organisms sometimes referred to as "nuisance" organisms. Although these organisms do not produce human disease, they can affect the taste, color, or odor of the water. Some also can form insoluble precipitates inside water pipes. Nuisance organisms include sulfur bacteria, iron bacteria, slime-forming bacteria, and algae. Among the sulfur bacteria are *Desulfovibrio,* which produces hydrogen sulfide, and *Thiobacillus,* which produces sulfuric acid that can corrode pipes. Iron bacteria deposit insoluble iron compounds that can obstruct water flow in pipes. Eukaryotic algae, diatoms, and cyanobacteria reproduce rapidly in water exposed to sunlight (as in reservoirs). They can become so numerous that they clog filters used in the purification process. Identifying these various nuisance organisms in water is a tedious task because different tests are required for each kind of organism. None of the tests are performed routinely but may be done when citizens complain of objectionable tastes, odors, or colors in water.

Water can also be contaminated with various organic and inorganic substances, especially when water supplies are drawn from rivers that have received upstream effluents of industrial wastes. Although it is at least theoretically possible to detect these substances through chemical analyses, such tests are rarely performed.

Sewage Treatment

Sewage is used water and the wastes it contains. It is about 99.9 percent water and about 0.1 percent solid or dissolved wastes. These wastes include household wastes (human feces, detergents, grease, and anything else people put down the drain or garbage disposal unit), indus-

➤ Figure 25.19   **The ONPG and MUG test.** A positive ONPG produces a yellow color; a positive MUG gives off a blue fluorescence on ultraviolet illumination. A sample without coliforms remains clear.

trial wastes (acids and other chemical wastes and organic matter from food-processing plants), and wastes carried by rainwater that enters sewers.

Sewage treatment is a relatively modern practice. Until recent years, many large U.S. cities dumped untreated sewage into rivers and oceans; many cities along the Mediterranean Sea still do! When small amounts of sewage are dumped into fast-flowing, well-oxygenated rivers, the natural activities of decomposers in the river purify the water. However, large amounts of sewage overload the purification capacity of rivers. Cities downstream are then forced to take their water supplies from rivers that contain wastes of the cities upstream. Fortunately, most U.S. cities now have some form of sewage treatment.

Complete sewage treatment consists of three steps: primary, secondary, and tertiary treatment (➢ Figure 25.20). In **primary treatment,** physical means are used to remove solid wastes from sewage. In **secondary treatment,** biological means (actions of decomposers) are used to remove solid wastes that remain after primary treatment. In **tertiary treatment,** chemical and physical means are used to produce an effluent of water pure enough to drink. Let us look at each of these processes in more detail.

➢ **Figure 25.20   An overview of a sewage treatment plant.** Primary, secondary, and tertiary treatment facilities are shown.

> Figure 25.21   **Trickling filters.** These filters are used in secondary sewage treatment.

(a)

Primary Treatment

As raw sewage enters a sewage treatment plant, several physical processes are used to remove wastes in primary treatment. Screens remove large pieces of floating debris, and skimmers remove oily substances. Water is then directed through a series of sedimentation tanks, where small particles settle out. The solid matter removed by these procedures accounts for about half the total solid matter in sewage. Flocculating substances can be used to increase the amount of solids that settle out and thus the proportion of solids removed by primary treatment. Sludge is removed from the sedimentation tanks intermittently or continuously, depending on the design of the treatment plant.

Secondary Treatment

The effluent from primary treatment flows into secondary treatment systems. These systems are of two types: trickling filter systems and activated sludge systems. Both systems make use of the decomposing activity of aerobic microorganisms. The BOD is high in secondary treatment systems, so the systems provide for continuous oxygenation of the wastewater.

In a **trickling filter system** (➤ Figure 25.21), sewage is sprayed over a bed of rocks about 2 m deep. The individual rocks are 5 to 10 cm in diameter and are coated with a slimy film of aerobic organisms such as *Sphaerotilus* and *Beggiatoa* (➤ Figure 25.22). Spraying oxygenates the sewage so that the aerobes can decompose organic matter in it. Such a system is less efficient but less subject to operational problems than an activated sludge system. It removes about 80 percent of the organic matter in the water.

In an **activated sludge system,** the effluent from primary treatment is constantly agitated, aerated, and added to solid material remaining from earlier water treatment. This **sludge** contains large numbers of aerobic organisms that digest organic matter in wastewater. However, filamentous bacteria multiply rapidly in such systems and cause some of the sludge to float on the surface of the water instead of settling out. This phenomenon, called **bulking,** allows the floating matter to contaminate the effluent. The sheathed bacterium *Sphaerotilus* (➤ Figure 25.22a), which sometimes proliferates rapidly on decaying leaves in small streams and causes a bloom, can interfere with the operation of sewage systems in this way. Its filaments clog filters and create floating clumps of undigested organic matter.

Sludge from both primary and secondary treatments can be pumped into **sludge digesters.** Here, oxygen is virtually excluded, and anaerobic bacteria partially digest the sludge to simple organic molecules and the gases carbon dioxide and methane. The methane can be used for heating the digester and providing for other power needs of the treatment plant. Undigested matter can be dried and used as a soil conditioner or as landfill (➤ Figure 25.23).

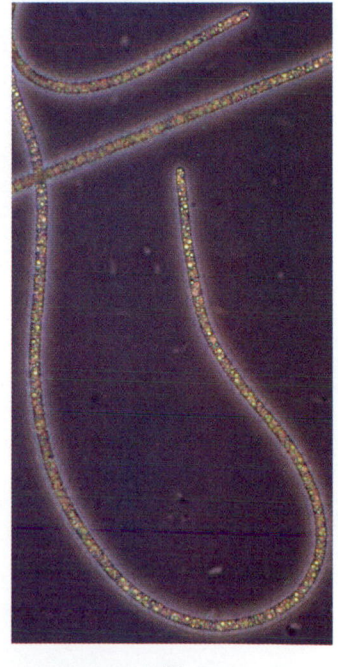

(b)

➤ Figure 25.22   **Two sheathed bacteria used in trickling filters.** (a) *Sphaerotilus* (29,300X); (b) *Beggiatoa* (400X).

➢ Figure 25.23 **Sludge digestion.** The injection of municipal sewage sludge into farmland is designed to return nutrients to the soil. But if treatment is not complete, it may also add pathogens to the soil.

Tertiary Treatment

The effluent from secondary treatment contains only 5 to 20 percent of the original quantity of organic matter and can be discharged into flowing rivers without causing serious problems. However, this effluent can contain large quantities of phosphates and nitrates, which can increase the growth rate of plants in the river. Tertiary treatment is an extremely costly process that involves physical and chemical methods. Fine sand and charcoal are used in filtration. Various flocculating chemicals precipitate phosphates and particulate matter. Denitrifying bacteria convert nitrates to nitrogen gas. Finally, chlorine is used to destroy any remaining organisms. Water that has received tertiary treatment can be released into any body of water without danger of causing eutrophication. Such water is pure enough to be recycled into a domestic water supply. However, the chlorine-containing effluent, when released into streams and lakes, can react to produce carcinogenic compounds that may enter the food chain or be ingested directly by humans in their drinking water. It would be safer to remove the chlorine before releasing the effluent, but this is rarely done today, although the cost is not great. Ultraviolet lights are now replacing chlorination as the final treatment of effluent. ➥ (Chapter 12, p. 334) They destroy microbes without adding carcinogens to our streams and waters. Likewise, especially in Europe, the treatment of effluent with ozone is replacing chlorination. Ozone generators are simple and not very costly, and they do not add carcinogens to natural waterways.

Septic Tanks

About 50 million rural families in the United States do not have access to city sewer connections or their treatment facilities. These homes rely on backyard **septic tank** systems (➢ Figure 25.24). Homeowners must be careful not to flush or put materials such as poisons and grease down the drain, as these might kill the beneficial microbes in the septic tank that decompose sludge solids that accumulate there. This would necessitate immediate pumping of the tank by a vehicle known as the "honey wagon" to prevent sewage from backing up into the house. Even with normal operation, it is occasionally necessary to pump the sludge from the tank and haul it to a sewage treatment plant.

Soluble components of the sewage continue out of the septic tank into the drainage (leaching) field. There they seep through perforated pipe, past a gravel bed and into the soil, which

CROSS SECTION OF PIPE

Soil

Gravel bed

Perforated drain pipe

Washing machine

Toilet

Sink

(a) SIDE VIEW

Sludge (must be pumped out periodically)

Septic tank   Cover

(b) TOP VIEW

Drainage field

Perforated drain pipe

> **Figure 25.24   Sewage disposal by means of a septic tank and drainage field system.**
(a) Solid materials settle out as sludge, which undergoes microbial decay, while soluble materials continue to the drainage field. There (b) they seep out into the soil, which filters out bacteria and some viruses and binds phosphate. Soil bacteria decompose organic materials.

filters out bacteria and some viruses and binds phosphate. Soil bacteria decompose organic materials. Drainage fields must be placed where they will not allow seepage into wells, a difficult problem on hills or in densely populated areas. Drainage fields cannot be used where the water table is too high or the soil is insufficiently permeable, such as in rocky areas. After 10 years or more, the average drainage field clogs up and can no longer be used.

Bioremediation

Bioremediation is a process that uses naturally occurring or genetically engineered microorganisms such as yeast, fungi, and bacteria to transform harmful substances into less toxic or nontoxic compounds. Microorganisms break down a variety of organic compounds in nature to obtain nutrients, carbon, and energy for growth and survival. Bioremediation promotes the growth of microorganisms to degrade contaminants by utilizing those contaminants as carbon and energy sources.

Bioremediation has been used since the late 1970s to degrade petroleum products and hydrocarbons. (See the box "Microbial Cleanup" in Chapter 5). ➡ (p. 128) In March 1989, the supertanker *Exxon Valdez* ran aground in Prince William Sound, Alaska. The leaking tanker flooded the shoreline with about 42 million liters (11 million gal) of crude oil. The gravel and sand beaches were saturated with oil more than half a meter thick. A massive cleanup program was organized. At first conventional cleanup techniques—such as booms, high-pressure hot-water sprays, skimmers, and manual scrubbers—were used. But the shore remained black and gooey, because these methods could not remove all the oil from beneath rocks and within beach sediments.

Scientists from the EPA decided to use bioremediation to enhance the cleanup. They sprayed the beach with fertilizer (nutrients) to stimulate the growth of the native microorganisms and promote the use of the oil wastes as the microbes' carbon source. The areas sprayed with fertilizer were soon nearly clean of oil to a depth of about one-third meter, while untreated areas were still coated with sticky oil. Studies of one treated beach show that 60 percent of total hydrocarbons and 45 percent of polycyclic aromatic hydrocarbons (PAHs), which are potentially toxic, were degraded by bacteria within 3 months. The *Exxon Valdez* cleanup was a bioremediation success story.

Another recent bioremediation success story involves the natural oil-detoxification actions of microbes. In 1995, scientists studying the Persian Gulf region, where lakes of oil had been left after pipelines and oil wells were damaged in the 1991 Gulf War, concluded that a natural recovery of the region's plants was occurring. Samir Radwan and his colleagues in Kuwait found that the roots of wildflowers in the oil-soaked desert were healthy and oil-free. Cultures made of bacteria and fungi from the sand revealed several types of known oil-eating microorganisms, such as the bacterium *Arthrobacter*. The researchers believe they may have found a cheap, safe, and natural method for cleaning up oil spills on land—cultivating plants whose roots recruit microbial oil-eaters.

The applications for bioremediation are rapidly expanding. It has already proved to be a successful means of decomposing wastes in landfills. It may be suitable for cleaning up soil and groundwater contaminated by leaks from underground tanks storing petroleum, heating oil, and other materials. The wood-preservative industry also appears to be a promising area for bioremediation. Each year the United States uses 450,000 tons of creosote, an oily liquid that is distilled from coal tar and used as a wood preservative. Creosote sometimes leaks from its holding tanks and seeps into the soil and underlying groundwater. The white-rot fungus *Phanerochaete chrysosporium* has been shown to degrade pentachlorophenol, the main contaminant at wood-preserving sites. This superfungus can also destroy other toxic compounds in the soil, including dioxins, polychlorinated biphenyls (PCBs), and PAHs.

At present only naturally occurring microbes are used in bioremediation. Scientists are experimenting with genetic engineering techniques, however, to develop microorganisms for use at hazardous waste sites. Before these bioengineered organisms can be used in the field, the EPA requires that they undergo a safety review in accordance with the Toxic Substance Control Act to evaluate any possible risks to human health or to the environment.

There are both advantages and disadvantages to the use of bioremediation. *Advantages:* Bioremediation is an ecologically sound, "natural" process; it destroys target chemicals at the contamination site instead of transferring contaminants from one site to another; and the process is usually less expensive than other methods used for cleaning up hazardous wastes. *Disadvantages:* Using bioremediation often takes longer than other remedial methods such as excavation or incineration, and bioremediation techniques are not yet refined for sites with mixtures of contaminants. More research is needed to perfect this technology. Nevertheless, bioremediation holds enormous promise for the future. As scientists develop more practical uses for bioremediation, this technology will become more important in cleaning up and protecting the environment.

✓ Define coliform bacteria. If they are not pathogens, why are they important?

✓ Compare primary, secondary, and tertiary sewage treatment as to what they remove.

✓ Why must you be very careful about what you flush down into a septic tank system?

✓ What are some of the advantages and disadvantages of bioremediation?

Putting Microbes to Work to Clean Our Environment

In 1991, manufacture of the East German auto the Trabant ceased. The two-cylinder, two-stroke engine couldn't compete with the Mercedes and BMWs of what was then West Germany. Imagine the pain, after perhaps 10 years of saving to buy one, of finding that what should have been a status symbol was now a costly dinosaur. Customers seeking to trade in their Trabants on a bigger car not only got no trade-in allowance but were charged $300 more to dispose of their old Trabant. Space in landfills is scarce and costly in Germany.

Biotechnology may soon have a solution to the problem by means of microbial *biodegradation,* or decomposition of humanmade materials by microbial activity. Trabants are made mostly of plastic. They are so light that two people can lift one into a dumpster. This plastic has a great deal of agricultural waste (cellulose) from the former Soviet Union incorporated into it, which should make it easily biodegradable by microbes. Scientists are searching the soil of landfills for microbes that are degrading the plastic of Trabants already buried there. These organisms could be used to disintegrate the glut of unwanted Trabants more rapidly. One Berlin biotechnology company claims to have developed a bacterium that will eat a Trabant in 20 days, leaving only a small pile of compost.

Plastics such as soda bottles and disposable diapers are an ever-increasing component of already overcrowded landfills. Some companies are producing biodegradable plastics meant to be broken down by microbes over time. In fact, the manufacture of those plastics depends on bacteria. Poly-β-hydroxyalkanes (PHAs) are biodegradable polyesters (polymers formed with ester bonds; Chapter 2) made by bacteria such as *Alcaligenes eutrophus.* ➡ (p. 37) These bacteria store PHAs in granules as a reserve of carbon. When carbon is readily available in the form of glucose, *A. eutrophus* produces large quantities of PHAs to store it for lean times. When carbon is scarce, the bacterium uses its own stores of carbon in the granules. Thus, this one bacterium can be used to manufacture PHAs—which are then formed into bottles, plastic bags, and such—as well as to degrade those products once their useful days have passed!

Although old Trabants and plastics are a nuisance, they aren't harmful, unlike contaminated soil and water. Dangerous chemical contaminants must be removed from soil and water.

The end of a dream: Dozens of discarded Trabant cars are stacked up in a Berlin junkyard. Scientists are looking for microbes that will eat these cars, which are made mostly of plastic. East Germans once saved for years to afford one of these "dream" cars.

Retracing Our Steps

Fundamentals of Ecology

- **Ecology** is the study of relationships among organisms and their environment.

The Nature of Ecosystems

- An **ecosystem** includes all the **biotic** and **abiotic factors** of an environment. It will always support life of **indigenous organisms** and sometimes supports **nonindigenous** ones. All the living organisms in an **ecosystem** make up a **community.**

The Flow of Energy in Ecosystems

- Energy in an ecosystem flows from the sun to **producers** to **consumers. Decomposers** obtain energy from digesting dead bodies and wastes of other organisms. The nutrients they release can be recycled.

Biogeochemical Cycles

- Although energy is continuously available, nutrients must be recycled from dead organisms and waste to make them available for other organisms.

The Water Cycle

- The **water cycle** is summarized in ➤ Figure 25.2.

The Carbon Cycle

- The **carbon cycle** is summarized in ➤ Figure 25.3.

The Nitrogen Cycle and Nitrogen Bacteria

- The **nitrogen cycle** is summarized in ➤ Figure 25.4.
- **Nitrogen fixation** is the reduction of atmospheric nitrogen to ammonia. It is accomplished by certain free-living aerobes and anaerobes but mostly by *Rhizobium*.
- *Rhizobium* cells accumulate around legume roots and change to **swarmer cells,** which invade root cells and become bacteroids. **Nitrogenase** in **bacteroids** catalyzes nitrogen-fixation reactions.
- **Nitrification** is the conversion of ammonia to nitrites and nitrates; nitrites are formed by *Nitrosomonas,* and nitrates by *Nitrobacter.*
- **Denitrification** is the conversion of nitrates to nitrous oxide and nitrogen gas. It is accomplished by a variety of organisms, especially in waterlogged soils.

The Sulfur Cycle and Sulfur Bacteria

- The **sulfur cycle** is summarized in ➤ Figure 25.7.
- Various bacterial species carry out **sulfate reduction, sulfur reduction,** or **sulfur oxidation.**

Other Biogeochemical Cycles

- The **phosphorus cycle** is summarized in ➤ Figure 25.9.
- All elements found in living organisms must be recycled.

Air

Microorganisms Found in Air

- Microorganisms are transmitted by air but do not grow in air. Air is analyzed by exposing agar plates to it and by drawing air over the surface of agar or into a liquid medium and then studying the organisms found.

Methods for Controlling Microorganisms in Air

- Microorganisms in air are controlled by chemical agents, radiation, filtration, and unidirectional airflow.

Soil

Components of Soil

- Soil contains inorganic rocks, minerals, water, gases, organic matter (**humus**), and many microorganisms.

Microorganisms in Soil

- Microbes of all major taxonomic groups are found in soil.
- Physical factors that affect soil microorganisms include moisture, oxygen concentration, pH, and temperature.
- Organisms also alter the characteristics of their environment as they use nutrients and release wastes.
- Soil microorganisms are important as decomposers in the carbon cycle and in all phases of the nitrogen cycle.

Soil Pathogens

- Soil pathogens affect mainly plants and insects.
- Various species of *Clostridium* are important human pathogens found in soil.

Water

Freshwater Environments

- Freshwater environments are characterized by low salinity and variability in temperature, pH, and oxygen concentration.
- Microorganisms from all major taxonomic groups are found in freshwater environments. Bacteria—aerobes where oxygen is plentiful, and anaerobes where oxygen is depleted—are especially abundant.

Marine Environments

- Marine environments are characterized by high salinity, smaller variability in temperature, pH, and oxygen concentration. As depth increases, pressure increases and sunlight penetration decreases. Microorganisms from all major taxonomic groups are also found in marine environments. Photosynthetic organisms are found nearest the surface, heterotrophs in surface and lower strata, and decomposers in bottom sediments.

Water Pollution

- Water is polluted if a substance or condition is present that renders the water useless for a particular purpose.
- Table 25.1 summarizes the effects of water pollution.
- Many human pathogens can be transmitted in water.
- Such pathogens are listed in Table 25.2.

Water Purification

- Water purification involves **flocculation** of suspended matter, **filtration,** and **chlorination.**
- Tests for water purity are designed to detect **coliform bacteria;** they include the **multiple-tube fermentation** and **membrane filter methods** and the **ONPG and MUG test.**

Sewage Treatment

- **Sewage** is used water and the wastes it contains.

Primary Treatment

- **Primary treatment** is the physical removal of solid wastes.

Secondary Treatment

- **Secondary treatment** is the removal of organic matter by the action of aerobic bacteria.

Tertiary Treatment

- **Tertiary treatment** is the removal of most organic matter, ni-trates, phosphates, and any surviving microorganism by physical and chemical means.

Septic Tanks

- Soil bacteria decompose soluble components of sewage in the drainage field of a **septic tank** system.

Bioremediation

- Naturally occurring or genetically engineered microbes transform harmful substances into less toxic or nontoxic compounds, by using the contaminants as carbon and energy sources.

Terminology Check

abiotic factor (p. 725)
activated sludge system (p. 749)
bacteroid (p. 731)
biogeochemical cycle (p. 727)
biological oxygen demand (BOD)
  (p. 743)
bioremediation (p. 751)
biosphere (p. 725)
biotic factor (p. 725)
bulking (p. 749)
carbon cycle (p. 727)
chlorination (p. 745)
coliform bacterium (p. 745)
community (p. 726)
completed test (p. 746)
confirmed test (p. 746)
consumer (p. 726)
decomposer (p. 726)
denitrification (p. 732)

ecology (p. 725)
ecosystem (p. 725)
eutrophication (p. 743)
filtration (p. 745)
flocculation (p. 745)
humus (p. 738)
hydrologic cycle (p. 727)
indicator organism (p. 745)
indigenous organism (p. 726)
membrane filter method (p. 746)
microenvironment (p. 726)
multiple-tube fermentation method
  (p. 746)
nitrification (p. 731)
nitrogen cycle (p. 729)
nitrogen fixation (p. 729)
nitrogenase (p. 730)
nonindigenous organism (p. 726)
ONPG and MUG test (p. 747)

phosphorus cycle (p. 734)
potable water (p. 743)
presumptive test (p. 746)
primary treatment (p. 748)
producer (p. 726)
secondary treatment (p. 748)
septic tank (p. 750)
sewage (p. 747)
sludge (p. 749)
sludge digester (p. 749)
sulfate reduction (p. 734)
sulfur cycle (p. 733)
sulfur oxidation (p. 734)
sulfur reduction (p. 734)
swarmer cell (p. 731)
tertiary treatment (p. 748)
trickling filter system (p. 749)
water cycle (p. 727)

See You on the Web

Don't forget I am expecting you on the web. The following is a preparedness checklist for your web explorations. If you have questions on any of these items, you can return to the end of Chapter 1 for more details.

- Questions We'll Explore
- Concept Check questions
- Flashcards
- Study Guide

If you are ready, come visit me at

http://www.prenhall.com/black

It will be a journey you'll never forget.

Applied Microbiology

GOLD—that precious metal we give in love and take in war—could it be a gift of bacteria? Bacterioform gold has been found in Alaska, China, Siberia, and South Africa, indicating that budding bacteria have probably played a role in the geochemistry of gold for more than half of the Earth's history.

The bacteria haven't manufactured the gold; they've collected it from nearby rocks containing gold ore. Certain soil bacteria can precipitate gold from ore dissolved by cold, waterlogged conditions. What's the advantage to the bacteria? Some scientists think that the golden armor acts like a spore coat, enabling the bacteria to survive as long as 1000 years!

We have now come full circle. As we enter this final chapter, I hope that you have been panning gold out of your book, fashioning it into a new way of seeing life.

Background Basics

T hroughout history, humans have used microorganisms. We have long found uses for microbes in making food and medicine, and today other industrial applications of microbes abound. And because biotechnology is advancing rapidly, the future of applied microbiology looks very bright. As we consider the many applications of microbiology—from food production to mining—we must also discuss food spoilage and methods of preservation.

Microorganisms Found in Food

Anything that people eat or drink can be used as food by microorganisms, too (➤ Figure 26.1). Most substances consumed by humans are derived from plants, which, of course, grow in soil, or from animals, which live in contact with soil and so have soil organisms on them. Although soil organisms usually are not human pathogens, many can cause food spoilage. The handling of foods from harvest or slaughter to human consumption provides many more opportunities for the foods to become contaminated with microorganisms. Unsanitary practices by food handlers and unsanitary working conditions frequently lead to contamination of foods with pathogens. Improper storage and preparation procedures at home and especially in restaurants can lead to further contamination with pathogens. Improper refrigeration of prepared foods is a major source of food poisoning.

Questions We'll Explore

A What kinds of microorganisms are found in different categories of food?

B How are diseases transmitted in food?

C How can food spoilage and disease transmission be prevented, and what standards relate to these problems?

D How are microorganisms used as food or in the making of food products?

E How are microbes used in the manufacture of beer, wine, and spirits?

F How can microbes be used in industry, and what problems are associated with their use?

G What is the role of microorganisms in the manufacture of simple organic compounds, antibiotics, enzymes, and other biologically useful substances?

H How are microorganisms used in mining?

I How are microorganisms used in waste disposal?

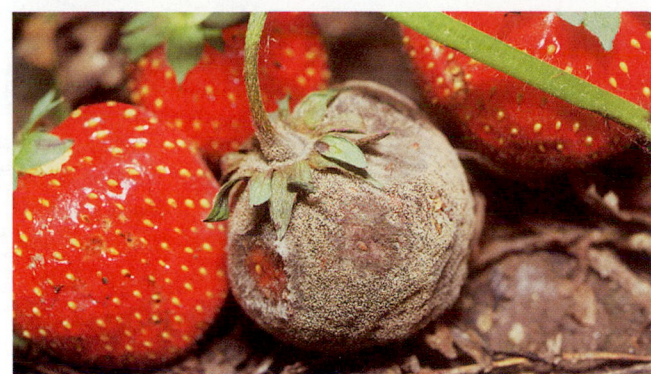

➤ Figure 26.1   **Microbes in food.** Microorganisms use the same foods that humans do, as can be seen from these strawberries covered with "gray mold."

Grains

When properly harvested, the various kinds of edible grains, such as rye and wheat, are dry. Because of the lack of moisture, few microorganisms thrive on them. However, if stored under moist conditions, grains can easily become contaminated with molds and other microorganisms. Insects, birds, and rodents also transmit microbial contaminants to grains.

Raw grains contaminated with the mold *Claviceps purpurea,* known as *ergot,* cause ergot poisoning, or ergotism. Compounds produced by this fungus are hallucinogenic and can alter behavior or even be deadly if eaten. Certain species of *Aspergillus* infecting peanuts and some grains produce *aflatoxins*. These toxic compounds have proved to be potent mutagens and carcinogens. ➥ (Chapter 22, p. 656)

Most grains are used in making breads and cereals. Humans have made breads for thousands of years, and some loaves of bread on display in the British Museum are 4000 years old. Although the first instance of leavening bread with yeast probably occurred by accident, special strains of *Saccharomyces cerevisiae* that produce large amounts of carbon dioxide are now added to bread dough to make it rise. We will describe this process in more detail later.

Like raw grain, bread is susceptible to contamination and spoilage by various molds. *Rhizopus nigricans* is the most common bread mold, but several species of *Penicillium, Aspergillus,* and *Monilia* also thrive in bread. Contamination of bread with *M. sitophila,* a pink bread mold, is especially dreaded by bakers because it is almost impossible to eliminate from a bakery once it has become established. Rye bread is particularly likely to become contaminated with *Bacillus* species, which hydrolyze proteins and starch and give the bread a stringy texture. If not killed by baking, *Bacillus* spores germinate and cause rapid and extensive damage to freshly baked bread.

Lactobacterium sanfrancisco adds lactic and acetic acid tartness to sourdough bread.

Fruits and Vegetables

Millions of commensal bacteria, especially *Pseudomonas fluorescens,* are found on the surfaces of fruits and vegetables. These foods also easily become contaminated with organisms from soil, animals, air, irrigation water, and equipment used to pick, transport, store, or process them. Pathogens such as *Salmonella, Shigella, Entamoeba histolytica, Ascaris,* and a variety of viruses can be transmitted on the surfaces of fruits and vegetables. However, the skins of most plant foods contain waxes and release antimicrobial substances, both of which tend to prevent microbial invasion of internal tissues.

Certain vegetables are particularly vulnerable to microbial attack and spoilage. Leafy vegetables and potatoes are susceptible to bacterial soft rot by *Erwinia carotovora*. The fungus *Phytophthora infestans* caused the Irish potato famine of 1846. Biotechnology is being called upon to help prevent such crop damage. Attempts are under way, for example, to insert into soybeans a gene for disease resistance, found in wild mustard plants, that could yield pathogen-resistant soybean crops. Similarly, a gene that protects plants against the bacterium *Pseudomonas syringae* may soon be introduced into corn, bean, and tomato plants.

Fruits are likewise susceptible to spoilage by microbial action. Tomatoes, cucumbers, and melons can be damaged by the fungus *Fusarium,* which causes soft rot and cracking of tomato skins. Fruit flies pick up the fungus from infected tomatoes and transmit it to healthy ones as they deposit eggs in surface cracks. Other insects pierce tomatoes to feed on them and at the same time introduce *Rhizopus,* which breaks down pectin and can turn a tomato into a bag of water. Fresh fruit juices, because of their high sugar and acid content, provide an excellent medium for growth of molds, yeasts, and bacteria of the genera *Leuconostoc* and *Lactobacillus.* Grapes and berries are damaged by a wide variety of fungi, and large numbers of stone fruits, such as peaches, can be destroyed overnight by brown rot due to *Monilia fructicola. Penicillium expansum,* which grows on apples, produces the toxin patulin, which can easily contaminate cider. Other *Penicillium* species produce blue and green mold on citrus fruits.

Meats and Poultry

Meat animals arrive at slaughterhouses with numerous and varied microorganisms in the gut and feces, on hides and hoofs, and sometimes in tissues. At least 70 pathogens have been identified among these microbes. Nearly all animal carcasses in slaughterhouses in the United States are inspected by a veterinarian or a trained inspector, and those found to be diseased are condemned and discarded (➤ Figure 26.2). Inspection, however, cannot guarantee that meat is parasite-free. The inspector examines the carcass surface and the heart (where helminths are often concentrated) but cannot look inside each cut of meat. Some of the most common diseases identified in slaughterhouses are abscesses, pneumonia, septicemia, enteritis, toxemia, nephritis, and pericarditis. Lymphadenitis, an inflammation of the lymph nodes, is especially common in sheep and lambs.

Even after animals are slaughtered and the carcasses are hung in refrigerated rooms to age, microorganisms sometimes spoil the meat. Several molds grow on refrigerated meats, and *Cladosporium herbarum* can grow on frozen meats. Mycelia of *Rhizopus* and *Mucor* produce a fluffy, white growth referred to as "whiskers" on the surfaces of hanging carcasses. The bacterium *Pseudomonas mephitica* releases hydrogen sulfide and causes green discoloration on refrigerated meat under low oxygen conditions. Several species of *Clostridium* cause putrefaction called **bone stink** deep in the tissues of large carcasses.

Ground meats sometimes contain helminth eggs and always contain large numbers of lactobacilli and molds. In most areas butcher shops are required to maintain two separate meat-grinding machines—one for pork and one for other meats. This practice is encouraged because it is difficult to clean a machine thoroughly enough to eliminate any possibility of transmitting raw pork bits, which may carry the helminth *Trichinella spiralis,* the cause of trichinosis, to other meats. ➡ (Chapter 11, p. 308) Lactobacilli in ground meats produce acids that retard the growth of enteric pathogens. Nevertheless, ground meats are subject to spoilage even when refrigerated and should be frozen if not to be used within a day or two. All meats, but especially ground meats, should be cooked thoroughly to kill pathogens.

More than 20 bacterial genera have been found on dressed poultry, and mishandling of poultry in restaurants accounts for many foodborne infections. Nearly half these infections have been traced to *Salmonella* and a fourth each to *Clostridium perfringens* and *Staphylococcus aureus.* Freezing fails to rid poultry of *Salmonella.* Pseudomonads and several other Gram-negative bacteria are common contaminants of poultry, where they cause slime and offensive odors.

Salmonella Vaccine on the Way

Millions of people around the world contract serious cases of salmonella food poisoning due to the improper handling of chicken, eggs, pork, and beef. A new vaccine—for use on animals—may drastically cut that number of poisonings. Dr. Roy Curtis, Washington University in St. Louis, Missouri, has developed a live salmonella vaccine that is given to baby chickens. The bacteria, no longer pathogenic due to the removal of the virulence genes, is given to the chickens in their drinking water. The baby chickens, who develop a lifelong immunity to salmonella, are protected a few hours after vaccination. The vaccine also protects the chickens' eggs. The vaccine for chickens is nearing U.S. governmental approval and should be available for use sometime in 1998. It is possible that a similar vaccine will be used for pigs, cattle, and eventually humans.

➤ Figure 26.2 **Meat inspection at a slaughterhouse.** Inspection helps provide, but cannot guarantee, a safe supply of meat for human consumption.

Beware of the Bad Egg!

Salmonella species are found in many animal reservoirs, especially poultry. Recent investigations of *Salmonella* in the poultry industry were not reassuring, and consumers can assume that poultry products will be heavily contaminated. What steps can you take to maneuver defensively in the sea of *Salmonella?*

First, assume that the shells of all eggs are contaminated, and wash your hands after handling them. (But do not wash the eggs—washing removes a protective surface coating that helps prevent microbes from entering the eggs.) *Salmonella* are part of the intestinal microflora of poultry, and eggs surely come in contact with feces, feathers, and contaminated surfaces. Cracked eggs cannot be sold for human consumption, but in some areas they can be sold as pet food. When organisms enter an egg, they find a superb, nutrient-rich culture medium in which they can rapidly multiply.

Does it make a difference whether you refrigerate eggs with the big end or the little end up? Big end up is the recommendation of food scientists. Why? The object is to keep the egg yolk and any embryo residing on it as close to the center of the egg as possible. This maximizes the distance an invading microbe would have to swim from the shell to the yolk. The egg white is filled with chemical hazards for the bacteria. Lysozymes attack the bacterial cell walls, killing many bacteria by rupturing their walls. Nutrients, vitamins, and metal ions of iron, copper, and zinc are wrapped tightly by proteins and other substances in the egg white and so are unavailable to the bacteria. Deprived of nutrients and chewed up by lysozymes, bacteria do not survive the swim; the embryo remains safe.

But why big end up? The natural tendency of the lipid-rich yolk is to rise, just as oil floats to the top of water. Bird eggs have two ropy cords (look for them next time you crack a raw egg) called chalaza, which act much like the ropes of a hammock to suspend the yolk from the shell lining. The larger chalaza is at the small end of the egg and can hold the yolk down better, preventing it from rising too close to the edge of the protective egg white.

Additionally, when baking, control your urge to lick raw batter if you've added any eggs to it. If the batter contains eggs, even powdered eggs in mixes, it might contain *Salmonella*. Pieces of shell dropped into batter, even when promptly retrieved, can inoculate it with organisms. Also, raw eggs can be contaminated internally if laid by an infected hen or if the eggs have undetected cracks or have been immersed in water.

Finally, handle raw poultry carefully. Poultry is naturally contaminated with fecal organisms, and contamination is exacerbated by the industry's practice of holding poultry in water baths during certain phases of the plucking and evisceration processes. The water becomes a *Salmonella* broth, and the surfaces of the birds become uniformly contaminated. Thus, use poultry promptly and discard contaminated wrappings and juices carefully. If you have laid poultry on a countertop or cutting board, scrub those surfaces thoroughly with hot soapy water before placing other foods on them. And don't touch other foods until you have washed your hands thoroughly.

Contaminated eggs being destroyed.

You might suppose that eggs, with their hard shells, would be free of microbial contamination. Most are, but the shells are porous, and pseudomonads and some other bacteria, as well as fungi such as *Penicillium, Cladosporium,* and *Sporotrichum,* grow on eggshells. These microbes can pass through pores in the shells to infect the inside of eggs. *Salmonella* also survives on eggshells and can enter broken eggs or be deposited with bits of shell in foods. Hens infected with *S. pullorum* lay infected eggs. The CDC reports that 1 out of every 10,000 eggs has *Salmonella* inside the shell. Any pathogens on eggshells or in eggs can be transmitted to humans unless eggs and foods that contain them are cooked thoroughly.

Fish and Shellfish

Fresh fish abound with microorganisms. Several species of enteric bacteria and clostridia, enteroviruses, and parasitic worms are commonly found on fresh fish. Many of these organisms survive shipment of fish packed in crushed ice, especially if the fish are packed too tightly together or are pressed against the slats of contaminated crates.

Fish contaminated with *Photobacterium phosphoreum* will glow. Although there have been no reports of this bacterium causing illness, it does indicate that the seafood is not fresh.

Shellfish, such as oysters and clams, carry many of the same organisms as fish. Raw oysters typically carry *Salmonella typhimurium* and sometimes *Vibrio cholerae*. Clams are especially likely sources of human infection because they are filter feeders—that is, they obtain food by filtering water and extracting microbes. If clams are exposed to increased levels of sewage, red tides, and other sources of large numbers of pathogens or toxin producers, clamming may be prohibited until the numbers of organisms decrease. Scallops are less likely to transmit diseases to humans because only the muscular part of the organism, and not its digestive tract, is eaten.

Among crustaceans, shrimp are extremely likely to be contaminated. Some studies have shown that over half the breaded shrimp on the market contain in excess of 1 million bacteria (and more than 5000 coliforms) per gram. Such high bacterial counts probably result from growth of bacteria during processing before the shrimp are frozen. Lobsters and crabs are even more perishable than shrimp and can carry a variety of enteric pathogens. Along the U.S. Gulf Coast, improperly cooked crabs have transmitted cholera. Crabs also carry *Clostridium botulinum* ➡ (Chapter 24, p. 712) and the pathogenic fungi *Cryptococcus* and *Candida*. Keep in mind, however, that the mere presence of microorganisms in seafoods—or in any other foods—does not necessarily mean that the foods are spoiled or contaminated with pathogens. In fact, *Lactobacillus bulgaricus,* which produces hydrogen peroxide, can be used to inhibit growth of other organisms found on seafood.

Milk

Modern mechanized milking and milk handling has greatly reduced the microbial content of raw milk (➤ Figure 26.3). However, breeding dairy cattle for increased milk production has resulted in exceptionally large udders and teats that easily admit bacteria. The first few milliliters of milk drawn from such cows can contain as many as 15,000 bacteria per milliliter, whereas the last milk drawn is free of microorganisms. Most microorganisms in freshly drawn milk are *Staphylococcus epidermidis* and *Micrococcus,* but *Pseudomonas, Flavobacterium, Erwinia,* and some fungi also can be present.

Microorganisms have many opportunities to enter milk before it is consumed. Hand milking, as opposed to mechanical milking, allows organisms from the body of the cow to enter the milk. These include *Escherichia coli,* which give milk a fecal flavor, and *Acinetobacter johnsoni* (formerly known as *Alcaligenes viscolactis*), which is especially abundant during summer months and causes a viscous slime to form in milk. Storing, transporting, and processing milk allow for contamination with any organisms in containers and for growth of those already present. Infectious organisms in milk usually come either from infected cows or from unsanitary practices of milk handlers. Diseased cattle can transmit *Mycobacterium bovis* and *Brucella* species to their milk. Dairy herds are tested for tuberculosis (with infected animals removed from the herd) and vaccinated against brucellosis (undulant fever); hence the risk of transmitting these diseases to humans is small. *Staphylococcus aureus, Salmonella* species, and other enteric bacteria can enter milk through unsanitary handling. Bacteria and molds will grow even in dried milk, which is made from pasteurized liquid milk but is unsterilized, if the water content of the powder reaches 10 percent.

➤ Figure 26.3   **A merry-go-round milking parlor.** Mechanized procedures for milking and milk handling have greatly reduced the microbial content of raw milk.

Some microorganisms, such as certain *Pseudomonas* species and some soil organisms, grow in refrigerated milk. These organisms are psychrophilic; although they normally grow at higher temperatures, they can grow at 5°C (refrigerator temperature). They also survive the concentration of chlorine normally used to purify drinking water.

Organisms that sour milk include *Streptococcus lactis* and species of *Lactobacillus*. When these microbes release enough lactic acid to bring the pH below 4.8, the milk proteins coagulate, and the milk is said to have soured. Souring of milk does not mean that the milk is unsafe for human consumption, but it does greatly alter the taste and appearance of the milk.

Various microorganisms that are involved in the spoilage of food are summarized in Table 26.1.

Other Edible Substances

Humans consume sugar, spices, condiments, tea, coffee, and cocoa—all of which are subject to microbial contamination—in addition to major nutrients. Fresh, dry, refined sugar is sterile, but raw sugar-cane juice supports growth of fungi such as *Aspergillus, Saccharomyces,* and *Candida* and several species of bacteria, including *Bacillus* and *Micrococcus*. Most are removed by filtration, and the remainder are killed by heat during evaporation of the juice. Foods to which sugar is added are especially susceptible to spoilage because the sugar is an excellent nutrient for many organisms. *Bacillus stearothermophilus,* a facultative anaerobe that grows best at 55° to 60°C, can multiply rapidly during processing of foods. Another thermophilic anaerobe, *Clostridium thermosaccharolyticum,* is often responsible for the gas production that causes cans to bulge. Conversely, sugar in high concentration acts as a preservative. The high sugar

> *Lactobacillus acidophilus adds lactic acid to milk, making that milk digestible by lactose-intolerant people.*

Bon Bon Bombs

Candy is generally safe from microbial contamination because its high sugar concentration creates an osmotic pressure too high for most organisms to survive. Occasionally, the cream filling in a batch of chocolates becomes contaminated with *Salmonella* or *Clostridium*. Chocolate lovers have been surprised to discover that the gas produced by clostridia has caused their favorite chocolates to explode.

➤ Table 26.1 Microorganisms involved in food spoilage

Food	Organism	Type of spoilage
Grains		
Bread	*Rhizopus nigricans*	Bread mold
	Penicillium species	
	Aspergillus species	
	Monilia sitophilia	
	Bacillus species	Protein and starch hydrolysis—stringy texture
Fruits and vegetables		
Leafy vegetables	*Erwinia carotovora*	Bacterial soft rot
Potatoes	*Phytophthora infestans*	Potato blight
Tomatoes, melons	*Fusarium* species	Soft rot; cracking skin
Grapes, berries, "stone" fruit	*Monilia fructicola*	Brown rot
Apples	*Penicillium expansum*	Patulin, a toxic cider contaminant
Meat		
	Rhizopus species	"Whiskers," white fluffy growth
	Mucor species	
	Pseudomonas	Green discoloration on refrigerated meat
	Clostridium species	Bone stink
Poultry		
	Pseudomonads	Sliminess
	Penicillium	Eggshell contamination
	Cladosporium species	
	Sporotrichum species	
Milk		
	Acinetobacter johnsoni	Sliminess
	Streptococcus lactis	Sour milk
	Lactobacillus species	

One Lump or Two?

When the coffee plantations of Ceylon (now Sri Lanka) were destroyed in the 1860s by coffee rust, the fields were replanted with tea. The British, who had depended on Ceylon for their coffee supplies, were forced to switch to tea. Thus, the lowly fungus played a major role in changing the British into a nation of tea drinkers. Two decades earlier, the Irish had not been as lucky. When a fungal disease destroyed the potato crop in the late 1840s, no replacement capable of feeding the population was available. A million people starved to death, and over a million emigrated.

✓ What organisms are most likely to contaminate milk?

✓ What dangers does honey present?

✓ What is bone stink?

---

concentration in such foods as jellies, jams, candies, and candied fruits creates sufficient osmotic pressure to inhibit microbial growth.

Maple trees are tapped, and sap is collected from them, in the early spring. The sap becomes increasingly contaminated as the weather becomes warmer. Organisms that thrive in maple sugar sap include species of *Leuconostoc, Pseudomonas,* and *Enterobacter*. Although these organisms may consume large quantities of sugar, they are killed as the sap is evaporated to syrup or sugar.

Honey can contain toxins if it is made of nectars from such plants as *Rhododendron* or *Datura*. It can also contain spores of *Clostridium botulinum*. Although the spores do not germinate in the honey, in infants they can germinate after the honey is ingested. Their toxins can cause "floppy baby syndrome."

Spices have been used in food preservation and embalming for centuries and have thereby gained a reputation as antimicrobials. This reputation is undeserved; spices more often mask odors of putrefaction than prevent spoilage. Leeuwenhoek, who first observed bacteria in spices, reported that water containing whole peppercorns teemed with them. Among the many microorganisms found in spices (Table 26.2), most are not pathogens. The small amounts of spices used in cooking are probably not a health hazard.

Condiments such as salad dressings, catsup, pickles, and mustard usually are markedly acidic. Although low pH prevents growth of many microorganisms, some molds are able to grow in these foods if they are not refrigerated.

Americans consume huge quantities of carbonated beverages and coffee and lesser quantities of tea and cocoa. The automated equipment in modern manufacturing plants can prepare carbonated beverages aseptically, but syrups can become contaminated with molds if there is a mechanical failure. Syrups sold in bulk to restaurants also can become contaminated. Freshly harvested coffee beans are subject to contamination by various molds and by microorganisms transmitted to them by insects. The coffee rust *Hemileia vastatrix,* a fungus, has devastated coffee plantations in Asia and is now a serious problem in South America. Tea leaves allowed to become moist are susceptible to contamination by *Aspergillus* and *Penicillium* molds, which impart an unpleasant aroma to tea brewed from them.

Microorganisms are useful in preparing coffee and cacao beans for marketing. The bacterium *Erwinia dissolvens* is used to digest pectin in outer coverings of coffee beans. Other bacteria are used to dissolve the covering of cacao beans, the beans from which cocoa and chocolate are made. These beans are subsequently treated with fermenting bacteria before they are roasted.

Preventing Disease Transmission and Food Spoilage

Diseases acquired from food are due mainly to the direct effects of microorganisms or their toxins (Table 26.3), but they can result from microbial action on food substances. Industrialization has increased the spread of foodborne pathogens. Large processing plants provide opportunities for contamination of great quantities of food unless sanitation is strictly prac-

---

> ## Table 26.2 Numbers of bacteria in spices

Type of spice	Number of microorganisms per gram of dry sample				
	Total aerobes	Coliforms	Yeasts and molds	Aerobic spores	Anaerobic spores
Bay leaves	520,000	0	3,300	9,200	<2
Cloves	3,000	0	18,005	<2	<2
Curry	>7,500,000	0	70	>240,000	>240,000
Marjoram	370,000	0	18,000	54,000	>24,000
Paprika	>5,500,000	600	2,300	>240,000	>620
Pepper	>2,000,000	0	15	>240,000	>24,000
Sage	6,800	0	10	7	>1,700
Thyme	1,900,000	0	11,000	160,000	>24,000
Turmeric	1,300,000	50	70	>110,000	>240,430

Source: Adapted from Karlson and Gunderson, 1965, as cited in *Food Technology* 1986, with permission from the Institute of Food Technologies.

> Table 26.3 | **Pathogenic organisms transmitted in food and milk**

Organism	Disease	Vector
Staphylococcus aureus	Food poisoning	Infected food handlers, unrefrigerated foods, milk from infected cows
Clostridium perfringens	Food poisoning	Unrefrigerated foods
Bacillus cereus	Food poisoning	Unrefrigerated foods
Clostridium botulinum	Botulism	Inadequately processed canned goods
Salmonella species	Salmonellosis	Infected food handlers, poor sanitation, contaminated seafood
Shigella species	Shigellosis	Infected food handlers, poor sanitation
Enteropathogenic *Escherichia coli*	Traveler's diarrhea and other diseases	Infected food handlers (sometimes asymptomatic), poor sanitation, contaminated meat
Campylobacter	Gastroenteritis	Undercooked chicken and raw milk
Vibrio cholerae	Cholera	Poor sanitation
Vibrio parahaemolyticus	Asian food poisoning	Undercooked fish and shellfish

ticed. In institutions that feed large numbers of people, contaminated food will cause many cases of disease. The increased popularity of convenience foods, especially fast foods, has also raised the risk of infection.

In addition to the enteric diseases described in Chapter 19, several other diseases can be transmitted in food. *Klebsiella pneumoniae* is commonly found in the human digestive tract. Although it is thought of mainly as a pathogen of the respiratory system, it can cause diarrhea in infants, abscesses, and nosocomial wound and urinary tract infections. Tuberculosis can be transmitted by fomites in food, unpasteurized milk and cheese, and meats from infected animals.

Several diseases can be transmitted to humans who eat infected meat. They include anthrax, brucellosis, Q fever, and listeriosis. Because of meat inspection procedures, animal and meat handlers are exposed to these diseases much more often than are consumers. Adirondack disease, caused by *Yersinia enterocolitica,* is thought to be transmitted by eating infected meat, but it also may be transmitted through milk and water. The disease has many forms, including mild to severe gastroenteritis, arthritis, glomerulonephritis, a fatal typhoidlike septicemia, and fatal ileitis, an inflammation of the small intestine that can be mistaken for appendicitis. People have also become infected with *Erysipelothrix rhusiopathiae* by eating infected pork. The resulting disease—called erysipelas (er-i-sip′e-las) in animals and erysipeloid in humans—infects swine, sheep, and turkeys. It is most likely to infect farmers and packing plant workers through skin wounds. The organism can infect skin, joints, and the respiratory tract.

Viruses are frequently transmitted through food. Enteroviruses are often spread during unsanitary handling of food, especially by asymptomatic food handlers. Droplet infection of food can transmit echovirus and coxsackie respiratory infections. Poliomyelitis viruses can be transmitted through milk and other foods. And the viruses responsible for hepatitis A can be transmitted through shellfish from contaminated waters. The virus that causes lymphocytic choriomeningitis, a flulike illness, can be spread to foods by mice.

Milk is an ideal medium for the growth of many pathogens. In addition to toxin producers and pathogens found in other foods, milk can contain organisms from cows. These organisms include *Mycobacterium bovis, Brucella* species, *Listeria monocytogenes,* and *Coxiella burnetii.* Spore-producing organisms such as *Bacillus anthracis* enter milk from infected cows or from soil. But milk also contains certain antibacterial substances, including lysozyme, agglutinins, leukocytes, and lactenin. Lactenin is a combination of thiocyanate, lactoperoxidase, and hydrogen peroxide. It is also present in human milk and other body secretions and may help prevent enteric infections in newborns. Fermented milk contains bacteria such as *Leuconostoc cremoris* that kill pathogens. A crucial factor in preventing spoilage and disease transmission in food and milk is cleanliness in handling. Other common-sense practices—prompt use of fresh foods, careful refrigeration, and prompt and adequate processing of foods to be stored—also help control disease transmission and spoilage.

Food Preservation

Many procedures used to preserve foods are based on practices begun early in human civilization. The ability to maintain a stable year-round food supply was essential in allowing humans to abandon a nomadic lifestyle, which followed the food supply, for a more settled village lifestyle. These practices were most likely based on simple observations like the following: Grain that was kept dry did not turn moldy. Dried and salted foods remained edible over a long period of time. And milk allowed to sour or made into cheese was usable over a much longer period of time than was fresh milk. Modern methods of food and milk preservation still utilize some of these early methods, but they also use heat and cold and other specialized procedures.

Many of the methods of food preservation have been described under antimicrobial physical agents in Chapter 12. They include canning using moist heat; refrigeration, freezing; lyophilization and drying; and the use of radiation. A number of chemical food additives also are used to retard spoilage.

Canning

The most common method of food preservation is **canning**—the use of moist heat under pressure. This method, analogous to laboratory autoclaving, is used to preserve fruits, vegetables, and meats in metal cans or glass jars (➤ Figure 26.4). If properly carried out, canning destroys all harmful spoilage microorganisms, including most heat-resistant endospores, prevents spoilage, and averts any hazard of disease transmission. Foods so treated may remain edible for years.

Some thermophilic anaerobic endospores, such as those of *Bacillus stearothermophilus,* may remain alive even after commercial canning. Hence canned food should not be stored in a hot environment, such as a car trunk. At elevated temperatures, the endospores can germinate, grow, and cause spoilage. Usually this produces gas, which causes the ends of cans to bulge so that the ends can be pressed up and down. Such spoilage also usually produces acid, which gives a sour flavor. Such changes are called **thermophilic anaerobic spoilage.** In some cases, however, spoilage due to growth of such spores does not cause cans to bulge with gas; such spoilage is called **flat sour spoilage.** Cans can also bulge due to **mesophilic spoilage,** which occurs when canning procedures were improperly followed or the seal has been broken. This type of spoilage can take place at room temperature, unlike flat sour and thermophilic spoilage, which occur only in properly processed and sealed cans that have been stored at high temperatures.

Because of the danger of botulism and other kinds of spoilage in improperly processed canned foods, people who do home canning should carefully follow instructions in a good,

Home Canning

Clostridium botulinum spores are present on most fresh food surfaces. The bacterial spores can be killed with a combination of acid and heat. It may be possible to acidify certain foods by the addition of lemon juice, citric acid, or vinegar. Low-acid foods, not safely processed with boiling water, must be sterilized using heat and pressurized steam. All low-acid foods should be sterilized at temperatures between 240°F to 250°F, attainable with a pressure canner operated at 10 to 15 PSIG. The time needed to sterilize the food depends on the kind of food being canned, the way it is packed in jars and the size of the jars. In addition the time for processing your home preserves will depend on the altitude where you live. Therefore it is important to carefully follow all directions.

➤ Figure 26.4 **Commercial canning equipment.** Canning destroys all microbes and their spores, thereby preventing food spoilage and disease transmission.

up-to-date canning handbook. The U.S. Department of Agriculture (USDA) now recommends that all low-acid foods be processed by pressure cooking. The USDA also recommends that jellies and jams, formerly packed hot and sealed with wax, be processed in a boiling water bath and sealed with lids like other canned goods to prevent accumulation of mold toxins. (See the box Home Canning in Chapter 12.) ➡ (p. 331) *Any can that has bulging ends—whether produced at home or commercially—should be discarded.*

Although many condiments are heat processed, some of their properties help preserve them. Sugar in jellies and jams increases osmotic pressure and retards growth of microorganisms. (Saccharin does not have this effect, so artificially sweetened products may require more stringent precautions to prevent spoilage than do those with a high sugar content.) Likewise, the high acidity of pickles and other sour foods helps prevent microbial growth.

Refrigeration and Freezing

Refrigeration at temperatures slightly above freezing (about 4°C) is suitable for preserving foods for only a few days. It does not prevent growth of psychrophilic organisms that can cause food poisoning. Freezing, another common method of food preservation, involves storage of foods at temperatures below freezing (about −10°C in most home freezers). All kinds of foods can be preserved by freezing for several months and some for much longer periods of time. An advantage of freezing is that it preserves the natural flavor of foods better than does canning. However, freezing has two disadvantages: (1) It causes some foods, especially fruits and watery vegetables, to become somewhat soft upon thawing and detracts from their appearance. (2) Although freezing prevents growth of most microorganisms, it does not destroy them. As soon as food begins to thaw, microorganisms begin to grow. In fact, freezing and thawing of foods actually promotes growth of microorganisms. Ice crystals puncture cell and plasma membranes and cell walls, allowing nutrients to escape from foods. These nutrients are then readily available to support microbial growth. Consequently, it is important never to thaw and refreeze foods.

Drying and Lyophilization

Drying (dessication, or dehydration) is one of the oldest methods used for food preservation. A certain level of water activity is necessary for microbial growth. Ideally, if more than 90 percent of the water is removed, the food can be stored in that state. Drying stops microbial growth but does not kill all microorganisms in or on food. Foods can be dehydrated by natural means, such as sun drying, or by artificial means by passing heated air over the food with controlled humidity. Addition of salt, high concentrations of sugar, or chemical preservatives that alter osmotic pressure and reduce water content are often included in the drying process.

Currently, **lyophilization** (freeze-drying) is employed in the food industry almost exclusively for the preparation of instant coffee and dry yeast for bread-making. (The technique is also used to preserve bacterial cultures; see Chapter 12.) ➡ (p. 331) Lyophilization involves drying frozen food in a vacuum. The process yields higher quality foods than are produced by ordinary drying methods.

Irradiation

Irradiation of food as a method of preservation is still new and quite controversial due to public concern about the hazards of radiation. There are two categories of radiation used to control microorganisms in food: nonionizing and ionizing.

Ultraviolet (UV) radiation, a form of *nonionizing radiation,* is limited by poor penetrating power. The radiation wavelength and exposure time also determine the effectiveness of this method of preservation. UV radiation is effective as a sanitizing agent for food-processing equipment and other surfaces. Microwaves, another form of nonionizing radiation, are useful for food preparation, cooking, and processing but not for preservation. Microwaves do not kill microorganisms directly, but the heat generated during cooking can be microbicidal. Uneven heating can interfere with this antimicrobial activity. Therefore, frequent rotation of food is necessary during microwaving.

Ionizing radiation, such as gamma rays, has great penetrating ability and is microbicidal. This type of radiation can be utilized either before or after packaging, depending on the foodstuff.

The Great Irradiation Debate: Are Gamma Rays Safe for Food?

What method of food preservation allows you to store onions sprout-free for up to three months, keeps refrigerated strawberries firm and fresh for three weeks, and virtually eliminates *Salmonella* from poultry? Irradiation—it may not be coming to a store near you. Despite its approval by the FDA and its endorsement by the World Health Organization, the American Medical Association, and even Julia Child, irradiation is practiced by only one major poultry company in the United States, and only one U.S. company is in the business of irradiating fruits and vegetables. Although advocates praise irradiation as an effective way to reduce bacterial and insect contamination of foods, critics charge that the practice may cause more harm than the organisms it destroys.

Irradiation is not a new technology; for over 20 years, selected food industries have used it to decontaminate foods. In 1963, the FDA approved irradiation for the elimination of mold from wheat and potatoes. In the 1980s, the FDA added dried spices, which attract insects, and pork, which carries the parasitic *Trichina* worm, to the list of foods approved for irradiation. In 1986, fruits and vegetables were approved, and poultry soon followed. The FDA has recently approved the irradiation of beef before it is shipped to markets.

The process of irradiating foods is simple. The food is passed on a conveyor belt through a lead-shielded, concrete chamber. Within the chamber, radioactive elements (usually cobalt-60 or cesium-137) bombard the food with gamma rays. The process does not make the food itself radioactive, any more than a dental X-ray makes your teeth radioactive. The gamma rays do, however, kill insects, molds, and bacteria and may slow the ripening of certain fruits and vegetables, extending their shelf life.

Proponents of irradiation think its time has come. They point to the frightening statistics gathered from poultry processing plants—that up to 60 percent of all birds are infected with *Salmonella* and that up to 100 percent are infected with *Campylobacter*. Both kinds of bacteria can cause serious gastrointestinal illnesses in humans if the poultry is not refrigerated or cooked properly. And although *Salmonella* and *Campylobacter* food poisonings are usually not fatal, other types are. In 1992, an outbreak of a deadly form of food poisoning caused by *Escherichia coli* strain 0157:H7 killed four children in the Pacific Northwest. The outbreak was traced to contaminated, undercooked beef served at a fast-food restaurant. Since then, 45 outbreaks of this kind of food poisoning have occurred throughout the United States. Supporters of irradiation say that the practice could have prevented many of these cases of food poisoning.

Tests show that irradiation is effective in destroying many unwelcome organisms. It kills 99.5 to 99.9 percent of *Salmonella* and *Campylobacter* bacteria in poultry, as well as fungi, parasites, and nonpathogenic bacteria that can cause food to spoil. And because irradiation kills insects, farmers would not have to use as many pesticides on their crops, reducing our exposure to these chemicals.

Opponents say that irradiation offers a false sense of security. Irradiation can't kill all the harmful bacteria that infect poultry and beef, and consumers who put their trust in irradiation may take short cuts when storing, handling, and cooking food. The food may look and smell fresh, but *Salmonella* and *Campylobacter*—which don't cause food to look or smell spoiled—may still be present in small amounts. In conditions favorable for bacterial growth, such as a warm kitchen in which a chicken is defrosting in the sink, these bacteria can multiply, with disastrous results. And irradiation doesn't touch toxins that many pathogenic

---

Gamma radiation from cobalt-60 or cesium-137 has been used in Japan and some European countries for food preservation for a number of years. The U.S. Food and Drug Administration (FDA) has declared such irradiation safe for preserving certain foods. Radiation has successfully been used to control microbial growth on fresh fish during transport to market and to kill insects in spices. It has also proved effective in reducing spoilage of fresh fruits and vegetables. Most recently the USDA proposed rules providing for the irradiation of fresh poultry. The public has much skepticism about irradiated foods. It should be emphasized that irradiation kills microorganisms but does not cause the food itself to become radioactive.

Chemical Additives

A large number of chemical compounds are added to various foods to kill microorganisms or retard their growth. We will describe a few examples and their uses.

Organic acids, some of which occur naturally in some foods, lower the pH of foods enough to prevent growth of human pathogens and toxin-producing bacteria. Acids such as benzoic, sorbic, and propionic inhibit growth of yeast and other fungi in margarine, fruit juices, breads, and other baked goods.

bacteria may have already deposited in the food. In many types of food poisoning, the toxins, not the bacteria, are what cause the nasty symptoms. Moreover, the most deadly food pathogen of all, *Clostridium botulinum,* is impervious to radiation.

These aren't the only drawbacks to irradiation, critics contend. Irradiation may create harmful chemicals in food. When gamma rays strike tissues, they disrupt chemical bonds. Living cells, such as bacteria, die, but some chemical bonds re-form in new configurations, creating chemicals that were not present before irradiation. Researchers have found minute amounts of formaldehyde and benzene—both cancer-causing chemicals—in samples of irradiated food. Irradiation supporters counter that many foods, such as eggs, naturally contain these chemicals and that some cooking techniques generate these chemicals in other foods. But along with benzene and formaldehyde, *unique radiolytic products (URPs)* form when the protein, fat, and carbohydrate molecules are rearranged during irradiation. It is not known whether these products are cancer-causing. Irradiation advocates contend that URPs are not new to preservation methods; they've only recently been identified by more sensitive detection techniques.

Another criticism is that irradiation depletes nutrients in food. The amounts of vitamins E, A, and C and the B vitamin thiamin are reduced during the irradiation process, sometimes by as much as 9 percent. Critics charge that the extension of shelf life irradiation promises can't make up for the loss of essential nutrients.

Some people think that irradiated food tastes "funny." One taste tester commented that irradiated poultry has the taste of "burnt hair." The "off" taste may result from the oxidation of fat, which produces by-products that create less-than-fresh flavors and odors. The same process occurs when food is kept too long in the freezer. Other tasters, however, could detect no difference in taste between irradiated foods and nonirradiated foods.

Are irradiated foods coming to your grocer's shelves soon? Not likely. Most poultry producers continue to rely on a disinfectant bath to rid their chickens of bacteria. A major spice manufacturer did irradiate its spices for a short time in the 1970s but has since reverted to chemical methods of decontamination. Until the perceived dangers are thoroughly studied and fears put to rest, Americans will continue to shun irradiated food.

Radiation can be used to sterilize food, as these two boxes of strawberries demonstrate. Irradiating food does not make it radioactive; it is safe to eat.

Alkylating agents such as ethylene oxide and propylene oxide are used only in nuts and spices. Sulfur dioxide, which is most effective at an acidic pH, is used in the United States only to bleach dried fruits and to eliminate bacteria and undesired yeasts in wineries. Ozone, a highly reactive form of oxygen, is used to kill coliform bacteria in shellfish and to treat water used in beverages. It has the advantage of leaving no residue. The disadvantages of ozone are that it tends to give foods a rancid taste by oxidizing fats, and it can damage molecules, especially lung lysozyme, if inhaled.

Sodium chloride, perhaps one of the first food additives, increases osmotic pressure in foods, keeping most microorganisms from growing. Salt used to cure meats is especially useful in preventing growth of clostridia deep in tissues, although fungi eventually grow on the surfaces of salted foods. Salt dehydrates bacteria and makes it difficult for them to take in water and nutrients. A recent discovery suggests that salt, in addition to increasing osmotic pressure, may create electric charges on the surface of meats, preventing bacteria from adhering to the surfaces.

Still other chemical additives have special uses. Halogen compounds, such as sodium hypochlorite, disinfect water and food surfaces. Gaseous chlorine prevents growth of microorganisms on food-processing equipment. Nitrates and nitrites suppress microbial growth in meats, especially ground meats and cold cuts. During cooking, however, they can be converted to nitrosamines, which are carcinogenic and toxic to the liver. We continue to use nitrates and nitrites because we have no good alternatives, particularly for sausage. The name botulism comes from the Latin word for sausage, so common was food poisoning from sausage in the days before the use of nitrites. Another reason nitrites are used is to keep colors bright, especially the red color of fresh meat. Carbon dioxide kills microorganisms, except for some fungi, in carbonated beverages. It also retards maturation of fruits and decreases spoilage during shipment. Finally, quaternary ammonium compounds (quats) can be used to sanitize many objects—utensils, cows' udders, fresh vegetables, and the surfaces of eggshells, which they do not penetrate.

Antibiotics

Antibiotics are added to foods and milk in some countries. In the United States only the anticlostridial agent *nisin,* a bacteriocin produced naturally during fermentation of milk by *Streptococcus lactis,* may be used. Antibiotic use in foods and milk is prohibited for the following good reasons:

- The antibiotics might be relied on instead of good sanitation.
- Pathogenic microorganisms might develop resistance to the antibiotics, so treatment of the diseases they cause would become difficult or impossible.
- Humans might be sensitized to the antibiotics and subsequently suffer allergic reactions.
- The antibiotics might interfere with the activities of microorganisms essential for fermenting milk and making cheese.

Pasteurization of Milk

Prevention of spoilage and disease transmission in milk begins with the maintenance of health in both dairy animals and in milk handlers. In the past, bovine tuberculosis was sometimes transmitted to humans from cows' milk. Many children were infected early in life and usually died by age 15. Establishment of mandatory testing of dairy herds for tuberculosis every 3 years (the time it takes an infection to progress to a transmissible stage) has greatly decreased incidence of the disease in the United States.

Milk is collected under clean, but not sterile, conditions and is usually subjected to pasteurization. Two methods of pasteurization are currently in use:

- In **high-temperature short-time (HTST) pasteurization,** or **flash pasteurization,** milk is heated to 71.6°C for at least 15 seconds.
- In **low-temperature long-time (LTLT) pasteurization,** or the **holding method,** milk is heated to 62.9°C for at least 30 minutes.

Both methods destroy vegetative cells of pathogens likely to be found in milk and decrease the numbers of organisms that can cause souring. Following pasteurization, milk is quickly cooled and refrigerated in sealed containers until it is used.

Milk can be preserved and kept safe to drink by means other than pasteurization. In Europe and increasingly in parts of the United States, milk may be sterilized by **ultra-high temperature (UHT) treatment** rather than simple pasteurization. UHT milk is heated to 87.8°C for 3 seconds. Such milk can be kept in sealed paper containers—called aseptic packaging—and stay unrefrigerated for about 6 months. Milk can also be preserved as canned condensed milk, which also is sterilized. It is reconstituted by adding an equal volume of water. Although sterilized milk is completely free of microorganisms, the heat treatment necessary to render it sterile alters its flavor.

Various chemical additives are sometimes used in milk. Addition of hydrogen peroxide to milk reduces the temperature required to destroy most pathogens. However, it fails to kill mycobacteria, and its use in milk has been banned in the United States for this reason.

Standards for Food and Milk Production

Because food and milk production are carefully regulated by federal, state, and local laws, consumers are better protected in the United States than in many other countries. Despite regulations, some hazards remain because of the use of food additives or heat treatment. The FDA regulates inspection of meat and poultry, accurate labeling, and quality standards for products being shipped across state lines. Other similar regulations are imposed by many state and local agencies within their own jurisdictions. And the USDA is now proposing that meat and poultry be inspected by microscope—a much stricter testing requirement than has ever been in place (➤ Figure 26.5).

Many food producers maintain quality checks on their own products. In canning factories, for example, counts of microorganisms in samples of foods are made during processing in an effort to minimize the number of organisms in the foods. Milk, because it is an extremely good growth medium for microorganisms, is subjected to several tests (Table 26.4). The use of these tests virtually guarantees high-quality milk to consumers.

➤ Figure 26.5   **Testing for trichinosis.** This new blood test kit can detect trichinosis infection in pigs.

Microorganisms as Food and in Food Production

Algae, Fungi, and Bacteria as Food

The rapid rise in world population is greatly increasing the demand for human food. Given present birthrates, the earth's population is expected to double in about 40 years, from 5 billion to 10 billion. To put these large numbers in perspective, consider that the world population increases by about 156 people per minute, 225,000 per day, or 9 million (the population of New York City) every 40 days. Even now, 25,000 people in developing countries die of starvation every day, and many more suffer from malnutrition. Clearly, this situation calls for expanding the human food supply.

Among microorganisms, yeasts show great promise for increasing our food supplies. Yeasts are a good source of protein and vitamins, and they can be grown on a variety of waste materials—grain husks, corncobs, citrus rinds, paper, and sewage. Each kilogram of yeast introduced into one of these media can produce 100 kg of protein—1000 times that obtained from 1 kg of soybeans and 100,000 times that obtained from 1 kg of beef. Yeast food products have been manufactured in Africa, Australia, Puerto Rico, Hawaii, Florida, and Wisconsin. Taiwan makes about 73,000 tons of yeast food per year, most of which is shipped to the United States for adding to processed foods. Dried yeast is sold as a nutritional supplement in health-food stores. Growing yeast on wastes could increase the food supply in the United States by 50 to 100 percent, but it has some drawbacks. Expensive equipment is required to begin production, and, more important, a way must be found to make people accept yeast as a desirable food. At

➤ Table 26.4	Tests for determining milk quality	
Test	**Description**	**Purpose and significance**
Phosphatase test	Detects the presence of phosphatase, an enzyme destroyed during pasteurization.	To determine whether adequate heat was used during pasteurization. If active phosphatase remains, pathogens also might be present.
Reductase test	Indirectly measures the number of bacteria in milk. The rate at which methylene blue is reduced to its colorless form is directly proportional to the number of bacteria in a milk sample.	To estimate the number of bacteria in a milk sample. High-quality milk contains so few bacteria that a standard concentration of methylene blue will not be reduced in $5\frac{1}{2}$ hours. Low-quality milk has so many that methylene blue is reduced in 2 hours or less.
Standard plate count	Directly measures viable bacteria. Diluted milk is mixed with nutrient agar and incubated 48 hours; colonies are counted, and the number of bacteria in original sample is calculated.	To determine the number of bacteria in a milk sample. The number per milliliter must not exceed 100,000 in raw milk before pooled with other milk or 20,000 after pasteurization.
Test for coliforms	Same as the test used for water (see Chapter 25).	To determine whether coliforms are present. A positive coliform test indicates contamination with fecal material.
Test for pathogens	Detects the presence of pathogens. Methods depend on the suspected pathogens.	To identify pathogens. Usually not required but can help locate the source of infectious agents that may appear in milk.

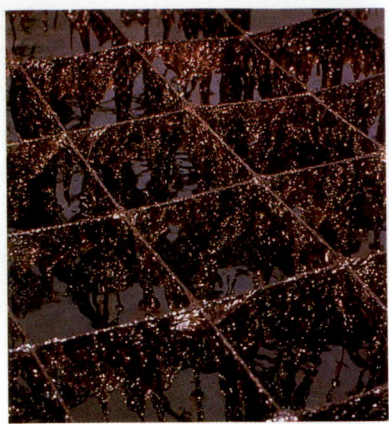

> **Figure 26.6 Increasing food production by algal culture.** In Asia, many red algae are grown in mariculture farms (agriculture in sea water). Nori, the dried sheets of pressed red algae used to wrap some sushi rolls, are produced in this way.

present, yeast is used chiefly in animal feeds. Only small amounts of single-celled foodstuffs (yeast or algal) can be handled well by the human digestive tract. Large quantities of nucleic acid derivatives can aggravate gout; therefore, yeast must be processed to reduce their levels.

Algal culture is another promising avenue for increasing human food supplies (➤ Figure 26.6). Algae such as *Scenedesmus* and *Chlorella* have been cultivated in Asia, Israel, Central America, several European countries, and even the western United States. Algae have also been used as ingredients in ice cream (as well as in nonfood consumer products such as diapers and cosmetics).

The use of algae as human food shortens the food chain. In other words, if humans eat algae directly rather than eating fish that have been nourished on algae, the algae will feed more people than will the fish. Approximately 100,000 kg of algae are required to make 1 kg of fish. Each acre of pond used to cultivate algae can produce 40 tons of dried algae—40 times the protein per acre from soybeans and 160 times that from beef. However, algal culture has so far proved economically feasible only in urban areas where large quantities of treated sewage on which to grow the algae are available. In addition to the problem of getting people to accept algal products as food, growing algae on sewage creates a potential health hazard because the products may contain viral pathogens.

Even some bacteria are used as food. The cyanobacterium *Spirulina* has been grown for centuries in alkaline lakes in Africa, Mexico, and by the Incas in Peru. The cyanobacteria are harvested, sun-dried, washed to remove sand, and made into cakes for human consumption.

If technical difficulties could be overcome and the products made acceptable as human food, yeasts, algae, and some bacteria could increase the world food supply. But, at best, the use of microorganisms as food can buy only a little time to allow humans to control their own numbers.

Food Production

The use of microorganisms in making bread, cheese, and wine is as old as civilization itself. Long before any microorganisms were identified, milk was being made into cheese and fermented beverages, and bread was being leavened by microbes. In modern food production, specific organisms are purposely used to make a variety of foods.

Bread

In bread, yeast is used as a **leavening agent**—that is, to produce gas that makes the dough rise. A particular strain of *Saccharomyces cerevisiae* is added to a mixture of flour, water, salt, sugar, and shortening. The mixture is allowed to ferment at about 25°C for several hours. During fermentation, yeast cells produce a little alcohol and much carbon dioxide. As the carbon dioxide bubbles become trapped in the dough, they cause the dough to increase in bulk and acquire a lighter, finer texture. When the dough is baked, the alcohol and carbon dioxide are driven off. The bread becomes light and porous because of spaces created by carbon dioxide bubbles. Home bakers often use activated dry yeast, a product that is prepared by the lyophilization of yeast cells.

A genetically engineered strain of fast-acting yeast cuts the rising time of most breads in half.

Dairy Products

Microorganisms are used in making a wide variety of dairy products. Cultured buttermilk, popular in the United States, is made by adding *Streptococcus cremoris* to pasteurized skim milk and allowing fermentation to occur until the desired consistency, flavor, and acidity are reached. Other organisms—*Streptococcus lactis, S. diacetylactis* and *Leuconostoc citrovorum, L. cremoris,* or *L. dextranicum*—make buttermilk with different flavors because of variations in fermentation products. Sour cream is made by adding one of these organisms to cream. Yogurt is made by adding *Streptococcus thermophilus* and *Lactobacillus bulgaricus* to milk. These organisms release still other products, and so yogurt has a different texture and flavor.

Fermented milk beverages have been made for centuries in various countries, especially Eastern European countries, by adding specific organisms or groups of organisms to milk. The products vary in acidity and alcohol content (Table 26.5). Acidophilus milk is made by adding *Lactobacillus acidophilus* to sterile milk. Sterilization prevents uncontrolled fermentation by organisms that might already be present in nonsterilized milk. Bulgarian milk is made by

Next time you buy yogurt "with active cultures," examine a sample with your microscope to see masses of rod-shaped *Lactobacillus bulgaricus* cells.

> **Table 26.5** **Fermented milk beverages**

Characteristics	Beverages and countries where made
Less than 1% lactic acid	Sour cream, cultured buttermilk (United States)
	Filmjolk (Finland)
2–3% lactic acid	Yogurt (United States); called leben in Egypt, matzoon in Armenia, naja in Bulgaria, and dahi in India
	Tarho (Hungary)
	Kos (Albania)
	Fru-fru (Switzerland)
	Kaimac (Yugoslavia)
	Acidophilus milk (United States)
Alcoholic (1–3%)	Koumiss, kefir, and araka (former Soviet Union)
	Fuli and puma (Finland)
	Taette (Norway)
	Lang (Sweden)

L. bulgaricus; it is similar to buttermilk except that it is more acidic and lacks the flavor imparted by the leuconostocs. The Balkan product kefir is made from the milk of cows, goats, or sheep, and the fermentation is usually carried out in goatskin bags. The Russian product koumiss is made from mare's milk. In kefir and koumiss, *Streptococcus lactis, L. bulgaricus,* and yeasts are responsible for the production of lactic acid, alcohol, and other products. These products are usually made by continuous fermentation—fresh milk is added as fermented product is removed.

Cheeses

The first step in making almost any cheese is to add lactic acid bacteria and either **rennin** (an enzyme from calves' stomachs) or bacterial enzymes to milk. The bacteria sour the milk, and the enzymes coagulate the milk protein casein. The solid portion—the **curd**—is used to make cheese, and the liquid portion—the **whey**—is a waste product of this process (➢ Figure 26.7). Sometimes lactic acid is extracted from whey. In separating the curd and the whey, different amounts of moisture are removed according to the kind of cheese being made. For soft cheeses, the whey is simply allowed to drain from the curd; for harder cheeses, heat and pressure are used to extract more moisture. Nearly all cheeses are salted. Salting helps remove water, prevents growth of undesired microorganisms, and contributes to the flavor of the cheese.

A few cheeses, such as cream cheese, cottage cheese, and ricotta, are unripened, but most cheeses are ripened. Ripening involves the action of microorganisms on the curd after it is pressed into a particular form. Soft cheeses are ripened by the action of microorganisms that occur naturally or are inoculated onto the surface of the pressed curd. Because the enzymes that ripen the cheese must diffuse from the surface into the center of the cheese, soft cheeses are relatively small. In contrast, hard cheeses are ripened by the action of microorganisms distributed through the curd. Because microbial action does not depend on diffusion, these cheeses can be quite large (➢ Figure 26.8). Cheese is ripened by microorganisms in a cool, moist environment. Many modern factories have environmentally controlled rooms, but some still use natural caves similar to those where cheeses were first ripened.

Cheeses can be classified by their consistency (soft to hard), by the kind of microorganisms involved in the ripening process, and by the length of time required for ripening (Table 26.6). The ripening period is shorter for soft cheeses (1 to 5 months) than for hard cheeses (2 to 16 months).

(a)

(b)

(c)

➤ Figure 26.7   **Making gouda cheese.** (a) Lactobacilli and the enzyme rennin are added to pasteurized milk. The bacteria sour the milk, and the rennin coagulates the milk protein casein. (b) The milk curdles into the solid curd and the liquid whey. Curds are drained and placed into "hoops," which are then squeezed in a press. (c) The pressed cheeses are removed from their hoops and floated in a tank of brine (salt solution). The high salt concentration extracts even more moisture from the cheese by osmosis (recall Chapter 4) and thus hardens it. ➡ (p. 99) The salt also adds flavor to the cheese and prevents the growth of undesired microorganisms.

➤ Figure 26.8   **Ripening gouda cheese.** The ripening process involves the action of microorganisms on the curd after it is pressed into its form. A plastic coating is added to the cheese to keep it from drying out. The cheese is then placed on shelves in the aging room, where the microorganisms that are distributed throughout the curd will ripen it. Because gouda is a hard cheese, it must ripen for a relatively long period—3 months to a year—depending on the size of the wheel.

➤ Table 26.6 Classification of ripened cheeses

Consistency and ripening period	Examples	Organisms associated with ripening
Soft (1–2 months)	Limburger	*Streptococcus lactis, S. cremoris, Brevibacterium linens*
Soft (2–5 months)	Brie and Camembert	*S. lactis, S. cremoris, Penicillium camemberti,* and *P. candidum*
Semihard (1–8 months)	Muenster and Brick	*S. lactis, S. cremoris, Brevibacterium linens*
Semihard (2–12 months)	Roquefort and Blue	*S. lactis, S. cremoris,* and *P. roqueforti* or *P. glaucum*
Hard (3–12 months)	Cheddar and Colby	*S. lactis, S. cremoris, S. durans,* and *Lactobacillus casei*
	Edam and Gouda	*S. lactis* and *S. cremoris*
	Gruyere and Swiss	*S. lactis, S. thermophilus, S. helveticus, Propionibacterium shermani,* or *L. bulgaricus* and *P. freudenreichii*
Hard (12–16 months)	Parmesan	*S. lactis, S. cremoris, S. thermophilus,* and *L. bulgaricus*

Several microbial actions, such as decomposition of the curd and fermentation, occur during the ripening of cheeses. Prior to ripening, the curd consists of protein, lactose, and if the cheese is made from whole milk, fat. As microorganisms act on the curd, they first break down the lactose to lactic acid and other products, such as alcohols and volatile acids. Proteolytic enzymes break down protein, more extensively in soft than in hard cheese. *Brevibacterium linens* and the mold *Penicillium camemberti* are especially adept at releasing proteolytic enzymes. Lipase, particularly in *Penicillium roqueforti,* releases short-chain fatty acids such as butyric, caproic, and caprylic acids. These acids and their oxidation products contribute significantly to the flavor of the cheeses. The effects of fermentation in cheese ripening are most easily seen in Swiss cheese. Bacteria of the genus *Propionibacterium* ferment lactic acid and produce propionic acid, acetic acid, and carbon dioxide. The acids flavor the cheese, and the carbon dioxide, which becomes trapped in the curd, produces the characteristic holes in the cheese.

Other Products

A great variety of fermented foods and food products are made throughout the world; here we consider only a few. (Beers, wines, and spirits are discussed later.)

Vinegar   Vinegar is made from ethyl alcohol by the acetic acid bacterium, *Acetobacter aceti* (➤ Figure 26.9), which oxidize the alcohol to acetic acid. Commercially produced vinegar contains about 4 percent acetic acid. Cider vinegar is made from alcohol in fermented apple cider; wine vinegar is made from alcohol in wine.

Sauerkraut   Sauerkraut was first made in Europe in the sixteenth century. Bacteria naturally present on cabbage leaves act on shredded cabbage placed in layers in large crocks. Enough dry salt is placed between the layers to make a 2 percent to 3 percent salt solution as the salt draws water out of the cabbage. The cabbage is firmly packed and weighted down to create an anaerobic environment. Although many organisms cannot tolerate such an environment, anaerobic, halophilic species of *Lactobacillus* and *Leuconostoc* can carry out fermentation under these conditions. These microbes produce lactic acid, acetic acid, carbon dioxide, alcohol, and small amounts of other substances. After 2 to 4 weeks at room temperature, the cabbage is changed by fermentation to sauerkraut, which can be refrigerated until eaten or preserved by canning.

Pickles   Pickles are made by essentially the same process as that used to make sauerkraut. Fresh cucumbers, either whole, sliced, or ground, are packed in brine and allowed to ferment for a few days to a few weeks. *Leuconostoc mesenteroides* is the major fermenting organism in low-salt brines (less than 5 percent salt), whereas species of *Pediococcus* are more active in high-salt brines (5 percent or higher salt concentration). A problem in pickle-making is to prevent a yeast film from forming on the surface of vats. Direct sunlight and ultraviolet light help solve this problem. After fermentation, vinegar and spices are added to sour pickles; sugar also is added to sweet pickles. Most pickles are pasteurized. Some pickles, such as sweet pickles and pickle relish, are made without fermentation. After the pickles have been soaked in brine for a

➤ **Figure 26.9   Fermented foods: vinegar.** The manufacture of vinegar makes use of the bacterium *Acetobacter aceti* (400X). This organism oxidizes ethyl alcohol to acetic acid.

(a)

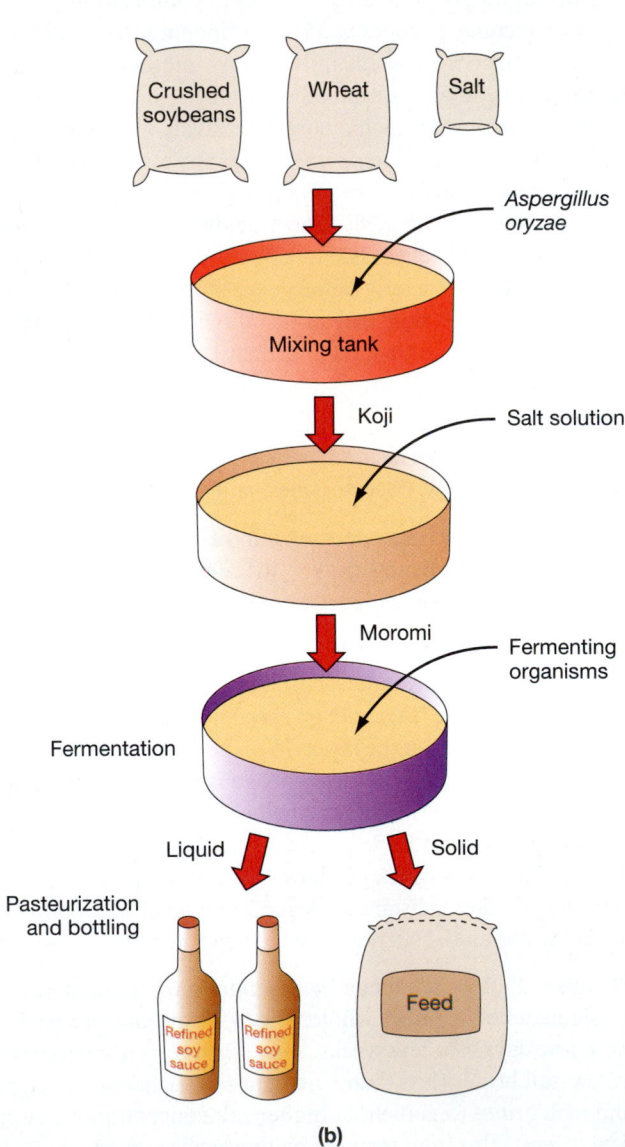

> Figure 26.10   **Making soy
sauce.** (a) The fermentation of
soy and wheat is carried out in
large steel tanks. (b) A diagram
of the soy fermentation process.

(b)

few hours, they are seasoned with vinegar and spices, heat-processed, and sealed. The entire operation can be done in a day in any kitchen.

Olives   Olives are treated with lye to hydrolyze oleuropein, a very bitter phenolic glucoside they contain that gives an undesirable flavor. Green olives are fermented in a 5 percent to 8 percent salt solution by *Leuconostoc mesenteroides* and *Lactobacillus plantarum*. They are packed in water and pasteurized. Ripe olives are picked when reddish in color but not fully ripe. They are oxidized with tannins to blacken them, fermented in dilute (less than 5 percent) salt solution, packed in water in cans, and processed at 116°C for 60 minutes.

Poi   Poi, a common food in the South Pacific, is made from ground roots of the taro plant. A paste of ground roots is allowed to ferment through the action of a succession of naturally occurring organisms. Pseudomonads and coliforms begin the fermentation process, and lactobacilli later take over. Yeasts add alcohol to the mixture.

Soy Sauce   Soy sauce is made in a stepwise process (➤ Figure 26.10). A salted mixture of crushed soybeans and wheat is treated with the mold *Aspergillus oryzae* to break down starch into fermentable glucose. The product, called *koji*, is mixed with an equal quantity of salt solution to make a mixture called moromi. Moromi is fermented for 8 to 12 months at low temperature and with occasional stirring. The fermenting organisms are mainly the bacterium *Pediococcus soyae* and the yeasts *Saccharomyces rouxii* and *Torulopsis* species. Lactic acid, other acids, and alcohol are produced. Upon completion of fermentation, the liquid and solid parts of the moromi are separated. The liquid part is bottled as soy sauce, and the solid part is sometimes used as animal food.

Soy Products   Other soy products include miso, tofu, and sufu. *Miso* is a fermented paste of soybeans made like soy sauce. *Tofu* is a soft curd of soybeans. It is made from soybeans ground to make a milk, which is boiled to inactivate enzymes. The curd is precipitated with calcium or magnesium sulfate and pressed into a soft mass similar in consistency to soft cheese. *Sufu* is made by action of a fungus on soy curd. Cubes of curd are dipped in a mixture of salt and citric acid, inoculated with *Mucor*, and incubated until coated with mycelia. The fungus-coated cubes are aged 6 weeks in a rice-wine brine.

Beer, Wine, and Spirits

Beer and wine are made by fermenting sugary juices; spirits, such as whiskey, gin, and rum, are made by fermenting juices and distilling the fermented product. **Distillation** separates alcohol and other volatile substances from solid and nonvolatile substances. Strains of *Saccharomyces* are the fermenters for all alcoholic beverages. Many different strains have been developed, each having distinctive characteristics. Both the organisms and how they are used are carefully guarded brewers' secrets.

To make beer, cereal grains (usually barley) are **malted** (partially germinated) to increase the concentration of starch-digesting enzymes that provide the sugar for fermentation (➤ Figure 26.11). Malted grain is crushed and mixed with hot water (about 65°C), producing **mash.** After a few hours, a liquid extract called **wort** is separated from the mix. *Hops* (flower cones from the hop plant) are added to the wort for flavoring, and the mixture is boiled to stop enzyme action and precipitate proteins. A strain of *Saccharomyces* is added. Fermentation produces ethyl alcohol, carbon dioxide, and other substances, including amyl and isoamyl alcohols and acetic and butyric acids, which add to the flavor of the beer. After fermentation, the yeast is removed, and the beer is filtered, pasteurized, and bottled.

Most wine is made from juice extracted from grapes (➤ Figure 26.12), although it can be made from any fruit—and even from nuts or dandelion blossoms. Juice is treated with sulfur dioxide to kill any wild yeasts that may already be present. Sugar and a strain of *Saccharomyces* are then added, and fermentation proceeds. Although ethyl alcohol is the main product of fermentation, other products similar to those in beer add to the flavor of the wine. In both beer and wine, the particular characteristics of the juice and the yeast strain determine the flavor of the final product. When fermentation is completed, liquid wine is siphoned to separate it from

✓ What is flat sour spoilage?

✓ How does thawing frozen foods make them more susceptible to microbial food spoilage?

✓ Why are there different methods of pasteurization?

✓ Describe the process of cheese production.

What Is Not in Your Beer

Fermentation wastes from breweries contain significant quantities of vitamins, proteins, and carbohydrates. Releasing these wastes into streams is prohibited in some areas because it enriches the waters, causing an overgrowth of algae. Some breweries dispose of these wastes by drying them and marketing them as supplements to animal feed.

Yeasts can also be grown specifically for the vitamins they produce. Certain *Candida* species can make and release into the medium up to 0.1 mg of riboflavin per gram of dry weight of yeast. This process has limited commercial value because the organisms are poisoned by traces of iron in the medium, and iron is nearly always present from equipment. Higher fungi, such as some that parasitize coffee and other plants, are not harmed by traces of iron and can be used commercially to produce riboflavin.

Saccharomyces uvarum produces ergosterol, a sterol that can be converted to vitamin D by UV radiation. This process is commercially feasible when carbon sources are adequate and vats are aerated.

Malted grain is crushed
and mixed with hot water
until enzymes degrade
the starch into sugars.

The resulting mash goes
to the mash tun for
further enzyme activity.

> ➤ Figure 26.11   **The beer fermentation
> process.**

yeast sediment, and if necessary, cleared with agents such as charcoal to remove suspended particles. Finally, it is bottled and aged in a cool place.

Spirits are made from the fermentation of a variety of foods, including malted barley (Scotch whiskey), rye (rye whiskey, gin), corn (bourbon), wine or fruit juice (brandy), potatoes (vodka), and molasses (rum). After fermentation, distilling separates alcohol and other volatile substances that impart flavor from the solid and nonvolatile substances. Because of distillation, the alcohol content of spirits ranges from 40 to 50 percent—much higher than the typical 12 per-

(a)　　　　　　　　　　　**(b)**　　　　　　　　　　　**(c)**

➤ Figure 26.12　**The fermentation of wine.** (a) The process begins with the rapid bubbling of the grape juice and yeast mixture, which is (b) stored in two-story-high stainless steel fermentation vats until the fermentation process is complete. (c) The wine is then transferred to wooden barrels, where it is aged, sometimes for many years. During this time, the flavor mellows and develops fully.

cent for wine and 6 percent for beer. (Wines do not contain more alcohol because when the alcohol concentration reaches 12 to 15 percent, it poisons the yeasts carrying out the fermentation. To produce fortified wines such as sherry and cognac, extra alcohol is added after fermentation.)

Industrial and Pharmaceutical Microbiology

Industrial microbiology deals with the use of microorganisms to assist in the manufacture of useful products or to dispose of waste products. **Pharmaceutical microbiology** is a special branch of industrial microbiology concerned with the manufacture of products used in treating or preventing disease. Today many industrial and pharmaceutical processes make use of genetic engineering, as we saw in Chapter 8. ➥ (p. 217)

Industrial microbiology, albeit in primitive form, had its beginnings more than 8000 years ago when the Babylonians fermented grain to make beer. However, little was understood about fermentation until Pasteur studied the process in the nineteenth century. Over the next several decades, various researchers studied fermentation and its products, but their findings were largely ignored until shortages of materials for making explosives in World War I created a use for them. Both glycerol and acetone are used to make explosives and other materials. The Germans developed a process for making glycerol; the British used acetone–butanol fermentation by *Clostridium acetobutylicum* to make acetone. An important sidelight of the acetone–butanol fermentation was the development of techniques to maintain pure cultures in industrial fermentation vats.

The serendipitous discovery by Fleming in 1928 that *Penicillium notatum* kills *Staphylococcus aureus* was the beginning of the antibiotic industry. ➥ (Chapter 13, p. 341) Today the manufacture of antibiotics is an immense component of pharmaceutical microbiology. Concurrent with the development of antibiotics was the development and industrial production of a variety of vaccines. The manufacture of antibiotics, vaccines, and many other pharmaceuticals all require pure-culture technology.

In recent years genetic engineering has been used to cause cells to synthesize products they otherwise would not make or to increase the yield of products they normally make. Genetic engineering may make it possible to program organisms to carry out specific industrial and pharmaceutical processes. Such processes might well be more efficient and more profitable than any others now available.

Today hundreds of different substances are manufactured with the aid of microorganisms. Various species of yeasts, molds, bacteria, and actinomycetes are used in manufacturing processes. The organisms themselves are sometimes useful, as when they can serve as a source of protein. Animal feed consisting of microorganisms is called **single-cell protein (SCP).** Single-cell protein is an important high-yield, relatively inexpensive source of protein-rich food. More often the valuable substance is a product of microbial metabolism.

Useful Metabolic Processes

The production of complex molecules and metabolic end products in amounts that are commercially profitable requires the manipulation of microbial processes. In nature, microbes have regulatory mechanisms, such as induction and repression, that cause them to make substances only in the amounts needed. ➥ (Chapter 7, p. 178) An important task in industry is to manipulate regulatory mechanisms so that the organisms continue producing large quantities of useful substances for humans. Industrial microbiologists accomplish this in several ways: (1) by altering nutrients available to the microbes, (2) by altering environmental conditions, (3) by isolating mutant microbes that produce excesses of a substance because of a defective regulatory mechanism, and (4) by using genetic engineering to program organisms to display particular synthetic capabilities. In some instances these efforts have been remarkably successful. The industrial strain of the mold *Ashbya gossypii* produces 20,000 times as much of the vitamin riboflavin as it uses. Industrial strains of *Propionibacterium shermanii* and *Pseudomonas denitrificans* produce 50,000 times as much cobalamin (vitamin B_{12}) as they use.

Problems of Industrial Microbiology

It is one thing to cause a microbe to carry out a useful process in a test tube. It is quite another to adapt the process so that it will be profitable in a large-scale industrial setting. In the past, most industrial processes have been carried out in large fermentation vats. Many new processes simply do not work in large vats, so smaller vats must be used (➤ Figure 26.13). Moreover, many of today's industrial microorganisms have been so extensively modified by selection of mutants or by genetic engineering that their products, which are useful to humans, may be useless or even toxic to the organisms.

(a)

➤ Figure 26.13   **Sealing up for fermentation.** Some fermentations do not progress satisfactorily (a) in large vats (at a Philadelphia beer-brewing plant) and therefore must be conducted (b) in small-scale containers.

(b)

Stirring
motor

pH controller

Continuous
addition of
fresh, sterile
medium

Continuous
removal of
product

Stirring blades
(impellers)

Cooling
water

Cooling
water

Sparger
(aerator)

Sterile air

> Figure 26.14   **A continuous
reactor.** Fresh culture medium is introduced on one side, and medium
containing product is withdrawn at
the other side in a continuous process.

Isolating and purifying the product, with or without killing the organisms, often presents technical difficulties. When the product remains within the cell, the plasma membrane must be
disrupted to obtain the product. Membranes can be disrupted by spraying a medium containing the organisms through a nozzle under high pressure or by putting the organisms in alcohol
or salt solutions that cause the product to form a precipitate. Product molecules from disrupted
cells sometimes can be collected on resin beads of appropriate size and electric charge. Secreted products can be collected relatively easily, sometimes without killing the organisms.
This can be done by using a **continuous reactor,** in which fresh medium is introduced on one
side and medium containing the product is withdrawn on the other side (> Figure 26.14).
Continuous reactors, which operate under strictly controlled temperature and pH conditions,
are used in many types of industrial fermentations.

Useful Organic Products

Simple Organic Compounds

Simple organic compounds such as solvents and organic acids can be manufactured with the
aid of microorganisms. The solvents include ethanol (ethyl alcohol), butanol, acetone, and glycerol. The acids include acetic, lactic, and citric acids. Although microorganisms are not now regularly used to make these substances, microbial synthesis will become economically feasible
as the cost of raw materials from petroleum increases, especially if organisms can be genetically programmed to increase their productivity.

Ethanol, an industrial chemical with annual sales of about $300 million, is used not only as
a solvent but also in the manufacture of antifreeze, dyes, detergents, adhesives, pesticides, explosives, cosmetics, and pharmaceuticals. In addition, ethanol is used as a fuel, either alone or
mixed with gasoline. Microbes are employed to produce ethanol for industrial purposes in
much the same way they are used to make alcoholic beverages, but microbiologists are developing new methods. In one method, cellulose is extracted from wood, digested to sugars, and
fermented by thermophilic clostridia. Because these organisms act at high temperatures, their
metabolic rates are higher than those of other fermenters, and they produce alcohol at a more
rapid rate. Also, because the effluent is already heated, less energy is needed to distill and purify the product. In another method, the bacterium *Zymomonas mobilis,* which ferments sugar
twice as fast as yeast can, is used (> Figure 26.15).

The yeast *Pachysolen tannophilus* produces relatively large quantities of alcohol from five-
carbon sugars. This is significant because grain products contain both five-carbon and six-

➤ Figure 26.15 **Alcohol production.** Equipment for the industrial production of ethanol, using *Zymomonas mobilis* as a catalyst.

carbon sugars. Thus, making alcohol with yeasts that use only six-carbon sugars fails to extract a significant amount of energy from the grain. Producing alcohol for fuel is currently almost a break-even activity; the energy obtained from the six-carbon sugars in the alcohol is approximately equal to the energy required to produce it. Extracting energy from both five- and six-carbon sugars would be much more economical.

Clostridium acetobutylicum acting on starch or *C. saccharoacetobutylicum* acting on sugar produces both butanol and acetone. Butanol is used in the manufacture of brake fluid, resins, and gasoline additives. Acetone is used mainly as a solvent.

Glycerol is made by adding sodium sulfite to a yeast fermentation. The sodium sulfite shifts the metabolic pathway so that glycerol, and not alcohol, is the main product. Glycerol is used as a lubricant and as an emollient in a variety of foods, toothpaste, cosmetics, and paper.

Of the organic acids made by microorganisms, acetic acid (vinegar) is used in the greatest volume in the manufacture of rubber, plastics, fabrics, insecticides, photographic materials, dyes, and pharmaceuticals. Acetic acid bacteria oxidize ethanol to make vinegar. But thermophilic bacteria can make it from cellulose, and *Acetobacterium woodii* and *Clostridium aceticum* can make it from hydrogen and carbon dioxide. Other acids made by microorganisms include lactic and citric acids. *Lactobacillus delbrueckii* metabolizes glucose to lactic acid, which is used to acidify foods, to make synthetic textiles and plastics, and in electroplating. The mold *Aspergillus niger* makes citric acid with great efficiency when molasses is used as the fermentation substrate. Citric acid is widely used to acidify and improve the flavor of foods.

Antibiotics

The antibiotic industry came into being in the 1940s with the manufacture of penicillin (➤ Figure 26.16a), and about 100 antibiotics have been manufactured in quantity since then. The market value of antibiotics worldwide now exceeds $5 billion annually.

Industrial microbiologists work diligently to find ways to cause organisms to make their particular antibiotic in large quantities. An effective way is to induce mutations and then screen the progeny to find strains that produce more antibiotic than the parent strain. Such methods, combined with improved fermentation procedures, have been quite successful. For example, a strain of *Penicillium chrysogenum* that once produced 60 mg of penicillin per liter of culture now produces 20 g/l. Genetic engineering may lead to more effective methods.

The reason only 2 percent of all known antibiotics have been marketed is that many antibiotics are either too toxic to use therapeutically or of no greater benefit than antibiotics already in use. Because pathogens continually develop resistance to antibiotics, industrial microbiologists not only seek new antibiotics but also try to make available antibiotics more effective. They look for ways to increase potency, improve therapeutic properties, and make antibiotics more resistant to inactivation by microorganisms. These efforts have led to the development of *semisynthetic antibiotics* (➤ Figure 26.16b), which are made partly by microorganisms and

(a)

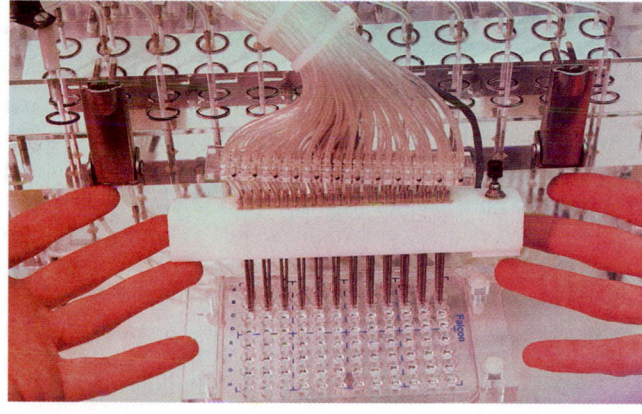

(b)

> Figure 26.16 **Penicillin production.** (a) The amber-colored droplets seen on the surface of this culture of the mold *Penicillium notatum* are the antibiotic penicillin, which it has produced. (b) Synthetic and semisynthetic antibiotics are now made in the laboratory for use alongside those produced by microorganisms.

partly by chemists. ➡ (Chapter 13, p. 341) This collaboration is illustrated by the way chemists have modified β-lactam antibiotics (penicillins and cephalosporins), which some microorganisms destroy with the enzyme β-lactamase. Binding a molecule such as clavulanic acid to the β-lactam ring prevents the enzyme from inactivating the antibiotic. The antibiotic augmentin, for example, is a semisynthetic antibiotic consisting of amoxicillin and clavulanic acid.

Enzymes

All enzymes used in industrial processes are synthesized by living organisms. With a few exceptions, such as the extraction of the meat tenderizer papain from papaya fruit, industrial enzymes are made by microorganisms. Enzymes are extracted from microorganisms rather than being synthesized in the laboratory because laboratory synthesis is as yet too complicated to be practical. Enzymes, like other proteins, consist of complex chains of amino acids in specific sequences. Organisms use genetic information to synthesize enzymes easily, but laboratory chemists find this a tedious and expensive process. Enzymes are especially useful in industrial processes because of their specificity. They act on a certain substrate and yield a certain product, thus minimizing problems of product purification.

The methods industrial microbiologists use to produce enzymes include screening for mutants and gene manipulation. Screening for mutants that produce large quantities of an enzyme has been an effective technique. Gene manipulation has been somewhat less effective. Compared with synthetic pathways for antibiotics, those for enzymes are simpler, and fewer genes are involved. Thus, gene manipulation shows great promise for increasing the yield of some enzymes and making industrial production of others feasible. Only about 200 of some 2000 known enzymes are now produced commercially, so there is room for significant progress in this area.

Of the commercially available enzymes, proteases and amylases are produced in greatest quantities. Proteases, which degrade proteins and are added to detergents to increase cleaning

Scientists are interested in the enzymes that allow extremophilic bacteria to function in harsh environments, because of their possible application to industrial environments.

power, are made industrially by molds of the genus *Aspergillus* and bacteria of the genus *Bacillus*. Amylases, which degrade starches into sugars, also are made by *Aspergillus* species. Other useful degradative enzymes are lipase from the yeast *Saccharomycopsis* and lactase from the mold *Trichoderma* and the yeast *Kluyveromyces*. Another important industrial enzyme is invertase (glucose isomerase) from *Saccharomyces;* it converts glucose to fructose, which is used as a sweetener in many processed foods.

Proteolytic Enzymes Pancreatic enzymes were first used more than 70 years ago to remove blood stains from butchers' aprons without weakening the cloth. These enzymes were tried as laundry aids, but it was found that they were inactivated by soap. In the 1970s proteolytic enzymes from bacteria, which retain their activity in the presence of detergents and hot water, were added to detergents. When this was first tried, workers in the detergent factories developed respiratory problems and skin irritations. The enzymes were removed from the detergents when the illnesses were traced to airborne enzyme molecules. Today, enzymes are incorporated in coated granules that dissolve in the wash instead of in the workers' environment.

Proteolytic enzymes have also been added to drain cleaners, where they are especially useful in degrading hair, which often clogs bathroom drains. A drain cleaner with a lipase should do a good job on kitchen drains.

Enzymes in Paper Production To make high-quality paper, much of the lignin, a coarse material in wood, is removed by expensive chemical means to produce a purer cellulose wood pulp. The fungus *Phanerochaete chrysosporium* secretes enzymes that digest both lignin and cellulose. If the enzymes can be separated and purified, one might be used selectively to digest lignin, leaving cellulose unchanged. If perfected, this process would provide a cheap way to prepare wood pulp for high-quality paper. A by-product of the research may benefit humans. Another lignin-digesting enzyme has been identified in a strain of *Streptomyces* bacteria. This enzyme modifies lignin into a molecule that enhances antibody production in mice. Might it someday do the same for humans?

Amino Acids

Microbial production of amino acids has become a commercially successful industry. Twenty different amino acids are utilized by animals in the production of proteins. Eight of the 20 are *essential amino acids*—they cannot be synthesized by the animal and therefore must be provided in the diet. Over 30,000 tons of lysine, an essential amino acid, are manufactured by microbial fermentation annually. Lysine is produced by a mutant strain of *Corynebacterium glutamicum* in which the biosynthetic pathway has been altered to promote the production of large quantities of the amino acid. Lysine is added to animal feed as a supplement and sold in health-food stores for human consumption.

Glutamic acid, another product of microbial fermentation, is also produced by a mutant strain of *C. glutamicum*. This mutant contains a high level of the enzyme glutamic acid dehydrogenase, which increases the yield of glutamic acid. The bacterium is grown in a culture medium deficient in biotin, which results in the formation of "leaky" plasma membranes. Such membranes permit the excretion of glutamic acid from the cell. Glutamic acid is used to make the flavor enhancer monosodium glutamate (MSG). About 200,000 tons is produced annually. Other microbially synthesized amino acids include phenylalanine, aspartic acid, and tryptophan.

Other Biological Products

Vitamins, hormones, and single-cell proteins are the major categories of other biologically useful products of industrial microbiology. The ability of microorganisms to make vitamin B_{12} and riboflavin was cited earlier as an example of highly successful amplification of microbial synthesis. Vaccines, of course, are exceedingly useful products (➤ Figure 26.17). The development of hepatitis vaccine was described in Chapter 8. ➡ (p. 215)

Microorganisms also are used in the production of steroid hormones. The process used is called **bioconversion,** a reaction in which one compound is converted to another by enzymes in cells. The first application of bioconversion to hormone synthesis was the use of the mold *Rhizopus nigricans* to hydroxylate progesterone. This microbial step simplified the chemical

➤ Figure 26.17   **Hepatitis B vaccine is being produced by recombinant DNA technology.** Technicians use a chromatography column to separate key proteins from batches of yeast cells.

synthesis of cortisone from bile acids from 37 to 11 steps. It reduced the cost of cortisone from $200 per gram to $6 per gram. (Subsequent improvements in procedures have brought the price below $0.70 per gram.) Other hormones that can now be produced industrially include insulin, human growth hormone, and somatostatin. They are made by recombinant DNA technology, using modified strains of *Escherichia coli*.

Single-cell proteins, as noted earlier, are whole organisms rich in protein. Currently used in animal feed, they may someday feed humans, too. An important advantage of single-cell protein is that it can be made from substances of less value than the protein produced. Certain *Candida* species make protein from paper pulp wastes, *Saccharomycopsis* makes it from petroleum wastes, and the bacterium *Methylophilus* makes it from methane or methanol.

Microbiological Mining

As the availability of mineral-rich ores decreases, methods are needed to extract minerals from less concentrated sources. This need spawned the new discipline known as **biohydrometallurgy,** the use of microbes to extract metals from ores. Copper and other metals originally were thought to be leached from the wastes of ore-crushing as a result of an inorganic chemical reaction such as those reactions used to extract metals from ores. It was then discovered that this leaching is due to the action of *Thiobacillus ferrooxidans*. This chemolithotrophic acidophilic bacterium lives by oxidizing the sulfur that binds copper, zinc, lead, and uranium into their respective sulfide minerals, with a resultant release of the pure metal. Copper in low-grade ores is often present as copper sulfide. When acidic water is sprayed on such ore, *T. ferrooxidans* obtains energy as it uses oxygen from the atmosphere to oxidize the sulfur atoms in sulfide ores to sulfate. The bacterium doesn't use the copper; it merely converts it to a water-soluble form that can be retrieved and used by humans (➤ Figure 26.18a).

Other minerals also can be degraded by microbes. *T. ferrooxidans* releases iron from iron sulfide by the same process (➤ Figure 26.18b). Combinations of *T. ferrooxidans* and a similar organism, *T. thiooxidans,* degrade some copper and iron ores more rapidly than either one does alone. Another combination of organisms, *Leptospirillum ferrooxidans* and *T. organoparus,* degrades pyrite (FeS_2) and chalcopyrite ($CuFeS_2$), although neither organism can degrade the minerals alone. Other bacteria can be used to mine uranium, and bacteria may eventually be used to remove arsenic, lead, zinc, cobalt, and gold.

Microbiological Waste Disposal

Sewage-treatment plants (Chapter 25) are prime examples of microbiological waste-disposal systems, but sewage disposal is a relatively simple problem compared with the problems associated with disposing of chemical pollutants and toxic wastes. ➥ (p. 747) Some wastes persist in the environment and contaminate water supplies for wildlife and humans. Bioremediation, the use of microorganisms to dispose of some chemical wastes, can help avert a monumental environmental disaster from accumulation of toxic wastes. ➥ (Chapter 25, p. 751)

(a)

(b)

> Figure 26.18   **Microbial mining.** Mining is made easier in many places by the activity of microorganisms. (a) Microbial leaching (extraction) helps concentrate copper from ore when conventional chemical approaches are not economically feasible. When acid solutions are added to dump surfaces, acidophilic bacteria such as *Thiobacillus ferrooxidans* quickly catalyze the oxidation of copper sulfate. (b) Microorganisms in these ponds reclaim metals from low-grade ores. Ions generated by microbial activity include ferric ions (orange) and ferrous ions (green).

✓ What are the origins and use of single-cell protein?

✓ List several enzymes and amino acids that are produced by industrial microbiology methods.

✓ How do microbes help in mining and in waste disposal?

Three strains of microorganisms have been found to deactivate Arochlor 1260, one of the most highly toxic of polychlorinated biphenyl (PCB) compounds. Other organisms have been shown to detoxify chemical substances such as cyanide and dioxin and to degrade oil spilled in the ocean. Genetic engineering has been used to develop a bacterium capable of detoxifying the defoliant Agent Orange, and work is under way to modify bacteria to detoxify other toxins.

One of the problems in developing microorganisms that can degrade toxic substances is the limited information available on the genetic characteristics of microorganisms found in wastes. Many researchers in this area focus their efforts on organisms found in wastes because these organisms probably already have degradatory capabilities. The researchers think it should be somewhat easier to modify them to metabolize other wastes than to use better-known organisms that have no known capacity to degrade wastes.

Retracing Our Steps

Microorganisms Found In Food

- Anything that humans eat is food for microorganisms, too.
- Many organisms in food are commensals; some cause spoilage, and a few cause human disease.

Grains

- Grains stored in moist areas become contaminated with various molds.
- Flour is purposely inoculated with yeasts to make bread.

Fruits and Vegetables

- Fruits and vegetables are subject to soft rot and mold damage.

Meats and Poultry

- Meats and poultry contain many kinds of microorganisms, some of which cause zoonoses. Poultry typically is contaminated with *Salmonella, Clostridium perfringens,* and *Staphylococcus aureus.*

Fish and Shellfish

- Seafood can be contaminated with several kinds of bacteria and viruses.

Milk

- Milk can contain organisms from cows, milk handlers, and the environment.

Other Edible Substances

- Sugars support growth of various organisms, which are killed during refining.
- Spices contain large numbers of microorganisms; condiments may support mold growth.
- Syrups for carbonated beverages can become contaminated with molds.
- Tea, coffee, and cocoa are also subject to mold contamination if not kept dry.

Preventing Disease Transmission and Food Spoilage

- Common diseases transmitted in food and milk are listed in Table 26.3.

- Good sanitation reduces the chance of getting foodborne diseases. The most important factor in preventing spoilage and disease transmission in food and milk is cleanliness in handling.

Food Preservation

- Methods of food preservation include **canning,** refrigeration, freezing, **lyophilization,** drying, ionizing radiation, and use of chemical additives.

Pasteurization of Milk

- Methods of preserving milk include pasteurization and sterilization.

Standards for Food and Milk Production

- Certain standards for food and milk production are maintained by federal, state, and local laws. Tests for milk quality are summarized in Table 26.4.

Microorganisms as Food and in Food Production

Algae, Fungi, and Bacteria as Food

- Rapid human population growth has created a need for new food sources.

- Yeasts can be grown on a variety of wastes and are good sources of cheap protein and vitamins. Equipment to grow them is expensive, and some effort will be needed to persuade humans to consider them acceptable as food.

- Algae can be grown in lakes and on sewage and also are good sources of protein. Problems with growing them include the danger of viral contamination and lack of acceptability as food.

Food Production

- Yeast is used to leaven bread.

- The fermentative capabilities of certain bacteria are used to make dairy products such as buttermilk, sour cream, yogurt, a variety of fermented beverages, and cheeses.

- In cheese-making, the **whey** of milk is discarded, and microorganisms ferment the **curd** and impart flavor and texture to a cheese.

- Other foods produced by microbial fermentation include vinegar, sauerkraut, pickles, olives, poi, soy sauce, and other soy products.

Beer, Wine, and Spirits

- Beer is made from **malted** grains; hops are added, and the mixture is fermented.

- Wine is made by fermenting fruit juices.

- Spirits are made by fermenting various substances and distilling the products.

Industrial and Pharmaceutical Microbiology

- **Industrial microbiology** deals with the use of microorganisms to assist in the manufacture of useful products or to dispose of waste products.

- **Pharmaceutical microbiology** deals with use of microorganisms in the manufacture of medically useful products.

Useful Metabolic Processes

- Modifications of microbial processes in industry include altering nutrients available to microbes, altering environmental conditions, isolating mutants that produce excesses of useful products, and modifying the organisms by genetic engineering.

Problems of Industrial Microbiology

- One problem in industrial microbiology is the development of small-scale processes; other problems concern techniques for recovering products.

Useful Organic Products

Simple Organic Compounds

- Microorganisms can make alcohols, acetone, glycerol, and organic acids. Microbial processes to make such products are likely to become more economically feasible in the future.

Antibiotics

- Antibiotics are derived from *Streptomyces, Penicillium, Cephalosporin,* and *Bacillus* species.

- Many antibiotics have a β-lactam ring; some semisynthetic antibiotics are made by modifying that ring so that microorganisms cannot degrade it.

Enzymes

- Enzymes extracted from microorganisms include proteases, amylases, lactases, lipases, and invertase. They are used in detergents, drain cleaners, enrichment of food, and manufacture of paper.

Amino Acids

- Several amino acids, such as lysine and glutamic acid, can be produced microbially.

Other Biological Products

- Vitamins and hormones are usually made by manipulating organisms so that they produce excessive amounts of these useful products.

- **Single-cell proteins** consist of whole organisms rich in proteins; they are used mainly as animal feeds.

Microbiological Mining

- Microbes are currently used to extract copper from low-grade ores. Other minerals that can be extracted by microorganisms include iron, uranium, arsenic, lead, zinc, cobalt, and nickel.

Microbiological Waste Disposal

- Sewage-treatment plants (Chapter 25) use microorganisms in waste disposal.

- A few organisms have been found that degrade toxic wastes, and research is under way to identify or develop others.

Terminology Check

bioconversion   *(p. 782)*
biohydrometallurgy   *(p. 783)*
bone stink   *(p. 758)*
canning   *(p. 764)*
continuous reactor
   (p. 779)
curd   *(p. 771)*
distillation   *(p. 775)*
flash pasteurization   *(p. 768)*
flat sour spoilage   *(p. 764)*

high-temperature short-time (HTST)
   pasteurization   *(p. 768)*
holding method   *(p. 768)*
industrial microbiology
   (p. 777)
leavening agent   *(p. 770)*
low-temperature long-time (LTLT)
   pasteurization   *(p. 768)*
lyophilization   *(p. 765)*
malted   *(p. 775)*

mash   *(p. 775)*
mesophilic spoilage   *(p. 764)*
pharmaceutical microbiology   *(p. 777)*
rennin   *(p. 771)*
single-cell protein (SCP)   *(p. 777)*
thermophilic anaerobic spoilage   *(p. 764)*
ultra-high temperature (UHT) treatment
   (p. 768)
whey   *(p. 771)*
wort   *(p. 775)*

See You on the Web

Don't forget I am expecting you on the web. The following is a preparedness checklist for your web explorations. If you have questions on any of these items, you can return to the end of Chapter 1 for more details.

- Questions We'll Explore
- Concept Check questions
- Flashcards
- Study Guide

If you are ready, come visit me at

http://www.prenhall.com/black

It will be a journey you'll never forget.

APPENDIX A

Metric System Measurements, Conversions, and Math Tools

Metric System Prefixes

pico (p)	=	10^{-12}
nano (n)	=	10^{-9}
micro (μ)	=	10^{-6}
milli (m)	=	10^{-3}
centi (c)	=	10^{-2}
deci (d)	=	10^{-1}
kilo (k)	=	10^{3}

Length

1 kilometer (km) = 0.62 mile
1 meter (m) = 39.37 inches = 3.281 feet
1 meter = 100 centimeters = 1000 millimeters
1 centimeter (cm) = 10 millimeters = 0.394 inch
1 millimeter (mm) = 0.0394 inch
1 micrometer (μm) = 10^{-6} meter
1 nanometer (nm) = 10^{-9} meter
1 angstrom (Å) = 10^{-10} meter

Volume

1 liter (l) = 1.057 quarts
1 liter = 1000 milliliters
1 milliliter (ml) = 1 cm^3 = 0.061 cubic inch
1 mm^3 = 10^{-3} cm^3 = 10^{-6} liter

Mass

1 kilogram (kg) = 1000 grams = 2.205 pounds
1 pound = 453.6 g
1 gram (g) = 1000 milligrams = 0.0353 ounce
1 ounce = 28.35 g
1 milligram (mg) = 10^{-3} g
1 microgram (μg) = 10^{-6} g

Temperature

degrees Fahrenheit (°F) = $\frac{9}{5}$(°C) + 32

degrees Celsius (°C) = $\frac{9}{5}$(°F − 32)

0°C = 32°F (freezing point of water)
100°C = 212°F (boiling point of water)
37°C = 98.6°F (normal body temperature)

Exponential (Scientific) Notation

Numbers that are either very large or very small are usually represented in *exponential* (scientific) *notation* as a number between 1 and 10 multiplied by a power of 10. In this kind of expression, the small raised number to the right of the 10 is the *exponent*.

Number	Exponential Form	Exponent
1,000,000	1×10^6	6
100,000	1×10^5	5
10,000	1×10^4	4
1,000	1×10^3	3
100	1×10^2	2
10	1×10^1	1
1		
0.1	1×10^{-1}	-1
0.01	1×10^{-2}	-2
0.001	1×10^{-3}	-3
0.000 1	1×10^{-4}	-4
0.000 01	1×10^{-5}	-5
0.000 001	1×10^{-6}	-6
0.000 000 1	1×10^{-7}	-7

Numbers greater than 1 have *positive* exponents, which tell how many times a number must be *multiplied* by 10 to obtain the correct value. For example, the expression 5.2×10^3 means that 5.2 must be multiplied by 10 three times:

$$5.2 \times 10^3 = 5.2 \times 10 \times 10 \times 10 = 5.2 \times 1000 = 5200$$

In so doing, we move the decimal point three places to the right:

$$5\ 2\ 0\ 0.$$
$$1\ 2\ 3$$

The value of a positive exponent indicates *how many places to the right the decimal point must be moved* to give the correct number in ordinary decimal notation.

Numbers less than 1 have *negative* exponents, which tell how many times a number must be *divided* by 10 (or multiplied by one-tenth) to obtain the correct value. Thus, the expression 3.7×10^{-2} means that 3.7 must be divided by 10 two times:

$$3.7 \times 10^{-2} = \frac{3.7}{10 \times 10} = \frac{3.7}{100} = 0.037$$

In so doing, we move the decimal point two places to the left:

$$0.\ 0\ 3\ 7$$
$$2\ 1$$

The value of a negative exponent indicates *how many places to the left the decimal point must be moved* to give the correct number in ordinary decimal notation.

Converting Decimal Numbers to Exponential Notation

To convert a number greater than 1 from decimal notation to exponential notation, first move the decimal point to the *left* until only a single digit is to the left of the decimal point. The *positive*

exponent needed for exponential notation is the same as *the number of places the decimal point was moved.*

$$6\ 3\ 5\ 7\ 8\ 1. = 6.35781 \times 10^5$$
$$5\ 4\ 3\ 2\ 1$$

To convert a number less than 1 from decimal notation to exponential notation, first move the decimal point to the *right* until a single *nonzero* digit is to the left of the decimal point. The *negative* exponent needed for exponential notation is the same as *the number of places the decimal point was moved.*

$$0.\ 0\ 0\ 0\ 4\ 2\ 6 = 4.26 \times 10^{-4}$$
$$1\ 2\ 3\ 4$$

Multiplying Exponential Numbers

To multiply two numbers in exponential form, *add* the exponents. For example:

$$
\begin{aligned}
(3.5 \times 10^3) \times (4.2 \times 10^4) &= 3.5 \times 4.2 \times 10^{(3+4)} \\
&= 14.7 \times 10^7 \\
&= 1.47 \times 10^8 = 1.5 \times 10^8 \\
&\text{(rounded off)}
\end{aligned}
$$

$$
\begin{aligned}
(5.2 \times 10^4) \times (4.6 \times 10^{-3}) &= 5.2 \times 4.6 \times 10^{[4+(-3)]} \\
&= 23.92 \times 10^1 \\
&= 2.392 \times 10^2 = 2.4 \times 10^2 \\
&\text{(rounded off)}
\end{aligned}
$$

Dividing Exponential Numbers

To divide two numbers in exponential form, *subtract* the exponents. For example:

$$
\begin{aligned}
\frac{4.1 \times 10^4}{6.2 \times 10^6} &= \frac{4.1}{6.2} \times 10^{(4-6)} = 0.6613 \times 10^{-2} \\
&= 6.613 \times 10^{-3} = 6.6 \times 10^{-3} \\
&\text{(rounded off)}
\end{aligned}
$$

$$
\begin{aligned}
\frac{6.6 \times 10^3}{8.4 \times 10^{-2}} &= \frac{6.6}{8.4} \times 10^{[3-(-2)]} = 0.7857 \times 10^5 \\
&= 7.857 \times 10^4 = 7.9 \times 10^4 \\
&\text{(rounded off)}
\end{aligned}
$$

pH Equation

The most convenient way to express the acidity or alkalinity of a solution is in terms of the *concentration of protons, or hydrogen ions* (H^+), present in the solution. Thus, to calculate the pH of a solution, you need to know the H^+ concentration—expressed as $[H^+]$—of that solution. Conversely, if you know the pH, you can determine $[H^+]$. A simple logarithmic equation can be used in either case:

$$pH = -\log_{10} [H^+]$$

In words, the pH of a solution is equal to the negative logarithm (to the base 10) of the hydrogen ion concentration, where $[H^+]$ is given in moles per liter. For example, water normally has

[H$^+$] = 10^{-7} moles/liter. The pH of water then is $-\log_{10}[10^{-7}] = -(-7) = 7$. Your stomach "juices" have [H$^+$] = 10^{-2} moles/liter. Therefore, the pH of your stomach juices is 2.

To calculate [H$^+$] from the pH equation when you know the pH, consider a structure in an animal cell with pH = 5. For this structure [H$^+$] must be 10^{-5} moles/liter. Antacids have a pH of 9, so for antacids [H$^+$] = 10^{-9} moles/liter.

Most Probable Number (MPN) Table

The following table can be used to determine the MPN of any organism in a test medium. ➡ (Chapter 6, p. 143) One of its most useful applications is in testing water purity. For example, the multiple-tube fermentation method provides an estimate for the MPN of coliforms. ➡ (Chapter 25, p. 746)

MPN Index for combinations of positive and negative results when five tubes are used per dilution (five each of 10 ml, 1ml, and 0.1ml)

Number of Tubes with Positive Results							
10 ml	1 ml	0.1 ml	MPN Index/100 ml	10 ml	1 ml	0.1 ml	MPN Index/100 ml
0	0	0	<2	4	3	1	33
0	0	1	2	4	4	0	34
0	1	0	2	5	0	0	23
0	2	0	4	5	0	1	30
1	0	0	2	5	0	2	40
1	0	1	4	5	1	0	30
1	1	0	4	5	1	1	50
1	1	1	6	5	1	2	60
1	2	0	6	5	2	0	50
2	0	0	4	5	2	1	70
2	0	1	7	5	2	2	90
2	1	0	7	5	3	0	80
2	1	1	9	5	3	1	110
2	2	0	9	5	3	2	140
2	3	0	12	5	3	3	170
3	0	0	8	5	4	0	130
3	0	1	11	5	4	1	170
3	1	0	11	5	4	2	220
3	1	1	14	5	4	3	280
3	2	0	14	5	4	4	350
3	2	1	17	5	5	0	240
4	0	0	13	5	5	1	300
4	0	1	17	5	5	2	500
4	1	0	17	5	5	3	900
4	1	1	21	5	5	4	1600
4	1	2	26	5	5	5	≥1600
4	2	0	22				
4	2	1	26				
4	3	0	27				

Source: A. E. Greenberg, L. S. Clesceri, and A. D. Eaton, eds. *Standard methods for the examination of water and wastewater.* 18th ed. Washington, D.C.: American Public Health Association, 1992.

Classification of Bacteria and Viruses

Classification of Bacteria According to *Bergey's Manual of Systematic Bacteriology*

The classification of bacteria changes as our understanding of them improves. The classification scheme currently accepted by microbiologists is set forth in the four volumes of *Bergey's Manual of Systematic Bacteriology* (Baltimore: Williams & Wilkins): Volume 1, 1984, R. Krieg; Volume 2, 1986, P. Sneath; Volume 3, 1989, J. Staley; and Volume 4, 1989, S. Williams.

This is a working model only, and changes were made even as the volumes were being published. For example, Section 26 in Volume 4 is an updated, slightly enlarged version of Section 17 in Volume 2. Some organisms present special difficulties and will, therefore, be listed in two different places until further research can determine their appropriate position. Bacteria not yet validated by publication in the *International Journal of Systematic Bacteriology* appear in quotation marks.

KINGDOM PROCARYOTAE

Division I.	**Gracilicutes**	Procaryotes with thinner cell walls, implying a Gram-negative type of cell wall
Division II.	**Firmicutes**	Procaryotes with thick and strong skin, indicative of a Gram-positive type of cell wall
Division III.	**Tenericutes**	Procaryotes of a pliable, soft nature, indicative of the lack of a rigid cell wall
Division IV.	**Mendosicutes**	Procaryotes having faulty cell walls, suggesting the lack of conventional peptidoglycan-type of cell wall

VOLUME 1

SECTION 1: The Spirochetes

Order I.	*Spirochaetales*
Family I.	*Spirochaetaceae*
Genus I.	*Spirochaeta*
Genus II.	*Cristispira*
Genus III.	*Treponema*
Genus IV.	*Borrelia*
Family II.	*Leptospiraceae*
Genus I.	*Leptospira*

Other Organisms
Hindgut spirochetes of termites and *Cryptocercus punctulatus* (wood-eating cockroach)

These helical, motile bacteria have a multilayered, membranous *outer sheath* that surrounds the cell wall, inside of which is the coiled *protoplasmic cylinder.* (➤ Figure B.1). Axial filaments are structurally similar to other bacterial flagella, but they lie between the outer sheath and the cell wall. Each axial filament is attached at one pole and permanently wound around the cylinder. Filaments vary in number from 2 in some species to as many

➤ **Figure B.1** A colorized TEM of *Leptospira interrogans.* Both the tightly coiled shape and axial filaments are characteristic of spirochetes. This organism causes leptospirosis in humans.

as 100 in others. In fluid environments, spirochetes display three kinds of movement: locomotion (straight-line motion), rotation on their longitudinal axis, and flexing (bending). Axial filaments enable spirochetes to move in relatively viscous (syrupy) fluids that prevent movement of bacteria with exposed flagella.

All spirochetes divide by binary fission, and division separates an elongated cell into two shorter ones. None form endospores. Among the spirochetes are aerobic, facultatively anaerobic, and anaerobic organisms. Free-living spirochetes inhabit a variety of aqueous environments, including sewage and muds containing hydrogen sulfide. Pathogenic spirochetes live mainly in body fluids.

The spirochetes include three genera that contain pathogens: *Treponema, Borrelia,* and *Leptospira,* all members of the order Spirochaetales. The pathogenic **treponemes** have not yet been grown on laboratory media. The type species of the treponemes is *Treponema pallidum,* the causative agent of the sexually transmitted disease syphilis. Other strains of this same species cause nonvenereal infections such as bejel (a mild rashlike infection of the skin) and yaws (deep, oozing skin lesions that can erode underlying bones). Treponemes such as *T. denticola* and *T. vincentii* are often found in the mouth. Although they usually are not pathogenic, under certain conditions they contribute to periodontal (gum) disease. Several other spirochetes can cause human disease. *Borrelia recurrentis,* which is carried by ticks, causes relapsing fever, in which the patient suffers repeated bouts of fever and chills. *B. burgdorferi,* also tick-borne, is responsible for Lyme disease. Several species of *Leptospira* can cause leptospirosis, a disease characterized by fever and liver and kidney damage in humans who ingest contaminated food or water. In laboratory cultures, leptospiras oxidize long-chain fatty acids, but borrelias do not.

SECTION 2: Aerobic/Microaerophilic, Motile, Helical/Vibrioid Gram-Negative Bacteria

Genus	*Aquaspirillum*
Genus	*Spirillum*
Genus	*Azospirillum*
Genus	*Oceanospirillum*
Genus	*Campylobacter*
Genus	*Bdellovibrio*
Genus	*Vampirovibrio*

Bacteria of this group are helical, or vibrioid (comma-shaped); as such they are similar to the spirochetes. The helix can have half a turn (comma) or many turns. All these bacteria have flagella and swim in a straight line with a corkscrew motion. Many are small spiral organisms found in soil and in fresh and stagnant water where oxygen to satisfy aerobes or microaerophiles is available; some grow in plant roots. All are Gram-negative.

Bacteria with a variety of spiral morphologies are common inhabitants of the gastrointestinal tracts of animals. The helical bacterium *Helicobacter pylori* has been found to lead to peptic ulcers in the human stomach and upper portion of the small intestine and has been linked to some cancers. Various species of *Campylobacter,* which are slender spiral rods, inhabit the reproductive and intestinal tracts and the mouth of humans. *C. jejuni* (➤ Figure B.2) causes enteritis (an inflammation of the digestive tract), resulting in diarrhea and sometimes vomiting. Bacteremia—bacteria in the blood—can lead to infections elsewhere in the body. Other *Campylobacter* species have been found to cause spontaneous abortion, stillbirth, and premature delivery in animals. In 1994 a species present in healthy and diarrhetic cats and dogs, *C. upsaliensis,* was found to infect and cause miscarriage in humans.

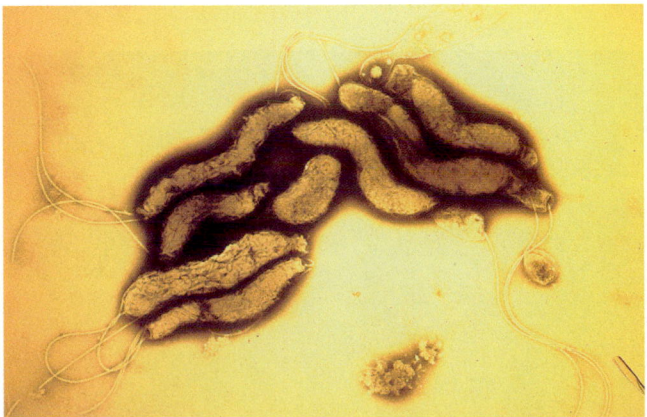

➤ **Figure B.2** *Campylobacter jejuni.* This colorized TEM (14,000X) shows spiral, Gram-negative organisms with flagella at one or both ends. *C. jejuni* is part of the normal intestinal microflora of birds and other animals. It is a common cause of enteritis in humans, causing fever, abdominal pain, diarrhea, and vomiting.

SECTION 3: Nonmotile (or Rarely Motile), Gram-Negative Curved Bacteria

Family I.	*Spirosomaceae*
Genus I.	*Spirosoma*
Genus II.	*Runella*
Genus III.	*Flectobacillus*
Other Genera	
Genus	*Ancyclobacter*
Genus	*Meniscus*
Genus	"*Brachyarcus*"
Genus	"*Pelosigma*"

Bacteria of this group are aerobic, free-living organisms found in soil, fresh water, and ocean water. These bacteria are a source of food for larger soil organisms.

SECTION 4: Gram-Negative Aerobic Rods and Cocci

Family I.	*Pseudomonadaceae*
Genus I.	*Pseudomonas*
Genus II.	*Xanthomonas*
Genus III.	*Frateuria*
Genus IV.	*Zoogloea*
Family II.	*Azotobacteraceae*
Genus I.	*Azotobacter*
Genus II.	*Azomonas*
Family III.	*Rhizobiaceae*
Genus I.	*Rhizobium*
Genus II.	*Bradyrhizobium*
Genus III.	*Agrobacterium*
Genus IV.	*Phyllobacterium*
Family IV.	*Methylococcaceae*
Genus I.	*Methylococcus*
Genus II.	*Methylomonas*
Family V.	*Halobacteriaceae*
Genus I.	*Halobacterium*
Genus II.	*Halococcus*
Family VI.	*Acetobacteraceae*
Genus I.	*Acetobacter*
Genus II.	*Gluconobacter*
Family VII.	*Legionellaceae*
Genus I.	*Legionella*
Family VIII.	*Neisseriaceae*
Genus I.	*Neisseria*
Genus II.	*Moraxella*
Genus III.	*Acinetobacter*
Genus IV.	*Kingella*
Other Genera	
Genus	*Beijerinckia*
Genus	*Derxia*
Genus	*Xanthobacter*
Genus	*Thermus*
Genus	*Thermomicrobium*
Genus	*Halomonas*
Genus	*Alteromonas*
Genus	*Flavobacterium*
Genus	*Alcaligenes*
Genus	*Serpens*
Genus	*Janthinobacterium*
Genus	*Brucella*
Genus	*Bordetella*
Genus	*Francisella*

Genus   *Paracoccus*
Genus   *Lampropedia*
Genus   *Cyclobacterium*

This group consists of a large number of bacteria—eight families and some genera that have not been assigned to families. Many of the genera include significant human pathogens.

The **pseudomonads** (su-do-mo′-nadz), members of the genus *Pseudomonas,* are generally aerobic motile rods with polar flagella (➤ Figure B.3). Many species synthesize a soluble yellow-green pigment that fluoresces under ultraviolet light. Most contain an oxidase enzyme and give positive oxidase test results. Free-living forms are found in soils and in freshwater and marine environments, where they are important decomposers of organic material. *Pseudomonas aeruginosa* is the main human pathogen; it is often seen in urinary tract infections and in wounds and burns.

Among the pseudomonads are a variety of animal and plant pathogens. One such animal pathogen causes glanders, a debilitating disease with ulceration of lymph nodes in horses. Another causes melioidosis, a disease similar to pneumonia, in both animals and humans in Southeast Asia. The plant pathogens cause rots, scabs, and wilts on a wide variety of plants, including some common house plants. Pseudomonads that infect plants require quite different growth conditions than those that infect humans, so you needn't worry about getting such an infection while caring for a sick plant.

Identifying and naming *Legionella* (le″jun-el′lah) *pneumophila* as the causative agent of Legionnaires' disease, which affected 200 veterans staying in the same hotel while attending an American Legion convention in Philadelphia in 1976, required that a new family (Legionellaceae) be established. *L. micdadei,* the causative agent of Pittsburgh pneumonia, and several other *Legionella* species have since been added to that family. These organisms had not been previously identified because fastidious nutritional requirements make them difficult to culture and isolate. They are pleomorphic (that is, they can have many different shapes), varying in shape from coccoid to filamentous. After **legionellas** were identified in the laboratory, they were found in water and soil, usually associated with other microorganisms, such as cyanobacteria, algae, protozoa, and other bacteria. They are often found living inside various species of amoebas.

Members of the family Neisseriaceae constitute another important group of medically important aerobic Gram-negative rods and cocci. Most are found on mucous membranes of humans and animals. *Neisseria gonorrhoeae* (ni-se′re-a gon-o-re′-e) thrives in the mucous membranes of the human urogenital tract and causes gonorrhea. *N. meningitidis* first colonizes nasopharyngeal mucous membranes, but it can invade blood and cerebrospinal fluid, where it causes meningitis, an inflammation of the membranes that cover the brain. Several species of the genus *Moraxella* can cause conjunctivitis, an inflammation of membranes of the eyes.

Several genera among the Gram-negative aerobic rods and cocci have been identified but not assigned to a family. Most members of the three genera *Brucella, Bordetella,* and *Francisella* are pathogens. All *Brucella* are *obligate intracellular parasites* (they can replicate only inside a living host) that usually multiply in phagocytic white blood cells. In humans they cause brucellosis, or undulant fever, in which the patient has daily episodes of fever and chills. They cause similar diseases in domestic animals. *Bordetella pertussis* causes whooping cough. One species, *Francisella tularensis,* causes tularemia in rabbits, some other wild animals, and humans. This pathogen's special need for the amino acid cysteine helps identify it in laboratory cultures. Because *F. tularensis* is highly infectious, specimens suspected of containing this bacterium should be sent to specially equipped laboratories.

Several other Gram-negative aerobic organisms are of interest in agriculture and environmental sciences. Some are important in nitrogen metabolism in plants (➤ Figure B.4); others cause some kinds of tumorlike growths called *galls* and other plant diseases. A few are specialized to metabolize methane, and many are important decomposers in mineral cycles. Another small group, the *halophiles,* tolerate high salt concentrations for

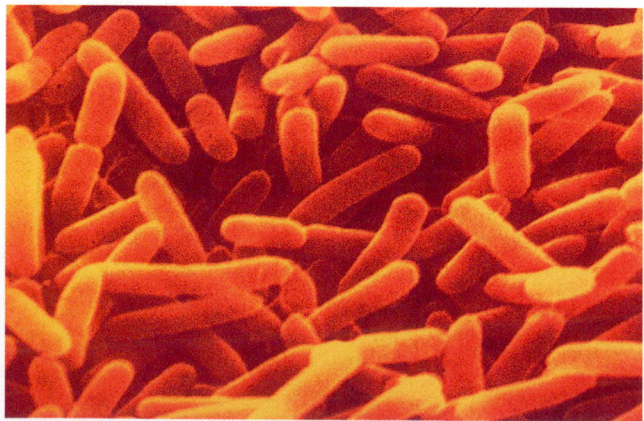

➤ **Figure B.3** SEM of *Pseudomonas aeruginosa.* Pseudomonads, aerobic motile rods, are widely distributed in nature. They are very hardy organisms, and some can even grow in surgical scrub solutions, where they can cause hospital-acquired infections.

➤ **Figure B.4** Nitrogen-fixing bacteria (*Rhizobium:* 1675X) inside nodules on alfalfa plant roots. These organisms capture N_2 gas from air and make it available to plants in more useful forms.

growth. Members of the Acetobacteraceae family produce acetic acid (vinegar) and are important in the food industry. They are found on fruits and vegetables and in alcoholic beverages. Some are a nuisance in the brewing industry because they change the alcohol produced by yeasts into acetic acid.

SECTION 5: Facultatively Anaerobic Gram-Negative Rods

Family I.	*Enterobacteriaceae*
Genus I.	*Escherichia*
Genus II.	*Shigella*
Genus III.	*Salmonella*
Genus IV.	*Citrobacter*
Genus V.	*Klebsiella*
Genus VI.	*Enterobacter*
Genus VII.	*Erwinia*
Genus VIII.	*Serratia*
Genus IX.	*Hafnia*
Genus X.	*Edwardsiella*
Genus XI.	*Proteus*
Genus XII.	*Providencia*
Genus XIII.	*Morganella*
Genus XIV.	*Yersinia*
Other Genera of Enterobacteriacae	
Genus	*Obesumbacterium*
Genus	*Xenorhabdus*
Genus	*Kluyvera*
Genus	*Rahnella*
Genus	*Cedecea*
Genus	*Tatumella*
Family II.	*Vibrionaceae*
Genus I.	*Vibrio*
Genus II.	*Photobacterium*
Genus III.	*Aeromonas*
Genus IV.	*Plesiomonas*
Family III.	*Pasteurellaceae*
Genus I.	*Pasteurella*
Genus II.	*Haemophilus*
Genus III.	*Actinobacillus*
Other Genera	
Genus	*Zymomonas*
Genus	*Chromobacterium*
Genus	*Cardiobacterium*
Genus	*Calymmatobacterium*
Genus	*Gardnerella*
Genus	*Eikenella*
Genus	*Streptobacillus*

This is another large and medically important group of bacteria. The group is divided into three families and includes several genera not assigned to families. Among them are many important human pathogens, including the enterics, the vibrios, and the *Pasteurella-Haemophilus* group. The distinguishing characteristics of these organisms are based on biochemical properties. These properties include kinds of fermentation reactions, percentages of bases in DNA, and the presence of certain enzymes.

The Enterics Members of the family Enterobacteriaceae are sometimes referred to as **enteric bacteria** because many of them inhabit the intestine of humans or other animals ("enteric" means pertaining to the intestine). In addition, some are free-living or-

ganisms found in soil and water. Others live in cooperation with, or at the expense of, their hosts; a few decompose dead organic matter. Enteric bacteria are small, morphologically similar Gram-negative rods. Some move with peritrichous flagella. They are facultative anaerobes that ferment glucose and other sugars and sometimes produce carbon dioxide and other gases.

Escherichia coli is an important enteric bacterium for three reasons:

1. *E. coli* is always present among the normal microflora of the human intestine. Finding it in water indicates that the water is contaminated with fecal matter and possibly with pathogenic enteric bacteria.

2. *E. coli* can be a pathogen in its own right; different strains produce various toxins and can cause opportunistic infections in a variety of body sites.

3. *E. coli* is an important organism in genetic engineering. Many important biological products can now be made by recombinant DNA technology, and *E. coli* frequently is the bacterium into which new genetic material is introduced. Consequently, this bacterium might be considered the workhorse of today's biotechnology.

Most members of the family Enterobacteriaceae, including *E. coli,* are found in the colon or large intestine and are sometimes referred to as "coliforms."

Members of the genera *Shigella* and *Salmonella* are closely related to *E. coli* in that all carry out mixed-acid fermentation (an anaerobic metabolic pathway that produces a combination of acids). *Shigella* can be distinguished from other enterics by its inability to produce gas during glucose fermentation. *Shigella* species cause a severe diarrhea called bacillary dysentery. *Salmonella* can be distinguished from *Escherichia* by its inability to produce acid and gas during lactose fermentation. Various species of *Salmonella* cause typhoid fever, enteritis, and food poisoning.

Another closely related set of genera are *Klebsiella, Enterobacter,* and *Erwinia,* all of which carry out butanediol fermentation. In laboratory cultures, the ability of *Klebsiella* to break down urea distinguishes that genus from *Enterobacter.* Not all members of the family Enterobacteriaceae cause enteric diseases. *Klebsiella pneumoniae* causes a bacterial pneumonia, and other *Klebsiella* species cause a variety of chronic respiratory diseases. *Enterobacter* species are opportunistic pathogens—that is, they can infect humans with lowered resistance but ordinarily do not cause disease. *Erwinia* species are of no medical significance, but they are plant pathogens and are thus important agriculturally.

Proteus, Providencia, and *Yersinia* constitute yet another related set of genera; all are mixed-acid fermenters, and they have a much lower percentage of G and C bases in their DNA than do other enterics. They can be distinguished in the laboratory by two enzyme reactions: *Proteus* has both enzymes phenylalanine deaminase and urease, but *Yersinia* lacks phenylalanine deaminase, and *Providencia* lacks urease.

Most species of *Proteus* and *Providencia* are associated with urinary tract infections and with burns and wounds. They are especially prominent as causative agents of nosocomial (hospital-acquired) infections. *Proteus mirabilis* is frequently found in

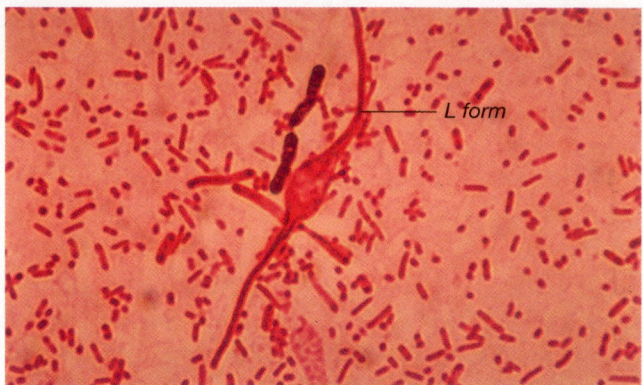

> **Figure B.5**   L forms of *Proteus vulgaris.*

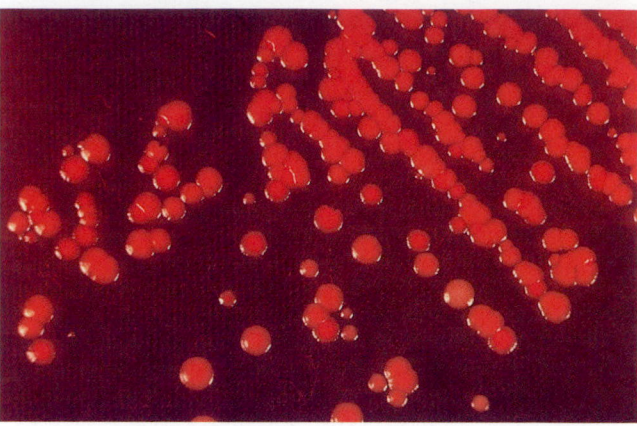

> **Figure B.6**   Colonies of *Serratia marcescens.* Notice the distinctive red pigment produced by this organism.

urine, blood, and other specimens. Some species of *Proteus* and a few other genera can lose their cell wall and swell into irregularly shaped cells called **L forms** (➢ Figure B.5). L forms can arise spontaneously or can be induced by lysozyme or penicillin, which remove the cell wall or prevent its formation. They can persist and divide repeatedly or spontaneously revert to normal-walled cells.

Yersinia pestis has a long history as an agent of human misery. When introduced into the body through the bite of an infected flea, *Y. pestis* causes bubonic plague, which is characterized by *buboes,* or abscesses of lymph nodes. When it reaches the lungs, it causes pneumonic plague, a pneumonia-like disease. Other species of *Yersinia* cause mesenteric lymphadenitis, an inflammation of lymph nodes in membranes that support abdominal organs, and a variety of other diseases.

Among enterics, *Serratia marcescens* seems to be in a class by itself. It is very different from common enteric pathogens and is also unusual in its ability to produce a characteristic pigment. Strains that grow at room temperature are generally harmless, free-living rods that usually produce the red pigment prodigiosin (➢ Figure B.6). Strains that grow at human body temperature fail to produce pigment and are less innocuous. In fact, they are avidly opportunistic! They seem to infect almost any system except the digestive tract; they often infect damaged heart valves and have been reported to cause pneumonia, urinary tract infections, meningitis, and wound infections. Many strains are resistant to several antibiotics. Because of their opportunism and antibiotic resistance, *S. marcescens* strains cause formidable problems among surgical patients, especially the immunodeficient, debilitated, and the elderly.

The Vibrios   The **vibrios** are curved or comma-shaped, Gram-negative facultative anaerobes with polar flagella. Many are found in fresh- or saltwater environments, but some inhabit the intestinal tract of humans and other animals. Among the vibrios are two important pathogens: *Vibrio cholerae* causes cholers, and *V. parahaemolyticus* causes a foodborne enteritis. Both are waterborne pathogens; the latter is common in coastal marine environments, and humans who eat contaminated shellfish often become infected. Also included with the vibrios are the luminescent marine bacteria of the genus *Photobacterium.*

The Pasteurella-Haemophilus Group   Members of the ***Pasteurella-Haemophilus* group** are very small Gram-negative bacilli and coccobacilli. They lack flagella and are nutritionally fastidious. Most are parasites of animals, and a few are human pathogens, usually attacking mucous membranes of the respiratory tract. The genus *Pasteurella* is named after Louis Pasteur, who identified *P. multocida* as the causative agent of fowl cholera. Most members of the *Pasteurella* genus infect animals. Humans become infected from cat and dog bites and can develop abscesses around the wound and septicemia (blood poisoning). Members of the genus *Haemophilus* ("blood-loving") received their name from the fact that they grow in the laboratory only in media enriched with blood (➢ Figure B.7). *Haemophilus* species sometimes inhabit the human respiratory membranes as normal microflora and can be responsible for opportunistic infections. *H. influenzae* does not itself cause influenza but at the time of its discovery was thought to do so, because the causative viruses could not be seen then. In patients with viral influenza, *H. influenzae* frequently causes a *secondary infection* (an infection that follows one already established). Some species of *Haemophilus* attack other membranes; *H. ducreyi* causes chancroid, a sexually transmitted disease, and *H. aegyptius* causes acute conjunctivitis ("pinkeye").

SECTION 6: Anaerobic Gram-Negative Straight, Curved, and Helical Rods

Family I.	*Bacteroidaceae*
Genus I.	*Bacteroides*
Genus II.	*Fusobacterium*
Genus III.	*Leptotrichia*
Genus IV.	*Butyrivibrio*
Genus V.	*Succinimonas*
Genus VI.	*Succinivibrio*
Genus VII.	*Anaerobiospirillum*
Genus VIII.	*Wolinella*
Genus IX.	*Selenomonas*
Genus X.	*Anaerovibrio*
Genus XI.	*Pectinatus*
Genus XII.	*Acetivibrio*
Genus XIII.	*Lachnospira*

➤ Figure B.7 *Haemophilus* species growing on blood agar. The clear areas surrounding the colonies are due to complete destruction of red blood cells (beta hemolysis).

These bacteria grow only in strictly anaerobic conditions and were cultured in the laboratory only after methods to maintain anaerobic conditions were developed. Although referred to as rods, many of these organisms are pleomorphic. Some have peritrichous flagella. Members of this group are found mainly in the intestinal and respiratory tracts of humans and other animals. Rod-shaped *Bacteroides* produce propionic acid or succinic acid. Spindle-shaped *Fusobacterium* and comma-shaped *Butyrivibrio* produce butyric acid. *Leptotrichia* produce lactic acid. Anaerobic Gram-negative rods cause a variety of opportunistic human infections not only in the intestine but also in the mouth, upper respiratory tract, and urogenital tract. They are especially likely to be found in deep puncture wounds. *Butyrivibrio* is important in digestive processes in cows and other animals that chew their cud. Among members of this section, species of *Bacteroides* are the most frequently encountered in human infections.

SECTION 7: Dissimilatory Sulfate- or Sulfur-Reducing Bacteria

Genus	*Desulfuromonas*
Genus	*Desulfovibrio*
Genus	*Desulfomonas*
Genus	*Desulfococcus*
Genus	*Desulfobacter*
Genus	*Desulfobulbus*
Genus	*Desulfosarcina*

These organisms, which give off sulfur-containing compounds as wastes, include the genus *Desulfovibrio* (➤ Figure B.8). They are anaerobes found mainly in water sediments, sewage, and polluted water. They are very important in the environmental sulfur cycle, and some members of the group can survive at extremes of temperature and high salinity. Although they can be found in the human intestine, they are not known to cause disease. They produce hydrogen sulfide gas, which is toxic and corrosive and smells like rotten eggs. In sufficiently high concentrations, it kills fish and plants and corrodes pipes.

SECTION 8: Anaerobic Gram-Negative Cocci

Family I.	*Veillonellaceae*
Genus I.	*Veillonella*
Genus II.	*Acidaminococcus*
Genus III.	*Megasphaera*

➤ Figure B.8 *Desulfovibrio gigas* (5000X). This anaerobic bacterium reduces sulfate to hydrogen sulfide.

Except for shape, the anaerobic Gram-negative cocci are quite similar to their rod-shaped cousins described earlier. Although several members of this group are found in animal intestines, only those of the genus *Veillonella* are of medical significance. Certain capnophilic (carbon dioxide-loving) species of *Veillonella* are found in the crevices between teeth and gums, where they often cause tooth abscesses and gum disease. Cocci occur in pairs, chains, or large masses, and they ferment lactic acid.

SECTION 9: The Rickettsias and Chlamydias

Order I.	*Rickettsiales*
Family I.	*Rickettsiaceae*
Tribe I.	*Rickettsieae*
Genus I.	*Rickettsia*
Genus II.	*Rochalimaea*
Genus III.	*Coxiella*
Tribe II.	*Ehrlichieae*
Genus IV.	*Ehrlichia*
Genus V.	*Cowdria*
Genus VI.	*Neorickettsia*
Tribe III.	*Wolbachieae*
Genus VII.	*Wolbachia*
Genus VIII.	*Rickettsiella*
Family II.	*Bartonellaceae*
Genus I.	*Bartonella*
Genus II.	*Grahamella*
Family III.	*Anaplasmataceae*
Genus I.	*Anaplasma*
Genus II.	*Aegyptianella*
Genus III.	*Haemobartonella*
Genus IV.	*Eperythrozoon*
Order II.	*Chlamydiales*
Family I.	*Chlamydiaceae*
Genus I.	*Chlamydia*

Rickettsias and **chlamydias** (➤ Figure B.9) are obligate intracellular parasites. Many are human pathogens. Although once thought to be viruses because they were first found in eukaryotic host cells, they are now known to have typical bacterial cell walls and to contain both DNA and RNA. ➥ (See Table 9.7). Most lack certain enzymes necessary for life outside host cells. These organisms probably arose from more typical bacteria by losing certain enzymes and other cell components. As they lost enzymes, they became more dependent on their hosts. These organisms also lack flagella and are nonmotile.

(a)

Chlamydia trachomatis

(b)

➤ **Figure B.9** Rickettsiae and Chlamydiae. (a) TEM of a *Rickettsia* species (95,000X), an obligate intracellular parasite. (b) TEM of *Chlamydia trachomatis* witin an infected cell (30,000X). Damage to the reproductive tract caused by this organism can eventually lead to sterility.

The rickettsias are small rods (1–2 μm), coccobacilli, or cocci that parasitize cells of mammals and arthropods. Many live in both human cells and cells of ticks, lice, fleas, and mites, which harbor and spread rickettsial diseases. Most human rickettsial pathogens belong to the genus *Rickettsia*. Variations in their characteristics are related to the diseases they cause. For example, members of the typhus fever group, which cause severe headache, chills, and fever, grow mainly in the nuclei of cells and have an optimum growth temperature of 35°C. In patients they cause **hemolysis,** the destruction of red blood cells. Members of the spotted fever group, which cause a more severe but similar disease, grow mainly in cytoplasm and have an optimum growth temperature of 32° to 34° C. They do not cause hemolysis in patients. Other rickettsias cause a variety of fevers. *Rickettsia tsutsugamushi* causes scrub typhus, a fever with rash and lymph node inflammation; it grows mostly in nuclei and has an optimum temperature of 35° C. *Rochalimaea quintana* causes

trench fever, which spreads especially among soldiers in trenches. *Rochalimaea* can be grown on artificial media and is an exception to the rule that all rickettsias are obligate intracellular parasites. The rickettsia *Bartonella henselae* causes bacillary angiomatosis, a complication of AIDS and other conditions that lessen the effectiveness of the immune system. Individuals with this infection experience fever, diarrhea, weight loss, and the formation of swellings (nodules) under the skin. *Bartonella bacilliformis* causes Oroya fever, which is accompanied by a wartlike rash. And *Coxiella burnetii* causes Q fever, a pneumonia-like disease.

The chlamydias are tiny (0.2–1.0 μm) rod-shaped or coccoid bacteria that invade host cells, where they are protected from host immune defense mechanisms. Like rickettsias, chlamydias also were once thought to be viruses because they are so small and because they multiply only in host cells. They have an unusual life cycle, in which the organisms appear in two different nonmotile forms (➤ Figure 9.24). When they are multiplying inside the host cell in membrane-enclosed cytoplasmic vacuoles, chlamydias consist of relatively large, metabolically active **reticulate bodies,** a noninfectious form with flexible cell walls. Aggregations of reticulate bodies are sometimes referred to as **inclusion bodies.** Reticulate bodies divide many times by binary fission, rupture, and release large numbers of small, dense **elementary bodies,** an infectious form with rigid cell walls. Elementary bodies are adapted for extracellular existence and are infectious; that is, they can attack other cells of the same host or be transmitted to new hosts.

Three species of chlamydia can cause human infections. *Chlamydia psittaci* causes severe pneumonia in humans and parrot fever in birds. A relatively new species, *C. pneumoniae,* is a major cause of human respiratory infections. *C. trachomatis,* long recognized as the cause of the eye infection trachoma, is also known to mimic *Neisseria gonorrhoeae* infections. One strain of *C. trachomatis* causes a severe form of sexually transmitted disease known as lymphogranuloma venereum. *C. trachomatis* can infect many different systems and organs. The ability of this pathogen to attack such a large number of different kinds of tissues and organs serves to illustrate an important concept: Although the "one organism, one disease" concept applies to many pathogens, it has many exceptions.

SECTION 10: The Mycoplasmas

Division *Tenericutes*
    Class I.   *Mollicutes*
    Order I.   *Mycoplasmatales*
       Family I.   *Mycoplasmataceae*
          Genus I.   *Mycoplasma*
          Genus II.   *Ureaplasma*
       Family II.   *Acholeplasmataceae*
          Genus I.   *Acholeplasma*
       Family III.   *Spiroplasmataceae*
          Genus I.   *Spiroplasma*

       Other Genera
       Genus   *Anaeroplasma*
       Genus   *Thermoplasma*
    Mycoplasma-like organisms of plants and invertebrates

The **mycoplasmas** are one of the few groups of bacteria for which an entire sequence of taxonomic categories from division to species has been defined. These organisms are so very small (0.1–0.8 μm) that they were once classified as viruses, but they are now known to have cell membranes, DNA, and RNA, all of which identify them as bacteria. They are sometimes said to be Gram-negative, not because they have the typical Gram-negative cell wall layers, but because they entirely lack a cell wall. Mycoplasmas are pleomorphic because of their flexible cell membrane, and they often form slender branched filaments. Most have sterols, such as cholesterol, in their cell membrane; in this respect they resemble fungi and protists. Mycoplasmas lack flagella. However, a few display a gliding movement on a wet surface. Most are facultatively anaerobic; a few are obligately anaerobic.

Although mycoplasmas can be grown in a special agar medium, where they form colonies with a "fried egg" appearance (➤ Figure B.10), in nature they have several modes of existence. Some must live at the expense of host organisms; some live on a host without damaging it; and others live on decaying organic matter. Human pathogens among the mycoplasmas are found in the genera *Mycoplasma* and *Ureaplasma*. *M. pneumoniae* causes primary atypical pneumonia, whereas other species are usually opportunists of the gums, oropharynx, and urogenital tract. *M. fermentans* may play a role in rheumatoid arthritis. *U. urealyticum* is unusual in that it metabolizes urea; it is found in the urogenital tract, where it sometimes causes sexually transmitted disease. Other mycoplasmas are plant pathogens. Some infect orange trees, which are treated by pumping solutions of the antibiotic tetracycline into the trees.

SECTION 11: Endosymbionts

A. Endosymbionts of Protozoa
    Endosymbionts of ciliates
    Endosymbionts of flagellates
    Endosymbionts of amoebas
    Taxa of endosymbionts:

Genus I.	*Holospora*
Genus II.	*Caedibacter*
Genus III.	*Pseudocaedibacter*
Genus IV.	*Lyticum*
Genus V.	*Tectibacter*

➤ **Figure B.10** The "fried egg" appearance of mycoplasma colonies (270X). This characteristic is unique to this group.

B. Endosymbionts of Insects
    Blood-sucking insects
    Plant sap-sucking insects
    Cellulose and stored grain feeders
    Insects feeding on complex diets
    Taxon of endosymbionts:

Genus	*Blattabacterium*

C. Endosymbionts of Fungi and Invertebrates Other than Arthropods
    Fungi
    Sponges
    Coelenterates
    Helminths
    Annelids
    Marine worms and mollusks

This is a diverse assortment of Gram-negative bacteria that live in the bodies of protozoa, fungi, algae, other bacteria, and insects. Little is known about their possible roles in the evolution of microorganisms. Certain endosymbionts of protozoa produce toxins. Some toxins pass from one protozoan to another during conjugation, killing the recipient. Other endosymbionts give toxin resistance to their hosts.

VOLUME 2
SECTION 12: Gram-Positive Cocci

Family I.	*Micrococcaceae*
Genus I.	*Micrococcus*
Genus II.	*Stomatococcus*
Genus III.	*Planococcus*
Genus IV.	*Staphylococcus*
Family II.	*Deinococcaceae*
Genus I.	*Deinococcus*
Other Genera	
Genus	*Streptococcus*

      Pyogenic Hemolytic Streptococci
      Oral Streptococci
      Lactic Acid Streptococci
      Anaerobic Streptococci
      Other Streptococci

Genus	*Enterococcus*
Genus	*Leuconostoc*
Genus	*Pediococcus*
Genus	*Aerococcus*
Genus	*Gemella*
Genus	*Peptococcus*
Genus	*Peptostreptococcus*
Genus	*Ruminococcus*
Genus	*Coprococcus*
Genus	*Sarcina*

The Gram-positive cocci represent one end of a spectrum of Gram-positive organisms distinguishable by shape and by whether they form spores (➤ Figure B.11). As we shall see when we consider the next few sections of bacteria, these organisms have other distinguishing properties.

The Gram-positive cocci constitute a relatively large and medically important group of bacteria. These cocci are divided into families on the basis of metabolic properties and cellular arrangements. The **micrococci** are aerobes or facultative anaerobes that

> ➤ Figure B.11 *Sporosarcina ureae,* Gram-positive cocci that produce spores. The cocci (57,600X) are arrqnged in cubical packets.

form irregular clusters by dividing in two or more planes. The **streptococci** are aerotolerant anaerobes that obtain energy from fermenting sugars to lactic acid. They form pairs or chains by dividing in one or two planes. Most lack the enzyme catalase. The **peptococci** are anaerobes that lack both catalase and the enzymes to ferment lactic acid. They also form pairs or irregular clusters.

Micrococci Many micrococci are free-living saprophytes (organisms that feed on dead matter) found in soil, fresh water, and marine environments. They are easily transmitted on the surfaces of plants and animals to meat, dairy products, and other foods. The micrococci include an aerobic genus, *Micrococcus,* and a facultatively anaerobic genus, *Staphylococcus.* Various species of *Micrococcus* are found on the skin and in the mouth and upper respiratory tract. *Staphylococcus aureus* is a common human pathogen. It is often responsible for skin abscesses and boils. If it invades the blood, it can travel to other tissues and cause pneumonia, meningitis, and osteomyelitis (infection of the marrow cavity of bones). *Staphylococcus epidermidis* normally inhabits skin and mucous membranes; it often causes opportunistic infections.

Streptococci The streptococci include a large number of species, which are differentiated by whether they have certain antigens on their cell surfaces and whether they carry out hemolysis. For example, some strains of *Streptococcus pyogenes* with so-called group-A antigens completely hemolyze red blood cells in laboratory cultures. Such complete hemolysis is called **beta hemolysis.** These strains are the causative agent of strep throat, scarlet fever, rheumatic fever, and a variety of other infections. In contrast, streptococci of the **viridans group,** which often infect the valves and lining of the heart, cause incomplete hemolysis, or **alpha hemolysis,** in laboratory cul-

tures. *S. agalactiae,* long known to cause mastitis (inflammation of the udders) in cattle, is now known to be sexually transmitted and responsible for female urogenital infections in humans. If transmitted to infants during the birth process, it can cause fatal illness.

Peptococci The peptococci are obligate anaerobes. Although the genus *Sarcina* consists of saprophytes found in soil and on plants, members of the genera *Peptococcus* and *Peptostreptococcus* are frequently found among the microbes that normally inhabit the digestive, respiratory, and urogenital tracts. These organisms are versatile opportunists; they can cause peritonitis (inflammation of the body cavity and its membranes), postpartum sepsis (blood poisoning after the birth of a child), accumulation of pus in deep wounds, osteomyelitis, vaginitis, and sinus and dental infections.

SECTION 13: Endospore-Forming Gram-Positive Rods and Cocci

Genus	*Bacillus*
Genus	*Sporolactobacillus*
Genus	*Clostridium*
Genus	*Desulfotomaculum*
Genus	*Sporosarcina*
Genus	*Oscillospira*

This group consists mainly of rod-shaped bacteria of the genera *Bacillus* and *Clostridium.* Cocci of the genus *Sporosarcina* are also included in this group, but they are soil bacteria of no medical significance. The ability to form endospores is the main distinguishing characteristic of these organisms.

All members of the genus *Bacillus* obtain nutrients from dead organic matter; some are obligate aerobes, and others are facultative anaerobes. They can be distinguished by the location of spores and by growth temperatures. For example, *B. cereus* and *B. subtilis* have centrally located endospores and grow at moderate temperatures. In contrast, *B. sterothermophilus* has terminal spores and grows at temperatures of 65° C or higher.

Several species of *Bacillus*—*B. subtilis, B. licheniformis, B. polymyxa,* and *B. brevis*—produce antibiotics. *B. anthracis* causes anthrax, a severe blood infection of cattle, sheep, and horses that can infect humans. *B. cereus* is sometimes implicated in food poisoning.

Members of the genus *Clostridium* are strictly anaerobic motile rods found in soil, water, and the intestinal tract of humans and other animals. Several species are important human pathogens that exert their effects mainly through the production of potent toxins. *C. tetani* causes tetanus (lockjaw), and *C. botulinum* causes botulism (a kind of food poisoning). *C. perfringens* also can cause food poisoning. Several species of *Clostridium* produce gas, predominantly hydrogen, under the anaerobic conditions of deep tissue wounds and cause gangrene, or tissue necrosis (death). Gangrene due to an infection is called *gas gangrene* to distinguish it from dry gangrene, which occurs when a tissue lacks an adequate blood supply. Other clostridia also infect wounds, and some can cause low-grade bacteremias, especially in debilitated patients.

SECTION 14: Regular, Nonsporing, Gram-Positive Rods

Genus	*Lactobacillus*
Genus	*Listeria*
Genus	*Erysipelothrix*
Genus	*Brochothrix*
Genus	*Renibacterium*
Genus	*Kurthia*
Genus	*Caryophanon*

These bacteria are called "regular" because they stain evenly and are of uniform shape rather than being pleomorphic. They are obligate or facultative anaerobes found in fermenting plant and animal products. They include three genera, *Lactobacillus, Listeria,* and *Erysipelothrix.* **Lactobacilli** (➤ Figure B.12) are found in a wide variety of foods. They are used in the production of cheeses, yogurt, sourdough, and many other fermented foods. *L. acidophilus* and *L. casei* are found among the natural microflora of the digestive and urogenital tracts of humans and are generally nonpathogenic. A few lactobacilli have been linked to tooth decay. *Listeria monocytogenes* causes listeriosis, an inflammation of the brain and the membranes that cover it, in humans and animals. *Erysipelothrix rhusiopathiae* causes erysipeloid lesions (red, painful skin lesions) in humans and more severe erysipelas in swine.

SECTION 15: Irregular, Nonsporing, Gram-Positive Rods

Genus	*Corynebacterium*
	Plant Pathogenic Species of Corynebacterium
Genus	*Gardnerella*
Genus	*Arcanobacterium*
Genus	*Arthrobacter*
Genus	*Brevibacterium*
Genus	*Curtobacterium*
Genus	*Caseobacter*
Genus	*Microbacterium*
Genus	*Aureobacterium*
Genus	*Cellulomonas*

Genus	*Agromyces*
Genus	*Arachnia*
Genus	*Rothia*
Genus	*Propionibacterium*
Genus	*Eubacterium*
Genus	*Acetobacterium*
Genus	*Lachnospira*
Genus	*Butyrivibrio*
Genus	*Thermoanaerobacter*
Genus	*Actinomyces*
Genus	*Bifidobacterium*

These bacteria are irregularly shaped and unevenly staining; they include club-shaped **corynebacteria,** pleomorphic **propionibacteria,** and filamentous **actinomycetes.** Nearly all are facultative anaerobes, but a few propionibacteria are strict anaerobes.

The genus *Corynebacterium* includes a large and diverse assortment of saprophytes, many of which are found in air, soil, and water or as pathogens in plants and a few animals. *C. xerosis* is a normal inhabitant of human conjunctiva; *C. pseudodiphtheriticum* is a normal inhabitant of the human pharynx. Neither produces damaging toxin. The most medically significant species is *C. diphtheriae*. It produces a potent toxin that causes diphtheria.

The family Propionibacteriaceae includes the genera *Propionibacterium* and *Eubacterium*. *P. freudenreichii* is found in dairy products. *P. acnes,* which is found in wounds and abscesses, commonly infects but does not cause acne lesions. Most species of *Eubacterium* are obligate anaerobes; they are found in soil and plant products and sometimes infect body cavities of humans and other animals. *E. foidans* is found in dental tartar and in a variety of infections.

Once classified as fungi, the order Actinomycetales, which means "ray fungus," includes mainly soil bacteria that tend to form branching filaments (➤ Figure B.13). Some of these bacteria provide nitrogen to plants; a few are normal inhabitants of the human mouth. The most important human pathogens in this group are members of the genus *Actinomyces*. *A. israelii* causes actinomycosis (jaw abscesses and sometimes lung disease), and several other species are found in dental caries (cavities) and

➤ **Figure B.12** *Lactobacillus acidophilus* (dark purple rods). This Gram-positive bacterium forms the predominant microflora in the vagina of sexually mature women. These Gram-positive bacteria produce lactic acid by fermenting the glycogen in the epithelial cells lining the vagina. The acidic environment that they produce and live in inhibits the growth of many other types of bacteria and thus helps protect the vagina from infection.

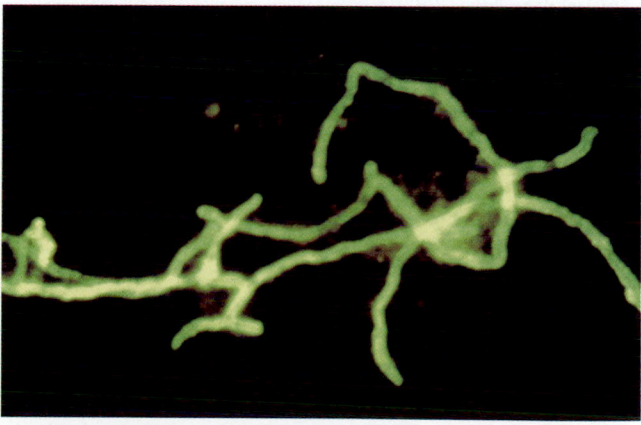

➤ **Figure B.13** *Actinomyces israelii* from a brain abscess. Notice the branching, filamentous type of growth, which formerly led the actinomycetes to be classified as fungi.

periodontal infections. *A. bovis* causes a disease called lumpy jaw in cattle.

SECTION 16: The Mycobacteria

Family	*Mycobacteriaceae*
Genus	*Mycobacterium*

The **mycobacteria** are unusual in that their cell walls contain large amounts of lipids. The lipids resist basic dyes, and so mycobacteria can be stained with carbolfuchsin in hot phenol and washed with acid alcohol. The red stain remains; thus, these organisms are said to be acid-fast. Mycobacteria are slender rods, usually without clubbed ends, and frequently form filaments. Most are soil saprophytes, but some are human pathogens. *Mycobacterium tuberculosis* causes tuberculosis, and *M. leprae* causes Hansen's disease, formerly known as leprosy. Another organism, *M. bovis,* causes tuberculosis in cattle and can be transmitted to humans. *M. avium-intracellulare,* or the *M. avium complex,* is recognized as an extremely important opportunist. It commonly causes lung and systemic infections in persons with AIDS and in others with poorly functioning immune systems. *M. avium-intracellulare* is widespread in soil, dust, and water.

SECTION 17: Nocardioforms

(These organisms are considered again in Section 26.)

Genus	*Nocardia*
Genus	*Rhodococcus*
Genus	*Nocardioides*
Genus	*Pseudonocardia*
Genus	*Oerskovia*
Genus	*Saccharopolyspora*
Genus	*Micropolyspora*
Genus	*Promicromonospora*
Genus	*Intrasporangium*

The **nocardioforms** comprise nine genera of aerobes, including *Nocardia*. They are closely related to corynebacteria, actinomycetes, and mycobacteria. Nocardioforms are Gram-positive, nonmotile, pleomorphic, aerobic, and generally acid-fast and filamentous. Although many species of *Nocardia* are found in soil, some are known pathogens. For example, *N. asteroides* causes skin abscesses and lung infections, and *N. brasiliensis* also causes skin abscesses.

VOLUME 3

SECTION 18: Anoxygenic Phototrophic Bacteria

I.	*Purple Bacteria*	
	Family I.	*Chromatiaceae*
	Genus I.	*Chromatium*
	Genus II.	*Thiocystis*
	Genus III.	*Thiospirillum*
	Genus IV.	*Thiocapsa*
	Genus V.	*Lamprobacter*
	Genus VI.	*Lamprocystis*
	Genus VII.	*Thiodictyon*
	Genus VIII.	*Amoebobacter*
	Genus IX.	*Thiopedia*
	Family II.	*Ectothiorhodospiraceae*
	Genus	*Ectothiorhodospira*

Purple nonsulfur bacteria	
Genus	*Rhodospirillum*
Genus	*Rhodopila*
Genus	*Rhodobacter*
Genus	*Rhodopseudomonas*
Genus	*Rhodomicrobium*
Genus	*Rhodocyclus*

II. Green Bacteria	
Green sulfur bacteria	
Genus	*Chlorobium*
Genus	*Prosthecochloris*
Genus	*Pelodictyon*
Genus	*Ancalochloris*
Genus	*Chloroherpeton*

Addendum to the green sulfur bacteria

Multicellular filamentous green bacteria	
Genus	*Chloroflexus*
Genus	*Heliothrix*
Genus	*"Oscillochloris"*
Genus	*Chloronema*

III. Genera Incertae Sedis	
Genus	*Heliobacterium*
Genus	*Erythrobacter*

As their name suggests, these organisms capture energy from light but do not release oxygen as photosynthetic plants do (➤ Figure B.14). These bacteria contain a special bacterial chlorophyll (called bateriochlorophyll) and a variety of other pigments: purple, red, orange, and brown. They include the purple sulfur bacteria, and the green nonsulfur bacteria. The sulfur varieties use sulfur rather than water in photosynthesis; the nonsulfur varieties use organic compounds. The anoxygenic phototrophic bacteria contribute to ecosystems by providing nutrients for other organisms.

SECTION 19: Oxygenic Photosynthetic Bacteria

Group I. Cyanobacteria
    Preface
    Taxa of the Cyanobacteria

Subsection I. Order Chroococcales

1.	Genus I.	*Chamaesiphon*
2.	Genus II.	*Gloeobacter*
3.		*Synechococcus*-group
4.	Genus III.	*Gloeothece*
5.		*Cyanothece*-group
6.		*Gloeocapsa*-group
7.		*Synechocystis*-group

➤ **Figure B.14** *Rhodospirillum rubrum*. This organism is an anoxygenic phototrophic bacterium.

Subsection II. Order Pleurocapsales

1.	Genus I.	*Dermocarpa*
2.	Genus II.	*Xenococcus*
3.	Genus III.	*Dermocarpella*
4.	Genus IV.	*Myxosarcina*
5.	Genus V.	*Chroococcidiopsis*
6.		*Pleurocapsa*-group

Subsection III. Order Oscillatoriales

Genus I.	*Spirulina*
Genus II.	*Arthrospira*
Genus III.	*Oscillatoria*
Genus IV.	*Lyngbya*
Genus V.	*Pseudanabaena*
Genus VI.	*Starria*
Genus VII.	*Crinalium*
Genus VIII.	*Microcoleus*

Subsection IV. Order Nostocales

Family I.	*Nostocaceae*
Genus I.	*Anabaena*
Genus II.	*Aphanizomenon*
Genus III.	*Nodularia*
Genus IV.	*Cylindrospermum*
Genus V.	*Nostoc*
Family II.	*Scytonemataceae*
Genus I.	*Scytonema*
Family III.	*Rivulariaceae*
Genus I.	*Calothrix*

Subsection V. Order Stigonematales

Genus I.	*Chlorogloeopsis*
Genus II.	*Fischerella*
Genus III.	*Stigonema*
Genus IV.	*Geitleria*

Group II. Order Prochlorales

Family I.	*Prochloraceae*
Genus	*Prochloron*

Other taxa

Genus	*"Prochlorothrix"*

The *cyanobacteria,* formerly called blue-green algae and classified with algae, are now known to be prokaryotes. Even so, they are remarkably similar to plants: Their chlorophyll is much like plant chlorophyll, and they release oxygen during photosynthesis, as do plants. Cyanobacteria may represent a step in the evolution of photosynthetic mechanisms.

The *prochlorophytes* are similar to the cyanobacteria, but they are placed in a different group because they have two distinctly plantlike chlorophylls found in structures much like chloroplasts. The only two currently known genera of this group are *Prochloron* (➤ Figure B.15) which lives with marine invertebrates called ascidians, and *Prochlorothrix*. Some scientists believe that *Prochloron* represents a "missing link" between photosynthetic prokaryotes and eukaryotic plants.

SECTION 20: Aerobic Chemolithotrophic Bacteria and Associated Organisms

Nitrifying Bacteria

Family	*Nitrobacteraceae*

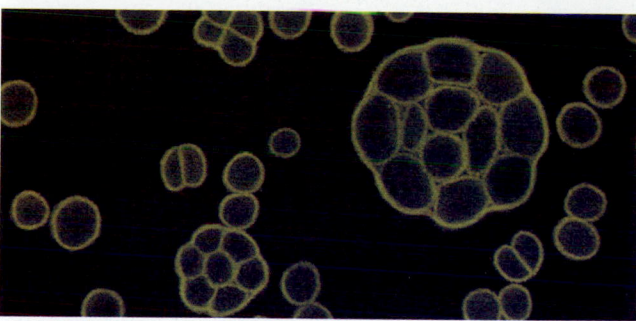

➤ **Figure B.15** *Prochloron,* a photosynthetic prokaryote with some similarities to plants. Each cell in a cluster contains chlorophyll arranged on membranous layers.

Nitrite-oxidizing bacteria

Genus I.	*Nitrobacter*
Genus II.	*Nitrospina*
Genus III.	*Nitrococcus*
Genus IV.	*Nitrospira*

Ammonia-oxidizing bacteria

Genus V.	*Nitrosomonas*
Genus VI.	*Nitrosococcus*
Genus VII.	*Nitrosospira*
Genus VIII.	*Nitrosolobus*
Genus IX.	*"Nitrosovibrio"*

Colorless Sulfur Bacteria

Genus	*Thiobacterium*
Genus	*Macromonas*
Genus	*Thiospira*
Genus	*Thiovulum*
Genus	*Thiobacillus*
Genus	*Thiomicrospira*
Genus	*Thiosphaera*
Genus	*Acidiphilium*
Genus	*Thermothrix*

Obligately Chemolithotrophic Hydrogen Bacteria

Genus	*Hydrogenobacter*

Iron- and Manganese-Oxidizing and/or Depositing Bacteria

Family	*"Siderocapsaceae"*
Genus I.	*"Siderocapsa"*
Genus II.	*"Naumaniella"*
Genus III.	*"Siderococcus"*
Genus IV.	*"Ochrobium"*

Magnetotactic Bacteria

Genus	*Aquaspirillum*
	(*A. magnetotacticum*)
Genus	*"Bilophococcus"*

This is another diverse group of bacteria that obtain energy from inorganic substances. They include nitrogen bacteria found in soil, such as *Nitrosomonas,* which oxidizes ammonia to nitrite, and *Nitrobacter,* which oxidizes nitrite to nitrate. Also in this group are *Thiobacillus,* which oxidizes sulfur, and certain other bacteria that reduce sulfur. Sulfur bacteria are found in soil, water, sewage, sulfur springs, and acid mine wastes. Still other chemolithotrophs oxidize iron and/or manganese and are responsible for large deposits of minerals on the ocean floor. A few of these organisms, known as *magnetotactic bacteria*, absorb iron into intracellular crystals. When placed in a magnetic field, they orient in a north-south direction. Chemolithotrophs are important in the cycling of nitrogen and sulfur. Exactly where the chemolithotrophs fit into the evolution of life is not yet clear; they present some interesting taxonomic problems.

SECTION 21: Budding and/or Appendaged Bacteria

Prosthecate Bacteria

  Budding bacteria

   Buds produced at tip of prostheca
   Genus   *Hyphomicrobium*
   Genus   *Hyphomonas*
   Genus   *Pedomicrobium*

   Buds produced on cell surface
   Genus   *Ancalomicrobium*
   Genus   *Prosthecomicrobium*
   Genus   *Labrys*
   Genus   *Stella*

  Bacteria that divide by binary transverse fission
   Genus   *Caulobacter*
   Genus   *Asticcacaulis*
   Genus   *Prosthecobacter*

Nonprosthecate Bacteria

  Budding bacteria

   Lack peptidoglycan
   Genus   *Planctomyces*
   Genus   *"Isophaera"*

   Contain peptidoglycan
   Genus   *Ensifer*
   Genus   *Blastobacter*
   Genus   *Angulomicrobium*
   Genus   *Gemmiger*

  Nonbudding, stalked bacteria
   Genus   *Gallionella*
   Genus   *Nevskia*

Other bacteria

  Nonspinate bacteria
   Genus   *Seliberia*
   Genus   *"Metallogenium"*
   Genus   *"Thiodendron"*

  Spinate bacteria

Bacteria in this group reproduce by *budding,* or dividing unequally. Small buds that contain DNA and a small amount of cytoplasm pinch off from the parent cell and grow to become parent cells themselves. Some of these bacteria, such as members of the genus *Caulobacter,* have an appendage, or stalk, that generally has a sticky holdfast at its tip (➤ Figure B.16. The holdfast is a device that allows a bacterium to anchor itself on a solid surface. These bacteria are found in soil and water. Both budding reproduction and development of stalks are examples of interesting evolutionary developments in this group.

➤ **Figure B.16** *Caulobacter.* This organism is a nonmotile stalked bacterium anchored to surfaces by a thin filament during part of its life cycle. However, it begins life as a free-swimming "swarmer" cell. Eventually the swarmer cell loses its flagellum and grows an attachment stalk at the site of its former flagellum. After settling down, it grows, elongates, and forms a new flagellum at the end opposite the attachment stalk. It then divides unequally by binary fission to produce a new swarmer cell plus the old attached cell.

SECTION 22: Sheathed Bacteria

   Genus   *Sphaerotilus*
   Genus   *Leptothrix*
   Genus   *Haliscominobacter*
   Genus   *"Lieskeella"*
   Genus   *"Phragmidiothrix"*
   Genus   *Crenothrix*
   Genus   *"Clonothrix"*

These bacteria form filaments and surround themselves with a secreted sheath made of lipoproteins and polysaccharides. If these Gram-negative organisms are grown in the presence of minerals such as iron or manganese, they incorporate the metals into their sheath. Sheathed bacteria reproduce by releasing flagellated swarmer cells, free-swimming cells that leave the filament and start new colonies. These water dwellers are responsible for the slimy coating on rocks in and along streams and for blooms (dense floating masses of organisms) when the nutrient supply suddenly increases. They provide evidence of the evolutionary trend toward multicellular structures.

SECTION 23: Nonphotosynthetic, Nonfruiting Gliding Bacteria

Order I.   *Cytophagales*
  Family I.   *Cytophagaceae*
    Genus I.   *Cytophaga*
    Genus II.   *Capnocytophaga*
    Genus III.   *Flexithrix*
    Genus IV.   *Sporocytophaga*

    Other Genera
    Genus   *Flexibacter*
    Genus   *Microscilla*
    Genus   *Chitinophaga*
    Genus   *Saprospira*
Order II.   *Lysobacterales*
  Family I.   *Lysobacteraceae*
    Genus I.   *Lysobacter*

Order III.	*Beggiatoales*
Family I.	*Beggiatoaceae*
Genus I.	*Beggiatoa*
Genus II.	*Thiothrix*
Genus III.	*Thioploca*
Genus IV.	*"Thiospirillopsis"*
Other Families and Genera	
Family	*Simonsiellaceae*
Genus I.	*Simonsiella*
Genus II.	*Alysiella*
Family	*"Pelonemataceae"*
Genus I.	*"Pelonema"*
Genus II.	*"Achroonema"*
Genus III.	*"Peloploca"*
Genus IV.	*"Desmanthus"*
Other Genera	
Genus	*Toxothrix*
Genus	*Leucothrix*
Genus	*Vitreoscilla*
Genus	*Desulfonema*
Genus	*Achromatium*
Genus	*Agitococcus*
Genus	*Herpetosiphon*

These organisms glide along a solid surface, leaving a slime trail behind them. They are said to be nonfruiting because they do not form cell aggregates called *fruiting bodies,* as do another group called the gliding fruiting bacteria. Gliding nonfruiting bacteria are important in ecological cycles. Members of the genus *Cytophaga* digest cellulose, chitin, agar, and other complex organic materials. *Beggiatoa* species oxidize hydrogen sulfide to sulfur, thereby contributing to the sulfur cycle.

SECTION 24: Fruiting Gliding Bacteria: The Myxobacteria

Order *Myxococcales*	
Family I.	*Myxococcaceae*
Genus	*Myxococcus*
Family II.	*Archangiaceae*
Genus	*Archangium*
Family III.	*Cystobacteraceae*
Genus I.	*Cystobacter*
Genus II.	*Melittangium*
Genus III.	*Stigmatella*
Family IV.	*Polyangiaceae*
Genus I.	*Polyangium*
Genus II.	*Nannocystis*
Genus III.	*Chondromyces*

Also referred to as the slime bacteria, these are Gram-negative aerobes. Found in well-oxygenated soil, these organisms are important decomposers. Their fruiting bodies consist of large, often highly colored, cell aggregates. Cells released from the fruiting bodies are released and migrate to form new colonies—more evidence for the evolutionary trend toward multicellular structures.

SECTION 25: Archaeobacteria

Group I. Methanogenic Archaeobacteria

Order I.	*Methanobacteriales*
Family I.	*Methanobacteriaceae*
Genus I.	*Methanobacterium*
Genus II.	*Methanobrevibacter*
Family II.	*Methanothermaceae*
Genus	*Methanothermus*
Order II.	*Methanococcales*
Family	*Methanococcaceae*
Genus	*Methanococcus*
Order III.	*Methanomicrobiales*
Family I.	*Methanomicrobiaceae*
Genus I.	*Methanomicrobium*
Genus II.	*Methanospirillum*
Genus III.	*Methanogenium*
Family II.	*Methanosarcinaceae*
Genus I.	*Methanosarcina*
Genus II.	*Methanolobus*
Genus III.	*Methanothrix*
Genus IV.	*Methanococcoides*
Other Taxa	
Family	*Methanoplanaceae*
Genus	*Methanoplanus*
Other Genus	*Methanosphaera*

Group II. Archaeobacterial Sulfate Reducers

Order	*"Archaeoglobales"*
Family	*"Archaeoglobaceae"*
Genus	*Archaeglobus*

Group III. Extremely Halophilic Archaeobacteria

Order	*Halobacteriales*
Family	*Halobacteriaceae*
Genus I.	*Halobacterium*
Genus II.	*Haloarcula*
Genus III.	*Haloferax*
Genus IV.	*Halococcus*
Genus V.	*Natronobacterium*
Genus VI.	*Natronococcus*

Group IV. Cell Wall-less Archaeobacteria

Genus	*Thermoplasma*

Group V. Extremely Thermophilic S⁰-Metabolizers

Order I.	*Thermococcales*
Family	*Thermococcaceae*
Genus I.	*Thermococcus*
Genus II.	*Pyrococcus*
Order II.	*Thermoproteales*
Family I.	*Thermoproteaceae*
Genus I.	*Thermoproteus*
Genus II.	*Thermofilum*
Family II.	*Desulfurococcaceae*
Genus	*Desulfurococcus*
Other Bacteria	
Genus	*Staphylothermus*
Genus	*Pyrodictium*
Order III.	*Sulfolobales*
Family	*Sulfolobaceae*
Genus I.	*Sulfolobus*
Genus II.	*Acidianus*

The *archaeobacteria* are a diverse group, mostly of anaerobes, that live in extremely harsh environments. These organisms share properties—such as unusual transfer RNA molecules, unusual lipids in their plasma membranes, and distinctive RNA polymerases—that are significantly different from those of the eubacteria. The archaeobacteria can be Gram-positive or Gram-negative, motile or nonmotile, and rods or cocci. Three distinct groups are currently recognized: methanogens, halobacteria, and thermoacidophiles. The **methanogens** produce methane and, in fact, are the sole source of natural methane, or marsh gas. They are found in soil, water, sewage, and even the digestive tract of humans and other animals. The **halobacteria** live in very concentrated salt solutions. In some places they can be seen from an airplane as pink patches on salt flats.

Thermoacidophiles metabolize sulfur and live in extremely hot, acidic environments. These bacteria and other organisms found near deep-sea "black smoker" vents (so named because of the black sulfur compounds that spew up from the sea bottom like plumes of black smoke) have been studied from the submersible vessel *Alvin*. The organisms are found in surprisingly dense populations. The bacteria provide food for many organisms, including mussels, giant clams big enough to provide 20 lb of meat per clam, and tube worms 3 m long and as thick as a human arm but lacking mouth, digestive tract, and anus (➤ Figure B.17). Some of the bacteria are free-living and form white, fluffy, tennis ball-sized masses in a bacterial mat. Others live symbiotically in the worm or clam tissues. There they contribute food that they have made with the use of energy from the sulfur compounds of the vent. These bacteria make these volcanic vents the most productive environments of all in terms of energy capture.

➤ Figure B.17 Thermoacidophilic archaeobacteria thrive in hot, acidic environments. In regions of the deep-ocean floor near volcanic vents they serve as the base of food chains, living symbiotically inside the tissues of animals such as these giant tube worms. The worms lack a mouth an an anus and thus rely on the bacterial capture of chemical energy to feed their tissues.

VOLUME 4

SECTION 26: Nocardioform Actinomycetes
(This is an updated, slightly enlarged version of Section 17 in Volume 2.):

Genus	*Nocardia*
Genus	*Rhodococcus*
Genus	*Nocardioides*
Genus	*Pseudonocardia*
Genus	*Oerskovia*
Genus	*Saccharopolyspora*
Genus	*Faenia*
Genus	*Promicromonospora*
Genus	*Intrasporangium*
Genus	*Actinopolyspora*
Genus	*Saccharomonospora*

The nocardioforms comprise nine genera of aerobes, including *Nocardia*. They are closely related to corynebacteria, actinomycetes, and mycobacteria. Nocardioforms are Gram-positive, nonmotile, pleomorphic, aerobic, and generally acid-fast and filamentous. Although many species of *Nocardia* are found in soil, some are known pathogens. For example, *N. asteroides* cause skin abscesses and lung infections, and *N. brasiliensis* also causes skin abscesses.

SECTION 27: Actinomycetes with Multilocular Sporangia

Genus	*Geodermatophilus*
Genus	*Dermatophilus*
Genus	*Frankia*

The actinomycetes that divide in more than one plane are similar to, but distinct from, the actinomycetes described earlier. These bacteria nearly always form masses of filaments. Some are free-living in soil; others live in plant root nodules where they oxidize molecular nitrogen. The medically important species belong to the genus *Dermatophilus,* which causes skin lesions.

SECTION 28: Actinoplanetes

Genus	*Actinoplanes*
Genus	*Ampullariella*
Genus	*Pilimelia*
Genus	*Dactylosporangium*
Genus	*Micromonospora*

Some *sporangiate actinomycetes* are soil bacteria that form relatively extensive filaments, complete with *sporangia* (spore sacs) filled with spores. Although these organisms do not infect humans, they are similar to the genera *Actinomyces* and *Dermatophilus,* which do.

Other *conidiate* genera have even more complex mycelial structures, including chains of spores borne on ends of hyphae, and more closely resemble fungi than other funguslike bacteria. The species *Micromonospora purpurea* produces the antibiotic gentamicin.

SECTION 29: Streptomycetes and Related Genera

Genus	*Streptomyces*
Genus	*Streptoverticillium*
Genus	*Kineosporia*
Genus	*Sporichthya*

Like the actinomycetes, the **streptomycetes** and their allies are soil-dwelling organisms that resemble fungi. They develop extensive branching filaments with spores of many different shapes. The characteristics of spores are used to separate the streptomycetes into genera. None are pathogenic, but many are medically significant because of the antibiotics they produce. More than 500 different antibiotic substances have been isolated from various species of the genus *Streptomyces*. Among these antibiotics are substances effective against bacteria, viruses, protozoa, and fungi.

SECTION 30: Maduromycetes

Genus	*Actinomadura*
Genus	*Microbispora*
Genus	*Microtetraspora*
Genus	*Planobispora*
Genus	*Planomonospora*
Genus	*Spirillospora*
Genus	*Streptosporangium*

SECTION 31: Thermomonospora and Related Genera

Genus	*Thermomonospora*
Genus	*Actinosynnema*
Genus	*Nocardiopsis*
Genus	*Streptoalloteichus*

SECTION 32: Thermoactinomycetes

Genus	*Thermoactinomyces*

SECTION 33: Other Genera

Genus	*Glycomyces*
Genus	*Kibdelosporangium*
Genus	*Kitasatosporia*
Genus	*Saccharothrix*
Genus	*Pasteuria*

Some *sporangiate actinomycetes* are soil bacteria that form relatively extensive filaments, complete with *sporangia* (spore sacs) filled with spores. Although these organisms do not infect humans, they are similar to the genera *Actinomyces* and *Dermatophilus,* which do.

ADDENDUM

Some novel actinomycete taxa that have been validly published or validated since Volume 4 of *Bergey's Manual of Systematic Bacteriology* was published.

Placement of certain genera has presented difficulties, as indicated by the following examples:

(a) The genus *Gardnerella*. The organisms of this genus have had a checkered taxonomic history and it is still not entirely clear whether they should be placed in Volume 1 with Gram-negative bacteria or in Volume 2 with Gram-positive bacteria.

(b) The genus *Butyrvibrio*. Although the cells stain Gram-negative, the ultrastructure of the cell wall is of the Gram-positive type. It is not clear whether the genus should be placed in Volume 1 or Volume 2.

(c) The genus *Xanthobacter*. The cells stain Gram-positive or Gram-variable, yet the cell wall structure and composition, as well as nucleic acid hybridization data, indicate that the organisms are of the Gram-negative type.

(d) The genus *Chromobacterium*. Although 80 percent of the strains attack glucose fermentatively and grow well anaerobically, the remainder attack glucose oxidatively and grow slowly under anaerobic conditions. It is consequently difficult to assign the organisms definitely to either Section 5 (Facultatively Anaerobic Gram-Negative Rods) or Section 4 (Gram-Negative Aerobic Rods and Cocci). Nucleic acid hybridization studies indicate a relationship to certain genera of aerobic rods.

(e) The genus *Zymomonas*. Although the organisms are facultatively anaerobic (a few obligately anaerobic), they are related genetically, phenotypically, and ecologically to the acetic acid bacteria, which are aerobic. Moreover, the occurrence of the Entner-Doudoroff pathway is typical of aerobic bacteria.

(f) The genus *Thermoplasma*. The lack of a cell wall makes this genus compatible with Section 10 (the Mycoplasmas); however, studies of the ribosomal RNA, as well as various phenotypic characteristics, indicate that the genus is related to the archaeobacteria, covered in Volume 3 of the *Manual*.

(g) The genera *Halobacterium* and *Halococcus*. Although these extreme halophiles are compatible with Section 4 (Gram-Negative Aerobic Rods and Cocci), nucleic acid studies and certain phenotypic characteristics indicate the genus is related to the archaeobacteria, covered in Volume 3.

As an interim solution to some of these problems, some taxa are described not only in Volume 1 but in an appropriate subsequent volume as well.

Classification of Viruses

The classification of viruses is undergoing greater changes than is bacterial taxonomy. Most viruses have not even been classified due to a lack of data concerning their reproduction and molecular biology. Estimates suggest that more than 30,000 viruses are being studied in laboratories and reference centers worldwide.

The classification and viral information presented here follows the outline given in Chapter 10 (Tables 10.1 and 10.2). Information also can be found in *Human Virology: A Text for Students of Medicine, Dentistry, and Microbiology* (L. Collier and J. Oxford, 1993, Oxford University Press), and in *Virology* (J. Levy, H. Fraenkel-Conrat, and R. Owens, 2d ed., 1994, Prentice Hall).

The 21 families of viruses listed here are primarily those that infect vertebrates. Thus, these families represent only a small part of the 71 families and more than 3000 viruses recognized in *Virus Taxonomy—Sixth Report of the International Committee on Taxonomy of Viruses* (F. A. Murphy, et al. [eds.], 1995, New York: Springer-Verlag Wien).

1. Family: Picornaviridae

Genera:

Enterovirus (gastrointestinal viruses, poliovirus, coxsackie viruses A and B, echoviruses)
Hepatovirus (hepatitis A virus)
Cardiovirus (encephalomyocarditis virus of mice and other rodents)
Rhinovirus (upper respiratory tract viruses, common cold viruses)
Aphthovirus >(foot-and-mouth disease virus)

Naked, polyhedral, positive-sense, ssRNA. Synthesis and maturation take place in the host cell cytoplasm. Viruses are released via cell lysis.

2. Family: Caliciviridae

Genus:

Calicivirus (Norwalk viruses and similar viruses causing gastroenteritis, hepatitis E virus)

Naked, polyhedral, positive-sense, ssRNA. Synthesis and maturation take place in the host cell cytoplasm. Viruses are released via cell lysis.

3. Family: Togaviridae

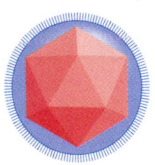

Genera:

Alphavirus (eastern, western, and Venezuelan equine encephalitis viruses, Semliki forest virus)
Rubivirus (rubella virus)
Arterivirus (equine arteritis virus, simian hemorrhagic fever virus)

Enveloped, polyhedral, positive-sense, ssRNA. Synthesis occurs in the host cell cytoplasm; maturation involves budding of nucleocapsids through the host cell plasma membrane. Viruses are released via cell lysis (*Arterivirus*). Many replicate in arthropods and vertebrates.

4. Family: Flaviviridae

Genera:

Flavivirus (yellow-fever virus, dengue fever virus, St. Louis and Japanese encephalitis viruses, tickborne encephalitis virus)
Pestivirus (bovine diarrhea virus, hog cholera virus)
Hepatitis C virus

Enveloped, polyhedral, positive-sense, ssRNA. Synthesis occurs in the host cell cytoplasm; maturation involves budding through host cell endoplasmic reticulum and Golgi apparatus membranes. Most replicate in arthropods.

5. Family: Coronaviridae

Genus:

Coronavirus (common cold viruses, avian infectious bronchitis virus, feline infectious peritonitis virus, mouse hepatitis virus)

Enveloped, helical, positive-sense, ssRNA. Synthesis occurs in the host cell cytoplasm; maturation involves budding through membranes of the endoplasmic reticulum and Golgi apparatus. Viruses are released via cell lysis.

6. Family: Rhabdoviridae

Genera:

Vesiculovirus (vesicular stomatitis-like viruses)
Lyssavirus (rabies and rabieslike viruses)
[Unnamed] (proposed, for bovine ephemeral feverlike viruses)
[Unnamed] (proposed, for many of the ungrouped rhabdoviruses of mammals, birds, fish, arthropods, and plants)

Enveloped, helical, negative-sense, ssRNA. Synthesis occurs in the host cell nucleus; maturation occurs via budding from the host cell plasma membrane. Many replicate in arthropods.

7. Family: Filoviridae

Genus:

Filovirus (Marburg and Ebola viruses)

Enveloped; long, filamentous forms, sometimes with branching, and sometimes U-shaped, 6-shaped, or circular; negative-sense, ssRNA. Synthesis occurs in the host cell cytoplasm; maturation involves budding from the host cell plasma membrane. Viruses are released via cell lysis. These viruses are "Biosafety Level 4" pathogens—they must be handled in the laboratory under maximum containment conditions.

8. Family: Paramyxoviridae

Genera:

Paramyxovirus (parainfluenza viruses 1–4, mumps virus, Newcastle disease virus)
Morbillivirus (measles and measleslike viruses, canine distemper virus)
Pneumovirus (respiratory syncytial virus)

Enveloped, helical, negative-sense, ssRNA. Synthesis occurs in the host cell cytoplasm; maturation involves budding through the host cell plasma membrane. Viruses are released via cell lysis. Morbilliviruses can cause persistent infections.

9. Family: Orthomyxoviridae

Genera:

Influenzavirus A and B (influenza viruses A and B)
Influenzavirus C (influenza C virus)

Enveloped, helical, negative-sense, ssRNA (eight segments). Synthesis occurs in the host cell nucleus; maturation takes place in the host cell cytoplasm. Viruses are released through budding from the host cell's plasma membrane. These viruses can reassort genes during mixed infections.

10. Family: Bunyaviridae

Genera:

Bunyavirus (Bunyamwera supergroup)
Phlebovirus (sandfly fever viruses)
Nairovirus (Nairobi sheep disease-like viruses)
Uukuvirus (Uukuniemi-like viruses)
Hantavirus (hemorrhagic fever viruses, Korean hemorrhagic fever, Sin Nombre hantavirus)

Enveloped, spherical, negative-sense, ssRNA (three segments; Phlebovirus ambisense ssRNA). Synthesis occurs in the host cell cytoplasm; maturation occurs within the Golgi apparatus. Viruses are released via cell lysis. Closely related viruses can reassort genes during mixed infections.

11. Family: Arenaviridae

Genus:

Arenavirus (Lassa fever virus, lymphocytic chorio-meningitis virus, Machupo virus, Junin virus)

Enveloped, helical, ambisense, ssRNA (two segments). Synthesis occurs in the host cell cytoplasm; maturation involves budding from the host cell plasma membrane. Virions contain ribosomes. The human pathogens Lassa, Machupo, and Junin viruses are "Biosafety Level 4" pathogens—they must be handled in the laboratory under maximum containment conditions.

12. Family: Reoviridae

Genera:

Orthoreovirus (reoviruses 1, 2, and 3)
Orbivirus (Orungo virus)
Rotavirus (human rotaviruses)
Cypovirus (cytoplasmic polyhidrosis viruses)
Coltivirus (Colorado tick fever virus)
Plant reovirus 1/3 (plant reoviruses subgroups 1, 2, and 3)

Each genus differs in morphology and physiochemical details. In general, virions are naked, polyhedral, dsRNA (10-12 segments). Synthesis and maturation take place in the host cell cytoplasm. Viruses are released via cell lysis. Virions contain ribosomes.

13. Family: Birnaviridae

Genera:

Birnavirus (infectious pancreatic necrosis virus of fish and infectious bursal disease virus of fowl)

Naked, polyhedral, dsRNA (two segments). Synthesis and maturation take place in the host cell cytoplasm. Viruses are released via cell lysis.

14. Family: Retroviridae

Genera:

MLV-related virus (spleen necrosis virus, mouse and feline leukemia viruses)
Mammalian type-B (mouse mammary tumor virus)
Type D (squirrel monkey retrovirus)
ALV-related virus (avian leukemia virus, rous sarcoma virus)
HTLV-BLV group (human T cell leukemia virus HTLV-I, HTLV-II, bovine leukemia virus)
Spumavirus (the foamy viruses)
Lentivirus (human, feline, simian, and bovine immunodeficiency viruses)

Enveloped, spherical, negative-sense, ssRNA (two identical strands). Synthesis occurs in the host cell cytoplasm; maturation involves budding through the host cell plasma membrane. These viruses contain the enzyme reverse transcriptase. The retroviruses (except the Spumavirus and Lentivirus genera) represent the RNA tumor viruses, causing leukemias, carcinomas, and sarcomas.

15. Family: Hepadnaviridae

Genera:

Orthohepadnavirus (hepatitis B virus)
Avihepadnavirus (duck hepatitis virus)

Enveloped, polyhedral, partially dsDNA. Synthesis and maturation take place in the host cell nucleus. Surface antigen production occurs in the cytoplasm. Persistence is common and is associated with chronic disease and neoplasia.

16. Family: Parvoviridae

Genera:

Parvovirus (feline leukopenia virus, canine parvovirus)
Dependovirus (adeno-associated viruses)
Densovirus (insect parvoviruses)
Erythrovirus (human erythrovirus B19)

Naked, polyhedral, negative-sense, ssDNA (*Parvovirus*) or positive-sense and negative-sense, ssDNA (other genera). Synthesis and maturation occur in rapidly dividing host cells, specifically in the host cell nucleus. Viruses are released via cell lysis.

17. Family: Papovaviridae

Genera:

Papillomavirus (wart viruses, genital condylomas, DNA tumor viruses)
Polyomavirus (human polyoma-like viruses, SV-40)

Naked, polyhedral, dsDNA. Synthesis and maturation take place in the host cell nucleus. Viruses are released via cell lysis.

18. Family: Adenoviridae

Genera:

Mastadenovirus (human adenoviruses A-F, infectious canine hepatitis virus)
Aviadenovirus (avian adenoviruses)

Naked, polyhedral, dsDNA. Synthesis and maturation take place in the host cell nucleus. Viruses are released via cell lysis.

19. Family: Herpesviridae

Subfamily:

Alphaherpesvirinae

Genera:

Simplexvirus (herpes simplex viruses 1 and 2)
Varicellovirus (varicella-zoster virus)

Subfamily:

Betaherpesvirinae

Genera:

Cytomegalovirus (human cytomegalovirus)
Muromegalovirus (murine cytomegalovirus)

Subfamily:

Gammaherpesvirinae

Genera:

Lymphocryptovirus (Epstein-Barr viruses)
Rhadinovirus (saimiri-ateles-like viruses)

Enveloped, polyhedral, dsDNA. Synthesis and maturation occur in the host cell nucleus, with budding through the nuclear envelope. Although most herpesviruses cause persistent infections, virions can be released by rupture of the host cell plasma membrane.

20. Family: Poxviridae

Subfamily:

Chordopoxvirinae

Genera:

Orthopoxvirus (vaccinia and variola viruses, cowpox virus)
Parapoxvirus (orf virus, pseudocowpox virus)
Avipoxvirus (fowlpox virus)
Capripoxvirus (sheep pox virus)
Leporipoxvirus (myxoma virus)
Suipoxvirus (swinepox virus)
Yatapoxvirus (yabapox virus and tanapox virus)
Molluscipoxvirus (molluscum contagiosum virus)

Subfamily:

Entomopoxvirinae

Genus:

Entomopoxvirus A/B/C (poxviruses of insects)
External envelope, large, brick-shaped (or ovoid), dsDNA. Synthesis and maturation take place in the portion of the host cell cytoplasm called viroplasm ("viral factories"). Viruses are released via cell lysis.

21. Family: Iridoviridae

Genera:

Iridovirus (small iridescent insect viruses)
Chloriridovirus (large iridescent insect viruses)
Ranavirus (frog viruses)
Lymphocystivirus (lymphocystis viruses of fish)

Enveloped (missing on some insect viruses), polyhedral, dsDNA. Synthesis occurs in both the host cell nucleus and cytoplasm. Most virions remain cell-associated.

Word Roots Commonly Encountered in Microbiology

a-, an- — <u>not, without, absence</u>   abiotic, not living; anaerobic, in the absence of air

acantho- — <u>thorn or spinelike</u>   *Acanthamoeba,* an amoeba with spinelike projections

actino- — <u>having rays</u>   *Actinomyces,* a bacterium forming colonies that look like sunbursts

aero- — <u>air</u>   aerobic, in the presence of air

agglutino- — <u>clumping or sticking together</u>   hemagglutination, clumping of blood cells

albo- — <u>white</u>   *Candida albicans,* a white fungus

amphi- — <u>around, doubly, both</u>   Amphitrichous describes flagella found at both ends of a bacterial cell.

ant-, anti- — <u>against, versus</u>   Antibacterial compounds kill bacteria.

archaeo- — <u>ancient</u>   Archaeobacteria are thought to resemble ancient forms of life.

arthro- — <u>joint</u>   arthritis, inflammation of joints

asco- — <u>sac, bag</u>   Ascospores are held in a saclike container, the ascus.

-ase — <u>denotes enzyme</u>   lipase, an enzyme attacking lipids

aureo- — <u>golden</u>   *Staphylococcus aureus* has gold-colored colonies.

auto- — <u>self</u>   Autotrophs, self-feeding organisms.

bacillo- — <u>rod</u>   bacillus, a rod-shaped bacterium

basid- — <u>base, foundation</u>   basidium, fungal cell bearing spores at its end

bio- — <u>life</u>   biology, the study of living things

blast- — <u>bud</u>   blastospore, spore formed by budding

bovi- — <u>cow</u>   *Mycobacterium bovis,* bacterium causing tuberculosis in cattle

brevi- — <u>short</u>   *Lactobacillus brevis,* a bacterium with short rod-shaped cells

butyr- — <u>butter</u>   Butyric acid gives rancid butter its unpleasant odor.

campylo- — <u>curved</u>   *Campylobacter,* a curved bacterium

carcino- — <u>cancer</u>   A carcinogen causes cancer.

caryo-, karyo- — <u>center, kernel</u>   Prokaryotic cells lack a true, discrete nucleus.

caseo- — <u>cheese</u>   caseous, cheeselike lesions

caul- — <u>stalk, stem</u>   *Caulobacter,* a stalked bacterium

ceph-, cephalo- — <u>of the head or brain</u>   encephalitis, inflammation of the brain

chlamydo- — <u>cloaked, hidden</u>   *Chlamydia* are difficult bacteria to detect.

chloro- — <u>green</u>   chlorophyll, a green pigment

chromo- — <u>colored</u>   Metachromatic granules stain various colors within a cell.

chryso- — <u>golden</u>   *Streptomyces chryseus,* a bacterium forming golden colonies

-cide — <u>to kill</u>   Fungicide kills fungi.

co-, con- — <u>with, together</u>   congenital, existing from birth

cocc- — <u>berry</u>   *Streptococcus,* spherical bacteria in chains

coeno- — <u>shared in common</u>   coenocytic, many nuclei not separated by septa

col-, colo- — <u>colon</u>   coliform bacteria, found in the colon (large intestine)

conidio- — <u>dust</u>   conidia, tiny dustlike spores produced by fungi

coryne- — <u>club</u>   *Corynebacterium diphtheriae,* a club-shaped bacterium

-cul — <u>little, tiny</u>   molecule, a tiny mass

cut-, -cut — <u>skin</u>   cutaneous, of the skin

cyan- — <u>blue</u>   cyanobacteria, formerly called the blue-green algae

cyst-, -cyst — <u>bladder</u>   cystitis, inflammation of the urinary bladder

cyt-, -cyte — <u>cell</u>   leukocyte, white blood cell

de- — <u>lack of, removal</u>   decolorize, to remove color

dermato- — <u>skin</u>   dermatitis, inflammation of the skin

di-, diplo- — <u>two, double</u>   diplococci, pairs of spherical cells

dys- — <u>bad, faulty, painful</u>   dysentery, a disease of the enteric system

ec-, ecto-, ex- — <u>outside, outer</u>   ectoparasite, found on the outside of the body

em-, en- — <u>in, inside</u>   encapsulated, inside a capsule

-emia — <u>of the blood</u>   pyemia, pus in the blood

endo- — <u>inside</u>   endospore, spore found inside a cell

entero- — <u>intestine</u>   enteric, bacteria found in the intestine

epi- — <u>atop, over</u>   epidemic, a disease spreading over an entire population at one time

erythro- — <u>red</u>   lupus erythematosus, disease with a red rash

etio- — <u>cause</u>   etiology, study of the causes of disease

eu- — <u>true, good, normal</u>   eukaryote, cell with a true nucleus

exo- — <u>outside</u>   exotoxin, toxin released outside of a cell

extra- — <u>outside, beyond</u>   extracellular, outside of a cell

fil- — <u>thread</u>   filament, thin chain of cells

flav- — <u>yellow</u>   flavivirus, cause of yellow fever

-fy — <u>to become, make</u>   solidify, to become solid

galacto- — <u>milk</u>   galactose, monosaccharide from milk sugar

gamet- — <u>marriage</u>   gamete, a reproductive cell, such as egg or sperm

gastro- — <u>stomach</u>   gastroenteritis, inflammation of the stomach and intestines

gel- — <u>to stiffen, congeal</u>   gelatinous, jellylike

gen-, -gen — <u>to give rise to</u>   pathogen, microbe that causes disease

-genesis — <u>origin, development</u>   pathogenesis, development of disease

germ, germin- — <u>bud</u>   germination, process of growing from a spore

-globulin — <u>protein</u>   immunoglobulins, proteins of the immune system

haem-, hem- — <u>blood</u>   hemagglutination, clumping of blood cells

halo- — <u>salt</u>   halophilic, organisms that thrive in salty environments

hepat- — <u>liver</u>   hepatitis, inflammation of the liver

herpes — <u>creeping</u>   herpes zoster, or shingles, in which vesicles erupt sequentially along a nerve pathway

hetero- — <u>different, other</u>   heterotroph, organism deriving nutrition from other sources

histo- — <u>tissue</u>   histology, the study of tissues

homo- — <u>same</u>   homologous, having the same structure

hydro- — <u>water</u>   hydrologic cycle, water cycle

hyper- — <u>over, above</u>   hyperbaric oxygen, higher than atmospheric pressure oxygen

hypo- — <u>under, below</u>   hypodermic, going beneath the skin

im-, in- — <u>not</u>   insoluble, cannot be dissolved

inter- — <u>between</u>   intercellular, between cells

intra- <u>inside</u> intracellular, inside a cell

io- <u>violet</u> iodine, element that is purple in gaseous state

iso- <u>same, equal</u> isotonic, having the same osmotic pressure

-itis <u>inflammation of</u> meningitis, inflammation of the meninges

kin- <u>moving</u> kinetic energy, energy of movement

leuko- <u>white</u> leukocyte, white blood cell

lip-, lipo- <u>fat, lipid</u> lipoprotein, molecule having both fatty and proteinaceous parts

-logy, -ology <u>study of</u> microbiology, study of microbes

lopho- <u>tuft</u> lophotrichous, having a tuft or group of flagella

luc-, luci- <u>light</u> luciferase, enzyme that catalyzes a light-producing reaction

luteo- <u>yellow</u> *Micrococcus luteus,* bacterium producing yellow colonies

lys-, -lysis <u>splitting</u> cytolysis, rupture of a cell

macro- <u>large</u> macroconidia, large spores

meningo- <u>membrane</u> meninges, membranes of the brain

meso- <u>middle</u> mesophile, organism growing best at medium temperatures

micro- <u>small, tiny</u> microbiology, study of tiny forms of life

mono- <u>one, single</u> monosaccharide, a single sugar unit

morph- <u>shape, form</u> pleiomorphic, having many different shapes

multi- <u>many</u> multicellular, having many cells

mur- <u>wall</u> muramic acid, a component of cell walls

muri-, mus- <u>mouse</u> murine, in or of mice

mut-, -mute <u>to change</u> mutagen, agent that causes genetic change

myc-, -myces <u>fungus</u> *Actinomyces,* a bacterium that resembles a fungus

myxo- <u>slime, mucus</u> myxomycetes, slime molds

necro- <u>dead, corpse</u> necrotizing toxin, causes death of tissue

nema-, -nema <u>thread</u> *Treponema,* nematode, threadlike organisms

nigr- <u>black</u> *Rhizopus nigricans,* a black mold

oculo- <u>eye</u> binocular, microscope with two eyepieces

-oid <u>like, resembling</u> toxoid, harmless molecule that resembles a toxin

-oma <u>tumor</u> carcinoma, tumor of epithelial cells

onco- <u>mass, tumor</u> oncogenes, genes that cause tumors

-osis <u>condition of</u> brucellosis, condition of being infected with Brucella

pan- <u>all, universal</u> pandemic, a disease affecting a large part of the world

para- <u>beside, near, abnormal</u> parainfluenza, a disease resembling influenza

patho- <u>abnormal</u> pathology, study of abnormal diseased states

peri- <u>around</u> peritrichous, flagella located all around an organism

phago- <u>eating</u> phagocytosis, cell eating by engulfing

philo-, -phil, -phile <u>loving, preferring</u> capnophile, organism needing higher than normal levels of carbon dioxide

-phob, -phobe <u>hating, fearing</u> hydrophobic, water-repelling

-phore <u>bearing, carrying</u> electrophoresis, technique in which ions are carried by an electric current

-phyte <u>plant</u> dermatophyte, fungus that attacks skin

pil- <u>hair</u> pilus, hairlike tube on bacterial surface

-plast <u>formed part</u> chloroplast, green body inside plant cell

pod-, -pod <u>foot</u> podocyte, foot cell of kidney

poly- <u>many</u> polyribosomes, many ribosomes on the same piece of messenger RNA

post- <u>afterward, behind</u> post-streptococcal glomerulonephritis, kidney damage following a streptococcal infection

pre-, pro- <u>before, toward</u> prepubertal, before puberty

pseudo- <u>false</u> pseudopod, projection resembling a foot, false foot

psychro- <u>cold</u> psychrophilic, preferring extreme cold

pyo- <u>pus</u> pyogenic, producing pus

pyro- <u>fire, heat</u> pyrogen, fever-producing compound

rhin- <u>nose</u> rhinitis, inflammation of nasal membranes

rhizo- <u>root</u> mycorrhiza, symbiotic growth of fungi and roots

rhodo- <u>red</u> *Rhodospirillum,* a large red spiral bacterium

-rrhea <u>flow</u> diarrhea, abnormal flow of liquid feces

rubri- <u>red</u> *Rhodospirillum rubrum,* a large red spiral bacterium

saccharo- <u>sugar</u> polysaccharide, many sugar units linked together

sapro- <u>rotten, decaying</u> saprophyte, organism living on dead matter

sarco- <u>flesh</u> sarcoma, tumor made up of muscle or connective tissue

schizo- <u>to split</u> schizogony, a type of fission in malarial parasites

-scope, -scopy <u>to see, examine</u> microscopy, use of the microscope to examine small things

sept-, septo- <u>partition, wall</u> septum, wall between cells

septi- <u>rotting</u> septic, exhibiting decomposition due to bacteria

soma-, -some <u>body</u> chromosome, colored body (when stained)

spiro- <u>coil</u> spirochete, spiral-shaped bacterium

sporo- <u>spore</u> sporocidal, spore killing

staphylo- <u>in bunches, like grapes</u> staphylococci, spherical bacteria growing in clusters

-stasis, stat- <u>stopping, not changing</u> bacteriostatic, able to stop the growth of bacteria

strepto- <u>twisted</u> *Streptobacillus,* twisted chains of bacilli

sub- <u>under, below</u> subclinical, signs and symptoms not clinically apparent

super- <u>above, more than</u> superficial mycosis, fungal infection of the surface tissues

sym-, syn- <u>together</u> symbiosis, living together

tact-, -taxis <u>touch</u> chemotaxis, orientation or movement in response to chemicals

tax-, taxon- <u>arrangement</u> taxonomy, the classification of organisms

thermo- <u>heat</u> thermophile, organism preferring or needing high temperatures

thio- <u>sulfur</u> *Thiobacillus,* organism that oxidizes hydrogen sulfide to sulfates

tox- <u>poison</u> toxin, a harmful compound

trans- <u>through, across</u> transduction, movement of genetic information from one cell to another

trich- <u>hair</u> monotrichous, having a single, hairlike flagellum

-troph <u>feeding, nutrition</u> phototroph, organism that makes its own food, using energy from light

uni- <u>one, singular</u> unicellular, composed of one cell

undul- <u>waving</u> undulant fever, disease in which fever rises and falls

vac-, vaccin- <u>cow</u> vaccine, disease-preventing product originally produced by inoculating it onto skin of calves

vacu- <u>empty</u> vacuole, empty-appearing structure in cytoplasm

vesic- <u>blister, bladder</u> vesicle, small blisterlike lesions

vitr- <u>glass</u> in vitro, grown in laboratory glassware

xantho- <u>yellow</u> *Xanthomonas oryzae,* bacterium producing yellow colonies

xeno- <u>strange, foreign</u> xenograft, graft from a different species

zoo- <u>animal</u> protozoan, first animal

zygo- <u>yoke, joining</u> zygote, fertilized egg

-zyme <u>ferment</u> enzymes, biological catalysts, some of which are involved in fermentation

Safety Precautions in the Handling of Clinical Specimens

Concern for maintaining safe conditions in school and hospital laboratories, in other work settings, and especially during patient interactions has led the federal government to formulate various regulations and recommendations. These are far too extensive to reproduce here in their entirety; several are nearly 200 pages long. As an introduction to the kinds of safety measures that should be taken, a few of the guidelines set forth in these publications are provided below. Some key references and sources are also listed, with the hope that they will stimulate the interested reader to investigate further.

In 1983 CDC published a document entitled "Guidelines for isolation precautions in hospitals" that contained a section headed "Blood and body fluid precautions." The recommendations in this section specified precautions to be observed regarding contact with the blood or body fluids of any patient known or suspected to be infected with bloodborne pathogens.

In August 1987 CDC published a document entitled "Recommendations for prevention of HIV transmission in health-care settings." In contrast with the 1983 document, the 1987 publication recommended that blood and body fluid precautions be consistently used for *all* patients regardless of their bloodborne infection status. These blood and body fluid precautions as they pertain to all patients are referred to as "Universal Blood and Body Fluid Precautions," or more simply as "Universal Precautions."

Following publication of this document, there were many requests for clarification—for example, as to which bodily fluids these precautions should apply. This led to a CDC publication on June 24, 1988, in *MMWR* entitled "Update: Universal precautions for prevention of transmission of human immunodeficiency virus, hepatitis B virus, and other bloodborne pathogens in health-care settings."

Copies of these two most recent reports (CDC August 1987 and June 1988) are available through the National AIDS Information Clearinghouse, P.O. Box 6003, Rockville, MD 20850.

In these two publications CDC makes the following recommendations regarding sharp instruments:

1. Take care to prevent injuries when using needles, scalpels, and other sharp instruments or devices; when handling sharp instruments after procedures; when cleaning used instruments; and when disposing of used needles. Do not recap used needles by hand; do not remove used needles from disposable syringes by hand; and do not bend, break, or otherwise manipulate used needles by hand. Place used disposable syringes and needles, scalpel blades, and other sharp items in puncture-resistant containers for disposal. Locate the puncture-resistant containers as close to the use area as is practical.

2. Use protective barriers to prevent exposure to blood, body fluids containing visible blood, and other fluids to which universal precautions apply. The type of protective barrier(s) should be appropriate for the procedure being performed and the type of exposure anticipated.

3. Immediately and thoroughly wash hands and other skin surfaces that are contaminated with blood, body fluids containing visible blood, or other body fluids to which universal precautions apply.

Glove use is recommended during phlebotomy (drawing blood samples) but cannot protect against penetrating injuries. Some institutions have relaxed recommendations for using gloves for phlebotomy procedures by skilled phlebotomists in settings where the prevalence of bloodborne pathogens is known to be very low (for example, volunteer blood-donation centers). Such institutions should periodically reevaluate their policy. Gloves should always be available to health care workers who wish to use them for phlebotomy. In addition, the following general guidelines apply:

1. Use gloves for performing phlebotomy when the health care worker has cuts, scratches, or other breaks in his/her skin.

2. Use gloves in situations where the health care worker judges that hand contamination with blood may occur, for example, when performing phlebotomy on an uncooperative patient.

3. Use gloves for performing finger and/or heel sticks on infants and children.

4. Use gloves when persons are receiving training in phlebotomy.

In March 1989 the Environmental Protection Agency (EPA) published a new set of "Standards for the tracking and management of medical waste," in part designed to prevent the deplorable pollution of our nation's beaches by medical wastes (*Federal Register,* March 24, 1989, pp. 12325–95). In May 1989 the Occupational Safety and Health Administration (OSHA) published a new set of rules for "Occupational exposure to bloodborne pathogens" (*Federal Register,* May 30, 1989, pp. 23041–139). Both publications provide very detailed information about procedures that must be followed.

Another useful and quite detailed publication, issued by CDC in February 1989 and reprinted in *MMWR* for June 23, 1989, is entitled "Guidelines for prevention of transmission of human immunodeficiency virus and hepatitis B virus to health-care and public-safety workers."

For those concerned primarily with the teaching laboratory, we recommend the publication "Handling infectious materials in the education setting" (G. Ballman, *American Clinical Laboratory,* July 1989, pp. 10–11). Other publications of interest include *Biosafety in Microbiological and Biomedical Laboratories* (CDC, U.S. Department of Health and Human Services, Public Health Service, 1988), which describes the four biosafety levels classified by CDC and recommended safety precautions for each, and "Labeling of microbial risks" (C. Robinson and T. H. Hatfield, *Journal of College Science Teaching,* May 1995, pp. 407–9), which evaluates that classification system of biosafety levels.

Metabolic Pathways

Stage 1

Confine substrate (glucose) inside the cell and form phosphorylated six-carbon units (fructose 1,6-diphosphate)

Step 1: Hexokinase transfers a phosphate group to glucose from ATP, forming glucose 6-phosphate.

Steps 2 & 3: The glucose molecule is rearranged to form fructose 6-phosphate, and a second phosphate group is added to give fructose 1,6-diphosphate. Phosphofructokinase (an allosteric enzyme) carries out this third step.

Step 4: The six-carbon sugar fructose is split into two different three-carbon sugar-phosphates by aldolase.

Step 5: Dihydroxyacetone phosphate is rearranged to form a second molecule of glyceraldehyde 3-phosphate.

Stage 2

A six-carbon unit is split into two three-carbon units, and two ATP molecules are produced. Two NAD molecules are reduced.

Step 6: Another phosphate group is added, and two hydrogen atoms, with their electrons, are transferred to NAD.

Step 7: Phosphoglycerate kinase catalyzes the transfer of a phosphate group to ADP, forming ATP.

Stage 3

Pyruvic acid is formed, and two more ATP molecules are produced.

Steps 8 & 9: The remaining phosphate group is moved from the end to the middle carbon atom by phosphoglycerate mutase, and a molecule of water is removed by enolase.

Step 10: The phosphate group is transferred to ADP by pyruvate kinase, forming another ATP molecule.

➤ Figure E.1 **Glycolysis (Embden-Meyerhof pathway).** Each of the 10 steps of glycolysis is catalyzed by a specific enzyme, which is indicated in a purple oval. (Refer to Chapter 5 for an explanation; Figure 5.11 shows a simplified version of the process. ➥ p. 115)

➤ Figure E.2   **Pentose phosphate pathway (phosphogluconate pathway).** This metabolic pathway occurs with glycolysis. It provides an alternative pathway for the breakdown of glucose as well as pentoses (five-carbon sugars). This pathway plays three important roles: (1) It provides intermediate pentoses, especially ribose, that the bacterial cell must use to synthesize nucleic acids. (2) This pathway's intermediates can be used to synthesize some amino acids. (3) The pentose phosphate pathway reduces NADP to NADPH. This coenzyme, like NADH, is an electron carrier and thus is a source of reducing power. The fates of several intermediates are indicated. For clarity, the specific enzymes catalyzing these reactions and the structural formulas of substrates have been omitted. (Refer to Chapter 5 for an explanation of this pathway. ➥ p. 116)

Stage A

Citrate synthase joins acetyl-CoA and oxaloacetic acid to form citric acid and coenzyme A (step 1).

Stage F

The last stage of the Krebs cycle involves three steps to regenerate oxaloacetic acid by the oxidation of succinic acid. First, succinate dehydrogenase generates an oxidation reaction, converting succinic acid to fumaric acid (step 6). The hydrogen acceptor for this reaction is FAD, which is reduced to FADH₂. Second, a water molecule is added to fumaric acid by the enzyme fumarase, forming malic acid (step 7). Third, malic acid participates in another oxidation reaction, catalyzed by malate dehydrogenase; oxaloacetic acid is reformed (step 8). In this reaction NAD is the hydrogen acceptor, being reduced to NADH .

Stage B

Further oxidation requires that citric acid be rearranged by aconitase into isocitric acid (step 2). This rearrangement involves two reactions, both catalyzed by aconitase.

Stage C

Isocitrate dehydrogenase brings about the oxidative decarboxylation (loss of CO_2) to form α–ketoglutaric acid (step 3). In this reaction NAD is reduced. The rate of this reaction is important in determining the overall rate of the Krebs cycle.

Stage D

A second oxidative decarboxylation, of α–ketoglutaric acid, forms succinyl-CoA (step 4). In this reaction NAD is reduced. The α–ketoglutarate dehydrogenase complex consists of three enzymes and works much like the pyruvate dehydrogenase complex that forms acetyl-CoA.

Stage E

Succinyl-CoA synthetase couples the cleavage of coenzyme A, which forms succinic acid (step 5), with the phosphorylation of guanosine triphosphate (GTP) from guanosine diphosphate (GDP). The terminal phosphate group of GTP can be transferred by nucleoside diphosphokinase to ADP, forming ATP.

➤ Figure E.3   **Krebs cycle (also called the citric acid cycle and the tricarboxylic acid cycle).** The reaction that converts pyruvic acid to acetyl-CoA precedes the Krebs cycle (see Figure 5.16.). This reaction is catalyzed by a pyruvate dehydrogenase complex, which contains three enzymes. Each of the eight steps of the Krebs cycle is also catalyzed by a specific enzyme, as indicated in a purple oval. (Refer to Chapter 5 for an explanation; Figure 5.17 shows a simplified version of the process. ➤ p. 120)

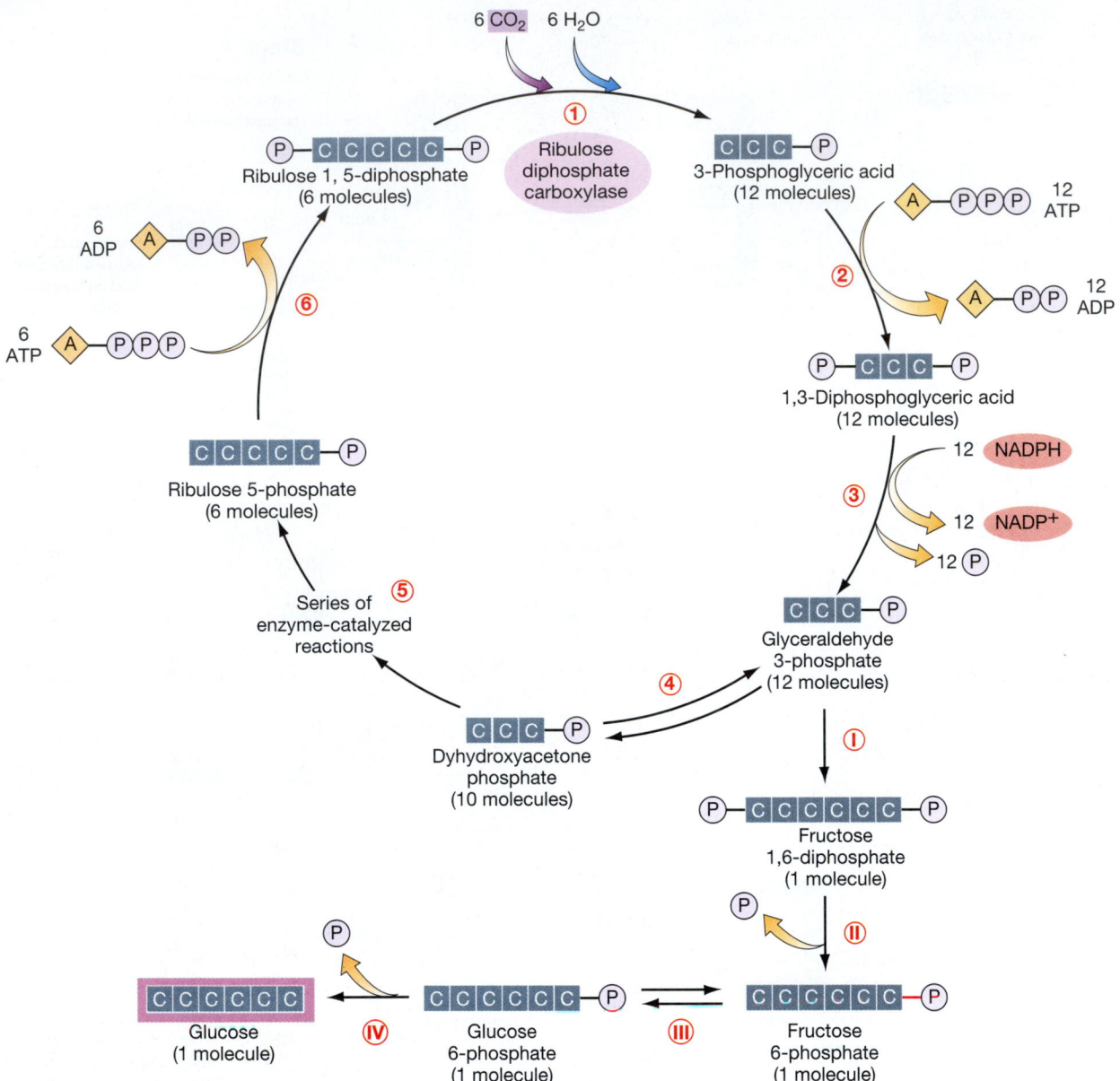

➤ **Figure E.4   Calvin-Benson cycle (dark reactions of photosynthesis).** Each step of the Calvin-Benson cycle is catalyzed by a specific enzyme, which for simplicity is not shown. Steps 1 through 3 produce 12 three-carbon intermediates. These three steps are dependent on photophosphorylation products (ATP and NADPH). Two of every 12 three-carbon molecules undergo chemical reactions (steps 1 through IV) to produce a six-carbon glucose molecule. The other 10 three-carbon molecules are recycled (steps 4 through 6), forming 6 five-carbon molecules. These are phosphorylated by ATP to ribulose-1,5-diphosphate. Each of these five-carbon molecules then combines with a CO_2 molecule, starting the process once again. The enzyme catalyzing this step is *ribulose diphosphate carboxylase,* the most prevalent enzyme in the biological world. (Refer to Chapter 5 for an explanation of the process. ➥ p. 129)

Glossary/Index

allergies, 485, 493–508
autoimmune disorders, 509
Genetic resistance, and antibiotics, 347
genetics The science of heredity, including the structure and regulation of
genes and how these genes are passed between generations.
in bacterial taxonomy, 244
genetic engineering, 6, 208, 211–219
genetic studies, types of, 211
gene transfer, 196–208
history of, 20
mutations, 182–191
plasmids, 82, 208–210
study of, 8
gene transfer Movement of genetic information between organisms by
transformation, transduction, or conjugation, 196–208
comparison of methods, 207–208
conjugation, 204–207, 208
significance of, 196–197
transduction, 199–203, 208
transformation, 197–199, 208
genital herpes. *See* **herpes simplex virus type 2,** 588
Genitalia, external, 567–568
genital wart An often malignant wart associated with sexually transmitted
viral disease having a very high association rate with cervical cancer
(also called **condyloma**), 547–548, 590
cervical carcinoma and, 590
progression of, 590
genome The genetic information in an organism or virus, 256–257
genotype The genetic information contained in the DNA of an organism, 182
Gentamicin, 360, 365
for brucellosis, 677
Gentian violet, as antimicrobial agent, 326
genus A taxon consisting of one or more species; the first name of an
organism in the binomial system of nomenclature—for example,
Escherichia in *Escherichia coli.*, 223
German measles. *See* **rubella.**
Germicide, 318
germination The start of the process of development of a spore or an
endospore.
of endospores, 84
of spore, 154–155, 298
germ theory of disease Theory that microorganisms (germs) can invade
other organisms and cause disease, 11–14
Germ tubes, 298
Gerstmann-Strassler disease, 716, 717
Geysers, and bacteria, 147
Giant tube worms, endosymbiont bacteria in, 97
Giardia, 96, 294, 368
Giardia intestinalis, 5
disease from, 654–655
virulence factor, 392
in water, 745
giardiasis A gastrointestinal disorder caused by the flagellated protozoan
Giardia intestinalis, 654–655
symptoms of, 654
transmission of, 654
treatment of, 655
gingivitis The mildest form of periodontal disease, characterized by
inflammation of the gums, 636
gingivostomatitis Lesions of the mucous membranes of the mouth, 589
Global warming, 416, 729
Globulins, 667
glomerulonephritis Inflammation of and damage to the glomeruli of the
kidneys (also called Bright's disease), 453, 571
glomerulus A coiled cluster of capillaries in the nephron, 566
Gloves, use of, 428, 429
Glucose, 35–36

complete oxidation of, 108
Glutamic acid, 782
Glutaraldehyde, as antimicrobial agents, 326
Glycerol, 36, 780
Glycerol production, 777
glycocalyx Term used to refer to all substances containing polysaccharides
found external to the cell wall, 89
capsule, 89
slime layer, 89
Glycogen, 35, 36, 83
Glycolipids, 81
glycolysis An anaerobic metabolic pathway used to break down glucose into
pyruvic acid while producing some ATP, 109, 114–116, 125
phosphorylation, 116
glycoprotein A long, spikelike molecule made of carbohydrate and protein
that projects beyond the surface of a cell or viral envelope; some viral
glycoproteins attach the virus to receptor sites on host cells, while
others aid fusion of viral and cellular membranes, 81, 257
glycosidic bond A covalent bond between two monosaccharides, 36
Gnats, 313, 554
Goblet cells, 536
Golden Age of Microbiology, 20, 21
Goldsmith, Timothy, 237
Golgi, Camillo, 296
Golgi apparatus An organelle in eukaryotic cells that receives, modifies,
and transports substances coming from the endoplasmic reticulum, 93
Gonococcal arthritis, 576
Gonococcal conjunctivitis, 554
Gonococcal infections, 453
gonorrhea A sexually transmitted disease caused by *Neisseria gonorrhoeae,*
89, 209, 439, 574–579
complications of, 576
diagnosis of, 577
drug-resistance, 577
organism in, 574–575
progression of, 575–579
treatment of, 577–579
Gonyaulax, 291, 292
Gorgas, William Crawford, 296
Gracilicutes, 248
Graft rejection, 513, 514–515
Grafts, 565
graft tissue Tissue that is transplanted from one site to another, 513
graft-versus-host (GVH) disease Disease in which host antigens elicit an
immunological response from graft cells that destroys host tissue, 513
Grains, organisms of, 757
Gram, Hans Christian, 64, 359
Gramicidin, 358
gram molecular weight. *See* **mole.**
Gram-negative bacteria
aerobic rods and cocci, 249
anaerobic cocci, 249
anaerobic straight/curved/helical rods, 249
cell wall of, 78–80
diseases from, 249
enzymes produced by, 153
facultatively anaerobic Gram-negative rods, 249
flagellum of, 86
staining of, 64, 66
Gram-nonreactive organisms, staining of, 64
Gram-positive bacteria
cell wall of, 78, 79
cocci, 250
endospore-forming rods and cocci, 250
enzymes produced by, 152
flagellum of, 86
irregular nonsporing rods, 250

major histocompatibility complex (MHC) A group of cell surface proteins that are essential to immune recognition reactions, 476, 513

Malachite green, 64, 67

malaria A severe parasitic disease caused by several species of the protozoan *Plasmodium* and transmitted by mosquitoes, 71, 291, 294–295, 296, 693–696

anemia as protection against, 694

causes of, 314, 693–694

diagnosis of, 695

prevention of, 418–420

progression of, 694–695

treatment/prevention of, 368, 695–696

Male cells, 205

male reproductive system The host system consisting of the testes, ducts, specific glands, and the penis.

components of, 567–568

infections of, 567, 568

malignant Relating to a tumor that is cancerous, 283

Malignant cells, 278, 473

Malignant diseases, 397–398

Kaposi's sarcoma, 264, 283, 520–521

Mallon, Mary, 648

Malnutrition, 518

Malta fever. *See* **brucellosis.**

Maltase, 153

malted Referring to cereal grains that are partially germinated to increase the concentration of starch-digesting enzymes, 775

mammary gland A modified sweat gland that produces milk and ducts that carry the milk to the nipple, 567

Mange, sarcoptic, 560

Manmade mineral fibers (MMMF), 736

Mannitol, 36

Mannitol-fermentation test, 118, 119

Mannose-binding protein (MBP), 453–455

Marble, bacterial action on, 33

Marburg disease, 263

Marburg virus, 690

Margulis, Lynn, 96–97, 226, 228, 248

Marine environments, 742–743

Marshall, Barry, 647

Mascara, 555

mash Malted grain that is crushed and mixed with hot water, 775

Mason, Dr. Arthur D., Jr., 430

mast cell A leukocyte that releases histamine during an allergic response, 437

mastigophoran A flagellate protozoan such as *Giardia*, 294

Mastitis, 567

mastoid area Portion of the temporal bone prominent behind the ear opening, 599

Mastoiditis, 599

Mastotermes darwiniensis, 96

Materia Medica (Dioscorides), 17

matrix Fluid-filled inner portion of a mitochondrion, 92

maturation The process by which complete virions are assembled from newly synthesized components in the replication process, 267, 268, 276

MBC. *See* Minimum bacterial concentration

measles A febrile disease with rash caused by the rubeola virus, which invades lymphatic tissue and blood (also called **rubeola**), 384, 542, 543–544

diagnosis and treatment of, 544

immunization against, 544

transmission of, 543–544

vaccine for, 544

measles encephalitis A serious complication of measles that leaves many survivors with permanent brain damage, 543

Meat

inspection of, 758

organisms of, 758–759, 763

mebendazole An antihelminthic agent that blocks glucose uptake by parasitic roundworms, 369, 370

mechanical stage Attachment to a microscope stage that holds the slide and allows precise control in moving the slide, 55

mechanical vector A vector in which the parasite does not complete any part of its life cycle during transit, 289, 414, 415

Medawar, Peter, 256

Media. *See* Culture media

Medical products, from recombinant DNA, 215–217

medium A mixture of nutritional substances on or in which microorganisms grow, 138

Mefloquine, 368

megakaryocyte Large cell normally present in bone marrow that gives rise to platelets, 436

meiosis Division process in eukaryotic cells that reduces the chromosome number by half, 92

Melarsoprol, 719

membrane attack complex A set of proteins in the complement system that lyses invading bacteria by producing lesions in their cell membranes, 452–453

membrane filter method Method of testing for coliform bacteria in water in which bacteria are filtered through a membrane and then incubated on the membrane surface in growth medium, 746–747

Membrane filters, pore sizes, 334–336

Membranes. *See also* **Cell membrane; Plasma membrane**

in cell walls, 77

Membrane transport, 97–102, 133

active transport, 98, 101, 133

phosphotransferase system, 133

Memory, in immune system, 466, 467

memory cell Long-lived B or T lymphocyte that can carry out an anamnestic or secondary response, 466, 470, 477

Mendel, Gregor, 20

Mendosicutes, 248

meninges Three layers of membrane that protect the brain and spinal cord, 701

Meningitis, 302, 396, 702–704

Meningitis belt, 417

Meningococcal infections, 453

Meningococcal meningitis, 703–704

progression of, 703

treatment of, 703–704

Meningoencephalitis, herpes, 589, 707

Mental disease, nature of, 383

Mental retardation from meningitis, 704

Mercuric naphthenate, 324

Mercurochrome, 324

Mercury, for treating syphilis, 17

Mercury compounds, as antimicrobial agents, 324

Merisopedia, 74

merozoite A malaria trophozoite found in infected red blood or liver cells, 294, 295

Merthiolate, 324

mesophile An organism that grows best at temperatures between 25° and 40°C, including most bacteria, 146

mesophilic spoilage Spoilage due to improper canning procedures or because the seal has been broken, 764

Mesosomes, 82

messenger RNA (mRNA) A type of RNA that carries the information from DNA to dictate the arrangement of amino acids in a protein, 167, 173, 174

properties of, 174

metabolic pathway A chain of chemical reactions in which the product of one reaction serves as the substrate for the next, 109, 116

metabolism The sum of all chemical processes carried out by living organisms, 106–136

aerobic, 119–125

obligate aerobe A bacterium that must have free oxygen to grow, 148
  culture of, 158–159
obligate anaerobe A bacterium that is killed by free oxygen, 148
  culture of, 159
obligate intracellular parasite An organism or virus that can live or multiply only inside a living host cell, 256
obligate parasite A parasite that must spend some or all of its life cycle in or on a host, 289
obligate psychrophile An organism that cannot grow at temperatures above 20°C, 146
obligate thermophile An organism that can grow only at temperatures above 37°C, 147
Oceans, 742–743
  characteristics of ocean water, 742
  organisms of, 743
Octet rule, 28
ocular lens Lens in microscope that further magnifies the image created by the objective lens, 55
ocular micrometer A glass disk with an inscribed scale that is placed inside the eyepiece of a microscope; used to measure the actual size of an object being viewed, 55–56
Oculinum A, 390
Oil spills, bioremediation for cleanup, 128, 751–752
Oil spraying, in control of vector-transmitted diseases, 415
Okazaki fragment One of the short, discontinuous DNA segments formed on the lagging strand during DNA replication, 169
Oleic acid, 38
Olives, production of, 775
Onchocerca volvulus, 309, 313, 369, 556
onchocerciasis An eye disease caused by the filarial larvae of the nematode *Onchocerca volvulus,* transmitted by blackflies; common in Africa and Central America (also known as river blindness), 369, 556
oncogene A cancer-causing gene, 283–284
Oncoviruses, 390–391
ONPG and MUG test Water purity test that relies on the ability of coliform bacteria to secrete enzymes that convert a substrate into a product that can be detected by a color change, 747
On the Nature of Things (De Rerum Natura), 8
Oomycota. *See* **water mold.**
Operator, 180
operon A sequence of closely associated genes that includes both structural genes and regulatory sites that control transcription, 180
Operon theory, 180
ophthalmia neonatorum Pyogenic infection of the eyes caused by *Neisseria gonorrhoeae* (also known as conjunctivitis of the newborn), 554
opportunist A species of resident or transient microbiota that does not ordinarily cause disease but can do so under certain conditions, 380–381
Opportunistic infections, in AIDS, 299, 302, 519, 520
opsonin An antibody that promotes phagocytosis when bound to the surface of a microorganism, 452, 472
opsonization The process by which microorganisms are rendered more attractive to phagocytes by being coated with antibodies (opsonins) and C3b complement protein (also called immune adherence), 450–452, 454, 472
optical microscope. *See* **compound light microscope.**
optimum pH The pH at which microorganisms grow best, 145
Oral cavity, 632–637
Oral cavity diseases
  dental caries, 633–635
  dental plaque, 632–633, 634, 635, 636, 637
  mumps, 637–638
  periodontal disease, 635–637
  thrush, 551, 638
  viral, 637–638
Oral contraceptives, 575
  rifampin and, 362

Oral hairy leukoplakia, 520
Oral rehydration therapy, 643
Oral sex and gonorrhea, 575
Oral Vaccination Program, 418
orbivirus Type of virus that causes Colorado tick fever, 264, 691
Orchids, fungi on roots of, 299
orchitis Inflammation of the testes; a symptom of mumps in postpubertal males, 637
Order, in taxonomy, 224
organelle An internal membrane-enclosed structure found in eukaryotic cells, 71–72, 96
Organic acids, 35
organic chemistry The study of compounds that contain carbon, 34
Organic compounds, 34–45
  carbohydrates, 35–37
  carbon compounds, 34
  industrial production of, 779–780
  lipids, 38–40
  nucleic acids, 43–45
  nucleotides, 42–43
  proteins, 41–42
Organic products, microorganisms in production of, 769–775
Organic wastes, 744
Organ transplant. *See* Transplantation
Oriental sore, 692
Origin of life, 95–97
Ornithodorus, 312
ornithosis Disease with pneumonia-like symptoms, caused by *Chlamydia psittaci* and acquired from birds (previously called psittacosis and parrot fever), 615
Oromandibular dystonia, 390
Oroya fever One form of bartonellosis; an acute fatal fever with severe anemia (also called Carrion's disease), 686
orthomyxovirus A medium-sized, enveloped RNA virus that varies in shape from spherical to filamentous and has an affinity for mucus, 262, 617
  characteristics of, 263
Orthopoxviruses, 265
Orthoreoviruses, 264
osmosis A special type of diffusion in which water molecules move from an area of higher concentration to one of lower concentration across a selectively permeable membrane, 75, 99–101
osmotic pressure The pressure required to prevent the net flow of water molecules by osmosis, 101
  as antibacterial method, 336
  microbial growth and, 149–150
Ossicles, 598
Osteoclasts, 438
otitis externa Infection of the external ear canal, 603
otitis media Infection of the middle ear, 603
Outer ear, 598
outer membrane A bilayer membrane, forming part of the cell wall of Gram-negative bacteria, 77
ovarian follicle An aggregation of cells in the ovary containing an ovum, 567
ovary In the female, one of a pair of glands that produce ovarian follicles, which contain an ovum and hormone-secreting cells, 567
Oxacillin, 355, 356
Oxaloacetic acid (oxaloacetate), 120
oxidation The loss of electrons and hydrogen atoms, 106
  events of, 107
  meaning of, 35
oxidative phosphorylation Process in which the energy of electrons is captured in high-energy bonds as phosphate groups combine with ADP to form ATP, 122
Oxidizing agents, as antimicrobial agents, 321, 325
Oxygen
  microbial growth and, 148–149
  structure of, 27

leishmaniasis, 291, 313, 314, 692–693
Lyme disease, 312, 314, 680–682
malaria, 71, 291, 294–295, 296, 693–696
parvovirus infections, 691
plague. *See* Plague
protozoan, 692–698
relapsing fever, 679–680
rickettsial and related, 682–687
rickettsialpox, 685
Rocky Mountain spotted fever, 312, 314, 684–685
toxoplasmosis, 291, 295, 696
trench fever, 314, 685–686
tularemia, 248, 312, 314, 677–678
typhus fever, 10, 314, 683–684, 685
viral, 687–692
yellow fever, 263, 266, 314, 688
systemic infection An infection that affects the entire body (also called generalized infection), 393
systemic lupus erythematosus A widely disseminated, systemic autoimmune disease resulting from production of antibodies against DNA and other body components, 453
characteristics of, 512–513
DNA and, 167
Systemic mycoses, 299

Taenia, 304, 307
Taenia saginata, 304, 307, 658
Taenia solium, disease from, 658
Tagament, 649
Tagged antibody tests, 529–531
Tampons, and toxic shock syndrome, 573
tapeworm Flatworm that lives in the adult stage as a parasite in the small intestine of animals., 289, 304, 369
characteristics of, 307–308
life cycle of, 307–308
types of, 307
Tapeworm infections, 658–659
diagnosis of, 659
sources of, 658–669
Taq DNA polymerase, 236
tartar Calcium deposition on dental plaque forming a very rough, hard crust, 636
Tate, P., 296
Tatum, Edward, contributions of, 20
taxonomy The science of classification, 222–254. *See also* Bacterial taxonomy
bacterial, 244–251
defined, 222
dichotomous key in, 224–225
evolutionary relationships, search for, 236–243
five-kingdom system, 227–231
genetic homology, 238–241
immunological reactions as basis, 242
Linnaeus taxonomy, 223–224
numerical taxonomy, 238
phage typing as basis, 243
problems related to, 225–226
prokaryotes, special methods for, 237–238
purpose of, 222
ribosomal properties as basis, 241
three-domain classification system, 231–236
viruses, classification of, 236, 259–261
taxon (plural: *taxa*) A category used in classification, such as species, genus, order, family, 222
T cell. *See* **T lymphocyte.**
T-dependent antigen Antigen requiring helper T cell (T$_H$2) activity to activate B cells.

Tears, 469
Teeth. *See also* Oral cavity diseases
components of, 630–631
tetracycline and staining of, 360, 361
teichoic acid A polymer attached to peptidoglycan in Gram-positive cell walls, 77
temperate phage A bacteriophage that does not cause a virulent infection; rather, its DNA is incorporated into the host cell chromosome, as a prophage, and replicated with the chromosome, 201, 270–272
Temperature
effects on enzymes, 113–114
microbial growth and, 146–148
template DNA used as a pattern for the synthesis of a new nucleotide polymer in replication or transcription, 167
temporary parasite A parasite that feeds on and then leaves its host (such as a biting insect), 289
Tenericutes, 248
teratogen An agent that induces defects during embryonic development, 278–279
teratogenesis The induction of defects during embryonic development, 278–279
Terbinafine, 366
terminator codon A codon that signals the end of the information for a particular protein (also called **nonsense codon** or stop codon), 173
Termites, 38, 294, 376
Terramycin, 360
tertiary structure The folding of a protein molecule into globular shapes, 41, 42
tertiary treatment Chemical and physical treatment of sewage to produce an effluent of water pure enough to drink, 748, 750
test A shell made of calcium carbonate and common to some protists. *See also* Diagnostic procedures and tests; specific tests
testis (plural: *testes*) One of a pair of male reproductive glands that produce testosterone and sperm, 567
tetanus Disease caused by *Clostridium tetani* in which muscle stiffness progresses to eventual paralysis and death (also called lockjaw), 710–711
progression of, 711
tetanus neonatorum, 711
vaccine, 485, 711
tetanus neonatorum Type of tetanus acquired through the raw stump of the umbilical cord, 711
Tetrachloroethylene, 660
tetracycline An antibacterial agent that inhibits protein synthesis, 209, 360, 361, 365, 382
diseases treated with, 360, 677
toxic effects, 360
types of, 360
tetrad Cuboidal groups of four cocci, 138
arrangement of, 74, 75
Tetrapeptides, 77
thallus The body of a fungus, 297
theca A tightly affixed, secreted outer layer of dinoflagellates that often contains cellulose, 293
Theophylline, 495
therapeutic dosage level Level of drug dosage that successfully eliminates a pathogenic organism if maintained over a period of time, 342
thermal death point The temperature that kills all the bacteria in a 24-old broth culture at neutral pH in 10 minutes, 327
thermal death time The time required to kill all the bacteria in a particular culture at a specified temperature, 327
Thermal hot springs, 740
thermoacidophile A member of one of the groups of the archaeobacteria that live in extremely hot, acidic environments, 228–229
extreme, 233
Thermoduric organisms, 146
Thermophila acidophilum, 146

PHOTO CREDITS

Halftitle Tom Van Sant/Geosphere Project/Science Photo Library/Photo Researchers, Inc.

About the Author/Owner's Manual Jacquelyn G. Black

Chapter 1 Chapter Opening Tom Van Sant/Geosphere Project/Science Photo Library/Photo Researchers, Inc. **1.1** Jacquelyn G. Black **1.2a** CNRI/Science Photo Library/Photo Researchers, Inc. **1.2b** R.B.Taylor/Science Photo Library/Photo Researchers, Inc. **1.2c** CBC/Phototake NYC **1.2d** Thomas Broker/CNRI/Phototake NYC **1.2e** Michael Abbey/Visuals Unlimited **1.2f** Cath Ellis, Department of Zoology, University of Hull/Science Photo Library/Photo Researchers, Inc. **1.3a, b, d, e** United States Department of Agriculture **1.3c** David Parker/Science Photo Library/Photo Researchers, Inc. **1.4** Pieter Brueghel the Elder (1528-1569), Flemish, *Trionfo della Morte.* Painting, Prado, Madrid, Spain/Scala/Art Resource NY **1.5** Corbis-Bettmann **Box, p. 10, bottom** Marcel Miranda **Box, p. 10, top** Beckman Coulter, Inc. **1.7** Charles O'Rear/Westlight **1.8** The Granger Collection **1.9** Science Photo Library/Photo Researchers, Inc. **1.10a** The Granger Collection **1.10b** National Library of Medicine/Science Photo Library/Custom Medical Stock Photo, Inc. **1.11** Georges-Gaston Melingue, *The First Vaccination,* 1796. The Granger Collection **1.12** Stock Montage, Inc./Historical Pictures Collection **1.13a** Omikron/Photo Researchers, Inc. **1.14** Corbis-Bettmann **1.15** UPI/Corbis-Bettmann **1.16** United States Department of Agriculture

Chapter 2 Chapter Opening Tom Sanders/Photri/The Stock Market **2.6b** Biophoto Associates/Photo Researchers, Inc. **Box, p. 33** Leonardo da Vinci, close-up of *The Last Supper.* Ultima cena-p. (tre apostoli a destra). Santa Maria della Grazie, Milano. Scala/Art Resource **2.18a, b, c** Jacquelyn G. Black

Chapter 3 Chapter Opening Dr. Tony Brain & David Parker/Science Library/Photo Researchers, Inc. **3.1b** John D. Cunningham/Visuals Unlimited **3.8** Southern Illinois University Biomedical Communications/Custom Medical Stock Photo, Inc. **Box, p. 53** Richard Megna/Fundamental Photographs **3.10c** Runk/Schoenberger/Grant Heilman Photography, Inc. **3.11** Carl Zeiss, Inc. **3.14a-b** Jim Solliday/Biological Photo Service **3.15** Biophoto Associates/Photo Researchers, Inc. **3.16** Biological Photo Service **3.17** Charles J. Wrobel, M.D., University of California at San Diego Medical Center **3.18a-d** David M. Phillips/Visuals Unlimited **3.19b** Jean Claude Revy/Phototake NYC **3.20a** Eric V. Grave/Photo Researchers, Inc. **3.20b** Biophoto Associates/Science Source/Photo Researchers, Inc. **3.21** John J. Cardamone Jr. & B.A. Phillips/Biological Photo Service **3.23** Biological Photo Service **3.24a** Ralph Slepecky/Visuals Unlimited **3.24b** R. Kessel & G. Shih/Visuals Unlimited **3.25a** Visuals Unlimited **3.25b** Dr. Heinz Schlesner, University of Kiel, Germany **3.25c** Dr. Anne Smith/Science Photo Library/Photo Researchers, Inc. **3.26** Courtesy of International Business Machine Corporation. Unauthorized use not permitted. **Table 3.2a, f, g** David M. Phillips/Visuals Unlimited **Table 3.2b** Jim Solliday/Biological Photo Service **Table 3.2c** Biophoto Associates/Photo Researchers, Inc. **Table 3.2d** Biological Photo Service **Table 3.2e** Charles J. Wrobel, M.D., University of California at San Diego Medical Center **3.27b** A.M. Siegelman/Visuals Unlimited **Table 3.3a** Michael Abbey/Photo Researchers, Inc. **Table 3.3b, e** George A. Wistreich, East Los Angeles College **Table 3.3c** John D. Cunningham/Visuals Unlimited **Table 3.3d** Jack Bostrack/Visuals Unlimited **3.28a-d** Raymond Otero/Visuals Unlimited **3.28e-f** George A. Wistreich, East Los Angeles College **3.29** John D. Cunningham/Visuals Unlimited **3.30** Jack Bostrack/Visuals Unlimited **3.31** George A. Wistreich, East Los Angeles College

Chapter 4 Chapter Opening (a) Sarah Jackson; Edifice/Corbis Corporation **Chapter Opening (b, d)** Kevin Schafer/Corbis Corporation **Chapter Opening (c)** Bob Krist/Corbis Corporation **Chapter Opening (e)** Dave Bartruff/Corbis Corporation **Chapter Opening (f)** Kevin Galvin/The Stock Market **4.2a-1** CNRI/Photo Researchers, Inc. **4.2a-2** David M. Phillips/Visuals Unlimited **4.2b** Cabisco/Visuals Unlimited **4.2c** R. Kessel & G. Shih/Visuals Unlimited **4.2d** Dr. Tony Brain/Science Photo Library/Photo Researchers, Inc. **4.2e** James T. Staley, University of Washington **4.2f** Dr. Heinz Schlesner, University of Kiel, Germany **4.2g** Photo by Dr. Walter Stoeckenius, courtesy of Dr. Hanna Stoeckenius **4.6a** Biological Photo Service **4.6b** Dr. T.J. Beveridge/Biological Photo Service **4.6c** Terrance J. Beveridge & T. Paul/Visuals Unlimited **4.9** CNRI/Science Photo Library/Custom Medical Stock Photo, Inc. **4.10** Paul W. Johnson/Biological Photo Service **Box, p. 84 (a)** D. Balkwill & D. Maratea/Visuals Unlimited **Box, p. 84 (b)** Richard Blakemore, University of New Hampshire **4.11** Institut Pasteur/CNRI/Phototake NYC **4.12a, c** E.C.S. Chan/Visuals Unlimited **4.12b** Jack M. Bostrack/Visuals Unlimited **4.12d** John D. Cunningham/Visuals Unlimited **4.12e** Fred Hossler/Visuals Unlimited **4.13b** Julius Adler/Visuals Unlimited **4.15a** CNRI/Science Photo Library/Photo Researchers, Inc. **4.15b** Dr. Max Listgarten, School of Dental Medicine, University of Pennsylvania, as published in *Journal of Bacteriology* 88:1087-1103, 1964 **4.16** Charles C. Brinton, Jr., University of Pittsburgh, & Judith Carnahan **4.17** Chris Bjornberg/Photo Researchers, Inc. **4.19a-b** Don Fawcett/Photo Researchers, Inc. **4.21b** Charles J. Flickinger, University of Virginia School of Medicine **4.22b** Dr. Kenneth R. Miller/Science Photo Library/Photo Researchers, Inc. **4.24a** Manfred Kage/Peter Arnold, Inc. **4.25b** Michael Abbey/Visuals Unlimited **4.26** Jian-Ming Li & Nancy Martin, University of Louisville School of Medicine **4.27** Micrograph by David G. Chase from *Early Life* by Lynn Margulies. (©1984 by Jones & Bartlett Publishers, Inc., Fig. 4-8, p. 90.)

Chapter 5 Chapter Opening Jacquelyn G. Black **5.5a** Clive Freeman, The Royal Institution/Science Photo Library/Photo Researchers, Inc. **5.12a-b** SuperStock, Inc. **5.12c** Mike Pares/Custom Medical Stock Photo, Inc. **5.15** George A. Wistreich, East Los Angeles College **Box, p. 128** Exxon Corporation **5.28a** John D. Cunningham/Visuals Unlimited **5.28b** Norbert Wu/Tony Stone Images

Chapter 6 Chapter Opening Jacquelyn G. Black **6.1b** George Musil/Visuals Unlimited **6.1c** Ralph A. Slepecky/Visuals Unlimited **6.2** G. Shih & R. Kessel/Visuals Unlimited **6.5** Wheaton Science Products, Inc. **6.8a** Society for Industrial Microbiology/Science/Visuals Unlimited **6.8b** Biological Photo Service **6.8c** Michael G. Gabridge/Visuals Unlimited **6.11** Raymond B. Otero/Visuals Unlimited **6.12** Spectronic Instruments, Inc. **6.13** Stephen Trimble/DRK Photo **6.16b** Michael Pasdzior/The Image Bank **Box, p. 152 (a)** Agricultural Research Service, USDA **(b)** David M. Phillips. Visuals Unlimited **6.18a** Soad Tabaqchali/Visuals Unlimited **6.18b** Alfred Pasieka/Science Photo Library/Photo Researchers, Inc. **6.19b** Christine Case/Visuals Unlimited **Box, p. 158** Becton Dickinson Microbiology Systems **Table 6.5 (a)** George A. Wistreich, East Los Angeles College **(b)** Runk Schoenberger/Grant Heilman Photography, Inc. **(c)** Cytographics, Inc./Visuals Unlimited **(d)** LeBeau/Custom Medical Stock Photo, Inc. **(e, g)** SuperStock, Inc. **(f)** Raymond B. Otero/Visuals Unlimited **6.21** Jack M. Bostrack/Visuals Unlimited **6.22** Shellab **6.23** Becton Dickinson Microbiology Systems **6.24** API/CounterPart Diagnostics/bioMerieux Vitek, Inc.

Chapter 7 Chapter Opening Information Staff/Agricultural Research Service, USDA **7.4** NIH/Kakefuda/Science Source/Photo Researchers, Inc. **7.9b** Tripos, Inc. **7.10a** O.L. Miller, Jr. & B.R. Beatty/*Journal of Cellular Physiology* 74 (1969). **7.10b** E. Kiselva & D. Fawcett/Visuals Unlimited **7.11** E. Kiselva & D. Fawcett/Visuals Unlimited **7.22a** Steven Frame/Stock Boston **7.22b** Ken Greer/Visuals Unlimited **Box, p. 189** Deneb Karentz, University of San Francisco **7.25a** Bruce N. Ames, University of California at Berkeley **Box, p. 192 (a)** Dr. Gerald Goldstein/Licking County Archaeological and Landmarks Society **(b)** Licking County Archaeological and Landmarks Society

PC1

19.11 World Health Organization 19.12 George A. Wistreich, East Los Angeles College 19.13a-b Kenneth Greer/Visuals Unlimited 19.14 Everett S. Beneke/Visuals Unlimited 19.15a-b Julius Kane/Michael R. McGinnis, University of Texas Medical Branch at Galveston 19.16 Kenneth Greer/Visuals Unlimited 19.16b Project Masters, Inc./The Bergman Collection 19.17a-b Ed Rottinger/Custom Medical Stock Photo, Inc. 19.18 National Medical Slide/Custom Medical Stock Photo, Inc. 19.19 Science VU/Visuals Unlimited 19.20 Carl Purcell/Photo Researchers, Inc. 19.21 Science VU/Armed Forces Institute of Pathology/Visuals Unlimited 19.22 Science VU/Visuals Unlimited 19.23 Ann Czapiewski/Animal and Plant Health Inspection Service, USDA

Chapter 20 Chapter Opening Jerome Duquesnoy I, *Mannekin-Pis,* 1619, bronze. Brussels, Belgium. Kavaler/Art Resource, New York. 20.4 Jon Meyer/Custom Medical Stock Photo, Inc. 20.6 CNRI/Science Photo Library/Photo Researchers, Inc. 20.7 CNRI/Phototake NYC 20.8 David M. Phillips/Photo Researchers, Inc. 20.9a A.M. Siegelman/Visuals Unlimited 20.9b Kwangshin Kim/Photo Researchers, Inc. 20.10a Kenneth Greer/Visuals Unlimited 20.10b Centers for Disease Control and Prevention CDC Box, p. 578 (a) Centers for Disease Control and Prevention CDC (b) Science VU/Visuals Unlimited 20.13a Centers for Disease Control and Prevention CDC 20.14a Southern Illinois University/Visuals Unlimited 20.14b Science VU/Visuals Unlimited 20.15a Science VU/Visuals Unlimited 20.15b Kenneth E. Greer, M.D., University of Virginia Health Sciences Center 20.15c Science VU/Visuals Unlimited 20.16 Centers for Disease Control and Prevention CDC 20.17a Science VU/Visuals Unlimited 20.17b Centers for Disease Control and Prevention CDC 20.18 Centers for Disease Control and Prevention CDC 20.20 CNRI/Phototake NYC 20.21 Centers for Disease Control and Prevention CDC 20.22 Kenneth E. Greer/Visuals Unlimited 20.23 Centers for Disease Control and Prevention CDC

Chapter 21 Chapter Opening Illustration by Andre Gill from Henry Murger's *Scenes de la Vie de Boheme*/The Granger Collection 21.3a Biophoto Associates/Photo Researchers, Inc. 21.3b Southern Illinois University/Peter Arnold, Inc. 21.4 G.W. Willis, M.D./Biological Photo Service 21.5 Science VU/Visuals Unlimited 21.8 Dr. S. Girod de Bentzmann 21.9a-b David M. Phillips/Visuals Unlimited 21.10 CDC/Science Source/Photo Researchers, Inc. 21.12a-b Biophoto Associates/Science Source/Photo Researchers, Inc. 21.13 National Museum of Health and Medicine/Armed Forces Institute of Pathology Box, p. 615 United States Department of Agriculture 21.14 Reproduced from T.F. McCaul & J.C. Williams, *Journal of Bacteriology* 147:1063-1076, Figure 1b, with permission of the American Society for Microbiology 21.15 R.J. Hawley & T. Imaeda/Visuals Unlimited 21.17 Iowa State University Library/University Archives 21.18b Omikron/Photo Researchers, Inc. 21.19a National Institute of Allergy and Infectious Diseases/NIH 21.19b Science VU/Visuals Unlimited 21.20 G.W. Willis, M.D./Biological Photo Service 21.21 A. M. Siegelman/Visuals Unlimited

Chapter 22 Chapter Opening Mary Teresa Giancoli 22.3a Fred E. Hossler/Visuals Unlimited 22.3b Dr. Ross P. Karlin 22.4 S. Rosen, Lenney, and O'Malley, "Dental caries in gnotobiotic rats," *Journal of Dental Research,* vol. 47, no. 3, p. 362, May-June 1968. 22.5 Dr. Ross P. Karlin 22.6a Dr. R. Gottsegen/Peter Arnold, Inc. 22.6b Science VU/Max Listgarten/Visuals Unlimited 22.8 John D. Cunningham/Visuals Unlimited 22.9 London School of Hygiene and Tropical Medicine/Science Photo Library/Photo Researchers, Inc. 22.10 Veronika Burmeister/Visuals Unlimited Box, p. 645 Michel Ulli/Reuters/Corbis-Bettmann 22.11 Veronika Burmeister/Visuals Unlimited 22.12 A.B. Dorsett/Science Photo Library/Photo Researchers, Inc. 22.13 K.G. Murti/Visuals Unlimited 22.14 Scott Camazine/Photo Researchers, Inc. Box, p. 653 AP/Wide World Photos 22.15a Jerome Paulin/Visuals Unlimited 22.16 Lauritz Jensen/Visuals Unlimited 22.17a Larry Jensen/Visuals Unlimited 22.18a John D. Cunningham/Visuals Unlimited 22.19a W. Ormerod/Visuals Unlimited 22.19b Forest W. Buchanan/Visuals Unlimited 22.20 Bruce Iverson/Science Photo Library/Photo Researchers, Inc. 22.21a Stanley Flegler/Visuals Unlimited 22.21b Bruce Iverson/Science Photo Library/Photo Researchers, Inc. 22.21c Larry Jensen/Visuals Unlimited 22.22 James Solliday/Biological Photo Service 22.23 R. Calentine/Visuals Unlimited 22.24 Dr. Daniel H. Connor/Armed Forces Institute of Pathology 22.25 Arthur M. Siegelman/Visuals Unlimited

Chapter 23 Chapter Opening Information Staff/Agricultural Research Service, USDA 23.2 Barts Medical Library/Phototake NYC Box, p. 669 Howard Shiang, D.V.M., *Journal of American Veterinary Medical Association* 163:981, Oct. 1973. 23.4a National Institute of Allergy and Infectious Diseases/NIH 23.4b Sinclair Stammers/Science Photo Library/Photo Researchers, Inc. 23.4c National Institute of Allergy and Infectious Diseases/National Institutes of Health 23.5a Institut Pasteur/Phototake NYC 23.5b Fred Marsik/Visuals Unlimited 23.6 Science VU/Charles W. Stratton/Visuals Unlimited 23.8 The Granger Collection 23.10a Eric Grave/Phototake NYC 23.10b National Medical Slide/Custom Medical Stock Photo, Inc. 23.12a Centers for Disease Control and Prevention CDC 23.12b Ken Greer/Visuals Unlimited 23.14a R. Calentine/Visuals Unlimited 23.14b Kenneth Greer/Visuals Unlimited 23.15 Armed Forces Institute of Pathology 23.16 George Musil/Visuals Unlimited 23.17 Science VU/Visuals Unlimited Box, p. 690 Patric Robert/Sygma 23.18a George A. Wistreich, East Los Angeles College 23.18b Science VU/Visuals Unlimited 23.20 Biophoto Associates/Science Source/Photo Researchers, Inc. 23.21 Moredun Animal Health Ltd./Science Photo Library/Custom Medical Stock Photo, Inc.

Chapter 24 Chapter Opening Marshall Gordon/PhotoDisc, Inc. 24.2 Edward J. Bottone, Mount Sinai School of Medicine 24.3 Science VU/Visuals Unlimited 24.5a CNRI/Phototake NYC 24.5b Ken Greer/Visuals Unlimited 24.6 Custom Medical Stock Photo, Inc. 24.7a Dr. T.J. Beveridge/Biological Photo Service 24.7b Biological Photo Service 24.8 Charles Bell (1774-1842), *Oposthotunus.* Reproduced by kind permission of The Royal College of Surgeons of Edinburgh 24.10a-b March of Dimes Birth Defects Foundation 24.12a-b Frederick C. Skvara, M.D. 24.13 D. Carleton Gajdusek, M.D. 24.14 United States Department of Agriculture 24.15 Nigel Dickinson/Still Pictures/Peter Arnold, Inc. 24.16a A.M. Siegelman/Visuals Unlimited 24.16b R. Ashley/Visuals Unlimited

Chapter 25 Chapter Opening SuperStock, Inc. 25.5a David Hall/Science Photo Library/Photo Researchers, Inc. 25.5b John D. Cunningham/Visuals Unlimited 25.5c A.M. Siegelman/Visuals Unlimited 25.6 Paul W. Johnson/Biological Photo Service 25.8 F. Widdel/Visuals Unlimited 25.10a Graseby Andersen, Inc. 25.10b Forma/Science VU/Visuals Unlimited 24.14 Dwight Kuhn Photography 25.15 Bruce F. Molnia/Biological Photo Service 25.16a Manfred Kage/Peter Arnold, Inc. 25.16b Paul W. Johnson/Biological Photo Service 25.18a Leon J. LeBeau/Biological Photo Service 25.18b Raymond B. Otero/Visuals Unlimited 25.19 Colilert photo courtesy of IDEXX Laboratories, Inc. 25.21 R.F. Ashley/Visuals Unlimited 25.22a Judith F.M. Hoeniger, University of Toronto/Biological Photo Service 25.22b Paul W. Johnson/Biological Photo Service 25.23 Larry Lefever/Grant Heilman Photography, Inc. Box, p. 753 AP/Wide World Photos

Chapter 26 Chapter Opening Corbis-Bettmann 26.1 R. Calentine/Visuals Unlimited 26.2 United States Department of Agriculture Box, p. 759 United States Department of Agriculture 26.3 Jane Latta/Photo Researchers, Inc. 26.4 Larry Lefever/Grant Heilman Photography, Inc. Box, p. 767 International Atomic Energy Agency 26.5 Tim McCabe/ United States Department of Agriculture 26.6 Biophoto Associates/Photo Researchers, Inc. 26.7a-c John Colwell/Grant Heilman Photography, Inc. 26.8 John Colwell/Grant Heilman Photography, Inc. 26.9 David M. Phillips/Visuals Unlimited 26.10a Bill Frantz Photography 26.12a Fred Lyon 26.12b Sylvain Grandadam/Photo Researchers, Inc. 26.12c Photo Researchers, Inc. 26.13a Joseph Nettis/Photo Researchers, Inc. 26.13b Science VU/Visuals Unlimited 26.15 Warren Gretz/National Renewable Energy Laboratory 26.16a A. McClenaghan/Science Photo Library/Photo Researchers, Inc. 26.16b Dan McCoy/Rainbow 26.17 Hank Morgan/Photo Researchers, Inc. 26.18a-b Corale L. Brierley/Visuals Unlimited

Appendix B B.1 CNRI/Science Photo Library/Photo Researchers, Inc. B.2 Heather Davies/Science Photo Library/Photo Researchers, Inc. B.3 David M. Phillips/Visuals Unlimited B.4 C.P. Venae/Visuals Unlimited B.5 Dr. Edward J. Bottone, Mount Sinai Medical Center B.6 Christine Case/Visuals Unlimited B.7 Bruce Iverson B.8 F. Widdel/Visuals Unlimited B.9a H. Pol/CNRI/Science Photo Library/Photo Researchers, Inc. B.9b Fred Hossler/Visuals Unlimited B.10 Michael Gabridge/Visuals Unlimited B.11 Dr. T.J. Beveridge/Biological Photo Service B.12 Eric V. Grave/Photo Researchers, Inc. B.13 Science VU/Visuals Unlimited B.14 Cabisco/Visuals Unlimited B.15 Centers for Disease Control and Prevention CDC B.16 J. Poindexter/Science VU/Visuals Unlimited B.17 D. Foster, Woods Hole Oceanographic Institution/Science VU/Visuals Unlimited

PATHOGENS AND THE DISEASES THEY CAUSE

VIRUSES

Virus	Group Family	Disease	Page	Virus	Group Family	Disease	Page
adenovirus	Adenoviridae	acute upper & lower respiratory tract distress, pharyngitis, pneumonia, follicular conjunctivitis, epidemic keratoconjunctivitis	257, 258 622	herpes simplex type 2	Herpesviridae	genital herpes, oral & whitlow	277 586
				herpesvirus	Herpesviridae	meningoencephalitis	264–5 586–90
arenavirus	Arenaviridae	Bolivian hemorrhagic fever	263, 691	human immunodeficiency (HIV)	Retroviridae	HIV disease, AIDS	399 518–23
	Arenaviridae	Lassa fever	263, 691				
bunyavirus	Bunyaviridae	encephalitis	263, 691	human papillomavirus	Papovaviridae	common warts (papillomas), genital warts (condylomas); associated with cervical cancer	283, 547–49
canine parvovirus	Paroviridae	severe vomiting & diarrhea	691				
Colorado tick fever	Reoviridae	encephalitis	312, 314, 691				
coronavirus	Coronaviridae	colds, GI disturbances	604	influenza	Orthomyxoviridae	influenza (flu)	263, 616–21
coxsackle	Picornaviridae	common cold syndrome & pharyngitis; severe systemic illness of newborn; muscle pain & damage; diabetes; meningoencephalitis	692	Marburg	Filoviridae	hemorrhagic fever	690
				measles	Paramyxoviridae	rubeola, sometimes subacute sclerosing panencephalitis (SSPE)	543–544
				parainfluenza	Paramyxoviridae	rhinitis, pharyngitis, bronchitis, pneumonia, croup	263, 604–605
cytomegalovirus	Herpesviridae	mononucleosis, congenital cytomegalic inclusion disease, severe birth defects	265, 590–2	paramyxovirus (mumps)	Paramyxoviridae	mumps	637–638
dengue	Flaviviridae	dengue fever (break-bone fever)	314, 687–88	poliovirus	Picornaviridae	poliomyelitis	262, 266, 604, 713
Eastern equine encephalitis	Togaviridae	encephalitis	405, 406, 706	polyomavirus: BK	Papovaviridae	associated with renal transplant infection, immunosuppressed patients	707–708
Ebola	Filoviridae	hemorrhagic fever	690				
enterovirus	Picornaviridae	acute hemorrhagic conjunctivitis	262	polyomavirus: JC	Papovaviridae	mild respiratory illness	707
Epstein-Barr	Herpesviridae	Burkitt's lymphoma, infectious mononucleosis, nasopharyngeal carcinoma	265 688–90	poxvirus group (unclassified)	?	molluscum contagiosum	265
				rabies	Rhabdoviridae	rabies	262, 263, 416, 704–6, 621–622
erythrovirus (B19)	Parvoviridae	aplastic crisis in sickle cell anemia, fifth disease (erythema infectiosum)	265–67, 691–2	respiratory syncytial	Paramyxoviridae	pneumonia in children under age 1, upper respiratory infection in older children & adults	
feline panleukopenia	Parvoviridae	decreased number of white blood cells with fever	691				
Hantaan	Bunyaviridae	Korean hemorrhagic fever	266	rhinovirus	Picornaviridae	common cold	604
hantavirus	Bunyaviridae	hantavirus pulmonary syndrome	266, 408, 691	Rift Valley fever	Bunyaviridae	fever & hemorrhage	691
				rotavirus	Reoviridae	enteritis	264
hepatitis A	Picornaviridae	infectious hepatitis	651–2	rubella	Togaviridae	German measles, 3-day measles	541–543
hepatitis B	Hepadnaviridae	serum hepatitis	651, 652–3				
hepatitis C	?	hepatitis C (non-A, non-B)	651, 652, 653	St. Louis encephalitis	Flaviviridae	encephalitis	314, 403, 404, 706
hepatitis D	?	hepatitis D (delta hepatitis)	265, 651, 652, 653	varicella-zoster	Herpesviridae	chickenpox, shingles	265, 544
hepatitis E	?	hepatitie E (enterically transmitted non-A, non-B non-C)	651, 652, 653	Venezuelan equine encephalitis	Togaviridae	encephalitis	706
				Western equine encephalitis	Togaviridae	encephalitis	314, 405, 406, 706
herpes simplex type 1	Herpesviridae	oral herpes, gingivostomatitis, herpes labialis (cold sores), keratoconjunctivitis, herpetic whitlow	277 586	yellow fever	Flaviviridae	yellow fever	263, 266,, 314, 688

FUNGI

Organism	Disease	Page	Organism	Disease	Page
Aspergillus sp.	aspergillosis, pneumonia in compromised patients, skin infections in burn patients, corneal & external ear infections	61	Epidermophyton sp.	ringworm (tinea)	549
			Filobasidiella neoformans	cryptococcosis	624
			Histoplasma capsulatum	histoplasmosis	624
Blastomyces dermatitidis	blastomycosis	550	Microsporum sp.	ringworm (tinea)	549
			Mucor sp.	zygomycosis	552, 758, 775
Candida albicans	candidiasis	551	Pneumocystis carinii	Pneumocystis pneumonia	520, 626
Claviceps purpurea	ergot poisoning	302, 656–57 757	Rhizopus sp.	zygomycosis	552
			Sporothrix schenckii	sporotrichosis	550
Coccidioides immitis	coccidioidomycosis (valley fever)	599	Trichophyton sp.	ringworm (tinea)	301–2, 549

BACTERIA

Organism	Gram Stain*	Basic Morphology	Diseases	Page	Organism	Gram Stain*	Basic Morphology	Diseases	Page
Actinomadura sp.	+	rod, some filamentous forms	Madura foot (maduromycosis)	552	Gardnerella vaginalis	−	small rod	bacterial vaginitis (nonspecific), urethritis	573
Actinomyces israelii	+	filamentous, diphtheroid, & coccal	actinomycosis, mouth & other lesions	AppB	Haemophilus aegyptius	−	coccobacillus	bacterial conjunctivitis	AppB
Afipia felis	−	rod	cat scratch fever	559	Haemophilus ducreyi	−	rod	chancroid	AppB, 583
Bacillus anthracis	+	rod, encapsulated	anthrax	131, AppB 389, 672–674, 740, 761	Haemophilus influenzae	−	coccobacillus, some strains form capsules	meningitis in children under 5, epiglottitis, ear infections, pneumonia in elderly or compromised patients	209 AppB, 601 603, 704
Bacillus cereus	+	rod, encapsulated	food poisoning	AppB, 389 639–640	Helicobacter pylori	−	curved rod	chronic gastritis, peptic ulcer	AppB, 647–649
Bacteroides sp.	−	small rod	mouth lesions, septicemia, abscesses, Vincent's angina	AppB, 668	Klebsiella pneumoniae	−	rod, encapsulated	pneumonia, infant diarrhea, urinary tract infections	118 AppB, 730 761
Bartonella bacilliformis	−	curved or coccoid	Oroya fever (systemic form), verruga peruana (cutaneous form)	AppB, 686	Legionella pneumophilia	−	coccoid rod	Legionnaires' disease (pneumonia)	AppB, 609–610, 745
Bartonella henselae	NA	coccobacillus	cat scratch fever	AppB, 559, 686	Leptospiria interrogans	−	spiral	leptospirosis	88
Bordetella pertussis	−	coccobacillus	whooping cough	AppB, 607	Listeria monocytogenes	+	rod	listeriosis, meningitis, abortion	AppB, 704 761
Borrelia burgdorferi	−	spiral	Lyme disease	AppB, 314 680–682	Mycobacterium avium	A-F	rod	chronic pulmonary disease, opportunistic infections in immunosuppressed patients	AppB
Borrelia recurrentis	−−	large spiral	epidemic relapsing fever	AppB					
Brucella sp.	−	coccobacillus	brucellosis (undulant fever or Malta fever)	AppB 678–679	Mycobacterium leprae	A-F	rod	Hansen's disease (leprosy)	66, AppB, 383, 708–710
Calymmatobacterium granulomatis	−	rod, encapsulated	granuloma inguinale (donovanosis)	586	Mycobacterium tuberculosis	A-F	rod, branching forms	tuberculosis	57, AppB, 377–378, 612
Campylobacter sp.	−	rod	gastroenteritis	148, AppB 646, 745	Mycoplasma pneumoniae	NA	too small to be visualized by light microscope	primary atypical bacterial pneumonia	152, AppB, 601, 609
Chlamydia psittaci	NA	coccoid, very tiny	ornithosis (psittacosis)	AppB, 615					
Chlamydia trachomatis	NA	coccoid, very tiny	conjunctivitis, trachoma, genital tract infection (nongonococcal urethritis), infant pneumonitis, lymphogranuloma venereum	AppB, 554–555 584, 585	Neisseria gonorrhoeae	−	cocci in pairs	gonorrhea, ophthalmia neonatorum, meningitis, arthritis, keratitis	89, 159 209, AppB, 574–579
					Neisseria meningitidis	−	cocci in pairs; capsules formed in young cells	meningitis, Waterhouse-Friderichson syndrome	AppB, 417 703–704
Clostridium botulinum	+	rod	food poisoning (botulism), wound infections, infant botulism	158, AppB, 330, 331 388, 389, 639, 711–713, 740, 758, 760	Nocardia sp.	+	rod, some filamentous forms	nocardiosis, Madura foot (maduromycosis)	213, AppB, 616
					Porphyromonas gingivalis	−	rod	periodontal disease	636
Clostridium difficile	+	rod	pseudomembranous colitis	359, 649	Propionibacterium acnes	+	rod	acne	AppB, 361, 541
Clostridium perfringens	+	rod	gas gangrene, food poisoning	83, AppB 389, 639, 740, 756	Providencia stuartii	−	rod	urinary tract infections, wound infections	AppB
Clostridium tetani	+	rod	tetanus	AppB, 389, 710–711, 740	Pseudomonas aeruginosa	−	rod	urinary tract infections, skin lesions, eye & ear infections, septicemia in immunocompromised patients	AppB, 389 430–431 541, 542 570 668
Corynebacterium diphtheriae	+	rod, clubshaped, pleomorphic, forms palisades	diphtheria: pharyngeal, laryngeal & cutaneous	AppB, 389 601–603					
					Rickettsia akari	NA	coccobacillus	rickettsialpox	685
Coxiella burnetii	NA	coccobacillus	Q fever pneumonia	761	Rickettsia prowazekii	NA	coccobacillus	epidemic typhus, Brill-Zinsser disease	683
Escherichia coli	−	rod	urinary tract infections, "traveler's diarrhea," nosocomial infections	5, 23, 61 76, 116, 133 161, 167, 179–180 189, 201 209, AppB, 389,644–645 570, 745, 758	Rickettsia rickettsii	NA	coccobacillus	Rocky Mountain spotted fever	685
					Rickettsia tsutsugamushi	NA	coccobacillus	tsutsugamushi fever	AppB, 314 684
					Rickettsia typhi	NA	coccobacillus	endemic or murine typhus	684
					Rochalimaea quintana	NA	coccobacillus	trench fever	AppB
					Salmonella enteritidis, S. paratyphi, S. typhimurium	−	rod	salmonellosis (food poisoning)	640–641, 758
Francisella tularensis	−	small rod (coccobacillus)	tularemia	AppB, 314, 677–678	Salmonella typhi	−	rod	typhoid fever	158, 159, 641, 745